Energy Conversion Engineering

This unique textbook equips students with the theoretical and practical tools needed to model, design, and build efficient and clean low-carbon energy systems. Students are introduced to thermodynamics principles, including chemical and electrochemical thermodynamics, moving on to applications in real-world energy systems, demonstrating the connection between fundamental concepts and theoretical analysis, modeling, application, and design. Topics gradually increase in complexity, nurturing student confidence as they build toward the use of advanced concepts and models for low- to zero-carbon energy conversion systems. The textbook covers conventional and emerging renewable energy conversion systems, including efficient fuel cells, carbon capture cycles, biomass utilization, geothermal and solar thermal systems, hydrogen, and low-carbon fuels. Featuring numerous worked examples, over 100 multi-component homework problems, and online instructor resources including a solutions manual, this textbook is the perfect teaching resource for an advanced undergraduate and graduate-level course in energy conversion engineering.

Ahmed F. Ghoniem is the Ronald C. Crane Professor of Mechanical Engineering, and Director of the Center for Energy and Propulsion Research and the Reacting Gas Dynamics Laboratory. He received his B.Sc. and M.Sc. degrees from Cairo University, and his Ph.D. from the University of California, Berkeley. His research covers computational engineering, turbulence and combustion, multiphase flow, and clean-energy technologies with a focus on CO_2 capture, renewable energy, and fuels. He has supervised more than 120 masters, doctoral, and post-doctoral students; published more than 500 articles in leading journals and conferences; and consulted for the aerospace, automotive, and energy industries. He is a fellow of the American Society of Mechanical Engineers, the American Physical Society, and the Combustion Institute, and an associate fellow of the American Institute of Aeronautics and Astronautics. He has received several awards, including the ASME James Harry Potter Award in Thermodynamics, the AIAA Propellant and Combustion Award, the KAUST Investigator Award, and the "Committed to Caring Professor" at MIT.

Energy Conversion Engineering

Ahmed F. Ghoniem
Massachusetts Institute of Technology

CAMBRIDGE
UNIVERSITY PRESS

University Printing House, Cambridge CB2 8BS, United Kingdom

One Liberty Plaza, 20th Floor, New York, NY 10006, USA

477 Williamstown Road, Port Melbourne, VIC 3207, Australia

314–321, 3rd Floor, Plot 3, Splendor Forum, Jasola District Centre, New Delhi – 110025, India

103 Penang Road, #05–06/07, Visioncrest Commercial, Singapore 238467

Cambridge University Press is part of the University of Cambridge.

It furthers the University's mission by disseminating knowledge in the pursuit of education, learning, and research at the highest international levels of excellence.

www.cambridge.org
Information on this title: www.cambridge.org/9781108478373
DOI: 10.1017/9781108777551

First published 2022

Printed in the United Kingdom by TJ Books Limited, Padstow, Cornwall

A catalogue record for this publication is available from the British Library.

Library of Congress Cataloging-in-Publication Data
Names: Ghoniem, Ahmed F., author.
Title: Energy conversion engineering / Ahmed Ghoniem, Massachusetts Institute of Technology.
Description: Cambridge, United Kingdom ; New York, NY, USA : Cambridge University Press, 2021. | Includes bibliographical references and index.
Identifiers: LCCN 2021028234 (print) | LCCN 2021028235 (ebook) | ISBN 9781108478373 (hardback) | ISBN 9781108777551 (epub)
Subjects: LCSH: Direct energy conversion. | BISAC: TECHNOLOGY & ENGINEERING / Mechanical | TECHNOLOGY & ENGINEERING / Mechanical
Classification: LCC TK2896 .G54 2021 (print) | LCC TK2896 (ebook) | DDC 621.31/24–dc23
LC record available at https://lccn.loc.gov/2021028234
LC ebook record available at https://lccn.loc.gov/2021028235

ISBN 978-1-108-47837-3 Hardback

Additional resources for this publication at www.cambridge.org/ghoniem

To the memory of my mother, Fatima Fouad Hussain
And to my wife, Elizabeth

Contents

Preface

In the early 2000s I started thinking about the next set of challenges to face us. After much reading and discussions with colleagues in my own and other disciplines (a short sabbatical was helpful), I concluded that it would be energy and its environmental impact. Not the lack of resources – conventional and non-conventional – but the rapid rise in consumption and increase of carbon dioxide concentration in the atmosphere. Not that it was a new challenge, either; many studies had concluded the same, over decades. I decided to expand my research portfolio to work on potential solutions, and introduced a senior undergraduate–graduate subject that we called "Fundamentals of Advanced Energy Conversion." The course focuses on energy conversion engineering, considering thermal, mechanical, chemical, and electrical energy forms, starting with fundamental principles and tools of analysis, and expanding toward systems. The course has not been solely about CO_2 reduction in energy and power systems, but this is an important theme that runs through the material. I cover material needed to evaluate the efficiency and CO_2 production in power plants, alternative and more efficient power plant designs including electrochemical systems, renewable concepts such as concentrated solar thermal, geothermal, and biomass, CO_2 capture-based power plants, and alternative fuels production and utilization, such as with hydrogen. The course includes an introduction to nuclear energy, wind, photovoltaics, and energy storage. I wrote lecture notes to help the students follow our coverage. These notes grew into this textbook, after many iterations that included help from students, associates, and colleagues.

While energy systems have relied mostly on burning fossil fuels to produce heat/thermal energy and converting this to work (then electricity by driving generators), things are changing, and the material in this book covers some of these changes. Newer energy conversion processes incorporate more machinery and other devices to improve efficiency, reduce CO_2 emission, capture CO_2, convert chemical energy to electricity directly, store energy in different forms, convert hydrocarbon to other fuels, produce hydrogen by reforming, thermolysis or electrolysis, and more. It is important to expose students studying energy conversion, at an early stage, to elements of general thermodynamics, including availability analysis and chemical and electrochemical thermodynamics, to broaden their perspectives as to what is possible, and to use examples of existing systems and systems under development to encourage them to think beyond systems covered in introductory courses. The application of this generalized treatment to power plants for efficient electricity production using chemical energy (fossil or renewable) and heat (from conventional or renewable sources) can follow. Beyond this, it is important to address the growing challenge of global warming by, besides emphasizing carbon-free energy, covering concepts related to carbon capture and storage. For this reason, introduction to gas separation processes and how they integrate with energy conversion in power plants is needed.

An underlying theme in the class, and in the coverage throughout this book, is energy conversion for electricity (power) production in a carbon-constrained world in which fossil fuels will continue to be used, renewable energy utilization will expand, nuclear energy may also grow, and integration/coproduction will be practiced. Thus, the overall theme is the conversion of thermal and chemical energies to work, efficiently and with reduced CO_2 release into the atmosphere. The focus of the coverage is on power plant fundamentals, or the power island part of the plant. In a carbon-constrained world, conversion efficiency is at a premium and hence all systems must be designed to operate near their ideal limit by reducing entropy production/exergy destruction. Thus, fundamental principles governing this near-ideal behavior are emphasized. These fundamental principles are very important and must be made clear first using simple systems, then discussed again within the context of complex systems. Efficiency can also be gained by integrating or combining different conversion subsystems, or using direct conversion – that is, skipping the production of mechanical work. Integrating power plants with chemical processes for gas separation is important for CO_2 capture designs. And a deeper understanding of the underlying chemistry is important for both, and in discussion of conversion of biomass to energy and fuels.

Chapter 1 frames the overall challenge of energy conversion by reviewing sources, consumption, and environmental concerns, especially as related to CO_2 emissions, accumulation in the atmosphere, and associated global warming. Evidence of growing consumption of energy, especially fossil fuels, and projections supporting the same trend in the intermediate future is shown. CO_2 emissions by fuel and sector (power generation, transportation, industry, etc.) are also used to prioritize intervention options, and international agreements to cap emissions show the urgency of action. Increased electrification of transportation and the economy overall adds more urgency to improving the efficiency and reducing the carbon intensity (CO_2 emission per unit of electricity generated) from power plants, the focus of this book.

While the book focuses on thermodynamics and chemistry (or thermochemistry) fundamentals used extensively in the analysis of existing energy systems and new higher-efficiency and low-CO_2 designs, it expands on system design to include recent studies of complex power plants that can capture CO_2 while burning gaseous and solid fuels. The coverage goes beyond the constraints of "heat-to-work" that most mechanical engineering students are familiar with in their undergraduate studies. Thus, fundamentals of chemical and electrochemical thermodynamics are treated in separate chapters which assume no prior introduction to these topics. A later chapter is dedicated to gas separation processes used in air separation, fuel reforming and hydrogen production, CO_2 separation from combustion gases, and more, and following chapters show how these processes are integrated in power plants, and their impact on efficiency. Solved examples are used to show how simplifying assumption can be used to make quick estimates, and very detailed analyses using computer simulations are used to show how to make more accurate calculations.

The book can be loosely divided into four parts: fundamentals in Chapters 2–4; analysis of power plant systems using conventional and renewable energy in Chapters 5–9; power plants for carbon capture and sequestration in Chapters 10–13; and biomass energy in Chapter 14. In each chapter, coverage gets progressively more specialized toward the end of the chapter, and coverage in the book overall gets more advanced toward the later parts. While in most chapters the emphasis is on using principles of thermodynamics to analyze the performance of simple and complex energy systems, in a couple of chapters I cover material related to kinetics and transport to highlight the need for non-equilibrium effects while estimating the performance of some systems (fuel cells in Chapter 7) and reactors (biomass gasification in Chapter 14). Some instructors can skip these sections (and others) without loss of continuity as desired. Especially in later chapters, I include some state-of-the-art systems and concepts, such as membrane reactors and chemical looping for combustion and gasification, some still at the research stage, to show recent progress and innovations in the field.

Chapters 2–4 cover the fundamentals of thermodynamics, chemical thermodynamics, and electrochemical dynamics, starting with basic concepts of mass and energy conservation, equilibrium, and entropy as a property of matter. These concepts are gradually generalized beyond thermal energy, to chemical energy to electrochemical energy. In the latter, charge conservation, direct conversion, and cell components are introduced.

Chapter 2 is essentially a review of the first and second laws, availability concepts, and their elementary applications. It starts with a quick review of thermodynamics with a formulation that focuses on evaluating the efficiency of heat-to-work conversion, that is availability or exergy analysis. It also reviews gas mixtures and entropy of mixing as an introduction to gas separation and its application in removing CO_2 from hydrocarbon combustion products, or in air separation for oxygen production. Special emphasis is given to the ideal energy of separation as an introduction to gas separation processes (discussed in detail in Chapter 10). It ends with a discussion of liquefaction and its application to hydrogen storage, among other topics. The chapter introduces mass transfer equilibrium and the concept of chemical potential. It reviews and extends the concept of thermodynamic efficiency beyond heat engines, an area revisited frequently in the rest of the book.

Chapter 3 covers chemical thermodynamics – that is, mass and energy conservation in chemical reactions and their application to combustion and thermochemical processes for the conversion of chemical energy to thermal energy and vice versa. Most students will have been introduced to chemical thermodynamics before, but will see its applications beyond combustion and in fuel reforming, hydrogen production, and water splitting in this chapter. Extension to the application of the second law is used to introduce equilibrium and the calculation of mixture composition under given constraints. Again, use of this powerful concept in new and conventional energy applications is explored using several examples and ongoing research. Extensions of equilibrium to define conditions under which maximum conversion efficiency of a “chemical engine” is achieved is also explored. This sets the stage for covering electrochemical thermodynamics. The chapter also seeks a deeper understanding of thermochemistry by discussing applications in syngas production and conversion to other chemicals, which is covered in more detail in later chapters.

Chapter 4 covers chemical reactions that involve charge exchange or transfer – that is, electrochemical reactions and reduction–oxidation or redox pairs. With the growing importance of fuel cells and "chemical engines" that convert chemical energy directly to work, it is important to expose students of energy conversion to fuel cells and batteries, where the whole operation is governed by electrochemical reactions. Conservation is extended beyond mass and energy and into charges in each chemical reaction. Work obtained from a chemical reaction, under equilibrium conditions, is the Gibbs free energy of the reaction. The same calculations are used to derive expressions for the voltage across the electrodes on an electrochemical cell. The chapter also introduces the concept of electrolysis – that is, converting electricity (work) to chemical energy via an electrochemical reaction. These are also related to energy storage (either chemical or electrical).

Chapters 5–9 discuss the "applications" of Chapters 2–4 to systems used in electricity generation (power plants), including fossil and renewable sources (as well as nuclear plant apart from the reactor itself). Several concerns, such as the temperature of the heat source, the overall conversion efficiency, and whether indirect conversion (chemical to heat to mechanical to electrical) or direct (chemical to electrical), are considered. Also addressed is combining conversion approaches or hybridizing different conversion technologies. Of special importance are systems designed to maximize conversion while using low-temperature sources that are typical of "waste heat" or renewable heat sources such as geothermal and solar. In these chapters, most examples and problem sets can be solved using hand calculations with the help of math software, but using equation solvers and packages such as EES (Engineering Equation Solver) should be encouraged.

In Chapters 5 and 6, we use material from Chapter 2 to develop expressions for efficiency of heat-to-power or thermomechanical conversion, first in high-temperature cycles using gas turbines then in intermediate cycles using Rankine cycles and working fluids undergoing phase change. In Chapter 5, simple and higher efficiency Brayton cycles for ideal gas are reviewed. The concept is extended to closed cycles, as well as those using unconventional working fluids such as CO_2, as well as how to enable the cycles to operate at even higher temperatures. Similar coverage is done for Rankine cycles using steam as a working fluid in Chapter 6. The focus is on how to minimize entropy generation (or exergy destruction) and improve efficiency. Pinch-point analysis is introduced and supercritical cycles are discussed for steam and CO_2 working fluids. Organic Rankine cycles play an important role in low-temperature applications and waste heat recovery. Finally, cooling and its impact on the cycle efficiency are discussed.

In Chapter 7, we go back to electrochemical conversion and cover elements of fuel cell operation, including electrochemical kinetics and charge transfer, and their impact of fuel cell efficiency. This is one of a few places in the book where we cover finite rate processes (kinetics and transport) and their impact on the operation of the conversion system, with sufficient detail so as to derive expressions for efficiency that account for these loss mechanisms. Relations are obtained between the finite current, cell voltage, and overall conversion efficiency. The chapter also summarize some of the most popular fuel cells and their performances. The emphasis is on conversion efficiency under different operating conditions and the impact of the cell design on losses.

Chapter 8 explores how to combine more than one cycle and/or a cycle with a fuel cell to improve the overall conversion efficiency. The traditional combined Brayton–Rankine cycle is covered, with the emphasis on how to use the pinch point to improve its efficiency. Another new concept considered here is oxy-combustion cycles in which fuel is burned in a mixture of oxygen and recycled CO_2 and/or water. These enable CO_2 capture, and while discussed in detailed in a later chapter, they are introduced here as an application of traditional cycle analysis. Two forms of oxy-combustion cycles are discussed: a combined cycle with the Brayton cycle operating on CO_2 as a working fluid, and a cycle with significant water recycling. Finally, hybridizing conventional cycles with fuel cells is also discussed.

Chapter 9 extends cycle analysis to renewable energy sources where low- and intermediate-temperature sources impose further constraints on the system design, starting with a brief characterization of each source and its geographic availability. We start with geothermal energy, where the source is at low temperature. In some cases, "collecting" the heat and the power island must be considered while maximizing the efficiency. Also discussed are approaches to hybridizing these sources and integrating them with fossil energy to overcome the intermittency problem and avoid the need for expensive storage. We describe interesting developments in integration involving solar reforming and its impact on efficiency.

Chapters 10, 11, and 13 focus on how to use fossil fuels for power generation while limiting CO_2 release into the atmosphere, essentially exploring carbon (dioxide) capture technologies (known as CCS for carbon capture and sequestration/storage, or CCUS for carbon capture use and storage) in ways compatible with the fuel and the cycle. Since a fuel must still be burned, these approaches have been classified into post-combustion, oxy-combustion, and pre-combustion, depending on where the CO_2 is separated. Since gas separation is an essential part of the system, the section starts with a chapter dedicated to reviewing these processes, and in particular their energy requirements, in order to assess how incorporating these processes in power cycles could affect their efficiency.

In Chapter 10, a new topic is introduced, namely gas separation, with the aim to discuss some of the related fundamentals before introducing power cycles that have been proposed for CO_2 capture (CCS or CCUS). Material in this chapter is likely to be new for mechanical engineering students, although most chemical engineering students should have been introduced to some of these concepts previously. The coverage is limited to some of the essential fundamentals used to model the separation process, which can be used to estimate the energy required under some operating conditions. The chapter covers absorption using liquid solvents, both physical and chemical, adsorption using solid sorbents, cryogenic separation, and membrane (porous and dense) separation. Some examples are given for how these processes are applied to different energy systems (including but also beyond power plants).

Chapters 11 discusses carbon (dioxide) capture from natural gas power plants, starting with classification of the different approaches followed by details for each. Simplified models that have been proposed are discussed and examples for the impact on efficiency penalty are solved. Many proposed power plant cycles that enable CO_2 capture via pre-combustion, post-combustion, and oxy-combustion are discussed. These include the "water" and Graz cycles, among others. Advanced materials on novel concepts including chemical looping combustion and membrane reactors in which air separation and combustion are integrated are also covered. Analytical treatment using simplified assumptions is used, and the results of analysis using computer models that relax these assumptions are also shown.

Chapters 12 and 13 focus on coal, one of the most widely used fuels for electricity production, and the one with the highest CO_2 emissions per unit of electricity produced. Chapter 12 starts with source characterization. Conventional systems, using boilers, and new plant designs that start with gasification in an integrated gasification combined cycle (IGCC) are reviewed, emphasizing the environmental performance of both. Because of its significance in both power plants and fuel and chemicals production, coal gasification is discussed in more detail, including types of gasifiers and processes to transform syngas to chemicals and fuels (these concepts extend beyond coal; any source of syngas can be used for the production of higher-value chemicals). In Chapter 13, CCS applications are discussed starting with post-capture using amines and membranes. Oxy-combustion of coal under atmospheric or higher pressure have received much attention and hence their results are reviewed. Pre-capture using gasification is also discussed, especially for applications using different feed technologies and with membrane separation of hydrogen. Finally, chemical looping combustion of coal is reviewed.

In Chapter 14 the topic of biomass is covered, starting with characterization, a quick review of organic chemistry, the composition of biomass and its derivatives, their constituents, and their energy contents. Conversion of biomass to thermal energy and biofuels is often characterized as biochemical or thermochemical, with subcategories in each. Mass and energy balances in bioconversion are shown in some detail, and the efficiency and products of each are presented. In thermochemical conversion, the dependency of the products on the process temperature (and heating rate) are shown, and reactor systems for low-temperature torrefaction and intermediate-temperature gasification are described. The chapter ends with a detailed gasification model for biomass in fluidized beds – an exercise combining single particle conversion with bed material heat and mass transport. This section, being the last in the book, involves the most complex modeling section, which is likely to go beyond what can be covered in a course. Nevertheless, it sets the stage for follow-up coverage.

Knowledge of college-level physical chemistry and introductory-level engineering thermodynamics is assumed, and the book can be used in senior-level undergraduate thermodynamics as well as energy conversion classes. More of the material can be covered in graduate courses. In my experience it is not possible to cover the material of the entire book in one semester, but one may choose to skip parts in each chapter, especially toward the end of the chapter, without losing continuity. It is also possible to skip entire chapters if they do not fit in the curriculum (e.g., Chapter 4 on equilibrium treatment of fuel cells or Chapter 7 on finite current performance, or Chapters 10, 11, and 13 on gas separation and carbon capture). There are plenty of solved examples to help the student, as well as homework problems for assignments. For doing the homework, familiarity with equation solvers packages is very helpful, and software such as EES will be very productive. Students familiar with ASPEN will find it useful while going through the examples in the later chapters, but it is not necessary or required.

The class at MIT has been taken by students mostly from mechanical engineering, but also from chemical engineering, nuclear engineering, material science and engineering, and occasionally by students from the school of science.

An early version of the PowerPoint slides used in teaching the class can be found on MIT OpenCourseWare. This will be replaced soon with a much-updated version that I used in 2019. In the class I teach at MIT, I ask the students to do a term project – some samples of these can also be found on the website.

Acknowledgments

A task like writing this book could not have been done without the help of many. I started teaching this class with late Prof. Mujid Kazimi, Profs. Jefferson Tester and Yang Shao-Horn, who helped shape the contents and reviewed some of the early notes that grew into this book. Prof. Alexandre Mitsos used the notes and made suggestions for improvement. Prof. Tarek Etchekki reviewed some of the early chapters, and Dr. Yousef Hazli edited some more and contributed to some of the examples and problems.

Several teaching assistants worked with me on the class and contributed problems and examples, and reviews, including Sunbae Park, Rory Monaghan, Simcha Singer, Reymond Speth, Gaurav Kewlani, Neerav Abani, Xiaoyu Wu, Nadim Chakroun, Aniket Patankar and Omar Labbani. Many undergraduate and graduate students who took the class provided valuable edits and comments on the notes and inspired changes and revisions.

I had the fortune of working with excellent graduate students and post-doctors whose research contributed to this material, including Akilesh Bakshi, Richard Bates, Cristina Botero, Nadim Chakroun, Lei Chen, Tianjaio Chen, Chuckwunwike Iloeje, Georgios Dimitrakopoulos, Jeffrey Hanna, Joungsup Hong, Katherine Hornbostel, Anton Hunt, Patrick Kirchen, Kevin Kung, Won Young Lee, Rory Monaghan, Michael Rutberg, Elysia Sheu, Santosh Shanbhogue, Yixiang Shi, Addison Stark, Mruthunjaya Uddi, Xiaoyu Wu, and Zhenlong Zhang. Without them this project would not have possible.

The book project received financial support from the MIT School of Engineering, MIT-ME Pappalardo Funds and the MIT-MITie Battelle Funds.

Nomenclature

A	Helmholtz free energy
a	Surface area
AF	Air-fuel ratio
$A_{f,b}$	Pre-exponential factor in the rate constant
A_s	Surface area
C_l	Speed of light
CV	Control volume
C	Molar concentration
c_p	Specific heat at constant pressure
d	Diameter
D	Diffusion coefficient
E	Total stored energy
E_a	Activation energy
e	Total specific internal energy
e^-	Electron
$\boldsymbol{F}$	Force
$\Im_a$	Faraday's number
FU	Fuel utilization
g_r	Gravitational acceleration
G	Gibbs free energy
H	Enthalpy
$\mathbf{H}$	Magnetic field
h	Specific enthalpy
HR	Heat rate
h_{conv}	Convective mass transfer coefficient
He	Henry's constant
I	Irreversibility
I	Number of different chemically distinct components
I	Total current
i	Current density, current per unit area
J	Flux of species, uncharged and charged
k_s	Elastic constant
KE	Total kinetic energy
K_p	Pressure-based equilibrium constant
k	Reaction rate constant
k	Isentropic index
K	Equilibrium constant

L	Length
$\dot{m}$	Mass flow rate
M	Molecular weight
MW	Molecular weight
m	Mass
N_a	Avogardo's number
N	Number of molecules
$\dot{N}$	Molar flux
n	Number of moles
$\vec{n}$	Unit vector
$\dot{n}$	Molar flow rate
n_e	Number of electrons in an electrochemical reaction
p	Pressure
PE	Potential energy
$\mathcal{P}$	Power
PP	Pinch point
$\mathcal{P}$	Power per unit area, or power density
Q	Heat transfer
$\tilde{P}_i$	$= D_i/(\mathfrak{R}Tt)$, Permeability (used mostly for porous membranes)
$\tilde{P}_i$	$= (D_i/\mathfrak{R}T)S_i$, Permeability coefficient (used for non-porous membranes)
$\mathfrak{R}$	Universal gas constant
R	Pressure ratio across membrane
$\mathbb{R}$	Resistance
S	Entropy
s	Specific entropy
S	Solubility
T	Temperature
TER	Thermal energy reservoir
t	Time
t	Membrane thickness
U	Internal energy
u	Specific internal energy
V	Velocity
$\forall$	Volume
V	Voltage
V	Convective velocity
v	Specific volume
W	Work
$\hat{w}_i$	Molar rate of formation/destruction
x	Distance
X	Mole fraction
Y	Mass fraction
z	Vertical distance, elevation
z_i	Charge number of a component

Greek

α Transfer coefficient in fuel cells
α Degraded efficiency in steam turbines
α_{ij} Separation factor between two gases, or permeability ratio, or selectivity
β Coefficient of performance
β Pressure ratio
$\tilde{\Gamma}$ Number of surface active sites
δ Thickness
δ Defect in stoichiometry
ε Electric potential
ε Porosity
η Efficiency
$\tilde{\eta}$ Overpotential
θ Fraction of occupied sites
ϑ Non-dimensional temperature
$\mathrm{\mu}$ Viscosity
μ Chemical potential
v Stoichiometric coefficient
π_P Pressure ratio
$\mathrm{\rho}$ Density
ξ Specific flow availability/exergy
Ξ Non-flow availability/exergy
λ Excess air ratio
ξ Flow availability per unit mass
v Stoichiometric coefficient
ϕ Equivalence ratio
φ Fuel utilization
τ Tortuosity
τ Time constant
χ Chemical symbol
ς Electrical charge

Subscripts

a Anodic
b Backward reaction
C Compressor
c Cathodic
ch Chemical
cr Critical conditions
d Diluent
el Electrical
$elas$ Elastic
f Flow
f Formation for enthalpy, entropy, Gibbs free energy.

f	Fuel
f	Forward reaction
fl	Fuel
g	Generation
g	Gas
i	Index indicating a component/species in a mixture
in	Inlet
L	Liquid
mag	Magnetic
mec	Mechanical
nuc	Nuclear
o	Environmental conditions
OC	Open circuit
ox	Oxidizer
out	Outlet
p	Product
r	Reactant
R	Reaction
s	Isentropic
T	Turbine
t	Turbine
th	Thermal
V	Vapor
∞	Bulk

Superscripts

o	Standard
^	Molar
·	Rate of change
‘	Reactants
“	Product
$*$	Cell conditions
o	Reference or standard conditions
‘	Reactants
“	Product
$*$	Cell conditions
o	Reference or standard conditions

1 Low-Carbon Energy, Why?

1.1 Introduction

Energy is one of the most important needs of humanity. Mobility, lighting, communications, heating, and air conditioning are all energy-intensive functions that are indispensable in modern life. Industrial production, food production, and clean water require energy.

Energy consumption correlates strongly with standards of living. The developed world has become accustomed to cheap and plentiful supplies, and recently more of the developing world is striving for the same. Competition over supplies of conventional resources is intensifying, and more challenging environmental problems, especially those related to carbon dioxide (CO_2), are looming. The evidence that atmospheric CO_2 concentration is correlated with the global temperature is strong, and models indicate that the century-old trend of rising temperatures could accelerate. Given the potential danger of such a scenario, steps must be taken to curb energy-related CO_2 emissions. Solutions include substantial improvements in conversion and utilization efficiencies, carbon capture and storage (CCS), and expanding use of zero-carbon sources, namely nuclear energy and renewable sources.

Recent energy consumption rates – total and per source – are summarized in Section 1.2. Nearly 80% of our energy is supplied from fossil resources. Global climate change has been shown to correlate with greenhouse gases. Evidence and trends are discussed in Section 1.3 [1]. In Section 1.4, CO_2 emissions by sector and fuel are reviewed. International agreements and targets for CO_2 reduction are summarized in Section 1.5. Technologies to address resources depletion and CO_2/climate change are discussed in Section 1.6. Conversion efficiency is at the forefront of the effort to conserve resources and reduce the environmental impacts, and is discussed in Section 1.7. Fossil fuels will remain as a major source of electricity generation for decades, and approaches to "decarbonize" power generation plants – including CO_2 capture – are discussed here. Zero-carbon energy includes nuclear energy and renewable sources, such as geothermal, wind, solar, and biomass. Transportation consumes a significant fraction of the total

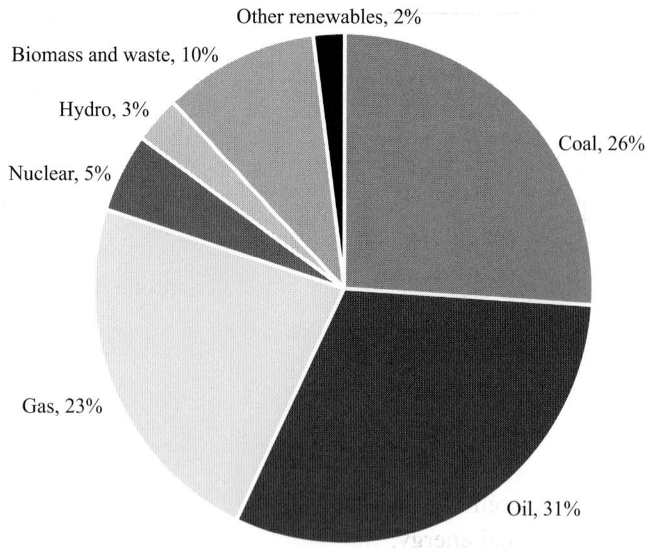

Figure 1.1 The breakdown of the world primary energy consumption in 2018. Except for hydropower, primary energy measures the thermal energy equivalent in the fuel that was used to produce a useful form of energy; e.g., thermal energy (heat), mechanical energy, electrical energy. When energy is obtained directly in the form of electricity, efficiency is used to convert it to equivalent thermal energy. 1 toe = ~42 GJ. IEA World Energy Outlook.

energy used worldwide. While heavily dependent on oil products,[1] the number of electrified and electric vehicles is growing, adding to the need for electricity production.

1.2 Consumption

The world consumed ~570 EJ (exajoule) in 2014 (up from ~440 EJ in 2004). According to the International Energy Agency (IEA), the world power capacity then was close to 18 TW, of which 6.1 TW was for electricity generation. The total capacity is expected to reach beyond 50 TW by the end of the twenty-first century, driven by population growth and rising living standards in developing countries. This is despite the anticipated improvement in energy intensity, defined as the energy used per gross domestic product (GDP), or J/GDP.

Figure 1.1 shows the breakdown of the world *primary* energy consumption in 2018.[2] Fossil-fuel use, measured by the total thermal energy equivalent, is dominated by oil, followed by coal and natural gas. Oil is used mostly in transportation. Coal is used mostly in electricity generation, where the consumption of natural gas has also been rising (in conventional energy units, 1 Mtoe = 41,868 TJ).[3] The total energy consumption in 2014 was 13,558 Mtoe (million tonne oil equivalent), up from 11,059 Mtoe in 2006. The contributions of nuclear and renewable sources such as hydropower, which produce electricity, are converted to thermal energy using First Law efficiencies (the First Law efficiency varies by source – for instance, for geothermal energy it is ~10%). The IEA uses 100% efficiency to

[1] According to the DoE/EIA, in 2007 the breakdown of transportation fuel in the USA was: petroleum 96.3%, natural gas 2.1%, biomass 1.2%, and electricity 0.3% (www.eia.doe.gov/oiaf/1605/gg04rpt/carbon.html).

[2] EJ is an exajoule, or 10^{18} J, and 1 TWh = 0.086 Mtoe.

[3] A megaton = 1×10^6 (metric) ton = 1×10^9 kg.

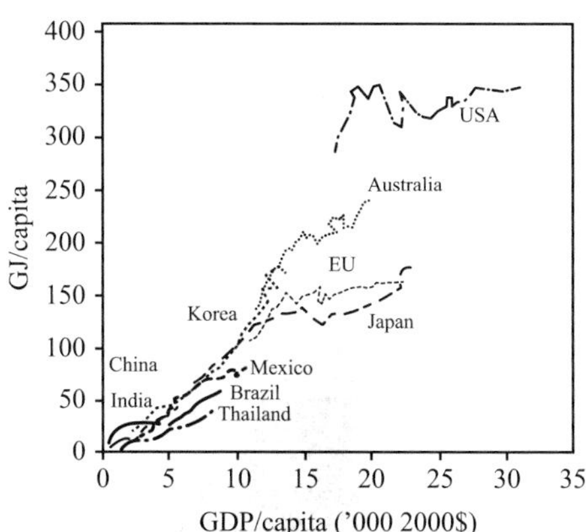

Figure 1.2 The per-capita energy consumption and the per-capita GDP for a number of developed and developing countries (1 BTU = 1.055 kJ). Energy use per capita is for the year 2003; GDP per capita is given for the year 2004 expressed in 2000 US dollars. Data downloaded from the United Nations Development Programme, Human Development Report (HDR) 2006, Table 1, pages 283–286, and Table 21, pages 353–356 (http://hdr.undo.org/hdr2006/report.cfm).

represent the energy content of electricity and 33% efficiency to convert nuclear electricity to thermal energy. Biomass sources[4] are used mostly in rural communities.

Our welfare depends on continuous and guaranteed supplies of energy at affordable rates. Per-capita GDP correlates well with per-capita energy consumption, with developed countries consuming energy at orders of magnitude greater than developing and poorer nations. One can in fact define an affluence index based on the per-capita energy consumption.

Increasing the per-capita GDP of a country goes hand-in-hand with rises in its per-capita energy consumption, especially during the early stages of development. This trend slows down as the economy matures and becomes more energy efficient. Figure 1.2 shows the rise of per-capita energy consumption against per-capita GDP for some developed and developing countries, and others undergoing rapid transition. Developed countries show significant improvement in energy efficiency as the per-capita energy use stabilizes while GDP continues to rise. This trend is enabled by investing in energy efficiency, adopting advances in technology that lead to energy savings, and citizens becoming more aware of their environmental impact. Rising energy prices often promote lower consumption, enabled primarily by switching to higher-efficiency systems, but the impact often persists even after energy prices fall back to more affordable levels. Some developing countries have started to take steps toward improving their economic conditions through industrialization, agricultural mechanization, and large-scale infrastructure improvement, causing their energy consumption to grow more rapidly.

Higher quality of life in developing countries could be achieved at energy intensity lower than the current standard in developed nations. For instance, it has been shown that the UN Human Development Index (HDI), which includes data that reflect the physical, social, and economic health and wellbeing of a population, such as per-capita GDP, education, longevity, use of technology, and gender development, rises steeply during the early stages of growth, alongside per-capita electricity consumption, before it levels

[4] Renewable sources are also non-exhaustable sources.

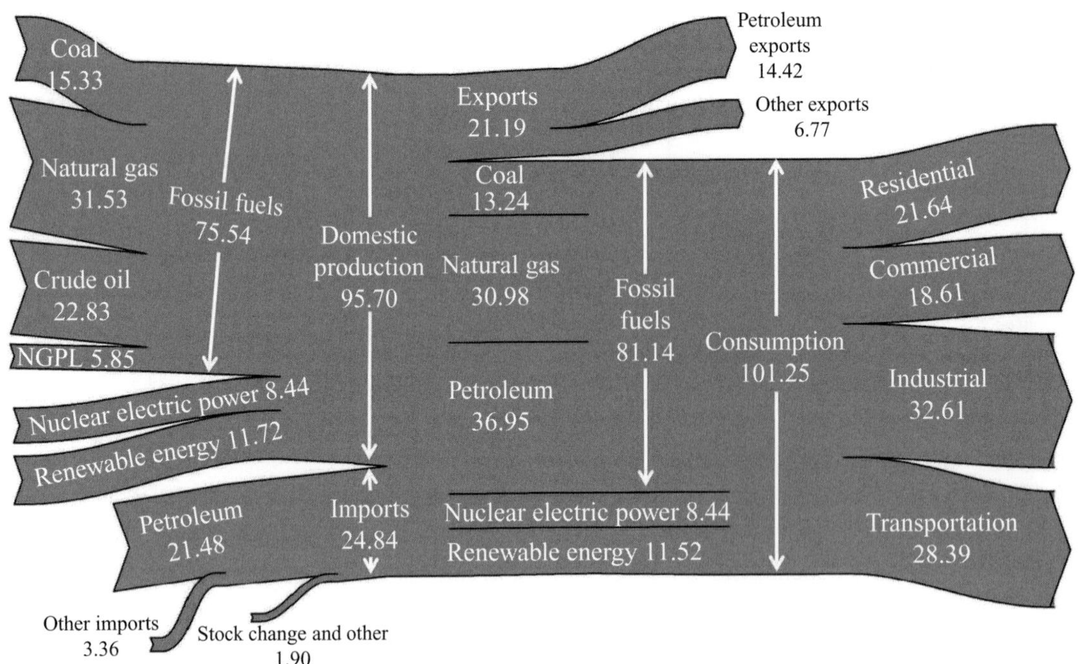

Figure 1.3 Energy sources and consumption patterns in the USA (2018 data), measured in quadrillion BTU (quad BTU or QBTU = 1.055 EJ = 1.055×10^{18} J). The units used here represent the thermal energy content of the fuel. In the case of nuclear and renewable energy, the latter of which is dominated by hydropower, where the energy output is electricity, an assumed First Law efficiency is used to convert the electricity to thermal energy. The efficiency used in assembling these data is an average over fossil-fuel power plants. Please note that some of the conversion factors used in the IEA and the EIA are different (www.eia.gov/totalenergy/data/monthly/pdf/flow/total_energy.pdf).

off [2]. That is, a "point saturation" of energy consumption is reached beyond which more energy use does not necessarily translate to a better standard of living. Meanwhile, nearly 20% of the world population does not have adequate access to electricity and rely on biomass as their primary source of energy.

Consumption patterns vary widely and depend on the economy, local weather, and population density, among other factors. The USA's energy consumption was ~100 EJ in 2018. The share of different sources and the utilization in different sectors is shown in Figure 1.3.[5] Currently, consumption is projected to rise over the next 25 years, with the fossil-fuel share being ~70%.

The share of different sources in energy production worldwide over the last 40 years is shown in Figure 1.4, along with the projection for the next 20 years. Continued growth in

[5] www.eia.doe.gov/cneaf/alternate/page/renew_energy_consump/rea_prereport.html.

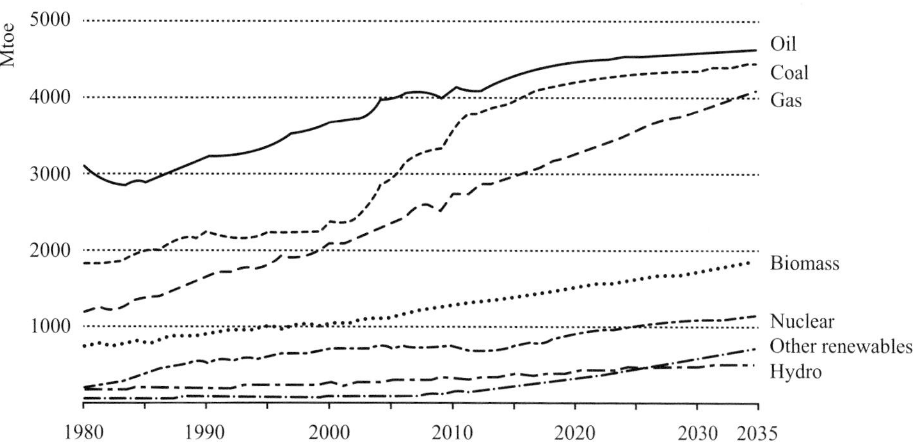

Figure 1.4 World primary energy consumption by fuel. Predicted based on the continuation of existing policies and measures as well as implementation of policies that have been announced by governments but are yet to be given effect (mid-2013). IEA World Energy Outlook 2013, p. 63.

natural gas, as more of its unconventional resources are tapped, will likely reduce the reliance on coal. Biomass contributes to heat and electricity, as well as biofuels such as ethanol. Its fast growth may contribute to slowing down oil consumption. Electricity from wind, solar, and geothermal (and other renewables) will continue to grow as their prices drop, satisfying a rising fraction of new demand. Nuclear energy must address waste storage, safety, and security in order to continue its expansion.

One challenge that energy production faces is related to possible limited supplies of fossil-fuel reserves (those that have been discovered and can be extracted economically using existing technologies) and resources (defined as those thought to exist but whose extraction may require advanced technologies and may not be presently economical), which some have estimated would last for only 100–300 years, depending on the fuel type and recovery.[6] Unconventional sources, such as oil shale and tar sands, could extend this period. For instance, it is estimated that while the proven reserves of oil are nearly 1 trillion barrels, Canadian oil sands could produce 1.7 trillion barrels, and oil shale in the USA could produce 2 trillion barrels. The environmental impact of producing light hydrocarbons from these resources could be significant. Other hydrocarbon resources include deep ocean methane hydrates. A case for the existence of abiogenic (non-organic) methane in deep, underground formations has been made, and if proven would be another vast resource.

A more serious challenge is global warming.

[6] For more on the subject, see [3], [4], and the ASPO website www.peakoil.net.

1.3 Global Warming

Data suggest that current temperatures are close to the highest ever reached in the past 400,000 years, and that CO_2 concentration is even higher than the highest level estimated during this same period. The records of temperature, CO_2 concentration, methane concentration, and solar insolation are shown in Figure 1.5. The temperature (in this case the temperature at Lake Vostok over the Antarctic) and atmospheric concentrations of CO_2 and CH_4 varied cyclically during this period. However, they remained well correlated. It is interesting to note that the time scales for the rise and fall of the quantities of interest are different. The figure shows that cyclic variation over geologic time scales is the norm. On the other hand, current CO_2 levels are higher than the peaks reached previously. Prior to the onset of the Industrial Revolution in the mid-1800s, natural causes were responsible for CO_2 concentration variation.

The correlation between the global temperature and CO_2 atmospheric concentration during the past 1000 years is shown in Figure 1.6. While reasonably stable at around

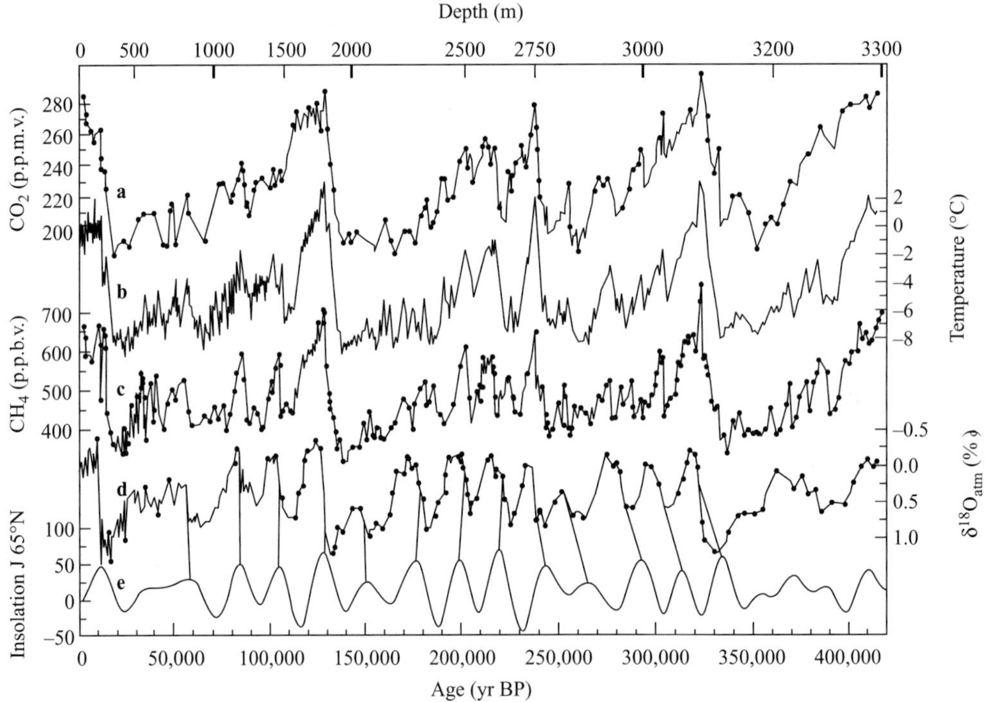

Figure 1.5 Time series of: (a) CO_2 concentration; (b) isotopic temperature of the atmosphere; (c) CH_4 concentration; (d) $\delta^{18}O$ atm (‰); and (e) mid-June insolation at the given location in W/m^2. The top axis shows the depth of the ice sample and the bottom axis shows the age. BP, before present. J.R. Petit, J. Jouzel, D. Raynaud, et al. "Climate and atmospheric history of the past 420,000 years from the Vostok ice core, Antarctica," *Nature*, vol. 399, pp. 429–436, 1999; figure 3, p. 431.

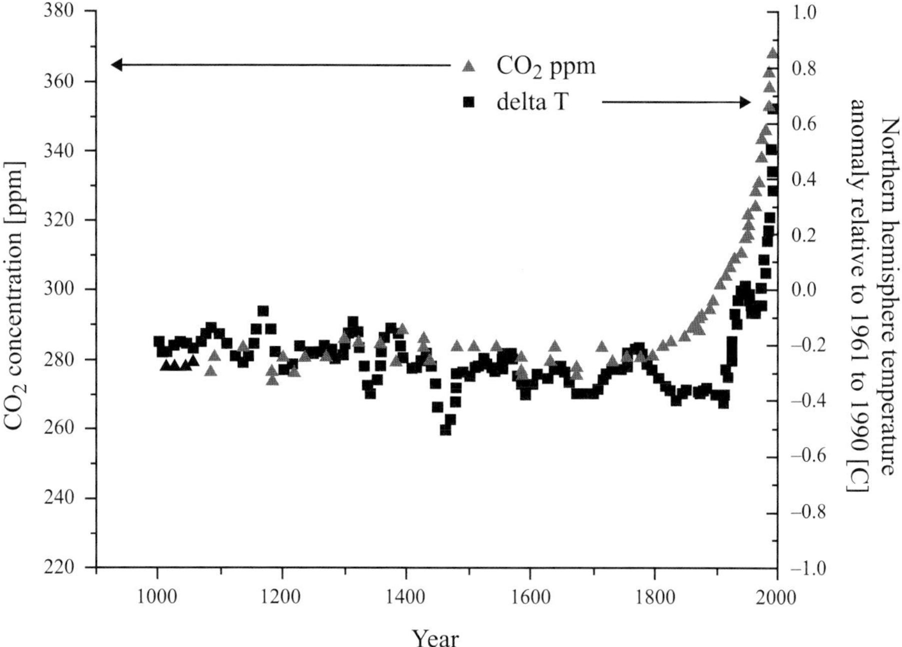

Figure 1.6 The rise in atmospheric concentration of CO_2 and temperature in the Northern Hemisphere over the past 1000 years. Data taken from the *IPCC Third Assessment Report 2001, Working Group I, Technical Summary*. The figure is a combination of data from figure 5, p. 29 (Millennial Northern Hemisphere (NH) temperature reconstruction) and figure 10b, p. 40 (CO_2 concentration in Antarctic ice cores for the past millennium). Recent atmospheric measurements (Mauna Loa) are shown for comparison. Triangles show CO_2 in ppm and squares show delta T in °C.

285 ppm before 1850, its rise around the onset of the Industrial Revolution, followed by a rise in the temperature, is evidence of "man-made" global warming.[7] The trend has intensified over the past ~50 years, as shown in Figure 1.7.[8] Recent measurements show CO_2 concentrations exceeding 400 ppm.

Greenhouse gases include water (H_2O), CO_2, methane (CH_4), nitrous oxide (N_2O), chlorofluorocarbons (CFCs), and aerosols. The greenhouse potentials of CO_2, CH_4, N_2O, and CFCs (taken as averages among different estimates) are 1:11:270:1300–7000. Because of

[7] For more data on the global mean temperature, the surface temperature anomaly (the difference from historical means), and the impact of solar irradiance variation on global mean temperature, see http://data.giss.nasa.gov/gistemp/2007. On a yearly average basis, the solar insolation (total energy received by an area perpendicular to a beam) at the outer edge of Earth's atmosphere is 1366 W/m^2. Despite a small decrease in solar irradiance recently, global temperature continues to rise, providing further evidence to the greenhouse gas mechanism.

[8] Arrhenius predicted an increase of the Earth surface temperature by 5–6 °C due to the doubling of CO_2 concentration in the atmosphere, more than 100 years ago.

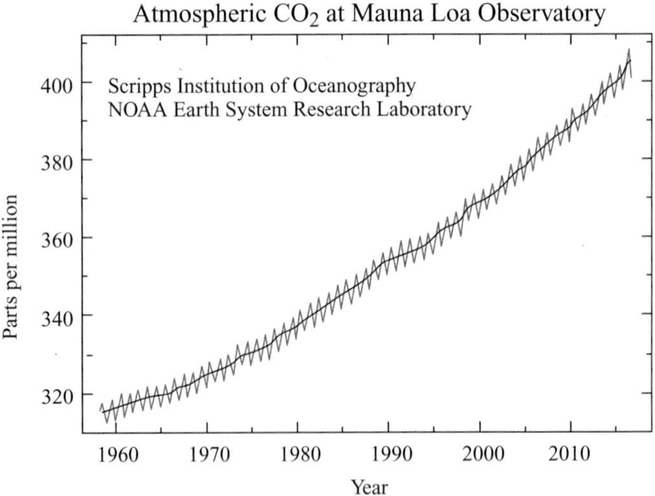

Figure 1.7 Close up of CO_2 concentration and temperature change over a 70-year period.

its much higher concentration, CO_2 is the most impactful. Most CO_2 anthropogenic emissions result from fossil-fuel combustion, with a smaller fraction from industrial processes.

1.3.1 Global Energy Balance

The energy fluxes to and from the Earth's atmosphere, and their change as radiation passes through the atmosphere, are shown in Figure 1.8. Solar radiation is concentrated at short wavelengths, within the visible range of 0.4–0.7 μm, because of the high temperature of the surface of the Sun, estimated to be ~6000 °C. Only a small fraction of solar radiation lies in the ultraviolet range, down to 0.1 μm, and in the infrared range, up to 3 μm. On average, 30% of the incoming solar radiation is reflected back by Earth's atmosphere and its surface (the albedo), 20% is scattered by the atmosphere at different altitudes, and the remaining 50% reaches the surface and is absorbed by the ground and the water. The fraction of the incoming radiation that is either absorbed or scattered while penetrating the atmosphere does so in a spectrally selective way, with the ultraviolet radiation absorbed by stratospheric ozone and oxygen, and infrared radiation absorbed by water, CO_2, ozone (O_3), N_2O, and CH_4 in the troposphere (lower atmosphere). Much of the radiation that reaches the ground goes into evaporating water from the oceans. Outgoing radiation from the cooler Earth's surface is concentrated at the longer wavelengths, in the range 4–100 μm [5].

Greenhouse gases in the atmosphere absorb part of the outgoing radiation, with water molecules absorbing in the 4–7 μm wavelength as well as at 15 μm, and CO_2 absorbing in the range 13–19 μm. A fraction of this energy is radiated back to Earth's surface and the remainder is radiated to outer space. The change of the energy balance due to this greenhouse gas radiation is known as the radiation forcing of these gases, and its contribution to Earth's energy balance depends on the concentration of greenhouse gases in the atmosphere. The net effect of absorption, radiation, and reabsorption is to keep Earth's surface warm, at

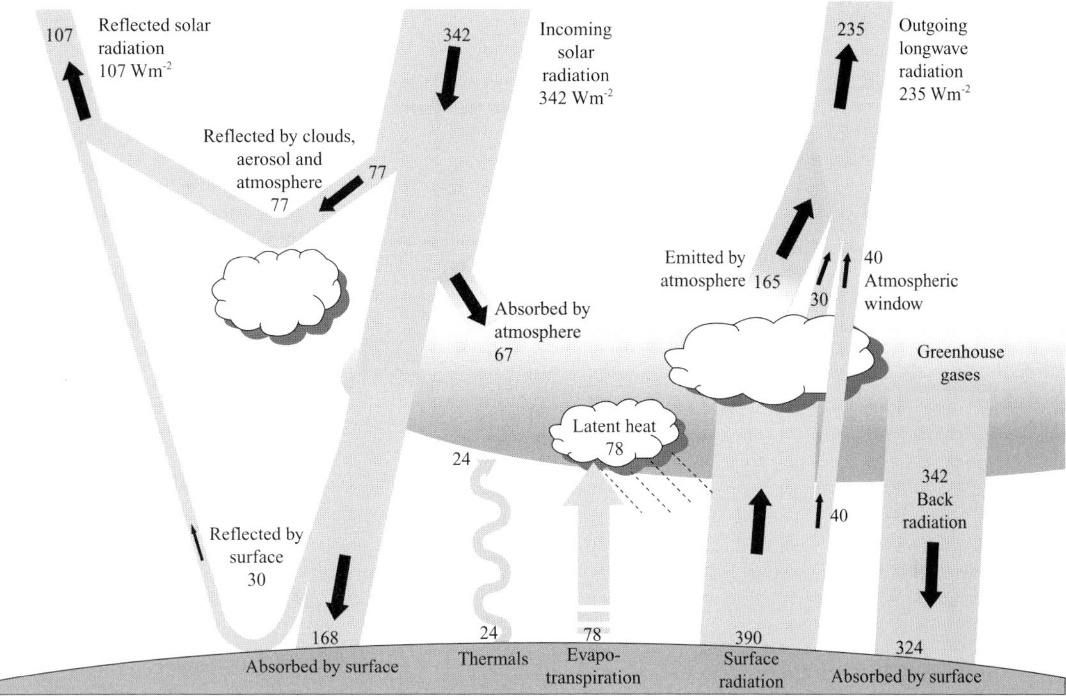

Figure 1.8 Solar energy flux and how much of it reaches Earth's surface; the radiation emitted by the ground and the balance that is re-radiated back to the surface. All numbers are given as averages over Earth's surface in units of Wm^{-2}. Adapted from Intergovernmental Panel on Climate Change, Working Group 1, *The Physical Basis of Climate Change*, p. 96, FAQ 1.1, figure 1 (2007) (http//ipcc-wg1.ucar.edu/wg1/Report/AR4WG1_Pub_Ch01.pdf).

average temperature close to 15 °C. In essence, Earth's atmosphere acts as a blanket; without it the surface temperature could fall to values as low as −19 °C. Because of its concentration CO_2 has the strongest radiation forcing among known greenhouse gases, except for water. However, water concentration in the atmosphere is the least controlled by human activities.

Increasing the concentration of greenhouse gases enhances the radiation forcing effect. Moreover, a number of feedback mechanisms, such as the melting of the polar ice (which reflects more of the incident radiation back to space) and the increase of water vapor in the atmosphere (due to the enhanced evaporation resulting from higher temperatures) are expected to accelerate the greenhouse contribution to the rise of the mean atmospheric temperature.

1.3.2 Carbon Balance

Current estimates indicate that fossil-fuel combustion produces ~6 GtC/y. This unit, gigaton carbon per year, is used to account for all forms of carbon injected into the atmosphere, with

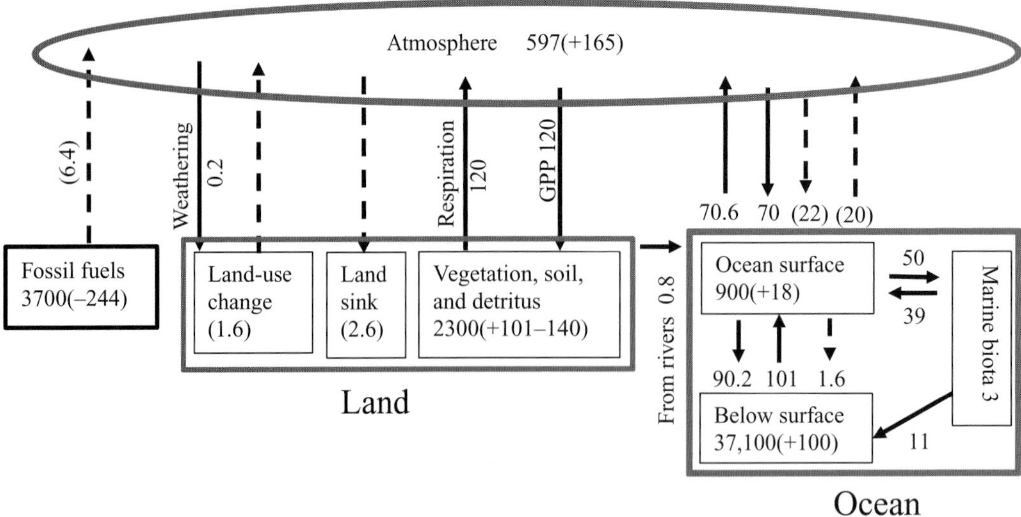

Figure 1.9 The global carbon cycle (in the 1990s), showing the primary annual fluxes in GtC/y: pre-industrial "natural" fluxes and added "anthropogenic" fluxes. Continuous arrows show natural fluxes and broken arrows show fluxes due to anthropogenic activity. Also shown are the reservoirs in GtC; numbers without brackets for natural reservoirs and numbers in brackets added to these reservoirs. Gross fluxes have uncertainty of more than ±20%. GPP is gross primary production during photosynthesis. Adapted from Intergovernmental Panel on Climate Change, Working Group 1, *The Physical Basis of Climate Change*, p. 514, Figure 7.3 (2007) (http//ipcc-wg1.ucar.edu/wg1/Report/AR4WG1_Pub_Ch07.pdf).

carbon accounting for 12/44 of CO_2 – that is, 1 GtC is equivalent to 44/12 = 3.667 $GtCO_2$. This should be compared with other sources/sinks that contribute to CO_2 concentration in the atmosphere, as shown in Figure 1.9. Carbon dioxide is injected into the atmosphere through respiration and the decomposition of waste and dead biomatter, and is removed by absorption during photosynthesis and by the phytoplankton living in the oceans. Respiration produces nearly 60 GtC/y, while photosynthesis removes nearly 61.7 GtC/y, with a balance of a sink of 1.7 GtC/y. The surfaces of the oceans act as a sink, contributing a net uptake of 2.2 GtC/y, a source–sink balance between production of 90 and consumption of 92.2 GtC/y. Changing land use (deforestation) and ecosystem exchange adds/removes 1.4/1.7 GtC/y, for a net balance of a sink of 0.3 GtC/y. The overall net gain of CO_2 in the atmosphere is estimated to be around 3.5 GtC/y. It is relative to these balances that the contribution of fossil-fuel combustion and related man-made sources appears significant. These numbers are somewhat uncertain, and there is 1–2 GtC/y unaccounted for in the overall balance, when all the uncertainties are traced. The uncertainty in the numbers is reflected in the different sources, and is demonstrated here by the different numbers.[9]

[9] It is estimated that for each 2.1 GtC introduced into the atmosphere, CO_2 concentration rises by 1 ppm, and that the average lifetime of CO_2 in the atmosphere is 100–200 years.

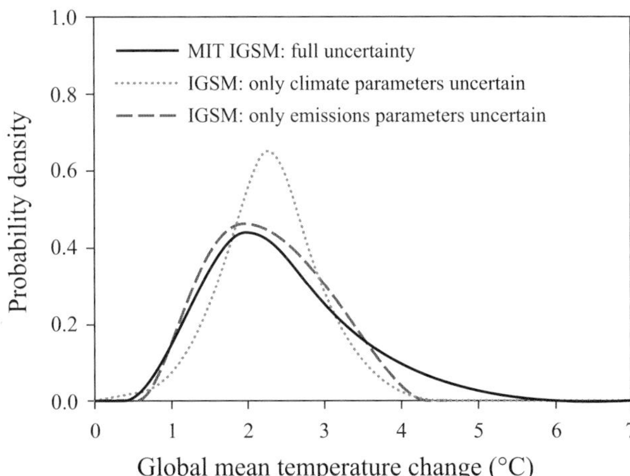

Figure 1.10 Probability distribution of global mean surface temperature change, 1990–2100, with all uncertain parameters (solid), only climate model parameters uncertain and emissions fixed (dotted), and only emissions uncertain with climate model parameters fixed (dashed). IGSM: Integrated Global System Model, a model with intermediate complexity for modeling global circulation coupled with an ocean circulation model, and supplemented with necessary emission models. M. Webster, C. Forest, J. Reilly, et al. "Uncertainty analysis of climate change and policy response," Global Change Science Policy Report 95, MIT, 2002.

Global climate or circulation models (GCMs) are complex computer models used to estimate the change in Earth's temperature and CO_2 concentration, among other things. These models are used to explore how different CO_2 emissions scenarios, other sources and sinks, and solar radiation could affect the atmosphere. Parametric studies are then used to construct ensemble probabilities for the different outcomes. An example of the results of such modeling is shown in Figure 1.10 in terms of the probability density function of the predicted temperature change with some sources of uncertainty. A rise of 2–3 °C is most probable by the end of the century.

Climate sensitivity, or the incremental change in the global mean climatological temperature resulting from the doubling of atmospheric CO_2 concentration, is still being debated, but most models estimate a range of 1.5–4.5 °C [6]. Cloud feedback is the largest source of uncertainty, with aerosols, non-CO_2 greenhouse gases, internal variability in the climate system, and land-use change being significant sources of uncertainty. Uncertainty in aerosols' radiative forcing remains large. Another source of uncertainty is the rate of heat diffusion into the deep oceans, given the sensitivity of the predictions to how much energy will be absorbed by this massive heat sink. Most predictions focus on CO_2-induced climate change, as CO_2 is the dominant source of change in Earth's radiative forcing in all the Intergovernmental Panel on Climate Change (IPCC) scenarios.[10]

[10] The IS92a scenario of the IPCC assumed that the carbon intensity of the source – that is, the carbon to energy (C/E) ratio of the fuel mix – would continue to drop monotonically well into the twenty-first century, reaching that of NG by 2030 but moving even lower as low- or zero-carbon sources, including nuclear and renewable sources, were introduced. In the *Second Assessment Report* in 1996, the IPCC predicted that CO_2 concentration of 360 ppm would rise to 750 ppm if annual emissions of near 24 $GtCO_2$ rose to more than 70 $GtCO_2$ by the end of the century, following the projected rise in energy consumption using sources used at that time. The corresponding global temperature, according to most climate models, would rise by 0.8–3.5 K by 2100. The same calculation showed that the annual CO_2 emissions would have to be limited to ~26 $GtCO_2$ (not far from the year 2000 levels) if CO_2 was to be stabilized close to 500 ppm by mid-century and stay fixed from there on. Carbon dioxide concentration and the

1.3.3 Impact

Temperature rise in colder climates may extend the growth seasons of some plants, especially in northern latitudes, but warmer temperatures may also support the spread and multiplication of pests that destroy crops. Dry seasons may become longer and droughts more frequent in areas already known for their hot climate and desert topography. Some animal habitats may become endangered, especially in colder climates. Melting of glaciers and icecaps would make more land available for agriculture where it is currently not possible, but water runoffs and heavier rainfalls would damage the soil. Several of these changes, especially drought and desertification, would impact poorer countries more severely, where adaptability and adjustment to substantial changes are less likely to be successful.

The impact of warmer temperatures on energy consumption is unclear. More cooling and air conditioning will become necessary during longer, warmer days, but less heat would also be required during the shorter, cool seasons. On the other hand, several major trends that could have strong impacts on life on Earth have been suggested with reasonable confidence, including sea-level rise, change of ocean acidity, and increased violent weather phenomena.

1.3.3.1 Sea-Level Rise

Sea level will rise because of the melting of polar icecaps, receding glaciers and the thermal expansion of the ocean surface waters. Records of different geologic periods confirm that the rise and fall of Earth's near-surface temperature is associated with the same trend. It is estimated that during the twentieth century, ~20 cm rise was observed over nineteenth-century levels. Interestingly, the melting of glaciers and icecaps may contribute the least to the rising sea level because of the balancing effect of increased evaporation. Most of the impact would result from the warming of the surface layer of the ocean waters and the resulting volumetric expansion. Combined, it has been estimated that by the end of the century, with 1–2 °C rise in temperature, a 30–50 cm rise in sea level should be expected [7]. This will have devastating impacts on coastal areas, especially agricultural land in the southern USA, India, Bangladesh, and Egypt.[11]

1.3.3.2 Change of Ocean Acidity

Carbon dioxide absorption in the ocean lowers its pH levels (making it less alkaline) and impacts aquatic life. Current average ocean water pH is 8.2. It is estimated that the rise in atmospheric CO_2 has already lowered the pH by 0.1 from pre-industrial levels. Ocean

temperature rise were revised upwards to 970 ppm and 1.4–5.8 °C, respectively, in the *Third Assessment Report* of 2001, with a 90% probability interval of 1.7–4.9 °C. Later models incorporated revised emission schedules and more accurate submodels for climate feedback, the radiative forcing of certain gases, and more accurate representation of atmosphere–ocean coupling.

[11] For more detail, see www.cresis.ku.edu/research/data/sea_level_rise/index_html.

circulation models predict a pH reduction of 0.7 units over the coming centuries if the current rise in CO_2 continues according to the business-as-usual scenario, and until fossil fuels are exhausted (leading to more than 1900 ppm in the atmosphere by 2300). Carbon dioxide solubility in water increases at lower temperatures and higher pressures, and hence CO_2-related acidity rise could increase at deeper water levels, affecting acidity-sensitive corals, in addition to the negative impact of rising water temperature alone. Calcareous plankton and other organisms whose skeletons or shells contain calcium carbonate may be endangered sooner [8].

1.3.3.3 Changes in Weather Phenomena

With warmer temperatures, on average, a more temperate climate will extend to higher latitudes, and extended periods of rainfall may occur due in part to the higher water concentrations in the warmer atmosphere. Hurricanes and typhoons, spawned by waters warmer than 27 °C within a band of 5–20 degrees north and south latitude, may occur more frequently. Ocean currents, such as the Gulf Stream and the Equatorial currents, driven by surface winds and density gradients, could also become more frequent and violent. Some of these currents can be accompanied by phenomena that cause strong weather perturbations. For instance, El Niño, which arises because of westward wind-driven surface water currents from the South American coast and sets up ocean circulation in which upwelling of colder water replaces the surface warmer waters, is known to increase the frequency of hurricanes and heavy storms.

1.4 CO_2 Emission by Fuel and Sector

Carbon dioxide concentration has grown from 280 ppm around 1850 to more than 400 ppm. Emission by fuel are shown in Figure 1.11. Oil produces ~40% of the primary energy because of its role in transportation, and is the largest source of CO_2. Coal, with a higher carbon content but lower overall share of energy consumption, produces nearly the same amount. Natural gas, with the lowest carbon content, contributes the least.

Worldwide CO_2 emissions from different sectors of the economy are shown in Figure 1.12. Electricity production has been, and a major source of CO_2, contributing 26% of total CO_2 emissions globally. It has been estimated that this number will grow to 43% by 2035 (from 30.2 billion metric tons in 2008 to 43.2 billion metric tons in 2035), with most of the increase contributed by developing economies (non-OECD countries), where most electricity production relies on coal. Sector-wise, electricity generation is followed by transportation, with industrial and residential following at lower rates. Electricity production plants use coal extensively, although the use of natural gas has been rising, alongside nuclear, hydro, geothermal, and more recently wind and solar. Other stationary sources of CO_2 include cement plants, oil refineries, iron and steel industries. As electricity-based transportation grows, so will emissions from power plants.

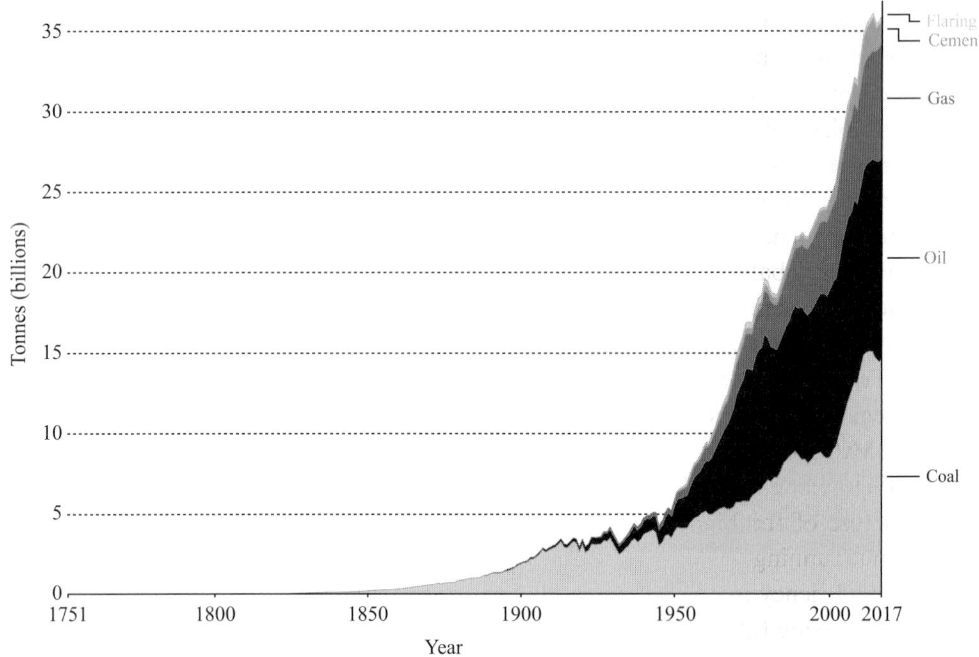

Figure 1.11 Worldwide emission by fuel, 1751–2017 (tonnes per year). H. Ritchie and M. Roser, "CO_2 and greenhouse gas emissions," 2002 (https://ourworldindata.org/co2-and-other-greenhouse-gas-emissions).

Figure 1.12 Global greenhouse gas emissions data (www.epa.gov/sites/production/files/2016-05/global_emissions_sector_2015.png).

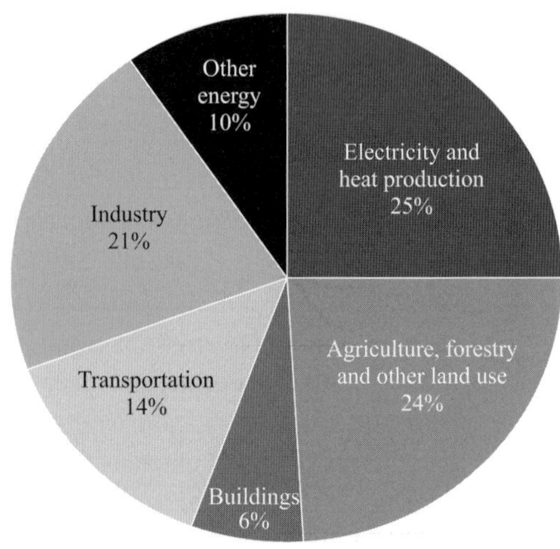

Estimates of CO_2 emissions must rely on careful assessments. For instance, an electric vehicle that charges its batteries from the grid is not a zero-emission vehicle, since grid power is currently mostly fossil-fuel based. Similarly, fuels produced from biomass, such as corn ethanol, cannot be considered as CO_2-neutral since growing corn and production of ethanol consume fossil fuels and hence emit CO_2. Such analysis is known as life-cycle analysis (LCA), and accounts for the contribution of emissions starting with the fuel source, refining, and conversion to useful energy. The methodology is based on conducting careful material and energy balances over a chain of events, and applies methodologies of process analysis in which a system boundary must be defined to assess the relationship between the input and output of the system.

To put some of the emissions numbers in perspective, a power plant rated at 500 MWe (megawatt electricity), running for 8000 h/y while burning coal, produces on average 4.8 $MtCO_2$/y (or 1.3 MtC/y). On average, coal-fired plants produce almost 1200 kg-CO_2/MWh, while natural gas-fired plants could produce as little as 400 kg-CO_2/MWh. This is because of the higher content of carbon in coal, and the higher efficiency of natural gas plants running on combined cycle rather than coal steam plant. Moreover, coal power plants lose efficiency in flue gas clean-up (to remove sulfur oxides, nitric oxides, and ash).

Predicting future trends in CO_2 emissions and the environmental impact must consider rise in energy consumption patterns worldwide. Carbon dioxide production in developing countries has exceeded that of developed countries because of their expanding economies and growing populations, and the fuels available domestically (Figure 1.13). China's dependence on coal for electricity generation, heating, and industrial production, and the production of liquid fuels, is growing rapidly. Recently it surpassed USA in total CO_2 production.

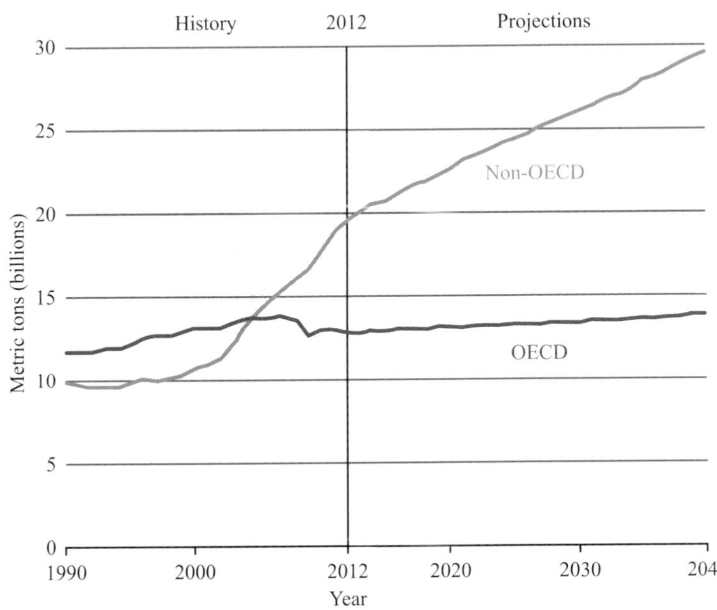

Figure 1.13 Worldwide energy-related CO_2 emissions in billion metric tons (2015). It is estimated that by 2040 non-OECD countries will contribute 69% of total CO_2 emissions (www.eia.gov/outlooks/ieo/pdf/emissions.pdf).

1.5 International Agreements

Scenarios have been proposed for stabilizing atmospheric CO_2 at the 450 ppm (or 550 ppm) level, arguing that the associated temperature rise would be acceptable. These scenarios rely on improving energy conversion and utilization efficiency, and using low- and zero-CO_2 sources [9]. Given the scale and scope of the problem, it is unlikely that a single approach can work, and multiple approaches must be considered in parallel, constrained by social and economic considerations. Action must be global to be effective, and must be sustainable over many decades.

Several actions have been suggested. Worth mentioning here is the United Nations Framework Convention on Climate Change (UNFCCC), which was signed in 1992, whose ultimate objective was to achieve stabilization of greenhouse gas concentration in the atmosphere at a level that would prevent dangerous anthropogenic interference with the climate system. Such a level should be achieved in such a time frame sufficient to allow ecosystems to adapt naturally to climate change, to ensure that food production is not threatened and to enable economic development to proceed in a sustainable manner. [10]

Several years later and following intensive debates and deliberations in the UN Conference on Climate Change, the Kyoto Agreement was proposed in 1997, which called for the reduction of CO_2 emissions to levels 5.2% below the 1990 level by 2008–2012, with most of the burden placed on developing countries to reduce their emissions according to a preset target for each country. The USA never ratified the agreement, and even countries that ratified the agreement did not enforce it.

The Cancún Agreement of 2010 was proposed during the UN Conference on Climate Change, and included both developed and developing countries. While recognizing that climate change represents an urgent and potentially irreversible threat to humans and the planet, the agreement included voluntary emission-reduction targets for 2020.

The Paris Agreement, or COP-21, was announced in 2015, promising the participation of all countries of the world, developed and developing, in efforts to decarbonize their economies. COP-21 aims to achieve peak emissions in the near future, reaching zero emissions by the second half of this century, with a target of less than 2 °C rise above the pre-industrial era temperature while pursuing the goal of less than 1.5 °C. While overall targets were announced, individual countries are given the freedom to develop their own plans on how to reach, collectively, these goals.

Figure 1.14 shows the growth in primary energy demand according to different scenarios, including current policies keeping things unchanged, a new policy scenario in which some measures are taken to curtail emissions but allowing continuing increase in emissions, and the 450 ppm scenario in which emissions peaks soon and start to fall rapidly soon after.

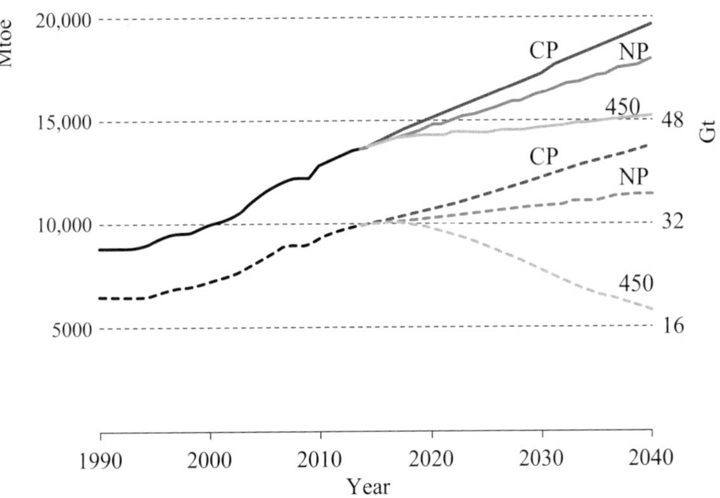

Figure 1.14 Primary energy demand and associated CO_2 emissions according to a number of policy scenarios: the current policies (CP) scenario, basically business as usual; the new policies (NP) scenario, which takes into account policies and implementation measures affecting energy markets that had been adopted as of mid-2015 (as well as the energy-related components of climate pledges in the run-up to COP-21); and the 450 ppm scenario, which depicts a pathway to the 2 °C climate goal that can be achieved by fostering technologies that are close to becoming available at commercial scales, and achieve a maximum of 450 ppm CO_2 concentration. Energy shown in continuous lines and energy-related emissions shown in broken lines. Primary energy is shown in millions of oil equivalent, and CO_2 in gigatons. IEA World Energy Outlook 2015, p. 55.

1.6 CO_2 Emissions Mitigation

While the fraction of renewable energy sources has risen significantly recently, the current energy infrastructure remains heavily dependent on fossil fuels, a trend that is unlikely to change for several decades. Approaches to reducing CO_2 emissions include:

1. improving conversion efficiency, such as for power plants, vehicle engines, and light bulbs;
2. improving demand-side efficiency, such as via building insulation, natural heating and cooling, public transportation, higher-efficiency appliances;
3. switching from coal to natural gas or other low C/H fuels, expanding the use of nuclear energy and renewables including solar, geothermal, biomass, and wind; and
4. CO_2 capture, utilization, and sequestration (CCUS) from power plants and industrial sources.

Given the scale of the issue, a portfolio of technological solutions is needed.

Figure 1.15 shows a scenario developed by the IEA that incorporates these solutions while assigning percentage contributions to different approaches. Each country must implement

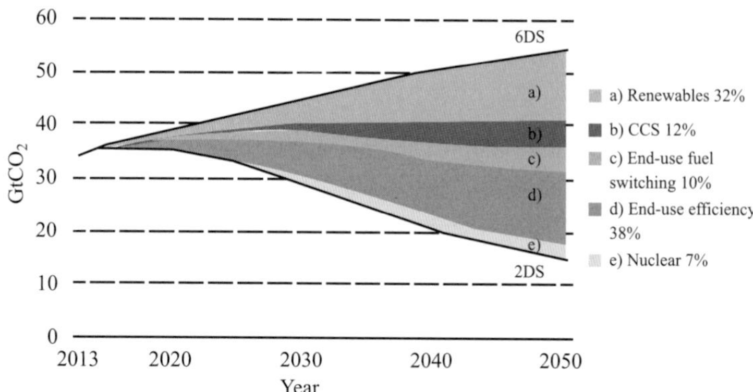

Figure 1.15 Reduction in global emissions using a portfolio of solutions including, in descending order from the top: renewable energy, CCS, end-use fuel switching, end-use efficiency improvement, and nuclear energy. T. Bryant, K. Burnard, P. Cazzola, et al. "Energy technology perspectives 2016: towards sustainable urban energy systems," technical report, International Energy Agency, 2016.

multiple solutions in parallel, depending on its resources, climate, and needs. The choice of solutions, and the percentage contribution of each approach, is likely to depend on the geographic location, including the cost of improved efficiency, available fuels, local renewables, and public acceptance (of nuclear energy in particular).

Given the existing expensive infrastructure, and the economic, social, political, and security concerns, gradual transition toward more efficient and less carbon-intensive energy options is more likely than sudden changes. Improving the efficiency of energy conversion and utilization is promising. Here, conversion refers to the production of useful forms of energy, such as thermal, mechanical, or electrical energy, from its original form (e.g., fuel). Utilization efficiency refers to how efficiently the final product is used, such as in insulation of heated spaces and reduction of aerodynamic and other forms of drag resistance in vehicles. Using thermodynamic terminology, efficiency-related improvements should reduce waste and hence entropy generation.

1.6.1 Fuels and Efficiency

Electricity generation is currently the largest CO_2 emitter because of the extensive use of coal, a trend that is likely to continue given its availability and low price. Developing countries such as China and India are rich in coal resources and are likely to exploit their natural resources to meet growing energy needs. While the efficiency of coal plants has been rising because of the implementation of supercritical cycles, and regulated emissions from these plants such as NO_x, SO_x, and particulates have fallen significantly, CO_2 emissions per unit of energy production from coal plants is the highest among all fuels. Further improvements are possible by, for example, using ultra-supercritical cycles and integrated gasification combined cycles (IGCC) with efficient gas turbines. Existing IGCC plants that do not incorporate a fuel cell have efficiencies of ~45%. With gasification, it is possible to incorporate a solid oxide fuel cell as a topping cycle for further improvements. Using a gasifier, solid and carbonaceous fuels are converted into syngas, a mixture of carbon monoxide and hydrogen. The syngas is cleaned to remove acidic and other contaminants and used in a high-temperature fuel cell, coupled with a combined cycle, with predicted efficiencies higher than 50%. See Chapter 12 for coal use.

The use of natural gas in electricity production has expanded significantly over the past two decades, in some cases replacing older, less efficient coal plants. Natural gas-fueled power plants have significant advantages because:

- they have higher efficiency, with combined cycles that can reach 60%;
- natural gas is easy to transport in pipelines;
- they produces less CO_2 per unit of energy due to the gas' higher hydrogen content; and
- gas is a clean-burning fuel, producing lower NO and CO.

Natural gas is considered an ideal "transition" fuel. Natural gas power plants can load follow – that is, respond to changes in load – and therefore act as effective backup for intermittent renewables. They can reach higher than 60% efficiency, which has proposed by incorporating a solid oxide fuel cell as a "topping cycle."

Distributed generation of electricity, where smaller units are installed to provide power at the level of an individual building up to a district, enables combined heat-and-power applications in which thermal energy in the exhaust of engines or turbines is used for heating. The introduction of more efficient direct-conversion devices such as low- and high-temperature fuel cells can improve the overall system efficiency further. However, low-temperature fuel cells use hydrogen, and unless it is generated efficiently and with low emissions, the advantage can be lost. Solid oxide cells can use methane or syngas, and can integrate well with power cycles. See Chapters 5 and 6 for discussion of efficient natural gas cycles.

1.6.2 CO_2 Capture Approaches

Efficiency improvement has limitations, and the next step is carbon (dioxide) capture and storage, or CCS, from fossil-fuel power plants, as well as other industrial processes using hydrocarbon fuels. While reducing the plant efficiency and raising its cost, it is possible to achieve zero emissions when using coal or gas in such plants. Maximizing the plant efficiency counters the efficiency penalty associated with CO_2 separation, but to a limited extent. Three general approaches have been proposed for power cycles with CCS, as shown in Figure 1.16: post-combustion capture, pre-combustion capture, and oxy-fuel combustion.

The first option is relatively simple to implement and can be retrofitted in existing power plants, although new designs should perform better. The second and third options need some special equipment in the power island, including a CO_2 turbine for oxy-fuel combustion and H_2 turbine when pre-combustion capture is used. In the case of coal, pre-combustion capture starts with gasification in an IGCC plant. In the case of natural gas, reforming is the first step. The efficiency penalty of CO_2 capture depends on the fuel, and the optimal design may not be the same for coal and natural gas. See also Chapters 10, 11 and 13.

Several underground storage sites for CO_2 have been proposed, including depleted oil and gas reservoirs, geological formations such as deep saline aquifers, coal seams, and solid mineral carbonates. Injection of CO_2 into oil wells is currently practiced at a small scale for enhanced oil recovery. Several projects/experiments have been conducted to test this concept (with total capacity of ~30 $MtCO_2/y$). A coal gasification power plant in North Dakota

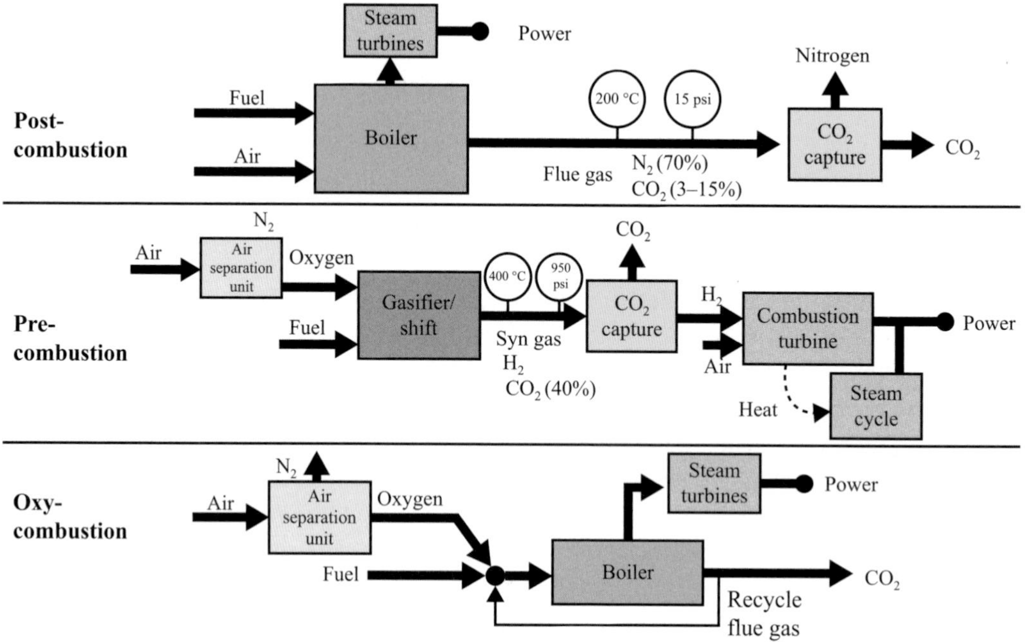

Figure 1.16 Different approaches to CO_2 capture from power plants, including post-combustion capture, oxyfuel combustion, and pre-combustion capture. The fuel can be solid, such as coal, or gaseous, such as natural gas. In the latter case, the gasifier in pre-combustion is a reformer.

injects 1.5 $MtCO_2$/y into the Weyburn field in Canada for enhanced oil recovery (EOR). Another project, where CO_2 is being separated from the outflow of a natural gas well and injected back deep underground, is the Sleipner field in Norway. This project targets less than 1 $MtCO_2$/y. Carbon dioxide injection is also used for enhanced methane recovery from coal beds in the Juan basin in the USA, and several more fields in Canada, where acid gas ($H_2S + CO_2$) is injected to recover sour natural gas. The IPCC estimates that the total cumulative emissions up to the year 2100 using the business-as-usual global energy consumption scenario (IS92a) is 1500 GtC. The carbon content of the remaining exploitable fossil fuels, excluding methane hydrates, is estimated to be 5000–7000 GtC. Sequestration in the form of solid carbonate minerals has also been proposed. One approach is to use the exothermic reaction between forsterite (Mg_2SiO_4) and CO_2, which is favored under ambient conditions, to form $MgCO_3$ (serpentine $Mg_3Si_2O_5(OH)_4$). Biological sequestration using reforestation and growing certain types of algae is another option.

1.6.2.1 Post-combustion Capture

Depending on the fuel used, CO_2 concentration in the products can be as low as 3% for lean-burning natural gas and much higher for coal plants. Gas separation processes are energy-intensive, and their integration into the power generation cycle lowers its overall efficiency. Chemical separation processes utilize a solvent to remove CO_2 from the exhaust gas stream of a conventional power plant, while thermal energy is needed to regenerate the solvent. Aqueous solutions of monoethanolamine have been used for this purpose. Carbon dioxide capture from coal plants using post-combustion capture could reduce their efficiency by 8–16 percentage points. In natural gas combined cycle (NGCC), the efficiency loss is less at 5–10 percentage points. Carbon dioxide compression for transport and storage add 2–4 more percentage points of efficiency loss. Ongoing research on different absorbents and other advanced separation techniques may reduce the efficiency penalty. In ideal separation work, expressed in terms of fuel chemical energy, the efficiency penalty is 2–3 percentage points. The technology required is broadly considered as conventional and "off-the-shelf." See Chapters 11 and 13 for related topics.

1.6.2.2 Oxy-fuel Combustion

In oxy-fuel combustion schemes, an air separation unit (ASU) is used to produce the pure oxygen needed in the combustion process. The products of oxy-fuel combustion – water and carbon dioxide – can be used in a steam-generation unit in a coal plant, or as the working fluid in a gas turbine plant. Following the power island, water can be condensed and carbon dioxide captured directly. Some CO_2 recycling is needed to moderate the combustion temperature. Large-scale distillation units are used for air separation, but membrane separation units are being considered. For coal-fired plants, estimates of the efficiency penalty are 5–7 percentage points for the ASU and 4 percentage points due to the recycling of the CO_2. Values for a net efficiency of 28–34% for optimized steam cycles with oxy-fuel combustion have been reported (without the CO_2 liquefaction energy). Natural gas-fired plants can reach 40–50% efficiency with CO_2 capture. The estimated reduction in efficiency for natural gas is 6–9 percentage points. Well-integrated combined cycles with membrane air separation, chemical recuperation, or chemical looping combustion using metal oxides have been proposed for natural gas plants, with efficiencies exceeding 50%, with CO_2 capture.

1.6.2.3 Pre-combustion Capture

In this approach, reforming or partial oxidation of the fuel to syngas followed by a gas–water shift reaction produces a mixture of CO_2 and H_2. Carbon dioxide is then separated and pure hydrogen is burned in air in the combustion chamber of the gas turbine. This approach reduces the load on the ASU since oxygen is required only for the partial oxidation. Moreover, gas clean-up in the case of coal is performed on smaller gas volumes. Because of the large difference in molecular weight between hydrogen and CO_2, gas separation can be performed using more efficient technologies, including membranes or physical absorption or adsorption. Special gas turbines that tolerate high moisture content in the working fluid are needed in pre-combustion capture plants. Estimated reduction in efficiency for coal (syngas) and natural gas are 7–13 and 4–11 percentage points, respectively. Gas separation technologies are introduced in Chapter 10.

1.6.2.4 Synthetic Fuel Production

Similar decarbonization[12] concepts can be applied to synthetic low-carbon fuel production, including hydrogen, from coal or other heavy hydrocarbons. Pre-combustion capture lends itself well to this application, since many synthetic fuel production processes start with synthetic gas, a mixture of hydrogen and carbon monoxide, using reforming or gasification. Following the clean-up of the synthetic gas, catalytic processes are used to combine the components of the gas at different ratios to produce hydrocarbons. In the case of hydrogen, carbon monoxide is reacted with water to form more hydrogen, which is then separated from CO_2. Carbon dioxide produced during reforming can be separated from the synthetic gas stream for underground storage. Higher pressures used in large-scale gasification and natural gas reforming are compatible with physical adsorption or membranes for H_2 separation from the shifted gas. The topic is covered in Chapters 3, 4 and 12.

Hydrogen can also be produced by water electrolysis without CO_2 emission if the source of electricity is carbon-free, or directly from high-temperature heat (~850 °C) using thermo-chemical cycles. Poly-generation plants can be optimized to maximize conversion efficiency and to deliver different products as needed.

1.6.3 Zero-Carbon Energy

Zero-carbon energy sources include nuclear energy and renewable sources such as hydro, geothermal, wind, solar, and biomass.[13] Nuclear energy is a scalable source that can supply a fraction of future energy needs. Concerns over waste management, proliferation, and the public perception of safety should be addressed before substantial expansion of nuclear power can be expected. Hydropower, which currently contributes a significant fraction of renewable electricity, has been widely exploited and has its own share of environmental concerns. Other sources of renewable energy have much lower energy and power density, and are characterized by varying degrees of intermittency (Chapter 9).[14] Biomass is used extensively in rural communities in developing countries, and for production of liquid transportation fuels, but the potential of biomass energy is limited by land and water resources. The most significant renewable sources are wind and solar energy, and geothermal sources. A source that is not necessarily renewable but could be considered non-exhaustible is waste. See Chapter 14 for more on biomass.

[12] "Decarbonization" and "carbon management" have become synonymous with the process of reducing CO_2 emissions.

[13] Other forms include ocean tidal waves and ocean thermal energy, which have not made much impact on energy resources yet. All forms of renewable energy originate in solar energy, except for geothermal energy (the original hot gases that formed the Earth) and ocean tidal waves (gravitational). It should be noted that the notion of zero-carbon power is relative, and for some forms, such as biomass, fossil fuels are still used in their production.

[14] Typically, fossil-fuel power flows though components in power and propulsion applications at ~100 kW/m^2, or higher for high-speed propulsion. Renewable sources have energy density flow rates 3–4 orders of magnitude lower, depending on the energy form. For instance, the average (total) solar power reaching Earth's surface is, on average, O (300 W/m^2).

Accelerated deployment of renewable energy systems can be achieved by: (1) improving their conversion efficiency; (2) reducing their cost; (3) raising the monetary incentives for those who wish to adopt renewable energy; and (4) significantly improving energy storage.

1.6.3.1 Nuclear Energy

Nuclear energy currently provides ~20% of the electricity needs of the USA, and more than 85% of that of France. Worldwide, it is estimated that nuclear energy supplies 6.4% of the primary energy (2.1% in the form of electricity), which amounts to nearly 17% of electricity supplies. Nuclear power plants, totaling ~500 worldwide, use uranium-235 (U-235), which is produced by enriching natural uranium. Light-water reactors, both the pressurized and boiling water types, represent the majority of current nuclear reactors, but some plants use gas-cooled graphite reactors. Progress has been made in designing passively safe reactors that reduce the chances of accidents, but current systems have yet to incorporate these designs at a large scale.

Current estimates for the ground-based reserves and ultimately recoverable resources of U-235 translate to 60–300 TW-year of primary power.[15] More uranium can be recovered from seawater, but extraction at large scales has not been attempted. Plutonium-239 is produced during the uranium reactions in power reactors, and can be separated from the spent fuel rods for use in sustained nuclear reaction for power generation, or for nuclear weapons. For this reason, reprocessing for spent fuels is currently banned in the USA and most other countries. Fast-breeder reactors, such as liquid-metal-cooled reactors, can be used to produce plutonium-239 and another fissile isotopes, such as thorium-233.

Fusion has been considered a promising technology that does not produce radioactive waste and is less prone to accidents, but efforts to achieve sustained power generation have evolved slowly, and it remains challenging. In fusion reactions, deuterium, which is abundant, reacts with itself, tritium, or helium to form helium. However, producing more energy from fusion reactions than that consumed to initiate them has been very difficult. Efforts to use Tokomak magnetic confinement of plasma to induce the fusion reaction, or high-powered strongly focused lasers to provide the energy for ignition, are underway.

1.6.3.2 Hydraulic Power

Currently an important source of renewable energy is hydraulic power plants built at natural waterfalls or behind river dams. There is ~ 0.7 TW capacity installed worldwide. Expansion possibilities are limited, and the 18 GW Three-Gorges Dam in China was one of the last large-scale projects. Overall, hydropower, when nearly fully utilized, is not expected to exceed 0.9 TW. The capacity might decrease if climate change leads to different rainfall patterns. Hydropower is seasonal, but large dams reduce seasonal fluctuations using high-

[15] If all current energy needs were to be met using nuclear fission energy using available uranium, these estimates would translate to a 5- to 25-year supply.

capacity reservoirs to regulate the flow of water into the power plants. Hydropower is not without negative ecological impacts, and the large reservoirs created behind dams can affect local ecosystems. Downstream of a dam, soil can become less fertile as silt that used to replenish its nutrients is no longer able to flow. River fish populations can be negatively impacted, and some dams have been recommended for removal to revive fish habitats. Large projects often lead to displacement of people.

1.6.3.3 Geothermal Energy

Geothermal energy is a scalable renewable source that relies on drilling deep wells in areas where ground sources of hot fluids are available, and thermal-electric power plants can take advantage of low-temperature heat. The efficiency of these plants is relatively low because of the low temperature of the reservoirs. Current installed capacity has grown from ~10 GW electricity (GWe) around 2010 to ~30 GWe recently (Figure 1.17). Most wells have a relatively small lifetime, five years on average, and new wells must be drilled to continue a plant's operation. The potential capacity of geothermal energy is large, ~10 TW worldwide. To reach its full potential, deeper wells reaching down 5–10 km will be needed. Drilling deeper wells allows for higher-temperature heat sources and higher efficiency, but is also more expensive. Shallow sources of geothermal energy have also been used for distributed heating and cooling systems. Newer concepts, called "heat mining" or "enhanced

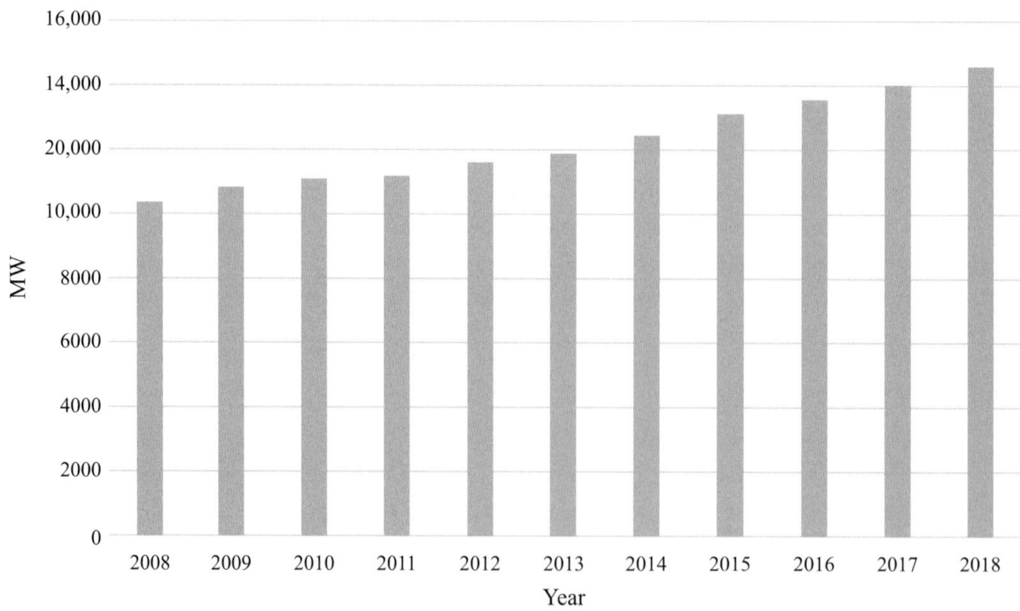

Figure 1.17 Cumulative installed geothermal plant capacity in GW (www.bp.com/content/dam/bp/business-sites/en/global/corporate/pdfs/energy-economics/statistical-review/bp-stats-review-2019-renewable-energy.pdf).

geothermal systems," rely on drilling deep wells and fractioning the hot rock at the well bottom. Fluids are then circulated between the power plant and the fractured rock to absorb the thermal energy and bring it to the surface. Concepts for hybridizing geothermal energy with fossil fuels or with solar energy are being considered to improve overall plant efficiency and extend plant lifetime. See Chapter 9 for more on geothermal energy.

1.6.3.4 Wind Energy

Although wind and solar contributions to total energy production currently represent a small fraction of the total, both have grown over the past decade at ~25–30% per year on average, and indications are that this trend will continue for some time. Figure 1.18 depicts the cumulative installed capacity in wind power, showing more than one order of magnitude increase since 2001. As the price of wind electricity has fallen, the technology has been adopted more widely. Larger turbines are being produced and installed, leading to further drops in price. Doubling the per-turbine capacity, actively controlled blade pitch for variable wind speed, and the use of sensors and actuators to protect against wind gusts and violent storms are significant technology innovations. Wind turbines with 5–10 MW capacity have been built and are favored especially in offshore installations where the wind is stronger and

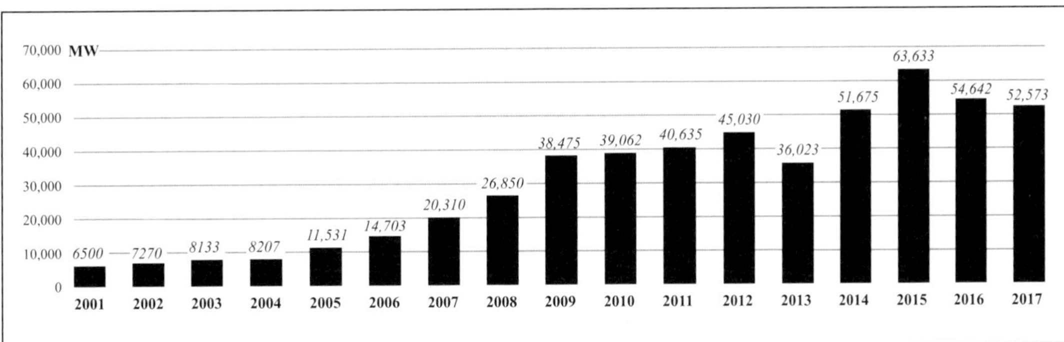

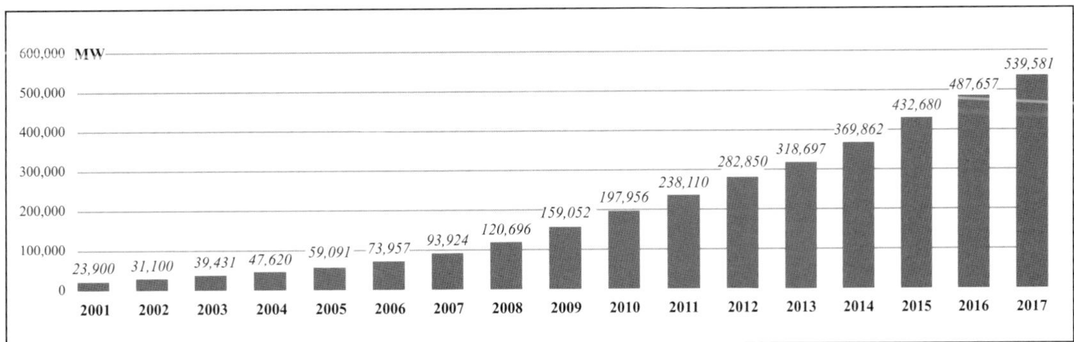

Figure 1.18 Global annual (a) and cumulative installed (b) wind capacity in GW (https://gwec.net/wp-content/uploads/vip/GWEC_PRstats2017_EN-003_FINAL.pdf).

less intermittent, and the impact on the local environment is minimized. The spread of offshore technology is enabled by progress in installation and maintenance technology. Depending on the turbine size and the extent of the wind farm, wind energy technology offers solutions for remote, off-grid applications, distributed power applications, and grid-connected central generation facilities. Integration into the grid, and into "micro-grids," with other sources of conventional and renewable energy generation is expanding as storage technologies evolve.

Total potential wind capacity that can be utilized practically is believed to exceed 10 TW worldwide, including offshore locations.

1.6.3.5 Solar Energy

Growth in solar energy utilization has been very robust recently (Figure 1.19). Solar thermal energy and solar electric conversion for heat and power applications are important for distributed utilization and centralized energy production. In solar thermal-electric conversion, intermediate-temperature heat (300–600 °C) collected and concentrated using troughs or towers is used to power steam cycles in plants ranging in capacity from 20 to 400 MW. The overall efficiency of concentrated solar power (CSP) is 20–30%, and some of the thermal energy can be stored for use after sunset. Hybrid solar–fossil operation can be beneficial in making the solar plant operable under cloudy conditions and at night, without the need for

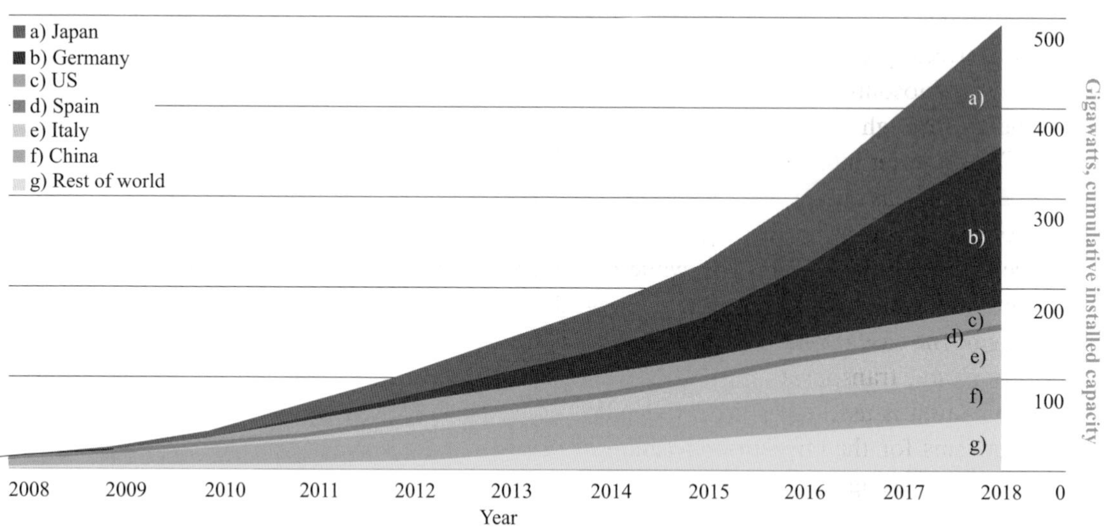

Figure 1.19 Global installed PV capacity in GW, shown for the following countries in descending order from the top: Japan, Germany, the USA, Spain, Italy, China, and the rest of the world (www.bp.com/content/dam/bp/business-sites/en/global/corporate/pdfs/energy-economics/statistical-review/bp-stats-review-2019-renewable-energy.pdf).

large-scale storage. Hybridization can also be used to retrofit existing fossil-fuel power plants by installing solar collectors in their surrounding space, if available.

Solar PV cells are solid-state devices that convert light to electricity directly, using semiconductors such as silicon, doped with small amounts of other elements that act as electron donors (n-type) or electron receptors (p-type). When two layers are joined, a potential difference is established when the cell is bombarded with photons of a particular wavelength (or energy exceeding the bandgap potential needed to move the electrons from the valence band to the conduction band). The freed electrons move from the donor to the receptor in an external circuit. The efficiency of silicon-type PV is 10–20%. Silicon-based PVs have been used extensively for distributed power applications, and more recently in central plants. Developments in nanostructured organic PV cells promise to lower the price further. Although these organic cells have lower efficiencies, they promise to be easier to fabricate, lighter in weight, and more adaptable. Efficiency improvements are pursued by blending the polymers with electron acceptors, while optimizing the cell to promote efficient excitation splitting and charge transport by reducing the bandgap so that a larger fraction of the solar spectrum can be absorbed. As in the case of wind, large-scale storage is necessary to overcome intermittency. Concentrated solar power is covered in Chapter 9.

1.6.3.6 Biomass

Biomass is the oldest and second largest source of renewable energy worldwide, following hydropower, contributing ~10% of primary energy production. Agricultural and silvicultural (trees) products and crops and their byproducts, as well as animal waste, have been used as biosources (unused animal parts have been used to produce fuels). Plants store energy through photosynthesis, converting radiation into chemical energy by combining CO_2 and water into carbohydrates, such as sugar, starch, and cellulose, in photon-energized reactions. This energy can be converted back to other forms through combustion (as most biomass is currently used), gasification, fermentation, or anaerobic digestion to produce fuels. During this process and while burning the biofuel, CO_2 is released, making biomass conversion carbon-neutral as long as no fossil fuels are used in its production. While this may be the case is rural economies, it is hardly the case when fossil fuels are usually used in agriculture, transportation, and conversion of biomass into biofuels [11], especially when corn starch is used to produce ethanol.[16] More recently, efforts have gone into producing organisms for the efficient conversion of cellulose, hemicellulose, and lignin into ethanol, hence increasing the yield beyond that attained from the grain alone. Thermochemical approaches, based on gasification and torrefaction, can use lower-value plant material and agri-waste in the production of biofuels. See Chapter 14 for more on biomass.

[16] Determining whether the yield is positive or negative requires complex calculations that start with the definition of the "system boundary" – that is, what is the input to the process producing the fuel – and whether, for instance, the chemical energy input to the production of the machinery used in agriculture should be included as input.

Similar to hydropower, bio-crops are not devoid of negative environmental impacts, such as the use of water, fertilizers, insecticides, and herbicides; soil erosion; and impacts on ecosystems, such as deforestation. Some agricultural products are used to reintroduce nutrients back into the soil, and if used in the production of biomass energy these will have to be replaced with synthetic fertilizers, or there is a threat of weakening the soil. Moreover, growing crops for the production of biofuels may have potential negative impacts on water resources and food supply. In general, the scalability of biomass is limited, given the low photosynthesis efficiency. Photosynthesis has power density less than 1 W/m^2 (in thermal power units), which is more than an order of magnitude lower than that of wind and solar power density (in electrical energy). Bio-energy produced from plant and agricultural waste is significantly more environmentally safe.

1.6.4 Storage

Expanding the use of renewables is very important in the effort to counter global warming, but renewables are intermittent – on the scale of hours to days for solar and wind; seasonal for biomass and hydro; and longer for geothermal. Without storage, backup power is needed for dispatchability – that is, for having access to continuous power as source availability is reduced and the load varies, adding to the cost and complexity.

Several options should be considered (Figure 1.20) [12]:

- high-capacity batteries for wind and photovoltaics, including lithium-ion and flow batteries – supercapacitors are more suitable for high power;
- thermal energy storage in molten salts or solids for concentrated solar thermal;
- high-capacity compressed-air storage containers or underground sites, include salt cavities, aquifers, and cavities with compensating surface reservoirs;
- high-capacity pumped-hydro storage facilities;
- large flywheels;
- electrolysis, to convert electricity to hydrogen and other fuels; and
- thermochemical production of fuels.

As an energy carrier, hydrogen can be used to generate electricity in polymer–electrolyte membrane (PEM) fuel cells. "Reversible" or two-way PEM fuel cell/electrolyzers have been designed for hydrogen production and utilization, thus reducing the hardware cost. The "roundtrip" efficiency of storage – that is, the overall efficiency from electricity (produced from solar or wind) back to electricity (produced in the fuel cell) – is rather low in these cells. The use of solar thermal-electric power plants simplifies short-term storage since these systems can store thermal energy in high heat capacity materials such as molten salt, which can be used later in running the same power plant. See Chapters 4 and 7 for more on fuel cells.

Large-scale higher-capacity storage options include pumped-hydro plants and compressed-air plants. Pressurized air can be stored in underground reservoirs in rock or salt cavities or in naturally contained porous aquifers. The system can be hybridized with fossil fuels – that is, fuel can be burned in the compressed air to raise its temperature and a

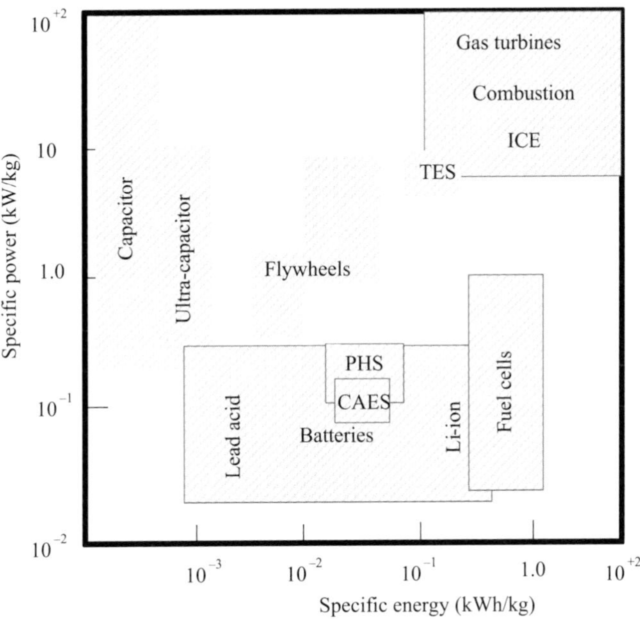

Figure 1.20 A schematic diagram showing the power and energy densities of some energy storage technologies, including batteries, supercapacitors, flywheels, and combustion engines. Data are orders of magnitude, and details inside the boxes are uncertain. Estimates change as technologies advance and alternative implementations are proposed. For instance, data for fuel cells depend on the operating temperature and fuel used. Other battery chemistries are still under development which can boost both energy and power density. PHS is pumped-hydro storage, whose energy density depends on the elevation of the storage tank. CAES is compressed-air energy storage, which is not yet used widely and hence data are scarce; it can also be hybridized with fuels such as hydrogen or natural gas. Estimates for combustion-based technologies, namely gas turbines and internal combustion engines (ICE), depend on the fuels used. Such diagrams are sometimes known as Ragone diagrams.

gas turbine can be used instead of the air turbine to produce more power. Pumped-hydro storage is used extensively because of its simplicity in places where natural or man-made large water reservoirs are available and where the natural topography assists. Some also consider biomass as an energy storage option (storing solar energy through photosynthesis in plant material). The potential for chemical storage (i.e., the production of fuels) is very high because of the large chemical energy density.

1.6.5 Transportation

Transportation consumes 27% of the total energy in the USA, and produces a proportional fraction of CO_2 emissions. While these fractions vary among countries, improved standards of living are likely to make transportation a significant cause of energy consumption and CO_2 emissions worldwide. The primary source of gasoline and diesel fuels, the most widely used transportation fuels, has been petroleum oil. Natural gas and propane have also been used for ground transportation on a small scale, and methane may find wider applications in this sector in the future. Chief among alternatives are biofuels such as ethanol and biodiesel. Decarbonizing the transportation sector is challenging because of the dependency on the "fueling" infrastructure.

Promising near-term solutions include: efficiency improvements achieved through efficient engines and transmissions, reducing vehicle weight, and improving aerodynamics; use of low-carbon fuels such as natural gas and biofuels; the use of nuclear- or renewable-generated hydrogen to fuel ICEs or fuel cells; and transitioning to hybrid and plug-in hybrids to raise gas mileage. Eventually pure electric cars, powered by low-carbon electricity, are the ultimate low-emission solution.

The efficiency of ICEs has improved steadily over the years. Vehicles that combine the advantages of gasoline engines and diesel engines, called homogeneous-charge-compression ignition (HCCI) or controlled ignition engines, achieve diesel engine efficiency with significantly fewer emissions. Other technologies have become widely available, such as: Stop and Start to reduce idle and part-load operation losses; variable valve timing (VVT), which controls the fraction of exhaust gas trapped in the cylinders to limit the power produced by the engine with reduced pumping loss; variable compression ratio, which makes it possible to increase efficiency at high power without the danger of knock; cylinder deactivation; supercharging; and turbocharging.

Optimized hybrid gas–electric powertrains in which the combustion engine is mated with one or more electric motor/generator and smaller batteries/supercapacitors further improve the efficiency, in some estimates by 30%. In the hybrid mode, the engine operates near its point of maximum efficiency most of the time, while the excess energy is stored, to be used when needed. Regenerative braking recovers and stores some of the braking energy using high-power-density storage devices such as supercapacitors. Several forms of hybridization are now widely available, including "mild hybrids" that use high-voltage batteries, parallel, series, and dual-mode hybrids, and plug-in hybrids that use larger batteries.

In the parallel hybrid, the wheels are driven by the engine and in parallel by a motor when more power is needed, which acts as a generator when the engine power exceeds what is needed. In series hybrids, the wheels are driven by an electric motor and electricity is supplied from a battery, or from a generator driven by a combustion engine. In dual-mode hybrids there are two motors; one drives the wheels directly and the other acts in parallel with the engine. This design takes advantage of the better features of both configurations, at the expense of more hardware complexity and an elaborate control system. Plug-in hybrids are equipped with larger and/or higher-energy-density batteries that can power the vehicle for longer distances on electricity drawn from the grid. Plug-in hybrids are considered as a transition technology to fully electric transportation.

A hybrid car with no engine is an electric vehicle equipped with a large high-performance battery pack, such as lithium-ion batteries, electric motors to drive the wheel, generators to take advantage of regen brakes, etc. The energy density of Li-ion batteries is close to twice that of its nearest competitor, the nickel–metal hydride battery, hence extending the driving range significantly. If using zero-CO_2 electricity, the electric car is a zero-emissions vehicle. Challenges are now mostly the price and charging infrastructure.

A low-temperature polymer exchange membrane fuel cell equipped car using hydrogen is another zero-emission electric vehicle, if the hydrogen is produced using zero-emission sources. Fuel cells are efficient direct-conversion devices. Wide deployment depends on the development of efficient, large-scale hydrogen production, distribution, and mobile storage technologies. Currently, hydrogen is produced mostly by methane-steam reforming without CCS. To make it emission-free, CCS must be introduced (using the technologies described earlier). Alternatively, we can use electrolysis to produce hydrogen from water splitting. In this case, electricity should be produced from nuclear energy or renewables.

1.6.6 Implementing Multiple Solutions

As shown in Figure 1.15, the key to achieving significant reduction in CO_2 emissions by mid-century is to implement multiple strategies in parallel. This concept was analyzed by Pacala and Socolow [13], where an approach was explored for achieving the goal of stabilizing atmospheric CO_2 concentration at the 550 ppm by 2050 (starting early in the twenty-first century). To reach this level, CO_2 emissions had to be held at 7 GtC/y (almost the same as the 1990 level) over 50 years. The business-as-usual rate of increase of 1.5% per year would have doubled the rate of emissions to 14 GtC/y in 50 years. A number of solutions were presented to achieve this reduction; each one would prevent the emission of 1 GtC/y by the end of the 50 years. For reference, 1 GtC/y is produced by a ~750 GW coal power plant at average efficiency of 34%, or a ~1500 GW natural gas power plant at an average efficiency of 46%. The overall strategy could be implemented using seven "wedges," each leading to the reduction of CO_2 emissions by 1 GtC/y by the end of the 50-year effort. Deploying seven solutions would lead to the desirable goal of stabilizing the CO_2 emission rate at 7 GtC/y. Each solution was expected to be deployed gradually, reaching full maturity in 50 years.

The solutions include improved conversion efficiency in power plants, improving efficiencies in building and in transportation vehicles, shifting to lower-carbon fuels, capturing CO_2 from fossil-fuel power plants and in hydrogen production plants that use fossil fuels, using nuclear energy for electricity and hydrogen production, and significantly expanding the use of renewable energy including wind, solar, geothermal, and biofuels. For instance, raising the fuel economy of 2B vehicles (light/medium trucks) from 30 to 60 mpg; reducing energy consumption of all buildings by 25%; doubling the efficiency of coal power plants to 60%; sequestering all the CO_2 emitted by 750 GW coal plants, etc. Scale matters: current rates of CO_2 injection underground, mostly for EOR, would have to be scaled up by two orders of magnitude to satisfy the needs for one wedge (that relies on CCS). For synthetic fuel production, the maximum capacity of some of the largest plants is ~165,000 bpd (barrel per day) of liquid fuel from coal. A wedge would require 200 Sasol-scale plants with CCS. The study also considered CO_2 sinks that can be expanded by reducing deforestation, or reforestation of clear-cut forests, especially in tropical areas. One half-wedge would be created by reforesting nearly 250 million hectares in the tropics or 400 million hectares in temperate zones (current areas of tropical and temperate forests are 1500 and 700 million hectares, respectively).

1.7 Conversion

We have summarized recent concerns regarding energy sources and consumption patterns, including the growing needs of a rising worldwide population that strives for better living standards and competes over resources, and the strong evidence that continuing use of fossil fuels without measures to reduce CO_2 emissions may lead to irreversible environmental damage resulting from global warming. The scale of energy consumption, with a positive time derivative, and the global nature of the problem are worth noting. A portfolio of approaches that adopt to local conditions must be implemented globally (Figure 1.21).

Carbon capture and sequestration from power plants, fuel production facilities, and other energy-intensive industries offer an opportunity to continue to use fossil fuels while mitigating their contribution to global warming. Given the supplies of fuels and their cheap prices, and the existing infrastructure, it is unlikely that a shift to alternatives will be sufficiently fast to avoid the predicted trends. A shift to low-carbon fuels such as natural gas and syngas (produced by coal gasification with partial CCS) are parts of an effective transition strategy.

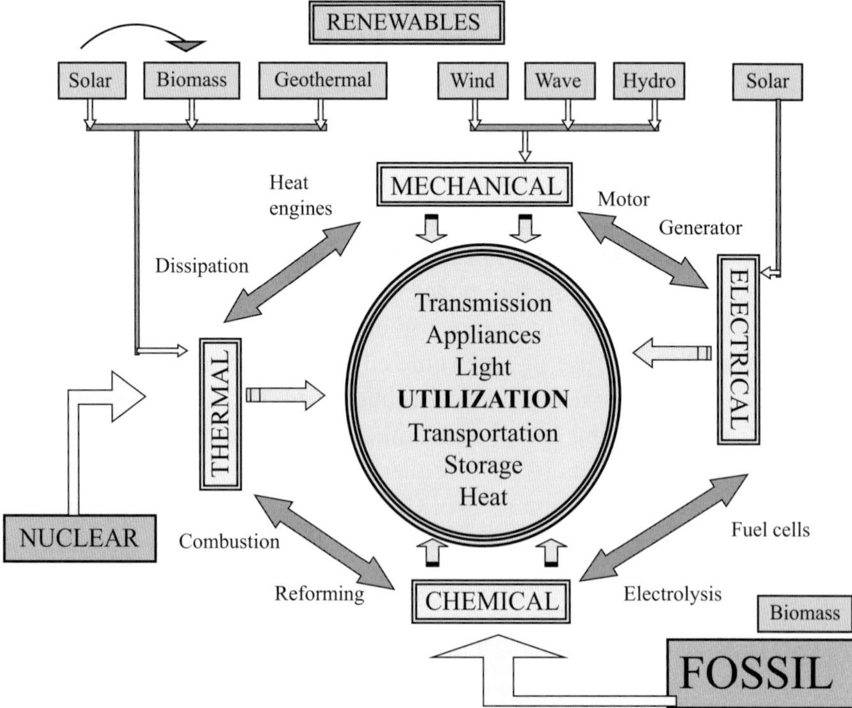

Figure 1.21 Energy sources (outer ring), useful forms, and conversion technologies.

Nuclear energy and renewable resources are necessary components. Nuclear power is a scalable resource that can satisfy a larger fraction of electricity-generation needs, but concerns over waste, safety, and security must be addressed. Biofuels contribute a small fraction of transportation fuel needs. Wind and solar are expanding rapidly, but large-scale storage and integration with conventional sources can lower costs and accelerate the transition.

Efficiency and biofuels will reduce CO_2 emissions from transportation. Hybridization and full electrification will eventually have a stronger impact, but will expand the need for low-carbon electricity generation.

REFERENCES

1. A. F. Ghoniem, "Needs, resources and climate change: clean and efficient conversion technologies," *Prog. Energy Combust. Sci.*, vol. 37, no. 1, pp. 15–51, 2011.
2. UN Development Programme, *Human development report*. New York: UNDP, 2006.
3. D. Goodstein, *Out of gas: the end of the age of oil*. New York: Norton, 2004.
4. K. S. Deffeyes, *Hubbert's Peak: the impending world oil shortage*. Princeton, NJ: Princeton University Press, 2001.
5. J. A. Fay and D. S. Golomb, *Energy and the environment*. Oxford: Oxford University Press, 2012.
6. K. Caldeira, A. K. Jain, and M. I. Hoffert, "Climate sensitivity uncertainty and the need for energy without CO2 emissions," *Science*, vol. 299, pp. 2052–2054, 2003.
7. S. C. B. Raper and R. J. Braithwaite, "Low sea level rise projections from mountain glaciers and icecaps under global warming," *Nature*, vol. 439, 2006.
8. K. Caldeira and M. E. Wickett, "Anthropogenic carbon and ocean pH," *Nature*, vol. 425, 2003.
9. T. M. L. Wigley, R. Richels, and J. A. Edmonds, "Economic and environmental choices in the stabilization of atmospheric CO2 concentrations," *Nature*, vol. 379, pp. 240–243, 1996.
10. United Nations, United Nations Framework on Climate Change, 1992.
11. T. W. Patzek, "Thermodynamics of corn–ethanol biofuelcycle," *Crit. Rev. Plant Sci.*, vol. 23, no. 6, pp. 519–567, 2004.
12. J. W. Tester, *Sustainable energy: choosing among options*. Cambridge, MA: MIT Press, 2012.
13. S. Pacala and R. Socolow, "Stabilizing wedges: solving the climate problem for the next 50 years with the current technologies," *Science*, vol. 305, 2004.

2 Thermodynamics

2.1 Introduction[1]

Thermodynamics is central to the analysis of energy conversion processes and systems. Although excluding rate processes, equilibrium thermodynamics' analysis can be used to examine the efficiency and specific work of a process or a series of processes executing work and heat transfer interactions with other systems, experiencing mass transfer, undergoing chemical and electrochemical reactions, or a combination of all of these events. Non-equilibrium and rate processes can indeed impact efficiency, and are necessary to determine the power as well as other performance measures such as size and emissions. Non-equilibrium effects will be examined in later chapters. In this chapter, the basic laws of equilibrium thermodynamics are reviewed, with an emphasis on some of the origins of the different statements, the meaning of the quantities appearing in these laws, the most relevant forms of the laws to be used in analysis of energy conversion, and some conclusions regarding how these systems should be designed. The early coverage is independent of the working fluid, and focuses on the energy conversion process. Pure substance, ideal gases, and mixtures of ideal gases and their equations of state are also mentioned.

We start with generalized forms of the First Law for closed and open systems, defining the different forms of energy storage and work transfer, and how they are impacted by heat and work transfer. This is followed by the Second Law, starting with the positivity of entropy generation for an isolated system, and extending to forms applicable to closed and open systems interacting with their environment. The combined form of the First and Second Laws is then used to define the availability, or maximum work during a process in which a system interacts with its environment. The availability equations for closed and open system analysis are then shown, along with the definitions of irreversibility and lost work. Implications for where loss of availability in a typical power cycle is, and where potential improvements can be expected, are discussed.

[1] Many thermodynamics textbooks have been written, both introductory and advanced, that contain detail on the topics covered in this chapter, see, e.g. [1–8]. More will be cited later.

Thermodynamics efficiency can be associated with a multitude of definitions that can be applied to the same process/system, depending on the focus of the analysis. An introduction to the conversion or First Law efficiency is given, where the output of an energy conversion process is compared to the input. Several other definitions are discussed. This is followed by discussion of the concepts of effectiveness, Second Law or availability/exergy efficiency, and their relations to the availability analysis. We also show that the definition of the efficiency depends on the source of energy and the energy conversion process. Efficiencies of individual processes, which are used extensively in later analyses, are also introduced. Much more on efficiency for different conversion processes and systems follows in later chapters.

Equilibrium is central to thermodynamic analysis of many phenomena and its principles are reviewed before several applications are introduced in the coming chapters. The condition of equilibrium of a system subject to a set of constraints is defined for several cases of interest, in terms of change of an extensive variable compatible with the constraints. These statements are used to define the direction of a process and set the stage for defining different types of equilibria.

Many energy systems analyses use ideal gas mixtures or multiphase pure substances as working fluids models. The property relations of both are reviewed, with an emphasis on the entropy of mixing in gases and the change in phase in pure substances undergoing heat or work transfer. The former is important for chemically reacting flows, but also in equilibrium of systems with complex composition.

Equilibrium is revisited, and conditions of thermal and mechanical equilibrium are defined in terms of the temperature and pressure. Conditions of equilibrium, applied under different constraints and in terms of extensive properties, are also presented. The concept of chemical potential is introduced to define mass transfer equilibrium in multi-component mixtures undergoing change.

The chapter ends with the analysis and discussion of two applications that illustrate the use of First and Second Law analysis in energy conversion. The first addresses the work of separation and its application to processes in energy systems, such as CO_2 capture from combustion or reforming products. The second addresses the work of liquefaction both for an idealized cycle and for a more practical system, such as the Linde system. Less idealized systems used in liquefaction and gas separation are discussed to determine more accurately the required work in both, and the efficiency or effectiveness of these systems.

2.2 The First Law of Thermodynamics

2.2.1 Work and Heat Transfer Interactions

The First Law of Thermodynamics, or simply the First Law, defines the **stored total energy**, E, of a system and relates its changes to the heat transfer and work transfer interactions between the system and its environment. A **system** is defined as a macroscopic region in space that has defined **boundaries** that separate it from its **surrounding** or **environment**. The system may have a fixed mass of material, also called **control mass** or a **closed system**. It also

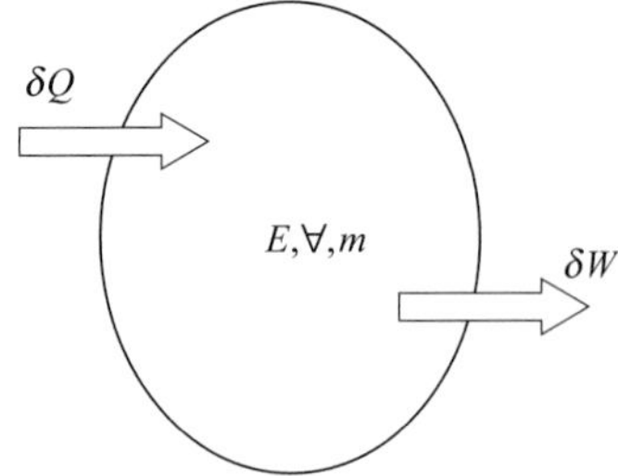

Figure 2.1 The First Law of Thermodynamics applied to a system.

can be primarily defined through its boundaries and allows mass to flow through them, also called a **control volume** or **open system**. Energy storage is associated with the presence of mass in the system. A system is characterized by a set of thermodynamic **properties**, which also characterize its thermodynamic **state**. These properties can be **extensive**, and therefore reflect the system size/mass, such as total mass or total energy storage. The properties can be **intensive**, and therefore do not characterize the system size. Examples of intensive properties include temperature and pressure, as well as extensive properties normalized by the system mass.

Conservation of energy, or the First Law of Thermodynamics for a closed system in differential form, states that the difference between the **heat** transfer to the system *from* its environment, δQ, and the net **work** transfer[2] *out* of the system to its environment, δW, equals the change in the stored energy:

$$\delta Q - \delta W = dE. \tag{2.1}$$

The balance is illustrated in Figure 2.1. The choice of δQ and δW are basic sign conventions that do not alter the physics. An *accumulation* of energy in the system is still due to a net transfer *into* the system. According to the "mechanical engineering" convention, heat transfer into the system is positive while work transfer out of the system is positive. Many texts in chemical engineering have adopted a different convention in which both heat and work transfer into the system are positive, since they both raise the stored energy. In this book, we will adopt the "mechanical engineering" convention.

In incremental form, we distinguish between small increments of heat and work transfer using the symbol δ versus the change in the stored energy denoted with a symbol d. Therefore, the symbol δ is used to emphasize that Q and W are not properties of the **state**, but are **interactions** between the system and its environment. In the simplest terms, heat interaction is associated with a transfer across the system boundary that increases its stored

[2] The first law distinguishes between interactions between the system and its environment, and changes in the internal or stored energy, with heat transfer and work transfer as interactions between the system and its environment and the quantities on the right-hand side as changes in the stored energies. Therefore, heat transfer and work transfer are interactions and should be characterized as such. In most textbooks, the word *heat* is used for heat transfer interaction and *work* for work transfer interaction.

energy, without work transfer. Heat conduction and radiation represent two modes of heat transfer interactions between the system and its surrounding. Work interaction involves the transfer of mechanical or electrical (or magnetic, etc.) energy across the system boundary, such that the net impact at the interaction across the boundary can be represented by, or is equivalent to, the change in the level of a weight outside the system boundary. In many cases, work interactions involve energy transfer due to forces acting on the boundaries and boundary displacement. Neither work nor heat transfer are system properties; they are interactions between the system and its environment (or another system). How the total stored energy change is manifested is due to a number of factors, including the thermodynamic properties of the substance contained within the system and its thermodynamic state. For example, heat addition can result in change in the temperature or **sensible heating**, change in phase (e.g., melting, evaporation) or **latent heating**, chemical or nuclear reaction, or a combination of some of these effects.

2.2.2 First Law for a System

Equation (2.1) can be integrated between two states labeled as 1 (start) and 2 (end):

$$Q - W = E_2 - E_1 = \Delta E. \tag{2.2}$$

The First Law statement simply relates the change of the stored energy to the work or heat transfer interactions. This statement of the First Law captures the observation that one can change the stored energy in the system through heat and work transfer interactions, without a distinction between the different modes of energy storage, or the separate impact of heat transfer or work transfer interactions on a particular mode.

The stored energy is a system property that depends on its state, as shown in the state plane in Figure 2.2. The state can be described by a number of **extensive properties** (proportional to the mass) and **intensive properties** (independent of the mass or measured per unit mass). The axes in Figure 2.2 are not labeled, but they refer to appropriate pairs of thermodynamic properties of the system. The path connecting the two states is defined as a **process**; different paths can connect states 1 and 2, depending on the constraints of

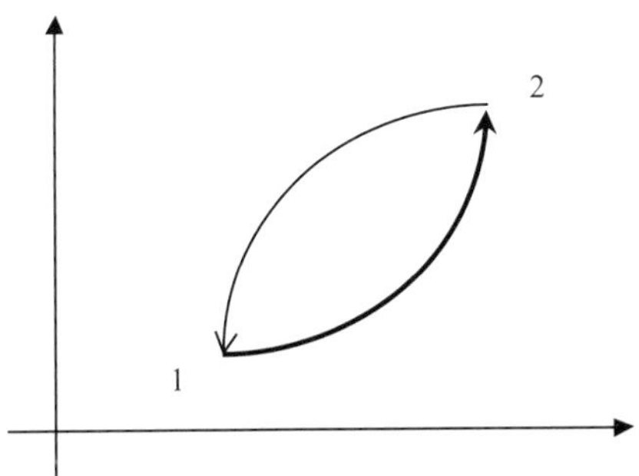

Figure 2.2 Process and cycle representation on a state plane.

the process. A series of connected processes that are implemented in a closed loop make up a **cycle**.

Cycles play a central role in energy conversion, as illustrated by the numerous applications in this book. Prominent among these cycles are the general class of **heat engines** or **power cycles**. These cycles produce mechanical work, which can be or is converted to electricity or propulsive power for transportation, while requiring heat transfer from a heat source (e.g., combustion, geothermal, solar). Cooling or refrigeration cycles transfer heat from a cool to a hot space, and are used widely, including in separation processes, cryogenic storage of gases, etc.

The stored energy can be expressed in terms of different modes or energy storage, such as [1]

$$E = KE + PE + U + E_{elas} + E_{el} + E_{mag} + E_{nuc} \tag{2.3}$$

In this equation:

1. $KE = \frac{1}{2}m\mathbf{V}^2$ is the **kinetic energy** of a mass, m, moving at a velocity $\mathbf{V}$; all elements in this mass have the same velocity.
2. PE is the **potential energy**; for a mass in a gravitational field in which the force exerted on the mass is $F = mg_r$, and g_r is the gravitational acceleration, $PE = mg_rZ$, while Z is the distance along but opposite to the direction of g_r, measured from a well-defined reference, assuming the mass is concentrated at a point.
3. U is the **internal** energy, which is the total energy associated with the kinetic and potential energy of the molecules (including the bond or chemical energy). For the purpose of our analysis, we will write: $U = U_{th} + U_{ch}$, where U_{th} is the **thermal energy**, including all the kinetic energy associated with the "random" motion of the molecules (the latent energy is part this as well), and U_{ch} is the **chemical energy** (to be defined in Chapter 3).
4. $E_{elas} = \frac{1}{2}k_s x^2$ is the **elastic energy** stored in a linear spring, where k_s is the spring constant and the distance x is measured with respect to a neutral state.
5. $E_{elect} = \varepsilon\varsigma$, is the **electrical energy** associated with the presence of an electric charge, ς, in an electric field potential, ε.
6. E_{mag} is the **magnetic energy**.
7. $E_{nuc} = mC_\ell^2$ is the **nuclear energy** stored within the nucleus of an atom, where C_ℓ is the speed of light in a vacuum.

Since the First Law is concerned with changes in the stored energy due to heat and work transfer interactions across the boundaries, in most cases, many of these terms do not appear in the analysis when the corresponding modes are not active or do not change during the process of interest. Moreover, some terms are neglected if their changes during a particular process are much smaller than others.

Similarly, work transfer can be expressed as a superposition of different modes. Under quasi-static conditions, when the system is internally in equilibrium and one can define thermodynamic properties such as pressure, p, temperature, T, etc., these work interactions can be expressed in terms of other thermodynamic state properties. Work interaction modes include the following:

1. **Mechanical work transfer** interactions, $-\delta W_{mech} = \vec{F} \cdot d\vec{x}$, where $\vec{F}$ is the force acting at the boundary causing a boundary displacement $d\vec{x}$. If the process is quasi-static and the system is in mechanical equilibrium at pressure p, then the force is $\vec{F} = -p\vec{n}\,da$, with $\vec{n}$ being the unit vector normal to the area and da being the surface area element, then $\delta W_{mech} = p\,d\forall$, where $\forall$ is the volume of the system.
2. **Electrical work** interactions, $-\delta W_{el} = \varepsilon d\varsigma$ (note that during discharge of an electrochemical cell, such as a battery or a fuel cell, the discharge is negative and hence the work transfer is positive).
3. **Magnetic work** interactions, $-\delta W_{mag} = \mathrm{H}dM_g$, where H is the external magnetic field and dM_g is the magnetization.

A work transfer is the product of an intensive property and the differential of an extensive property (sometimes called the generalized force and generalized displacement, respectively). It should be clear from the above discussion of stored energies and work modes that there are correspondences between some work interaction modes and energy storage modes, the clear exception being the thermal part of the internal energy.

The state of a system is defined by $(N + 1)$ properties, where N is the number of the relevant quasi-static work transfer modes that can impact its stored energy, or the state properties that determine these energy storage modes. The extra 1 is often taken as the internal energy. Equations of state are used to define the relation among these thermodynamic state properties. For example, if the substance within the system is subject to only mechanical work, then only two independent thermodynamic properties are needed to determine its thermodynamic state. Such a substance is referred to as a **simple compressible substance**.

Example 2.1 Electrical Work Transfer in a Battery

A common example of a system interacting with its environment through heat transfer and electrical work transfer is **the battery** shown in Figure E2.1. A battery is an electrochemical cell that stores energy in the form of chemical energy. It is an example of the **direct conversion** of chemical energy to electricity. When connected to a resistance, internal electrochemical reactions force electrons to flow externally through a resistance by establishing an electric potential across the terminals of the battery (accompanied by an internal current of positively charged ions). However, a battery also has an internal resistance, which converts part of the work to heat. This resistance, denoted as r_{el}, and the load resistance, denoted by R_{el}, operate in series, and the total effective resistance is $R_{el} + r_{el}$. The chemical reaction and the internal flow of the charge raise the temperature of the battery through ohmic resistance, leading to heat transfer to the environment. A rechargeable or secondary battery undergoes the reverse processes during charging – that is, electrical work is transferred into the battery and the reverse chemical reactions restore chemical state and chemical energy inside the battery to its original level (or close to it). Some heat transfer is also associated with the charging process. Thus, losses occur during charging and discharging.

Example 2.1 (cont.)

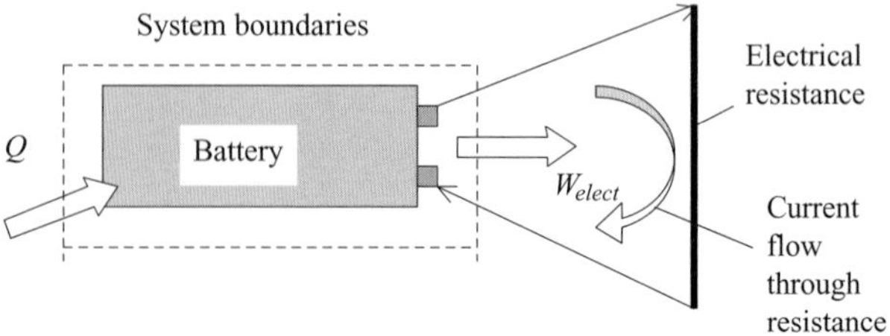

Figure E2.1 The application of the First Law to a battery discharging in a resistance.

1. Derive an expression for the electric work transfer from a battery at steady-state operation and with sufficient insulation.
2. Determine the rate of change in stored chemical energy in a battery attached to an electrical resistance with a load resistance of 2 Ω and an internal resistance of 0.2 Ω. Under open-circuit conditions (i.e., with no current flow) the measured voltage across the (multi-cell) battery is $V_{\text{emf}} = 12$ V.

Solution

In this case, during a typical discharge, the First Law shows that

$$Q - W_{el} = (U_2 - U_1)_{th} + (U_2 - U_1)_{ch}$$

where Q is the heat transfer "to" the battery and W_{el} is the electrical work out of the battery via the current flowing through an external resistance. At steady state, the battery temperature is constant and $\Delta U_{th} = 0$:

$$Q - W_{el} = (U_2 - U_1)_{ch}.$$

1. The electric current running through the internal and load resistances may be expressed as $I = V_{emf}/(R_{el} + r_{el})$. The change of chemical energy stored in the battery, $(U_1 - U_2)_{ch}$, over a period of Δt is $I^2(R_{el} + r_{el})\Delta t$. Of that, part is converted to thermal energy that is transferred out as heat, $Q = I^2 r_{el}\Delta t$, and the work transferred to the external resistance is $W_{el} = I^2 R_{el}\Delta t$.
2. For the given conditions, the electric current is $I = 12/(2 + 0.2) = 5.45$ A. So, the rate of change in stored chemical energy is obtained as follows: $\Delta U_{ch}/\Delta t = (5.45)^2 \times 2.2 = 65.3455$ W.

2.2.3 First Law for a Control Volume

For an open system or control volume, shown in Figure 2.3, an additional energy transfer mechanism is associated with the mass transfer across the system boundaries. In energy

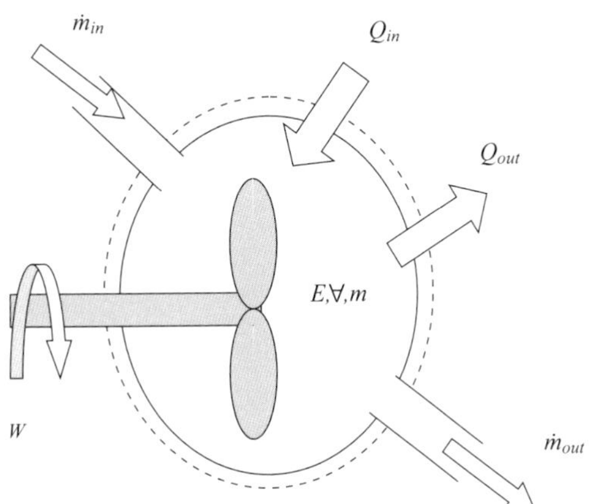

Figure 2.3 Control volume analysis of the First Law.

balance, this transfer is manifested in two forms: (1) transfer of stored energy associated with the transfer of mass; and (2) flow work, which corresponds to work done across boundaries by the volume displacement, $d\forall$, of mass dm entering or existing the system, such that $dm = d\forall/v$, where v is the specific volume of the mass crossing the control volume. In infinitesimal form, the incremental change in energy stored in the control volume may be expressed as

$$dE = \delta Q - \delta W + e\,dm + pd\forall, \tag{2.4}$$

where the last two terms correspond to the energy transfer via mass transfer, $e\,dm$, and the flow work through mass, $pd\forall$. e is the specific stored energy of the mass element introduced into the system, $e = E/m$, m is the total mass, and δQ and δW are the heat and work transfer during this process, respectively. The flow work term and the internal energy can be combined into a single thermodynamic variable, the **enthalpy**,

$$h = u + pv, \tag{2.5}$$

where v is the specific volume, which is defined as $v = \forall/m$. Equation (2.4) can be written in the differential form as follows:

$$dE = \delta Q - \delta W + (h + ke + pe + \cdots)\,dm. \tag{2.6}$$

We added the kinetic and potential energies of the mass introduced into the control volume. W is often called the **shaft work** when the boundaries are rigid and the work transfer interaction is through a rotating shaft only. Examples of shaft work include the work transferred to turbine blades or from compressor blades. If the boundaries move (without mass transfer), an extra work term, the boundary work, is added to account for the work transfer due to boundary motion. All the lower-case symbols stand for per unit mass properties – that is, they are intensive properties. Equation (2.6) shows that a flowing mass, under steady-state conditions, which does not exchange heat or work with its environment, has a constant value of total stored energy and flow work.

Integrating (2.6) over an arbitrary number of mass elements within the system boundaries, we get [9]

$$E_2 - E_1 = Q - W + \sum_{in} m_{in}(h + ke + pe + \cdots)_{in} - \sum_{out} m_{out}(h + ke + pe + \cdots)_{out}. \tag{2.7}$$

For an unsteady process,

$$\frac{dE}{dt} = \dot{Q} - \dot{W} + \sum_{in} \dot{m}_{in}(h + ke + pe + \cdots)_{in} - \sum_{out} \dot{m}_{out}(h + ke + pe + \cdots)_{out}. \tag{2.8}$$

The dot over the symbol indicates a time rate of change. In (2.7) and (2.8), the energy balance indicates that energy storage in the form of internal, kinetic, and potential energies can change due to heat and mass transfers, the common energy transfer mechanism with a closed system, and mass transfer (carrying energy stored in mass and flow work). In a steady-flow system, the mass fluxes in and out of the system are fixed in time, and there is no accumulation of mass inside the control volume. In steady-state conditions, there is no storage/change of energy within the system and the time derivative is zero.

While the system's representation of the First Law is self-contained, because the mass of the system is fixed, the control volume representation must be complemented with the statement of conservation of mass:[3]

$$\frac{dm_{CV}}{dt} = \sum_{in} \dot{m}_{in} - \sum_{out} \dot{m}_{out}. \tag{2.9}$$

Example 2.2

An example for the application of control volume analysis to energy conversion systems is the diesel generator shown in Figure E2.2, which is used for an auxiliary power unit in a hospital. The control volume is taken to surround the fuel tank, the diesel engine, and the electrical generator. The application of (2.7) between two arbitrary states yields

$$(U_2 - U_1) = Q - W_{elect} + m_{air}h_{air} - m_{exhaust}h_{exhaust} \tag{2.10}$$

and

$$(m_2 - m_1)_{fuel} = m_{air} - m_{exhaust}. \tag{2.11}$$

[3] This equation does not account for "mass loss" in the case when nuclear energy is converted to thermal energy. Given the value of the speed of light, 2.998×10^8 m/s, this is not unreasonable, even with such conversion.

Example 2.2 (cont.)

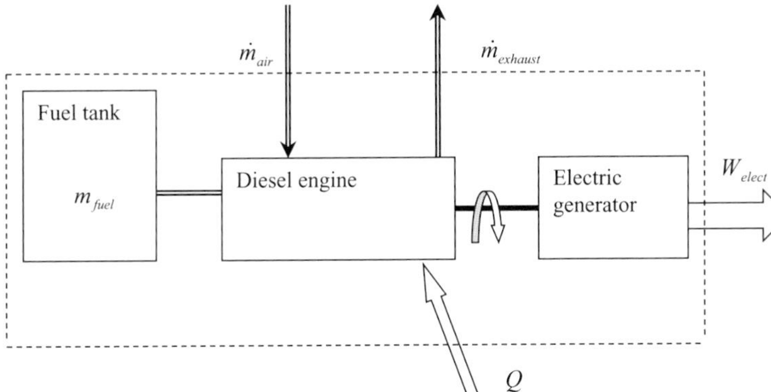

Figure E2.2 Control volume analysis of electric power generation using a diesel engine, a fuel tank, and an electric motor.

Assuming that the temperature of the components inside the control volume remains constant during the period of analysis, at least under steady operation, then $(U_2 - U_1)_{ch} = (m_2 - m_1)u_{ch}$, where u_{ch} is the chemical internal energy per unit mass of fuel (to be defined in detail in Chapter 3). This system is an example of **indirect conversion** from chemical energy to electrical work. First, the chemical energy in the fuel is converted to thermal energy by combustion, which is then converted by the engine into mechanical work that drives the electric generator.

2.3 The Second Law of Thermodynamics

In the previous discussion, we reviewed the First Law, which deals primarily with energy conservation. However, energy bookkeeping is not the only constraint that a process or a cycle needs to satisfy. Some additional constraints are inherently intuitive, but they represent a manifestation of the Second Law. For example, we expect spontaneous heat transfer from a hotter substance to a colder substance and not the opposite (naturally). We also expect that, when we puncture a balloon, all the gas that escapes cannot spontaneously return to the balloon. These processes are denoted as "irreversible" and represent some embodiments of the Second Law. The Second Law has other manifestations, and powerful tools to address the limitations of thermodynamic processes through the concepts of entropy and exergy or availability can be developed using the Second Law.

2.3.1 Entropy

The form of the Second Law that is most useful for our development, written for a control mass or closed system for which the total internal energy and mass are fixed, is as follows: There exists a variable/property of state called the **entropy**, S, such that

$$dS = (dS)_f + (\delta S)_g. \tag{2.12}$$

The generation term, which is process-dependent, must satisfy the following inequality [10]:

$$(\delta S)_g \geq 0. \tag{2.13}$$

Subscript *f* and *g* correspond to entropy flow across the system boundary, associated with heat transfer interaction with the environment, and entropy generation inside the system, respectively. Entropy generation is zero in reversible processes. For a **closed system** or a **control mass** undergoing a process, entropy flow is associated with heat transfer only and is given by

$$(dS)_f = \frac{\delta Q}{T}, \tag{2.14}$$

where δQ is the heat transfer *to* the system and T is the system temperature,[4] and more specifically the temperature at which heat is transferred to the system. Entropy flows into, or is transferred to, a system during a heat transfer interaction, and there is no entropy flow due to work transfer. The contribution of irreversibility is manifested in the entropy generation term. Entropy is a thermodynamic property of matter. It can be related to other properties of the system through equation-of-state relations. In later sections, we will illustrate one of them, based on the Gibbs equation. It is important to emphasize that entropy can increase or decrease during a process, based on the balances of transfer and entropy generation. The Second Law simply states that entropy generation is positive and is at best zero. When entropy generation is zero, the process is called reversible.

Entropy is generated during irreversible processes, such as:

1. heat transfer across a finite temperature gradient, from the higher to the lower temperatures;
2. free expansion, or expansion against a pressure other than the system pressure – the escape of a gas from a punctured balloon is an example of free expansion;
3. friction;
4. mixing of different substances, or mixing substances at different temperatures, or stirring; and
5. chemical reactions (under some conditions).

These **irreversible** processes cannot be reversed spontaneously (with finite probability) to bring the system back to its original state, and hence generate entropy. Entropy generation or irreversibility, as shown later, reduces the available work transfer interaction between a given system at a given state and its environment.

Note that (2.13) carries the important implication that at **equilibrium**, with no further change in the isolated system for which the total internal energy and volume are fixed, the

[4] If the system and the environment temperatures are different, then the entropy flow is calculated using this equation, but there is also an entropy generation terms associated with the heat transfer between the two different temperatures. This entropy generation terms will be shown later.

entropy must reach a maximum. This statement has important implications in determining the final state of a system once perturbed from an equilibrium state, and in defining the direction of that change. It also implies that with no heat transfer interactions, and under ideal, reversible conditions, the entropy remains constant.

As stated above, entropy is a property of the state of the system. It can be defined in extensive form, as indicated by the above equations for entropy balance, or intensive form, $s = S/m$.

2.3.2 Second Law for a Closed System

The statement of the Second Law for a closed system, in differential form, is

$$dS = \frac{\delta Q}{T} + \delta S_g. \tag{2.15}$$

The finite change form takes on the integrated version:

$$S_2 - S_1 = \int_1^2 \frac{\delta Q}{T} + S_g. \tag{2.16}$$

The last equation is used to determine the change in the system's entropy as a function of its properties by carrying out reversible processes during which these properties change and the entropy generation is zero. Note that the temperature T is the temperature at which the heat transfer increment δQ enters the system, as shown in Figure 2.4. A fixed entropy process or **isentropic** process corresponds to a process that is adiabatic (no heat interactions) and reversible (no entropy generation).

In an integral form, we have a system receiving heat transfer at K areas, where the system temperature at these points is T_k, and heat transfer at these points is Q_k; we write the total change of entropy of the system as

$$S_2 - S_1 = \sum_{k=1}^{K} \frac{Q_k}{T_k} + S_g. \tag{2.17}$$

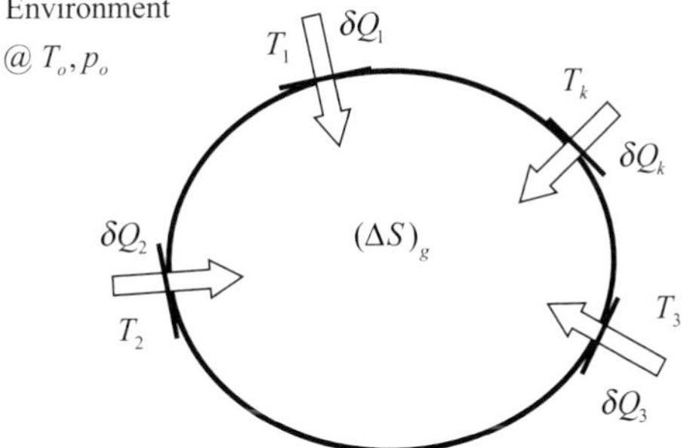

Figure 2.4 Entropy flow/transfer into the system, associated with heat transfer at multiple locations. The temperature is that of the control surface where heat is transferred, and entropy is generated inside the control surface.

In this equation, the entropy flow or transfer is (Q_k/T_k), while all the entropy generation is within the system.

2.3.3 Second Law for a Control Volume

For an open system, an additional mechanism for entropy transfer is present in the entropy balance equation. It corresponds to mass transfer. In infinitesimal form, the entropy balance for a control volume is written as

$$dS = \frac{\delta Q}{T} + s\,dm + \delta S_g, \tag{2.18}$$

where s is the entropy of the mass element flowing into the system. In integral form, assuming multiple ports for mass flow in or out of the control volume:

$$(S_2 - S_1)_{CV} = \int_1^2 \frac{\delta Q}{T} + \sum_{in} s_{in} m_{in} - \sum_{out} s_{out} m_{out} + S_g. \tag{2.19}$$

For finite heat transfer at fixed boundary temperatures,

$$(S_2 - S_1)_{CV} = \sum_{k=1}^{K} \frac{Q_k}{T_k} + \sum_{in} s_{in} m_{in} - \sum_{out} s_{out} m_{out} + S_g. \tag{2.20}$$

For an unsteady control volume, the rate form of the Second Law is written as

$$\frac{dS_{cv}}{dt} = \frac{\dot{Q}}{T} + \sum_{in} s_{in} \dot{m}_{in} - \sum_{out} s_{out} \dot{m}_{out} + \dot{S}_g. \tag{2.21}$$

Example 2.3 Entropy Generation in a Mixing Chamber

Liquid water at 15 °C (stream 1) is mixed in a mixing chamber with superheated steam at 200 °C (stream 2). The mass flow rates of the liquid and the superheated steam are 0.5 and 1 kg/s, respectively. The mixing chamber operates at a pressure of 200 kPa. The chamber rejects heat to the environment at 25 °C at the rate of 500 kW. Determine the entropy production rate within the mixing chamber.

Solution

We assume steady-state and steady-flow operation, no work transfer, and negligible changes in kinetic and potential energies. The fluid used is water. For liquid water, we consider saturated liquid properties at the prescribed temperature. For saturated liquid-vapor properties, we use saturated liquid and saturated vapor properties at the prescribed temperature,

Example 2.3 (cont.)

or pressure-weighted by the quality. For superheated steam, we use superheated steam tables at the prescribed temperature and pressure.

The energy balance across the mixing chamber is

$$\dot{m}_1 h_1 + \dot{m}_2 h_2 = \dot{m}_3 h_3 + \dot{Q}_{out} = (\dot{m}_1 + \dot{m}_2)h_3 + \dot{Q}_{out},$$

where state 3 corresponds to the exit stream. Solving for the enthalpy in the exit stream,

$$h_3 = \frac{\dot{m}_1}{\dot{m}_1 + \dot{m}_2} h_1 + \frac{\dot{m}_2}{\dot{m}_1 + \dot{m}_2} h_2 - \frac{\dot{Q}_{out}}{\dot{m}_1 + \dot{m}_2}.$$

State 1 is compressed liquid at 15 °C and 200 kPa. Its enthalpy may be approximated as the saturated liquid enthalpy at 15 °C: $h_1 = h_{f@15°C} = 62.98$ kJ/kg, $s_1 = s_{f@15°C} = 0.22447$ kJ/kg·K. State 2 is superheated steam at 200 °C and 200 kPa. Using superheated steam tables, $h_2 = 2870.4$ kJ/kg, $s_2 = 7.5081$ kJ/kg·K.

Solving for h_3,

$$h_3 = \frac{0.5}{0.5+1}(62.98) + \frac{1}{0.5+1}(2870.4) - \frac{500}{0.5+1} = 1601.26 \text{ kJ/kg}.$$

This state is also saturated, since the enthalpy value is between the saturated liquid enthalpy $h_{f@200kPa} = 504.71$ kJ/kg and the saturated vapor $h_{g@200kPa} = 2706.3$ kJ/kg at 200 kPa. The quality of this mixture is $x_3 = (h_3 - h_f)/(h_g - h_f) = 0.498$. The corresponding entropy is: $s_3 = x_3 s_g + (1 - x_3)s_f = (0.498)(1.5302) + (0.502)(7.1270) = 4.3394$ kJ/K.

The entropy balance across the mixing chamber is

$$\dot{m}_1 s_1 + \dot{m}_2 s_2 + \dot{S}_g = \dot{m}_3 s_3 + \frac{\dot{Q}_{out}}{T_0}.$$

Substituting the properties and rates into the entropy generation equation,

$$\dot{S}_g = (0.5+1)(4.3394) - (0.5)(0.22447) - (1)(7.5081) + \frac{500}{25+273} = 0.5666 \text{ kW/K}.$$

2.4 The Combined Statement and Availability

2.4.1 Maximum Work and Availability

Combining the First and Second Laws for a system/control mass, (2.1) and (2.15), we can show the impact of entropy generation on work interaction between the system and its environment during a given process:

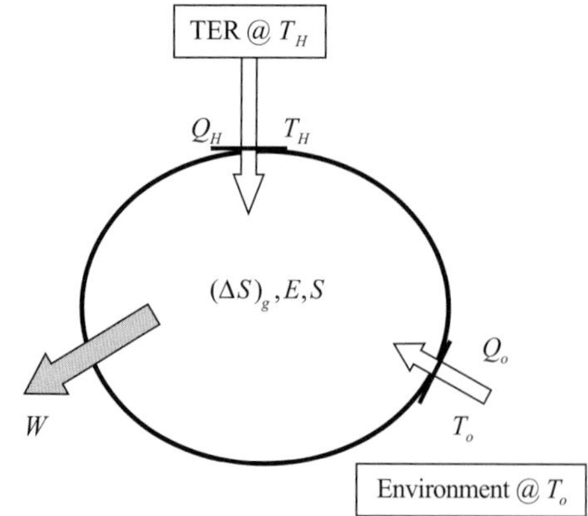

Figure 2.5 Heat and entropy flow/transfer into the system from a high-temperature TER, and from the environment, work interaction with the environment, and entropy generation inside the control mass.

$$\delta W = (T\,dS - dE) - T\,\delta S_g. \tag{2.22}$$

There is a loss of work equal to $T\delta S_g$, which is also called the **irreversibility**, or the **lost work** due to entropy generation. According to the Second Law, this term is always positive, and it is larger when entropy generation occurs at higher temperatures. We also define the **availability**, or the maximum work transfer interaction, between a system and its environment, given the conditions of the system and the environment, as they reach a state of equilibrium. This quantity is also named "essergy" – that is, "essence of energy" – or more commonly **exergy**. Note that the availability is determined using the combined statements of the First and Second Laws. For simplicity, we take a system interacting with two constant-temperature thermal energy reservoirs, depicted in Figure 2.5, at temperatures T_H and T_o (this analysis will be generalized in the next section). A **thermal energy reservoir** (TER) is an idealized system whose intensive state (T_o, p_o)[5] does not change because of its heat and work transfer interactions with a system (later we will add the chemical potential, μ_i^o, of its components to this list). Examples of systems that approximate such TERs include the atmosphere and large bodies of water (e.g., lakes, rivers, and oceans). If the system undergoes a process 1–2, during which there is heat transfer Q_H from the high-temperature reservoir and Q_o from the low-temperature reservoir, and work transfer W to the environment, we can write:

$$Q_H + Q_o - W = E_2 - E_1 \tag{2.23}$$

and

$$S_2 - S_1 = \frac{Q_H}{T_H} + \frac{Q_o}{T_o} + S_g. \tag{2.24}$$

[5] To stay within the accepted convention, T_o and p_o designate the temperature and pressure of the environment. Reference quantities used to define the state will have a superscript, e.g., p^o, T^o, h^o.

After few manipulations to eliminate the heat transfer interactions with the environment, we find that the net work interaction during this process is

$$W = Q_H\left(1 - \frac{T_o}{T_H}\right) + \left[(E_1 - T_oS_1) - (E_2 - T_oS_2)\right] - T_o\,S_g. \tag{2.25}$$

This expression is useful for any system. A simplified form of it is particularly important for thermodynamic cycles where different components of the cycle may draw heat from a hot TER at T_H at a given stage or component of the cycle and reject heat to a "cold" reservoir at T_o at another stage or component of the same cycle. For a cycle, the initial and final states are identical; the second term, $(E_1 - T_oS_1) - (E_2 - T_oS_2)$, is zero. From the above expression, it is clear that entropy generation reduces the available work, and the last term, $T_o\,S_g$, is a "lost work" term. The available work consists of the first two terms on the right-hand side. The available work is due to:

1. the net heat transfer interaction, as captured by the first term, the so-called **exergy transfer** due to heat transfer; and
2. the net change of the term $(E - T_oS)$.

If the system experiences change of volume during this process, it must transfer work to the environment that equals $p_0(\forall_2 - \forall_1)$, where p_0 is the atmospheric pressure. In this case, the useful work, also called **available exergy** [11], is

$$W_{use} = Q_H\left(1 - \frac{T_o}{T_H}\right) + (E_1 - T_oS_1 + p_o\forall_1) - (E_2 - T_oS_2 + p_o\forall_2) - T_oS_g. \tag{2.26}$$

The term **non-flow availability function** is used to describe the quantity:

$$\Xi_f = (E + p_o\forall - T_oS). \tag{2.27}$$

In this definition, $(T_o,\ p_o)$ are, respectively, the temperature and pressure of the environment. The **non-flow exergy function** is a function of both the state and the environment. The last term in (2.26) is the **destroyed exergy** through entropy generation. The implications of this expression are as follows:

1. If the system undergoes heat transfer interactions with the environment only, then the first term is zero and the maximum work as it passes from state 1 to state 2 is the difference of the availability at these two end states, with no internal entropy generation:

$$W_{max} = (E_1 - T_oS_1 + p_o\forall_1) - (E_2 - T_oS_2 + p_o\forall_2). \tag{2.28}$$

2. If states 1 and 2 of the system/control mass are identical, while it interacts with a TER whose temperature is T_H, and with the environment at T_o, then the maximum work transfer is

$$W_{max} = W_{Carnot} = \left(1 - \frac{T_o}{T_H}\right)Q_H. \tag{2.29}$$

Recall that a TER is a special control mass whose state does not change with heat transfer; it is a very large reservoir. Equation (2.29) is precisely the work obtained from a **Carnot** engine operating between the two TERs.

As expected, a Carnot cycle delivers the maximum work while interacting with two fixed-temperature TERs, and is the most efficient cycle operating between two given fixed-temperature reservoirs. In this cycle, heat transfer occurs **isothermally** and work transfer occurs **isentropically**, that is adiabatically and reversibly, as shown in Figure 2.6. The expression in (2.29) shows that the exergy flow associated with a heat transfer Q_H from the high-temperature reservoir at T_H is $\left(1 - \frac{T_o}{T_H}\right) Q_H$. Isothermal heat transfer is essential to reversibility, or maximum work transfer.

One can estimate the work loss, $T_o S_g$, in many cases by comparing the actual work interaction in an energy conversion process to the maximum possible value determined by the available temperatures. Take the simple example in which a Carnot engine is operated between T_L and T_M, where $T_L > T_o$ and $T_M < T_H$ (to allow for finite heat transfer rates between the engine and the TERs), as shown in Figure 2.7. In this case, the work transfer is

$$W = \left(1 - \frac{T_L}{T_M}\right) Q_H.$$

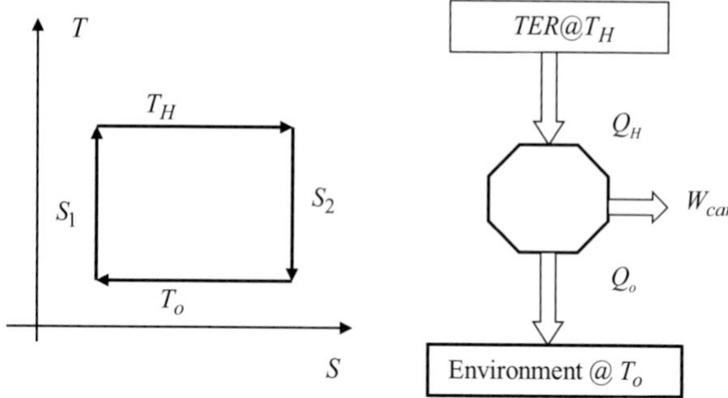

Figure 2.6 The T–S diagram of a Carnot cycle, and the corresponding heat interactions with two TERs and the work interactions with the environment.

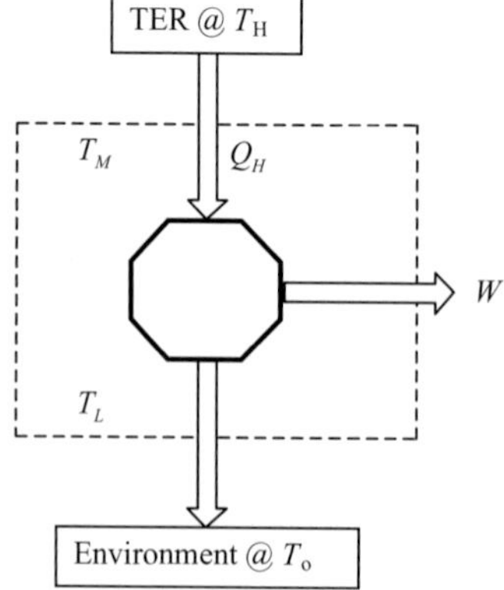

Figure 2.7 Schematic of a Carnot engine operating between two TERs but with a small temperature difference between the high and low temperatures of the engine and those of the energy reservoirs.

The maximum work is given by (2.29). The difference is the lost work, which is the difference between these expressions:

$$T_o S_g = \left(\frac{T_L}{T_M} - \frac{T_o}{T_H}\right) Q_H \geq 0.$$

The entropy generation $S_g = \left(\frac{T_L}{T_M} - \frac{T_o}{T_H}\right)\frac{Q_H}{T_o} \geq 0$ is indeed the entropy generated in the two heat transfer processes across the temperature difference at the high-and low-temperature ends (remember that entropy transfer at a fixed temperature is Q/T).

Using the availability or maximum work equations is convenient in determining the "best case scenario" in energy conversion and estimating the loss in actual processes and system. This is especially true in complex conversion systems where many processes and components interact, and when it is important to isolate the sources where most losses occur. This will be shown using different examples throughout this chapter and in the following chapters.

Example 2.4 Maximum Work/Irreversibility of a Heat Engine

A proposed heat engine cycle receives heat at the rate of 200 kW from a hot source at 2000 K and delivers a power output of 100 kW. Heat is rejected to the environment at 300 K.

1. Is this engine thermodynamically feasible?
2. If it is feasible, determine the rate of irreversibility of the cycle.

Solution

1. The cycle operates between $T_o = 300$ K and $T_H = 2000$ K. The maximum power output is that of a Carnot engine: $\dot{W}_{\max} = \dot{Q}_H\left(1 - \frac{T_o}{T_H}\right) = 170$ kW, which is higher than the power output of the proposed engine $\dot{W} = 100$ kW. Therefore, both the First and Second Laws are satisfied.
2. The irreversibility may be expressed as the difference between the maximum power output and the actual power output, which is $\dot{W}_{\max} - \dot{W} = 70$ kW.

2.4.2 System Availability

The origin of the availability or exergy can be further understood by evaluating the maximum work transfer interaction during a process in which a control mass at an arbitrary initial state, defined by $(E, S, \forall)$, is allowed to reach equilibrium with its environment, at the so-called **restricted dead state** (T_o, p_o) while experiencing heat and work interactions with the environment [12]. For the system shown in Figure 2.8, applying (2.26) with $Q_H = 0$ and $S_g = 0$, we get

$$\begin{aligned} W_{\max} &= (E - U_o) + p_o(\forall - \forall_o) - T_o(S - S_o) \\ &= \Xi. \end{aligned} \tag{2.30}$$

Figure 2.8 Maximum work and the availability analysis of a system interacting with an environment at the restricted dead state.

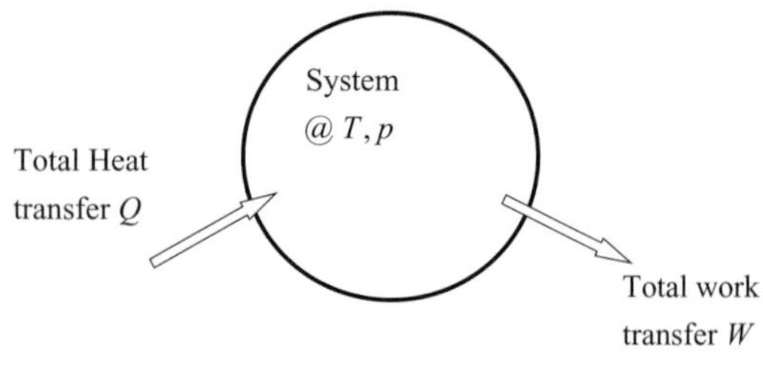

In this expression, U_o, $\forall_o$, and S_o are, respectively, the total internal energy, volume, and entropy of the control mass when it reaches equilibrium with the environment, and Ξ is the total exergy or availability of the control mass at the given state.

It should be clear now that maximum work can be obtained only when the control mass reaches the restricted dead state, or gets in equilibrium with the environment. When the control system changes its state from state 1 to state 2 while experiencing heat and work transfer interactions with the environment, then

$$W = \Xi_1 - \Xi_2 - T_o S_g. \tag{2.31}$$

These definitions capture the important notion that availability is measured with respect to the environment. Now (2.31) can be written in terms of the change in the system's availability:

$$W_{\max} = \Xi_1 - \Xi_2. \tag{2.32}$$

Exergy, like stored energy, decreases as the system experiences (positive) work transfer interaction with the environment. However, contrary to energy, exergy is not conserved, it is destroyed due to internal irreversibility in the system.

Using (2.26) and (2.16), we can show that for a control mass interacting with the environment and with multiple TERs [12],

$$\dot{W} = \int \left(1 - \frac{T_o}{T_s}\right) \dot{q}_s da - \left(\frac{d\Xi}{dt} - p_o \dot{\forall}\right) - \dot{I}_{ir}. \tag{2.33}$$

In this equation, $\dot{q}_s$ is the heat transfer flux per unit area of the control mass surface, defined by the element da, and the irreversibility rate is $\dot{I}_{ir} = T_o \dot{S}_g$. If the heat transfer interactions occur with a finite number of TERs defined by their temperature, T_i, and environment, then

$$\dot{W} = \sum_{TERs} \left(1 - \frac{T_o}{T_i}\right) \dot{Q}_i - \left(\frac{d\Xi}{dt} - p_o \dot{\forall}\right) - \dot{I}_{ir}. \tag{2.34}$$

Equations (2.33) or (2.34) define the maximum work interaction as the control mass changes its state and experiences heat transfer interactions with multiple sources when $\dot{S}_g = 0$.

Example 2.5 Entropy Generation/Exergy Loss During Free Expansion

Air contained initially in a rigid vessel of volume 0.2 m^3, pressure 200 kPa, and temperature 300 K is allowed to expand to an identical initially empty vessel, doubling its volume. Both vessels are insulated from the environment and there are no work transfer interactions with the environment. Determine the entropy generation and irreversibility or exergy loss during the free expansion process.

Solution

Assume air is an ideal gas, gas constant $R = 0.287$ kJ/kg·K, and a constant volume specific heat of $c_v = 0.708$ kJ/kg·K. The change in internal energy is $U_2 - U_1 = m\,c_v(T_2 - T_1)$. Entropy change for an ideal gas with constant specific heats is $\Delta S = m\,c_v \ln \frac{T_2}{T_1} + m\,R \ln \frac{v_2}{v_1}$. The ideal gas law $p\,V = mR\,T$ also applies.

Take the two vessels as a single control volume.

$$\text{First Law}: \; \Delta U = Q - W = 0 = mc_v(T_2 - T_1).$$

Heat and work transfer interactions are zero, the initial and final temperatures are equal: $T_2 = T_1$

$$\text{Second Law}: \Delta S = \int_1^2 \frac{\delta Q}{T} + S_g = S_g.$$

For this adiabatic process, $\Delta S = S_g$. Substituting ΔS for an ideal gas and using the constraint of an isothermal process, $S_g = mR \ln \frac{v_2}{v_1}$.

The volume ratio $\frac{v_2}{v_1}$ is 2. The mass based on the initial state and the ideal gas equation may be written as $m = \frac{p_1 V_1}{R\,T_1} = \frac{(100)(0.2)}{(0.287)(300)} = 0.232$ kg.

The entropy generation is $S_g = m\,R \ln \frac{v_2}{v_1} = (0.232)(0.287) \ln 2 = 0.020$ kJ/K.

The exergy loss is $T_o S_g = (300)(0.020) = 6.02$ kJ.

This is also the work lost in this free expansion process.

2.4.3 Availability Equation for a Control Volume

For another demonstration of the maximum work in an interaction, we look at an open flow system. Here, too, we consider a system interacting with two heat reservoirs, one of them the atmosphere, and limit the stored energy to internal, kinetic, and potential energy. To simplify, we will use the "methalpy" $\tilde{h} = h + ke + pe$. The two relevant equations are

$$\frac{dE}{dt} = \dot{Q}_H + \dot{Q}_o - \dot{W} + \sum_{in} \dot{m}_{in}\tilde{h}_{in} - \sum_{out} \dot{m}_{out}\tilde{h}_{out} \tag{2.35}$$

and

$$\frac{dS}{dt} = \frac{\dot{Q}_H}{T_H} + \frac{\dot{Q}_o}{T_o} + \sum_{in} \dot{m}_{in}s_{in} - \sum_{out} \dot{m}_{out}s_{out} + \dot{S}_g. \tag{2.36}$$

We may combine the equations into a single expression for the work interaction for the case in which the control volume has a *fixed volume*:

$$\begin{aligned}\dot{W} = &\left(1 - \frac{T_o}{T_H}\right)\dot{Q}_H + \sum_{in} \dot{m}_{in}\left(\tilde{h}_{in} - T_o s_{in}\right) \\ &- \sum_{out} \dot{m}_{out}\left(\tilde{h}_{out} - T_o s_{out}\right) - \frac{d}{dt}(E - T_o S) - T_o\dot{S}_g.\end{aligned} \tag{2.37}$$

Thus, maximum work in this open flow system can be attained only under reversible conditions for which $S_g = 0$; otherwise, the last term again is the lost work. The term $\left(\tilde{h} - T_0 s\right)$ is called the **thermomechanical or physical flow availability function** or **flow exergy function**. At steady state, the work transfer is due to two components: the heat transfer across the boundaries (the first term in (2.37)); and the difference between the incoming and the outgoing flow exergy. The useful work can be recovered fully as the irreversibilities inside the control volume are eliminated, and the system is allowed to reach equilibrium with its environment.

Example 2.6

Gas enters a steady-flow, steady-state device at state 1 and is accelerated to supersonic speeds at state 2. The device can exchange heat with the environment at temperature T_0. Derive an expression for the maximum velocity at state 2, V_2, in terms of the gas properties at states 1 and 2. You may assume that the inlet velocity is negligible compared to that of state 2.

Solution

The energy and entropy balance equations for this device, assuming no work interactions, negligible potential energy changes, and kinetic energy at state 1 are:

$$\text{First Law}: \ \dot{m}\,h_1 = \dot{m}\left(h_2 + \frac{V_2^2}{2}\right) + \dot{Q}.$$

$$\text{Second Law}: \ \dot{m}\,s_1 + \dot{S}_g = \dot{m}\,s_2 + \frac{\dot{Q}}{T_0}.$$

Example 2.6 (cont.)

Combining to eliminate the heat transfer term, and solving for the maximum velocity:

$$\frac{V_2^2}{2} = (h_1 - T_0 s_1) - (h_2 - T_0 s_2) - T_0 \frac{\dot{S}_g}{\dot{m}}.$$

The maximum velocity occurs when the last term is zero:

$$V_2 = \sqrt{2[(h_1 - T_0 s_1) - (h_2 - T_0 s_2)]}.$$

We can now draw some useful conclusions from the application of the combined statement to a control volume, and for simplicity we restrict the analysis to steady-state operations:

1. For maximum work, the control volume components interacting with the environment must reach thermal and mechanical equilibrium with the environment. Moreover, the mass flowing out must also reach equilibrium with the environment.
2. If the control volume is allowed heat transfer interaction with the environment only, then the first term in (2.37) drops out. Furthermore, at steady state the unsteady term disappears and the maximum work is the difference between the availabilities at the two end states, with no internal entropy generation:

$$\dot{W}_{max} = \dot{m}[(h_{in} - T_o s_{in}) - (h_{out} - T_o s_{out})]. \tag{2.38}$$

3. In the case when the system interacts with a "thermal energy reservoir" at a fixed temperature T_H, and with the environment at T_o, while states 1 and 2 are identical, the maximum work transfer is given by

$$\dot{W}_{max} = \dot{W}_{Carnot} = \left(1 - \frac{T_o}{T_H}\right)\dot{Q}_H. \tag{2.39}$$

Example 2.7 Entropy Generation/Irreversibility in a Heat Exchanger

A liquid "C" with a specific heat of 4 kJ/kg·K is heated in a counter-flow heat exchanger by another "H" with a specific heat of 6 kJ/kg·K. The cold liquid enters the heat exchanger at 10 °C at a rate of 0.5 kg/s and exits at 60 °C. The hot liquid enters at 150 °C and exits at 40 °C. The heat exchanger is well insulated, and only internal heat exchange is allowed. The environment is at 10 °C. Determine the entropy generation rate and irreversibility in the system.

Example 2.7 (cont.)

Solution

We assume constant specific heats for both liquids with specific heats $C_C = 4\,\text{kJ/kg·K}$, $C_H = 6\,\text{kJ/kg·K}$. We also use $h_2 - h_1 = C\ (T_2 - T_1)$ and $s_2 - s_1 = C\ln\left(\frac{T_2}{T_1}\right)$.

The energy balance around the heat exchanger is considered an adiabatic process with no work interactions. Inlets are labeled as 1 and outlets as 2:

$$m_C C_C (T_2 - T_1)_C = m_H C_H (T_1 - T_2)_H.$$

Solving for the hot liquid flow rate, we get

$$\dot{m}_H = \dot{m}_C \frac{C_C(T_2 - T_1)_C}{C_H(T_1 - T_2)_H} = (0.5)\frac{(4)(60 - 10)}{(6)(150 - 40)} = 0.2\,\text{kg/s}.$$

The entropy balance across the boundary of the heat exchanger is

$$\dot{m}_C(s_1 - s_2)_C + \dot{m}_H(s_1 - s_2)_H + \dot{S}_{\text{gen}} = 0.$$

Substituting the parameters of the two streams, we get:

$$\dot{S}_{\text{gen}} = (0.5)(4)\ln\left(\frac{60 + 273}{10 + 273}\right) + (0.2)(5)\ln\left(\frac{40 + 273}{140 + 273}\right) = 2.09 \times 10^{-2}\,\text{kW/K}.$$

The rate of irreversibility in the system is

$$T_0 \dot{S}_{\text{gen}} = (10 + 273) \times 2.09 \times 10^{-2} = 5.917\,\text{kW}.$$

Example 2.8

Water at 200 kPa and 100 °C is expanded in an adiabatic throttle valve to a final pressure of 20 kPa. The process does not involve any work transfer interactions. An inventor claims to have designed a device that generates work of 10 kJ/kg of water while maintaining the same inlet and outlet conditions of the throttle and exchanging heat with the environment at 25 °C. Is this claim feasible?

Solution

We assume steady operation and neglect changes in the kinetic and potential energies. At 200 kPa and 100 °C, $h_1 = h_{f@100°\text{C}} = 419.17\,\text{kJ/kg}$ and $s_1 = s_{f@100°\text{C}} = 1.3072\,\text{kJ/kg·K}$. The energy balance across an adiabatic throttle is: $h_2 = h_1 = 419.17\,\text{kJ/kg}$. The final state is

Example 2.8 (cont.)

determined by knowing the final pressure, p_2, and the final enthalpy, h_2. Since the enthalpy falls between the saturated liquid and the saturated vapor values at 20 kPa, $h_{f@20\text{kPa}} = 251.42$ kJ/kg and $h_{g@20\text{kPa}} = 2608.9$ kJ/kg, the quality of the mixture is $x_2 = \left(h_2 - h_f\right)/h_{fg} = 0.0712$ and the entropy is $s_2 = s_f + x_2\ s_{fg} = 0.8320 + 0.0712 \times 7.9073 = 1.3354$ kJ/kg·K. The maximum work is the difference between the availability functions between the initial and final states:

$$w_{\max} = (h_1 - T_o s_1) - (h_2 - T_o s_2) = T_o(s_2 - s_1).$$

$$w_{\max} = T_o(s_2 - s_1) = (25 + 273) \times (1.3354 - 1.3072) = 8.4174 \text{ kJ/kg}.$$

The work output claimed by the inventor is higher than the maximum value, and the proposed design is not possible.

Now we define the **specific flow thermomechanical exergy** with respect to an environment at (T_o, p_o) as

$$\xi = \left(\tilde{h} - h_o\right) - T_o(s - s_o). \tag{2.40}$$

Equation (2.37) can be written in terms of the specific flow availability, while keeping the definition of the control volume exergy as before:

$$\begin{aligned}\dot{W}_{cv} = &\int\left(1 - \frac{T_o}{T_s}\right)\dot{q}_s da - \left(\frac{d\Xi_{cv}}{dt} - p_o\frac{d\forall_{cv}}{dt}\right)\\ &+ \sum_{in}\dot{m}_{in}\xi_{in} - \sum_{out}\dot{m}_{out}\xi_{out} - \dot{I}_{ir}.\end{aligned} \tag{2.41}$$

The irreversibility is $I_{ir} = T_o S_g$ and the control mass availability is given by (2.30):

$$\Xi = (E - U_o) + p_o(\forall - \forall_o) - T_o(S - S_o). \tag{2.42}$$

Quantities with a subscript o are evaluated at the conditions of the environment for the mass of the control volume. While interacting with a finite number of heat reservoirs at fixed temperatures, and transferring a finite amount of heat, (2.41) is written in an integral form as follows:

$$\begin{aligned}\dot{W}_{cv} = &\sum_{TERs}\left(1 - \frac{T_o}{T_i}\right)\dot{Q}_i - \left(\frac{d\Xi_{cv}}{dt} - p_o\dot{\forall}_{cv}\right)\\ &+ \sum_{in}\dot{m}_{in}\xi_{in} - \sum_{out}\dot{m}_{out}\xi_{out} - \dot{I}_{ir}.\end{aligned} \tag{2.43}$$

Equation (2.43) shows that for a steady-state process with a single stream interacting with a high-temperature TER and the environment, the work is

$$\dot{W}_{cv} = \left(1 - \frac{T_o}{T_H}\right)\dot{Q}_H + \dot{m}(\xi_{in} - \xi_{out}) - \dot{I}_{ir}. \tag{2.44}$$

Thus, for instance, in an adiabatic process or system, a decrease in availability inside the control volume that is not balanced by work transfer leads to irreversibility.

It is important to remember the strong connection between availability loss, entropy generation, irreversibility, and work loss. While these quantities are strongly related and describe similar effects, they are often used in different contexts depending on convenience and how a process/system is described. Entropy generation in a process, S_g, implies work loss or irreversibility equal to T_oS_g. The change in the availability, however, is more than the work loss if the process in non-adiabatic, and/or involves work transfer interactions and mass transfer across the system boundaries. The availability changes when the system interacts with the environment through heat transfer and work transfer. When following a sequence of processes, using availability is often an effective way to determine where the losses are, and to focus attention where improvement opportunities exist. It is also important to mention that in evaluating availability and its change, the only properties of the environment required are the temperature and pressure (when mass transfer is not allowed).

To summarize, availability/exergy is the maximum work that can be obtained from the combined system and its environment as the system reaches the restricted dead state of the environment. At this final equilibrium state, the availability of the system relative to its environment is zero, although the total energy of both is not. The contribution of the kinetic and potential energy of the system availability relative to the environment is 100%, while that of heat transfer is at the ratio of $\left(1 - \frac{T_o}{T_s}\right)$. All the analysis presented in this chapter pertains to the thermomechanical availability; that is, assuming that the chemical state of the system and environment do not change during the process of work and heat transfer. Thermochemical availability will be discussed in detail in the next chapter.

As an application of this development, we calculate the minimum compression work required to raise the pressure of an ideal gas. The minimum compression work is achieved in isothermal processes, operating in equilibrium with the environment. It is given by (2.44), with $I_{ir} = 0$ and $T_H = T_o$ – that is, for a unit mass of an ideal gas:

$$\hat{w} = \left(\hat{\xi}_{in} - \hat{\xi}_{out}\right) = \left(\hat{h}_{in} - \hat{h}_{out}\right) - T_o(\hat{s}_{in} - \hat{s}_{out}) = \Re T_o \ln\left(\frac{p_o}{p}\right). \tag{2.45}$$

Keeping the temperature constant is difficult in practical processes, and multistage adiabatic compression with intercooling is used to approach this ideal isothermal process, as shown in Figure 2.9.

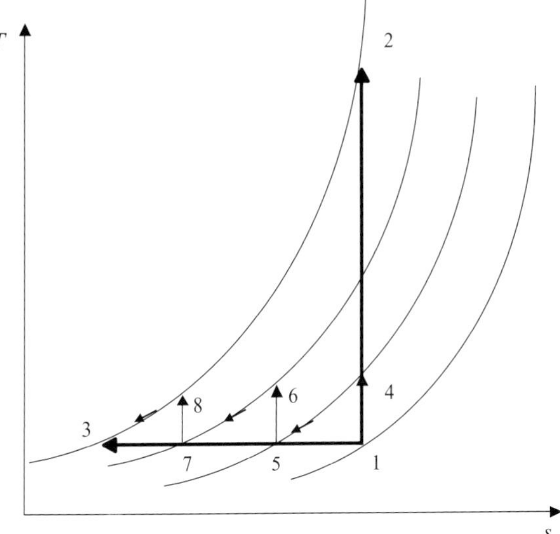

Figure 2.9 Isentropic single-stage compression, process 1–2; isothermal single-stage compression, process 1–3; and multistage isentropic compression with intercooling, process 1–4–5–6–7–8–3.

Example 2.9 A Gas Turbine Cycle

Moran [12] gives a nice example for insight gained from availability analysis. A closed-cycle gas turbine power plant, shown in Figure E2.9, operates with air as a working fluid. Air is first compressed from 1 atm and 300 K to a pressure of 20 atm. The air exits the heat exchanger at a temperature of 1600 K. It is expanded in the turbine to atmospheric pressure and a temperature of 816.4 K. Table E2.9a summarizes the properties of air at 4 different states of the cycle.

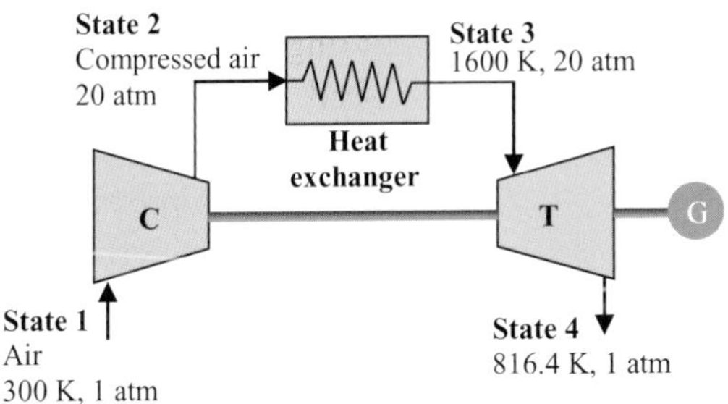

Figure E2.9 Schematic of gas turbine operation in Example 2.0.

Table E2.9a State properties for Example 2.9

State	T(K)	p (atm)	h (kJ/kg·K)	ξ (kJ/kg)
1	300	1	0	0
2	808.3	20	510.4	469.8
3	1600	20	1305.2	1058.9
4	816.4	1	519.4	217.5

Example 2.9 (cont.)

Determine the availability from the heat exchanger, the compressor, and the turbine, and the Second Law efficiency of the cycle.

Solution

Assume air is an ideal gas, the compressor and turbine are adiabatic, and the kinetic and potential energy changes in the different devices are negligible. Furthermore, the system experiences no pressure drop. The environmental conditions are taken to be the same as that of the air at state 1.

Table E2.9b shows the enthalpy and availability change between different states; the latter is obtained from the definition in (2.40). Energy and availability are added to the system in the heat exchanger. Note that the changes in the availability in processes 1–2 and 3–4 are not the same as the changes in the enthalpy during the same processes because of the irreversibility in the compression and the expansion processes. They would have been the same if the processes were ideal (isentropic). The second column shows the results of the energy analysis of the system, while the third column shows the availability analysis. Different information can be gleaned from both.

Table E2.9b Enthalpy and availability changes between states in Example 2.9

	Enthalpy change (kJ/kg)	Availability change (kJ/kg)
Increase through heat exchanger	$h_3 - h_2 = 794.8$	$\xi_3 - \xi_2 = 589.1$
Compressor	$W_c = 510.4$	–469.8
Turbine	$W_t = 785.8$	841.4
Net work	$(h_3 - h_4) - (h_2 - h_1) = 275.4$	
Air out at 4	$h_4 - h_1 = 519.4$	217.5
Irreversibility		
Compressor		$\frac{\dot{I}}{\dot{m}} = -\frac{\dot{W}}{\dot{m}} + \xi_1 - \xi_2 = 40.6$
Turbine		$\frac{\dot{I}}{\dot{m}} = -\frac{\dot{W}}{\dot{m}} + \xi_3 - \xi_4 = 55.6$

The conservation of energy shows that the enthalpy increase in the heat exchanger equals the sum of the net work out and the enthalpy in the exiting stream. Energy analysis also shows that the energy input is through the heat exchanger, and the exhaust stream enthalpy is rather high – that is, most of the input energy is leaving with the exhaust stream. The availability added in the heat exchanger is 589.1 kJ/kg. The maximum work of the system is the sum of the availability change across the compressor and turbine, plus the availability out, 589.1 kJ/kg (the same as the availability added in the heat exchanger).

Example 2.9 (cont.)

Moreover, the sum of the network output, the irreversibility in the turbine and the compressor, and the availability in the exhaust stream equals the availability added to the system in the heat exchanger. The thermodynamic First Law efficiency of this system, defined by the net work divided by the heat input, is 34.6%. The availability analysis shows that the turbine and compressor losses are relatively low, and that the availability loss associated with the exhaust stream is high. The overall performance can be improved significantly if the exhaust availability could be used in another process to generate more work. A Second Law efficiency, which will be defined later, is the work produced by the system as a fraction of the maximum work, or the availability added. For this system, the Second Law efficiency is 275.4/589.1 = 0.467.

2.5 Thermodynamic Efficiency

Thermodynamics is used to evaluate the performance of heat engines and other energy conversion devices such as reformers, fuel cells, and storage devices. Heat engines or power cycles convert heat, from combustion or other sources, to work. Thermal energy transferred to the engine may be converted to mechanical work (in engines) or electrical work such as in a thermoelectric device. Energy can also be available in the form of chemical energy stored in a fuel, which is converted into thermal energy during a combustion process, and the thermal energy is then converted to work. This is a combustion engine. The inverse uses mechanical or electrical work to transfer heat out of a system, as in a refrigerator, or from a low temperature to a high temperature, as in a heat pump. Heat engines for power generation nominally operate between two temperature limits; one is often that of the environment and the other is the highest temperature available. The high temperature is determined by fuel combustion, the nuclear reactor operating conditions, a solar collector, or the geothermal source, for example.

There are two types of efficiencies that we use; one relates to the conversion efficiency and the other to the quality or effectiveness of the process. **Conversion efficiencies** are also known as First Law efficiencies since they involve processes that convert one form of energy to another – for example, heat to work, thermal energy to chemical energy, chemical energy to electrical work, one form of chemical energy to another. The second category, **effectiveness**, is known as the Second Law efficiency as they most often compare the outcome of a process with that obtained in the corresponding ideal process, or the output of a system executing a series of processes to the output of the same system if all processes were ideal. In the following we define a number of these efficiencies, and more will be defined in later chapters.

2.5.1 Heat Engine Conversion Efficiency

One important performance parameter of heat engines used in power generation and propulsion cycles is the thermodynamic or **thermomechanical** or simply thermal efficiency. For a heat engine interacting with two heat reservoirs, as shown in Figure 2.10, the thermodynamic efficiency or First Law efficiency is defined as

$$\eta_I = \frac{W_{net}}{Q_H}, \tag{2.46}$$

where W_{net} is the net work output or the algebraic sum of all work interactions during the cycle, and Q_H is the sum of all heat transfer from the high-temperature TER(s) to the engine. The fact that these two quantities – the heat in and the work out – are positive is convenient. The efficiency definition implicitly states that the network output in a heat engine is always less than the required heat input. This is one of the statements of the Second Law that requires a heat engine to reject heat to the environment to operate as a cycle. Using energy balance on the entire cycle operating under steady-state condition yields a balance between the net energy transfers in and out of the cycle:

$$Q_L + W_{net} = Q_H \text{ or } W_{net} = Q_H - Q_L,$$

where Q_L is the absolute value of the sum of all heat transfer out of the engine. Therefore, the thermomechanical efficiency can be alternatively written as

$$\eta_I = 1 - \frac{Q_L}{Q_H}. \tag{2.47}$$

Despite its convenience, this expression is limited to this special case, and other expressions should be developed according to the system or process.

It is often useful to express the efficiency of a power cycle in terms of the terminal temperatures available for the heat engine. This depends on the processes or cycle performed by the engine during the heat and work interactions. Efficiency expressions for many power cycles will be developed in the following chapters. The ideal power cycle, with the highest

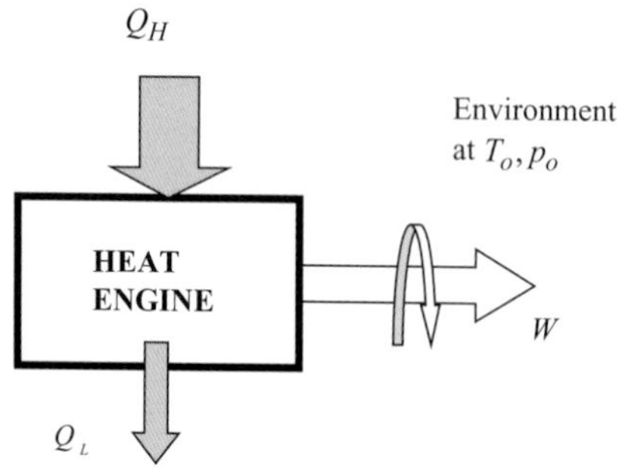

Figure 2.10 A model for a thermomechanical energy conversion system. A heat engine uses a hot reservoir in a power cycle.

possible thermal efficiency, is the Carnot cycle. For this engine, heat is transferred isothermally at two temperatures, and the thermal efficiency is

$$\eta_{Carnot} = 1 - \frac{T_L}{T_H}. \tag{2.48}$$

It should be clear from the maximum work expressions in the previous section that the Carnot cycle extracts the maximum work from a given amount of heat by interacting with two thermal energy sources at two different temperatures. The implications of this expression are generic: for maximum energy conversion efficiency, heat transfer to the engine must be at the highest possible temperature, while it should be at the lowest possible temperature for heat transfer out. Practical limitations related to material tolerance and the availability of a thermal energy source/sink at the requisite temperatures often limit what is possible, and reaching the highest/lowest possible temperature to transfer heat in/out often requires more complex hardware and hence higher capital cost for the energy conversion device.

The Carnot cycle operating between two "infinite" TERs, that is between a TER at a "fixed" temperature, T_H, and the environment at T_o, achieves the theoretical maximum cycle efficiency $\eta_{Carnot} = (1 - T_o/T_H)$. As an example, consider the case of combustion-generated high-temperature products. Typical combustion processes reach T_H/T_o of the order of 6–8, depending on the fuel and the fuel–air ratio. In such a case, this model shows that the theoretical efficiency of an engine operating between combustion products and the environment is high, with values of η_{Carnot} in the range of 84–88%.

However, this model suffers from one important shortcoming: the assumption that the high-temperature TER's temperature remains constant. This may be unreasonable since the thermal energy utilized during the heat transfer–work transfer process is likely to be associated with a temperature drop.

In another model that addresses this shortcoming, a high-temperature stream is supplied to the heat engine at T_H and is cooled down to T_o using an infinite series of Carnot cycles, as shown in Figure 2.11, whose maximum temperature varies such that $T_0 \leq T \leq T_H$. The total work transfer is obtained by summing over the work done be all the differential engines operating between T and T_o:

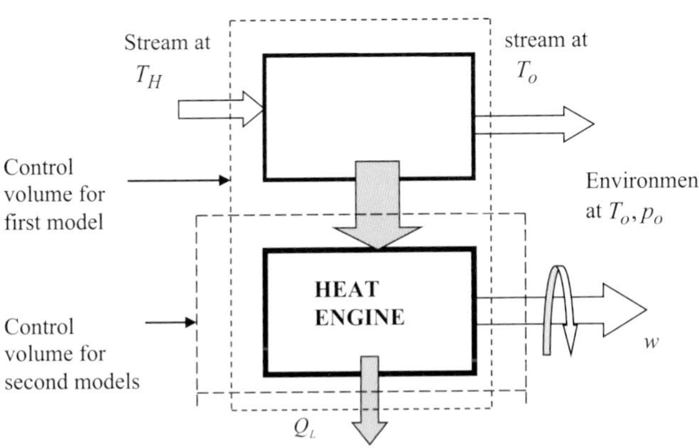

Figure 2.11 Models for a heat engine operation with a hot stream being cooled to the environmetal temperature and the environment (as a cool reservoir).

$$W_{max} = \int_{T_o}^{T_H} \left(1 - \frac{T_o}{T}\right) \delta Q_{in}. \tag{2.49}$$

The temperature difference available for the work transfer decreases and hence the work transfer decreases while the stream is flowing between the inlet and outlet of this engine. Writing the heat transfer in terms of the absolute temperature scale, $dQ_{in} = CdT$,

$$W_{max} = \int_{T_o}^{T_H} \left(1 - \frac{T_o}{T}\right) CdT = C\left[(T_H - T_o) - T_o \ln \frac{T_H}{T_o}\right]. \tag{2.50}$$

The total heat transfer is $Q_{in} = C(T_H - T_o)$, and the efficiency of this heat engine is

$$\eta^*_{Carnot} = 1 - \frac{\ln \dfrac{T_H}{T_o}}{\left(\dfrac{T_H}{T_o} - 1\right)}. \tag{2.51}$$

The same expression for the maximum work can be obtained from the availability expression in (2.37), assuming an ideal gas as a working fluid, and achieving thermal and mechanical equilibrium with the environment:

$$W_{max} = (H_H - H_o) - T_o(S_H - S_o) = C\left[(T_H - T_o) - T_o \ln \frac{T_H}{T_o}\right]. \tag{2.52}$$

Typical values of this efficiency can be obtained using the same estimates for an ideal combustion temperature as before. For instance, for $T_o = 300$ K and $T_H = 2400$ K, the efficiency is 70%. As expected, this is lower than the ideal Carnot cycle efficiency operating at a single high temperature, T_H. It is interesting to note that our first model leads to the same thermal efficiency as the second model if we used the log mean temperature between the high-temperature TER and the environment. Figure 2.12 shows a comparison between the two efficiencies as the ratio of the high temperature to the low temperature changes.

2.5.2 Conversion Efficiencies of Other Systems

The definition of process, cycle, or system efficiency in applications other than simple heat engines depends on the detail of the energy conversion process. Other efficiencies must be designed to emphasize the success of a process in delivering its design objectives. Generically, we can write the efficiency as

$$\eta_I = \frac{\text{Work/energy/heat out}}{\text{Heat/energy/work in}}. \tag{2.53}$$

Several other efficiencies will be defined later, as appropriate, to determine the "goodness" of an energy conversion system. Especially when energy conversion involves chemical energy,

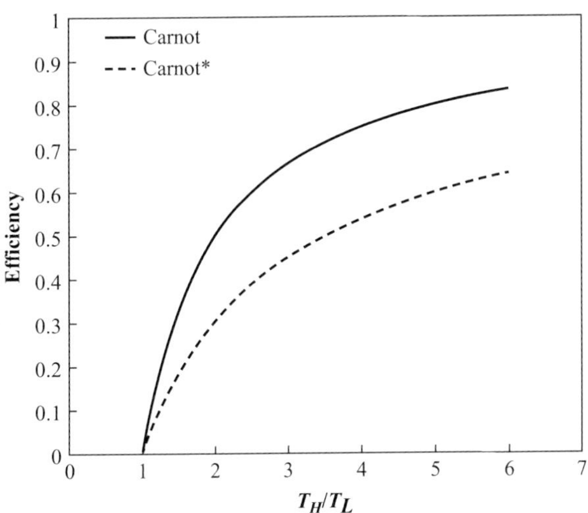

Figure 2.12 The dependence of the simple Carnot efficiency, and the Carnot efficiency for an engine operating by a cooling stream on the ratio of the high temperature to that of the environment, Carnot*.

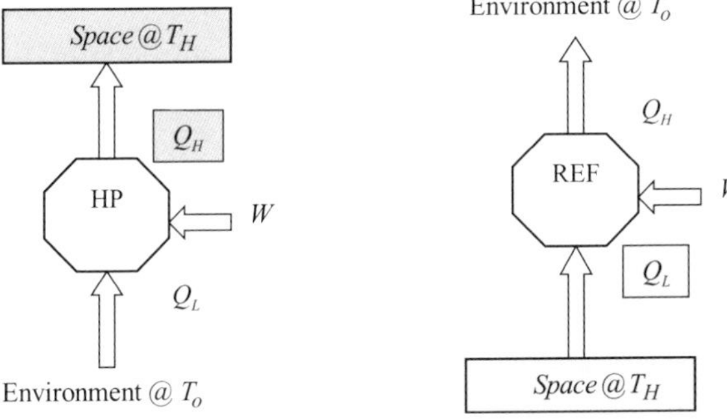

Figure 2.13 Heat pump HP used to heat up a space on the left-hand side, and refrigerator REF used to cool down a space on the right-hand side.

several consequences will be discussed and quantified through efficiencies. Note, however, that in all cases both the numerator and denominator are forms of work, heat, or energy, and the transition between them is an energy conversion process.

It is important to think of conversion efficiency as a ratio between the intended output and what we must supply to make it happen. One way of writing this definition is

$$\eta_I = \frac{\text{All that you get}}{\text{All that you pay}}. \tag{2.54}$$

Note that "all" indicates that in some systems multiple inputs are possible, as well as multiple outputs. This makes it necessary sometimes to use different names for this "efficiency" in order not to confuse it with heat engine efficiency. For example, in the refrigeration process shown in Figure 2.13, the input is likely to be work, and the output is heat transfer at the low-temperature end. The ratio of these two quantities is known as the coefficient of performance, and designated as β. In the case of a refrigerator, it is

$$\beta_{ref} = \frac{Q_L}{W}. \tag{2.55}$$

On the other hand, a heat pump is a device used to transfer heat from a low-temperature source to a high-temperature environment, using work as input also. The coefficient of performance of a heat pump is

$$\beta_{HP} = \frac{Q_H}{W}. \tag{2.56}$$

Other efficiencies and performance measures will be defined as needed.

2.5.3 Second Law Efficiency

Building heat engines that execute a Carnot cycle or other forms of idealized processes/systems is nearly impossible, either because of the difficulties in executing isothermal heat transfer processes, or isentropic expansion and compression, or because of the presence of other forms of irreversibility, such as the need for a finite temperature gradient for a finite rate of heat transfer, friction, pressure losses due to flow processes, etc. More practical cycles, for instance, are designed to transfer heat under approximately constant-pressure or constant-volume conditions. Some of these cycles will be discussed in the coming chapters. Another efficiency, the Second Law efficiency of heat engines, or the exergy efficiency, or the effectiveness, is often used to measure the efficiency of these cycles relative to that of the corresponding ideal cycles that operate under similar conditions but execute ideal processes:

$$\eta_{II} = \frac{W}{W_{max}} = \frac{\eta_I}{(\eta_I)_{ideal}}. \tag{2.57}$$

Consider, for instance, an engine operating between two fixed-temperature heat reservoirs, and compare the engine efficiency to that of the Carnot cycle:

$$\eta_{II} = \frac{W_{net}/Q_H}{1 - T_L/T_H}. \tag{2.58}$$

The numerator, which is the thermal efficiency of the cycle, must be evaluated according to the heat engine design, while the denominator refers to the highest and lowest available temperatures. Clearly this term must be lower than unity.

Using a heat engine, work transfer is achieved by interacting with either:

1. one high-temperature thermal reservoir and the environment, which is a low-temperature reservoir (the reverse is also possible); or
2. a hot stream, which is cooled to the temperature of the environment while it is used in the system.

Thus, other more general Second Law efficiencies measure the performance of the cycle with respect to the maximum possible work; for example,

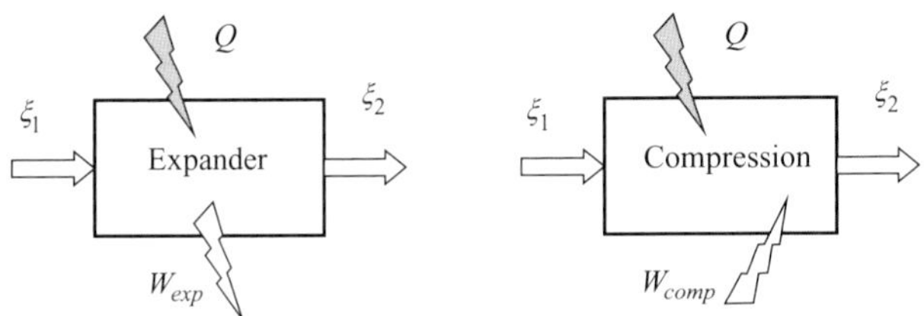

Figure 2.14 Availability analysis for an expansion process and a compression process.

$$\eta_{II} = \frac{W_{net}}{\Delta \Xi} = \frac{W_{net}}{W_{max}}. \tag{2.59}$$

The maximum work is calculated from the combined statements of the First and Second Laws, or the difference in availability (2.32).

Second Law efficiencies can be used to measure the performance of an individual process, or system, with respect to the maximum possible work in that process, or to an idealized form of that process. For instance, if the process is work-producing, then

$$\eta_{II} = \frac{W}{W_{max}}. \tag{2.60}$$

Note that in this expression we may allow heat transfer with the environment. For example, using (2.44) for the steady expansion process/system interacting only with the environment, shown in Figure 2.14, we get

$$\eta_{II,\,exp} = \frac{W_{exp}}{\Delta \xi_{exp}}. \tag{2.61}$$

In this expression, and per unit mass of the expanding stream, (2.44) shows that $w_{max} = \Delta \xi_{exp}$, with $I_{ir} = 0$, and $Q_H = 0$. The inverse, under the same conditions, is used in the case of a compression process/system:

$$\eta_{II,com} = \frac{\Delta \xi_{comp}}{\left| W_{comp} \right|}. \tag{2.62}$$

Second Law efficiencies for flow mixing and heat transfer processes can also be defined. Note that these efficiencies are not the "adiabatic" or isentropic efficiencies used in cycle analysis, which are defined next.

For energy conversion devices other than heat engines, different definitions must be used. The ratio of the actual to the ideal coefficient of performance is the "figure of merit" of a refrigerator or a heat pump:

$$\eta_{II} = \frac{(\beta_{HP})_{act}}{(\beta_{HP})_{ideal}}. \tag{2.63}$$

If the heat pump operates between T_L and T_H, its ideal performance is: $\beta_{HP} = \frac{T_H}{T_H - T_L}$.

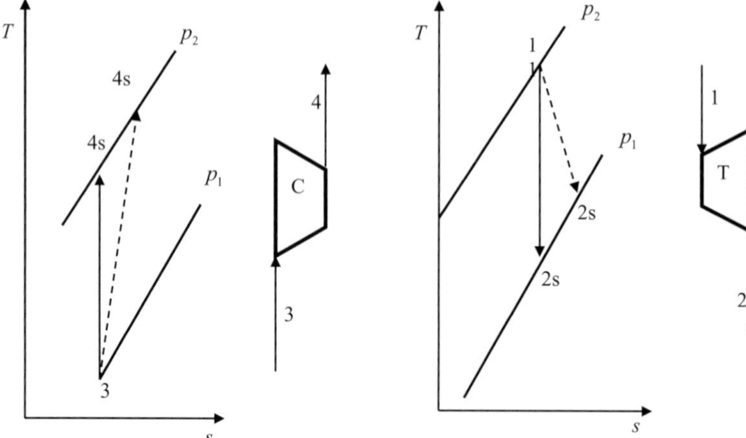

Figure 2.15 Adiabatic expansion (right) and compression (left) processes on T–s diagram. Solid arrows: ideal processes; broken arrows: actual processes.

2.5.4 Isentropic Efficiency

In many cases, other process efficiencies must be defined that are different from the Second Law efficiencies defined above. For example, adiabatic reversible or isentropic efficiencies are used for turbines, pumps, and compressors, which are defined as follows:

$$\eta_{is,\,tur} = \frac{W}{W_{is}}. \tag{2.64}$$

In this expression, the actual work is the adiabatic expansion work, while the ideal or isentropic work is the ideal adiabatic expansion work. Similarly, for a compressor, we define the isentropic efficiency as follows:

$$\eta_{is,\,comp} = \frac{W_{is}}{W}. \tag{2.65}$$

A similar expression may be given for the isentropic efficiency of a pump. These processes are shown schematically in Figure 2.15, which illustrates compression and expansion processes on a T–s diagram between two pressures.

Figure 2.15 shows that under adiabatic conditions, non-ideal behavior results in higher entropies than the ideal isentropic (adiabatic and reversible work) behavior. Isentropic processes on a T–s diagram are vertical. Similar definitions based on kinetic energies may also be defined for nozzles and diffusers.

Example 2.10

An industrial plant requires a large quantity of high-temperature heat, which it generates by burning kerosene. After extracting the "useful" high-temperature heat from the combustion products, the plant discharges gases at 950 K and 1 atm. The flow rate of combustion gases is

Example 2.10 (cont.)

2.0 kg/s. A waste heat recovery system (WHRS), shown in Figure E2.10 is proposed for the utilization of the energy in the hot exhausted gases. Waste heat recovery is an important approach to improve the efficiency of many industrial processes. The WHRS consists of a steam generator called the heat recovery steam generator (HRSG) and a steam turbine as shown in the figure. In this problem we examine the performance of the HRSG. Assume that the HRSG is adiabatic.

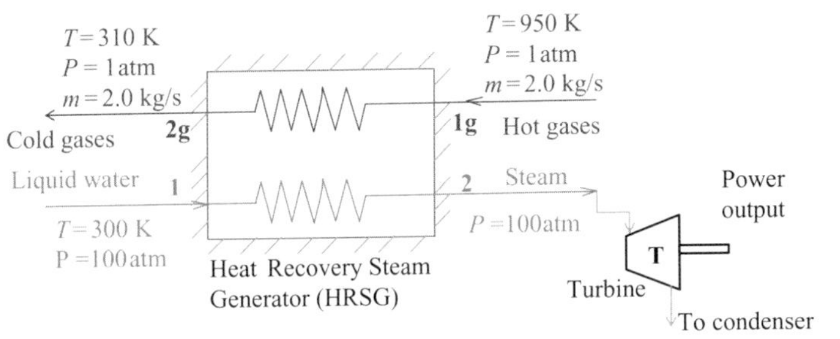

Figure E2.10a Schematic of a WHRS.

The overall efficiency of the WHRS depends, among other things, on the performance of the HRSG. The temperature profiles of the two streams in the HRSG are shown schematically in Figure E2.10. The pinch point (PP) determines the minimum temperature difference in the HRSG. The PP is where the cold stream starts to evaporate. It is determined by points 2g and 1c. The PP gives rise to larger temperature difference on the hot side, resulting in lower turbine output power.

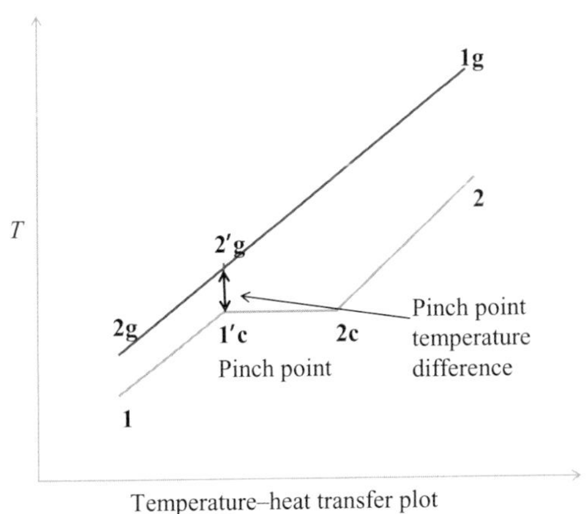

Figure E2.10b Schematic showing temperature–heat transfer plot in the HRSG. 1g–2g shows the temperature profile of the gas and 1–2 shows the temperature profile of the steam.

Example 2.10 (cont.)

Assume the PP temperature difference in the HRSG is 10 °C. The atmospheric conditions are at 1 atm and 300 K. You may assume the hot combustion products as an ideal gas with the same properties as air. The specific heat of air is $c_{p,GAS} = 1.048$ KJ/kg·K. The inlet and exit conditions of the hot and cold streams in the HRSG are shown in Figure E2.10a.

a. Calculate the maximum work that can be obtained from the hot gases leaving the industrial plant.
b. Given the conditions of the gases leaving the HRSG, calculate the maximum work again.
c. Evaluate the mass flow rate of the water.
d. Evaluate the exit conditions of the steam leaving the HRSG.
e. Calculate the increase in the availability of water in the HRSG
f. Calculate the irreversibility in the HRSG.
g. The steam expands in the turbine to a pressure that corresponds to saturation temperature of 40 °C. The isentropic efficiency of the turbine is 94%. Calculate the change of availability in the turbine and the condenser. You can assume that the exit of the condenser is saturated liquid at 40 °C. After the condenser, there is an additional heat exchanger that supercools water further before it gets pumped to the entry of the HRSG. This completes the Rankine cycle.

Solution

Consider the control volume as the boundary of the heat exchanger.

a. The maximum work can be obtained using the availability equation for the hot gases:

$$\text{Maximum work} = \dot{\Xi}_{GASES} = \dot{\Xi}_{1g} = \dot{m}_{1g}\left[\left(h_{1g} - h_{0g}\right) - T_0\left(s_{1g} - s_{0g}\right)\right]$$

$$= \dot{m}_{1g}\left[\left(h_{1g} - h_{0g}\right) - T_0\left(c_{p,GAS}\ln\left(\frac{T_1}{T_0}\right)\right)\right]$$

$$= 2[(995.6 - 314.4) - 300(1.048^* \ln(590/300))] = 638.1\,\text{kW}.$$

b. Change in the availability of gases from inlet to exit:

$$\Delta\dot{\Xi}_{GASES} = \dot{\Xi}_{1g} - \dot{\Xi}_{2g} = \dot{m}_{1g}\left[\left(h_{1g} - h_{2g}\right) - T_0\left(s_{1g} - s_{2g}\right)\right]$$

$$= \dot{m}_{1g}\left[\left(h_{1g} - h_{2g}\right) - T_0\left(c_{p,GAS}\ln\left(\frac{T_1}{T_2}\right)\right)\right]$$

$$= 2[(995.6 - 324.9) - 300(1.048\ln(950/310))] = 637.3\,\text{kW}.$$

c. Mass flow rate of water:

We have information on the cold side of the HRSG and PP states. Using the energy balance between the two streams from the cold side of the HRSG to the PP, we can solve for the mass flow rate of water:

Example 2.10 (cont.)

$$\dot{m}_{1g}(h_{2'g} - h_{2g}) = \dot{m}_w(h_{1'c} - h_1)$$
$$\dot{m}_{1g}C_{P,GAS}(T_{2'g} - T_{2g}) = \dot{m}_w(h_{1'c} - h_1).$$

Looking at the steam tables to determine the specific enthalpy of steam as saturated liquid at a pressure of 100 atm (10132.5 kPa), we get $h_{1'c} = 1413.0\,\text{kJ/kg}$. Looking at the enthalpy of water at $T = 300\,\text{K}$ and $p = 101{,}325\,\text{kPa}$, we get $h_1 = 121.8\,\text{kJ/kg}$.

From the steam tables, the saturation temperature $T_{1'c}$ at a pressure of 101,325 kPa is **585.2 K**.

The PP temperature difference is 10 K. $T_{2'g} = T_{1'c} + 10 = \mathbf{595.2\ K}$.

Therefore, the mass flow of water is:

$$\dot{m}_w = \frac{\dot{m}_{1g}C_{P,GAS}(T_{2'g} - T_{2g})}{(h_{1'c} - h_1)} = \frac{2.0\,1.048\,(595.2 - 310)}{(1413 - 121.8)} = \mathbf{0.4629 kg/s}.$$

d. Exit conditions of steam:

Now we know the mass flow rate of the water, we can apply the energy equation for the whole of the HRSG as a control volume

$$\dot{m}_{1g}(h_{1g} - h_{2g}) = \dot{m}_w(h_2 - h_1)$$
$$\dot{m}_{1g}C_{P,GAS}(T_{1g} - T_{2g}) = \dot{m}_w(h_2 - h_1)$$
$$2\,1.048\,(950 - 310) = 0.4629(h_2 - 121.8).$$

This gives $h_2 = 3020\,\text{kJ/kg}$. Using $h_2 = 3020\,\text{kJ/K}$ and $p_2 = 100\,\text{atm}$, looking in the steam tables, we get $T_2 = 650.7\,\text{K}$. The exit temperature is **650.7 K**.

e. Increase in the availability of water coming out of the HRSG:

The increase in the availability of the water in the HRSG can be evaluated in a similar way to how we calculated the availability in part (b) for gases:

$$\Delta\dot{\Xi}_{STEAM} = \dot{\Xi}_2 - \dot{\Xi}_1 = \dot{m}_w[(h_2 - h_1) - T_0(s_2 - s_1)].$$

Using the steam tables, at exit conditions of $T_2 = 650.7$ K and $p_2 = 100$ atm, we find $s_2 = 6.091$ [kJ/kg·K]. Similarly, at inlet conditions, $T_1 = 300.0$ K and $p_2 = 100$ atm, and $s_1 = 0.3928$ [kJ/kg·K]. The increase in the availability of water is:

$$\Delta\dot{\Xi}_{STEAM} = 0.4629[(3020 - 112.8) - 300(6.091 - 0.3928)] = \mathbf{549.8\ [kJ/s]}.$$

f. Irreversibility in the HRSG:

Using the exergy balance equation for the HRSG, we can evaluate the exergy destroyed in the HRSG. The exergy destroyed will provide the irreversibility in the HRSG. Alternatively, we can also solve using the entropy balance equation to calculate irreversibility (entropy generation multiplied by the ambient temperature):

Example 2.10 (cont.)

$$\dot{\Xi} = 0 = \sum\left(1 - \frac{T_0}{T_j}\right)\dot{Q}_j - \dot{W}_{CV} + \dot{\Xi}_{1g} - \dot{\Xi}_{2g} + \dot{\Xi}_1 - \dot{\Xi}_2 - \dot{\Xi}_{DESTRUCTION}.$$

Because the HRSG is adiabatic and there is no net work done by it, heat transfer and work terms are zero.

$$\dot{\Xi} = 0 = \sum\left(1 - \frac{T_0}{T_j}\right)\dot{Q}_j - \dot{W}_{CV} + \dot{\Xi}_{1g} - \dot{\Xi}_{2g} + \dot{\Xi}_1 - \dot{\Xi}_2 - \dot{\Xi}_{DESTRUCTION}$$

$$\dot{\Xi} = 0 = \Delta\dot{\Xi}_{GASES} - \Delta\dot{\Xi}_{STEAM} - \dot{\Xi}_{DESTRUCTION}.$$

Irreversibility $= \dot{\Xi}_{DESTRUCTION} = 637.7 - 549.8 = 87.48$ kW.

g. Change of availability in the turbine:

Steam expands to pressure corresponding to $T_{sat} = 273 + 40 = 313\,\text{K}$.

From steam tables, we can find the saturation pressure at this temperature is 7.323 kPa.

The turbine inlet condition of steam equals the exit conditions from the HRSG: $T_2 = 650.7\,\text{K}$, $p_2 = 100\,\text{atm}$ or $101{,}325\,\text{kPa}$, $h_2 = 3020\,\text{kJ/kg}$, and $s_2 = 6.091\,\text{kJ/kg·K}$

Let us give 3 as the state of the steam exiting the turbine. The isentropic conditions of steam exiting the turbine are: $p_3 = 7.323\,\text{KPa}$, $s_{3s} = s_2 = 6.091\,\text{kJ/kg·K}$. Using these two values in the steam tables and interpolating, we get the isentropic enthalpy as $h_{3s} = 1895\,\text{kJ/kg}$. The actual conditions (enthalpy) of the steam exiting the turbine can be found as

$$\eta_T = \frac{(h_2 - h_3)}{(h_2 - h_{3s})} \Rightarrow 0.94 = \frac{(3020 - h_3)}{(3020 - 1895)}.$$

This gives $h_3 = 1962\,\text{kJ/kg}$. Using h_3 and P_3, we find from the steam tables after using interpolation:

$s_3 = 6.307\,\text{kJ/kg·K}$, $T_3 = 313\,\text{K}$. The actual conditions of steam are a two-phase flow mixture. The change of availability in the turbine is:

$$\dot{\Xi}_{Change,T} = \dot{m}_w[(h_2 - h_3) - T_0(s_2 - s_3)] = 0.4629\,[(3020 - 1962) - 300(6.091 - 6.307)]$$

$\dot{\Xi}_{Change,T} =$ **519.4 kW** (this exergy change is due to turbine power output, 489.5 kW, and exergy destroyed in the turbine, **29.88 kW**).

Loss of Availability in the Condenser

Water exits the condenser and is pumped to a higher pressure that corresponds to the inlet conditions of water at the HRSG. We can assume that the isentropic efficiency of the pump is 100%. Let 4 be the state of the water exiting the condenser.

$s_4 = s_1 = 0.3928\,\text{KJ/kg·K}$. Using s_4 and $p_4 = p_3 = 7.323\,\text{KPa}$, we can interpolate values in the steam tables to find:

Example 2.10 (cont.)

$$h_4 = 111.6 \text{ KJ/kg·K}, T_4 = 299.8 \text{ K}.$$

Loss of availability in the condenser is

$$\dot{\Xi}_{LOSS,C} = \dot{m}_w[(h_3 - h_4) - T_0(s_3 - s_4)] = 0.4629[(1962 - 111.6) - 300(6.307 - 0.3928)]$$
$$= \mathbf{35.05\ kW}.$$

Now it is possible to compare the exergy loss in the different components to determine which is responsible for the maximum loss in the system; at times the result is surprising!

2.6 Equilibrium

Equilibrium states can be described in terms of intensive properties. Two systems in thermal equilibrium must have the same temperature; otherwise they undergo heat transfer interactions when brought into contact until thermal equilibrium is reached. Similarly, two systems in mechanical equilibrium must have the same pressure; otherwise, when brought into contact they will experience work transfer interactions until their pressures are the same. These experience-supported statements, which are also consistent with rigorous definitions of thermodynamic temperature and pressure, will be revisited. The temperature difference is the potential that drives heat transfer, and the pressure difference is the potential that drives mechanical work transfer. Other potentials, such as the chemical potential that drives mass and charge transfer, determine whether mass transfer of particular species can occur when two systems of different compositions are brought into contact. Chemical potential is a useful property in mixing and separation, and in the treatment of chemical reactions. As will be seen, the chemical potential is defined to express the equilibrium state in multi-component systems, and to express the driving force for mass transfer, the law of mass action, etc.

2.6.1 Conditions of Equilibrium (Extensive Properties)

In its general form, the Second Law *states that for a closed system with fixed mass and internal energy, equilibrium is reached when entropy generation is maximum.* It is often desirable to determine the condition of equilibrium under constraints other than fixed internal energy and volume. The corresponding conditions are often described in terms of free energies. *For simplicity, we restrict the stored energy to the internal energy and the work transfer interaction to the mechanical work mode only* (extension to more generalized cases, such as when electrical work modes are allowed, will be addressed later). To derive the conditions of equilibrium, we first recall that the First and Second Law expressions for a closed system or a control mass undergoing an arbitrary process under quasi-static conditions are

$$dU = \delta Q - p d\forall \tag{2.66}$$

and

$$\delta S_g = dS - \frac{\delta Q}{T} \geq 0. \tag{2.67}$$

The quasi-static equilibrium assumption is important for expressing the work transfer interaction in terms of state variables $\delta W = p d\forall$. Combining these two equations to eliminate the heat transfer term gives the **general condition for equilibrium for a control mass**:

$$T\delta S_g = T\,dS - dU - p d\forall \geq 0. \tag{2.68}$$

Entropy generation must be greater than zero, as stated originally in the Second Law, reaching a maximum at the equilibrium state. For example, a process seeking to balance temperature (thermal equilibrium), pressure (mechanical equilibrium), composition of two phases (phase equilibrium), or reaction (chemical equilibrium) proceeds according to the relation above. Next we seek to express the same condition in terms of the **properties** of the system, under external constraints, instead of the entropy generation.

At fixed volume and internal energy, or when the system is constrained by a given volume and internal energy (closed system), (2.68) shows that $dS = \delta S_g$ and the equilibrium state corresponds to the point of maximum entropy, as shown graphically in Figure 2.16:

$$dS \geq 0 \quad \text{at constant } \ (U, \forall). \tag{2.69}$$

Writing (2.68) in terms of the enthalpy, $H = U + p\forall$, yields the following general condition for equilibrium:

$$T\delta S_g = T\,dS - dH - \forall dp \geq 0. \tag{2.70}$$

Thus, at constant p and H, or when the system is constrained to a given p and H, it reaches the equilibrium state when

$$dS \geq 0 \quad \text{at constant } (p, H). \tag{2.71}$$

Therefore, at equilibrium, when conditions of constant pressure and enthalpy are applied, the total entropy is a maximum, as shown in Figure 2.16.

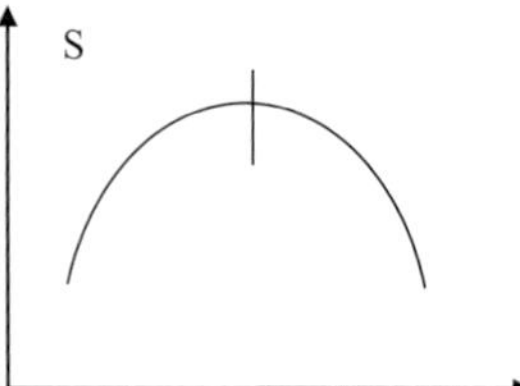

Figure 2.16 Equilibrium at constant internal energy and volume, and constant enthalpy and pressure.

2.6.2 The Free Energies

Before proceeding to find the condition of equilibrium when the system is constrained at a given temperature and pressure, we introduce another thermodynamic function, G, the **Gibbs free energy** or the **Gibbs function**:

$$G \equiv H - TS. \tag{2.72}$$

The role of the Gibbs free energy becomes clear when deriving the condition of equilibrium for a system undergoing a process at constant p and T. To find that, we differentiate (2.72) and combine it with the general condition, (2.68), to find that at equilibrium,

$$T\delta S_g = -dG + \forall dp - SdT \geq 0. \tag{2.73}$$

Now we constrain the system to constant T and p, and find that $dG = -\delta S_g$; hence, at equilibrium under these conditions, the system satisfies

$$dG \leq 0 \quad \text{at constant } (p, T), p\forall \text{ work only.} \tag{2.74}$$

Therefore, G must reach a minimum when the state is defined by the pressure and temperature, as depicted graphically in Figure 2.17. This condition is widely used because it is often convenient to define a system in terms of (p, T).

To determine the condition of equilibrium when the system is constrained to constant T and $\forall$, it is convenient to use A, the **Helmholtz free energy** or the **Helmholtz function**:

$$\mathrm{A} \equiv U - TS. \tag{2.75}$$

To derive the condition of equilibrium for a system undergoing change at constant T and U, we differentiate (2.75) and combine it with (2.68):

$$T\delta S_g = -d\mathrm{A} - p\,d\forall - S\,dT \geq 0. \tag{2.76}$$

Now we constrain the system to constant T and $\forall$, and we find that at equilibrium under these conditions the system must satisfy $d\mathrm{A} = -dS_g$, and we must have

$$d\mathrm{A} \leq 0 \quad \text{at constant } (T, \forall). \tag{2.77}$$

That is, A must reach a minimum.

Given a system, the appropriate forms of the equations of state, and a set of constraints in terms of some thermodynamic properties and mass conservation, one can use (2.69), (2.71), (2.74), or (2.77) to determine the equilibrium state, depending on the constraints.

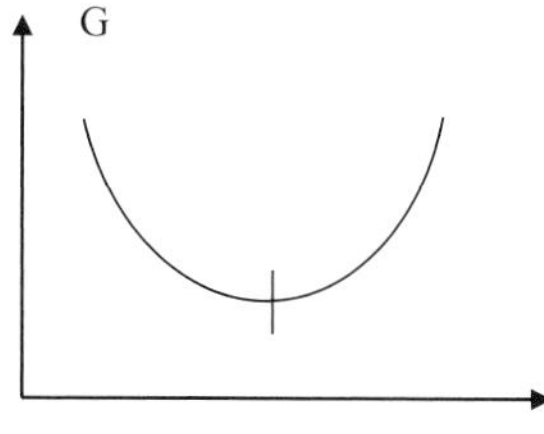

Figure 2.17 Equilibrium at constant temperature and pressure (the figure applies to equilibrium at constant temperature and volume if Helmholtz free energy is used instead of Gibbs free energy).

2.7 The Fundamental Equation

The combined statement of the First and Second Laws, written for a reversible process in which $\delta Q = T dS$ and $\delta S_g = 0$, is a special case of (2.68), in which

$$T\,dS = dU + p\,d\forall. \tag{2.78}$$

This is the **Gibbs fundamental equation** (or simply the **Gibbs equation**) of thermodynamics applied to a control mass, with a single work interaction mode and a single internal energy storage mode. This equation will be generalized to describe systems undergoing change in its chemical composition – that is, in the amount of each component in the system, or even a chemical reaction. Since this equation is applicable to reversible processes in equilibrium, it can also be used to describe the relationship between the properties of the substance (i.e. an equation of state), assuming that the composition or the phase of the fluid does not change:

$$T\,ds = du + p\,dv, \tag{2.79}$$

where the intensive properties (e.g., s) are defined per unit mass of the corresponding extensive property, $s = S/m$. An alternative form of the Gibbs equation is

$$ds = \frac{dh}{T} - v\frac{dp}{T}.$$

Therefore, the Gibbs equation relates changes in entropy to changes in internal energy and specific volume or to changes in enthalpy and pressure. For a simple compressible substance for which there is only one single work transfer mode, the state is defined by two variables:

$$S = S(U, \forall). \tag{2.80}$$

Differentiating this expression using the chain rule,

$$dS = \left(\frac{\partial S}{\partial U}\right)_{\forall} dU + \left(\frac{\partial S}{\partial \forall}\right)_{U} d\forall. \tag{2.81}$$

From (2.78) we have the following result:

$$\frac{1}{T} = \left(\frac{\partial S}{\partial U}\right)_{\forall} \tag{2.82}$$

and

$$\frac{p}{T} = \left(\frac{\partial S}{\partial \forall}\right)_{U}. \tag{2.83}$$

Equations (2.82) and (2.83) are often recognized as the fundamental definitions of **temperature** and **pressure**, respectively. Thermodynamic arguments can be used to show that these definitions are compatible with the experiential-based definitions of temperature and pressure. Two systems in thermal contact can reach equilibrium only when their temperatures are the same, which maximizes the total entropy. If two systems at different temperatures are brought into contact, heat is spontaneously transferred from the higher-temperature system to the lower-temperature system until the entropy of the combined system is maximized. When the systems are in mechanical contact, they reach mechanical equilibrium only when their pressures are the same, also maximizing the total entropy. In the process, work is transferred from the higher- to the lower-pressure system.

Example 2.11 Isentropic Processes for Ideal Gas

We can use the Gibbs equation to derive equations relating changes in the state properties of an ideal gas during an isentropic process for constant and variable specific heats.

Solution

The ideal gas equation of state is $pv = RT$ and $h = h(T)$ or $u = u(T)$.

Using the definitions for specific heats at constant volume and constant pressure respectively, we get $c_p = dh/dT$ and $c_v = du/dT$. Substituting these relations into the Gibbs equation, we get

$$ds = c_v \frac{dT}{T} + R\frac{dv}{v} = c_p \frac{dT}{T} - R\frac{dp}{p}.$$

Since the process is isentropic, $ds = 0$, the above equations can be integrated between an initial state "1" and a final state "2" to yield

$$0 = \int_1^2 c_v \frac{dT}{T} + \int_1^2 R\frac{dv}{v} = \int_1^2 c_v \frac{dT}{T} + R\ln\left(\frac{v_2}{v_1}\right)$$

and

$$0 = \int_1^2 c_p \frac{dT}{T} - \int_1^2 R\frac{dp}{p} = \int_1^2 c_p \frac{dT}{T} - R\ln\left(\frac{p_2}{p_1}\right).$$

Hence,

$$\frac{v_2}{v_1} = \exp\left[-\frac{1}{R}\int_1^2 c_v \frac{dT}{T}\right] \text{ and } \frac{p_2}{p_1} = \exp\left[\frac{1}{R}\int_1^2 c_p \frac{dT}{T}\right].$$

For the special case of constant specific heats, the integrals yield: $\frac{v_2}{v_1} = \left(\frac{T_1}{T_2}\right)^{\frac{\alpha-1}{\alpha}}$ and $\frac{p_2}{p_1} = \left(\frac{T_1}{T_2}\right)^{\alpha-1}$, and $\alpha = (k-1)/k$ and $k = c_p/c_v$.

For the general case, we may define a reduced specific volume and a reduced pressure as follows: $v_r = \exp\left[-\frac{1}{R}\int_{T^\circ}^{T} c_v \frac{dT}{T}\right]$ and $p_r = \exp\left[\frac{1}{R}\int_{T^\circ}^{T} c_p \frac{dT}{T}\right]$ where T° is a reference temperature. Both v_r and p_r are functions of temperature. Under isentropic conditions, two states of an ideal gas with variable specific heats are related with the relationships $\frac{v_2}{v_1} = \frac{v_{r2}}{v_{r1}}$ and $\frac{p_2}{p_1} = \frac{p_{r2}}{p_{r1}}$.

2.8 Describing the State

The applications of the First and Second Laws to energy systems involve a number of extensive state properties, including the internal energy, enthalpy, and free energies and their relation to other intensive properties such as the pressure, entropy, and specific enthalpy. Not all of these properties are independent and equations of state are used to relate them. Equations of state depend on the nature of the substance, whether its chemical composition changes during the process, and whether it undergoes phase transition. In the next two subsections, we review briefly the equations of state of a nonreacting mixture because of their significance in many energy conversion processes.

2.8.1 Nonreacting Mixtures

When different, chemically distinct substances are thoroughly mixed, they constitute a **multi-component** mixture. Chemically distinct substances such as oxygen (O_2), nitrogen (N_2), argon (Ar), hydrogen (H_2), carbon dioxide (CO_2), carbon monoxide (CO), water (H_2O), methane (CH_4), propane (C_3H_8), and benzene (C_6H_6) are also called components, species, or constituents. A **homogeneous** mixture is made of a single phase, in contrast to a **heterogeneous** mixture, such as gas and liquid, or gas and solid. In a nonreacting mixture, the mass of each component in a unit volume of the mixture remains constant while the mixture undergoes different processes; this makes it possible to develop simple equations of state for the mixture (as long as none of the components condenses during the process). The mixture composition changes if mixing with another stream is allowed. Prominent examples of nonreacting mixtures are dry air, made up primarily of O_2, N_2, and Ar, moist air, made up of dry air components and H_2O, and products of combustion, consisting mostly of N_2, CO_2, and H_2O for hydrocarbon fuel combustion.

2.8.1.1 Mole and Mass Fraction

The composition of a mixture can be described by either the mass of the different components, m_i, or the number of **moles** of these components, n_i, in the mixture, where

$$n_i = m_i / M_i. \tag{2.84}$$

In this expression, M_i is the molecular weight of component i, and $i = 1, 2, \ldots, N$, where N is the number of components; the molecular weight is also called the molar mass. In terms of fractions, we describe the mixture using **mass fractions**, which for a given component i correspond to the ratio of its mass to the total mass of the mixture:

$$Y_i = m_i / m, \tag{2.85}$$

where m is the total mass: $m = \sum_{i=1}^{N} m_i$. Composition also can be expressed in terms of **mole fractions**, which, for a component i may be expressed as the ratio of its number of moles to the total number of moles in the mixture:

$$X_i = n_i/n, \tag{2.86}$$

where n is the total number of moles: $n = \sum_{i=1}^{N} n_i$. Invariably, the sum of all mole or mass fractions must be unity: $\sum_{i=1}^{N} X_i = 1$ and $\sum_{i=1}^{N} Y_i = 1$. The mole fraction plays an important role in defining other thermodynamic properties, especially for chemically reacting mixtures. Note that knowing the components and their mass fractions, we can find the mole fractions, and vice versa:

$$X_i = \frac{(Y_i/M_i)}{\sum_{j=1}^{N} (Y_j/M_j)}, \tag{2.87}$$

$$Y_i = \frac{(X_i M_i)}{\sum_{j=1}^{N} (X_j M_j)}. \tag{2.88}$$

A useful quantity to define here is the **average molecular weight** of the mixture, also known as the molar mass of the mixture, $M = m/n$, such that

$$M = \frac{1}{n}\sum_{i=1}^{N} n_i M_i = \sum_{i=1}^{N} X_i M_i. \tag{2.89}$$

Mixing "independent" gases preserves their number of moles. For example, "pure" air is a mixture of approximately 3.76 moles of nitrogen for each mole of oxygen, and hence the mole fractions are: $X_{N_2} = 3.76/4.76 = 0.79$ and $X_{O_2} = 0.21$. Based on these mole fractions, the average molecular weight of air is $M_{air} = 0.79 \times 28.016 + 0.21 \times 32.00 = 28.86$ kg/kgmol. Air contains traces of other gases, such as argon and carbon dioxide, making its standard molecular weight $M_{air} = 28.97$ kg/kgmol.

2.8.1.2 Mixtures of Independent Substances

We focus this analysis on a mixture of compressible gases, for which the only significant reversible work interaction mode is through volume change. Moreover, we consider ideal mixtures of **independent** substances – that is, those substances for which the presence of other substances does not affect their own properties. For a gas mixture of independent substances occupying a volume $\forall$, and at temperature T, *all components have the same volume and temperature*. The components share the same volume because the molecules of each component are free to move throughout the available volume as if the other constituents do not exist. Note that the total volume occupied by the molecules of a gas at relatively low pressures and high temperatures (compared to its critical pressure and temperature) is miniscule compared to the available volume. The components have the same temperature because of thermal equilibrium.

For mixtures of independent components, extensive properties such as the internal energy (U), enthalpy (H), and entropy (S) are simply the sum of those of the components, each multiplied by its own mass or number of moles:

$$U = \sum_{i=1}^{N} n_i \hat{u}_i(T, \forall) = \sum_{i=1}^{N} m_i u_i(T, \forall), \tag{2.90}$$

$$H = \sum_{i=1}^{N} n_i \hat{h}_i(T, \forall) = \sum_{i=1}^{N} m_i h_i(T, \forall), \tag{2.91}$$

$$S = \sum_{i=1}^{N} n_i \hat{s}_i(T, \forall) = \sum_{i=1}^{N} m_i s_i(T, \forall). \tag{2.92}$$

A quantity with a "hat" is measured per mole (e.g., $\hat{h}_i = M_i h_i$). Moreover, mixture average specific quantities can be obtained in terms of the fractions; for example:

$$u = U/M = \sum_{i=1}^{N} Y_i u_i, \tag{2.93}$$

and

$$\hat{h} = H/n = \sum_{i=1}^{N} X_i \hat{h}_i. \tag{2.94}$$

The specific or intensive quantities of individual components in these expressions are evaluated *at the total volume and mixture temperature.*

2.8.1.3 Mixture of Ideal Gases

So far we have not talked about the pressure. While it is possible to continue the development for an arbitrary gas, we limit the following to perfect gases. The equation of state of the mixture of perfect gases is

$$p\forall = n\Re T. \tag{2.95}$$

The universal gas constant is $\Re = 8.3143$ kJ/kgmol·K. Since each component occupies the same total volume and is at the same temperature as the entire mixture, and since the different components behave independently as perfect gases, then

$$p_i \forall = n_i \Re T. \tag{2.96}$$

The *partial pressure*, p_i, of each component is simply

$$p_i = X_i p. \tag{2.97}$$

The total pressure is $p = \sum_{i=1}^{N} p_i$. This is **Dalton's law of additive pressures**. This description is consistent with the mechanical definition of the pressure as the force generated by the impact of the molecules on an elementary surface inside the gas per unit area as they travel through this surface. In a mixture, different molecules contribute different values to the pressure, while the total pressure is simply the sum over the contributions of all the different components. As given above, the ideal gas law applies to both the entire mixture as well as the individual mixture components if the proper pressure and number of moles are used.

2.8.1.4 Enthalpy of a Mixture

Substituting from the equation of state of individual components, we write the intensive or extensive properties of the mixture as the sum over the individual components evaluated at the *mixture temperature and their own partial pressure*:

$$H = \sum_{i=1}^{N} m_i h_i(T, p_i) = \sum_{i=1}^{N} n_i \hat{h}_i(T, p_i). \tag{2.98}$$

For ideal gases, the enthalpy is a function of temperature only, and the mole-based and mass-based enthalpies may be expressed as

$$\begin{aligned} h_i &= h_i^o + \int_{T^o}^{T} c_{p,i}(T)\, dT \\ \hat{h}_i &= \hat{h}_i^o + \int_{T^o}^{T} \hat{c}_{p,i}(T)\, dT, \end{aligned} \tag{2.99}$$

and $\hat{h}_i^o = \hat{h}_i(T^o)$ is the "reference" molar enthalpy of that component, evaluated at the reference state (T^o, p^o), and c_p is the specific heat at constant pressure. In all applications in which we have a single component, a mixture of fixed composition, or when we mix or separate nonreacting substances, there is no need to use the reference enthalpy since we are always interested in the difference of enthalpies of the same gases (and the same masses as well) between two end states. The choice of the reference enthalpy is important in the analysis of reacting mixtures since some components of the mixture disappear, and others appear during the process. If the mixture composition is fixed, we can define mixture average specific heat at constant pressure:

$$\bar{c}_p(T) = \sum_N Y_i c_{p,i}(T) \tag{2.100}$$

and

$$\bar{\hat{c}}_p(T) = \sum_N X_i \hat{c}_{p,i}(T). \tag{2.101}$$

Figure 2.18 shows the molar constant-pressure specific heat of typical gaseous substances versus temperature. For ideal gases $\hat{c}_p = \hat{c}_v + \mathfrak{R}$, where $\hat{c}_v$ is the specific heat at

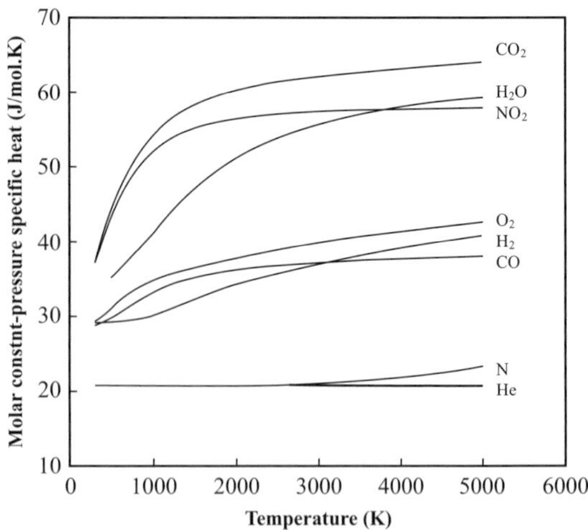

Figure 2.18 Variation of constant-pressure specific heats of some important species in combustion products (including monatomic, diatomic, and triatomic gases) with temperature. Note the substantial increase in $\hat{c}_p$ as the number of atoms/molecules increases. Source: NIST data.

constant volume. The dependence of the specific heat on temperature for high-temperature applications contributes significantly to the internal energy and enthalpy of the mixture. Specific heats for monatomic gases such as Ar, O, N, and helium (He), diatomic gases such as O_2 and N_2, and triatomic gases such as CO_2 and H_2O are different because of the way they store energy: The larger the number of atoms per molecule, the more modes of internal energy storage are available. These modes include translational (associated with the motion of molecules in space), electronic (associated with the motion of electrons around the nucleus), vibrational (associated with the motion of atoms within a bond relative to their center of mass), and rotational (associated with the rotation of a molecule relative to its center of mass) modes. According to the principle of equipartition of energy, the constant-volume specific heat of an ideal gas equals the number of degrees of freedom times a constant proportional to $(\mathfrak{R}/2)$, being the kinetic energy/mode. This is responsible for increasing the specific heat of complex molecules as the temperature rises, as seen in Figure 2.18.

The specific heat ratio of a mixture, or its isentropic index, is

$$k = \frac{\hat{c}_p}{\hat{c}_v} = \frac{\hat{c}_p}{\hat{c}_p - \mathfrak{R}}. \tag{2.102}$$

The isentropic index of monatomic, diatomic, and triatomic gases at ambient temperature conditions and moderately higher temperatures are, respectively, 1.667, 1.4, and 1.285.

2.8.1.5 Entropy of a Mixture

For the specific entropy, we first recall that for a single ideal gas we have

$$\hat{s}(T,p) = \hat{s}^o(T) - \mathfrak{R} \ln \frac{p}{p^o}, \tag{2.103}$$

where

$$\hat{s}^o(T) = \int_{T^o}^{T} \frac{\hat{c}_p(T)}{T}\, dT, \tag{2.104}$$

where T° and p° are the reference temperature and pressure, respectively. The specific entropy of a mixture per mole is given by

$$\hat{s} = \sum_{i=1}^{N} X_i \hat{s}_i(T, p_i) = \sum_{i=1}^{N} X_i \hat{s}_i^o(T) - \Re \sum_{i=1}^{N} X_i \ln \frac{p_i}{p^o}. \tag{2.105}$$

Note that the reference temperature and pressure (T^o, p^o) are the same for all components in the mixture. Similar to the enthalpy, it is not necessary to carry a nonzero value for the entropy at the reference state since for a mixture with fixed composition, or in the mixing or separation of nonreacting gases, we only need the difference between the entropy of the same gases (with the same masses) at different states. Nevertheless, the numerical value of the entropy at the reference state is important in reacting mixtures. The second term on the right-hand side of (2.105) can be written in terms of the total pressure, p, and the mole fractions, showing that

$$\hat{s}(T, p) = \sum_{i=1}^{N} X_i \hat{s}_i^o(T) - \Re \ln \frac{p}{p^o} - \Re \sum_{i=1}^{N} X_i \ln X_i. \tag{2.106}$$

This is an important result in chemical thermodynamics. Another convenient definition that will also be used extensively in chemical calculations is

$$\hat{s}(T, p) = \sum_{i=1}^{N} X_i \hat{s}_i(T, p, X_i), \tag{2.107}$$

where

$$\hat{s}_i(T, p, X_i) = \hat{s}_i(T, p) - \Re \ln X_i. \tag{2.108}$$

This is the entropy of an ideal gas in a gas mixture, written in terms of the temperature, total pressure, and mole fraction. Note that the contribution of each species in its mole fraction is nonlinear.

2.8.1.6 Volume Fraction

One more definition often used in gas mixtures is the **volume fraction**, $\forall_i$, which is defined as the volume occupied by a particular component if its pressure was the same as the total pressure of the mixture:

$$\forall_i = \frac{n_i \Re T}{p}. \tag{2.109}$$

Using the equation of state of mixture, we can show that

$$\forall_i/\forall = X_i. \tag{2.110}$$

Therefore, the volume fraction, also known as the **partial volume**, *is the same as the mole fraction*, and the mass fractions can be found from (2.88). This is convenient because mixtures of perfect gases are often specified by their partial volumes or volume fractions. The partial volumes can be obtained from laboratory analysis by gas absorption at constant pressure and temperature, or from knowledge of the chemical reactions used to form these mixtures.

2.8.2 Mixing of Perfect Gases

The definition of the volume fraction is useful in determining the properties of a gaseous mixture formed by adding known volumes of gases at the same pressure and temperature. In this case, all the original components have the same pressure, $p^{(1)}$, and temperature, but each occupies its own volume, $\forall_i$. After being allowed to mix adiabatically, the mixture volume becomes

$$\forall = \sum_N \forall_i. \tag{2.111}$$

This is the **Amagat–Luduc law of additive volumes**. Noting that

$$\forall = \sum_{i=1}^{N} \forall_i = \sum_{i=1}^{N} n_i \Re T/p^{(1)} = n\Re T/p^{(1)},$$

the volume fractions of the different components in the mixture are given by

$$\frac{\forall_i}{\forall} = \frac{n_i \Re T/p^{(1)}}{n\Re T/p^{(1)}} = \frac{n_i}{n} = X_i.$$

That is, the volume fraction is equal to the mole fractions,

$$X_i = n_i / \sum_{i=1}^{N} n_i,$$

consistent with the above definition of the volume fraction.

Since the mixing process does not involve heat or work interactions, the total and all the individual internal energies remain the same (First law application), and hence the temperature remains the same after mixing (as the number of moles of each constituent remains fixed). Moreover, mixing allows the volume of each constituent to expand from its original volume, $\forall_i$, to the total volume, $\forall$. Therefore, the pressure of each constituent changes such that, originally, $p^{(1)}\forall_i = n_i\Re T$, and finally, $p_i^{(2)}\forall = n_i\Re T$.

At the final state, the total pressure is the sum of the partial pressures:

$$p_t^{(2)} = \sum_{i=1}^{N} p_i^{(2)} = \frac{\Re T}{\forall}\sum_{i=1}^{N} n_i = \frac{n\Re T}{\forall} = p^{(1)}. \tag{2.112}$$

Note that $p^{(1)} = n_i \Re T/\forall_i = n\Re T/\forall$, showing that the total pressure in the final state is the same as the pressure of each component in the initial state. Moreover, the partial pressures of the components in the final state are

$$\frac{p_i^{(2)}}{p^{(2)}} = \frac{n_i \Re T/\forall}{n\Re T/\forall} = \frac{n_i}{n} = X_i. \tag{2.113}$$

Therefore, the partial pressure of each component in the final state is determined by its mole or volume fraction.

Example 2.12 Composition and Properties of an $O_2/H_2/CO_2$ Mixture

Given a gas mixture of O_2, H_2, and CO_2, with volume fractions of 0.15, 0.6, and 0.25, respectively, at $p = 4.08$ atm and $T = 26.0\,^\circ C$, determine the mole fractions and the mass fractions of the different components in the mixture and the average molecular weight, the mixture gas constant, and the mixture density. The molecular weights of O_2, H_2, and CO_2 are 32, 2.016, and 44.01 g/mol, respectively.

Solution

The composition of the mixture is given in volume fractions, which also correspond to mole fractions. The mixture molecular weight is given by $M = \sum_{j=1}^{N} X_j M_j$. The mass fractions are given by $Y_i = X_i M_i/M$. Table E2.12 summarizes the mole and mass fractions of the three components of the mixture. The mixture molecular weight is 17.01 g/mol.

Table E2.12 Nominal composition of gas mixture

Component	Mole fraction	Mass fraction
O_2	0.15	0.282
H_2	0.60	0.071
CO_2	0.25	0.647

Now we can calculate the gas constant of the mixture, R = 8.314462/17.01 = 0.489 J/kg·K. The density of the mixture is then obtained as 2.825 kg/m^3. As an additional step to validate the calculations of the mole and mass fractions, these fractions must add up to unity.

Note that if the mixture is kept in a vessel that contains an absorbent for one of the gases, and the vessel temperature is kept constant, the vessel pressure will drop by an amount equal to the partial pressure of that gas. On the other hand, if the temperature and pressure are kept constant during the absorption process, the volume will decrease by an amount equal to the partial volume of this gas.

2.8.3 Entropy of Mixing

Mixing processes are irreversible; try to "spontaneously" separate salt and pepper after you have mixed them, or for that matter try to separate pure oxygen from air without doing some work! To demonstrate this fact quantitatively, we evaluate the entropy change due to the mixing of N individual gaseous species, which are at identical pressure and temperature. The entropy of N species is

$$\hat{S}_1 = \sum_{i=1}^{N} n_i \hat{s}_i(T,p) = \sum_{i=1}^{N} n_i \left[\hat{s}_i^o(T) - \Re \ln \frac{p}{p^o} \right],$$

or, per total mole of species $n_{tot} = \sum_{i=1}^{N} n_i$, we have

$$\hat{s}_1 = \sum_{i=1}^{N} X_i \hat{s}_i(T,p) = \sum_{i=1}^{N} X_i \left[\hat{s}_i^o(T) - \Re \ln \frac{p}{p^o} \right]. \tag{2.114}$$

The specific entropy *after mixing* is (see (2.106))

$$\hat{s}_2 = \sum_{i=1}^{N} X_i \hat{s}_i(T,p_i) = \sum_{i=1}^{N} X_i \hat{s}_i^o(T) - \Re \ln \frac{p}{p^o} - \Re \sum_{i=1}^{N} X_i \ln X_i. \tag{2.115}$$

Thus, entropy generation due to this adiabatic, constant-temperature, constant-pressure mixing process, measured per mole of mixture, is given by

$$\hat{s}_g = (\hat{s}_2 - \hat{s}_1) = -\Re \sum_{i=1}^{N} X_i \ln X_i. \tag{2.116}$$

This is the entropy of mixing, which is a positive quantity since the mole fractions of all species are less than 1. The physical origin of the entropy of mixing can be understood by realizing that the gases, originally at the same pressure but different volume, must expand to fill the total volume after being allowed to mix. That is, the partial pressures of individual components must drop to a value corresponding to the partial volume times the original pressure. This expansion occurs from the original high pressure to a lower pressure, without work interaction to the environment, leading to entropy generation.

2.9 Mass Transfer Equilibrium

2.9.1 The Chemical Potential

Similar to the use of T and p to define the condition of thermal and mechanical equilibrium between two interacting but isolated systems, we need an intensive property to define the mass transfer equilibrium between two isolated systems that can exchange mass of a particular chemical species or component. This property is the chemical potential.

For a mixture of different chemical species maintained in mechanical and thermal equilibrium, (2.78) and (2.80) must be modified to reflect the contribution of the mixture constituents or species to the total entropy. First, the Gibbs fundamental relation is extended to allow for the change of entropy due to incremental changes in the moles of different species due to mass transfer at constant volume and internal energy:

$$T\,dS = dU + p\,d\forall - \sum_{i=1}^{N} \mu_i dn_i. \tag{2.117}$$

The new property, μ_i, is chemical potential (to be defined below) and the minus sign is chosen by convention. The entropy function now depends on the number of moles of each species, n_i, such that

$$S = S(U, \forall, n_i). \tag{2.118}$$

Following the steps taken before to define T and p, we write

$$dS = \left(\frac{\partial S}{\partial U}\right)_{\forall, n_i} dU + \left(\frac{\partial S}{\partial \forall}\right)_{U, n_i} d\forall + \sum_{I} \left(\frac{\partial S}{\partial n_j}\right)_{\forall, U} dn_j. \tag{2.119}$$

The new variable, the chemical potential, $\hat{\mu}_i$, is defined such that

$$-\frac{\mu_i}{T} = \left(\frac{\partial S}{\partial n_i}\right)_{U, \forall, n_{j \neq i}}. \tag{2.120}$$

Similar to the roles played by the temperature and pressure in defining the conditions of thermal and mechanical equilibrium between two thermally and mechanically interacting systems, respectively, the chemical potential defines the condition of mass transfer equilibrium between two subsystems within an isolated system that can exchange mass of chemically distinct species. Using (2.69) and (2.117), we can show that at mass transfer equilibrium (i.e., in the absence of mass transfer between the two subsystems), the chemical potential of the component of interest in the two subsystems is the same. Also, similar to the roles of temperature and pressure differences, ΔT and Δp, being the thermal and mechanical potentials or the driving forces for heat transfer and work transfer between the subsystems, respectively, the difference in the chemical potential, $\Delta\mu_i$, is the driving force for **mass transfer** of this particular chemical species between two subsystems in which the chemical potential is different. This is shown next in a more generalized form.

Consider an isolated system composed of two phases, A and B, each containing two substances, 1 and 2 [2]. The total volume, internal energy, and masses of 1 and 2 are those for the combined system, which is shown schematically in Figure 2.19. In each phase, the pressure and temperature are the same, satisfying internal equilibrium within each phase. The two phases are in mutual stable equilibrium while they are allowed to exchange mass *only* with each other (they do not exchange mass with another system). For the isolated system,

$$dS = dS_A + dS_B. \tag{2.121}$$

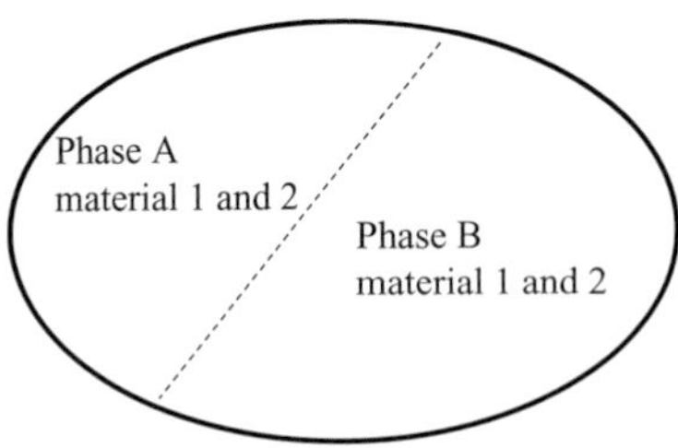

Figure 2.19 Two subsystems at different phases, which are in mechanical and thermal equilibrium, exchange mass until chemical equilibrium is achieved.

Using (2.117):

$$dS = \left(\frac{dU_A}{T_A} + \frac{p_A}{T_A} d\forall_A - \frac{\mu_{1A}}{T_A} dn_{1A} - \frac{\mu_{2A}}{T_A} dn_{2A}\right) + \left(\frac{dU_B}{T_B} + \frac{p_B}{T_B} d\forall_B - \frac{\mu_{1B}}{T_B} dn_{1B} - \frac{\mu_{2B}}{T_B} dn_{2B}\right), \tag{2.122}$$

where μ_{1A} and μ_{1B} denote the chemical potential of substance 1 in phases A and B, respectively. For the isolated system, the total energy, total volume, and total mass for each substance are constant; that is; $dU_A + dU_B = 0$, $d\forall_A + d\forall_B = 0$, $dn_{1A} + dn_{1B} = 0$, and $dn_{2A} + dn_{2B} = 0$. Thus, (2.122) can be written as

$$dS = \left(\frac{1}{T_A} - \frac{1}{T_B}\right) dU_A + \left(\frac{p_A}{T_A} - \frac{p_B}{T_B}\right) d\forall_A - \left(\frac{\mu_{1A}}{T_A} - \frac{\mu_{1B}}{T_B}\right) dn_{1A} - \left(\frac{\mu_{2A}}{T_B} - \frac{\mu_{2B}}{T_B}\right) dn_{2A}. \tag{2.123}$$

Maximizing S with respect to U_A, V_A, n_{1A}, and n_{2A} (i.e., $\frac{\partial S}{\partial U_A} = \frac{\partial S}{\partial \forall_A} = \frac{\partial S}{\partial n_{1A}} = \frac{\partial S}{\partial n_{2A}} = 0$) shows that $T_A = T_B$, $p_A = p_B$, $\mu_{1A} = \mu_{1B}$, and $\mu_{2A} = \mu_{1B}$.

The analysis can be extended to systems with more than two phases and with any number of components to give the following conditions for equilibrium:

Thermal equilibrium: the temperature of every phase is the same.
Mechanical equilibrium: the pressure of every phase is the same.
Chemical equilibrium: the chemical potential of each component has the same value in every phase.

Consider again the isolated system consisting of phases A and B. Let the two phases be in thermal and mechanical equilibrium, as shown in Figure 2.19. Also, let the chemical potential of substance 2 be the same in both phases. The entropy change is now

$$dS = \frac{1}{T}(\mu_{1B} - \mu_{1A}) dn_{1A}, \tag{2.124}$$

where dn_{1A} represents the number of moles of substance 1 transferred from phase B to phase A. Entropy increases in a spontaneous process of the isolated system: $dS > 0$. Accordingly, when $\mu_{1B} > \mu_{1A}$, $dn_{1A} > 0$; substance 1 passes from phase B to phase A. When $\mu_{1B} < \mu_{1A}$, $dn_{1A} < 0$;

substance 1 passes from phase A to phase B. Therefore, the chemical potential is a measure of the **escaping tendency** of a substance. Any substance will try to move from the phase having the higher chemical potential for that substance to the phase having lower chemical potential.

Equation (2.120) shows that the chemical potential is proportional to the increase in the total entropy of the mixture due to the change in the number of moles of a particular species at constant total volume and internal energy. Other definitions follow by rewriting (2.117) in terms of other thermodynamic variables – for example, if (H, p) are used instead of $(U, \forall)$, it can be shown that

$$-\frac{\mu_i}{T} = \left(\frac{\partial S}{\partial n_i}\right)_{H,p,n_{j\neq i}}. \tag{2.125}$$

Furthermore, (2.117) can be rearranged such that the **Gibbs fundamental equation** for a mixture can be written in energy terms:

$$dU = TdS - pd\forall + \sum_I \mu_i dn_i. \tag{2.126}$$

Now expressing the internal energy as $U = U(S, \forall, n_i)$, and differentiating, another important definition of the chemical potential is obtained:

$$\mu_i = \left(\frac{\partial U}{\partial n_i}\right)_{S,\forall,n_{j\neq i}}. \tag{2.127}$$

That is, the chemical potential of a particular species, or the driving force for mass transfer involving this particular species, is the incremental increase in the internal energy of the mixture at constant total entropy, volume, and number of moles of all other species, as the number of moles of this species changes. Using the above definitions, it is easy to show that

$$\mu_i = \left(\frac{\partial H}{\partial n_i}\right)_{S,p,n_{j\neq i}}, \tag{2.128}$$

$$\mu_i = \left(\frac{\partial \mathrm{A}}{\partial n_i}\right)_{T,\forall,n_{j\neq i}}, \tag{2.129}$$

$$\mu_i = \left(\frac{\partial G}{\partial n_i}\right)_{T,p,n_{j\neq i}}. \tag{2.130}$$

These definitions are used to express the conditions of chemical equilibrium in terms of the chemical potential.

2.9.2 Chemical Potential and Entropy Generation

Now we obtain an important result in setting up the analysis of mass transfer equilibrium using the chemical potential. Using the general expression for entropy generation (2.68) and the Gibbs fundamental equation (2.117), we can show that for a control mass, at constant $(U, \forall)$,

$$T\,\delta S_g = -\sum_{i=1}^{N} \mu_i dn_i. \tag{2.131}$$

The entropy generation in this equation is due to mass transfer only. At equilibrium,

$$\sum_{i=1}^{N} \mu_i dn_i \leq 0. \tag{2.132}$$

One can verify that this is a general condition for mass transfer equilibrium – that is, besides its applicability for constant (U,V), it also applies for constant (H,p), constant (T,p), and constant $(T, \forall)$ using (2.71), (2.74), or (2.77), the definitions of H, G, and A, and (2.117). For instance, using the definitions of G and (2.117), we find

$$dG = -S\,dT + \forall\,dp + \sum_{i=1}^{N} \mu_i dn_i. \tag{2.133}$$

At constant T and p, (2.74) shows that, at equilibrium,

$$\sum_{i=1}^{N} \mu_i dn_i \leq 0. \tag{2.134}$$

This is the convenience of the chemical potential and why it plays an important role in chemical thermodynamic calculations. In the next section, we find the dependence of the chemical potential on other thermodynamic properties of state.

2.9.3 Dependence of Chemical Potential on *p* and *T*

The Gibbs fundamental equation for multi-component systems (2.117) is

$$T\,dS = dU + p\,d\forall - \sum_{i=1}^{N} \mu_i dn_i.$$

It is important to determine the dependence of the chemical potential on other thermodynamic variables. For this, we choose a system composed of a number of components in thermal and mechanical equilibrium. Next, we make infinitesimal perturbations of the components' composition in the mixture, such that $dn_i = n_i d\varepsilon$, without changing the pressure or the temperature. The form involving $d\varepsilon$ indicates that all fractional changes are implemented the same way for all species, such that $d\varepsilon = dn_i/n_i$. Since all the extensive variables are additive, (e.g., $U = \sum_N n_i \hat{u}_i$, $S = \sum_N n_i \hat{s}_i$), a simple calculation shows that $dU = \sum_{i=1}^{N} \hat{u}_i dn_i = \left(\sum_{i-1}^{N} \hat{u}_i n_i\right) d\varepsilon = U d\varepsilon$, $dS = \sum_{i=1}^{N} \hat{s}_i dn_i = \left(\sum_{i-1}^{N} \hat{s}_i n_i\right) d\varepsilon = S\,d\varepsilon$, etc. Similarly, we can show that $d\forall = \forall d\varepsilon$. Because the constituents are all in thermal and mechanical equilibrium, substituting in (2.117) and eliminating the differential change, we get the important relation

$$U = TS - p\forall + \sum_{I} \mu_i n_i. \tag{2.135}$$

Moreover, since $H = U + p\forall$,

$$H = TS + \sum_{i=1}^{N} \mu_i n_i. \tag{2.136}$$

By definition, $G = H - TS$, and hence

$$G = \sum_{i=1}^{N} \mu_i n_i. \tag{2.137}$$

The last result allows us to calculate the chemical potential. The total Gibbs free energy is an extensive property:

$$G = \sum_{i=1}^{N} \hat{g}_i n_i. \tag{2.138}$$

Comparing (2.137) and (2.138), we find that in a mixture, the chemical potential of a constituent is the same as the molar Gibbs free energy of that constituent evaluated at its partial pressure in the mixture:

$$\mu_i = \hat{g}_i(T, p_i). \tag{2.139}$$

The Gibbs free energy of individual species in a mixture is

$$\hat{g}_i(T, p_i) = \hat{h}_i - T\hat{s}_i(T, p_i). \tag{2.140}$$

The chemical potential depends on the mixture temperature and pressure, and the partial pressure of the species (similar to the entropy of a mixture, which depends on the temperature and pressure and the molar concentrations or partial pressure of the component species in the mixture).

Equation (2.137) can be used to derive an important condition of equilibrium at constant temperature and total pressure. Differentiating this expression to yield $dG = \sum_i \mu_i dn_i + \sum_i n_i d\mu_i$, and using (2.134) and (2.74), we show that at equilibrium,

$$\sum_{I} n_i d\mu_i = 0. \tag{2.141}$$

Equation (2.141) is the Gibbs–Duhem equation at constant p and T, and is used extensively in equilibrium calculations. For a mixture with two components, the change of the chemical potentials of the component are related as follows: $d\mu_1 = -\dfrac{n_2}{n_1} d\mu_1$.

2.9.4 The Chemical Potential in a Mixture of Ideal Gases

We have seen before that in a **mixture of ideal gases**, the entropy of each component is calculated at the component partial pressure, and hence the Gibbs free energy of each constituent can be written as follows:

$$\hat{g}_i(T,p_i) = \hat{g}_i^o(T) + \Re T \ln \frac{p_i}{p^o}. \tag{2.142}$$

In terms of the total pressure and the mole fraction, the above expression takes the following form:

$$\hat{g}_i(T,p_i) = \hat{g}_i^o(T) + \Re T \ln \frac{p}{p^o} + \Re T \ln X_i. \tag{2.143}$$

Another useful form is that obtained by the Gibbs free energy in the mixture in terms of the Gibbs free energy at the total pressure, $\hat{g}_i(T,p) = \hat{g}_i^o(T) + \Re T \ln \frac{p}{p^o}$:

$$\hat{g}_i(T,p_i) = \hat{g}_i(T,p) + \Re T \ln X_i. \tag{2.144}$$

The Gibbs free energy of a pure substance at the reference pressure p^o is a function of temperature only,

$$\hat{g}^o(T) = \hat{h} - T\ \hat{s}^o(T), \tag{2.145}$$

and is often evaluated at atmospheric pressure. At an arbitrary pressure, the Gibbs free energy of a pure substance is

$$\hat{g}(T,p) = \hat{g}^o(T) + \Re \ln \frac{p}{p^o}. \tag{2.146}$$

The Gibbs free energy of a pure substance can be written as

$$\hat{g}^o(T) = g^{oo} + \Im n(T), \tag{2.147}$$

where $\hat{g}^{oo} = \hat{g}^o(T^o, p^o = 1\ \text{atm})$ and $\Im n(T)$ is a function of T that can be obtained from (2.145). In a mixture, the Gibbs free energy is modified to reflect the partial pressure of the species in the mixture, as shown in (2.142), (2.143), or (2.144).

Similarly, the chemical potential of a component in a mixture must also be computed at the partial pressure,

$$\mu_i = \hat{g}_i^o(T) + \Re T \ln \frac{p_i}{p^o}, \tag{2.148}$$

or

$$\mu_i = \hat{g}_i^o(T) + \Re T \left[\ln \left(\frac{p}{p^o} \right) + \ln X_i \right]. \tag{2.149}$$

It is important to emphasize that the chemical potential of a component in a mixture is a thermodynamic property of the component, evaluated at the mixture conditions, and is affected by the concentration of the component in the mixture.

2.9.5 Equilibrium Across a Semipermeable Membrane

Semipermeable membranes are common in some energy conversion systems, such as some air separation units (ASUs) and fuel cells, and are likely to be more ubiquitous in future systems. These systems may include processes of CO_2 separation from combustion products for the purpose of sequestration, or separation of hydrogen from fuel reforming products (mostly made up of hydrogen and carbon dioxide). Physically, both polymer and ceramic membranes allow certain ionic species to pass from one side to another without allowing others to pass through. As will be seen in the analysis of fuel cells (Chapter 4) and in gas separation (Chapter 10), some membranes allow protons and others allow negative ions.

Consider the case in which on one side of a membrane a mixture exists, and on the other side a component of the same mixture exists [12]. Under conditions of equilibrium, the chemical potential of the component in the mixture equals the chemical potential of the same component in its pure form:

$$\mu_{k,mixture} = \mu_{k,pure}. \tag{2.150}$$

In the case of perfect gases, and because the temperature is the same across the membrane, (2.150) translates to equal pressure of the pure component and the partial pressure of the same component in the mixture, $p_{k,pure} = p_{k,mixture} = X_k\, p_{mixture}$, where $p_{k,pure}$ is the pressure of the pure component on the side where it exists by itself, $p_{k,mixture}$ is the partial pressure of that component in the mixture, X_k is the mole fraction of the component in the mixture, and $p_{mixture}$ is the pressure of the mixture. Clearly, in this case, the membrane must support a pressure differential between the two sides because $X_i < 1$. Under non-equilibrium conditions, the difference in the chemical potential of that component across the membrane is the driving force of the flux of this component from one side to the other. The subject of membrane separation will be covered in detail in Chapters 10, 11 and 13.

2.9.6 System Availability with Mass Transfer

We now revisit the maximum work produced while a control mass attains equilibrium with its environment, and is allowed to exchange heat, work, and mass with the environment. Similar expressions to those obtained in (2.30) and (2.38) will be derived, but modified to account for the work required to transfer mass from the control mass to the environment. Consider the control mass shown in Figure 2.20 with total energy E, total volume $\forall$, total entropy S, temperature T, pressure p, number of moles of different species n_i and chemical potentials μ_i. The mole fractions are X_i. The control mass is to attain equilibrium with its environment, which is at pressure p_o, temperature T_o, and chemical potentials μ_i^o. These conditions define the restricted dead state for this case. Note that we assume that the same species in the control mass exist in the environment, but at different concentrations, X_i^o.

The maximum work transfer is still given by the combined statement of the First and Second Laws, with zero entropy generation; when the system reaches the pressure, temperature, and chemical potentials of the environment,

Figure 2.20 Maximum work or availability analysis of a system exchanging heat and mass with its environment, which defines the restricted dead state.

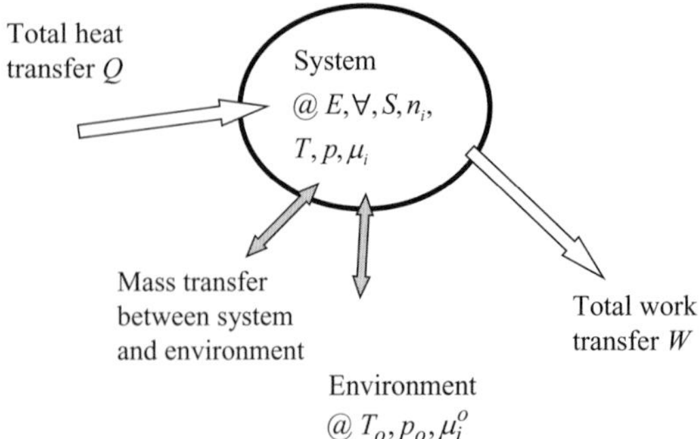

$$W_{max} = (E - U_o) + p_o(\forall - \forall_o) - T_o(S - S_o) = \Xi. \tag{2.151}$$

The total energy, volume, and entropy of the control mass at equilibrium are calculated at the conditions of the environment, including the mole fractions or the chemical potentials. This expression can be reorganized by adding and subtracting $(\tilde{U}_o, \tilde{\forall}_o, \tilde{S}_o)$, which are calculated at (T_o, p_o), but for the composition (or mole factions, or chemical potential) of the original control mass:

$$\begin{aligned} W_{max} = & \left[(E - \tilde{U}_o) + p_o(\forall - \tilde{\forall}_o) - T_o(S - \tilde{S}_o)\right] \\ & + \left[(\tilde{U}_o - U_o) + p_o(\tilde{\forall}_o - \forall_o) - T_o(\tilde{S}_o - S_o)\right]. \end{aligned} \tag{2.152}$$

The expression in the first line is the **thermomechanical or physical availability**, as if the control mass did not experience any mass transfer interactions with the environment, while the second line can be written in terms of the change in the chemical potential as the control mass, having achieved thermal and mechanical equilibrium with its environment, now achieves chemical equilibrium as well:

$$\begin{aligned} W_{max,\,tot} &= \Xi_{th} + \left[(\tilde{H}_o - H_o) - T_o(\tilde{S}_o - S_o)\right] \\ &= \Xi_{th} + \sum_I n_i(\mu_{io} - \mu_i^o). \end{aligned} \tag{2.153}$$

In this expression, the chemical potential μ_{io} is that of the components at their concentrations in the original mixture but evaluated at (T_o, p_o), while μ_i^o is the chemical potential of the same components in the environment. The second term in (2.153) is the total maximum work expression associated with the **chemical availability**, or the change of the concentrations in the mixture as it reaches chemical equilibrium with the environment through mass transfer:

$$W_{max,\,tot} = \Xi_{th} + \Xi_{ch}. \tag{2.154}$$

Thus, for a control mass in mechanical and thermal equilibrium with its environment, the maximum work is the **chemical exergy**:

$$W_{max,\,ch} = \sum_I n_i(\mu_{io} - \mu_i^o) = \Xi_{ch}. \tag{2.155}$$

For a mixture of ideal gases, the total chemical availability can be obtained using (2.149) to evaluate the chemical potential at the different states:

$$\Xi_{ch} = \Re T_o \sum_{i=1}^{I} n_i \ln\left(\frac{X_i}{X_i^o}\right), \tag{2.156}$$

where X_i^o is the concentration in the environment. If the control system changes state from 1 to 2, then the maximum work in the process is

$$W_{max,tot} = (\Xi_{th,1} - \Xi_{th,2}) + (\Xi_{ch,1} - \Xi_{ch,2}). \tag{2.157}$$

Example 2.13

Gaseous methane is completely burned in a combustor with 200% theoretical air. The gaseous products leave the combustor at 1600 K. Determine (a) the chemical exergy, and (b) the total flow exergy of the products per mole of fuel. Treat air and the combustion product as ideal gas mixtures. Assume $T_0 = 300$ K and $p_0 = 1$ atm. Take the mole fractions of nitrogen, oxygen, water vapor, and carbon dioxide in the environment as 0.7567, 0.2035, 0.0303, and 0.0003, respectively. The total flow exergy is defined bu (2.158) following this example.

Solution

The stoichiometric of combustion of methane with air is (see Chapter 3)

$$CH_4 + 4(O_2 + 3.76N_2) \rightarrow CO_2 + 2H_2O + 2O_2 + 15.04N_2.$$

The mole fractions of CO_2, H_2O, O_2, and N_2 in a mixture of the combustion products are 0.0499, 0.0998, 0.0998, and 0.7505, respectively. The total specific flow exergy is the sum of thermomechanical flow exergy and chemical exergy. The chemical exergy of the combustion products per mole of fuel is calculated using (2.156):

$$\Xi_{ch} = \Re T_o \sum_{i=1}^{I} n_i \ln\left(\frac{X_i}{X_i^o}\right) = \Re T_o \left[n_{CO_2} \ln\left(\frac{X_{CO_2}}{X_{CO_2}^o}\right) + n_{H_2O} \ln\left(\frac{X_{H_2O}}{X_{H_2O}^o}\right) + n_{O_2} \ln\left(\frac{X_{O_2}}{X_{O_2}^o}\right) + n_{N_2} \ln\left(\frac{X_{N_2}}{X_{N_2}^o}\right) \right].$$

$$= 8.314 \times 300 \left[\ln\left(\frac{0.0499}{0.0003}\right) + 2\ln\left(\frac{0.0998}{0.0303}\right) + 2\ln\left(\frac{0.0998}{0.2035}\right) + 15.04\ln\left(\frac{0.7505}{0.7567}\right) \right]$$

$$= 14{,}838.75 \text{ J/mol of } CH_4.$$

Example 2.13 (cont.)

The thermomechanical flow exergy of the products per mole of the fuel as follows:

$$\Xi_{th} = \sum_{i=1}^{I} n_i[h_i - h_o - T_o(s_i - s_o)]$$

$$= \left[h_{CO_2} - h_{CO_2,o} - T_o\left(s^o_{CO_2} - s_{CO_2,o}\right)\right] + 2\left[h_{H_2O} - h_{H_2O,o} - T_o\left(s^o_{H_2O} - s_{H_2O,o}\right)\right] +$$
$$2\left[h_{O_2} - h_{O_2,o} - T_o\left(s^o_{O_2} - s_{O_2,o}\right)\right] + 15.04\left[h_{N_2} - h_{N_2,o} - T_o\left(s^o_{N_2} - s_{N_2,o}\right)\right].$$

The values of specific enthalpy and entropy of the four constituents at 1600 K and 300 K (environment temperature) can be obtained from thermodynamic tables. Hence,

$$\Xi_{th} = [76{,}944 - 9431 - 300(295.91 - 213.92)] + 2[62748 - 9966 - 300(253.51 - 188.93)] +$$
$$2[{,}52{,}961 - 8736 - 300(260.33 - 205.21)] + 15.04[{,}50{,}571 - 8723 - 300(244.03 - 191.68)]$$
$$= 558{,}300.72\ \text{J/mol of } CH_4$$

The total flow exergy of the combustion products is

$$\Xi_{tot} = 558{,}300.72 + 14{,}838.75 = 573{,}139.47\ \text{J/mol of } CH_4.$$

The contribution of chemical exergy to total flow exergy is a small fraction of 2.6%.

2.9.7 Flow Availability with Mass Transfer

A similar analysis leads to the definition of the flow availability of a stream in the presence of mass transfer as an extension of the definition in (2.40) as follows:

$$\hat{\xi} = \left[\left(\hat{h} - \hat{h}_o\right) - T_o(\hat{s} - \hat{s}_o)\right] + \sum_{i=1}^{I} X_i(\mu_{io} - \mu_i^o)$$
$$= \hat{\xi}_{th} + \hat{\xi}_{ch}. \tag{2.158}$$

Using this expression, (2.41)–(2.44) can be applied to calculate maximum work, irreversibility, etc. in a flow process or a control volume. For ideal gases, the chemical availability, or the maximum work interaction between a system and its environment whose component mole fractions are X_i and X_i^o in the system and environment, when the system is at thermal and mechanical equilibrium with the environment, is

$$\hat{\xi}_{ch} = \Re T_o \sum_{i=1}^{I} X_i \ln\left(\frac{X_i}{X_i^o}\right). \tag{2.159}$$

Substituting in (2.158), another convenient expression for the total availability is obtained:

$$\hat{\xi} = \sum_{i=1}^{N} X_i \left\{ \left[\hat{h}_i(T) - \hat{h}_i(T^o) \right] - T^o \left[\hat{s}_i^o(T) - \hat{s}_i^o(T^o) \right] + \Re T^o \ln \left(\frac{X_i}{X_i^o} \right) \right\} + \Re T^o \ln \frac{p}{p^o}. \tag{2.160}$$

These expressions will be used in the analysis of separation processes. In the next chapter, the same expressions will be applied to compute chemical availability in mixtures undergoing chemical reactions. Equation (2.159) shows that the minimum work required to extract gases from the environment, in which the mole fractions are X_i^o, into a mixture in which the mole fractions are X_i is $W_{\min} = -\hat{\xi}_{ch} = -\Re T_o \sum_{i=1}^{I} X_i \ln \left(\frac{X_i}{X_i^o} \right)$.

2.10 Gas Separation

An important application of the availability analysis in multi-component mixtures is the separation of one component from a gas mixture. Detail of gas separation technologies is covered in Chapter 10; here we only cover an ideal process in the limit of thermodynamic equilibrium. In many energy conversion applications, we may need to separate the components of a gas mixture into their pure forms. Examples of where we encounter such need include the following:

1. Separating air into oxygen and nitrogen for the purpose of "oxyfuel" combustion – that is, burning fuel in oxygen instead of air such that the products of combustion are carbon dioxide and water. This is one approach for CO_2 capture. The separation of air into oxygen and nitrogen is done in the air separation unit (ASU) in coal- or gas-fired power plants built to capture CO_2.
2. Separating the products of burning hydrocarbons under rich conditions – that is, with excess fuel – into different streams of hydrogen and carbon dioxide for the generation of hydrogen (hydrocarbon fuel-rich combustion, which will be explained in more detail in Chapter 3, starts out with much less oxygen than required to burn all its hydrogen to water and all its carbon to CO_2).
3. Separating CO_2 from the rest of the combustion products for the purpose of CO_2 capture and sequestration in power plants (where the products are a mixture of N_2, CO_2, and H_2O).

It is possible to evaluate the minimum work required for separation using the chemical availability expression. However, we will derive the equations from first principles to illustrate the procedure. We start by recalling that mixing processes are irreversible, and that the entropy generated during mixing is always positive. Therefore, work is required to separate gases into their individual components [3,13].

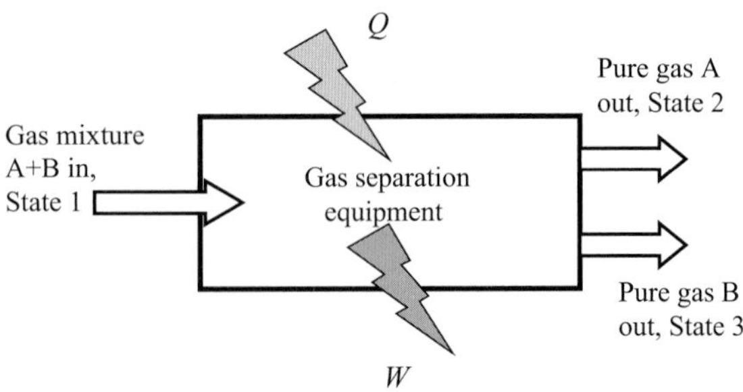

Figure 2.21 Conceptual representation of gas separation process in an open flow system.

Consider the process shown schematically in Figure 2.21. Applying the First and Second Laws to this steady-state flow system, and assuming that the system interacts only with the atmosphere, we get the following expressions:

$$\dot{Q} - \dot{W} + \dot{n}_1 \hat{h}_1 - \left(\dot{n}_a \hat{h}_{a2} + \dot{n}_b \hat{h}_{b3}\right) = 0 \tag{2.161}$$

and

$$\frac{\dot{Q}}{T_0} + \dot{n}_1 \hat{s}_1 - (\dot{n}_a \hat{s}_{a2} + \dot{n}_b \hat{s}_{b3}) + \dot{S}_g = 0, \tag{2.162}$$

where $\dot{n}$ is the molar flow rate of a stream, and T_0 is the temperature of the environment. Next, eliminate the heat transfer term between the two expressions:

$$-\dot{W} = \left[\dot{n}_a\left(\hat{h}_{a2} - T_0\hat{s}_{a2}\right) + \dot{n}_b\left(\hat{h}_{b3} - T_0\hat{s}_{b3}\right)\right] - \dot{n}_1\left(\hat{h}_1 - T_0\hat{s}_1\right) + T_o\dot{S}_g. \tag{2.163}$$

The first two terms on the right-hand side of (2.163) are the total flow availability in the outgoing and incoming streams. Dividing both sides of (2.163) by the total molar flow rate using $w = \dot{W}/\dot{n}_1$, $X_a = \dot{n}_a/\dot{n}_1$, $X_b = \dot{n}_b/\dot{n}_1$, and $S_g = \dot{S}_g/\dot{n}_1$, we get

$$-\hat{w} = \left[X_a\left(\hat{h}_{a2} - T_0\hat{s}_{a2}\right) + X_b\left(\hat{h}_{b3} - T_0\hat{s}_{b3}\right)\right] - \left(\hat{h}_1 - T_0\hat{s}_1\right) + T_oS_g. \tag{2.164}$$

Note that this is the work required per unit mole of the original mixture to separate it into two constituents. Substituting $\hat{h}_1 = X_a\hat{h}_{a1} + X_b\hat{h}_{b1}$ and $\hat{s}_1 = X_a\hat{s}_{a1} + X_b\hat{s}_{b1}$ in (2.164) yields

$$\begin{aligned} -\hat{w} = {} & X_a\left[\left(\hat{h}_{a2} - \hat{h}_{a1}\right) - T_0(\hat{s}_{a2} - \hat{s}_{a1})\right] \\ & + X_b\left[\left(\hat{h}_{b3} - \hat{h}_{b1}\right) - T_0(\hat{s}_{b3} - \hat{s}_{b1})\right] + T_0S_g. \end{aligned} \tag{2.165}$$

Ideal gas relations can be used to express the enthalpy difference and entropy difference for both components, assuming constant specific heats: $\Delta\hat{h}_i = \hat{c}_{pi}\Delta T_i$, $\Delta\hat{s}_a = \hat{c}_{pa}\ln\frac{T_{a2}}{T_1} - \Re\ln\frac{p_{a2}}{p_{a1}}$ and $\Delta\hat{s}_b = \hat{c}_{pb}\ln\frac{T_{b3}}{T_1} - \Re\ln\frac{p_{b3}}{p_{b1}}$.

To minimize the work input, we set entropy generation to zero. The best we can do is $S_g = 0$:

$$-\hat{w} = X_a\left[\left(\hat{h}_{a2} - \hat{h}_{a1}\right) - T_0(\hat{s}_{a2} - \hat{s}_{a1})\right] + X_b\left[\left(\hat{h}_{b3} - \hat{h}_{b1}\right) - T_0(\hat{s}_{b3} - \hat{s}_{b1})\right]. \quad (2.166)$$

Recall that, in this and previous expressions, the inlet and exit conditions of the streams are arbitrary. To further reduce the work, the exit conditions must be in equilibrium; thermally: $T_{a2} = T_{b3} = T_1$. In this case, (2.166) reduces to

$$-\hat{w} = T_0 X_a\left(\Re \ln \frac{p_{a2}}{p_{a1}}\right) + T_0 X_b\left(\Re \ln \frac{p_{b3}}{p_{b1}}\right).$$

Using the partial pressures of the two gases at the inlet, the work expression reduces to

$$\frac{\hat{w}_{\min}}{\Re T_0} = X_a \ln\left(X_a \frac{p_1}{p_{a2}}\right) + X_b \ln\left(X_b \frac{p_1}{p_{b2}}\right). \quad (2.167)$$

Finally, if we also assume *mechanical equilibrium* between the inlet and outlet streams (i.e., $p_{a2} = p_{b3} = p_1$), and generalize to the case in which we are separating a mixture of N components to their individual components, the minimum work required is given by

$$\begin{aligned}\frac{\hat{w}_{\min}}{\Re T_0} &= X_a \ln X_a + X_b \ln X_b \\ &= \sum_N X_i \ln X_i.\end{aligned} \quad (2.168)$$

This, as expected, is the same expression as in (2.159). Less work is required for separation at low temperatures. We recall that X_i is the mole fraction of the components in the original mixture. Equation (2.168) can be used for the case of separating a single stream into its components. This expression is similar to the entropy of mixing at constant pressure and temperature (should not it be!). Note that this work is negative; work is done on the gas stream to separate the two components. Also, since the temperature is kept constant and we are dealing with ideal gases whose enthalpies are functions of temperature only, this is the same as the heat rejected during the separation process (from the application of the First Law). A conceptual membrane-assisted ideal separation process is shown in Figure 2.22.

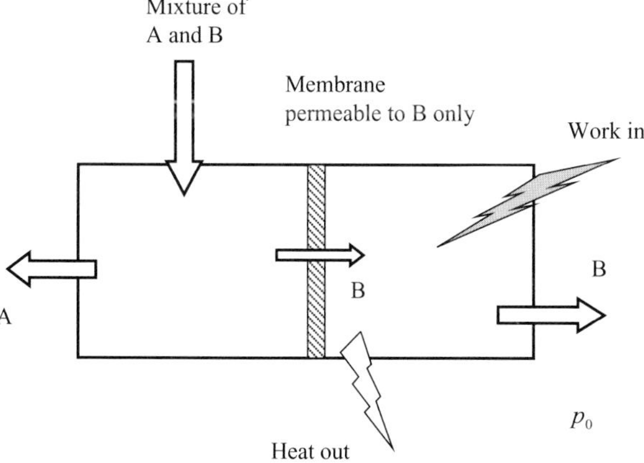

Figure 2.22 A gas separation process using a permeable membrane. For more detail, see Chapter 10.

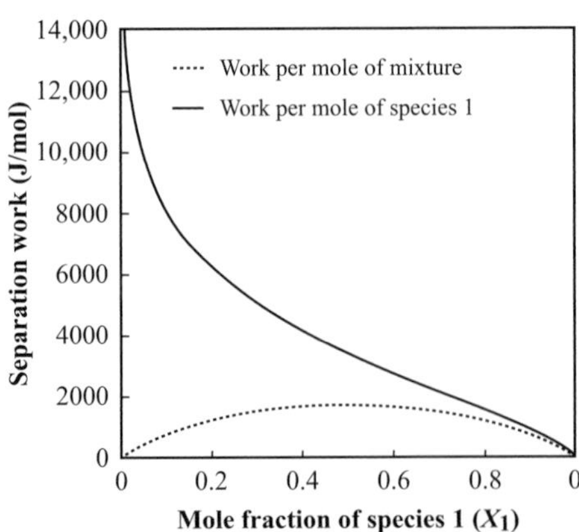

Figure 2.23 Separation work for a two-component mixture, shown per mole of mixture and per mole of X_1, evaluated at $T = 300$ K.

For a mixture of two components, X_1 and $(1 - X_1)$, or if X_1 is the mole fraction of the component to be separated and $(1 - X_1)$ is what is left, the minimum work per mole of mixture is

$$\begin{aligned} \hat{w}_{mole\ of\ mixture} &= -\Re T_o[X_1 \ln X_1 + (1 - X_1)\ln(1 - X_1)] \\ &= -\Re T_o\left[X_1 \ln \frac{X_1}{1 - X_1} + \ln(1 - X_1)\right]. \end{aligned} \tag{2.169}$$

The minimum work per mole of component 1 is

$$\hat{w}_{mole\ of\ X_1} = \frac{\hat{w}_{mole\ of\ mixture}}{X_1} = -\Re T_o\left[\ln \frac{X_1}{1 - X_1} + \frac{\ln(1 - X_1)}{X_1}\right]. \tag{2.170}$$

The two expressions in (2.169) and (2.170) are plotted in Figure 2.23. For X_1, it takes less work per mole of this component to separate it as the concentration of this component increases in the mixture.

Example 2.14 Work Requirement in an Air Separation Unit

Consider separating air into oxygen and nitrogen at 1 atm and 25 °C. The mole fractions of oxygen and nitrogen are 0.21 and 0.79, respectively. Substituting in the minimum work expression, (2.168), the work required is 1.273 MJ/kmol of air. Given the mole fractions of nitrogen and oxygen, the work requirement can be expressed in terms of kmol produced of each component using (2.170) as follows: 1.612 MJ/kmol of nitrogen, or 6.067 MJ/kmol of oxygen (or 0.19 MJ/kg-O_2). One possible realization of this process is using two pistons, each being permeable to either gas only. You may recall from your earlier studies of thermodynamics that this isothermal compression process requires the minimum work transfer, as was also demonstrated in the discussion of chemical availability.

Another interesting example is the production of argon by separating it from air, which contains 0.9% argon and 99.1% nitrogen and oxygen, by volume; that is:

Example 2.14 (cont.)

$X_{Ar} = 0.009$ and $X_{O_2+N_2} = 0.991$. Using the minimum work equation, we find that to separate argon, we need $w = \Re T_0(X_{Ar} \ln X_{Ar} + X_{O_2+N_2} \ln X_{O_2+N_2})$. Substituting, at 25 °C, we get −0.127 MJ/kmol of air, or −14.3 MJ/kmol of argon. Note that these values are ideal, assuming isothermal, frictionless processes that minimize the work.

Example 2.15 Ideal Work for CO_2 Separation

In this example, we consider the complete combustion of benzene in air at stoichiometric proportions, where the only products considered are CO_2, H_2O, and N_2. The products of stoichiometric combustion of benzene in air are found from atom conservation (more detail on the atom and mass balance in a chemical reaction will be given in Chapter 3):

$$C_6H_6 + 7.5(O_2 + 3.76\ N_2) \Rightarrow 6\ CO_2 + 3\ H_2O + 28.2\ N_2.$$

We will assume that water exists in the vapor phase at all the temperatures of interest (note that water concentration is low anyway, and the products are dominated by nitrogen; therefore, if most of the water is condensed before CO_2 separation, the change in the result is small). Calculate the minimum work required to separate CO_2 from the combustion products for the purpose of sequestering it in a zero-emission power plant when ambient conditions are 300 K and 1 atm.

Solution

Based on the reaction balance, the concentrations of CO_2 and the rest of the products in this mixture are $X_{CO_2} = 0.16$ and $X_{H_2O+N_2} = 0.84$. Substituting in the minimum work equation, assuming all streams are at the same pressure and temperature, and in equilibrium with the environment:

$$\hat{w}_{mole\ of\ X_1} = -\Re T_o\left(\ln \frac{X_1}{1-X_1} + \frac{\ln(1-X_1)}{X_1}\right) = 8.3143\ T_o\left(\ln \frac{0.16}{0.84} + \frac{\ln 0.84}{0.16}\right).$$

We find that, per mole of benzene, the minimum work to separate CO_2 in the gaseous form at atmospheric pressure is $w = -0.137\ T_0\,\mathrm{MJ/kmol}$ of benzene, where T_0 is the temperature of the environment. At $T_0 = 300$ K, the work is −41.1 MJ/kmol of benzene. The enthalpy of reaction of benzene is −3171 MJ/kmol, and taking a 20% efficient cycle, the "useful" work produced by this cycle is 634.2 MJ/kmol of benzene. Therefore, there is a penalty of 6.5% for the separation of CO_2 at $T = 27$ °C. Another way to express the loss is in terms of the separation work per fuel chemical energy – that is, the efficiency penalty in absolute percentage points. For the case discussed here, the efficiency penalty is 1.3%. Actual separation processes require more work, with higher energy penalty, due to irreversibility. Note also that the work is linearly proportional to the equilibrium temperature.

Example 2.15 (cont.)

If the gas stream is at a higher temperature than the environment, (2.166) should be used. In this case, the work transfer may be positive if the temperature difference between the gas stream and the environment is sufficiently large to allow us to use a power cycle to produce work. For instance, if $T_1 = 500\,\text{K}$, the $w_{net} = 17.1$ MJ/kmol of benzene, which is a positive quantity. Obviously, without separation, one could have used a cycle to extract even more work.

2.11 Liquefaction and Its Application

Liquefaction is used extensively in energy systems, such as storage and transportation of gaseous fuels like methane and hydrogen. The ideal liquefaction work can be obtained from (2.38). Under steady-state conditions,

$$\frac{\dot{W}_{min}}{\dot{m}} = (h_{in} - T_o s_{in}) - (h_{out} - T_o s_{out}), \tag{2.171}$$

where the *in* conditions are for the gas, and *out* conditions are for a saturated liquid at thermal and mechanical equilibrium with the environment. This minimum work system can be realized using two processes:

1. Isothermal gas compression from the original pressure, p_1, to a higher pressure, p_2. At the end of compression, the entropy is the same as that of a saturated liquid at the original pressure (see Figure 2.24).

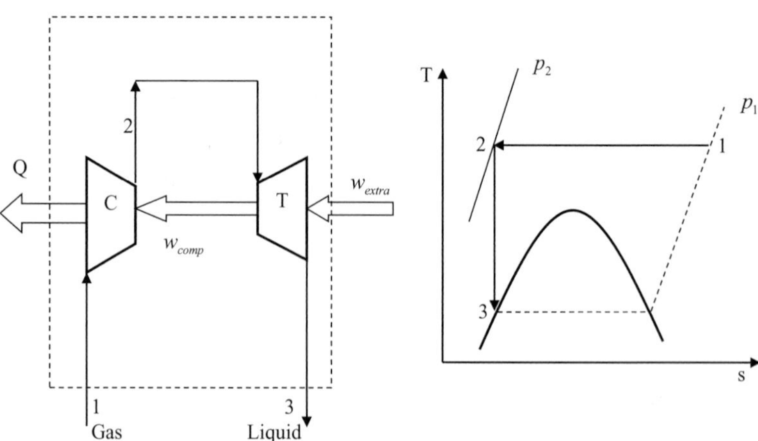

Figure 2.24 Ideal liquefaction cycle with isothermal compression and isentropic expansion.

2. Isentropic expansion of the gas from the high pressure, p_2, to the original pressure, p_1, that is, to a liquid state in mechanical equilibrium with the environment.

Figure 2.24 shows the system and the corresponding T–s diagram.

It can be shown [4,13] that the work required by these two ideal processes is given by (2.172), which defines the minimum work required to achieve the final state:

$$w_{min} = -[(h_3 - h_1) - T_1(s_3 - s_1)]. \tag{2.172}$$

Example 2.16 Minimum Work for Hydrogen Liquefaction

Using the ideal liquefaction cycle shown in Figure 2.24, determine the minimum work per kilogram of hydrogen in the cycle. At the cycle inlet, hydrogen is a superheated vapor at 1 atm and 298 K. At the outlet, it becomes saturated liquid at 1 atm. The following properties are provided: (1) the superheated vapor properties at the inlet are: $s_1 = 53.44$ kJ/kg·K and $h_1 = 3929.6$ kJ/kg; (2) the outlet (saturated vapor) properties at 1 atm are $s_3 = -0.02$ kJ/kg·K and $h_3 = -0.44$ kJ/kg.

Solution

We determine the minimum work per kilogram of hydrogen using (2.172):

$$w_{min} = [298(-0.02 - 53.436) - (-0.44 - 3929.6)] = 11.870 \text{ MJ/kg}.$$

The critical point of hydrogen is 33.3 K and 1.3 MPa. The lower heating value of hydrogen is 120.9 MJ/kg. Hence, the ideal liquefaction work is less than 10% of the chemical energy stored in hydrogen. As we will see next, more realistic values of the energy of liquefaction are higher than these ideal numbers, making the energy requirements for hydrogen storage in the liquid form a large fraction of the energy it carries.

Example 2.17 Minimum Work for Nitrogen and Methane Liquefaction

For comparison, repeat the previous example by replacing hydrogen with nitrogen or methane. The following properties are provided for nitrogen: (1) The properties of nitrogen at 1 atm and 298 K (state 1) are $h_1 = 311.1$ kJ/kg and $s_1 = 6.84$ kJ/kg·K; (2) the properties of nitrogen at 1 atm at saturated liquid conditions (state 3) are $h_3 = -122.1$ kJ/kg, $s_3 = 2.83$ kJ/kg·K, and $T_3 = 79$ K.

Solution

Substituting in (2.172), we find

$$w_{min} = [(298)(2.83 - 6.84) - (-122.1 - 311.1)] = -769.4 \text{ kJ/kg}.$$

Example 2.17 (cont.)

The minimum liquefaction work for nitrogen is much lower than that for hydrogen, which is not surprising given the higher liquefaction temperature for nitrogen. The critical point of nitrogen is 126.2 K and 3.4 MPa.

For methane, the **initial conditions** are: pressure p_i ¼ 1 atm, temperature $T_i = 293\,\text{K}$. The **final conditions** are: pressure p_f ¼ 2.6 atm and temperature $T_i = 110$ K.

The minimum work required to achieve the liquid state from the gas state is again calculated by the following equation (note that the critical point of methane is 126.2 K and 3.4 MPa):

$$w_{min} = -\left[\left(h_f - h_i\right) - T_0\left(s_f - s_i\right)\right].$$

From the NIST website, the thermodynamic properties of methane in the initial and final states is

$$\begin{array}{ll} h_i = 854.5\,\text{kJ/kg} & s_i = 6.48\,\text{kJ/kg·K} \\ h_f = -5.55\,\text{kJ/kg} & s_f = -0.053\,\text{kJ/kg·K} \end{array}.$$

Therefore, the minimum work required to liquefy methane is

$$w_{min} = -[(-5.55 - 854.5) - 273(-0.053 - 6.48)] = -923.46\,\text{kJ/kg of methane.}$$

If the plant has a Second Law efficiency of 25%, the actual work required to liquefy the natural gas is

$$w^{actual} = \frac{w^{ideal}}{25} \cdot 100 = \frac{-923.46}{25} \cdot 100 = -3693.84\,\text{kJ/kg of methane.}$$

The energy required can be expressed as a fraction of the lower heating value (LHV) of methane (see Chapter 3):

$$\%\text{energy required} = \frac{w^{actual}}{LHV_{CH_4}} \cdot 100 = \frac{3.693}{50.1} \cdot 100 = 7.37\%.$$

In smaller liquefaction plants, the expander is often replaced with a Joule–Thompson or throttle valve. The constant enthalpy process of gas expansion through this valve introduces irreversibility that reduces the performance of the system, and reduces the amount of liquid following expansion. To maximize the amount of liquid following expansion, the gas is cooled before it is expanded, using the cooled gas at the end of the throttling process, as shown in Figure 2.25. This regenerative cooling improves the performance of the Linde–Hampson liquefaction process.

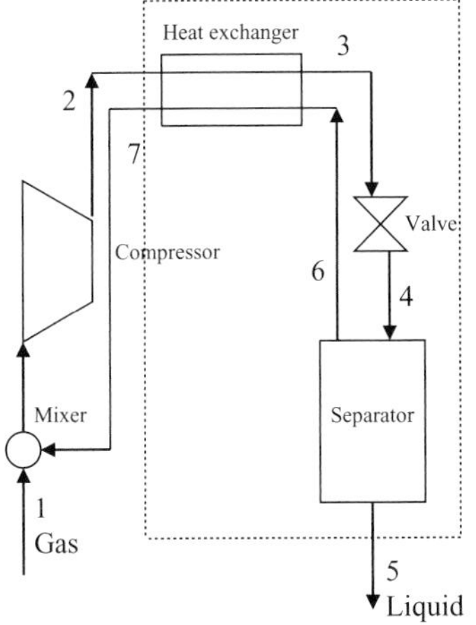

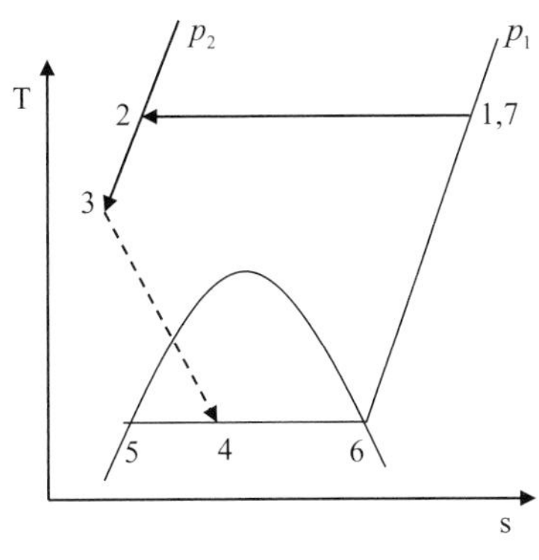

Figure 2.25 Linde–Hampson liquefaction cycle, with isothermal compression, regeneration, and throttle valve, and the corresponding T–s diagram.

The mass fraction of liquid produced after expansion per unit mass of compressed gas, $Y_l = m_5/m_2$, can be obtained by applying the First Law to the control volume shown by the broken lines in Figure 2.25:

$$m_2 h_2 = m_5 h_5 + (m_2 - m_5)h_1. \tag{2.173}$$

Rearranging (2.173) gives

$$Y_l = \frac{m_5}{m_2} = \frac{h_1 - h_2}{h_1 - h_5}. \tag{2.174}$$

The isothermal compression work can be obtained by applying the First and Second Laws to that process:

$$Q - W = h_2 - h_1, \tag{2.175}$$

and since the process is isothermal,

$$Q = T_1(s_2 - s_1). \tag{2.176}$$

Now combine the two statements to eliminate the heat transfer and obtain the compression work:

$$W = T_1(s_2 - s_1) - (h_2 - h_1). \tag{2.177}$$

The work done per unit mass of liquid produced is thus

$$W_l = \frac{W}{Y_l} = \frac{h_1 - h_5}{h_1 - h_2}[T_1(s_2 - s_1) - (h_2 - h_1)]. \tag{2.178}$$

The impact of the irreversibility associated with the Joule–Thomson valve can be seen by comparing the two work expressions in (2.172) and (2.178). For nitrogen initially at standard temperature and pressure (STP) that is compressed to 20 MPa, with $h_2 = 279.0$ kJ/kg and $s_2 = 5.16$ kJ/kg·K, calculations show that the work requirement is

$w = -6354.3\ \text{kJ/kg}$ per unit mass of liquid nitrogen. The Second Law efficiency of the Linde–Hampson liquefaction system is $w_{min}/w = 0.121$.

Isothermal compression is an idealization, and in most cases a series of adiabatic compression with intermediate cooling is used to approximate isothermal compression. In some cases, it may be necessary to add extra cooling before or after compression to increase the percentage of liquid following the throttle. This is particularly true in the case of hydrogen liquefaction because throttling below a certain temperature raises the temperature of the gas. In cases when cooling is used, the following Linde–Hampson liquefaction system is more representative [10]. It consists of:

1. an isothermal compression to raise the gas pressure;
2. a heat exchanger to cool the pressurized gas using a cooling medium, such as liquid nitrogen;
3. a Joule–Thompson valve used to throttle the high-pressure gas adiabatically to pressures below its own critical pressure – that is, to within the gas-vapor dome; and
4. a separation chamber used to separate the liquid stream from the gas stream, which is recycled for another round of compression, cooling, and throttling. Figure 2.26 shows the components used in this cycle.

The corresponding T–s diagram for this cycle is shown in Figure 2.27. Note that extra gas is constantly supplied at state 0, at atmospheric conditions. At steady state the mass flow rate of the liquid leaving the separator equals that of the gas added. A coolant at atmospheric conditions can be used following compression. If cooling is used before compression, a cold agent should be used, and the energy required to cool this agent should be added to the total energy required to liquefy the gas. Instead of the throttle valve, it is possible to use an expander, such as a turbine or a reciprocating device, to take advantage of the expansion work and hence improving the overall efficiency. However, this increases the cost and complexity of the plant. Note also that the mass flow rates in the heat exchanger are not equal, the "hot stream" is the total mass flow of the gas, while the "cold stream" is only part

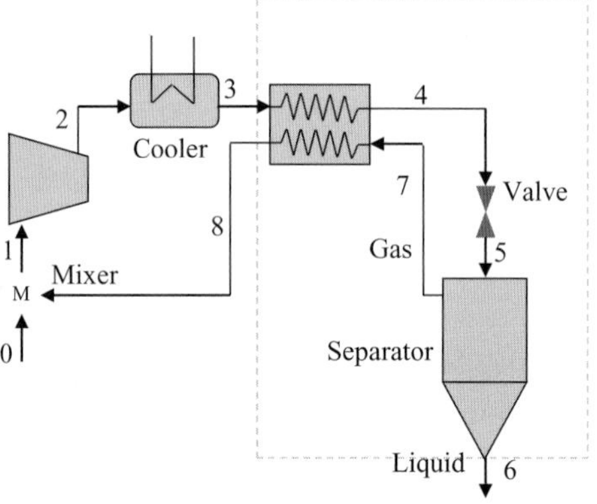

Figure 2.26 Liquefaction cycle with adiabatic compression and a separate cooler.

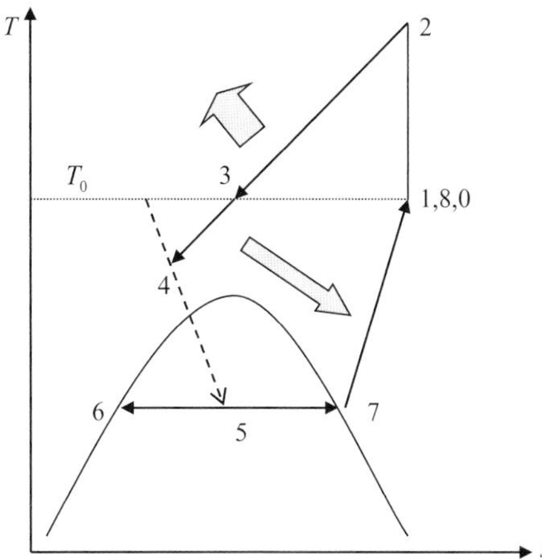

Figure 2.27 The T–s diagram of the adiabatic compression liquefaction cycle shown in Figure 2.26, utilizing a cooler to reduce the gas temperature between 2 and 3, and a heat exchanger to further reduce the gas temperature between 3 and 4 (while heating up the separated vapor between 7 and 8). The thick arrows show cooling between 2 and 3, and internal heat exchange between 2–3 and 7–8.

of it that is not liquefied. Despite its simplicity, the performance of this cycle is far from ideal because of the heat exchanger and the throttling process.

2.12 Summary

In this chapter we reviewed some of the fundamental laws of thermodynamics posed in forms most useful for the analysis of energy conversion processes and systems, with a focus on heat, work, and thermal internal energy. Essentially the First Law postulates that the stored internal energy of a system can change as the system experiences heat and work transfer with its environment, and the Second Law postulates that the entropy of a closed system must increase, or that entropy generation in this system must be positive. Both statements led to the definition of the availability/exergy and a methodology for evaluating the performance of energy conversion components and systems and identification of related losses and inefficiencies. Because of the focus on heat and work, this leads naturally to the concept of a heat engine, its absolute (First Law) and relative (Second Law) efficiency. The concept of efficiency was extended to other energy conversion devices and to individual processes.

Most energy conversion systems operate continuously and hence steady (or unsteady) control volume analysis is a more useful tool for evaluating their performance. In such analysis, mass conservation must also be satisfied. The combined statement of the First and Second Laws can be used to demonstrate the relation between entropy generation and availability loss clearly. Examples are used to show that some of the observations leading to the Second Law can be quantified – for example, heat transfer across a temperature difference leading to entropy generation. We demonstrated also that the Carnot engine is the most efficient device operating between two fixed-temperature heat reservoirs. Another expression is obtained in the high-temperature reservoir temperature changes during the heat transfer process. In that case, the log mean temperature must be used.

In thermodynamics it is important to define the state of a system and, appealing to equilibrium, one can define a generic form for relating the changes in state properties. A widely used equation of state, the ideal gas law, was also reviewed for a pure component and a mixture. Because of the significance of gas (or other states such as liquid) mixing and separation, the concepts of mass transfer and chemical potential were introduced. An important application is ideal work for gas separation and the relation between entropy generation/availability loss during mixing and work required for separation. Several examples are used to quantify the ideal work of air separation for oxygen production and for CO_2 separation from combustion products. Both will be revisited in later chapters, when actual processes are discussed. Another important process is liquefaction (e.g., for hydrogen storage or natural gas transport), which is also discussed. Both the ideal approach of liquefaction (most efficient) and processes using more practical devises are discussed. An important take-away is the inevitable inefficiency associated with reducing the complexity (and cost) of components.

In the following two chapters, the concepts developed in this chapter will be extended to systems undergoing chemical and electrochemical reactions, under equilibrium, by extending the definitions of internal energy to include bond energies and mode of work transfer to include electrical work transfer.

Problems

2.1 An air compressor with isentropic efficiency of 82% is used to pressurize air at 298 K and 1 atm. Exiting air with a pressure of 8 atm cools in a heat exchanger to 345 K. Water is used as a coolant in the heat exchanger with inlet/outlet temperature of 298/312 K, as shown in Figure P2.1. The volumetric flow rate of air entering the compressor is 0.4 m^3/s. Treat air as an ideal gas and assume no pressure drop in the heat exchanger; determine:

a. the required mass flow rate of the coolant; and

b. irreversibility within the compressor and the heat exchanger.

Figure P2.1

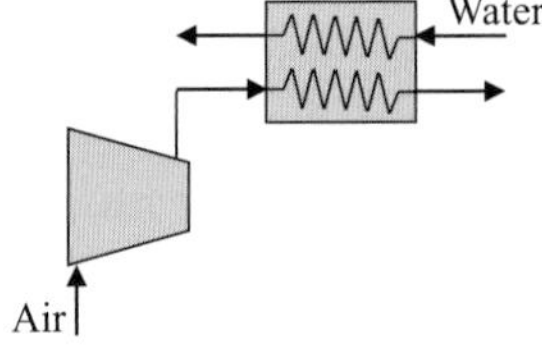

2.2 A three-stage compressor is used to raise the pressure of an ideal gas, initially at p_1 and T_1, to a certain pressure, p_2. The pressurized gas is cooled in an intercooling heat exchanger after each stage, as shown in Figure P2.2. Ideally, the gas temperature after each intercooler drops to the initial temperature, T_1. Assume that there is no pressure drop across the process path of the gas and the isentropic efficiency of the compression process is identical in all three stages, η_c.

a. Develop a relationship for minimum work requirement for this three-stage compression process.

b. Develop an expression for the compression work if the ideal gas is compressed isothermally from the initial pressure, p_1, to the desired pressure, p_2, and compare it with the result of part (a).

c. Using the result of part (a), compare the minimum compression work per unit mass of the gas versus pressure ratio p_2/p_1 for carbon dioxide, nitrogen, air, and oxygen.

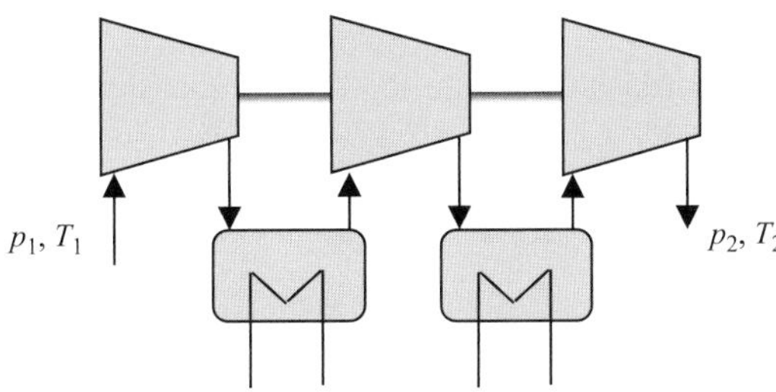

Figure P2.2

2.3 Consider the simple open gas turbine power plant shown in Figure P2.3, which uses air supplied from the atmosphere as a working fluid. The isentropic efficiencies of compressor and turbine are denoted by η_C and η_T, respectively. The pressure drop within the combustor is Δp_{Com}. The idea is to determine the optimum performance of the power plant based on various criteria. The temperature and pressure of air at the compressor inlet are fixed since it is supplied from the atmosphere. The maximum temperature (i.e., turbine inlet temperature), is also fixed due to the limitation of turbine blades' durability. Assume that air behaves like an ideal gas; the mass flow rate of fuel, $\dot{m}_f$, burned in the combustor is negligible compared to that of air, $\dot{m}_a$; and specific heat of air is constant throughout the process.

a. Derive a relationship for power output of the plant, P_{net}.

b. Derive a relationship for thermal efficiency of the plant, η_{th};

c. Derive a relationship for entropy generation rate within the power plant, $\dot{S}_g$.

d. Plot the normalized power output $(P_{net}/\dot{m}_a)$, thermal efficiency, and normalized entropy generation rate $(\dot{S}_g/T_1\dot{m}_a)$ versus the pressure ratio across the compressor, p_2/p_1 for $\eta_C = 0.80$, $\eta_T = 0.90$, $\Delta P_{Com} = 0.05P_2$, and $T_3/T_1 = 4$. Identify the optimum pressure ratios corresponding to maximum power output, maximum thermal efficiency, and minimum entropy generation rate.

e. Using the relationships obtained in parts (a) and (b), derive explicit expressions for the optimum pressure ratio leading to maximum thermal efficiency and maximum power output.

f. Derive a relationship for the Second Law efficiency of the power plant.

Figure P2.3

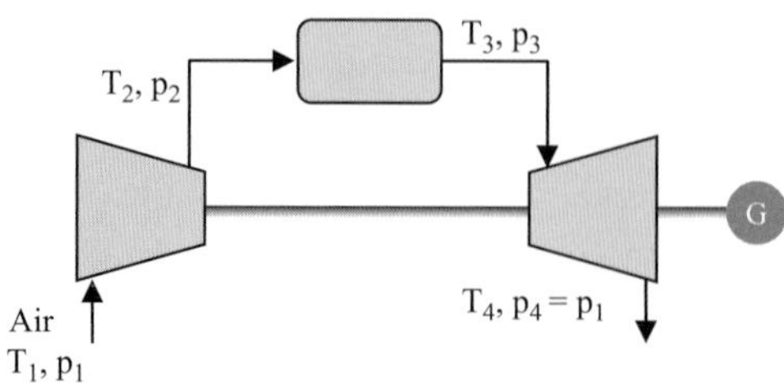

2.4 A gaseous mixture composed of equal mass of substance A and substance B enters a separation unit, as shown in Figure P2.4.

a. Show that the minimum work per unit mass for separation of A from this mixture is obtained from the following equation:

$$W_{min} = T_0 R \ln\left[\frac{(r+1)^2}{r}\right],$$

where r is the ratio of the molecular weight of substance B to the molecular weight of substance A.

b. Determine the value of r for which W_{min} is minimal.

Figure P2.4

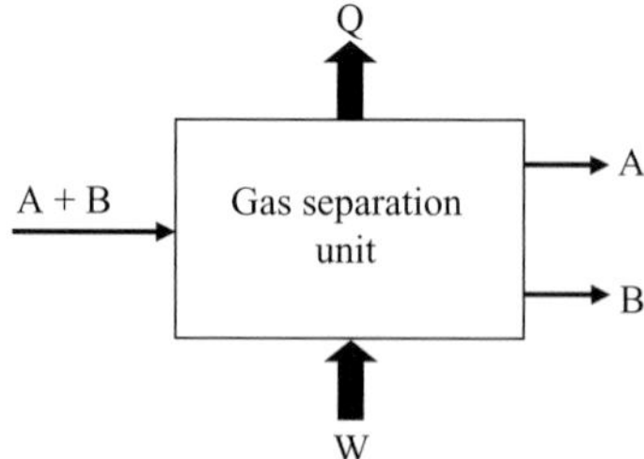

2.5 The volumetric composition of light gases released from the pyrolysis (chemical decomposition in an inert environment) of beech wood is reported to be CO (33.5%), CO_2 (56.7%), H_2 (0.6%), and CH_4 (9.2%). Three separation units are operating in series for separation of the above individual substances from the gaseous mixture. As shown in Figure P2.5, CO_2 is separated from the mixture in the first unit. In the second unit, CO is separated and the mixture of H_2 and CH_4 is directed to the third unit, which separates the mixture into H_2 and CH_4. Determine the minimum work requirement per unit mole of the initial mixture of the light gases.

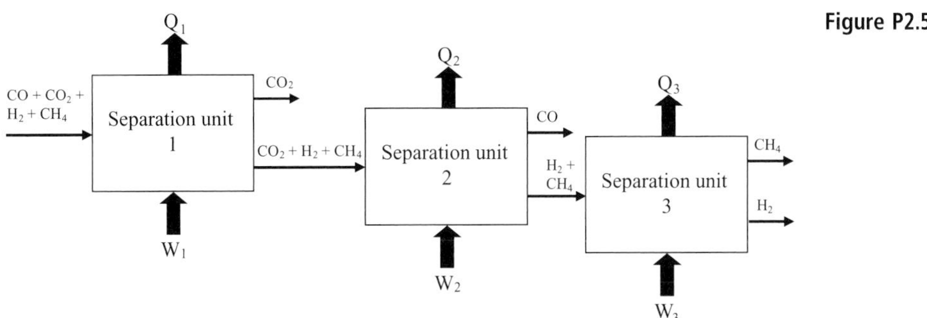

Figure P2.5

2.6 Air flows from a source at p_s (N/m^2) and T_s (K) into an insulated rigid tank, as shown in Figure P2.6. During the filling process, the pressure inside the tank is maintained at p_o (N/m^2) by discharging air through a valve from the tank in a controlled manner. Assume that air is an ideal gas and that its initial temperature in the tank is T_1. The air flowrate into the tank, $\dot{m}_s$ (kg/s) is kept constant. Derive an expression for the variation of the air temperature in the tank with time.

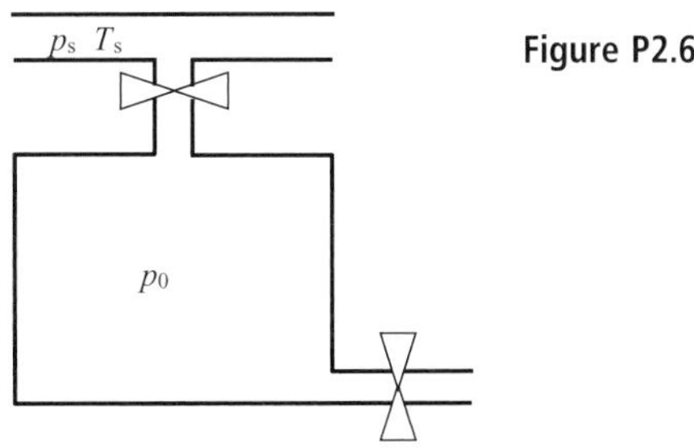

Figure P2.6

2.7 Assume that the USA consumes approximately 400 million gallons of gasoline per day, mainly in vehicles. A proposal has been made to reduce the consumption of gasoline by 100 million gallons per day by using either liquid hydrogen (LH) or liquid natural gas (LNG). A four-cylinder, 2-liter gasoline engine, at cruising speed, has the engine shaft rotating at 2500 rpm, producing a torque of 80 Nm. The engine thermal efficiency is 25%. Assume that the same engine can be modified to use LH or LNG without any change in the thermal efficiency. The average speed of the vehicle is 65 miles per hour.

a. Find the power generated by the engine.

b. Compute the average distance covered by the car per gallon of gasoline consumed.

c. Determine the distance covered per gallon of LH and LNG.

d. Determine the daily requirement of LH and LNG to replace 100 million gallons of gasoline.

e. Hydrogen can be produced by reacting coal with steam and oxygen according to the following reaction: $C + H_2O + O_2 \rightarrow CO_2 + H_2$. The required oxygen can be obtained from an ASU. Considering air to be a mixture of O_2:N_2 = 1:3.76, obtain the minimum work required by an ASU, operating at 298 K and 1 atm to produce 1 kg of oxygen. If the Second Law efficiency of the ASU is 30%, how much power is needed to produce oxygen for the proposal?

f. To use LH as the fuel in transportation vehicles, work is needed in the liquefaction process. Assuming the Second Law efficiency of liquefaction is 15%, calculate the power required to produce LH for the proposal. The specific enthalpy and entropy at the initial and final states are 3975 kJ/kg, 53.53 kJ/kg K, and −3.57 kJ/kg, −0.178 kJ/kg·K, respectively.

g. Now, consider that the natural gas (assume CH_4) is obtained from an offshore gas field. The gas leaving the well contains 10% CO_2 and 90% CH_4 by volume, and is at a temperature of 0 °C. How much work is required to separate CH_4 from the mixture under ideal conditions, per kilogram of CH_4 produced? What is the power required for the proposal if the Second Law efficiency of the process is 20%?

h. After separation of CO_2, the next step is to liquefy the natural gas. In the LNG plant, this is accomplished by decreasing the temperature of CH_4 to −163 °C while increasing the pressure to 2.6 atm. The initial pressure of CH_4 is assumed to be 1 atm. If the Second Law efficiency of the plant is 25%, calculate the power required to liquefy CH_4 per kilogram for the proposal. The specific enthalpy and entropy at the initial and final states are 854.5 kJ/kg, 6.48 kJ/kg·K, and −5.55 kJ/kg, −0.053 kJ/kg K, respectively.

You may use the data in Table P2.7 for the calculations.

Table P2.7

Fuel	*Gasoline*	*Liquid hydrogen*	*LNG*
Density, kg/m^3	719	70	475
Lower heating value, MJ/kg	42	120.9	48.5

2.8 Compressed air storage (CAS) has been used to store energy (electricity generated by renewable or other sources) in the form of compressed air in underground caverns (Figure P2.8a).

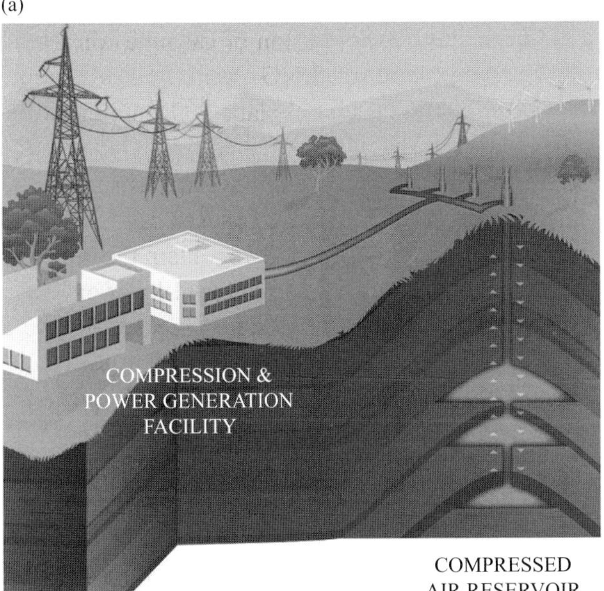

Figure P2.8a Illustration of a compressed air energy storage system (courtesy of Pacific Northwest National Laboratory (CC)).

In order to increase the mass of high-pressure air stored in the available volume of the cavern, air may be cooled after compression using a liquid. The heated liquid is then stored in a separate tank. This is the charging process. During discharging, the high-pressure air from the air cavern and the hot liquid from the tank are used to generate work (electricity) by expanding the air and extracting heat from the salt tank, respectively, using heat engines interacting with the environment. The overall system (air tank, molten salt tank, and heat engines) acts as a giant "thermomechanical" battery. In this problem we use an insulated tank to stored high-pressure air, and another insulated tank of molten salt to store the thermal energy, and generic heat engines to generate work from the stored energy.

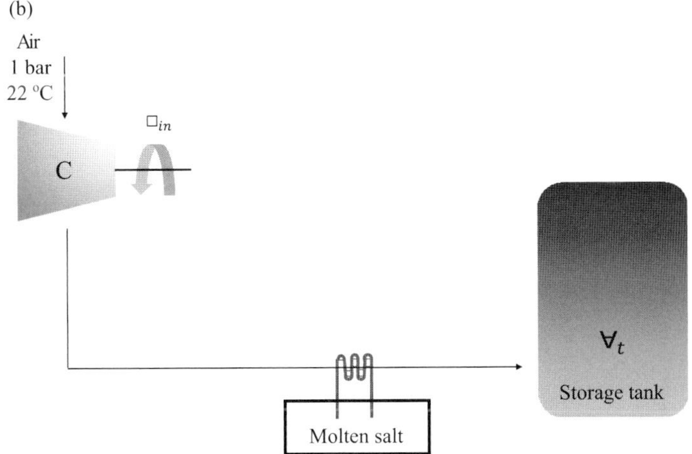

Figure P2.8b Schematic of the proposed CAS system.

The air tank volume is 1000 m^3. The molten salt mass in the tank is 100 ton (10^5 kg). The specific heat of the salt is 1,500 J/kg·K. The conditions of the gas in the tank when fully discharged is $p = 1$ bar and $T = 22$ °C. The temperature of the salt in the tank when fully discharged is 300 °C (to maintain the salt in the liquid phase). When fully charged, the conditions in the air tank are: $p = 100$ bar and $T = 600$ °C. When fully charged, the conditions in the salt tank are: $T = 600$ °C. Atmospheric conditions are $p = 1$ bar and $T = 22$ °C.

Calculate:

a. The energy stored in both tanks between the fully charged state and discharged state.

b. The work required by the compressor train to charge the system of the air tank and the salt tank. The compressor train operates adiabatically.

c. The maximum work that can be extracted from the molten salt tank starting with the fully charged state.

d. The maximum work that can be extracted from the gas tank starting with the fully charged state.

e. Assume that the Second Law efficiency of the machinery used to extract work from the gas tank is 70% and that of the machinery used to extract work from the salt tank is 60%. What is the roundtrip efficiency of this storage system?

f. How long does it take to charge the system using a wind turbine operating at 1 MW?

2.9 A non-conducting piston having an area of 10^{-3} m^2 and negligible mass is free to move in an insulated cylinder, as shown in Figure P2.9. Initially, compartment A of the cylinder with a volume of 10^{-4} m^3 contains helium at 315 K and 15 bar, while the piston is held in position by a pin and compartment B is evacuated. The environment is at 298 K. The piston's movement is resisted by a spring with a spring constant of 0.9 kN/m, which is initially uncompressed. The pin is now removed, allowing the helium gas to expand until the piston hits a second pin at a distance of 0.18 m from the first pin.

a. Calculate the work done by helium and the process irreversibility.

b. If the helium undergoes free expansion without the spring, calculate the quantities listed in (a) for the new arrangement.

c. If the expansion is carried out in a quasi-static manner, compute the quantities listed in (a).

Hint: Work done by a spring that extends from x_0 to x_1 may be obtained from

$$W_{x_0}^{x_1} = \frac{1}{2}k\left(x_1^2 - x_0^2\right).$$

Figure P2.9

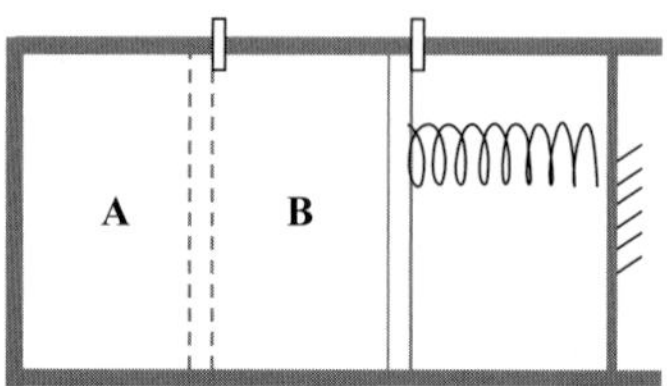

2.10 Two rigid tanks A and B are connected by a valve, as shown in Figure P2.10. Tank A is insulated and contains 0.2 m^3 of steam at 400 kPa and 80% percent. Tank B is not insulated and contains 3 kg of steam at 200 kPa and 250 °C. The valve is now opened and steam flows from tank A to tank B until the pressure in tank A drops to 300 kPa. During the process, 600 kJ of heat is transferred from tank B to the surroundings that is at 0 °C. Assume that the steam inside tank A undergoes a reversible adiabatic process.

a. Determine the final temperature in each tank.

b. Determine the irreversibility of the process.

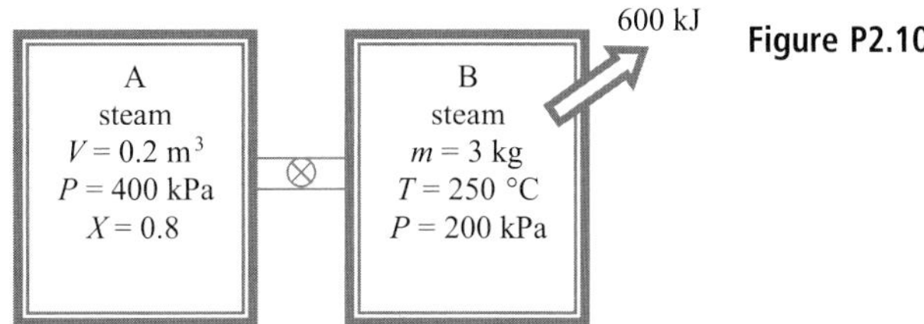

Figure P2.10

2.11 (Adopted from Baron and Baron, *Cryogenic Systems*.) A Linde–Hampson cycle for hydrogen liquefaction is shown schematically in Figure P2.11. Hydrogen is isothermally compressed, cooled at a constant pressure, and then expands adiabatically across a throttle valve. To cool down the hydrogen through the throttle value, the temperature at the inlet of the throttle valve should be, at least, lower than its inversion temperature of ~200 K (to ensure that the Joule–Thomson coefficient is positive). Moreover, to liquefy hydrogen as it expands through the valve, the inlet temperature should be lowered as close as possible to the critical temperature of ~30 K. Liquid nitrogen is used to cool the gaseous hydrogen before it enters the throttle valve. Assume that all the heat exchangers and the liquid bath are perfectly insulated, and streams from the make-up gas and from the exit of heat exchanger 1 have the same temperature and pressure. The states are shown in Table P2.11.

a. Draw the T–s diagram of the cycle.

b. Calculate the mass of liquid H_2 produced in the hydrogen liquid separator per 1 kg of H_2 at state 2. Note that the conditions of state 3 and 7 are given.

c. Determine the pressure and temperature of H_2 at states 1–7.

d. Calculate the work required in the compressor per 1 kg of H_2 at state 2. Assume the isothermal compressor is reversible.

e. Calculate the mass of liquid N_2 in kilograms required to cool 1 kg of H_2 in the N_2 liquid bath. Assume the liquid bath is perfectly insulated.

f. Assuming a Second Law efficiency of 30% for the N_2 liquefaction plant, determine the work required to reliquefy the nitrogen needed in part (d).

g. Determine the total work required to produce 1 kg of liquid hydrogen.

h. Calculate the irreversibility (per kilogram of H_2 at state 2) in heat exchanger 1, heat exchanger 2, and the throttle valve.

Table P2.11

State	Temperature (K)	Pressure (atm)
1	298	1
2	298	100
3	64	100
7	63.3	1
8	298	1

Figure P2.11

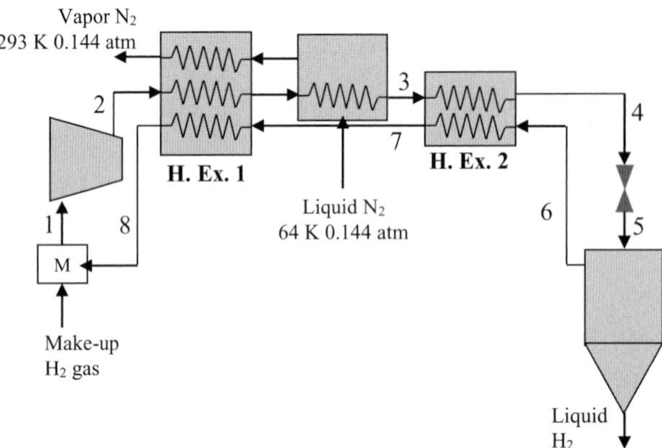

2.12 A schematic of the Claude cycle is shown in Figure P2.12 (adapted from Baron and Baron, *Cryogenic Systems*). It is identical to the Linde–Hampson cycle except that the high-pressure hydrogen stream is split after heat exchanger 1. A fraction of the high-pressure hydrogen is expanded to the low pressure in a turbine producing some of the work needed by the compressor. This also makes it possible to eliminate the liquid N_2 bath. Following expansion, this stream is mixed with the gas exiting heat exchanger 3 to provide the cooling in heat exchanger 2. The portion of the high-pressure hydrogen stream bypassing the turbine is cooled in heat exchangers 2 and 3, and expanded through the valve. The liquid is passed to the liquid receiver, and the vapor is recycled to provide cooling for the incoming high-pressure stream of the gas. Assume that

1. all the heat exchangers are perfectly insulated;
2. the two streams from the make-up gas and the exit of heat exchanger 1 have the same temperature and pressure;
3. the streams exiting the turbine and heat exchanger 3 have the same temperature and pressure; and
4. 50% of the high-pressure hydrogen stream leaving heat exchanger 1 is sent through the turbine, and the turbine isentropic efficiency is 90%.

The states are shown in Table P2.12.

a. Draw schematically the *T–s* diagram of the Claude cycle.

b. Calculate the mass of liquid H_2 produced in the hydrogen liquid separator per 1 kg of H_2 at state 2.

c. Determine the pressure and temperature of H_2 at states 1–9.

d. Calculate the work produced by the turbine per 1 kg of the produced liquid H_2.

e. Determine the total work required to produce 1 kg of liquid hydrogen.

f. Determine the Second Law efficiency of the Claude cycle.

Table P2.12

State	Temperature (K)	Pressure (atm)
1	298	1
2	298	100
3	160	100
10	125	1

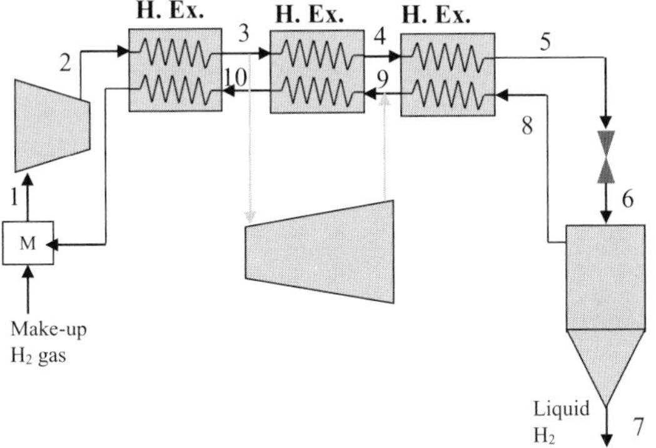

Figure P2.12

2.13 Helios, a slow-flying, remotely controlled, electric propeller plane that set the altitude record of 96,500 ft, is driven by photovoltaic cell-generated electricity during daytime, and by a hydrogen-powered proton-exchange membrane fuel cell at night. The solar cells with efficiency of 19% are mounted on the wide span wing and can produce up to 35 kW at midday. The total power required for the plane to fly is 15 kW. The current version of the plane carries hydrogen in pressurized tanks to power the fuel cells at night and uses oxygen from the atmosphere. This limits how long the plant can fly by the hydrogen storage capacity of the tanks, and work is underway to replace the compressed tanks with liquid hydrogen tanks. This will extend the range, but will still be limited. Another redesign of the propulsion system will use a closed loop regenerative system that generates hydrogen and oxygen by cracking water using extra electricity from the photovoltaic cells during the day. This will be done in an electrolyzer, which is actually the fuel cell operated in reverse. The two gases will be stored separately, and will be used at night to power the fuel cell (see Figure P2.13). Combining hydrogen and oxygen to produce electricity in the fuel cell also produces water, which is stored for further processing during the day.

Assume that the storage tanks can operate up to 500 bar. The fuel cell efficiency defined as $\eta_{FC} = P_{out}/(\dot{m}_{H_2} LHV_{H_2})$ is 45%, where P_{out} is the power produced by the fuel cell, $\dot{m}_{H_2}$ is the mass flow rate of hydrogen, and LHV_{H_2} is the lower heating value of hydrogen. Also, assume that the 24-hour cycle consists of 12 hours of daytime and 12 hours of night.

a. Calculate the mass of hydrogen required to fly the plane during the night.

b. Calculate the mass of oxygen required, noting that for each mole of hydrogen, 0.5 mole of oxygen is consumed in the fuel cell.

c. Treating hydrogen and oxygen as ideal gases, calculate the volume of the tanks in which the gases are stored if the tank temperature is 25 °C.

d. Calculate the mass of the hydrogen storage tank if the gravimetric efficiency of hydrogen storage, defined as the ratio of the mass of hydrogen in the tank to the total mass of the hydrogen and the tank, is 5.5%.

e. If both tanks operate at the same maximum pressure, estimate the gravimetric efficiency of the oxygen tank, and its mass.

f. During the daytime, the electrolyzer produces separate streams of hydrogen and oxygen at 1 bar and 25 °C (each). Calculate the compressor work per unit mass required to raise the pressure of hydrogen and oxygen to 500 bar, assuming steady-state operation. The Second Law efficiencies of the hydrogen and oxygen compressors are 60% and 80%, respectively.

g. Calculate the total power required by the photovoltaic cells during the daytime, assuming that the electrolyzer efficiency, defined as $\eta_{elec} = (\dot{m}_{H_2} LHV_{H_2})/P_{in}$, is 80%.

h. A smart engineer thinks that liquefying hydrogen will reduce the size of its storage tank, but it will require higher photocell power. Determine the additional power required to store the hydrogen in liquid form at 1 bar compared to part (g). The Second Law efficiency of that plant is 50%.

i. In part (f), it is assumed that the compressors operate steadily to raise the pressure from 1 bar to 500 bar. This, however, is unnecessary, since early in the morning the pressure inside the tank is close to 1 bar, and only late in the day does it rise to 500 bar. Calculate the ideal work required to compress the hydrogen in its tank, starting with 1 bar and ending with 500 bar, and compare it with the result in part (f). (The compressor exit pressure is the pressure inside the tank, which is increasing from 1 bar to 500 bar during the process).

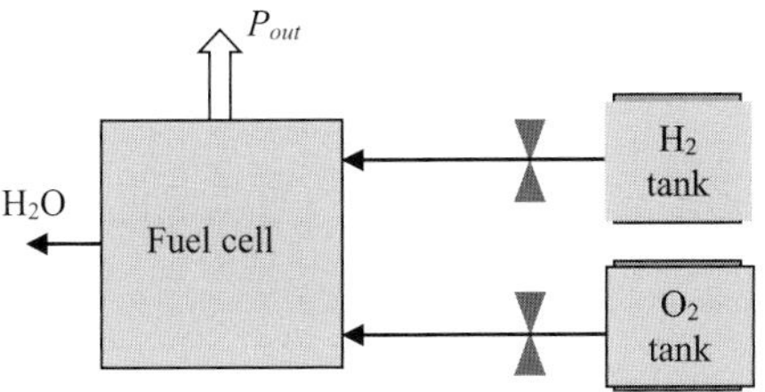

Figure P2.13a Operation during the night.

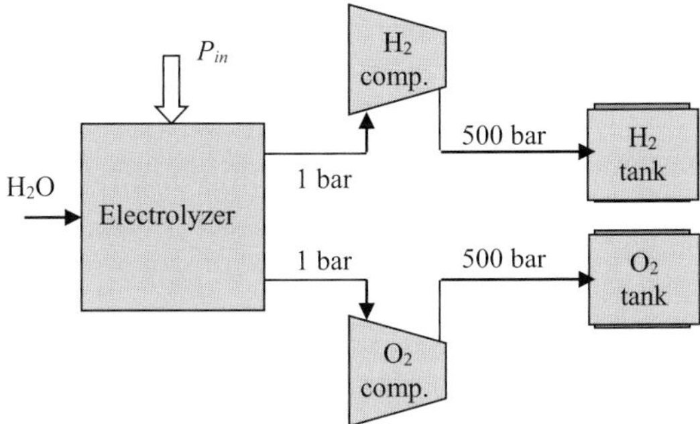

Figure P2.13b Operation during the day.

2.14 Desalination – that is the removal of a significant fraction of the salt (NaCl) – provides a potential solution to water scarcity. At its core, desalination is a separation process that aims to remove salt and produce a stream of pure water using thermal or mechanical energy.

Consider the desalination of seawater at standard temperature and pressure. Seawater is modeled as a 1% concentration by mass of NaCl. We can assume that:

1. the solution is an incompressible liquid with ideal entropy of mixing and zero enthalpy of mixing;
2. the molar heat capacity of seawater is constant and equal to the molar heat capacity of pure water; and
3. the density of seawater is constant and equal to the pure water density.

We wish to produce 1 kg/s of pure water from 2 kg/s of seawater. The two outlet streams of the desalination process are pure water and brine or concentrate (the remaining solution).

a. Calculate the brine flow rate and composition.
b. Calculate the ideal work of separation expressed in kJ/m^3 of product.

Note that in a liquid mixture, the entropy of component i is approximately $s_i = s_i^\circ - R \ln X_i$.

Hint: For an incompressible liquid $h = c(T - T_0) + pv$.

Now consider the following two options for the technology:

Technology I: Once-Through Boiling

Once-through boiling is the most basic thermal desalination method, which heats the seawater feed to the boiling point, bringing about freshwater vaporization. You can ignore the boiling point elevation and any change in the enthalpy of vaporization associated with the addition of salt.

Technology II: Reverse Osmosis (RO)

Osmosis refers to the movement of water across a membrane from a region of low salt concentration to that of a higher salt concentration. This transport of water is caused by the buildup of a pressure difference between the two sides of a membrane due to a difference in the chemical potential of water. For more information on osmosis, please refer to *Physical Chemistry* by P.W. Atkins or https://en.wikipedia.org/wiki/Osmosis.

While desalination has historically been a thermal process, the invention of RO membranes in the 1960s revolutionized the process. In contrast to osmosis, RO produces freshwater by pumping the seawater to a pressure, P_H, higher than the osmotic pressure, as shown in Figure P2.14. The osmotic pressure is given by the Morse equation as

$$\Pi = 2\hat{\rho}_{sol}RT,$$

where $\hat{\rho}_{sol}$ is the molar density of the solute in the brine, R is the ideal gas constant, and T is the temperature.

c. Draw a flow diagram for the once-through boiling process.

d. Calculate the required heat transfer rate.

e. What work transfer rate could this heat transfer rate produce in an ideal heat engine, assuming that the heat source is at the boiling temperature of water? What is the ratio of this work transfer rate to the minimal work transfer rate?

f. Propose an improvement for this design.

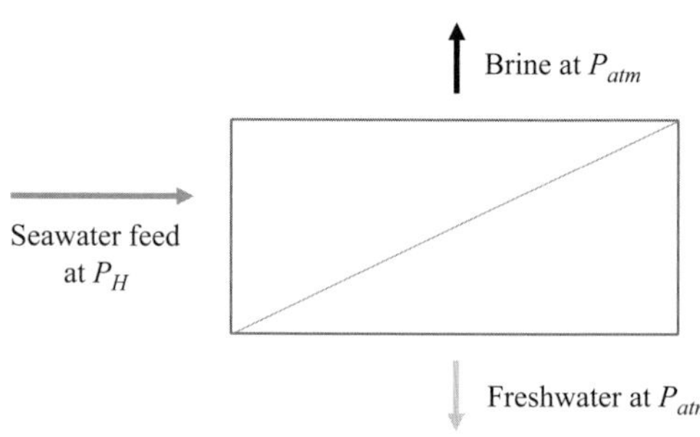

Figure P2.14 Schematic of the reverse osmosis process.

Consider a basic RO process. This consists of a high-pressure pump increasing the pressure of the feed solution from ambient P_0 to the desired operating pressure P_H, and a perfect semipermeable RO membrane through which water can pass, but dissolved ion species cannot.

g. Draw a flow diagram for the RO process.

h. Calculate the minimum pressure for the outlet of the pump and the corresponding work transfer rate. How does this compare to the minimum work transfer rate and why?

i. Calculate the Second Law efficiency of an RO pump operating at a pressure of 30 bar.

j. Propose an improvement for this design.

REFERENCES

1. G. N. Hatsopoulos and J. H. Keenan, *Principles of general thermodynamics*. Huntington, NY: R.E. Krieger Pub. Co., 1981.
2. A. Bejan, *Advanced engineering thermodynamics*. Hoboken, NJ: Wiley, 2006.
3. Y. A. Çengel and M. A. Boles, *Thermodynamics: an engineering approach*. New York: McGraw-Hill Education, 2015.
4. J. M. Smith, H. C. Van Ness, and M. M. Abbott, *Introduction to chemical engineering thermodynamics*. Boston, MA: McGraw-Hill, 2005.
5. M. J. Moran and H. N. Shapiro, *Fundamentals of engineering thermodynamics*. Chichester: Wiley, 2006.
6. E. P. Gyftopoulos and G. P. Beretta, *Thermodynamics: foundations and applications*. Mineola, NY: Dover Publications, 2005.
7. H. B. Callen, *Thermodynamics: an introduction to the physical theories of equilibrium thermostatics and irreversible thermodynamics*. New York: Wiley, 1960.
8. W. C. Reynolds and H. C. Perkins, *Engineering thermodynamics*. New York: McGraw-Hill, 1977.
9. E. G. Cravalho and J. L. Smith, *Engineering thermodynamics*. Boston, MA: Pitman, 1981.
10. W. G. Vincenti and C. H. Kruger, *Introduction to physical gas dynamics*. New York: Wiley, 1965.
11. T. J. Kotas, *The exergy method of thermal plant analysis*. London: Butterworths, 1985.
12. M. J. Moran, *Availability analysis: a guide to efficient energy use*. New York: ASME Press, 1989.
13. F. F. Huang, *Engineering thermodynamics: fundamentals and applications*. New York: Macmillan, 1976.

3 Chemical Thermodynamics and Thermochemical Calculations

3.1 Introduction

Energy conversion systems, in which the energy source is a fuel or a chemical energy carrier such as hydrogen, very often involve reacting mixtures. Reactions among species in these mixtures result in the conversion of their chemical bond energy into other forms, such as thermal energy or electrical energy. In **exothermic** reactions, of which **combustion** reactions are an important subset, the stored chemical energy is converted into thermal energy, which raises the temperature of the mixture. In jet engines and rockets, the thermal energy is converted into kinetic energy in the nozzle, which produces thrust. Exothermic chemical reactions are also used in industrial operations – where thermal energy delivered at fast rates is required – and domestic and commercial heating systems. **Endothermic** reactions, which absorb energy, are also important. Examples of practical applications of endothermic reactions include some fuel reforming processes, the production of synthetic fuels and other chemicals, chemical storage of hydrogen, and the formation of nitric oxides in combustion. In endothermic reactions, thermal energy is supplied through heat transfer to energize the reactions. Reactions involving the conversion between chemical and thermal energies are generally called **thermochemical reactions**. Such reactions are the principal scope of this chapter.

In other reactions, the chemical energy is converted directly to electrical energy, as in the case of a battery or a fuel cell. In these **electrochemical** processes, the reactions may be isothermal and in equilibrium with their environment, and intermediate steps in the overall reaction can free an electric charge that moves across a finite potential into an external resistance, transferring work to the environment. Although the process may ideally be isothermal, some of the chemical energy may be converted to thermal energy either directly or indirectly through internal resistances. Many **biochemical** processes are thermochemical or electrochemical processes or a combination of both. Another class of reactions depends on the availability of light, or results in the release or absorption of photons; these are **photochemical** reactions encountered in, for example, chemical lasers, the formation of photochemical smog in the atmosphere, and photosynthesis. The energetics of the aforementioned reactions can be addressed in similar ways, and such methodologies are illustrated in this chapter.

A reacting mixture undergoing a chemical reaction converts the mixture of "fuel" and "oxidizer" into a mixture of products. Fuels can be stored either in gas, liquid, or solid phases. These phases are dictated by their states at standard atmospheric conditions. Surface reactions are important for solid fuels, such as coal and biomass, although some of the heat release occurs during combustion in the gas phase after devolatization. For liquid fuels, combustion occurs mostly after evaporation. Fuels come from different sources, including fossil fuels (e.g., crude oil, coal), biomass (e.g., vegetation, animal fat, vegetable oils), and synthesis reactions (e.g., syngas from coal gasification, biofuels, gas-to-liquid fuels). Invariably, the most common fuels in practical use are hydrocarbons. The most common oxidizer is oxygen, and air is the most common oxidizer for hydrocarbon or hydrogen combustion. There are different ways of characterizing fuel–air mixtures. These include heat of combustion, which indicates the amount of thermal energy available from the combustion process, and the adiabatic flame temperature, which indicates the highest temperature achieved during a process. Both quantities are important, and their importance will be discussed when these concepts are introduced.

When burning hydrocarbons in air, the products are mostly carbon dioxide, water, and nitrogen. During such a reaction, some of the mixture components undergo a chemical conversion during which chemical bonds between atoms are broken and other chemical bonds are formed. The difference between the bond energies before and after the reaction causes an increase in the thermal energy of the mixture at the expense of the bond energy. In electrochemical reactions, the bond energy is that of the electrons, which flow out of the system. The objectives of the following analysis are to review some basic concepts in chemistry, and to extend the equations of nonreacting mixtures to reacting mixtures. In particular, we wish to develop the forms of the internal energy and enthalpy, which can be used in conjunction with the First Law written in its familiar form, to calculate heat and work interactions when the system undergoes chemical reactions. The Second Law will be used to determine the final state and the equilibrium composition of a reacting mixture, and will be used to define the direction of a chemical reaction. Application of the Second Law requires the knowledge of the entropy of a reacting mixture.

The first step in analyzing a chemical reaction is to satisfy mass conservation at the overall as well as the elemental levels. This determines conditions for stoichiometric mixtures of fuel and oxidizer and defines fuel-lean and fuel-rich mixtures. Examples for the application of lean and rich burn in combustion, reforming, gasification, and fuel production are discussed. There are other definitions for the state of the mixture, including the equivalence ratio, air-to-fuel ratio, percent theoretical air or percent excess air.

Reactions energetics are concerned with the conversion of the chemical bond energy into thermal energy during a chemical reaction. Careful accounting starts with writing the internal energy as a sum of the thermal and chemical energy, and defining the energy of formation of a molecule. These concepts are then applied to define the enthalpy and internal energy of reaction, the fuel heating values, and the adiabatic flame temperature. Chemical reactions involve the change of entropy as reactants form products; to calculate this change, the reference entropy for each component of the mixture is defined. Another important state property is the Gibbs free energy, and we use its relation to enthalpy and entropy to determine its change in a reaction.

Chemical equilibrium is used to determine the composition of a reacting mixture. Equilibrium conditions are stated in terms of change in extensive properties of the mixture, or in terms of the chemical potential of the participating components. This expression defines the condition of equilibrium for a single reaction, or the equilibrium constant in terms of the Gibbs function of the reaction. The equilibrium constant is then used to determine the mixture composition.

We conclude the chapter with a discussion of the maximum work expressions for a chemical reaction, the chemical reaction availability, fuel availability, and the efficiency of processes in which chemical energy and work interactions are important.

3.2 Stoichiometry and Mass Conservation

A chemical reaction follows a chemical balance equation in which the total number of the atoms of each distinctive atomic species is conserved. On the left-hand side of a chemical reaction equation, we have the reactants. On the right-hand side, we have the products, which define the end state of the process. A simple example for a chemical reaction is the hydrogen–oxygen reaction, often encountered in some liquid fuel rocket engines, such as space shuttle engines, in polymer electrolyte membrane fuel cells, etc. The chemical balance equation for this reaction is shown schematically in Figure 3.1, a schematic representation of a continuous combustion process in which the chemical reaction between oxygen and hydrogen produces water. The number of moles is for a stoichiometric combustion process:

$$2H_2 + O_2 \Rightarrow 2H_2O. \tag{3.1}$$

The balance reaction lists two groups of species within the mixture, which correspond to two states, reactants on the left-hand side and products on the right-hand side. Two hydrogen molecules react with a single oxygen molecule to form two molecules of water. Note that the reaction may go in the opposite direction if certain conditions prevail. However, for the purpose of converting chemical energy into thermal energy, or for the

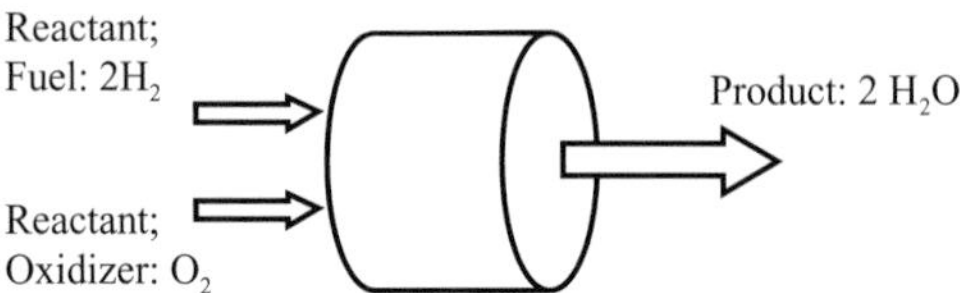

Figure 3.1 A schematic representation of a continuous combustion process in which the chemical reaction between oxygen and hydrogen produces water. The number of moles is for a stoichiometric combustion process.

purpose of producing electric work, the reaction should proceed in the direction indicated in (3.1). The coefficients in front of the molecules are required to balance the number of hydrogen and oxygen atoms on both sides of the equations. These coefficients are called **stoichiometric coefficients** of the reaction, which are often denoted by the symbol ν_i, where the subscript corresponds to the chemical species or chemical component in the mixture. Note also that this chemical reaction equation can be interpreted as indicating that two moles of hydrogen react with one mole of oxygen to form two moles of water. Recall that a mole of any substance contains the same number of molecules; it is the Avogadro's number, 6.022×10^{23} molecule/gmol. Moreover, a mole of any ideal gas, at the same pressure and temperature occupies the same volume, $\hat{v} = \Re T/p$. This is useful in practical calculations.

Typically, in energy conversion applications involving combustion reactions, the chemical reaction equation has an oxidizer and a fuel. The oxidizer may be air, in which nitrogen does not participate in the reaction (except at negligible quantities, but with significant consequences with regards to the emissions of pollutants), but only acts as an added thermal capacity. We should note that the reactants and the products are two distinct states of the mixture. The total number of moles and the mole fractions of the two states are different. Only the total mass is the same on both sides of the chemical equation, as well as the mass of distinct chemical elements. We also note that, besides nitrogen in fuel–air combustion, some other species may appear on both sides of the equation as reactants and products.

A **stoichiometric mixture** is a mixture in which the ratio between the reactants in the mixture (i.e., fuel and oxidizer) is the same as that given by the chemical reaction equation. **Stoichiometric combustion** is one in which the ratio between the available fuel and the available oxygen matches the stoichiometric ratio of their chemical reaction equation. For instance, in a rocket engine using hydrogen and oxygen, combustion is stoichiometric if the ratio of the mass flow rate of oxygen to that of hydrogen into the combustion chamber is 8 (i.e., $32/(2 \times 2)$).

While it is trivial to balance a reaction, such as the hydrogen reaction, we discuss next a simple procedure for balancing reactions based on methane and a mixture of syngas. The balance is based on atom/element conservation between reactants and products. Consider the **complete combustion** of methane, CH_4, in air. Complete combustion is a conceptual reaction that requires adequate proportions of fuel and oxidizer so that only CO_2, H_2O, and N_2 (if the oxidizer is air) are assumed in the products. In reality, even with the same proportions of fuel and oxidizer, other products, albeit potentially in smaller proportions, may be present. The complete combustion reaction of methane with air may be written as follows:

$$CH_4 + a\,(O_2 + 3.76\,N_2) \rightarrow b\,CO_2 + c\,H_2O + d\,N_2.$$

The reaction is written per mole of fuel consumed. Balancing the reaction requires the determination of the coefficients a, b, c, and d. These coefficients are denoted as the stoichiometric coefficients for O_2, CO_2, H_2O, and N_2, respectively. The stoichiometric coefficients for CH_4 and N_2 are 1 and 3.76 a, respectively. The elements represented in the reactants and products are C, H, O, and N. Balancing the elements,

$$\begin{aligned} \text{C}: &\quad 1 = b \\ \text{H}: &\quad 4 = 2c \\ \text{O}: &\quad 2a = 2b + c \\ \text{N}: &\quad (3.76)(2)\, a = 2d, \end{aligned}$$

where the values on the left-hand side of the equations corresponds to the number of atoms on the reactants side, and the right-hand side of the equations corresponds to the number of atoms on the products side. The atom balance results in the following stoichiometric coefficients: $a = 2$, $b = 1$, $c = 2$, and $d = 7.52$. Substituting the values of these stoichiometric coefficients, the complete combustion reaction may be written as

$$CH_4 + 2\,(O_2 + 3.76\,N_2) \rightarrow CO_2 + 2\,H_2O + 7.52\,N_2. \tag{3.2}$$

The reaction of complete combustion, albeit conceptual, serves to define the stoichiometric composition of a fuel and an oxidizer. Based on the above reaction balance, the stoichiometric mixture of methane and air requires 2 moles of O_2 per mole of fuel and $2 \times (1 + 3.76)$ or 9.52 moles of air. The definition of stoichiometric proportions of fuel and oxidizer can be defined to also characterize rich and lean burn conditions. If more than 2 moles of O_2 are used to burn 1 mole of CH_4, the mixture is fuel-lean; proportions of O_2 of less than 2 per mole of fuel indicates fuel-rich burning.

Example 3.1 Complete Combustion of Syngas

A coal sample is gasified to yield a syngas composition by volume including 14% H_2, 5% CO_2, 27% CO, 1% O_2, 3% CH_4, and 50% N_2. The syngas is burned at stoichiometric proportions with air. Determine the products' composition, assuming complete combustion.

Solution

The reaction balance may be written as follows:

$$\begin{aligned} &(0.14\,H_2 + 0.05\,CO_2 + 0.27\,CO + 0.01\,O_2 + 0.03\,CH_4 + 0.50\,N_2\,) + a\,(O_2 + 3.76\,N_2) \\ &\quad \Rightarrow b\,CO_2 + c\,H_2O + d\,N_2. \end{aligned}$$

To balance the reaction, we need to determine the stoichiometric coefficients a, b, c, and d using the atom balance:

$$\begin{aligned} \text{C}: &\quad 0.05 + 0.27 + 0.03 = b \Rightarrow b = 0.35 \\ \text{H}: &\quad 2 \times 0.14 + 4 \times 0.03 = 2c \Rightarrow c = 0.20 \\ \text{O}: &\quad 2 \times 0.05 + 0.27 + 2 \times 0.01 + 2a = 2b + c \Rightarrow a = b + c/2 - 0.195 = 0.255 \\ \text{N}: &\quad 2 \times 0.5 + (3.76)(2)\, a = 2d \Rightarrow d = 1.4588. \end{aligned}$$

Example 3.1 (cont.)

Rewriting the reaction balance:

$$(0.14\ H_2 + 0.05\ CO_2 + 0.27\ CO + 0.01\ O_2 + 0.03\ CH_4 + 0.50\ N_2) + 0.255\ (O_2 + 3.76\ N_2)$$
$$\Rightarrow 0.35\ CO_2 + 0.20\ H_2O + 1.4588\ N_2.$$

The concept of complete combustion is used in other contexts such as the determination of the fuel–air mixture stoichiometry and the heat of combustion. Under fuel-lean conditions, the products of hydrocarbon combustion also contain excess air, and under fuel-rich conditions the products may include CO and H_2, plus other minor species that are present at low concentrations measured in parts per million (ppm). With excess air in fuel-lean conditions, the atom balance is sufficient to determine the make-up of the products. However, sometimes additional equations, such as equilibrium conditions, may be needed to determine the make-up of the products.

As suggested by the reaction balances discussed above, a reaction relates states of reactants and products. They are written as if they were **irreversible** – that is, the reaction proceeds from left to right, or from reactants on the left-hand side to products on the right-hand side of the chemical equation, as participating chemical components change states while passing through a device such as a combustor, a chemical reactor, or a fuel cell. Symbolically, a chemical reaction is written as

$$\sum_{i=1}^{I} v_i{}'\chi_i \Rightarrow \sum_{i=1}^{I} v_i{}''\chi_i, \tag{3.3}$$

where χ_i stands for a chemical substance, $v_i{}'$ and $v_i{}''$ are the stoichiometric coefficients of the reactants and products, respectively; the index i defines a distinct molecule, and the total number of distinct molecules in this reaction is I. Some, but not all, of the species may appear on both sides, such as nitrogen in the case of burning a fuel in air; see (3.2). To express mass conservation symbolically, we define for each participating species/component/molecule a new two-dimensional array that defines the number of atoms in this molecule. This two-dimensional array is given the symbol v_{ij}^*, in which i stands for the molecule and j stands for the atom in this molecule, $i = 1 - I$ (which indicates that the index i runs from 1 to I) and $j = 1 - J$, and J is the total number of atoms participating in this reaction. For instance, for the reaction in (3.2), we write $i = 1$ for methane, $i = 2$ for oxygen, $i = 3$ for water, and $i = 4$ for carbon dioxide. Moreover, we define $j = 1$ for the carbon atom, $j = 2$ for the hydrogen atom, and $j = 3$ for the oxygen atom. Then, $v_{11}^* = 1$, $v_{12}^* = 4$, $v_{13}^* = 0$, and $v_{21}^* = 0$, $v_{22}^* = 0$, and $v_{23}^* = 2$, and $v_{31}^* = 0$, $v_{32}^* = 2$, and $v_{33}^* = 1$. Atom conservation for (3.3) can be written as

$$\sum_{react} v_i{}'v_{ij}^* = \sum_{prod} v_i{}''v_{ij}^* \quad \text{for } j = 1, J \tag{3.4}$$

or

$$\sum_{react} \nu_i \nu_{ij}^* = 0 \quad \text{for } j = 1, J, \tag{3.5}$$

where $\nu = \nu'' - \nu'$ is the net stoichiometric coefficient of a substance participating in the reaction. Equation (3.5) constitutes a number of equations equal to the number of atoms participating in a chemical reaction. The solution of these equations yields the stoichiometric coefficients in the chemical reaction equation.

3.2.1 Lean and Rich Burn

It is important to quantify the amount of air required to burn a fuel. Many combustion systems run on stoichiometric mixtures (e.g., most spark-ignition internal combustion engines). The **stoichiometric mole (or volume) ratio of air to fuel**, $\widehat{AF}$, is defined as the number of moles of air required to burn one mole of the fuel:

$$\widehat{AF}_{ST} = \left(\frac{n_{air}}{n_{fuel}}\right)_{ST} = \left(\frac{n_{oxygen} + n_{nitrogen}}{n_{fuel}}\right)_{ST} = \frac{\nu'_{oxygen} + \nu'_{nitrogen}}{\nu'_{fuel}}. \tag{3.6}$$

For the methane–air combustion in (3.2), for instance, $\widehat{AF}$ is $2 \times (1 + 3.76)/1 = 9.52$; that is, 9.52 unit volumes of air are required to burn a unit volume of methane, both measured at the same pressure and temperature, or equivalently 9.52 moles of air are required to burn 1 mole of methane. The stoichiometric mass air-to-fuel ratio is the ratio of the mass of air to that of fuel:

$$AF_{ST} = \left(\frac{m_{air}}{m_{fuel}}\right)_{ST} = \frac{\nu'_{oxygen} M_{oxgyen} + \nu'_{nitrogen} M_{nitrogen}}{\nu'_{fuel} M_{fuel}}. \tag{3.7}$$

For the case of methane–air combustion, we have $AF_{ST} = 2 \times (32 + 3.76 \times 28)/(1 \times 16) = 17.16$. Typically, for most hydrocarbons, the mass AF ratio is 16–18. The actual air-to-fuel ratio is simply whatever is used, and the corresponding definitions are

$$\widehat{AF} = \left(\frac{n_{air}}{n_{fuel}}\right) = \left(\frac{n_{oxygen} + n_{nitrogen}}{n_{fuel}}\right) \tag{3.8}$$

and

$$AF = \left(\frac{m_{air}}{m_{fuel}}\right) = \left(\frac{m_{oxygen} + m_{nitrogen}}{m_{fuel}}\right). \tag{3.9}$$

Many other combustion systems run with **excess air** (e.g., diesel engines and gas turbine engines) to reduce the products temperature, limit the maximum allowable pressure, ensure complete combustion, control the power output, etc. In this case, more air than that required by the stoichiometric balance is used. Excess air, or **fuel-lean burning**, may be used because it is difficult to ensure complete mixing between the fuel and air at fast rates, or because low-temperature combustion produces less nitric oxide, a regulated pollutant. With excess air, $AF > AF_{ST}$, and we define the excess-air ratio, λ, such that

$$\lambda = \frac{AF}{AF_{ST}}. \tag{3.10}$$

Excess air is usually given in percentage of stoichiometric or theoretical air, and the percentage refers to the extra volume, or moles, of air added to the mixture. For instance, for the methane–air combustion shown in (3.2), 20% excess air means that we add 0.4(4.76) = 1.904 moles of air to the reactants – that is, the total air used in the combustion is 2.4(4.76) = 11.424 moles. The chemical reaction equation in this case is

$$CH_4 + 2.4(O_2 + 3.76N_2) \Rightarrow CO_2 + 2H_2O + 0.4O_2 + 9.024N_2. \tag{3.11}$$

In this example, it is assumed that all the carbon in the fuel is converted to carbon dioxide, and all the hydrogen is converted to water. We call this type of combustion **complete combustion** since all the fuel has been consumed. If insufficient oxygen is available, CO may form, as well as CO_2, and some hydrogen may leave unburned. This is called **incomplete combustion**. The terminology **excess-air ratio** is used extensively in the automotive industry.

The opposite of lean combustion is **fuel-rich combustion**. This is the case when more fuel is supplied than required by the stoichiometric balance equation. This mode of combustion is used, for example, in the primary stage in gas turbine engines to ensure flame stability. It is also used in the production of carbon monoxide and hydrogen in methane reforming, or in coal gasification in a mixture of oxygen and steam. The process of converting a heavy hydrocarbon to synthetic gas, or syngas, which is a mixture of CO and H_2, is called **fuel reforming**. **Gasification** is the production of gaseous fuels from solid fuels. The ratio between the actual fuel–air ratio and the stoichiometric fuel–air ratio is known as the **equivalence ratio**. The terminology "equivalence ratio" is used extensively in the gas turbine industry. The equivalence ratio is larger than one for rich combustion and less than one for lean combustion. It is given the symbol ϕ, which is defined as

$$\phi = \frac{1}{\lambda}. \tag{3.12}$$

For instance, in the presence of an appropriate catalyst (often a metal that participates in the reaction only as an intermediate) and at a temperature determined by equilibrium limitation (as discussed later), a "heavy" hydrocarbon burning in rich mode produces carbon monoxide and hydrogen, in addition to the nitrogen gas. For example, in the case of benzene:

$$C_6H_6 + 3(O_2 + 3.76N_2) \Rightarrow 6CO + 3H_2 + 11.28N_2. \tag{3.13}$$

Note that this **partial oxidation** reaction uses 50% less air than is required for complete combustion of benzene (stoichiometric burning of benzene requires 4.5×4.76 moles of air).

Following partial oxidation, the product gases may be separated into their individual components with hydrogen used in a fuel cell to produce electricity and carbon monoxide combusted to form carbon dioxide in a second-stage combustor. In other implementations, water may be used to oxidize carbon monoxide according to the homogeneous **water–gas shift reaction**:

$$CO + H_2O \Rightarrow CO_2 + H_2. \tag{3.14}$$

This process, when combined with reforming, produces hydrogen in the products gas stream.

Example 3.2 Determining the Reactants' Composition

In a lean propane (C_3H_8) combustor, 3% (by volume) oxygen is measured in the products. The propane–air mixture is fuel-lean, and therefore, excess air is expected in the products. Assuming "complete" combustion without dissociation such that the products contain only CO_2, H_2O, O_2, and N_2, determine: (1) the mass-based air–fuel ratio supplied to the combustor; (2) the equivalence ratio of the propane–air mixture; and (3) the excess-air ratio.

Solution

The combustion reaction is written as follows:

$$C_3H_8 + a\ (O_2 + 3.76N_2) \rightarrow b\ CO_2 + c\ H_2O + d\ O_2 + e\ N_2.$$

In terms of the theoretical value for a, the reaction may be balanced using atom conservation across the reaction:

$$C_3H_8 + \frac{a_{th}}{\phi}\ (O_2 + 3.76N_2) \rightarrow 3\ CO_2 + 4\ H_2O + a_{th}\left(\frac{1}{\phi} - 1\right)O_2 + 3.76\frac{a_{th}}{\phi}\ N_2,$$

where, $a_{th} = 5$ and ϕ is the equivalence ratio. The mole fraction of O_2 in the products is given. Hence,

$$X_{O_2} = \frac{5\left(\frac{1}{\phi} - 1\right)}{3 + 4 + 5\left(\frac{1}{\phi} - 1\right) + 3.76\frac{5}{\phi}} = 0.03 = \frac{\frac{5}{\phi} - 5}{2 + \frac{23.8}{\phi}}.$$

Solving for the equivalence ratio, we get $\phi = 0.84704 < 1$, which confirms that the mixture is fuel-lean. The corresponding excess-air ratio is $\lambda = \frac{1}{\phi} = \frac{1}{0.84704} = 1.181$.

The air–fuel ratio is expressed on the reactants side as

$$AF_{ST} = \frac{\frac{5}{0.84704}(32 + 3.76 \times 28)}{44} = 18.42.$$

Chemical reaction equations can be added and subtracted to form more complex reactions. These operations follow the Lavoisier and Laplace's law as well as Hess' Law, which are important for reaction energetics as well. Lavoisier and Laplace's law states that the heat added reversing a reaction is equal and opposite in sign to the heat added for the forward reaction. Hess' law states that reactions and associated heat addition can be formed through a linear combination of reactions (usually simpler) and associated heat additions. These laws are very useful for computing reaction energetics as well as for equilibrium calculations, as discussed below. For example, we can add six times the water–gas shift reaction to the rich benzene reaction above to get an overall reaction equation for production of hydrogen through **partial oxidation and steam reforming** of benzene that produces a hydrogen-rich stream:

$$C_6H_6 + 3(O_2 + 3.76N_2) + 6H_2O \Rightarrow 6CO_2 + 9H_2 + 11.28N_2. \quad (3.15)$$

Example 3.3 Reforming and Hydrogen Fuel Cells

Figure E3.3 shows the arrangement in which a heavy hydrocarbon fuel (defined generically as a fuel with a large number of carbon molecules) is used in the production of electricity. The primary conversion device here is the fuel cell. Low-temperature fuel cells run on hydrogen, which can be produced first by reforming a hydrocarbon. This is done in the fuel reformer, which combines the hydrocarbon and air to produce a mixture of CO and H_2, and a shift reactor in which CO is converted to CO_2. Electrochemical reactions in which electrons and ions are exchanged during the reactions will be discussed in detail in the next chapter.

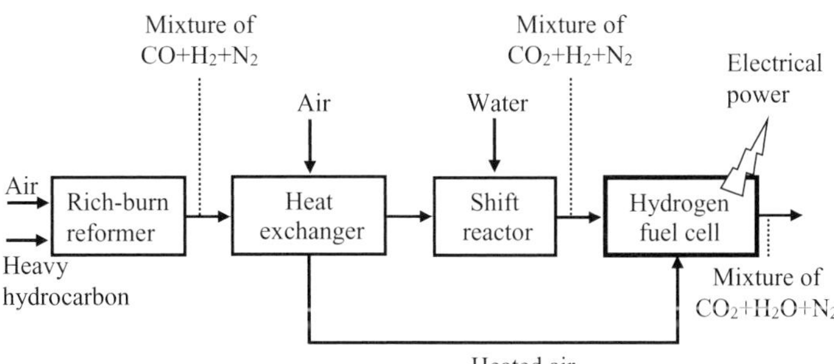

Figure E3.3 A schematic diagram showing the operation of a hydrogen fuel cell using a hydrocarbon. A fuel reformer and a shift reactor are used to produce a hydrogen-rich gas for the cell. The gas produced by the reformer is first cooled in a heat exchanger before it is shifted in the shift reactor. The hot air leaving the heat exchanger is used as an oxidizer in the fuel cell.

3.3 Reaction Energetics

As discussed in Chapter 2, concepts of energy conservation and energy conversion can be extended to apply to systems in which chemical energy is stored in molecular bonds. If these systems undergo chemical reactions that change the molecular structure of their components, chemical energy may be converted into thermal energy or transferred as heat or work. To account for these changes, we need to define measures for the chemical energy stored in molecular bonds in simple and complex molecules. What will follow from these definitions is calculations of the heat transferred between the system and its environment as it undergoes chemical reactions, isothermally or adiabatically, or under arbitrary conditions. Very useful definitions follow from this analysis.

3.3.1 Reference Enthalpy and Enthalpy of Formation

When applying the First Law to reacting mixtures, it is important to choose a common reference state for measuring the enthalpy of the different compounds that participate in the reaction [1,2]. Since chemical transformations involve the consumption and production of chemical compounds, and the breaking and formation of bonds, with the associated change in the bond energy, the expressions for the internal energy and enthalpy (and entropy and free energies) of chemical species participating in chemical reactions must account for these bond energies in an explicit form. These bond energies can be considered as part of the internal energy of the molecules. To extend the applicability of the familiar forms of the First Law to reacting mixtures, we revisit the original expressions of enthalpy and internal energy of an ideal gas, written here on a per mole basis:

$$\hat{h} = \hat{h}^o + \int_{T^o}^{T} \hat{c}_p(T)dT \tag{3.16}$$

and

$$\hat{u} = \hat{u}^o + \int_{T^o}^{T} \hat{c}_v(T)dT, \tag{3.17}$$

where, $\hat{h}^o$, and $\hat{u}^o$ (which are used to account for the bond energy) are the enthalpy and internal energy at the reference state, (T^o, p^o), and for an ideal gas,

$$\hat{u}^o = \hat{h}^o - \Re T^o. \tag{3.18}$$

In addition to the reference temperature and pressure, there is a need to account for the reference chemical state as well, since the chemical energy stored in chemical bonds can be converted to sensible energy. The most common convention is to use species representing different elements at their stable phase at the standard reference state, $T^o = 25\,°\mathrm{C}$ and $p^o = 1\,\mathrm{atm}$. For example, for the C, H, O, and N atoms, respectively, the following species are considered reference chemical species, C(s), H_2(g), O_2(g), and N_2(g), where (s) and (g) correspond to solid and gas phases of the substance. The **reference enthalpies** of O_2(g), H_2(g), N_2(g), mercury in liquid form, and carbon in solid form are taken to be zero. These values are then used to determine the reference enthalpies of molecules that result from chemical reactions between these elemental substances.

To define the reference enthalpy of a chemical species/compound, we need to know about its **enthalpy of formation** [3]. The enthalpy of formation of a compound, $\Delta\hat{h}_i^o$, is defined as the enthalpy of the formation reaction of this compound from the reference chemical states. For example, the heat of formation of CO_2 is the heat added at constant pressure to form one mole of CO_2 from reference chemical species, C(s) and O_2(g) in the following formation reaction: $C(s) + O_2(g) \Rightarrow CO_2$. The enthalpy of formation represents effectively the chemical contribution to the enthalpy and internal energy of a species involved in a reaction.

The enthalpy of reaction is the total heat transfer during the constant-pressure reaction process, in which the reactants enter at the reference pressure and temperature, and the products leave at the reference pressure and temperature, (T^o, p^o). Consider a compound that is formed by a chemical reaction between v_i moles (molecules) of elemental substances, χ_i:

$$\sum_i v_i \chi_i \Rightarrow \chi_{comp}. \tag{3.19}$$

The enthalpy of formation of the compound, $\hat{h}_{comp}^o$, is evaluated from the following expression, where $\Delta\hat{h}_R^o$ is the enthalpy of reaction per mole of the compound:

$$\Delta\hat{h}_R^o = \hat{h}_{f,comp}^o - \sum_i v_i \hat{h}_i^o. \tag{3.20}$$

Since the enthalpy of all elemental substances is defined to be zero at the standard reference state, then

$$\hat{h}_{f,comp}^o = \Delta\hat{h}_R^o \Big|_{\text{formation reaction}}. \tag{3.21}$$

Formation reactions are not necessarily isothermal (i.e., they are not in equilibrium with the environment); what is important is that the initial state of the reactants and the final state of the products are in equilibrium with an environment at the reference state. In many texts, the enthalpy of formation is often written as $\Delta h_{f,i}^o$. For an ideal gas, the enthalpy at an arbitrary temperature is expressed by a modified equation (3.16) as

$$\hat{h} = \hat{h}_f^o + \int_{T^o}^{T} \hat{c}_p(T)dT. \tag{3.22}$$

Values of the enthalpy of formation or the reference enthalpy of many compounds have been tabulated, and examples are shown in Table 3.1. For instance, the enthalpy of formation for methane is the heat transfer in an isobaric formation reaction of methane:

$$C_{(s)} + 2H_{2(g)} \Rightarrow CH_{4(g)}. \tag{3.23}$$

Table 3.1 The standard molar enthalpy and Gibbs free energy of formation of a number of energy-relevant molecules at 25 °C and 1 atm [3]

Substance	Formula	$\Delta\hat{h}_f^o$ MJ/kmol	$\Delta\hat{g}_f^o$ MJ/kmol
Hydrogen	H_2	0	0
Hydrogen atom	H	218.0	203.3
Oxygen	O_2	0	0
Oxygen atom	O	249.2	231.8
Hydroxyl	OH	39.5	34.3
Water	H_2O	−242.0	−228.8
Nitrogen	N_2	0	0
Nitric oxide	NO	90.4	86.8
Ammonia	NH_3	−45.7	−16.2
Carbon	C	0	0
Carbon monoxide	CO	−110.6	89.7
Carbon dioxide	CO_2	−393.8	2.9
Methane	CH_4	−74.9	−50.9
Ethane	C_2H_6	−84.7	−33.0
Ethylene	C_2H_4	52.3	68.2
Acetylene	C_2H_2	226.9	209.3
Propane	C_3H_8	−103.9	−23.5
Benzene	C_6H_6	83.0	−156.9
N-octane	C_8H_{18}	−208.6	16.4
Isooctane	C_8H_{18}	−224.3	13.7
Methanol	CH_3OH	−201.3	−162.6
Ethanol	C_2H_5OH	−235.0	−168.4

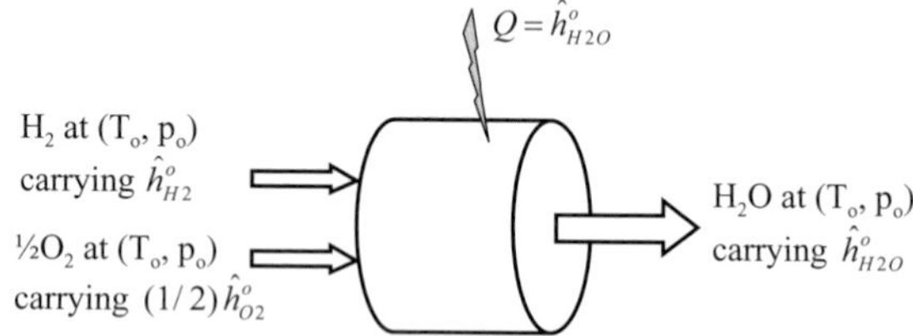

Figure 3.2 The application of the First Law to a constant-pressure, isothermal reactor used to define the enthalpy of formation of water. Heat interaction with the environment is the enthalpy of formation.

According to the First Law, negative enthalpy of formation means that the formation reaction is **exothermic**. The enthalpy of formation of water is the heat transfer during its formation reaction,

$$H_{2(g)} + \frac{1}{2}O_{2(g)} \Rightarrow H_2O_{(l)}, \tag{3.24}$$

at standard pressure and temperature (Figure 3.2), and is equal to −286 kJ/mol of water in the liquid form (the normal phase at the standard state), or −242 kJ/mol of water if it is produced in the gaseous form. Clearly, this reaction is exothermic. This value is also the enthalpy of water in standard conditions, that is $\hat{h}^o_{H_2O} = -242\,\text{kJ/mol} = -242\,\text{MJ/kmol}$.

The enthalpy of formation of carbon dioxide, and its reference enthalpy −393.8 kJ/mol, is the heat transfer during the formation reaction of the compound:

$$C_{(s)} + O_{2(g)} \Rightarrow CO_{2(g)} \tag{3.25}$$

from its elemental compounds – solid carbon and gaseous oxygen – at the standard temperature and pressure. This is also an exothermic reaction, and $\hat{h}^o_{CO_2(g)} = -393.8\,\text{MJ/kmol}$.

3.3.2 Enthalpy of Reaction

Having defined the enthalpy of a compound in terms of the sum of the reference or chemical enthalpy and the thermal or sensible energy, we go back and apply the familiar forms of the First Law to calculate the heat interaction involved in any process during which a chemical reaction occurs. Let us first consider the case in which this reaction occurs at constant pressure and constant temperature, at the reference state. For a given reaction, the heat transfer, Q, per mole of fuel is

$$\begin{aligned} Q &= H_{out} - H_{in} \\ &= \sum_{prod} n_i \hat{h}_i - \sum_{react} n_i \hat{h}_i. \end{aligned} \tag{3.26}$$

In the case of a stoichiometric reaction, the numbers of moles on the right-hand side are the same as the stoichiometric coefficients of the reaction, $n = v$. At standard pressure and temperature, the "heat" of reaction becomes the standard enthalpy of reaction:

$$\Delta H^o_R = \sum_{prod} v_i'' \hat{h}^o_i - \sum_{react} v_i' \hat{h}^o_i. \tag{3.27}$$

Note that for exothermic reactions, $\Delta H^o_R < 0$. The evaluation of the enthalpy of reaction is shown schematically in Figure 3.3, where the variations of the products' and reactants' total

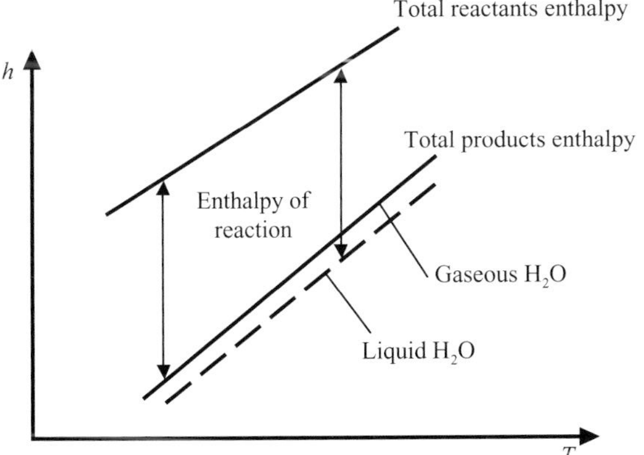

Figure 3.3 Enthalpy–temperature diagram showing the enthalpy of reaction as a function of the reaction temperature. Note that the increase of products' specific heat with temperature makes the slope of the products curve higher than that of the reactants – that is, it decreases the enthalpy of reaction at higher temperatures. Note also the impact of the state of water in the products.

enthalpies with temperature for both cases of water in the liquid phase and the gaseous phase in the products are depicted. The difference between the total enthalpies of the reactants and the products at any temperature is the enthalpy of reaction at that temperature:

$$\Delta H_R(T) = \sum_{prod} \nu_i''\hat{h}_i(T) - \sum_{react} \nu_i'\hat{h}_i(T) = \sum_{species} \nu_i\hat{h}_i(T). \tag{3.28}$$

Example 3.4. Heat of Reaction for CO Oxidation

Evaluate the heat interaction of the following constant-pressure CO oxidation reaction at standard conditions:

$$CO + \frac{1}{2}O_2 \Rightarrow CO_2. \tag{3.29}$$

Solution

Since we are at the standard state, then

$$Q = \hat{h}^o_{CO_2} - \left(\hat{h}^o_{CO} + \frac{1}{2}\hat{h}^o_{O_2}\right) = -393.8 - (-110.6 + 0) = -282.2\,\text{kJ/gmol of CO.}$$

This reaction transfers 282.2 kJ of heat per mole of CO out of the combustor when burning occurs under isothermal, isobaric conditions. This is an exothermic reaction, and CO is indeed a fuel.

Since hydrocarbons, such as methane CH_4, propane, C_3H_8, n-butane C_4H_{10}, benzene, C_6H_6, octane, C_8H_{18}, etc., are used extensively in power and propulsion industries, we will write the reaction for a generic hydrocarbon fuel, C_nH_m, burning in air. The stoichiometric oxidation reaction with air for this generic fuel is

$$C_nH_m + \left(n + \frac{m}{4}\right)(O_2 + 3.76N_2) \Rightarrow nCO_2 + \frac{m}{2}H_2O + \left(n + \frac{m}{4}\right)3.76N_2. \tag{3.30}$$

The enthalpy of reaction for this complete combustion at the reference state is given by

$$\Delta\widehat{H}^o_{R,C_nH_m} = n\hat{h}^o_{CO_2} + \frac{m}{2}\hat{h}^o_{H_2O} - \hat{h}^o_{C_nH_m} = \Delta\hat{h}^o_{R,C_nH_m}. \tag{3.31}$$

The lower-case symbol in the second line indicates per mole of the hydrocarbon. The enthalpy of reaction of hydrogen is −242 MJ/kmol of hydrogen when water appears in the vapor phase in the products.

The enthalpy of reaction per unit mass of the fuel, $\Delta h^o_R = \Delta\hat{h}^o_R/M$, of several fuels are shown in Table 3.2. The values shown in this table correspond to water appearing in the

Table 3.2 The standard enthalpy and Gibbs free energy of stoichiometric combustion in air for a number of fuels at $T_o = 25\,°C$, and $p_o = 1\,atm$ (water leaving as vapor). All gases are assumed to be ideal gases [5]

Fuel	Formula	M kg/kmol	Δh_R^o MJ/kg	Δg_R^o MJ/kg
Hydrogen	H_2	2.016	−120.0	−113.5
Carbon (graphite)	C	12.011	−32.8	−32.9
Methane	CH_4	16.043	−50.0	−49.9
Acetylene	C_2H_2	26.038	−48.3	−47.1
Ethylene	C_2H_4	28.054	−47.2	−46.9
Ethane	C_2H_6	30.07	−47.5	−48.0
Propylene	C_3H_6	42.081	−45.8	−45.9
Propane	C_3H_8	44.097	−46.4	−47.1
N-butane	C_4H_{10}	58.12	−45.8	−46.6
N-pentane	C_5H_{12}	72.15	−45.4	−46.3
Benzene	C_6H_6	78.114	−40.6	−40.8
N-hexane	C_6H_{14}	86.18	−45.1	−46.1
N-heptane	C_7H_{16}	100.21	−45.0	−45.9
N-octane	C_8H_{18}	114.232	−44.8	−45.8
Isooctane	C_8H_{18}	114.232	−44.7	−45.8
N-nonane	C_9H_{20}	128.26	−44.7	−45.7
N-decane	$C_{10}H_{22}$	142.29	−44.6	−45.7
Carbon monoxide	CO	28.01	−10.1	−9.2
Methanol	CH_3OH	32.042	−21.1	−21.5
Ethanol	C_2H_5OH	46.069	−27.8	−28.4

products in vapor form, which is called the lower enthalpy of reaction (similar to the lower heating value that will be defined shortly). The highest value belongs to hydrogen, at 120 MJ/kg. For most hydrocarbons, the lower value of the enthalpy of reaction is ~45 MJ/kg [4]. "Oxygenated" hydrocarbons, that is, those with oxygen atoms in their molecules, have lower enthalpy of reaction, mostly below 30 MJ/kg. Carbon monoxide has a particularly low value at 10.1 MJ/kg.

3.3.3 The Heating Values

The enthalpy of reaction is negative for all fuels; its absolute value is called the **heat of combustion** or **heating value** of the fuel. Depending on whether water in the combustion products appears in the gas or liquid phase, the heating value is called the **lower heating value (LHV)** or **higher heating value (HHV)**. Heating values are often quoted per unit mass of the fuel, and are used for comparative purposes. As an example, we consider methane, for which $n = 1$ and $m = 4$. The enthalpy of reaction of methane at the reference state is

$$\Delta\hat{h}^o_{CH_4} = -393.8 + 2(-242) - (-74.9) = -802.9\,\mathrm{MJ/kgmol}\text{ of methane.}$$

The LHV per unit mass of methane is: 802.9/16.042 = 50.050 MJ/kg of methane, since we use the enthalpy of water in a gaseous form to obtain the value of the enthalpy of reaction. The HHV is obtained by adding the enthalpy of condensation of water – that is, adding 2 × (−44 MJ/kmol), to 802.9 MJ/kmol, resulting in −890.9 MJ/kmol, or 55.54 MJ/kg of methane. Interestingly, for most hydrocarbons the higher enthalpy of reactions is around 50 MJ/kg of fuel. For hydrogen, the HHV is much higher, 141 MJ/kg. Consider an engine having a thermodynamic efficiency of 10%, which uses a modest 0.1 kg of a typical hydrocarbon. The work production of the engine is 0.5 MJ (= 0.1 kg × 50 MJ/kg × 0.1). At a consumption rate of 10 watts, you can power your power-hungry laptop for nearly 14 hours! This has been the attraction of micro engines (however, a substantial amount of thermal energy is rejected; try to calculate the exit temperature of the products and how hot your laptop will get!).

The enthalpy of reaction (3.30) at an arbitrary temperature can be evaluated easily from the following equation:

$$\Delta\hat{h}^T_{R,C_nH_m} = n\hat{h}_{CO_2}(T) + \frac{m}{2}\hat{h}_{H_2O}(T) - \left[\hat{h}_{C_nH_m}(T) + \left(n + \frac{m}{4}\right)\hat{h}_{O_2}(T)\right]. \tag{3.32}$$

The enthalpy of reaction does not change significantly with temperature at temperatures below 1000 K. For instance, the enthalpy of reaction of methane drops from 802.9 MJ/kmol

Table 3.3 The LHV, HHV, and standard Gibbs free energy of reaction (ΔG_R^{oo}) of some fuels at 25 °C and 1 atm. The Gibbs free energy assumes water leaving in the liquid phase [6–8]

Fuel (phase)[b]	LHV (kJ/mol)	HHV (kJ/mol)	$-\Delta G_R^{oo}(T^o, p^o)$ (kJ/mol)
Hydrogen (g), H_2	241.8	285.9	228.6
Carbon (s), C	393.5	393.5	394.4
Methane (g), CH_4	802.3	890.4	818
Ethane (g), C_2H_6	1427.9	1559.9	1467.5
Propane (g), C_3H_8	2044	2220	2108.4
Butane (g), C_4H_{10}	2658.5	2878.5	2747.8
Pentane (l), C_5H_{12}	3245.5	3509.5	3385.8
Hexane (l), C_6H_{14}	3855.1	4163.1	4022.8
Heptane (l), C_7 H_{16}	4464.9	4816.9	4660
Octane (l), C_8H_{18}	5074.6	5470.7	5297.2
Ethylene (g), C_2H_4	1323	1411	1331.3
Propylene (g), C_3H_6	1926.5	2058.5	1957.3
Butene (g), C_4H_8	2542.6	2718.6	2598.3
Pentene (g), C_5H_{10}	3155.8	3375.9	3236.5
Benzene (g), C_6H_6	3169.5	3301.6	3207.5
Toluene (l), C_7H_8	3771	3947	3834

g, gas; l, liquid; and s, solid.

at standard temperature to 801.52 MJ/kmol at 1100 K. Nevertheless, it becomes more sensitive to temperature at temperatures above 1200 K due to the increase of the specific heat of the products and the dissociation of products at even higher temperatures. The constant-pressure specific heats of combustion products increase with temperature at higher temperature. Thus, it is important to consider their temperature dependence when evaluating the enthalpy of reaction and the products' temperature at high temperatures. Note that combustion products except nitrogen are primarily triatomic gases whose specific heats are higher than those of diatomic gases; see Figure 2.18.

The LHVs and the HHVs of several fuels per mole of the fuel are shown in Table 3.3. As expected, the bigger the molecule, the higher the per mole heating value, but this also depends on the hydrocarbon family of the fuel (classification of hydrocarbons will be discussed in later chapters).

Example 3.5 Higher and Lower Heating Values of Natural Gas

A sample of natural gas is made up of the following, by volume: 95% methane (CH_4), 2.5% ethane (C_2H_6), 1.5% nitrogen (N_2), 1% carbon dioxide (CO_2), and other trace elements and lower concentrations of propane (C_3H_8), butane (C_4H_{10}), and pentanes (C_5H_{12} and $C_{10}H_{22}$). Determine the HHV of the combustion of this mixture of natural gas.

Solution

To find the heat of reaction of the natural gas mixture, we first write the reaction equation for its complete combustion. Hence,

$$\begin{aligned}&(0.95\ CH_4 + 0.025\ C_2H_6 + 0.01\ CO_2 + 0.015\ N_2\ ) + a\ (O_2 + 3.76\ N_2)\\&\quad \Rightarrow b\ CO_2 + c\ H_2O + d\ N_2.\end{aligned}$$

To balance the reaction, we need to determine the stoichiometric coefficients a, b, c, and d using the atom balance:

$$\begin{aligned}
&C: && 0.95 + 2 \times 0.025 + 0.01 = b && \Rightarrow b = 1.01\\
&H: && 4 \times 0.95 + 6 \times 0.025 = 2c && \Rightarrow c = 1.975\\
&O: && 2 \times 0.01 + 2a = 2b + c && \Rightarrow a = 1.9875\\
&N: && 2 \times 0.015 + (3.76)(2)\ a = 2d && \Rightarrow d = 7.488.
\end{aligned}$$

Substituting the values of the stoichiometric coefficients into the reaction balance yields

$$\begin{aligned}&(0.95\ CH_4 + 0.025\ C_2H_6 + 0.01\ CO_2 + 0.015\ N_2\ ) + 1.9875\ (O_2 + 3.76\ N_2)\\&\quad \Rightarrow 1.01\ CO_2 + 1.975\ H_2O + 7.488\ N_2.\end{aligned}$$

Example 3.5 (cont.)

The heat of reaction is now obtained from (3.27):

$$\begin{aligned}
\Delta H_R^o &= \sum_{prod} \nu_i'' \hat{h}_i^o - \sum_{react} \nu_i' \hat{h}_i^o \\
&= \left(1.01\,\hat{h}_{CO_2}^o + 1.975\,\hat{h}_{H_2O}^o + 7.488\,\hat{h}_{N_2}^o\right) \\
&\quad - \left(0.95\,\hat{h}_{CH_4}^o + 0.025\,\hat{h}_{C_2H_6}^o + 0.01\,\hat{h}_{CO_2}^o + 1.9875\,\hat{h}_{O_2}^o + 7.488\,\hat{h}_{N_2}^o\right) \\
&= 1.00\,\hat{h}_{CO_2}^o + 1.975\,\hat{h}_{H_2O}^o - 0.95\,\hat{h}_{CH_4}^o - 0.025\,\hat{h}_{C_2H_6}^o - 1.9875\,\hat{h}_{O_2}^o \\
&= 1.00 \times (-393.8) + 1.975 \times (-242.0) - 0.95 \times (-74.9) - 0.025 \times (-84.7) - 1.9875 \times (0) \\
&= 798.5\ \text{MJ/kmol fuel.}
\end{aligned}$$

In many applications, we can simply apply the First Law while using the appropriate expressions for the internal energy or enthalpy, including the enthalpy of formation and the thermal enthalpy, to compute the heat and work interactions during the process involving a chemical reaction. For the open flow reacting system depicted in Figure 3.4, we write

$$\frac{dE}{dt} = \dot{Q} - \dot{W} + \left(\sum_{react} \dot{m}_i\left(h_i + \mathbf{V}_i^2 + g_v z_i\right) - \sum_{prod} \dot{m}_i\left(h_i + \mathbf{V}_i^2 + g_v z_i\right)\right). \tag{3.33}$$

In this expression, $\dot{Q}$ is the rate of heat transfer across the boundaries of the control volume, $\dot{W}$ is the rate of work interaction across the boundary, and h is the total enthalpy.

In a closed system, the First Law may be written as

$$\begin{aligned} Q - W &= \Delta U \\ &= U_p - U_r. \end{aligned} \tag{3.34}$$

The relation between the enthalpy and internal energies of formation will be discussed shortly. The energy balance is not different from that of a nonreacting pure substance, except that the chemical enthalpy terms and the mixture composition in the reactants and products mixture now have to be explicitly evaluated.

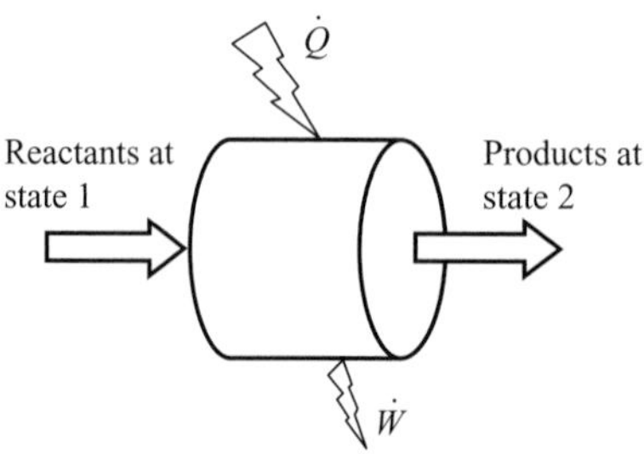

Figure 3.4 Application of the First Law to an open flow reacting system.

Example 3.6 Maximum Temperature in a Gas Turbine Engine

Gas turbine engines cannot operate at temperatures higher than a certain maximum because of stringent materials requirements, which depends on the alloy used to build the turbine blades, the rotational speed of the shaft, and the cooling technology. Consider the case of a turbine whose maximum inlet temperature cannot exceed 950 °C. The gas turbine cycle shown in Figure E3.6 operates at a pressure ratio of 4. The compressor isentropic efficiency is 82%. The atmospheric air at 27 °C and 1 atm enters the compressor. Methane used as a fuel is fed to the combustor at 25 °C. Find the excess air required for safe operation of this plant. Use constant specific heat assumption.

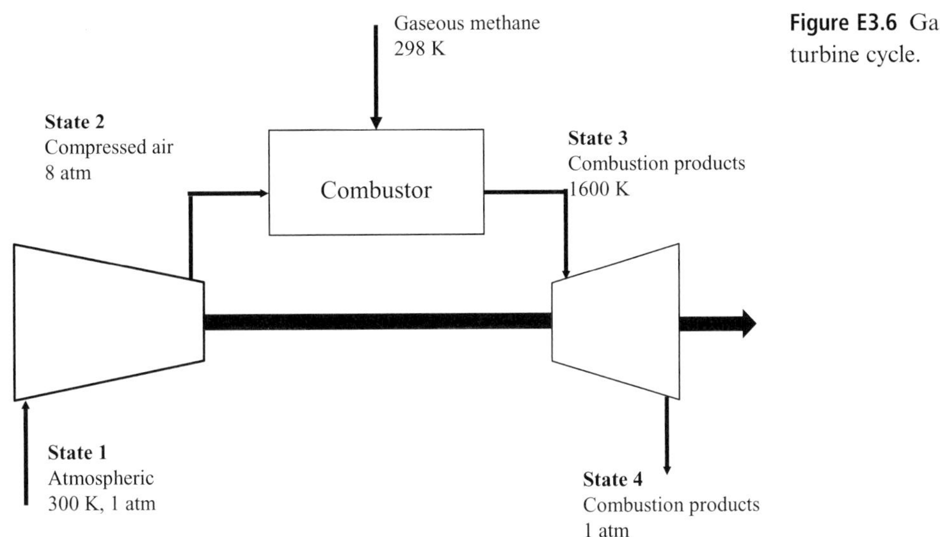

Figure E3.6 Gas turbine cycle.

Solution

Assume constant specific heats and a specific heat ratio of $k = 1.4$ for air behaving like an ideal gas. The temperature at the compressor exit (state 2) is obtained using the isentropic compression from the inlet of the compressor (state 1) to its outlet (state 2) and the compressor isentropic efficiency.

$$T_{2s} = T_1 (p_2/p_1)^{k-1/k} = (27 + 273) \times 8^{0.4/1.4} = 543.4\,\text{K}$$

Using the definition of the isentropic efficiency, the exit temperature of the compressor is obtained as follows:

$$T_2 = T_1 + \frac{T_{2s} - T_1}{\eta_c} = 300 + \frac{543.4 - 300}{0.82} = 596.8\,\text{K}.$$

Example 3.6 (cont.)

Methane combustion is determined by the following chemical equation:

$$CH_{4_{(g)}} + \alpha(O_2 + 3.76N_2) \rightarrow CO_2 + 2H_2O + \beta O_2 + 3.76\alpha N_2.$$

We have assumed complete fuel combustion. Oxygen atom balance shows that $\beta = \alpha - 2$.

Applying the First Law over a CV surrounding the combustor, assuming adiabatic conditions, we get

$$\hat{h}_{CH_4}^{298\,K} + \alpha\left(\hat{h}_{O_2}^{596.8\,K} + 3.76\hat{h}_{N_2}^{596.8\,K}\right) = \hat{h}_{CO_2}^{1600\,K} + 2\hat{h}_{H_2O}^{1600\,K} + (\alpha - 2)\hat{h}_{O_2}^{1600\,K} + 3.76\alpha\hat{h}_{N_2}^{1600\,K}.$$

The values of all the enthalpies can be calculated using (3.22):

$$\hat{h}_{CH_4}^{298\,K} = -74{,}900\,kJ/kmol$$

$$\hat{h}_{O_2}^{596.8\,K} = \hat{h}_{O_2}^{298\,K} + \hat{c}_p \int_{298}^{596.8} dT = 0 + 29.38 \times (596.8 - 298) = 8779\,kJ/kmol.$$

Similarly,

$$\hat{h}_{N_2}^{596.8\,K} = \hat{h}_{N_2}^{298\,K} + \hat{c}_p \int_{298}^{596.8} dT = 8692\,kJ/kmol,$$

$$\hat{h}_{CO_2}^{1600\,K} = -345{,}340\,kJ/kmol,$$

$$\hat{h}_{H_2O}^{1600\,K} = -202{,}484\,kJ/kmol,$$

$$\hat{h}_{O_2}^{1600\,K} = 38{,}253\,kJ/kmol,$$

$$\hat{h}_{N_2}^{1600\,K} = 37{,}875\,kJ/kmol.$$

Hence,

$$-74900 + \alpha(8779 + 3.76 \times 8692) = -345340 - 2 \times 202484 + (\alpha - 2) \times 38253 + 3.76 \times 37875 \times \alpha.$$

Solving the above equation yields $\alpha = 5.4$.

Thus, $5.4 \times 4.76 = 25.704$ moles of air are used to burn a mole of methane. The reaction equation is: $CH_{4_{(g)}} + 5.4(O_2 + 3.76N_2) \rightarrow CO_2 + 2H_2O + 3.4O_2 + 20.304N_2$.

At the condition of combustion of methane with 100% theoretical air, the theoretical stoichiometric coefficient of air is 2; see (3.2). Thus, the equivalence ratio is $\phi = \frac{2}{5.4} = 0.3704$.

The excess-air ratio is then obtained using (3.12): $\lambda = \frac{1}{0.3704} = 2.7$; 270% theoretical air is required to keep the maximum temperature at 1600 K.

3.3.4 Adiabatic Flame Temperature

The adiabatic flame temperature is defined as the temperature of the reaction products if the combustion occurs in an adiabatic process. Two important quantities are discussed here: the adiabatic flame temperature at constant pressure and the adiabatic flame at constant volume. We first establish the condition for adiabatic flame temperature at constant pressure. Applying the First Law for an adiabatic flow process, with reactants at temperature T_r (Figure 3.5), and considering the enthalpy only, we have

$$\sum_{react} n_i \hat{h}_i(T_r) = \sum_{prod} n_i \hat{h}_i(T_p). \tag{3.35}$$

The only unknown in this equation is the products temperature, T_p. Clearly, the flame temperature depends on the stoichiometry of the mixture and the fuel used. When the mixture is stoichiometric, the numbers of moles in this equation are the stoichiometric coefficients, $n_i = \nu_i$.

To illustrate the calculations of the adiabatic flame temperature, let us use the stoichiometric methane–air mixture governed by (3.2), and calculate the products temperature assuming adiabatic constant-pressure combustion, starting with reactants at 298 K. In this case, the left-hand side of (3.35) is

$$\sum_{react} \nu_i \hat{h}_i(298) = \hat{h}^o_{methane} = -74{,}831\ \mathrm{kJ/kgmol}.$$

The methane enthalpy at the reference state is the same as its enthalpy of formation. The right-hand side is

$$\hat{h}_{CO_2}(T_p) + 2\hat{h}_{H_2O}(T_p) + 7.52\hat{h}_{N_2}(T_p) = \sum_{i=1}^{3} n_i \hat{h}^o_i + \sum_{i=1}^{3} n_i \int_{T_o}^{T_p} \hat{c}_{p,i} dT.$$

The standard enthalpy of formation of the species is given in Table 3.1. The polynomial expressions for the specific heats are integrated, and substituted in the above equation. The solution is found by trial and error; the result for this case is 2340 K. (A more accurate value of 2247 K is obtained after accounting for dissociation of product, as will be shown later.)

Note that the flame temperature depends on the specific heat of the products only, since the starting temperature is the reference temperature. An approximate value for the flame temperature can be obtained by choosing average values for the specific heats at an intermediate temperature between the reactants temperature and the flame temperature,

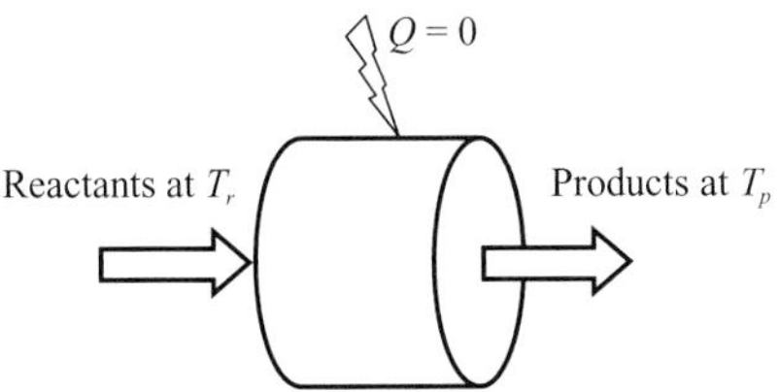

Figure 3.5 Constant-pressure adiabatic combustion used in evaluating the adiabatic flame temperature.

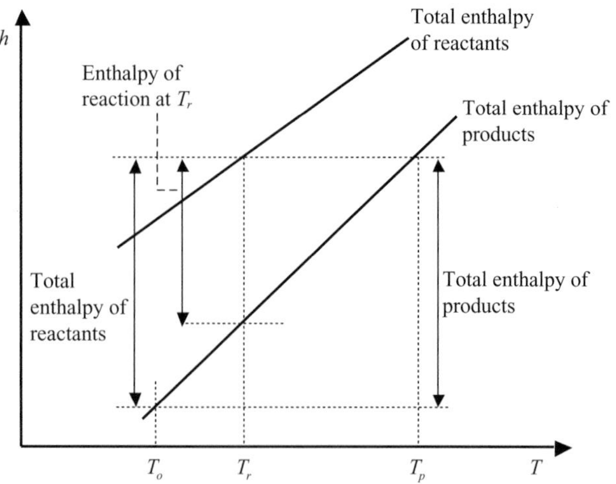

Figure 3.6 Enthalpy–temperature diagram showing the approach for evaluating the adiabatic flame temperature.

say 1200 K since the products temperature is expected to be close to 2200 K. The average specific heat is defined as

$$\bar{c}_{p,i} = \frac{1}{T_2 - T_1} \int_{T_2}^{T_1} c_{p,i} dT \approx c_{p,i}\left(\frac{T_2 + T_1}{2}\right).$$

The second expression is obtained using the mean value theorem. The specific heats of CO_2, H_2O, and N_2 at 1200 K are, respectively, 56.21, 43.8,7 and 33.71 kJ/kmol·K. Substituting in the right-hand side of (3.35), we get the constant specific heat-based flame temperature. The resulting adiabatic flame temperature is 2318 K. This is a reasonable approximation of the more accurate value obtained using the detailed variation of the specific heats, which requires much less effort.

The adiabatic flame temperature for most hydrocarbon fuels burning in air, starting with reactants at 298 K and 1 atm, falls in the range 2300–2600 K. In fuel-oxygen reactions (known as **oxyfuel combustion**), such as in rocket engines, some industrial processes, and technologies for carbon capture described later in the book, the adiabatic flame temperature is higher, reaching above 3000 K for hydrogen. Figure 3.6 shows the enthalpy–temperature diagram describing the procedure used in evaluating the adiabatic flame temperature. A note of warning: Calculations presented so far are based on the assumption of complete combustion of the hydrocarbon to CO_2 and H_2O. Complete combustion is not guaranteed by the Second Law, as will be shown later.

Example 3.7 Adiabatic Flame Temperature of a Lean Methane–Air Mixture

One mole of methane is burned adiabatically at constant atmospheric pressure with 110% theoretical air. The reactants temperature is 298 K. The major products of combustion (i.e., without dissociation) include H_2O, CO_2, O_2, and N_2.

Example 3.7 (cont.)

1. Balance the reaction including only the major products of combustion by determining the unknown coefficients, *a, b, c, d*, and *e*:

$$CH_4 + a\,(O_2 + 3.76N_2) \rightarrow b\,CO_2 + c\,H_2O + d\,O_2 + e\,N_2.$$

2. Determine the air-to-fuel ratio, AF, and the equivalence ratio, φ.
3. Obtain the adiabatic flame temperature of the mixture for a constant-pressure combustion process, considering only the major products of combustion.

Solution

1. To balance the reaction with a prescribed amount of theoretical air, we need to first balance the stoichiometric reaction:

$$CH_4 + 2\,(O_2 + 3.76N_2) \rightarrow CO_2 + 2\,H_2O + 7.52\,N_2.$$

The stoichiometric coefficient for air, a_{th}, is 2. Therefore, $a = 1.1 \times a_{th} = 2.2$. With excess air, CO_2 and H_2O are the same, but d and e are different. Balancing O and N, we get $d = 0.2$, $e = 8.272$. Therefore, the reaction equation reads:

$$CH_4 + 2.2(O_2 + 3.76N_2) \rightarrow CO_2 + 2H_2O + 0.2O_2 + 8.272\,N_2.$$

2. The air-to-fuel ratio, AF, is expressed as

$$AF = \frac{m_{\text{air}}}{m_{\text{fuel}}} = \frac{2.2 \times (32 + 3.76 \times 28)}{16} = 18.876.$$

The equivalence ratio, ϕ, may be expressed as

$$\phi = \frac{a_{th}}{a} = \frac{2}{2.2} = \frac{100}{\text{percent theoretical air}} = \frac{100}{110} = 0.909.$$

3. The adiabatic flame temperature of the mixture for a constant-pressure combustion and considering only the major products of combustion is obtained from

$$H_r = H_p \Rightarrow \sum_{reac} n_i \hat{h}_i = \sum_{prod} n_i \hat{h}_i.$$

Substituting the expression for the enthalpies of the reactants and the products,

$$n_{CH_4}\Delta\hat{h}^{\circ}_{f,CH_4} = n_{CO_2}\left[\left(\hat{h}_{T_p,CO_2} - \hat{h}_{298,CO_2}\right) + \Delta\hat{h}^{\circ}_{f,CO_2}\right] + n_{H_2O}\left[\left(\hat{h}_{T_p,H_2O} - \hat{h}_{298,H_2O}\right) + \Delta\hat{h}^{\circ}_{f,H_2O}\right]$$
$$+ n_{O_2}\left[\left(\hat{h}_{T_p,O_2} - \hat{h}_{298,O_2}\right) + 0\right] + n_{N_2}\left[\left(\hat{h}_{T_p,N_2} - \hat{h}_{298,N_2}\right) + 0\right].$$

Example 3.7 (cont.)

Now, substituting the values of the enthalpies and the numbers of moles, we get

$$-74{,}900 = \left[\left(\hat{h}_{T_p,\mathrm{CO_2}} - \hat{h}_{298\,K,\,CO_2}\right) - 393{,}800\right] + 2\left[\left(\hat{h}_{T_p,\mathrm{H_2O}} - \hat{h}_{298\,K,\mathrm{H_2O}}\right) - 242{,}000\right]$$
$$+ 0.2\left[\left(\hat{h}_{T_p,\mathrm{O_2}} - \hat{h}_{298\,\mathrm{K},\mathrm{O_2}}\right) + 0\right] + 8.272\left[\left(\hat{h}_{T_p,\mathrm{N_2}} - \hat{h}_{298\,K,\mathrm{N_2}}\right) + 0\right].$$

Note that the enthalpy of formation for water is that of water vapor, because we expect the final products of adiabatic combustion to be hot enough to only include water in the vapor phase. With excess oxidizer, we expect the adiabatic flame temperature to be lower than the stoichiometric value. We guess two values, 2000 K and 2200 K, and interpolate or extrapolate a next guess. At 2000 K,

$$\begin{aligned} H_p(2000\,\mathrm{K}) &= [(100{,}804 - 9364) - 393{,}800] + 2[(82{,}593 - 9904) - 242{,}000] \\ &\quad + 0.2[(67{,}881 - 8682) + 0] + 8.272[(64{,}810 - 8669) + 0] \\ &= -164{,}744\ \mathrm{kJ/kmol}. \end{aligned}$$

The above value for H_p is much less than the enthalpy of the reactants (i.e., –74,900 kJ/kmol).
At 2200 K,

$$\begin{aligned} H_p(2200\,\mathrm{K}) &= [(112{,}939 - 9364) - 393{,}800] + 2[(92{,}940 - 9904) - 242{,}000] \\ &\quad + 0.2[(75{,}484 - 8682) + 0] + 8.272[(72{,}040 - 8669) + 0] \\ &= -70{,}588\ \mathrm{kJ/kmol} \end{aligned}$$

Comparing with the enthalpy of the reactants, it can be seen than H_p at 2200 K is slightly higher than –74,900 kJ/kmol. This means that the adiabatic flame temperature is slightly below 2200 K. Therefore, an interpolation for the first guess of the adiabatic flame temperature is implemented as follows:

$$\frac{T_p - 2000}{2200 - 2000} = \frac{H_r - H_p(2000\,\mathrm{K})}{H_p(2200\,\mathrm{K}) - H_p(2000\,\mathrm{K})}.$$

Solving the above equation for T_p yields

$$T_p = 2000 + (2200 - 2000)\frac{-74{,}900 - (-1647{,}44)}{-70{,}588 - (-164{,}744)} = 2190.8\,\mathrm{K}.$$

In more accurate calculations of the flame temperature, the Second Law must be invoked to determine the actual products composition, given the constraints imposed on the reaction. This is the subject of chemical equilibrium, which will be used to determine the concentration of different species that can exist in the products and the corresponding temperature. As will be shown, equilibrium at high temperature acts against complete combustion, and some CO and H_2 may be present in the products. In this case, the flame temperature is somewhat lower than the values obtained from the complete combustion assumptions.

Figures 3.7 and 3.8 show results of equilibrium calculations of the flame temperature and mixture composition, at different equivalence ratio for benzene–air combustion and methane–air combustion, respectively. These calculations account for equilibrium constraints that will be discussed shortly, but are given here to show the impact of stoichiometric fractions on temperature and composition. We note that the flame temperature reaches a maximum value at slightly rich mixtures because of dissociation, as will be shown in detail later in this chapter. Lean and rich mixtures burn at lower temperatures than those corresponding to stoichiometric combustion. As expected, under fuel-rich conditions, there is more carbon monoxide and hydrogen in the products.

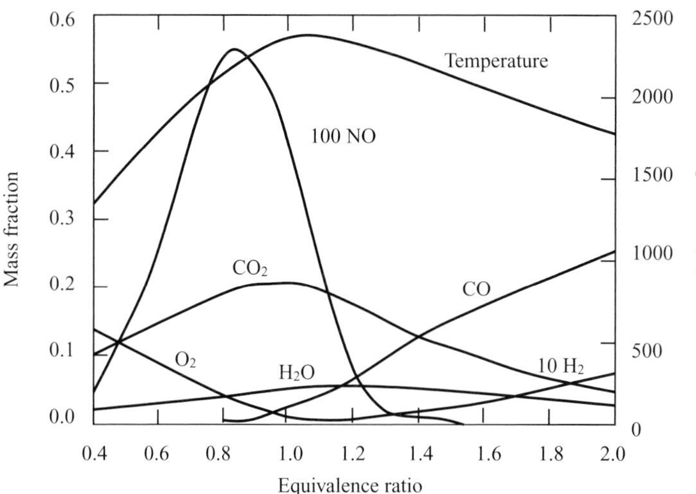

Figure 3.7 Flame temperature and equilibrium mass fractions of species released during combustion of benzene (C_6H_6) in air combustion at different equivalence ratios under adiabatic constant-pressure conditions, with initial temperature of 300 K. Calculations were performed using equilibrium, assuming the existence of the major species shown, as well as the following minor species: OH, H, and O, which are not shown in the figure.

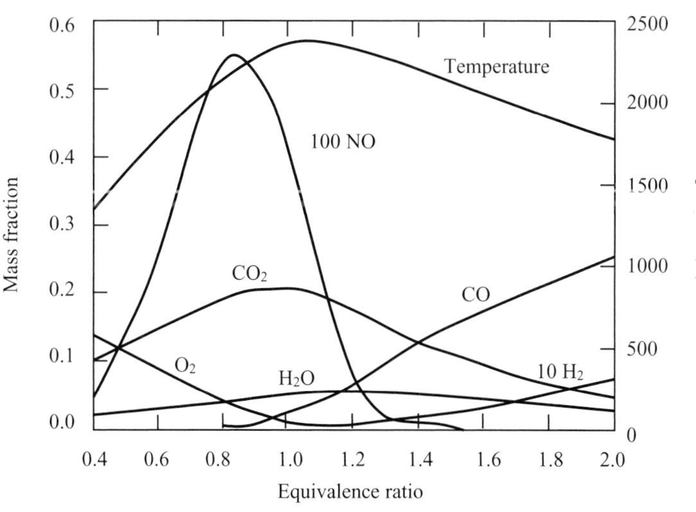

Figure 3.8 The figure shows the flame temperature and equilibrium mass fractions of species released during methane–air combustion at different equivalence ratios under adiabatic constant-pressure conditions, with initial temperature of 300 K. Calculations were performed using equilibrium, assuming the existence of the major species shown as well as the following minor species: OH, H, and O, which are not shown in the figure.

3.3.5 Internal Energy of Reaction

In many applications, such as in internal combustion engines, non-flow processes are utilized. In these cases, the application of the First Law shows that

$$\begin{aligned} Q - W = \Delta U &= U_p - U_r \\ &= H_p - H_r - \left((p\forall)_p - (p\forall)_r\right) \\ &= \sum_{prod} n_i \hat{h}_i(T_p) - \sum_{react} n_i \hat{h}_i(T_r) - \left(n_p \Re T_p - n_r \Re T_r\right). \end{aligned} \tag{3.36}$$

The subscripts r and p are used for the reactants and products, respectively, $n_p = \sum_{prod} n_i$ and $n_r = \sum_{react} n_i$, and ideal gas mixtures are assumed. Equation (3.36) shows that we can use the enthalpy expressions in computing interactions in closed systems, since the change in the number of moles can be calculated from the chemical reaction equations.

If the initial and final temperatures are the same and are equal to the reference temperature, and the volume remains constant, the internal energy of reaction evaluated under stoichiometric conditions (equivalent to the definition of the enthalpy of reaction) is given by the following equation:

$$\begin{aligned} \Delta \widehat{U}_R^o &= \sum_{prod} \nu_i \hat{h}_i(T^o) - \sum_{react} \nu_i \hat{h}_i(T^o) - \left(\nu_p - \nu_r\right)\Re T_r \\ &= \Delta \widehat{H}_R^o - \left(\nu_p - \nu_r\right)\Re T^o, \end{aligned} \tag{3.37}$$

where $\nu_p = \sum_{prod} \nu_i''$ and $\nu_r = \sum_{react} \nu_i'$.

Clearly, if the number of moles remains constant during the reaction, as in the case of stoichiometric methane–air combustion, the enthalpy of reaction and the internal energy of reaction are the same.

3.3.6 Adiabatic Flame Temperature at Constant-Volume Reaction

Application of the internal energy of combustion is illustrated by evaluating the constant-volume adiabatic flame temperature for stoichiometric methane–air combustion starting with 298 K and 1 atm. Using the First Law statement given above, with no heat and work interactions, we have "conservation of the total internal energy":

$$\sum_{prod} \nu_i \hat{h}_i(T_p) - \sum_{react} \nu_i \hat{h}_i(T_r) - \left(\nu_p \Re T_p - \nu_r \Re T_r\right) = 0. \tag{3.38}$$

The details of calculations are given in Example 3.8.

Example 3.8 Adiabatic Flame Temperature at Constant Volume

For stoichiometric methane–air at constant volume and assuming complete combustion (i.e., the products are CO_2, H_2O, and N_2), the initial mixture is at 298 K and 1 atm. Determine the adiabatic flame temperature of the mixture. Start with *t* values of $T_{p1} = 2800\,\text{K}$ and $T_{p2} = 2900\,\text{K}$.

Solution

We use (3.36) with the stoichiometric reaction:

$$CH_4 + 2\,(O_2 + 3.76\,N_2) \Rightarrow CO_2 + 2\,H_2O + 7.52\,N_2.$$

The internal energy of the reactants mixture based on one mole of methane consumed is

$$U_r(T_R) = U_r(298.15\,\text{K}) = H_r(T_r) - n_r\mathfrak{R}T_r = \sum_{i=1}^{N} n_i\left[\hat{h}_f^{\circ}\right]_i - n_r\mathfrak{R}\,T_r$$
$$= n_{CH_4}\hat{h}^{\circ}_{f,CH_4} - (n_{CH_4} + n_{O_2} + n_{N_2}) \times \mathfrak{R} \times (298.15) = -100{,}977\,\text{kJ}.$$

For the products, we make two guesses for T_P at 2800 and 3000. The results of the these are shown in Table E3.8.

Table E3.8

	Sensible enthalpy (kJ/kmol)			Chemical enthalpy (kJ/kmol)		
T [K]	CO_2	H_2O	N_2	CO_2	H_2O	N_2
2800	140,435	115,463	85,323	−393,800	−242,000	0
3000	152,853	126,548	92,715	−393,800	−242,000	0

$$U_p(2800\,\text{K}) = H_p(2800\,\text{K}) - n_p\mathfrak{R}\,T_p = \sum_{i=1}^{N} n_i\left[\hat{h}_f^{\circ} + \left(\hat{h} - h_{298.15}\right)\right]_i - n_p\mathfrak{R}\,T_p$$
$$= 1 \times (-393{,}800 + 140{,}435) + 2 \times (-2420{,}00 + 115{,}463)$$
$$+ 7.52 \times (0 + 85{,}323) - 10.52 \times 8.314 \times 2800$$
$$= -109{,}707\,\text{kJ},$$
$$U_p(3000\text{K}) = 1 \times (-393{,}800 + 152{,}853) + 2 \times (-242{,}000 + 126{,}548)$$
$$+ 7.52 \times (0 + 92{,}715) - 10.52 \times 8.314 \times 3000$$
$$= -37{,}024\,\text{kJ}.$$

Interpolating for the temperature, since: $U_P(3000\,\text{K}) < U_R < U_P(2800\,\text{K})$, we get

$$T_P = 2800 + (3000 - 2800)\frac{-100{,}977 - (-109{,}707)}{-37024 - (-109{,}707)} = 2824\,\text{K}.$$

Example 3.8 (cont.)

We can also compute the final pressure from the equation of state, $\frac{p_p}{p_r} = \frac{n_p T_p}{n_r T_r}$, which leads to $p_p = 9.47\,\text{atm}$. In spark-ignition engines, this pressure rise in addition to the pressure rise resulting from the compression stroke can result in a high pressure in the cycle, of the order of 100 atm for a compression ratio of 10.

3.4 Combustion and Combustion Engine Efficiency

The concept of combustion efficiency is complex and must be approached carefully. In the literature, it has many definitions and is assigned many meanings. We will introduce several definitions and make a distinction between them. In hydrocarbon combustion, while most of the fuel is converted to water and carbon dioxide, some escapes in the form of unburned hydrocarbon fragments,[1] carbon monoxide, and hydrogen (and in some cases pure carbon or soot). This incomplete combustion of the hydrocarbon is due to a number of factors, including the finite-rate reaction within the combustion chamber and local chemical quenching processes. We define the combustion efficiency as the chemical energy converted to thermal energy during a particular reaction divided by the maximum chemical energy in the fuel, or the fuel heat value. That is

$$\eta_{comb} = \frac{\sum_{prod} n_i \hat{h}_i - \sum_{react} n_i \hat{h}_i}{-\Delta \hat{h}_{R,f}}. \tag{3.39}$$

In this equation, the change in enthalpy is evaluated per mole of fuel. This definition ensures that the value computed here determines how "complete" the combustion of the fuel is, without reference to how the thermal energy is utilized later to produce work. A Second Law efficiency will be defined in terms of the availability shortly.

The combustion efficiency defined in (3.39) should not be confused with the **fuel utilization efficiency** of engines using fuels, which is defined as the net work produced in a power cycle utilizing the fuel as a source of chemical energy divided by the chemical energy in the fuel. When fuels are burned to convert their chemical energy into thermal energy, Q_H in the thermodynamic cycle efficiency is replaced with the "heating value" of the fuel, HV_{fuel}. In this case, the thermal efficiency is replaced by the **fuel utilization efficiency**:

$$\eta_{FU} = \frac{\dot{W}_{net}}{\dot{m}_{fuel} HV_{fuel}}. \tag{3.40}$$

[1] The expression *hydrocarbon fragments* refers to hydrocarbons with the number of carbon and hydrogen atoms in their molecules mostly smaller than those in the original hydrocarbon molecule.

The heating value in this equation can be either the HHV or the LHV, but it must be stated unambiguously. It is more appropriate to use the LHV for IC engines, since water in the products is expected to be in vapor phase. One can also express the fuel utilization efficiency in terms of Second Law efficiency and the combustion/fuel conversion efficiency as follows:

$$\eta_{FU} = \frac{W_{net}}{W_{max}} \frac{W_{max}}{HV_{fuel}} = \eta_{II} \cdot \eta_{max.chem.eng}. \tag{3.41}$$

In the second expression, the first term is a Second Law efficiency and the second term is the efficiency of the most efficient chemical engine.

Other efficiencies are often defined for other systems utilizing the fuel's chemical energy in multiple tasks. For instance, in a cogeneration cycle, one has work interaction as well as "useful" heat transfer to a lower-temperature surrounding for heating purposes. One useful definition is

$$\eta_{co-gen} = \frac{\dot{W}_{net} + \dot{Q}_{heating}}{\dot{m}_{fuel} HV_{fuel}}. \tag{3.42}$$

The efficiencies described here are First Law efficiencies since they are essentially based on the energy input to the system. Second Law efficiencies based on the availability input will be discussed after chemical reaction equilibrium has been covered.

3.5 Absolute Entropy and the Third Law of Thermodynamics

The application of the Second Law of Thermodynamics in chemical reaction calculations requires the evaluation of the entropy of a mixture using a common reference state, and knowledge of the entropy at that state. This is because during the course of a chemical reaction chemical components are constantly breaking up and recombining to form new species, and hence it is important that the entropies of all components are properly referenced to a common state. The situation is similar to that encountered in the application of the First Law to a reacting system, where we found that a reference enthalpy was needed. The Third Law of Thermodynamics allows us to define such reference state using the following statement: *The entropy of any pure substance in thermodynamic equilibrium approaches zero as the absolute temperature approaches zero*. This statement has been supported by a large body of experimental data and through quantum statistical calculations of the entropy. A pure substance at absolute zero temperature is in the pure crystalline state.

Since absolute zero temperature is not a convenient reference state, it is customary to write the entropy for a perfect gas as

$$\hat{s}(T,p) = \hat{s}^o(T) - \Re \ln \frac{p}{p^o}, \tag{3.43}$$

where

$$\hat{s}^o(T) = \hat{s}^{oo} + \int_{T^o}^{T} \frac{\hat{c}_p(T)}{T} dT \tag{3.44}$$

and $\hat{s}^{oo}(T) = \hat{s}(T^o, p^o)$ is evaluated at the standard temperature and pressure, and $\hat{s}^{oo} = \hat{s}^o(T^o)$. Similar to the reference enthalpy, the standard conditions are 298 K and 1 atm. Values for the reference entropy are provided in tables, along with the reference enthalpy and the reference Gibbs free energy, $\hat{g}^{oo} = \hat{h}^o - T^o\hat{s}^{oo}$, which is used extensively in chemical calculations. Values of $\hat{s}_i^{oo}$ are also expressed as Δs_f^o which are evaluated at standard temperature and pressure. Remember that for a mixture, the entropy is the sum over the entropies of the components, scaled to their mass or mole fractions, evaluated at their partial pressures in the mixture.

Gibbs free energy of formation and entropy of formation can be defined in a way similar to how the enthalpy of formation was defined. Thus, the Gibbs free energy at the reference conditions is not zero, except for the elemental substances whose Gibbs free energy is arbitrarily set to zero.[2] The absolute entropy at the standard reference condition, or the entropy of formation, is not zero. However, the entropy of formation of the elemental substances is set to zero. The entropy of formation at the reference conditions is $\Delta s_f^o = \left(\Delta h_f^o - \Delta g_f^o\right)/T^o$. Table 3.1 shows the values of the entropy of formation $\Delta\hat{s}_f^o \equiv \hat{s}^{oo}$, and the Gibbs free energy of formation, $\Delta\hat{g}_f^o \equiv \hat{g}^{oo}$, at the standard conditions for a number of elements and molecules.

There are several important applications of the Second Law in reacting mixture calculations, such as confirming that a particular reaction can proceed in a certain direction to transform reactants into products; evaluating the actual equilibrium products composition under certain conditions; and determining the maximum work transfer possible from a process involving a chemical reaction. We review some of these applications using examples next, and leave the detail of the equilibrium calculations for the next section.

3.5.1 Entropy Generation in a Chemical Reaction

The statement of the Second Law for a control volume holds true for a reacting mixture as well, that is:

$$\frac{dS}{dt} = \sum_k \frac{\dot{Q}_k}{T_k} + \sum_{react} \dot{n}_i \hat{s}_i - \sum_{prod} \dot{n}_i \hat{s}_i + \dot{S}_g. \tag{3.45}$$

[2] Measuring the Gibbs free energy of formation requires running the formation reaction isothermally and at equilibrium with the environment to avoid entropy generation. The work generated in this case can be shown to be $-\Delta g_f^o$. The voltage established by a fuel cell under open-circuit conditions could be used to measure this Gibbs free energy, as will be discussed in detail in Chapter 4.

The entropy of individual species must include their entropy of formation at the same reference temperature and pressure, and $\dot{S}_g \geq 0$ (since time change is always positive). Moreover, the properties of each stream must be evaluated at its own conditions; that is, the entropy of the reactants is evaluated at the pressure and temperature of the reactants stream and the entropy of the products is evaluated at the pressure and temperature of the products stream. The entropy of each species depends on the partial pressure of the species in the mixture in the corresponding stream. In an adiabatic constant-pressure reaction, the entropy of reaction is

$$\Delta S_R = \Delta S_g = \sum_{prod} v_i'' s_i(T, p, X_i) - \sum_{react} v_i' s_i(T, p, X_i). \tag{3.46}$$

In this expression, the entropy of each species is

$$\hat{s}_i\,(T, p, X_i) = \hat{s}_i^o(T) - \Re \ln \frac{p_i}{p^o} = \left(\hat{s}_i^{oo} + \int_{T^o}^{T} \frac{\hat{c}_{p,i}(T)}{T} dT \right) - \left(\Re \ln \frac{p}{p^o} + \Re \ln X_i \right). \tag{3.47}$$

The Gibbs free energy of reaction, or the change in the Gibbs free energy in the isothermal combustion of a number of fuels in air at the standard conditions, is shown in Table 3.2. Also shown is the change in the Gibbs free energy for the same combustion reactions. Entropy generation in these isothermal reactions is small. When measured in terms of the work lost during the reaction, $T_o \Delta\hat{s}_R^o / \Delta\hat{g}_R^o$, it is below 2% for most fuels, with higher values for hydrogen and carbon monoxide. Remember that the change in the Gibbs free energy is the same as the change in availability at the standard conditions, and hence it signifies the maximum work interaction with the environment during this process. Work interaction during chemical reactions will be discussed in detail later in this chapter.

Is a reaction possible? Let us determine the entropy generation in the steady-state, adiabatic methane–air combustion at standard conditions discussed above, in which separate streams of methane and air are admitted at atmospheric pressure and 298 K into a combustor. In this case, the entropy generation per mole of methane burned is

$$\begin{aligned}\hat{s}_g = {} & \hat{s}_{CO_2}(T_p, p_{CO_2}) + 2\hat{s}_{H_2O}(T_p, p_{H_2O}) + 7.52\hat{s}_{N_2}(T_p, p_{N_2}) \\ & -\left(\hat{s}_{CH_4}(T_r, p_{CH_4}) + 2\hat{s}_{O_2}(T_r, p_{O_2}) + 7.52\hat{s}_{N_2}(T_r, p_{N_2})\right).\end{aligned}$$

Note that computing the entropy generation requires knowledge of the adiabatic flame temperature, and the partial pressures of all the components in the reactants and the products mixtures. The former is found as $T = 2330$ K with the method explained in Example 3.6. For the reactants, the partial pressures of oxygen and nitrogen are, respectively, 0.21 and 0.79 atm. For the products, the partial pressure of CO_2, H_2O, and N_2 are 0.095, 0.19, and 0.715 atm, respectively. Using the standard entropy of formation for each species, and integrating the specific heats divided by temperature for the products, we evaluate the component's entropy. Substituting in the above expression, we find that entropy generation for this system is 358.7 kJ/K. Since the entropy generation is positive, this reaction can proceed in the direction assumed. It should be noted that an increase in the total entropy of the mixture does not guarantee that the reaction has proceeded to equilibrium.

As discussed earlier, the "complete" combustion of a fuel – that is, the conversion of all the fuel carbon into carbon dioxide and all its hydrogen to water – under arbitrary conditions is not always guaranteed. In most cases, some dissociation of the products, in particular dissociation of some CO_2 to CO and some H_2O to H_2, may occur at high temperatures. The calculation of the products concentration of a hydrocarbon–air mixture must, in general, allow for the possibility of existence of at least CO_2, CO, H_2O, H_2, O_2, and N_2 in the products stream. Carrying out these calculations requires the careful application of the Second Law to determine the conditions necessary for chemical equilibrium – a subject that we will tackle in the next section.

3.6 Equilibrium

Chemical thermodynamics is concerned with evaluating the state of a reacting mixture undergoing a single or multiple reactions under equilibrium conditions. So far, we have analyzed chemical reactions based on some knowledge of the composition of the products. This has been enabled by the assumption of complete combustion at stoichiometric fuel–oxidizer mixtures or fuel-lean combustion where excess oxidizer is recovered in the products. However, neither assumption is exact and, in fact, these assumptions are not valid for fuel-rich mixtures, where there is not enough oxygen to burn all the carbon to CO_2 and all the hydrogen to H_2O, and instead the products contain CO and H_2. Moreover, at high temperatures, dissociation of CO_2 into CO and H_2O into H_2 is expected. As illustrated in Example 3.9, the atom balance is not sufficient to estimate all possible products of combustion, and additional analysis based on equilibrium calculation may be useful to determine equilibrium products concentrations.

Chemical equilibrium is about the balance of all possible reactions in the products. For this, we enable reactions to occur in both the **forward** and **backward** directions. Equilibrium is reached when the forward and backward reactions proceed at equal rates. For example, if we expect that H_2 and H are in the products, then the following reaction must be equilibrium: $H_2 \Leftrightarrow 2\,H$. Equally, if we expect dissociation in the products of hydrocarbon combustion from CO_2 and H_2O to CO and H_2, then we can write a balance reaction in the products: $H_2O + CO \Leftrightarrow H_2 + CO_2$. This is known as the water–gas shift reaction. Invariably, we can write a multitude of combinations of equilibrium reactions. However, it is important to write reactions that only involve species that we are interested in. For example, if we do not expect H to be a major species in a reaction, then we may not want to consider $H_2 \Leftrightarrow 2\,H$. The water–gas shift reaction is always important for hydrocarbons under stoichiometric and rich conditions.

In this chapter, we discuss in detail the conditions of equilibrium as applied to chemical reactions under different constraints, and the tools used to define the equilibrium state. The starting point here is to admit the notion that equilibrium chemical reactions are reversible and, hence, the reaction equation is written with double arrows pointing in both directions. For example, a reaction involving H_2, O_2, and H_2O can be written as

$$2H_2 + O_2 \Leftrightarrow 2H_2O. \tag{3.48}$$

The Second Law of Thermodynamics defines the equilibrium state. In its general form, it states that: "for a closed system with fixed mass, volume, and internal energy, equilibrium is reached when the total entropy generation is maximum."

The equilibrium condition expressed in terms of the extensive properties of the mixture is constraints-dependent, as was shown in the Chapter 2. At constant volume and total internal energy, or at constant pressure and total enthalpy, one maximizes the entropy to determine the equilibrium state. Equilibrium can also be expressed in terms of the intensive variable. At constant pressure and temperature, one minimizes the total Gibbs free energy. At fixed temperature and specific volume, one minimizes the Helmholtz free energy to determine the equilibrium state. Since chemical reactions involve transfer of mass internally between subsystems – that is, the possible exchange of mass between the chemical components – the chemical potential is often used to define the equilibrium state in terms of mole fractions. This leads to the formulation of the equilibrium constant of a chemical reaction, and the law of mass action.

3.6.1 Conditions of Equilibrium

At constant volume and internal energy, or when the system is constrained by a given volume and internal energy (closed system), we showed that $dS = dS_g$, and the equilibrium state corresponds to the point of maximum entropy:

$$dS \geq 0 \quad \text{at constant } (U, \forall). \tag{3.49}$$

Thus, in order to determine the state of equilibrium of a mixture in terms of the number of moles, or mole fractions, of the species that can coexist within the mixture, we use the following equation of state (for an ideal gas):

$$S = \sum_I n_i \hat{s}_i(T, p_i) = \sum_N n_i \left[\hat{s}_i(T) - \Re \ln \frac{p_i}{p^o} \right],$$

$$U = \sum_I n_i \hat{u}_i,$$

$$\forall = \sum_I n_i \hat{v}_i$$

in (3.49) while holding the total mass of all the chemical elements in the mixture constant. The problem then becomes maximizing S while holding the known U, V and the total mass of the elemental components constant. The solution of this constraint maximization problem is the mole fractions of the coexisting species and the other thermodynamics properties of state. This can be done efficiently using computer code. This approach is known as the element potential method to equilibrium calculations.

At constant p and H (in an open flow system), or when the system is constrained to a given p and H, the equilibrium state is reached when

$$dS \geq 0 \quad \text{at constant } (p, H). \tag{3.50}$$

Therefore, at equilibrium, when conditions of constant pressure and enthalpy are applied, the total entropy is maximal. The solution steps are similar to those described following (3.49).

Constraining the system to constant T and p, we find that $dG = -dS_g$, and at equilibrium we have

$$dG \leq 0 \quad \text{at constant } (p, T). \tag{3.51}$$

Therefore, at equilibrium, G must reach a minimum when the state is defined by the pressure and temperature. Finally, constraining the system to constant T and $\forall$, we find that $d\mathrm{A} = -dS_g$, and at equilibrium we must satisfy

$$d\mathrm{A} \leq 0 \quad \text{at constant } (T, \forall). \tag{3.52}$$

That is, A must reach a minimum at equilibrium.

For a given system, using the appropriate forms of the equations of state, and a set of constraints in terms of some thermodynamic properties and mass conservation, one can employ (3.49)–(3.52) to determine the equilibrium state, depending on the constraints. As described before, this requires writing the extensive variables as a sum over the number of moles times the intensive property, imposing the corresponding constraints and solving the maximization/minimization problem iteratively. These steps have been implemented in widely available computer codes. On the other hand, these procedures do not always lend themselves well to analytical treatment.

In the next section, the process of determining the equilibrium state/composition is presented using the concept of chemical potential. It will be shown that a universal condition can be found. This condition is then applied to determine the equilibrium state in a mixture whose constituents can interact internally by exchanging/transferring mass among the chemical species through a number of simultaneous chemical reactions.

3.6.2 Chemical Potential and Entropy Generation

Using the general expression for entropy generation, we showed that

$$T\,\delta S_g = -\sum_N \mu_i dn_i. \tag{3.53}$$

The entropy generation in this equation is due to mass transfer in a chemical reaction. At equilibrium,

$$\sum_N \mu_i dn_i \leq 0. \tag{3.54}$$

This expression shows the convenience of using the chemical potential in thermodynamic calculations. The chemical potential of the species, μ_i, depends on the mixture temperature and pressure, and the partial pressure of the species. This dependence of the chemical potential on the state properties and partial pressure is inherited from the entropy of a mixture, which depends on the temperature and pressure and the molar concentrations, or partial pressure of the components/species in the mixture. Using the relation between the chemical potential and the Gibbs free energy of a component in a mixture, $\mu_i = \hat{g}_i((T, p_i)$, we write:

$$\mu_i = \hat{g}_i^o(T) + \Re T \ln \frac{p_i}{p^o} \tag{3.55}$$

or

$$\begin{aligned} \mu_i &= \hat{g}_i^o(T) + \Re T(\ln p + \ln X_i) \\ &= \hat{g}_i(T, p) + \Re T(\ln X_i). \end{aligned} \tag{3.56}$$

It is important to emphasize that the chemical potential of a component in a mixture is a thermodynamics property of the component, evaluated at the mixture conditions, and the concentration of the component in the mixture. Now we are ready to define the condition or reaction of equilibrium in terms of the chemical potential.

3.6.3 Chemical Equilibrium of a Single Reaction

Using (3.54), equilibrium is reached when

$$\sum_N \mu_i dn_i = 0. \tag{3.57}$$

With further manipulation, this expression can conveniently be transformed into an explicit expression for a relationship between the partial pressures/concentrations of the species in the mixture, and the Gibbs free energy of the reaction governing the equilibrium in the mixture. To impose the conservation of mass constraint on the above equation for the following generic equation:

$$v_a' A + v_b' B \Leftrightarrow v_c'' C + v_d'' D. \tag{3.58}$$

Writing a chemical reaction with two opposing arrows indicates that the reaction can proceed in both directions, and reactants and products can coexist, with different concentrations that satisfy the condition of equilibrium. This is an important degree of freedom that allows us to satisfy the Second Law (e.g., to allow the species participating in a chemical reaction to exchange mass until the equilibrium condition is satisfied).

According to the reaction, (3.58), changing the number of moles of one component, n_i, by a small amount, dn_i (it is negative for a reactant), should cause all other concentrations to change by dn_i/v_i. This ratio is often called the reaction progress variable, $d\xi$. Remember that for any component, $\nu_i = \nu_i'' - \nu_i'$. Using this relation in (3.57) reveals that the condition of equilibrium, also called the **law of mass action**, for a chemical reaction is

$$\sum_{prod} \nu_i''\mu_i - \sum_{react} \nu_i'\mu_i = 0$$

or (3.59)

$$\sum_{species} \nu_i\mu_i = 0,$$

where μ_i is the chemical potential at *equilibrium*. One can proceed to find the concentrations, or partial pressures, of the reactants and products in the mixture that satisfy (3.59). Expressing the chemical potential of an ideal gas using (3.55), rearranging and setting $p^o = 1\,\text{atm}$, we get the important result that, under equilibrium, the species in a mixture participating in a particular chemical reaction must satisfy the following relation:

$$\frac{\prod_{prod} p_i^{\nu_i''}}{\prod_{react} p_i^{\nu_i'}} = K_p(T). \tag{3.60}$$

The pressure $p_i = X_i p$ is the partial pressure (measured in atm) of a species in the mixture *at equilibrium*, while K_p is the pressure-based **equilibrium constant** of the chemical reaction. The equilibrium constant is a function of temperature only:

$$K_p(T) = \exp\left(-\frac{\Delta G_R^o(T)}{\Re T}\right). \tag{3.61}$$

The Gibbs free energy of the reaction at T is given by

$$\begin{aligned}\Delta G_R^o(T) &= \sum_{prod} \nu_i''\hat{g}_i^o(T) - \sum_{react} \nu_i'\hat{g}_i^o(T) \\ &= \sum_{species} \nu_i\hat{g}_i^o(T).\end{aligned} \tag{3.62}$$

The Gibbs free energy of a reaction can also be written in terms of the enthalpy and entropy at $p = 1\,\text{atm}$:

$$\Delta G_R^o(T) = \sum_{species}^{reacting} \nu_i\hat{h}_i^o(T) - T\sum_{species}^{reacting} \nu_i\hat{s}_i^o(T). \tag{3.63}$$

The summations in this equation are over the species that participate in a particular reaction of interest, and not over all the species in the mixture. The result in (3.60) shows that the ratio of the product of the partial pressures of the products to that of the reactants is only a function of temperature. So, the equilibrium constant is a property of the reaction only. The Gibbs free energies and the corresponding equilibrium constants of many important

Table 3.4 The values of the equilibrium constant, $\log_{10} K_P(T)$, for eight reactions in ideal gas mixtures in equilibrium [3]

T (K)	$\frac{1}{2}H_2 \Leftrightarrow H$	$\frac{1}{2}O_2 \Leftrightarrow O$	$\frac{1}{2}N_2 \Leftrightarrow N$	$H_2O \Leftrightarrow H_2 + \frac{1}{2}O_2$	$H_2O \Leftrightarrow OH + \frac{1}{2}H_2$	$\frac{1}{2}O_2 + \frac{1}{2}N_2 \Leftrightarrow NO$	$CO_2 \Leftrightarrow CO + \frac{1}{2}O_2$	$CO_2 + H_2 \Leftrightarrow CO + H_2O$
100	−110.954	−126.73	−243.583	−123.6	−143.8	−46.453	−143.2	−19.6
298	−35.612	−40.604	−79.8	−40.048	−46.137	−15.171	−45.066	−5.018
500	−20.158	−22.94	−46.336	−22.886	−26.182	−8.783	−25.025	−2.139
1000	−8.646	−9.807	−21.528	−10.062	−11.309	−4.062	−10.221	−0.159
1200	−6.707	−7.604	−17.377	−7.899	−8.811	−3.275	−7.764	+0.135
1400	−5.315	−6.027	−14.406	−6.347	−7.021	−2.712	−6.014	+0.333
1600	−4.266	−4.842	−12.175	−5.18	−5.677	−2.290	−4.706	+0.474
1800	−3.448	−3.918	−10.437	−4.27	−4.631	−1.962	−3.693	+0.577
2000	−2.790	−3.178	−9.046	−3.54	−3.793	−1.699	−2.884	+0.656
2200	−2.251	−2.571	−7.905	−2.942	−3.107	−1.484	−2.226	+0.716
2400	−1.800	−2.065	−6.954	−2.443	−2.535	−1.305	−1.679	+0.764
2600	−1.417	−1.636	−6.149	−2.021	−2.052	−1.154	−1.219	+0.802
2800	−1.089	−1.268	−5.457	−1.658	−1.637	−1.025	−0.825	0.833
3000	−0.803	−0.949	−4.858	−1.343	−1.278	−0.913	−0.485	+0.858

The values of K_p are for the reactions moving from left to right. The equilibrium constant is $K_P(T) = \frac{X_3^{\nu_3} X_4^{\nu_4}}{X_1^{\nu_1} X_2^{\nu_2}} \left(\frac{p}{p_o}\right)^{-\nu_1-\nu_2+\nu_3+\nu_4}$, in which $p_o = 1$ atm. The equilibrium constant is written for the chemical equation: $\nu_1 A_1 + \nu_2 A_2 \Leftrightarrow \nu_3 A_3 + \nu_4 A_4$.

reactions have been well tabulated or approximated as polynomials. The Gibbs free energies for the combustion of a number of fuels at standard conditions are given in Table 3.2. JANAF tables [3] are a comprehensive source of data on equilibrium constants. Table 3.4 provides a sample of these data. The procedure outlined above is used extensively to determine the equilibrium composition of a mixture, which is governed by single or multiple reactions. Note that the reactions listed in JANAF tables are listed in their elementary forms, and the unit of pressure is *atmosphere*.

Equations (3.59)–(3.61) show important properties of chemical reactions. When the Gibbs free energy of a reaction is a large negative quantity (recall that $\Delta G = \Delta H - T\Delta S$), as it is often the case when the reaction is strongly exothermic, the equilibrium constant is very large and, *under appropriate conditions* that will be discussed later in the present chapter, the reaction proceeds to near completion – that is, the products' concentrations are so much higher than those of the reactants that the latter exist in negligible fraction. This is not the case for reactions with small Gibbs free energy, especially at higher temperatures when the equilibrium constant becomes of order one, O(1). In this case, the coexistence of both reactants and products is likely.

At this stage, it is worth mentioning some of the properties of the equilibrium constant, $K_p(T)$. First, the equilibrium constant for the reverse of a given reaction is the inverse of the equilibrium constant for that reaction. For example, the equilibrium constant for the dissociation reaction of O_2, $O_2 \Rightarrow 2\,O$, is equal to the inverse of the formation reaction of O_2, $2\,O \Rightarrow O_2$. Second, if a reaction is multiplied by a given factor, then the equilibrium constant of the new reaction is the equilibrium reaction of the original reaction to the power of the given factor. For example, the equilibrium constant of $O \Rightarrow 1/2O_2$ is the square root of the formation reaction of O_2, $2O \Rightarrow O_2$. Similarly, summing (subtracting) two reactions results in the multiplication (division) of their corresponding equilibrium constants.

From the above observations, it is possible to construct equilibrium constants for any reaction using combinations of formation reactions. For example, the equilibrium constant for the reaction $CO + 0.5O_2 \Rightarrow CO_2$ can be expressed in terms of the formation reactions of CO and CO_2: $K_p(T) = K_{p,CO_2} \times K_{p,CO}^{-1} = \frac{K_{p,CO_2}}{K_{p,CO}}$. In this expression, the equilibrium constants for the formation reactions of CO_2 and CO are defined as the equilibrium constants for $C + O_2 \Rightarrow CO_2$ and $C + 0.5\,O_2 \Rightarrow CO$, respectively. The power associated with the above equation corresponds to the net stoichiometric coefficients of the species in the reactions, $\nu'' - \nu'$. Expressing the equilibrium constant of a given reaction in terms of the equilibrium constants of formation reactions is very convenient, since it is easier to tabulate the specific reactions (i.e., the total number of tables is the same as the number of species minus the number of reference chemical species involved in the reaction) than tabulating equilibrium constants for a much higher number of possible reactions. Finally, if the equilibrium constant is very small/high, then the reactants are more/less dominant than the products of the reaction at equilibrium conditions.

Example 3.9 Equilibrium of Water, Hydrogen, and Oxygen Reaction

To demonstrate the equilibrium calculation in a hydrogen, oxygen, and water mixture, with known total mass of hydrogen and oxygen, the equilibrium composition at a given temperature and pressure can be found using the reaction

$$H_2O \Leftrightarrow H_2 + \frac{1}{2}O_2. \tag{3.64}$$

The Gibbs free energy of the reaction is: $\Delta G^o_{R,H_2O} = \left(\hat{g}^o_{H_2} + \frac{1}{2}\hat{g}^o_{O_2}\right) - \hat{g}^o_{H_2O}$.

The equilibrium constant, $K_{p,H_2O} = \exp\left(-\frac{\Delta G^o_{R,H_2O}(T)}{\Re T}\right)$, defines the relation between the partial pressures of the species:

$$\frac{p_{H_2} p_{O_2}^{1/2}}{p_{H_2O}} = K_{p,H_2O}. \tag{3.65}$$

The last expression can be written more conveniently in terms of the mole fractions:

$$\frac{X_{H_2} X_{O_2}^{1/2}}{X_{H_2O}} = \frac{K_{p,H_2O}}{\sqrt{p}}. \tag{3.66}$$

Example 3.9 (cont.)

Starting with one mole of H_2O, the composition of the dissociated mixture is made up of

$$\left((1-\alpha)\ \mathrm{H_2O}+\alpha\,\mathrm{H_2}+\frac{\alpha}{2}\,\mathrm{O_2}\right).$$

This composition satisfies the conservation of mass of oxygen and hydrogen. The mole fractions of the species in the mixture are: $X_{\mathrm{H_2O}}=\frac{1-\alpha}{1+\alpha/2}$, $X_{\mathrm{H_2}}=\frac{\alpha}{1+\alpha/2}$, and $X_{\mathrm{O_2}}=\frac{\alpha/2}{1+\alpha/2}$.

Substituting in (3.66), we get: $\dfrac{\alpha}{1-\alpha}\left(\dfrac{\alpha}{2+\alpha}\right)^{1/2}=\dfrac{K_{p,\mathrm{H_2O}}}{\sqrt{p}}$.

As an example, at 3000 K, Table 3.4 shows that $K_{p,\mathrm{H_2O}}=0.0454$, and the solution of the above equation for $p=1$ atm, yields: $X_{\mathrm{H_2O}}=0.794, X_{\mathrm{H_2}}=0.137$, and $X_{\mathrm{O_2}}=0.069$. Therefore, not all the hydrogen has reacted to form water at this temperature.

This example shows that both the reaction equilibrium condition and conservation of mass of the participating species must be used together to determine the equilibrium composition of a chemically reacting mixture.

Example 3.10 Formation of NO in Hot Air

When air is heated, NO is formed. Assume air is made up of 21% O_2 and 79% N_2. Determine the concentration of NO at 1500 K and 2000 K at 1 atm.

Solution

The equilibrium reaction for NO formation is: $\mathrm{N_2}+\mathrm{O_2}\Leftrightarrow 2\,\mathrm{NO}$, or $1/2\,\mathrm{N_2}+1/2\,\mathrm{O_2}\Leftrightarrow\mathrm{NO}$. We will use the first equilibrium reaction to show how one can relate the equilibrium constant of this reaction to that of the formation reaction. According to mass balance: $\mathrm{O_2}+3.76\,\mathrm{N_2}\Leftrightarrow a\,\mathrm{O_2}+b\,\mathrm{N_2}+c\,\mathrm{NO}$. The atom balance yields:

$$\begin{aligned}\mathrm{O}:&\quad 2=2a+c\\ \mathrm{N}:&\quad 3.76\times 2=2b+c.\end{aligned}$$

The equilibrium constant relates the partial pressures of the species in the products:

$$K_p(T)=\frac{p_{\mathrm{NO}}^2}{p_{\mathrm{N_2}}p_{\mathrm{O_2}}}=\frac{X_{\mathrm{NO}}^2p^2}{(X_{\mathrm{N_2}}p)(X_{\mathrm{O_2}}p)}=\frac{X_{\mathrm{NO}}^2}{X_{\mathrm{N_2}}X_{\mathrm{O_2}}}.$$

Example 3.10 (cont.)

The mole fractions are expressed as: $X_{O_2} = \frac{a}{a+b+c}$, $X_{N_2} = \frac{b}{a+b+c}$, $X_{NO} = \frac{c}{a+b+c}$, which after substitution of b and c in terms of a from the above elemental balance yields

$$X_{O_2} = \frac{a}{4.76}, \quad X_{N_2} = \frac{2.76 + a}{4.76}, \quad X_{NO} = \frac{2(1-a)}{4.76}.$$

Substituting these mole fractions into the expression for the equilibrium constant above yields

$$K_p(T) = \frac{4(1-a)^2}{a(2.76+a)}.$$

We can formulate the expression for the unknown a in terms of a quadratic equation:

$$(4 - K_p)\,a^2 - (8 + 2.76\,K_p)\,a + 4 = 0,$$

with the only valid root for a that corresponds to a positive value. The equilibrium constants at 1500 K and 2000 K are: 1.0617×10^{-5} and 3.9945×10^{-4}, respectively. Increasing the equilibrium constant between 1500 K and 2000 K indicates a higher amount of dissociation of heated air at the higher temperature. Solving the above quadratic equation at 1500 K gives $a = 0.997$, $b = 3.769$, and $c = 6.306 \times 10^{-3}$. On the other hand, at 2000 K, we get $a = 0.981$, $b = 3.741$, and $c = 3.831 \times 10^{-3}$. Thus, the concentrations of NO at 1500 K and 2000 K are 0.13% and 0.8%, respectively.

3.6.4 The van't Hoff Equation

One can carry out further manipulations to derive the **van't Hoff** equation for the equilibrium constant:

$$\frac{d(\ln K_p)}{d(1/T)} = -\frac{\Delta H_R}{\Re}, \tag{3.67}$$

where ΔH_R is the enthalpy of reaction at a given temperature T, as defined in (3.28). Because the enthalpy of reaction of a mixture is often a weak function of temperature over a small temperature range, (3.67) can be approximated as follows:

$$\begin{aligned} K_p &\approx \left[K_p(T^0)\exp\left(\frac{\Delta H_R^0}{\Re T^0}\right)\right]\exp\left(-\frac{\Delta H_R^0}{\Re T}\right) \\ &= \bar{K}_p^o \exp\left(-\frac{\Delta H_R^0}{\Re T}\right). \end{aligned} \tag{3.68}$$

This equation reveals that for an exothermic reaction, with negative enthalpy of reaction, increasing the temperature reduces the equilibrium constant – that is, it lowers the

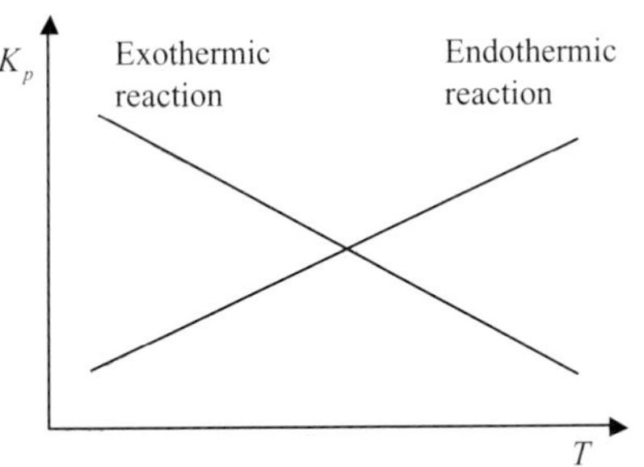

Figure 3.9 The dependence of the equilibrium constant of exothermic and endothermic chemical reactions on temperature.

concentration of the products in the mixture – and vice versa (see Figure 3.9). Thus, some exothermic reactions, such as the oxidation reactions of carbon, carbon monoxide, and hydrogen, may not reach complete consumption of the fuel in order to satisfy equilibrium conditions. For example, in the hydrogen–oxygen–water mixture examined in Example 3.8, the higher the temperature, the lower the water concentration in the mixture since the reaction in (3.64) is strongly exothermic. In other words, *adiabatic or near-adiabatic exothermic reactions proceed in the direction that reduces complete conversion*, thus leaving some of the reactants in their original form. Moreover, the higher the temperature of the products, the more dissociation is expected. The opposite is also true; endothermic reactions that absorb thermal energy to break the bonds of reactants have a higher equilibrium constant at high temperatures, and more products are therefore expected.

Example 3.11 Dissociation of Carbon Dioxide to Carbon Monoxide and Oxygen

Consider a mixture of carbon monoxide, carbon dioxide, and oxygen, whose composition is governed by the following reaction:

$$CO_2 \Leftrightarrow CO + \frac{1}{2}O_2. \tag{3.69}$$

This reaction is strongly endothermic when proceeding to the right. The higher the operating temperature, the more CO is found in the mixture. The equilibrium constant is used to determine the ratio of reactants and products in this mixture.

A stream of CO_2 is heated from 298 K and 6 atm to 2800 K and 5 atm in a steady-state process, shown schematically in Figure E3.11. Determine the amount of heat transfer and the composition of the mixture at the exit of the heater.

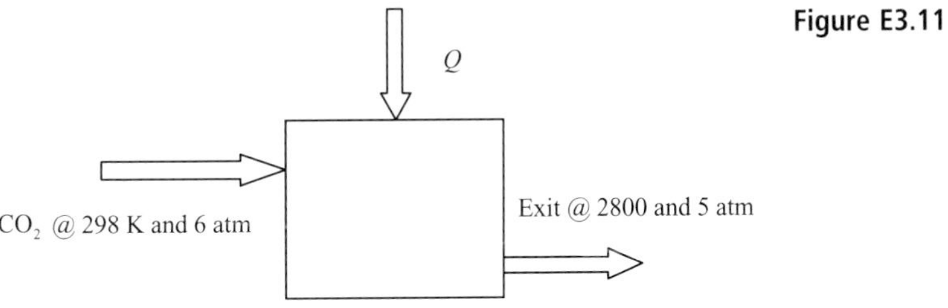

Figure E3.11

Example 3.11 (cont.)

Solution

Applying the First Law to this steady-state process, we have

$$Q = \hat{h}_2 - \hat{h}_1.$$

In this expression, $\hat{h}$ is the enthalpy per unit mole of CO_2, and states 1 and 2 are the inlet and exit of the heater, respectively. The gas stream enters the heater as CO_2. At elevated temperatures it is expected to dissociate according to (3.69), and a mole of CO_2 becomes $\left((1-\alpha)CO_2 + \alpha CO + \frac{\alpha}{2}O_2\right)$. The choice of the number of moles of each species satisfies the conservation of mass of carbon and oxygen. Using the definition of the enthalpy,

$$Q = \left[(1-\alpha)\hat{h}^{2800}_{CO_2} + \alpha\hat{h}^{2800}_{CO} + \frac{\alpha}{2}\hat{h}^{2800}_{O_2}\right] - \hat{h}^{298}_{CO_2}.$$

The enthalpies are all known, but we need to determine α. Since this mixture is in equilibrium, it must satisfy the law of mass action:

$$\frac{p_{CO}p^{1/2}_{O_2}}{p_{CO_2}} = \frac{X_{CO}X^{1/2}_{O_2}}{X_{CO_2}}p^{1/2} = K_{p,CO_2}(T). \tag{3.70}$$

The mole fraction can be expressed in terms of α such that $X_{CO} = \frac{\alpha}{1+\alpha/2}$, $X_{O_2} = \frac{\alpha/2}{1+\alpha/2}$, $X_{CO_2} = \frac{1-\alpha}{1+\alpha/2}$, and the equilibrium equation becomes

$$\frac{K_p^2}{p} = \frac{\alpha^3}{(1-\alpha)^2(2+\alpha)}.$$

The equilibrium constant can be found from Table 3.4. At 2800 K, the equilibrium constant is $K_{p,CO_2} = 0.15$. Note that the equilibrium constant is given in atm, so the pressure should be substituted in the same unit. A solution of the above equation at $p = 5$ atm gives $\alpha = 0.1867$, that is 18.7% of carbon dioxide dissociates back to carbon monoxide and oxygen. The mixture composition at the heater exit is $X_{CO} = 0.171$, $X_{O2} = 0.086$, and $X_{CO2} = 0.743$. Substituting in the First Law, we find that $Q = 139.4$ MJ/kmol CO_2. If we had assumed no dissociation, we would have found that $Q = 140.4$ MJ/kmol CO_2.

We note here that the equilibrium constant for CO_2 dissociation reaction, at atmospheric pressure, increases from 9×10^{-46} at 300 K, to 6.34×10^{-11} at 1000 K, to 1.37×10^{-3} at 2000 K, and to 0.34 at 3000 K, showing the significance of dissociation at high temperatures.

3.6.5 Pressure Dependence

Another interesting result of the law of mass action is the dependence of the equilibrium composition of a mixture, as governed by a particular reaction, on the total pressure of the mixture. Equation (3.60) can be written in terms of the mole fractions, the total number of moles, and the total pressure as follows:

$$\frac{\prod_{prod} X_i^{\nu_i''}}{\prod_{react} X_i^{\nu_i'}} = \frac{K_p(T)}{p^\sigma}, \tag{3.71}$$

where $\sigma = \sum_{prod} \nu_i'' - \sum_{react} \nu_i'$. The right-hand side of this expression is the mole fraction-based equilibrium constant. Therefore, if the sum of the moles on the products side of the reaction is greater than that on the reactants side, $\sigma > 0$, then increasing the total pressure reduces the products concentration in the mixture, and the opposite is obviously true. On the other hand, if the reaction leads to a lower number of moles of products, increasing the pressure forces the reaction to move toward the products side. For example, increasing the total pressure increases the water concentration in a mixture of hydrogen, oxygen, and water since the equilibrium of this mixture is governed by (3.66).

Equilibrium constants have been tabulated for most "elementary" reactions. One can algebraically add several reactions to derive the equilibrium constant of a more complex reaction. Take, for example, the homogeneous water–gas shift reaction (3.14):

$$H_2O + CO \Leftrightarrow H_2 + CO_2.$$

We will call this reaction R3, the water dissociation reaction (3.64) will be R1, and the carbon dioxide dissociation reaction (3.69) will be R2. Clearly, R3 = R1 – R2, $\Delta G_{R,R3} = \Delta G_{R,R1} - \Delta G_{R,R2}$, and hence the equilibrium constant is $K_{p,R3} = K_{p,R1}/K_{p,R2}$.

3.6.6 Equilibrium and Gibbs' Free Energy

Equation (3.51) states that at constant pressure and temperature the Gibbs free energy of the mixture must decrease on its way to equilibrium, $\Delta G \leq 0$. This expression embodies both principles of energy minimization and entropy maximization during change toward equilibrium, $\Delta G = \Delta H - T\Delta S$. This principle applies to any change, including mixing (which experiences little energy change), phase change (including the latent enthalpy), and chemical reactions. The same equation is used to determine whether a change, under constant pressure and temperature, can occur spontaneously or not. Thus, the criterion for a reaction to proceed from the reactants to products is

$$\Delta \bar{G}_{mixture} = \sum_{prod} n_i \hat{g}_i - \sum_{react} n_i \hat{g}_i \leq 0. \tag{3.72}$$

The overbar emphasizes the fact that the number of moles in this expression is that of a mixture undergoing a reaction. Equilibrium is reached when the change of the total Gibbs' free energy reaches zero:

$$\Delta\bar{G}_{mixture} = 0, \quad \text{reacting mixture at equilibrium.} \tag{3.73}$$

In this equation, the number of moles, or mole fractions, are those found in the actual reaction, which can change during the reaction. The mixture may undergo one or a number of spontaneous reactions. Clearly, for the equilibrium condition to be satisfied, the number of moles of the reacting species must adjust to satisfy the Gibbs free energy minimization condition. Highly exothermic reactions, like combustion reactions, have large negative values of ΔH_R and hence tend to proceed toward the products side spontaneously. With positive change in entropy, higher temperatures favor a further reduction in Gibbs free energy. (This statement does not imply that the rate of reaction is high, only that the reaction does proceed. The reaction rate is determined by the kinetics, as will be shown later, and under some conditions the reaction may proceed so slowly that the mixture, in effect, does not change its composition.) In the case of reactions with low exothermicity, one must account for the entropy change for determining the reaction direction.

Positive value of the change in the Gibbs' free energy indicates that the reaction will proceed in the opposite direction. The difference $(\Delta G - \Delta G_R)$ is a departure from equilibrium, or a driving force toward equilibrium. Change in the Gibbs free energy in a spontaneous reaction implies irreversibility, or loss of availability, such as in adiabatic combustion processes in which the chemical energy is converted into thermal energy. If this same conversion is managed such that it occurs reversibly and in equilibrium with the environment, work can be produced, as in the case of ideal electrochemical reactions.

It is instructive to express the results in the preceding paragraphs in terms of Gibbs free energy. For instance, the change in Gibbs free energy during a chemical reaction is given by

$$\Delta G_R(T,p) = \Delta G_R^o(T) + \Re T \frac{\prod\limits_{prod} p_i^{\nu_i''}}{\prod\limits_{react} p_i^{\nu_i'}}. \tag{3.74}$$

Moreover, (3.61) can be rewritten in the following form:

$$\Delta G_R^o(T) = -\Re T \ln K_p(T). \tag{3.75}$$

Recall that ΔG_R^o is the change in the Gibbs free energy during the reaction if all the partial pressures of the reactants and products are at atmospheric pressure. This equation will be useful in later developments.

3.7 Hydrocarbon Combustion Reactions

Because of the tendency of combustion products to dissociate back to their more elementary components at higher temperature, which impact the energy budget and hence the temperature, it is important to include the equilibrium of CO_2 (and its elementary components) and H_2O (and its elementary components) when working with hydrocarbon oxidation. Moreover, higher temperatures drive the equilibrium of endothermic reactions such as NO formation.

3.7.1 Accurate Calculation of the Adiabatic Flame Temperature

As we saw from previous examples, products of combustion dissociate at high temperatures, and some "fuel" in the form of hydrogen and carbon monoxide (in the case of burning a hydrocarbon) can remain unburned in the mixture. This reduces the products or flame temperature, and leaves some of the chemical energy in the products species. The presence of H_2 and CO, a form of incomplete combustion, often called dissociation, is associated with the breakup of H_2O and CO_2. Example 3.12 shows the impact of the initial mixture temperature on the post-combustion temperature and products composition.

Example 3.12

To proceed with the calculation of the flame temperature while accounting for incomplete combustion of a pure hydrocarbon fuel, we write the chemical reaction in the following form (extension to oxygenated hydrocarbon is straight forward):

$$\begin{aligned} C_nH_m + \left(n + \frac{m}{2}\right)(O_2 + 3.76N_2) \Rightarrow\ & \alpha_{CO_2}CO_2 + \alpha_{CO}CO + \alpha_{H_2O}H_2O \\ & + \alpha_{H_2}H_2 + \alpha_{O_2}O_2 + \alpha_{N_2}N_2. \end{aligned} \tag{3.76}$$

In this equation, $\alpha_{N_2} = 3.76\left(n + \frac{m}{2}\right)$, since the nitrogen does not participate in the reaction. To compute all the unknowns, we start by writing mass conservation equations for all the chemical elements:

Carbon

$$n = \alpha_{CO_2} + \alpha_{CO}. \tag{3.77}$$

Hydrogen

$$m = 2(\alpha_{H_2O} + \alpha_{H_2}). \tag{3.78}$$

Oxygen

$$\left(n + \frac{m}{2}\right) = \alpha_{CO_2} + \frac{1}{2}\alpha_{CO} + \frac{1}{2}\alpha_{H_2O} + \alpha_{O_2}. \tag{3.79}$$

Next, we assume that the products on the right-hand side of (3.76) are in chemical equilibrium. The mole fractions of the different species in these products can then be computed as $X_i = \alpha_i / \sum_N \alpha_i (N = CO_2, CO, H_2O, H_2, O_2, N_2)$.

Example 3.12 (cont.)

For the mixture on the right-hand side of (3.76), the reactions (3.64) and (3.69) are sufficient to calculate the equilibrium composition for this mixture. One more equation is needed to close this system: the First Law for an adiabatic system at constant pressure:

$$\sum_{react} \nu_i \hat{h}_i(T_1) = \sum_{prod} \alpha_i \hat{h}_i(T_p). \tag{3.80}$$

The resulting six equations – two from equilibrium, three from mass conservation, and the First Law – are solved for the six unknowns. These are nonlinear equations because of the dependence of the chemical potential on the mole fraction, the exponential dependence of the equilibrium constant on temperature, and the dependence of the specific heats on temperature. They must be solved by iterations.

Table E3.12 shows the results of a stoichiometric methane–air mixture for different values of $T_1 = 300$ K, 500 K, and 700 K for three different cases:

1. complete combustion;
2. dissociation of CO_2 and H_2O; and
3. dissociation of CO_2 and H_2O and formation of NO.

Table E3.12 Post-combustion equilibrium temperature and products composition of methane–air combustion

			Mole fraction (final)						
Case no.	$T_{initial}$ (K)	T_{final}(K)	CO_2	H_2O	O_2	N_2	CO	H_2	NO
1	300	2329	0.0951	0.1901	0.0000	0.7148	N/A	N/A	N/A
1	500	2474	0.0951	0.1901	0.0000	0.7148	N/A	N/A	N/A
1	700	2628	0.0951	0.1901	0.0000	0.7148	N/A	N/A	N/A
2	300	2248	0.0856	0.1854	0.0062	0.7104	0.0089	0.0036	N/A
2	500	2354	0.0808	0.1831	0.0093	0.7081	0.0133	0.0053	N/A
2	700	2458	0.0749	0.1801	0.0132	0.7054	0.0189	0.0075	N/A
3	300	2241	0.0853	0.1852	0.0053	0.7091	0.0091	0.0037	0.0022
3	500	2344	0.0805	0.1829	0.0079	0.7064	0.0136	0.0054	0.0033
3	700	2445	0.0747	0.1799	0.0110	0.7029	0.0191	0.0077	0.0047

Instead of solving the nonlinear equations, the results shown in the table are obtained using the equilibrium program. The results show the following:

1. As the temperature of the reactants increases, the impact of dissociation on the flame temperature, or T_{final} in the table, becomes more significant.

Example 3.12 (cont.)

2. In all cases, the mole fraction of CO is higher than that of H_2 by more than a factor of 2, although the mole fraction of CO_2 is lower than that of H_2O by more than a factor of ½; that is, more CO_2 dissociates than H_2O.
3. The mole faction of NO is smaller than that of hydrogen, and its formation has a very weak effect on the temperature. The formation of NO is endothermic, which therefore leads to a small reduction of temperature.

In this example, the possible presence of the original fuel in the products, or any further dissociation to the level of C, O, H, OH, etc., is neglected. The reasons for these assumptions are as follows:

1. The equilibrium concentration of methane that participates in these extremely exothermic reactions at the products temperature is extremely small at the temperatures of interest (traces of hydrocarbon can often be found in combustion products due to other mechanisms, such as flame quenching leading to incomplete combustion of the fuel).
2. Below 3000 K, the concentrations of O, H, and OH are too small to impact the energy and mass balance significantly, and are often ignored.
3. At temperatures above 3000 K, the dissociation of H_2 and O_2 can lead to finite but small concentrations of O, H, and OH. They can, however, be important in high-temperature reacting gas dynamic applications, or in plasma dynamics, for example.
4. Equilibrium tends to estimate much higher concentration of NO than found in actual combustion processes. This is because NO formation in actual combustion processes is kinetics limited – a subject that will be discussed later.

The equilibrium concentrations of the products of combustion depend strongly on the mixture and the fuel–air ratio. Rich mixtures produce more CO and H_2.

Figures 3.7 and 3.8 were used to show the dependence of the concentrations of the products species on the fuel–air equivalence ratio for benzene and methane. Note that the maximum flame temperature is reached at fuel concentrations slightly higher than stoichiometric because of dissociation. Note also the presence of carbon monoxide in the products even in fuel-lean conditions. "Radical" species such as O, H, and OH are not shown because they appear at very small concentrations, in the order of magnitude of 100–2000 ppm, with OH being the highest, and have negligible effect on the mass and energy balances of the reaction.

The adiabatic flame temperature for constant-pressure combustion for a stoichiometric mixture of hydrocarbon and standard air, with reactants starting at standard temperature and pressure, range from 2230 K to 2300 K, depending on the fuel and the assumption regarding the composition of the products. Higher values for any fuel correspond to complete combustion without dissociation of products. It is necessary to assume the presence of H_2 and CO, as well as H_2O and CO_2, in order to get more accurate but lower values than obtained under the assumption of complete combustion. Exceptions to this range are acetylene, whose corresponding flame temperature is ~2600 K, and pure hydrogen at ~2450 K.

The adiabatic flame temperature for constant-volume combustion for a stoichiometric mixture of hydrocarbon and standard air, with reactants starting at standard temperature and pressure, range from 2650 K to 2800 K, with acetylene at ~3000 K and hydrogen around 2850 K. As expected, the flame temperature is higher in the constant-volume case (no expansion work transfer), and the pressure rises for all cases to values around 10 atm. The higher temperature of constant-volume combustion is associated with higher concentration of dissociated products, H_2 and CO. The concentrations of both species are nearly doubled in the constant-volume case compared to the constant-pressure case. It should be mentioned that many combustion systems operate under lean conditions – that is, with excess air – in order to reduce the temperature and achieve more complete burning of the fuel – that is, reduce H_2 and CO in the products. Reducing the temperature also reduces unwanted products such as NO. With twice as much air available for combustion, or excess-air ratio of 2 (or equivalence ratio of 0.5), the adiabatic constant-pressure flame temperature of most hydrocarbons is ~1500 K.

3.8 Maximum Work in a Chemical Reaction

An interesting application of reacting mixture energy and entropy analysis is to find the maximum work interaction between a reactive stream starting with reactants at (T_1, p_1) and ending with products at (T_2, p_2), while interacting with the environment at (T_o, p_o)[8,9]. To evaluate the work interaction in a reactive flow process, we combine the statements of the First and Second Laws to eliminate the heat transfer term (this is identical to the analysis done to calculate the maximum work from nonreacting streams in Chapter 2). Assuming steady-state operation for the system shown in Figure 3.10, and per mole of the fuel:

$$\begin{aligned} W = &\sum_{react} n_i\left(\hat{h}_i(T_1) - T_o\hat{s}_i(T_1, p_{i1})\right) \\ &- \sum_{prod} n_i\left(\hat{h}_i(T_2) - T_o\hat{s}_i(T_2, p_{i2})\right) - T_o\Delta S_g. \end{aligned} \tag{3.81}$$

The partial pressures of the components are $p_{i1} = X_i p_1$ and $p_{i2} = X_i p_2$, while the mole fractions of the different components are evaluated for the corresponding streams. The first two terms on the right-hand side of (3.81) represent the difference between the flow availability in the incoming (reactants) stream and the outgoing (products) stream. The last

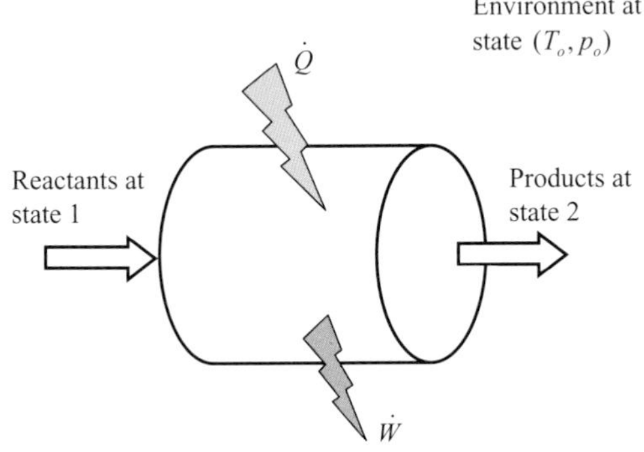

Figure 3.10 Application of the First and Second Laws to an open flow reactive system, used to derive expressions of chemical reaction availability.

term is the entropy generation inside the control volume due to processes other than the chemical reaction. The enthalpies and entropies in this expression are evaluated at the inlet conditions of the reactants stream(s) and the exit conditions of the products stream(s), and include both the thermal and the chemical terms. Note that we may choose to introduce the reactants individually, and separate the products inside the control volume so that the different components leave separately. The enthalpy and entropy include a chemical term and a thermal term.

To maximize the work interactions:

1. entropy generation due to processes inside the control volume, other than the chemical reaction, must be zero; and
2. the products stream must be (at least) in thermal and mechanical equilibrium with the environment (chemical or mass transfer equilibrium will be discussed later).

This maximum work is the **availability** of the process, starting with a fuel–air mixture (or two separate streams of fuel and oxidizer), and ending with a products stream (or products in different streams). Chemical or mass transfer equilibrium with the environment, in which the concentrations in the products stream match the concentrations of the same species in the environment, is not assumed here, and will be discussed in more detail later. (If chemical equilibrium is achieved, we truly have the maximum possible work.) The maximum work, or chemical reaction availability (without chemical equilibrium with the environment) starting at conditions (T_1, p_1) and arbitrary concentrations of fuel and oxidizer, is given by

$$W_{max} = \sum_{react} n_i\left(\hat{h}_i(T_1) - T_o\hat{s}_i(T_1, p_{i1})\right) - \sum_{prod} n_i\left(\hat{h}_i(T_o) - T_o\hat{s}_i(T_o, p_{io})\right). \quad (3.82)$$

In this equation, the pressures at the exit are $p_{io} = X_i p_o$, and the stream of products leaves at thermal and mechanical equilibrium with the environment at (T_o, p_o).

Example 3.13 Maximum Work during Hydrogen Oxidation

One mole of H_2 is burned with ½ mole of O_2 to form one mole of H_2O (g) at 25 °C and 1 atm: $H_2 + \frac{1}{2}O_2 \Rightarrow H_2O$ at standard conditions. Determine the maximum work.

Solution

The reaction balance assumes complete combustion where the only product is water. The maximum work is expressed as

$$W_{max} = \sum_{react} n_i\left(\hat{h}_i(T_1) - T_o\hat{s}_i(T_1, p_{i1})\right) - \sum_{prod} n_i\left(\hat{h}_i(T_o) - T_o\hat{s}_i(T_o, p_{io})\right)$$
$$= n_{H_2}(h_{H_2} - T_o s_{H_2}) + n_{O_2}(h_{O_2} - T_o s_{O_2}) - n_{H_2O}(h_{H_2O} - T_o s_{H_2O}).$$

The first two terms on the left-hand side are zero, since they correspond to the Gibbs functions of two chemical reference species at standard temperature and pressure. Therefore, the maximum work can be simply expressed as

$$W_{max} = -n_{H_2O}\Delta g^{\circ}_{f,H_2O} = -(1) \times (-228.8) = 228.8\,\text{MJ/kmol of } H_2O.$$

3.8.1 Standard Chemical Reaction Availability

Equation (3.82) can be used to define the standard chemical reaction availability, or the maximum work possible from converting the chemical energy in fuel during a chemical reaction while interacting with the environment [8]. In this case, both the reactants and products are in thermomechanical equilibrium with the environment. For the generic reaction in (3.3), the maximum work is obtained as

$$W_{max} = \sum_{react} \nu_i'\left(\hat{h}_i(T_o) - T_o\hat{s}_i(T_o, p_{io})\right) - \sum_{prod} \nu_i''\left(\hat{h}_i(T_o) - T_o\hat{s}_i(T_o, p_{io})\right). \tag{3.83}$$

Assuming that the environment is also at the standard conditions, $T_o = T^o$ and $p_o = p^o$, the resulting expression, after some manipulations, is the **standard chemical reaction availability**, ΔG_R^{AV}:

$$W_{max} = \sum_{react} \nu_i'\left(\hat{h}_i(T^o) - T^o\hat{s}_i(T^o, p_i^o)\right) - \sum_{prod} \nu_i''\left(\hat{h}_i(T^o) - T^o\hat{s}_i(T^o, p_i^o)\right)$$
$$= -\Delta G_R^{AV}, \tag{3.84}$$

where

$$\Delta G_R^{AV} = G_p(T^o, p^o) - G_r(T^o, p^o) \tag{3.85}$$

is the Gibbs free energy of the reaction, and $G(T^o, p^o) = \sum_I \nu_i^{\#} \hat{g}_i(T^o, p_i^o)$, is the Gibbs free energy of the reactants, $\nu_i^{\#} = \nu_i'$ for the reactants and $\nu_i^{\#} = \nu_i''$ for the products, evaluated for each species at the pressure and temperature of the environment, and its concentrations in the mixture.[3]

Table 3.2 shows the standard Gibbs free energy of reaction for the stoichiometric combustion of a number of fuels at 25 °C and 1 atm, where Δg_R^o is presented per kilogram of the fuel. Similar to the enthalpy of reaction, the largest value of the Gibbs free energy of reaction is that of hydrogen. Most hydrocarbon fuels have values in the range 45–50 MJ/kg. For the isothermal reaction, in thermomechanical equilibrium with the environment, the Gibbs free energy of reaction is very close to the enthalpy of reaction, and the entropy generation or loss of available work is a small fraction of enthalpy of reaction.

For an ideal gas mixture, with $p^o = 1$ atm, we write

$$\begin{aligned} G(T^o, p^o) &= \sum_I \nu_i^{\#} \left(\hat{g}_i^{oo} + \Re T^o \ln X_i\right) = \sum_I \nu_i^{\#} \hat{g}_i^{oo} + \Re T^o \ln\left(\prod_N X_i^{\nu_i^{\#}}\right) \\ &= G^{oo} + \Re T^o \ln\left(\prod_N X_i^{\nu_i^{\#}}\right). \end{aligned} \tag{3.86}$$

The molar Gibbs free energy, $\hat{g}_i^{oo} = \hat{g}_i^o(T^o)$, is evaluated at the standard temperature and pressure, and total Gibbs free energy $G^{oo} = \sum_I \nu_i^{\#} \hat{g}_i^{oo}$ is evaluated as if all the species were at the same pressure: 1 atm. In these terms, the Gibbs free energy of reaction is written as follows:

$$\begin{aligned} \left(-\Delta G_R^o\right) &= \left(-\Delta G_R^{oo}\right) + \Re T^o \ln\left(\frac{\prod_{react} X_i^{\nu_i'}}{\prod_{prod} X_i^{\nu_i''}}\right) \\ &= \left(-\Delta G_R^{AV}\right), \end{aligned} \tag{3.87}$$

where

$$\Delta G_R^{oo} = G_P^{oo} - G_R^{oo} \tag{3.88}$$

and $\left(\Delta G_R^o\right)$ was also defined in (3.62). The second term in (3.87) is often much smaller than the first term, so

$$\Delta G_R^o \approx \Delta G_R^{oo}. \tag{3.89}$$

Assuming no entropy generation and equilibrium with the environment implies **reversibility** of the reaction, and hence maximum work (except that the product stream is not in chemical

[3] If the air and fuel enter the reactor separately, they both have pressure 1 atm in this expression, and the partial pressure of oxygen and nitrogen are those in air.

equilibrium with the environment). For the complete combustion of the generic hydrocarbon reaction in (3.76) with oxygen,

$$C_nH_m + \left(n + \frac{m}{4}\right)O_2 \Rightarrow nCO_2 + \frac{m}{2}H_2O_{(v)}, \tag{3.90}$$

the standard Gibbs free energy of the reaction is

$$\Delta G^{oo}_{R,C_nH_m} = n\hat{g}^{oo}_{CO_2} + \frac{m}{2}\hat{g}^{oo}_{H_2O(v)} - \left[\hat{g}^{oo}_{C_nH_m} + \left(n + \frac{m}{4}\right)\hat{g}^{oo}_{O_2}\right]. \tag{3.91}$$

Values of the standard chemical reaction availability of a number of fuels, ΔG^{oo}_R, calculated using (3.91), are shown in Table 3.3 [7]. The Gibbs free energy of the reaction is calculated assuming that water appears in the liquid phase in the exit stream; this is the "higher Gibbs free energy." To get the lower values we subtract the absolute value of Gibbs free energy for evaporation of water, which is approximately 40.7 kJ/mol from the absolute values shown. The Gibbs free energy of a reaction is less than the HHV. These values can be used in calculations of maximum work interaction during a chemical reaction. In calculating the actual Gibbs free energy of a reaction in a particular mixture, this value should be used in (3.87), and the logarithm term should be evaluated for the components in the mixture. Moreover, care must be taken if either stream is not in thermomechanical equilibrium with the environment, in which case the difference in the thermomechanical availability of the stream must be added (see examples for detail). However, as noted previously, the contribution of the logarithmic terms is small.

3.8.2 Availability of Fuels

In the previous section, we introduced the concept of fuel availability as the maximum work interaction between a system undergoing a chemical reaction while experiencing heat transfer with an environment, and maintaining thermal and mechanical equilibrium with that environment, at the standard pressure and temperature: (3.87). Here, we extend the analysis by assuming chemical equilibrium with the environment for all reactants and products, except for the fuel [8–10] (Figure 3.11).

For this purpose, we assume that the concentrations of all the components at the inlet and exit streams are the same as those in the environment, except for the fuel, which enters separately at the temperature and pressure of the environment. Using (3.84), we write the maximum work, or the fuel total availability, as follows:

$$\begin{aligned}\hat{\xi}_{fuel} &= \hat{g}_{fuel}(T^o,p^o) + \sum_{react\neq fuel} \nu_i'\left(\hat{h}_i(T^o) - T^o\hat{s}_i(T^o,p_i^o)\right) \\ &\quad - \sum_{prod} \nu_i''\left(\hat{h}_i(T^o) - T^o\hat{s}_i(T^o,p_i^o)\right) \\ &= \hat{g}_{fuel}(T^o,p^o) + \sum_{react\neq fuel} \nu_i'\mu_i^o - \sum_{prod} \nu_i''\mu_i^o.\end{aligned} \tag{3.92}$$

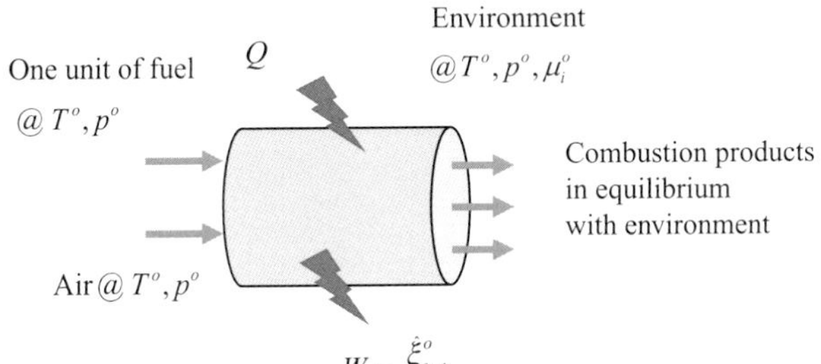

Figure 3.11 A schematic representation to help define fuel availability. A unit of fuel enters and the stoichiometric air enters separately. Products of combustion leave separately, all in equilibrium with the environment.

In this equation, μ_i^o is the chemical potential of the components at their concentrations in the environment. This equation assumes that all components, except the fuel(s), exist in the environment at known concentrations. These include oxygen, nitrogen, water, and carbon dioxide. Because the mixture composition changes during the reaction, using the definition of the chemical potential in (3.55), we obtain the following expression for the fuel chemical availability:

$$\hat{\xi}^o_{fuel} = \left(-\Delta G_R^{oo}\right) + \Re T_o \ln \left(\prod_{react \neq fuel} \left(X_i^o\right)^{\nu_i'} \Big/ \prod_{prod} \left(X_i^o\right)^{\nu''} \right). \tag{3.93}$$

In this expression, X_i^o is the concentration of the species in the environment (assuming that H_2O, CO_2, etc., exist in the environment). We note that the fuel availability in (3.93) is mostly dominated by the first term – that is, the **chemical reaction availability** of the stoichiometric combustion of the fuel at standard conditions (see (3.88)) – and the logarithm terms are relatively very small. The fuel availability evaluated in (3.93) is higher than the chemical reaction availability value evaluated in the previous section since in the calculation of the fuel availability we assume chemical equilibrium with the environment, as well as the thermal and mechanical equilibrium. The extra work results from the expansion of the individual products species from their partial pressures in the products to that of the environment. Typical deviations are O(1%).

If the fuel enters the reaction zone as part of a mixture, with concentration X_f, the value of $\hat{\xi}_f$ in (3.93) should be reduced by $\Re T^o \ln X_f$; that is

$$\hat{\xi}^o_{fuel, X_{fuel}} = \hat{\xi}^o_{fuel} + \Re T^o \ln X_{fuel}. \tag{3.94}$$

For the generic hydrocarbon reaction in (3.90), assuming complete combustion, the total availability of the fuel in (3.93) can be written as

$$\hat{\xi}^o_{C_nH_{\{m\}}} = \left(-\Delta G_R^{oo}\right)_{fuel} + \Re T^o \ln \left[\frac{\left(X^o_{O_2}\right)^{n+m/4}}{\left(X^o_{CO_2}\right)^n \left(X^o_{H_2O}\right)^{m/2}} \right], \tag{3.95}$$

$\hat{\xi}^o_{C_nH_m} > -\Delta G^{oo}_{R, C_nH_m}$, with difference between these two depending on the concentration of the products in the environment; that is, the expansion of the individual combustion products from their concentrations in the products stream to their concentrations in the environment. The difference is, however, rather small. Both values are also lower than the HHV of the fuel.

3.8.3 Entropy Generation and Loss of Availability

Given the preceding discussion on the availability and the maximum work transfer interaction between a system undergoing a chemical reaction and its environment, it is instructive to look at the evaluation of entropy generation and loss of availability associated with a chemical reaction. If the reaction occurs adiabatically at constant pressure p_o in equilibrium with the environment, the entropy generated during the reaction is simply

$$S_g = S_p(T_F, p_o) - S_r(T_o, p_o). \tag{3.96}$$

We assume that the reactants temperature is the same as that of the environment, T_O. The availability loss as a fraction of the maximum available work $\frac{T_o S_{irr}}{(-\Delta g_R^o)}$ for most hydrocarbons combusting adiabatically and stoichiometrically in air at constant pressure is 25–30%, and a few percentage points lower for constant-volume combustion. While the change in the lost work in the isothermal reactions is a small fraction of the Gibbs free energy of the reaction, the corresponding values in the adiabatic reactions are much higher. These results have significant implications for the efficiency of combustion engines, which will be discussed later in this chapter.

3.9 Fuel Reforming Efficiency

In some energy conversion systems it is necessary to break down a complex hydrocarbon into simpler components before further processing, or to convert a solid fuel such as coal or biomass to a gaseous stream, as in gasification to syngas (CO and H_2). A common example for the first scenario is in the production of hydrogen for hydrogen fuel cells when hydrogen is not available and a hydrocarbon is all that is available as a fuel source locally (hydrogen storage and transportation are difficult because it is very light, and very high pressures or very low temperatures are required to store sufficient amounts of it). Hydrogen can be produced using a hydrocarbon as a source centrally where hydrogen is produced in large quantities and shipped for distribution. The production of hydrogen can also be done locally in stationary application or on board in mobile applications. It can also take place internally inside a fuel cell if the temperature is sufficiently high and steam is available.

3.9.1 Methane Reforming for H_2 Production

A common process of hydrogen production from hydrocarbons is **steam reforming** (SR), in which the hydrocarbon is reacted with water to form hydrogen and carbon monoxide. An example of SR is that of methane:

$$CH_4 + H_2O \Rightarrow CO + 3H_2. \tag{3.97}$$

This is an endothermic reaction, with $\Delta H_R^o = 206.4\,\text{kJ/gmol}$ methane. It is also a "reversible" reaction that can move in both directions, depending on the temperature of the mixture. The actual composition of the mixture is a combination of the "reactants" and "products"; that is, CH_4, CO, H_2, H_2O, as well as CO_2. The composition depends strongly on the pressure and temperature, and the composition of the feed gas: methane and water. Being strongly endothermic, increasing the temperature by heating the reactor favors the forward reaction and hydrogen formation. Since the number of moles increases in the forward direction, lowering the pressure also favors hydrogen production. Figure 3.12 plots the equilibrium composition of the methane–steam mixture as a function of temperature. It shows that maximum hydrogen concentration is reached at $T \sim 800\,°\text{C}$ (this is a relatively low temperature and the chemical kinetic rates are likely to be low for fast production, and hence a catalyst is used to speed up the reaction). These results were obtained for reforming one mole of methane with one mole of water (in the liquid phase), at 1 atm, starting at 300 K. Raising the temperature of the mixture requires adding heat externally. Steam reforming is the most widely used process to generate hydrogen on an industrial scale.

The SR reaction becomes more endothermic at higher temperatures, requiring more heat to be added from external sources (such as burning extra methane in oxygen or air in an external combustor or within the same reactor), but the conversion ratio of methane to hydrogen is also better at higher temperatures. At 1100 K, the reaction requires about 10 MJ/kg mixture to produce a mixture with over 70% hydrogen (there is still a trace of methane at this temperature). Note how CO_2 reaches higher concentrations at intermediate temperatures, but dissociates back to lower concentrations as the temperature rises.

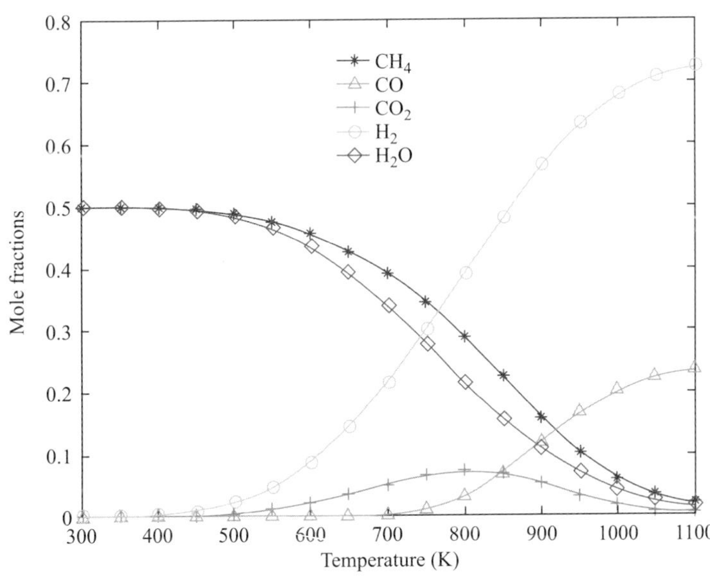

Figure 3.12 The equilibrium composition of a mixture of one mole of methane and one mole of water as a function of temperature, at 1 atm. The original mixture is at 300 K.

Following the methane reforming reaction (or in parallel to it) more water is added to convert CO to CO_2 and produce more hydrogen, according to the **water–gas shift** reaction:

$$H_2O + CO \Rightarrow H_2 + CO_2. \tag{3.98}$$

This is a weakly exothermic reaction with $\Delta H_R^o = -42\,\text{kJ/gmol}$ CO, when water is introduced in the gaseous state (it is endothermic otherwise). It becomes more endothermic at higher temperatures. Similar to the SR reaction, the water–gas shift reaction is reversible and its products composition is strongly temperature-dependent. Figure 3.13 shows the change in the concentration of hydrogen with temperature in a mixture of one mole of CO and one mole of H_2O at a pressure of 1 atm. Water–gas shift favors conversion to hydrogen at lower temperatures, with CO and H_2O concentrations reaching very small values at 300 K. For this reason and to reach high concentrations of hydrogen, it is necessary to perform the shift reaction separately from the SR reaction, after cooling the products of the reforming reaction, as shown schematically in Figure 3.14. Figure 3.13 shows a negligible CO concentration at 300 K. It is often necessary to use two water–gas shift reactors to achieve maximum conversion to hydrogen, where the second reactor runs at an even lower temperature than the first one. Running the gas shift reactions at lower temperatures but at sufficiently high rates requires a metal catalyst.

The combined methane–water reforming and water–gas shift reactions, obtained by adding reactions (3.97) and (3.98), is endothermic. Although the overall reaction can be

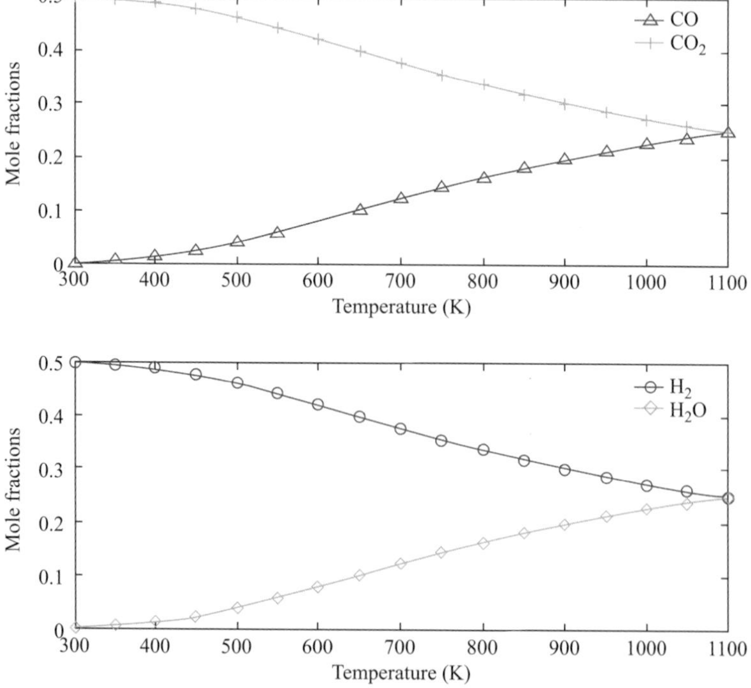

Figure 3.13 The equilibrium composition of one mole of water and one mole of CO as function of temperature, at 1 atm. At 1100 K. The mixture has equal concentrations of CO, CO_2, H_2 and H_2O.

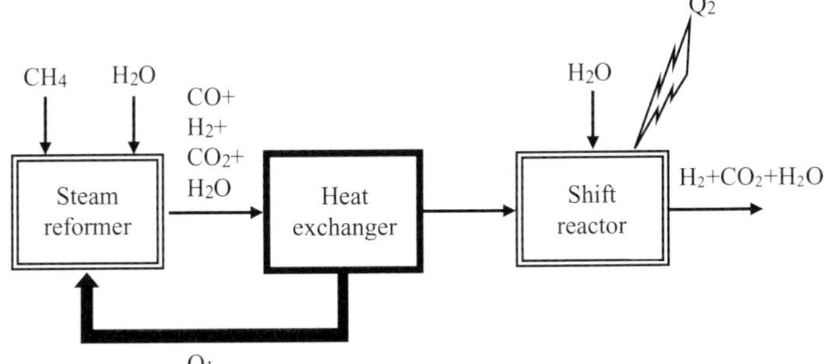

Figure 3.14 Schematic representation of a two-reactor system for production of hydrogen from methane using water reforming. A heat exchanger between the reactors is used to cool the reformer exhaust before it is introduced in the shift reactor. Some of the heat can be used in the reformer itself.

performed in the same reactor, at different sections, it is more effective to perform the two steps separately. Either extra fuel is burned outside the reforming reactor to provide the heat necessary for the endothermic reforming reaction in (3.97) to proceed, or heat is supplied from other parts of the system to the reforming reactor to improve the overall efficiency of the system. For instance, the exit stream of the reformer can be used to heat up the inlet stream. Two moles of water are needed to reform a single mole of methane to hydrogen and CO_2, although more water is often added to ensure the complete reforming of methane and the complete conversion of all its carbon to CO_2. It is important to avoid soot (almost pure carbon molecules) formation during these reactions. Soot can form via the pyrolysis reaction of methane – $CH_4 \rightarrow C + 2H_2$ $\left(\Delta H^o_R = 75\,\mathrm{kJ/mol}\right)$ – or through the Boudouard reaction: $2CO \rightarrow C + CO_2$. Soot can poison the catalyst and renders it ineffective.

To speed up the reaction kinetics, reduce the reactor size and improve the overall efficiency of the process, a nickel-based catalyst is used in both the reforming and the shift reactors. As mentioned above, carbon or soot formation should be avoided since it tends to deposit on catalyst surfaces and poison them by plugging catalytic "sponges."

Because of the conflicting requirements of the methane reforming and water–gas shift reactions with respect to the operating temperature, the two processes are performed separately in separated reactor vessels with a cooling step in between, as shown in Figure 3.15. For maximum conversion to hydrogen, two separate shift reactors may be used, a high-temperature and a low-temperature shift reactor operating at two different temperatures, with an intercooler in between. With proper system integration, one can use the heat removed between the SR reactor and the first water–gas shift reactor to heat up the incoming stream into the SR reactor (in combined cycle applications, this heat can be used to raise production of steam as well), but extra fuel is still required to achieve the working temperature of the SR reactor. It is also possible to use the heat removed between the high- and low-temperature reactors for the same purpose, although system integration becomes too complex and some of this heat is often wasted. Following the second low-temperature shift, hydrogen is separated. Separation energy is required for this step, and this is an extra penalty in the hydrogen generation process.

Figure 3.15 shows a schematic diagram for a process using three reactors, a combustor to heat up the incoming reactors, and an expander to take advantage of the hot stream exiting

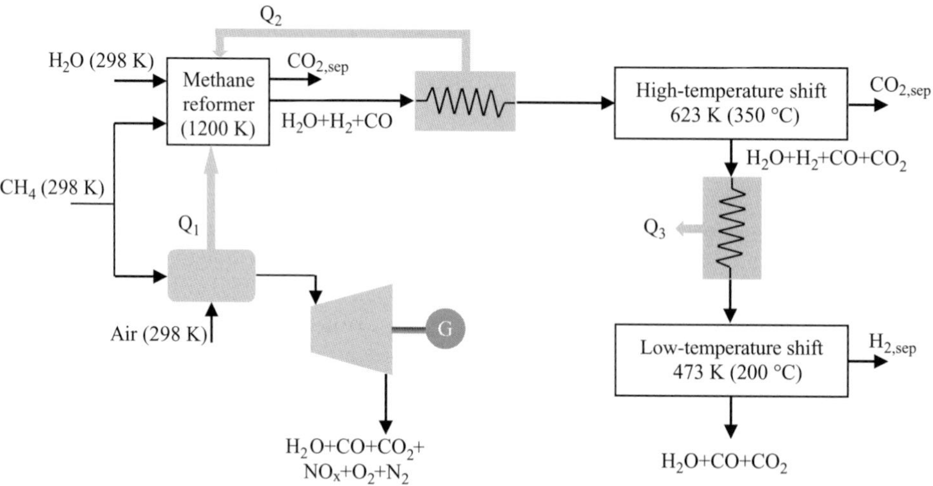

Figure 3.15 A detailed flow diagram for SR of methane followed by two water–gas shift reactors for hydrogen production. A separate combustor is used to generate heat for the steam reformer.

the combustor. Heat integration is represented on this diagram. The diagram shows the possibility of separating CO_2 during the reforming process.

It is possible to steam-reform higher hydrocarbons, whose molecules have numbers of carbon atoms larger than one, using nickel-based catalysts. The products of reforming are mostly methane and syngas ($CO + H_2$). The composition of the products of the pre-reformer and the conditions at which it must operate depend strongly on the fuel being reformed.

Other reforming processes [11] involve burning a fuel-rich mixture of hydrocarbon and oxygen to produce syngas, known as **partial oxidation (POX)**. In the case of methane, the reaction is

$$CH_4 + \frac{1}{2}O_2 \Rightarrow CO + 2H_2. \tag{3.99}$$

This is a slightly exothermic reaction, with $\Delta H_R^o = -24\,\text{kJ/gmol}\;\; CH_4$, and is often carried out at high temperatures in the range 1200–1500 °C. As the reaction is exothermic, it does not need extra heat or a catalyst to proceed at sufficiently fast rates. POX reactions are self-sufficient energy-wise. At lower temperatures, typically less than 1000 K, partial oxidation can be performed on a catalyst, depending on the temperature range: mostly platinum-based for low temperatures or nickel-based for intermediate temperatures. In the case that a catalyst is used, the reaction is known as catalytic partial oxidation (CPO). The reaction utilizes pure oxygen and produces less hydrogen per mole of methane than SR. It can also be used with air as an oxidizer, but this reduces the partial pressure of hydrogen in the products, as well as the overall temperature, and may lead to NO formation. The absence of water in POX is convenient in mobile applications (unless water is condensed from the combustion products and recycled for SR). Heavier hydrocarbons can be partially oxidized as well, but they often require higher temperature, for both catalytic and non-catalytic processes. The formation of CO_2 in partial oxidation can be minimized by keeping the temperature relatively low or by injecting steam to quench the reaction.

When water is added to the partial oxidation reaction, the process is called **autothermal reforming (ATR)**, as shown in the following reaction:

$$CH_4 + \frac{1}{2}O_2 + H_2O \Rightarrow CO_2 + 3H_2.$$

This is a mildly exothermic reaction at lower temperatures, but becomes endothermic at higher temperatures (and depending on the composition of the feed gas). In the case of an endothermic reaction, some heat may be added externally, and some generated internally by adding extra methane and oxygen. As mentioned before, reducing the temperature by employing a catalyst improves the overall efficiency and increases the production of CO_2.

Another example is the reforming of methanol, which has been proposed as a practical fuel for polymer electrolyte membrane (PEM) fuel cells, where moderate-temperature SR can be used to produce hydrogen according to the following reaction:

$$CH_3OH + H_2O \rightarrow 3H_2 + \cdots + CO_2 \qquad \Delta H_R^o = 49.7\,\mathrm{kJ/mol}.$$

Around 250 °C, this reaction can proceed kinetically at a reasonable rate over a catalyst with moderate activity, such as copper supported over zinc oxide. The products of reforming has small fractions of CO, which is desirable for the fuel cell (e.g., at 200 °C around 0.1% CO is found in the products).

In fuel reforming processes, in which we start with a "complex" fuel such as methane, and produce a simpler one such as hydrogen, an appropriate efficiency should be defined to determine the performance of the process. The efficiency may be the chemical energy in a unit mass of hydrogen divided by the chemical energy in the mass of the original fuel used to produce a unit mass of hydrogen. Some of the energy in the original fuel may be used to raise the temperature of the mixture being gasified to a value that promotes shifting the composition of the products gas to the desired values, and some thermal energy may be lost due to heat transfer across the boundaries of the reformer. In some cases, other forms of energy are used besides the original fuel chemical energy, such as thermal energy from other sources or work for compression. Energy may also be required for separating one product gas, say hydrogen, from the products of reforming in order to produce a pure stream of fuel. Furthermore, some of the thermal energy generated in the reformer could be used directly in another process (for instance, producing steam for a local power plant). In this case, the efficiency must account for all the energies in and all the energies out. A First Law efficiency of a reformer can thus be defined as follows:

$$(\eta_{reform})_I = \frac{\text{Energies out}}{\text{Energies in}} = \frac{\text{Chemical} + \text{thermal energy out}}{\text{Chemical} + \text{thermal} + \text{separation energy in}}. \qquad (3.100)$$

Because some of the chemical energy in the original fuel is often converted to thermal energy while raising the temperature inside the reformer, leading to entropy generation, availability loss should be expected in these processes. Thus, a better measure of the reformer performance may be the availability or the "Second Law" efficiency, which is defined as the ratio between the availability in the original fuel stream and the availability in the exiting stream. For both streams, one must account for the chemical and thermal availabilities:

$$(\eta_{reform})_{II} = \frac{\text{Chemical} + \text{thermal availability out}}{\text{Chemical} + \text{thermal availability in}}. \tag{3.101}$$

In practice, the temperature of the stream leaving the reformer is often low, its thermal energy is wasted, and only the chemical energy is utilized. Furthermore, the overall reforming efficiency depends on the level of system integration, how much heat is recovered between the different components, and the separation energy needed to produce pure fuel stream. The practical First Law reformer efficiency of methane to hydrogen is 70–85%.

3.9.2 Coal and Biomass Gasification Availability Efficiency

Coal gasification may play a growing role in energy conversion due to a number of advantages it has over direct coal combustion. While the advantages of and systems used in coal gasification will be described in more detail in the next chapters, it is worth mentioning some here, including: (1) easier and more effective gas clean-up leading to lower overall emissions from the power plant; (2) better potential for carbon dioxide capture and sequestration, and lower efficiency penalty; and (3) potential for higher overall power plant efficiency when used in an integrated gasification combined cycle. Gasification is also used in synthetic fuel production using coal or other solid hydrocarbons. In this case, the synthetic gas (CO and H_2) produced during gasification is combined in a catalytic reactor to form liquid fuels for transportation and other applications. The carbon monoxide of the synthetic gas can be used to produce hydrogen by reacting it with steam to oxidize CO to CO_2 and produce more hydrogen. Carbon monoxide and hydrogen can then be separated chemically or using a membrane to produce a pure stream of hydrogen. Although here we discuss the gasification reactor, similar analysis and overall merits apply to biomass gasification for the production of power or synthetic fuels.

In gasification, pulverized coal or other solid fuels such as biomass, oxygen, and steam react to produce synthetic gas. If the gasifier runs adiabatically, the overall energy balance shows that the total (chemical and thermal) enthalpy of the inlet streams is equal to the total enthalpy of the exit streams. However, there is availability loss. We note here that the temperature inside the gasifier must be maintained relatively high for fast kinetics and to shift the equilibrium composition toward the desired synthetic gas composition. Coal gasification reactions resemble methane reforming reactions. For instance, steam gasification and partial oxidation are given by

$$C + H_2O \rightarrow CO + H_2 \tag{3.102}$$

and

$$C + \frac{1}{2}O_2 \rightarrow CO. \tag{3.103}$$

These reactions can be combined in one gasifier:

$$2C + H_2O + \frac{1}{2}O_2 \rightarrow 2CO + H_2. \tag{3.104}$$

Water–gas shift can then be used to produce more hydrogen and convert CO to CO_2. Example 3.14 shows how to perform availability analysis for coal gasification.

Example 3.14

Biomass is a potentially large source of renewable energy and fuel. Biomass refers to living and recently decomposed biological materials that can be burned for heating, electricity generation, or to produce biofuels. If used in its raw form, it is CO_2-neutral since it is part of the carbon cycle. We will analyze woody biomass in this problem. Its chemical formula can be assumed to be $C_6H_{12}O_6$.

Technology One: Combustion

One way to utilize biomass is to combust it in air. In this configuration, biomass is burned in an adiabatic combustion chamber and the products are used in a steam cycle to produce work. Biomass and air enter the combustion chamber at atmospheric conditions.

1. If the combustion of biomass is stoichiometric, calculate the temperature of products assuming complete combustion.
2. The combustion products are cooled down to 50 °C and 1 atm. The maximum temperature of the steam cycle is 550 °C. Calculate the maximum possible efficiency for the system, and maximum efficiency of the steam cycle (both based on the logarithmic mean temperature).
3. If the Second Law efficiency of the steam cycle is 50%, calculate the work produced by the plant per mole if biomass.
4. What is the overall fuel utilization efficiency of this configuration?

Technology Two: Gasification

Another way to utilize biomass is to reform it via a gasification process. In this process, one mole of biomass and eight moles of H_2O enter the steam reformer at atmospheric conditions to produce a mixture of CO, CO_2, H_2, and H_2O at 500 °C. Next, the mixture goes through a water–gas shift reactor (WGS) to convert CO to CO_2. In the WGS, one mole of H_2O (at 25 °C) is added for every mole of CO originally in the mixture. The mixture leaves the WGS reactor at 100 °C. The resulting mixture is used in the fuel cell to generate work. A schematic is shown in Figure E3.14.

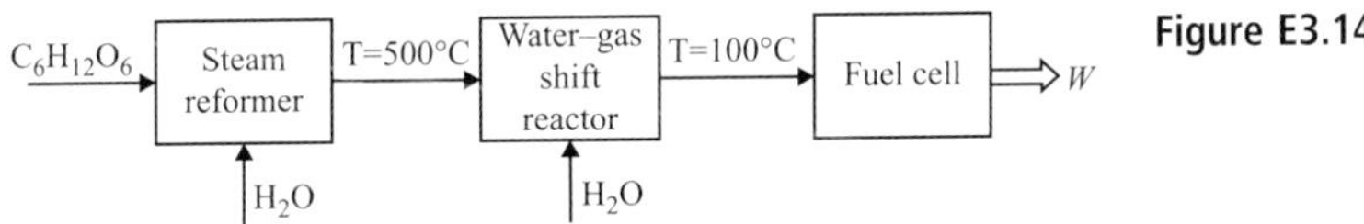

Figure E3.14

Example 3.14 (cont.)

5. What is the composition of the gas leaving the steam reformer if the temperature of the mixture is 500 °C. Assume that this mixture of CO, CO_2, H_2, and H_2O is at equilibrium.
6. What is the heat transfer required in the steam reformer?
7. If CO concentration leaving the WGS is negligible, what is the hydrogen concentration in the gas leaving the WGS reactor?
8. Evaluate the overall process efficiency, including the reforming and WGS process, if no waste heat is recuperated.

The following information might be useful in your calculations:

- the LHV of woody biomass is 21 MJ/kg;
- thermodynamic properties (Table E3.14a); and
- thermodynamic equilibrium constant for WGS reaction (Table E3.14b).

Table E3.14a Thermodynamic properties

Enthalpy of formations	Specific heat
$\hat{h}^o_{f,H_2O(g)} = -242$ kJ/mol	$\hat{c}_{p,O_2} = 33.4$ J/mol·K
$\hat{h}^o_{f,H_2O(l)} = -286$ kJ/mol	$\hat{c}_{p,N_2} = 31.1$ J/mol·K
$\hat{h}^o_{f,CO} = -110.6$ kJ/mol	$\hat{c}_{p,CO_2} = 50.6$ J/mol·K
$\hat{h}^o_{f,CO_2} = -393.8$ kJ/mol	$\hat{c}_{p,H_2} = 30.0$ J/mol·K
$\hat{h}^o_{f,C_6H_{12}O_6} = -1267.1$ kJ/mol	$\hat{c}_{p,CO} = 29.3$ J/mol·K
	$\hat{c}_{p,H_2O} = 38.2$ J/mol·K
	$\hat{c}_{p,C_6H_{12}O_6} = 75.6$ J/mol·K

Table E3.14b Equilibrium constant values for water–gas shift reaction $CO_2 + H_2 \Leftrightarrow CO + H_2O$

$\text{Log}_{10}\, Kp(T)$	T (K)
−19.6	100
−5.018	298
−2.139	500
−0.159	1000
0.135	1200
0.333	1400
0.474	1600
0.577	1800

Example 3.14 (cont.)

Solution

Technology One: Combustion

1. If the combustion of biomass is stoichiometric, then the reaction has the form

$$C_6H_{12}O_6 + 6(O_2 + 3.76N_2) \rightarrow 6CO_2 + 6H_2O + 22.56N_2$$

Assuming complete combustion occurs in an adiabatic chamber, we can write the First Law as follows:

$$n_{C_6H_{12}O_6}\hat{h}_{C_6H_{12}O_6,in} + n_{O_2}\hat{h}_{O_2,in} + n_{N_2}\hat{h}_{N_2,in} = n_{CO_2}\hat{h}_{CO_2,out} + n_{H_2O}\hat{h}_{H_2O,out} + n_{N_2}\hat{h}_{N_2,out}.$$

Biomass and air enter the combustion chamber at atmospheric conditions; therefore:

$$n_{C_6H_{12}O_6}\hat{h}_{C_6H_{12}O_6,in} = n_{CO_2}\hat{h}_{CO_2,out} + n_{H_2O}\hat{h}_{H_2O,out} + n_{N_2}\hat{h}_{N_2,out},$$

where for material i,

$$\hat{h}_i(T) = \hat{h}^o_{f,i} + \int_{T_o}^{T} \hat{c}_{p,i}dT,$$

$$\hat{c}_{p,i} = a_i \Rightarrow \hat{h}_i(T) = \hat{h}^o_{f,i} + a_i(T - T_o).$$

Using this, we can determine the enthalpy values as

$$\begin{aligned}\hat{h}_{C_6H_{12}O_6,in} &= -1267.1\,\text{kJ/mol}\\ \hat{h}_{CO_2,out} &= -393.8 + 50.6(T_{out} - 298)\,\text{kJ/mol}\\ \hat{h}_{H_2O,out} &= -242 + 38.2(T_{out} - 298)\,\text{kJ/mol}\\ \hat{h}_{N_2,out} &= -31.1(T_{out} - 298)\,\text{kJ/mol}.\end{aligned}$$

Hence, per mole of biomass,

$$\begin{aligned}-1267.1 = {} & 6(-393.8 + 50.6(T_{out} - 298)) + 6(-242 + 38.2(T_{out} - 298))\\ & + 22.56(-31.1(T_{out} - 298)).\end{aligned}$$

Solving, we get $T_{out} = 2359.7\,\text{K}$.

2. Now, we know that

$$\eta_{car} = 1 - \frac{\ln\left(\frac{T_F}{T^*}\right)}{\frac{T_F}{T^*} - 1}.$$

Example 3.14 (cont.)

For the steam cycle, $T^* = T_{ambient} = 298\,\text{K}$ and $T_F = T_{turb\_max} = 823\,\text{K}$. In addition, for the overall system, $T^* = T_C = 323\,\text{K}$ and $T_F = T_{products} = 2359.7\,\text{K}$. Substituting, we get the maximum possible efficiencies of the system and steam cycle:

$$\eta_{sys} = 1 - \frac{\ln\left(\frac{2359.7}{323}\right)}{\frac{2359.7}{323} - 1} = 0.684 = 68.4\%,$$

$$\eta_{steam\_cycle} = 1 - \frac{\ln\left(\frac{823}{298}\right)}{\frac{823}{298} - 1} = 0.423 = 42.3\%.$$

3. First, we have to find the amount of heat addition to the steam cycle:

$$Q_H = \sum n_i \hat{h}_{i,out} - \sum n_i \hat{h}_{i,in}$$

$$\Rightarrow Q_H = n_{CO_2}\hat{h}_{CO_2,out} + n_{H_2O}\hat{h}_{H_2O,out} + n_{N_2}\hat{h}_{N_2,out} - \left(n_{CO_2}\hat{h}_{CO_2,in} + n_{H_2O}\hat{h}_{H_2O,in} + n_{N_2}\hat{h}_{N_2,in}\right).$$

Hence, per mole of biomass, we get

$$Q_H = 6(50.6(2359.7 - 323)) + 6(38.2(2359.7 - 323)) + 22.56(31.1(2359.7 - 323))$$
$$\Rightarrow Q_H = 2514\,\text{kJ}.$$

Since we calculated the First Law efficiency of the steam cycle, we have

$$W_{ideal} = \eta_{steam\_cycle} \cdot Q_H = 1063\,\text{kJ}.$$

The Second Law efficiency of the cycle is given as 50%. Therefore:

$$W_{actual} = 0.5 \times W_{ideal} = 531.7\,\text{kJ}.$$

Thus, the actual work produced by the system is 533.5 kJ per mole of biomass.

4. The overall fuel utilization efficiency of the configuration can be calculated (per mole of biomass) as

$$\eta_{conf} = \frac{W_{actual}}{LHV_{C_6H_{12}O_6}} = \frac{531.7}{21{,}000 \times 0.18} = 0.141 = 14.1\%.$$

You can also determine the LHV for $C_6H_{12}O_6$ and use it here.

Technology Two: Gasification

5. The reaction in the steam reformer has the following form:

$$C_6H_{12}O_6 + 8H_2O \rightarrow a\text{CO} + bH_2 + cH_2O + dCO_2.$$

Example 3.14 (cont.)

Mass balance for C, H, and O gives

$$a + d = 6$$
$$b + c = 14$$
$$a + c + 2d = 14.$$

We also have the expression for product equilibrium:

$$CO_2 + H_2 \leftrightarrow CO + H_2O.$$

At 500 °C, K_p of the equilibrium is 0.24 (this can be determined by interpolating the K_p values (not $\log K_p$) using a polynomial expression of order 3 from the range 300–1200 K). Therefore:

$$0.24 = \frac{ac}{bd}.$$

Solving the four equations simultaneously, we have

$$a = 2.169$$
$$b = 9.831$$
$$c = 4.169$$
$$d = 3.831.$$

Therefore, the composition of the gas leaving the steam reformer (in terms of mole fractions) is:

CO: 0.1085
H_2: 0.4915
H_2O: 0.2084
CO_2: 0.1916.

6. Writing the First Law for the steam reformer, we have

$$Q_{reform} = a\hat{h}_{CO,out} + b\hat{h}_{H_2,out} + c\hat{h}_{H_2O,out} + d\hat{h}_{CO_2,out} - \left(\hat{h}_{C_6H_{12}O_6,in} + 8\hat{h}_{H_2O,in}\right),$$

where for material i,

$$\hat{h}_i(T) = \hat{h}^o_{f,i} + \int_{T_o}^{T} \hat{c}_{p,i} dT,$$

$$\hat{c}_{p,i} = a_i \Rightarrow \hat{h}_i(T) = \hat{h}^o_{f,i} + a_i(T - T_o),$$

Example 3.14 (cont.)

and

$$T_{in} = 298\,\mathrm{K}$$
$$T_{out} = 773\,\mathrm{K}.$$

We get the enthalpy values as

$$\hat{h}_{C_6H_{12}O_6,\,in} = -1267.1\,\mathrm{kJ/mol}$$
$$\hat{h}_{H_2O,\,in} = -286\,\mathrm{kJ/mol}$$
$$\hat{h}_{CO,\,out} = -96.7\,\mathrm{kJ/mol}$$
$$\hat{h}_{H_2,\,out} = 14.25\,\mathrm{kJ/mol}$$
$$\hat{h}_{H_2O,\,out} = -223.9\,\mathrm{kJ/mol}$$
$$\hat{h}_{CO_2,\,out} = -369.8\,\mathrm{kJ/mol}.$$

We already know the coefficient values. Therefore, we get

$$Q_{reform} = 1135.3\,\mathrm{kJ}.$$

Heat required for steam reaction is 1135.3 kJ per mole of biomass.

7. H_2O is added to match the moles of CO. We assume that the CO concentration leaving the WGS is negligible. Therefore, the WGS reaction has the form

$$a\mathrm{CO} + b\mathrm{H_2} + (a + c)\mathrm{H_2O} + d\mathrm{CO_2} \rightarrow e\mathrm{H_2} + f\mathrm{CO_2} + g\mathrm{H_2O}.$$

The atom balances are

$$a + d = f$$
$$a + b + c = g + e$$
$$2a + 2d + c = 2f + g.$$

We can solve for the three unknowns to get $e = 12$.

Therefore, the hydrogen concentration leaving the WGS reactor is

$$X_{H_2} = \frac{12}{22.169} = 0.5413 = 54.13\%.$$

8. We can define the overall reformer efficiency as

$$\eta_{reform} = \frac{e \times LHV_{H_2}}{LHV_{C_6H_{12}O_6} + Q_{reform}} = \frac{12 \times 242}{21{,}000 \times 0.18 + 1135.3} = 0.591 = 59.1\%.$$

Example 3.15

Figure E3.15 shows a schematic diagram of a coal gasification plant used to produce hydrogen. The chemical composition of coal used is $CH_{0.8}$. For each mole of coal, 1 mole of water and Y moles of oxygen enters the gasifier. Oxygen is supplied by an air-separation unit (ASU). The reactants are supplied at 298 K and 40 bar and the gasifier operates adiabatically and at constant pressure. Inside the gasifier, coal reacts with water and oxygen to form a mixture of H_2O, H_2, CO, and CO_2 (and other contaminants that we will neglect here for the sake of simplicity). The gas exits the gasifier at 1200 °C and 40 bar. Next, the gas enters the high-temperature shift reactor (HTSR) where ½ mole of water is added to raise the hydrogen concentration in the mixture. Water is added in the liquid form at 298 K and 40 bar. The new mixture leaves the HTSR at 600 °C, entering a separation unit to remove all the CO_2 in the mixture at this stage. The mixture of H_2O, H_2, and CO enters a low-temperature shift reactor (LTSR) where an extra ½ mole of water is added. Water is added in the liquid form at 298 K and 40 bar. The gas exits the LTSR at 200 °C. The newly formed CO_2 is again removed from the mixture.

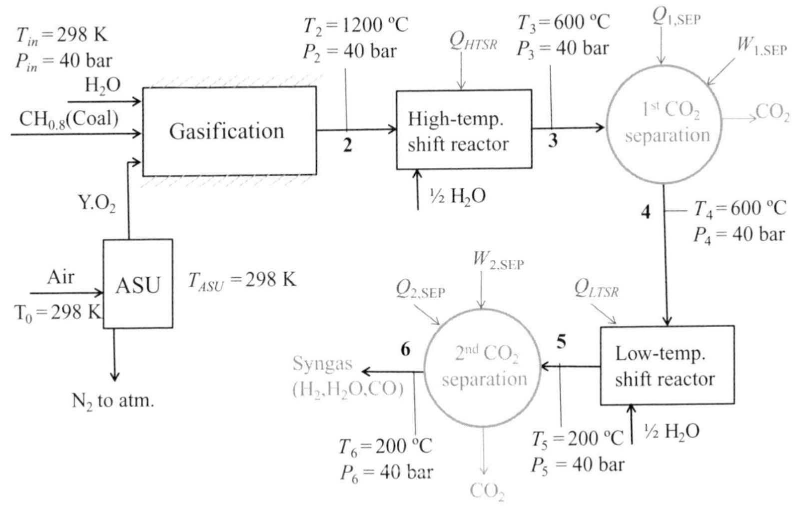

Figure E3.15 Schematic of coal gasification plant for hydrogen production.

1. Calculate the enthalpy of formation of the coal. The LHV of this coal is 25 MJ/kg.
2. If the gas mixture existing the adiabatic gasifier is at equilibrium and 1200 °C, calculate the number of moles of oxygen per mole of coal needed in the gasification process, and the mixture composition leaving the gasifier.
3. Calculate the mixture composition leaving the HTSR and the heat transfer from the reactor. Assume that the mixture is H_2O, H_2, CO, and CO_2, and at equilibrium.

Example 3.15 (cont.)

4. Calculate the mixture composition leaving the LTSR and the heat transfer from the reactor. Assume that the mixture is H_2O, H_2, CO, and CO_2, and at equilibrium.
5. Calculate the cold gas efficiency of this process. The cold gas efficiency is defined as the ratio of chemical energy of the hydrogen as a fraction of the chemical energy used in the gasifier, both evaluated at STP.
6. Calculate the work required to separate CO_2 from the two CO_2 separators. Assume that the Second Law efficiency is 25% in both cases.
7. Calculate the heat transfer from the two CO_2 separators.

Use the following data and Table E3.14a:

$$\hat{c}_{p,\mathrm{H_2O}} = 43.24\,\mathrm{J/mol{\cdot}K}$$
$$\hat{c}_{p,\mathrm{H_2}} = 30.56\,\mathrm{J/mol{\cdot}K}$$
$$\hat{c}_{p,\mathrm{CO}} = 50.8\,\mathrm{J/mol{\cdot}K}$$
$$\hat{c}_{p,\mathrm{CO_2}} = 57.6\,\mathrm{J/mol{\cdot}K}$$
$$\hat{c}_{p,\mathrm{O_2}} = 30.56\,\mathrm{J/mol{\cdot}K}.$$

Enthalpy of formation:

$$\hat{h}^0_{f,\mathrm{H_2O(g)}} = -242{,}000\,\mathrm{J/mol}$$
$$\hat{h}^0_{f,\mathrm{H_2O(l)}} - 286{,}030\,\mathrm{J/mol}$$
$$\hat{h}^0_{f,\mathrm{CO}} = -110{,}600\,\mathrm{J/mol}$$
$$\hat{h}^0_{f,\mathrm{CO2}} = -393{,}800\,\mathrm{J/mol}.$$

Entropy of formation:

$$\hat{s}^0_{f,\mathrm{H_2O}}(g) = -44.4\,\mathrm{J/mol{\cdot}K}$$
$$\hat{s}^0_{f,\mathrm{H_2O}}(l) = -163.3\,\mathrm{J/mol{\cdot}K}.$$

Table E3.15 Equilibrium constant values for water–gas shift reaction

$\mathrm{Log}_{10}\, K_p(T)$ for $CO_2 + H_2 \Leftrightarrow CO + H_2O$	T (K)
−19.6	100
−5.018	298
−2.139	500
−0.159	1000
0.135	1200
0.333	1400
0.474	1600
0.577	1800

Example 3.15 (cont.)

$$\hat{s}^0_{f,\mathrm{CO}} = 89.55\,\mathrm{J/mol{\cdot}K}$$
$$\hat{s}^0_{f,\mathrm{CO_2}} = 3.02\,\mathrm{J/mol{\cdot}K}$$
$$\hat{s}^0_{f,\mathrm{H_2}} = 130.6\,\mathrm{J/mol{\cdot}K}$$
$$\hat{s}^0_{f,\mathrm{O_2}} = 205.0\,\mathrm{J/mol{\cdot}K}.$$

Solution

1. The LHV of coal is 25 MJ/kg. We can consider complete combustion of coal with oxygen or air at 298 K and consider the products also leave at 298 K. The heat of reaction in this case is equal to the LHV of the coal. Because the enthalpy of reaction is equal to the LHV of the fuel, the water appears in the form of gas in the product side. We will consider coal reacting with oxygen at 298 K. The coal reaction is

$$\mathrm{CH_{0.8}} + a\mathrm{O_2} \rightarrow x\mathrm{CO_2} + y\mathrm{H_2O(g)}$$

H balance: $0.8 = 2y \Rightarrow y = 0.4$
C balance: $1 = x \Rightarrow x = 1$
O balance: $2a = 2x + y \Rightarrow a = 1.2$
the coal–oxygen combustion reaction is

$$\mathrm{CH_{0.8}} + 1.2\,\mathrm{O_2} \rightarrow \mathrm{CO_2} + 0.4\,\mathrm{H_2O(g)}$$

the enthalpy of reaction is

$$\Delta H_R = H_R - H_P = h^0_{f,\mathrm{CH0.8}} - \left(h^0_{f,\mathrm{CO2}} + 0.4 \times h^0_{f,\mathrm{H_2O}}\right) = -LHV_{\mathrm{CH0.8}} \text{ on per mol basis}$$

Note: You use negative sign because it is an exothermic reaction.

$$LHV_{\mathrm{CH0.8}}\text{per kg basis} = 25 \times 10^6\,[\mathrm{J/kg}]$$
$$\begin{aligned} LHV_{\mathrm{CH0.8}}\text{ per mol basis} &= 25 \times 10^6 \times (\text{mol wt of CH0.8})/1000\,[\mathrm{J/mol}] \\ &= 25 \times 10^6 \times 12.8/1000\,[\mathrm{J/mol}] \\ &= 320{,}000\,[\mathrm{J/mol}] \end{aligned}$$
$$\begin{aligned} h^0_{f,\mathrm{CH0.8}} &= \left(h^0_{f,\mathrm{CO2}} + 0.4 \times h^0_{f,\mathrm{H_2O}}\right) + LHV_{\mathrm{CH0.8}} \\ &= -393{,}800 - 0.4 \times 242{,}000 + 320{,}000 \\ &= -170{,}600\,[\mathrm{J/mol}]. \end{aligned}$$

Example 3.15 (cont.)

2. The mixture leaving the gasifier has composition of CO_2, CO, H_2, and H_2O. We also do not know how much oxygen is needed to produce the mixture leaving gasifier at $T = 1200\,°C$.

   Hence, we need five equations for five unknowns (the concentrations of CO_2, CO, H_2, H_2O leaving the gasifier and O_2 entering the gasifier). The coal gasification reaction is $CH_{0.8} + H_2O + yO_2 \rightarrow aH_2O + bH_2 + cCO + dCO_2$

   H balance: $0.8 + 2 = 2a + 2b$
   C balance: $1 = c + d$
   O balance: $1 + 2y = a + c + 2d$.

   We make use of the equilibrium reaction for the product species. Interpolating the value of $\log_{10}(K_p)$ from the table provided at 1200 °C (1473 K), we get $K_p = 2.441$

$$K_p = 2.441 = (a \times c)/(b \times d).$$

   The fifth equation is using the First Law for the adiabatic gasifier, which gives

$$H_R = H_P$$

$$h^0_{f,CH_{0.8}} + h^0_{f,H_2O(l)} + y h^0_{f,O_2} = a\left[h^0_{f,H_2O(g)} + c_{p,H_2O}(T - T_0)\right] + b\left[h^0_{f,H_2} + c_{p,H_2}(T - T_0)\right]$$
$$+ c\left[h^0_{f,CO} + c_{p,CO}(T - T_0)\right] + c\left[h^0_{f,CO_2} + c_{p,CO_2}(T - T_0)\right]$$

   where $T = 1473$ K and $T_0 = 298$ K. Using enthalpies for formation and specific heat of gases given, we get our fifth equation. Solving the above five equations for a, b, c, d, and y, we get

   $a = 1.178$
   $b = 0.2221$
   $c = 0.3151$
   $d = 0.6849$
   $y = 0.\ 9143$.

3. Mixture composition leaving the HTSR:

$$T_2 = 1200 + 273 = 1473\,\text{K}$$
$$T_3 = 600 + 273 = 873\,\text{K}$$
$$a_2 = a; b_2 = b; c_2 = c; d_2 = d.$$

   The products leave the HTSR at 873 K and are in equilibrium. The HTSR reaction is

$$a_2H_2O + b_2H_2 + c_2CO + d_2CO_2 + \tfrac{1}{2}H_2O \rightarrow a_3H_2O + b_3H_2 + c_3CO + d_3CO_2$$

   H balance: $2a_2 + 2b_2 + \tfrac{1}{2} \times 2 = 2a_3 + 2b_3$
   C balance: $c_2 + d_2 = 2c_3 + 2d_3$
   O balance: $a_2 + c_2 + 2d_2 + \tfrac{1}{2} = a_3 + c_3 + 2d_3$.

Example 3.15 (cont.)

Equilibrium reaction:
At 873 K, $\log 10(K_p) = -0.6619$, $K_p = 0.2178$. The fourth equation is

$$K_p = 0.2178 = (a_3 \times c_3)/(b_3 \times d_3).$$

Solving the above four equations, we get the following mixture composition:

$$\begin{aligned} a_3 &= 1.43 \\ b_3 &= 0.4703 \\ c_3 &= 0.06687 \\ d_3 &= 0.9331. \end{aligned}$$

The heat transfer from the reactor can be found by applying the First Law to the HTSR:

$$\begin{aligned} Q_{HTSR} &= H_P - H_R \\ &= \sum_P n_i \hat{h}_i - \sum_R n_i \hat{h}_i \\ &= a_2\left(\hat{h}^0_{f,\mathrm{H_2O(g)}} + \hat{c}_{p,\mathrm{H_2O}}(T_3 - T_0)\right) + b_2\left(\hat{h}^0_{f,\mathrm{H_2}} + \hat{c}_{p,\mathrm{H_2}}(T_3 - T_0)\right) \\ &\quad + c_2\left(\hat{h}^0_{f,\mathrm{CO}} + \hat{c}_{p,\mathrm{CO}}(T_3 - T_0)\right) + d_2\left(\hat{h}^0_{f,\mathrm{CO_2}} + \hat{c}_{p,\mathrm{CO_2}}(T_3 - T_0)\right) \\ &\quad - \left\{ \begin{array}{l} a\left(\hat{h}^0_{f,\mathrm{H_2O(g)}} + \hat{c}_{p,\mathrm{H_2O}}(T_2 - T_0)\right) + b\left(\hat{h}^0_{f,\mathrm{H_2}} + \hat{c}_{p,\mathrm{H_2}}(T_2 - T_0)\right) \\ + c\left(\hat{h}^0_{f,\mathrm{CO}} + \hat{c}_{p,\mathrm{CO}}(T_2 - T_0)\right) \\ + d\left(\hat{h}^0_{f,\mathrm{CO_2}} + \hat{c}_{p,\mathrm{CO_2}}(T_2 - T_0)\right) + \frac{1}{2}\left(\hat{h}^0_{f,\mathrm{H_2O(l)}}\right) \end{array} \right\}. \end{aligned}$$

Solving the above equations gives
$Q_{HTSR} = -44{,}525$ J/mol-syngas; 44.52 KJ/mol-syngas of heat is lost from the HTSR.

4. Mixture composition leaving the LTSR:

$$T_4 = T_3 = 600 + 273 = 873\,\mathrm{K}$$

$$T_5 = 300 + 273 = 473\,\mathrm{K}$$

$$a_4 = a_3;\, b_4 = b_3;\, c_4 = c_3.$$

The CO_2 has already been removed in the previous step.
The products leave the LTSR at 473 K and are in equilibrium. The HTSR reaction is

$$a_4\mathrm{H_2O} + b_4\mathrm{H_2} + c_4\mathrm{CO} + \tfrac{1}{2}\mathrm{H_2O} \rightarrow a_5\mathrm{H_2O} + b_5\mathrm{H_2} + c_5\mathrm{CO} + d_5\mathrm{CO_2}$$

H balance: $2a_4 + 2b_4 + \tfrac{1}{2} \times 2 = 2a_5 + 2b_5$
C balance: $c_4 = 2c_5 + 2d_5$
O balance: $a_4 + c_4 + 2d_4 + \tfrac{1}{2} = a_5 + c_5 + 2d_5$.

Example 3.15 (cont.)

Equilibrium reaction:
At 673 K, $\log10(K_p) = -2.524$, $K_p = 0.002992$. The fourth equation is

$$K_p = 0.002992 = (a_5 \times c_5)/(b_5 \times d_5).$$

Solving the above four equations, we get the following mixture composition:

$$\begin{aligned} a_5 &= 1.863 \\ b_5 &= 0.5372 \\ c_5 &= 5.765 \times 10^{-5} \\ d_5 &= 0.06681. \end{aligned}$$

The heat transfer from the reactor can be found by applying the First Law to the LTSR:

$$\begin{aligned} Q_{LTSR} &= H_P - H_R \\ &= \sum_P n_i \hat{h}_i - \sum_R n_i \hat{h}_i \\ &= a_5\left(\hat{h}^0_{f,H_2O(g)} + \hat{c}_{p,H_2O}(T_5 - T_0)\right) + b_5\left(\hat{h}^0_{f,H_2} + \hat{c}_{p,H_2}(T_5 - T_0)\right) \\ &\quad + c_5\left(\hat{h}^0_{f,CO} + \hat{c}_{p,CO}(T_5 - T_0)\right) + d_5\left(\hat{h}^0_{f,CO_2} + \hat{c}_{p,CO_2}(T_5 - T_0)\right) \\ &\quad - \left\{ \begin{array}{l} a_4\left(\hat{h}^0_{f,H_2O(g)} + \hat{c}_{p,H_2O}(T_4 - T_0)\right) + b_4\left(\hat{h}^0_{f,H_2} + \hat{c}_{p,H_2}(T_4 - T_0)\right) \\ + c_4\left(\hat{h}^0_{f,CO} + \hat{c}_{p,CO}(T_4 - T_0)\right) + \frac{1}{2}\left(\hat{h}^0_{f,H_2O(l)}\right) \end{array} \right\}. \end{aligned}$$

Solving the above equation gives
$Q_{LTSR} = -8858$ J/mol-syngas; 8.58 KJ/mol-syngas of heat is lost from the LTSR.

Note: the water coming out of the LTSR at 200 °C and 40 bar is liquid and hence a better estimate is obtained using the enthalpy of formation of liquid water in the products side. Using $\hat{h}^0_{f,H_2O(l)}$ in the products side will give $Q_{LTSR} = -90.879$ KJ/mol-syngas.

5. Cold gas efficiency:
$LHV_{H2} = 120$ MJ/kg. We need to find out LHV_{H2} on a per mol basis:

$$LHV_{H_2} = 120 \times (\text{mol.wt.}H_2)/1000 = 120 \times 2/1000 = 0.24 \text{ MJ/mol} = 2.4 \times 10^5 \text{ J/mol}.$$

Similarly,

$$\begin{aligned} \text{LHV}_{\text{CH}0.8} &= 25 \times (\text{Mol.Wt.CH}_{0.8})/1000 = 25 \times 12.8/1000 = 0.32\,\text{MJ/mol} \\ &= 3.2 \times 10^5\,\text{J/mol}. \end{aligned}$$

$$\eta_{cold} = \frac{b_5 \times LHV_{H_2}}{LHV_{CH_{0.8}}} = \frac{0.5372 \times 2.4 \times 10^5}{3.2 \times 10^5} = 0.4029 = 40.29\%.$$

Example 3.15 (cont.)

6. Work required to separate CO_2 from the mixture in the first CO_2 separator:

$$T_3 = 873\,\text{K}$$
$$X_{CO_2} = d_3/(a_3 + b_3 + c_3 + d_3) = 0.9331/2.9 = 0.3218$$
$$X_{REST} = 1 - X_{CO_2} = 0.6782.$$

The minimum work required to separate the using Second Law is

$$\begin{aligned} W_{min} &= RT[X_{CO_2}\ln(X_{CO_2}) + X_{Rest}\ln(X_{Rest})] \\ &= 8.314 \times 873[0.3218 \times \ln(0.3218) + 0.6782 \times \ln(0.6782)] \\ &= -4560\,\text{J/mol of mixture.} \end{aligned}$$

For every mol of coal, 2.9 moles of mixture enter the HTSR. Thus, the minimum work required to separate CO_2 for every mole of coal is

$$W_{min,total} = -4560 \times n_{total} = -13{,}223\,\text{J/mol of coal.}$$

Using the Second Law efficiency of 0.25, work required to separate CO_2 is

$$W_{1,SEP} = W_{min,total}/\eta_{II} = -13{,}223/0.25 = -52{,}890\,\text{J/mol of coal.}$$

The work required to separate CO_2 from the mixture in the second CO_2 separator:

$$T_5 = 473\,\text{K}$$
$$X_{2,CO_2} = d_5/(a_5 + b_5 + c_5 + d_5) = 0.06681/2.467 = 0.0271$$
$$X_{2,REST} = 1 - X_{2,CO_2} = 0.9729.$$

The minimum work required to separate using the Second Law is

$$\begin{aligned} W_{2,min} &= RT_5[X_{2,CO_2}\ln(X_{2,CO_2}) + X_{2,Rest}\ln(X_{2,Rest})] \\ &= 8.314 \times 473[0.0271 \times \ln(0.0271) + 0.9721 \times \ln(0.9721)] \\ &= -489.6\,\text{J/mol of mixture.} \end{aligned}$$

For every mol of coal, 2.467 moles of mixture enter the HTSR. Thus, the minimum work required to separate CO_2 for every mole of coal is

$$W_{2,min,total} = -489.6 \times n_{total} = -489.6 \times 2.467 = -1208\,\text{J/mol of coal.}$$

Using the Second Law efficiency of 0.25, the work required to separate CO_2 is

$$W_{2,SEP} = W_{2,min,total}/\eta_{II} = -1208/0.25 = -4832\,\text{J/mol of coal.}$$

7. Heat transfer from the first CO_2 separator. Using the energy balance for the first CO_2 separator, we have

$$\Delta E_{1,SEP} = 0 = Q_{1,SEP} - W_{1,SEP} + \sum_{in} m_i h_i - \sum_{out} m_i h_i.$$

Example 3.15 (cont.)

Because the species leaving the first CO_2 separator have the same temperature and pressure conditions as the species entering the CO_2 separator, the inlet and outlet flow enthalpies cancel each other. Therefore, the heat transfer across the separator is equal to the work required to separate CO_2 from the mixture. Hence,

$$Q_{1.SEP} = W_{1.SEP} = -52{,}890\,\mathrm{J/mol\ of\ coal\ or} - 5289\,\mathrm{KJ/mol\ of\ coal.}$$

Similarly,

$$Q_{2.SEP} = W_{2.SEP} = -4832\,\mathrm{J/mol\ of\ coal\ or} - 5289\,\mathrm{KJ/mol\ of\ coal.}$$

3.9.3 Chemical Reaction Second Law Efficiency

We learned that $\Delta G_R^{AV} = \Delta H_R^o - T^o \Delta S_R^o$, where $\Delta S_R^o = S_p^o - S_r^o$ is the entropy of reaction. At standard conditions, the entropy of reaction may be positive or negative. For most fuels of interest; $|T^o \Delta S_R^o| \ll |\Delta H_R^o|$, and using the standard Gibbs free energy of reaction,

$$\Delta G_R^{\circ\circ} \approx \Delta H_{R,LHV}^{\circ}, \tag{3.105}$$

where $\Delta H_{R,LHV}^{\circ}$ is the LHV. This conclusion is supported by the numerical values in Table 3.3. For cases in which $|\Delta G_R^{av}| < |\Delta H_R^o|$, one defines an efficiency that describes the best-case scenario of converting chemical energy to work as follows:

$$\eta_{chem.eng} = \frac{\Delta G_R^{\circ}}{\Delta H_R^{\circ}}. \tag{3.106}$$

For example, the "efficiency" of the hydrogen–oxygen reaction is 95% with water exiting in the liquid phase, and 83% with water exiting in the vapor phase. Extracting this maximum work requires an ideal conversion device in which the chemical energy is converted to work during the reaction. *It should be emphasized that extracting the maximum work requires the system to undergo a series of quasi-equilibrium processes from the "reactants" to the "products," during which it remains at the same temperature and pressure as the environment, while transferring heat and work to that environment*. Only a special device can do that, running perhaps slowly to keep that quasi-equilibrium all the time. An electrochemical device such as a fuel cell running at very low current may approach this ideal limit.

An indirect conversion of the chemical energy to mechanical work is done by first converting the chemical energy to thermal energy. A heat engine can then be used to get work from the thermal energy. We call this device a **combustion engine**. In both cases of direct conversion and indirect conversion, we write

$$
\begin{aligned}
W_{comb.eng.} &= W_{max} - T^o S_g \\
&= \left(-\Delta G_R^o\right) - T^o S_g.
\end{aligned} \tag{3.107}
$$

The entropy generation is due to the chemical reaction in the combustion process operating away from equilibrium with the environment, and other irreversibilities in the energy conversion systems, such as heat transfer across temperature gradients, work producing/consuming component irreversibility, friction, etc. Examples will be discussed later.

3.10 Indirect Conversion, Combustion Engines

In an indirect chemical to mechanical/electrical energy conversion process (Figure 3.16), work is extracted from a chemical reaction by first converting the chemical energy to thermal energy in a combustion device, then running a heat "engine" in which the thermal energy is utilized to enable work transfer (for electrical work, a generator is then used, but we assume 100% conversion efficiency for the mechanical to electrical energy conversion). If a stoichiometric mixture is used, and the environment is at the standard conditions, the maximum work equals the fuel chemical availability or Gibbs free energy of reaction, depending on whether chemical equilibrium with the environment is assumed, as shown in (3.87) and (3.92), and given numerically in Table 3.3. The maximum work can be extracted if the reaction occurs isothermally and in equilibrium with the environment. However, combustion does not transfer work to the environment directly, and a cyclical heat engine must be incorporated for this purpose. Moreover, for the latter to operate efficiently, it must operate between a high-temperature source, provided by the combustion process in the case that a fuel is used, and a low temperature – that is, the environment. But, as we saw already, adiabatic combustion generates entropy and hence lowers the availability, as it raises the temperature of the combustion products.

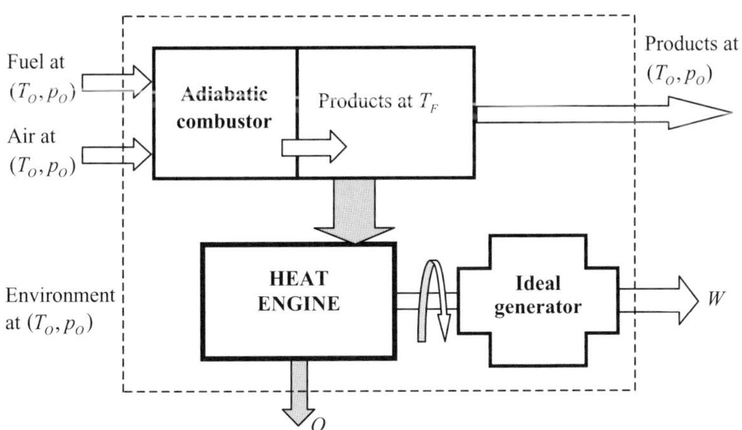

Figure 3.16 A model for an indirect chemical energy conversion system. A heat engine uses combustion products internally in a power cycle.

To calculate the work transfer interaction and the efficiency of a system that uses a combustion process to provide heat to a power cycle, we must know the cycle processes, and how the thermal energy is utilized. This will be done in more detail in later chapters.

3.10.1 Best Efficiency of Combustion Engines

Combustion engines operate on open cycles, in which the working fluid is a mixture of a fuel and an oxidizer that undergoes a reaction converting its chemical energy into thermal energy and providing the high temperature and pressure necessary to operate the engine. The power cycles themselves may use the combustion products as a working fluid or another working fluid, such as is the case in steam cycles. The power cycle may use a rotating machine, such as a gas turbine, or a reciprocating machine, such as an internal combustion engine. The maximum allowable temperature in rotating engines is lower since the rotating machines are continuously exposed to the hot gases. In reciprocating machines, one can tolerate much higher combustion temperatures since hot combustion gases are only present for a small fraction of the cycle, and it is often possible to cool the walls of the cylinder as well. Power cycles that use combustion gases to heat the working fluid, similar to those used in steam power plants, may suffer from another source of irreversibility – that is, the heat transfer between the combustion gases and the working fluid. On the other hand, gas turbine cycles that use the products of combustion directly can suffer from limitations imposed by the properties of the working fluid, as will be shown in detail later. In modeling combustion engines using thermodynamic cycles, the adiabatic flame temperature corresponding to the reactants' temperature and composition prior to burning can be regarded as the maximum cycle temperature.

Before examples of practical combustion engine cycles are analyzed in detail in the following chapters, here, we give some estimates for the theoretical maximum efficiency of systems utilizing a combustion process and a heat engine for the conversion of chemical energy to thermal energy to mechanical energy – that is, indirect conversion of chemical energy to mechanical energy. The models we develop here are merely illustrative; they are attractive because of their simplicity. However, they may not reflect the actual performance of a power cycle, since the efficiency of a power cycle depends strongly on the working fluid and the cycle detail.

As shown in Chapter 2, a heat engine may interact with a hot stream and the environment to produce work. Combustion products can be used to provide the high-temperature stream; the initial temperature of the products is the flame temperature, T_F. The temperature of the environment is designated here as T_o. As the gases cool down to T_o, using an infinite series of Carnot cycles, the thermal efficiency of this heat engine is

$$\eta^*_{Carnot} = \frac{W_{net}}{Q_{in}} = 1 - \frac{\ln \dfrac{T_F}{T_o}}{\left(\dfrac{T_F}{T_o} - 1\right)}. \tag{3.108}$$

In this equation, Q_{in} is the heat input to the heat engine, which is generated in a combustion process. Per unit mass of fuel, the heat input to the engine is $Q_{in} = |\Delta h_R|$. This model shows that achieving a high combustion temperature is necessary for achieving high overall efficiency. Values of this efficiency can be estimated using typical values for the flame temperature. For instance, for $T_o = 300\,\mathrm{K}$, and $T_F = 2400\,\mathrm{K}$, the efficiency is 70%. Modern natural gas combined cycle power plants, which will be discussed later, have $T_F \approx 1700\,\mathrm{K}$, for which $\eta^*_{Carnot} \approx 64\%$. In practice, these plants reach efficiencies close to 58%.

Now we use an availability-based model to show the impact of combustion-generated irreversibility on the efficiency of a combustion engine. As shown before, if all internal processes are reversible, the maximum work is the change in the availability of the reactive mixture across the control volume:

$$\begin{aligned} W_{max.chem.eng} &= (H_1 - T_o S_1) - (H_2 - T_o S_2) \\ &= \Delta G^\circ_R. \end{aligned} \tag{3.109}$$

We use $W_{max.chem.eng}$ to indicate the maximum work generated in a chemical engine operating reversibly (i.e., without internal entropy generation). On the other hand, if the engine uses an adiabatic combustion process as a heat source, we must subtract the work lost because of the entropy generation associated with the adiabatic chemical reaction, expressed before as loss of availability or entropy generation associated with adiabatic combustion (other processes assumed to be ideal):

$$\begin{aligned} W_{max.comb.eng.} &= \Delta G^o_R - T_o \Delta S_{ad.comb} \\ &= (H_1 - T_o S_1) - (H_2 - T_o S_2) - T_o S_g \\ &= (H_1 - T_o S_1) - (H_2 - T_o S_2) - T_o (S_F - S_1) \\ &= (H_F - T_o S_F) - (H_2 - T_o S_2). \end{aligned} \tag{3.110}$$

For maximum work, the exit stream must be at equilibrium with the environment, $T_2 = T_o$ and $p_2 = p_o$. Assuming constant specific heat and ideal gases, $H_F - H_o = \Delta H_R = \bar{C}(T_F - T_o)$, the maximum combustion engine efficiency is

$$(\eta_{best.comb}) \approx \frac{\left[\bar{C}(T_F - T_o) - T_o (S_F - S_o)\right]}{\left[\bar{C}(T_F - T_o)\right]}. \tag{3.111}$$

If we assume constant specific heat and ideal gases, $S_F - S_o = \bar{C} T_o \ell n(T_F/T_o)$. Thus, when using an ideal heat engine operating between the flame temperature and the environment to convert the fuel chemical energy into work, the source of loss is associated with the adiabatic combustion process. In actual cycles, the heat engine efficiency is lower than that predicted by (3.111) due to other forms of irreversibility, further reducing the work produced by the combustion engine below that predicted by (3.108) or (3.111).

Another result can be obtained using availability analysis. One can define a Second Law efficiency of a combustion engine as follows:

$$\eta_{comb.II} = \frac{W_{max.comb.eng}}{\Delta G_R^o} = 1 - \left|\frac{T_o \Delta S_{ad.comb}}{\Delta G_R^o}\right|, \tag{3.112}$$

where the second term is the loss of availability, or irreversibility due to adiabatic combustion. Values of the irreversibilities generated during adiabatic stoichiometric combustion of different fuels burning in air, respectively, normalized by the Gibbs free energy of reaction for the fuel were shown to be 26–28% for constant-pressure processes and 22–26% for constant-volume processes. Constant-volume combustion shows smaller irreversibility than constant-pressure combustion, and hydrogen has the least irreversibility of all the fuels shown. On average, there is a loss of available work of almost 25%, leading to a Second Law efficiency of almost 75% [12].

Lean burning produces more irreversibility. The reduction of Second Law efficiency of a combustion engine associated with increasing the air-to-fuel ratio is shown in Figure 3.17. It can also be inferred that it is possible to reduce the combustion-associated irreversibility by preheating the air before combustion.

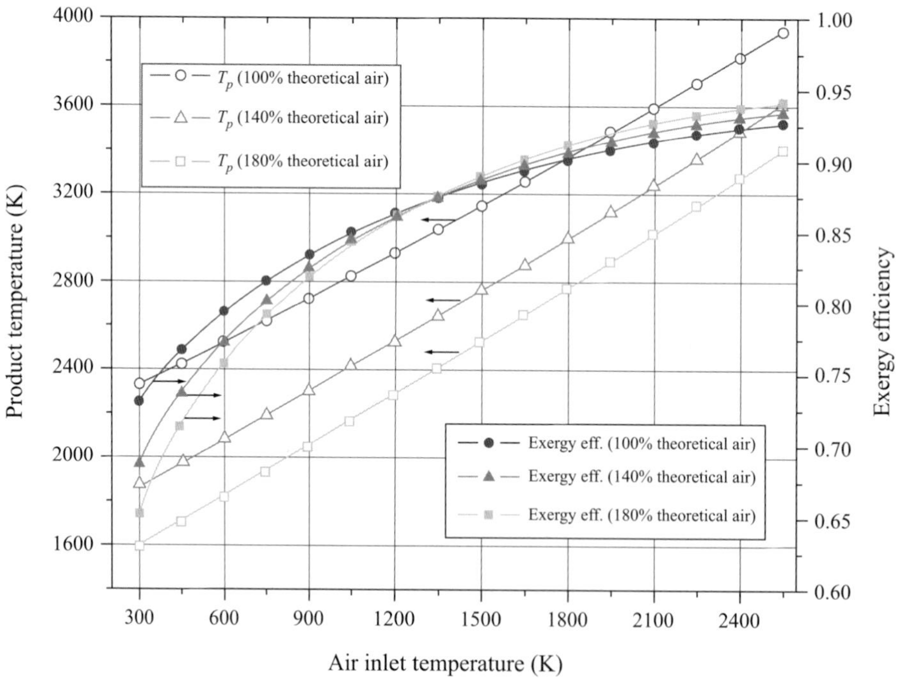

Figure 3.17 Impact of reactants temperature and the percentage of air on the products temperature and the Second Law efficiency of the process. The Second Law efficiency is defined as the change of availability across the combustion process as a fraction of the fuel chemical availability. Complete combustion is assumed.

Another commonly used efficiency in combustion engines is the **fuel–air cycle efficiency,**

$$\eta_{FU} = \frac{W}{\Delta H^{\circ}_{R,f}}, \tag{3.113}$$

which can be manipulated to read

$$\eta_{FU} = \frac{W}{W_{max.comb.eng}} \frac{W_{max.comb.eng}}{W_{max.chem.eng}} \frac{\Delta G^o_R}{\Delta H^o_{R,f}} \approx \frac{W}{W_{max.comb.eng}} \left(1 - \left|\frac{T_o \Delta S_{ad.comb}}{\Delta G^o_R}\right|\right). \tag{3.114}$$

In the second line, we use the approximation $\Delta G^o_R \approx \Delta H^o_{R,LHV}$. According to the above discussion, $W_{max.comb.eng} = \eta^*_{car} \Delta H^o_R$, and the ideal efficiency is calculated from (3.108), with T_F being the flame temperature in the combustion engine.

3.11 About Fuels and Their Production

Refining refers to the separation of crude hydrocarbon liquids (e.g., petroleum and coal tar) into various fractions by: (1) **distillation**, which is chemical processing used to change the structure or size of the molecule; and (2) purification to reduce nitrogen and sulfur species concentration. Distillation is a physical process that relies on the difference between the boiling points of the different compounds in a mixture, which allows us to boil off the mixture as the temperature is raised gradually. In general, the boiling point of an organic compound increases as the carbon number of the compound increases. Table 3.5 shows typical petroleum fractions, their carbon numbers, boiling ranges, and use [13].

Figure 3.18 shows the different components extracted from crude oil by distillation at different points in the column, as determined by the distillation temperature.

Thermal decomposition or **pyrolysis** is performed in the absence of oxygen, and is used to crack larger molecules to smaller, lighter compounds. **Tar** refers to the liquid obtained during the pyrolysis of coal, wood, or shale at intermediate temperature, around

Table 3.5 Carbon number range (CN), boiling temperature (TB), range, and some uses of products of petroleum refining

Fraction	CN	TB (°C)	Uses
Gases	1–4	<0	Fuels (mostly heating)
Light naphtha	5–7	26–93	Industrial solvents
Heavy naphtha	6–10	93–177	Transportation fuels
Kerosene	9–15	149–260	Jet fuels
Light gas oil	13–18	204–343	Fuel oil (power generation)
Heavy gas oil	>22	385–565	Heavy fuel oil (power and ships)
Lube oils and waxes	>22	385–565	Lubricants
Residuum	>44	>565	Asphalt tar for roads/pavements

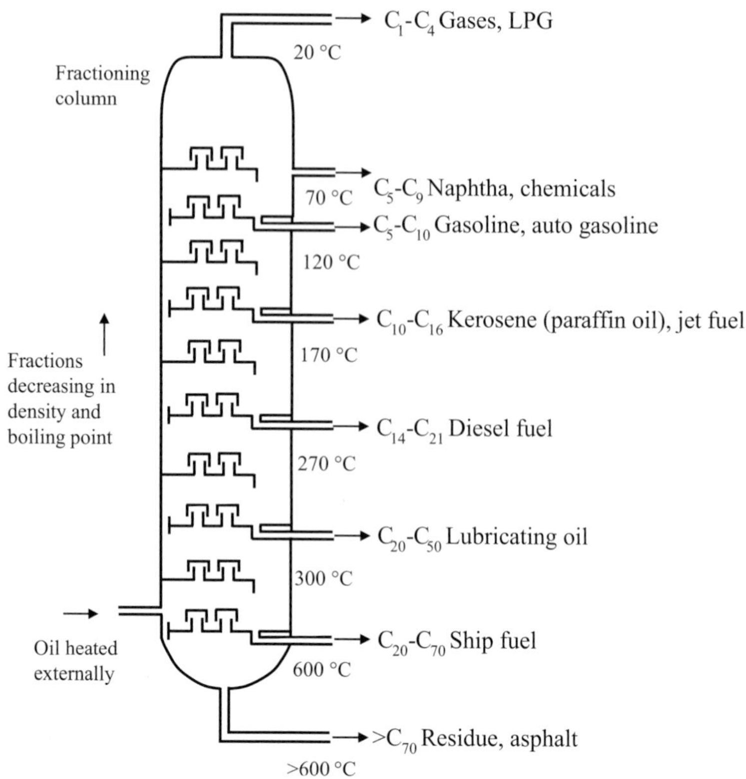

Figure 3.18 The distillation process of crude oil, showing the temperature at which different components are recovered from the distillation column, and the range of the number of carbon molecules found within each range.

300–600 °C for coal. The higher the temperature, the less tar is formed. **Pitch** refers to the solid that is left on the distillation of tar. Other common words used to refer to liquids left on the processing of feedstock are: **bitumen**, which is a substance soluble in carbon disulfide, and often obtained from tar sands, and **kerogen**, which is the main organic obtained on the processing of oil shale. Petroleum is a **bituminous** liquid, and some coals are also bituminous.

As mentioned above in defining refining, chemical processing often follows distillation. Chemical processing to change the molecular structure refers to the decomposition or **cracking** of large molecules, combination or **polymerization** of small molecules, and the **reforming** of molecules. Cracking can be accomplished thermally by raising the temperature of heavy oils above 360 °C, and may be assisted by a catalyst. For instance, thermal cracking of *n*-butane can be used to produce methane and propylene according to the following reaction:

$$CH_3CH_2CH_2CH_3 \rightarrow CH_4 + (CH_3CH = CH_2).$$

The opposite reaction to pyrolysis is **coking** or polymerization of the smaller molecules to larger molecules, sometimes called coke. Coking is often used to remove some of the carbon in the original hydrocarbon to produce hydrocarbons with lower carbon content.

Another important process in fuel production is reforming, in which the size of the molecule does not change by much, but its structure does. For instance, reforming can be used to remove hydrogen from the original molecule – that is, dehydrogenation. Another reforming process involves isomerization.

Another opposite process to decomposition is **synthesis**, in which a larger molecule is formed by combining smaller molecules, often in the presence of a catalyst. A well-known example is the Fischer–Tropsch synthesis used to produce gasoline from synthesis gas:

$$nCO + 2nH_2 \rightarrow C_nH_{2n} + nH_2O,$$

$$nCO + (2n+1)H_2 \rightarrow C_nH_{2n+2} + nH_2O.$$

As shown by these equations, a fraction of the products are alkenes, with the remainders being alkanes and aromatics.

3.12 Summary

In this chapter we introduced the concepts of mass and energy balances in systems undergoing chemical or, more precisely, thermochemical reactions in which conversion of the chemical bond energy into thermal energy, and vice versa, takes place alone or while the system is undergoing other changes. Several important definitions were introduced to standardize the applications of these concepts, such as the enthalpy of formation, and derived quantities used extensively in applications such as the enthalpy of reaction, fuel heating value, and adiabatic flame temperature were demonstrated. While in many cases these concepts are applied to combustion calculations, they are equally applicable to other applications, such as fuel reforming and hydrogen production, as shown by several examples. Most importantly, it was shown that the familiar forms of the First Law are still applicable as long as the internal energy or enthalpy includes, besides the sensible component, the formation component.

As shown in Chapter 2, description of a thermodynamic, or in this case thermochemical, process often requires the application of the Second Law in which entropy balance is used to determine the equilibrium state. It is shown that, depending on the imposed constraints (or what properties are used to define the state), equilibrium is reached when entropy is maximized or the free energy is minimized. When applied to a single reaction, and with the aid of a chemical potential, the condition of chemical equilibrium, or the composition of the reacting mixture, can be determined using the law of mass action. In this case, either the Gibbs free energy of reaction or the equilibrium constant can be used to determine the mixture composition as the temperature changes. Applications of these tools to calculate the composition of combustion products, or the result of fuel reforming, water or CO_2 splitting (into hydrogen and oxygen, or carbon monoxide and oxygen, respectively) are shown.

Since combustion is used extensively to operate many engines, it is reasonable to ask about the maximum work produced by burning a fuel, or the fuel availability. The process in this case must remain in (thermal and chemical) equilibrium with the environment, a requirement which is not compatible with practical (fast) engines. Nevertheless, the analysis and results are useful in defining the relative (Second Law) efficiency of an engine using fuels as the energy source. It also suggests that achieving higher chemical energy to work conversion requires other devices that will be covered in Chapter 4. As shown in Chapter 2, efficiency definitions depend on the process/engine used in the energy conversion process, and when using a chemical reaction to provide the heat, a fuel utilization efficiency should be used.

Other efficiency definitions are more useful when thermochemical reactions are used to produce fuels, similar to the case of methane reforming for hydrogen production. This application illustrates clearly how equilibrium limitations must be considered when converting chemical compounds to other compounds while operating at certain temperatures. It was shown that two reactors operating at two different temperatures must be used to achieve conversion of a significant fraction of methane into hydrogen. Another example is used to introduce the concept of coal (or other heavy hydrocarbons) gasification, which will be covered in more detail later, and the associated conversion efficiency.

Problems

3.1 Methane at 298 K and 1 atm is burned with 400% theoretical air at 298 K and 1 atm.

- **a.** Compute the adiabatic flame temperature.
- **b.** Compute the minimum work for CO_2 separation considering that the combustion products include CO_2, water vapor, N_2, and O_2.
- **c.** Methane is combusted with 400% pure oxygen. Calculate the adiabatic flame temperature.

3.2 Consider an internal combustion engine that uses liquid octane (C_8H_{18}) as a fuel at a rate of 0.015 kg/s. The heat loss from the engine is 100 kJ. Before the combustion process, the fuel is mixed with air in a stoichiometric ratio. Both fuel and air are supplied at 298 K. Only 93% of the carbon of the fuel is converted to carbon dioxide, and the remaining 7% burns to form carbon monoxide. Also, the combustion products include hydrogen in addition to water vapor and nitrogen. The temperature of the combustion products leaving the engine is 923 K. Calculate the thermal efficiency of the engine.

3.3 An internal combustion engine operating on the basis of the Otto cycle uses gaseous propane (C_3H_8) as a fuel at a rate of 0.02 kg/s, and 150% theoretical air as a working fluid. During the intake, both fuel and air are supplied at 300 K and 1 atm, which are mixed and compressed up to 8 atm through a compression process with isentropic efficiency of 80%. The mixture is then ignited, resulting in a constant-volume combustion process. The products of combustion undergo an expansion process with isentropic efficiency of 88% down to the atmospheric pressure, after which they are discharged to the atmosphere. Only 95% of the carbon is converted to carbon dioxide and the remaining carbon burns to form carbon monoxide. All the hydrogen of the products burns, forming water vapor. Assuming ideal gas behavior of the air–fuel mixture and the combustion products, determine the thermal efficiency of the engine.

3.4 Syngas at 600 K and 30 atm is used as a fuel in a gas turbine power plant that operates on an open Brayton cycle, as shown in Figure P3.4. The air supplied from the atmosphere at 298 K and 1 atm is pressurized in the compressor up to 30 atm. The composition of syngas is 1 mole hydrogen, ½ mole carbon monoxide, and ½ mole carbon dioxide. It is mixed with air in the combustor, where it undergoes an adiabatic complete combustion process. The combustion products leaving the combustor at 1500 K and 30 atm are expanded in the turbine down to the atmospheric pressure. The products are then passed through a heat exchanger, where they are cooled down to 298 K so the water content of the products is condensed and collected. The carbon dioxide is separated from the mixture of the remaining products in a separator located downstream of the heat exchanger. The work requirement for separation process is 580 kJ per kilogram of CO_2. The isentropic efficiencies of the compressor and turbine are 85% and 90%, respectively. Determine:

a. the number of moles of air for each mole of syngas used in the combustor. Use $\gamma_{mix} = c_{p,mix}/c_{v,mix} = 1.359$ to determine the compressor exit temperature;

b. the work requirement of the compressor per mole of syngas;

c. the work production of the turbine per mole of syngas. Use $\gamma_{mix} = c_{p,mix}/\, c_{v,mix} = 1.327$ to determine the turbine exit temperature;

d. the CO_2 separation work per one mole of syngas;

e. the overall efficiency of the power plant based on the LHV of the fuel;

f. the total flow availability at the combustor inlet; and

g. the total flow availability at the combustor outlet.

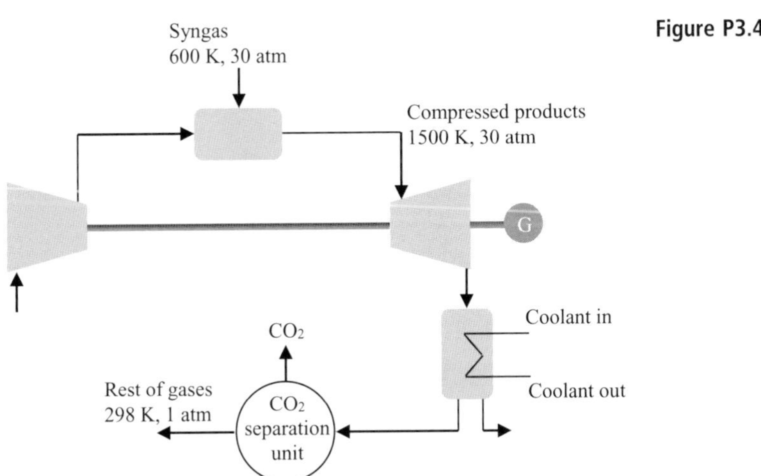

Figure P3.4

3.5 Consider the gas turbine power plant that is now operating in an oxyfuel combustion configuration, as shown in Figure P3.5. The system uses a separation unit where oxygen of the air supplied from the atmosphere is separated by spending 0.27 kWh per kilogram of oxygen. A mixture of oxygen and carbon dioxide is compressed up to 30 atm in the compressor. The hot combustion products leaving the turbine at 1 atm are cooled in a heat exchanger so the water content of the products is condensed and collected. Some of the carbon dioxide leaving the heat exchanger is recycled using a control valve and the rest is captured and stored.

a. Calculate the number of moles of CO_2 per mole of syngas needed to operate the gas turbine cycle under stoichiometric conditions while limiting the turbine inlet temperature to 1500 K.

b. Calculate the work production of the turbine per mole of syngas. You may use $T_{4S} = 909.1$ K as the isentropic exit temperature of the turbine.

c. Calculate the compression work of the mixture of oxygen and carbon dioxide per mole of syngas.

d. Calculate the O_2 separation work per one mole of syngas.

e. Determine the overall efficiency of the gas turbine plant based on the LHV of the fuel, taking into account the work required by the ASU.

f. Compare the overall efficiency of the gas turbine power plant with that of the power plant in P.3.22. Why is one lower than the other?

g. If 40% of the heat transferred in the heat exchanger can be converted to useful work, recalculate the overall efficiency. What are possible ways to utilize the heat transfer from the heat exchanger into useful work? Your answer to this part should be realistic based on the exit temperature of the turbine.

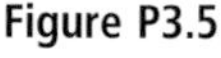

Figure P3.5

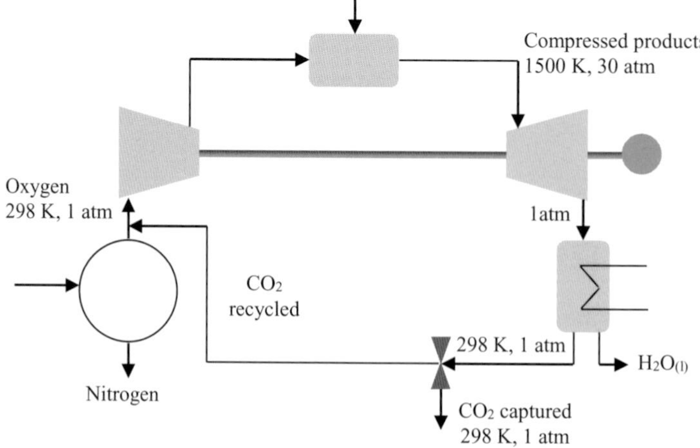

3.6 Figure P3.6 shows a schematic diagram of a hydrogen production plant using electrolysis. Water at 25 °C and 1 bar is pumped to 10 bar. Some of the pressurized water is injected directly into a combustor. The rest of the water is heated to 80 °C in heat exchanger 1 and then it enters an isothermal electrolyzer that operates at the same temperature. Oxygen produced by the electrolyzer is heated in heat exchanger 2 to 750 °C prior to entering the combustor. Methane enters the combustor at 25 °C and 10 bar. The equivalence ratio of the methane–oxygen combustion is 1 and the combustion is complete. The combustion products leaving the combustor at 1400 °C are expanded in a turbine to 1 bar. The isentropic efficiency of the turbine is 87%. The electricity generator has an efficiency of 97%. The electricity produced by the plant powers the pump and partially powers the electrolyzer. Neglect the pressure drop inside the heat exchangers. You can treat all gases as ideal and assume constant specific heats. Liquid water is pure and incompressible. The temperature rise across the pump can also be neglected.

a. Calculate the total moles of water injected into the combustor per mole of methane burned in the combustor.

b. Calculate the turbine exit temperature and the electric work generated per mole of methane.

c. Calculate the work requirement of the electrolyzer per mole of methane if its Second Law efficiency is 55%.

d. Determine the heat transfer between the electrolyzer and its environment per mole of methane.

e. Determine the total flow availability added to the plant and the total flow availability leaving the plant.

f. As the plant is designed for hydrogen production, define an appropriate efficiency for the plant and calculate it.

g. The plant proposes to utilize the high-temperature exhaust stream to raise the overall efficiency of hydrogen production. The Second Law efficiency of the heat engine utilizing this stream to produce mechanical work is 45%. If the exhaust temperature is 780 °C and the ambient is at 25 °C and 1 bar, calculate the mechanical work generated by the heat engine per mole of methane.

h. Recalculate the efficiency defined in part (f) after adding the heat engine in part (g).

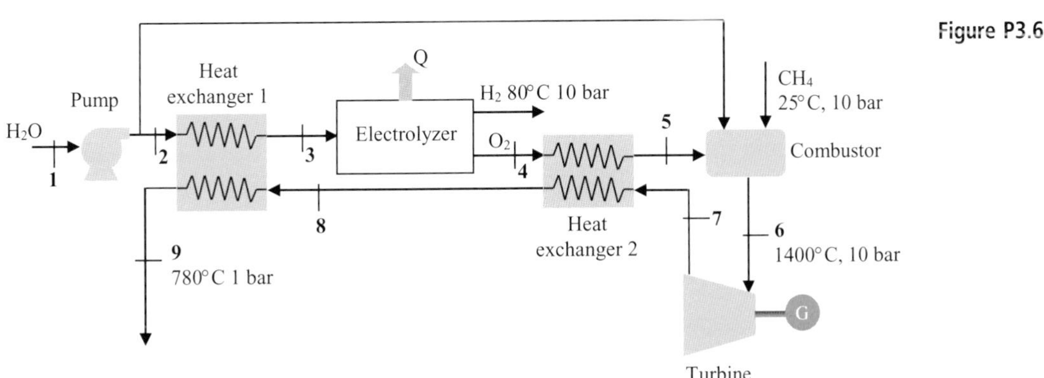

Figure P3.6

3.7 A schematic of a hydrogen production plant using water reforming of methane followed by shift reaction is shown in Figure P3.7. For both reactions, the outlet temperature is equal to the operating temperature. All streams and processes are at ambient pressure. Note that the reforming reactor operates at the given temperature, and the water–gas shift reactor is adiabatic. Assume that only the species given at the system outlet will be formed.

- **a.** Calculate the composition out of the steam reformer assuming complete reforming occurs.
- **b.** Determine the equilibrium composition out of the steam reformer and compare with your result from part (a).
- **c.** Calculate $\dot{Q}_1$ and $\dot{Q}_2$.
- **d.** Calculate the equilibrium composition out of the water–gas shift reactor using the answer from part (b).

Figure P3.7

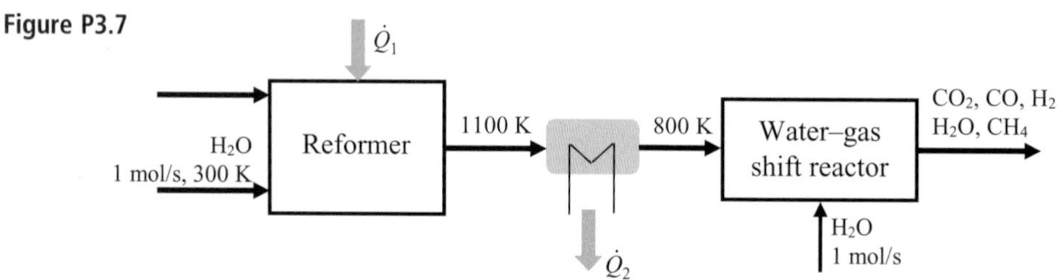

3.8 Power is generated from a stream of 1 kg/s of H_2 available at ambient temperature and pressure. Ambient air is used to oxidize the fuel. The processes considered in this problem take place at atmospheric pressure. The enthalpy and entropy of each species may be evaluated assuming a constant heat capacity.

- **a.** Calculate the maximal work that can be obtained assuming a stoichiometric mixture and complete combustion, and that the flue gases leave the power generation at 110 °C.
- **b.** Consider now a combustion engine. The fuel and air are mixed and fed into an adiabatic combustor, and then the flue gases are cooled to 110 °C by a perfect heat engine. Assume complete consumption of fuel or oxygen (whichever is limiting). Plot the combustion temperature in the process, the lost work rate of the adiabatic combustion, and the power generation as a function of the molar excess-air ratio in the range 0.5–10. If the maximal temperature in the process is 1000 K, what is the minimum molar excess-air ratio?

3.9 In the oil and gas industry, hydrocarbons such as ethane (C_2H_6), propane (C_3H_8), and butane (C_4H_{10}) are derived as byproducts in different amounts. Assume that one of the natural gas fields in Canada contains significant amounts of propane and, after separating propane as a byproduct, instead of selling it the gas company wants to use it as fuel in a simple gas turbine cycle to generate electricity. Perform your calculations per kilogram of propane. Propane is mixed with air first. The mixture is compressed and burned in a combustor (assume adiabatic complete combustion). Next, the products of combustion are expanded in the gas turbine.

Assume that the air–propane mixture can be treated as an ideal gas. The initial conditions of the mixture before compression are 298 K and 1 atm. The compression ratio is 10 and the isentropic efficiency of the compressor is 80%. In the turbine, the mixture is expanded to atmospheric pressure and the isentropic efficiency of the turbine is 85%. The mixture composition remains frozen during expansion.

a. Evaluate the air–fuel ratio of the propane–air mixture if the turbine inlet temperature is 1500 K.

b. Determine the equivalence ratio for combustion.

c. Calculate the net work produced by the gas turbine cycle.

d. Evaluate the thermal efficiency of the gas turbine cycle.

3.10 An autonomous underwater vehicle (AUV) uses a Stirling engine for power and propulsion. Hydrogen and oxygen are stored on board in liquid form, heated and fed to a combustor at 298 K. After combustion, the product gas is fed to the hot end of the Stirling engine, where it cools. After this it is vented to the sea.

a. At a depth of 10 m in seawater (density = 1020 kg/m^3), what is the minimum required pressure in the combustor if the exhaust stream does not need compression prior to venting. Please note that a Stirling engine is an external combustion engine, which does not employ its heat source as a working fluid. As a result, there is no pressure drop of product gases across the engine.

b. Assuming complete combustion in an adiabatic chamber, how many moles of O_2 per mole of H_2 are required to keep the combustor outlet temperature at 3000 K?

c. Assuming the products of combustion reach equilibrium at 3000 K in an adiabatic combustor, find the inlet and outlet compositions. Use the inlet temperature from part (b). The temperature-dependent equilibrium constant for the $H_2O \leftrightarrow H_2 + 0.5O_2$ reaction is $K_p(T) = \exp(6.7085 - 29562/T)$, where T is in kelvin.

d. The product gas enters the Stirling engine at 3000 K and leaves at 800 K. If the Stirling engine converts 70% of the maximum available energy in the product gas stream to mechanical work, find the mechanical energy output per mole of product gas. Assume that the temperature of the ambient seawater is 288 K.

e. Define an overall efficiency for this process and determine it.

3.11 A schematic of a typical steam–methane reforming plant is shown in Figure P3.11. For each mole of methane, three moles of water enter the methane reformer, both at 25 °C. Methane reforming is performed at 900 °C. The heat required for methane reforming is supplied from the exhaust of the reformer, which exits the reformer at a high temperature of 900 °C, and from the heat exchange with the products of stoichiometric methane–air combustion in a separate combustor. Due to the heat loss to the environment, only 90% of the heat generated in the combustor is supplied to the methane reformer. The mixture leaving the methane reformer, after being cooled in the heat exchanger to a temperature of 350 °C, enters the high-temperature shift reactor (HTSR) that operates at 350 °C. The products of the HTSR are cooled in another heat exchanger to 200 °C. This is followed by a CO_2-separation process. The mixture then enters the low-temperature shift reactor (LTSR), operating at 200 °C. The mixture exiting the LTSR is cooled in a third heat exchanger to 25 °C and then hydrogen is separated and transported to the location where it is stored. The mixture composition is frozen except when it passes through the methane reformer, HTSR, and LTSR. Assume average and constant values for the specific heats of the species.

a. Find the composition of the mixture at the exit of the methane reformer, assuming that the mixture is in thermodynamic equilibrium and all methane is consumed. Consider the equilibrium of the species CO_2, CO, H_2, and H_2O.

b. Calculate the amount of heat required by the methane reformer.

c. Calculate the ratio of moles of methane burned in the combustor to the moles of methane entering the methane reformer. Assume complete combustion of methane in the combustor.

d. Find the composition of the mixture at the exit of HTSR.

e. Calculate the amount of heat rejected between the HTSR and the CO_2 separator in J/mol CH_4 entering the methane reformer.

f. Calculate the separation work required for the CO_2 separator in J/mol CH_4 entering the methane reformer if it has a Second Law efficiency of 35%. How much heat is rejected from the separator?

g. Find the composition of the mixture at the exit of the LTSR.

h. Calculate the amount of heat rejected between the LTSR and the hydrogen separator in J/mol CH_4 entering the methane reformer.

i. Calculate the separation work required for the hydrogen separator in J/mol CH_4 entering the methane reformer if it has a Second Law efficiency of 25%.

j. Define and calculate the efficiency of this system.

k. Recalculate the efficiency if all the heat rejected/lost in this plant is recovered to supply heat to the methane reformer.

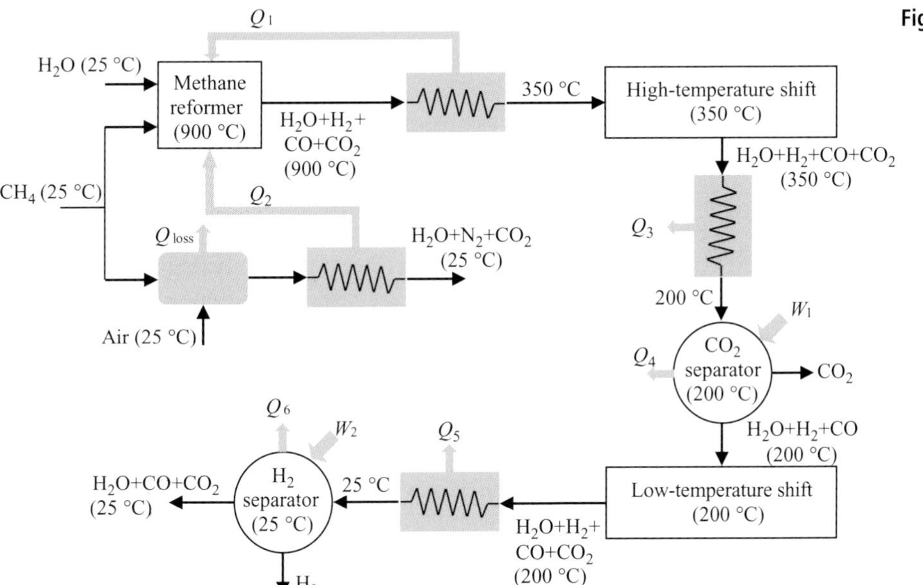

Figure P3.11

3.12 The power requirement of a vehicle is provided using a four-stroke SI engine, which runs on a stoichiometric methane–air mixture. The mixture enters the cylinder at 25 °C and 1 atm. The compression ratio of the engine is 10. The combustion products, CO_2, CO, H_2O, H_2, O_2, and N_2, are in thermodynamic equilibrium at the end of the combustion process. The product mixture is frozen during the expansion process. The isentropic index during the compression and expansion processes are 1.4 and 1.3, respectively. All the gases can be treated as ideal.

a. Calculate the mixture composition at the end of the combustion process, assuming that the temperature at the end of combustion is 2750 K. The equilibrium constants of the reactions $H_2O \leftrightarrow H_2 + 0.5O_2$ and $H_2O \leftrightarrow H_2 + 0.5O_2$ are given in Table 3.3.

b. Determine the amount of heat lost during combustion as a percentage of the HHV of the fuel.

c. Calculate the work produced in the expansion stroke if 5% of the ideal work is lost due to heat loss to the environment and other sources of irreversibility. Irreversibilities are usually small compared to the heat loss.

d. Calculate the fuel utilization efficiency of the cycle based on the HHV of methane.

3.13 Oxy-combustion is a promising technology for use in carbon capture systems. This involves burning a fuel in pure oxygen where the CO_2 in the products can be easily separated out by condensing the H_2O and then stored. Due to the high flame temperatures of oxy-combustion, a diluent is needed to moderate the temperatures in the combustor. Common diluents used are CO_2 and H_2O.

Sour gas is a type of fuel that consists of $CH_4 + CO_2 + H_2S$, and is a raw form of natural gas that is extracted from a well. Sour gas–oxy-combustion cycles are a novel way of combining the environmental benefits of oxy-combustion with the economic advantage of using sour gas [14]. In this problem we will examine two types of cycles and evaluate them based on their efficiency. The two cycles are:

1. methane–oxy cycle; and
2. sour gas–oxy cycle.

The type of cycle we will consider is a Brayton cycle (as shown in the Figure P3.13). An ASU is used to produce the pure oxygen stream required for oxy-combustion. Since the separation process is expensive and energy-intensive, oxy-combustion cycles commonly operate very close to stoichiometric conditions in the combustor.

Givens and assumptions:

- Specific power required to produce O_2 in the ASU = 810 kJ/kg-O_2;
- assume complete combustion of fuel;
- sour gas composition (by volume): 70% CH_4, 15% CO_2, 15% H_2S;
- assume that the LHV of methane is 50 MJ/kg-CH_4 and of hydrogen sulfide 15 MJ/kg-H_2S;
- H_2S combustion reaction: $H_2S + 3/2\,O_2 \rightarrow SO_2 + H_2O$;
- assume that the combustion process is stoichiometric ($\varphi = 1$);
- isentropic efficiencies of the turbine and compressor are 92% and 85%, respectively;
- combustor is adiabatic (no heat loss) and there is no pressure drop;
- inlet fuel and oxygen conditions are: $T = 25\,°C$, $P = 40$ bar;
- CO_2 is used as the diluent to keep the combustor exit temperature fixed at 1200 °C for both cycles;
- mole flow rate of fuel (methane or sour gas) for both cases = 0.1 kmol/s; and
- cycle conditions (for both cycles) are as given in Table P3.13.

Table P3.13

State #	T (°C)	P (bar)	Gas composition
1	25	1	100% CO_2
2	?	40	100% CO_2
3	1200	40	?
4	?	1	?

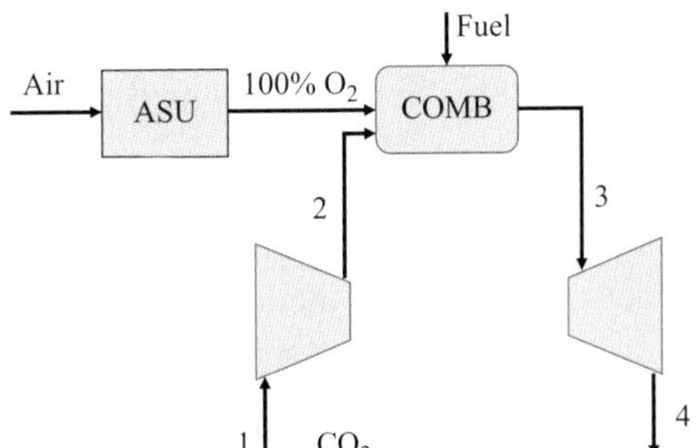

Figure P3.13 Simple oxy-fuel cycle.

Calculate the following for **each cycle**:

a. The required mole flow rate of O_2 and CO_2 (kmol/s) in order to maintain the given combustor temperature.
b. temperatures (°C) and gas compositions (mole fractions) at all four states (see Table P3.13).
c. The power required by the ASU (MW).
d. The **net** power of the cycle (MW).
e. The **net** cycle efficiency (based on LHV).
f. In practical combined cycles, the exit stream at state 4 is then sent to a heat exchanger to produce steam to power a steam cycle, thus generating more power and increasing the combined cycle efficiency. For the sour gas cycle, cooling the stream at state 4 down to low temperatures will result in corrosion of the heat exchanger due to sulfuric acid formation. Therefore, it is suggested to cool the stream down to a temperature that is just above the dew point of the mixture, in order to prevent condensation.

   Assume that the steam cycle efficiency is 40% and that the exit stream at state 4 is cooled to produce that steam. That stream is cooled to:

   - 30 °C in the methane–oxy cycle; and
   - 160 °C for the sour gas–oxy cycle.

   Calculate the new net power and net cycle efficiency of each combined cycle.
g. Comment on the performance of each cycle.

   Treat all of the gases as ideal

3.14 Consider a power plant with a power rating of 100 MW that can use lignite, methane, or octane as fuel. The environment is at standard conditions. Suppose first that the plant has a First Law efficiency of 30%, defined based on the LHV of the fuels.

a. For each fuel, calculate the required fuel flow rate and CO_2 emissions. Explain their relative values, and comment on which is the best fuel and why.
b. Is it reasonable to assume the same efficiency for all fuels? Why or why not?

Additional tasks for the case of octane:

c. Calculate the adiabatic flame temperature, assuming an air excess of 20% and that the reaction is complete.
d. Based on the adiabatic flame temperature, calculate a maximum First Law efficiency using the standard Carnot efficiency.
e. Correct the efficiency to account for the fact that the heat source has finite flow rate and to allow for a 10 K temperature difference for the removal of heat to the environment from the heat engine. Is it the same as part (d)?
f. Why is heat removal to the environment necessary and why do we account for a temperature difference?

3.15 Gaseous propane (C_3H_8) undergoes combustion with 20% excess air in a constant-pressure process at 200 kPa. The equilibrium temperature of the products of combustion is 2800 K. The CO_2 and H_2O in the products undergo dissociation reactions. The products of combustion contain only CO_2, CO, O_2, H_2O, H_2 and N_2.

Using EES or other solvers (CHEM_EQUIL Libraries or other similar equilibrium libraries are not allowed):

a. calculate the degree of dissociation of CO_2;
b. calculate the mole fraction of H_2 in the products;
c. plot the degree of dissociation of CO_2 as excess air varies from 20% to 60% (temperature is maintained at 2800 K).

You may have to provide guess values to get the solution. Also, you may use the following as the dissociation reactions:

$$\mathrm{CO_2} \leftrightarrow \mathrm{CO} + \frac{1}{2}\mathrm{O_2},$$

$$\mathrm{H_2O} + \mathrm{CO} \leftrightarrow \mathrm{H_2} + \mathrm{CO_2}.$$

The equilibrium constants at 2800 K are given by

$$\frac{P_{\mathrm{CO_2}}}{P_{\mathrm{CO}} P_{\mathrm{O_2}}^{1/2}} = 6.6\,\mathrm{bar}^{-1/2},$$

$$\frac{P_{\mathrm{CO_2}} P_{\mathrm{H_2}}}{P_{\mathrm{CO}} P_{\mathrm{H_2O}}} = 0.15.$$

3.16 A particular bituminous coal has the following composition by mass: 75% carbon, 12% water, 3% sulfur, and 10% ash (modeled as SiO_2). The coal enters a combustor at 25 °C and is burned in air with an equivalence ratio of 50% (with respect to C as the fuel). Ash is separated by gravity (with no work expenditure) from the products prior to leaving the combustor.

Figure P3.16

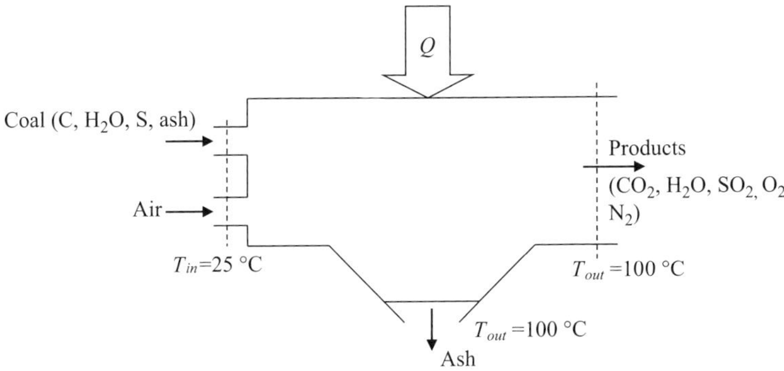

a. What is the heat transfer required to operate the combustor at these conditions if all oxidation reactions proceed to completion, and the gaseous products and ash leave at 100 °C? Express your answer in terms of kilojoules per mole of coal.

b. After combustion, SO_2 is separated from the product stream at 100 °C. Calculate the separation work if the separation device has a Second Law efficiency of 20%. Express your answer in terms of kilojoules per mole of coal.

Table P3.16

Species	Molecular weight (g/mol)	$\hat{h}_f^0$ (J/mol)	$\hat{c}_p$ (J/mol·K)
C	12	0	10.7
H_2O	18	−242,000	30.0
S	32	0	22.7
SiO_2	60	−910,860	72.1
O_2	32	0	33.4
N_2	28	0	31.1
CO_2	44	−393,790	50.6
SO_2	64	−296,840	44.5

REFERENCES

1. R. J. Silbey, R. A. Alberty, and M. G. Bawendi, *Physical chemistry*. Hoboken, NJ: Wiley, 2005.
2. P. W. Atkins and J. De Paula, *Atkins' physical chemistry*. Oxford: Oxford University Press, 2006.
3. D. R. Stull and H. Prophet, United States National Bureau of Standards, *JANAF thermochemical tables*. Washington, DC: U.S. Dept. of Commerce, National Bureau of Standards, 1971.
4. R. C. Weast, *CRC handbook of chemistry and physics*. 67th ed. Boca Raton, FL: CRC Press, 1986.
5. L. L. Turns, and S. R. Pauley, *Thermodynamics, concepts and application*, 2nd ed. Cambridge: Cambridge University Press, 2020.

6. I. Rossini, F., Wagman, D., Evans, W., Levine, S. and J. Affe, "Selected values of chemical thermodynamic properties," NBS Circular 500. Washington, DC: National Bureau of Standards, 1952.
7. M. Sussman, "Steady-flow availability and the standard chemical availability," *Energy*, vol. 5, pp. 793–808, 1980.
8. M. J. Moran, *Availability analysis: a guide to efficient energy use*. New York: ASME Press, 1989.
9. A. Bejan, *Advanced engineering thermodynamics*. Hoboken, NJ: Wiley, 2006.
10. J. Szargut, *Exergy method: technical and ecological applications*. Southampton: WIT Press, 2005.
11. G. Hoogers, *Fuel cell technology handbook*. Boca Raton, FL: CRC Press, 2003.
12. E. P. Gyftopoulos and G. P. Beretta, *Thermodynamics: foundations and applications*. Mineola, NY: Dover Publications, 2005.
13. R. F. Probstein and R. E. Hicks, *Synthetic fuels*. Mineola, NY: Dover Publications, 2006.
14. N. W. Chakroun and A. F. Ghoniem, "High-efficiency low LCOE combined cycles for sour gas oxy-combustion with CO2 capture," *Int. J. Greenhouse Gas Contr.*, vol. 41, pp. 163–173, 2015.

4 Electrochemical Thermodynamics

4.1 Introduction

Adiabatic combustion raises the temperature of the working fluid in a power cycle and provides the source of "high-temperature heat" to the heat engine. Analysis in the previous chapter showed that adiabatic combustion reactions are irreversible, and lead to entropy generation and hence loss of availability. Isothermal reactions that operate at equilibrium with the environment avoid this loss mechanism. If carefully executed, these can lead to more efficient use of the chemical energy. One practical way to directly convert chemical energy to electricity under nearly isothermal conditions is in a fuel cell, where reactions occur in the form of an electrochemical pair, or a redox pair.

Electrochemical systems, such as batteries and fuel cells, convert chemical energy stored directly to electrical energy by moving electrons across a potential difference created by charge separation at two different electrodes. The electrodes are separated by an electrolyte, which allows ions to migrate across without permitting neutral molecules or electrons to move across. These electrochemical devices are not combustion or heat engines. In combustion engines, the chemical energy is converted to thermal energy prior to its conversion to work. In heat engines, heat interaction with a high-temperature source is converted to work by, for example, changing the volume of the working fluid against the pressure differential. If combustion reactions were more "controllable" so that one could isothermally convert chemical energy gradually to thermal energy while allowing the system to maintain a constant temperature, an equivalent reversible process producing a maximum work would be possible. (Even if it was possible, this would be a slow process requiring catalysis to enhance the chemical reaction rate at or near the low temperature of the environment; in other words, it would have been impractical.)

In this chapter, we introduce electrochemical reactions as described by a reduction–oxidation pair, or a redox pair. The essential elements of a fuel cell are discussed in connection with different common electrochemical reactions. The same reaction (e.g., a hydrogen–oxygen reaction) can be executed in multiple pairs of redox reactions, depending on the type of the electrolyte used to separate the two electrodes. We will also see that, theoretically, many reactions can be performed electrochemically. We will define the rules governing the ideal potential produced in these reactions. The inverse of a fuel cell is an electrolyzer, and the reactions occurring in these devices are also described. The thermodynamics of direct or electrochemical conversion are then used to show the relation between the ideal work, the electrochemical potential of a cell, and the Gibbs free energy of a reaction. The dependence of the cell potential on the reaction, temperature, pressure, and species concentrations is expressed in the form of the Nernst equation, and the implications of the resulting expressions are discussed. The concept of fuel utilization is introduced and examples of the impact of fuel utilization "overpotential" on the cell performance are discussed.

Ideal, open-circuit fuel cell efficiency is defined as the ratio of the Gibbs free energy to the enthalpy of reaction. However, this is only the open-circuit potential, and a cell operating at a finite current suffers potential loss due to a number of processes that will be described in detail in Chapter 7. Nonetheless, we define a number of efficiencies that are used to relate the performance of an operating cell to that of the ideal cell, including the relative or Second Law efficiency and the Faradic efficiency.

For convenience, the half-cell potential, defined for each of the two halves of a redox reaction, is often used to characterize electrochemical reaction. This is particularly true for reactions in primary and secondary batteries, some of which are introduced later in this chapter. The concept of chemical potential is expanded to that of an electrochemical potential. The relation between the chemical potential and the electrochemical potential is reviewed.

Under ideal conditions, a fuel cell extracts the maximum work from the fuel chemical energy. We show that the maximum work is the Gibbs free energy of the reaction. Under these ideal conditions, the heat rejected is the difference between the enthalpy of reaction and the Gibbs free energy of reaction. If the temperature of the fuel cell is higher than that of the environment, this heat transfer can be used for other purposes, such as for space heating (known as combined heat and power, or CHP, applications) or to power a heat engine. The fact that the fuel cell, ideally, produces work equal to the Gibbs free energy of reaction gives another meaning of that quantity: it is the maximum (non-$p\forall$) work that can be produced from a system at constant pressure and temperature, in this case converting the chemical energy into electrical work.

4.2 Electrochemical Reactions

Electrochemical reactions in most energy conversion devices occur in a **simple cell**, in which two **electrodes** are immersed in the same **electrolyte**, which allows one type of ion to move through, but does not allow electrons or neutral molecules to cross. An electrode is an electron-conducting plate. Electrochemical reactions occur at the interface between the electrode and the electrolyte. During this reaction, a reactant loses an electron in an **oxidation reaction**, or it gains one in a **reduction reaction**. Thermodynamic analysis allows us to calculate the work interaction between the electrochemical cell and its environment, and the cell ideal energy conversion efficiency. Before introducing the thermodynamic analysis of electrochemical cells, it is important to introduce some definitions and concepts.

Electrochemical reactions involve charged species, and occur at the interface between an **electrode** and an **electrolyte**. They involve **electron transfer** in **oxidation–reduction reactions**; that is, the two halves of a **redox** reaction, which occur in different compartments of the cell isolated by the electrolyte. By definition, in **a reduction reaction** the reactant **gains electrons**, and a reducing agent is an electron donor. In an **oxidation** reaction, the reactant **loses electrons**, and an oxidizing agent is an electron acceptor. For example, consider the following reaction between hydrogen and oxygen, producing water:

$$H_2 + \frac{1}{2}O_2 \Rightarrow H_2O. \tag{4.1}$$

This reaction involves the transfer of two electrons per molecule of hydrogen consumed or molecule of water produced, and can take place in several redox pairs. One example is the following **redox pair**, which occurs across an electrolyte that is permeable to positive ions only – that is, an **acidic electrolyte**:

$$\frac{1}{2}O_2 + 2H^+ + 2e^- \Rightarrow H_2O \qquad \text{Reduction}, \tag{4.2}$$

$$H_2 \Rightarrow 2H^+ + 2e^- \qquad \text{Oxidation}. \tag{4.3}$$

The positive ion in these reactions is H^+. In electrochemical reactions, these two reactions occur at the interface between a metallic/catalytic surface (phase) and an aqueous or solid solution (phase) that acts as an electrolyte, or an ion-conducting medium, as shown in Figure 4.1. Local spots on the interface form the **triple phase boundary** (TPB). The electrode

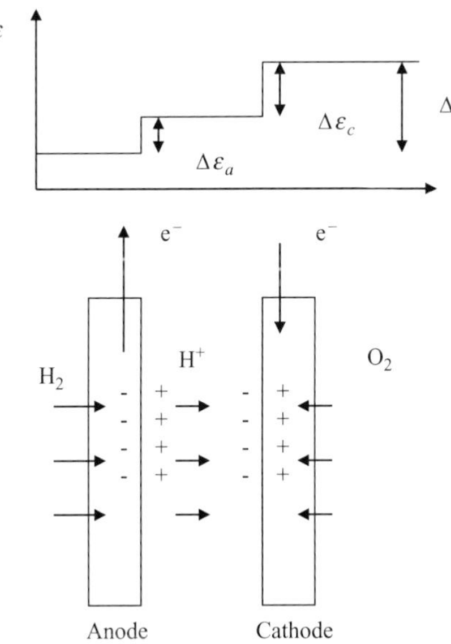

Figure 4.1 The electrochemical reactions at the anode–electrolyte interface and cathode–electrolyte interface, and the potential jump across each interface because of charge separation. The electric double layer formed at each interface is assigned relative charges to indicate the change in the potential. The overall potential difference across the cell (i.e., between the two electrodes) is the total jump. According to the convention, the anode reaction in this figure, that is the hydrogen oxidation reaction, has $\Delta\varepsilon_a = 0$.

on which oxidation (where the reactant is losing electrons) occurs is the **anode**. The electrode on which reduction (where the reactant is gaining electrons) occurs is known as the **cathode**. In reaction (4.3), hydrogen on the electrode surface (anode) leaves the surface in the form of positive ions while losing two electrons to the electrode. The electrode surface becomes negatively charged positively charged ions leave the surface and electrons move through it. This is the oxidation **half-reaction**. Electrons travel through an external circuit, reaching the other electrode (the cathode) on the other half of the cell. Meanwhile, hydrogen ions diffuse through the electrolyte and deposit on the other electrode surface (cathode), making it a positively charged. Hydrogen ions gain the two electrons coming from the external circuit while reacting with oxygen to form water on the electrode surface, in accordance with (4.2).

On the electrode surface, an **electric double layer** consisting of electrons on the surface and ions immediately adjacent to it forms between the electrode and the electrolyte. Ionic species that transfer the charge internally are the **charge carriers**, and it depends on the properties of the cell and the electrolyte. In the case of the redox reaction in (4.2) and (4.3), the charge carrier is the positively charged hydrogen atom. The pair of redox reactions shown above occur on the two sides of the polymer electrolyte of a proton exchange membrane fuel cell (PEMFC). In this cell, positive ions are the internal charge carriers, and hence a PEMFC is an **acidic cell**. Energetically, most of the chemical energy in each half-reaction is converted to the electrical energy associated with charge separation at the interface, establishing a potential "jump" across the electrical double layer between the electrode and the electrolyte. As shown in Figure 4.1, there is a potential jump at each electrode, and the total potential difference across the cell (i.e., the potential difference between the two electrodes) is the sum of the potential jumps across the two electrodes.

Because the two surfaces of the electrodes are electrically separated by the electrolyte, neither electrons nor neutral hydrogen are allowed to move through the electrolyte, and an external charge transfer from the anode to the cathode can be maintained. In the case of a cell running on reactions (4.2) and (4.3), the hydrogen electrode is the **negative electrode** because of the accumulation of electrons on its surface. The oxygen electrode is the **positive electrode** because of the accumulation of positive ions on its surface. Electrons passing through an external circuit perform work outside the **electrochemical cell** and hence directly convert the chemical energy into electrical work. Internal charge transfer in the form of ions between the electrodes must occur without mixing hydrogen and oxygen.

4.2.1 Redox Reactions and Electrolytes

The properties of the electrolyte determine to a large extend the electrochemical reactions at the electrodes. The hydrogen–oxygen reaction in (4.1) occurs differently if an **alkaline electrolyte** that conducts negatively charged oxygen with negative ions is used. In this case, the following reactions occur:

$$O_2 + 4e^- \Rightarrow 2O^{2-} \qquad \text{Reduction at cathode,} \tag{4.4}$$

$$H_2 + O^{2-} \Rightarrow H_2O + 2e^- \qquad \text{Oxidation at anode.} \tag{4.5}$$

Note that the overall reaction and the energetic characteristics of this redox reaction pair are the same as those described by (4.2) and (4.3). The redox reaction pair in (4.4) and (4.5) occurs in high-temperature **solid oxide fuel cells** (SOFC). Note that in this case, water forms at the anode side, where hydrogen is introduced, while the charge carrier is the negatively charged oxygen atom.

The charge carried by a charged species, which can be positive or negative, is quantified by the **charge number**, z_i, which also reflects the sign of the charge. The charge number is the absolute value of an electron charge carried by the species/component and its sign. The total charge of a component in a chemical reaction is the charge number multiplied by the absolute value of the electron charge. In the example in (4.2) and (4.3), the charge number of the hydrogen ion is $z_{H^+} = 1$. In the examples in (4.4) and (4.5), the charge number of the oxygen ion is $z_{O^{2-}} = -2$. In both cases, the charge number of electrons is $z_{e^-} = -1$.

An electrochemical half-reaction must conserve mass by conserving the number of atoms of each elemental species, like other chemical reactions (see Chapter 3), that is:

$$\sum_{react} \nu_i' \chi_i = \sum_{prod} \nu_i'' \chi_i \tag{4.6}$$

or

$$\sum_{allspecies} \nu_i \chi_i = 0, \tag{4.7}$$

where the stoichiometric coefficient of a reactant is ν_i', the stoichiometric coefficient of a product is ν_i'', and the net stoichiometric coefficient is $\nu_i = \nu_i'' - \nu_i'$. The electrochemical reaction must also satisfy charge conservation, such that

$$\sum_{react} \nu_i' z_i = \sum_{prod} \nu_i'' z_i \tag{4.8}$$

or

$$\sum_{allspecies} \nu_i z_i = 0. \tag{4.9}$$

Thus, we must extend the definition of the stoichiometric coefficient to account for charge as well as atomic species. In this case, an electron is considered a component with zero mass, $M_{e^-} = 0$, but with finite charge, and hence its stoichiometric coefficient is determined solely by the conservation of charge.

Another pair of reactions occurs in a hydrogen–oxygen cell if an OH-ion conducting alkali electrolyte is used. In this case, the charge carrier is the negatively charged hydroxyl radical OH^-. This pair, which occurs in **alkaline fuel cells** (AFCs), is

$$H_2 + 2OH^- \rightarrow 2H_2O + 2e^- \tag{4.10}$$

and

$$\frac{1}{2}O_2 + H_2O + 2e^- \rightarrow 2OH^-. \tag{4.11}$$

Not all redox reactions are hydrogen–oxygen reactions. One important redox reaction for high-temperature fuel cells is the oxidation reaction of carbon monoxide:

$$CO + \frac{1}{2}O_2 \rightarrow CO_2. \tag{4.12}$$

The redox pair is

$$\frac{1}{2}O_2 + 2e^- \rightarrow O^{2-} \tag{4.13}$$

and

$$CO + O^{2-} \rightarrow CO_2 + 2e^-. \tag{4.14}$$

High-temperature SOFCs may be fueled with a mixture of hydrogen and carbon monoxide, also known as synthetic gas or syngas. In this case, the conducting ion or charge carrier is still the negatively charged oxygen atoms traveling from the cathode to the anode, where hydrogen and carbon monoxide are oxidized, and water and carbon dioxide form as a result of the electrochemical reactions in (4.5) and (4.14), respectively. It is possible that water forming as a product at the anode can oxidize CO to CO_2 in a water–gas shift reaction, forming more hydrogen. The hydrogen then participates in the electrochemical reaction.

If a high-temperature SOFC is supplied with methane instead of a carbon monoxide–hydrogen mixture, methane oxidation reaction may take place:

$$CH_4 + 2O_2 \rightarrow CO_2 + 2H_2O. \tag{4.15}$$

This reaction produces eight electrons for each methane molecule consumed (remember that the charge of O is -2 so producing four negatively charged oxygen atoms requires eight electrons). This can also be verified by adding the reactions in (4.1) (or (4.5)) and (4.12) (or (4.14)) with proper stoichiometry. The details of the electrochemical reactions that consume methane depend on the conditions, and several routes are possible. Another representation of the methane electrochemical oxidation is as follows. In the direct oxidation of methane, at the anode, the following reduction reaction occurs:

$$CH_4 + H_2O \rightarrow 6H^+ + 6e^- + CO. \tag{4.16}$$

In addition to the six electrons produced in this reaction, two more electrons are produced at the anode in the electrochemical oxidation of CO that forms in this reaction, according to (4.14). Alternatively, methane may first be reformed internally to hydrogen and carbon monoxide on the anode side through reactions with the water that forms as a product in the electrochemical reactions. In this case, methane is first electrochemically reformed in the following reaction:

$$CH_4 + O^{2-} \rightarrow CO + 2H_2 + 2e^-, \tag{4.17}$$

producing two electrons and forming carbon monoxide and hydrogen, which are then oxidized according to (4.14) and (4.5), producing six more electrons. Another possible route is a water–gas shift reaction following the reforming reaction, in which CO is converted to CO_2 and H_2 is produced. Hydrogen then reacts electrochemically. The net result of all these alternative routes is the same for the overall oxidation reaction.

In **direct methanol fuel cells**, a mixture of methanol and water is used as a fuel, and oxygen or air is used as an oxidizer. The overall reaction in this case is

$$CH_3OH + \frac{3}{2}O_2 \rightarrow CO_2 + 2H_2O. \tag{4.18}$$

The corresponding electrochemical reactions are

$$CH_3OH + H_2O \rightarrow 6H^+ + 6e^- + CO_2 \tag{4.19}$$

and (4.2).

A rule that emerges from these reactions is that for every oxygen molecule consumed in the overall reaction, four electrons are consumed on the cathode side (or two electrons for each oxygen atom). Thus, four electrons must be produced on the anode side for each oxygen molecule consumed to maintain electroneutrality.

4.2.2 Types of Electrochemical Cells

A redox reaction is written symbolically as the following pair [1–3]:

$$\mathrm{Red_b} \Rightarrow \mathrm{Ox_b} + \nu_e e^- \qquad \text{Oxidation (anode)}, \tag{4.20}$$

$$\mathrm{Ox_a} + \nu_e e^- \Rightarrow \mathrm{Red_a} \qquad \text{Reduction (cathode)}. \tag{4.21}$$

Electrochemical reactions must individually satisfy charge conservation and overall neutrality, and therefore Red and Ox can be charged. In more complex electrochemical cells the oxidizing agent in the reduction reaction, $\mathrm{Ox_a}$, may be different than that produced in the oxidation reaction, $\mathrm{Ox_b}$. The same is true for the reducing agent. In such cases, there are two different electrolytes for the two electrodes, and the electrolytes are separated. In this section, we discuss three types of electrochemical cells:

1. A cell at equilibrium, which produces zero net current while maintaining a finite voltage across the two electrodes, which is equal to the ideal or open-cell potential;
2. A galvanic cell producing a finite (electronic) current from the anode to the cathode while maintaining a voltage between the two terminals that is less than the ideal value; and

3. An electrolytic cell, in which a power source is attached to the two terminals/electrodes that imposes a voltage larger than the ideal cell voltage and drives a current through the cell.

The definitions of the anode – where the oxidation of a reactant occurs and it loses electrons – and the cathode – where the reduction of another reactant occurs and it gains electrons, are independent of whether the cell is galvanic or electrolytic.

At equilibrium, the two halves of the electrochemical reactions reach equilibrium locally, resulting in zero net current between the two electrodes. The currents across the anode–electrolyte interface, produced by the forward and backward reactions of the oxidation half-reactions (i.e., (4.20)), are equal and opposite, so the net local current is zero. Furthermore, the currents across the cathode–electrolyte interface, produced by the forward and backward reactions of the reduction half-reaction (i.e., (4.21)), are equal and opposite, and the net current is zero. As will be shown in later sections, each half-reaction produces a potential jump at the interface where it takes place, supported by the charge separation at the electrode surface (see Section 4.7). That potential jump is related to the Gibbs free energy of the half-reaction. The relationship between the Gibbs free energy and the local potential jump will be derived shortly. At equilibrium, the voltage of the cathode is higher than the voltage of the anode. The ideal potential difference is called the open-circuit potential, $\Delta\varepsilon$. The balance between the chemical energy of each half-reaction and the potential difference across the interface keeps the cell at equilibrium, unless a finite current is drawn from the cell, thereby resulting in deviation of the voltage jump from its equilibrium value.

In a **galvanic** cell, shown in Figure 4.2, the cathode has a higher potential than the anode, but the difference is smaller than that achieved under equilibrium. The potential losses occur

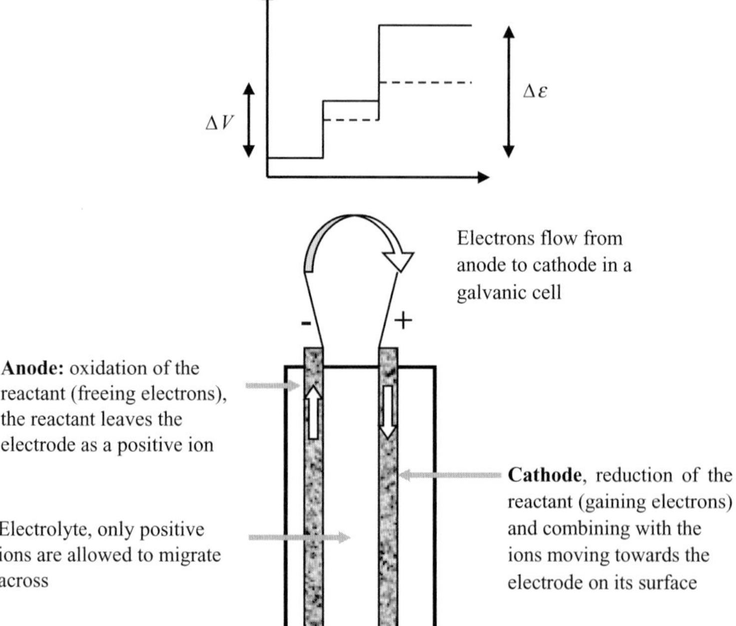

ΔV
Δε
Electrons flow from anode to cathode in a galvanic cell
-
+
Anode: oxidation of the reactant (freeing electrons), the reactant leaves the electrode as a positive ion
Cathode, reduction of the reactant (gaining electrons) and combining with the ions moving towards the electrode on its surface
Electrolyte, only positive ions are allowed to migrate across

Figure 4.2 A simple galvanic cell. The small arrows indicate the direction of motion of electrons. The positively charged ions move inside the cell in the opposite direction to that of the electrons. The solid lines in the potential diagram show the equilibrium potential differences, and the broken lines show the case under finite current operation, in which case losses reduce the available external potential.

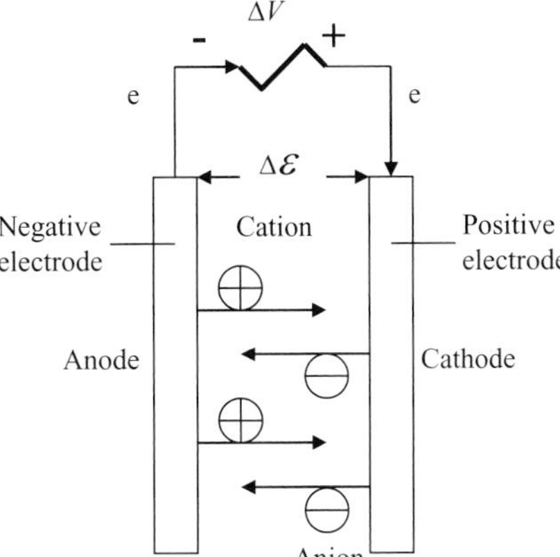

Figure 4.3 Galvanic or voltaic cell. When a finite current is drawn, the voltage difference across the external resistance, $\Delta V < \Delta\varepsilon$, where $\Delta\varepsilon$ is the equilibrium potential of the reaction. The polarity of this galvanic cell is such that the anode (supplying the electrons) is the negative electrode and the cathode is the positive electrode.

because of the finite current. Inside the cell, negative ions or **anions**, flow toward the anode, and positive ions or **cations**, move toward the cathode, as shown in Figure 4.3. The opposite flow of equal amounts of anions and cations establishes the conditions of **electric neutrality** inside the cell. Given these definitions, the positive ions flow internally toward the cathode, and negative ions flow toward the anode. In the galvanic or voltaic cell, electrons produced in the electrochemical reaction flow externally as a result of the internal spontaneous reaction, which also establishes an external voltage that increases from the anode side (–) to the cathode side (+). Electrons flow from the negatively charged anode, where they are produced, to the positively charged cathode, where they are consumed. In actual operations, because of internal losses that will be discussed later, the production of a current affects the voltage at the cell terminals such that $\Delta V \leq \Delta\varepsilon$, where ΔV is the actual voltage difference across the external resistance, and $\Delta\varepsilon$ is the ideal equilibrium voltage difference across the cell under zero current conditions.

The opposite of a galvanic cell is an **electrolytic** cell, in which an external voltage is supplied to drive a redox reaction. In the case shown in Figure 4.4, the cell electrolyzes water into hydrogen and oxygen. *The anode is still where oxidation occurs* (by definition, where electrons are produced). However, in this case, electrons are drawn from the anode using the external power source, leaving the anode positively charged with the hydrogen ions (only neutral oxygen molecules leave the electrode in this case) according to the reverse of the reaction in (4.2); that is:

$$H_2O \Rightarrow \frac{1}{2}O_2 + 2H^+ + 2e^-. \tag{4.22}$$

Electrons flow toward the cathode, where they accumulate, making it negatively charged. Hydrogen ions migrate toward the anode, where they recombine, forming hydrogen molecules that leave the cell. *The reaction at the cathode side is a reduction reaction, where electrons are consumed*, and it is the reverse of reaction (4.3); that is:

$$2H^+ + 2e^- \Rightarrow H_2. \tag{4.23}$$

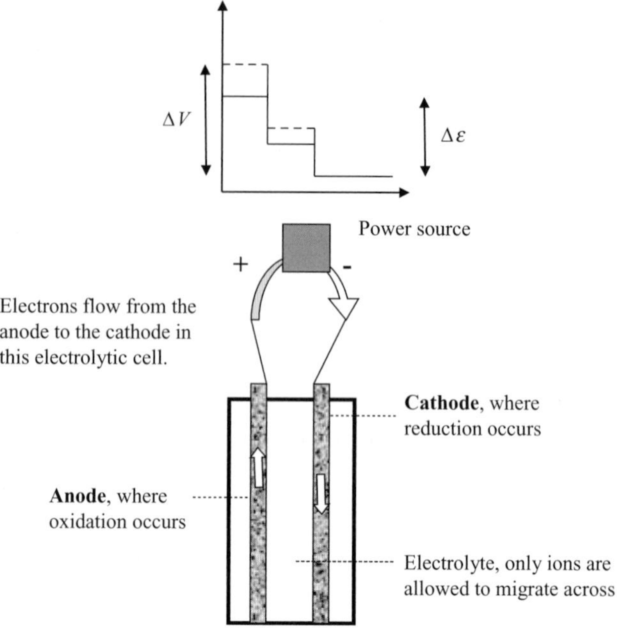

Figure 4.4 A simple electrolytic cell. The small arrows indicate the direction of motion of electrons. The positively charged ions move inside the cell in the opposite direction. Often, neutral species are removed from both electrodes. The solid lines in the potential diagram show the equilibrium potential differences, and the broken lines show the case under finite current operation.

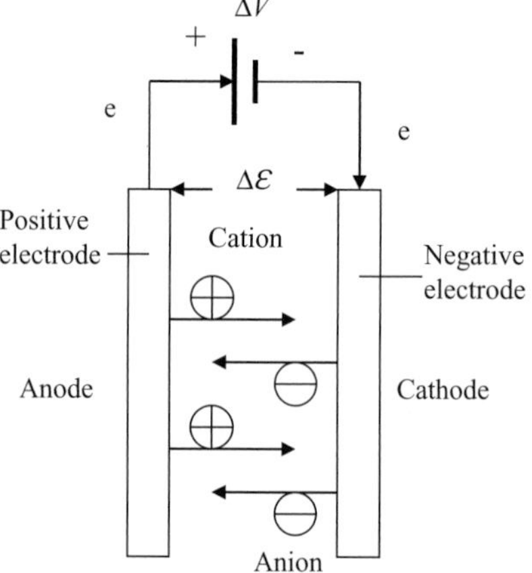

Figure 4.5 Electrolytic cell. In this case, a battery is connected to supply a potential to overcome the equilibrium potential of the reaction, $\Delta V > \Delta \varepsilon$. The cathode is now negatively charged, supplied externally with electrons, while the anode is positively charged.

In an electrolysis cell, an external voltage is applied to energize an internal reaction that breaks up water molecules (or other molecules such as CO_2). In this case, electrons flow externally from the positively charged anode to the negatively charged cathode. Internally, and similar to the galvanic cell, oxidation or loss of electrons occurs at the anode, and reduction or gaining electrons occurs at the cathode. The internal charges are carried by anions moving toward the anode and cations flowing toward the cathode, as shown in Figure 4.5. Under actual operating conditions, and because of internal irreversibility that will be discussed later, a finite current affects the voltage at the cell terminals such that $\Delta V \geq \Delta \varepsilon$.

4.3 Thermodynamics of Fuel Cells

One system that may come close to operating under ideal thermodynamic conditions, producing the maximum possible work from a chemical reaction, is a fuel cell, when operating close to equilibrium. In this electrochemical cell, separate streams of oxygen (or air) and hydrogen (or other fuels) are introduced into a "reactor." The fuel and oxidizer react inside the cell to produce electricity, and water exits according to the reaction described in (4.1).

The fuel and the oxidizer flow into the cell in two separate streams. Products formed inside the cell flow out in a different stream, or they are mixed with either the fuel or oxidizer stream. All streams are at the temperature and pressure of the environment. The maximum work produced per mole of hydrogen (or the fuel) can be calculated using the application of the First and Second Laws of Thermodynamics to the reversible open system shown in Figure 4.6. The system undergoes an isothermal change at the temperature of the environment, and hence:

$$\begin{aligned} Q - W &= H_{out} - H_{in} \\ &= \Delta H_R \end{aligned} \tag{4.24}$$

and

$$\begin{aligned} \frac{Q}{T^*} &= S_{out} - S_{in}, \\ &= \Delta S_R \end{aligned} \tag{4.25}$$

where T^* is the temperature of the electrochemical cell, which is maintained at the temperature of its immediate environment. It is the same as the temperature of the reactants entering and the products leaving the cell. Solving for the heat and work interactions, we get

$$Q = T^* \Delta S_R((T^*, p^*)) \tag{4.26}$$

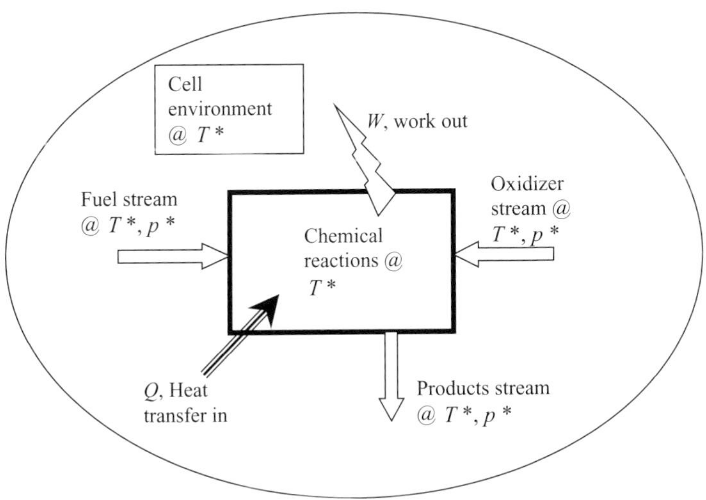

Figure 4.6 Mass and energy balances of a typical fuel cell.

and

$$\begin{aligned}-W &= (H - T^* S)_{out} - (H - T^* S)_{in} \\ &= \Delta G_R(T^*, p^*).\end{aligned} \tag{4.27}$$

Equation (4.27) shows the meaning of the Gibbs free energy of reaction: It is the maximum (non-$p\forall$) work done by a system undergoing a chemical reaction. For isothermal reactions at equilibrium with their environment, reversible heat and work transfer interactions should be expected. Both depend on the temperature due to the dependence of the enthalpy of reaction and entropy of reaction on temperature (which is relatively weak for low and intermediate temperatures).

The Gibbs free energy of reaction at the cell state is given by (see Chapter 3)

$$\begin{aligned}-\Delta G_R(T^*, p^*) &= \sum_{react} \nu_i' \left[\hat{h}_i(T^*) - T^* \hat{s}_i(T^*, p^*)\right] \\ &\quad - \sum_{prod} \nu_I'' \left[\hat{h}_i(T^*) - T^* \hat{s}_i(T^*, p^*)\right],\end{aligned} \tag{4.28}$$

where p^* is the partial pressure calculated in the corresponding streams (i.e., reactants and products). The Gibbs free energy and the enthalpy of reaction are evaluated at the conditions of the cell, (T^*, p^*). The concentrations of the reactants and products are calculated in the corresponding reactants and products streams, noting that the fuel and oxidizer are introduced in separate streams. The products concentration must be evaluated in that stream. As shown in the introduction to electrochemical reactions, depending on the type of electrolyte used, products may mix with the oxygen or the fuel streams. Moreover, depending on the application, oxygen can be used or can be part of air.

The "thermodynamic" or galvanic efficiency of the cell is defined as the ratio of the open-circuit (zero current) work interaction divided by the enthalpy of reaction:

$$\eta_{OC} = \frac{W_{max}}{\Delta H^*_{R,H_2O}} = \frac{\Delta G^*_{R,H_2O}}{\Delta H^*_{R,H_2O}}. \tag{4.29}$$

The efficiency calculated from this equation is called the "open-circuit" or maximum work efficiency, since it assumes no current flow, which corresponds to equilibrium conditions. The choice of the enthalpy of reaction is arbitrary; the higher heating value (HHV) is used, except when otherwise indicated. Thus,

$$\eta_{OC} = \frac{-\Delta G^*_{R,H_2O}}{HHV_{H_2O}}. \tag{4.30}$$

Actual efficiencies are lower because of several internal irreversibilities associated with the resistance to ion flow, finite chemical reaction rates, transport of chemical species within the channels, and incomplete consumption of fuel. Under the conditions of finite currents, the concentrations change within the cell as the fuel and oxygen are consumed and products are formed and mixed in the reactants streams. These lead to a change in the value of the Gibbs free energy locally, as will be discussed in Chapter 7.

With oxygen and hydrogen (or any other fuel) flowing into the cell in separate streams, both in the pure form (i.e., $(X_{O_2})_{oxygen\ stream} = 1$ and $(X_{H_2})_{fuel\ stream} = 1$), at the pressure and temperature of the environment, and water (or other products of reaction) leaving in the vapor phase (i.e., $(X_{H_2O})_{products\ stream} = 1$), the Gibbs free energy of reaction is given by

$$\Delta G^o_R(T) = \sum_{react} \nu'_i \left[\hat{h}^o_i(T) - T\hat{s}^o_i(T)\right] - \sum_{prod} \nu''_i \left[\hat{h}^o_i(T) - T\hat{s}^o_i(T)\right]. \tag{4.31}$$

In this case of individual pure reactants streams and a pure products stream, at the standard conditions for pressure and temperature, $T = 300$ K and $p = 1$ atm, and with water leaving in the vapor phase, $\Delta H_{R,H_2O} = -242$ MJ/kmol, $\Delta G_{R,H_2O} = -228$ MJ/kmol, the open-circuit efficiency based on the low heating value (LHV) is 94%. With water leaving in the liquid phase, $\Delta H_{R,H_2O} = -286$ MJ/kmol, $\Delta G_{R,H_2O} = -237$ MJ/kmol, the open-circuit efficiency based on the HHV is 83%. The efficiency is lower because we used the higher value of enthalpy of reaction in the definition of the efficiency, although the maximum work is also higher.

It is sometimes desirable to run the fuel cell at higher temperatures. High temperatures accelerate chemical reactions and reduce the need for catalysts. High temperature can also reduce the ionic resistance of certain electrolytes. To determine the impact of temperature on the open-circuit efficiency, we perform the following calculations. Using the definitions of the enthalpy of reaction and the Gibbs free energy of reaction in (4.31), we can show that at 500 K and 1 atm, with water in the vapor phase, $\Delta H_{R,H_2O} = -242$ MJ/kmol, $\Delta G_{R,H_2O} = -219$ MJ/kmol, and the conversion efficiency is 76.5%. At 1000 K and 1 atm, the enthalpy of reaction is $\Delta H_{R,H_2O} = -245$ MJ/kmol, the Gibbs free energy is $\Delta G_{R,H_2O} = -192.6$ MJ/kmol, and the conversion efficiency is 67.3%. The enthalpy of reaction changes slightly within this range because of a slight change in the specific heat. The change in entropy is notable because of the logarithmic dependence on temperature, and the change in the Gibbs free energy with temperature is remarkable because of the $(-T\Delta S_R)$ term. At higher temperatures, the cell rejects more of the fuel chemical energy in the form of heat transfer to its environment, and its efficiency decreases. That high-temperature "waste heat" could be used in a heat engine to generate work. This is the idea of the hybrid fuel cell gas turbine cycle that will be discussed later.

Figure 4.7 shows the essential elements in a fuel cell. Note that the reactants (i.e., fuel and oxidizer) are admitted in separate streams – they do not mix. Instead, charged species are formed separately at the **electrode–electrolyte** interfaces, and they migrate across the electrolyte to form products. The products leave on either side of the cell with the fuel or the oxidizer stream, depending on the design of the cell. An **ideal** electrolyte (liquid or solid) is a medium that acts as a perfect electrical insulator – that is, does not allow **electrons** to pass through – a perfect conductor of a certain type of **ions**, depending on the electrolyte material, and a perfect insulator of uncharged species. Therefore, in these electrochemical reactors, uncharged species do not mix or react spontaneously. Instead, a controlled reaction is allowed on the two different interfaces between the electrolyte and the electrodes, resulting in freeing electrons and their flow outside the cell. Ideally, thermochemical reactions are not allowed to take place in these cells,

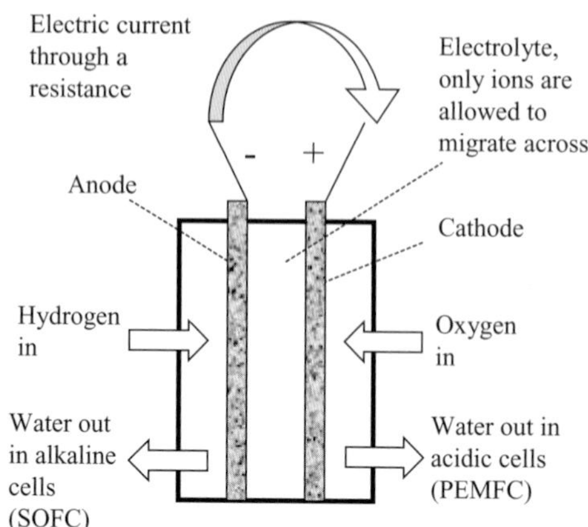

Figure 4.7 A fuel cell using pure oxygen and hydrogen to produce water. In an acidic cell, similar to PEMFC, in which the electrolyte conducts the positive ions, water forms on the cathode side and leaves on the side of the oxidizer. In an alkaline cell, in which the electrolyte conducts the negative ions, water forms on the anode side and leaves on the side of the fuel.

4.3.1 Electrochemical Cell Voltage

The electric potential is another characteristic parameter of a fuel cell, which can be obtained using the expression of the maximum work. The electrical work can be written in terms of the potential difference between the electrodes, $\Delta\varepsilon$, measured in volts, and the charge, ς, measured in coulombs:

$$W_{el} = W_{max} = \Delta\varepsilon\, \varsigma. \tag{4.32}$$

The chemical reactions that free up the electrons depend on the fuel cell type. In a hydrogen–oxygen cell (4.1), per mole of water forming or hydrogen consumed, the total charge produced is $2|\varsigma_{e^-}|N_a$, where 2 is the number of electrons generated for each molecule reacting and

$$N_a = 6.023 \times 10^{23}\ \text{mole}^{-1} \tag{4.33}$$

is Avogadro's number, or the number of molecules per mole, and the charge carried by an electron, e^-, is

$$\varsigma_{e^-} = -1.602 \times 10^{-19}\ \text{coulombs}. \tag{4.34}$$

The combination

$$\Im_a = |\varsigma_{e^-}|N_a = 9.6485 \times 10^4\ \text{coulombs/mole} \tag{4.35}$$

is Faraday's number, or the charge of a mole of electrons.

Substituting in the maximum work equation (4.31), the ideal electric field potential at 300 K is

$$\Delta\varepsilon = \frac{-(-228{,}000)}{2 \times 96{,}485} = 1.18\ \text{volts if water leaves in vapor form,}$$

$$\Delta\varepsilon = \frac{-(-237{,}000)}{2 \times 96{,}485} = 1.23\ \text{volts if water leaves in liquid form.}$$

Example 4.1

Determine the standard open-circuit potential for the carbon monoxide–oxygen reaction in (4.12), the methane reaction in (4.15), and methanol–oxygen reaction in (4.18).

Solution

The Gibbs free energy of the carbon monoxide oxidation reaction at standard conditions is $\Delta G^o_{R,\mathrm{CO}} = -257.2$ kJ/mol. The reaction produces two electrons. Hence,

$$\Delta\varepsilon^o = \frac{-(-257{,}200)}{2 \times 96{,}485} = 1.33 \text{ volts.}$$

The Gibbs free energy of the methane oxidation reaction at standard conditions is $\Delta G^o_{R,\mathrm{CH4}} = -803.4$ kJ/mol. The reaction produces eight electrons. So,

$$\Delta\varepsilon^o = \frac{-(-803{,}400)}{8 \times 96{,}485} = 1.04 \text{ volts.}$$

The Gibbs free energy of the methanol–oxygen reaction at standard conditions is $\Delta G^o_{\mathrm{R,CH_3OH}} = -698$ kJ/mol. The number of electron exchanges for each molecule consumed is six. Hence, the standard cell potential is

$$\Delta\varepsilon^o = \frac{-(-698{,}000)}{6 \times 96{,}485} = 1.21 \text{ volts.}$$

The inverse of a fuel cell is an electrolyzer, in which water is introduced and an external electric field is applied to the electrodes. The chemical reaction inside an electrolyzer is described by (4.22). It is the reverse of the reaction that takes place inside a fuel cell. The equations derived for the fuel cell still apply, but the reactants and products are now reversed. Thus, for water electrolysis, water is the reactant (in) and oxygen and hydrogen are the products (out). The elements of an electrolyzer are shown in Figure 4.8.

Unlike fuel cells, whose maximum electrical work is the Gibbs free energy of the reaction taking place within the cell, in the case of an electrolyzer the Gibbs free energy is in fact the minimum work required to split water into hydrogen and oxygen. On the other hand, the ideal potential of a fuel cell is the maximum voltage that can be obtained in the absence of losses (to be discussed in Chapter 7), whereas the ideal potential of an electrolyzer calculated from the Gibbs free energy is the minimum voltage required to run the electrolyzer for splitting water. The comparison of fuel cell and electrolyzer is somewhat similar to the comparison of heat engine and refrigerator. If we operate a heat engine (a work-producing device) in a reverse direction, it would be a refrigerator (a work-consuming device), and vice versa. Likewise, if a fuel cell (a work-producing device) is operated inversely, it would be an electrolyzer (a work-consuming device), and vice versa. Based on these considerations, one may use the equations derived for a fuel cell to model the performance of an electrolyzer.

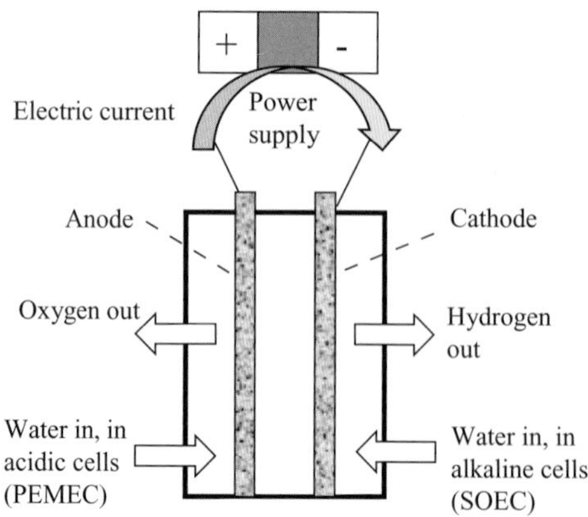

Figure 4.8 An electrolyzer cell, in which pure oxygen and pure hydrogen are produced by electrochemically splitting water using an electric current.

4.4 Nernst Equation

Direct energy conversion, in which chemical energy is converted into electrical energy directly without intermediate steps, was described in Chapter 3. There, we obtained expressions for the maximum work and efficiency and related them to "chemical reaction or fuel availability." Now we apply these concepts to fuel cells. Consider the case in which the fuel and oxidizer streams flow into and out of the cell at (p^*, T^*). The ideal work done by the cell is the Gibbs free energy of reaction (see (3.74)):

$$(-\Delta G_R(p^*, T^*)) = \left(-\Delta G_R^o(T^*)\right) - \Re T * \ln\left(\frac{\prod\limits_{prod}(p_i^*)^{\nu_i''}}{\prod\limits_{react}(p_i^*)^{\nu_i'}}\right). \tag{4.36}$$

In this equation, p_i^* is the partial pressure of the species. The change in the Gibbs free energy is for the overall reaction. Recall that $\Delta G_R^o(T^*)$ is evaluated using (3.75):

$$\Delta G_R^o(T^*) = -\Re T \ln K_p(T^*), \tag{4.37}$$

where K_p is the partial pressure-based equilibrium constant. Since we assume that the cell and its "environment" are at the same temperature and total pressure, then for ideal gas streams, (4.36) reduces to

$$(-\Delta G_R(p^*, T^*)) = \left(-\Delta G_R^o(T^*)\right) - \sigma \Re T * \ln(p^*) - \Re T * \ln\left(\frac{\prod\limits_{prod} X_i^{\nu_i''}}{\prod\limits_{react} X_i^{\nu_i'}}\right). \tag{4.38}$$

In this expression, $\sigma = \sum_{prod} \nu_i'' - \sum_{react} \nu_i'$; all other parameters, were defined earlier. The first term on the right-hand side is given in (4.31). Recall that the reactants enter in different streams, and each reactant may be mixed with other components in its own stream.

For instance, air could be used as an oxidizer. In this case, oxygen is mixed with nitrogen. The reactants meet within the membrane electrode assembly (MEA) to undergo the electrochemical reaction. As will be shown later, the MEA, made of an electrolyte sandwiched between two electrodes, is very thin, and it is reasonable to assume that all the reactants are at the same temperature. Determining the exact concentration of each reactant as the streams flow along the electrodes requires a detailed analysis of transport within the MEA; a subject that will be tackled in detail later. Here, equilibrium is assumed. In most cases, the products leave in either the fuel stream or the oxidizer stream (in special cases products can leave in a separate stream).

4.4.1 Dependence on Pressure and Temperature

Equation (4.38) shows the important impacts of the cell pressure and temperature, and the concentrations of the fuel, oxidizer, and products on the work produced by the cell. As the pressure within the cell rises, the maximum work increases/decreases if the number of moles in the reaction decreases/increases from the reactants side to the products side in (4.6). The work also increases as the concentrations of the fuel and oxidizer rise.

The maximum work is $(-\Delta G_R) = n_e \Im_a \ \Delta\varepsilon$, where n_e is the number of electrons produced/consumed by the electrochemical reaction at the electrodes. The ideal open-circuit EMF (electromotive force) or voltage change across the electrodes of the cell can be expressed in terms of the partial pressures of the reactants and products using (4.36):

$$\Delta\varepsilon(p^*, T^*) = \Delta\varepsilon^o(T^*) - \frac{\Re T^*}{n_e \Im_a} \ln \left(\frac{\prod\limits_{prod} \left(p_i^*\right)^{v_i''}}{\prod\limits_{react} \left(p_i^*\right)^{v_i'}} \right). \tag{4.39}$$

The first term on the right-hand side is the ideal potential with all species at 1 bar. It can be expressed as

$$\Delta\varepsilon^o(T^*) = \frac{-\Delta G_R^o(T^*)}{n_e \Im} = \frac{\Re T^*}{n_e \Im_a} \ln K_p(T^*). \tag{4.40}$$

The partial pressure of each reactive species in reactants or products, in its own stream, is $p_i^* = X_i p*$. Written explicitly in terms of the pressure of the cell/environment and the mole fractions of the species in their respective streams, the equilibrium potential across the cell is

$$\begin{aligned} \Delta\varepsilon_{max}(p^*, T^*) &= \Delta\varepsilon^o(T^*) - \frac{\sigma \Re T^*}{n_e \Im_a} \ln (p^*) - \frac{\Re T^*}{n_e \Im_a} \ln \left(\frac{\prod\limits_{prod} X_i^{v_i''}}{\prod\limits_{react} X_i^{v_i'}} \right) \\ &= \Delta\varepsilon^o(T^*) + \Delta\varepsilon_p(p^*, T^*) + \Delta\varepsilon_{conc}(X_i, T^*). \end{aligned} \tag{4.41}$$

Equation (4.41) is the **Nernst equation**. It is used to predict the dependence of the theoretical, maximum open-circuit voltage or potential, or the Nernst voltage, of an electrochemical reaction, on the fuel, temperature and pressure, and concentrations of the fuel and oxidizer at the electrodes. In the second line of (4.41), $\Delta\varepsilon^o(T^*)$ shows the dependence of the open-circuit voltage on the operating temperature when all the streams are at 1 bar and the fuel, oxidizer, and products are at unity concentration; $\Delta\varepsilon_p(T^*, p^*)$ shows the dependence on the cell pressure, and $\Delta\varepsilon_{conc}(X_i, T^*)$ shows the dependence on the concentrations of the fuel, oxidizer, and products. Clearly, higher pressures and higher reactants concentrations raise the EMF of a hydrogen–oxygen fuel cell, especially at higher temperatures. Recall that the prefactor $\Re/\Im_a = 8.6 \times 10^{-5}$ volt/K. The concentrations of the fuel, oxygen, and products in (4.39) and (4.41) are the local values where the chemical reaction takes place. In calculating the products concentration, (4.41) assumes complete consumption of reactants. In a working cell, the concentrations change along the electrode surfaces as fuel and oxygen are consumed and products are formed. The analysis will be extended to the case of incomplete consumption of reactants later.

As an example, consider a hydrogen–oxygen fuel cell, in which $\sigma = -½$. The governing overall reaction is given in (4.1). Under reversible, ideal conditions, and with thermal and mechanical equilibrium with the environment, the pressure and concentration dependence of the electrical potential is

$$\Delta\varepsilon_{max}(p^*, T^*) = \Delta\varepsilon^o_{max}(T^*) + \frac{\Re T^*}{4\Im_a} \ln(p^*) + \frac{\Re T^*}{2\Im_a} \ln\left(\frac{X_{H_2}\sqrt{X_{O_2}}}{X_{H_2O}}\right). \tag{4.42}$$

This is the Nernst equation of a hydrogen–oxygen fuel cell. The dependence of the open-circuit potential on the temperature is shown in Figure 4.9 (see Figure 4.10 as well). The dependence on temperature is a result of the change in the Gibbs free energy and, as is typical of many reactions, the Gibbs free energy of reaction decreases as the temperature increases. For the hydrogen–oxygen reaction, at atmospheric pressure and with pure hydrogen and oxygen streams and the products stream at one atmosphere and unity mole fractions, the Nernst voltage drops from 1.185 V at 25 °C to 0.960 V at 1000 °C [4]. The change in the Nernst potential with temperature is much larger for the CO reaction, dropping from 1.333 V at 25 °C to 0.896 V at 1000 °C. On the other hand, for methane

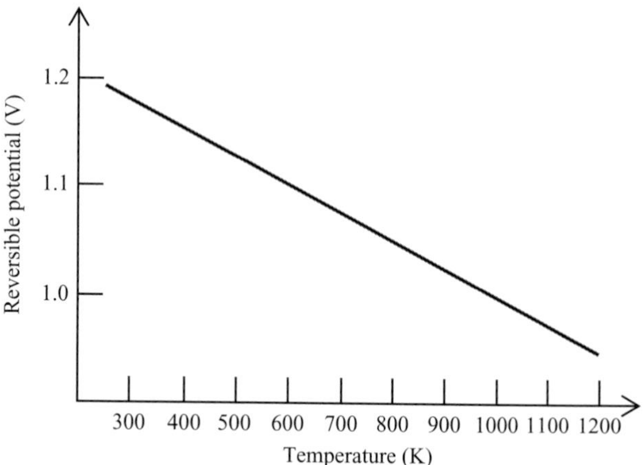

Figure 4.9 Dependence of the ideal potential in a hydrogen–oxygen fuel cell on the temperature of the cell.

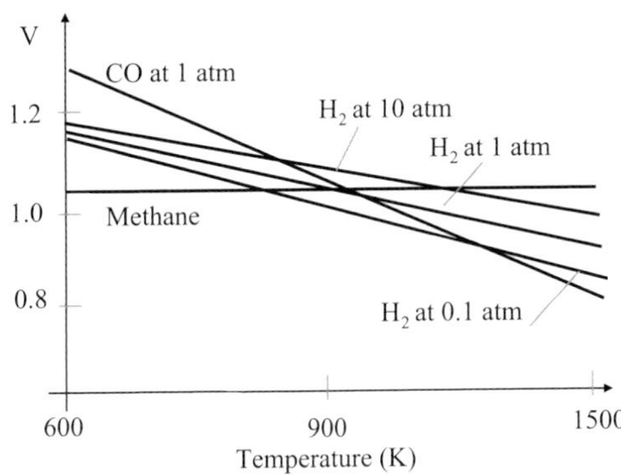

Figure 4.10 Dependence of the cell potential on the pressure and temperature for CO, shown at 1 atm, hydrogen shown at three pressures, and methane shown at any pressure.

reactions, there is almost no change in the open-circuit volt within this temperature range, as shown in Figure 4.10.

The dependence of the open-circuit potential on the cell pressure is depicted by the second term in (4.41) in general, and in (4.42) for hydrogen–oxygen reactions. This dependence is affected strongly by the change in the number of moles during the reactions, as expressed by σ. For negative values of σ, as in the cases of hydrogen–oxygen and the carbon monoxide–oxygen reactions, the open-circuit potential increases with pressure (note that mechanical equilibrium means that the cell and all the streams are at the same pressure as the environment). Moreover, $\Delta\varepsilon_{max}$ increases almost linearly with the operating temperature, as shown in Figure 4.10. Similar trends are observed for carbon monoxide electrochemical oxidation for which $\sigma = -1/2$. On the other hand, because $\sigma = 0$ for methane oxidation, the open-circuit potential is pressure-independent. Note how the decrease in the cell potential with temperature contrasts that of the heat engine (a Carnot engine for simplicity), where the work and efficiency increase with temperature.

4.4.2 Dependence on Concentrations

The dependence of the open-circuit potential on the reactants and products concentrations is captured by the last term on the right-hand side in the Nernst equation (4.41). For the case of hydrogen–oxygen reaction, (4.42), we can rewrite this term as follows:

$$\Delta\varepsilon_{conc} = \frac{\Re T^*}{2\Im_a}\left[\ln\left(X_{H_2}\right)_{fl} + \frac{1}{2}\ln\left(X_{O_2}\right)_{ox} - \ln\left(X_{H_2O}\right)_{sep/fl/ox}\right]. \tag{4.43}$$

In this equation, subscript "fl" and "ox" refer to the fuel and oxidizer channels, and "sep/fl/ox" indicate that the products may be found in a separate channel – the fuel or the oxidizer channel, respectively – depending on the design of the cell. *In principle, this equation represents the operation of a cell at equilibrium where the concentrations of the fuel, oxygen, and products have the same value everywhere along the electrodes surfaces. It also assumes that the fuel and oxidizer have been consumed and the products have been formed.* Equation (4.43) shows that $\Delta\varepsilon_{conc}$ becomes more significant as the operating temperature increases.

Example 4.2

Determine the change in the open-circuit potential of a hydrogen–oxygen cell if air is used as an oxidizer instead of oxygen.

Solution

In the air stream is $X_{O_2} = 0.21$, following (4.43), using air reduces the cell potential by $\Delta\varepsilon = \frac{\Re T^*}{4\Im_a}\left[\ln\left(\frac{X_{O_2}}{1}\right)_{ox}\right]$. At 300 K, $\Delta\varepsilon = -0.01$ V. The cell potential drops 0.01 V if air is used compared to the case with pure oxygen.

If both fuel and oxygen are diluted or are introduced as part of a carrier gas, then the concentration potential is

$$\Delta\varepsilon_{conc} = \frac{\Re T^*}{2\Im_a}\left[\ln\left(X_{H_2}\right)_{fl} + \frac{1}{2}\ln\left(X_{O_2}\right)_{ox} - \ln\left(X_{H_2O}\right)^{**}_{sep/fl/ox}\right]. \tag{4.44}$$

In this equation, the products concentration in the stream, $\left(X_{H_2O}\right)^{**}_{sep/fl/ox}$, is evaluated using the assumption that all the fuel and oxygen have been consumed.

In an operating fuel cell with finite-length electrode, as the streams flow along the electrodes from the inlet to the exit, oxygen and fuel are consumed and their concentrations decrease continuously. Moreover, products form and their concentrations increase along the length of the electrodes. Depending on the design, products diffuse toward the oxygen channel (as in the case of a PEMFC) or the fuel channel (as in a SOFC), where their concentrations rise as they move from the inlet side to the exit side, further reducing the concentration of the fuel or oxygen. The reduction of the reactants concentrations reduces the "local" potential. This reduction in the open-circuit reduction is known as the "fuel utilization" effect. For instance, oxygen utilization reduces the value of $\left(X_{O_2}\right)_{ox}$ between the inlet and the exit, and hence increases the negative value of $\Delta\varepsilon_{conc}$ and reduces $\Delta\varepsilon_{max}$. The lowest value of the potential is encountered at the exit. Since the electrodes are designed to be almost perfect conductors, the lowest value of the potential is often taken as the characteristic value of the cell potential. The reduction in the cell potential because of fuel consumption is known as "**utilization overpotential**" and is one of the "loss" mechanisms in fuel cells.

Example 4.3

Determine the change in the open-circuit voltage of a fuel cell operating at 250 °C if a mixture of 50% hydrogen and 50% carbon dioxide is used instead of the case with pure hydrogen.

Example 4.3 (cont.)

Solution

The change in ideal voltage due to a change in concentration of hydrogen is $\Delta\varepsilon = \frac{\Re T^*}{2\mathfrak{I}_a}\left[\ln\left(\frac{X_{2,H_2}}{X_{1,H_2}}\right)\right]$. For the case of pure hydrogen, $X_{1,H_2} = 1$, whereas for a mixture of H_2 and CO_2 with equal mole fractions, $X_{2,H_2} = 0.5$. Hence, $\Delta\varepsilon = -0.016$ V. The cell voltage will drop by 0.016 V if a mixture of 50% hydrogen and 50% carbon dioxide is used compared to the case with pure hydrogen as a fuel.

4.4.3 Impact of Fuel/Oxygen Depletion on Cell Potential

Equation (4.41) is used to calculate the cell potential at a point where the concentrations of the fuel and oxidizer are given. However, concentrations change between the inlet and exit of along the channels. In an operating cell, the fuel stream enters the cell on the anode side, and the oxidizer stream enters on the cathode side, on the opposite side of the electrolyte. The streams flow along the surface of the anode/cathode in a long, narrow, winding channel, and the fuel and oxidizer diffuse toward the electrode–electrolyte interface in a direction normal to the electrodes. The products diffuse back toward either the fuel or oxidizer channel, depending on the design. Figure 4.11 shows schematically the operation of a PEM fuel cell using a mixture of oxygen and a diluent and a mixture of hydrogen and diluent. Products

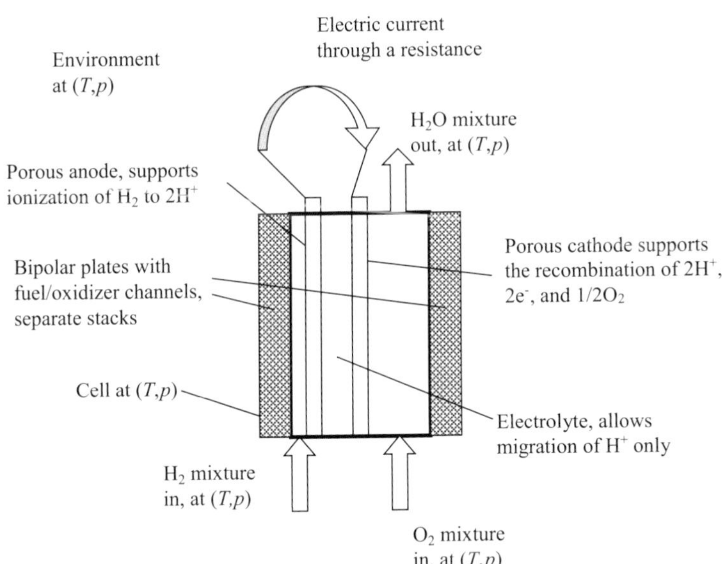

Figure 4.11 A PEM fuel cell using an oxygen mixture and a hydrogen mixture. All components are at the same temperature and pressure as the environment.

form on the oxygen side and flow out on the oxygen stream. All components are at the same temperature and pressure as the environment.

In the case of incomplete fuel utilization (i.e., some of the fuel leaves before being consumed within the cell), the potential is lower than that corresponding to complete fuel utilization or consumption. It is often assumed that the actual cell potential in this case is that corresponding to the concentrations of fuel and oxidizer at the exit.[1] These concentrations must be calculated based on the cell design, inlet concentrations, and stoichiometry. Consider the following reaction:

$$\nu'_{fl}\chi_{fl} + \nu'_{ox}\chi_{ox} \rightarrow \nu''_p\chi_p. \tag{4.45}$$

The calculations are carried out for two different configurations, depending on whether the products form on the fuel (anode) side or the oxidizer (cathode) side. If the products form on the *oxidizer or cathode side*, we have the following: For one mole of fuel, assume the presence of n_{d1} moles diluent, $n_{ox1} = \nu'_{ox}$ moles oxygen (stoichiometric conditions at inlet), $n_{n1} = 3.76\nu'_{ox}$ moles of nitrogen (taking air as an oxidizer). The fuel utilization is defined as

$$\varphi = \frac{n_{fl1} - n_{fl2}}{n_{fl1}}. \tag{4.46}$$

Thus, $\varphi = 0$ corresponds to no fuel utilization, that is, all the fuel leaves the cell, while $\varphi = 1$ is the case in which all the fuel is consumed. Using this definition for the fuel utilization, for one mole of fuel fed to the cell we write $n_{fl2} = 1 - \varphi$ and $n_{d2} = n_{d1}$. Mass conservation requires that

$$\frac{n_{i2} - n_{i1}}{\nu''_i - \nu'_i} = \varphi. \tag{4.47}$$

Thus: $n_{xo2} = (1 - \varphi)\nu'_{ox}$ and $n_{p2} = \nu''_p\varphi$. The mole fractions of the reactants at the inlet of the cell are $X_{fl1} = \frac{1}{1+n_{d1}}$ and $X_{ox1} = 0.21$. The mole fractions of the reactants at the exit are $X_{fl2} = \frac{1-\varphi}{1-\varphi+n_{d1}}$ and $X_{ox2} = \frac{(1-\varphi)\nu'_{ox}}{3.76\nu'_{ox}+(1-\varphi)\nu'_{ox}+\nu''_p\varphi}$. Note that the mole fractions of the fuel and the oxidizer at the exit depend on fuel utilization. For the purpose of calculating the cell potential corresponding to the mole fractions at the inlet or at the exit, complete fuel consumption is assumed, and the products mole fraction is $X_p = \frac{\nu''_p}{n_n+\nu''_p}$. Thus, while computing the open-circuit voltage corresponding to the inlet conditions or the exit conditions, the products mole fraction is the same.

[1] This should be taken as an assumption. In operating cells with a finite current flowing through the electrodes, there may be a finite potential difference between the inlet and the outlet. However, for a well-designed cell, this potential drop should be small.

If the *products form on the fuel/anode side*, the concentrations in the exit stream depend on the fuel utilization. For one mole of fuel, we have n_{d1} moles diluent, $n_{ox1} = v'_{ox}$ moles oxygen (stoichiometric mixture at inlet), $n_{n1} = 3.76v'_{ox}$ moles of nitrogen (again air is used as an oxidizer). With φ fuel utilization, we have $n_{fl2} = 1 - \varphi$ and $n_{d2} = n_{d1}$. Mass conservation is imposed using (4.47). The mole fractions of the reactants at the inlet are $X_{fl1} = \frac{1}{1+n_{d1}}$ and $X_{o1} = 0.21$. The mole fractions of the reactants at the exit are $X_{fl2} = \frac{1-\varphi}{1-\varphi+n_{d1}+v''_p\varphi}$ and $X_{ox2} = \frac{(1-\varphi)v'_{ox}}{3.76v'_{ox}+(1-\varphi)v'_{ox}}$. The products concentration at the inlet and the exit are taken to correspond to complete fuel consumption, $X_{p1} = \frac{v''_p}{n_{d1}+v''_p}$. Similar to the previous case, the concentrations at the exit depend on the fuel utilization. For $n_e = 2$ and $T = 300$ K, the prefactor: $\frac{\Re(8.314\text{J/mol.K})\ T(300\text{ K})}{n_e(2\text{ mol e}^-/\text{mol fuel})\ \Im_a(96{,}487\text{ coul/mol e}^-)} = 0.01293$ V.

4.4.4 Fuel Utilization

The reduction of the reactants concentrations between the inlet and the exit of the fuel and oxidizer channels is known as "fuel utilization." As shown by the Nernst equation, a decrease in the EMF as the reactants concentrations are reduced is relatively small, especially at low temperatures. However, significant fuel utilization can lead to a large drop in the cell potential if the fuel is introduced as part of a mixture, air is used as an oxidizer, and the cell is operating at a high temperature. As mentioned above, because the electrodes are good conductors, the open-circuit potential calculated on the basis of the concentrations at the cell exit is used as the open-circuit potential of the cell. Figure 4.12 shows the results of a calculation that emphasize the effect of fuel utilization

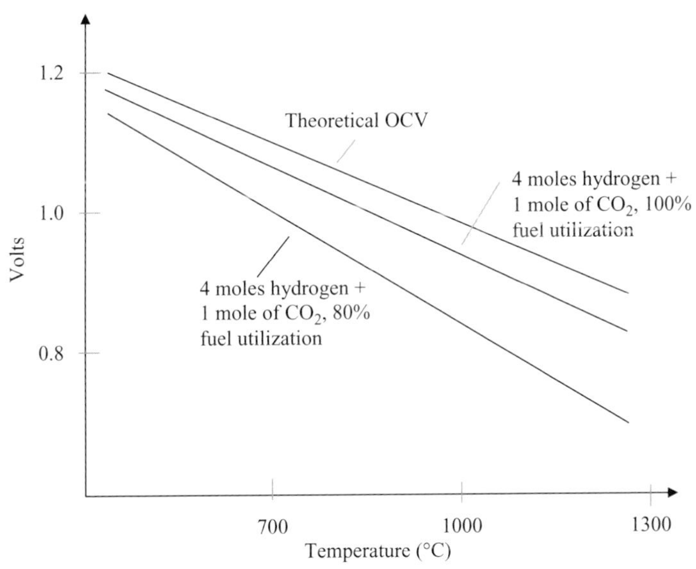

Figure 4.12 Open-circuit potential for a hydrogen–air fuel cell. The top line is the theoretical open-circuit voltage (OCV). The second line is the OCV for a mixture of four moles of hydrogen and 1 mole of CO_2. The third line is the OCV for the same mixture with 80% fuel utilization.

on the open-circuit potential [5]. The top line in this figure shows the dependence of the standard cell potential on the temperature, which is simply a reflection of the fact that the Gibbs free energy of reaction for the hydrogen–oxygen reaction decreases almost linearly with temperature. To obtain this curve, the cell is assumed to operate on pure hydrogen and oxygen, and the open-circuit voltage corresponds to the inlet conditions. The second curve shows the impact of mixing the fuel with CO_2, with four moles of hydrogen and one mole of carbon dioxide (the result of methane steam reforming $CH_4 + 2H_2O \rightarrow CO_2 + 4H_2$) and air is used on the oxidizer side. This curve also shows the cell potential at the inlet condition – that is, without utilization loss. The third and fourth curves correspond to the same case of mixed hydrogen and carbon dioxide, but show the voltage at exit condition with 80% and 90% fuel utilization, both with 50% oxygen utilization. Note how increasing the fuel utilization leads to a substantial drop in the cell potential at high temperatures.

Reducing fuel utilization within the cell means some fuel is left over in the "exit" fuel stream that should be recycled back into the cell after replenishing the stream with extra fuel, or used in another "bottoming cycle," as will be discussed in the hybrid cycle section at the end of this chapter. The penalty associated with oxygen utilization can be minimized by increasing the flow rate of oxygen/air.

As shown in (4.46), fuel utilization is defined as the change in the flow rate of the fuel across the cell divided by its value at the inlet:[2]

$$\varphi = \frac{n_{fl1} - n_{fl2}}{n_{fl1}} = \frac{\dot{n}_{fl1} - \dot{n}_{fl2}}{\dot{n}_{fl1}} = \frac{\dot{m}_{fl,in} - \dot{m}_{fl,out}}{\dot{m}_{fl,in}}. \tag{4.48}$$

To demonstrate the role that fuel utilization plays in changing the performance of a fuel cell, we show an example of a SOFC [6] for a cell using different fuels: hydrogen, methane, and butane, all with air as the oxidizer. Air is assumed to remain at the same concentration between the inlet and the exit of the cell, but fuel is being consumed and is replaced by products. Products of reaction mix with the fuel stream. Thus, for the reaction $\sum_{react} \nu_i' \chi_i \rightarrow \sum_{prod} \nu_i'' \chi_i$, with $\nu_f' = 1$ and φ as the fraction of fuel utilized, $(1 - \varphi)$ of fuel is left in the stream and $\varphi\left(\sum_{products} \nu_i''\right)$ moles of products are formed that mix with the same stream. The mole fraction of fuel at the exit stream can be calculated as $X_{fl}^{out} = (1 - \varphi)/\left((1 - \varphi) + \varphi \sum_{prod} \nu_i''\right)$.

[2] The mole fractions of the participating reactants in this equation are those at the electrode surface where the electrochemical reaction takes place. These values, especially the mole concentrations/mole fractions, vary inside the small confines of the cell, across the electrodes, and possibly across the electrolyte as well, due to finite rate chemical kinetic rates and the transport processes, leading to changes to the potential from that of the open-circuit potential expressed by the equilibrium expression. These effects, which lead to a further drop of the cell voltage from that predicted based on the equilibrium assumption (zero current), produce more "overpotential" losses that will be analyzed in detail in Chapter 7.

Take, for instance, butane as the fuel for which complete reaction is given by

$$C_4H_{10} + \frac{13}{2}O_2 \rightarrow 4CO_2 + 5H_2O. \tag{4.49}$$

With φ fuel utilized, the mixture on the anode side (fuel) is composed of $(1-\varphi)$ moles of butane, 5φ moles of H_2O, 4φ moles of CO_2, with total number of moles being $(1+8\varphi)$. The mole fractions of fuel, H_2O, and CO_2 are $X^{out}_{C_4H_{10}} = (1-\varphi)/(1+8\varphi)$, $X^{out}_{H_2O} = (5\varphi)/(1+8\varphi)$, and $X^{out}_{CO_2} = (4\varphi)/(1+8\varphi)$, respectively. For this reaction, the number of electrons produced for each butane molecule is $n_e = \frac{13}{2} \times 4 = 26$. The Nernst potentials for three fuel–air systems are shown in Figure 4.13 [6]. The dependence of the open-cell potential on fuel utilization is strong because of the high operating temperature.

The following example illustrates the impact of fuel utilization on a molten carbonate fuel cell (MCFC). The reaction in this case is

$$H_2 + \frac{1}{2}O_2 + CO_2 \rightarrow H_2O + CO_2. \tag{4.50}$$

The carbonate ion, CO_3^{2-}, is the charge carrier, and it is produced and consumed in the following pair of redox reactions:

$$H_2 + CO_3^{2-} \rightarrow H_2O + CO_2 + 2e^- \qquad \text{Anode,} \tag{4.51}$$

$$\frac{1}{2}O_2 + CO_2 + 2e^- \rightarrow CO_3^{2-} \qquad \text{Cathode reduction.} \tag{4.52}$$

The open-circuit potential in this case is given by

$$\Delta\varepsilon_{OC} = \Delta\varepsilon^o + \frac{\Re T}{2\Im_a}\left(\frac{1}{2}\ln p + \ln\frac{X_{H_2,C}X^{1/2}_{O_2,C}X_{CO_2,C}}{X_{H_2O,A}X_{CO_2,A}}\right).$$

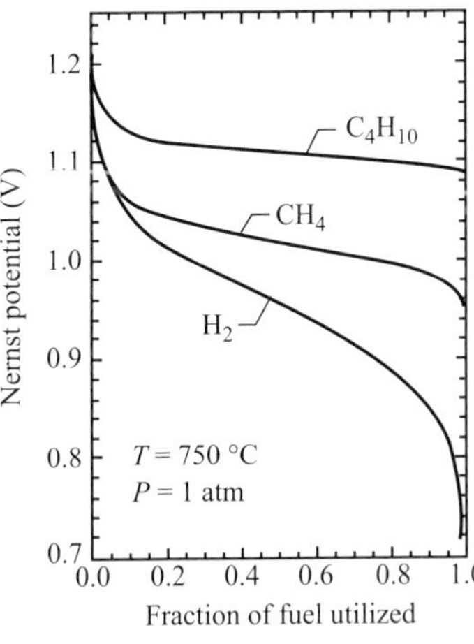

Figure 4.13 Dependence of ideal cell potential on the fraction of fuel utilized for a number of fuels. A solid oxide fuel cell is used, in which the products mix with the fuel stream, while air composition is assumed to remain constant [6].

The designation A and C indicate anode and cathode, respectively. Using the mole fractions at the inlet and assuming complete consumption of fuel would give the maximum cell potential. Using exit values corresponding to a given fuel utilization lowers the value of the last term in the open-circuit potential (Nernst) equation. Fuel utilization lowers the fuel and oxygen concentrations in the two reactants streams.

Example 4.4

Methane is used as a source of fuel in a hydrogen fuel cell. It is first reformed in a reactor producing hydrogen and carbon monoxide. The mixture is fed to a secondary reactor where the shift reaction takes place, converting all carbon monoxide and steam to carbon dioxide and hydrogen. Air is used as an oxidizer. Derive appropriate relationships for the mass flow rates of methane and air in terms of power and actual voltage of the fuel cell.

Solution

The overall methane reforming and shift reactions can be represented as

$$CH_4 + 2H_2O \rightarrow CO_2 + 4H_2.$$

Four moles of hydrogen are produced for each mole of methane. In a hydrogen fuel cell, two electrons are released for each mole of hydrogen spent. If all the fuel is converted electrochemically, the molar flow rate of hydrogen is obtained from $\dot{n}_{H_2} = \frac{I}{2\mathfrak{F}_a}$. In terms of power and actual voltage, the flow rate is $\dot{n}_{H_2} = \frac{P}{2\mathfrak{F}_a V_{act}}$, where P is the power output and V_{act} denotes the actual cell potential/voltage. For a fuel cell with fuel utilization less than unity (i.e., $\varphi < 1$), $\dot{n}_{H_2} = \frac{P}{2\mathfrak{F}_a \varphi V_{act}}$.

Based on the overall reforming and shift reaction, the molar flow rate of methane is $\dot{n}_{CH_4} = \frac{P}{8\mathfrak{F}_a \varphi V_{act}}$ or $\dot{m}_{CH_4} = 2.073 \times 10^{-8} \frac{P}{\varphi V_{act}}$ kg/s.

Similarly, the mole flow rate of the utilized oxygen is $\dot{n}_{O_2} = \frac{I}{4\mathfrak{F}_a}$.

Note that four electrons are released for each mole of oxygen used in the fuel cell. Taking into account the mole fraction of oxygen in air (i.e., 0.21), the molar flow rate of oxygen required for the cell operation is $\dot{n}_{air} = \frac{I}{0.21 \times 4\mathfrak{F}_a}$, or $\dot{m}_{air} = 3.574 \times 10^{-7} \frac{P}{V_{act}} \left[\frac{\text{kg}}{\text{s}}\right]$. In the case of higher air utilization, represented by a stoichiometry factor $\Lambda > 1$, $\dot{m}_{air} = 3.574 \times 10^{-7} \Lambda \frac{P}{V_{act}}$ kg/s.

4.5 Fuel Cell Efficiencies

We have defined the thermodynamic or **open-circuit efficiency** of a fuel cell (similar to that of a combustion engine) as the ratio of the Gibbs free energy to the enthalpy of reaction of the fuel:

$$\eta_{OC} = \frac{\Delta G_R}{\Delta H_R}. \tag{4.53}$$

Note the difference between this efficiency and that of the Carnot engine efficiency. In (4.53), we take the ratio of the maximum work to the chemical energy input. This is in contrast to the heat engine efficiencies, where we take the ratio of the work out to the heat in. The definition of the fuel cell open-circuit efficiency resembles that used for a combustion engine. A comparison between the three devices is shown in Figure 4.14.

The open-circuit efficiency, known also as the Nernst efficiency, is the ratio of the chemical free energy or chemical availability to the chemical energy in the reactants. It should be noted that, in most cases, $|\Delta G_R| < |\Delta H_R|$, both being negative for an exothermic reaction capable of doing work on its environment ($W = -\Delta G_R$), and the efficiency is less than 100%. The difference between the Gibbs free energy and the enthalpy of reaction is the "reversible" isothermal heat transfer, $T\Delta S_R = \Delta H_R - \Delta G_R$, which is negative, corresponding to heat transfer out of the cell for the case of less than 100% efficiency.

Another efficiency often used to characterize the performance of an operating cell, in which a finite current is drawn and a finite power is delivered, is the ratio between the actual measured potential to the ideal Nernst open-circuit voltage, and is referred to as relative efficiency:

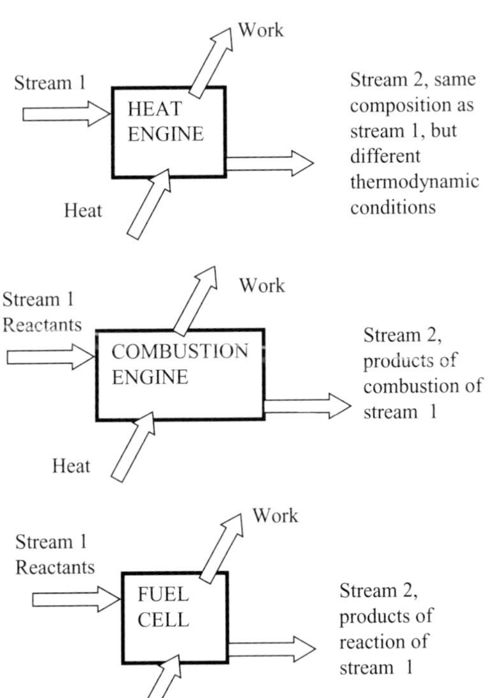

Figure 4.14 The operations of a heat engine, a combustion engine, and a fuel cell, all shown to justify the definitions of their efficiency.

$$\eta_{rel} = \frac{V_{act}}{V_{OC}}, \tag{4.54}$$

where V_{OC} is the open-circuit voltage at standard conditions. This is also called the relative, utilization, or Second Law efficiency, since it measures the work as a ratio of the maximum work. Internal voltage drop due to fuel utilization, finite rate chemical reactions, transport, and ohmic resistance ensures that this ratio is less than unity. Multiplying the numerator and denominator by the current, this efficiency defines the power delivered as a percentage of the maximum power at standard conditions that could be delivered at the same current in a cell operated under ideal conditions.

A third efficiency often quoted is the **Faradic** or **current efficiency**, used to describe the fraction of the reactants that actually participate in the electrochemical reaction. The current efficiency is given by

$$\eta_{Far} = \frac{I}{n_e \Im_a \dot{n}_f}, \tag{4.55}$$

where I is the total current and $\dot{n}_f$ is the molar fuel flow rate into the cell. If all of the fuel is consumed in the electrochemical reaction within a fuel cell, this fraction is unity, according to Faraday's Law. The difference $(1 - \eta_{Far})$ is the fraction of the fuel that does not react electrochemically. The fuel that does not participate in the electrochemical reaction may react thermochemically, because of fuel leakage across the membrane, or simply leave the cell unreacted. The combination $\eta_{rel}\eta_{Far} = \frac{IV_{act}}{(n_e\Im_a\dot{n}_f)V_{OC}} = \frac{I\,V_{act}}{\dot{n}_f \Delta\hat{g}^o_f}$ is the fraction of the chemical free energy or availability in a molar fuel flow rate of $\dot{n}_{fuel}$ that is converted to electrical power producing a total current I across a voltage of V_{act}.

Similar to other energy conversion systems, we define the overall **fuel utilization efficiency** as the ratio of the power out of the cell to the rate of chemical energy into the cell:

$$\eta_{fuel-utilization} = \frac{\text{Power out}}{\text{Rate of chemical energy in}} = \frac{IV_{act}}{\dot{n}_f \Delta\hat{h}^o_{R,f}}. \tag{4.56}$$

Substituting using the above definitions, we find that

$$\eta_{fuel-utilization} = \eta_{OC}\eta_{rel}\eta_{Far}. \tag{4.57}$$

Example 4.5

A schematic of a 500 W fuel cell operating at 80 °C and 1 bar is shown in Figure E4.5. It uses pure hydrogen as a fuel and air as an oxidizer. The actual cell voltage is 0.7 volts. Compute: (a) the open-circuit voltage; (b) the fuel utilization efficiency; and (c) the rate of heat generation by the fuel cell.

Example 4.5 (cont.)

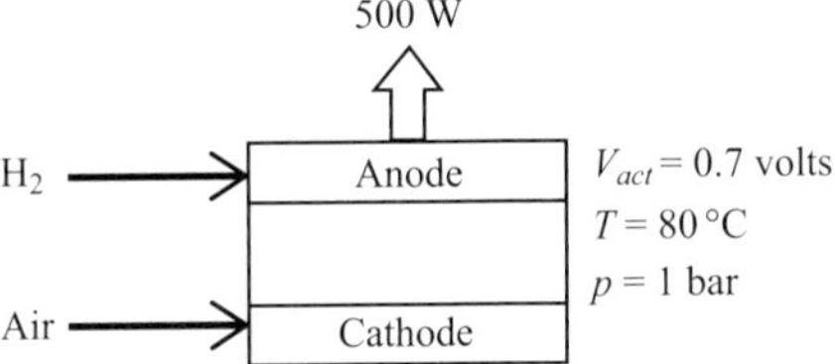

Figure E4.5

Solution

a. The open-circuit voltage for a hydrogen–oxygen fuel cell at atmospheric pressure is obtained from $\Delta\varepsilon = -\frac{\Delta g_{80^\circ C}}{2\Im_a}$. The Gibbs free energy of hydrogen oxidation reaction at 80 °C is $\Delta g_{80^\circ C} = -228$ kJ/mol. Substituting, we find $\Delta\varepsilon = 1.182$ volts. As the oxidizer is air, we need to calculate the voltage drop as explained in Example 4.2. Hence,

$$\Delta\varepsilon = \frac{\Re T^*}{4\Im_a}\left[\ln\left(\frac{X_{O_2}}{1}\right)_{ox}\right] = \frac{8.314 \times 353}{4 \times 96{,}485}\left[\ln\left(\frac{0.21}{1}\right)_{ox}\right] = -0.012$$

Therefore, the open-circuit potential of the fuel cell is $\Delta\varepsilon = 1.182 - 0.012 = 1.17$ volts.

b. The fuel utilization efficiency is the ratio of the power output to the chemical energy spent. At 80 °C and 1 bar, the water leaving the cell is in liquid phase. Hence, $\eta_{fuel-utilization} = \frac{P}{\dot{n}_{H_2} HHV_{H_2}}$. In Example 4.4, we showed that the molar flow rate of hydrogen is $\dot{n}_{H_2} = \frac{P}{2\Im_a V_{act}}$. Hence, $\dot{n}_{H_2} = \frac{500}{2\times 96{,}485\times 0.7} = 0.0037\ \frac{\text{mol}}{\text{s}}$.

The HHV of hydrogen is 285.8 kJ/mol, and the fuel utilization efficiency is obtained as follows. $\eta_{fuel-utilization} = \frac{500}{0.0037\times 285{,}800} \times 100 = 47.3\%$.

c. The fraction of fuel chemical energy that is not converted to electrical energy is wasted (due to various voltage losses such as ohmic resistance) in the form of heat. Hence,

$$\dot{Q} = \dot{n}_{H_2} HHV - P = P\left(\frac{1}{\eta_{fuel-utilization}} - 1\right) = 557\ \text{W}.$$

4.6 Open-Circuit Potential and Thermodynamic Equilibrium

As mentioned before, the maximum potential E_{rev} between the fuel and oxidizer streams is the Nernst potential. Assuming that chemical equilibrium exists in the anode channel, $\prod_k p_{k,a}^{\nu_{k,a}} = K_p = \exp(-\Delta G^o/\Re T)$, where K_p is the equilibrium constant, and the Nernst potential can then be written in terms of the oxygen partial pressures in the anode and cathode channels:

$$E_{rev} = |\nu_{O_2}| \frac{\Re T}{n_e \mathfrak{F}_a} \ln \left(\frac{p_{O_2,c}}{p_{O_2,a}} \right).$$

The anodic partial pressure $p_{O2,a}$ is determined by equilibrium chemistry in the anode fuel channel. Regardless of the oxidation reaction, the number of electrons per mole of oxygen is constant – that is four electrons per oxygen molecule – because the stoichiometry is set by the cathodic reduction reaction. The influence of the anodic chemistry and electrochemistry appears indirectly through the anode oxygen partial pressure $p_{O2,a}$. The overall half-cell reduction–oxidation (redox) reactions for four typical fuels (on the anode side, a) and oxygen (on the cathode side) can be written as follows:

$$H_2(g) + O^{2-}(el) \rightleftarrows H_2O(g) + 2e^-(a)$$

$$CO(g) + O^{2-}(el) \rightleftarrows CO_2(g) + 2e^-(a)$$

$$CH_4(g) + 4O^{2-}(el) \rightleftarrows CO_2(g) + 2H_2O(g) + 8e^-(a)$$

$$C(s) + 2O^{2-}(el) \rightleftarrows CO_2(g) + 4e^-(a).$$

The cathode half-cell reduction reaction is

$$O_2(g) + 4e^-(a) \rightleftarrows 2O^{2-}(el).$$

Figure 4.15 shows the calculated reversible cell potential of different fuels using the equations above. The results show that the reversible cell potential from the oxidation of H_2 (top-left panel) or CO (top-right) decreases at higher temperature and increases with higher pressure. On the other hand, the reversible cell potential from the full oxidation of CH_4 (bottom-left, straight lines) or solid carbon (bottom-right, slight negative-sloping line) is pressure-independent. The influence of controlling the oxidation reaction is illustrated in the bottom two panels. For both, the curves are calculated assuming all anode-side reactions are in partial equilibrium, so the fuel stream is in a state of full chemical equilibrium and the reversible potential is calculated from the equation above, using the equilibrium anode-side oxygen partial pressure. It is clear that full oxidation (straight lines) of CH_4 determines the open-circuit voltage at lower temperatures, while its partial oxidation (not shown) is correlated to the positive slope at higher temperatures and higher methane partial pressures. Moreover, oxidation of H_2 contributes to decreasing open-circuit voltage at higher temperatures and higher product partial pressures. For the electrochemical oxidation of solid carbon, the competing oxidation mechanisms and their influence on the reversible cell potential are very clear that at lower temperatures full oxidation to CO_2 is favored, while partial oxidation to CO is preferred at higher temperatures.

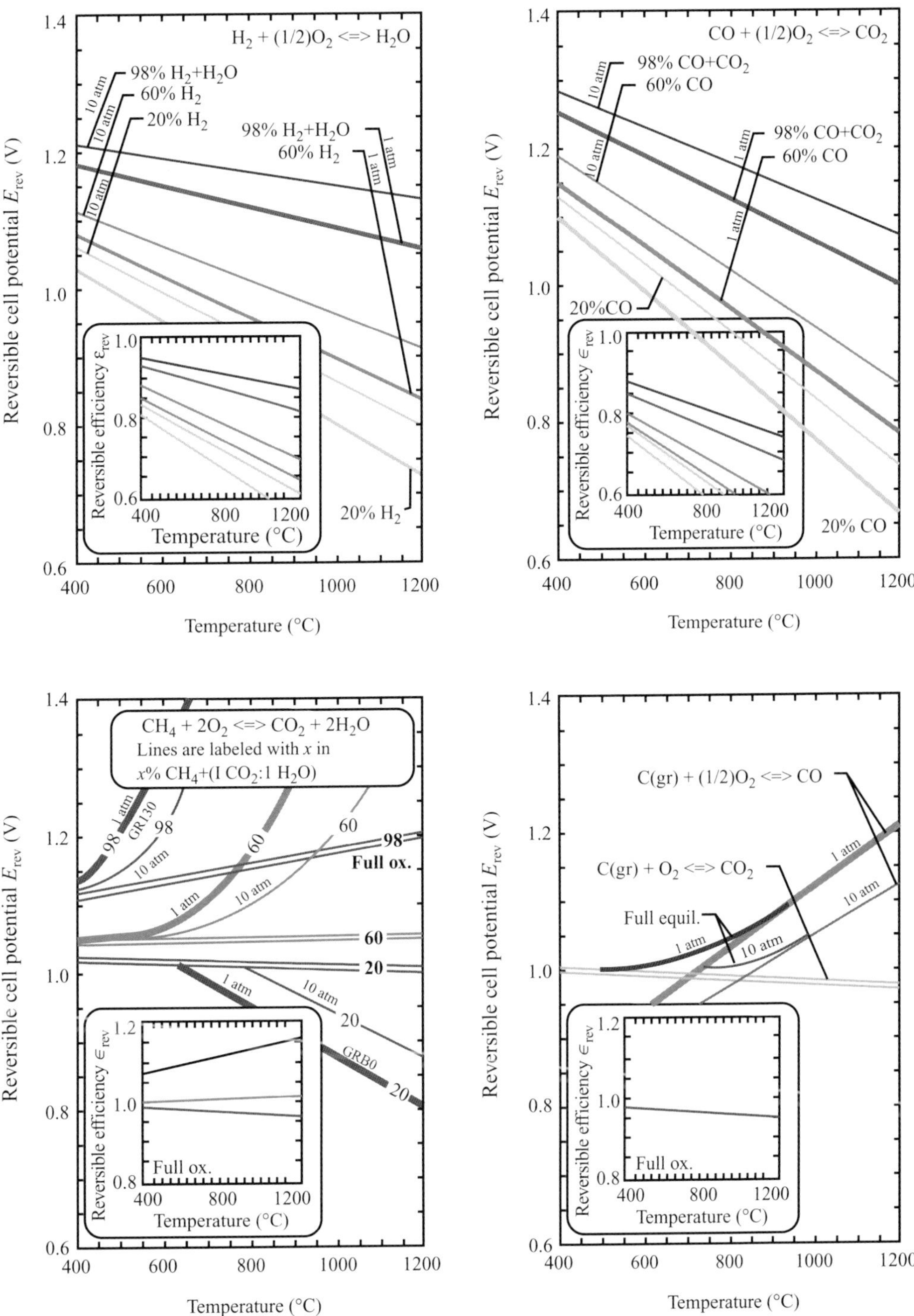

Figure 4.15 Open-circuit voltage (OCV) and thermodynamic efficiency (inset) for different fuels at pressures of 1 atm (thick lines) and 10 atm (thin lines). The fuel streams are equilibrated mixtures of the indicated ratios for fuel and oxidation products as the fuel becomes more diluted (think of this as being representative of moving along the flow channel). The OCV is calculated assuming air as the oxidant.

Another important application of thermodynamic equilibrium in determining the results of partial fuel oxidation in the fuel cell anode is in carbon deposition potential [7].

4.7 Half-Cell Potential

The standard cell potential is a characteristic quantity of a particular redox reaction pair. It depends on the standard Gibbs free energy of the overall reaction and the number of electrons exchanged in the half-cell reactions at the electrodes' surfaces. This is the reason why the standard cell potentials are tabulated. However, it has become standard to tabulate the half-cell reactions separately according to the standard Gibbs free energy and the number of electrons exchanged in each half-cell reaction separately. For this purpose, it has become the standard to set the enthalpy and Gibbs free energy of reaction of the hydrogen molecule to hydrogen ions to zero. Accordingly, the standard half-cell potential for the following hydrogen ionization reaction is, by definition, set to zero:

$$H_{2(g)} \Rightarrow 2H^+ + 2e^-. \tag{4.58}$$

All other half-cell reactions are measured against this half-cell reaction (note that the enthalpy and Gibbs free energy of the hydrogen dissociation reaction are not zero). Lewis and Latimer (see [9]), two of the pioneers of electrochemistry, tabulated half-cell reaction potentials for many reactions with electrons written on the right-hand side of that reaction – that is, for oxidation reaction. Reverse half-cell reactions have the opposite sign for the half-cell potential. For example, the oxidation half-cell reaction of copper,

$$Cu_{(s)} \rightarrow Cu^{2+}_{aq} + 2e^-, \tag{4.59}$$

has a standard potential $\Delta\varepsilon^o = -0.337$ volts. The corresponding reduction reaction,

$$Cu^{2+}_{aq} + 2e^- \rightarrow Cu_{(s)}, \tag{4.60}$$

Figure 4.15 (*cont.*) The reversible efficiency is calculated from $\Delta G^o/\Delta H^o = |\nu_{O_2}|(\Re T/\Delta H^o)$. $\ln\left(p_{O_2,a}/P_{O_2,c}\right) = -n_e \Im_a E_{rev}/\Delta H^o$, where ΔH^o is the temperature-dependent standard-state enthalpy change associated with the global oxidation reaction. The top-left panel is for H_2, top-right for CO, bottom-left for CH_4, and bottom-right for of solid graphitic carbon C(gr). For methane oxidation, the straight lines give the OCV calculated assuming full oxidation of methane to CO_2 and H_2O (pressure-independent), and the curves give the OCV from a full equilibrium calculation using the GRI30 mechanism. Reversible efficiencies are calculated based only on the full oxidation of methane. For carbon oxidation, the two straight lines assume oxidation to CO only (positive slope) or CO_2 only (slight negative slope and independent of pressure) over the entire temperature range, whereas the curved line accounts for the thermodynamically favored oxidation product (which shifts from CO_2 at lower temperatures to CO at higher temperatures) [8].

has $\Delta\varepsilon^o = 0.337$ volts. The corresponding values for lithium Li(s), sodium Na(s), and water (l) are, respectively, 3.05, 2.71, −1.23, all being single-electron reactions.

A high negative cell potential corresponds to a positive Gibbs free energy, $\Delta\varepsilon^o(T^*) = \frac{-\Delta G_R^o(T^*)}{n_e\mathfrak{I}}$ and small reaction equilibrium constant, $K_p(T) = \exp\left(\frac{-\Delta G_R^o(T^*)}{\Re T}\right)$. Thus, reactions at the top of the list have the strongest tendency to move in their forward direction, while those at the bottom would rather move in the opposite direction.

4.7.1 Cell Diagram

An example of an electrochemical cell in which the two electrodes are immersed in different electrolytes separated by an ion-conducting membrane [10] is the Daniel cell depicted in Figure 4.16.

Electrochemical reactions are often described using a cell diagram, which shows the two different half-cell reactions taking place on the two different electrodes separated by a double line indicating the ion transfer medium [1]. For the reaction in the Daniel cell, the cell diagram is

$$Zn_{(s)} \mid ZnSO_{4(aq)} \| CuSO_{4(aq)} \mid Cu_{(s)}. \tag{4.61}$$

A single bar indicates a phase boundary, and a double bar indicates a junction between electrolytes. The convention is to write the oxidation reaction (reactant losing electrons) on the left-hand side of the double bar and the reduction reaction (reactant gaining electrons) on the right-hand side. The same cell diagram can be expanded to indicate the presence of ionized species in the electrolyte:

$$Zn_{(s)} \mid Zn^{2+}, SO_{4(aq)}^{2-} \| Cu^{2+}, SO_4^{2-} \mid Cu_{(s)}. \tag{4.62}$$

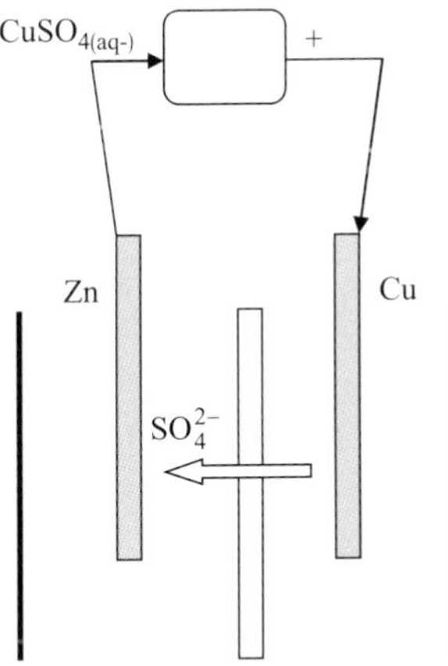

Figure 4.16 In a Daniel cell, two different compartments and two different electrolytes, $CuSO_{4(aq)}$ forming $Cu_{(aq)}^{2+}$ on the right-hand side and $ZnSO_{4(aq)}$ forming $Zn_{(aq)}^{2+}$ on the left-hand side, separated by an ion permeable membrane, are used. The zinc anode loses electrons that travel in the outside circuit to the copper cathode.

In the presence of a salt bridge, this cell diagram becomes

$$Zn_{(s)} \mid ZnSO_{4(aq)} \mid \text{KCLbridge} \mid CuSO_{4(aq)} \mid Cu_{(s)}. \tag{4.63}$$

For the hydrogen–oxygen reaction in a fuel cell, the cell diagram is

$$H_{2(g)} \mid H^+ \parallel H_2O \mid O_{2(g)}. \tag{4.64}$$

4.8 The Electrochemical Potential

Another approach to analyzing the equilibrium of electrochemical reactions is to rely on a potential for mass transfer among species undergoing electrochemical reactions [11]. This approach is similar to that used in analyzing equilibrium of thermochemical reactions. In the case of an electrochemical reaction, the transfer between components involves both mass and charge – that is, transfer among charged species. Moreover, the electrical work interaction associated with charge transfer must be considered when writing the combined statement of the First and Second Laws. First, by adding an electrical work (non-$p\forall$) term $dW_{elec} = \varepsilon d\varsigma$, the fundamental equation becomes

$$dU = TdS - pd\forall + \varepsilon d\varsigma + \sum_I \mu_i dn_i. \tag{4.65}$$

The electrical work interaction can be written in terms of the change in the charge during a reaction. For this purpose, recall that the charge number of a component, z_i, was defined as the multiple of electron charges carried by a species and it is taken to be positive if the charge is positive and negative if the charge is negative. The net change in the total charge during a reaction, associated with a mole change of dn_i in the participating components, is

$$d\varsigma = \sum_I z_i \Im_a dn_i. \tag{4.66}$$

Substituting (4.66) into (4.65), we find that

$$\begin{aligned} dU &= TdS - pd\forall + \varepsilon \sum_I z_i \Im_a dn_i + \sum_I \mu_i dn_i \\ &= TdS - pd\forall + \sum_I (\mu_i + z_i \Im_a \varepsilon) dn_i \\ &= TdS - pd\forall + \sum_I \bar{\mu}_i dn_i. \end{aligned} \tag{4.67}$$

The **electrochemical potential** is defined as

$$\bar{\mu}_i = \mu_i + z_i \Im_a \varepsilon. \tag{4.68}$$

For a single electrochemical reaction, $\sum_{react} \nu_i' \chi_i \Rightarrow \sum_{prod} \nu_i'' \chi_i$, changing the number of moles of one component, n_i, by a small amount, dn_i (it is negative for a reactant and positive for a product, and $\nu_i = \nu_i'' - \nu_i'$) should cause all other components participating in the reaction to change by $d\xi = dn_i/\nu_i$. In this case, the net change in internal energy is

$$dU = TdS - pd\forall + \left(\sum_{I} \nu_i \bar{\mu}_i\right) d\xi. \tag{4.69}$$

Moreover, the net charge transferred in an electrochemical reaction defined in (4.8) or (4.9) is determined by

$$\sigma_e = n_e = \sum_{prod} \nu_i'' z_i - \sum_{react} \nu_i' z_i. \tag{4.70}$$

Equation (4.69) can be written in terms of other thermodynamic functions, such as in terms of enthalpy, Helmholtz, and Gibbs free energies, as the equivalent expressions are, respectively,

$$dH = TdS - \forall dp + \left(\sum_{N} \nu_i \bar{\mu}_i\right) d\xi, \tag{4.71}$$

$$dA = -SdT - pd\forall + \left(\sum_{N} \nu_i \bar{\mu}_i\right) d\xi, \tag{4.72}$$

$$dG = -SdT + \forall dp + \left(\sum_{N} \nu_i \bar{\mu}_i\right) d\xi. \tag{4.73}$$

Now, using the conditions of equilibrium obtained in Chapter 2, the condition of equilibrium of an electrochemical reaction is

$$\left(\sum_{N} \nu_i \bar{\mu}_i\right) d\xi \le 0. \tag{4.74}$$

Or, at equilibrium,

$$\sum_{N} \nu_i \bar{\mu}_i = 0. \tag{4.75}$$

Finally, expanding the electrochemical potential in terms of the chemical potential, we get

$$\sum_{N} \nu_i \mu_i = -\sigma_e \Im_a \varepsilon. \tag{4.76}$$

Writing the chemical potential in terms of the temperature, pressure, and concentrations, we get the Nernst equation.

4.9 Thermodynamics of Hybrid Systems

In this chapter we have focused on the ideal performance of fuel cells – that is, the open-circuit voltage, and its dependence on the cell temperature, pressure, reactants concentration, and fuel utilization, assuming equilibrium of the cell with its environment. The actual performance of an operating fuel cell, under finite current, is different. The cell voltage decreases as the current increases because of a number of loss mechanisms, including those associated with the finite rates of the electrochemical reactions, the concentration gradients generated by the finite fluxes of species through the electrodes, and the electrical resistance across different elements of the cell structure, which will be described in the next chapter. As such, at finite current, the impact of the temperature on the cell performance may be different from that predicted at zero current, in part because higher temperatures promote faster kinetics and may also change the resistance of the cell material to the transport of uncharged and charged species. Higher temperatures accelerate the kinetic rates of the electrochemical reactions without the need for expensive catalysts. Moreover, operating at high temperatures allows for internal fuel reforming, allowing complex fuels, such as mixtures of hydrogen and carbon monoxide, or methane, to be used. The thermodynamic performance is degraded at high temperatures. In this section, we analyze "hybrid" operation of an ideal fuel cell and an ideal heat engine (i.e., a Carnot cycle) to show that the drop of the efficiency while operating at high temperature can be compensated for by operating the cell in series with a heat engine.

Figure 4.17 shows the ideal efficiency of a hydrogen fuel cell operating at different temperatures, while maintaining the "external" environment at 100 °C. These results are similar to those shown in Figure 4.10, and show a continuous drop in the cell efficiency as

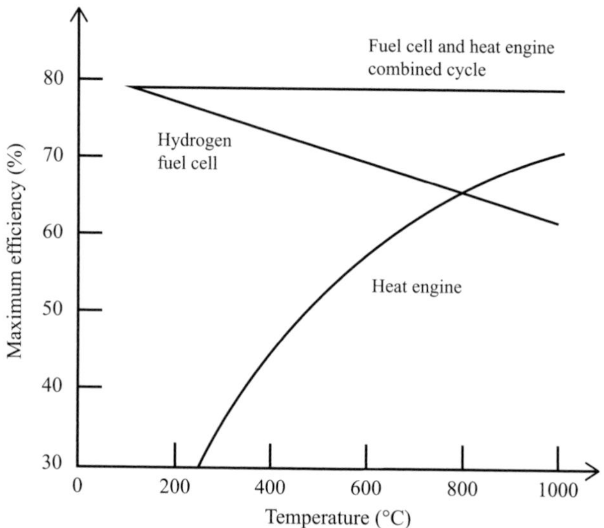

Figure 4.17 Ideal efficiency of a hydrogen fuel cell as a function of the fuel cell temperature, a Carnot engine operating between the given temperature and a lower temperature of 100 °C, and a hybrid system combining a fuel cell and a Carnot engine operating between the cell temperature and the environment.

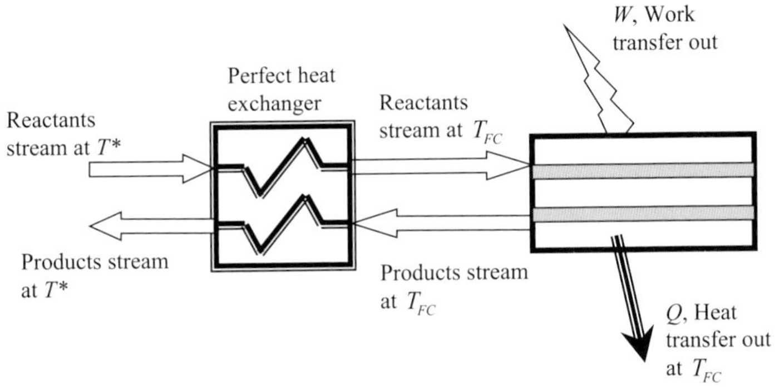

Figure 4.18 A block diagram showing a fuel cell operating isothermally at a temperature higher than the environment temperature. An ideal heat exchanger is used to heat the reactants using the exhaust thermal energy.

the temperature of the cell rises. The system required to operate the fuel cell at these conditions is shown schematically in Figure 4.18. A heat exchanger is used to heat up the reactants to the cell temperature, using the exhaust stream. The Carnot efficiency of a heat engine operating between the temperature of the cell and the environment at 100 °C is shown in Figure 4.17. The figure also shows that adding the two efficiencies algebraically results is a constant overall efficiency of the combined fuel cell and Carnot engine, independent of the operating temperature of the cell. The analysis validating these plots is shown next.

Analysis of the fuel cell performance so far has been based on the assumption that the cell temperature is the same as that of the environment, T^*. In this section, we analyze the impact of running a cell at a temperature $T_{FC} > T^*$. In this case, the reactants stream must be heated up to the cell temperature. The thermal energy in the exhaust gases is used to heat up the fuel and air streams flowing into the cell to raise their temperature to that of the cell. As the cell operates at temperatures higher than the environment, it rejects heat at its operating temperature. To mitigate the negative impact of rejecting "waste" heat at temperatures higher than that of the environment, the heat rejected by the cell at this temperature should be used in a **bottoming cycle** to reduce the efficiency penalty. The heat generated during the electrochemical reactions, at T_{FC}, can be used in a power cycle across the temperature difference (T_{FC}, T^*) to generate more work. This "bottoming" cycle can be a steam cycle if the gas pressure is low, or a gas turbine cycle if the cell operates at higher pressures (more details on these cycles will be given in later chapters). The hybrid operation of a fuel cell and a thermal power cycle is an important development route for high-temperature fuel cells in order to maintain the overall system efficiency.

Figure 4.19 shows a schematic diagram of a fuel cell operating at a temperature higher than that of its environment [12]. In this case, it is necessary to heat the reactants to the fuel cell temperature using the products stream. With the assumption that the average specific heats of both streams are the same, and using a perfect heat exchanger, it is possible to raise the reactants temperature from T^* to T_{FC} while cooling the products from T_{FC} to T^*, without using an extra heat source. The electrical work produced by the cell is the Gibbs free energy of the reactants at the cell temperature:

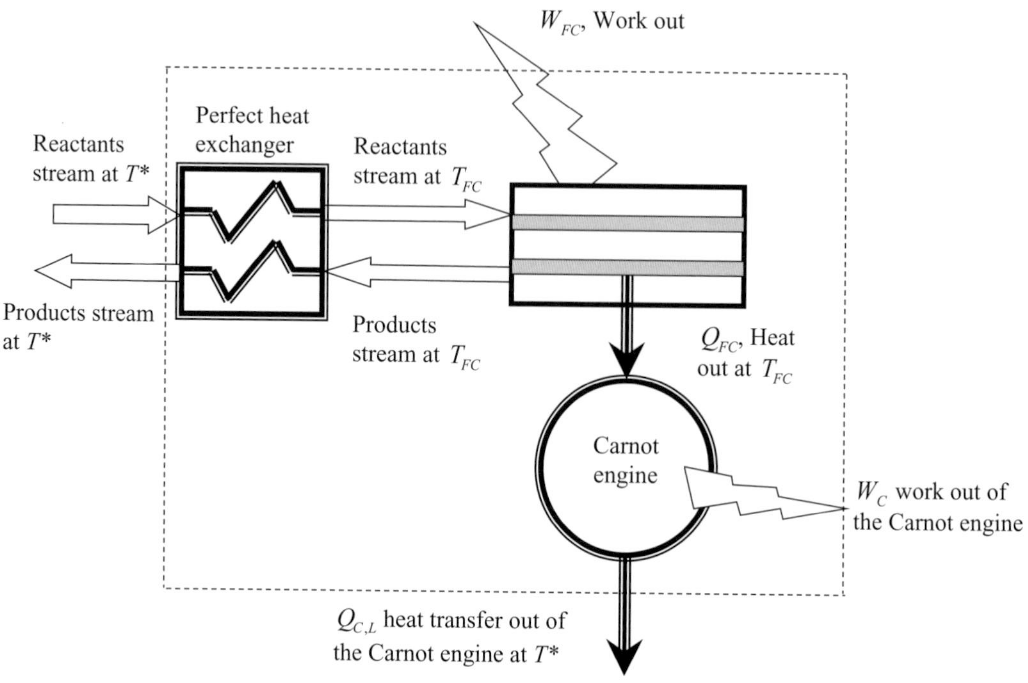

Figure 4.19 A block diagram showing a fuel cell operating isothermally but at temperature higher than the environment temperature, with a Carnot engine operating between the fuel cell temperature and T^* while taking advantage of the heat rejected by the cell. An ideal heat exchanger is used to heat the reactants using the exhaust thermal energy.

$$\Delta G_R(T_{FC}) = \Delta H_R(T_{FC}) - T_{FC}\Delta S_R(T_{FC}). \tag{4.77}$$

The heat rejected by the cell at this temperature is

$$Q_{FC} = T_{FC}\Delta S_R(T_{FC}). \tag{4.78}$$

The electrical work delivered by the cell alone is

$$W_{FC} = -\Delta G_R(T_{FC}). \tag{4.79}$$

If the cell operates at the temperature of the environment, the maximum work it produces is given by

$$W^* = -\Delta G_R(T^*) = -[\Delta H_R(T^*) - T^*\Delta S_R(T^*)], \tag{4.80}$$

and the heat rejected is

$$Q^* = T^*\Delta S_R.$$

The difference between the work in both cases – that is the difference between W^* in (4.80), and W_{FC} in (4.79) – is lost because of the irreversible heat transfer between the cell and the environment across a temperature gradient, $T_{FC} > T^*$.

We can use the heat rejected from the cell at its higher temperature to the environment at the lower temperature in a heat engine. If we make another assumption that the enthalpy and entropy of reaction are both constant within the temperature range of interest, it is clear that the electrical work produced by the cell at T_{FC} is lower than that at T^* by $(T_{FC} - T^*)\Delta S_R$, which is the extra heat transfer from the cell when it operates at T_{FC}. If the heat rejected at T_{FC} is used in a Carnot cycle operating between the cell and environment temperatures, then the extra work gained from this heat engine is

$$W_{car} = -(T_{FC}\Delta S_R)\left(1 - \frac{T^*}{T_{FC}}\right) = -(T_{FC} - T^*)\Delta S_R. \tag{4.81}$$

This is the work lost by operating the cell at the higher temperature. The total work of the fuel cell operating at high temperature and the Carnot engine operating between the fuel cell temperature and the environment is the same as that produced by the fuel cell operation at the low temperature – that is, the environment temperature. This result is not surprising, because the conversion process is still performed reversibly inside the cell, both isothermally and in equilibrium with its own environment. Moreover, the Carnot engine operates reversibly as well. The combined system is reversible and in equilibrium with the environment. If we draw a box over the combined cell and Carnot engine, and since there are no irreversibility inside, the new hybrid system looks like the original fuel cell working at the temperature of the environment.

This is an interesting and useful result despite the fact that it is obtained using a number of simplifying assumptions. It demonstrates that operating a cell at high temperature to take advantage of lower overpotential loss or other desirable features need not impair the overall efficiency of the plant if a bottoming cycle is used to take advantage of the "waste" heat in the exhaust. It shows also that for such operation to be successful, heat and mass integration between the different components is necessary. This integration will be discussed in later chapters.

Example 4.6

A schematic diagram of a 200 kW *internal reforming* fuel cell system is shown in Figure E4.6. Methane is the source of fuel, which is first reformed with superheated steam, producing hydrogen and carbon monoxide. The mixture then undergoes the water–gas shift reaction. The byproducts leaving the shift reactor are fed to the fuel cell. No disassociation in reformer and shift reactor is assumed. Air is utilized as an oxidizer with a stoichiometric ratio of 4. The fuel cell operates with an average actual voltage of 0.68 volts. The exit flows of anode and cathode are well mixed and leave the fuel cell at a thermal equilibrium. The thermal energy of the fuel cell exit stream is used in a heat engine with 40% thermal efficiency. Calculate: (a) the operating temperature of the fuel cell; and (b) the net power production of the combined fuel cell/heat engine system.

Example 4.6 (cont.)

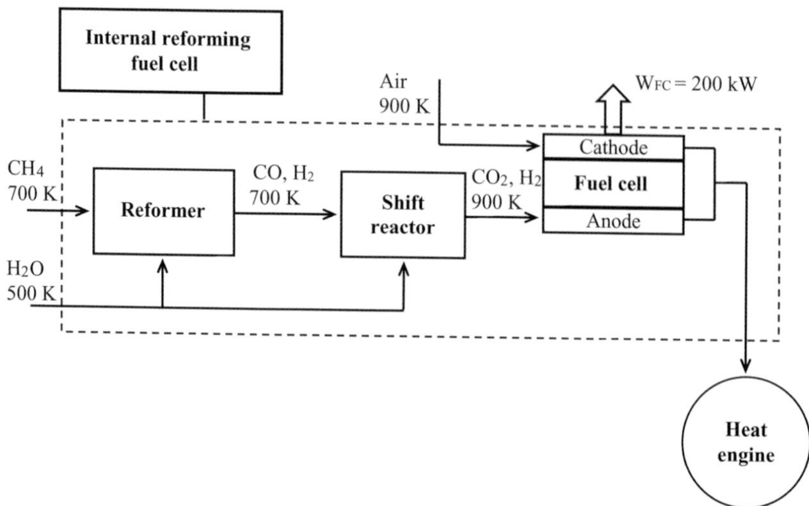

Figure E4.6

Solution

The net reforming and shift reactions can be written as

$$CH_4 + 2H_2O \rightarrow CO_2 + 4H_2.$$

The molar flow rate of hydrogen is determined using the results of Example 4.4:

$$\dot{n}_{H_2} = \frac{P}{2\mathfrak{F}_a V_{act}} = \frac{200{,}000}{2 \times 96{,}485 \times 0.68} = 1.524 \; \frac{\text{mol}}{\text{s}}.$$

The rate of heat generation is calculated as follows:

$$\dot{Q} = \dot{n}_{H_2} LHV_{H_2} - P = 1.524 \times 242{,}000 - 200{,}000 = 168{,}808 \text{ W}.$$

The molar flow rate of methane is calculated based on the flow rate of hydrogen determined above, $\dot{n}_{CH_4} = \frac{\dot{n}_{H_2}}{4} = 0.381 \; \frac{\text{mol}}{\text{s}}$.

We now compute the heat requirement of the reforming and shift reactions:

$$\dot{Q}_{R+S} = \dot{n}_{CH_4}(\hat{h}_{CO_2} + 4\hat{h}_{H_2O} - 2\hat{h}_{H_2O} - \hat{h}_{CH_4}).$$

The enthalpy of methane is obtained using the following relationship:

$$\hat{h} = \hat{h}_f^o + \int_{298}^{T} c_p dT,$$

Example 4.6 (cont.)

where $\hat{h}^{o}_{f,\mathrm{CH_4}} = -74{,}900$ kJ/kmol and $c_{p,\mathrm{CH_4}} = 1.1126T^{0.6053}$ J/mol·K. This relationship is valid for the temperature range 298–1400 K. The enthalpy of methane at 700 K is calculated as $\hat{h}_{\mathrm{CH_4}}(700\ \mathrm{K}) = -55{,}809$ J/mol. The enthalpies of other species are obtained from the thermodynamic tables. Hence, $\hat{h}_{\mathrm{H_2O}}(500\ \mathrm{K}) = -234{,}904$ J/mol, $\hat{h}_{\mathrm{CO_2}}(900\ \mathrm{K}) = -365{,}492$ J/mol, $\hat{h}_{\mathrm{H_2}}(900\ \mathrm{K}) = 17{,}657$ J/mol.

Substituting in the equation for $\dot{Q}_{R+S}$, we find

$$\dot{Q}_{R+S} = 0.381(-365,492 + 4 \times 17,657 + 2 \times 234,904 + 55{,}809) = 87{,}917\ \mathrm{W}.$$

The operating temperature of the fuel cell can be obtained by applying the first Law to the fuel cell and assuming that the exit temperature is the same as the cell temperature:

$$(\dot{n}LHV)_{\mathrm{H_2}} + (\dot{n}h)_{\mathrm{CO_2}} + (\dot{n}h)_{\mathrm{O_2},in} + (\dot{n}h)_{\mathrm{N_2},in} - \dot{Q}_{R+S} - P$$
$$= (\dot{n}h)_{\mathrm{O_2},out} + (\dot{n}h)_{\mathrm{N_2},out} + (\dot{n}h)_{\mathrm{H_2O}} + (\dot{n}h)_{\mathrm{CO_2}}.$$

The molar flow rate of air at the inlet of the cell cathode is obtained (see Example 4.4):

$$\dot{n}_{air} = \frac{P}{(0.21)4\Im_a V_{act}}\Lambda = \frac{200{,}000}{(0.21)4 \times 96{,}485 \times 0.68} \times 4 = 14.516\ \frac{\mathrm{mol}}{\mathrm{s}}.$$

The molar flow rate of nitrogen is 79% of $\dot{n}_{air}$, which does not participate in the electrochemical reaction. So,

$$\dot{n}_{\mathrm{N_2},in} = \dot{n}_{\mathrm{N_2},out} = 0.79 \times 14.516 = 11.468\ \frac{\mathrm{mol}}{\mathrm{s}}.$$

The molar flow rate of oxygen at the cathode inlet is $\dot{n}_{O_2,in} = 0.21 \times 14.516 = 3.048\ \frac{\mathrm{mol}}{\mathrm{s}}$. The difference between $\dot{n}_{O_2,in}$ and the rate of oxygen consumption in the fuel cell is the molar flow rate of oxygen at the cathode outlet:

$$\dot{n}_{O_2,out} = \dot{n}_{O_2,in} - \frac{P}{4\Im_a V_{act}} = 3.048 - \frac{200{,}000}{4 \times 96{,}485 \times 0.68} = 2.286\ \frac{\mathrm{mol}}{2}.$$

The molar flow rate of water at the exit stream is the same as the molar flow rate of the hydrogen (for each mole of hydrogen, one mole of water is produced): $\dot{n}_{\mathrm{H_2O}} = 1.524$ mol/s.

The left-hand side of the energy equation of the fuel cell is the rate of energy that is carried out of the fuel cell:

$$(1.524 \times 242{,}000)_{\mathrm{H_2}} + (0.381 \times 849.7 \times 44)_{\mathrm{CO_2}} + (3.048 \times 872.75 \times 32)_{\mathrm{O_2},in}$$
$$+(11.468 \times 960.25 \times 28)_{\mathrm{N_2},in} + 87{,}917 - 200{,}000 = 3 = 488{,}600\ \mathrm{W}.$$

The molar flow rates of gaseous species at the exit stream are known. So, we find

$$T_{out} = 1000\ \mathrm{K}.$$

Example 4.6 (cont.)

This is the operating temperature of the fuel cell. Note that this value is obtained assuming 100% fuel utilization factor; i.e., $\varphi = 1$. In practice, not all the fuel is converted in the cell. The reported values for φ are usually in the range 0.80–0.85.

It is obvious that utilizing a bottoming power cycle to use the energy of the exhaust stream of the fuel cell for further power production can improve the thermal efficiency. If this amount of energy is used in a power plant with 40% efficiency, the additional power production is

$$\dot{W}_{power\ plant} = 0.4 \times 488{,}600 = 195{,}440\ \text{W}.$$

The total power produced by the combined system is therefore

$$\dot{W}_{CP} = 195{,}440 + 200{,}000 = 395{,}440\ \text{W}.$$

4.10 Summary

In electrochemical reactions, the charge must be conserved, along with mass and energy conservation. Electrochemical reactions occur in a redox (reduction–oxidation) pair that when added together constitute the original reaction among uncharged species. They are encountered in fuel cells, electrolysis cells, and batteries, among other applications. In these cells, two electrodes, the anode and cathode, are separated by an electrolyte. Under equilibrium – that is at zero current – the application of the First and Second Laws shows that the electrical work produced by the reaction is the same as the Gibbs free energy of the reaction. The difference between the enthalpy of reaction and Gibbs free energy of reaction is heat transfer with the environment. Given the work and charge separated during the reaction, we can calculate the potential difference between the electrodes. Both the work and potential depend on the operating temperature and pressure of the reaction. They also depend on the concentrations of the fuel and oxidizer in their corresponding channels, as given by the Nernst equation. The depletion of the fuel and oxidizer along their corresponding channels reduces the open-circuit voltage.

The open-circuit efficiency of a fuel cell, as measured by the ratio of the Gibbs free energy divided by the enthalpy of reaction, is quite high. Losses responsible for the reduction of this efficiency at finite current will be discussed in detail in Chapter 7.

Problems

4.1 A fuel cell operating at 100 °C and 2 atm uses pure hydrogen and air. Determine (a) the open-circuit potential; and (b) the relative efficiency for an actual voltage of 0.74 volts.

4.2 Derive a relationship for the Nernst potential of a fuel cell operating with methanol as a fuel. Determine the change in open-circuit potential of the cell maintained at 200 °C if the operating pressure is increased 50%.

4.3 Methanol is used as a fuel source in a fuel cell system that operates at 90 °C and 1 atm with relative efficiency of 58%. Methanol is reformed with steam and the byproducts (i.e., H_2 and CO_2) are fed to the cell system, which produces 2 kW of power. The reforming efficiency is 90%. Determine the required mass flow rate of the methanol.

4.4 Ammonia is used as a source of hydrogen for a fuel cell operating at 90 °C. The dissociation reaction of ammonia is $2NH_3 \rightarrow N_2 + 3H_2$. The mixture of nitrogen and hydrogen is fed to the fuel cell instead of pure hydrogen. Determine: (a) the change in the open-circuit potential of the cell when the above mixture is fed to the cell, compared to the case with pure hydrogen; (b) the open-circuit potential of the cell using this mixture; and (c) the relative efficiency of the cell whose measured actual voltage is 0.7 volts.

4.5 In practice, the byproducts of methane reforming and water–gas shift reactions are composed of CH_4, H_2O, CO, CO_2, H_2. The composition of the mixture strongly depends on the temperature. Consider a fuel cell that operates at 1100 K and atmospheric pressure and uses the above mixture and air as fuel and oxidizer, respectively. The reforming process proposed here is a combined steam reforming and gas shift in one step. Gaseous methane and water enter the reformer at 300 K. The reformer operates at atmospheric pressure. From Figure 3.12, the composition of the mixture exiting the reformer at 1100 K is 16% CO, 4% CO_2, 64% H_2, and 16% H_2O. The system is shown schematically in Figure P4.5.

a. Calculate the heat requirement of the reformer per unit mass of the mixture.

b. Determine the maximum electrical work per unit mass of the mixture.

c. Calculate the open-circuit potential of the fuel cell knowing that two electrons are produced for each mole of hydrogen or carbon monoxide consumed.

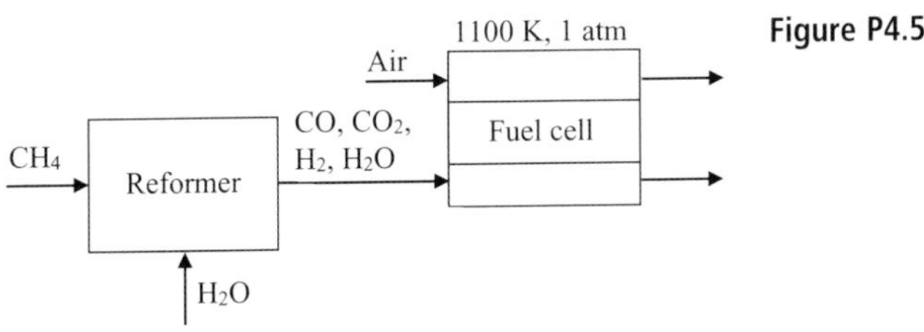

Figure P4.5

4.6 The power output of an internal reforming fuel cell operating at 1100 K is 50 kW. Methane at 750 K is fed to the fuel cell. Superheated steam at 600 K is used for reforming and shift reaction, both of which are assumed to take place completely without any dissociation. The byproducts, hydrogen and carbon dioxide at 800 K, are then fed to the anode. A hot stream of air at 850 K is used as the oxidizer. The fuel cell operates at an average voltage of 0.71 volts, with a fuel utilization of 80%.

a. Determine the stoichiometric ratio of the air required for the operation of the fuel cell system.

b. Compute the overall efficiency of the system. For this, you need to take into account the energy spent for heating the air stream and production of steam assuming that both air and water are supplied from an environment at 298 K and 1 atm.

4.7 A fuel cell operates at 1000 °C and 1 atm, as shown in Figure P4.7. The cell uses methane and air. Both streams enter and exit the cell at 1000 °C and 1 atm. The flow rate of air at the cathode inlet is such that the methane/oxygen ratio is stoichiometric. The fuel utilization is 75%, therefore some CH_4 exits the anode with the products, and some oxygen leaves unutilized. All gases can be treated as ideal.

a. Assume that the products stream on the anode side is CO_2, H_2O, and CH_4. Determine the maximum work per kmol of CH_4 that can be obtained from this fuel cell under these conditions.

b. Now assume that the products exiting at the anode consist of a mixture of CO, H_2, CO_2, and H_2O. In other words, all of the unutilized CH_4 has reformed internally to form a mixture of CO, H_2, CO_2, and H_2O. If there are 0.4094 moles of CO on the exit side per mole of CH_4 entering the fuel cell, find the mole fractions of the four species at the exit of the anode.

c. Assume that all the utilized oxygen reacts electrochemically and the total flow rate of methane is 20 mg/s. Calculate the current produced by the cell.

d. If the cell operates at a voltage of 0.65 V, determine the heat transfer rate out of the fuel cell.

e. Show that the composition calculated in part (b) at the anode exit is the equilibrium composition for that stream, assuming that all the methane has been consumed.

Figure P4.7

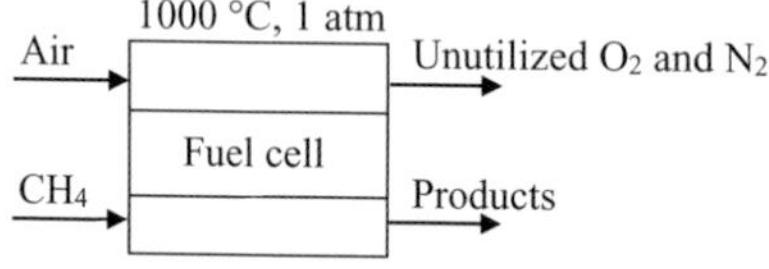

4.8 A fuel cell system is proposed as shown in **Figure P4.8**. The system utilizes methane and air. All stream conditions are given in the figure. Across the fuel cell, assume that only hydrogen is consumed in the electrochemical reactions with 100% hydrogen utilization. Carbon monoxide passes through the fuel cell without change. Within the cell, O^{2-} ions move through the electrolyte from the cathode to the anode. The flow rate of air at the cathode inlet is determined such that the ratio of the flow rate of the fuel stream (1) to the air stream is stoichiometric. All gases can be treated as ideal gases.

Assume that stream (1) at the exit of the reformer/water–gas shift reactor consists of CO, CO_2, H_2, and H_2O. The products are at thermodynamic equilibrium at the given temperature and pressure. At equilibrium, there are 0.049 kmol/s of CO_2 and 0.284 kmol/s of H_2O within the product stream (1).

a. Determine the molar flow rate of methane and water fed to the reformer. Note that the partial pressure-based equilibrium constant of the water–gas shift reaction at 800 °C is $K_p = 0.8879$.
b. Calculate the flow rate of hydrogen and the current produced by the fuel cell.
c. Determine the heat transfer rate across the reformer/water–gas shift reactor.
d. Calculate the flow rate of air into the fuel cell.
e. Assuming that the product stream (2) of the fuel cell consists of CO, CO_2, and H_2O, determine the maximum work transfer of the fuel cell.
f. Determine the actual work transfer from the fuel cell, assuming a Second Law efficiency of 70% for the fuel cell.
g. Calculate the operating voltage of the fuel cell.

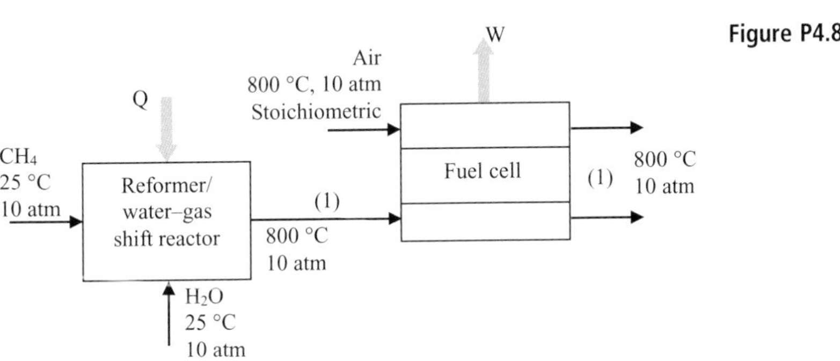

Figure P4.8

4.9 The gas mixture leaving the water–gas shift reactor of the biomass gasification plant in Figure P4.9 enters a low-temperature fuel cell operating at 100 °C, which uses air heated to 100 °C using some of the reformer rejected heat. Assume that the fraction of CO in the syngas is negligible and unimportant to the fuel cell reactions.

a. Assuming stoichiometric ratios of fuel and air at the inlet, and 50% fuel utilization, find the gas compositions at the outlets of the fuel cell anode and cathode.
b. Determine the open-circuit voltage ($\Delta\varepsilon_{max}$) for this fuel cell.

c. For a voltage efficiency of 50% and Faradic efficiency of 100%, determine the work per mole of biomass supplied to the reformer.
d. Calculate the overall First Law efficiency of the reformer + shift + fuel cell system.
e. Verify the assumption that CO concentration in the products of the WGS is indeed negligible.
f. Calculate the availability loss for the reformer and shift reactor.

Figure P4.9

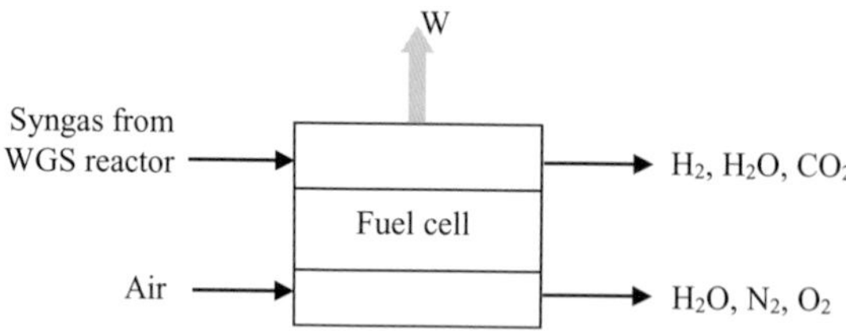

4.10. The steam–methane reforming process shown in Figure P4.10 produces syngas, which consists of three moles of H_2 and one mole of CO per mole of methane reformed. The syngas is used in a fuel cell operating at 700 °C to produce electrical work. The fuel is supplied on the anode side and air is supplied on the cathode side. The gas streams enter and exit the anode and the cathode at 10 atm and 700 °C. The fuel utilization efficiency for both H_2 and CO is 75%. At the entrance, enough air is supplied to the fuel cell for a stoichiometric reaction. All the gases can be treated as ideal. Specific heats can be assumed constant.

a. Calculate the mixture composition at the inlet and exit of the anode in mole fractions.
b. Calculate the mixture composition at the inlet and exit of the cathode in mole fractions.
c. Determine the maximum (reversible) work that the fuel cell can produce per mole of fuel consumed.
d. Calculate the cell voltage corresponding to the maximum work.
e. Calculate the corresponding fuel cell efficiency if the relative efficiency is 50%.
f. Determine the molar flow rate of the fuel entering the anode if the allowable internal current between the anode and the cathode is 1 A/cm^2 and the stack is 1 m^2.
g. Plot the reversible cell voltage as a function of fuel utilization efficiency between 0 and 1.

Figure P4.10

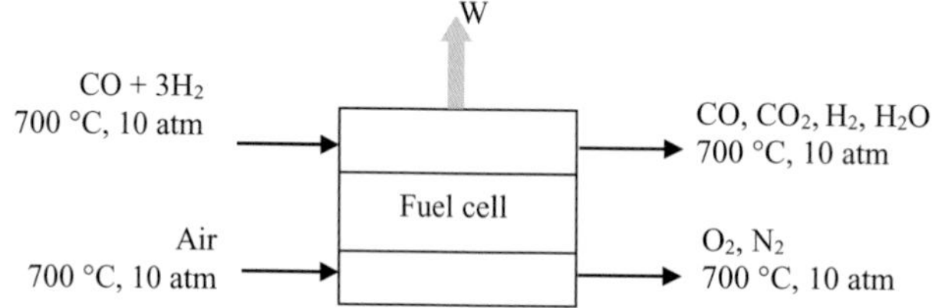

4.11 Water management is critical for the operation of proton exchange membrane fuel cells (PEMFC). These cells use sulphonate fluoropolymers as an electrolyte (membrane). The most well-known membrane brand is Nafion. For the electrolyte to be a good proton conductor, it should be well hydrated. However, too much water can flood the electrodes, blocking gas diffusion through the pores. Therefore, water produced by the electrochemical reaction on the cathode side should be removed by airflow. Even though both the fuel and air streams are mixed with water vapor at the inlet, humidity level is controlled primarily by air stream.

The following figure shows a schematic diagram for the operation of a fuel cell operating on hydrogen produced by steam reforming of methane.

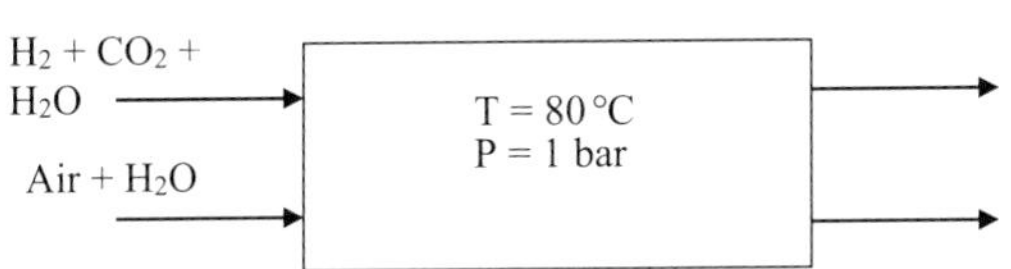

Figure P4.11 PEMFC schematics.

We will make following assumptions

- All streams are at 1 bar and 80 C.
- The fuel consists of H_2 and CO_2 with a molar ratio 4:1. The relative humidity of the fuel steam is 20%, where the relative humidity is defined as $\phi = \frac{p_{H_2O}}{p_{H_2O,sat}(T)}$.
- CO, which is poisonous to Pt catalyst, is completely removed from the fuel stream.
- Air stoichiometry, defined as the ratio of inlet oxygen flow rate to the oxygen consumption rate $\lambda = \frac{\dot{n}_{O_2,inlet}}{\dot{n}_{O_2,usage}}$ is 2 at the inlet, a very typical value.
- The inlet air relative humidity is 40%.
- The electrolyte is well hydrated and the level of hydration is constant and steady. In other words, there is no incoming or outgoing water from and to the electrolyte.
- Water produced electrochemically is in the form of vapor.
- All the species are considered as an ideal gas.

Values given at 353 K, saturated steam pressure: 0.4708×10^5 Pa

c_p (Temperature range [298 K~353 K])	Enthalpy of formation (J/mol)	Entropy of formation (J/mol·K) @ 1 bar
$\hat{c}^0_{p,O_2} = 29.55$	$\hat{h}^0_{f,H_2O(g)} = -241{,}826$	$\hat{s}^0_{f,H_2O(g)} = 188.835$
$\hat{c}^0_{p,H_2} = 28.96$		$\hat{s}^0_{f,H_2} = 130.68$
$\hat{c}^0_{p,H_2O} = 33.59$		$\hat{s}^0_{f,O_2} = 205.152$
$\hat{c}^0_{p,N_2} = 29.15$		$\hat{s}^0_{f,N_2} = 191.609$

Answer the following questions:

a. Determine the mole fractions at the anode and cathode. What is the theoretical open-circuit voltage of this cell?

b. What is the open-circuit (thermodynamic) efficiency based on the LHV of H_2? (LHV of H_2 is 120.1 MJ/kg.)

A fuel cell stack is used to power a small vehicle that requires 80 kW. Each cell operates at 0.6 V, and a typical current density is 1 A/cm^2. Assume that the surface area of each individual cell is 650 cm^2, the Faradaic efficiency is 100%, and the fuel utilization is 90%.

c. How many individual cells are needed to supply the required power for the vehicle?

d. What is the total molar flow rate of oxygen?

e. What is the composition of the air-side stream at the exit of the stack, expressed in terms of the mole fractions of N_2, O_2, and H_2O?

f. When the design exit air relative humidity is 90%, do the current operating conditions satisfy the design target?

g. What is the composition at the anode-side exit stream?

h. What is the molar flow rate of the fuel stream at the inlet?

i. What is the First Law efficiency of the cell?

$$\eta_I = \frac{\text{generated power}}{\text{rate of chemical energy in}}.$$

j. What is the cooling rate required to keep the fuel cell at 80 °C?

k. Derive an expression for the exit relative humidity in terms of the air stoichiometry, λ, the relative humidity of the inlet air stream, ϕ_{inlet}, the water saturation pressure, $p_{H_2O,sat}$, and the pressure p^0.

l. What should be the relative humidity of the inlet air if the relative humidity of the exit air is 90%?

4.12 Figure P4.12 shows a schematic diagram of a hydrogen generation plant using electrolysis. Water available at 25 °C and 1 bar is pumped to 10 bar. Some of the pumped water is injected directly into a combustor. The rest of the water is heated to 80 °C in heat exchanger 1 and then enters the isothermal electrolyzer, which operates at the same temperature. Oxygen produced by the electrolyzer is at 80 °C and is heated in heat exchanger 2 to 750 °C prior to entering the combustor. Methane enters the combustor at 25 °C and 10 bar. The equivalence ratio of the methane–oxygen combustion is 1 and the combustion is complete. The turbine inlet temperature is fixed at 1400 °C and the mixture is expanded in the turbine to 1 bar. The isentropic efficiency of the turbine is 87%. The electricity generator has an efficiency of 97%. The electricity produced by the plant powers the pump and partially powers the electrolyzer. Neglect the pressure drop inside the heat exchangers.

You can treat all gases as ideal and assume constant specific heats. Liquid water is pure and incompressible. You can neglect the temperature rise across the pump.

Answer the following questions:

a. Calculate the total moles of water injected into the combustor per moles of methane burned.

b. Calculate the turbine exit temperature (T_7) and the electric work generated per moles of methane burned. The isentropic index in the turbine is 1.245.

c. Calculate the additional work required by the electrolyzer in J/mol methane burned if its Second Law efficiency is 55%.

d. Calculate the total heat transfer between the electrolyzer and its environment in J/mol methane burned.

e. Given that the plant is designed for hydrogen production, define an appropriate efficiency for the plant and calculate it.

f. The plant proposes to utilize the high-temperature exhaust stream to raise the overall efficiency of hydrogen production. The Second Law efficiency of the heat engine utilizing this stream to produce mechanical work is 45%. If the exhaust temperature (T_9) is 780 °C and the ambient is 25 °C and 1 bar, calculate the mechanical work generated by the heat engine per mole of methane burned.

g. Recalculate the efficiency defined in part (e) after adding the heat engine in part (f).

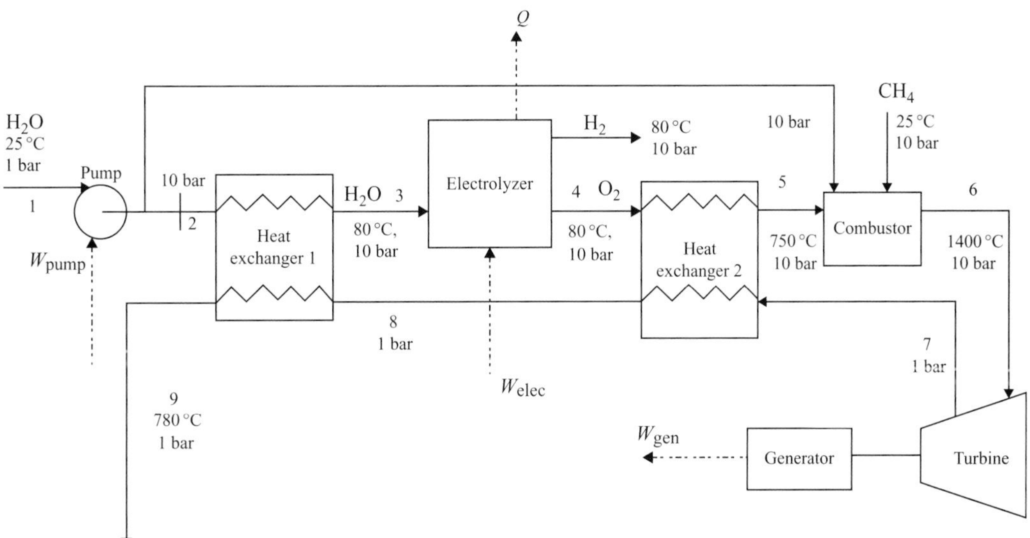

Figure P4.12 Hydrogen generation plant using electrolysis.

Useful Data

Standard Enthalpy of Formation of Relevant Species

$$\hat{h}^o_{f,H_2O(l)} = -286{,}030 \text{ J/mol}$$

$$\hat{h}^o_{f,H_2O(g)} = -242{,}000 \text{ J/mol}$$

$$\hat{h}^o_{f,CO_2} = -393{,}790 \text{ J/mol}$$

$$\hat{h}^o_{f,CH_4} = -74{,}920 \text{ J/mol}$$

Standard Entropy of Relevant Species

$$\hat{s}^o_{H_2O(l)} = 69.95 \text{ J/mol·K}$$

$$\hat{s}^o_{H_2O(g)} = 188.85 \text{ J/mol·K}$$

$$\hat{s}^o_{CO_2} = 213.83 \text{ J/mol·K}$$

Constant Specific Heats

$$\hat{c}_{p,H_2O(g)} = 40.9 \text{ J/mol·K}$$

$$\hat{c}_{p,CO_2} = 54.0 \text{ J/mol·K}$$

$$\hat{c}_{p,O_2} = 31.4 \text{ J/mol·K}$$

$$\rho_{H_2O(l)} = 1000 \text{ kg/m}^3 \text{ (density of liquid water)}$$

$$M_{H_2O} = 18 \text{ kg/kmol (molecular weight of liquid water)}$$

$$\mathfrak{R} = 8.314 \text{ J/mol·K (universal gas constant)}$$

For the reaction $H_2O \rightarrow H_2 + \frac{1}{2}O_2$
Enthalpy of reaction at 80 °C, 10 bar = 284,290 J/mol
Entropy of reaction at 80 °C, 10 bar = 132.45 J/mol·K

REFERENCES

1. D. R. Crow, *Principles and applications of electrochemistry*. London: Blackie, 1994.
2. J. S. Newman and K. E. Thomas-Alyea, *Electrochemical systems*. Hoboken, NJ: Wiley, 2004.
3. A. J. Bard and L. R. Faulkner, *Electrochemical methods: fundamentals and applications*. New York: Wiley, 2001.
4. S. C. Singhal and K. Kendall, *High temperature solid oxide fuel cells: fundamentals, design, and applications*. Oxford: Elsevier, 2003.
5. J. Larminie and A. Dicks, *Fuel cell systems explained*. Chichester: Wiley, 2003.
6. H. Zhu and R. J. Kee, "A general mathematical model for analyzing the performance of fuel-cell membrane-electrode assemblies," *J. Power Sources*, vol. 117, no. 1, pp. 61–74, 2003.
7. A. F. Lee, W. Y., Hanna, and J. Ghoniem, "On the prediction of carbon deposition on the nickel anode of a SOFC and its impact on open-circuit conditions," *J. Electrochem. Soc.*, vol. 160, no. 2, pp. F94–F105, 2013.

8. J. Hanna, W. Y. Lee, Y. Shi, and A. F. Ghoniem, "Fundamentals of electro- and thermochemistry in the anode of solid-oxide fuel cells with hydrocarbon and syngas fuels," *Prog. Energy Combust. Sci.*, vol. 40, pp. 74–111, 2014.
9. S. W. Angrist, *Direct energy conversion*. Boston, MA: Allyn and Bacon, 1965.
10. A. P. H. Warn and J. R. W. Peters, *Concise chemical thermodynamics*, 2nd ed. Boca Raton, FL: CRC Press, 2000.
11. G. N. Hatsopoulos and J. H. Keenan, *Principles of general thermodynamics*. Huntington, NY: R.E. Krieger Pub. Co., 1981.
12. US DOE, Office of Fossil Energy, *Fuel cell handbook*, 7th ed. Washington, DC: US DOE, Office of Fossil Energy, 2004.

5 Thermomechanical Conversion

Gas Turbine Cycles

5.1 Introduction

Thermomechanical energy conversion is concerned with the conversion of heat and thermal energy to mechanical work or mechanical energy. The latter may be used directly (e.g., for propulsion) or to generate electricity. Currently, the major source of thermal energy is the combustion of fossil fuels and biomass, followed by thermonuclear reactions, with geothermal energy and solar energy at much smaller scales. The conversion efficiency is directly related to the temperature of the heat source or the "quality" of the thermal energy. The temperature of the source and that of the environment determine the maximum efficiency of energy conversion cycles.

When the source of the thermal energy is combustion, one can control the maximum temperature with reasonable flexibility. For instance, most hydrocarbons burning adiabatically and stoichiometrically in air at constant pressure reach temperatures around 2250 K. The corresponding value for hydrogen is 2380 K. Air preheat or compression prior to combustion can raise the temperature further if a higher temperature is desired. Constant-volume combustion reaches even higher temperatures. Another option to raise the maximum temperature is to burn fuel in pure oxygen, where the flame temperature of most hydrocarbons, as well as hydrogen, is O (3000 K). Lower combustion temperatures can be achieved through lean burning, which has emissions and combustion efficiency advantages, and by mixing the combustion products with extra air downstream of the primary combustion zone. Exhaust gas recirculation can also be used to lower the combustion temperature without adding more air. Other factors involved in determining the appropriate combustion temperature, such as the kinetics of the chemical reactions or the overall burning rate, must be considered. Furthermore, except for reciprocating engines, continuous-flow energy conversion devices, such as gas turbines, cannot operate anywhere near stoichiometric combustion temperatures because of metallurgical constraints.

Another important temperature in energy conversion cycles is that of the heat sink or the environment. This is often taken as the atmospheric temperature, or that of nearby lake or ocean water. In some implementations, it is the lowest pressure that determines some of the characteristics of the heat engine, because it limits the work produced during the expansion

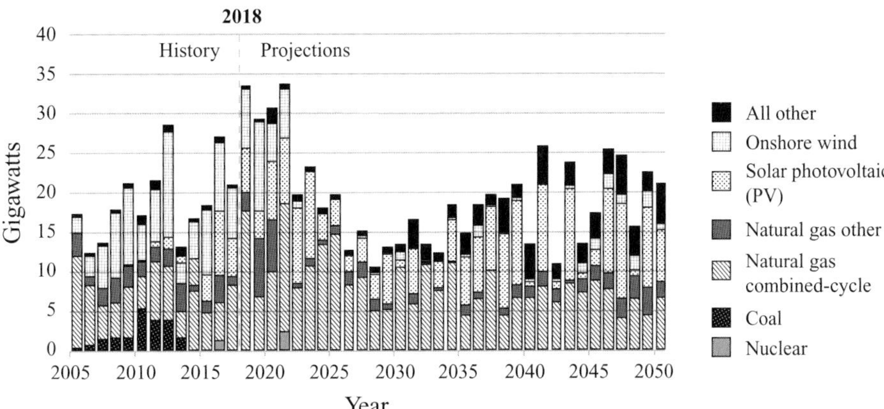

Figure 5.1 US electricity generation capacity additions by fuel [1].

of the working fluid. In open cycles, the lowest pressure is the atmospheric pressure. In closed cycles, such as two-phase cycles and closed gas cycles, lower pressures can be reached by incorporating a condenser or a heat exchanger to cool the gas, respectively.

In this chapter and the next one, we are concerned primarily with the thermodynamics of energy conversion cycles used in electricity generation. This chapter focuses on gas turbine cycles. We start with a review of heat engines and their maximum efficiency, as well as some other definitions of the efficiency, and an overview of thermal energy sources. All the systems and cycles discussed here are those used in electricity generation. Next, we discuss the Brayton cycle in the ideal form and when practical components are used, with a focus on factors determining the efficiency and the specific work of the cycles. We compare open and closed cycles, and determine the impact of the working fluid. Next, we introduce cycle modifications used to improve the efficiency, such as recuperation, intercooling, and reheating. We conclude with a summary of the state of the art of practical gas turbines.

Figure 5.1 shows the recent capacity addition of power generation equipment in the USA, and projections for the next ~20 years, highlighting the significant role of gas turbines, mostly powered by natural gas in the USA [1].

5.2 Heat Engines

Heat engines convert heat or thermal energy into mechanical energy or work. These thermomechanical devices are used widely in power and propulsion and deserve a closer look when considering the effective use of resources. Heat engines come in continuously operating forms, such as steam and gas turbines, and intermittently operating forms, such as internal combustion engines. Steam turbine cycles have been the most extensively used form of electricity generation, with gas turbines coming in second place and diesel engines being at a smaller scale. The operation of these engines can be described thermodynamically using one of a small number of cycles, which we review here.

The choice of the engine, or cycle, is determined by many requirements, such as the power:weight ratio, the part-load performance, the startup/shutdown time, the emission characteristics when fossil fuels are involved, the fuel or thermal energy source, the cooling requirements, the initial vs. running cost of the plant, and more. For large-scale electricity production, steam turbines and gas turbines are used extensively. We will start our coverage of advanced thermomechanical energy conversion by reviewing gas turbine cycles. The emphasis here is on the aspects of the cycle operation that control the overall efficiency and specific work, the compromise between simplicity and efficiency, and measures that can be taken to improve both. Some subtle effects, such as the complexity of the hardware required to improve efficiency, will also be discussed.

This chapter is concerned primarily with power production or electricity generation, and hence the emphasis is on turbine engines and their operating cycles. Current gas turbine engines are external combustion engines in which the chemical energy of the fuel is converted to thermal energy in a combustor, and the hot products of combustion are expanded in a turbine to produce work. Gas turbines have not been used in the nuclear energy industry because the reactor core temperature is too low. For the same reason, they have not been used in solar or geothermal plants.

5.2.1 Heat Engine Efficiency

In most cases, the source of thermal energy is the chemical energy of a fuel, thermonuclear energy, geothermal energy, or solar energy. The conversion of chemical energy to thermal energy – that is, combustion – was discussed in Chapter 3. The conversion of thermal energy or heat into mechanical work is discussed here. Continuous conversion or cyclical operation is performed in thermodynamic cycles, and is limited by the requirements of the Second Law. In the cycle analysis, the thermal efficiency of a work-producing cycle is defined as

$$\eta_I = \frac{|\text{net work out}|}{\text{total heat input}} = \frac{|\sum \text{work out}|}{\text{total heat input}}. \tag{5.1}$$

In the second expression, the algebraic sign of the work must be used, with work out being positive. As we have seen before, this is also called the First Law efficiency, the cycle efficiency η_{cycle}, or the thermodynamic efficiency, η_{th}. As we saw in Chapter 2, the best-performing heat engine cycle is that running according to a Carnot cycle for which the efficiency is given by one of the two following expressions depending on the source of heat or thermal energy, where T_o is the temperature of the environment and T_H is the maximum temperature available for heat transfer:

$$\eta_{car} = 1 - \frac{T_o}{T_H} \tag{5.2}$$

or

$$\eta_{car}^* = 1 - \frac{\ln \dfrac{T_H}{T_o}}{\left(\dfrac{T_H}{T_o} - 1\right)}. \tag{5.3}$$

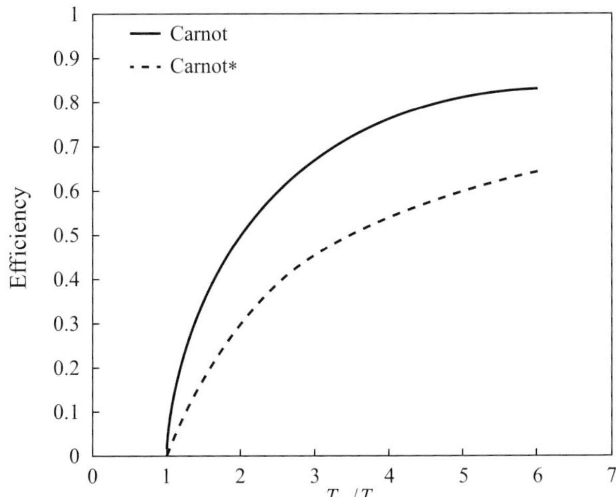

Figure 5.2 The efficiencies of a Carnot engine operating between two reservoirs at fixed temperatures, and a Carnot* engine operating by extracting heat from a stream starting at the high temperature until it is cooled to the lower temperature. The efficiency of the Carnot engine is used to define the quality of the heat.

In the second expression, the heat source temperature drops between the high temperature and the low temperature. The second expression also applies to a case in which a hot stream is used as a source of heat or thermal energy. The two expressions for the heat engine efficiency are shown in Figure 5.2.

Another important efficiency measures how well the system is performing with respect to the best possible system, and uses the availability or maximum work as a reference. This is the Second Law efficiency, also known as the utilization efficiency (not to be confused with the fuel utilization efficiency, which is defined next):

$$\eta_{II} = \frac{\text{net work out}}{\text{maximum work}} = \frac{\text{net work out}}{\text{available energy}}. \tag{5.4}$$

The Second Law efficiency is often used to determine the performance of a given cycle relative to that delivering the maximum work, or to a Carnot cycle operating between the maximum and minimum temperatures. Practical cycles, even in their idealized forms, cannot perform as well as the Carnot cycle because of the difficulties of performing isothermal processes. The Second Law efficiency shows how far the performance of a given energy conversion system is from the operation of the Carnot cycle.

When the source of the thermal energy is the fuel chemical energy, then a more relevant definition is the fuel utilization efficiency, defined mostly in terms of the power and the fuel flow rate times its availability:

$$\eta_{fuel-exergy-utilization} = \frac{\text{net power out}}{\text{fuel mass flow rate} \times \text{fuel availability}} = \frac{\wp}{\dot{m}_f \cdot \xi_f^o}. \tag{5.5}$$

As we saw before, the numerical value of the fuel availability is close to the lower heating value (LHV) of the fuel. Thus, a common expression for the fuel utilization efficiency is

$$\eta_{fuel-utilization} = \frac{\wp}{\dot{m}_f \cdot LHV}. \tag{5.6}$$

The relations between these efficiencies were discussed in detail in Chapters 2–4. Another source of irreversibility, or loss of availability, is the heat transfer between the hot combustion gases and the working fluid, which for all practical purposes occurs across a finite temperature difference. (The same notion applies to systems operating on different sources of "primary" energy, such as nuclear plants, in which there must be a finite temperature difference between the nuclear reactor coolant and the working fluid.) Note that the fuel utilization efficiency accounts for both the cycle efficiency and the combustion efficiency, where the combustion efficiency is defined as

$$\eta_{comb} = \frac{(\text{thermal enthalpy out} - \text{thermal enthalpy in})_{combustor}}{\text{fuel mass flow rate} \times \text{fuel heating value}}. \tag{5.7}$$

The fuel utilization efficiency is then

$$\eta_{fuel-utilization} = \eta_{comb}\eta_I. \tag{5.8}$$

The use of gas turbines in power generation is growing rapidly with the expanding utilization of natural gas in this sector, both in simple cycles and combined cycles. This trend is expected to continue, even if fuels other than natural gas are used. For instance, with the expansion of coal and biomass gasification, the natural gas can be replaced with synthetic gas produced from coal. Gas turbine cycles analysis lends itself more to analytical study when the ideal gas assumption is incorporated, and one can obtain analytical expressions for cycle performance. Many of the efficiency improvement techniques suggested for gas turbine cycles carry over to steam turbine cycles as well.

Improving the thermal efficiency of the cycle requires modifications that make the cycle look as close as possible to the Carnot cycle – that is, heat transfer with external sources should be as close as possible to isothermal, and work processes should be as close as possible to isentropic. Heat transfer processes, either with external sources or between internal components, should be done with the least temperature difference between the streams exchanging heat.

5.2.2 Working with Different Sources of Thermal Energy

Different heat engine cycles, including Brayton and Rankine cycles, and Otto and Diesel cycles, can be used for power generation. The choice of a power cycle depends on several factors; chief among them is the source of thermal energy and the maximum temperature of the source. Power cycles operating with fossil fuels or biomass have the potential to reach high temperatures, often much higher than rotating machines can tolerate. Rotating machines, such as gas turbines and steam turbines used in large-scale electricity generation, spin at a high speed and sustain large mechanical stresses while being consistently exposed to high temperature and often high-pressure working fluids. High-temperature gases, mostly in the form of combustion products, can operate gas turbines directly – that is, they can be expanded through a gas turbine or used as working fluids. Combustion products are also used in combined cycle machines. In these cycles, combustion gases are first expanded though the gas turbine. Next, the gas turbine exhaust is used in steam generation and superheating. Because of the combination of the continuous exposure to high temperatures and spinning at high speeds, gas turbine cycles have severe limitations on the maximum temperature. Higher temperatures are tolerated in reciprocating machines, which are used in smaller power plants or for "peak shaving," as well as in transportation power trains. This is because the cylinders of these engines are exposed to hot gases only part of the time, and are also regularly cooled.

Thermonuclear energy and renewable sources such as solar thermal energy and geothermal sources provide lower-temperature heat. In these cases, two-phase cycles are much more suitable for thermomechanical conversion. Intermediate temperature heat sources are compatible with steam Rankine cycles. Low-temperature geothermal plants often use other working fluids whose evaporation temperatures are lower than water (e.g., organic fluids).

Although not usually defined in a formal way, the grade of the thermal energy can be defined according to the highest temperature available from the source. This energy grade is related to the best heat engine efficiency – that is, the Carnot efficiency – that the source can provide when used to run a heat engine. For instance, for $T_L \sim 300\,\text{K}$, the Carnot efficiency is $\eta^*_{car} \leq 30\%$ when $T_H \sim 600\,\text{K}$, and $\eta^*_{car} \leq 50\%$ when $T_H \sim 1000\,\text{K}$. These values set the upper limits on the efficiencies achieved from heat engine cycles using sources at these temperatures. Temperatures higher than 1000 K allow reaching theoretical efficiencies above 50%.

Currently, because of the concern over the contribution of carbon dioxide to global warming, another important measure of power plant performance is the amount of CO_2 emitted per unit of useful energy produced. Carbon dioxide generation per unit of useful energy produced – that is, per unit of electricity – is related to the efficiency of the plant and can be estimated from the following expression:

$$CO_2 \text{ generation} = \frac{\nu_{CO_2} M_{CO_2}}{\eta_e \left|\Delta \hat{h}_{R,f}\right|} \,[\text{kg-CO}_2/\text{MJ}_\text{e}], \tag{5.9}$$

where ν_{CO_2} is the number of moles of CO_2 per mole of fuel burned (= 1 for coal or methane), M_{CO_2} is the molecular weight of CO_2, η_e is the overall chemical-to-electrical conversion efficiency, and $\Delta \hat{h}_{R,f}$ is the fuel enthalpy of reaction per mole. For methane (with 800 MJ/kgmol), when used in a combined gas–steam cycle with 55% efficiency, CO_2 production is close to 0.1 kg-CO_2/MJ. For coal (with 360 MJ/kgmol), when used in a simple steam cycle with 40% efficiency, CO_2 production is close to 0.3 kg-CO_2/MJ. In practice, numbers vary because of the variation of the fuel properties.

Electricity generation using fossil fuels takes advantage of the abundance and low prices of coal and, in recent years, of the convenience and higher efficiencies afforded by natural gas. Natural gas-burning plants emit cleaner exhaust, and hence it is easier to site and operate them in urban areas. Natural gas turbine plants can be built to operate on simple cycles (SCs) or combined cycles (CCs). The use of higher inlet temperature aero-derivative gas turbines can boost the conversion efficiency beyond what is achievable with the larger size, heavy-duty turbines. In the future, natural gas CCs are expected to be hybridized with a high-temperature fuel cell for further efficiency improvements. Coal plants operate mostly on steam cycles, running at subcritical but more recently at supercritical conditions. When gasified, coal can be used to operate a CC power plant, which is more efficient than a subcritical steam plant. In an integrated gasification CC (IGCC) plant, it is also possible to incorporate a high-temperature fuel cell.

Thermal energy produced in nuclear reactors is delivered at temperatures that depend on the reactor design and the coolant [2]. The following types of nuclear reactors have been used in power generation:

- pressurized heavy water reactors (PHWR) operate at lower temperatures of 260–280 °C;
- boiling water reactors (BWR) provide heat in the range 280–290 °C;
- pressurized water reactors (PWR) operate around 300–350 °C;
- metal-cooled reactors operate at 550 °C; and
- compressed gas reactors (CGR or GCR, which stands for gas-cooled reactors) run at 700 800 °C.

As will be shown in the discussion of gas turbine thermodynamic cycles, the inlet rotor temperature or the highest temperature of the cycle is critical in determining the cycle efficiency, and temperatures below 700 °C are not recommended for gas turbine cycles. Nuclear power plants currently operate on steam Rankine cycles, both subcritical and supercritical. For the same reason, most solar thermal electric power plants, which use concentrators to heat the working fluid to temperatures in the range of 350 °C (for parabolic or trough concentrators) to 550 °C (for tower concentrators), operate on steam cycles. More recent designs, in which a dish concentrator is used to reach 750 °C, incorporate gas Stirling cycles.

Geothermal sources are often in the 150–200 °C range, and typical pressures are above 1 MPa. Except for a few plants in which steam is directly expanded in a turbine, geothermal water pressure is reduced in a flash unit until steam is generated, and subsequently expanded in a steam turbine. Ocean thermal energy is a low-temperature source. Closed and open two-phase Rankin cycles for this type of energy operate on the temperature difference between the warmer surface water and the cooler deep water. This temperature difference is low, and at peak values is in the range of 20–25 °C. Therefore, the thermal efficiency of these cycles is low.

5.3 Gas Turbines

There has been a substantial growth in the use of gas turbines in electricity generation power plants, especially those burning natural gas or oil, but mostly natural gas. This is supported in part by improvements in gas turbine cycle components, which made it possible to raise the pressure ratio across the turbine and compressor as well as the maximum turbine inlet temperature to high values. Another contributing factor to the overall conversion efficiency of these plants is the improvement in the aerodynamic efficiency of the rotating components. Some of these components were originally developed for application in aerospace propulsion (when converted to power generation, these engines are known as aero-derivative turbines). Gas turbines designed to sustain temperatures $> 1400\,^{\circ}\mathrm{C}$, in which special materials are used in the turbine blades and advanced cooling techniques are applied to maintain the conditions safe for these materials, have become available for power generation. Implementing other measures described in the next several sections, such as regeneration and reheating, have also contributed significantly to improving the cycle efficiency.

Using gas turbines in electricity generation became more attractive with the introduction of high-efficiency natural gas CC plants. In these plants, gas turbines are incorporated as topping cycles for steam turbines, thus taking advantage of the higher combustion temperatures and the clean exhaust of natural gas in the gas turbines, while using the higher temperature of the gas turbine exhaust to raise steam for a bottoming Rankine cycle. With the spread of coal gasification technology, CC plants will find even more utilization. In these gasification-based CCs, synthetic gas is produced in a coal gasifier, thoroughly cleaned by removing most of the sulfur compounds and particulate, and then used as a gas turbine combustor fuel. Other advantages of gas turbines include their compact size, their short startup and shutdown times, and the absence of a need for cooling when simple cycles are used.

5.3.1 Brayton Cycle

Gas turbines used in power generation run on the **Brayton cycle**. Ideally, the cycle consists of an isentropic compression, a constant-pressure heat addition, an isentropic expansion, and a constant-pressure heat rejection [3,4]. The cycle efficiency depends strongly on the pressure ratio. In practice, the efficiency suffers from the non-ideal compression and expansion processes. Under non-ideal conditions, the cycle efficiency depends on the maximum cycle temperature as well. Practical Brayton cycles suffer from irreversibility associated with non-isothermal heat transfer processes at higher and lower temperatures, and, as will be seen in the calculations, the Second Law efficiency can be low. When modified through reheat and recuperation, substantial efficiency gains can be achieved.

There are two types of gas turbine cycles, the **open cycle** and the **closed cycle**. In the open or combustion turbine cycle, air is compressed and delivered to the combustor, where fuel is burned to raise the temperature. The maximum temperatures sustainable by the gas turbine blades are currently close to $1400\,^{\circ}\mathrm{C}$ for power generation. Combustion products are expanded in the turbine and exhausted into the atmosphere. The air temperature at the combustor inlet depends on the compression ratio and the atmospheric temperature. Limiting the combustion products to a temperature that can be tolerated by the turbine means that the overall fuel–air ratio in the combustor is much below the stoichiometric,

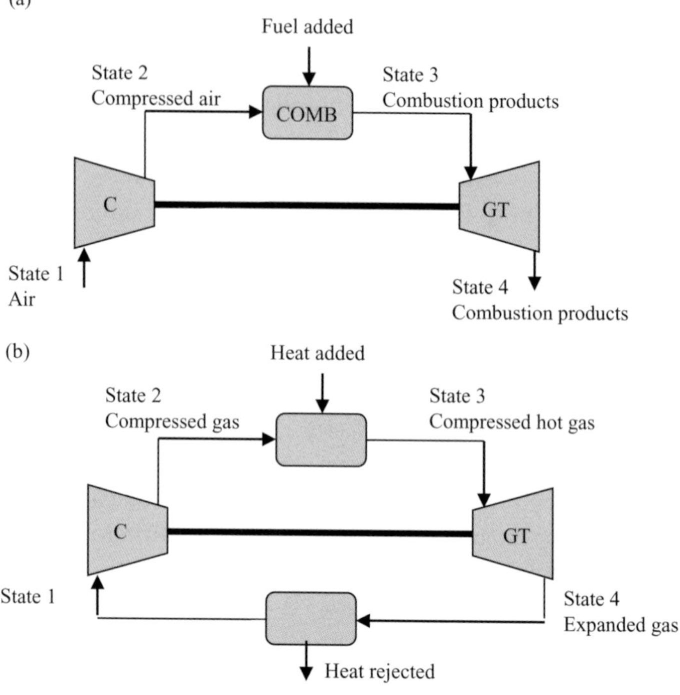

Figure 5.3 The Brayton cycle layout and its components in an open-cycle combustion turbine plant (a) and a closed cycle (b). The numbers correspond to the states in Figure 5.4. The fuel introduced into the combustor must be compressed to at least the pressure of the air entering the combustor. Note, however, that the mass of the fuel is much smaller than that of air.

often below 1:30. In the closed cycle, a working fluid circulates around the cycle without ever being exhausted. The working fluid is heated in a heat exchanger using an available source of heat, and it is cooled in another heat exchanger using a cooling medium.

The schematics of an open Brayton cycle and a closed cycle are shown in Figure 5.3. The minimum pressure of the open cycle is the atmospheric pressure, while it is arbitrary in the case of a closed cycle.

The T–s diagrams of an ideal cycle, in which the compression and expansion are isentropic, and an irreversible cycle in which the work interaction processes are irreversible, are shown in Figure 5.4. Under reversible conditions, the area bound by a curve and the s-axis on the T–s diagram indicates the heat transfer interaction, $q = \oint_{curve-s\_axis} Tds$, with a rightward pointing arrow indicating heat addition to the working fluid. Moreover, under these conditions, the area bound by the closed cycle on the T–s diagram, $q = \oint_{closed\_curve} Tds$ is the net heat transfer and also the net work transfer with the environment. The constant-pressure lines on the T–s planes of an ideal gas with constant and fixed isentropic index do not diverge as T or s increase (contrary to commonly shown hand drawings). Instead, they are exponential curves whose slope rises as the temperature or entropy increase:

$$\left(\frac{T}{T_0}\right)_p = \exp\left(\frac{\hat{s} - \hat{s}_o}{\mathfrak{R}}\right). \tag{5.10}$$

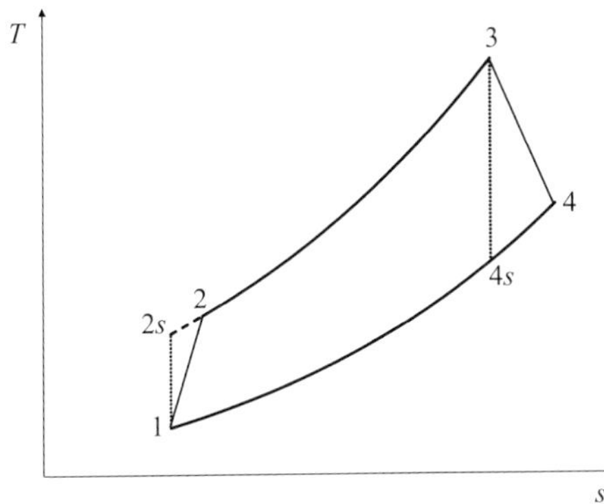

Figure 5.4 The T–s diagram of a simple Brayton cycle operating between two pressures, with isentropic compression and expansion shown in continuous curves, and with adiabatic compression and expansion shown in broken curves the sequence of states: 1-2s(2)-3-4s(4).

An important implication here is that during a constant-entropy process between two pressures, the temperature change ΔT, and hence the enthalpy change Δh, increase as the initial temperature in the processes rises. The enthalpy difference, Δh, is the work interaction in an ideal isentropic process. This result has many important applications in the analysis of gas turbine cycles.

Irreversibility in gas turbine cycle components, in particular in the work transfer components, can have significant negative impacts on the overall efficiency and specific work of the cycle. To quantify the impact of this irreversibility, it is necessary to define the **isentropic efficiencies** of the compressor, η_C, and the turbine, η_T, as follows (see Figure 5.4):

$$\eta_C = \frac{\text{isentropic compression work}}{\text{actual work required}} = \frac{h_{2s} - h_1}{h_2 - h_1} \tag{5.11}$$

and

$$\eta_T = \frac{\text{actual work output}}{\text{isentropic work output}} = \frac{h_3 - h_4}{h_3 - h_{4s}}. \tag{5.12}$$

The change in the kinetic energy of the fluid across work transfer components is often negligible with respect to the change in enthalpy, and hence the work transfer in adiabatic processes is the difference between the fluid enthalpies across the machine. Because of the aerodynamic effects, the compressor efficiency is often lower than that of the turbine. In fact, it was the complexity of building high-efficiency compressors that had plagued the development of gas turbine engines for many years, originally for propulsion then later for energy conversion purposes. While there is still room for improvement, there have been substantial strides in this endeavor, and this is no longer a showstopper.

5.3.2 Efficiency and Specific Work

The calculations of the Brayton cycle efficiency and work output involve a standard application of the First Law to the individual processes [3,5]. Assuming that the working fluid behaves as an ideal gas with constant specific heat and isentropic index, $k = c_p/c_v$, where c_p and c_v are the specific heats at constant pressure and constant volume, respectively, we get the following equations that define the states at the end of the compression and expansion processes under ideal conditions, where sub C corresponds to compressor:

$$\frac{T_{2s}}{T_1} = \left(\frac{p_2}{p_1}\right)^{\frac{k_C-1}{k_C}} \tag{5.13}$$

and

$$\frac{T_3}{T_{4s}} = \left(\frac{p_3}{p_4}\right)^{\frac{k_T-1}{k_T}} = \left(\frac{p_2}{p_1}\right)^{\frac{k_T-1}{k_T}}. \tag{5.14}$$

and sub T corresponds to turbine. In most analytical treatment, the isentropic indices are assumed to be the same for the compression and expansion processes. To determine the per unit mass flow heat transfer into the cycle, q_{in}, and the total work out per unit mass, w_{net}, we use the following equations, assuming constant specific heats:

$$q_{in} = h_3 - h_2 = c_p(T_3 - T_2) \tag{5.15}$$

and

$$\begin{aligned} w_{net} &= w_t - w_c \\ &= (h_3 - h_4) - (h_2 - h_1), \end{aligned} \tag{5.16}$$

while

$$w_{net} = c_p[(T_3 - T_4) - (T_2 - T_1)]. \tag{5.17}$$

The last equation is obtained assuming the same value for the specific heat during the compression and expansion. In the case of non-ideal components, and for an ideal gas, (5.11) and (5.12) are used to calculate the temperatures at the end of the compression and expansion processes, while replacing the enthalpies with the temperatures:

$$T_2 = T_1 + \frac{T_{2s} - T_1}{\eta_C} \tag{5.18}$$

and

$$T_4 = T_3 - \eta_T(T_3 - T_{4s}). \tag{5.19}$$

For an ideal gas with constant specific heat ratio, the isentropic compressor work per unit mass of working fluid, across a pressure ratio π_p, is given by

$$w_c = c_p T_1 \left[\pi_p^{(k-1)/k} - 1\right]. \tag{5.20}$$

Under the same conditions, the ideal turbine work per unit mass of the working fluid is

$$w_t = c_p T_3 \left[1 - \frac{1}{\pi_p^{(k-1)/k}} \right]. \tag{5.21}$$

The cycle efficiency is the First Law efficiency:

$$\eta = \frac{w_{net}}{q_{in}}. \tag{5.22}$$

Using (5.15), (5.20), and (5.21), we get the expression for the simple Brayton cycle efficiency under ideal conditions:

$$\begin{aligned} \eta &= 1 - \left(\frac{1}{\pi_p} \right)^{(k-1)/k} \\ &= 1 - \frac{1}{\vartheta_{2s}}, \end{aligned} \tag{5.23}$$

where $\vartheta_{2s} = T_{2s}/T_1$ and $\pi_P = p_2/p_1$. The thermal efficiency increases monotonically with the pressure ratio. As the pressure ratio is raised, the compressor exit temperature increases and the heat required to reach the same turbine inlet temperature decreases. Thus, for the same net work output, the thermal efficiency also increases with the pressure ratio. This result is illustrated schematically in Figure 5.5. In this figure, two Brayton cycles that produce the same net work output and operate between the same minimum and maximum temperatures are drawn on a T–s diagram. The heat added in the high-pressure cycle is lower than the heat added in the low-pressure cycle.

Another important cycle performance parameter is the specific work, or the net work output per unit mass of the working fluid. In gas turbines, the maximum turbine rotor inlet temperature (TRIT or TIT), T_3, or normalized as $\vartheta_3 = T_3/T_1$, is another important design parameter. Using (5.20) and (5.21), it can be shown that for an ideal air standard cycle:

$$\tilde{w}_{net} = (\vartheta_2 - 1)\left(\frac{\vartheta_3}{\vartheta_2} - 1 \right), \tag{5.24}$$

where $\tilde{w}_{net} = w_{net}/c_p T_1$ is a normalized work output. The specific work of the cycle depends on the pressure ratio and the maximum cycle temperature. The maximum cycle temperature depends strongly on the material used to build the gas turbine and how the

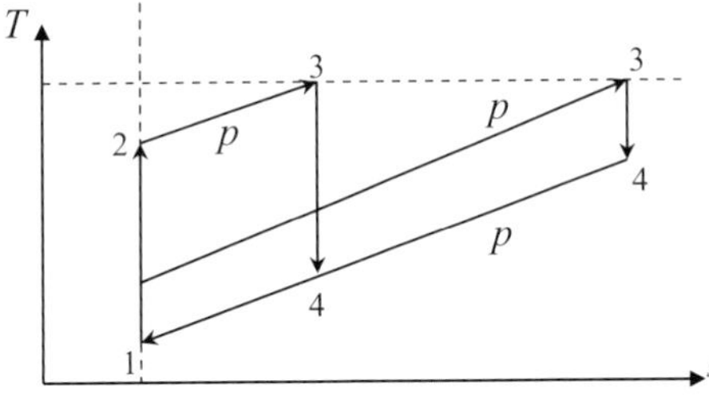

Figure 5.5 A schematic T–s diagram comparing high-pressure ratio and low-pressure ratio ideal Brayton cycles for the same work output (the area within the cycle) and the same turbine inlet temperature. The heat added is higher for the low-pressure ratio cycle.

blades are cooled. The value of this temperature has been steadily rising, and so are the efficiency and specific work of gas turbines. Note that while the maximum temperature has no direct impact on the simple cycle ideal efficiency, raising it allows one to use higher pressure ratios without reducing the specific work. Conventional designs limit the maximum temperature to less than 1000 °C. Modern designs that use special coatings for turbine blades and advanced cooling techniques such as steam cooling of blades have demonstrated the ability to push the maximum temperature to about 1400 °C for electricity generation applications.

Since the maximum temperature is fixed by the turbine design, the above expression can be differentiated with respect to the pressure ratio to determine the condition of maximum specific work. Written in terms of the temperature ratio across the compressor, the value required for **maximum specific work** is

$$\vartheta_{2,w_{max}} = \sqrt{\vartheta_3}, \tag{5.25}$$

and the maximum specific work is

$$\tilde{w}_{max} = \left(\sqrt{\vartheta_3} - 1\right)^2. \tag{5.26}$$

The dependence of the thermal efficiency and specific work on the compressor temperature ratio and the corresponding pressure ratio is shown in Figure 5.6 for two values of the isentropic index, $k = 1.4$ (air) and $k = 1.67$ (helium). Higher efficiency is realized at larger pressure ratios, as expected. However, under ideal operating conditions, there is little improvement in the cycle efficiency by increasing the pressure ratio beyond a certain point, say 30. On the other hand, the specific work peaks at a certain, rather low, pressure ratio. Lower-pressure cycles deliver larger specific work, which corresponds to a higher power: weight ratio, but they suffer from lower efficiency. Mobile applications, especially in aerospace propulsion, can benefit more from this design point. Other considerations such as real gas properties show benefits from using a higher pressure ratio, and modern designs reach pressure ratios of 30 and higher. Multistage axial compressors are often used to achieve these high pressures, both for propulsion applications and for mechanical and electrical energy generation applications.

Figure 5.6 shows the impact of the normalized turbine inlet temperature, ϑ_3, on the efficiency and the net work of the cycle, $w_{net}/\left(c_p T_1\right)$, for different pressure ratios. Under ideal conditions, the cycle efficiency is independent of the turbine inlet temperature, but the specific work produced by the cycle depends strongly on this design parameter. A substantial increase in the specific work is obtained as the turbine inlet temperature is raised. It is interesting to note that while the efficiency increases monotonically with the pressure ratio, the specific work reaches a maximum at a certain, relatively low, pressure ratio. Reaching high pressures requires more compressor work and reduces the work output of the cycle. The choice of whether to operate at the condition of maximum work or at the condition of higher efficiency depends on the application. For instance, a gas turbine can be designed to operate at the maximum efficiency most of the time to deliver the base load. It can also be designed to operate at maximum work if it is used to meet peak load only.

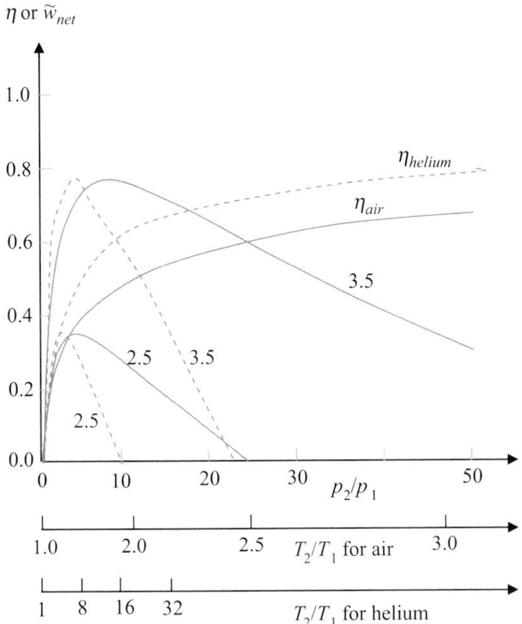

Figure 5.6 Dependence of ideal simple Brayton cycle efficiency on the pressure ratio and the corresponding temperature ratio across the compressor for two different ideal gases: air with $k = 1.4$ shown in continuous lines, and helium with $k = 1.667$ shown in broken lines. The lines showing the efficiency are labeled η_{air} or η_{helium}. The figure also shows the impact of the maximum cycle temperature on the specific work. The work lines are labeled with the normalized maximum temperature.

Figure 5.6 also compares the performance obtained using two different working fluids, air and helium. While air is a good surrogate for combustion products, which are dominated by nitrogen and oxygen (under the typical lean combustion conditions used in a combustion gas turbine), helium could be used in a closed cycle. At the same pressure ratio, the cycle efficiency is higher when helium is used because of the higher isentropic index of this gas. However, the specific work drops with increasing pressure faster after reaching its maximum in the case of helium than in the case of air. Moreover, at the same pressure ratio, the operating temperatures within the cycle are much higher for helium than for air, and special materials would have to be used in the turbine design.

5.3.3 Impact of the Isentropic Efficiency

An expression for the cycle thermal efficiency that accounts for the isentropic efficiencies of the compressor and the turbine can be obtained using (5.13), (5.19), and the work and heat interaction expressions [6]. Writing $\vartheta_{max} = \vartheta_3 = T_3/T_1$, the thermal efficiency in this case is

$$\eta = \frac{\eta_T \vartheta_{max}\left(1 - \dfrac{1}{\pi_P}\right) - \dfrac{1}{\eta_C}(\pi_P - 1)}{\vartheta_{max} - \left(1 + \dfrac{\pi_P - 1}{\eta_C}\right)}. \tag{5.27}$$

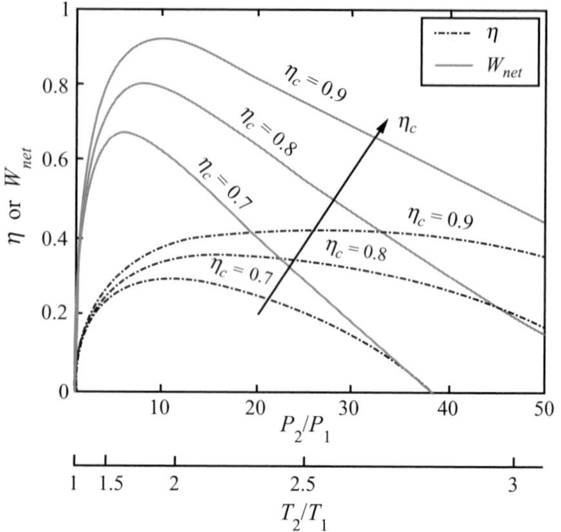

Figure 5.7 The impact of the compressor isentropic efficiency on the Brayton cycle efficiency shown in broken lines and specific work shown in continuous lines, for $\vartheta_3 = 4.5$, $\eta_T = 90\%$, $\beta = 1$. The last symbol, β, is the ratio between the pressure at the end of the heat addition process and that at the beginning of the same process – that is, it accounts for the pressure drop within the combustor.

The efficiency is no longer a monotonic function of the pressure ratio, and there exists a complex dependence on the pressure ratio, maximum temperature ratio, and component efficiencies. The pressure ratio that maximizes the specific work, and the specific work are given by

$$\pi_{P,w_{max}} = (\eta_T \eta_C \vartheta_{max})^{\frac{k}{2(k-1)}} \tag{5.28}$$

and

$$\tilde{w}_{net} = \vartheta_{max}\eta_T\left(1 - \frac{1}{\pi_P}\right) - \frac{1}{\eta_C}(\pi_P - 1). \tag{5.29}$$

Figure 5.7 shows the impact of the pressure ratio and the turbine inlet temperature on the efficiency and the specific work at three different values of the compressor isentropic efficiency. The figure shows that there is a significant efficiency penalty associated with using low-isentropic-efficiency compressors, especially in high-pressure ratio cycles. There is a corresponding penalty in the specific work as well. Moreover, the cycle efficiency is no longer monotonic in the pressure ratio; as the pressure ratio rises, the fraction of the turbine work used in compression increases rapidly and impacts the net work output. Under these conditions, and according to (5.27) and (5.29), the impact of the turbine inlet temperature is also significant. This has motivated continuous effort to build turbines that can sustain higher temperatures, recently exceeding 1400 °C, by employing better blade-cooling techniques and utilizing materials that can withstand high temperatures.

The effects of the design parameters on the cycle performance, measured in terms of the thermal efficiency and the net work, are also illustrated in the following example. Consider a gas turbine cycle, in which we assume that the working fluid is air throughout, operating at

Table 5.1 Results of Brayton cycle calculations for a pressure ratio of 4 and a turbine inlet temperature of 800 °C, for three different cases. Air is used as a working fluid, and work and heat are given in kJ/kg

Case	W_C	W_T	W_{net}	Q_H	η	T_4 (K)
$\eta_C = \eta_T = 1$	143.0	352.3	209.3	640.2	0.327	722.1
$\eta_T = 0.90$, $\eta_C = 0.85$	168.2	317.1	148.9	614.9	0.242	757.2
$\eta_T = 0.90$, $\eta_C = 0.65$	219.9	317.1	97.2	563.2	0.173	757.2
With regeneration $\eta_c = 0.85$					0.412	

Table 5.2 Results of Brayton cycle calculations for a pressure ratio of 8 and a turbine inlet temperature of 800 °C, for three different cases. Air is used as a working fluid, and work and heat are given in kJ/kg

Case	W_C	W_T	W_{net}	Q_H	η	T_4 (K)
$\eta_C = \eta_T = 1$	238.7	482.6	243.9	544.4	0.448	592.3
$\eta_T = 0.90$, $\eta_C = 0.85$	280.8	434.3	153.5	502.3	0.306	640.4
$\eta_T = 0.90$, $\eta_C = 0.65$	367.2	434.3	67.1	415.9	0.161	640.4
With regeneration, $\eta_c = 0.85$					0.345	

two different pressure ratios, $\pi_P = p_2/p_1 = 4$ and 8. The inlet conditions are 20 °C and 1 atm, and the temperature exiting the combustor is relatively low, 800 °C. We solve the problem for three cases,

1. isentropic compression and expansion; i.e., $\eta_C = \eta_T = 1$;
2. the compressor isentropic efficiency is $\eta_C = 0.85$ and turbine isentropic efficiency is $\eta_T = 0.9$; and
3. the compressor efficiency is $\eta_C = 0.65$ and the turbine efficiency is $\eta_T = 0.9$.

Results of the calculations for these three cases are shown in Tables 5.1 and 5.2. The isentropic index is 1.4, and the specific heat is 1.004 kJ/kg·K. The units for the heat and work in these tables are kJ/kg. Reducing the compressor isentropic efficiency from 85% to 65% lowers the cycle efficiency by a factor close to 2, at both pressure ratios. Clearly the compressor efficiency plays a very important role in determining the cycle efficiency. Raising the pressure ratio improves the cycle efficiency when the compressor efficiency is high. However, in the case of low compressor efficiency, the cycle efficiency drops as the pressure ratio increases because of the large compression work required in this case.

The Carnot efficiency of this cycle, when operated between the minimum compressor inlet temperature and the maximum inlet turbine temperature, is 62.5%, independent of the pressure ratio. This value is used to define the Second Law efficiency of the cycle, $\eta_{II} = \eta_{Brayton}/\eta_{Carnot}$. Since the minimum and maximum temperatures are fixed, the Second Law efficiency is proportional to the First Law efficiency.

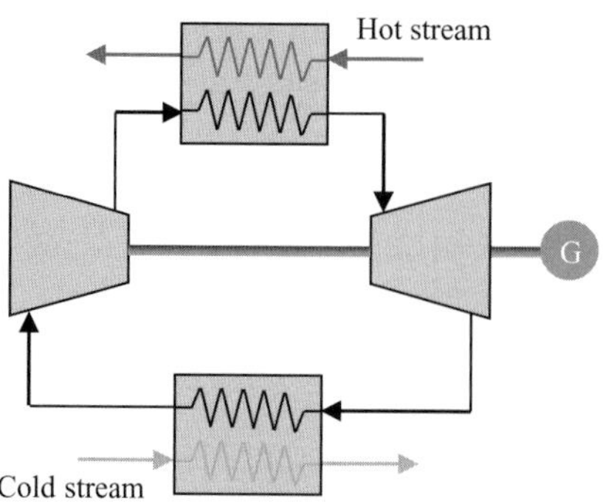

Figure 5.8 Schematic of the closed Brayton cycle.

5.3.4 The Closed Cycle

In the closed cycle engines, shown in Figure 5.8, the working fluid is simply circulated around the cycle between the different components: It is heated in a heat exchanger using the available heat source, such as a nuclear reactor, and cooled using a heatsink, such as cooling water from a nearby lake/river, in a second heat exchanger. The availability of a high-temperature source compatible with the high efficiency requirements of Brayton cycles is assumed.

The advantages of the closed cycle are:

1. We can use a working fluid with the properties that can improve the cycle performance – that is, a working fluid with high isentropic efficiency.
2. We can use a lower pressure at the turbine exhaust with a turbine that does not discharge the working fluid to the atmosphere.

The disadvantages include:

1. The added hardware is complex.
2. There is a need for heat exchangers at the high-temperature side and the low-temperature side. The heat exchanger on the high-temperature side may require special material, given the prevailing temperature there. The heat exchanger on the low side is not needed in the open cycle, and also requires a cooling medium.

Using low turbine exhaust pressure increases the overall pressure ratio of the cycle, and hence the efficiency and specific work, without elevating the cycle maximum pressure. In the case of a closed gas turbine cycle, the limiting operating condition at the low-pressure side is the cycle's lowest temperature. This is in contrast to the open cycle in which the limiting operating condition is the cycle's lowest pressure.

Since gas turbine engines require high temperatures at the turbine inlet to achieve high specific power and efficiency, they have not been used extensively in nuclear power plants, where the temperature is limited by the reactor core temperature. Next-generation reactors

Table 5.3 Results of Brayton cycle calculations for a pressure ratio of 4 and a turbine inlet temperature of 800 °C, for three different cases. Helium is used as a working fluid, and work and heat are given in kJ/kg

Case	W_C	W_T	W_{net}	Q_H	η	T_4 (K)
$\eta_C = \eta_T = 1$	1128	2372	1244	2922	0.426	616.2
$\eta_T = 0.90$, $\eta_C = 0.85$	1327	2135	808	2723	0.297	661.9
$\eta_T = 0.90$, $\eta_C = 0.65$	1735	2135	408	2315	0.173	661.9
With regeneration $\eta_c = 0.85$					0.363	

Table 5.4 Results of Brayton cycle calculations for a pressure ratio of 8 and a turbine inlet temperature of 800 °C, for three different cases. Helium is used as a working fluid, and work and heat are given in kJ/kg

Case	W_C	W_T	W_{net}	Q_H	η	T_4 (K)
$\eta_C = \eta_T = 1$	1975	3147	1172	2076	0.565	466.9
$\eta_T = 0.90$, $\eta_C = 0.85$	2323	2832	509	1727	0.295	528
$\eta_T = 0.90$, $\eta_C = 0.65$	3038	2832	−205.8	1012		528

using helium as a reactor coolant, and running at higher temperatures, could be a better match. Tables 5.3 and 5.4 show the results for the simple Brayton cycle analyzed above, but using helium as a working fluid. Again, two pressure ratios and two compressor efficiencies are used to demonstrate the impact of the pressure ratio and the significance of the compressor efficiency in determining the cycle efficiency. Note, in particular, that in the high-pressure ratio case, the cycle cannot produce positive power when the compressor efficiency is low!

Before leaving the topic of the closed cycle, we should mention that the choice of **helium** has other advantages and disadvantages, besides the higher value of k that, as already shown, leads to a better cycle efficiency compared to a cycle using air for the same pressure ratio. Helium has a much higher specific heat capacity than air; at atmospheric conditions, it is 5.193 kJ/kg·K. Thus, per unit mass flow rate, it holds more thermal energy. However, helium also has a lower density than air, and higher velocities are necessary to deliver the same mass flow rate or same thermal energy rates. Higher velocities lead to larger pressure drop in the components. Another advantage of helium is the relatively lower expansion ratio through the work transfer machine for the same temperature ratio, since the temperature ratio is lower for the same efficiency.

The results obtained from the analysis so far show that:

1. higher pressure ratio across the compressor leads to higher cycle efficiency;
2. at low compressor efficiency, most of the turbine work is used to drive the compressor, and the net work is small;

Table 5.5 Advancement enabling higher pressure ratios and higher turbine inlet temperatures. The impacts on the SC efficiency and the CC efficiency are rather dramatic [7]. All temperatures are in °C, and power in MW; SC is simple cycle and CC is combined cycle

	501A	501B	501D	501D5	501DA	501F	501G	ATS
Commercial year	1968	1973	1976	1982	1994	1992	1997	2000
Power SC	45	80	95	107	120	160	230	290
Pressure ratio	7.5	11.2	12.6	14.0	15.0	15.0	19.2	28.0
Rotor inlet T	879	993	1096	1132	1177	1277	1417	1510
Exhaust T	474	486	513	527	540	584	593	593
Efficiency – SC	27.1	29.4	31.2	34.0	34.5	35.5	38.5	–
Efficiency – CC	37.9	46.4	46.4	48.4	48.6	53.1	48.0	60.0

3. increasing the pressure ratio, in the case of low compressor efficiency, may reduce the cycle thermal efficiency; and
4. working fluids with a higher isentropic index, such as helium, can lead to higher efficiency, if used in closed cycles.

Table 5.5 shows the impact of the maximum cycle temperature on the net work output of the cycle. This table shows values obtained by a particular turbine manufacturer. However, these values demonstrate the historic trends reported by other manufacturers as well. Note that the turbine designation, such as D, F, or G is used, and is related to the turbine inlet temperature. As shown in the analysis of cycles with isentropic efficiencies less than unity, the efficiency of the cycle depends strongly on the turbine inlet temperature. This temperature is limited in practice by what the gas turbine can tolerate, and can be increased using high-temperature material and special blade-cooling techniques. Given that both the thermal efficiency and the net work of the cycle depend strongly on the TRIT, there has been extensive effort to build gas turbines that can withstand higher temperatures, with most recent designs reaching beyond 1400 °C, and with efforts to boost this value even higher. In the most recent designs, air or steam is injected to keep the blade temperature several hundred degrees below that of the gas. More on the material used and the cooling techniques will be discussed shortly.

At the pressure ratio of four, the ideal Brayton cycle Second Law efficiency is 52.5%, and hence there is plenty of room to improve its performance and to design cycles that perform closer to the Carnot cycle. Cycle improvements include reheating, intercooling, and regeneration. These modifications will be explored shortly.

5.3.5 Effect of Pressure Drop across the Combustor

Pressure drop across the combustor reduces the turbine work and hence the cycle efficiency. In a closed cycle, there is pressure drop across the low-temperature heat exchanger as well, as

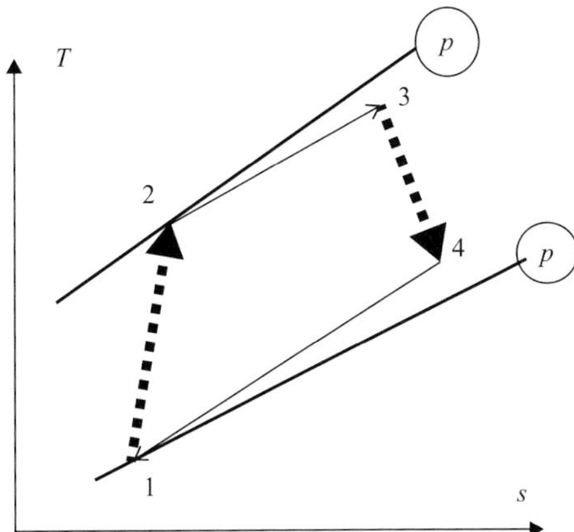

Figure 5.9 The temperature entropy diagram of a simple Brayton cycle, with isentropic efficiencies for the work transfer components, and pressure drop across the heat transfer components.

shown in Figure 5.9. Denoting the pressure ratio across the gas heater/combustor as $\beta_H = p_3/p_2$, and in the case of a closed cycle, the ratio across the gas cooler as $\beta_L = p_1/p_4$, the expression for the efficiency in (5.27) is modified such that [6]

$$\eta = \frac{\eta_T \vartheta_{max}\left(1 - \dfrac{1}{\beta^* \pi_P}\right) - \dfrac{1}{\eta_C}(\pi_P - 1)}{\vartheta_{max} - \left(1 + \dfrac{\pi_P - 1}{\eta_C}\right)}, \tag{5.30}$$

where $\beta^* = (\beta_L \beta_H)^{\frac{k-1}{k}}$.

The pressure ratio at maximum efficiency is found by solving $\partial\eta/\partial\pi_P = 0$:

$$\pi_{P,\eta_{max}} = \left(\frac{-b - \sqrt{b^2 - 4ac}}{2a}\right)^{\frac{k}{k-1}}, \tag{5.31}$$

where: $a = \frac{\eta_T}{\eta_C}\vartheta_{max} - \frac{\vartheta_{max}-1}{\eta_C}$, $b = -\frac{2\eta_T\vartheta_{max}}{\eta_C\beta^*}$, and $c = \frac{\eta_C}{\beta^*}\vartheta_{max}\left(\vartheta_2 - 1 + \frac{1}{\eta_C}\right)$.

In the case of finite pressure drop across the heat interaction components, the expressions for the normalized net work of the cycle and the optimum pressure ratio maximizing the net work given in (5.29) and (5.28), respectively, are modified as follows:

$$\tilde{w}_{net} = \vartheta_{max}\eta_T\left(1 - \frac{1}{\beta^* \pi_P}\right) - \frac{1}{\eta_C}(\pi_P - 1) \tag{5.32}$$

and

$$\pi_{P,w_{max}} = \left(\vartheta_{max}\frac{\eta_T\eta_C}{\beta^*}\right)^{\frac{k}{2(k-1)}}. \tag{5.33}$$

Note that the pressure ratios corresponding to maximum efficiency and maximum specific work are not the same.

5.4 Regenerative/Recuperative Cycles

Analysis so far shows one of the most important characteristics of gas turbine cycle operation. Besides the large compressor work and the possible perils of low compressor isentropic efficiency, the exit temperature from the turbine is rather high. Turbine inefficiency adds to the high turbine exit temperature. Higher turbine exhaust temperature lowers the efficiency of the cycle, even under ideal conditions. As mentioned before, the Second Law efficiency of these cycles can be low, while the availability of the exit stream can be high. One way to overcome this limitation and improve the cycle thermal efficiency is to use **recuperation** or **regeneration**. In this case, a heat exchanger is utilized to transfer some of the thermal energy of the turbine exhaust to the stream exiting the compressor, as shown in Figures 5.10 and 5.11. Utilization of a recuperative heat exchanger results in using less thermal energy from the source, or less fuel in the combustor, to achieve a desired maximum cycle temperature.

Counter-flow heat exchangers are employed as recuperators. Except for the hardware cost and the added weight, the improvement of the cycle efficiency associated with recovering a fraction of the exhaust thermal energy and using it to heat the gases leaving the compressor can be impressive. Take, for example, the cases analyzed previously, whose results are shown

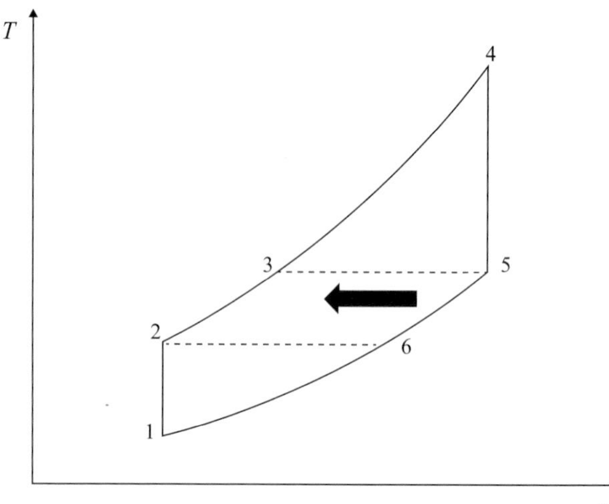

Figure 5.10 The T–s diagram for an ideal recuperative Brayton cycle. The sequence of states: 1-2-3-4-5-6, thick arrow shows internal heat transfer.

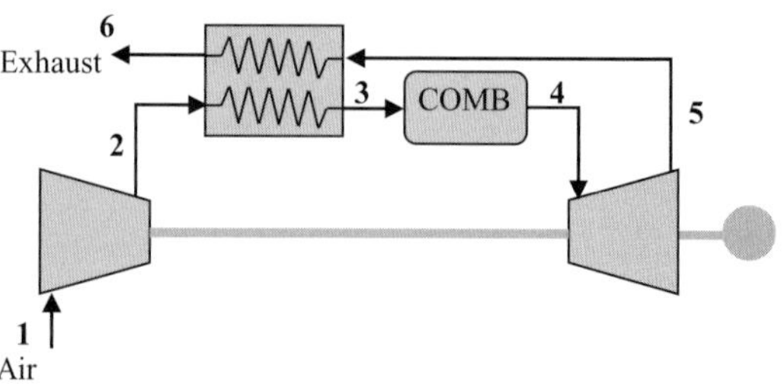

Figure 5.11 Brayton cycle with recuperation.

in Figures 5.12 and 5.13. Assuming 85% temperature recovery from the turbine exhaust – that is, the actual drop in the turbine exhaust stream temperature through the recuperator is 0.85 of the difference between the turbine exhaust temperature and the compressor exit temperature – we calculate the improvement in the cycle efficiency for the case with the higher compressor efficiency of 85%. The regenerative cycle efficiency for $\pi_P = 4$ is 41.2%, while the simple cycle efficiency is 24.2%. For the high-pressure ratio cycle, $\pi_P = 8$, the regenerative cycle efficiency is 34.5% instead of 30.6% for the simple cycle. It is clear that regeneration is very beneficial, especially for low-pressure cycles, where $T_5 \gg T_2$, which

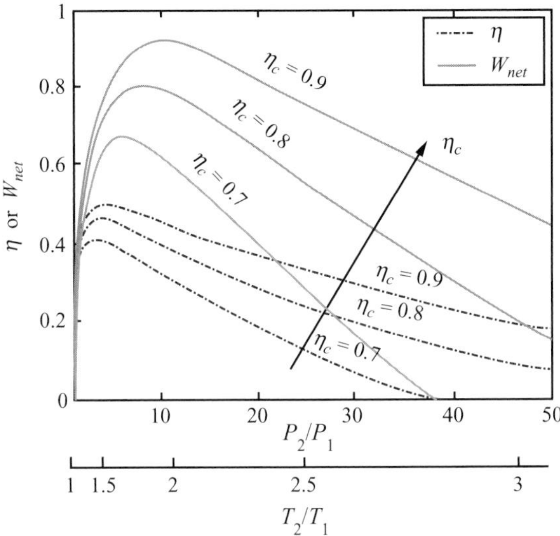

Figure 5.12 The impact of recuperation on the thermal efficiency and normalized specific work of a Brayton cycle, $\eta_T = 0.9$, $\vartheta_3 = 4.5$, $\beta = 1$, and $\eta_{reg} = 85\%$.

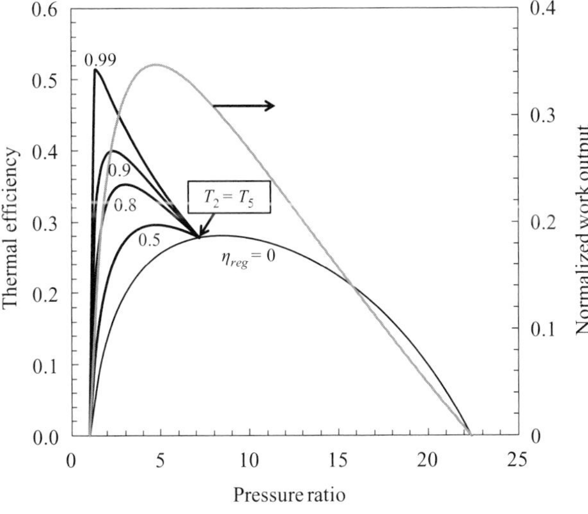

Figure 5.13 Impact of regeneration efficiency on the Brayton cycle efficiency, $\eta_T = \eta_C = 0.90$, $\vartheta_{\max} = 3$, $\beta = 1$. Numbers on the lines show the regeneration efficiency, defined as $(T_5 - T_2)/(T_5 - T_6)$.

allows for the recovery of a larger fraction of the exhaust thermal energy. This may not be the case for high-pressure cycles, in which the turbine exhaust temperature is closer to the compressor exit temperature. Regeneration is also used in small gas turbine engines in mobile applications to boost their fuel economy.

The thermal efficiency of an ideal simple recuperative cycle, with perfect expander and compressor and 100% recuperation (i.e., for the cycle shown in the *T*–*s* diagram in Figure 5.10), is

$$\eta = 1 - \left(\frac{1}{\vartheta_4}\right)\left(\frac{p_2}{p_1}\right)^{(k-1)/k}. \tag{5.34}$$

Thus, the thermal efficiency depends on the cycle maximum temperature and the pressure ratio. For a fixed turbine inlet temperature, the efficiency decreases as the pressure ratio rises, as seen in the numerical examples. Taking into consideration other factors, such as isentropic efficiencies and pressure drops, can change this conclusion.

More complex expressions for the cycle efficiency can be obtained for the case with turbine and compressor isentropic efficiencies lower than 100%, and with pressure drop across the two sides of the regenerator and across the combustor/heater. Call the pressure ratio across the hot side of the regenerator β_{Hreg}, across the cold side β_{Lreg}, and across the heater/combustor β_H, then $\bar{\beta}^* = (\beta_H \beta_{Hreg} \beta_{Lreg})^{\frac{k-1}{k}}$. Moreover, call the regenerator effectiveness η_{reg}, such that

$$\eta_{reg} = \frac{\text{actual temperature rise across regenerator}}{\text{maximum available temperature rise}} = \frac{\text{regenerator exit temperature} - \text{compressor exit temperature}}{\text{turbine exit temperature} - \text{compressor exit temperature}}.$$

The normalized specific work is given by [6]

$$\tilde{w}_{net} = \vartheta_{max}\eta_T\left(1 - \frac{1}{\bar{\beta}^*\pi_P^*}\right) - \frac{1}{\eta_C}\left(\pi_P^* - 1\right), \tag{5.35}$$

where $\pi_P^* = (\pi_P)^{\frac{k-1}{k}}$. Maximizing this expression with respect to the pressure ratio across the compressor, we find the pressure ratio required for maximum specific work:

$$\pi_{P,w_{max}} = \left(\vartheta_{max}\frac{\eta_T\eta_C}{\bar{\beta}^*}\right)^{\frac{k}{2(k-1)}}. \tag{5.36}$$

Meanwhile, we can calculate the heat transfer into the cycle, $\tilde{q}_{in} = \frac{q_{in}}{c_p T_1}$,

$$\tilde{q}_{in} = \vartheta_{max} - \left(1 + \frac{\pi_P^* - 1}{\eta_C}\right) - \eta_{reg}\left\{\vartheta_{max}\left[1 - \eta_T\left(1 - \frac{1}{\bar{\beta}^*\pi_P^*}\right)\right] - \left[1 + \frac{\pi_P^* - 1}{\eta_C}\right]\right\}. \tag{5.37}$$

Dividing the specific work by the heat transfer into the cycle, we get the efficiency. The efficiency and normalized specific work of a gas turbine engine utilizing a recuperator with

$\eta_{reg} = 85\%$ are shown in Figure 5.12. Pressure drops are all neglected, the turbine efficiency is 90%, maximum normalized cycle temperature is 4.5, and the compressor isentropic efficiency is varied between 0.7 and 0.9. Comparing Figures 5.7 and 5.12, it is clear that there is a substantial gain in efficiency, especially at lower pressure ratios, when recuperation is utilized in gas turbine cycles.

The maximum efficiency can be obtained by solving $\partial\eta/\partial\pi_P = 0$. The solution is

$$\pi_{P,max\ \eta} = \left(\frac{-b + \sqrt{b^2 - 4ac}}{2a}\right)^{\frac{k}{k-1}}, \tag{5.38}$$

with the following coefficients: $a = \frac{\bar{\beta}^*}{\eta_C}\left[\left(1 - \eta_{reg}\right)(\vartheta_{max} - 1) + \vartheta_{max}\eta_T\left(2\eta_{reg} - 1\right)\right]$, $b = -\frac{2\eta_T\vartheta_{max}\left(1 - 2\eta_{reg}\right)}{\eta_C}$, and $c = \eta_T\vartheta_{max}\left(1 - \eta_{reg}\right)(1 - \vartheta_{max}) + \frac{\vartheta_{max}\eta_T}{\eta_C}\left(2\eta_{reg} - 1\right)$.

These equations, although obtained assuming that the working fluid is an ideal gas with constant specific heat ratio throughout the cycle, are useful design equations, and their results reveal some important characteristics of gas turbine cycles. The impact of the regeneration efficiency on the cycle improvement is shown in Figure 5.13 for a Brayton cycle with isentropic efficiencies of 0.9 for the turbine and compressor, and $\vartheta_4 = 3$.

Example 5.1

Air at 298 K and 100 kPa enters the compressor of a regenerative gas turbine power plant, where it is compressed to 900 kPa. The regenerator efficiency is 85%, and the turbine inlet temperature is 1373 K. Assume that the fuel mass flow rate is negligible compared to that of air. The isentropic efficiencies of turbine and compressor are 90% and 85%, respectively. The pressure drops on the cold and hot sides of the regenerator and within the combustor are 2%, 2%, and 3%, respectively.

a. Determine the specific work output.
b. Determine the thermal efficiency.
c. Calculate the exhaust temperature (see Figure 5.11).
d. What would be the thermal efficiency without the regenerator?

Solution

Compressor: $T_2 = T_1\left\{1 + \frac{1}{\eta_C}\left[\left(r_{p_1}\right)^{\frac{k-1}{k}} - 1\right]\right\} = 596.8\,\text{K}$

$$W_C = c_{p,c}(T_2 - T_1) = 1.021 \times (596.8 - 298) = 305.07\ \text{kJ/kg}.$$

Note that the specific heat of air is determined at $T_{ave} = (T_1 + T_2)/2$.

Turbine: $p_4 = 0.97p_3 = 0.97 \times 0.98p_2 = 0.97 \times 0.98\pi_P p_1 = 855.54\ \text{kPa}$

$$p_5 = p_6/0.98 = 102.04\ \text{kPa}$$

Example 5.1 (cont.)

$$T_5 = T_4\left\{1 - \eta_T\left[1 - \left(\frac{P_5}{P_4}\right)^{\frac{k-1}{k}}\right]\right\} = 867.3\ \text{K}$$

$$W_T = c_{p,t}(T_4 - T_5) = 586.11\ \text{kJ/kg}.$$

Combustor: $T_3 = T_2 + \eta_{reg}(T_5 - T_2) = 826.7\,\text{K}$

$$Q_{in} = c_{p,com}(T_4 - T_3) = 633.16\ \text{kJ/kg}.$$

a. The specific work output is $W_{net} = W_T - W_C = 281.04\ \text{kJ/kg}$.

b. The thermal efficiency is $\eta_{th} = \frac{W_{net}}{Q_{in}} = 0.444$.

c. The exhaust temperature is obtained by applying the First Law to the regenerator, $c_{p,HX_C}(T_3 - T_2) = c_{p,HX_H}(T_5 - T_6)$, where the specific heat at the cold side and the hot side of the regenerator are denoted by c_{p,HX_C} and $c_{p,HXH}$, respectively. Rearranging for T_6, we find

$$T_6 = T_5 - \frac{c_{p,HX_C}}{c_{p,HX_H}}(T_3 - T_2) = 867.3 - \frac{1.078}{1.088}(826.7 - 596.8) = 639.5\,\text{K}.$$

d. In the absence of the regenerator, the combustor inlet temperature would be equal to the compressor exit temperature: $T_3 = T_2 = 596.8\,\text{K}$. Thus, the heat added in the combustor would be

$$Q_{in} = 1.141 \times (1373 - 596.8) = 885.64\,\text{kJ/kg}.$$

Thus, $\eta_{th} = 0.317$. Note that the work output is not influenced by the regenerator.

5.5 Reheating and Intercooling

High-efficiency, large gas turbine installations utilize multiple **intercooling** and **reheating**, with or without recuperation to improve the cycle performance. The cycle layout with intercooling and reheating is shown in Figure 5.14. A supercharger, essentially a high-speed fan, may be used before the first stage of compression to deliver high velocity air with a small pressure rise. The two compressors are responsible for most of the pressure rise. As shown in the *T–s* diagram in Figure 5.15, intercooling lowers the compressor work, which reaches its minimum when the number of intercooling stages is very large and the process is nearly isothermal. However, intercooling increases the required heat transfer into the cycle between the last compressor stage and the turbine inlet to achieve the same

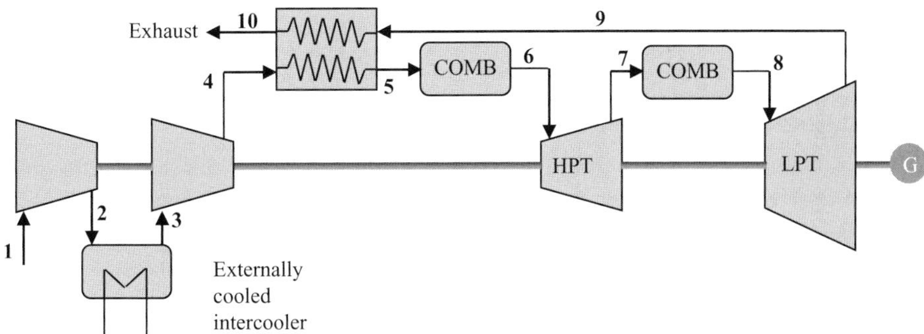

Figure 5.14 A Brayton cycle equipped with two-stage compression, intercooling, two-stage expansion, and reheating; see Figure 5.15 for the T–s diagram.

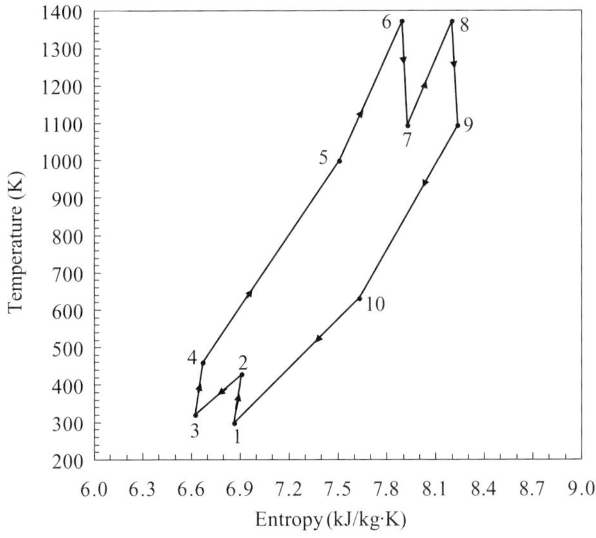

Figure 5.15 The T–s diagram of a regenerative Brayton cycle with intercooling and reheating; see Example 5.2.

maximum cycle temperature. Thus, it may lead to lower overall cycle efficiency unless regeneration between the turbine exhaust and the fluid exiting the compressor is also used to heat the compressed gas. With real components, intercooling has been shown to boost the efficiency of the cycle in most applications. Furthermore, intercooling raises the specific work of the cycle since it reduces the compression work. Intercooling also increases the potential for regeneration since it lowers the gas temperature at the end of the compression process.

Reheating between expansion stages, on the other hand, boosts the total turbine work, hence increasing the specific work of the cycle and its overall efficiency. Adding heat to the cycle at the high-temperature side, between expansion steps, improves the conversion efficiency, as shown before in reviewing basic principles of thermodynamics. Again with a large number of reheat stages, the expansion process in the turbine approaches isothermal expansion, which maximizes the work and cycle efficiency. Reheating increases the temperature of the turbine exhaust, making regeneration more effective. Integrating intercooling, reheating, and recuperation leads to a high-efficiency **intercooled recuperated cycle** (ICR).

It is also possible to boost the cycle power by employing supercharging – that is, by using a fan upstream of the compressor to increase the flow of air into the compressor and raise its pressure by 10–15 % over the pressure of its environment. Further increase in the power can be attained if the air leaving the surpercharger is also cooled before the compression. Raising the pressure and lowering the temperature increases the air density and hence the net power. Both may also lower the efficiency of the plant because of the extra work required by the supercharger, and because of the temperature rise before compression if intercooling is not used (note that the ideal compression power is linearly proportional to the inlet temperature to the compressor). Thus, care must be exercised when optimizing the cycle design. In some cycles, evaporative cooling – that is, using direct water injection into the streams – is used instead of a surface heat exchanger. Water addition to the compressor air streams before the

Example 5.2

The gas turbine power plant of Example 5.1 is modified by compressing air in two stages, with an intercooler between the compressors, and by expanding the working fluid in two turbines with an additional combustor between the turbines (see Figure 5.14). The pressure ratio across the compressors is the same. The inlet temperature of the second compressor is 320 K. The pressure drop within the intercooler is 3%. The pressure drop in each combustor is 2%. The outlet temperature of the second combustor is 1373 K. The pressure ratio across the turbines is also the same.

a. Determine the specific work output of the power plant.
b. Calculate the thermal efficiency of the power plant.

Solution

The pressure ratio across the compressors is the same. Hence, $\frac{p_2}{p_1} = \frac{p_4}{p_3}$. Note that $p_3 = 0.97p_2$. Hence, $p_3 = 0.97 \times 304.6 = 295.5\,\text{kPa}$.

The pressure ratio across the turbines is also identical. Hence, $p_8 = 0.97p_7$, $p_6 = 0.97\,p_5 = 855.54\,\text{kPa}$, and $p_9 = p_{10}/0.98 = 102.04\,\text{kPa}$.

Substituting, we find that $p_8 = 291\,\text{kPa}$.

$$T_2 = T_1\left\{1 + \frac{1}{\eta_C}\left[\left(\frac{p_2}{p_1}\right)^{\frac{k-1}{k}} - 1\right]\right\} = 428.5\,\text{K},$$

Compressor 1:

$$W_{C_1} = c_{p,c1}(T_2 - T_1) = 1.01 \times (428.5 - 298) = 131.8\,\text{kJ/kg}.$$

$$T_4 = 320\left\{1 + \frac{1}{0.85}\left[\left(\frac{900}{295.5}\right)^{\frac{1.395-1}{1.395}} - 1\right]\right\} = 459.6\,\text{K},$$

Example 5.2 (cont.)

Compressor 2:

$$W_{C_2} = 1.01 \times (459.6 - 320) = 141 \text{ kJ/kg}$$

$$T_7 = T_6\left\{1 - \eta_T\left[1 - \left(\frac{P_7}{P_6}\right)^{\frac{k-1}{k}}\right]\right\} = 1094.6\,\text{K},$$

Turbine 1:

$$W_{T_1} = c_{p,t}(T_6 - T_7) = 328.2\,\text{kJ/kg}.$$

$$T_9 = 1373 \times \left\{1 - 0.9 \times \left[1 - \left(\frac{102.04}{291}\right)^{\frac{0.322}{1.322}}\right]\right\} = 1094.6\,\text{K}.$$

Turbine 2:

$$W_{T_2} = c_{p,t}(T_8 - T_9) = 1.179 \times (1373 - 1094.6) = 328.2\,\text{kJ/kg}.$$

Combustors:

$$T_5 = T_4 + \eta_{reg}(T_9 - T_4) = 999.4\,\text{K}$$

$$Q_{in,1} = c_{p,com}(T_6 - T_5) = 439\,\text{kJ/kg}$$

$$Q_{in,2} = c_{p,com}(T_8 - T_7) = 327.1\,\text{kJ/kg}.$$

a. Specific work output: $W_{net} = W_{T_1} + W_{T_2} - W_{C_1} - W_{C_2} = 383.6\,\text{kJ/kg}$.
b. Thermal efficiency: $\eta_{th} = \frac{W_{net}}{Q_{in,1} + Q_{in,2}} = 0.501$.

The T–s diagram of the cycle is shown in Figure 5.15.

turbine is discussed in more detail for the humid air cycles. Adding water has other advantages that will be described in the humid air cycle, such as reducing nitric oxide emissions in the combustion process.

5.5.1 Ericsson Cycle

In theory, with an "infinite" number of intercoolers and reheaters, compression and expansion are performed isothermally; the heat transfer into the cycle takes place at the highest possible temperature, and the heat transfer out is achieved at the lowest temperature, fulfilling the Second Law requirement for reversible heat transfer. The constant-pressure processes in between the isothermal processes are used in a regenerative arrangement – that is, heat is transferred from the expander exhaust to the compressor outflow. The resulting cycle is the **Ericsson cycle**, which has the same efficiency as the Carnot cycle! The T–s

Figure 5.16 The T–s diagram of the Ericsson cycle in which external heat transfer processes are performed isothermally during the compression and expansion processes, and the constant-pressure processes in which heat regeneration between the turbine exhaust and the fluid exiting the compressor is performed internally.

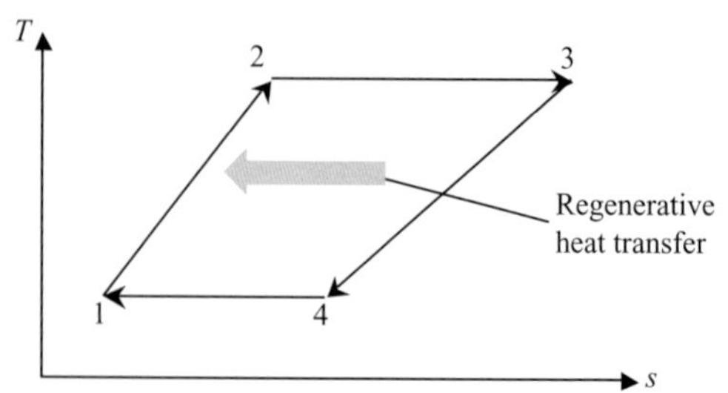

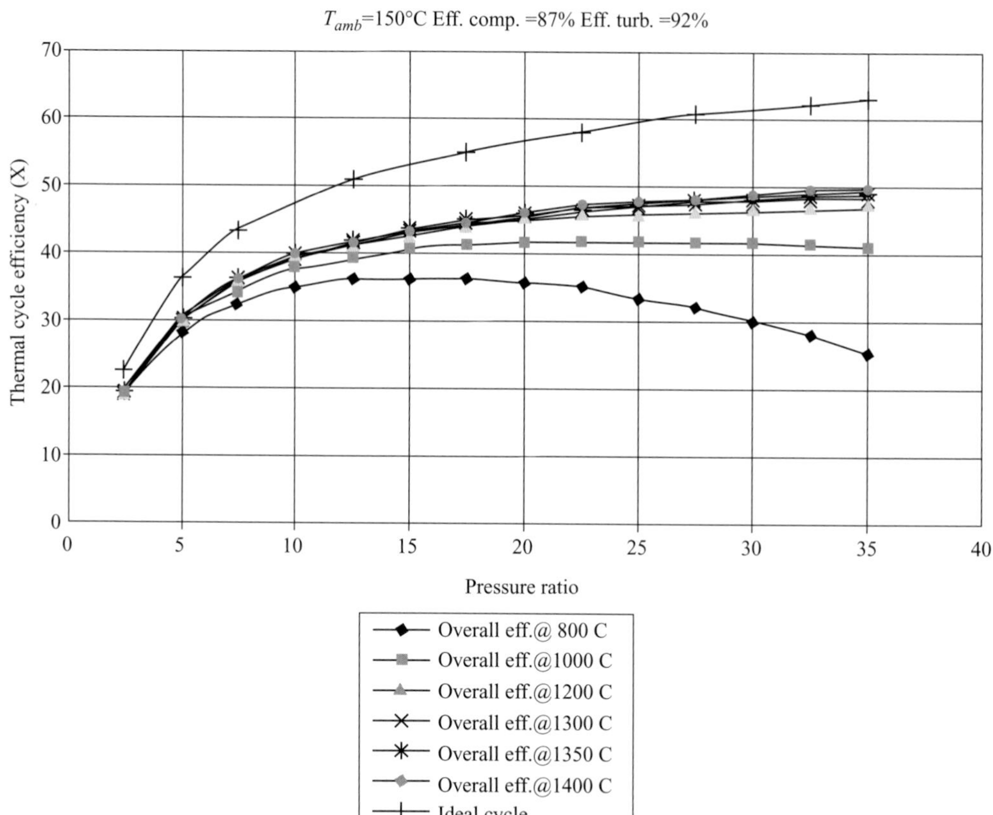

Figure 5.17 Impact of pressure ratio and turbine inlet temperature on overall cycle efficiency [8].

diagram for the Ericsson cycle is shown in Figure 5.16. Practical engines utilizing this cycle have not been demonstrated yet, as it would be costly to achieve a large number of reheat and intercooling steps.

Regenerative cycles with multiple heat exchangers can be expensive because of the need for high-temperature material. Raising the turbine inlet temperature and overall cycle pressure remain the most effective ways to improve gas turbine efficiency. Figure 5.17 shows the impact of the firing temperature and pressure ratio on the overall cycle efficiency.

As mentioned before, the efficiency increases monotonically with temperature, and with pressure up to values around 20 [8].

Other potential improvements in gas turbine performance include the use of wave energy exchangers, which have been under development [9].

5.6 Impact of Working Fluid Properties

The analysis so far has been based on an ideal gas assumption to drive analytical expressions for the cycle efficiency and specific work. More detailed cycle analysis must consider the real properties of the gas through the different components. For instance, in the compressor, air at low temperature is compressed and (5.20) is a good approximation for the ideal work. However, the gas stream through the turbine in an open cycle is made of a mixture of gases. Assuming complete combustion under fuel-lean conditions, the combustion products are a mixture of nitrogen, oxygen, carbon dioxide, and water (N_2, O_2, CO_2, H_2O). The mass flow rate through the turbine is that of air plus the fuel. The fuel required for raising the gas temperature to the TRIT must also be calculated using the correct thermochemical model, as described in Chapter 3. Since the inlet turbine rotor temperature is well below the typical stoichiometric flame temperature, combustion in gas turbines is well below stoichiometric, with an equivalence ratio in the range of 0.2–0.3. Moreover, with TRIT ~1400 °C, the properties of the gas are no longer those of low-temperature air.

In open-cycle gas turbine operation, which is currently the most widely used in practice, air, at mass flow rate of $\dot{m}_a$, is compressed in a single- or multistage compressor, then fuel is added in the combustor at a mass flow rate of $\dot{m}_f$ to raise the temperature to that tolerated by the turbine rotor. The products of combustion expand in the turbine before they are exhausted to the atmosphere or sent to a heat recovery steam generator in combined cycles. Assuming ideal gas properties, and taking average values for the specific heats of the reactants and the products, and for a simple cycle, the compressor power, turbine power, and cycle fuel utilization efficiency are given by

$$\dot{W}_C = \dot{m}_a c_{pa} (T_2 - T_1), \tag{5.39}$$

$$\dot{W}_T = \left(\dot{m}_a + \dot{m}_f\right) \bar{c}_p (T_3 - T_4), \tag{5.40}$$

and

$$\eta_{fuel-utilization} = \frac{\dot{W}_T - \dot{W}_C}{\dot{m}_f \cdot LHV}. \tag{5.41}$$

The high specific heat of the hot combustion products lowers the turbine work.

5.6.1 Carbon Dioxide as a Working Fluid in Closed and Semi-Closed Cycles

Closed cycles can be designed to use CO_2 as a working fluid. For instance, the working fluid in this case can be heated using a nuclear reactor. Even open cycles may be designed in the future to operate with CO_2 as a working fluid, or a mixture of CO_2 and stream. The products of combustion of pure oxygen and hydrocarbon have a high concentration of CO_2. Thus, cycles operating in the oxy-combustion mode – that is, burning either coal, synthetic gas (synfuel) produced from oxy-gasification of coal, or natural gas in pure oxygen – would use a working fluid with a high concentration of CO_2. The concentration of CO_2 in these cases can be made as high as desired by recycling a fraction of the flue gas after water has been condensed. Recycling CO_2 in oxy-combustion or oxy-gasification applications is necessary to keep the maximum combustion temperature below the allowable gas turbine temperature. These oxy-combustion-based semi-closed cycles are not only of theoretical interest, but they also present a good alternative if CO_2 capture and sequestration is part of the design criteria used in building new power plants.[1] Brayton cycles can be used with pure CO_2 or with $CO_2 + H_2O$, with water remaining in the vapor form during the expansion of the gas throughout the turbine.

In analyzing CO_2 Brayton cycles, it is important to remember that CO_2 has different characteristics to air or combustion products, which under lean combustion conditions are dominated by nitrogen. Some of the important characteristics of CO_2 are as follows:

1. The critical temperature of CO_2 is 31 °C and the critical pressure is 73.9 atm.
2. CO_2 is a triatomic gas with an isentropic exponent close to 1.3. If used as a working fluid, it results in a higher turbine exit temperature and a lower efficiency (for the same pressure ratio) than in the case of air.
3. The heat capacity of CO_2 is higher than that of air at a higher temperature (9% higher at 100 °C).
4. It is denser and has a lower kinematic viscosity than air, and hence its flow within the turbine is at a higher Reynolds number (for the same velocities).

These changes lead to the following:

1. CO_2 turbines have lower specific expansion work than air turbines for the same pressure ratio. A CO_2 gas turbine must process higher mass flow rate than an air turbine to deliver the same power.
2. The gas turbine exit temperature is higher when operating on CO_2 than when operating on air for the same expansion ratio, because of the isentropic index (see Figure 5.18 for a comparison between the exit turbine temperature when different working fluids are used).
3. Achieving the maximum efficiency of the cycle requires higher pressure ratios for CO_2 than for air.
4. Special cycle designs may be required to overcome some of these limitations.

Figure 5.19 shows the efficiency vs the specific power measured in kJ/kg fluid for air and CO_2 simple Brayton cycles, calculated for different pressure ratios. In both cases, a constant

[1] Different schemes have been suggested for CO_2 capture from power cycles, including post-combustion separation of CO_2, pre-combustion separation of CO_2 following steam reforming and water–gas shift reactions to convert all CO to CO_2, and oxy-combustion. These schemes will be described in detail in Chapters 10–13.

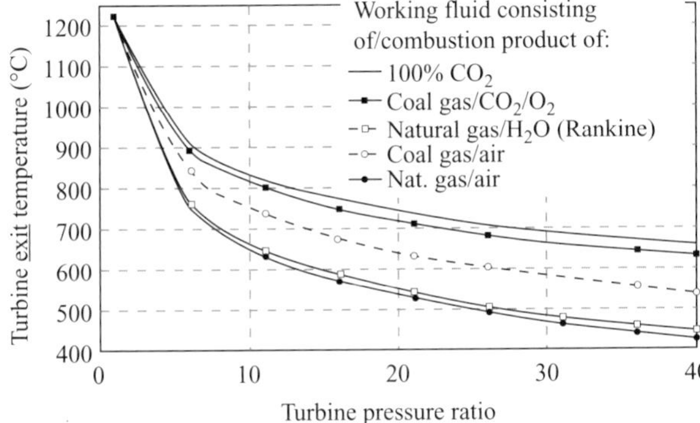

Figure 5.18 The gas turbine exit temperature calculated for different working fluids and pressure ratios across the turbine, for a turbine inlet temperature of 1200 °C. The working fluid is either pure CO_2, or the combustion product of the fuel and the oxidizer as listed in the figure, with stoichiometry adjusted to give the specified inlet temperature. The essential difference between the different gases is the isentropic index [10].

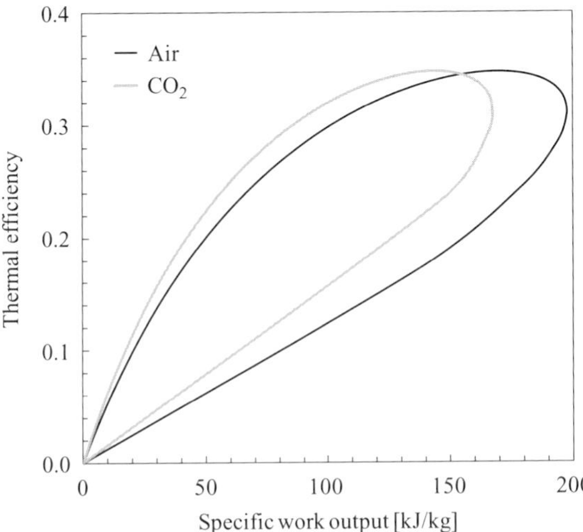

Figure 5.19 The thermal efficiency of a Brayton cycle plotted against the specific net work, measured in kJ/kg, at different pressure ratios [10]. The plot shows the results for different working fluids: air (black) and CO_2 (light gray). Note that the maximum efficiency for an air cycle is reached at a pressure ratio near 30, while it is near 80 for CO_2. In both cases, the working fluid is assumed to be an ideal gas with constant specific heats.

specific heat was used in the cycle calculation. The isentropic efficiencies of the turbine and compressor are 90% and 85%, respectively. The figure is constructed for the same minimum and maximum cycle temperatures of 298 K and 1192 K for both working fluids, and the same compressor and turbine characteristics. The maximum cycle efficiency for both working fluids is almost the same at 34.8%. However, it is reached at a pressure ratio of 30 for CO_2 and 15 for air. Furthermore, the maximum work output of the cycle with air is 197.4 kJ/kg, which occurs at a pressure ratio of 6.7, whereas the maximum work output of the cycle with carbon dioxide is 167.2 kJ/kg, which takes place at a pressure ratio of 13.6.

Gas turbine machinery compatible with CO_2 as a working fluid, including both the compressor and the expander, will be different in size, number of stages, and blade shapes, and the standard air-based air turbine cannot be used without a large performance penalty. More about the use of CO_2 or mixtures of CO_2 and other fluids in power generation cycles will be covered in Chapters 10–13.

Example 5.3

A gas turbine power plant, shown in Figure E5.3, operates with oxyfuel combustion (for carbon capture and storage applications, a subject that will be described in more detail in upcoming chapters) and uses syngas (a mixture of one mole of hydrogen and one mole of carbon monoxide) as a fuel. Air at 25 °C and 1 atm is compressed adiabatically to 8 atm within an air-separation unit (ASU), which produces oxygen at 1 atm. The oxygen stream is cooled to 30 °C within a heat exchanger before mixing with the recycled CO_2. The mixture of oxygen and carbon dioxide enters the compressor, where its pressure rises to 28 atm. The syngas undergoes a complete and adiabatic combustion. The pressure drop within the combustor is 5%. The combustor exit temperature is 1600 K. The combustion products expand in the turbine, whose isentropic efficiency is 90%. The turbine exhaust enters a condenser, where it is cooled to 30 °C so the water is separated from CO_2. The CO_2 stream passes through a valve, which allows recycling of CO_2. Assume an isentropic efficiency of 80% for the compressors. How much CO_2 recycle is needed? Calculate the net power production and the thermal efficiency of the power plant.

Figure E5.3

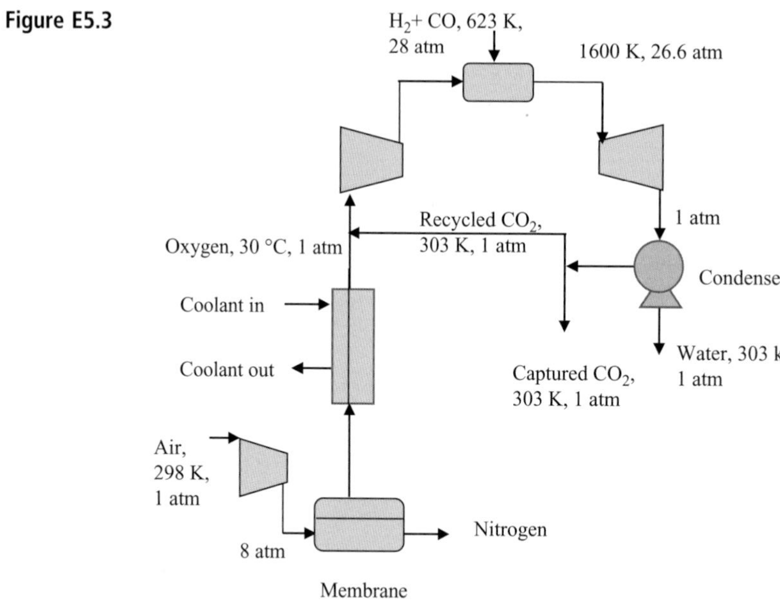

Solution

We begin the analysis from the first compressor, where air is compressed for separation within the ASU. The air temperature at the compressor outlet is

Example 5.3 (cont.)

$$T_1' = T_0\left[1 + \frac{(p_1'/p_0)^{k-1/k} - 1}{\eta_c}\right] = 298 \times \left[1 + \frac{8^{0.4/1.4} - 1}{0.8}\right] = 600.3\,\text{K}.$$

Next, we calculate the temperature of the oxygen and carbon dioxide mixture at the exit of the gas turbine cycle compressor. We assume that the specific heat ratio of the O_2–CO_2 mixture is that of carbon dioxide and that they are at the same temperature. This will be verified later. Hence,

$$k_{mix} = k_{CO_2} = 1.289,$$

$$T_2 = T_1\left[1 + \frac{(p_2/p_1)^{k-1/k} - 1}{\eta_c}\right] = 303 \times \left[1 + \frac{28^{0.342/1.342} - 1}{0.8}\right] = 723.7\,\text{K}.$$

The combustion reaction can be written as

$$\underbrace{H_2 + CO}_{\text{Syngas}} + n_{O_2}O_2 + n_{CO_2,r}CO_2 \rightarrow \left(n_{CO_2,r} + 1\right)CO_2 + H_2O,$$

where $n_{CO_2,r}$ is the number of CO_2 moles recycled.

From the oxygen balance, we find $n_{O_2} = 1$. Applying energy conservation to the adiabatic combustor,

$$\hat{h}_{H_2}^{623\text{K}} + \hat{h}_{CO}^{623\text{K}} + \hat{h}_{O_2}^{723.7\text{K}} n_{CO_2,r}\hat{h}_{CO_2}^{723.7\text{K}} = \left(n_{CO_2,r} + 1\right)\hat{h}_{CO_2}^{1600} + \hat{h}_{H_2O}^{1600}.$$

The enthalpies of gases are calculated as follows.

$$\hat{h}_{H_2}^{623\text{K}} = \hat{h}_{f,H_2}^0 + \hat{c}_{p,H_2}(623 - 298) = 0 + 28.6x325 = 9295\,\text{kJ/kmol}.$$

$$\text{Similarly}: \hat{h}_{CO}^{623\text{K}} = \hat{h}_{f,CO}^0 + \hat{c}_{p,CO}(623 - 298) = -101{,}072.5\,\text{kJ/kmol}$$

$$\hat{h}_{O_2}^{723.7\text{K}} = 12{,}517\,\text{kJ/kmol},\ \hat{h}_{CO_2}^{809.7\text{K}} = -377{,}963\,\text{kJ/kmol}$$

$$\hat{h}_{H_2O}^{1600} = 202419.2\,\text{kJ/kmol},\ \hat{h}_{CO_2}^{1600} = 345{,}365.6\,\text{kJ/kmol}.$$

Substituting into the equation above and solving for $n_{CO_2{'}r}$, we find $n_{CO_2{'}r} = 14.4$.

Next, we calculate the exit temperature of the turbine:

$$T_4 = T_3\left\{1 - \eta_t\left[1 - (p_4/p_3)^{k-1/k}\right]\right\} = 1600\left\{1 - 0.9\left[1 - (1/26.6)^{\frac{0.289}{1.289}}\right]\right\} = 850.1\,\text{K}.$$

The specific heat ratio of CO_2 is used because over 90% of the mixture is carbon dioxide. The work of the gas turbine is obtained as follows:

$$W_t = \left(n_{CO_2}\hat{c}_{p,CO_2} + n_{H_2O}\hat{c}_{p,O_2}\right)(1600 - 850.1) = (15.4 \times 37.2 + 1 \times 30.4)(1600 - 850.1)$$
$$= 451.668\,\text{kJ}.$$

Example 5.3 (cont.)

Similarly, the work of the O_2–CO_2 compressor is calculated:

$$W_c = \left(n_{CO_2,r}\hat{c}_{p,CO_2} + n_{O_2}\hat{c}_{p,O_2}\right)(723.7 - 303) = (14.4 \times 37.2 + 1 \times 29.4) \times 420.7$$
$$= 237.333 \text{ kJ}.$$

The work requirement of the ASU compressor is

$$W_{c,ASU} = (29.4 + 3.76x29.1)(600.3 - 303) = 41.265 \text{ kJ}.$$

The net work produced by the power plant is therefore

$$W_{net} = W_t - W_c - W_{c,ASU} = 173.070 \text{ kJ}.$$

The thermal efficiency of the power plant is $n_{th} = \frac{W_{net}}{LHV_{H_2}+LHV_{CO}} = \frac{173,070}{242,000+283,270} = 0.33.$

In more detailed analysis, one must also account for additional energy penalties due to liquefaction of carbon dioxide captured and pumping the coolants through the condenser and the heat exchanger. It should also be mentioned that in actual implementation, one would use a combined cycle to raise the efficiency, as shown in forthcoming chapters.

5.7 Bottoming Cycles

Another practical way to use the thermal energy in the gas turbine exhaust, besides regeneration, is to add a **bottoming cycle**. Given the high temperature of the gas turbine exhaust, a Rankine cycle is most suitable for taking advantage of the thermal energy in the exhaust to maintain high thermodynamic efficiency (see Chapter 2). In this case, the gas turbine exhaust is used in a heat recovery steam generator (HRSG). The steam is then expanded in a steam turbine to extract more work, followed by condensation, pumping the condensate back to the high pressure of the Rankine cycle. The two-phase cycle, or steam Rankine cycle when water is used as a working fluid, will be discussed in detail in the Chapter 6, and the detailed analysis of the combined cycle will be discussed in Chapter 8. With proper design, the combined cycle can achieve impressive efficiencies much higher than either the Brayton cycle or Rankine cycle alone. Many current natural gas power plants are built to use CCs. As will be shown, CCs can also be used with coal, if a gasifier is used first to generate synthetic gas that can then be burned in the gas turbine combustor. Other ways to use the thermal energy of the exhaust include the utilization of a humid air cycle and thermochemical recuperation. Both will be discussed in the Chapter 8.

5.8 Cogeneration

Another way to utilize the thermal energy in the gas turbine exhaust stream is cogeneration. In this arrangement, the exhaust energy is utilized in the form of "process heat." From the Second Law point of view, this is not necessarily an efficient utilization of energy since more work can be extracted from the high-temperature exhaust stream. However, from a practical and economic standpoint, it is a convenient and widely utilized approach.

There are two ways to define the efficiency of cogeneration plants; the First Law efficiency, also known as **the energy utilization factor** (EUF):

$$\eta_{FirstLaw} = \text{EUF} = \frac{\text{work out} + \text{process heat}}{\text{heat transfer in}}, \tag{5.42}$$

and the Second Law efficiency of cogeneration plants, in which the fuel availability is used:

$$\eta_{SecondLaw} = RC = \frac{\text{work out} + \text{process heat}}{(\text{work out} + \text{process heat})_{rev-plant}}. \tag{5.43}$$

This is also known as the rational criterion of performance (RC). For a given "work out" and "process heat," the RC is the ratio between the fuel chemical energy (or total thermal energy) required by a reversible plant to that required in the actual cogeneration plant.

Other parameters are often quoted for cogeneration or combined heat and power (CHP) plants, such as the heat to power ratio (HPR):

$$\text{HPR} = \frac{\text{process heat rate}}{\text{power output}}. \tag{5.44}$$

Cogeneration often achieves overall fuel utilization efficiencies or energy utilization factors exceeding 80%.

5.9 Gas Turbines Operating at High Temperature

Gas turbines achieve high power density by processing large mass flow rates while spinning at high RPM, and hence they must be constructed of high-strength material that can withstand the high temperatures of the expanding gases and the high stress generated in the spinning rotors. Gas turbine blades are often coated with thermally insulating and corrosion-resistant materials. Advanced turbines are manufactured using composite materials and "superalloys" of nickel (Ni) and cobalt (Co), mixed with molybdenum (Mo), tungsten (W), titanium (Ti), aluminum (Al), and chromium (Cr). The blades are hollowed for cooling. Recent efforts to manufacture gas turbine blades using oxide-dispersion-strengthened and ceramic material could raise the TRIT even further. Figure 5.20 shows the impact of the turbine blade material and cooling technologies on the turbine inlet temperature [11]. Significant material improvements have been made over the years, including adding a thermal barrier, enabling significant increase in the gas temperature flowing

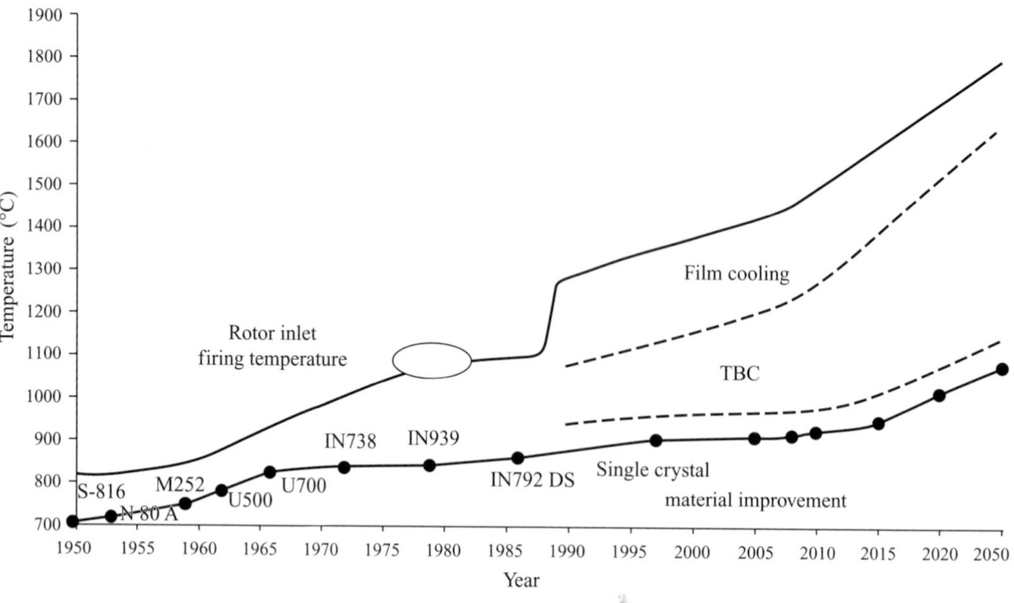

Figure 5.20 Impact of turbine blade metal, thermal barrier coating (TBC), and film cooling on the turbine inlet temperature [11].

over the turbine blade. However, as shown in the figure, improvements in blade-cooling technology have had an even greater impact.

A combination of high-temperature and oxygen-rich gases make gas turbine blade vulnerable to corrosion. The blades are coated with chromium, or at higher temperatures with XCrAlY, where X stands for cobalt or nickel and Y is yttrium, mixed in a dense aluminum oxide layer on the blade surface. This is part of the thermal barrier coating (TBC) applied to the blade surface, which is often a ceramic layer of zirconia (ZrO_2) stabilized with yttria and a bonding of a metallic layer of XCrAlY. The ceramic layer has low thermal conductivity. Advanced manufacturing techniques, including physical vapor deposition or plasma vapor deposition, are used in applying these coats [12].

Besides fabricating the turbine blades using corrosion- and heat-resistant materials, advanced cooling techniques are used to keep the blades at temperatures much lower than that of the gas. These include air and steam cooling using jet impingement, inner extended surfaces, and cooling films on the surface. The latest generation of gas turbine engines used in electricity generation have turbine inlet temperatures in the range of 1250–1350 °C and pressure ratios of 14–30, with exit temperatures ~550 °C. The simple cycle efficiency of these plants is 35–38% with power. These plants use large heavy-duty gas turbines. Smaller-scale aero-derivative turbines, originally designed for jet engine applications but modified for use in electricity generation plants, can operate at higher temperatures (1350–1400 °C) and higher pressure ratios (>30), boosting their efficiency to 39%. Because of the high-pressure ratio of these turbines, the turbine exit temperature is relatively low and the overall thermal efficiency is high.

5.9.1 Impact of Ambient Conditions

Gas turbine engines are designed for inlet conditions of 15 °C and 1 bar, and their specifications are given as ISO conditions. As the outside temperature increases, at a fixed maximum turbine inlet temperature, the compression work increases and the cycle efficiency decreases. Furthermore, with higher ambient temperature (or lower pressure) the mass flow rate at the compressor inlet decreases and the net power declines at fixed compressor inlet velocity. These considerations are especially important when installing these machines in warm environments.

5.10 Gas Turbines in Power Generation

Gas turbines have many advantages in the electricity generation sector, including the following:

1. They operate at high speed and utilize compact equipment.
2. They can be started, turned down for part-load operation, and stopped relatively easily and within a short period of time, and hence are capable of following load oscillation and meeting peak-load demands.
3. Gas turbine engines are compact and easy to operate, and they take advantage of ongoing developments in aerospace as well as sea and some ground propulsion applications.
4. Gas turbine cycles operate at relatively low pressures compared to steam turbines, and this simplifies the plumbing of the plant.
5. Many installations, for a wide range of loads, have been built and operated over the past couple of decades, especially those burning natural gas, or in dual-fuel mode burning natural gas and oil.
6. Gas turbines do not handle wet gases like steam turbines do, and are not as vulnerable to corrosion as steam turbines.
7. Open-cycle or combustion gas turbines do not require heat transfer equipment on the low-temperature side, and nor do they need a coolant, and hence they can be built and operated in hot, dry climates.

However, gas turbines also suffer from several limitations:

1. Their thermodynamic efficiency can be low since their maximum temperature is limited by what the turbine blade material can handle, even with cooling.
2. Their Second Law efficiency is rather low because of the irreversibility of the heat transfer processes, the high compressor work, and the low efficiency of compressors.

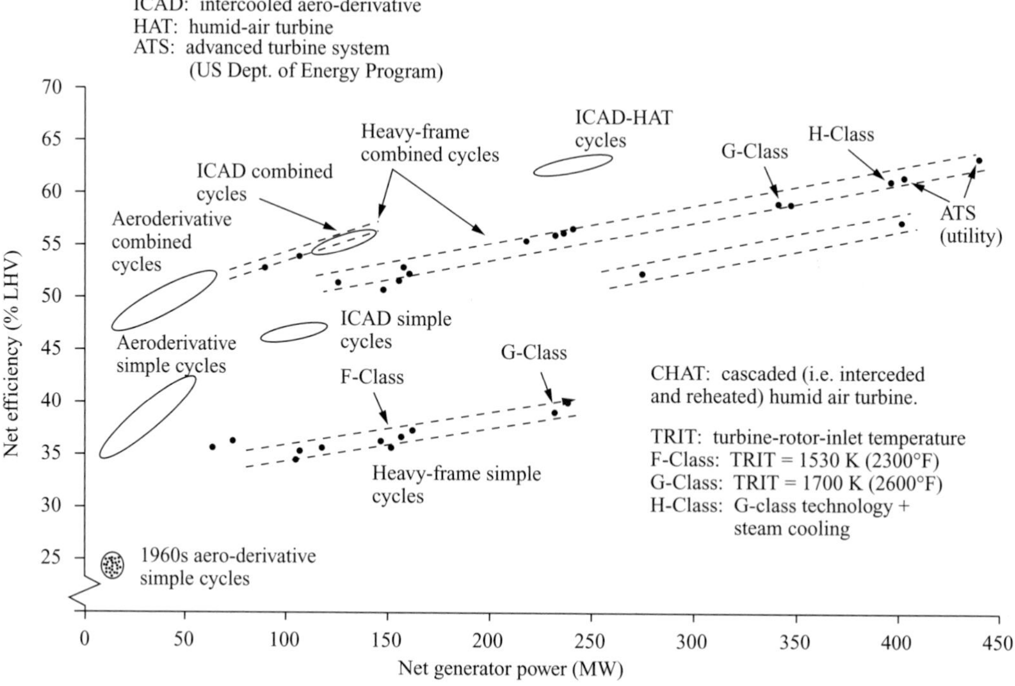

Figure 5.21 Compilation of simple and combined gas turbine cycle efficiency as a function of the total power output. The diagram includes data for heavy-duty designs as well as aero-derivative based designs. The different class turbines: F, G, and H classes, are designations for the inlet temperature, with the most recent being the advanced turbine system (ATS) [13]. Other advanced cycles, such as the HAT cycle, will be described later.

3. Open-cycle turbines are limited by the relatively high exhaust pressure compared to that of steam cycles, which limits the work transfer of the turbine.
4. They cannot be used with "dirty" fuels such as coal without special treatment, since sulfur oxides can damage turbine blades.

Steam or vapor Rankine cycles overcome some of these limitations, and hence have been popular in electric power generation. We discuss some of their characteristics in the next chapter.

Figure 5.21 shows a compilation of the net thermal efficiency of gas turbine power generation plants as a function of the power generation capacity. Shown are the efficiencies of simple cycle, combined cycle, aero-derivative, and heavy-duty turbines [13]. The turbines are classified as F, G, or H according to their inlet temperature. Smaller aero-derivative turbines operate at higher efficiency when used in smaller installations, but the efficiency of heavy-duty turbines increases as the capacity increases. Shown also are conceptual designs, such as the humid air turbine (HAT).

5.11 Performance Criteria

Gas turbine power plants are rated according to their **net power output**, measured in kW or MW of electricity, which is determined by the specific turbine work times the total flow rate of gases through the turbine, subtracted from the specific compressor work times its mass flow rate:

$$\wp = \dot{m}_T w_T - \dot{m}_C w_C. \tag{5.45}$$

The compressor mass flow rate is that of air, $\dot{m}_C = \dot{m}_a$, and the difference between the two mass flow rates is that of the fuel added in the combustor, $\dot{m}_f = \dot{m}_T - \dot{m}_C$. Gas turbine engines run under fuel-lean conditions due to the limitations on the TRIT, and hence $\dot{m}_f/\dot{m}_a = \mathrm{O}(1/30)$. Another characteristic quantity used to define the performance of gas turbine systems is the **net heat rate**, HR, defined as the total rate of heat input per unit of power output:

$$HR = \dot{Q}_{in}/\wp = \dot{m}_f \cdot HV/\wp. \tag{5.46}$$

The HV can be the LHV or the HHV.

Since the performance of gas turbines depends on the outside pressure and temperature, these values are usually quoted at the ISO (International Standards Organization) conditions of 15 °C, 1.013 bar and 60% humidity. Advanced, third-generation, heavy-duty turbine engines burning natural gas are rated up 250 MW (depending on the frequency of the generator), with high-pressure ratios in the range of 15–30, TIT > 1300 °C, and efficiency of 35–39%. First-, second-, and third-generation turbines were distinguished by the TRIT. A more common terminology to classify gas turbines is the F, G, and H classes, with the F class having TRIT > 1200 °C and the G class having TIT > 1400 °C. The H class is a G class with blade cooling whose TRIT > 1500 °C. Figure 5.21 summarizes the performance of these turbine systems when operating alone or as part of a combined cycle.

As mentioned before, another class of gas turbines is known as aero-derivative engines, which are modified jet engines made of lighter material and equipped with advanced blade-cooling techniques. Their pressure ratios often exceed 30, and TRIT is >1300 °C, with efficiencies reaching 39%. They are smaller engines with power in the range of 50 MW. When used for power generation, they are often equipped with intercoolers and recuperators for further efficiency improvements.

The **plant capacity factor** (CF) is defined as the average power output over the course of a year as a fraction of its rated power. It is also the total energy produced during a year of operation, measured in kWh, divided by the rated power times 8760 hours:

$$\begin{aligned} \mathrm{CF} &= \frac{\text{yearly average power}}{\text{rated power}} \\ &= \frac{\text{total energy produced over a year in kWh}}{\text{rated power} \times 8760 \text{ hrs}}. \end{aligned} \tag{5.47}$$

Base load plants, such as coal-fired and nuclear power plants, have a CF > 68%. These plants operate for more than 6000 hours/year at full load. These are large, high-capital-cost plants with low operating costs. Coal, nuclear, and natural gas combined cycle (NGCC) plants are used mostly for base load, but some SC natural gas plants have been recently introduced for the same purpose [14]. Intermediate-load plants operate at 28% < CF < 68%. Peaking or peak shaving units operate with CF < 28% and tend to be cheaper plants but with high operating cost. Natural gas combustion turbines and diesel engines are easy to start and stop, and can be used for peak shaving [8].

5.12 Compressed Air Energy Storage System

Energy storage can take many forms, including electrical energy storage in batteries, mechanical energy storage in flywheels, potential energy storage in hydraulic pumped storage systems, and potential energy storage in compressed air. Energy storage systems are very important when any form of renewable energy is used, because of the intermittency of the source. Air can be stored in deep underground reservoirs, such as caverns or salt domes. High-pressure cool air can be kept in these reservoirs for several hours until the energy is needed. This system is suitable for operation with renewable intermittent sources such as wind power or solar power, for overnight storage. Wind- or solar-generated electricity can be used to drive a motor that powers a compressor for pumping compressed air in the available underground space. Compressed air can also be used to store some of the energy generated by gas turbine engines running at optimum conditions, whose power might exceed the demand during part of the day. This excess power is used to drive the air compressors directly or indirectly (through a generator-motor), and hence is translated to energy storage.

The stored high-pressure air is used, when needed, to run the gas turbine directly, or as high-pressure air for the gas turbine combustor, where fuel is burned and extra power is produced. Figure 5.22 shows the layout of a compressed air energy storage (CAES) system. The gas turbine compressor is used to charge the underground reservoir during times when power is not needed. An intercooler is used to lower the air temperature and increase its density. When needed, air is discharged from the reservoir to power the gas turbine.

CAES efficiency, defined as the ratio of the energy recovered from the stored air to the energy required to compress the gas originally, is typically below 75%. The losses are due to the compressor losses, air cooling in the reservoir, and friction losses in the system. Storage reservoir pressures are typically 40–60 bar. The energy recovered can be increased if fuel is used in the turbine combustor, and some heat regeneration is applied to take advantage of the hot turbine exhaust.

5.13 Summary

Thermomechanical systems convert heat/thermal energy to work using a power cycle. These are known as heat engines, irrespective of the source of heat. However, given that the

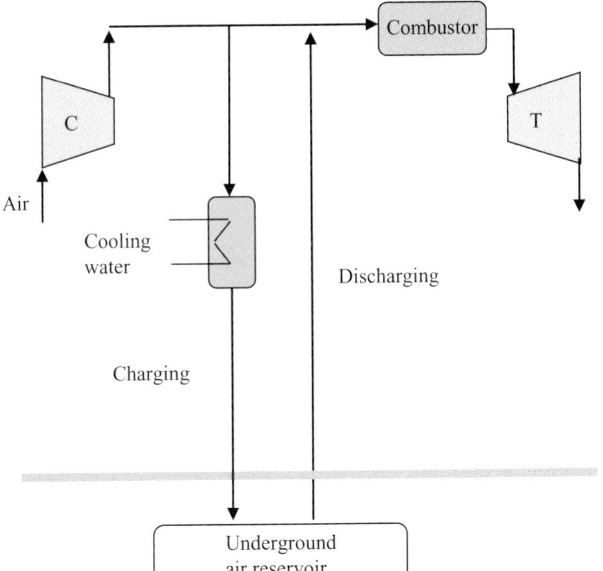

Figure 5.22 A compressed air storage system used in connection with a gas turbine power plant. During normal operation, the power plant operates without communication with the underground air reservoir. At times of excess power, some of the compressed gas is sent to the underground storage, after being cooled, to charge the storage system. When needed, stored pressurized air flows directly to the combustor, discharging the storage system.

conversion efficiency depends strongly on the temperature of the heat source, careful selection of the power cycle is necessary. Moreover, given the concern over CO_2 production per unit of work produced when fossil fuels are used to produce the heat, other criteria such as CO_2 production per unit of thermal energy and the impact of the working fluid on efficiency must be used. The materials used in the cycle components play a role in limiting the operating conditions, as well as other practical and economic concerns.

In this chapter we reviewed systems using high-temperature and moderate-pressure working fluids in the gas phase to perform a series of processes for converting heat to work. A simple Brayton cycle is very popular, as well as several improvements including recuperation, reheat, and intercooling. Raising the compressor pressure ratio increases the cycle efficiency, as well as using higher quality turbomachinery (higher isentropic efficiency). Recuperation and multiple reheat and intercooling bring the Brayton cycle efficiency closer to the ideal Carnot efficiency (in the Ericsson cycle). In the open cycle, the working fluid is combustion products, and their efficiency is limited by the exhaust pressure of the gas turbine. Closed cycles can improve the efficiency if a different working fluid is selected, but they also require cooling (heat rejection).

Besides the cycle pressure ratio, the turbine inlet temperature plays an important role in the cycle efficiency. Raising this value by building turbine blades using composites and superalloys with thermal barrier coating, and improving blade cooling (where the gas temperature can be higher than the blade surface temperature) continues to increase gas turbine systems efficiency. On the other hand, addressing the major availability loss in gas turbines (see Chapter 2) – that is the availability loss in the turbine exhaust to the environment – requires methods for waste heat recovery. Steam plants can be used for that purpose, and are discussed in detail in the next chapter. These two-phase power cycles find broad and extensive applications because of their compatibility with lower-temperature heat sources.

Problems

5.1 A gas turbine power plant operating on a simple Brayton cycle produces 20 MW power. Atmospheric air at 101.3 kPa and 293 K enters the compressor, whose pressure ratio is 15. The amount of heat added in the combustor is 1000 kJ/kg. Assume that the fuel mass flow rate is negligible compared to that of air. The isentropic efficiencies of turbine and compressor are 90% and 85%, respectively. Determine: (a) the turbine inlet temperature; (b) the mass flow rate of air; and (c) the thermal efficiency of the power plant.

5.2 Air at 298 K and 1 atm enters the compressor of a regenerative gas turbine power plant, where it is compressed to 11 atm. The regenerator efficiency is 80%, and the turbine inlet temperature is 1500 K. Assume that the fuel mass flow rate is negligible compared to that of air. The isentropic efficiencies of turbine and compressor are 90% and 85%, respectively.

a. Determine the thermal efficiency of the power plant.
b. Calculate the exhaust temperature.
c. What would be the thermal efficiency without the regenerator?

5.3 The pressure ratio across the compressor of a regenerative gas turbine power plant, shown in Figure P5.3 is 12. Air, initially at 290 K and 100 kPa, is used as the working fluid, and 60% of the power produced by the gas turbine is consumed by the compressor. The isentropic efficiencies of the turbine and the compressor are 90% and 85%, respectively. The efficiency of the regenerator is 80%.

a. Determine the turbine inlet temperature.
b. Calculate the specific work output of the plant.
c. Determine the thermal efficiency of the power plant.

Figure P5.3

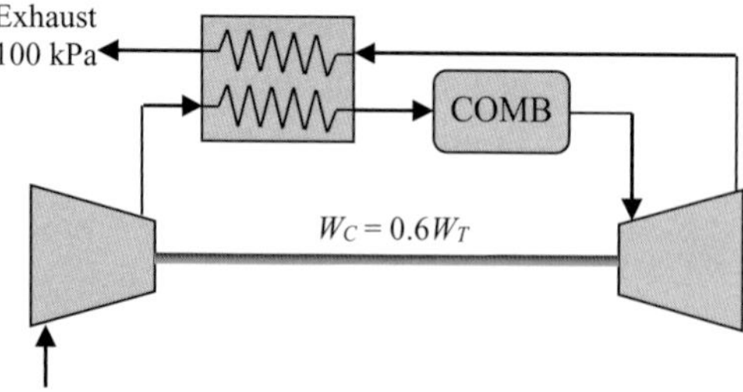

5.4 An intercooler is used between two compressors, as shown in Figure P5.4. Air at 1 atm and 288 K enters the first compressor. It is then compressed to pressure p_2 before entering the intercooler. The temperature of the air at the outlet of the intercooler is

310 K. The pressure at the outlet of the second compressor is 12 atm. Assume that there is no pressure drop within the intercooler. Determine (a) the optimum value of p_2 that would lead to a minimum total compression work; (b) the quantity of heat transferred from the intercooler per unit mass of air; and (c) the minimum work requirement of the compressors per unit mass of air.

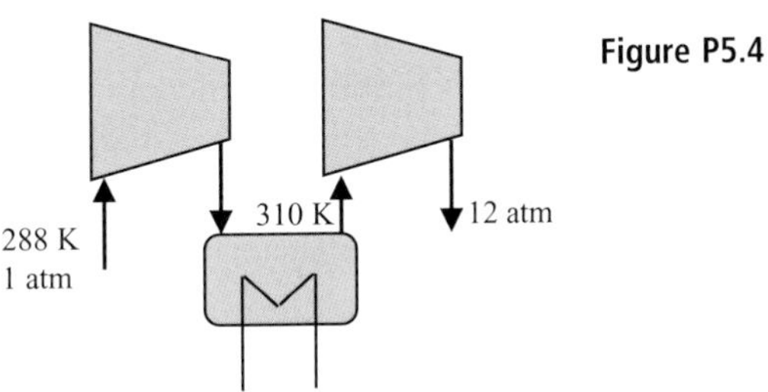

Figure P5.4

5.5 A regenerative gas turbine power plant with single intercooling and single reheat is shown in Figure P5.5. Air at 298 K and 1 atm enters the first compressor. The inlet temperature of the second compressor is 323 K. The pressure ratio across each compressor is 4. The pressure drop on the cold and hot sides of the regenerator is 4%. The pressure drop in each combustor is 2%. The pressurized air is heated to 1423 K in the first combustor. The outlet temperature of the second combustor is 1373 K. The isentropic efficiency of the compressors is 82%, the isentropic efficiency of the turbines is 90%. The regenerator operates with 80% efficiency. The pressure ratio across the turbines is optimized, yielding a maximum work production.

a. Determine the air temperature at the entry of the first combustor.
b. Calculate the exhaust temperature.
c. Determine the specific work output of the power plant.
d. Calculate the thermal efficiency of the power plant.
e. What would be the efficiency of the power plant with only one compressor with a pressure ratio of 16?
f. What would be the efficiency of the power plant with only one turbine?

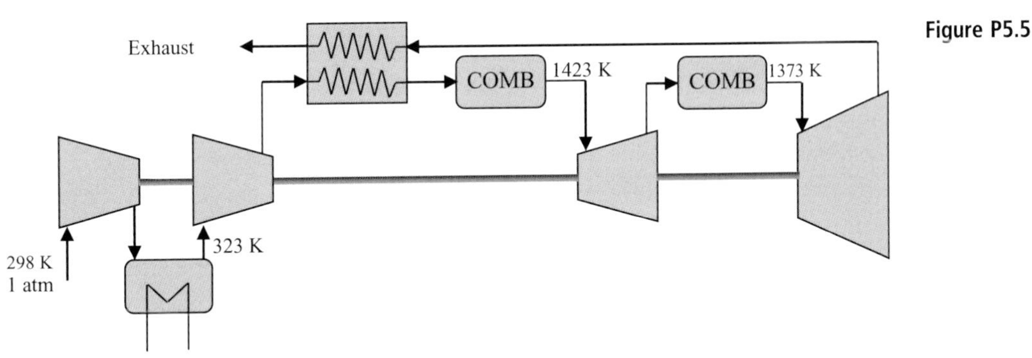

Figure P5.5

5.6 A schematic of a gas turbine power plant operating with methane (LHV = 50050 kJ/kg) as the fuel is shown in Figure P5.6. Air, initially at 288 K and 100 kPa, is compressed in three stages, with one intercooler between two compressors. The pressure ratio across each compressor is 2.3:1. The inlet temperatures of the second and third compressors are 303 K and 323 K, respectively. The pressure drop in each intercooler is 1%. The isentropic efficiency of the compressors is 85%. At full-load operation, the ratio of the air-to-fuel mass flow rates is 40:1, and the plant produces a net power of 150 MW. The pressure drops within the combustor, and on the cold and hot sides of the regenerator are 2.5%, 4%, and 3%, respectively. The exhaust pressure is 100 kPa. The isentropic efficiency of the turbine is 88%. The regenerator operates with 85% efficiency.

a. Calculate the mass flow rates of the air and fuel.

b. Determine the total power consumption of the compressors.

c. Determine the power production of the turbine.

d. Calculate the thermal efficiency of the power plant.

e. Part-load operation of the power plant is possible by reducing the turbine inlet temperature, hence reducing the mass flow rate of the fuel. Determine the turbine inlet temperature at 70% of full-load operation.

f. Determine the thermal efficiency at 70% full-load operation.

Figure P5.6

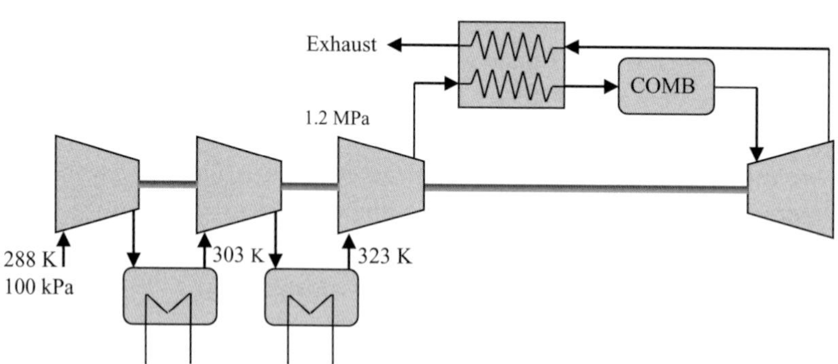

5.7 A regenerative gas turbine power plant operating on the basis of a Brayton cycle is shown in Figure P5.7. Air, initially at 300 K and 1 atm, enters the compressor, where its pressure rises to 12 atm. The pressurized air is then heated up in the heat exchanger before entering the combustor. Methane at 300 K and 12 atm is fed at the rate of 0.5 kg/s to the combustor, where it is mixed with 400% theoretical air and burned completely. The combustion products, including carbon dioxide, water vapor, oxygen, and nitrogen, are then expanded in the turbine to the atmospheric pressure. The products are directed to the heat exchanger to heat the pressurized fresh air. The flue gas is finally discharged to the atmosphere. Take an isentropic efficiency of 80% for the compressor, an isentropic efficiency of 88% for the turbine, and an efficiency of 85% for the heat exchanger. Also, assume that all components operate adiabatically.

a. Determine the net power production and the thermal efficiency of the power cycle.

b. Calculate the irreversibility in each component of the power cycle.

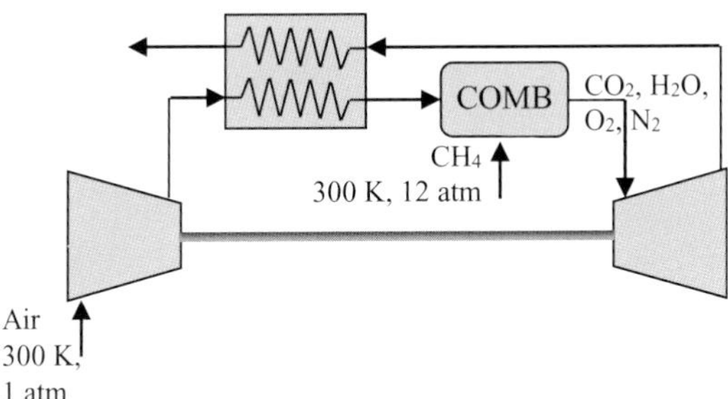

Figure P5.7

5.8 Consider a compressed air energy storage (CAES) system that uses off-peak electricity to compress 100 kg/s of atmospheric air ($T_{atm} = 298$ K, $P_{atm} = 1$ atm) in two stages, first to 5 bar and then to 30 bar for underground storage. The air is cooled back to ambient temperature between compression stages. Both compressors have isentropic efficiency of 80% and operate at steady state. Air can be modeled as an ideal gas. The air is stored in an adiabatic cavern at 30 bar and the temperature corresponding to that at the exit of the first compressor until electrical power is needed. When power is required, the stored air is expanded back to atmospheric pressure in a turbine with an isentropic efficiency of 85%.

a. Determine the power required for the entire compression process.

b. Determine the power produced by the turbine.

c. Calculate the roundtrip efficiency of this CAES system.

5.9 The CAES system of Problem 5.8 is modified so that when power is required, the stored air is fed to a combustor along with natural gas at 298 K (model as CH_4). The mixture is burned to completion at 30 bar and the products at 1500 K enter a turbine with an isentropic efficiency of 85%, where they are expanded to atmospheric pressure.

a. Determine the number of moles of O_2 and N_2 required per mole of CH_4.

b. Determine the equivalence ratio for combustion.

c. Determine the power produced by the turbine.

d. Calculate the roundtrip efficiency of this CAES configuration. Neglect the power required to compress CH_4.

5.10 Consider a CAES system that is similar to the one in Problem 5.9, but CO_2 emissions from the turbine are captured. Assume that the capture equipment has a Second Law efficiency of 20%.

a. Determine the power needed to separate CO_2 from the exhaust gas stream using the stated capture equipment if both the inlet and outlets are at the turbine outlet temperature.

b. Determine the power needed to separate CO_2 from the exhaust gas stream using the stated capture equipment if the inlet is at the turbine outlet temperature and the outlets are at atmospheric temperature.

c. Determine the roundtrip efficiency of this CAES configuration for the conditions given in parts (a) and (b).

REFERENCES

1. EIA, *EIA annual energy outlook*. Washington, DC: EIA, 2013.
2. M. M. El-Wakil, *Powerplant technology*. New York: McGraw-Hill, 1984.
3. E. G. Cravalho and J. L. Smith, *Engineering thermodynamics*. Boston, MA: Pitman, 1981.
4. R. W. Haywood, *Analysis of engineering cycles: power, refrigerating, and gas liquefaction plant*. Oxford: Pergamon Press, 1991.
5. W. C. Reynolds and H. C. Perkins, *Engineering thermodynamics*. New York: McGraw-Hill, 1977.
6. F. F. Huang, *Engineering thermodynamics: fundamentals and applications*. New York: Macmillan, 1976.
7. Westinghouse Electric Co., *Advanced natural gas fired turbine system utilizing thermochemical recuperation and/or partial oxidation for electricity generation, greenfield and repowering applications*. Morgantown, WV; Department of Energy, 1997.
8. M. P. Boyce, *Gas turbine handbook*, 2nd ed. Houston, TX: Gulf Professional Publishing, 2002.
9. R. Decher, *Energy conversion: systems, flow physics, and engineering*. New York: Oxford University Press, 1994.
10. G. Gottlicher, *The energetics of carbon dioxide capture in power plants*. Washington, DC: U.S. Department of Energy, Office of Fossil Energy, NETL, 2004.
11. A. Rao, "Advanced Brayton cycles," in NETL (ed.), *Gas turbine handbook*, Washington, DC: NETL, 2006.
12. N. V. Kharchenko and V. M. Kharchenko, *Advanced energy systems*. Boca Raton, FL: CRC Press, 2014.
13. D. G. Wilson and T. Korakianitis, *The design of high-efficiency turbomachinery and gas turbines*. Cambridge, MA: Massachusetts Institute of Technology, 2014.
14. M. P. Boyce, *Gas turbine engineering handbook*. Amsterdam: Elsevier/Butterworth-Heinemann, 2012.

6 Thermomechanical Conversion

Two-Phase Cycles

6.1 Introduction

In Chapter 5 we covered gas turbine cycles and discussed conditions under which they are expected to achieve high efficiency or high specific work. With high-temperature energy sources, such as combustion of clean fuels, gas turbines offer many advantages in electricity generation, as well as high-speed propulsion. They are not, however, compatible with intermediate-temperature sources, such as lower-temperature nuclear reactors and concentrated solar thermal energy, or low-temperature sources such as concentrated solar power or geothermal energy. Two-phase Rankine cycles can be designed to operate at high efficiency while utilizing these sources. This is the subject of this chapter.

Rankine cycles take advantage of the low pumping work required to raise the pressure of an incompressible fluid and the isothermal heat transfer during evaporation and condensation to improve the cycle efficiency. Closed Rankine cycles that use a condenser also take advantage of the low saturation pressure of many fluids at a temperature close to the environment temperature to expand the working fluid between the turbine inlet pressure and pressures much lower than the atmosphere pressure. While this makes it necessary to use a condenser, it raises the cycle efficiency significantly. Cycle modifications similar to those used in Brayton gas cycles are implemented in Rankine cycles to improve their efficiency, such as reheating and regeneration. Similarly, two-phase cycles achieve better efficiencies when high-temperature turbines are utilized. These modifications are discussed in detail in this chapter.

In two-phase cycles, the working fluid is heated externally and pinch point analysis is used to reduce the temperature difference between the source and cycle maximum temperature. In many modern plants, cycles use superheated steam to reduce heat transfer irreversibility and raise the efficiency further. When using alternative working fluids, such as carbon dioxide, whose critical pressure and temperature are smaller than those of water, it is possible to operate a "hypercritical" cycle in which the lower pressure is closer to the critical pressure of the cycle. With modifications such as regeneration and split flow compression, these cycles can reach impressive efficiency. On the other hand, when working with lower-temperature heat sources, such as geothermal energy and "waste heat," fluids with low critical temperature such as organic fluids and their mixtures are used.

In two-phase cycles, the condenser cooling plays an important role in determining the cycle efficiency. Different condenser cooling options are discussed close to the end of the chapter, as well as their impact on water consumption and cycle efficiency.

6.2 Two-Phase Power Cycles

Steam-based Rankine cycles have been used in power plants for many years, long before gas turbines were invented. Modern steam turbine engines grew out of the original piston-driven steam engines, when rotating machines became available. Currently, two-phase turbine cycles are used extensively in fossil fuel, nuclear, solar thermal, and geothermal power plants (geothermal power plants use working fluids other than water, as will be shown later, to match the low temperature of the heat source). Steam Rankine cycles have many desirable characteristics; the most important one is that they can be used at much lower maximum temperature than that required by gas turbines while delivering high efficiency close to 50% when operating at supercritical conditions and with multiple reheat and regeneration. Their relatively high efficiency at these lower temperatures is due to the very small pumping power required to pressurize the working fluid following the condensation process, in contrast to the high compression work needed for gas turbine cycle compressors. Two-phase Rankine cycles are also used in combined cycle configurations, as a bottoming cycle for gas turbine engines, with impressive overall plant efficiency.

Steam turbine plants operate on the two-phase Rankine cycle, in which the fluid undergoes a phase change as it flows within and in between components. For water, the critical conditions are: $p_{cr} = 221.2$ atm $= 22$ MPa and $T_{cr} = 374.14\,°\mathrm{C}$ [1]. The T–s diagram for water is shown in Figure 6.1 [2]. Thus, in cycles operating at high pressures below 220 bar, the working fluid undergoes phase change in the boiler and condenser. This phase change endows the cycle with some of its most important advantages, such as small pumping work, large heat transfer, and isothermal heat transfer during parts of the cycle. The essentials of Rankine cycles are reviewed next, with an emphasis on how to improve the cycle efficiency.

6.3 The Rankine Cycle

The simple Rankine cycle has two isentropic processes, for compression and expansion, and two constant-pressure processes, for heat transfer interactions, as shown in the T–s diagram in Figure 6.2. The working fluid is compressed as a liquid, and hence the compression work is much smaller than the expansion work. During the expansion, the working fluid is mostly a gas or a two-phase medium with high vapor concentration. This makes the cycle less dependent on the pump isentropic efficiency. Furthermore, most of the heat interactions

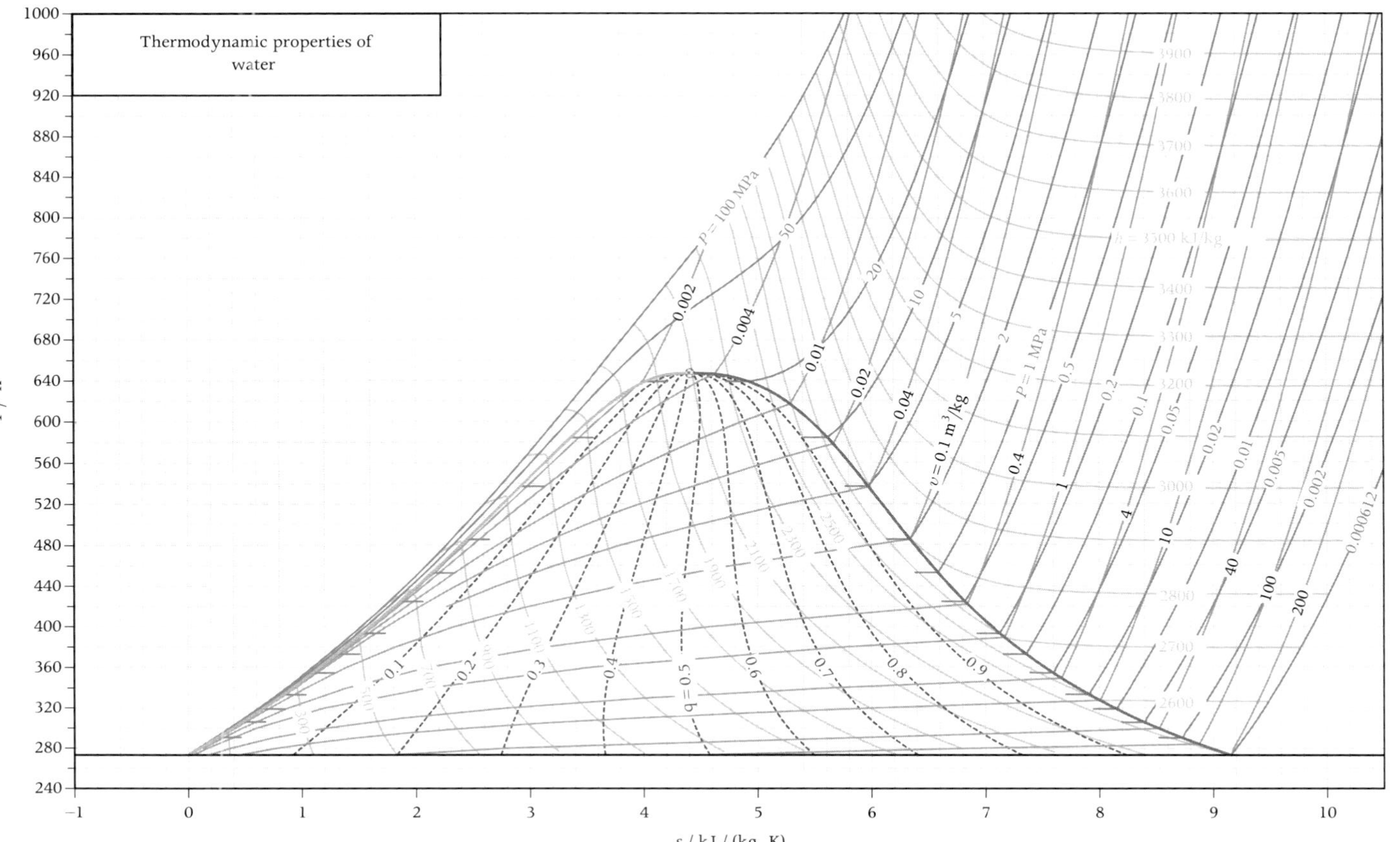

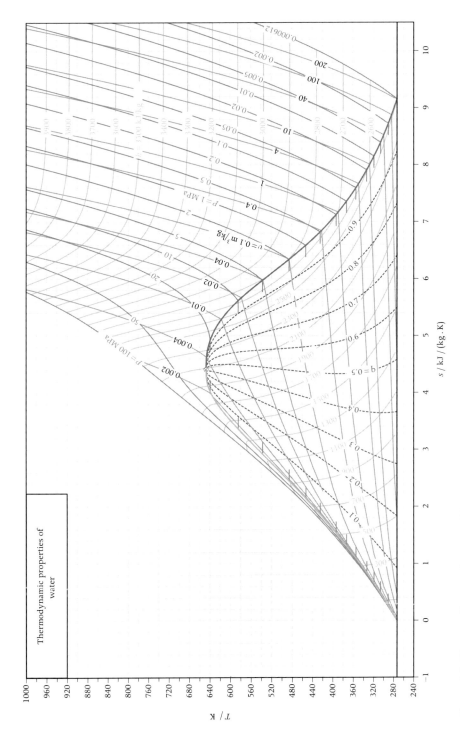

Figure 6.1 The T–s diagram for water [2].

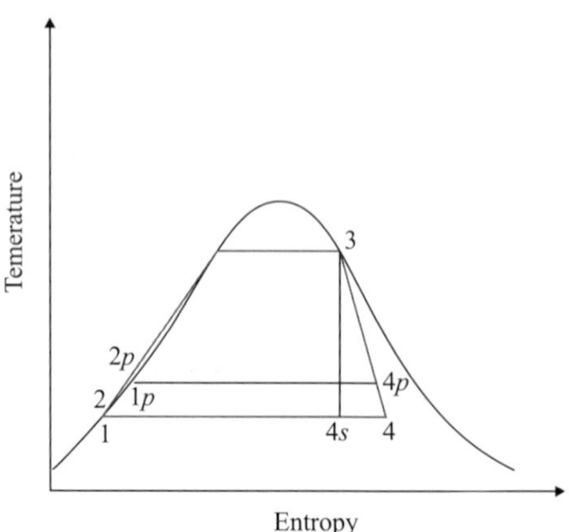

Figure 6.2 The T–s diagram of an open Rankine cycle, $1p$–$2p$–3–$4p$, operating between the high pressure of state 3 and the atmospheric pressure of state $1p$, and a closed Rankine cycle, 1–2–3–4, with sub-atmospheric pressure at state 1.

occur during phase transition, when the temperature of the fluid is constant. Both of these factors help raise the efficiency of Rankine cycles. Since the two-phase cycle transfers heat during evaporation and condensation, they take advantage of the large latent enthalpy of evaporation of the working fluid to increase the specific work.

The Rankine cycle performance depends strongly on the working fluid properties inside and around the vapor dome, and hence they are sometimes called **fluid-property limited cycles** [3]. Contrary to gas turbine cycles, which are known as **temperature-limited cycles** because the material maximum allowable temperature limits the performance, in these vapor cycles (at least when water is used as a working fluid) it is the maximum pressure that is of concern to the plant design and hence they are **pressure-limited** [3,4].

Cycles whose maximum pressure is below the critical pressure of the working fluid, p_c, are called **subcritical** cycles; otherwise they are called **supercritical** cycles. Figure 6.2 shows the T–s diagram of a subcritical cycle. A pump is used to raise the liquid pressure to a desired value. Next, the liquid is evaporated in a boiler, where fuel is burned to provide the necessary thermal energy, or in an evaporator in which the working fluid is heated using the coolant of a nuclear reactor, a solar collector, or the geothermal fluid. Saturated steam is then expanded in a turbine to a lowest allowable pressure. If the exhaust of the turbine is vented into the atmosphere (i.e., in an open cycle configuration), the lowest allowable pressure of the cycle is the atmospheric pressure. All current Rankine cycle plants operate in a closed cycle mode since this allows expansion of the working fluid to sub-atmospheric pressures, and hence much higher turbine output and better overall efficiency. The lowest pressure corresponds to the vapor pressure of the working fluid near or at the available coolant temperature, or the temperature of the environment. In this case, a **condenser** is used to cool the working fluid while condensing the remaining vapor in the turbine exhaust. The components required for the simple Rankine cycle are shown in Figure 6.3. Note that for water, at 15 °C, the vapor pressure is 5.63 kPa, which is much lower than the atmospheric pressure.

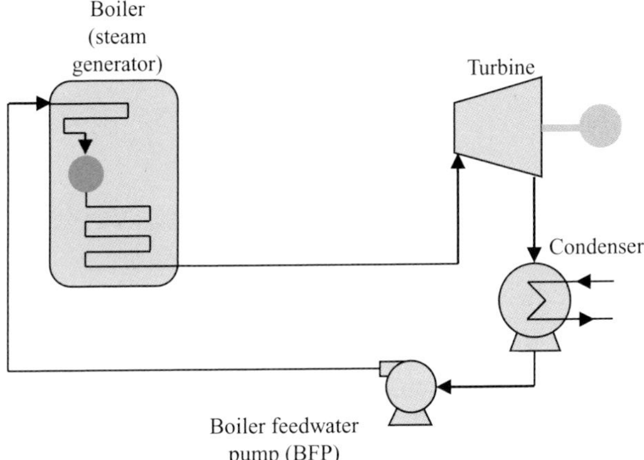

Figure 6.3 The layout and the components used in a simple Rankine cycle.

The cycle analysis follows the standard procedure of applying the First Law to the different processes. We note here that the **ideal (adiabatic reversible) pump work** per unit mass of the working fluid for an incompressible liquid is given by

$$w_{pump,ideal} = h_{2s} - h_1 = v(p_2 - p_1). \tag{6.1}$$

The inlet and exit pressures to the pump are p_1 and p_2, respectively, and the specific volume of the liquid is v. The actual pump work is obtained from the definition of the pump isentropic efficiency:

$$\eta_{pump} = \frac{w_{pump,ideal}}{w_{pump,actual}} = \frac{h_{2,s} - h_1}{h_2 - h_1}. \tag{6.2}$$

This relation also allows calculation of the enthalpy of the liquid at the pump exit. The ideal turbine work per unit mass of the working fluid, neglecting the small changes in the kinetic energy of the fluid, is

$$w_{T,ideal} = h_3 - h_{4s}. \tag{6.3}$$

The actual work is obtained from the definition of the turbine isentropic efficiency:

$$\eta_T = \frac{h_3 - h_4}{h_3 - h_{4s}}. \tag{6.4}$$

The heat transfer to the cycle per unit mass of the working fluid is

$$q_{in} = h_3 - h_2. \tag{6.5}$$

During this heat transfer process, the fluid temperature increases at a constant pressure until it reaches the saturation temperature at that pressure, then it remains constant during the phase transition. The heat transfer out of the cycle occurs almost always entirely at constant temperature unless the turbine exhaust is a superheated steam. The constant pressure lines in the liquid phase are very close to the two-phase dome, practically on the dome line itself. The definition of the net work out of the cycle and the thermal efficiency are the same as before:

$$w_{net} = w_t - w_c = (h_3 - h_4) - (h_2 - h_1) \tag{6.6}$$

and

$$\eta = \frac{w_{net}}{q_{in}}. \tag{6.7}$$

Figure 6.2 shows a subcritical closed cycle, with a sub-atmospheric pressure condenser, and the corresponding open cycle without a condenser, in which the turbine exhaust pressure is atmospheric. The following numbers show the significant improvement of the specific work and efficiency when a condenser is used in the cycle. Consider a case with atmospheric pressure of 10^5 Pa, atmospheric temperature of 20 °C, the pump efficiency is 65% and turbine efficiency is 90%, and the pressure ratio across the pump/turbine is 8, with negligible pressure drop in the boiler and condenser. We note here that it is generally easier to build high-efficiency pumps for incompressible fluid than high-efficiency compressors for gases. It is also easier to build high-efficiency turbines than high-efficiency compressors because of the smaller flow losses during expansion processes than during compression processes. We also neglect the kinetic energy of the fluid in calculating the work transfer in the turbine and pump. We calculate the pump work, the turbine work, and the specific work transfer for the open and closed cycles, the cycle efficiency in which pump and turbine losses are finite, the ideal cycle with ideal components, and the Carnot cycle. The results are shown in Table 6.1.

Adding a condenser to this cycle almost doubles the efficiency because the turbine work doubles as well. It is clear that the condenser plays a very important role in improving the

Table 6.1 Rankine cycle calculations for a number of designs and operating conditions. In the first column, a conventional cycle operating with vacuum at the low-pressure side, or venting to the atmosphere. The second column shows that case with vacuum but with superheat of 100 °C. The third column is also with vacuum and for three different values of reheat. The last column shows the calculations for a regenerative cycle, with 100 °C superheat

	Conventional		Superheating	Reheat cycle			Regenerative
	$T_{min} = 20$	$P_{min} = 1$ atm	+100 °C	+100 °C	+200 °C	+300 °C	cycle +100 °C
w_{pump} (kJ/kg)	1.23	1.12	1.23	1.23	1.23	1.23	1.26
w_t (kJ/kg)	736	316	818	947.2	1086	1400	774
w_{net} (kJ/kg)	735	315	817	946	1085	1398	773
η	27.4 %	13.4%	28.1%	28.1%	30.3%	35.5%	29.4%
η_{ideal}	30.4%	14.9%					
η_{car}	33.9%	15.8%	46.0%	46.0%	54.4%	60.6%	46.0%
X_4	0.794	0.8856	0.8517	0.9583	Vapor	Vapor	N/A

Rankine cycle performance. This is because having a lower backpressure at the turbine exhaust increases the work from that component, while the extra work required by the pump remains a very small fraction of the turbine work. Further advantages of the closed cycle include the continuous circulation of the working fluid inside the cycle; theoretically there is no loss of working fluid. This is important since the working fluid must be purified to improve the turbine reliability. The Carnot efficiencies of these cycles are 16% and 33.8%, also reflecting the improvement gained by adding a condenser. The Second Law efficiency of the cycle with a condenser is 81%, better than that of the gas turbine cycle with the same pressure ratio, as expected.

One problem with employing a condenser, however, is the air leakage from the environment through the seals into the condenser. The mixing of air with water raises the condenser pressure and changes the properties of the working fluid throughout the cycle. It is thus necessary to equip the condenser with a vacuum pump to remove the leaked air and maintain the pressure there at the design condition. As with most efficiency improvement measures, the condenser adds to the initial cost and some to the running cost of the plant, and requires pumps to circulate coolant. The latter is often water from nearby large sources such as a lake or a river. In the absence of water, it is necessary to build large cooling towers for cooling the condenser coolant. In a large installation, the cost of the condenser, coolant pumps, etc., is offset by the fuel saving resulting from the efficiency improvement. As large quantities of thermal energy, equivalent to the latent enthalpy of water, must be removed, we expect the need for large quantities of coolant to flow through the heat exchange passages of the condenser, requiring extra pumping power. Note the interesting compromise between:

1. using high flow rates of the coolant – which leads to higher pressure drops and requires large pumping power – to keep the coolant temperature from rising too much; and
2. reducing the coolant flow rates to minimize the coolant power requirements.

In the former case, a small temperature rise in the coolant keeps the difference between the coolant and the condensing steam temperature low, and hence reduces heat transfer irreversibility. Also, limiting the coolant temperature rise may minimize the danger posed to local wildlife in the coolant source (e.g., a lake) near the location of the coolant circulation.

Figure 6.4 shows the impact of the maximum pressure/temperature of a simple Rankine cycle on its efficiency and specific work. The isentropic efficiency of the pump is 65%, the turbine isentropic efficiency is 90%, and the condenser temperature is 30 °C. In this calculation, the steam at the turbine inlet is saturated. Contrary to the gas turbine cycle, in which the specific work peaks at a relatively low pressure ratio where the efficiency is rather low, in the steam cycle, the specific work peaks at much higher pressures, where the efficiency is high. The reason for the correlation between the efficiency and the net work is the very small pumping work associated with the liquid phase. Any further improvement in efficiency must utilize superheating of the steam beyond its saturation temperature, using supercritical steam, regeneration, and/or reheating. All these improvements will be addressed shortly.

The large pressure and temperature variation between the front and the back ends of the steam turbine means that the flow passages must expand. Indeed, the size and outer

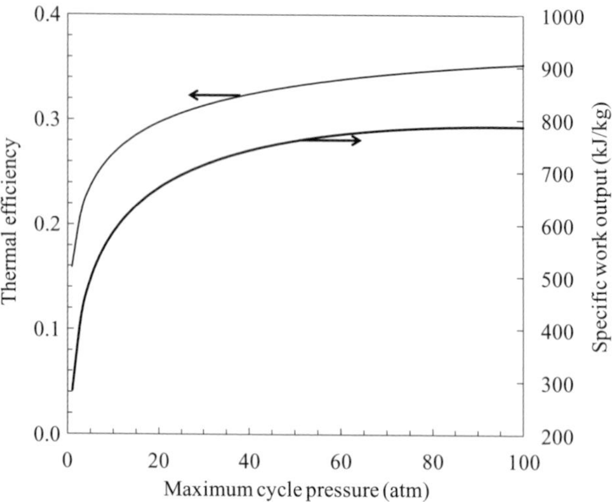

Figure 6.4 Impact of the maximum pressure/temperature in a saturated Rankine cycle on the efficiency and the specific net work.

diameters of the later stages are much bigger than those of the early stages. Whether several turbine stages are mounted on the same turbine shaft or several turbines are used depends on the design and the extra cycle complexity, as described next. Modern power plants often utilize several turbine units, each producing up to 1 GW. The size of these very large steam turbines makes their startup and turn down rather slow and complex. Moreover, boilers supplying steam to these large turbines have high thermal inertia. Steam turbines generally run with small load variation – that is, they are used to deliver the base load. The problem is compounded by the high thermal inertia of other components in the cycle, such as the boiler. In general, steam power plants are more capable of delivering the "base" load, and other cycles, such as gas turbines and diesel engines, are used for peak loads.

6.4 Superheat

Increasing the net specific work and the cycle efficiency of steam turbine cycles requires superheating the steam, and using one or more reheat steps. There is another reason for superheating and reheating. When the backpressure of the turbine is low, the liquid content of the steam can be rather high at the turbine latter stages. Low steam quality, or high moisture content, can cause blade damage. It is often desirable to keep the steam quality inside the turbine higher than 90%, which can limit the back pressure of the turbine, or force the reduction of the highest pressure of the cycle (see the vapor dome in the *T*–*s* diagram in Figure 6.1; the diagram shows that the higher the maximum turbine pressure, the lower the steam quality at the end of isentropic expansion, for the same condenser pressure). A better solution is **superheating**, which has the added advantage of raising the specific work and the cycle efficiency. Before exiting the boiler, in another chamber called the superheater, steam is heated above the saturation temperatures before entering the turbine. In the case that a

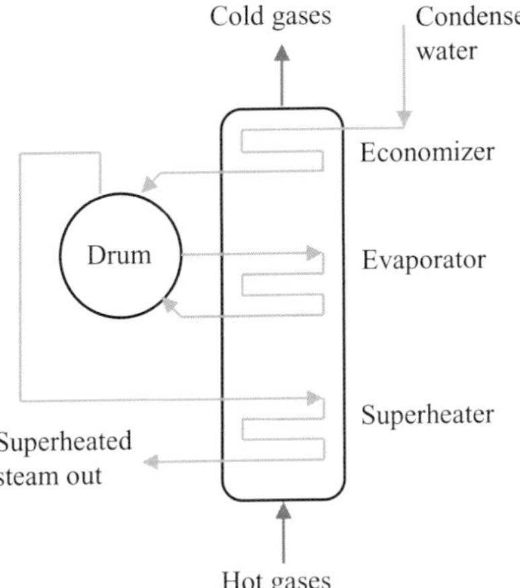

Figure 6.5 The typical arrangement used in a superheated Rankine cycle: an economizer heats up water in the liquid phase, an evaporator allows steam to leave the boiler in the gas phase, keeping the liquid in the vessel, and a superheater in which the steam temperature is raised above the saturation temperature. The hot combustion gases enter at the bottom.

nuclear reactor is used, more reactor cooling fluid is used to superheat the steam. As shown on the *T–s* diagram, with superheating, more work is generated by the turbine, and the steam enters the condenser at a higher quality.

The above example is reworked with a turbine inlet temperature being 100 °C higher than the saturation temperature at 0.8 MPa, and the results are shown in Table 6.1. As can be seen, the efficiency improvement is small, but the specific work is much higher, and the steam quality in the turbine exhaust is also higher. The Carnot efficiency of the superheated cycle is 46%, and the Second Law efficiency is 61%. Superheating involves more irreversibilities since the heat addition process occurs over a wider temperature change.

Superheating requires more complex steam generators in which a superheater is added that raises the temperature of the saturated steam, leaving the boiler at the required temperature. Most often, an economizer is used to bring the feedwater temperature to the saturation temperature before it enters the evaporator. The evaporator, or boiler, separates steam from water. A typical arrangement used in superheated Rankine cycles is shown in Figure 6.5.

6.5 Reheat

Yet another way to increase the specific work and improve the steam quality in the late stages of the turbine is **reheat**. The steam exiting the first-stage turbine, at an intermediate pressure between the highest and lowest cycle pressures, is brought back to a superheater to raise its temperature to the maximum cycle temperature. In the example shown in Table 6.1, the intermediate pressure is taken as the geometric average of the highest and lowest pressures. Reheat is applied between the end of expansion in the high-pressure turbine and the inlet of the low-pressure turbine. The maximum temperatures for the high-pressure and low-pressure turbines are the same. The *h–s* diagram of a reheat Rankine cycle is shown in

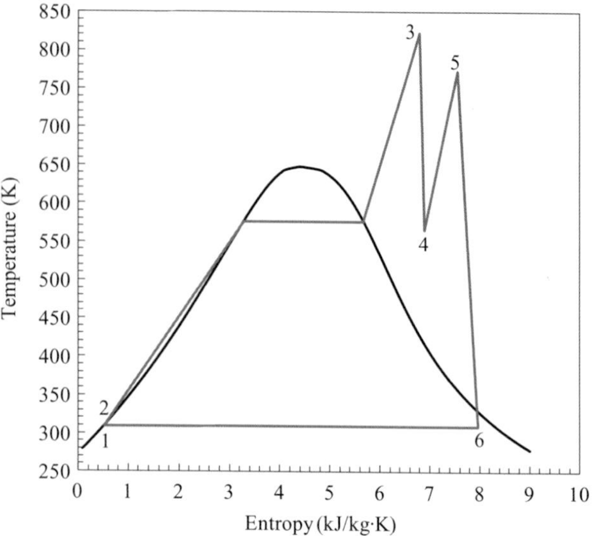

Figure 6.6 The T–s diagram of a Rankine cycle with a single-stage reheat and superheating; see Example 6.1. The working fluid states change in the following sequence: 1, 2, 3, 4, 5, 6, 1.

Figure 6.6. The total heat input and the turbine work, per unit mass of the working fluid flowing through the boiler, are

$$q_{in} = (h_3 - h_2) + (h_5 - h_{4r}) \tag{6.8}$$

and

$$w_T = h_3 - h_{4r} + (h_5 - h_6). \tag{6.9}$$

Note that the specific work increases and the steam quality at the backpressure of the low-pressure turbine rises. With higher superheat and reheat temperatures, a measurable increase in efficiency is observed. In cycles with large pressure ratios, several stages of reheat are often used. Reheat is easily attainable in fossil fuel plants, since the combustion temperature, O (1200 °C), is much higher than the critical temperature of steam, and the maximum reheat temperature is determined by what is allowable by the turbine material. In nuclear plants, the reactor coolant temperature may limit the superheat/reheat temperature. Nevertheless, the reactor coolant temperature is sufficient, especially in gas-cooled reactors. Multistage reheat reduces some of the cycle irreversibility by making the heat addition and the working fluid expansion occur at nearly isothermal conditions.

The improvement of the cycle efficiency as the turbine inlet pressure and temperature are raised, and as the pressure in the condenser is lowered, are shown in Figure 6.7 for the case of coal-fueled power plants [5]. Notice the improvement associated with moving from one-stage to two-stage reheat as well. The figure quotes values for high-efficiency supercritical cycles, whose detail will be described shortly. Raising the steam pressure and temperature at the turbine inlet leads, as expected, to important gains in efficiency. Similar gains are achieved by lowering the condenser pressure.

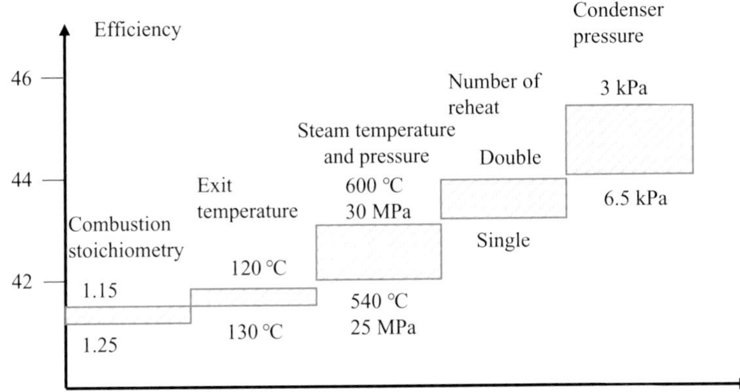

Figure 6.7 Improvement in the efficiency of pulverized coal-fired steam power plants as the cycle design parameters change. Note that the diagram shows supercritical cycles, but also shows a large change when the temperature and pressure of the steam at the turbine inlet are increased. There is also improvement with reheating and with decreasing the pressure of the condenser [5].

Example 6.1

In a steam power plant (Figure E6.1) with a single reheat stage, steam at 550 °C and 9 MPa enters a high-pressure turbine, where it is expanded to 1.5 MPa. It is then reheated up to 500 °C, and further expanded to 6 kPa in a low-pressure turbine. The isentropic efficiencies of the high- and low-pressure turbines are 92% and 89%, respectively. The isentropic efficiency of the pump is 72%. Determine: (a) the steam quality at the exit of the low-pressure turbine; (b) the specific work output of the power plant; (c) the amount of heat transfer in the boiler; and (d) the cycle efficiency.

Solution

The first step is to evaluate the enthalpy at all states:

$$\text{State 1: } \left.\begin{array}{l} p_1 = 6\,\text{kPa} \\ \text{Saturated liquid} \end{array}\right\} \Rightarrow h_1 = 151.5\,\text{kJ/kg};\, v_1 = 0.001\,\text{m}^3/\text{kgtw.}$$

$$\text{State 2: } h_{2s} = h_1 + v_1(p_2 - p_1) = 151.5 + 0.001 \times (9000 - 6) = 160.5\,\text{kJ/kg}$$

$$h_2 = h_1 + \frac{h_{2s} - h_1}{\eta_P} = 151.5 + \frac{160.5 - 151.5}{0.72} = 164\,\text{kJ/kg.}$$

$$\text{State 3: } \left.\begin{array}{l} T_3 = 550\,^\circ\text{C} \\ p_3 = 9\,\text{MPa} \end{array}\right\} \Rightarrow h_3 = 3511\,\text{kJ/kg};\, s_3 = 6.814\,\text{kJ/kg·K.}$$

$$\text{State 4: } \left.\begin{array}{l} p_4 = 1.5\,\text{MPa} \\ s_{4s} = s_3 \end{array}\right\} \Rightarrow h_{4s} = 2979\,\text{kJ/kg}$$

$$h_4 = h_3 - \eta_{T_{HP}}(h_3 - h_{4s}) = 3511 - 0.92 \times (3511 - 2979) = 3021.6\,\text{kJ/kg.}$$

$$\text{State 5: } \left.\begin{array}{l} T_5 = 500\,^\circ\text{C} \\ p_3 = 1.5\,\text{MPa} \end{array}\right\} \Rightarrow h_5 = 3473\,\text{kJ/kg};\, s_5 = 7.569\,\text{kJ/kg·K.}$$

$$\text{State 6: } \left.\begin{array}{l} p_6 = 6\,\text{kPa} \\ s_{6s} = s_5 \end{array}\right\} \Rightarrow h_{6s} = 2332\,\text{kJ/kg}$$

Example 6.1 (cont.)

$$h_6 = h_5 - \eta_{T_{LP}}(h_5 - h_{6s}) = 3473 - 0.89 \times (3473 - 2332) = 2457.5\,\text{kJ/kg}.$$

a. To determine the quality of steam at state 6, we note that $h_f < h_6 < h_v$, where

$$h_f = h_{p=6\,kPa,x=0} = h_1 = 151.5\,\text{kJ/kg},$$
$$h_v = h_{p=6\,kPa,x=1} = 2566\,\text{kJ/kg}.$$

Hence, $x = \dfrac{h_6 - h_f}{h_v - h_f} = \dfrac{2457.5 - 151.5}{2566 - 151.5} = 0.955.$

b. The specific work produced by the turbines is

$$W_T = W_{HPT} + W_{LPT} = (h_3 - h_4) + (h_5 - h_6) = 1504.9\,\text{kJ/kg}.$$

The specific work consumption by the feedwater pump is $W_P = h_2 - h_1 = 12.5\,\text{kJ/kg}$. The specific net work production is therefore $W_{net} = W_T - W_P = 1492.4\,\text{kJ/kg}$.

c. The heat input per unit mass of the steam is calculated as follows:

$$q_{in} = (h_3 - h_2) + (h_5 - h_4) = 3798.4\,\text{kJ/kg}.$$

d. The thermal efficiency of the power plant is $\eta_{th} = \frac{W_{net}}{q_{in}} = 0.393$ or 39.3%.

The operation of the cycle on a T–s diagram is depicted in **Figure 6.7**.

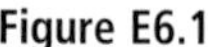
Figure E6.1

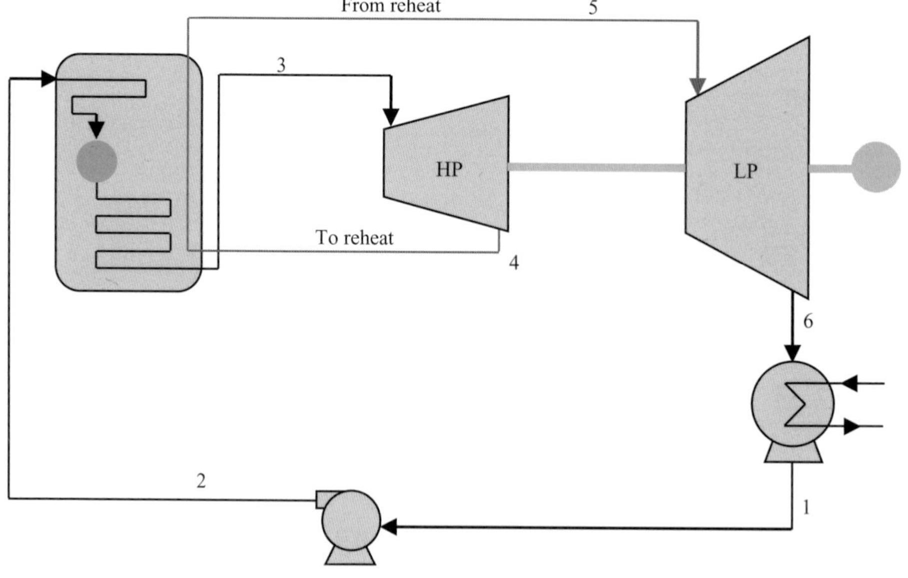

6.6 Regeneration

Regeneration is also widely used in steam turbine cycles. Regeneration is utilized in single or multiple stages, depending on the overall pressure ratio and turbine size. Similar to gas turbine cycles, regeneration offers a good opportunity to improve the cycle efficiency. With multistage regeneration, and under idealized conditions, the efficiency can reach that of the Carnot cycle. One major source of irreversibility in Rankine cycles, in this case an external irreversibility associated with the heat transfer between the external source and the cycle, is heating the feedwater from the condenser temperature to that of the saturation temperature inside the boiler. Internal regeneration, or heating up the feedwater from the condenser temperature to the boiler temperature internally and without using an external heat source, overcomes this problem. If the feedwater is pumped gradually while it is being heated by steam extracted at higher temperatures from the turbine, one can achieve "maximum" regeneration between the two saturation temperatures. Figure 6.8 shows the T–s diagram of a cycle with an infinite number of regeneration steps. It can be shown that this cycle achieves the Carnot efficiency. Note that only a small fraction of the steam in the turbine is required for heating the feedwater since the latent heat is much higher than the sensible heat. Thus, the work penalty is small while the efficiency gain is high [6].

To implement regeneration in a Rankine cycle, a fraction of the thermal energy of the steam that is expanding through the turbine is transferred to the feedwater downstream of the condenser, before or following the pumping stage. There are three different schemes for regeneration implemented using one of the following scenarios [6]:

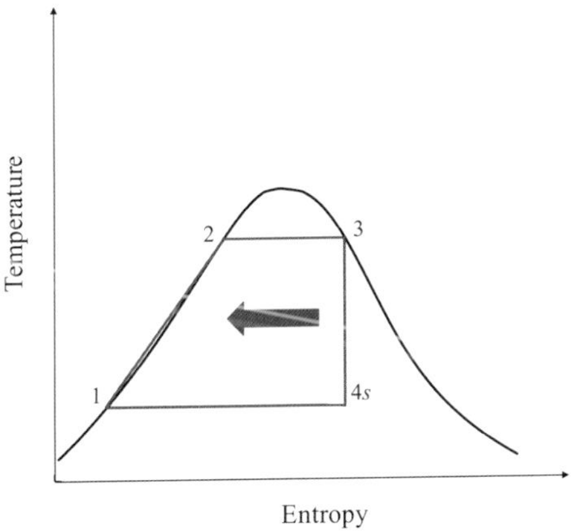

Figure 6.8 Rankine cycle with infinite number of regeneration steps, in which steam extracted from the turbine is used to heat the feedwater from the condenser temperature to the boiler saturation temperature. The completely regenerative cycle has the same efficiency as the Carnot cycle operating between the condenser saturation temperature and the boiler saturation temperature. The big arrow in the diagram indicates heat transfer between the expanding steam in the turbine and the feedwater heater. During process 1–2, the feedwater is gradually being pumped while it is heated. The boiler inlet is now in state 2.

1. An open feedwater heater in which steam is extracted from an intermediate stage in the turbine is mixed with the condensed water after it has been pumped to the same pressure. The heated water is then pumped to the boiler pressure for the next stage of regeneration.
2. A closed feedwater heater uses the steam extracted from an intermediate stage in the turbine to heat the feedwater after it has been pumped to the boiler pressure. The cooled steam is then throttled back to the condenser pressure and mixed with the feedwater leaving the condenser.
3. A closed feedwater heater in which the steam extracted from an intermediate stage in the turbine is used to heat the feedwater after it has been pumped to the boiler pressure. The cooled steam is then pumped up to the boiler pressure and feedwater leaving the heater.

The detailed implementation of these three schemes is described next, along with the advantages and disadvantages of each. In most power plants, several stages of regeneration are used, with possibly different types of feedwater heaters in each stage.

6.6.1 Open Feedwater Heaters

In the first form of regeneration, a fraction of the steam flowing through the turbine is bled off at an intermediate pressure, and mixed with the water flowing between a two-stage pump to raise its enthalpy in an open or direct contact feedwater heater, as shown in Figure 6.9. In this case, steam bled from the turbine at state 6 is mixed with water at state 2 to produce saturated water at state 3.

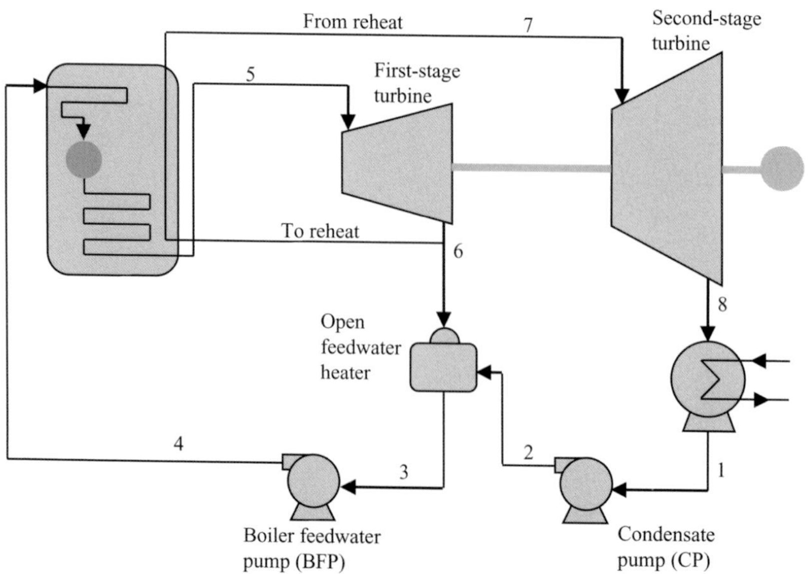

Figure 6.9 The layout and components of a reheat cycle with regeneration using an open feedwater heater.

The pressures of all these states are the same, and the fraction of steam required in this regeneration step is obtained from the following energy balance, where α is the fraction of steam bled from the turbine:

$$\alpha h_6 + (1 - \alpha)h_2 = h_3. \tag{6.10}$$

Note that regeneration reduces the heat rejected from the cycle in the condenser, since only $(1 - \alpha)$ of the steam that enters the high-pressure turbine passes through the condenser. Furthermore, this regeneration step raises the starting temperature in the heat addition process (boiler) from T_2 to T_3; that is, to the saturation temperature of the feedwater heater. By raising the temperature at which external heat is added to generate steam, some of the irreversibility in the cycle is reduced. The heat added externally is

$$q_{in} = (h_5 - h_4) + (1 - \alpha)(h_7 - h_8). \tag{6.11}$$

The exit state from the feedwater heater must be saturated or slightly subcooled liquid, since the pump must not handle two-phase mixtures. Clearly, the turbine work is reduced since only $(1 - \alpha)$ of the steam passes through the second stage of the turbine, per unit mass of the water passing through the boiler:

$$w_T = (h_5 - h_6) + (1 - \alpha)(h_7 - h_8). \tag{6.12}$$

Using a direct feedwater heater guarantees that the exit temperature of the two streams is the same, as mixing occurs between two streams with the same pressure. Thus, it does not lead to extra losses. By repeating this step several times at higher and higher pressures, the temperature of the water approaches the temperature at which external heat is added in the boiler, before entering the steam generator. Note that with a very large number of recuperation stages, the pumping line follows the left side of the vapor dome, and the cycle efficiency approaches that of the Carnot efficiency since all the external heat is added isothermally during evaporation, while heat is rejected isothermally in the condenser. Open feedwater heaters, however, require an extra large pump for each feedwater heater, as shown in Figure 6.10. While this is the most efficient regeneration, the requirement for a large pump following each open feedwater heater limits the use of this form of recuperation to one line or to one pressure stage. Other recuperation stages in the cycle utilize closed feedwater heaters, as shown next. In practice, the incremental improvement in efficiency diminishes as more regeneration steps are used.

Figure 6.10 shows the T–s diagram of this regenerative cycle. In the example shown in Table 6.1, if the intermediate pressure at which regeneration is applied is chosen to be the same as the reheat pressure, the cycle efficiency improvement is about 10% over that of the simple cycle. Studies show that reheat and regeneration applied at the same pressure yields best results for efficiency improvement. The choice of the pressures at which multiple reheat and multiple regeneration are applied, together or separately, depends on the conditions and cycle design, and requires optimization analysis for individual cycles.

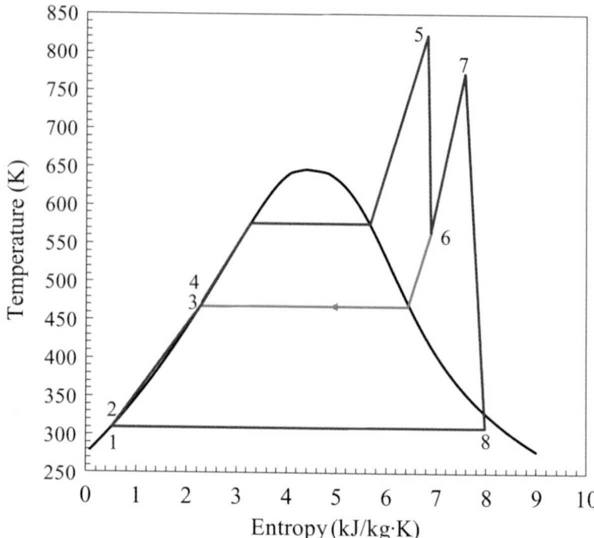

Figure 6.10 The temperature–entropy diagram of a regenerative reheat Rankine cycle with an open feedwater heater. The layout of this cycle is shown in Figure 6.9.

Example 6.2

Consider the power plant of Example 6.1 that is now equipped with an open feedwater heater, so its layout is the same as shown in Figure 6.9. The isentropic efficiency of the condensate pump is 75%. A fraction of steam leaving the first-stage turbine is used in the open feedwater heater, and the rest is reheated. Determine: (a) the fraction of steam used in the heater; (b) the specific work output of the power plant; and (c) the thermal efficiency of the power plant.

Solution

States 1, 5, 6, 7, and 8 are known from Example 6.1. Hence,
State 2: $h_{2s} = h_1 + v_1(p_2 - p_1) = 151.5 + 0.001 \times (1500 - 6) = 153$ kJ/kg,

$$h_2 = h_1 + \frac{h_{2s} - h_1}{\eta_{P_1}} = 151.5 + \frac{153 - 151.5}{0.75} = 153.5 \text{ kJ/kg}.$$

State 3: $h_3 = h_{sat.liq}(p_3) = 844.9$ kJ/kg, $v_3 = v_{sat.liq}(p_3) = 0.00115$ m^3/kg.
State 4: $h_{4s} = h_3 + v_3(p_4 - p_3) = 853.5$ kJ/kg, $h_4 = h_3 + \frac{h_{4s} - h_3}{\eta_{P_2}} = 856.8$ kJ/kg.

a. The fraction of steam extracted from the first-stage turbine that is used in the open feedwater heater is calculated using (6.10): $\alpha = \frac{h_3 - h_2}{h_6 - h_2} = 0.241$.

b. $W_T = (h_5 - h_6) + (1 - \alpha)(h_7 - h_8) = 1260.2$ kJ,

$$W_P = (h_4 - h_3) + (1 - \alpha)(h_2 - h_1) = 13.4 \text{ kJ/kg}.$$

The specific net work production is $W_{net} = W_T - W_P = 1246.8$ kJ/kg.

Example 6.2 (cont.)

c. The amount of heat transfer in the boiler per unit mass of the steam is

$$q_{in} = (h_5 - h_4) + (1 - \alpha)(h_7 - h_6) = 2996.8 \text{ kJ/kg}.$$

We now calculate the thermal efficiency of the power plant as $\eta_{th} = \frac{W_{net}}{q_{in}} = 0.416$ or 41.6%.

6.6.2 Closed Feedwater Heater

In another regenerative cycle design, the steam bled from the turbine at an intermediate pressure is used in a closed feedwater heater, or a shell and tube heat exchanger, to raise the feedwater temperature to a value near the saturation temperature of the bled steam, as shown in Figure 6.11. The cooled steam is then throttled down to the pressure of the condenser and mixed with the condensing steam. In the case of multistage regeneration using this type of closed feedwater heater, the bled steam is throttled back to the pressure of the lower-pressure closed feedwater heater and used in the lower-pressure feedwater

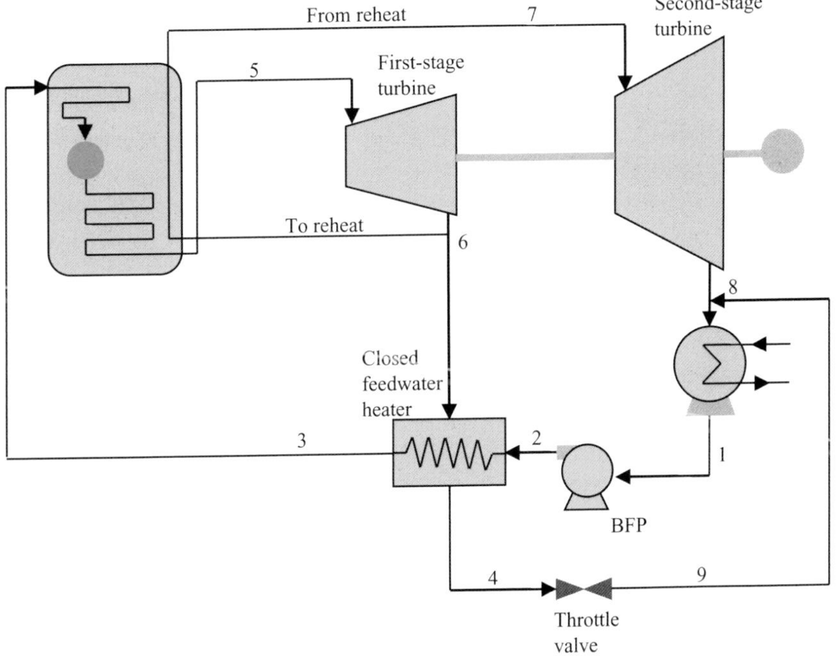

Figure 6.11 The layout and components of a reheat cycle with regeneration using a closed feedwater heater cascading backward.

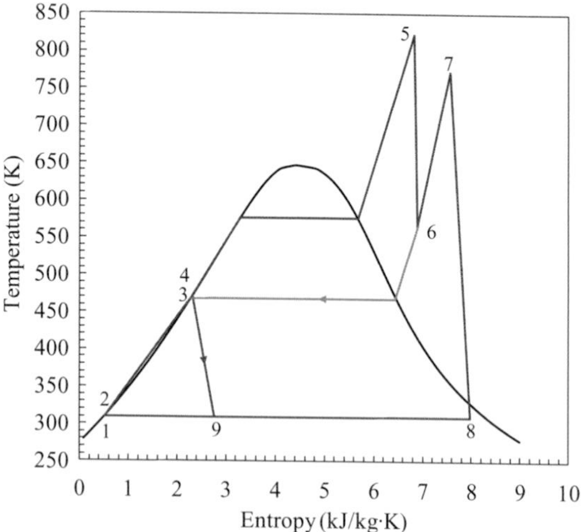

Figure 6.12. The temperature–entropy diagram of the cycle shown in Figure 6.11 – that is, a regenerative reheat Rankine cycle employing a closed feedwater heater cascading backward.

heater again. This **cascading backward feedwater heater** does not require the extra large pump in the feedwater line that is required in the open feedwater heater case. However, it requires a large surface heat exchanger, and it suffers more irreversibility than the open feedwater heater because of the pressure loss in the throttling process and the temperature difference across the heat exchanger. As such, more heat is rejected in the condenser, as can be seen in the cycle layout. Figure 6.12 shows the *T*–*s* diagram for a cascading backward feedwater heater [6].

In most applications with a closed feedwater heater, a single open feedwater heater is employed along with a number of closed feedwater heaters. The open heater acts both as a regenerator and as an aerator to rid the steam of air that leaks into the working fluid at different stages, especially in the condenser, where a strong vacuum is created.

Several backward-cascading closed feedwater heaters can be used without increasing the number of required pumps, and the bled steam is mixed at the lower pressure heaters or the condenser. These closed feedwater heaters are less efficient than the open type since they require a relatively larger temperature difference between the two streams, and the exit water stream temperature cannot reach the temperature of the inlet steam stream, or in most cases even the saturation temperature of the bled steam. Moreover, some of the enthalpy of the bled steam is eventually rejected in the condenser and the throttling process.

The difference between the maximum temperature of the feedwater water exiting the heater and the saturation temperature of the bled steam is known as the "**pinch point**" or PP. As shown in Figure 6.13, the pinch point is the smallest temperature difference between the steam/condensate and the feedwater, or between the stream being cooled and that being heated. Large values of PP lead to large heat transfer irreversibility, while small values require large heat transfer area within the heater. Feedwater heaters use a PP of 3–5 °C. Pinch point analysis, common in heat transfer between single-phase and multiphase streams, is important in minimizing heat exchanger size while keeping heat transfer irreversibility

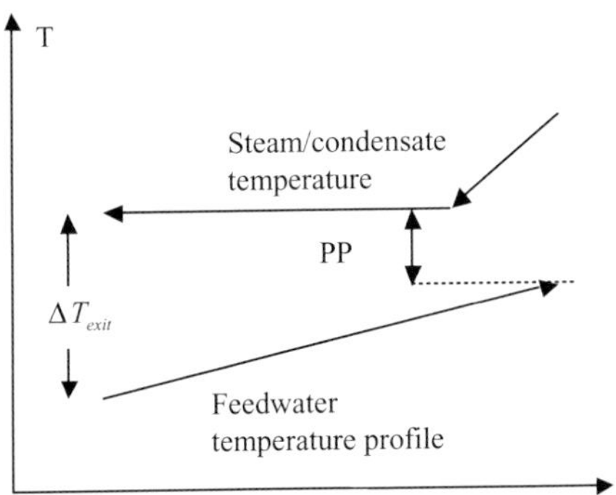

Figure 6.13 Temperature profiles of the bled steam and the feedwater in a counter-flow closed feedwater heater. Arrows indicate the directions of the temperature change. PP is the smallest temperature difference inside the heat exchanger, and ΔT_{exit} is the largest.

small, and will be discussed again in comparing subcritical and supercritical cycles, and in the design of combined cycles.

Example 6.3

Consider the power plant of Example 6.1 that is now equipped with a closed feedwater heater, so its layout is the same as that shown in Figure 6.12. A fraction of steam leaving the first-stage turbine is used in the open feedwater heater, and the rest is reheated. The pinch point is 5 °C. Determine: (a) the fraction of steam used in the heater; (b) the specific work output of the power plant; and (c) the thermal efficiency of the power plant.

Solution

States 1, 2, 5, 6, 7, and 8 are known from Example 6.1.
State 4: $h_4 = h_{sat.liq}(p_4) = h_{sat.liq}(1.5\,MPa) = 844.9$ kJ/kg,

$$T_4 = T_{sat.liq}(p_4) = 471.5\text{ K}.$$

State 3: $\left.\begin{array}{l} p_3 = p_2 = 9\,\text{MPa} \\ T_3 = T_4 - PP = 466.5\,\text{K} \end{array}\right\} \Rightarrow h_3 = 825.9\text{ kJ/kg}.$

a. The fraction of steam extracted from the first-stage turbine that is used in the closed feedwater heater is obtained by applying the First Law to the heater:

$$\alpha(h_6 - h_4) = h_3 - h_2 \Rightarrow \alpha = \frac{h_3 - h_2}{h_6 - h_4} = \frac{825.9 - 164}{3021.6 - 844.9} = 0.304.$$

b. $W_T = (h_5 - h_6) + (1 - \alpha)(h_7 - h_8) \;= 1196.2\text{ kJ}$

$$W_P = h_2 - h_1 = 164 - 151.5 = 12.5\text{ kJ/kg}$$

$$W_{net} = W_T - W_P = 1196.2 - 12.5 = 1183.7\text{ kJ/kg}.$$

Example 6.3 (cont.)

c. $q_{in} = (h_5 - h_3) + (1 - \alpha)(h_7 - h_6) = 2996.3\ \text{kJ/kg}$

$$\eta_{th} = \frac{W_{net}}{q_{in}} = \frac{1196.2}{2999.3} = 0.399 \quad \text{or} \quad 39.9\%.$$

Comparing with the efficiency of the power plant of Example 6.1, including a closed feedwater heater improves raises the efficiency by 0.6 percentage point. The operation of the cycle on a T–s diagram is depicted in Figure 6.13.

Another type of closed feedwater heater is known as the **cascading forward closed feedwater heater**, in which following the heat exchange in the heater, a pump is used to raise the pressure of the bled stream to that of the boiler pressure before mixing the two streams downstream of the heater. The layout of this scheme is shown in Figure 6.14. Because a small fraction of the total flow rate through the turbine is bled for regeneration, only a small pump is required to raise the pressure of the bled stream to that of the high-pressure water stream. This arrangement is more efficient than the cascading backward heater arrangement since no pressure loss is encountered, and the extra enthalpy in the bled steam is added to the water stream instead of rejecting it in the condenser. However, the efficiency improvement comes at the cost of adding a small pump for each feedwater heater. Similar to the other closed feedwater heater case, one open heater is often

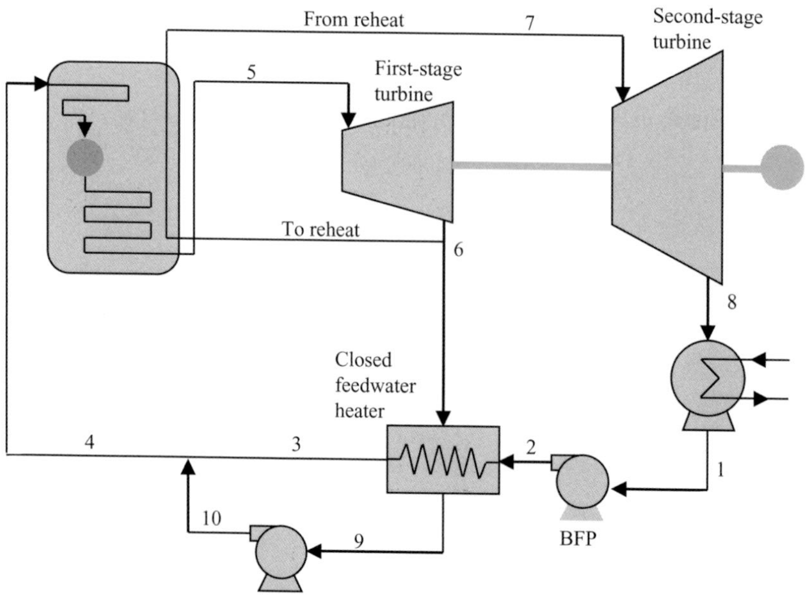

Figure 6.14 The layout and components of a reheat cycle with regeneration using a closed feedwater heater cascading forward.

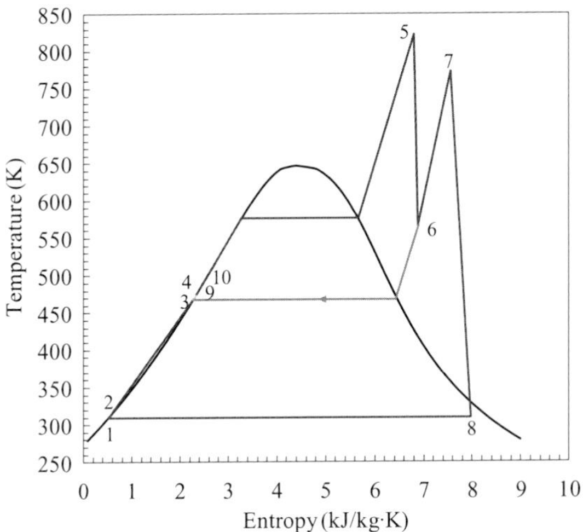

Figure 6.15 The temperature–entropy diagram of a recuperative reheat Rankine cycle employing a closed feedwater heater cascading forward. Note how close states 3, 4, 9, and 10 are on the T–s diagram; see Example 6.4.

employed for aeration purposes, along with a number of closed heaters. The T–s diagram of the regenerative cycle is shown in Figure 6.15.

The difference in the efficiency between the closed forward and the closed backward-cascading heaters is small, and the choice is often made using a careful compromise between the added cost and hardware complexity, and the cost of the extra fuel. The backward-cascading heaters are employed more frequently because of convenience, except for the lowest pressure heater, which is mostly a forward-cascading heater. The latter avoids rejecting the thermal energy accumulated in the bled streams, and instead adding it to the water stream.

Example 6.4

Consider Example 6.3, in which the closed feedwater heater is a cascading forward heater (see Figure 6.15). The pinch point is 5 °C, and the isentropic efficiency of the second pump is 75%. Determine: (a) the fraction of steam used in the heater; (b) the specific work output of the power plant; and (c) the thermal efficiency of the power plant.

Solution

States 1, 2, 3, 5, 6, 7, 8, and 9 are known from Example 6.3.

a. The fraction of steam extracted is obtained by applying the First Law to the heater:

$$\alpha(h_6 - h_9) = (1 - \alpha)(h_3 - h_2) \Rightarrow \alpha = (h_3 - h_2)$$

$$\alpha = \frac{h_3 - h_2}{(h_6 - h_9) + (h_3 - h_2)} = \frac{825.9 - 164}{(3021.6 - 844.9) + (825.9 - 164)} = 0.233.$$

b. $W_T = (h_5 - h_6) + (1 - \alpha)(h_7 - h_8) = 1268.3$ kJ.

Example 6.4 (cont.)

To determine the work requirement of the pumps, we need the enthalpy at state 10:

$$h_{10s} = h_9 + v_9(p_{10} - p_0) = 844.9 + 0.00115 \times (9000 - 1500) = 853.5 \text{ kJ/kg},$$

$$h_{10} = h_9 + \frac{h_{10s} - h_9}{\eta_{P_2}} = 844.9 + \frac{853.5 - 844.9}{0.75} = 856.4 \text{ kJ/kg},$$

$$W_P = (1 - \alpha)(h_2 - h_1) + \alpha(h_{10} - h_9) = 12.3 \text{ kJ/kg},$$

$$W_{net} = W_T - W_P = 1268.3 - 12.3 = 1256 \text{ kJ/kg}.$$

c. The enthalpy of water at state 4:

$$h_4 = (1 - \alpha)h_3 + \alpha h_{10} = (1 - 0.233) \times 825.9 + 0.233 \times 856.4 = 833 \text{ kJ/kg},$$

$$q_{in} = (h_5 - h_4) + (1 - \alpha)(h_7 - h_6) = 3024.2 \text{ kJ/kg},$$

$$\eta_{th} = \frac{W_{net}}{q_{in}} = \frac{1256}{3024.2} = 0.415 \quad \text{or} \quad 41.5\%.$$

Comparing with the efficiency of the power plant of Example 6.3, we see that the cascading forward heater is better than the cascading backward heater.

Reheat and recuperation are standard technologies in large steam power plants. Using a modest number of stages of reheat and recuperation, the efficiency improvement and hence fuel cost saving more than offsets the extra cost of the hardware. The impact of superheating, reheating, and regeneration on the efficiency are significant, although the difference between the different types of feedwater heaters is small. Supercritical cycles have a strong advantage in efficiency compared to subcritical cycles.

Steam turbine cycles, or in general two-phase Rankine cycles, have several advantages over gas turbine cycles, including their fuel flexibility:

1. Rankine cycles operate in a closed cycle mode, and reject heat isothermally at temperatures close to that of the environment.
2. In a Rankine cycle, there is no contact between the working fluid and the combustion products; there is no need to purify the fuel or the combustion products.
3. They achieve high efficiency even when using cooler high-temperature sources such as solar thermal energy, since the pumping work of the liquid is much smaller than the expansion work of the gas.
4. They utilize relatively low working fluid flow rate, since heat is added in part to evaporate the fluid.
5. Because they work well with cooler high-temperature sources, they are suitable to intermediate-grade thermal energy sources, such as nuclear reactors, concentrated solar thermal (300–550 °C), and lower-grade geothermal sources (200 °C). The latter often uses organic fluids as working fluids because their critical temperatures are low.

However, steam cycles and their components tend to have large thermal and mechanical inertia, and more significantly, they require large amount of cooling in the condenser.

6.7 Supercritical Two-Phase Rankine Cycles

Many steam Rankine cycle plants used in electricity generation burn coal to provide the necessary heat for the cycle, and many others utilize thermonuclear energy as a source. To raise the cycle efficiency, running the steam cycle of these plants under supercritical conditions has become more popular. In a supercritical cycle, the pressure and temperature of steam at the inlet of the turbine are set above their critical values, to raise the cycle efficiency. Thus, it is worth spending some time analyzing the advantages and costs of these cycles.

Supercritical cycles operate at maximum pressures higher than the critical pressure of the working fluid and raise the temperature above the critical temperature. The lower pressure remains at very low values, compatible with the condenser cold temperature. In steam power plants, the pump must raise the water pressure above 220.88 atm, and the steam temperature above the critical temperature, 648 K, before the first turbine stage. To minimize the water content in the last-stage turbine exhaust, substantial superheat is also used in supercritical cycles (see Figure 6.1 for the T–s diagram of water). Supercritical cycles also employ several reheat stages to further improve the efficiency and increase the steam quality at the exit of the final turbine stage, as well extensive regeneration using feedwater heating. Pressures at the early turbine stages can exceed 300 bar, with temperatures above 500 °C requiring special attention in the design of the turbine components, especially in the early stages.

Supercritical cycles achieve higher efficiency not only because of the higher pressure and temperature of the working fluid, but also by minimizing "external irreversibilities" associated with heat transfer from the heat source to the steam in the steam generator, or the heat recovery steam generator in the case of combined cycles. The external irreversibilities refer to losses "outside" the steam cycle itself, and associated with the process of heat addition to the steam cycle. Because the heat source is a stream of hot combustion products or a stream of a nuclear reactor coolant, the process of steam generation involves availability loss due to the temperature difference between the two streams. Minimizing the temperature difference between the hot stream and the cold stream all along their paths is necessary to minimize the associated availability losses. It is also important to minimize the temperature difference between the two streams at the inlet and exit of the steam generator.

Figure 6.16 shows schematically the temperature profiles in a subcritical cycle and in a supercritical cycle during the steam generation processes. The figure shows schematically the temperature profile throughout the steam generation process for the heat source, in this case combustion products, and for the stream being heated, in this case water, two-phase mixture, and superheated steam. The temperature profiles in the subcritical case show a much larger temperature difference between the cooling exhaust stream and that of the

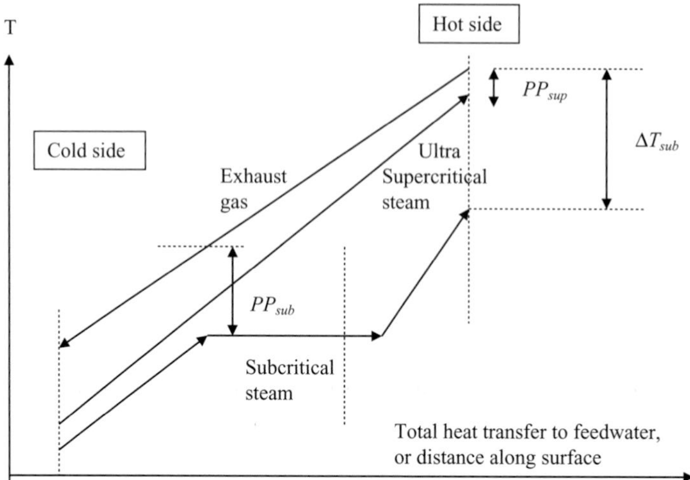

Figure 6.16 Schematic diagram for the temperature plotted against the heat transfer to the feedwater stream, or the distance along the heat transfer surface, for the cooling exhaust gases and the warming water/steam in subcritical and ultra-supercritical cycles, in a counter-flow heat transfer process. The arrows indicate the direction of temperature change for each stream. The pinch point is identified in each case. The profile for supercritical steam lies between the two curves shown in the figure. Here PP_{sup} is the temperature difference between the steam exiting the steam generator and the hot gases under supercritical conditions, while ΔT_{sub} is the same temperature difference in the case of a subcritical cycle. In the latter case, that difference results from a PP_{sub} temperature difference at the pinch point.

water–steam stream on the hot section, because of the constant-temperature phase-transition region. The temperature difference between the two streams, in a nonwork-producing device, amount to an availability loss, and the higher the difference is the higher this loss becomes. Moreover, the large temperature difference between the exiting steam and the entering hot gases, indicated on the diagram by ΔT_{sub}, leads to a large loss of availability for the cycle, since the temperature of the steam entering the turbine is much lower that the highest available temperature in the hot products.

The need for minimizing the heat transfer-associated availability loss within the steam generator is stronger when using low-temperature sources, since the overall cycle efficiency is low to begin with. In this case, choosing working fluids with relatively low critical temperatures, which makes it possible to operate a supercritical cycle, is important. These choices will be discussed later in the chapter.

6.7.1 Pinch Point Analysis

The local enthalpy flux, $\dot{H}_g$, for the hot combustion gas or the hot source in general (it could be liquid in the case of a nuclear reactor or a solar concentrator) is given by

$$\dot{H}_g = \dot{m}_g c_{pg} \left(T_g - T^o \right), \tag{6.13}$$

where the mass flow rate of the hot gas is $\dot{m}_g$, its specific heat is c_{pg}, the local temperature is T_g, and T^o is a reference temperature. For the steam, the enthalpy flux, $\dot{H}_{st}$, is given by

$$\dot{H}_{st} = \dot{m}_{st} h_{st}, \tag{6.14}$$

where the mass flow rate of steam is $\dot{m}_{st}$ and its local enthalpy is h_{st}. Assuming an ideal gas behavior outside the dome, then

$$\dot{H}_{st} = \dot{m}_{st} c_{ps}(T_{st} - T^o). \tag{6.15}$$

Equations (6.13)–(6.15) show the linear relation between the enthalpy flux and the temperature for constant mass flow rates and constant heat capacity. The convective heat transfer flux between the two streams is also proportional to the temperature difference, since

$$\dot{q}_{ht} = \kappa(T_g - T_{st}). \tag{6.16}$$

In this relation, κ is the overall convective heat transfer coefficient between the two streams. Clearly, a finite temperature difference must be maintained between the two streams to develop a sufficiently high heat flux from the hot stream to the cold stream. A compromise must hence be found between the size of the heat exchanger and the overall cycle efficiency. The total heat transfer flux is related to the enthalpy fluxes as follows:

$$\dot{q}_{ht} L_{ht} = -\frac{d\dot{H}_g}{dx} = -\dot{m}_g c_{pg} \frac{dT_g}{dx} \tag{6.17}$$

and

$$\dot{q}_{ht} L_{ht} = \frac{d\dot{H}_{st}}{dx} = \dot{m}_{st} \frac{dh_{st}}{dx}, \tag{6.18}$$

where L_{HT} is an effective length for the heat transfer area measured in the direction normal to the flux – that is, the differential element of the heat transfer element is $dA_{HT} = L_{HT} dx$, and dx is the differential length in the flow direction. For subcooled water and superheated steam,

$$\dot{q}_{ht} L_{ht} = \dot{m}_{st} c_{pst} \frac{dT_{st}}{dx}. \tag{6.19}$$

Equations (6.17) and (6.18) or (6.19) show that the temperature profiles of the gas and steam are strongly related, and once the mass flow rates are fixed, the temperature profiles are determined only by their boundary condition – that is, their value at the inlet of the heat exchanger.

In subcritical cycles, the smallest temperature difference, the PP, occurs at the point where evaporation starts, as shown in Figure 6.16. Away from this point, especially on the side of evaporation, the temperature difference increases substantially to the value indicated by ΔT_{sub} because of the constant temperature of the two-phase fluid during evaporation. The higher ΔT_{sub}, the lower the steam temperature at the exit of the heat recovery steam generator (HRSG), and the lower the steam cycle work becomes. On the other hand, in the supercritical cycle case, the smallest temperature difference occurs at the hot side of the heat transfer process, PP_{sup}, keeping the exit steam temperature much closer to the hot gas temperature on the hot side of the HRSG. Because of the nature of both curves in the supercritical case, and as seen from (6.17) and (6.19), the temperature difference remains nearly the same all along the heat transfer zone. Since $\Delta T_{sub} > PP_{sup}$, the availability loss in the subcritical cycle case is higher than that in the supercritical cycle case. We note here that even if the subcritical steam plant has the same maximum temperature as the supercritical plant, the supercritical plant achieves a higher efficiency because of the higher pressure and the lower losses in the heat transfer process inside the HRSG.

One approach to improve the efficiency of the pulverized coal plants is to raise the steam pressure and temperature. Subcritical steam cycles for coal applications reach efficiencies up to 38.5%. Supercritical cycles, which reach up to 45% efficiency, are particularly attractive in cases when pulverized coal is used as a fuel.

Despite the technical challenges and expense of constructing supercritical and ultra-supercritical steam cycles, especially for coal and natural gas, several of these power plants have been built, operating at 250–300 bar, and up to 600 °C, and achieved impressive efficiencies up to 49%, using mostly coal but some with natural gas.

6.8 Hypercritical Cycles

Brayton cycles operate at temperatures much higher than the critical temperature of working fluid, and pressures mostly lower than the critical pressure, in a regime in which the working fluid behaves essentially as an ideal gas. That is, Brayton gas cycles operate entirely above the supercritical region, to the far right of the critical point and the vapor dome on the T–s diagram. Gas cycles are not called supercritical cycles; this name is reserved in the literature for two-phase cycles. Supercritical cycles are two-phase cycles whose higher pressure is above the critical pressure, while the lower pressure is much lower than the critical pressure, and most often much below the atmospheric pressure. On the other hand, essentially gas cycles whose lower pressure is close to the critical pressure are called "hypercritical" cycles. The lower pressure can either be just below or just above the critical pressure of the working fluid, while the upper pressure is much higher than the critical pressure. Thus, hypercritical cycles are essentially gas cycles in which the lower pressure is very close to the critical pressure. Two examples of these hypercritical cycles are the Feher cycle and the Gohstjejn cycle, as shown in Figure 6.17. In both cases, the compressor work is lower than in a regular Brayton cycle since the working fluid is within the liquid phase or in the early stages of the supercritical fluid phase – that is, in a region where the constant pressure lines on the T–s or h–s diagram are close to each other. The characteristics of the constant pressure lines region are shown in Figure 6.18 for carbon dioxide.

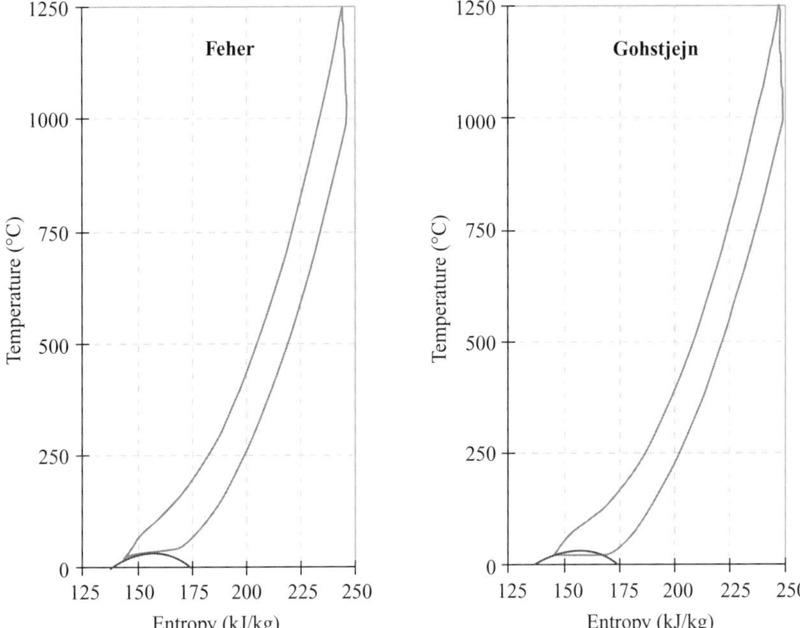

Figure 6.17 Two hypercritical CO_2 cycles: the "Feher" cycle, in which the lower pressure is slightly above the critical pressure, and the "Gohstjejn" cycle, in which the lower pressure is just below the critical pressure [7]. These cycles bridge the gap between supercritical Rankine cycles and Brayton gas cycles, and when regeneration is implemented their efficiency can potentially exceed 50% (without the complexity of combined cycles). Further developments on these cycles, which utilizes CO_2 as a working fluid under oxy-combustion conditions (for carbon dioxide capture), have been proposed and will be discussed in Chapters 8 and 11, such as the GRAZ and MATIANT cycles, among others.

6.8.1 Using CO_2 for Hypercritical Cycles

Hypercritical cycles are not practically possible when water is used as a working fluid because the critical pressure of water is high. They may, however, be practically possible for CO_2 because the critical pressure of carbon dioxide is a rather moderate 73 atm, and it is possible to design a 4:1 pressure ratio cycle above the critical pressure. Clearly, such cycles are meant to be closed cycles, and CO_2 will have to be cooled or condensed at the lower pressure before it is compressed to the higher pressure. In most planned applications, the working fluid, CO_2, would be heated in a nuclear reactor core. Because the critical temperature of CO_2 is 31 °C, the environmental conditions are almost compatible with the cooling needs. Hypercritical cycles, such as those shown in Figure 6.17, are excellent candidates for regeneration, across a wide temperature range, thus enabling heat addition near the higher temperatures while rejecting heat closer to ambient conditions. Note that the maximum temperatures attained in these cycles are much higher than those attained in steam cycles, while the maximum pressures are within the same range. Because of these special conditions, hypercritical cycles require special turbines.

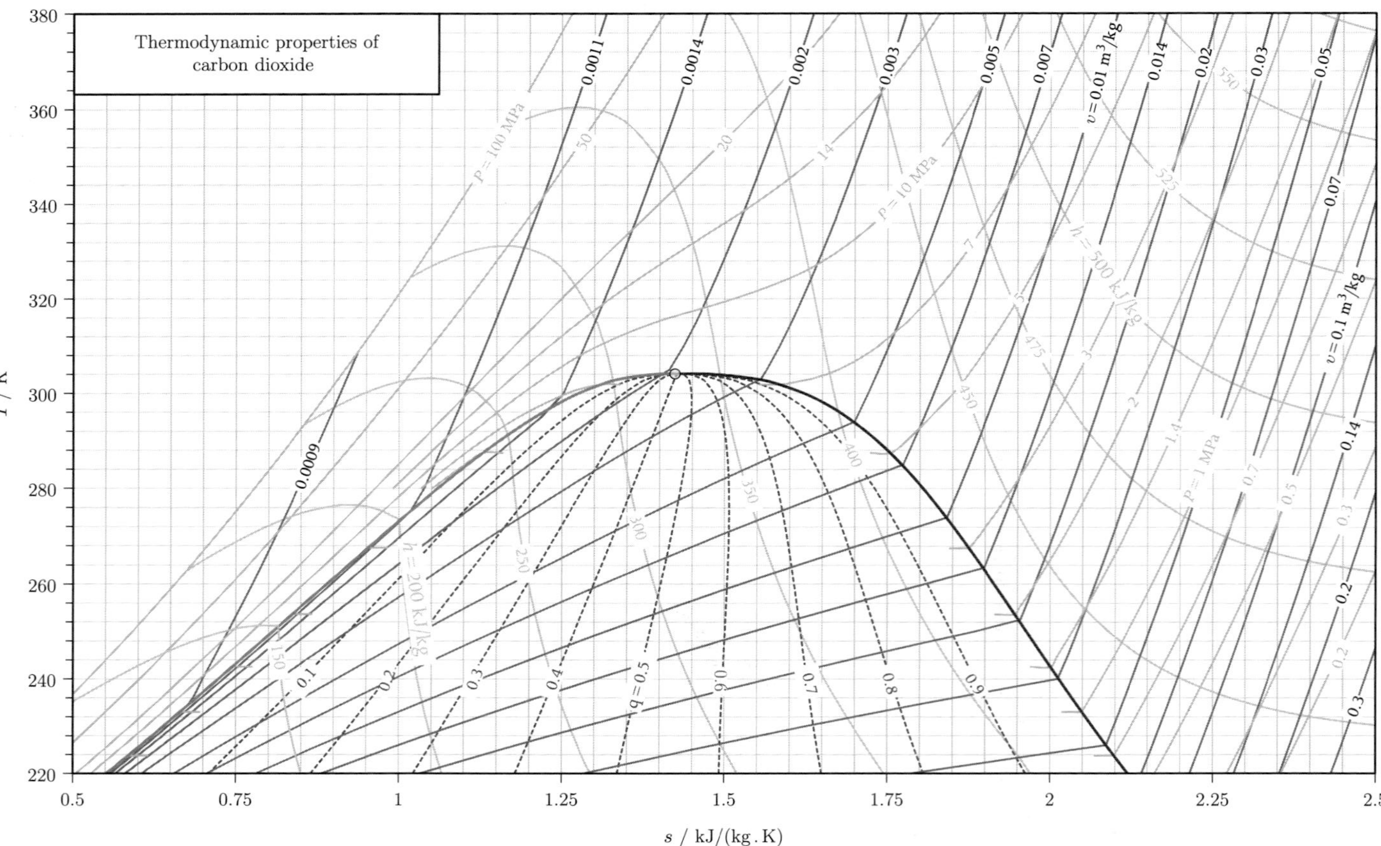

Figure 6.18 The T–s diagram of carbon dioxide [2].

Another improvement in the efficiency of supercritical cycles can be gained by employing split flow compression, as shown in Figure 6.19. To understand the benefit of this technique, remember that the specific heat of the fluid on the high-pressure side of the regenerator, immediately leaving the compressor and when the fluid is close to the critical point, is much higher than the specific heat of the fluid passing on the lower pressure of the regenerator as it leaves the turbine and enters the regenerator. The specific heat of the fluid is given by

$$c_p = \frac{T}{(\partial T/\partial s)_p}. \tag{6.20}$$

Isobars go through an inflection point on the T–s diagram at the critical point – that is $\left\{(\partial T/\partial s)_p\right\}_{critical} \to \infty$ – which boosts the value of the specific heat around the critical state, as shown in Figure 6.20 for CO_2. Therefore, in order to maintain a nearly fixed temperature

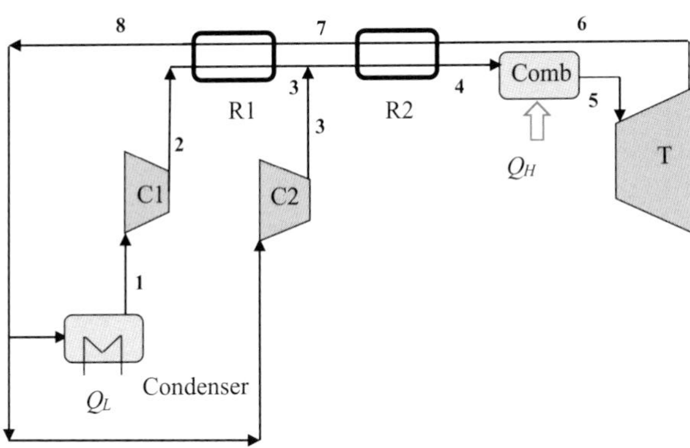

Figure 6.19 A hypercritical CO_2 cycle using split flow compression between two compressors and two regenerators; R1 and R2 are recuperators. Heat is supplied by a combustor (COMB.) or external heat source (nuclear reactor).

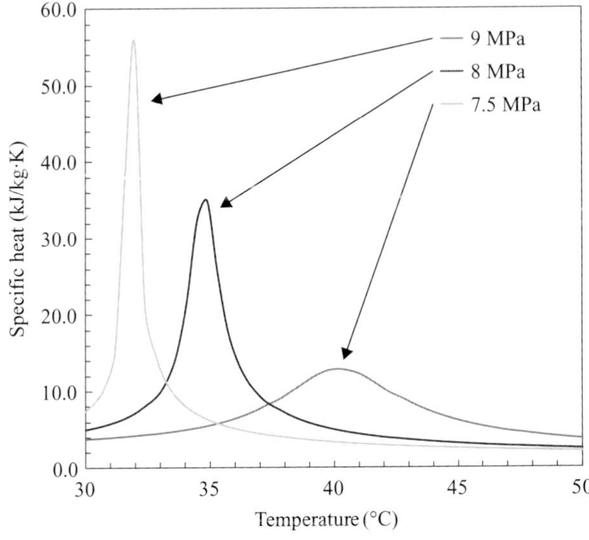

Figure 6.20 Constant-pressure specific heat of carbon dioxide at temperatures in the vicinity of the critical temperature (31 °C) and critical pressure, 79.9 atm. The specific heat reaches a maximum near the critical pressure and temperature.

difference between the two streams in the regenerator, between its inlet and its exit sections, a lower mass flow should be used on the high-pressure side, where the specific heat is higher than the mass flow rate on the low-pressure side, where the specific heat is lower. To achieve that, about one-third of the mass flow that passes through the lower-pressure side of the regenerator bypasses the high-pressure side by being diverted to a second compressor. Reducing the temperature difference in the heat exchanger lowers the irreversibility and improves the overall efficiency. The split cycle is described in more detail in Example 6.5.

6.8.2 CO_2 Transcritical Cycle for Nuclear Applications

Hypercritical or transcritical (hypercritical at lower pressure ratios) CO_2 cycles have been suggested for next-generation nuclear power plants that will employ higher-temperature nuclear reactor coolant [8]. While superheated and supercritical steam cycles show slow improvements in thermal efficiency above 550 °C, these hypercritical CO_2 cycles show continued measurable improvement beyond this point, as seen in Figure 6.21. Although labeled as supercritical on the diagram, these cycles satisfy the same conditions as hypercritical cycles defined here (i.e., their lower pressure is very close to the critical pressure of the working fluid and their high pressure is above the critical pressure). Because they operate in the hypercritical regime, they can reach higher efficiencies even at the relatively low temperatures of 600–800 °C. Given the critical pressure of CO_2, these cycles must operate in the range of 80–200 bar, mostly with lower pressure ratio of 2–4. They also operate within the range of 30–800 °C. Because of the high pressure and low temperature – that is, working near the critical point – the density of CO_2 is high, near 600 kg/m^3 (at 77 bar and 32 °C) at the compressor inlet, and close to 125 kg/m^3 (at 200 bar and 550 °C) at the turbine inlet. Such high density makes it possible to reduce the size of the gas turbine well below what is used in steam turbines, and even below combustion gas turbines that operate at higher temperatures and lower pressures.

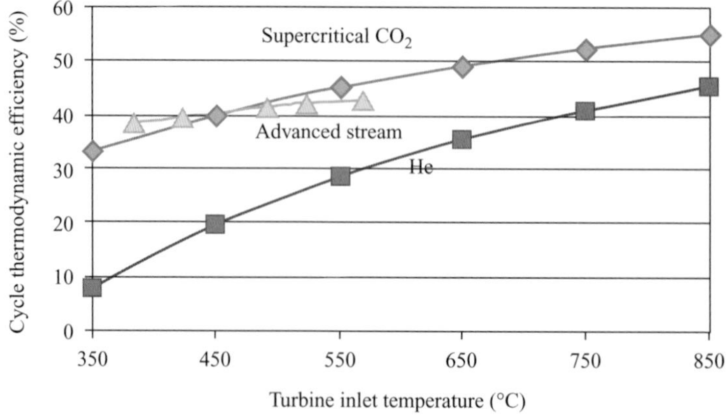

Figure 6.21 Thermal efficiency of a number of cycles within the low temperature range. Helium cycles are Brayton cycles, which can only achieve low efficiency at these low temperatures. Advanced steam cycles are superheated or supercritical steam cycles. Supercritical CO_2 cycles are actually hypercritical CO_2 cycles, similar to those shown in [8].

Both supercritical and hypercritical cycles benefit from the low compression work of a liquid or a supercritical fluid, and the large turbine work of a gas-like fluid. They also benefit from regeneration to improve the efficiency. Hypercritical cycles bridge the gaps between Brayton gas cycles and supercritical Rankine cycles.

Example 6.5

Consider a hypercritical cycle with CO_2 as the working fluid. The critical pressure and temperature of CO_2 are $P_c = 73.8$ bar and $T_c = 30.4\,°C$. The lowest temperature in the cycle (after heat rejection and before compression) is $T_L = 22\,°C$, and the highest temperature of the working fluid (after heat addition, before expansion) is $T_H = 800\,°C$. The low-pressure side runs at $P_L = 75$ bar, and the pressure ratio is 3.5. The isentropic efficiency of all compressors/pumps (the working fluid may be liquid or supercritical) is 80%, and the turbines have isentropic efficiencies of 90%. This applies to all three cycles below.

Please answer the corresponding questions for each cycle below.

Cycle 1: Simple Cycle (Figure E6.5a)

a. The net power produced per unit mass flow;
b. the heat addition to the cycle per unit mass flow; and
c. the thermal efficiency (First Law efficiency).

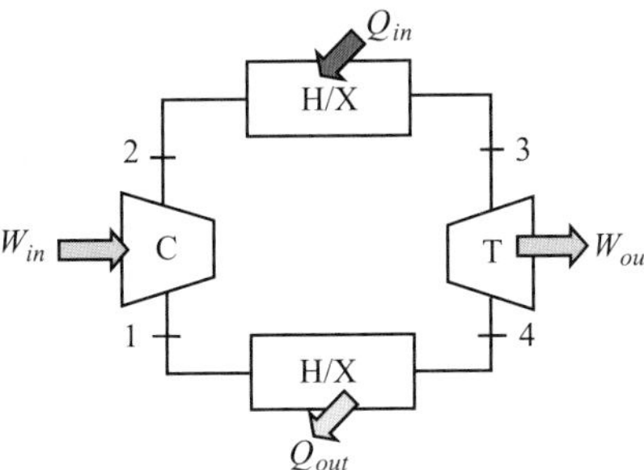

Figure E6.5a Simple Rankine cycle.

Cycle 2: Regenerative (Figure E6.5b)

The turbine exhaust in gas cycles is typically still hot in the simple cycle, and thus provides a good opportunity for regeneration. A regenerative cycle is shown in Figure E6.5b. Please calculate:

d. the net power produced per unit mass flow;
e. the heat addition to the cycle per unit mass flow; and
f. the thermal efficiency (First Law efficiency).

Note: The regenerator efficiency is 75% in this case, and is defined as

$$\eta_{regen} = \frac{T_{cold\_side\_out} - T_{cold\_side\_in}}{T_{hot\_side\_in} - T_{cold\_side\_in}} = \frac{T_x - T_2}{T_4 - T_2}.$$

Example 6.5 (cont.)

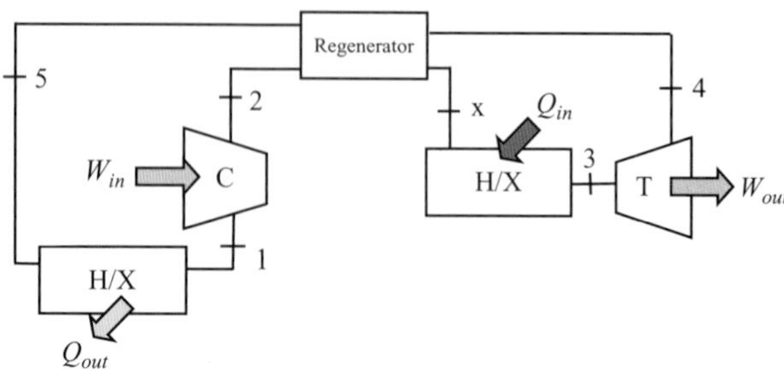

Figure E6.5b Regenerative Brayton cycle.

Cycle 3: Split Flow Compression Cycle (Figure E6.5c)

Finally, consider the split flow compression cycle shown in Figure E6.5c. This cycle improves efficiency by reducing the stream-to-stream temperature difference in the regenerators. Since the fluid leaving the compressor/pump at state 2 is near the critical point, it has a much higher specific heat, c_p, than the fluid on the low-pressure side of the regenerator (see Figure 6.22 regarding the variation of specific heat with temperature at pressures near the critical pressure). Therefore, the low-pressure side experiences a larger change in temperature than does the high-pressure side, and the heat exchanger is unbalanced.

For this problem, we want to maintain a stream-to-stream temperature difference of 10 °C (at both inlet and exit) in regenerator #1 and therefore, we must have a smaller mass flow rate, m_x, through the high-pressure side of this regenerator. For this reason, some of the mass flow is diverted at state 5 (shown as X) and is sent directly to its own compressor. Before entering regenerator #2, the two high-pressure streams join up at the point marked with the circle.

g. What is the percentage of the mass flow, m_x, that goes through the high-pressure side of regenerator #1?

h. What is the net power produced per unit mass flow?

i. What is the heat addition to the cycle per unit mass flow?

j. What is the thermal efficiency (First Law efficiency)?

k. What are the resulting regenerator efficiencies for regenerators #1 and #2?

l. Calculate and compare the entropy generation (kJ/kg·K) in the regenerator in cycle 2 with the total entropy generated in both regenerators in cycle 3.

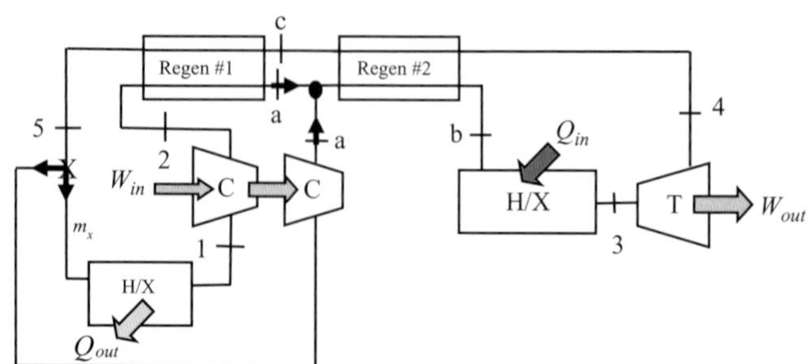

Figure E6.5c Split flow compression cycle.

Example 6.5 (cont.)

Solution

Cycle 1: Simple cycle (Figures E6.5a and E6.5d)

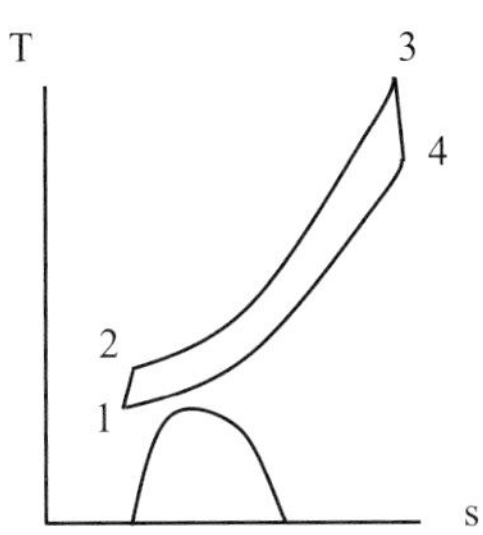

Figure E6.5d Simple cycle T–s diagram.

Once all the states are fixed, the calculations for heat, work, and efficiency are straightforward. Two independent properties are required to fix the state. Some states are already known, and some will have to be determined. All properties will be obtained using the NIST tables of thermophysical properties of fluid systems for CO_2.

T_1 and T_3 are given. $P_L = 75$ bar, and the pressure ratio is 3.5, which yields the high pressure as 262.5 bar. Since $P_L > P_{crit}$, there are no saturated states. We assume there are no pressure drops in the heat exchangers. The values are given in the Table E6.5.

Table E6.5 Data for states in Example 6.5

	T (°C)	p (bar)	h (kJ/kg)	s (kJ/kg·K)
State 1	22	75	254.78	1.1764
State 2		262.5		
State 3	800	262.5	1349.8	3.0206
State 4		75		

Given states 1 and 3, we can look up the corresponding enthalpies and entropies. To fix state 2, we use the given compressor efficiency of 0.8.

For an isentropic compressor, $s_{2,s} = s_1$ and $p_{2,s} = p_2 = 262.5$ bar, so we can look up $h_{2,s} = 277.07$ kJ/kg. Now we use the isentropic efficiency to find h_2:

$$\eta_C = 0.8 = \frac{h_{2,s} - h_1}{h_2 - h_1} \Rightarrow h_2 = 282.64\,\text{kJ/kg}.$$

Using h_2 and p_2, we can look up $T_2 = 45.26\,°C$ and $s_2 = 1.194$ kJ/kg·K.

We do a similar thing to fix state 4. For isentropic turbine operation, $s_{4,s} = s_3$ and $p_{4,s} = p_4 = 75$ bar, so we can look up $h_{4,s} = 1111.9$ kJ/kg. Now we use the isentropic efficiency to find h_4:

Example 6.5 (cont.)

$$\eta_T = 0.9 = \frac{h_3 - h_4}{h_3 - h_{4,s}} \Rightarrow h_4 = 1135.69.$$

Using h_4 and p_4, we look up $T_4 = 624.6\,°\text{C}$ and $s_4 = 3.0473\,\text{kJ/kg·K}$.

Knowing all the states, we can now calculate the network output, heat input, and efficiency.

a. $\frac{\dot{W}_{out}}{\dot{m}} - \frac{\dot{W}_{in}}{\dot{m}} = (h_3 - h_4) - (h_2 - h_1) = 214.1 - 27.86 = 186.24\,\text{kJ/kg}.$

b. $\frac{\dot{Q}_{in}}{\dot{m}} = h_3 - h_2 = 1067.16\,\text{kJ/kg}.$

c. $\eta_{Th} = \frac{\dot{W}_{out} - \dot{W}_{in}}{\dot{Q}_{in}} = 0.1745.$

Cycle 2: Regenerative Cycle (Figure E6.5e)

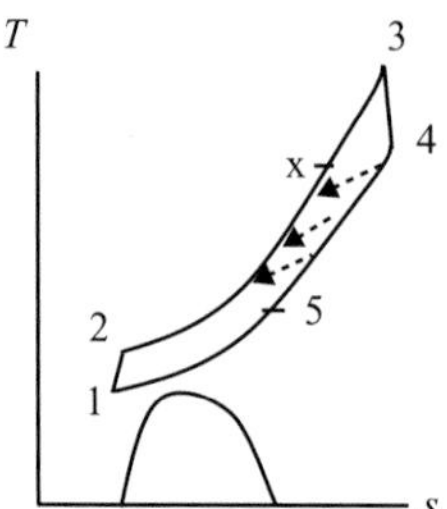

Figure E6.5e Regenerative cycle T–s diagram.

In this problem, states 1 and 3 are given as the same as in cycle 1. States 2 and 4 are also the same as in cycle 1 because the compressor/turbine efficiencies are the same and the pressure ratio is the same. Therefore, we only have to determine states 5 and x. We are given:

$$\eta_{regen} = \frac{T_{cold\_side\_out} - T_{cold\_side\_in}}{T_{hot\_side\_in} - T_{cold\_side\_in}} = \frac{T_x - T_2}{T_4 - T_2} = 0.75.$$

Using T_2 and T_4 as determined above, we can solve for $T_x = 479.8\,°\text{C}$. Since $p_x = 262.5$ bar, we have two properties and we can look up $h_x = 943$ kJ/kg and $s_x = 2.5708$ kJ/kg·K.

To solve for state 5 we use the First Law around the adiabatic regenerator to get h_5:

$$0 = h_2 + h_4 - h_x - h_5 \Rightarrow h_5 = 475.33\,\text{kJ/kg}.$$

Using $p_5 = 75$ bar, we look up $T_5 = 66.1\,°\text{C}$ and $s_5 = 1.8893$ kJ/kg·K.

Since states 1, 2, 3, and 4 are the same as cycle 1, the net power produced is the same, but the heat input is much smaller and the efficiency is higher:

d. $\frac{\dot{W}_{net}}{\dot{m}} = 186.24\,\text{kJ/kg}.$

e. $\frac{\dot{Q}_{in}}{\dot{m}} = h_3 - h_x = 406.8\,\text{kJ/kg}.$

f. $\eta_{Th} = \frac{\dot{W}_{out} - \dot{W}_{in}}{\dot{Q}_{in}} = 0.4578.$

Example 6.5 (cont.)

Cycle 3: Split Flow Compression Cycle (Figure E6.5f)

In Figure E6.5f the dashed arrows indicate internal heat transfer and the solid arrows indicate mass flow.

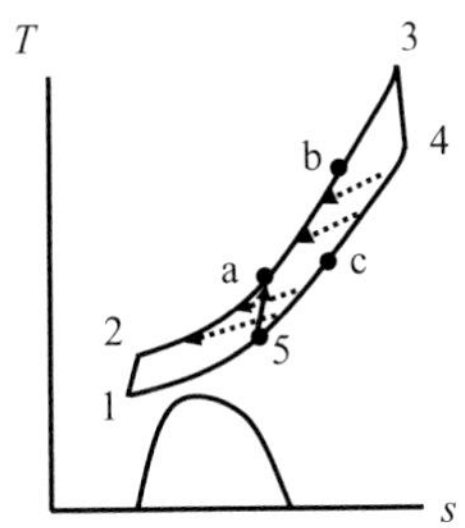

Figure E6.5f Split flow compression cycle T–s diagram.

Again, based on the information given, states 1, 2, 3, and 4 are the same as in cycles 1 and 2. We are told that the stream-to-stream temperature difference is 10 °C at both ends of regenerator #1; therefore $T_5 - T_2 = 10\,°\text{C}$ and $T_c - T_a = 10\,°\text{C}$.

Thus, $T_5 = 55.26\,°\text{C}$ and since $p_5 = 75\,\text{bar}$, we can look up $h_5 = 456.34\,\text{kJ/kg}$ and $s_5 = 1.8322\,\text{kJ/kg·K}$.

To fix state a, we examine the compressor that compresses the gas from state 5 to state a. For isentropic operation $s_{a,s} = s_5$ and $p_{a,s} = p_a = 262.5\,\text{bar}$, so we can look up $h_{a,s} = 519.9\,\text{kJ/kg}$ and $T_{a,s} = 162.1\,°\text{C}$. Now we use the isentropic efficiency to find h_a:

$$\eta_C = 0.8 = \frac{h_{a,s} - h_5}{h_a - h_5} \Rightarrow h_a = 535.8\,\text{kJ/kg}.$$

Using h_a and p_a, we can look up $T_a = 171.7\,°\text{C}$ and $s_2 = 1.8683\,\text{kJ/kg·K}$.

Since $T_c - T_a = 10\,°\text{C}$, $T_c = 181.7\,°\text{C}$ and $p_c = 75\,\text{bar}$, we can look up $h_c = 620.8\,\text{kJ/kg}$ and $s_c = 2.2611\,\text{kJ/kg·K}$.

g. To determine the mass fraction m_x going through the high-pressure side of regenerator #1, we use the First Law on the adiabatic regenerator:

$$0 = \dot{m}_x(h_2 - h_a) + 1(h_c - h_5) \Rightarrow \dot{m}_x = 0.6496.$$

h. Now we need to fix state b, which is the final unknown state. We use the First Law on regenerator #2:

$$0 = (h_a - h_b) + (h_4 - h_c) \Rightarrow h_b = 1050.7\,\text{kJ/kg}.$$

Using this with $p_b = 262.5\,\text{bar}$, we look up $T_b = 565.6\,°\text{C}$ and $s_b = 2.7062\,\text{kJ/kg·K}$.

Now we can answer all the other questions, since all states are known.

$$\frac{\dot{W}_{comp\_Total}}{\dot{m}} = \dot{m}_x(h_2 - h_1) + (1 - \dot{m}_x)(h_a - h_5) = 45.88\,\text{kJ/kg},$$

Example 6.5 (cont.)

$$\frac{\dot{W}_T - \dot{W}_C}{\dot{m}} = 214.1 - 45.88 = 168.22\,\text{kJ/kg}.$$

i. $\frac{\dot{Q}_{in}}{\dot{m}} = h_3 - h_b = 299.1\,\text{kJ/kg}$.

j. $\eta_{Th} = \frac{\dot{W}_T - \dot{W}_C}{\dot{Q}_{in}} = 0.5624$.

k. $\eta_{regen\,\#1} = \frac{T_a - T_2}{T_c - T_2} = 0.927$ and $\eta_{regen\,\#2} = \frac{T_b - T_a}{T_4 - T_a} = 0.869$.

l. For cycle 2, applying the Second Law, we have the entropy generated in the regenerator as

$$0 = 0 + s_2 + s_4 - s_x - s_5 + \frac{\dot{S}_{gen}}{\dot{m}} \Rightarrow \frac{\dot{S}_{gen}}{\dot{m}} = 0.2188\,\text{kJ/kg·K}.$$

For cycle 3, applying the Second Law to each regenerator individually, we get:

$$\dot{S}_{gen\_R\#1} = \dot{m}_x(s_a - s_2) + (s_5 - s_c) = 0.00912\ \text{kJ/kg·K}$$
$$\dot{S}_{gen\_R\#2} = (s_b - s_a) + (s_c - s_4) = 0.0517\ \text{kJ/kg·K}.$$

Thus, the total entropy generated is $\frac{\dot{S}_{gen}}{\dot{m}} = 0.0608\,\text{kJ/kg·K}$.

This could also be determined by applying the Second Law to a single control volume surrounding both regenerators, and we would get the same answer.

The entropy generated in the regenerators is less in cycle 3 than in cycle 2 because the split compression cycle reduces the temperature difference between the streams, which is the source of entropy generation and irreversibility. Thus, higher regenerator efficiencies lead to lower entropy generation.

The example shows that the basic hypercritical CO_2 gas cycle has low efficiency. This is because of the higher turbine outlet temperature. The efficiency of the hypercritical CO_2 gas cycle with regeneration clearly improves by recovering some of the heat from the turbine outlet stream. Introduction of split flow compression also increases the efficiency, since it lowers the irreversibility by reducing the stream-to-stream temperature difference in the regenerators. However, the improvement is rather small. Given the cost of the extra equipment, split flow compression may not be favorable.

6.9 Binary Two-Phase Cycles

Binary vapor cycles have also been suggested to improve the performance of Rankine cycle power plants. In this case, two different cycles with two different fluids are used. The heat rejected from the "**topping**" or high-temperature cycle is added to the "**bottoming**" or low-temperature cycle (Figure 6.22). Exotic fluids, such as mercury, potassium, and sulfur, have been proposed for the bottoming cycle. Although theoretical studies have demonstrated the higher efficiency of these cycles, the complexity of handling these fluid has so far prevented

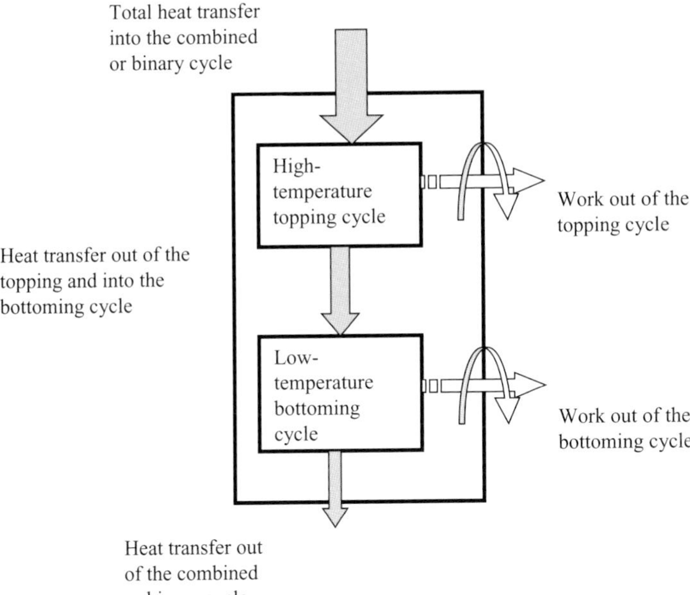

Figure 6.22 A schematic diagram showing the layout of a binary or combined cycle with a single heat source, with a fuel stream or a nuclear/renewable heat source, and two power-producing cycles.

their utilization (it is unlikely that either mercury or sulfur would ever be a working fluid in a neighborhood power plant any time soon!). Binary cycles are a subset of combined cycles, which will be discussed in detail in the next chapter, that use two-phase fluids in both cycles.

6.10 Organic Rankine Cycles

Water is the primary working fluid for Rankine cycles for many applications, including fossil fuels, nuclear, and solar thermal. But the critical properties, especially the critical pressure of steam, are high and operating under near-critical or supercritical conditions for high efficiency requires high-temperature sources and operating at high pressures. Moreover, steam plants are not easy to downscale if the available thermal power is low. To operate under these conditions while maintaining reasonable efficiency, organic fluids with low critical properties are used as working fluids, forming organic Rankine cycles (ORCs). In an ORC, the working fluid is selected to match the low temperature of the heat source and the low available power, and careful optimization is necessary.

6.10.1 ORC Systems and Applications

Organic Rankine cycles have been used with biomass combustion, in which the flame temperature resulting from burning raw or partially dried biomass is low, and the associated power density is also low, making it uneconomical to use steam plants. These are particularly attractive when used for combined heat and power (CHP), where the exhaust thermal energy is used for space or process heating. A similar scenario is often encountered in geothermal energy where the typical temperature of the source (brines) is 150–200 °C, and with relatively low flow from each geo-well the available thermal power is also low for steam Rankine cycle applications. Another range of applications for ORCs come under the important category of waste heat recovery (WHR) in industrial applications, such as small cement production plants, metals production plants such as iron and steel, and glass production, where heat sources at 200–400 °C are available but at low thermal energy rates [9]. Waste heat recovery is also possible in internal combustion engine applications, in which the exhaust temperature is 200–500 °C, depending on the type of engine, and the engine coolants are at ~100 °C, but the exhaust and coolant flows are typically low (Figure 6.23). For instance, in a 300 kW_e engine running at 40% efficiency, the waste heat rate is 450 kW_{th}. An ORC with 8% efficiency could recover 36 kW_e, which raises the overall system efficiency by 12%.

Smaller-scale solar thermal power plants (CSP) that employ low-temperature concentrators are also good candidates for ORCs (solar thermal plants are described in detail in Chapter 9). In CSP plants, the working fluid is heated in a solar collector instead of a combustor, as shown in Figure 6.24. In this case it may be possible to choose an organic fluid that does not evaporate as it flows through the solar collector (that is, a fluid whose boiling temperature at the high-pressure side of the cycle is higher than the concentrator temperature), and hence the hot liquid fluid can be stored for later use in the power production cycle

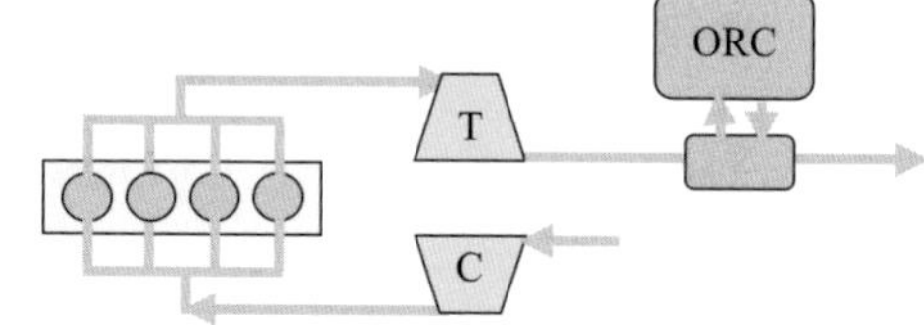

Figure 6.23 Using an ORC to improve a turbocharged diesel engine efficiency by converting some of the waste heat in the exhaust gases to electricity. In the ORC, the organic fluid is heated by the engine exhaust following expansion in the turbocharger.

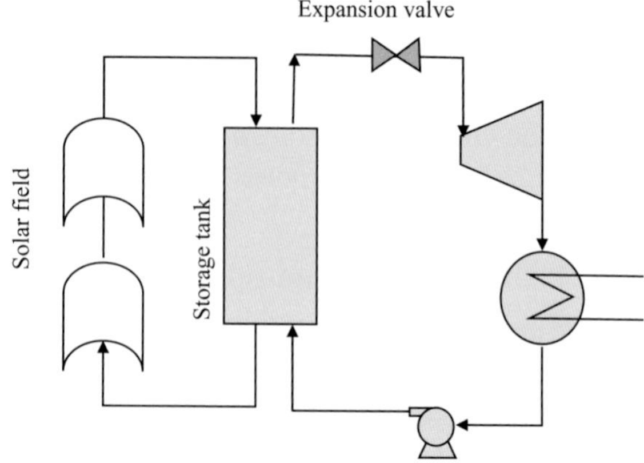

Figure 6.24 Using some organic fluids that do not evaporate while being heated in a low-temperature solar concentrator makes it possible to store some of the solar energy (in the form of hot liquid) for later use. The valve following the concentrator/storage tank is used to expand the high-pressure hot organic fluid at constant enthalpy, thus causing it to evaporate before the turbine. The fluid leaving the turbine is condensed and pumped back to the tank pressure. Some of the fluid can be stored in the cold side of the tank if not needed (when the solar energy is not sufficient to raise a high flow rate of organic fluid to a sufficient temperature).

(after sunset), providing integrated thermal storage and power [10]. In order to use the hot fluid in the turbine, it is first evaporated by expanding at a constant enthalpy value. Following expansion in the turbine, a recuperator is used to preheat the pumped liquid. Finally, the fluid is condensed and pumped back up to the maximum pressure (which is the same as the storage tank/solar collector pressure). Thus, these are three-pressure cycles: condenser pressure (lowest pressure), turbine pressure (intermediate pressure), and collector/tank pressure (highest pressure). Using a dry fluid, described next, is compatible with the needs of this system.

The low temperature of these heat sources sets a limit on the maximum efficiency of the cycles. Operating with superheated or supercritical two-phase cycles is preferred because of their higher efficiency. Therefore, organic fluids with low critical temperature are used. Recently, some organic fluid mixtures have also been suggested because of their unique properties.

6.10.2 ORC Fluids and Mixtures

Selecting organic fluids with molecular weights higher than that of water, it is possible to design compact ORC engines, and hence integrate them in mobile applications. Heavier fluids can also have more favorable heat transfer characteristics. Besides their thermodynamic properties, which will be described next, organic fluids used in ORCs must have other characteristics including low flammability and low freezing temperature, low toxicity and high chemical stability, low corrosion potential, and low viscosity. Important environmental properties include low toxicity, low global warming potential (GWP), and low ozone depletion potential (OPD). Working fluid for ORCs can be: (1) pure, such as hydrofluorocarbons (HFCs), hydrochlorofluorocarbons (HCFCs), perfluorocarbons (PFCs), hydrocarbons (HCs), siloxanes, hydrofluoroolefines (HFOs), chlorofluorocarbons (CFCs), and hydrofluoroethers (HFEs); or (2) zeotropic mixtures of organic fluids that have different boiling temperatures, such as benzene, ethanol, toluene, and isobutane. It should also be affordable and easy to dispose of after it has been used. Some candidates are shown in Table 6.2.

Table 6.2 Properties of some ORC fluids, including the boiling temperature and the critical temperature and pressure

	Boiling temperature (°C)	Critical temperature (°C)	Critical pressure (MPa)
R123	27.8	183	3.7
R134a	26	101	4.06
R152a	−24	113	4.5
R245fa	25.2	154	3.6
R290	41.1	97	4.25
R600	−0.5	152	3.8

Another desirable but not necessary property of ORC fluids is that their saturation curve on the *T–s* diagram should be "overhanging" – that is, the slope of that curve on the vapor side should be positive, similar to that on the liquid side. Fluids with this property are known as dry fluids. For these fluids, for any subcritical cycle and for supercritical cycles, expansion within the turbine does not lead to condensation without having to go to very high pressures and temperatures at the turbine inlet, even for isentropic expansion. Examples of these fluids and their *T–s* diagram are given in Figure 6.25. This excludes fluids such as water and methane, which are known as wet fluids, for which the slope of the saturation curve on the right-hand side of the critical point is negative. Fluids with higher molecular weight and molecular complexity tend to be dry fluids. These same fluids tend to have higher heat capacity and low specific heat ratio.

Using zeotropic mixtures of two fluids whose boiling temperatures in the pure state are different (for instance, fluids with very different molecular weights and hence different critical temperatures), the boiling and condensation curves of the mixture are no longer isothermal and hence one can avoid the large temperature difference between the two streams in the heat exchangers during heat addition and rejection in the cycle. The difference between the temperature of the dew line and bubble line of the mixture is known as the temperature glide, and it determines the mixture temperature change during boiling and condensation (Figure 6.26). Using mixtures is still limited in applications because of challenges associated with potential heat transfer degradation and mixture composition change if one of the two components leaks.

Examples of results showing the dependency of the ORC efficiency on the choice of the working fluid and operating conditions are shown in Figure 6.27. In that study, an organic Rankine cycle operated on diesel engine exhaust, assumed to be available at 250 °C. This is one of the lower exhaust temperatures for internal combustion engines; diesel engines

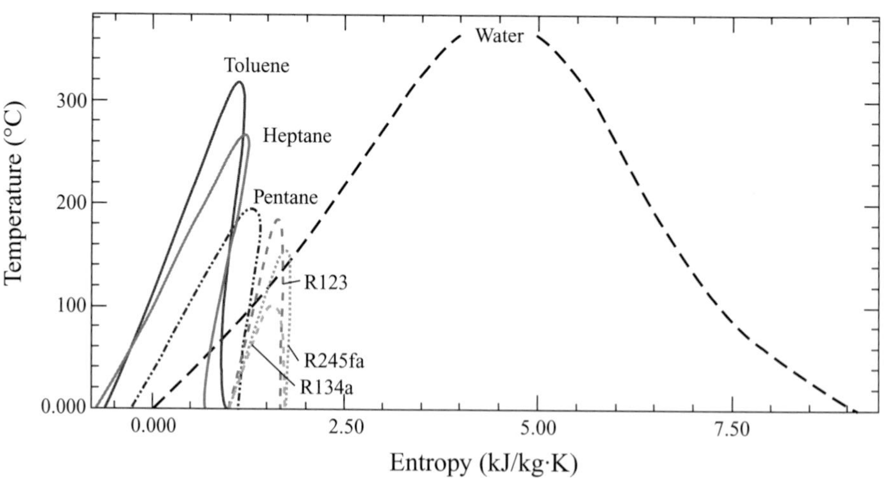

Figure 6.25 *T–s* diagram for a number of hydrocarbons and refrigerants with an overhang saturation curve, compared with water.

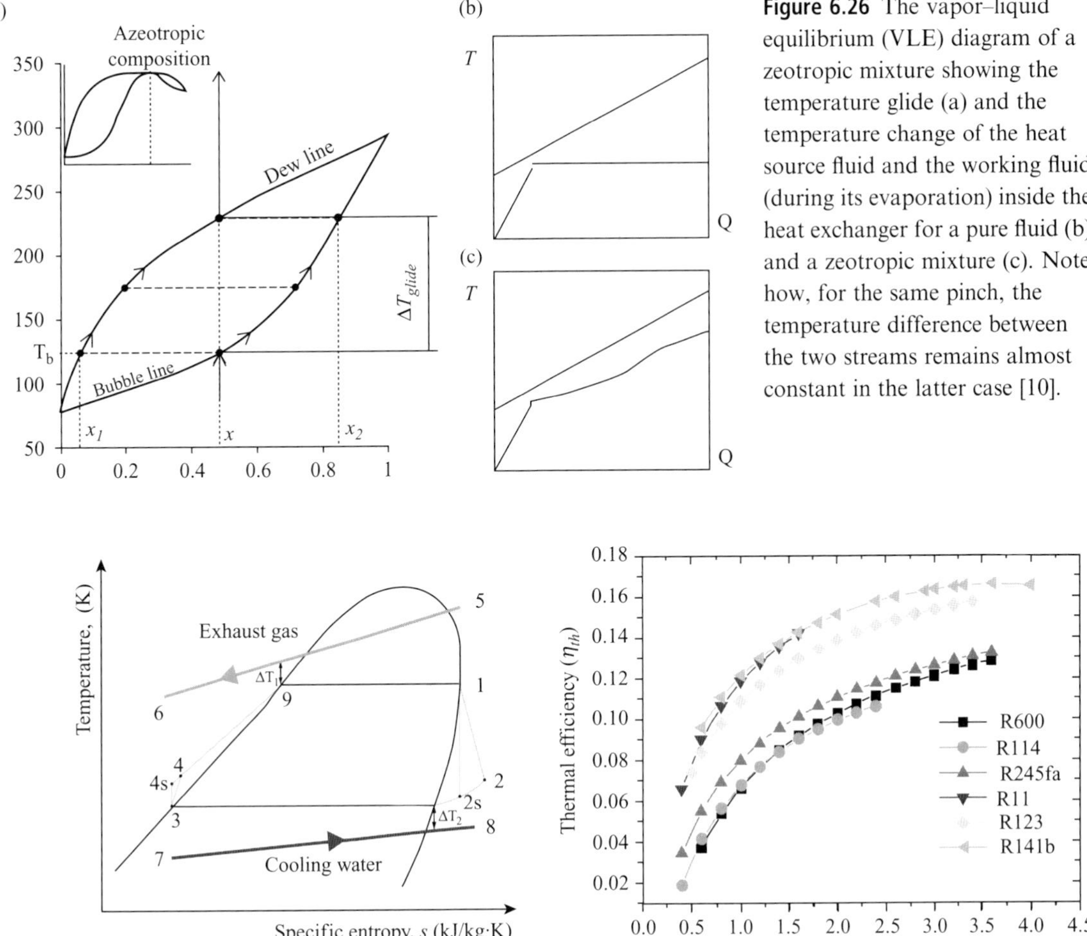

Figure 6.26 The vapor–liquid equilibrium (VLE) diagram of a zeotropic mixture showing the temperature glide (a) and the temperature change of the heat source fluid and the working fluid (during its evaporation) inside the heat exchanger for a pure fluid (b) and a zeotropic mixture (c). Note how, for the same pinch, the temperature difference between the two streams remains almost constant in the latter case [10].

Figure 6.27 (a) The T–s diagram of an ORC using a dry fluid and (b) a simple ORC showing schematically the change of temperature of the heat source (exhaust fluid temperature shown as R114) and of the condenser cooling fluid (cooling water shown as R11) [9]. The ORC efficiency depends on the high cycle pressure using a number of working fluids. The model used to produce these results applied realistic models for the different cycle components and working fluid equation of state (engine exhaust fluid with an inlet temperature of 250 °C) [11].

operate with lean fuel–air mixtures and with high compression/expansion ratios, resulting in low exhaust temperatures. Spark-ignition engine exhaust temperature is typically much higher because the engine operates on stoichiometric mixtures and with lower expansion ratios (to avoid knocking, see Chapter 4). The efficiency of the ORC is strongly dependent on the maximum pressure of the cycle or fluid evaporation pressure, but plateaus at higher pressures. It also depends on the choice of the fluid, with significant improvement seen as we switch from R141b to R600.

6.11 Cooling

Cooling is a necessary and integral part of Rankine cycle power plants; without being able to bring the working fluid back to the condensed state at or near the temperature of the environment, this cycle loses a significant part of its efficiency advantage. Moreover, given the significant latent heat of condensation of most working fluid, the heat rejected is typically a large fraction of the heat input, as determined by the thermodynamic efficiency of the plant. This is the reason why, traditionally, steam power plants were built close to a large body of water (watershed) to facilitate the cooling of the power plant condensers. However, growing concerns regarding the ecological impact of warming up the water in rivers, lakes, and oceans, and the negative impact on marine animal life caused by the cooling equipment and the biological health of these important resources, as well as the related evaporation and changes in the regional weather conditions, has prompted development of alternative cooling technologies and policies that discourage or ban the use of natural water resources in power production. These alternative cooling technologies are also used when power plants are built in dry areas. In this section, cooling needs and approaches are discussed.

Power plants cooling impacts the efficiency and cost of power in several ways. As shown before, the exit temperature of the steam turbine affects the efficiency of the plants by impacting the exit pressure [12]. Figure 6.28 shows the impact of the condenser temperature on the plant output for different steam turbine designs, with a 7% drop in output (and hence plant efficiency) associated with ~16 °C change in the condenser temperature (which may correspond to change between winter to summer conditions). It also shows the impact of the cooling technology, whether wet or dry, as will be described next, on this change. Of course, changing the efficiency impacts the economics of the plant as more fuel must be used to generate the same power at lower efficiency. A compromise must be made between investing in cooling technology to improve the efficiency or using more fuel. Thus, the choice of

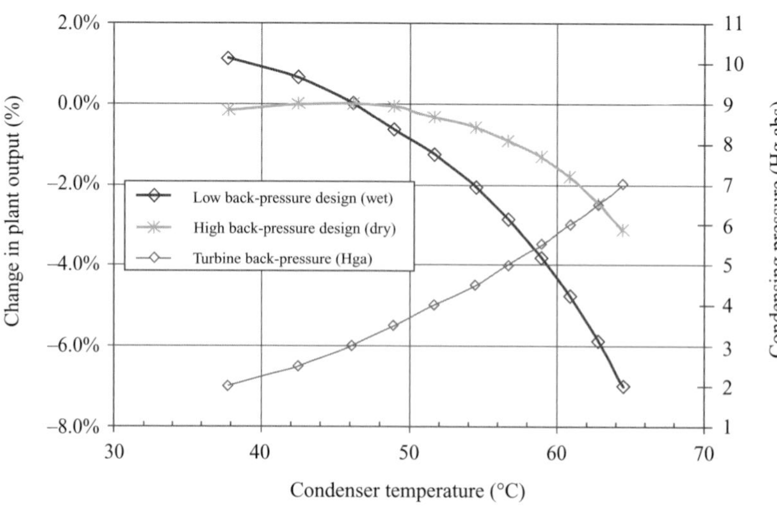

Figure 6.28 Change in a steam power plant output with the condenser temperature for two types of turbine designs connected with condensers that are cooled by wet towers and by air or dry cooling [13].

cooling must be considered because of its environmental impact and also because of the cost, as well as the local available water resources and regulations on water use.

6.11.1 Cooling Types

Water is used in power plants either by withdrawing it from a nearby watershed or by consuming a given supply. In withdrawal, water is taken from a watershed or aquifer, and is returned or discharged to the same watershed but at higher temperature. Consumption in this case typically refers to the water evaporated in the watershed because of the higher temperature, or water evaporation when cooling towers are used. While water is used for more than condenser cooling (e.g., in coal plants water is used in desulfurization of the combustion products, in ash ponds, in gasification, etc.), cooling typically dominates water use in power plants.

Figure 6.29 shows the energy flow in a steam Rankine cycle plant, presented in the form of a Sankey diagram where the total input thermal energy is divided among essentially three part: the mechanical energy or electricity output; the thermal energy in the exhaust or flue gases (in the case of combustion or fossil fuel-based plants); and energy leaving through the cooling system. The fraction leaving in the form of electricity depends on the thermodynamic efficiency of the plant. The fraction leaving in the flue gases depends on fuel, cycle, and components. For instance, the boiler flue gases in a coal power plant typically leave at ~100 °C. In nuclear plants and solar thermal plants there are no flue gases. The difference

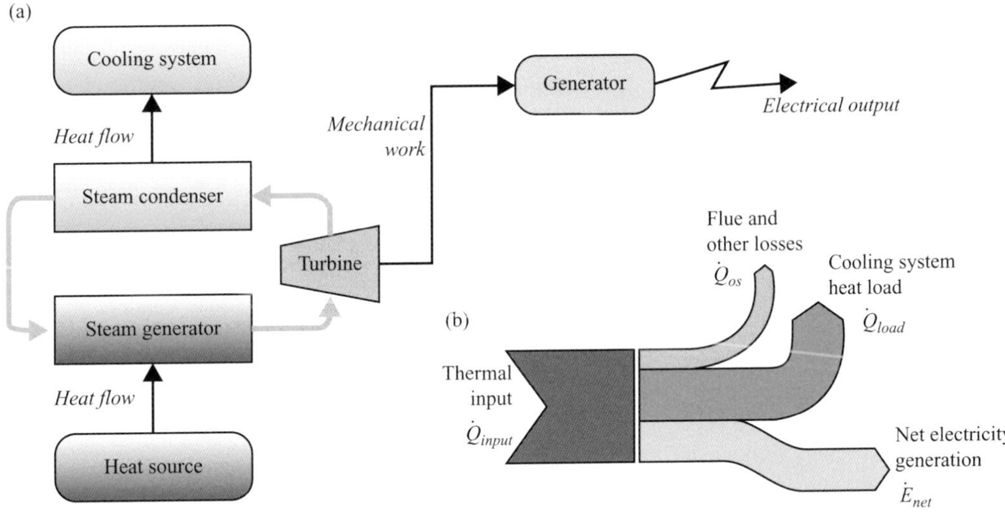

Figure 6.29 Block diagram showing a typical Rankine cycle steam power plant (a) and a Sankey diagram showing the energy flow in the same power plants (b). Flue losses are associated with combustion gases in fossil fuel power plants, and are zero in nuclear, solar thermal, and geothermal plants [13].

between the input energy and the turbine out plus the exhaust energy must be rejected through the condenser.

Power plant cooling systems are categorized in terms of the cooling medium and how water is used during the cooling process as discussed below.

Once-through cooling, also called open-loop cooling, involves withdrawing water from the ocean, river, lake, or large ponds to cool the condenser, and discharging it to the same water body at higher temperature. In this case, water consumption is via evaporation due to the higher temperature of the discharged water. While simple and cheap when a large body of water is available nearby, it endangers aquatic life directly and indirectly by raising the temperature. It is now considered environmentally unsafe.

Wet tower cooling uses a recirculating loop of water to cool the condenser (Figure 6.30). Hot water used to cool the condenser is sprayed down through the cooling tower over the fill. This is a matrix or lattice-like material used to increases contact area between the hot water and cooling air driven upward through the fill by a fan or natural draft. The hot water is cooled by direct contact with air, and by the evaporation of a small fraction that acts to cool the remaining water. The cooled water is collected at the bottom of the tower and sent back to the condenser. Some of the cooling water is consumed by evaporation within the tower and must be replaced. In addition, smaller amounts of water are purged from the cooling water circuit to avoid buildup of harmful contaminants, known as blowdown. Wet towers withdraw about two orders of magnitude less water than once-through systems, but consume more. They are also more expensive, are bulky, and produce significant vapor plumes.

Dry cooling, also called air cooling, does not consume any water, but circulates air through the condenser. Because of the lower heat capacity of air, the surface area of the

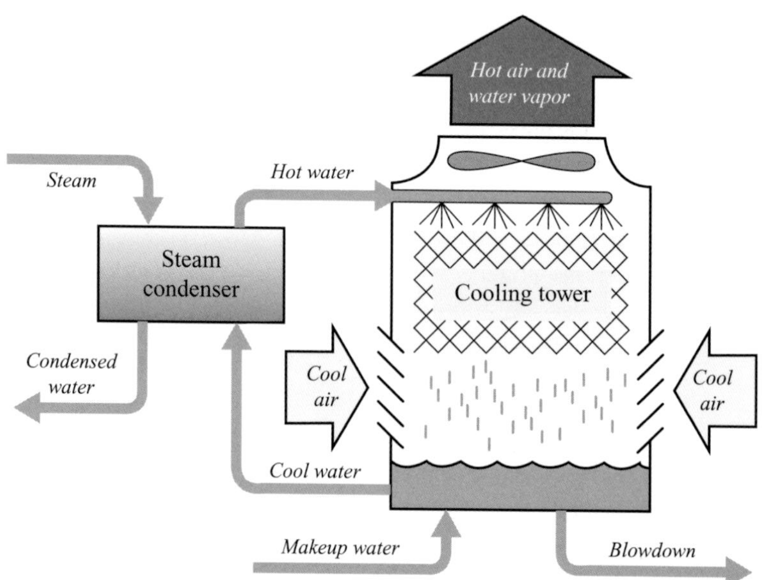

Figure 6.30 Schematic drawing of a wet cooling tower using mechanical draft to flow cooling air upward while the cooling water is sprayed at the top to flow counter-current to air. Some water evaporates and flows out with the warmed air (constituting water consumption). Cooled water is collected at the bottom of the tower and circulated back to the plant steam condenser. Some of the collected water is removed in a blowdown process to remove excess contaminants [13].

condenser must be very large and hence they can be three or four times as expensive as an equivalent wet tower cooling system. Moreover, dry cooling is less effective and reduces the plant efficiency and output, especially on hot days (when there is often more demand for power for air-conditioning).

Hybrid cooling systems combine wet and dry cooling with the typical trade-offs between wet and dry systems in terms of cost, performance, and water use.

Pond cooling uses a system of ponds instead of a cooling tower. Hot water from the condenser is discharged to the pond, where it is cooled through convection and evaporation. This form of cooling can be considered as a compromise between open-loop and wet tower cooling.

6.11.2 Withdrawal and Consumption Rates

Estimates of water withdrawal and consumption in wet towers and once-through cooling can be made based on semi-empirical models that account for the needs of the plant, local conditions, and best practices. Recently a comprehensive model based on energy conservation and mass balance was suggested to estimate these quantities. For instance, in cooling towers, water withdrawal intensity was given by [14]

$$I_{ww} = 3600\frac{(1-\eta_{net}-k_{os})}{\eta_{net}}\frac{(1-k_{sens})}{\rho_w h_{fg}}\left(1+\frac{1}{n_{cc}-1}\right)+I_{proc},$$

where I_{ww} is the water withdrawal intensity in L/MWh; η_{net} is the net plant efficiency; k_{os} is the thermal energy fraction lost to non-cooling sinks (such as desulfurization, mirror washing in solar thermal plants, etc.), taken as 0.2 for combined cycle natural gas plants, 0.12 for steam fossil plants, and 0 otherwise; k_{sens} is the thermal energy input fraction lost through convection, given by: $k_{sens} = 1 - k_{evap}\frac{h_{fg}}{c_p\Delta T}$, and $k_{evap} = \frac{\text{evaporation loss}}{\text{water circulation rate}} \sim \Delta T(\text{in}\,^\circ\text{C})/7$, where ΔT is the water temperature difference between the inlet and exit of the tower (but more accurate correlations have been developed for k_{sens} based on cooling tower design and local conditions); ρ_w is water density; h_{fg} is the latent heat evaporation of water; I_{proc} is the water use intensity for non-cooling purposes, taken as 200 L/MWh for oil and coal plants and 10 L/MWh for gas and nuclear plants; and n_{cc} is the number of cycles of cooling before blowdown (removal for purification),which is typically 2–10.

Water consumption through evaporation and blowdown is given by

$$I_{ww} = 3600\frac{(1-\eta_{net}-k_{os})}{\eta_{net}}\frac{(1-k_{sens})}{\rho_w h_{fg}}\left(1+\frac{1-k_{bd}}{n_{cc}-1}\right)+I_{proc},$$

where k_{bd} is the fraction of blowdown that is treated and discharged.

In the case of once-through cooling, the withdrawal is given by

$$I_{wo} = 3600\frac{(1-\eta_{net}-k_{os})}{\eta_{net}}\frac{1}{\rho_w c_p \Delta T}+I_{proc},$$

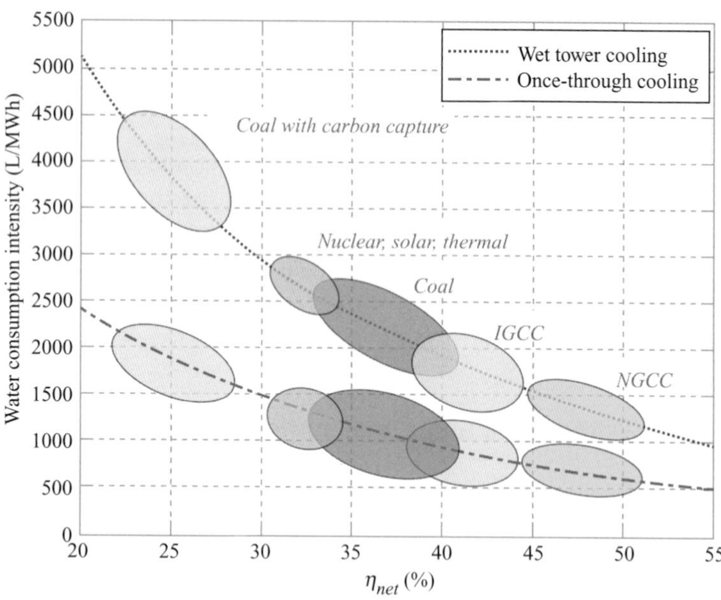

Figure 6.31 Water consumption intensity measured in L/MHh in electric power plants using different technologies as a function of First Law efficiency, in cooling towers and once-through systems [13]. Natural gas combined cycle (NGCC) plants have the highest efficiency and hence the lowest water consumption, followed by integrated gasification combined cycle (IGCC) coal plants, and pulverized coal combustion plants. Nuclear, solar thermal, and geothermal plants have lower efficiency and do not reject heat in the form of flue gases, and hence the higher cooling water consumption. With carbon capture technology added to coal plants, the efficiency penalty translates to more water consumption.

and consumption is

$$I_{co} = 3600\frac{(1-\eta_{net}-k_{os})}{\eta_{net}}\frac{k_{de}}{\rho_w c_p \Delta T} + I_{proc},$$

where k_{ed} is the fraction of discharge that evaporates, O (1%).

Clearly the plant thermal efficiency is the most critical factor determining water consumption. The efficiency depends strongly on the plant type and fuel. The equations presented above have been calibrated against data available from power plants. Results are shown in Figure 6.31. Natural gas combined cycle (NGCC) plants are very efficient and the flue gases leave the HRSG near the environment temperature, hence the low consumption. This is followed by integrated gasification combined cycle (IGCC) plants for coal, which have high efficiency, although they also use water for coal gasification. Coal steam plants (especially the subcritical type) have lower efficiency and use water for other purposes as well. Nuclear and solar thermal plants have much higher water use because of the lower efficiency and lack of exhaust. The lowest are geothermal plants because of the lower-temperature heat source.

6.12 Modeling Software

Computer simulations of thermodynamics has advanced significantly over the past 20 years and have become necessary for accurate modeling of cycles and systems of different types. It is almost indispensable now for systems in which the working fluid undergoes changes that cannot be modeled using ideal gases or steam tables. This is especially true for capturing the thermodynamics of gases with temperature-dependent specific heats and their mixtures, vapor–liquid equilibrium, reacting gas mixtures at equilibrium, reacting mixtures undergoing phase change, etc. Even for relatively simple systems, using computer software for modeling enables parametric studies over a wide range of conditions, optimization, and adding some complexity in modeling components such as turbomachinery and heat exchangers. Fortunately, many packages have been written for this purpose, and some are even open source, allowing users to adapt the software to their purposes. Commercial software packages are also available for professional use, and for academic use mostly at discounts. Research groups have also written their own and some are willing to share their packages.

While analytical solutions are still very useful in establishing trends and building intuition and insight, they are limited. Solutions using hand calculations in which one must search for properties and solve equations are also very useful to gain appreciation for the limitations of analytical models and for the value of simplifications and impact of different assumptions. Software saves time in searching for and assembling property data, which can be tedious and time-consuming, and let the user focus on the modeling and examining the results. With object-oriented programming it is possible to simply and at times graphically translate a model into equations that can be solved using built-in algebraic and differential equation solvers. Many packages include tools for sensitivity and uncertainty analysis and optimization, and can present the results graphically using built-in intuitive interfaces. Some packages allow the user to add data and functionality to extend the capability.

6.13 Summary

Converting low- and intermediate-temperature heat sources, such as geothermal, solar, and nuclear energy sources, to work is efficiently done using a two-phase power cycle. In the absolute majority of applications, the working fluid is water/steam. With even lower-temperature sources, such as waste heat (from industrial plants) and low-quality geothermal sources (to be discussed in detail in Chapter 9), it is possible to use an organic fluid (with lower critical pressure) instead of water as a working fluid. Rankine cycles are closed cycles in which the working fluid is externally heated and cooled in a cyclical fashion, and a pump (instead of a compressor) is used to raise the pressure of the liquid (instead of a gas) following condensation, which contributes to improving the cycle efficiency even with relatively lower maximum temperature (compared to the case of gas turbine cycles). Because the working fluid is isolated, it is possible to use lower-quality fuels (such as coal of different grades [see Chapter 12] or biomass [see Chapter 14]) as the heat source. On the other hand, maintaining the efficiency requires lowering the lower (condenser) pressure by using low-temperature coolant, as well as raising the maximum pressure of the cycle. Similar to Brayton cycles, reheat and regeneration (internal transfer of heat between the expansion side and the liquid heating side) are used to further improve the efficiency.

Cycles whose maximum pressure is higher than the critical pressure of the working fluid are called supercritical cycles, and gain efficiency improvements by reducing the temperature difference between the heat source and the working fluid (and the corresponding irreversibility). If the minimum pressure is close to or slightly higher than the critical pressure, these are called hypercritical cycles; these have demonstrated remarkable efficiencies. Cycles using CO_2 as a working fluid mostly fall in that latter category (more is discussed on these cycles in Chapters 8 and 11).

Because of the role cooling plays in determining the cycle efficiency, and because of concerns about water supplies, cooling technologies that minimize or eliminate water consumption should be used, keeping in mind the potential reduction of the plant efficiency as the turbine exit pressure may rise.

Problems

6.1 In a two-stage turbine, steam at the entrance of the first stage is at 550 °C and 4 MPa. Following expansion to 800 kPa, it is reheated and expanded in the second stage. Steam leaves the second stage as a statured vapor, and condenses at 40 °C. The heat transfer rate in the condenser is 15 MW. The isentropic efficiencies of the first and the second stages of the turbine are 88% and 84%, respectively. The feedwater pump operates with an isentropic efficiency of 75%. Determine: (a) the steam temperature after reheat; (b) the heat transfer rate in the boiler; and (c) the thermal efficiency of the power plant.

6.2 In a regenerative steam power cycle with one closed feedwater heater, steam at 580 °C and 10 MPa expands to 6 kPa in a turbine whose isentropic efficiency is 88%. The steam required for preheating the water in the closed heater is extracted from the turbine at 2 MPa. The extracted steam completely condenses within the heater and becomes saturated water. It is then pumped to the main water line before entering the boiler. Take a pinch point of 6 °C, and an isentropic efficiency of 75% for both pumps. Determine: (a) the fraction of the extracted steam; (b) the temperature of the feedwater at the entrance of the boiler; (c) the temperature at the exit of the heater; and (d) the thermal efficiency of the cycle.

6.3 Consider the power plant of Figure 6.10. The boiler produces superheated steam at 550 °C and 8 MPa. It is then expanded to 2 MPa in the high-pressure turbine. A fraction of the steam leaving the high-pressure turbine is used in the open feedwater heater, and the rest is reheated to 485 °C, which is then expanded in the low-pressure turbine. The condenser operates at 45 °C. The state of the flow at the exit of the condenser and the open heater is saturated liquid. The rate of heat transfer in the boiler is 100 MW. The isentropic efficiencies of the high-pressure and low-pressure turbines are 89% and 80%, respectively. The condensate pump and the feedwater pump operate with 70% and 65% isentropic efficiency, respectively.

a. Determine the steam mass flow rate though the first turbine.
b. Determine the steam mass flow rate required in the feedwater heater.
c. Calculate the net power output of the power plant.
d. Calculate the thermal efficiency of the power plant.

6.4 A schematic of a steam power plant is shown in Figure P6.4. The boiler produces 10,000 kg/h superheated steam at 540 °C and 10 MPa. The power plant is equipped with a reheat, an open feedwater heater, a low-pressure feedwater heater, and a high-pressure feedwater heater. Heat is transferred to the feedwater in the three heaters by extracting steam from three different points. The water leaving the open heater is saturated liquid, which is pumped to the boiler. The condenser operates at 7.5 kPa. The efficiencies of the turbines and pumps are the same as in Problem 6.3. The pinch point for the low-pressure and high-pressure heaters are 4 °C and 6 °C, respectively. Other process parameters are given in the figure.

a. Determine the mass flow rate of steam at the extraction points.
b. Determine the feedwater temperature at the inlet of the boiler.
c. Calculate the power output of the power plant.
d. Calculate the thermal efficiency of the power plant.

Figure P6.4

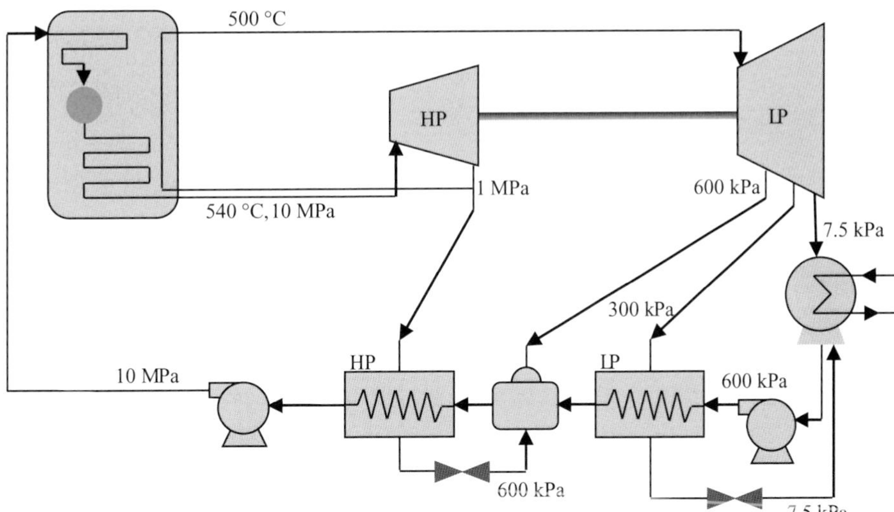

6.5 A schematic of a steam power plant equipped with a reheat, an open feedwater heater, a low-pressure feedwater heater, an intermediate-pressure heater, and a high-pressure feedwater heater is shown in Figure P6.5. The boiler produces superheated steam at 600 °C. The amount of power transmitted to the generator is 100 MW. The condensate of the low-pressure heater is returned to the condenser. The condensate of the other two closed heaters is returned to the open heater. The isentropic efficiencies of the condensate and boiler feedwater pumps are 65% and 70%, respectively. The isentropic efficiencies of the high- and low-pressure turbines are 89% and 92%, respectively. Other process parameters are given in the figure. The pinch point for the high-, intermediate-, and low-pressure heaters is 6 °C, 8 °C, and 4 °C, respectively.

a. Determine the mass flow rate of steam required in the cycle.

b. Determine the mass flow rate of steam at the extraction points.

c. Determine the specific work production by the turbines.

d. Calculate the rate of heat transfer in the boiler.

e. Calculate the mass flow rate of the cooling water required in the condenser.

f. Calculate the thermal efficiency of the power plant.

Figure P6.5

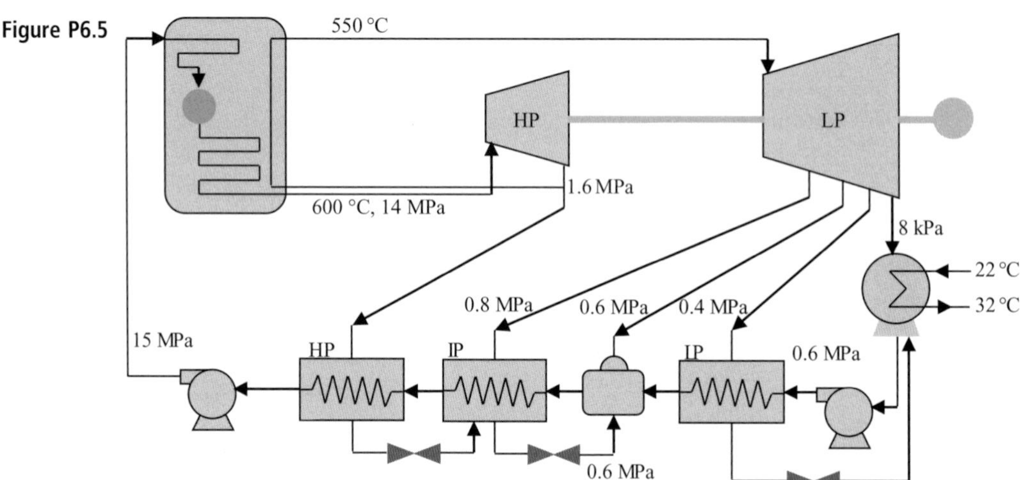

6.6 Superheated steam at 250 °C and 15 bar is supplied to the turbine of a CHP system with a net work output of 3000 kW, as shown in Figure P6.6. Some of the steam is extracted from the turbine at a location where the pressure is 1.4 bar and the rest expand to the condenser pressure of 0.05 bar. Part of the extracted steam is then sent to an open feed heater while the rest is sent to a heating system with a heat load of 8000 kW. The condensate leaving the heating system at 55 °C enters the open feed heater. The feedwater leaves the feed heater as a saturated liquid at 1.4 bar and is then pumped to the boiler. The isentropic efficiency of the turbine is 80%. The pumps operate with isentropic efficiency of 75%.

a. Calculate the steam flow rate through the boiler.

b. Determine the overall thermal efficiency of this CHP plant.

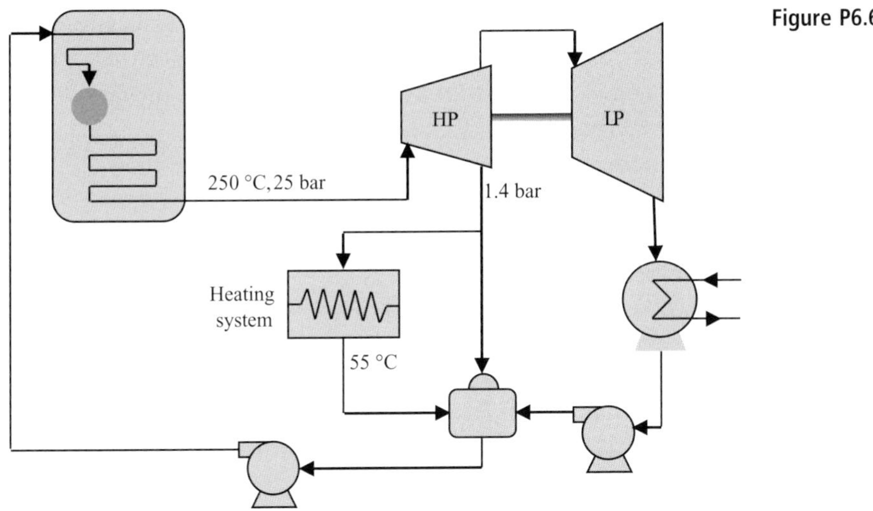

Figure P6.6

6.7 A steam power plant with reheat, one open heater, and one closed heater is schematically shown in Figure P6.7. The net power output of the power plant is 100 MW. Steam enters the first turbine at 8.0 MPa and 480 °C, and expands to 0.7 MPa. The steam is reheated to 440 °C before entering the second turbine, where it expands to the condenser pressure of 0.008 MPa. Steam is extracted from the first turbine at 2 MPa and fed to the closed feedwater heater. Feedwater leaves the closed heater at 205 °C and 8.0 MPa, and the condensate exits as saturated liquid at 2 MPa. The condensate is throttled down into the open feedwater heater, which operates at 0.3 MPa. The steam exiting the open feedwater heater is saturated at 0.3 MPa. The isentropic efficiencies of turbines and pumps are 88% and 75%, respectively.

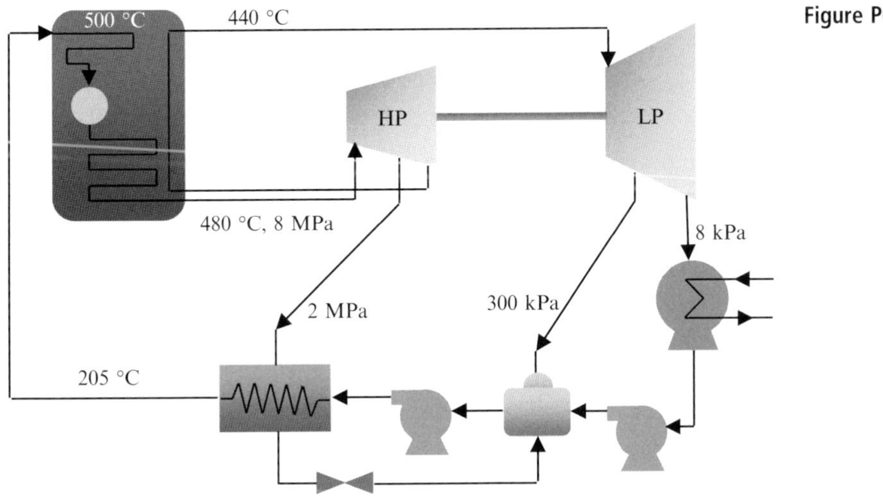

Figure P6.7

a. Calculate the fractions of steam used in the closed heater and the open heater.
b. Determine the mass flow rate at the inlet of the high-pressure turbine.
c. Calculate the power produced by the high-pressure and low-pressure turbines.
d. Calculate the power required by pumps 1 and 2.
e. Calculate the rate of heat added in the boiler.
f. Determine the thermal efficiency of the power plant.
g. If the water enters and exits the condenser at 15 °C and 35 °C, respectively, and at 1 bar, what is the mass flow rate of the cooling water through the condenser.
h. Calculate the total irreversibilities in the closed feedwater heater, the open feedwater heater, the throttle valve, and condenser when the environment temperature is 25 °C. Which device produces the largest irreversibility?

6.8 As the hypercritical CO_2 cycle examined in Example 6.5 rejects heat at a high temperature, it is a good candidate for regeneration. The high-temperature post-turbine fluid is passed through a heat exchanger to preheat the fluid before the heat addition, thus enabling heat addition near the higher temperature while rejecting heat closer to ambient conditions. The schematic of the hypercritical CO_2 cycle with regeneration is depicted in Figure P6.8. The operating conditions are the same as in Example 6.5. The regenerator efficiency is 80%.

a. Determine the temperature of the fluid between the heat exchangers.
b. Calculate the heat addition to the cycle per unit mass flow of CO_2.
c. Determine the thermal efficiency of the cycle.

Figure P6.8

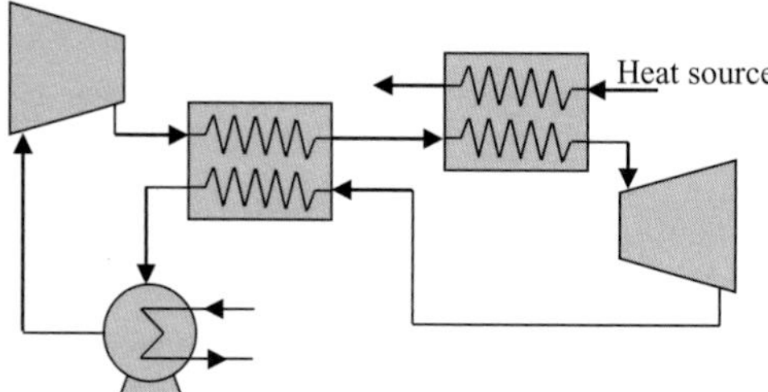

6.9 Consider the split flow compression cycle shown in Figure P6.9. Since the fluid leaving the compressor at state 2 is near the critical point, it has a much higher specific heat than the fluid on the low-pressure side of the regenerator (see Figure 6.22). Therefore, the low-pressure side experiences a larger change in temperature than the high-pressure side, and the heat exchanger is unbalanced. Assume that we want to maintain a stream-to-stream temperature difference of 10 °C (at both inlet and exit) in regenerator 1 and therefore we must have a smaller mass flow rate, m_x, through the high-pressure side of this regenerator. For this reason, some of the mass flow is diverted at point X and is sent directly to its own compressor. Before entering regenerator 2, the two high-pressure streams join up. The process conditions are the same as in Figure P6.8.

a. Calculate the percentage of the mass flow that goes through the high-pressure side of regenerator 1.
b. Determine the net work produced per unit mass flow of CO_2.
c. Evaluate the thermal efficiency of this CO_2 cycle.

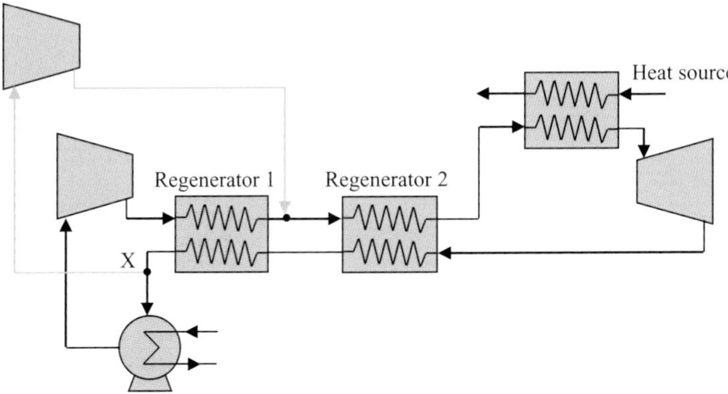

Figure P6.9

REFERENCES

1. W. C. Reynolds, *Thermodynamic properties in SI units*. Stanford, CA: Department of Mechanical Engineering, Stanford University, 1979.
2. W. C. Reynolds and P. Colonna, *Thermodynamics: fundamentals and engineering applications*. Cambridge: Cambridge University Press, 2018.
3. R. Decher, *Energy conversion: systems, flow physics, and engineering*. New York: Oxford University Press, 1994.
4. F. F. Huang, *Engineering thermodynamics: fundamentals and applications*. New York: Macmillan, 1976.
5. J. M. Beér, "High efficiency electric power generation: the environmental role," *Prog. Energy Combust. Sci.*, vol. 33, no. 2, pp. 107–134, 2007.
6. M. M. El-Wakil, *Powerplant technology*. New York: McGraw-Hill, 1984.
7. G. Gottlicher, *The energetics of carbon dioxide capture in power plants*. Washington, DC: U.S. Department of Energy, Office of Fossil Energy, NETL, 2004.
8. P. Dostal, V., Driscoll, and M. J. Hejzlar, "A supercritical carbon dioxide cycle for next generation nuclear reactors," Advanced Nuclear Power Technology Program, 2004.
9. A. T. Hoang, "Waste heat recovery from diesel engines based on organic Rankine cycle," *Appl. Energy*, vol. 231, pp. 138–166, 2018.
10. M. Astofli, "An innovative approach for the techo-economic optimization of organic Rankine cycles," Politecnico di Milano, 2014.
11. L. Tian, H., Shu, G., Wei, et al., "Fluids and parameters optimization for the organic Rankine cycles (ORCs) used in exhaust heat recovery of internal combustion engines (ICE)," *Energy*, vol. 47, pp. 125–136, 2012.

12. C. F. Turche, C. S., Wagner, and M. J. Kutscher, *Water use in parabolic trough power plants: summary results from Worley Parsons' analysis*. Boulder, CO: NREL 2010.
13. M. J. Rutberg, "Modeling water use at thermoelectric power plants," MIT, 2012.
14. M. J. Rutberg, A. Delgado, H. J. Herzog, and A. F. Ghoniem, "A system level generic model of water use at power plants and its application to regional water use estimation," in ASME 2011 Int. Mechanical Eng. Congress & Exposition IMECE 2011, November 11–17, 2011, Denver, CO, 2011, p. 11.

7 Fuel Cells at Finite Current

7.1 Introduction

The performance of fuel cells at finite current, or finite power, is presented in this chapter, focusing on the sources of loss, how each loss mechanism is modeled, and how the design parameters and operating conditions contribute to each. In particular, we examine the role of chemical kinetics and transport processes in fuel cell efficiency. At finite current, fuel cells cannot achieve the ideal thermodynamic efficiency, corresponding to the maximum work or the Gibbs free energy of the overall reaction, due to a number of intrinsic loss mechanisms. These include: (1) non-electrochemical, or thermochemical, reactions, occurring on the surfaces or within the fuel channel; (2) potential loss associated with finite-rate electrochemical reactions; (3) decrease in reactants concentrations because of finite-rate transport processes; and (4) losses associated with the transport of ions and electrons across different elements. All of these mechanisms depend on the current drawn from the cell. Some small losses are observed even at open-circuit conditions, mostly due to electron and fuel leakage across the electrodes. Modeling these losses is tackled in some detail in this chapter.

Before discussing the mechanisms leading to losses, we review briefly types of fuel cells and their typical components. The objective is to explain the differences between different types of fuel cells, the materials used in their construction, and how these may impact their performance. Essentially five types of fuel cells are available or under development: proton exchange membrane cells, alkaline cells, phosphoric acid cells, molten carbonate cells, and solid oxide cells. The electrochemical reactions occurring in the different cells and the electrode and electrolyte materials used in them are also listed.

The first significant loss mechanism encountered as the current increase is electrochemical kinetic loss. To understand the physical origin of this mechanism, a quick review of chemical reaction kinetics, first in homogeneous and then heterogeneous environments is reviewed. This is followed by a theory describing the kinetics in an electrochemical reaction, or discharge separation reactions. Charge separation at equilibrium is related to the voltage difference at the electrode–electrolyte interface and the generation of exchange currents. The role of a finite net current in disturbing the equilibrium is then used to quantify a relation between the voltage drop, or kinetic overpotential, and the current. The Butler–Volmer equation and its simplification are then derived.

The transport of uncharged species is associated with the establishment of finite species concentration gradients across the electrodes and the electrolyte. The relation between the transport flux and the species concentration gradient is governed by Fick's law. Simple forms of Fick's law are used to determine the change in concentration across the electrodes for a given current, and to define the concentration at the electrolyte surfaces, where the reactions occur. A lower concentration at the interfaces corresponds to lower potential, according to the Nernst equation, and the difference between the voltage evaluated using the concentrations at the fuel channels and the concentration at the interface is known as the concentration overpotential.

Other concentration overpotentials are generated by the change of the species concentration inside the flow channels between the bulk and the electrode surface, associated with mass transfer boundary layers inside the fuel/oxidizer supply channels. This is quantified in terms of simple mass transport expressions and the mass transport overpotential expressions are extended to incorporate this effect.

The resistance of the cell material to the flow of charged species causes a potential drop that reduces the measured cell potential below its ideal value as the current increases. This voltage drop is determined by Ohm's law and the proportionality constant is the electrical conductivity of the medium.

We end the chapter with a review of the different types of fuel cells, their reactions and polarization characteristics. For more detail on the modeling and operation of different types of fuel cells, *the reader can consult with* [1]. For background material of the physical chemistry of electrochemical reactions, see [2] and [3]. For comprehensive coverage of electrochemistry, see [4] and [5].

7.2 Fuel Cell Components

Fuel cells are often built by assembling a stack of individual cells, connected in series to build up the voltage and deliver the necessary potential for a particular application. Several stacks can be connected in parallel if higher currents are required.

A single cell consists of two thin electrodes sandwiching an electrolyte, forming the membrane–electrode assembly (MEA). The electrodes are made of electronic conducting material. The electrolyte is a perfect electronic insulator and a good ionic conductor. The interfaces between the electrodes and the electrolyte are coated with a metallic catalyst, used to speed up the electrochemical reaction. On the outside of each electrode, opposite the electrolyte, there is a fuel or an oxidizer supply channel carved in the interconnect between neighboring cells. The interconnect is a bipolar plate since, in a stacking arrangement, it is connected to the anode of one cell and the cathode of its neighboring cell. This positive-to-negative stacking of cells allows the accumulation of the voltage across the stack. The components of a typical fuel cell are shown in Figure 4.11, in which a polymer electrolyte

Table 7.1 Electrochemical reactions at the anode and cathode of different types of fuel cells, as described is more detail in Chapter 4

Fuel cell, fuel	At the anode	At the cathode
PEM, hydrogen	$H_2 \rightarrow 2H^+ + 2e^-$	$\frac{1}{2}O_2 + 2H^+ + 2e^- \rightarrow H_2O$
Alkaline, hydrogen	$H_2 + 2OH^- \rightarrow 2H_2O + 2e^-$	$\frac{1}{2}O_2 + H_2O + 2e^- \rightarrow 2OH^-$
Molten carbonate, H_2, CO	$H_2 + CO_3^{2-} \rightarrow H_2O + CO_2 + 2e^-$ $CO + CO_3^{2-} \rightarrow 2CO_2 + 2e^-$	$\frac{1}{2}O_2 + CO_2 + 2e^- \rightarrow CO_3^{2-}$
Hydrogen solid oxide, CO, methane	$H_2 + O^{2-} \rightarrow H_2O + 2e^-$ $CO + O^{2-} \rightarrow CO_2 + 2e^-$ $CH_4 + 4O^{2-} \rightarrow 2H_2O + CO_2 + 8e^-$	$\frac{1}{2}O_2 + 2e^- \rightarrow O^{2-}$

membrane fuel cell (PEMFC) is used as an example (other stacking arrangements have been used). The electrode material is porous to allow gas diffusion from the fuel/oxidizer channel to the electrode–electrolyte interface and the diffusion of the products in the opposite direction. The gas here is a fuel, an oxidizer, or a product of reaction, which can flow separately or as part of a gas mixture made up of the reacting or active species and some inert or carrier gas.

Fuel cells differ in their construction, materials, and operating temperatures. The construction of a cell, and in particular the material used in the electrolyte, determine the electrochemical reactions taking place at the anode–electrolyte and cathode–electrolyte interfaces. The electrolyte material allowing the migration of positive ions is an acidic electrolyte. If the material permits negative ions to pass through, it is as alkaline electrolyte. Table 7.1 shows the electrochemical reactions in different acidic and alkaline cells. In acidic cells (the first row in the table) the charge carrier is a hydrogen proton. In the other three cases, negative ions – that is hydroxyl, carbonate, or oxygen ions – act as charge carriers across the electrolyte. Depending on the electrolyte material and hence the charge carrier, the electrochemical reactions on the anode and cathode sides are different, even when the same fuel and oxidizer are used. The electrolyte also determines on which side the product forms, and how it is removed from the cell. In acidic cells, water forms on the cathode side where oxygen is supplied, mostly in an air stream, and is removed from the exit stream of the cathode. In alkaline cells, water forms on the anode side, where fuel hydrogen is supplied.

7.3 Fuel Cell Types and Materials

Several types of fuel cells are shown in Table 7.2, including the polymer electrolyte membrane (PEMFC), alkaline (AFC), phosphoric acid (PAFC), molten carbonate (MCFC), and solid oxide fuel cells (SOFC). PEMFCs are also known as proton exchange membrane fuel cells. Despite the variety, the number of overall reactions and the number

Table 7.2 Fuel cell designs and their operating conditions; except for molten carbonate and solid oxide cells, all others use hydrogen

Fuel cell	Proton exchange	Alkaline	Phosphoric acid	Molten carbonate	Solid oxide
Electrolyte	Polymer ion exchange membrane	Potassium hydroxide in asbestos	Liquid phosphoric acid in SiC	Liquid molten carbonate in $LiAlO_2$	Perovskites
Electrode	Carbon	Transition metals	Carbon	Nickels and nickel oxides	Perovskites/ metal cermet
Catalyst	Platinum	Platinum	Platinum	Same as electrode	Transition metal
Temperature	Less than 100 °C	80–220 °C	~ 200 °C	600–700 °C	700–950 °C
Charge carrier	H^+	OH^-	H^+	CO^{3-}	O^-

of electrochemical redox pairs are rather limited, as shown in Table 7.1. Most fuel cells use hydrogen and air as fuel and oxidizer, respectively. A number of high-temperature fuel cells use a mixture of hydrogen and carbon monoxide, or methane. Furthermore, the number of internal change carriers is limited. For these reasons, it is possible to analyze fuel cell systems rather generically, although the types of electrolyte used and the nature of the electrode material may affect the performance. Table 7.2 shows the electrolyte material, the electrode and the catalyst deposited on its surface, the internal charge carrier, and the interconnect material of several fuel cells. It also shows the operating temperature, and whether a reformer is needed when using hydrocarbon fuels. A reformer is used to break down a complex hydrocarbon fuel to hydrogen and carbon monoxide, or hydrogen and carbon dioxide, when the hydrocarbon is used to power a cell built for hydrogen consumption. Thus, the need for a reformer is determined by the tolerance of the cell, in particular the catalyst, to carbon and carbon monoxide and hydrocarbons. Water, and heat management – that is, how to remove water and maintain the cell temperature at the desired values, are also important in some cases. Some of the construction detail will be summarized in later sections, after a detailed compilation of the different loss mechanisms and approaches to model them, is covered.

7.4 Polarization Curves

Under working conditions – that is, beyond the condition of equilibrium in which the cell current is zero and its potential corresponds to the thermodynamic limit – fuel cells produce a finite current. Under these conditions, the power, $\mathcal{P}$, is the current, I, times the voltage, V:

$$\mathcal{P} = IV. \tag{7.1}$$

The current per unit area of the MEA, i, and so the power, $\mathcal{P}$:

$$\mathcal{P} = iV. \tag{7.2}$$

As more current per unit area is drawn, the potential difference across the electrodes drops due to a number of mechanisms, including the finite-rate kinetics at the electrode–electrolyte interfaces, called activation overpotential, the finite transport fluxes of species internally, called the mass transfer overpotential, and the ohmic losses associated with the flow of electrons in the electrodes and ions through the electrolyte. Higher currents decrease the voltage efficiency, defined in (4.54), as well as the overall efficiency.

The relation between the current density and the operating voltage is known as the polarization curve or the characteristic or performance curve. At zero current, corresponding to open-circuit conditions, the maximum voltage allowed at the thermodynamic equilibrium limit is expected. At this point the power drawn from the cell is also zero. Beyond that, as shown in Figure 7.1, the voltage drops rather quickly at low currents, reaching a lower slope for intermediate current density, then experiencing a precipitous drop at higher currents. As the current increases, the power density increases gradually, reaching a maximum at current densities observed at low voltages, closer to the limiting current density, and drops sharply afterwards. Maximum power does not necessarily coincide with high efficiency (or high voltage), and depending on the operating conditions and application, one may consider operating the fuel cell at maximum power and a lower efficiency, or at a higher efficiency and a lower power density.

As shown in Chapter 4, the fuel utilization efficiency of a fuel cell is defined as follows:

$$\eta_{FU} = \frac{\wp}{\dot{n}_f \Delta \hat{h}_{R,f}} = \frac{IV}{\dot{n}_f \Delta \hat{h}_{R,f}} = \frac{I}{n_e \Im_a \dot{n}_f} \frac{V}{V_{OC}} \frac{\varsigma V_{OC}}{\Delta \hat{h}_{R,f}} = \eta_{far}\ \eta_{rel}\ \eta_{OC}, \tag{7.3}$$

where the three terms stand for the Faradic or current efficiency, the Second Law or voltage efficiency, and the open-circuit efficiency. The total molar fuel rate into the cell is $\dot{n}_f$, and $\varsigma = n_e \Im_a$. The current or Faradic efficiency measures the fraction of the total fuel supply that participates in the electrochemical reaction, (i.e., that produces useful current). The dependence of the voltage efficiency, η_{rel}, on the current density is shown in Figure 7.1. The contribution of the different loss mechanisms, measured by "overpotentials," will be explored shortly.

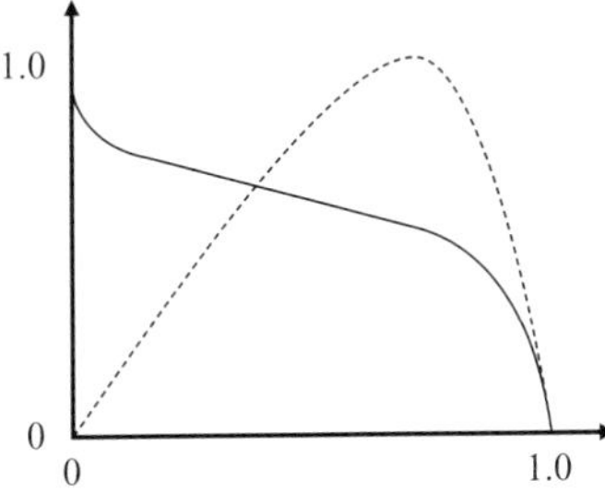

Figure 7.1 The dependence of the fuel cell voltage efficiency, shown in solid line, and the power as a percentage of the maximum power, on the current density as a percentage of the maximum current density.

7.4.1 Faraday's Law

Faraday's law defines the relation between the current and the rate of fuel consumption in an electrochemical reaction, when all the fuel is converted in the reaction, that is, under complete consumption conditions. Since the current density, measured in amperes per unit area, is the charge rate, measured in coulomb/s, then in an ideal electrochemical reaction, where all the fuel participates in the reaction:

$$i = n_e \Im_a j_f, \tag{7.4}$$

where n_e is the number of electrons produced/consumed per molecule of fuel, $\Im_a$, Faraday's number, is the number of molecules per mole, i is the current density, and j_f is the fuel flux, defined as the fuel flow rate per unit cell area. The quantity $n_e \Im_a$ is the total charge produced per mole of fuel consumed. The same law can be written for other reactants participating in an electrochemical reaction, as long as the number of electrons produced for each molecule or mole of reactants is used. For a typical cell operating at $i = 1\,\mathrm{A/cm^2}$, using hydrogen as a fuel, the required hydrogen flux is $j = 5.2\ 10^{-6}\,\mathrm{mol/cm^2s} = 10.4\ 10^{-6}\,\mathrm{g/cm^2s}$. The hydrogen density at standard temperature and pressure is $0.089\ 10^{-3}\,\mathrm{gm/cm^3}$, and hence the corresponding volume flux is $0.117\,\mathrm{cm^3/cm^2 \cdot s}$.

7.5 Kinetics of Electrochemical Processes at Surfaces

Homogeneous thermochemical reactions in gas phase between molecules or atoms involve collision between two or more bodies that possess sufficient energy for the reaction to proceed. Energy is required to either break a chemical bond or form another bond. In these reactions, the rate or the probability of a chemical reaction is determined by the rate or probability of molecular collision times the rate or probability that the colliding molecules possess the requisite energy (the Boltzmann factor). Directionality may also be important and one needs to multiply by the steric factor, which determines the probability that the colliding molecules are properly oriented. The collision rate depends on the concentration of the gas molecule, and so does the reaction rate. Since the collision rate and the collision energy depend on the temperature, most gas reaction rates are strong functions of temperature (the Arrhenius dependence). Most chemical reactions are reversible and can proceed in the forward and backward directions simultaneously, at different rates, unless equilibrium is reached, at which point both rates are equal and no changes in the overall concentrations are observed. The energy required for a reaction to proceed is called the activation energy, and it is different from the energy of reaction. The activation energy is the energy barrier that the reactants must overcome during collision, or the total energy they must acquire before they can form products. It is the energy of the reactants at their activated states before they form products and give up some of their energy (in an isothermal reaction). This is the essence of the "activated state or transition state" theory of chemical reactions, shown schematically in Figure 7.2.

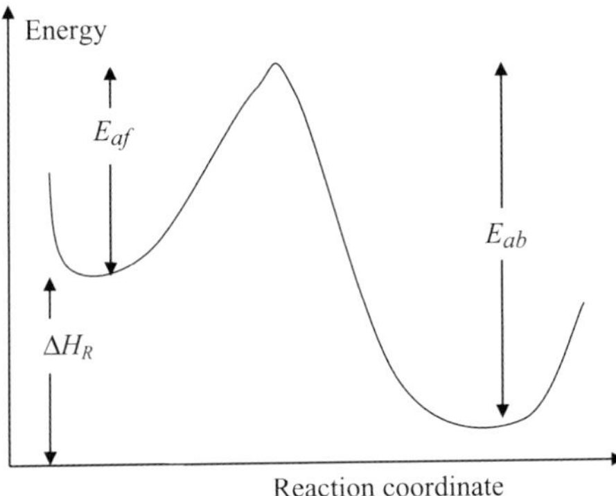

Figure 7.2 A schematic diagram showing the forward and backward reactions in an exothermic thermochemical reaction. In the case of an electrochemical system, the activation energy for the forward and backward reactions are replaced with an activation Gibbs free energy.

For the elementary reversible homogeneous thermochemical reaction between A and B, forming C and D, where v_j is the stoichiometric coefficient:

$$v'_A A + v'_B B \Leftrightarrow v''_C C + v''_D D. \tag{7.5}$$

The reaction rate is the difference between the forward and the backward rates:

$$R_r = k_f(T) C_A^{v'_A} C_B^{v'_B} - k_b(T) C_C^{v''_c} C_D^{v''_D}. \tag{7.6}$$

The dependence of the reaction rate on the concentrations reflects its impact on the collision rate. The reaction rate constants in (7.6) are written in such a way that they capture the dependence of the reaction rate on the temperature,

$$k_f(T) = A_f \, e^{-\hat{E}_{af}/\Re T} \tag{7.7}$$

and

$$k_b(T) = A_b \, e^{-\hat{E}_{ba}/\Re T}. \tag{7.8}$$

The activation energies for the forward and backward rates are $\hat{E}_{af}$ and $\hat{E}_{ab}$, while the pre-exponential factors, A_f or A_b, are also known as the frequency factors. The transition state theory shows that the difference between the forward and backward activation energies is the enthalpy of reaction, $\left(\hat{E}_{af} - \hat{E}_{ab} = \Delta H_R\right)$, and the ratio of the forward and backward reaction rate constants is $k_f(T)/k_b(T) = K_C(T)$, which is the concentration-based equilibrium constant. The exponential term, in the general case, is the activation Gibbs free energy, $G_a = E_a - TS_a$, and changes the pre-exponential factor to absorb the entropy; that is,

$$k_{f,b}(T) = \tilde{A}_{f,b} \, e^{-\hat{G}_{af,b}/\Re T}. \tag{7.9}$$

Reactions between complex molecules (global reactions) occur in a series of "elementary" steps that define the global mechanism. In homogeneous reactions, we first see dissociation or decomposition of large stable molecules (i.e., the reactants) to form smaller and elementary molecules or atoms, and radicals (atoms or molecules with a strong affinity to react), followed by radicals reacting among themselves or with larger molecules to form other molecules and/or more radicals, and finally recombination reactions to form the new molecules (i.e., the products). The presence of radicals and intermediate species is important for the progress of the overall reaction, and often complicates the analysis. Chemical reactions must satisfy mass and energy conservation. At equilibrium, each reversible reaction step is also in equilibrium, that is, the rates of forward and backward reactions are equal.

Heterogeneous and most catalytic reactions occur between a gas and a solid surface, and gases adsorbed on the surface. In heterogeneous reactions, gas molecules are first adsorbed on active sites of the solid surface. Adsorbed molecules hop between neighboring active sites, react with other adsorbed molecules, and get desorbed back to the gas phase. The overall rate of reaction, or the probability that a surface reaction occurs, is the product of the probabilities that the molecules collide with the surface on an empty active site. The collision with the surface has the requisite direction, and the collision has the energy required for adsorption. Other events include reactions between the adsorbed molecules, and desorption of the product molecule from the surface back to the gas.

Figure 7.3 shows the series of events required for a catalytic heterogeneous reaction to occur: the adsorption of the gas molecules by the surface, followed by surface reactions among mobile molecules, then desorption of other gas molecules.

Adsorption and desorption reactions may be chemisorption (a molecule forming a chemical bond with the surface) or physisorption (a molecule forming a physical bond with the surface), depending on the surface and the molecules. The sequence of events that follows is complicated, and many reaction steps that constitute the global mechanism may occur in series or simultaneously, depending on the conditions. This chain of events occurs between molecules very close to the surface and the surface itself – that is, reactants molecules must first diffuse from the bulk of the gas to the surface before chemical reactions can occur, and products molecules must diffuse away from the surface after they form on the surface and get desorbed. Faster rates of surface reactions are enabled by impregnating the surface with

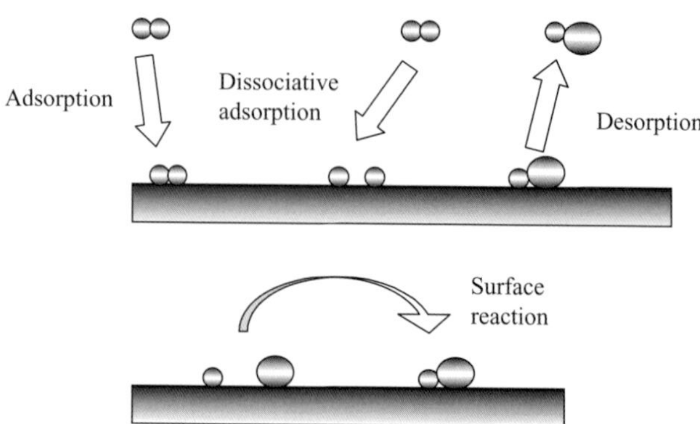

Figure 7.3 Processes at the atomistic level in surface/catalytic reactions.

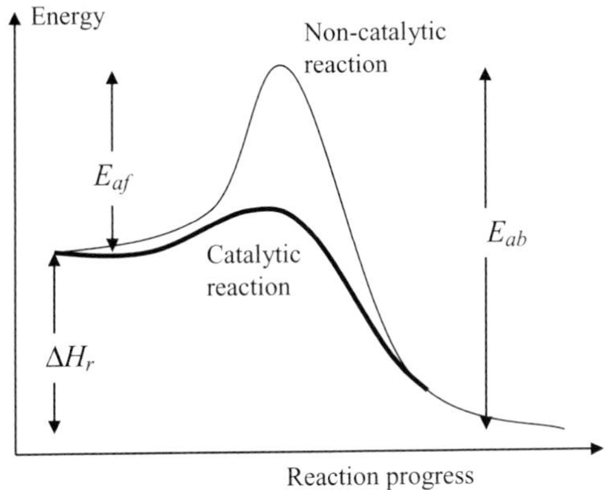

Figure 7.4 The activation energy required for non-catalytic and catalytic reactions.

catalytic particles, often metallic particles (mostly noble metals or transition metals), which reduce the energy barrier, or the activation energy required for the reaction to occur. Catalytic surfaces are used to reduce the reaction temperature without reducing the reaction rates. Noble metals, such as platinum, are excellent low-temperature catalysts (used extensively in the catalytic converters of combustion engines to remove pollutants such as nitric oxides and carbon monoxide before the exhaust is emitted). Transition metals like nickel and copper are used for intermediate temperature reactions such as reforming. Figure 7.4 shows schematically the impact of catalysis on the activation energy in the forward and backward directions of a chemical reaction.

Electrochemical reactions involve **charged particles** in the presence of an **electric field**. They occur in bulk and on the surface of an **electrode** in homogeneous or heterogeneous form. In some of the latter cases, uncharged particles are adsorbed on surface sites of an electrode, and electrochemical reactions result in the stripping/adding of electrons and the formation of charged species. The reverse reaction can also happen, in which a positive ion is adsorbed onto a surface, where it combines with an electron on a surface to form a neutral product that gets desorbed into the gas. The overall rate of these processes is governed by a combination of diffusive transport (transport due to molecular motion) and migration (transport of charged particles in an electric field) processes that bring the species to the electrode surface, and **electrokinetic** reactions at the surface. The surface electrokinetic processes include adsorption, surface reactions and charge separation/recombination, and desorption. Some of these processes involve neutral particles, such as the adsorption dissociation of a hydrogen molecule, and others involve the ionization of neutral atoms and the formation of an ion and an electron.

Electrochemical (like many thermochemical) reactions are reversible. A "half-cell" reaction is one of the two reactions constituting a redox reaction. Half-cell reactions occur on either side of the electrolyte, and involve charge transfer (separation) that creates a potential difference across a very thin zone next to the surface. The **potential** plays a role in determining the kinetic rate. Here, we review single-step reaction kinetics.

Under equilibrium, a charge-transfer reaction proceeds between the reactants at the electrode and the electrolyte surfaces, causing a local charge separation and the formation of an electric double layer. The layer thickness is in the order of nanometers, and satisfies charge neutrality – that is, the charge is almost equal on both sides of the layer (exactly equal under equilibrium conditions). In an electrochemical cell, an electric double layer is established between each electrode and the electrolyte, and is supported by a reaction on each side – that is, a half-cell reaction. The total potential difference across the cell is the sum of the two potential differences on the two sides of the electrolyte. Under open-circuit equilibrium conditions, the potential difference at each electrode is the equilibrium value, and no net current flow results (local equilibrium can be interpreted as equal and opposite currents within each electric double layer).

A reversible electrochemical reaction that takes place at the electrode can be expressed in the following form, where the reactant or product can be charged:

$$\mathrm{R} \underset{k_b}{\overset{k_f}{\Leftrightarrow}} \mathrm{P} + e, \tag{7.10}$$

where k_f and k_b are the forward and backward reaction rate constants. For instance, in the case of an oxidation reaction at the anode, the forward reaction describes an oxidation reaction in which an element loses an electron (e.g., a hydrogen atom losing an electron to the anode to form a hydrogen proton). The backward reaction is a reduction reaction in which the element gains an electron. The forward reaction rate is proportional to the surface concentration of R, and the backward reaction rate is proportional to the surface concentration of P. Equation (7.10) describes the electrochemical reaction on a single electrode – the anode in this case. However, since it is a reversible reaction, it consists of a forward oxidation reaction and a backward reduction reaction, both occurring at the same surface, and both producing their own currents, as shown in Figure 7.5. Under equilibrium conditions, the two currents are equal and opposite; that is, the net current is zero.

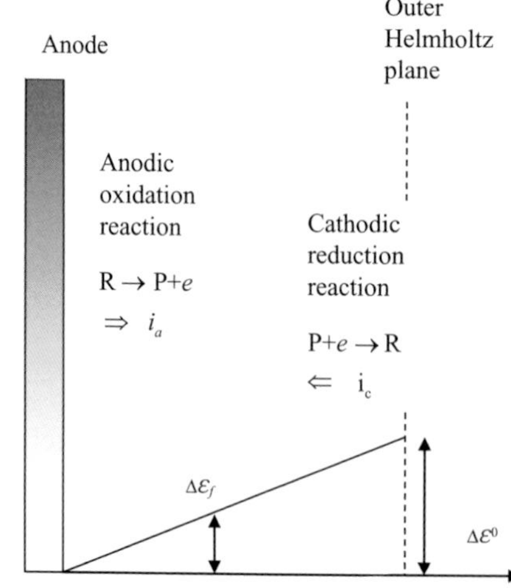

Figure 7.5 The electrochemical reactions at the anode surface, showing the forward and backward reactions, producing the two different currents, and the resulting potential difference between the electrode surface and the outer Helmholtz layer.

Energetically, there is a Gibbs free energy of reaction, $\Delta\hat{G}_R$, associated with the reaction, which proceeds in the forward direction, and its negative value is associated with the backward reaction. Moreover, with the reaction moving forward, there is a potential difference established across an electric double layer, which under thermodynamic equilibrium conditions is given by $\Delta\hat{G}_R = -n_e\Im_a\Delta\varepsilon^o$, where $\Delta\varepsilon^o$ is the equilibrium potential of (7.10). The potential difference associated with a single electrode is known as the **Galvani potential**, and is measured experimentally with respect to a reference electrode (similar to the case of many other energy levels that are not measured absolutely, but with respect to a reference value).

The reaction rate, which is proportional to the rate of formation of charged species, is the current produced at the electrode. The **oxidation** reaction in (7.10) is an **anodic** reaction, producing an anodic **current** proportional to the rate of reaction, with the current density given by

$$i_a = n_e\Im_a\left(k_f C_R\right). \tag{7.11}$$

The surface concentration of the reactant is C_R. The backward reaction is a **reduction** or **cathodic** reaction, producing a cathodic current given by

$$i_c = n_e\Im_a(k_b C_P). \tag{7.12}$$

The net current density, taken as positive if it is an anodic current (positive charge) flowing out of the anode toward the cathode, is the difference between these two current densities in (7.11) and (7.12):

$$i = i_a - i_c. \tag{7.13}$$

A model of the reaction rate constant is constructed using the transition state theory described briefly in previous discussions, and it is extended in the following. There are several ways to construct this model. Assume that the electrochemical reaction takes place between the surface molecules, and another layer of molecules, which, in effect, are adjacent to the first layer of molecules, as shown schematically in Figure 7.5. The extreme extent of the second layer is called the **outer Helmholtz plane**, and the two layers are known as the **electric double layer**.

Recall that the reaction rate constant depends on the activation energy – that is, the energy required to form an activated complex or a transition state, and the energy exchanged in an electrochemical reaction is the Gibbs free energy. For charged species in an electric field, the total Gibbs free energy is the chemical Gibbs free energy plus the energy associated with placing the charge in the electric field $\Delta G^{\#}$. Now, we write the reaction rate constant in the same form as before:

$$k = \widehat{A}\,\exp\left(\frac{-\Delta\hat{G}_a^{\#}}{\Re T}\right). \tag{7.14}$$

We introduce the notation $\Delta\hat{G}^{\#}$ to indicate that the Gibbs free energy is augmented by the energy of the charged species in the electric field – that is $\Delta\hat{G}^{\#} = \Delta\hat{G} + n_e\Im_a\Delta\varepsilon$, where $n_e\Im_a$ is the total charge in a mole of reactants and $\Delta\varepsilon$ is the electrical potential (see Chapter I, where the "entropy-free" stored energy in a charge placed in an electric field is defined). Equation (7.14) applies to both the forward and backward reactions:

$$k_{f/b} = \widehat{A}_{f/b}\ \exp\left(\frac{-\Delta\hat{G}^{\#}_{af/ab}}{\Re T}\right). \tag{7.15}$$

Before the onset of the chemical reaction at the electrode surface leads to charge separation across the double layer, the Gibbs free energies of activation are chemical energies only, $\Delta G^{\#}_{af,ab} = \Delta G_{af,ab}$, and as before, $\Delta G_{af} - \Delta G_{ab} = \Delta G_R$. After charge separation, and under equilibrium conditions, the local value of the Gibbs free energy changes by the electrical energy, such that

$$\Delta G^{\#0}_{af} = \Delta G_{af} + \Im_a \Delta\varepsilon_f \tag{7.16}$$

and

$$\Delta G^{\#0}_{ab} = \Delta G_{sb} + \Im_a \Delta\varepsilon^o = \Delta G_{sf} - \Delta G_R + \Im_a \Delta\varepsilon^o, \tag{7.17}$$

where in this reaction $n_e = 1$, and $\Delta\varepsilon_f < \Delta\varepsilon^o$ is the electric potential change at the anode surface (see Figure 7.5 for a schematic representation of this derivation, and note that the distribution of the potential is purely notional), which causes the Gibbs free energy of the forward reaction to change (the peak of the activation energy is reduced by $|\Im_a \Delta\varepsilon_f|$ and hence the forward Gibbs free energy of activation is reduced by the same amount). The values of $\Delta G^{\#0}_{af,ab}$ indicate the electrochemical Gibbs free energies under equilibrium conditions.

7.5.1 Reaction Rates at Equilibrium

At equilibrium, the anodic and cathodic currents are equal, the concentrations of the reactants and products are the same, $C^*_R = C^*_P$, and hence $k^0_f = k^0_b$. In this case, the net current drawn from the electrode is zero, and the chemical energy supports the separation of charge at the electrode surface without producing a finite net current. The potential difference between the electrode surface and the electrolyte, across the double layer, corresponds to the equilibrium value of the potential, $\Delta\varepsilon^O$:

$$-n_e \Im_a \Delta\varepsilon^o = \Delta\hat{G}_R. \tag{7.18}$$

Moreover, at equilibrium, the forward and backward reaction rates, and the two corresponding currents are equal; that is

$$i_o = k^0_f C^*_R = k^0_b C^*_P, \tag{7.19}$$

which is known as the exchange current density. It is the current, in both directions, exchanged between the electrode surface and the outer Helmholtz plane under equilibrium

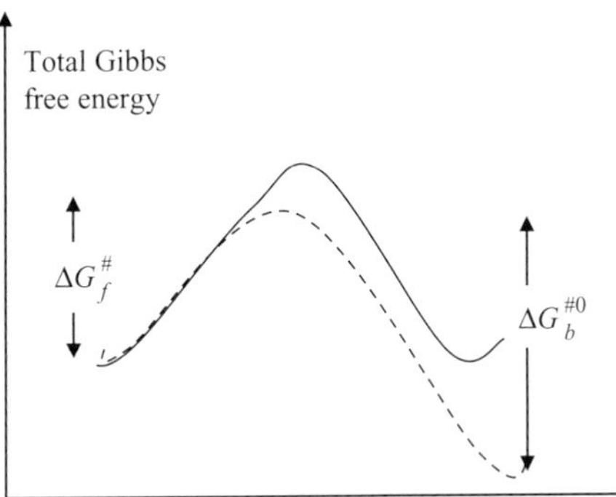

Figure 7.6 The reaction progress diagram under conditions of electrochemical equilibrium, where the total activation Gibbs energy is the same for both directions. The reaction progress diagram in the presence of an activation overpotential is shown by the broken line.

conditions. All quantities in (7.19) are evaluated under equilibrium conditions, including the concentrations and the energies in the rate constants. Higher values for the exchange current density indicate fast dynamics (low activation energy). Figure 7.6 shows the transition state theory representation of an electrochemical reaction at equilibrium.

The **exchange current density** plays an important role in determining the performance of the fuel cell electrode under finite current non-equilibrium conditions. It can be written, for a reaction generating/consuming n_e electrons, as follows:

$$i_o = (n_e \Im_a) \hat{A}_f C_O^* e^{-\Delta G_{af}^{\#0}/\Re T}. \tag{7.20}$$

This expression shows that the exchange current density at the electrode surface depends on the reactants concentration, temperature, and activation energy barrier. The activation energy can be reduced by depositing an active electrocatalyst on the surface. Increasing the catalyst surface density raises the number of active sites on the electrode surface where the reaction may occur. Using nano-structured porous electrodes provides the requisite large surface area where the catalyst material can be deposited and the extra contact surface between the catalytic material, the gas, and the electrons, called the triple phase boundary (TPB), can be extended. Increasing the area of the TPB can also be accomplished through special design of the electrode surface where the catalyst is deposited. The ratio of the active surface area to the geometrical area, or the surface roughness, can reach values of 500–1000. Raising the temperature and pressure also increases the exchange current density through the exponential term and the concentration.

The exchange current density, at 25 °C, varies from 10^{-3} A/cm^2 (measured per unit real surface area of metal catalyst) for hydrogen oxidation on platinum electrodes and acidic electrolyte, to 10^{-4} A/cm^2 for the same but in alkaline electrolyte. The value drops by three orders of magnitude if iron is used as a catalytic metal on the electrode, and by nine orders of magnitude if mercury is used. For oxygen reduction, typical values are 10^{-9} A/cm^2 for platinum surfaces in acidic electrolyte, and drops by two orders of magnitude if other metals are used, such as lead and iridium. For alkaline electrolytes, the value varies between 10^{-9} and 10^{-8} A/cm^2 for pure platinum to platinum deposited on other metals such as iron and cobalt [6,7].

7.5.2 Reaction Rates at Finite Current

Under finite current conditions, the potential difference between the two planes of the double layer is no longer the equilibrium potential. The net current, or the difference between the forward and the backward currents, creates a potential difference between the two planes within the electric double layer. The difference between the ideal potential and the finite net current potential is known as the **activation overpotential**. According to (7.15), the two reaction rate constants change from their equilibrium values because the total free energies of the two planes are no longer the same. The following model is developed to relate the net current to the perturbation in the potential defined as the difference between the ideal potential and the potential under finite current conditions, or the **kinetic overpotential**.

Now we go back to (7.15)–(7.17) for the reaction rate constant and the Gibbs free energy (for activation). If the total potential difference across the double layer is changed (see Figure 7.7) from $\Delta\varepsilon^0$ to $\Delta\varepsilon$ by the overpotential $\tilde{\eta}_{act}$, which is negative if the potential difference is a loss, defined as

$$\tilde{\eta}_{act} = \Delta\varepsilon - \Delta\varepsilon^0, \tag{7.21}$$

then the new total Gibbs free energies of the forward and backward reactions change such that

$$\Delta\hat{G}^{\#}_{af} = \Delta\hat{G}^{\#0}_{af} - \alpha\Im_a\tilde{\eta}_{act} \tag{7.22}$$

and

$$\Delta\hat{G}^{\#}_{ab} = \Delta\hat{G}^{\#0}_{ab} + (1-\alpha)\Im_a\tilde{\eta}_{act}. \tag{7.23}$$

The constant α is known as the **transfer coefficient**, $0 \geq \alpha \geq 1$, and its value depends on the energy barrier in both reaction directions. In many cases, it is taken as $\alpha \approx 0.5$. Substituting the free energies of the forward and backward reactions, from (7.22) and (7.23), into the

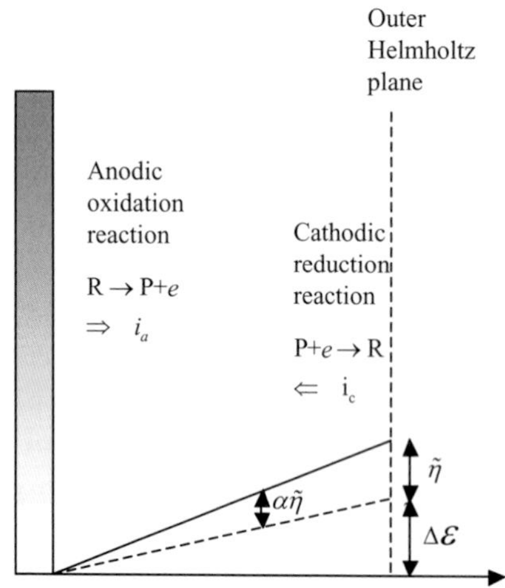

Figure 7.7 An overpotential at the electrode surface lowers the overall potential of the cell below its value under equilibrium conditions. The solid line shows the Galvani potential of the electrode under equilibrium, and the broken line shows the same in the presence of an overpotential, at finite current.

reaction rate constant equation (i.e. (7.15)), we get the form of the rate constants for the forward and backward reactions:

$$k_f = \widehat{A}_f \exp\left(\frac{-\Delta \hat{G}_{af}^{\#0}}{\Re T}\right) \exp\left(\alpha \frac{\Im_a \tilde{\eta}_{act}}{\Re T}\right), \tag{7.24}$$

$$k_b = \widehat{A}_b \exp\left(\frac{-\Delta \hat{G}_{ab}^{\#0}}{\Re T}\right) \exp\left(-(1-\alpha)\frac{n\Im_a \tilde{\eta}_{act}}{\Re T}\right). \tag{7.25}$$

Substituting in the current density equation (7.13), we get the net anodic current density of an electrode. Assuming that the concentrations do not change much from their values at equilibrium, and using the definition of the exchange current density, we get after generalizing for a reaction generating/consuming more than one electron:

$$i = i_0\left\{\exp\left(\alpha \frac{n_e \Im_a \tilde{\eta}_{act}}{\Re T}\right) - \exp\left(-(1-\alpha)\frac{n_e \Im_a \tilde{\eta}_{act}}{\Re T}\right)\right\}. \tag{7.26}$$

This equation is known as the **Butler–Volmer** equation, and relates the kinetic overpotential to the current density. The overpotential of an anode is positive and so is the anodic current. A cathodic current is negative and so is the cathodic overpotential. The total overpotential of an electrochemical cell is given by

$$\begin{aligned} \tilde{\eta} &= \Delta\varepsilon_{cell} - \Delta\varepsilon^o_{cell} = (\Delta\varepsilon_{cathode} - \Delta\varepsilon_{anode}) - \left(\Delta\varepsilon^o_{cathode} - \Delta\varepsilon^o_{catode}\right) \\ &= \tilde{\eta}_{cathode} - \tilde{\eta}_{anode} = \sum |\tilde{\eta}_{electrode}|. \end{aligned} \tag{7.27}$$

Equation (7.26) is sometimes written in terms of two transfer coefficients:

$$i = i_0\left\{\exp\left(\alpha_a \frac{n_e \Im_a \tilde{\eta}_{act}}{\Re T}\right) - \exp\left(-\alpha_c \frac{n_e \Im_a \tilde{\eta}_{act}}{\Re T}\right)\right\}. \tag{7.28}$$

Equation (7.26) assumes that the concentrations of P and R do not change from their equilibrium values. This assumption can be relaxed by rewriting (7.28) in the following form:

$$i = i_0\left\{\frac{C_R}{C_R^*}\exp\left(\alpha_a \frac{n_e \Im_a \tilde{\eta}_{act}}{\Re T}\right) - \frac{C_P}{C_P^*}\exp\left(-\alpha_c \frac{n_e \Im_a \tilde{\eta}_{act}}{\Re T}\right)\right\}, \tag{7.29}$$

where the asterisk denotes the equilibrium value. The total potential in (7.27) against the current density is shown in Figure 7.8 for different values of exchange current density. As shown in this figure, the kinetic overpotential drops rapidly as the current increases, but it levels off at a high exchange current density. High exchange current density indicates fast kinetics and low overpotential at high net current.

The magnitude of the activation overpotential depends on the rates at which the reaction proceeds – that is, the current produced. The overpotential times the current represents lost electrical power that is converted into thermal energy. In a working cell at finite current, kinetic overpotential losses occur at the anode and the cathode, but at different magnitudes

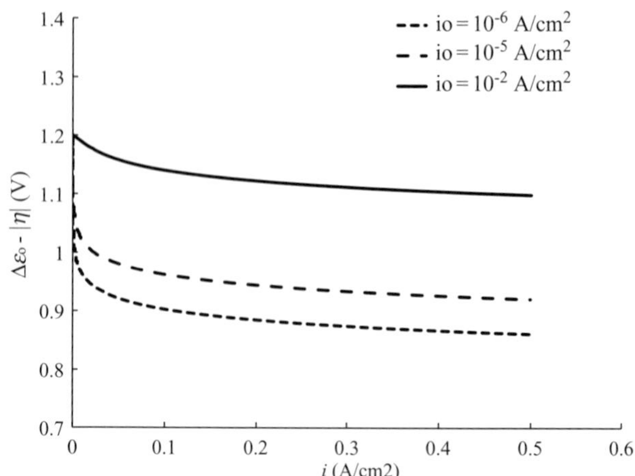

Figure 7.8 Effect of activation overpotential on fuel cell voltage, as a function of exchange current density of the electrode ($\alpha = 0.5$, $n_e = 2$, at $T = 25\,°C$).

that depend on the exchange current density of the electrode, transfer coefficient, electrode design, reaction, and catalyst used in each electrode. The exchange current density depends on the reactant and its concentration, the catalyst and its surface distribution. The dependence of the exchange current density on the electrochemical kinetics and electrode properties, $i_0 = A_e \Im k^0 C_O$, shows the impact of increasing the temperature, surface area, and catalytic properties on the activation losses.

In general, hydrogen chemistry is much faster than oxygen chemistry, making kinetic overpotential of the hydrogen electrode smaller than that of the oxygen electrode. When air is used as an oxidizer, the overpotential is increased because of the lower concentration of oxygen compared to the pure oxygen case. It is often the case that the overpotential on the hydrogen anode is negligible compared to that on the oxygen cathode, in the case of acidic electrolyte. More detail are shown in Section 7.9.

Reducing activation polarization is achieved by using electrocatalysis and increasing the surface area of the TPB. This is done through the utilization of small pores, O (10) μm diameter, that create a large surface area for the reactions between the gases – which must first be adsorbed onto the solid surfaces – to proceed. A catalyst layer is usually deposited at the interface to facilitate the electrochemical reaction. The ratio of the effective area of the TPB to the geometric area of the electrode is an important design parameter, sometimes called the roughness factor. The choice of the material for the electrode and electrolyte are critical for the operation of the cell; the electrode should be a good electrical conductor, while the electrolyte should be a good ionic conductor. The electrode should also be porous to allow the flow of the fuel/oxygen gas to the interface.

7.5.3 Simplifications of the Butler–Volmer Equation

While the Butler–Volmer equation applies at any current, it is often simplified at the limit of low and high current densities, corresponding to low and high kinetic overpotentials, respectively. At low overpotentials, the exponential terms can be expanded and combined to show that the relation between the anodic current and the anodic overpotential is

$$i = i_0\left(\frac{n_e\mathfrak{I}_a}{\mathfrak{R}T}\right)\tilde{\eta}_{act}. \tag{7.30}$$

Within this range, the current is linearly proportional to the overpotential. This relation is valid for $\tilde{\eta}_{act} < 0.015\,\text{V}$.

On the other hand, at high overpotentials, one of the exponential terms drops with respect to the other terms and we are left with one term:

$$i = i_0 \exp\left(\alpha\frac{n_e\mathfrak{I}_a}{\mathfrak{R}T}\tilde{\eta}_{act}\right). \tag{7.31}$$

This equation is known as the **Tafel equation**. The form of the Tafel equation most often used is obtained by taking the logarithm:

$$\tilde{\eta}_{act} = \frac{\mathfrak{R}T}{\alpha n_e\mathfrak{I}_a}\ln\frac{i}{i_o} = \frac{\mathfrak{R}T}{\alpha n_e\mathfrak{I}_a}\ln i - \frac{\mathfrak{R}T}{\alpha n_e\mathfrak{I}_a}\ln i_o. \tag{7.32}$$

The parameters (α, i_o), being the transfer coefficient and the exchange current density, are kinetic parameters related to the reaction and the electrode design, which can be measured experimentally by fitting the high-current data to (7.32). The coefficients are known as the **Tafel slope**. Experimental measurements support this relation, and such measurements are used to determine the parameters of the equation.

Poisoning of the catalyst, clogging the electrode pores, development of pinholes in the electrolyte, the unintended rise of the concentration of certain components such as water that may flood the cell, etc. must all be avoided to limit the negative impact of the kinetic overpotential.

Example 7.1

An experimental test is conducted to determine the coefficients of the Tafel equation. The activation overpotential at two current densities of 400 mA/cm^2 and 600 mA/cm^2 is measured to be 0.397 V and 0.414 V, respectively. Determine the values of exchange current density and α, assuming $n_e = 2$ and $T = 300\,\text{K}$.

Solution

We first derive an equation for a linear line connecting [ln(0.4 A/cm^2), 0.397 V] and [ln(0.6 A/cm^2), 0.414 V]. Hence,

$$\frac{\tilde{\eta}_{act} - 0.397}{0.414 - 0.397} = \frac{\ln i - \ln 0.4}{\ln 0.6 - \ln 0.4} \Rightarrow \tilde{\eta}_{act} = 0.4354 + 0.0419\ln i\,\cdot$$

Example 7.1 (cont.)

Comparing the above equation with (7.32), we find that

$$\frac{\Re T}{\alpha n_e \Im_a} = 0.0419$$

$$-\frac{\Re T}{\alpha n_e \Im_a} \ln i_o = 0.4354.$$

The value of α is obtained from the first expression:

$$\alpha = \frac{8.314 \times 300}{0.0419 \times 2 \times 96,485} = 0.308.$$

Substituting it into the second expression, the exchange current density is

$$i_o = \exp\left(-\frac{0.4354 \times 0.308 \times 2 \times 96,485}{8.314 \times 300}\right) = 3.1 \times 10^{-5}\,\mathrm{A/cm^2}.$$

7.5.4 Voltage Distribution Inside a Fuel Cell

Figure 7.9 shows schematically the voltage distribution across different components. Note that there is no voltage change across the anode or the cathode, although, as we will show

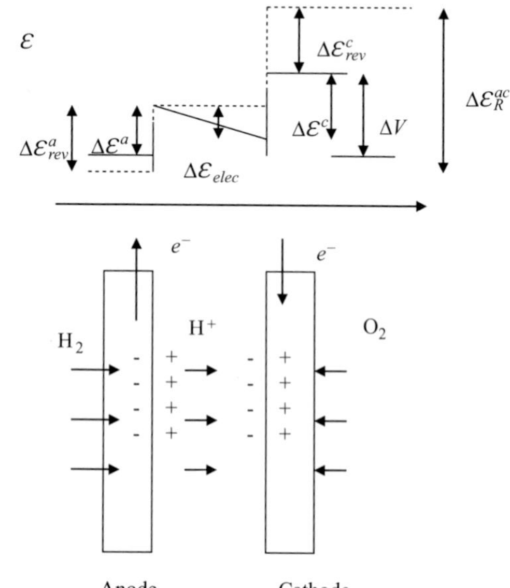

Figure 7.9 Voltage distribution inside the fuel cell, showing the impact of the kinetic overpotentials at the two electrodes, as well as the charge transport overpotential, or ohmic overpotential within the electrolyte. The kinetic overpotential is shown as positive at the anode and negative at the cathode. The charge transport overpotential creates a voltage drop between the anode–electrolyte interface and the cathode–electrolyte interface.

next, the concentrations of species change across these electrodes due to the finite current/ species fluxes, where:

- $\Delta\varepsilon_R^{ac}$ is the open-circuit potential between the two electrodes, which depends on the concentrations of the fuel and oxidizer at the interfaces between the electrodes and the electrolyte, and the current. $\Delta V < \Delta\varepsilon_R^{ac}$, where ΔV is the actual potential difference at finite current;
- $\Delta\varepsilon_{rev}^{a}$ is the equilibrium anode potential, and the actual anode voltage change is $\Delta\varepsilon^{a} < \Delta\varepsilon_{rev}^{a}$, where $\Delta\varepsilon^{a}$ is the potential difference across the anode–electrolyte interface at finite current;
- $\Delta\varepsilon_{rev}^{c}$ is the reversible potential difference of the cathode, and the actual cathode voltage is $\Delta\varepsilon^{c} < \Delta\varepsilon_{rev}^{c}$, where $\Delta\varepsilon^{c}$ is the potential difference across the anode–electrolyte interface at finite current; and
- $\Delta\varepsilon_{elec}$ is the ohmic overpotential within the electrolyte.

Besides kinetic overpotentials, there is an ohmic overpotential loss within the electrolyte associated with the electric migration of the positive charge from the anode to the cathode that will be described in a later section. Before that, we describe the transport of uncharged particles across the electrodes, which contributes another loss mechanism, known as the transport overpotentials. This loss mechanism is associated with the drop of the fuel and oxidizer concentrations as they flow through the respective electrodes. The drop in the concentration lowers the available potential at the electrodes.

7.6 Transport of Uncharged Species

Mass transport in fuel cells is driven by several mechanisms, including diffusion, electric migration, and convection. Diffusion is associated with molecular motion of species within other species, driven by concentration gradients in multi-component mixtures. Electric migration is the motion of charged species in an electric field. Convection corresponds to the bulk motion of a fluid, driven by a pressure gradient. Diffusion is the primary mode of transport across the concentration boundary layer in the supply channels and across the porous electrodes. Transport across the electrode is also driven by a small pressure drop between the supply channel and the electrode–electrolyte interface (this pressure drop is too small to impact the cell thermodynamics, though). Electric migration contributes to transport across the electrolye. Convection is the transport mode through the flow channel. In many cases, mass transport occurs via more than one mode; we focus, however, on the primary mode in each case.

Mass transport in multi-component mixtures by diffusion is driven by the concentration gradient, according to Fick's law. Given that j_i is the molar flux, C is the total molar concentration, C_i is the molar concentration of a particular component, and $X_i = C_i/C$ is the mole fraction of the same component, Fick's law states that

$$j_i = -D_i C \frac{dX_i}{dx}, \tag{7.33}$$

with x taken along the direction of diffusion. The mass diffusion coefficient, D_i, accounts for the diffusive transport of a species in the bulk of the gas mixture. In the case of diffusion in porous medium, it is replaced by the effective diffusion coefficient, modified by the effects of porosity and tortuosity:

$$D_i^{eff} = D_i \frac{\varepsilon}{\tau}, \tag{7.34}$$

where ε is the porosity or the fraction of the pore volume to the total volume and τ is the tortuosity or the average length of a pore divided by the thickness of the electrode. Depending on the nature of the electrode, the diffusion coefficient can be the diffusion coefficient in the bulk gas (or the binary diffusion coefficient in the case of a two-component mixture) or the bulk diffusion coefficient plus the Knudsen diffusion coefficient if the average pore diameter of the porous medium is on the same order of magnitude as the mean free path of the gas, or less. For simplicity in writing the expressions, from now on we will drop the "eff" superscript, but it is understood that the effective diffusivity is being used.

When the change in the total molar concentration is weak, it is possible to write Fick's law for mass diffusion in a multi-component mixture in terms of the concentration gradient of the component concentration:

$$j_i = -D_i \frac{dC_i}{dx}. \tag{7.35}$$

For more accurate analysis, Fick's law is replaced by the Stefan–Maxwell relations of multispecies diffusion [8]. In more generalized analysis, the diffusion flux is related to gradients of the chemical potential.

Transport of a single component gas through a porous medium is driven by the pressure gradient. Moreover, the molar concentration of a gas is proportional to its pressure, $C = p/(\Re T)$, and the pressure gradient is associated with a concentration gradient. Thus, the flux of a single component gas through a porous medium is proportional to the concentration gradient.

Figure 7.10 shows schematically different mass transport processes for a hydrogen–oxygen fuel cell in which product-formation reactions occur on the cathode side (an acidic electrolyte cell). Hydrogen is transported from the fuel supply channel, on the backside of the anode, through the porous anode. Hydrogen flows from the inlet to the exit of the fuel cell through a hydrogen supply channel that remains in contact with the anode. Oxygen is supplied to the fuel cell as part of a mixture of oxygen and nitrogen (i.e., air). Air flows in an air supply channel on the backside of the cathode. Oxygen is transported from the air supply channel to the cathode–electrolyte interface.

Water forming on the same interface is transported back to the air supply channel, within the same space when oxygen is transported forward. That establishes a binary mixture of oxygen and water in which the diffusion flux of each component is driven by the concentration gradient of that component. Hydrogen protons forming on the anode–electrolyte

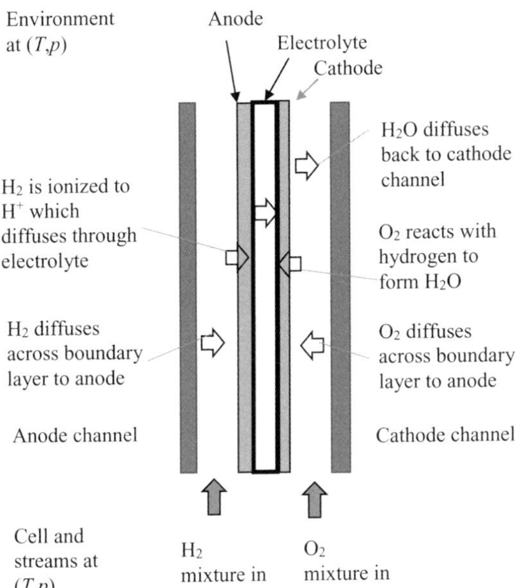

Figure 7.10 Elements of a hydrogen–oxygen acidic fuel cell (such as a PEM) and the transport of different components through these elements. Vertical arrows show flow in the anode and electrode channels, while horizontal arrows indicate flow of fuel, oxidizer, and products between channels and electrodes, and ions across the membrane. Each transport mechanism, including the transport of uncharged and charged species, leads to a certain amount of potential loss due to the drop of the concentration of the transported component.

interface migrate through the electrolyte membrane across the potential field established between the two electrodes.

7.6.1 Concentration Overpotential in Electrodes

To understand the origin of the concentration or mass transfer overpotential, recall that the cell potential depends on the standard potential and the concentration of the species participating in the electrochemical reaction at the electrolyte surfaces. The values of the concentrations at the electrode–electrolyte interfaces are affected by the transport of species across the electrodes, which establishes gradients due to diffusion of uncharged species from the supply channels. To illustrate the impact of the concentration gradients within the membrane electrolyte assembly on the overpotentials, consider again the hydrogen–oxygen cell in the case of a PEM fuel cell where the product-formation reaction occurs at the cathode surface. For this cell, the open-circuit potential, at zero current, is given by

$$\Delta\varepsilon(T,p_i) = \Delta\varepsilon^o(T) + \frac{\Re T}{2\Im_a}\left(\frac{1}{2}\ln p + \ln\left(\frac{X_{H_2}X_{O_2}^{1/2}}{X_{H_2O}}\right)\right). \tag{7.36}$$

The mole fractions or molar concentrations in the last expression should be calculated at the anode–electrolyte and cathode–electrolyte interfaces, where the electrochemical reactions take place. These values are different than those in the supply channel, and depend on the transport fluxes, or the current. Because of the internal transport processes that establish gradients between the fuel and oxygen channels and the electrode–electrolyte interfaces, the concentrations at the reactive interfaces are not the same as those in the fuel/oxidizer channels. The impact of the lower concentration on the potential constitutes a loss.

When a pure gas is supplied and products do not mix with that gas, such as hydrogen in the acidic cell, its concentration remains fixed at $X_i = 1$ in the supply channel. The gas concentration is slightly lower at the electrode–membrane interface due to the pressure drop associated with the transport across the electrode. On the other hand, the concentration of a component could be less than unity in the supply channel, such as in the case when air is used as an oxidizer. Moreover, when a product forms at the cathode–membrane interface and diffuses back toward the supply channel, it competes with the gas diffusing in the opposite direction. In this case, a binary mixture of oxygen and products form within the electrode, and the concentrations of oxygen and products vary strongly between the two sides of the electrode (see Figure 7.11).

To illustrate the impact of changing the concentration on the electrode potential, let us consider a simple case in which the concentration of a single species at the electrode surface, in this case the cathode, determines the potential:

$$\Delta\varepsilon = \Delta\bar{\varepsilon}^o + \frac{\Re T}{n_e \Im_a} \ln C_i. \tag{7.37}$$

The standard cell potential, $\Delta\varepsilon^o$, has been modified to absorb all other "constants" in the problem, $\Delta\bar{\varepsilon}^o$. We will use this equation to estimate the concentration overpotential in cases when the concentration C_i changes across the electrode, between the supply or backside and the membrane or the surface side. The concentration C_i at the electrode–electrolyte interface is C_{is}. At zero current, the concentration at the interface is the same as that at the flow channel (fuel or oxidizer), which we call C_{ib}:

$$\Delta\varepsilon = \Delta\bar{\varepsilon}^o + \frac{\Re T}{n_e \Im_a} \ln C_{ib}. \tag{7.38}$$

The fact that the concentration is depleted by the reaction at the interface, and hence the concentration is lower, means that the potential is different:

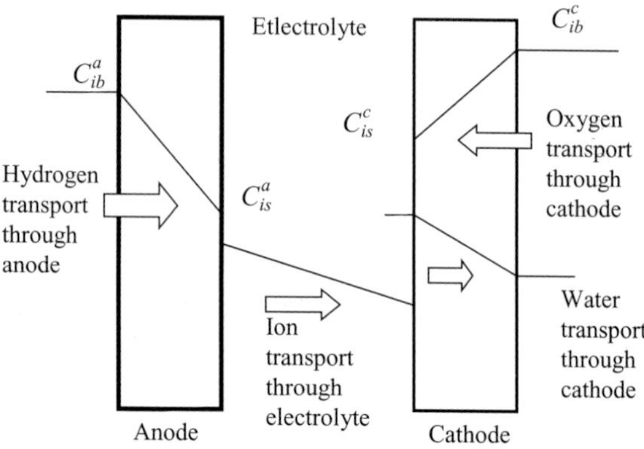

Figure 7.11 Distributions of uncharged and charged species within a fuel cell, in the case when positive ions migrate from the anode to the cathode, similar to the case of a hydrogen–oxygen PEM fuel cell.

$$\Delta\varepsilon = \Delta\bar{\varepsilon}^o + \frac{\Re T}{n_e \Im_a} \ln C_{is}. \tag{7.39}$$

We note that $C_{ib} > C_{is}$ under conditions of finite current. The difference,

$$\Delta\varepsilon(C_{is}) - \Delta\varepsilon(C_{ib}) = \tilde{\eta}_{el,conc}, \tag{7.40}$$

is the concentration overpotential that we are trying to estimate. In terms of the concentrations, the concentration overpotential is

$$\tilde{\eta}_{conc} = \frac{\Re T}{n_e \Im_a} \ln \frac{C_{is}}{C_{ib}}. \tag{7.41}$$

The presence of a gradient within the electrode forms a flux according to Fick's law (7.35). Given the values of the concentration in the bulk and at the electrode–electrolyte interface, the molar flux can be estimated as

$$j_i = D_i \frac{C_{ib} - C_{is}}{\delta_{el}}. \tag{7.42}$$

The diffusion coefficient D_i is used to characterize this diffusion process, and δ_{el} is the thickness of the electrode. The "anodic" current density associated with this flux is given by Faraday's law:

$$i = n_e \Im_a j_i = n_e \Im_a D_i \frac{C_{ib} - C_{is}}{\delta_{el}}. \tag{7.43}$$

The maximum or limiting current density corresponds to $C_{is} = 0$, and is given by

$$i_{\lim} = n_e \Im_a D_i \frac{C_{ib}}{\delta_{el}}. \tag{7.44}$$

The limiting current is a property of the cell design, and it is a function of the concentration of the reactant in the supply channel, the diffusivity within the electrode, the operating pressure, and temperature, as well as the number of charges exchanged in the electrochemical reaction at the electrode–electrolyte interface. Most cell designs have $i_{\lim} = O(1\,\text{A/cm}^2)$. This is based on: $n_e = 1 - 2$, $\Im_a = 98,485\,\text{coulombs/mol}$, $D_i \sim 10^{-5} - 10^{-6}\,\text{m}^2/\text{s}$, $C_{ib} \sim 10\,\text{mol/m}^3$, and $\delta_{el} = 10^{-4}\,\text{m}$.

The limiting current density, an important parameter for the operation of the cell, can be raised by increasing the effective diffusivity and the concentrations in the supply channels, and by decreasing the electrode thickness.

A fast electrochemical reaction at the electrode surface consumes the reactants as they arrive at the anode–electrolyte interface, leading to lower concentration of reactants on that surface: $C_{is} \to 0$. This is consistent with the high flux and large current density expressions in (7.42) and (7.43), respectively. A zero concentration at the reactive interface is approached at a very fast electrochemical reaction rate. Since our objective is to relate the concentration overpotential to the net current density, we use (7.44) in (7.43) to express the current density in terms of the bulk concentration and the concentration at the electrode–electrolyte interface:

$$\frac{C_{is}}{C_{ib}} = \left(1 - \frac{\delta_{el}}{n_e \Im_a D C_{ib}}\right) = 1 - \frac{i}{i_{\lim}}. \tag{7.45}$$

Next, we substitute (7.45) in the relation between the overpotential and the concentration ratio; that is, (7.41):

$$\tilde{\eta}_{conc} = \frac{\Re T}{n_e \Im_a} \ln\left(1 - \frac{i}{i_{\lim}}\right). \tag{7.46}$$

At zero current/flux, the concentration overpotential is zero. As the current/flux increases, the absolute value of the concentration overpotential rises and the gradients within the electrode become steeper. Note that this overpotential is the difference between the potential corresponding to the actual concentration at the electrode surface and that which would have been obtained if the concentration had been the same as the bulk concentration. This is also known as the Nernstian loss.

We should note that the drop in the concentration due to the diffusion through the electrode affects the kinetic overpotential, since the latter depends on the concentration of the reactive species at the electrolyte surface, according to (7.20) or (7.29). The activation overpotential should be calculated using C_{is} instead of C_{ib}. The difference between the kinetic overpotential calculated on the basis of C_{ib} and that calculated using C_{is} is known as the "reaction loss." Using the Tafel equation – (7.31) – to estimate the kinetic overpotential (at the high current limit), this difference can be shown to be

$$\tilde{\eta}_{act,conc} = \frac{\Re T}{\alpha n_e \Im_a} \ln \frac{C_{is}}{C_{ib}} = \frac{\Re T}{\alpha n_e \Im_a} \ln\left(1 - \frac{i}{i_{\lim}}\right), \tag{7.47}$$

where the ratio between the concentrations on both sides of the electrode is given by (7.46). This extra mass transfer loss is now added to the original kinetic activation overpotential calculated on the basis of the concentration in the flow channel – (7.32) – to estimate the total activation loss. Alternatively, we can add up the concentration loss in (7.46) and the activation loss due to the concentration gradient in (7.47) to get the total loss due to the concentration gradient:

$$\tilde{\eta}_{conc,tot} = \frac{\Re T}{n_e \Im_a}\left(1 + \frac{1}{\alpha}\right) \ln\left(1 - \frac{i}{i_{\lim}}\right). \tag{7.48}$$

The concentration loss in (7.48) should be added to the kinetic overpotential calculated on the basis of C_{ib}. The results in Figure 7.12 show typical values of the total concentration overpotential for different values of the limiting current density. The impact of this loss is rather small for small current densities, and increases rapidly at a higher current density to a point where the cell voltage drops to zero. Experimental measurements show that while the trend is accurate, the constant in front of the logarithmic term in (7.48) is under-predicted.

There are two concentration overpotentials associated with the two electrodes.

7.6.2 Concentration Overpotential in the Channel Layer

In cases in which a reactant concentration in less than unity in its own stream, that is, when a carrier gas is used to introduce a reactant into the fuel cell or when products diffuse back to the flow channel and mix with the supply, another type of concentration overpotential arises. This is because, in this case, the reactants concentration decreases between the free stream in the flow channel and the boundary between the flow channel and the electrode, as shown in Figure 7.13. For example, if air is used on the cathode side, oxygen concentration will

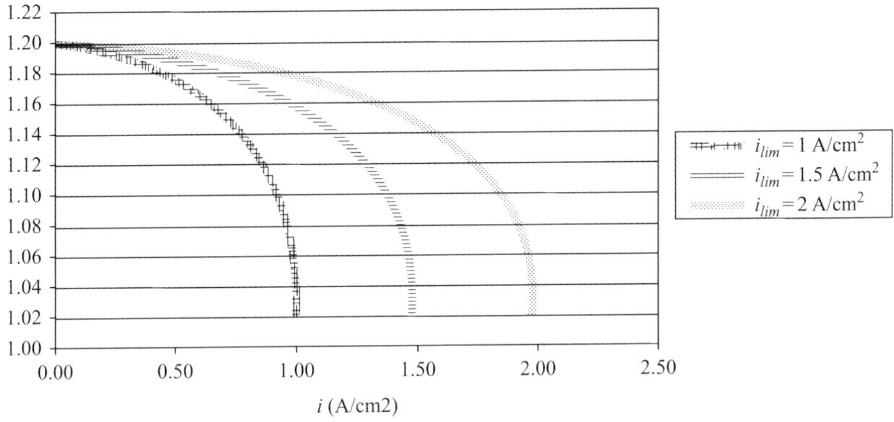

Figure 7.12 Concentration overpotential calculated using (7.48) for the values in the caption. Curves are obtained for three values of the limiting current density, 1.0, 1.5, and 2, all for $\alpha = 0.5$, $n_e = 2$, $T = 300\,\text{K}$.

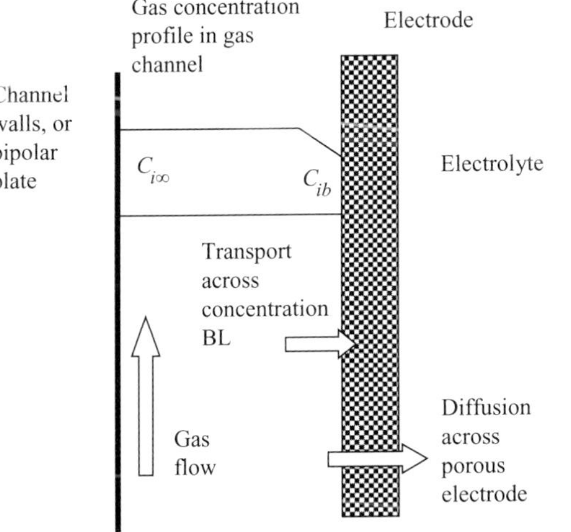

Figure 7.13 The change in the reactants concentration in the flow channel.

decrease between the bulk flow and the backside of the electrode. Momentum and mass diffusion boundary layer theories should be used to determine this change of the concentration in the direction normal to the stream. The transport flux of the reactants between the bulk flow in the flow channel, where its concentration is given as $C_{i\infty}$, and the electrode–channel boundary, where the concentration is $C_{ib} < C_{i\infty}$, is related to their difference – that is, $C_{i\infty} - C_{ib}$. Given that h_{conc} is the convective mass transfer coefficient within the flow channel, the flux equation is

$$j_i = h_{conc}(C_{i\infty} - C_{ib}). \tag{7.49}$$

The convective mass transfer coefficient depends on the flow characteristics within the flow channel. It is often given in terms of the Sherwood number, Sh, a convective mass transfer coefficient normalized with respect to the characteristic length scale and the mass diffusivity. The Sherwood number is a function of the Reynolds number, Re, the Schmidt number, Sc, and wall conditions:

$$\mathrm{Sh} = \frac{h_{conv} D_h}{D_i} = \mathrm{Sh}(\mathrm{Re}, \mathrm{Sc}), \tag{7.50}$$

where D_h is the hydraulic diameter of the channel. The Reynolds number is defined as $\mathrm{Re} = \frac{\bar{U} D_h}{\nu_f}$, and the Schmidt number is defined as $\mathrm{Sc} = \frac{\nu_f}{D_i}$. Under laminar flow conditions, the Sherwood number is O (3–5), depending on the geometry of the flow channel.

Combining (7.42) and (7.49), we rewrite (7.43) and (7.44) in the following forms:

$$i = n_e \Im_a \dot{n}_i = n_e \Im_a \frac{C_{si} - C_{ib}}{\left(\frac{\delta_{el}}{D} + \frac{1}{h_{conv}}\right)} \tag{7.51}$$

and

$$i_{\lim} = n_e \Im_a \frac{C_{i\infty}}{\left(\frac{\delta_{el}}{D_i} + \frac{1}{h_{conv}}\right)}. \tag{7.52}$$

In fact, we have replaced the diffusion resistance in the limiting current equation, $\left(\delta_{el}/D_i^{eff}\right)$, with a resistance that accounts for the concentration boundary layer within the flow channel, $\left(\delta_{el}/D_i^{eff} + 1/h_{conc}\right)$. All other expressions, such as (7.45) and (7.47) remain the same, with $C_{i\infty}$ used in place of C_{ib}. Again, if (7.48) is used to calculate the total concentration overpotential using $C_{i\infty}$, then the activation overpotential should also be based on $C_{i\infty}$. Even when the flow resistance in the electrode is small, the drop in the concentration of the reactant between the "free" stream in the supply channel, away from the electrode surface, and the electrode surface, which establishes a concentration boundary layer within the gas channel, can lead to limiting current conditions. Note that the concentration drop depends on the concentration boundary layer thickness and the diffusivity of the reactive gas in the reactants mixture.

It should be noted here that the reactants concentrations in both supply channels decreases along the flow direction, causing the local current density to decrease as well. A detailed procedure to calculate the local current density at each point along the electrode can be developed.

Example 7.2

A hydrogen fuel cell operating at 85 °C and 1 atm uses circular channels with a diameter of 2 mm. The anode electrode with porosity of 40% is 0.1 mm thick. Determine the limiting current density and the total overpotential on the anode side for an operating current density of 0.3 A/cm^2. Assume a binary diffusion coefficient of 10^{-6} m^2/s, Sh = 3.6, and $\alpha = 0.5$.

Solution

The limiting current density can be calculated using (7.52). First, we need to determine hydrogen concentration at infinity, the effective binary diffusion, and mass transfer coefficients:

$$C_{i\infty} = \frac{p}{\Re T} = \frac{101,300[\text{pa}]}{8.314\left[\dfrac{\text{J}}{\text{mol}\cdot\text{K}}\right] \times (85+273)[\text{K}]} = 34\,\frac{\text{mol}}{\text{m}^3},$$

$$D_i^{eff} = \varepsilon.D_i = 0.4 \times 10^{-6} = 4 \times 10^{-7}\,\text{m}^2/\text{s},$$

$$h_{conv} = \frac{\text{Sh}\cdot D_i}{D} = \frac{3.6 \times 10^{-6}}{0.002} = 0.0018\,\frac{\text{m}}{\text{s}}.$$

Substituting these values in (7.52),

$$i_{\lim} = 2 \times 96,485 \frac{34}{\left(\dfrac{0.0001}{4x10^{-7}} + \dfrac{1}{0.0018}\right)} = 8145\,\frac{\text{A}}{\text{m}^2}.$$

The total concentration overpotential at the anode side is obtained from equation (7.48):

$$\tilde{\eta}_{conc,tot} = \frac{8.314 \times 358}{2 \times 96,485}\left(1 + \frac{1}{0.5}\right)\ln\left(1 - \frac{0.3}{0.8145}\right) = -0.0213\,\text{V}.$$

7.6.3 Electric Resistance Polarization and Ohmic Overpotential

The **migration** of ions across the potential field within the electrolyte and the flow of electrons along the electrodes cause a further loss of potential. In cases in which multiple species carry charge, the generalized Faraday's law, or the relation between charged species' molar flux, j_i, and the electric current density, i, is given by

$$i = \sum z_i \mathfrak{I}_a j_i. \tag{7.53}$$

Note that the charge number, z_i, has the same sign as the charge – it is negative for electrons and negatively charged ions and positive for protons. The summation in this equation is taken over all mobile charged species.

The relation between the current density and the electric potential gradient is

$$i = -\sigma \frac{dV}{dx}, \tag{7.54}$$

where σ is the electrical conductivity. This relation resembles Fick's law of uncharged species diffusion, and Fourier's law of heat conduction. The transport of charged species is similar in nature to other diffusion processes. The current density is measured in terms of amperes/unit area, and the current density is the charge flux, where ampere = coulomb/s, and coulomb is the unit of charge. Thus, similar to the mass or mole flux, the current density is a charge flux. The electrical conductivity is measured in $(\text{ohm}\cdot\text{m})^{-1}$, where the unit of electrical resistance is ohm = volt/ampere.

Electrical conductivity is determined by the molar concentration of the charge-carrying species, C_i, and the mobility of the charge carrier, u_i:

$$\sigma_i = (|z_i|\mathfrak{I}_a) C_i u_i. \tag{7.55}$$

Fundamental expressions for the charge carrier mobility depend on the charge (electrons vs. ions) and the nature of the medium. For instance, in crystalline solid electrolytes, the relation between the mobility and the diffusivity of the charge carrier is given by the Nernst–Einstein relation:

$$u_i = \frac{|z_i|\mathfrak{I}_a D_i}{\mathfrak{R}T}. \tag{7.56}$$

Moreover, the diffusivity depends on the temperature in an Arrhenius fashion:

$$D = D_o e^{-\Delta G_{act}/\mathfrak{R}T}, \tag{7.57}$$

where ΔG_{act} is the Gibbs free energy barrier for the ion-hopping process. These expressions apply to ceramic electrolytes in SOFC.

Combining (7.55) and (7.56), we obtain the following expression for ion conductivity in terms of the diffusivity and the concentration:

$$\sigma_i = \frac{(|z_i|\mathfrak{I}_a)^2 C_i D_i}{\mathfrak{R}T}. \tag{7.58}$$

The exponential terms describing the temperature dependence of the diffusivity in (7.57) dominate over the algebraic dependence in the last equation, raising the diffusivity and mobility of ions with temperature, and reducing the associated potential losses.

For constant charge flux or current density through the electrolyte, (7.54) is written in the form of Ohm's law as follows:

$$i = -\sigma \frac{\Delta V}{\delta_{electolyte}}, \tag{7.59}$$

where $\delta_{electolyte}$ is the thickness of the electrolyte. Rearranging this equation, we get

$$\Delta V = -\frac{\delta_{electrolyte}}{\sigma} i. \tag{7.60}$$

Thus, the ohmic overpotential (potential drop) resulting from an ionic flux in an electric field is proportional to the electrolyte thickness and the current density, as given by the following expression using the total resistance and the total current:

$$\tilde{\eta}_{oh} = -R_{electrolyte} I. \tag{7.61}$$

In this equation, the total absolute value of the current is $I = iA$, where A is the surface area of the electrolyte in the direction normal to the current flow, and the total resistance is the ionic resistance, $R_{electrolyte}$:

$$R_{electrolyte} = \frac{\delta_{electrolyte}}{A_{electrolyte}\,\sigma_{electrolyte}}. \tag{7.62}$$

The total electrical resistance depends on the surface area in the direction normal to the current, A, and the thickness of the conductor, δ. The electric conductivity is the inverse of the electric resistivity, $\sigma = 1/\rho$, which accounts for the resistance of the "conductor" to charge motion within the medium (similar to the resistance of a medium to heat flow in the case of heat conduction). The drop in electric potential is associated with a current flux in the direction of a lower potential, which is the case within the electrolyte, as shown in Figures 7.9 and 7.11. A similar expression to that in (7.61) can be written for the resistance to the electron flow along each electrode, where in this case δ is the length of the electrode and A is the cross-sectional area of the electrode:

$$R_{electrode} = \frac{\delta_{electrode}}{A_{electrode}\,\sigma_{electrode}}. \tag{7.63}$$

Finally, the total ohmic overpotential is the sum of the two expressions:

$$\tilde{\eta}_{ohic-total} = -\left(R_{anode} + R_{cathode} + R_{electrolyte}\right) I. \tag{7.64}$$

Expressions for the electrical conductivity can be written for ion mobility through the electrolyte, and for electron mobility through the electrode media. Mobility can be continuous motion of the ions, or the hopping of ions between neighboring charge sites can be responsible for charge transport (similar to other conduction and diffusion mechanisms). Solid electrolytes are often doped with metals to improve their ion conductivity (or the diffusion/hopping of ions between neighboring vacancies within the lattice). More complex ion transport mechanisms are found in aqueous electrolytes and polymer electrolytes. In the former, a salt dissolved in water is ionized, and the ions transport the charge by moving in the solvent under the impact of the potential difference. In polymer electrolytes, fixed charge sites within the polymer provide centers where the moving ions can reside. Moreover, free volume within the polymer provides another opportunity for mobile ions, carried by a vehicular species (such as water), to be transported across the electrolyte. This mechanism resembles that of aqueous electrolytes, making polymers a better ionic conductor than solid electrolytes, especially at low temperatures. The presence of water in polymer electrolyte, like Nafion, is necessary to take advantage of this extra ion transport mechanism [6]. However, too much water can have an adverse effect since water can diffuse in the direction opposite to that desired for ion transport (see Figure 7.12 for the distribution of different species in a PEM fuel cell).

Typical values for electric conductivity of fuel cell material are in the range of 10^{-2} to $10^{2}\,(\text{ohm·m})^{-1}$. Ion diffusivities of polymer electrolytes are $\sim 10^{-8}\,\text{m}^2/\text{s}$, while they are $10^{-11}\,\text{m}^2/\text{s}$ for ceramic electrolytes at high temperatures. Ion carrier concentration is in the range of 10^{2} to $10^{3}\,\text{mol/m}^3$. Clearly, higher carrier concentration is needed in ceramic electrolytes to achieve the desired electric conductivity. For an electrolyte thickness of $10^{-4}\,\text{m}$, current density of $10^{4}\,\text{A/m}^2$, and electric conductivity of $10\,(\text{ohm·m})^{-1}$, the ohmic overpotential is 0.1 V.

The ion concentration gradient within the electrolyte gives rise to an ion diffusion flux, which is proportional to the ion concentration gradient, independent of the electric potential difference across the electrolyte. The concentration gradient-driven ion flux is given by an expression similar to that in (7.35). This flux increases the total ion flux such that

$$j_i = -D_i \frac{dC_i}{dx} - \frac{\sigma_i}{(|z_i|\Im_a)} \frac{dV}{dx}, \tag{7.65}$$

where D_i is ion diffusivity. Moreover, the current density associated with this total ion flux is given by

$$i = (|z_i|\Im_a)\left(-D_i \frac{dC_i}{dx} - \frac{\sigma_i}{|z_i|\Im_a} \frac{dV}{dx}\right). \tag{7.66}$$

In most cases, the electrical driving force expressed by the second term in (7.65) and (7.66) dominates, and the impact of the concentration gradient on the ion flux/current density is neglected. As such, the impact of the ion transport through the concentration gradient on the ohmic overpotential is neglected with respect to that caused by the ion transport due to the electrical potential difference. Integration of (7.66) for constant current density can be used if the impact of the gradient-driven ion transport is required (note that the electric conductivity depends on the concentration, which should be taken into consideration in the integration) [9,10].

The ion flux lowers the concentration of the ions from the side of the electrolyte membrane where they are formed, to the side where they are consumed. Thus, the ion flux also contributes another form of overpotential associated with (1) the lower concentration of ions, and (2) the lower electrochemical reaction rates and hence the higher kinetic overpotential on the side of the membrane where ions are consumed. *In most approximate models, this form of overpotential is often neglected, given that the electrolyte layer is thin and the drop of the ion concentration is small.*

7.7 Fuel Utilization and Gas Consumption Losses

The drop in the cell potential due to **fuel utilization** is measured by the difference between the cell potential based on the inlet fuel and oxygen molar concentrations and that based on the exit values (Figure 7.14). Depending on the type and design of the cell, the fuel or oxygen concentrations drop continuously along the surface of the electrode as both reactants are consumed and products are produced and mixed with one of the reactants streams. In the case when the fuel/oxygen is supplied as part of a mixture, such as when hydrogen is produced by steam reforming methane to $H_2 + CO_2$, the concentration of H_2 drops between the inlet and exit of the cell. The drop of the concentration affects the cell potential according to the Nernst equation:

$$\tilde{\eta}_{FU} = \frac{\Re T}{2\Im_a}\left(\ln\frac{X_{H_2}^{(2)}}{X_{H_2}^{(1)}} + \frac{1}{2}\ln\frac{X_{O_2}^{(2)}}{X_{O_2}^{(1)}} - \ln\frac{X_{H_2O}^{(2)}}{X_{H_2O}^{(1)}}\right). \tag{7.67}$$

Another loss mechanism comes from the fact that all overpotentials depend on the local concentrations of the fuel and oxidizer, which have been so far assumed to be constant

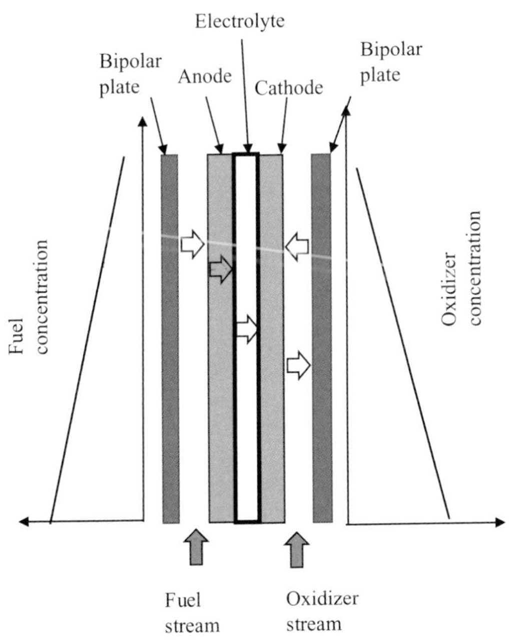

Figure 7.14 A fuel cell using a mixture of oxygen and a diluent in the cathode channel and a mixture of hydrogen and a diluent in the anode channel. Vertical arrows indicate the major direction of flow of different components, while horizontal arrows indicate the flow of fuel and oxidizer from the channels to the electrodes, products from the electrode to the flow channel, and charged species across the electrolyte (membrane).

within the flow channels as we travel along the channel from the inlet to the exit of the fuel cell. However, this is not realistic as both the fuel and oxidizer are consumed, and hence, depending on the flow in both channels, their concentrations change along the flow channel: $C_{i\infty} = C_{i\infty}(y)$. If the fuel stream is a pure fuel, the concentration remains constant. However, if the oxidizer stream is air, then the oxygen concentration will drop as we move along the channel, and will affect all the overpotentials. Careful mass conservation modeling is needed to calculate the local concentrations of the fuel and oxygen concentrations, and relate them to the local current produced within the cell. We also note that the overall potential difference remains the same along the electrode, since the electrodes are made of electrically conducting material, but the current changes. Since the concentrations decrease, and the overpotential increases, the current density decreases from the inlet to the exit of the fuel cell.

7.8 Fuel Crossover/Internal Currents

Even without a finite current drawn from a fuel cell, the potential difference across the cell electrodes may be lower than that predicted by the Nernst open-circuit potential, especially at low temperature. This is often observed in the form of an initial steep drop of the polarization curve at very small currents, as shown in Figure 7.15. The reason for this drop is the unintended conduction of a very small number of electrons across the electrolyte, and some hydrogen fuel as well. Even though the electrolyte is supposed to be a perfect electrical insulator, perfect uncharged species insulator, and perfect ion conductor, in practice it often allows some electrons and uncharged hydrogen to pass through. The voltage drop associated with these internal electron currents is known as crossover overpotential. Moreover, the transfer of fuel across the electrolyte without the associated electrochemical reaction, and the transfer of electrons internally (instead of externally) add to the potential losses. Some

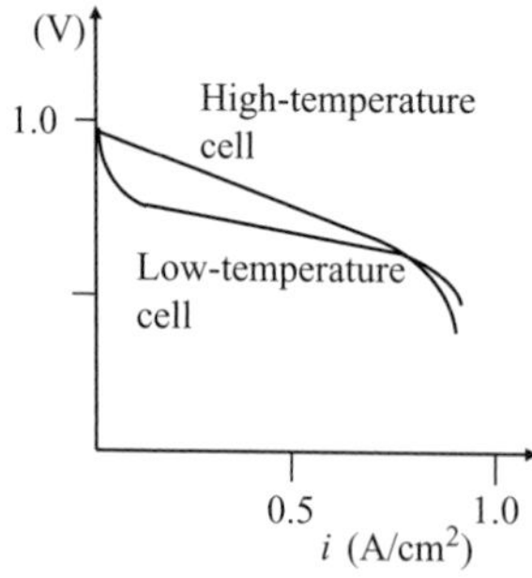

Figure 7.15 Schematics of polarization curves typical for a low-temperature fuel cell, such as a PEMFC, and a high-temperature cell, such as an SOFC. The thermodynamic potential of the low-temperature cell is higher, but it suffers a sharp drop from the open-circuit voltage due to charge leakage and fuel crossover. Moreover, due to the low-temperature operation, kinetic losses are higher and tend to contribute to the higher potential loss at low currents. Higher-temperature cells start out with lower open-circuit voltage, but their voltage decreases more gradually as the current increases.

research has suggested that the overpotential associated with internal currents can be modeled by adding a term i_n to the Tafel equation, such that

$$\tilde{\eta}_{act} = \frac{\Re T}{\alpha n_e \Im_a}\left(\ln \frac{i + i_n}{i_o}\right), \tag{7.68}$$

with $i_n = O(2 - 4\,\text{mA/cm}^{-2})$. Under typical operation of O(1000 mA/cm^2), this is a small current that can be neglected.

Because the electrical conductivity increases with temperature, these losses decrease in high-temperature operation. Moreover, high-temperature operation speeds up the chemical reactions at the electrode surfaces and further contributes to the reduction of the difference between the ideal potential and actual operating potential near zero current.

Figure 7.15 shows the polarization curves of a low-temperature cell (e.g., a PEMFC) and a high-temperature cell (e.g., an SOFC), which are compared to highlight the roles of the different mechanisms in determining the impact of finite current on the performance. For the low-temperature cell, following the sharp drop associated with open-circuit leakage, another rapid drop is seen because of the activation overpotential. The slope decreases as the current increases until transport polarization becomes important, leading to another fast drop in the cell potential. In the high-temperature cell, the open-circuit potential in lower but the initial drop in voltage is very small.

Example 7.3

A hydrogen PEM fuel cell stack consisting of 100 cells each with a surface area of 200 cm^2 operates at 80 °C and 1 atm. Determine the internal fuel crossover leakage that would cause a loss of voltage from 0.95 V to 0.94 V. Assume that the fuel cell operates at the regime of low current density at which only activation losses dominate. Assume $i_0 = 0.04\,\text{mA/cm}^2$ and $\alpha = 0.5$.

Solution

The Gibbs free energy of hydrogen oxidation at 80 °C and 1 atm, with water leaving in liquid form at −228 kJ/mol. Therefore, the open-circuit voltage is

$$\Delta\varepsilon^o = -\frac{-228,000}{2 \times 96,485} = 1.18\,\text{V}.$$

Due to the activation losses, the voltage drops from 1.18 V to 0.95 V. Thus, the current density can be determined using (7.31) as follows:

$$i = i_0 \exp\left(a\frac{n_e \Im_a}{\Re T}\hat{\eta}_{act}\right) = 0.04 \exp\left[0.5\frac{2 \times 96,485}{8.314 \times 353}(1.18 - 0.95)\right] = 76.9\,\text{mA/cm}^2.$$

Example 7.3 (cont.)

Now we calculate the equivalent current density, i_n, which causes a voltage drop from 0.95 V to 0.92 V:

$$\Delta\varepsilon = \Delta\varepsilon^o - \frac{\Re T}{\alpha n_e \Im_a} \ln\left(\frac{\mathrm{i} + \mathrm{i_n}}{\mathrm{i_0}}\right)$$

$$0.94 = 1.18 - \frac{8.314 \times 353}{0.5 \times 2 \times 96,485} \ln\left(\frac{76.9 + \mathrm{i_n}}{0.04}\right).$$

Solving the above equation for i_n, we find

$$i_n = 27\,\mathrm{mA/cm^2}.$$

The internal fuel crossover leakage for one cell is obtained as follows:

$$\dot{n}_{H_2} = \frac{i_n A_{cell}}{n_e \Im_a} = \frac{0.027 \times 200}{2 \times 96,485} = 2.8 \times 10^{-5}\,\mathrm{mol/s}.$$

The equivalent fuel crossover leakage for 100 cells is

$$(\dot{n}_{H_2})_{tot} = (2.8 \times 10^{-5}) \times 100 = 2.8 \times 10^{-3}\,\mathrm{mol/s}.$$

7.9 Overall MEA Model

The above analysis points to the significance of finite-rate chemical kinetics and transport processes in establishing the current–voltage relationship in fuel cells and other electrochemical systems such as batteries and electrolyzers. This relationship is a complex function of the pressure, temperature, electrode material and surface area, catalyst used, thicknesses of the electrodes and electrolyte, and their conductivities.

Fuel cell losses are described in terms of the **polarization** or **overpotential**; that is, the difference between the ideal potential and the actual potential at a given current at any point within the cell, which is lower than the ideal value. Not to be confused by the nomenclature, while "overpotential" in fuel cell applications is always **negative**, it is always either added with its own sign, or subtracted in absolute value. The following expression accounts for these losses, starting with the ideal thermodynamic potential, which is calculated based on the concentration of reactants at the entrance of the cell, and subtracting the overpotential (note that the concentration overpotential across the electrolyte is neglected):

$$\Delta\varepsilon = \Delta\varepsilon^o + \left(\tilde{\eta}_{a,act} + \tilde{\eta}_{a,conc} + \tilde{\eta}_{a,FU}\right) + \tilde{\eta}_{electrolyte,oh} + \left(\tilde{\eta}_{c,act} + \tilde{\eta}_{c,conc} + \tilde{\eta}_{c,FU}\right). \qquad (7.69)$$

The different overpotentials are listed below. The parameters in each overpotential are different for the anode and the cathode:

$$\begin{aligned}
\tilde{\eta}_{act} &= \frac{\Re T}{\alpha n_e \Im_a}\left(\ln \frac{i+i_n}{i_o}\right) && \begin{cases} i_o = \text{exchange current density} \\ \alpha = \text{charge transfer coeff.} \\ i_n = \text{crossover current} \end{cases} \\
\tilde{\eta}_{conc} &= \frac{\Re T}{n_e \Im_a}\left(1+\frac{1}{\alpha}\right)\ln\left(1-\frac{i}{i_{\lim}}\right) && i_{\lim} = n_e \Im_a D \frac{C_{i\infty}}{\left(\frac{\delta_{el}}{D}+\frac{1}{h_{conv}}\right)} \\
\tilde{\eta}_{ohm} &= -\left(\sum_k \frac{\delta_k}{\sigma_k}\right)(i+i_n) && \{\sigma_k = \text{conductivity in } a, c, el \\
\tilde{\eta}_{FU} &= \frac{\Re T}{n_e \Im_a}\left(\ln \frac{X_i^{(exit)}}{X_i^{(inlet)}}\right) && \{X_i^{(exit)} < X_i^{(inlet)}, \text{ channel flow.}
\end{aligned} \tag{7.70}$$

We note that each overpotential results in the conversion of some of the chemical energy of the fuel into thermal energy, at a rate equal to $I^2\tilde{\eta}$. Under isothermal operation, this amounts to heat transfer out of the system equal to the same amount. This is besides the heat generated by (unwanted) fuel that reacts thermochemically and the reversible part ($T\Delta S$).

Figure 7.16 shows the polarization curve of a SOFC using hydrogen (50% H_2 in H_2O). Both the overall cell potential, different overpotentials associated with different

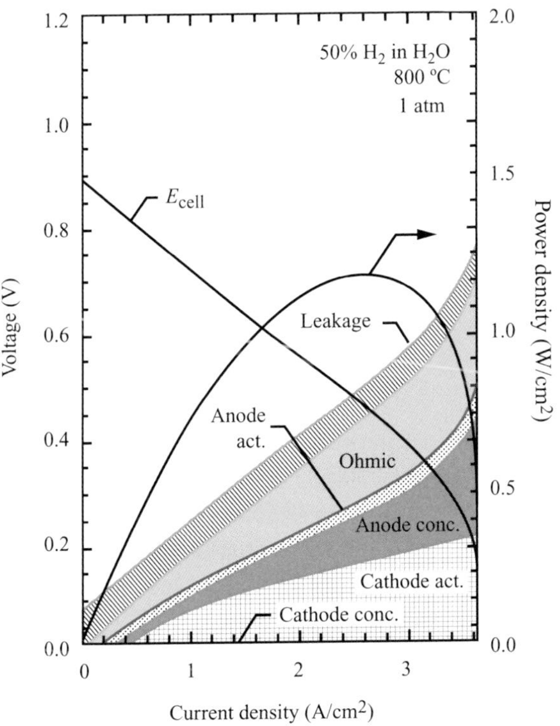

Figure 7.16 Dependence of cell voltage, overpotentials, and power density on current density [12].

loss mechanisms, and the cell power are shown. Calculations were performed using a detailed (differential equation-based) model of the MEA in which the transport through the porous electrodes was described by the "dusty gas" model, and multistep electrochemical mechanisms were used for oxygen reduction and hydrogen oxidation reactions [8, 11]. The results are shown here to illustrate the role of the different loss mechanisms in determining the voltage as a function of current, and the corresponding power curve. The shaded areas show the contributions of the various overpotentials plotted in an additive format. Thus, the top of the shaded region is a quantitative measure of the total available chemical potential that is used to overcome internal losses, decreasing the cell potential from its open-circuit value. Because the cathode used here was very thin, its concentration overpotential is essentially zero. While the detail depends on the construction of the MEA and the operating conditions (p,T), overall the trends are similar. At open circuit, the cell potential is given by the Nernst potential, and as the current density increases, so do the overpotentials, until the total losses equal the Nernst potential, where the cell voltage and power go to zero. This is the limiting current density. Generally, at low currents, the cell response is dominated by charge-transfer reaction kinetics quantified by the activation overpotentials. The linear central portion of the cell voltage curve is mostly dominated by ohmic losses, while the concentration overpotentials affect the shape of the curve at high current densities (and are therefore typically responsible for limiting the maximum attainable current from the cell). At high current densities, reactants are quickly consumed in charge-transfer reactions and new reactants cannot be sufficiently supplied through the phases of the porous electrodes.

Each of the overpotential curves is illustrated in a cumulative fashion. While shown here for illustration, these curves were computed using a more detailed model of SOFC performance based on 1D formulation of transport within the different elements, as well as more detailed modeling of the thermochemical and electrochemical reactions.

Example 7.4

The specifications of an SOFC are listed in Table E7.4. In theoretical modeling of fuel cells, the conductivity σ is usually represented as $\frac{1}{\sigma_i} = \mathrm{A}_i \exp\left(\frac{\mathrm{B}_i}{\mathrm{T}}\right)$, where i stands for electrolyte, cathode electrode, anode electrode, and parameters A_i and B_i denote the resistivity constants of the material. If the operating current density is 400 mA/cm^2, determine the actual voltage and the electrical power production of the SOFC.

Example 7.4 (cont.)

Table E7.4

Temperature	1073 K
Pressure	1 atm
Fuel	Hydrogen
Oxidizer	Air
Cell surface area	270 cm^2
Number of cells	200
Limiting current density	900 mA/cm^2
Exchange current density (anode)	650 mA/cm^2
Exchange current density (cathode)	250 mA/cm^2
Electrolyte thickness	0.001 cm
Electrolyte resistivity constants	$A_e = 0.00294$ Ωcm; $B_e = 10,350$
Cathode thickness	0.005 cm
Cathode resistivity constants	$A_c = 0.00814$ Ωcm, $B_c = 600$
Anode thickness	0.05 cm
Anode resistivity constants	$A_a = 0.00298$, $B_a = -1392$
Charge-transfer coefficient	0.5

Solution

The actual voltage is the difference between the Nernst potential and the voltage losses due to various sources shown in (7.69). We first calculate the Nernst potential:

$$\Delta\varepsilon^o = \frac{-\Delta G^o}{n_e F} + \frac{RT}{n_e \Im_a}\ln\left(\frac{P_{H_2}\sqrt{P_{O_2}}}{P_{H_2O}}\right) = -\frac{(-188{,}600)}{2\times 96{,}485} + \frac{8.314\times 1073}{2\times 96{,}485}\ln\left(\frac{1\times 0.21^{0.5}}{1}\right) = 0.833\,\text{V}.$$

Activation overpotential is obtained using the Butler–Volmer equation (7.26), with $\alpha = 0.5$. Hence:

$$i = i_0\left[\exp\left(0.5\times\frac{2\times 96,485\tilde{\eta}_{act}}{8.314T}\right) - \exp\left(-0.5\times\frac{2\times 96,485\tilde{\eta}_{act}}{8.314T}\right)\right],$$

$$\tilde{\eta}_{act} = \frac{T}{11,605}\sinh^{-1}\left(\frac{i}{2i_0}\right).$$

The activation overpotential is calculated as follows:

$$\tilde{\eta}_{act} = \tilde{\eta}_{act,a} + \tilde{\eta}_{act,c} = \frac{T}{11,605}\left[\sinh^{-1}\left(\frac{i}{2i_{o,a}}\right) + \sinh^{-1}\left(\frac{i}{2i_{o,c}}\right)\right]$$

$$= \frac{1073}{11605}\left[\sinh^{-1}\left(\frac{400}{2\times 650}\right) + \sinh^{-1}\left(\frac{400}{2\times 250}\right)\right]$$

$$= 0.028 + 0.068 = 0.096\,\text{V}.$$

Example 7.4 (cont.)

The ohmic overpotential is determined, taking into account the ohmic losses in electrolyte and cathode and anode electrodes. Hence:

$$\begin{aligned}\tilde{\eta}_{ohm} &= i\left(\frac{\delta_e}{\sigma_e}+\frac{\delta_a}{\sigma_a}+\frac{\delta_c}{\sigma_c}\right)\\ &= \left[0.4\times 0.001\times 0.00294\exp\left(\frac{10350}{1073}\right)+0.005\times 0.00298\exp\left(-\frac{1392}{1073}\right)\right.\\ &\quad\left.+0.05\times 0.00811\exp\left(\frac{600}{1073}\right)\right]\\ &= 0.018\,\mathrm{V}.\end{aligned}$$

We now calculate the concentration overpotential as follows:

$$\tilde{\eta}_{conc} = -\frac{8.314\times 1073}{2\times 96,485}\left(1+\frac{1}{0.5}\right)\ln\left(1-\frac{400}{900}\right) = 0.081\,\mathrm{V}.$$

The actual operating voltage of the cell is therefore determined as

$$\Delta\varepsilon = \Delta\varepsilon^o - (\tilde{\eta}_{act}+\tilde{\eta}_{ohm}+\tilde{\eta}_{conc}) = 0.833-(0.096+0.018+0.081) = 0.638\,\mathrm{V}.$$

The electric power produced by the fuel cell stack with 200 cells is

$$P_{dc} = (iA_{cell}n_{cell})\Delta\varepsilon = (0.4\times 270\times 200)\times 0.638 = 13,780.8\,\mathrm{W}.$$

Note that the above calculated power is dc (direct current), which is converted to ac (alternative current) using a dc–ac converter. To obtain the value of ac power, P_{dc} should be multiplied by the converter efficiency, $\eta_{converter}$.

7.10 Fuel Cell Types

In this section, we review the basic construction of some of the commonly known fuel cells, and how their construction and internal chemical and electrochemical reactions affect their performance. The focus of this review is fuel performance and how modifying the internal structure in ways that impacts the internal flows may influence the polarization curve.

7.10.1 Solid Oxide Fuel Cells

Solid oxide fuel cells are high-temperature units that operate on hydrogen, syngas, or methane (often mixed with water) fuels. All the components of these cells are ceramic solids, with the anode typically made of a porous nickel mixed with yttria-stabilized zirconia (YSZ), and the cathode of lanthanum manganese doped with strontium (LSM). The electrolyte is also solid, and is often made of zirconia oxide (ZrO_2) doped with yttrium oxide (Y_2O_3), which has good oxide ions conductivity at temperatures above 800 °C. These cells do not require noble metals as catalysts on the electrode surfaces since they operate at higher temperatures, sufficient to accelerate the electrochemical reactions and reduce the corresponding overpotential. Moreover, the ionic conductivity of the electrolyte increases with temperature, improving the overall performance of the cell. Both the high temperature of the anode and the availability of water on the anode side (as a product of the reactions) enable it to shift carbon monoxide to hydrogen (and CO_2), and/or reform methane to syngas, before the electrochemical reactions. Solid oxide fuel cells have been developed as independent power sources, combined heat and power (CHP) units, auxiliary power units for automotive applications (AUP), and other applications. They can play a role in increasing the efficiency of power plants using hybrid combined cycles (fuel cell + steam turbine for lower-temperature cells, and fuel cell + gas turbine + steam turbine for higher-temperature cells).

Operating at high temperatures is advantageous for the finite current efficiency of the cell. This, however, has a limit since raising the temperature also lowers the open-circuit potential, as shown before. Moreover, higher temperatures increase the internal thermal stresses and the overall thermal inertia, which is not desirable during startup and load variations.

Schematic representations of the operation of an SOFC are shown in Figures 7.17 and 7.18. Depending on the fuel, and the ability of the anode to reform carbon monoxide to carbon dioxide, reactions may vary. In all cases, the cathode reduces oxygen according to the reaction:

$$\frac{1}{2}O_2 + 2e^- \rightarrow O^{2-}. \tag{7.71}$$

The oxide ions are conducted through the electrolyte and hydrogen is oxidized to water:

$$H_2 + O^{2-} \rightarrow H_2O + 2e^-. \tag{7.72}$$

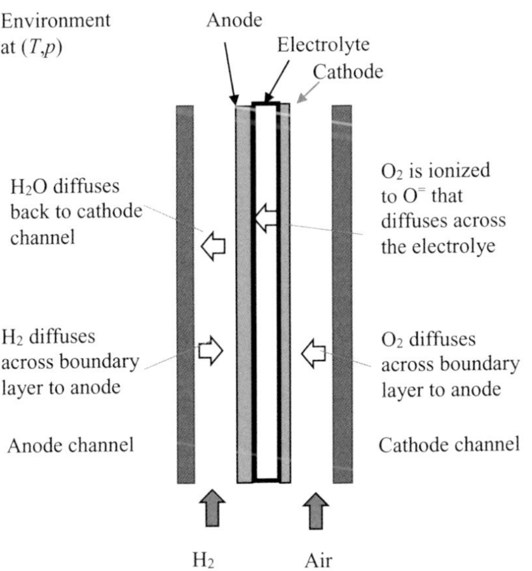

Figure 7.17 Schematic diagram of an SOFC operation.

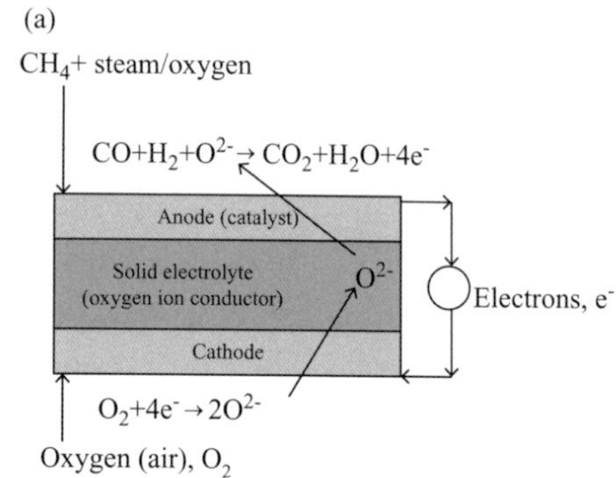

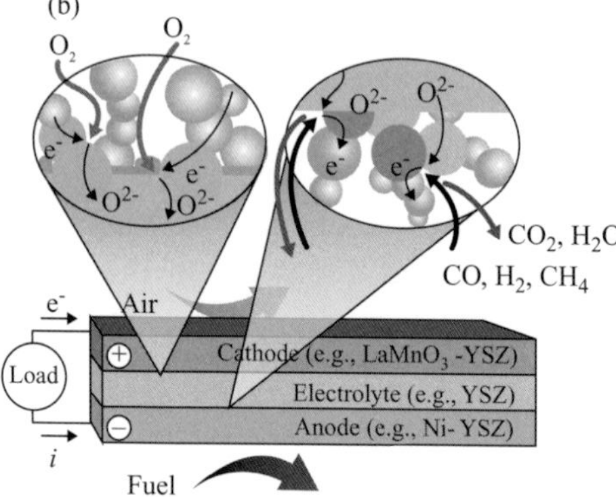

Figure 7.18 (a) An SOFC powered by methane (or syngas) in which both CO and H_2 participate in electrochemical reactions. Depending on the operating conditions, methane may get thermochemically reformed and carbon monoxide shifted first, followed by the electrochemical oxidation of hydrogen. (b) Detailed operation of the membrane–electrode assembly. In the left-hand bubble, oxygen molecules are reduced at the cathode and its ions are conducted through the electrolyte, moving through the electrolyte into the anode (right-hand bubble), where they are used for electrochemical oxidation of fuel at the triple-phase boundary (TPB). Electrons released in the charge-transfer reactions are conducted through the anode (metal) to the external circuit.

At high operating temperatures, the availability of water at the anode side can reform carbon monoxide to produce more hydrogen:

$$CO + H_2O \rightarrow CO_2 + H_2. \tag{7.73}$$

Thus, for each mole of hydrogen and half-mole of oxygen, one mole of water is produced and a charge of $2\mathfrak{F}_a$ coulombs. The water produced in this process can be used to reform a mole of carbon monoxide to produce another mole of hydrogen, which when oxidized can produce another $2\mathfrak{F}_a$ coulombs. Alternatively, carbon monoxide may also be electrochemically oxidized according to the reaction:

$$CO + O^{2-} \rightarrow CO_2 + 2e^-. \tag{7.74}$$

With a methane–water mixture in the anode, internal reforming according to the following reaction can be used to produce hydrogen and carbon monoxide:

$$CH_4 + H_2O \rightarrow 3H_2 + CO. \tag{7.75}$$

Following the reforming, the electrochemical reactions given in (7.70) and (7.71) occur. Accordingly, the open-circuit voltage of the cell can be estimated using the Nernst equation for hydrogen–oxygen reaction:

$$\varepsilon(T, p_i) = \varepsilon^o(T) + \frac{\Re T}{2\Im_a} \ln\left(\frac{p_{H_2} p_{O_2}^{1/2}}{p_{H_2O}}\right). \tag{7.76}$$

In an SOFC, the concentrations or partial pressures of hydrogen and oxygen should be evaluated at the anode side, where water formation occurs. Equation (7.75) shows that the pressure has a positive effect on the Nernst voltage.

The open-circuit potential decreases with temperature (e.g., from 1.18 V at 80 °C to 0.92 V at 1100 °C) because of the change in the Gibbs free energy of the hydrogen–oxygen reaction with temperature. In fact, over the range of temperature of interest, $300 < T < 1200\,\text{K}$, the changes in ΔH_R and ΔS_R are weak enough to be negligible and

$$\frac{\partial \Delta G_R}{\partial T} = -\Delta S_R. \tag{7.77}$$

Changes in the Gibbs free energy and entropy are negative during the reaction, and the ideal potential of the cell decreases at the rate of almost 0.05 kJ/K·mol, and $0.05/2\Im_a$ volts/K, respectively, as shown in Figure 7.19, close to zero current. However, as shown in experimental investigations, the actual performance of the cell, at current densities typically used in practice, improves at higher temperatures. The improved performance is because of the drop in the ohmic losses due to the drop in the resistivity of the ceramic components used to build the cell, as well as the reduction of the activation overpotentials.

The partial pressures in the Nernst equation are those for the reactants participating in the reactions at the electrode–electrolyte interfaces. The partial pressures, the concentrations, or in general the activity coefficients must be calculated at the site where the reaction occurs. As discussed, many factors contribute to these concentrations, including gas transport in the channel and across the electrodes, and the charged species transport. The impact of the total

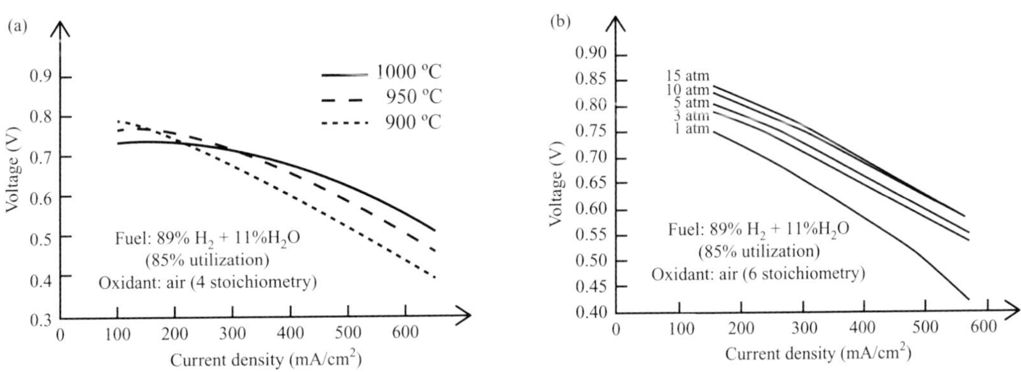

Figure 7.19 The effect of temperature at 1 bar (a) and pressure at 1000 °C (b) on the polarization curve [1].

pressure on the cell performance is shown in Figure 7.19 for the case of hydrogen, when both the open-circuit potential and the reaction rates increase with the total pressure (the total pressure affect the kinetics because if affects the molar concentration).

The cell efficiency dependence on the current largely resembles that of the cell voltage dependence on current – the polarization curve. It is better to operate at low current density if high efficiency is the goal. However, the power can be low at this point and far from the maximum power that can be drawn from the cell stack. This is demonstrated in Figure 7.20, where experimental results (and computational modeling) are shown for the performance of the same cell operating on CO (left) or H_2 (right). For both fuels, maximum power is achieved at currents closer to the limiting current, where the voltage efficiency is rather low. The figures also show that the current density is higher when using H_2 because of its faster kinetics and transport. This also leads to a slower reduction of the voltage at higher current, and slower reduction with decreasing fuel concentration on the anode side.

In general, not all the fuel entering the cell is consumed electrochemically as it flows along the MEA. As the fuel concentration decreases, the reaction rates also decrease, and it may not be economical to make the MEA long enough for 100% fuel consumption. The fuel left in the stream can be burned externally for other applications. While the cell voltage increases

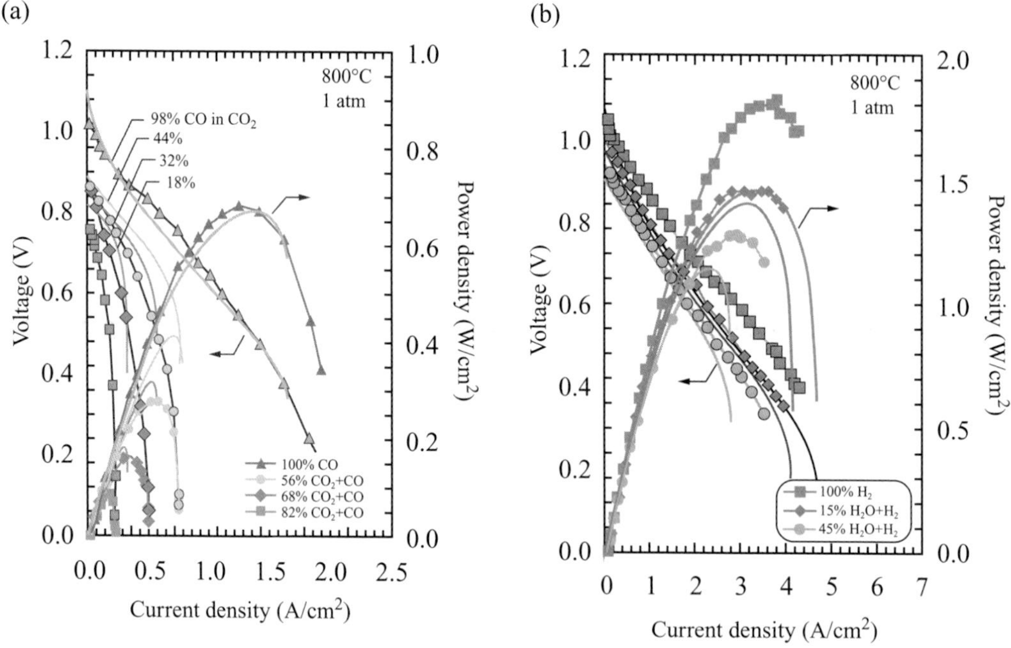

Figure 7.20 Comparison of measured (symbols) and computed (solid lines) MEA performance operating on CO–CO_2 (a) and H_2–H_2O (b) mixtures at 800 °C and 1 atm. The symbols are experimental measurements of button-cell performance [13] using an Ni–YSZ porous anode. The solid lines are the computed performance of the same button cell using a detailed computational framework [8,11] to model the MEA. For details, see [12].

with temperature, it decreases with fuel utilization. In most cases, fuel utilization near 80–85% is found to be an optimal operating point.

7.10.2 Polymer Electrolyte Membrane Fuel Cell

Polymer electrolyte membrane fuel cells (PEMFCs) have been used for automotive applications as well as power sources. In PEMFCs, a solid phase polymer membrane is used as an electrolyte and hence the cell operates at relatively low temperatures, below 100 °C, allowing for faster startup and response to load changes. These cells operate on pure hydrogen, and fuel reformers are necessary if hydrocarbons are to be used. They produce higher current densities, 1–2 A/cm^2. Operating with hydrogen and air (with three times stoichiometric air), and fuel utilization more than 85%, they produce voltages of ~0.7 V and power of ~400–600 mW/cm^2. Most cells operate at pressure less than 3 bar. Catalyst (mostly platinum) loading of ~1 mg/cm^2 is often used. Under similar conditions, the same cell produces power around 50 mW/cm^2 at 0.7 V while operating on reformate of hydrocarbon (CO_2 and N_2, plus hydrogen).

Polymer electrolyte membrane fuel cells utilize platinum as an electrocatalyst in the form of either pure platinum or as carbon or graphite-supported platinum catalyst. When reformates are used, the catalyst on the anode side is often an alloy of platinum and ruthenium. Platinum contributes to the high cost of PEMFCs, and to reduce this cost small nanoparticles of ~3.5 nm are used on a support to optimize their reactivity. Platinum loading, ~1–4 mg Pt/cm^2 on both the anode and cathode are used. The electrodes can be cast as thin films and attached to the membrane, or applied to the membrane [1].

One significant limitation of PEMFCs is the need for almost pure hydrogen. The use of carbon monoxide-containing fuel streams in these low-temperature cells causes significant performance deterioration, because carbon deposits on the platinum catalyst sites, reducing their activity (catalysis poisoning). One way to minimize CO poisoning of the catalyst is to raise the operating temperatures above 120 °C. However, this dries out the membrane, causing reduction in chemisorption and proton transport. To avoid this negative effect, only a few ppm of CO can be tolerated at 80 °C. Therefore, after reforming and the water–gas shift, which can leave up to 1% CO in the fuel stream, a mechanism to eliminate CO is needed. This can be done using preferential oxidation (PROX) that selectively oxidizes CO over H_2 using a precious metal catalyst. The low operating temperature also means that little, if any, heat is available from the fuel cell for endothermic reforming. Another approach to eliminate CO content is to add some oxygen to the fuel stream, which, while consuming some of the fuel, allows the performance to remain the same.

Water management is critical for the operation of PEMFCs. For the electrolyte to maintain its high proton conductivity, it should be well hydrated. On the other hand, too much water can flood the electrodes, blocking gas diffusion through the pores. Water produced by the electrochemical reaction on the cathode side should be removed by the airflow. Even though the fuel and air streams may be mixed with water vapor at the inlet, the humidity level is controlled primarily by the air stream, where water is produced by the electrochemical reaction. The cells of PEMFCs use sulfonate fluoropolymers as an electrolyte (membrane). The most well-known membrane brand is Nafion.

In PEMFCs, water forms on the cathode side (see Figure 7.16). Air flows over the cathode and, apart from supplying the necessary oxygen, it dries out excess water. Because the membrane electrolyte is so thin, water diffuses from the cathode side to the anode side, and throughout the whole electrolyte a suitable state of hydration is achieved. During the operation of the cell, H^+ ions moving from the anode to the cathode "drag" water molecules with them (typically, five water molecules per proton). Thus, especially at high current densities, the anode side of the electrolyte may dry out, even if the cathode side is well hydrated. Operating at temperatures above 60 °C, which may be beneficial for the kinetics, the electrodes may dry out faster than the rate of water production by electrochemical reactions. It is common to humidify air or hydrogen, or both, before they enter the cell to reduce the impact of this high-temperature operation. Moreover, unless pure oxygen is used, which is rare, air is fed at a rate higher than that required by stoichiometry to supply the necessary oxygen [7].

As already indicated, PEMFCs require tight control of fuel and air feed, humidity level, and temperature to operate properly, and hence heat exchangers, humidifiers, and condensers form important components in the fuel cell system. Hydrogen can be stored at high pressure (up to 700 bar), in liquid form or in metal hydrides (which require heating to release the gas), depending on the application. Hydrocarbon-based PEMFC systems avoid the complexities of pure hydrogen storage, but need a reformer, a water–gas shift reactor, and a reformate purifier. Besides the extra weight, reforming can reduce the system efficiency (fuel reforming efficiency typically ranges from 75% to 90%, defined as the ratio between the chemical energy in the produced hydrogen divided by the chemical energy of the original fuel).

Direct methanol fuel cells (DMFC) are specially designed PEMFCs that can operate on methanol, a liquid fuel with a relatively high energy density. Direct methanol fuel cells using air can generate up to 250 mA/cm^2 with cell voltages in the range of 0.25–0.4 V, and hence the power density is 40–100 mW/cm^2. The low cell voltage is caused by the crossover of methanol without participating in the electrochemistry. Moreover, large quantities of water are transported across the membrane, requiring complex water-recovery systems [1].

Example 7.5

Figure E7.5 shows a schematic diagram of a fuel cell operating on hydrogen produced by steam reforming of methane. All streams are at 1 bar and 80 °C. The fuel consists of H_2 and CO_2, with a molar ratio 4:1. The relative humidity of the fuel steam is 20%, where the relative humidity is defined as $\phi = \frac{p_{H_2O}}{p_{H_2O,sat}(T)}$. Carbon monoxide, which is poisonous to Pt catalysts, is completely removed from the fuel stream. Air stoichiometry, defined as the ratio of inlet oxygen flow rate to the oxygen consumption rate, is 2 at the inlet. The inlet air relative humidity is 40%. The electrolyte is well hydrated and the level of hydration is constant and

Example 7.5 (cont.)

steady. In other words, there is no incoming and outgoing water from and to the electrolyte. We assume that water produced at the cathode side is in the form of vapor, and all species behave like an ideal gas. The fuel cell stack is used to power a small vehicle that requires 80 kW. Each cell operates at 0.6 V, and a typical current density is 1 A/cm^2. Assume that the fuel utilization is 90%. Determine (a) the mole fractions at the inlet of the anode and cathode; (b) the composition at the exit streams of anode and cathode; and (c) the relative humidity at the exit streams.

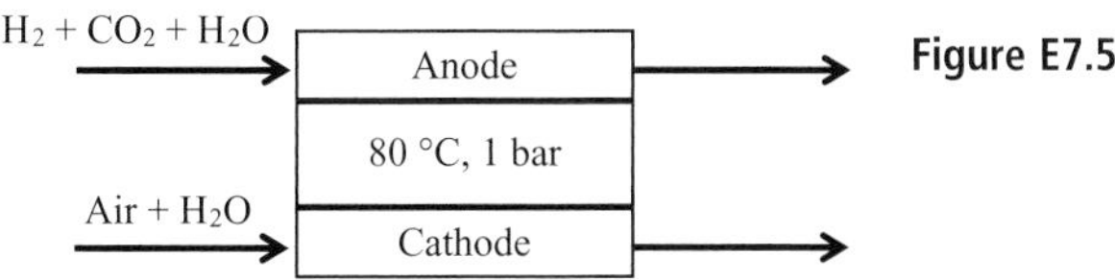

Figure E7.5

Solution

a. We first calculate the composition at the inlet of the anode channel. The saturation pressure of water at 80 °C is $P_{H_2O,sat}(T) = 0.4739$ bar. Thus, the partial pressure and the mole fraction of water are

$$(P_{H_2O,a})_{in} = 0.2 \times 0.4739 = 0.0948\,\text{bar},$$

$$(X_{H_2O,a})_{in} = \frac{0.0948\text{ bar}}{1\text{ bar}} = 0.0948.$$

The mole fractions of CO_2 and H_2 are determined, noting that $X_{H_2}/X_{CO_2} = 4$ and $X_{H_2} + X_{CO_2} = 1 - X_{H_2O} = 0.9052$. Hence,

$$(X_{CO_2,a})_{in} = 0.1810,$$

$$(X_{H_2,a})_{in} = 0.7242.$$

Similarly, the mole fraction of water at the cathode channel inlet is obtained as follows:

$$(X_{H_2O,c})_{in} = \frac{0.4 \times 0.4739\text{ bar}}{1\text{ bar}} = 0.1896.$$

The mole fractions of oxygen and nitrogen are found from $X_{N_2}/X_{O_2} = 3.76$ and $X_{N_2}/X_{O_2} = 1 - 0.1896 = 0.8104$. Hence,

$$(X_{N_2,c})_{in} = 0.6402,\ (X_{O_2,c})_{in} = 0.1702.$$

Example 7.5 (cont.)

b. To find the composition at the channel exit, we need to calculate the molar flow rates of individual species. The molar flow rate of hydrogen consumed in the fuel cell is

$$\dot{n}_{H_2,consumed} = \frac{P}{2FV} = \frac{80,000}{2 \times 96,485 \times 0.6} = 0.691\frac{\text{mol}}{\text{s}}.$$

The fuel utilization factor is 0.9. So, the inlet hydrogen molar flow rate is $(\dot{n}_{H_2,a})_{in} = 0.691/0.9 = 0.768\,\text{mol/s}$. Hence,

$$(\dot{n}_{H_2,a})_{exit} = (\dot{n}_{H_2,a})_{in} - \dot{n}_{H_2,consumed} = 0.768 - 0.691 = 0.077\,\text{mol/s}.$$

The molar flow rates of water and carbon dioxide at the anode channel exit are the same as their corresponding values at the anode channel inlet. Hence,

$$(\dot{n}_{H_2O,a})_{exit} = (\dot{n}_{H_2O,a})_{in} = \frac{(X_{H_2O,a})_{in}}{(X_{H_2,a})_{in}}(\dot{n}_{H_2,a})_{in} = \frac{0.0948}{0.7242} \times 0.768 = 0.101\,\text{mol/s},$$

$$(\dot{n}_{CO_2,a})_{exit} = (\dot{n}_{CO_2,a})_{in} = \frac{(X_{CO_2,a})_{in}}{(X_{H_2,a})_{in}}(\dot{n}_{H_2,a})_{in} = \frac{0.1810}{0.7242} \times 0.768 = 0.192\,\text{mol/s}.$$

Therefore, the total molar flow rate at the exit stream of the anode channel is

$$\dot{n}_{a,exit} = 0.077 + 0.101 + 0.192 = 0.37\,\text{mol/s}.$$

Thus, the mole fractions of species at the anode channel exit are

$$(X_{H_2,a})_{exit} = \frac{0.077}{0.37} = 0.2081,$$

$$(X_{H_2O,a})_{exit} = \frac{0.101}{0.37} = 0.2730,$$

$$(X_{CO_2,a})_{exit} = \frac{0.192}{0.37} = 0.5189.$$

To determine the composition at the exit of the cathode channel, we first determine the molar flow rate of oxygen:

$$\dot{n}_{O_2,consumed} = \frac{\dot{n}_{H_2,consumed}}{2} = \frac{0.691}{2} = 0.3455\,\text{mol/s},$$

$$(\dot{n}_{O_2,c})_{in} = \Lambda.\dot{n}_{O_2,consumed} = 2 \times 0.3455 = 0.691\,\text{mol/s},$$

$$(\dot{n}_{O_2,c})_{exit} = (\dot{n}_{O_2,c})_{in} - \dot{n}_{O_2,consumed} = 0.691 - 0.3455 = 0.3455\,\text{mol/s}.$$

The molar flow rate of nitrogen at the cathode channel inlet and exit are the same. Hence,

$$(\dot{n}_{N_2,c})_{exit} = (\dot{n}_{N_2,c})_{in} = 3.76 \times 0.691 = 2.598\,\text{mol/s},$$

Example 7.5 (cont.)

$$(\dot{n}_{H_2O,c})_{exit} = (\dot{n}_{H_2O,c})_{in} + \dot{n}_{H_2O,produced} = \frac{(X_{H_2O,c})_{in}}{(X_{O_2,c})_{in}}(\dot{n}_{O_2,c})_{in} + \dot{n}_{H_2O,produced}$$

$$= \frac{0.1896}{0.1702} \times 0.691 + 0.691 = 1.461 \text{ mol/s}.$$

Thus, the total molar flow rate at the exit of the cathode channel is

$$\dot{n}_{c,exit} = 0.3455 + 2.598 + 1.461 = 4.4045 \text{ mol/s}.$$

Thus, the mole fractions of species at the cathode channel exit are

$$(X_{O_2,c})_{exit} = \frac{0.3455}{4.4045} = 0.0784,$$

$$(X_{N_2,c})_{exit} = \frac{2.598}{4.4045} = 0.5899,$$

$$(X_{H_2O,c})_{exit} = \frac{1.461}{4.4045} = 0.3317.$$

c. The relative humidity at the exit streams is determined as follows:

$$\phi_{c,exit} = \frac{(X_{H_2O,c})_{exit} P \left(= \frac{0.3317 \times 1}{0.4729} \right)}{P_{sat}} = 0.7,$$

$$\phi_{a,exit} = \frac{(X_{H_2O,a})_{exit} P \left(= \frac{0.2730 \times 1}{0.4729} \right)}{P_{sat}} = 0.576.$$

7.10.3 Alkaline Fuel Cells

Alkaline fuel cells (AFCs) operate at temperatures of 80–200 °C and are fueled by hydrogen. Precious metals are used as part of the catalyst at the lower end of the temperature range. In these units, the electrolyte conducts OH^- from the cathode side, where water, oxygen, and electrons react to form the hydroxyl ions. Thus, water forms on the anode (hydrogen supply) side, and some (one-half) of it must be supplied to the cathode (oxygen supply) side according to the reactions shown in Table 7.1. The electrolyte is potassium hydroxide (KOH). In static designs, the liquid electrolyte is immobilized in an asbestos matrix between the electrodes. In mobile electrolyte designs, KOH is circulated between the cell and a heat exchanger used to keep the temperature down. In both cases, it is necessary to keep the MEA at a low temperature by cooling the (recycled) hydrogen stream and the electrolyte (in the mobile case). Humidity management is also important since water is produced on the anode side and must be supplied at the cathode side. The other components are typically an Ni porous electrode and a lithiated NiO porous anode.

These cells operate at high pressures, up to 50 bar, and at the higher end of their temperature they can deliver high performance, with 0.4 A/cm^2 at 0.78 V. They were used in the space program, but have not been used extensively since; PEMFCs are used more extensively [1,7].

Example 7.6

A hydrogen fuel cell operating at 70 °C delivers a direct current power of 10 W at an average cell voltage of 0.4 V. Humidified air at 60 °C and 1.2 bar with 80% relative humidity is fed to the cathode. Assume a 5% pressure drop within the stack. Determine (a) the air stoichiometry in order to maintain a relative humidity of 90% at the cathode exit; and (b) the portion of water at the exit stream that should be recycled back to the inlet air stream to maintain the inlet air relative humidity.

Solution

a. It can be shown that the relationship between the inlet relative humidity and the exit relative humidity is (see Problem 7.2)

$$\phi_{exit} = \frac{\Lambda\phi_{in} + 0.42(\pi - \phi_{in})}{\Lambda\pi + 0.21(\pi - \phi_{in})}\pi,$$

where Λ is the air stoichiometry and $\pi = P_{exit}/P_{sat}$.

There is a 5% pressure drop within the stack, so the exit pressure is

$$P_{exit} = 0.95P_{in} = 0.95 \times 1.2 = 1.14\,\text{bar}.$$

The saturation pressure of water at 70 °C is obtained from thermodynamic tables. Hence, $P_{sat}(70°C) = 0.3119\,\text{bar}$.

So, $\pi = 0.2736$.

Substituting in the above relationship, we get

$$0.9 = \frac{0.8\Lambda + 0.42(0.2736 - 0.8)}{0.2736\Lambda + 0.21(0.2736 - 0.8)} \times 0.2736.$$

Solving for Λ, we find $\Lambda = 1.42$.

b. The mass flow rate of (dry) air at the cathode inlet is determined using the relationship derived in Example 4.4:

$$\dot{m}_{air} = 3.574 \times 1.42 \times 10^{-4}\frac{10}{0.4} = 0.0127\,\text{g/s}.$$

The mass flow rate of water in the inlet stream is obtained as follows:

$$\left(\frac{\dot{m}_{H_2O}}{\dot{m}_{air}}\right)_{in} = \frac{M_{H_2O}}{M_{air}}\left(\frac{\dot{n}_{H_2O}}{\dot{n}_{air}}\right)_{in} = \frac{18}{28.97}\frac{P_{H_2O}}{P_{air}} = 0.622\frac{P_{H_2O}}{P_{in} - P_{H_2O}},$$

Example 7.6 (cont.)

where: $P_{H_2O} = \phi_{in} P_{sat}(60\,°C) = 0.8 \times 0.1595\,\text{bar}$ and $P_{in} = 1.2\,\text{bar}$.

Substituting these values, we find

$$\dot{m}_{H_2O} = 0.622 \frac{0.1595}{1.2 - 0.1595} \times 0.0127 = 0.0012\ \text{g/s}.$$

Next, we calculate the amount of water production due to the electrochemical reaction within the stack. Noting that for each mole of hydrogen, one mole of water is formed, we have

$$\dot{n}_{H_2O} = \dot{n}_{H_2} = \frac{P}{2\mathfrak{F}_a V_{act}} = \frac{10}{2 \times 96{,}485 \times 0.4} = 0.0001296\,\text{mol/s},$$

or $\dot{m}_{H_2O} = M_{H_2O}\dot{n}_{H_2O} = 18 \times 0.0001296 = 0.0023\,\text{g/s}$.

Thus, $(\dot{m}_{H_2O})_{exit} = 0.0012 + 0.0023 = 0.0035\,\text{g/s}$.

The ratio of $(\dot{m}_{H_2O})_{in}/(\dot{m}_{H_2O})_{exit}$ is 0.343 (= 0.0012/0.0035). In other words, 34.3% of the exit water needs to be recycled to the air inlet stream to maintain a relative humidity of 80%.

7.10.4 Molten Carbonate Fuel Cell

Molten carbonate fuel cells (MCFCs) utilize a mixture of molten alkali metal carbonates, such as lithium–sodium or lithium–potassium carbonate, as an ion-conducting electrolyte. The molten carbonate is embedded in a ceramic porous matrix of lithium alumina ($LiAlO_2$). The alkali metal carbonate is conductive to carbonate ions, CO_3^{2-}, in the temperature range of 600–700 °C. Moreover, at these temperatures, and similar to SOFCs, transition metals such as Ni can be used to catalyze electrochemical reactions at the electrolyte surfaces, even when using natural gas or other hydrocarbons. The half-cell reactions in an MCFC are shown Table 7.1. These equations show that charge transport across the electrolyte is enabled by the presence of CO_2, which moves from the cathode to the anode in the form of a carbonate ion. Clearly there is a need for a continuous CO_2 supply at the cathode with the oxidizer stream. If the cell is fueled by a hydrocarbon or syngas, some of the CO_2 from the products stream (produced at the anode) can be mixed with air/oxygen and introduced with the cathode stream. If the cell is fueled by hydrogen, an external supply of CO_2 is needed. This CO_2 will need to be separated from the anode products and recycled back to the cathode inlet (which adds to the complexity of these cells). Similar to SOFCs, MCFCs tolerate fuels beyond hydrogen.

The anodes in MCFCs are often made of porous sintered Ni–Cr/Ni–Al alloys (~0.4–0.8 mm thick), while the cathode is made of lithiated nickel oxide. Typical MCFCs generates up to 400 mA/cm^2 at 0.750–0.900 V.

7.11 Summary

At finite current, fuel cells suffer from a number of loss mechanisms that reduce their voltage and efficiency beyond their values at zero current. These are measured by voltage overpotentials beyond the open-circuit potential, and include activation, ohmic, and concentration overpotentials, which result from finite-rate electrokinetics at the interface between the electrode and electrolyte, the charge transport through different elements, and the drop of ionized species concentration across the electrodes, respectively. Losses occur on both electrodes and are additive. The losses depend on the operating conditions – that is, the temperature, pressure, and fuel and oxidizer concentrations, as well as the materials/catalysts used to construct the electrodes and electrolyte, their thicknesses, and the current density. Other losses occur because of the neutral species (oxygen and fuel) diffusion across the mass transfer boundary layers in the supply channels and associated concentration drop. Some small losses are attributed to fuel crossover in pinholes in the electrode or electron leakage across the electrolyte.

Approximate models for these losses were developed in this chapter, which can be used to describe the I–V curve of a fuel cell. Reaction kinetics at the interfaces depend on the operating temperature and activation potential, while the electrochemical rate reaction is proportional to the local concentration and the current. Species diffusion is approximated by Fick's law relating the concentration gradient to the flux, which is proportional to the current. Ohm's law is used to model the resistance. It is shown that activation overpotential dominates the losses at low current, while concentration overpotential is the most significant at high current. The limiting current is found when one of the reactants' concentration reaches zero at one of the reactive interfaces.

Problems

7.1 Repeat Example 7.3 but for an SOFC operating at 900 °C; all other data remain the same.

7.2 Consider a PEMFC operating at temperature T and pressure p. Air is fed to the cathode side with a relative humidity ϕ_{in} and air stoichiometry (the actual air–fuel ratio to the theoretical air–fuel ratio) Λ.

a. Show that the relative humidity at the cathode exit can be obtained from the following relationship:

$$\phi_{exit} = \frac{\Lambda\phi_{in} + 0.42(\pi - \phi_{in})}{\Lambda\pi + 0.21(\pi - \phi_{in})}\pi \qquad \pi = \frac{p}{p_{sat}}$$

b. Air is supplied at 20 °C, 1.2 bar, and 50% relative humidity. Determine the required air stoichiometry to maintain a relative humidity of 80% at the cathode exit of a fuel cell operating at 60 °C.

7.3 A hydrogen fuel cell that uses air as an oxidizer operates at 70 °C and 1 atm. The thickness of the anode is 300 μm, and the mass transfer coefficient in the anode channel is $h_{conv} = 0.003$ m/s. Assuming an effective diffusion coefficient of 10^{-7} m²/s and $\alpha = 0.5$, determine the concentration overpotential at the anode side at an operating current density of 200 mA/cm².

7.4 Consider an SOFC operating at 750 °C and 1 atm. The exchange current density at the anode and cathode sides is 400 mA/cm² and 140 mA/cm², respectively. The fuel cell uses hydrogen as a fuel and air as an oxidizer. All the transfer coefficients are $\alpha = 0.5$.

a. Determine the total activation overpotential at a current density of 150 mA/cm².

b. The actual voltage of the cell is measured to be 0.85 V. If the total ohmic resistance is 0.0565 Ω and the concentration overpotential is negligible, find the surface area of the fuel cell.

c. Plot the activation overpotential versus current density.

7.5 Using a computing tool such as MATLAB or EES, generate voltage vs. current density and power vs. current density graphs for the SOFC examined in Example 7.4.

7.6 Consider the SOFC of Example 7.4. Investigate the effect of the operating temperature on the performance of the fuel cell.

a. Derive a relationship for the open-circuit voltage as a function of temperature.

b. Derive a relationship for the total overpotential (including activation, ohmic, and concentration overpotentials) as a function of temperature.

c. Derive an equation for the actual cell voltage as a function of temperature.

d. Plot the cell voltage as a function of temperature.

7.7 In Example 7.4, we assumed that all the hydrogen supplied to the fuel cell was consumed; that is, the fuel utilization was 100%. Using the data given in Example 7.4:

a. plot the actual cell voltage versus the fuel utilization factor ranging from 50% to 100%;

b. investigate the effect of the operating pressure on the actual voltage and the power production of the fuel cell at a fuel utilization factor of 80% and 100%.

7.8 Electrochemical cells are used extensively as concentration sensors. For instance, as the oxygen sensor of an internal combustion engine. In this case, the products of combustion from the engine are introduced along one electrode of the cell, and air is introduced along the other electrode. The sensor is used to measure the concentration of oxygen in the products given that the oxygen concentration in air is known (0.21). The electrolyte of this cell conducts oxygen ions (O^{2-}). The open-circuit potential difference between the two electrodes, which depends on the ratio of oxygen concentration in the two streams, is used to determine the concentration of oxygen in the products.

a. Derive a relation between the open-circuit potential of the cell and the oxygen concentration in the products stream. Plot this relation when the cell temperature is 25 °C, 100 °C, and 400 °C.

b. A similar concept can be used as an isothermal expander (electric work-producing machine). In this case, pure hydrogen at high pressure is introduced along one electrode of the cell while the hydrogen concentration is maintained at much lower values along the opposite electrode. The electrolyte conducts hydrogen protons (H^+). Derive an expression for the open-circuit potential and the ideal work of expansion in this case in terms of the hydrogen partial pressure ratio across the electrolyte. Compare this expression with the isothermal mechanical work of expansion across the same pressure ratio. Comment on this result. Calculate the open-circuit potential at $T = 30\,°C$, and hydrogen pressure ratio across the electrolyte of 100 and 10,000.

c. It has been proposed to construct a power cycle using (1) isothermal compression between pressures p_1 and p_2 at the initial temperature T_1; (2) constant pressure heating at p_2 to T_3; and (3) expansion back to p_1. Derive an expression for the efficiency of this cycle and compare it to that of a conventional Bryton cycle between the same two pressures. Calculate both efficiencies for pressure ratio of 30, T_1=300 K and $T_3 = 1600$ K.

Figure P 7.8

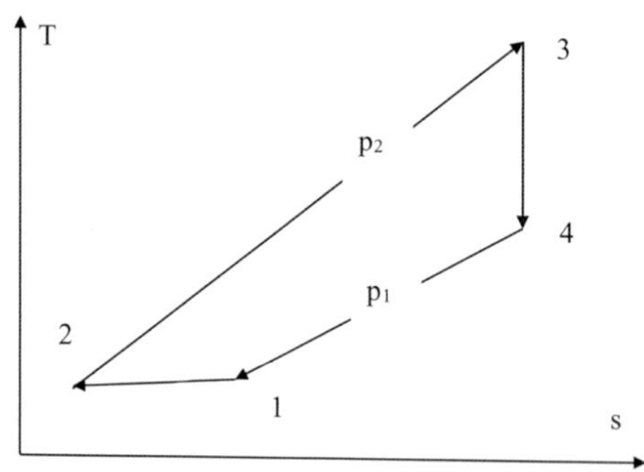

d. How can you improve the efficiency of the cycle proposed in part (c), and what is the new efficiency?

e. Explain how the reverse of the setup described in part (b) can be used as an isothermal electrochemical compressor, in which a voltage is applied to pump the gas (hydrogen in the case of part (b), but oxygen for this part) from the low-pressure side to the high-pressure side. What is the open-circuit voltage required to produce an oxygen stream at 10 bar from air using this cell?

7.9 Consider a high-temperature fuel cell operating on 95% hydrogen/5% steam and air. The fuel utilization factor is 0.8 and the stoichiometric ratio of air is 4. The exchange current densities at the anode and cathode electrodes are 530 mA/cm^2 and 200 mA/cm^2, respectively. Both electrodes are 40% porous with tortuosity of 6, and each is 50 μm thick. The electrolyte thickness is 500 μm. The operating pressure and temperature are 1 atm and 900 °C, respectively. Estimate the concentration overpotential at the anode side. The concentration overpotential at the cathode side needs to be determined using the relationships derived in the text. The total ohmic resistance is 0.0565 Ω and the gas diffusivity through the electrodes is 10^{-6} m^2/s.

a. Find the current density and the corresponding cell voltage and power density at maximum power.
b. Calculate the thermal efficiency of the fuel cell at maximum power density.
c. Determine the heat generated within the cell at maximum power density.
d. Repeat parts (a) and (b) for operating temperatures of 800 °C and 1000 °C.

7.10 Hydrators are used on the anode and cathode feedstreams in a fuel cell system. The fuel and air streams must be hydrated as water is needed in the electrolyte to improve proton conductivity. The hydrator is heated with an 80% efficient electrical heater. Assuming that the water production within the fuel cell has a negligible contribution to the humidity of the cathode stream flow, determine:
a. the electric power required to support a feed of 1 mole/s of hydrogen to the anode or 1 mole/s of air to the cathode;
b. the water supply rate;
c. the partial pressures of the reactants at the exit;
d. the mole fraction of water vapor in the exit stream; and
e. the effect of including the water produced within the cell on the mole fraction of the water and the partial pressures of the reactants at the exit.

Use the data in Table P 7.11.

Table P 7.11

	Anode	Cathode
Feed	Dry hydrogen at 25 °C at 1.5 atm	60% RH air at 25 °C at 1.5 atm
Water supply	25 °C	25 °C
Hydrator exit	100% RH at 80 °C, 1.5 atm	100% RH at 80 °C, 1.5 atm

7.11 Consider an air-cooled fuel cell operating at 500 watts and an average voltage of 0.68 volts. Water is supplied to the hydrogen stream. We need to determine the air supply to the stack for both the cathode (i.e., oxygen in the air) and for cooling the fuel cell. The air blower operates at 16 kPa above atmospheric pressure. Hydrogen supplied to the stack is humidified to 100% relative humidity (RH), and air is fed with a stoichiometric ratio of 2.2 at 25 °C and 25% RH. The cell operates at 80 °C and 105 kPa.
a. Determine the mass flow rate of air required for cooling and for oxidation.
b. Determine the RH of air at the cathode exit.
c. Determine the RH of exhaust air if all the air required for cooling and reaction flows through the cathode.

7.12 Consider the PEMFC of Example 7.6. The surface area of each individual cell is 650 cm^2, and the Faradaic efficiency is 100%.
a. Determine the theoretical open-circuit voltage of this cell.
b. Calculate the efficiency of the fuel cell.
c. Determine the cooling rate required to keep the fuel cell at 80 °C.

7.13 Consider a single PEMFC operating at 80 °C with pure H_2 and pure O_2 introduced at 150 kPa. The anode and cathode electrode areas are 5 cm^2. The total ohmic resistance is 0.005 ohms. The exchange current densities for the rate-determining steps of oxygen reduction and hydrogen oxidation are 5×10^{-11} A/cm^2 and 1×10^{-3} A/cm^2, respectively.

a. Calculate the open-circuit voltage.

b. Develop an analytical expression that relates the operating voltage to the current density by considering the ohmic and activation overpotentials only. Plot the ohmic and activation overpotentials as a function of current density in the range 0–2 A/cm^2.

c. Plot the power density in W/cm^2 as a function of current density.

REFERENCES

1. US DOE, Office of Fossil Energy, NETL, *Fuel cell handbook*, 7th ed. Washington, DC: US DOE, Office of Fossil Energy, NETL, 2004.
2. R. J. Silbey, R. A. Alberty, and M. G. Bawendi, *Physical chemistry*. Hoboken, NJ: Wiley, 2005.
3. P. W. Atkins and J. De Paula, *Atkins' physical chemistry*. Oxford: Oxford University Press, 2006.
4. A. J. Bard and L. R. Faulkner, *Electrochemical methods: fundamentals and applications*. New York: Wiley, 2001.
5. J. S. Newman and K. E. Thomas-Alyea, *Electrochemical systems*. Hoboken, NJ: Wiley, 2004.
6. R. P. O'Hayre, S.-W. Cha, W. G. Colella, and F. B. Prinz, *Fuel cell fundamentals*. Hoboken NJ: Wiley, 2016.
7. J. Larminie and A. Dicks, *Fuel cell systems explained*. Chichester: Wiley, 2003.
8. W. Y. Lee, D. Wee, and A. F. Ghoniem, "An improved one-dimensional membrane–electrode assembly model to predict the performance of solid oxide fuel cell including the limiting current density," *J. Power Sources*, vol. 186, no. 2, pp. 417–427, 2009.
9. R. F. Probstein, *Physicochemical hydrodynamics, an introduction*. London: Butterworth, 1989.
10. X. Li, *Principles of fuel cells*. New York: Taylor & Francis, 2006.
11. H. Zhu, R. J. Kee, V. M. Janardhanan, O., Deutschmann, and D. G. Goodwin, "Modeling elementary heterogeneous chemistry and electrochemistry in solid oxide fuel cells," *J. Electrochem. Soc.*, vol. 152, pp. A2427–A2440, 2005.
12. J. Hanna, W. Y. Lee, Y. Shi, and A. F. Ghoniem, "Fundamentals of electro- and thermochemistry in the anode of solid-oxide fuel cells with hydrocarbon and syngas fuels," *Prog. Energy Combust. Sci.*, vol. 40, pp. 74–111, 2014.
13. A. V. Jiang, and Y. Virkar, "Fuel composition and diluent effects on gas transport and performance of anode-supported SOFCs," *J. Electrochem. Soc.*, vol. 150, no. 7, pp. A942–A951, 2003.

8 Combined, Oxy-Combustion, and Hybrid Cycles

8.1 Introduction

In Chapters 5 and 6, we discussed gas turbine cycles and Rankine cycles used in power generation, with a focus on how to improve the cycle efficiency and recover the maximum availability from the primary energy source. We also discussed the conditions under which one chooses to build a plant running on a gas cycle or a two-phase cycle, and related these to the characteristics of the primary energy source. High-temperature energy sources can be effectively utilized in a gas turbine cycle, which exhausts its stream at atmospheric pressure (unless the cycle is closed, but this is not currently practiced). Lower-temperature sources must use two-phase Rankine cycles, and must be operated in a closed cycle mode to achieve the desired efficiency. In both cases, reheat and regeneration are effective approaches to raising the thermal efficiency.

In this chapter we cover extensions in which more complex cycle design and integration can further improve the efficiency. These include the humid air cycle, in which the hot exhaust of the gas turbine is used in a heat exchanger to raise steam, which is then added to the products of combustion before they are expanded in a gas turbine. Thus, a fraction of the thermal energy extracted from the turbine products is recycled back in the form of superheated steam. The cycle achieves higher efficiency, but the higher humidity of the expanding products may result in a corrosive environment inside the gas turbine. Another approach to recovering some of the thermal energy in the gas turbine exhaust is thermochemical recuperation – that is, steam reforming of the fuel in a reactor, which is heated by the gas turbine products. If steam reforming is used, the steam is also raised using some of the thermal energy in the exhaust. The partially reformed fuel is then burned in the combustor. By virtue of capturing some of the exhaust thermal energy, these cycles are more efficient than their simple counterpart. Moreover, burning partially reformed fuels has some combustion advantage as well.

Combined cycles simply integrate a gas turbine cycle and a steam turbine cycle such that the gas turbine exhaust is used as the heat source for the steam turbine cycle. Currently, combined cycles (CCs) achieve the highest possible overall conversion efficiency when using natural gas. Combined cycles have also been used with coal gasification, which will be discussed in a later chapter.

Currently the majority of power cycles use fuels as an energy source, mostly fossil fuels with a smaller fraction using biomass (otherwise nuclear, solar thermal, or geothermal energy is used). To reduce CO_2 emissions, carbon (dioxide) capture and (underground) storage could be used. This is a subject covered in Chapters 10–13, in which several approaches for natural gas and coal are discussed in detail. As an early introduction, natural gas oxy-combustion cycles in which the fuel is burned in a stream of oxygen diluted with recycled carbon dioxide and/or steam is discussed using examples with different compositions of the recycled stream.

"Hybrid" cycles in which a fuel cell and a gas/steam turbine or a CC are used in series are also discussed. The motivation behind such integration is described and some modes of integrating a fuel cell and a heat engine are shown. A fuel cell operating at temperature higher than the environment rejects heats at that temperature and hence a heat engine can be operated between the cell temperature and that of the environment. Moreover, if the fuel utilization of the cell is not 100%, the stream leaving the anode side in the case of a solid oxide fuel cell (SOFC) contains fuel that can be burned in the combustor of a heat engine cycle. In the case that the fuel cell is operating at higher pressures, a gas turbine can also be used to expand the existing streams and produce more work.

8.2 Humid Air Cycles

In simple Brayton cycles, the turbine exit temperature, which is determined by the turbine inlet temperature and pressure and the pressure ratio, tends to be high, even for high-pressure ratio cycles. These temperatures are typically in the range of 450–650 °C. Regeneration, in which the turbine exhaust is used to heat the compressed air before combustion, is an effective way to take advantage of the hot gas turbine exhaust and to improve the cycle efficiency. Combined cycles in which the turbine exhaust is used as the heat source for a steam cycle is another approach that will be described later in this chapter. Combined cycles, however, require a steam turbine, a large heat exchanger, and a condenser, which can be costly. **Humid air turbine** (HAT) cycles apply a different approach for regeneration. Humid air cycles utilize the high temperature and high enthalpy of the gas turbine exhaust to evaporate a pressurized water stream and to raise steam. The steam is mixed with the compressed air in the combustor, and the mixture of combustion products and steam is expanded in the turbine. Adding steam increases the total mass flow rate and the total enthalpy flow rate into the gas turbine, and hence the gas turbine work, without increasing the compressor work. Adding steam to the combustor also lowers the flame temperature [1]. Although the turbine work increases, lowering the turbine rotor inlet temperature (TRIT) or simply turbine inlet temperature (TIT) may negatively impact the cycle efficiency. Moreover, adding steam may lower the gas turbine isentropic efficiency and care must be taken in optimizing the cycle operation.

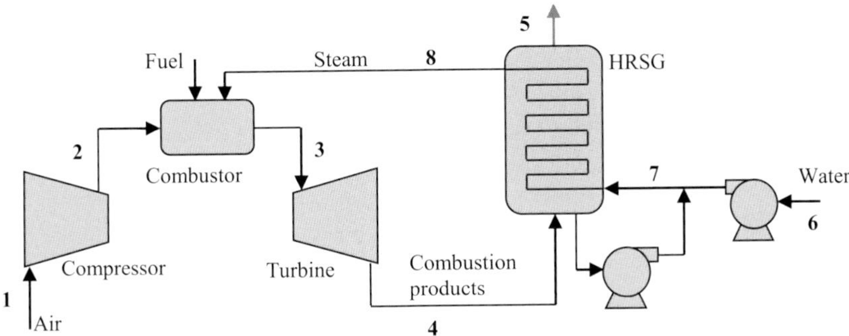

Figure 8.1 Steam injected gas turbine cycle. The gas turbine exhaust is used in a heat recovery steam generator to evaporate feedwater, which is then injected into the combustor to augment the mass flow rate and enthalpy of the turbine stream. Some of the water is condensed in the HRSG, and is recycled back as feedwater.

One form of humid air cycles is the steam injection and heat recovery cycle (known broadly as steam injection gas turbine, or STIG). In these cycles, the turbine exhaust is cooled in a heat recovery steam generator (HRSG) and the steam raised in the HRSG is injected into the combustor to increase the total mass flow rate and the enthalpy flow rate of the turbine inlet stream. Water is pumped to the combustor pressure before it is evaporated in the HRSG. The cycle diagram is shown in Figure 8.1.

The amount of water circulating in the cycle is estimated as follows. Call the water mass flow rate in the HRSG $\dot{m}_{st}$, its latent heat of evaporation at the turbine pressure Δh_L, the total gas (products of combustion) mass flow rate $\dot{m}_g$, its specific heat c_p, the turbine exit temperature T_4, and the HRSG exit temperature T_e. Energy balance across the HRSG shows that

$$\dot{m}_s \Delta h_L = \left(\dot{m}_s + \dot{m}_g\right) c_p (T_4 - T_e). \tag{8.1}$$

Solving for the ratio of the added water mass flow to the combustion products mass flow, we get:

$$\frac{\dot{m}_s}{\dot{m}_g} = \frac{1}{\dfrac{\Delta h_L}{c_p(T_4 - T_e)} - 1}. \tag{8.2}$$

For typical values of the parameters on the right-hand side of (8.2), this ratio is O(20%).

Some of the water added to the combustor is recovered by condensation in the HRSG, and is pumped up to the combustor pressure and recirculated back as feedwater. However, more water must still be added to make up for the amount leaving with the gaseous exhaust of the HRSG as steam. Humid air cycles that rely on recirculating some of the water can be thought of as being semi-closed cycles because some of the working fluid is circulated. A technological challenge for this cycle is the gas turbine design. The gas turbine is exposed to very high-temperature moist combustion products, an environment that can be corrosive. The amount of water added should be limited by the saturation pressure of water at the turbine exit temperature to prevent condensation inside the turbine. Moreover, a cooled condenser may be needed to recover as much water as possible from the gas turbine exhaust, adding to the cost and complexity of the system.

Example 8.1

Consider the STIG cycle shown in Figure 8.1. Air at 298 K and 100 kPa enters the compressor. The pressure ratio is 9. The TIT is 1423 K. The flue gases leaving the turbine flow through a HRSG, which produces superheated steam at 700 K. The exhaust gases leave the HRSG at 400 K. Water at 298 K and 100 kPa is pumped to the HRSG. No water recycling from the flue gas is considered. The compressor, turbine, and pump operate with isentropic efficiency of 90%, 85%, and 70%, respectively. The fuel burned in the combustor is methane with a lower heating value (LHV) of 50.05 MJ/kg. The properties of the flow at states 3, 4, and 5 may be assumed to be those of air + steam mixture. Determine (a) the amount of steam injection and fuel consumption per unit mass of the air; (b) the network; and (c) the thermal efficiency.

Solution

Compressor: $T_2 = T_1\left\{1 + \frac{1}{\eta_C}\left[(r_p)^{\frac{k-1}{k}} - 1\right]\right\} = 298\left\{1 + \frac{1}{0.85}\left[9^{\frac{1.39-1}{1.39}} - 1\right]\right\} = 596.8\,\text{K}.$

Combustor: The First Law is applied to the combustor:

$$\alpha{\cdot}LHV_{CH_4} + h_2 + \beta{\cdot}h_8 = (1+\alpha)h_{a,3} + \beta h_{s,3},$$

where α is the fuel-to-air mass ratio, β is the steam-to-air mass ratio, and $h_{a,3}$ and $h_{s,3}$ denote the enthalpies of air and steam at state 3. Note that in writing this equation, we assume that the thermal properties of the combustion products can be approximated by that of air, and hence the difference between the total enthalpy of the products of combustion and the fraction of original air that participated in combustion and fuel is $\alpha{\cdot}LHV$. This is because we expect the fuel/air in this case to be very small:

$$h_2 = 604.8\text{ kJ/kg}; \quad h_{a,3} = 1545\text{ kJ/kg}; \quad h_{s,3} = 5019\text{ kJ/kg}; \quad h_8 = 3322\text{ kJ/kg}.$$

Turbine: $T_4 = T_3\left\{1 - \eta_T\left[1 - (r_p)^{\frac{1-\gamma}{\gamma}}\right]\right\} = 1423 \times \left[1 - 0.9 \times \left(1 - 9^{-\frac{0.29}{1.29}}\right)\right] = 923.8\text{ K}.$

Pump: $h_6 = 104.2\text{ kJ/kg}; v_6 = 0.001\text{ m}^3/\text{kg},$

$$h_{7s} = 104.2 + 0.001 \times (900 - 100) = 105\text{ kJ/kg},$$

$$h_7 = 104.2 + \frac{105 - 104.2}{0.7} = 105.3\text{ kJ/kg}.$$

HRSG: Applying the First Law, we have: $(1+\alpha)(h_{a,4} - h_{a,5}) + \beta(h_{s,4} - h_{s,5}) = \beta(h_8 - h_7)$

Example 8.1 (cont.)

$h_{a,4} = 956$ kJ/kg; $h_{a,5} = 401.3$ kJ/kg; $h_{s,4} = 3809$ kJ/kg; $h_{s,5} = 2730$ kJ/kg.

a. Solving the enthalpy balance equations of the combustor and HRSG yields

$$\alpha = 0.0287 \text{ and } \beta = 0.267.$$

b. $W_C = h_2 - h_1 = 604.8 - 298.3 = 306.5$ kJ/kg

$$\begin{aligned} W_T &= (1+\alpha)(h_{a,3} - h_{a,4}) + \beta(h_{s,3} - h_{s,4}) \\ &= (1 + 0.0287)(1545 - 956) + 0.267 \times (5019 - 3809) \\ &= 929 \text{ kJ/kg} \\ W_P &= \beta(h_7 - h_6) = 0.267(105.3 - 104.2) = 0.3 \text{ kJ/kg} \\ W_{net} &= W_T - W_C - W_P = 929 - 306.5 - 0.3 = 622.2 \text{ kJ/kg.} \end{aligned}$$

c. $\eta_{th} = \dfrac{W_{net}}{\alpha.LHV_{CH_4}} = \dfrac{622.2}{0.0287 \times 50{,}050} = 0.433.$

Humid air cycles of different forms have been suggested to increase the specific work and maximize the efficiency, especially when aeroderivative gas turbines are used. For instance, if multistage compression and intercooling are used to reduce the compression work, the cooling water used in these intercoolers can be added to the air stream exiting the high-pressure compressor, before or after the combustor, to increase the enthalpy and the mass flow rate into the gas turbine. This is known as the **recuperated water injected cycle (RWI).** In the cycle shown in Figure 8.2, the intercooler water is heated further in an economizer using the exhaust of the recuperator, and then mixed with the high-pressure compressed air. The economizer is used to preheat water between the intercooler and the mixer, so as to extract the most thermal energy from the turbine exhaust stream. Following the mixing stage, the air–steam mixture is heated in the recuperator. Instead of heating the combustion air in the recuperator, as in the case of a regenerative gas turbine cycle, the air–steam stream is heated to take advantage of the higher thermal capacity of the mixture of steam and air. Fuel can also be heated in the economizer and the recuperator before it is injected into the combustor, but the mass flow of fuel is usually much smaller than that of the air or that of the cooling water and hence this step may not contribute much to the regeneration. The humidified air is then admitted to the combustor. The efficiency of this cycle is improved over the regular intercooled recuperated cycle because most of the thermal energy is recovered and fed into the gas turbine. Furthermore, compression work is reduced by intercooling. Using an economizer and a recuperator maximizes the thermal energy recovery from the turbine exhaust stream.

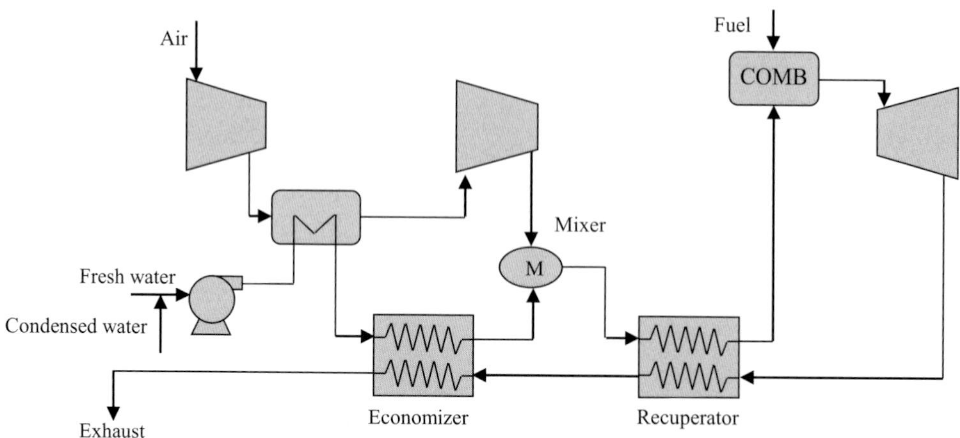

Figure 8.2 Steam injected gas turbine cycle with an economizer, regeneration, and water condensation for recycling. The cycle uses an intercooler to reduce the compression work, and the enthalpy recovered by the cooling water in the intercooler is added to the turbine. This is an example of a humid air turbine (HAT) cycle.

Several variations exist, including a modification in which more than one water stream is used in the intercoolers and aftercooler, and several loops are used between the different components to minimize the temperature difference in the heat transfer equipment and in the mixer. When the maximum amount of water is added, the stream leaving the mixer is saturated; in this case the mixer is called the saturator. This is the **HAT cycle**. Analysis shows that this humid air cycle is the most efficient gas turbine cycle – even more efficient than the combined gas–steam cycles. However, to achieve this high efficiency, most of the enthalpy in the exhaust stream must be recovered, and hence most of the water must be condensed before leaving the economizer. As mentioned before, the steam content in the gas turbine stream is 15–25% by mass, depending on the pressure and the TRIT. In most cases, increasing the steam content should raise the cycle efficiency. Because these cycles are suggested for use with aeroderivative turbines they operate at high pressure ratios and TRIT. Another reason for the higher efficiencies for these humid air cycles is that steam is generated in the gas stream through flash evaporation, hence avoiding large temperature differences between the heating stream and the evaporating steam. Efficiencies in the range of 45–50% have been predicted for HAT cycles. Table 8.1 shows a comparison between performance predictions for different cycles [1–3].

As mentioned already, a potential disadvantage of humid air cycles is that the combined effect of humid air, higher pressure, and high temperature could lead to extensive corrosion of the turbine blades. Moreover, care must be taken to avoid condensation of water in the later stages of the turbine, where high moisture content and low temperature may lead to the formation of water droplets. Fast-spinning rotors can be impacted negatively by such formation. Another disadvantage of humid air cycles is the need for purified water, and

Table 8.1 Performance of some water injection cycles [4]

Performance data	Simple cycle	CC	HAT
Gas turbine type	Heavy duty	AD	AD
Pressure ratio	30	46	48
TIT (°C)	1250	1500	1500
Water consumption (kg/kWh)	0.892	0.74	0.72
Efficiency (%)	39.7	55.5	57.0

CC, combined cycle (described later in the chapter); AD, aeroderivative

the need to purify the recirculated water. On the other hand, these HAT cycles use a single set of turbines, and with technological development of turbine technology, they may present an opportunity for higher efficiency with reduced equipment cost.

8.3 Thermochemical Recuperation

As an alternative to the HAT-type cycles, thermochemical recuperation (TCR) has been suggested. In this case, the thermal energy recovered from the turbine exhaust is used to steam reform a hydrocarbon fuel to a mixture of hydrogen and carbon monoxide (syngas). As shown in previous chapters, steam reforming of hydrocarbons into a hydrogen-rich gas mixture is mostly endothermic and involves reactions with water. For instance, for methane:

$$CH_4 + H_2O \rightarrow CO + 3H_2 \qquad \Delta H_R = 226\ \text{kJ/mole of methane.} \tag{8.3}$$

To maximize the hydrogen content in the reformed gas, the water–gas shift reaction can be used in addition to the methane reforming reaction:

$$CO + H_2O \rightarrow CO_2 + H_2 \qquad \Delta H_R = -41\ \text{kJ/mole.} \tag{8.4}$$

As shown in previous discussions on reforming in Chapter 3, equilibrium determines the composition of the reformate, and in most cases some carbon dioxide is present. The thermal energy needed for reforming is extracted from the turbine exhaust stream. The resulting fuel stream – that is, the mixture of $CO + H_2 + N_2 + CO_2$ – is burned in the gas turbine combustor. Burning these lighter fuels has some combustion and emissions advantages. The composition of the reformate is a strong function of the temperature and pressure of reforming. At typical high-pressure gas turbine conditions, the exit temperature of the turbine is around 600 °C, and only partial reforming is expected, and some hydrocarbons will be left in the stream. Furthermore, at these low temperatures, the kinetic reaction rates can be slow and equilibrium may not be reached within the reformer. Special metal-based (mostly nickel) catalytic reactors are required to speed up the reforming reactions while conducting them at low temperatures that favor hydrogen formation.

In a modeling study to assess the potential of TCR, a Westinghouse 501F heavy-duty gas turbine, spinning at 3600 RPM, was used. The study used methane as a fuel. The air flow rate was 436 kg/s, TRIT = 1316 °C, turbine exhaust T = 606 °C, the temperature of the reformed fuel was 579 °C. The results showed that 38% of the methane was reformed, given the amount of water added and the conditions of the reformer. The thermal energy in the turbine exhaust gases is used partially in the thermochemical recuperator, or the reformer, and partially to raise steam in the HRSG, as shown in Figure 8.3. In this example, the exhaust gas leaves the reformer at 427 °C to raise steam at 207 °C and 21.6 atm. Table 8.2 lists the major performance parameters for the cycle.

Clearly, TCR raises the cycle efficiency, but not as much as a CC. However, this study did not attempt to optimize the process, and better overall performance should be expected using advanced higher-temperature G and H class turbines.

There are other thermochemical cycles in which the exhaust gas is used directly for the partial oxidation of the fuel, but studies show that steam reforming cycles are more efficient [4]. While the concept of TCR has not been used much in practice, similar concepts have

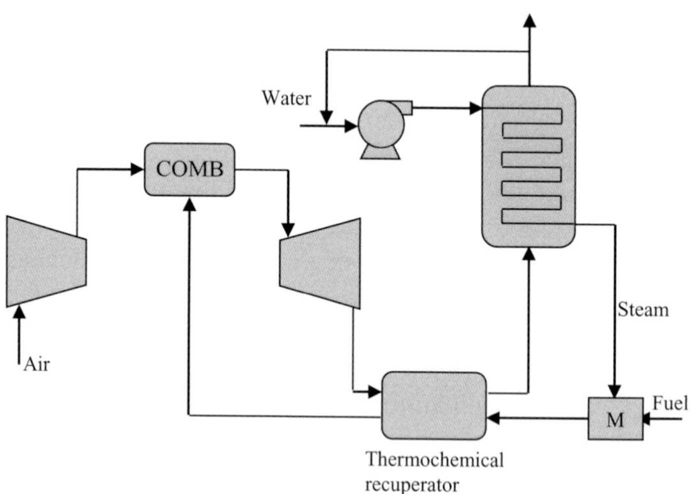

Figure 8.3 Chemically recuperated gas turbine cycle using a HAT. Fuel is methane reformed using the steam raised in the HRSG. The high-temperature gas turbine exhaust is used to supply thermal energy for reforming, and then to raise steam for reforming in the HRSG.

Table 8.2 Performance of a thermochemical recuperation cycle (TCR) compared to conventional simple cycle (SC) and conventional combined cycle (CC) using NG (natural gas) [4]

	TCR	SC	CC
Steam to NG ratio by mass	6.6	NA	NA
Air to NG ratio by mass	43	42.7	42.7
Makeup water (kg/kWh)	1	0	0.02
Stack gas temperature (°C)	126	590	129
Net cycle power (MW)	216	166	264
Cycle efficiency (%)	48.9	35.7	56.8

been suggested in carbon capture cycles (Chapters 11 and 13) and cycles that integrate solar and fuel energy sources (Chapter 9).

8.4 Combined Cycles

Combined gas and steam cycles have been used to improve the overall thermal efficiency of power plants [5,6]. As mentioned before, although supercritical steam cycles can achieve high efficiency, they require high-temperature and high-pressure steam turbines. High-pressure superheated steam turbines can be bulky, and hence have high inertia. Meanwhile, while gas turbine cycles operate at much higher maximum temperatures, the turbine exhaust temperature is also high, and hence plenty of thermal energy and availability are wasted. A CC uses the gas turbine as the high-temperature **topping cycle**, and the steam plant as the low-temperature **bottoming cycle**, thus recovering some of the thermal energy and availability in the gas turbine exhaust and using that thermal energy to operate a Rankine cycle. The matching between the two cycles is an optimization process in which some parameters, such as the pressure ratio of one or both cycles, the intermediate temperature, etc., are varied until efficiency is maximized.

Figure 8.4 shows the layout of a CC. The boiler/steam generator is replaced by a heat exchanger in which the gas turbine exhaust is used to raise steam. All other components remain the same.

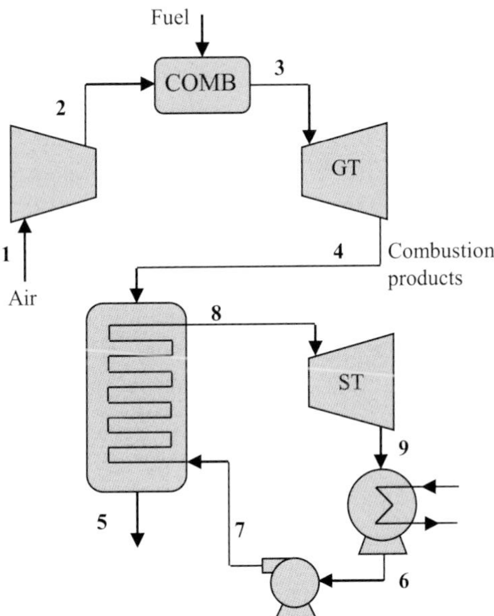

Figure 8.4 A typical arrangement for a steam–gas (STAG) CC, with the gas turbine exhaust used to generate steam for the steam cycle.

Combined steam and gas turbine plants are known as **STAGs**. Even with recuperation, the exhaust temperature of the plant remains high – higher than the temperature at the exit of the compressor. With sufficiently high pressure ratio, and practical values for compressor efficiency, the exhaust gas temperature is sufficiently high to power a steam cycle. The overall utilization of the fuel chemical energy increases substantially when a bottoming cycle is added. Higher specific work can be achieved in a CC with **supplementary firing**; that is, using extra fuel to further raise the gas turbine exhaust gas temperature before using it to generate steam. This, however, does not guarantee better efficiency.

An expression for the CC efficiency, written in terms of the efficiencies of the individual cycles, can be used to demonstrate their benefits. Call the topping, high-temperature, or gas turbine cycle efficiency η_{GT}, and that of the bottoming, low-temperature, or steam turbine cycle η_{ST}. The total work of the plant is the sum of the gas turbine plant and the steam turbine plant [1],

$$W = W_{GT} + W_{ST}. \tag{8.5}$$

The total heat added to this cycle is the heat added to the gas turbine plant by burning fuel in the gas turbine cycle combustor, $Q_{in} = Q_{GT}$. The gas turbine cycle work is $W_{GT} = \eta_{GT} Q_{GT}$. The heat added to the steam turbine cycle and the work it generates are given by

$$Q_{ST} = (1 - \eta_{GT}) Q_{GT}, \tag{8.6}$$

$$W_{ST} = \eta_{ST}(1 - \eta_{GT}) Q_{GT}. \tag{8.7}$$

The CC efficiency is then

$$\eta_{CC} = \frac{W_{GT} + W_{ST}}{Q_{GT}} = \eta_{GT} + \eta_{ST}(1 - \eta_{GT}). \tag{8.8}$$

Thus, for $\eta_{GT} = 0.25$ and $\eta_{ST} = 0.4$, the CC efficiency is $\eta_{CC} = 0.55$, a substantial improvement over both. Studies have shown that the CC efficiency increases as the gas turbine cycle efficiency is raised by increasing the TRIT and its pressure ratio. It should be mentioned, however, that changing the pressure ratio and other design parameters that raise the gas turbine cycle efficiency can impact the steam cycle efficiency negatively, and care must be exercised in matching the two cycles for the best efficiency.

Equation (8.8) shows that the CC efficiency is more strongly impacted by that of the gas turbine cycle. Increasing the gas cycle efficiency is bound to raise the overall cycle efficiency, even if the steam cycle efficiency decreases slightly. This is because while the overall cycle efficiency increases directly with η_{GT}, the impact of a small drop in η_{ST} is reduced by the reduction in $(1 - \eta_{GT})$ as η_{GT} is raised. For example, for $\eta_{GT} = 0.3$, and $\eta_{ST} = 0.28, \eta_{CC} = 0.5$, but a substantial increase in the gas turbine cycle and a concomitant decrease in the steam turbine cycle such that: $\eta_{GT} = 0.38$, and $\eta_{ST} = 0.25$, yields a significant increase in the overall cycle efficiency, $\eta_{CC} = 0.535$. Natural gas CCs (NGCCs) reach 55% efficiency using available components. Similar performance has been observed when using liquid fuels. Combined cycles that use gasified coal products have

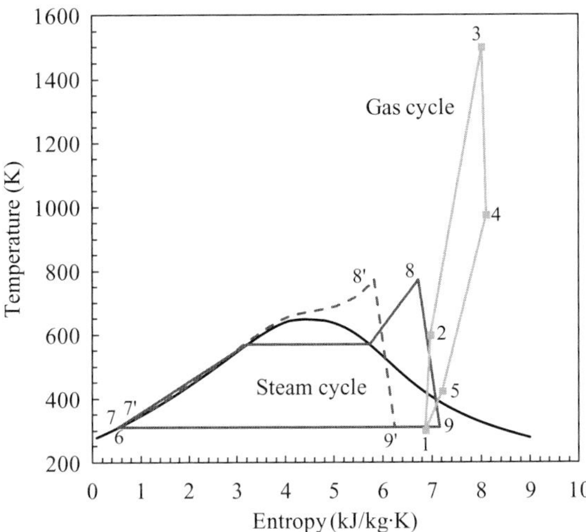

Figure 8.5 A supercritical (6–7′–8′–9′) and a subcritical (6–7–8–9) steam cycle, and a "matching" gas turbine cycle for use in a CC plant. Note that points 7 and 7′ are not the same since the pressures are different, although they appear to overlap on the diagram; see Example 8.2.

also been built, but their overall efficiency is lower because of issues connected with the gasification process, gas clean-up, and the integration of different units, as will be shown later. Analysis of Oxy-combustion CC is described later and in Chapter 9.

The *T–s* diagram of a CC is shown in Figure 8.5 for two cases: when the steam turbine cycle is subcritical, as indicated by the states 6–7–8–9–6; and when it is supercritical, as marked by the states 6–7′–8′–9′–6. The maximum pressure of the subcritical and supercritical steam cycles is 8 MPa and 29 MPa, respectively. Other process data are given in Example 8.2. The heat transfer process between the gas turbine exhaust and the steam turbine cycle in the HRSG depends on the steam turbine cycle conditions. Higher steam TIT can be achieved when using a supercritical cycle since supercritical cycles avoid the flat temperature zone within the evaporation dome. Moreover, in a supercritical cycle, the temperature difference between the two streams, the gas turbine exhaust and the steam, is smaller all along, and higher overall cycle efficiency is achievable. Supercritical cycles reduce the irreversibility associated with heat transfer across a large temperature difference.

8.4.1 Supplementary Firing

If there is a need for more power, extra fuel is burned in the steam generator using the excess oxygen in the gas turbine exhaust stream. This is the supplementary fuel cycle. As mentioned before, gas turbines run under lean to very lean conditions because of the gas turbine material's considerations, and the exhaust stream of the topping cycle gas turbine contains extra oxygen that can be used in a combustion process in the steam generator if more fuel is used. While supplementary firing raises the CC total work, it does not necessarily raise the overall efficiency because the extra heat is transferred to the cycle at a temperature lower than the cycle maximum temperature. The loss of efficiency can be determined by driving the overall cycle efficiency for the case with supplementary firing. Given the same definitions as before, the heat transferred to the steam turbine cycle with supplementary firing is

$$Q_{ST} = (1 - \eta_{GT})Q_{GT} + Q_{sup} = \left[(1 - \eta_{GT}) + f_{sup}\right]Q_{GT}, \tag{8.9}$$

where $f_{sup} = Q_{sup}/Q_{GT}$ is the ratio of the supplementary heat added to the steam cycle to the heat added to the gas turbine cycle. The CC efficiency is the total work out divided by the total heat transferred in, and is given by

$$\eta_{CC} = \frac{\eta_{GT} + \eta_{ST}(1 - \eta_{GT}) + \eta_{ST} f_{sup}}{1 + f_{sup}}. \tag{8.10}$$

Since $\eta_{ST} < 1$, as f_{sup} increases, the overall cycle efficiency decreases.

Example 8.2

Consider the CC power plant shown in Figure 8.4. Air at 300 K and 100 kPa enters the compressor, whose isentropic efficiency is 80%. It is then compressed to 800 kPa and heated to 1500 K. The hot stream leaving the gas turbine, which operates with an isentropic efficiency of 88%, flows through the HRSG, where it is cooled to 423 K. Superheated steam at 723 K and 8 MPa enters the steam turbine and is expanded to 6 kPa. The mass of fuel burned in the combustor is negligible compared to that of the air. The isentropic efficiencies of the steam turbine and pump are 90% and 70%, respectively. Determine: (a) the amount of steam produced in the HRSG per unit mass of air; (b) the work output of the gas and steam cycles; (c) the thermal efficiency of gas and steam cycles if they would operate separately; and (d) the thermal efficiency of the CC.

Solution

Gas cycle:

$$T_2 = T_1\left\{1 + \frac{1}{\eta_C}\left[(r_p)^{\frac{k-1}{k}} - 1\right]\right\} = 300\left\{1 + \frac{1}{0.8}\left[8^{\frac{1.39-1}{1.39}} - 1\right]\right\} = 597.1\ \text{K},$$

$$W_C = h_2 - h_1 = 605 - 300.3 = 304.7\ \text{kJ/kg},$$

$$T_4 = T_3\left\{1 - \eta_{GT}\left[1 - (r_p)^{\frac{1-k}{k}}\right]\right\} = 1500 \times \left[1 - 0.88 \times \left(1 - 8^{-\frac{0.323}{1.323}}\right)\right] = 974.5\ \text{K},$$

$$W_{GT} = h_3 - h_4 = 1637 - 1017 = 620\ \text{kJ/kg},$$

$$q_{in} = h_3 - h_2 = 1637 - 605 = 1032\ \text{kJ/kg}.$$

Example 8.2 (cont.)

Steam cycle:

$$\left.\begin{array}{l} p_6 = 6\ \text{kPa} \\ \text{Saturated liquid} \end{array}\right\} \Rightarrow h_6 = 151.5\ \text{kJ/kg};\ v_6 = 0.001\ \text{m}^3/\text{kg},$$

$$h_{7s} = h_6 + v_6(p_7 - p_6) = 151.5 + 0.001 \times (8000 - 6) = 159\ \text{kJ/kg},$$

$$h_7 = h_6 + \frac{h_{7s} - h_6}{\eta_P} = 151.5 + \frac{159 - 151.5}{0.7} = 162.2\ \text{kJ/kg},$$

$$\left.\begin{array}{l} T_8 = 773\ \text{K} \\ p_8 = 8\ \text{MPa} \end{array}\right\} \Rightarrow h_8 = 3398\ \text{kJ/kg};\ s_8 = 6.724\ \text{kJ/kg·K},$$

$$\left.\begin{array}{l} p_9 = 6\ \text{kPa} \\ s_{9s} = s_8 \end{array}\right\} \Rightarrow h_{9s} = 2070\ \text{kJ/kg},$$

$$h_9 = h_8 - \eta_{ST}(h_8 - h_{9s}) = 3398 - 0.9 \times (3398 - 2070) = 2202.8\ \text{kJ/kg}.$$

a. To determine the mass of the steam per mass of the air, α, we apply the First Law to the HRSG. Hence,

$$h_4 - h_5 = \alpha(h_8 - h_7),$$

$$\alpha = \frac{h_4 - h_5}{h_8 - h_7} = \frac{1017 - 424.7}{3398 - 162.2} = 0.183.$$

b. The work production of the gas cycle is

$$W_{net,GC} = W_{GT} - W_C = 620 - 304.7 = 315.3\ \text{kJ/kg}.$$

The net work produced by the steam cycle is calculated as follows:

$$W_{ST} = \alpha(h_8 - h_9) = 0.183 \times (3398 - 2202.8) = 218.7\ \text{kJ/kg},$$

$$W_P = \alpha(h_7 - h_6) = 0.183 \times (162.2 - 151.5) = 2\ \text{kJ/kg},$$

$$W_{net,SC} = W_{ST} - W_P = 218.7 - 2 = 216.7\ \text{kJ/kg}.$$

The net work output of the CC is

$$W_{net} = W_{net,GC} + W_{net,SC} = 315.3 + 216.7 = 532\ \text{kJ/kg}.$$

c. The thermal efficiency of the gas turbine cycle alone is

$$\eta_{th,GC} = \frac{W_{net,GC}}{q_{in}} = \frac{315.3}{1032} = 0.306.$$

Example 8.2 (cont.)

The efficiency of the steam cycle if it operated separately would be

$$\eta_{th,GC} = \frac{W_{net,SC}}{\alpha(h_8 - h_7)} = \frac{216.7}{0.183 \times (3398 - 162.2)} = 0.366.$$

d. The thermal efficiency of the CC is

$$\eta_{th,CC} = \frac{W_{net}}{q_{in}} = \frac{532}{1032} = 0.516.$$

The efficiency of the CC is much higher than that of the simple cycles. The *T–s* diagram of the CC is shown in Figure 8.5.

Analysis of CCs must consider the difference between the circulating mass flow rate of water/steam in the bottoming cycle and the air and combustion exhaust flow rate in the topping cycle. The *T–s* diagram of the CC shows that a bottoming supercritical steam cycle is a better match for a topping gas turbine cycle because the heat transfer between the gas turbine exhaust and the steam generator is less irreversible than that between the gas turbine exhaust and a subcritical steam generator. However, in most cases, CC power plants utilize subcritical steam cycles and special attention should be given to the heat transfer process between the exhaust gases and the steam to minimize the irreversibility. These are discussed next.

8.4.2 Pinch Point Analysis

A CC with a single steam turbine operating on a single stream of steam is called a single-pressure CC plant. If the steam turbine operates under subcritical conditions, the temperature–heat transfer diagram inside the HRSG exhibits the features shown in Figure 8.6 – namely the discontinuous slope of the steam temperature line. The point of minimum temperature difference between the two streams is the **pinch point (PP)**, and it occurs at the transition between the economizer and the evaporator. From there on, the temperature difference between the two streams increases, as the evaporation temperature remains constant while the temperature of the hot stream decreases. The largest temperature difference between the two streams is found at the hot section, between the entering hot gas stream and the exiting superheated steam. Large values of PP increase this temperature difference – that is, the difference between T_4 and T_9. The former is the gas turbine exit temperature, which is determined by the gas TIT and the gas turbine cycle pressure ratio. A large temperature difference inside the HRSG lowers the value of T_9, which is the steam TIT. This leads to less work available from the steam turbine cycle. In most designs, PP = 10–15 °C.

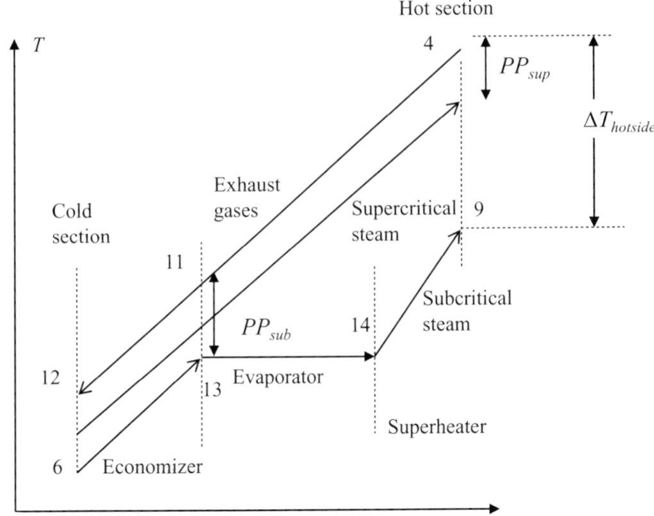

Figure 8.6 A schematic diagram showing the temperature–heat transfer plot in a HRSG in a single-pressure CC power plant, showing the pinch point in both cases of a subcritical steam cycle and a supercritical steam cycle.

The diagram in Figure 8.6 uses the same numbers to define the states of the working fluid as in the T–s diagram in Figure 8.5, with extra numbers identifying the intermediate stages within the HRSG. According to this diagram,

$$PP = T_{11} - T_{13}. \tag{8.11}$$

Several considerations are used in the design of the HRSG, and in defining the different states. To facilitate the discussion, we recall that the convective heat flux, $\dot{q}_{HT}$, between two streams is given by

$$\dot{q}_{ht} = \kappa(T_g - T_{st}). \tag{8.12}$$

In this relation, κ is the overall heat transfer coefficient between the two streams, T_g and T_{st} are the exhaust gas and steam temperatures. A finite temperature difference must be maintained between the two streams to develop a finite heat flux. The temperature of the gas turbine exhaust gases is kept relatively high inside the HRSG to limit water condensation on the outside of the tubes, the possible formation of sulfuric acids, and the potential corrosion of the heat exchange pipes. These considerations are used to determine the lowest temperature of the gas turbine exhaust stream, T_{12}.

The feedwater temperature at the entry of the steam generator, that is the economizer section, which is T_6 in Figure 8.6, is kept low for the best efficiency of the steam cycle. Essentially, T_6 is the saturation temperature corresponding to the condenser pressure, and it is kept as close as possible to the coolant temperature. This determines the temperature difference between the two streams at the cold end of the HRSG, $(T_{12} - T_6)$. Meanwhile, the convective heat transfer coefficient at the hot end of the

HRSG is relatively low because of the poor heat transfer between two gaseous streams compared to that between a gaseous and a liquid stream. To overcome this limitation, the temperature difference $(T_4 - T_9)$ is kept relatively high at the hot end of the HRSG. The temperature difference rises within the evaporator section, where the evaporating steam temperature stays constant while the exhaust products temperature is decreasing. Thus, the minimum temperature difference between the two streams within the HRSG is indeed the value marked as PP, and we write:

$$T_{11} = T_{13} + PP. \tag{8.13}$$

The cycle calculations may proceed as follows. Given the steam turbine inlet pressure, which is fixed by the coolant temperature plus a difference O(10–20 °C), we determine h_{13} and T_{13}. Equation (8.13) is then used to determine T_{11}. Calling the water/steam flow rate in the HRSG $\dot{m}_s$, and that of the exhaust gas $\dot{m}_g$, and assuming ideal gas behavior, we write the following relation between the two flow rates over the economizer section:

$$\dot{m}_g c_{pg}(T_{11} - T_{12}) = \dot{m}_s(h_{13} - h_6). \tag{8.14}$$

Given the gas mass flow rate of the exhaust gas, this expression can be used to determine $\dot{m}_s$. Finally, the exit conditions of the steam from the HRSG is determined from the energy balance over the evaporator/superheater section:

$$\dot{m}_g c_{pg}(T_4 - T_{11}) = \dot{m}_s(h_9 - h_{13}). \tag{8.15}$$

Alternatively, instead of specifying the exit temperature of the exhaust gas, or the stack temperature, T_{12}, one can specify the inlet temperature to the steam turbine, T_9 and recast these equations to calculate the rest. The energy balance over the evaporator–superheater, (8.15), can now be used to calculate the steam mass flow rate since the gas flow rate is known:

$$\dot{m}_{st} = \dot{m}_g \frac{c_{pg}(T_4 - T_{11})}{(h_9 - h_{13})}. \tag{8.16}$$

In this equation, the exhaust gas mass flow rate and its temperature at the HRSG inlet are known, T_{13} is given by the steam turbine pressure, and T_{11} is given by (8.13). Next, (8.14) is used to calculate the exhaust exit temperature.

8.4.3 Multi-pressure HRSG

The temperature difference between the exhaust gas stream and the water–steam stream inside the HRSG can be reduced by employing a dual- or triple-pressure steam plant, with all pressures under subcritical conditions. The steam generator/turbine part of the plant and the corresponding temperature heat flux diagrams are shown for a dual-pressure turbine design in Figures 8.7 and 8.8. In this case, two pumps are used, with the low-pressure stream being heated at the cold end of the HRSG and the high-pressure

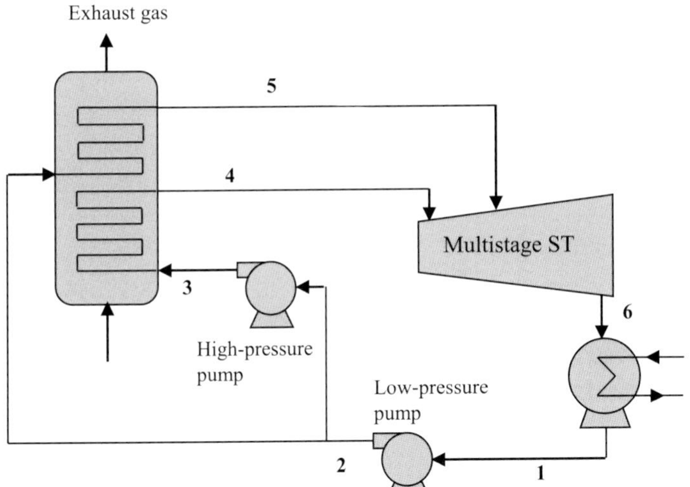

Figure 8.7 The steam cycle section of a dual-pressure CC power plant, showing the two-stage pump, steam generation, and steam turbine.

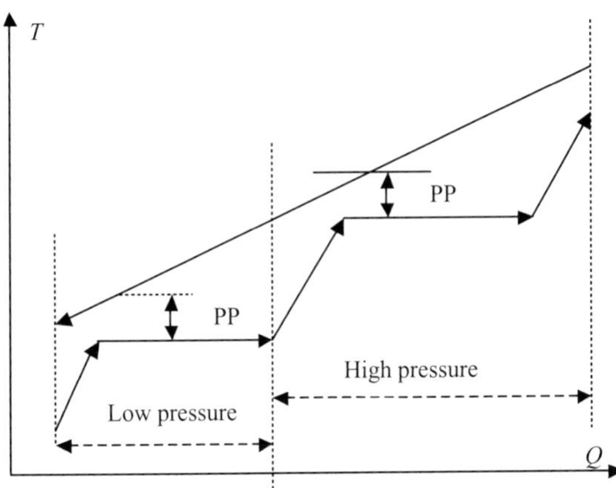

Figure 8.8 A schematic diagram showing the temperature–total heat transfer in the HRSG in a dual-pressure CC power plant, showing the two pinch points. The cycle components are shown in Figure 8.7.

stream at the hot end. While the PP is kept the same as before, O (10–15 °C), the water/steam temperature stays closer to the gas stream temperature than in the case of a single-pressure cycle. Given the two pressures, one can calculate the saturation temperatures, and using the PP condition in two places, one can evaluate the corresponding gas temperatures.

Calculations show that the CC efficiency increases by 2–3 percentage points – that is, from, say, 50% to 52–53%, with each extra pressure stage in the steam turbine cycle. Up to three stages have been used to achieve the high efficiencies reported for CCs. These cycles often utilize all reheat and regeneration steps used in conventional steam power plants to achieve maximum efficiency. In particular, reheat back to the maximum temperature available in the exhaust gases, 550–600 °C, is used in many

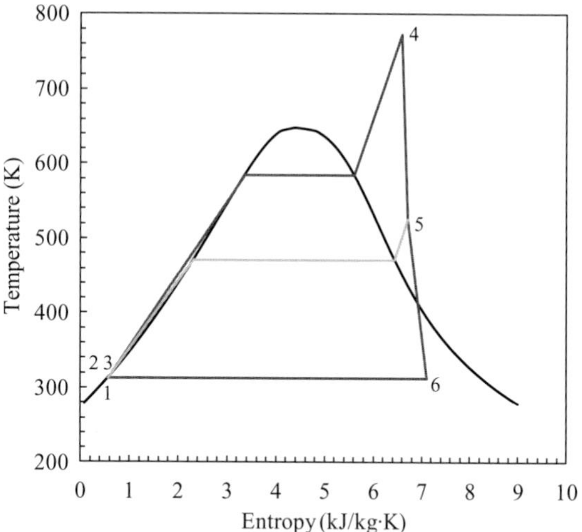

Figure 8.9 The T–s diagram of the steam cycle section of a CC plant with a dual-pressure HRSG. This part of the cycle is shown in Figure 8.7, and the corresponding temperature profiles are shown in Figure 8.8.

modern plants. Intercooling in the compression stage is also used to reduce the compressor work. The T–s diagram of the steam cycle for a dual-pressure HRSG is shown in Figure 8.9.

Another approach to minimizing heat transfer-caused loss of work potential in the steam turbine is to operate the steam cycle under supercritical conditions. In this case, and as shown in Figure 8.6, it is possible for the superheated steam exiting the HRSG to reach a temperature much closer to that of the exhaust gases.

Most CC power plants built so far burn natural gas, and can be designed to reach 55% efficiency. The maximum efficiency is largely controlled by the TRIT, and values more that 60% are expected as advanced turbine systems are introduced, whose maximum inlet temperature is 1500 °C. Combined cycle plants burning oil have also been built. Note that since combustion products are expanded in the gas turbine, the fuel must be cleaned up before entering the gas turbine if it contains harmful impurities such as sulfur, to safeguard against corrosion. Maximizing the plant efficiency requires, besides raising the gas TIT and the cycle pressure ratio, employing multistage reheat and intercooling in the gas turbine cycle.

Despite their complexity, CCs have gained popularity because of their high efficiency, especially when used in natural gas-fired electricity generation plants. Natural gas is a clean-burning fuel and produces the least amount of carbon dioxide per unit of electricity generated of any other hydrocarbon. On the other hand, it can be relatively expensive, especially when compared to coal. Even when used in lower-efficiency steam power plants, coal can prove to be economical because of the low price of the fuel. Coal is also abundant in the USA and many other countries, like Australia, China, and India, the last two have substantially increased their use of coal in electricity generation during the past few years. The use of coal in CC plants is described in Chapters 12 and 13, without and with carbon capture.

8.5 Oxy-Combustion Cycles

Oxy-combustion cycles have been proposed for carbon dioxide capture power plants, and have been demonstrated at the pilot scale. In these cycles, an air-separation unit (ASU) is used to separate oxygen from air, and an almost pure stream of oxygen is used to burn the fuel. A diluent is used in oxyfuel combustion to moderate the flame temperature (replacing nitrogen), and in CC applications this diluent is a recycled stream of CO_2 or CO_2 with a small concentration of water. Following the condenser of the steam cycle, liquid water is condensed and separated, some CO_2 is removed and sent to a CO_2 purification and liquefaction unit, and the rest is recycled to the gas turbine cycle. Examples 8.3–8.5 illustrate the operation of an oxy-combustion CC. Chapters 11 and 13 describe in detail many alternative oxy-combustion cycles.

Example 8.3

A gas turbine power plant operating with oxyfuel combustion and methane generates 100 MW power. Air at 298 K and 1 bar is compressed up to 6 bar within an ASU. The oxygen produced is cooled to 310 K and is then pressurized to 35 bar in another compressor located upstream of the combustor. Methane initially at 298 K and 1 bar is supplied as a fuel and its pressure is increased to 35 bar using the fuel compressor. The combustion products comprising carbon dioxide and steam leave the combustor at 1473 K. Assume a 5% pressure drop within the combustor. The products are expanded in the turbine down to 1 bar.

In an initial design, the flue gases leaving the gas turbine are cooled to 320 K in a condenser and the water is separated from the mixture. The remaining carbon dioxide is separated into two streams using a control valve; most of the CO_2 is recycled and sent to a compressor, where it is pressurized up to 35 bar, and the rest is captured. Assume an isentropic efficiency of 80% for all the compressors, and isentropic efficiency of 88% for the gas turbine. Determine the mass flow rates of the air and methane, and the thermal efficiency.

In an improved design, the turbine exhaust passes through a HRSG, where a fraction of its thermal energy is transferred to pressurized water at 3 MPa, resulting in the production of superheated steam at 798 K. The temperature of the turbine exhaust gases reduces to 473 K at the HRSG outlet. The superheated steam is directed to a steam turbine, where it is expanded for production of additional power. The saturated steam leaves the turbine and condenses in a condenser maintained at 0.08 bar. The condensate is circulated back to the HRSG using a pump. Assuming isentropic efficiencies of 89% and 70% for the steam turbine and the pump, determine the net power production and the thermal efficiency of the CC.

Example 8.3 (cont.)

Figure E8.3

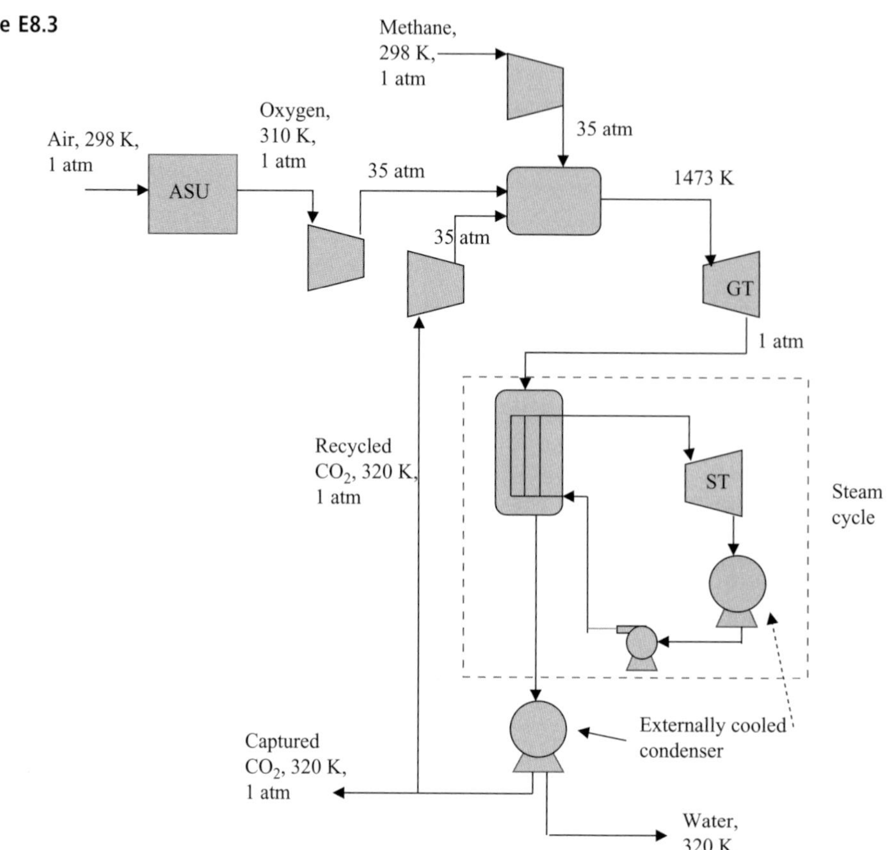

Solution

Initial design: gas turbine power plant with oxyfuel combustion and CO_2 capture

Air compressor: The outlet temperature of the air compressor within the ASU is obtained from

$$T_{air,o} = T_{air,in}\left[1 + \frac{\left(p_{air,o}/p_{air,in}\right)^{\frac{k_{air}-1}{k_{air}}} - 1}{\eta_c}\right].$$

Example 8.3 (cont.)

The power requirement of the air compressor is therefore determined as follows:

$$\dot{W}_{air,C} = \dot{m}_{air}(h_{air,o} - h_{air,in}).$$

Oxygen compressor: $$T_{O_2,o} = T_{O_2,in}\left[1 + \frac{(p_{O_2,o}/p_{O_2,in})^{\frac{k_{O_2}-1}{k_{O_2}}} - 1}{\eta_c}\right],$$

$$\dot{W}_{O_2,C} = \dot{m}_{O_2}(h_{O_2,o} - h_{O_2,in}).$$

Carbon dioxide compressor: $$T_{CO_2,o} = T_{CO_2,in}\left[1 + \frac{(p_{CO_2,o}/p_{CO_2,in})^{\frac{k_{CO_2}-1}{k_{CO_2}}} - 1}{\eta_c}\right]$$

$$\dot{W}_{CO_2,C} = \dot{m}_{CO_2,r}(h_{CO_2,r,o} - h_{CO_2,r,in}).$$

Fuel compressor: $$T_{CH_4,o} = T_{CH_4,in}\left[1 + \frac{(p_{CH_4,o}/p_{CH_4,in})^{k_{CH_4}-1/k_{CH_4}} - 1}{\eta_c}\right]$$

$$\dot{W}_{CH_4,C} = \dot{m}_{CH_4}(h_{CH_4,o} - h_{CH_4,in}).$$

Gas turbine: $$\mathrm{TOT} = \mathrm{TIT}\left[1 - \eta_t\left(1 - \left(\frac{(p_{TO})}{p_{TI}}\right)^{\frac{k_{CO_2}-1}{k_{CO_2}}}\right)\right],$$

where TOT is the turbine outlet temperature, TIT is the turbine inlet temperature, p_{TO} is the turbine outlet pressure, and p_{TI} is the turbine inlet pressure. Due to the 5% pressure drop in the combustor, we have $p_{TI} = 0.95 p_{CO_2,o}$. Note that $p_{CO_2,o} = p_{O_2,o} = p_{CH_4,o} = 35\,\text{bar}$.

$$\dot{W}_T = \dot{m}_{CO_2}(h_{CO_2,\mathrm{in}} - h_{CO_2,o}) + \dot{m}_{H_2O}(h_{H_2O,in} - h_{H_2O,o}).$$

The net power production of the power plant is

$$\dot{W}_T - \dot{W}_{air,C} - \dot{W}_{O_2,C} - \dot{W}_{CO_2,C} - \dot{W}_{CH_4,C} = \dot{W}_{net} = 100{,}000\ \text{kW}.$$

Combustor:

The chemical reaction equation in the combustor can be written as

$$CH_4 + aCO_2 + bO_2 \rightarrow cCO_2 + 2H_2O.$$

Example 8.3 (cont.)

From the carbon and oxygen balance, we find

$$a + 1 = c$$
$$b = 2.$$

Notice that a is the number of moles of carbon dioxide recycled, whereas c denotes the number of moles of carbon dioxide leaving the combustor.

The energy balance equation for the combustor takes the following form (assuming adiabatic combustion):

$$\hat{h}_{CH_4} + a\hat{h}_{CO_2,r} + 2\hat{h}_{O_2} = (a+1)\hat{h}_{CO_2} + 2\hat{h}_{H_2O}.$$

Upon determination of the enthalpies of individual species using the method described in Examples 3.5 and 10.2, the number of moles of CO_2 recycled is obtained from the combustor energy balance equation: a = 15.59.

Relationship between mass flow rates: Based on the chemical reaction equation above, the mass flow rates of the different gaseous species can be related to the fuel mass flow rate:

$$\dot{m}_{O_2} = 4\dot{m}_{CH_4}$$

$$\dot{m}_{CO_2,r} = \frac{a \times 44}{16}\dot{m}_{CH_4}$$

$$\dot{m}_{H_2O} = 2.25\dot{m}_{CH_4}$$

$$\dot{m}_{CO_2} = \frac{(a+1) \times 44}{16}\dot{m}_{CH_4}$$

$$\dot{m}_{air} = 4.76\dot{m}_{O_2} = 19.04\dot{m}_{CH_4}.$$

The formulation described above can be implemented using engineering software such as EES. The results are summarized in Table E8.3a. The required air and fuel mass flow rates for the operation of the gas turbine power plant are 149.4 kg/s and 8.669 kg/s.

The thermal efficiency of the gas turbine power plant is simply obtained from

$$\eta_{th} = \frac{\dot{W}_{net}}{\dot{m}_{CH_4} LHV_{CH_4}} = \frac{100{,}000}{8.669 \times 50{,}050} = 0.2305.$$

Example 8.3 (cont.)

Table E8.3a

Mass flow rates (kg/s)	
Air	149.4
Fuel	8.669
Recycled CO_2	371.8
Water	19.51
Temperatures (K)	
Air compressor outlet	544.6
O_2 compressor outlet	894.5
CO_2 compressor outlet	698.5
Fuel compressor outlet	654
Turbine outlet	916.3
Power consumption (kW)	
Air compressor	37518
O_2 compressor	20250
CO_2 compressor	140,784
Fuel compressor	8679
Turbine power production (kW)	307,231

Improved design: gas turbine combined cycle

Pump: $\dot{W}_{Pump} = \dot{m}_{steam}\left(h_{Pump,o} - h_{Pump,in}\right)$,

where

$$h_{Pump,o} = h_{Pump,in} + \frac{h_{Pump,os} - h_{Pump,in}}{\eta_P}$$

$$h_{Pump,in} = h_{sat}^{0.08\ \text{bar}}$$

$$h_{Pump,os} = h_{s=s_{in}}^{30\ \text{bar}}.$$

HRSG: Applying the energy conservation equation for the HRSG assuming no heat loss, we have

$$\dot{m}_{CO_2}\left(h_{CO_2,HRSG,in} - h_{CO_2,HRSG,out}\right) + \dot{m}_{H_2O}\left(h_{H_2O,HRSG,in} - h_{H_2O,HRSG,out}\right)$$
$$= \dot{m}_{steam}\left(h_{steam}^{723\text{K},30\ \text{bar}} - h_{Pump,o}\right).$$

From this equation, the mass flow rate of steam is calculated.

Steam turbine: $\dot{W}_{ST} = \dot{m}_{steam}(h_{ST,in} - h_{ST,o})$,

where

$$h_{ST,in} = h_{steam}^{798\text{K},30\ \text{bar}}$$

$$h_{ST,o} = h_{steam}^{798\ \text{K},30\ \text{bar}} - \eta_{ST}\left(h_{stam}^{798\ \text{K},30\ \text{bar}} - h_{ST,os}\right)$$

$$h_{ST,os} = h_{s=s_{in}}^{0.08\ \text{bar}}.$$

Example 8.3 (cont.)

The net power production and the efficiency of the CC are obtained from

$$\dot{W}_{CC} = \dot{W}_{net} + \dot{W}_{SteamCycle} = \dot{W}_{net} + \left(\dot{W}_{ST} - \dot{W}_{Pump}\right)$$

$$\eta_{th} = \frac{\dot{W}_{CC}}{\dot{m}_{CH_4} LHV_{CH_4}}.$$

The results for the steam cycle and the combined gas and turbine cycle are given in Table E8.3b.

Table E8.3b

Steam mass flow rate (kg/s)	64.14
Steam turbine power (kW)	70,069
Pump power (kW)	276.5
Steam cycle net power (kW)	69,792.5
Combined cycle net power (kW)	169,792.5
Combined cycle thermal efficiency (%)	39.1

Example 8.4

The power plant cycle of Example 8.3 is modified in a way that 88% of the flue gases are cooled in a condenser, and the liquid water is recycled to the combustor using a pump (see Figure E8.4). The rest of the flue gases are sent to another condenser where water and carbon dioxide are separated. Determine the composition of the combustion products, the mass flow rates of air and methane, and the thermal efficiency of the gas turbine cycle.

Solution

Combustor:

The chemical reaction equation in the combustor can be written as

$$CH_4 + aCO_2 + bO_2 + cH_2O \rightarrow dCO_2 + eH_2O.$$

Carbon balance: $a + 1 = d.$

Hydrogen balance: $c + 2 = e.$

Oxygen balance: $a + b + 0.5c = d + 0.5e.$

From these three equations, we find $b = 2$.

Since 89% of the combustion products are recycled, we can write an additional equation:

$$\dot{m}_{CO_2,r} + \dot{m}_{H_2O,r} = 0.89(\dot{m}_{CO_2} + \dot{m}_{H_2O}).$$

Example 8.4 (cont.)

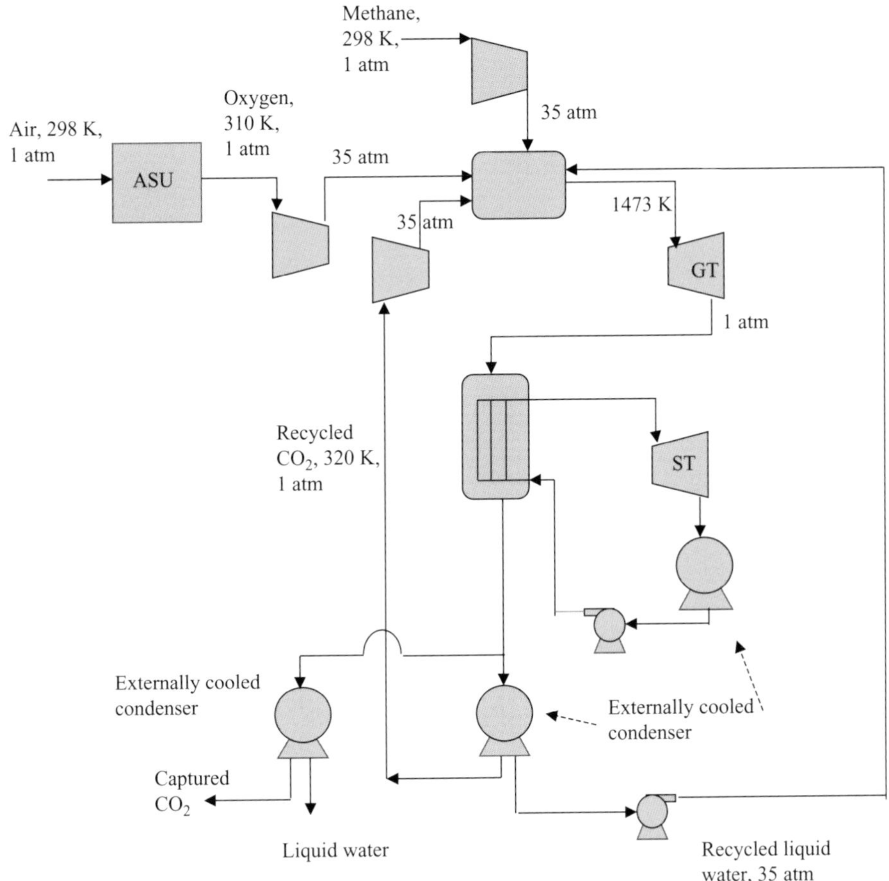

Figure E8.4

Based on the combustion reaction equation, we can relate mass flow rates to that of methane. Hence:

$$\dot{m}_{O_2} = 4\dot{m}_{CH_4}$$

$$\dot{m}_{air} = 4.76\dot{m}_{O_2} = 19.04\dot{m}_{CH_4}$$

$$\dot{m}_{CO_2,r} = \frac{a \times 44}{16}\dot{m}_{CH_4}$$

$$\dot{m}_{H_2O_r} = \frac{c \times 18}{16}\dot{m}_{CH_4}$$

$$\dot{m}_{CO_2} = \frac{d \times 44}{16}\dot{m}_{CH_4}$$

$$\dot{m}_{H_2O} = \frac{e \times 18}{16}\dot{m}_{CH_4}.$$

Example 8.4 (cont.)

Substituting, we find an additional equation:

$$\left(\frac{44a}{16}\right)\dot{m}_{CH_4} + \left(\frac{18c}{16}\right)\dot{m}_{CH_4} = 0.89 \times \left[\left(\frac{44d}{16}\right)\dot{m}_{CH_4} + \left(\frac{18e}{16}\right)\dot{m}_{CH_4}\right].$$

Simplifying, we get

$$44a + 18c = 39.16d + 16.02e.$$

The fourth equation can be established from the combustor energy conservation equation. Hence,

$$\hat{h}_{CH_4} + a\hat{h}_{CO_2,r} + 2\hat{h}_{O_2} + c\hat{h}_{H_2O(l),r} = d\hat{h}_{CO_2} + e\hat{h}_{H_2O},$$

where

$$\hat{h}_{H_2O(l),r} = -286{,}000 + \left[h_{Pump1,o} - h_{H_2O(l)}^{298\,K}\right].$$

Notice that recycled water is in liquid phase, whose enthalpy is obtained as follows.

$$h_{Pump,in} = h_{320K}^{1bar} = 196.2\,\text{kJ/kg}$$

$$h_{Pump,os} = h_{s=s_{in}}^{35bar} = 199.7\,\text{kJ/kg}$$

$$h_{Pump,o} = 196.2 + \frac{199.7 - 196.2}{0.7} = 201.1\,\text{kJ/kg}.$$

Substituting the enthalpies of the individual species in the energy equation of the combustor, we find $a = 14.49$.

The chemical reaction equation within the combustor is therefore

$$CH_4 + 14.49CO_2 + 2O_2 + 0.546H_2O_{(l)} \rightarrow 15.49CO_2 + 2.546H_2O_{(v)}.$$

The composition of the combustion products is

$$X_{CO_2} = \frac{15.49}{18.036} = 0.8588$$

$$X_{H_2O} = \frac{2.546}{18.036} = 0.1412.$$

To calculate the turbine exit temperature, we need to first determine the specific heat ratio for the mixture of CO_2 and H_2O:

$$\hat{c}_{p,mix} = X_{CO_2}\hat{c}_{p,CO_2} + X_{H_2O}\hat{c}_{p,H_2O} = 0.8588 \times 53.2 + 0.1412 \times 41.2 = 51.5\,\text{kJ/kmol·K},$$

$$k_{mix} = \frac{\hat{c}_{p,mix}}{\hat{c}_{p,mix} - \bar{R}} = 1.193,$$

Example 8.4 (cont.)

$$TOT = TIT = \left[1 - \eta_t \left(\frac{p_{TO}}{p_{TI}}\right)^{\frac{k_{mix}-1}{k_{mix}}}\right] = 913\ \text{K}.$$

The solution for the rest of the cycle is the same as described in Example 10.3. The results of the cycle analysis are given in Table E8.4. The mass flow rates of air and fuel are 149.4 kg/s and 8.667 kg/s, which are nearly the same as the corresponding values obtained in Example 8.3 for a gas turbine cycle with CO_2 recycling only. The calculated thermal efficiency of the simple cycle is 31.1%.

Table E8.4

Mass flow rates (kg/s)	
Air	149.4
Fuel	8.667
Recycled CO_2	345.3
Recycled water	5.328
Temperatures (K)	
Air compressor outlet	544.6
O_2 compressor outlet	894.5
CO_2 compressor outlet	698.5
Fuel compressor outlet	654
Turbine outlet	913
Power consumption (kW)	
Air compressor	37,509
O_2 compressor	20,245
CO_2 compressor	130,766
Fuel compressor	8677
Water recycling pump	26
Turbine power production (kW)	297,223
Gas turbine cycle efficiency (%)	23.05

It can be shown that the thermal efficiency of the CCs is 38.4%.

More accurate simulation of a CC with oxy-combustion will be shown in Chapter 11. An alternative to CCs that has been proposed and built at a prototype scale is the oxy-combustion water cycle, in which the working fluid is primarily steam mixed with a small fraction of CO_2. The cycle is essentially a Rankine cycle with high-temperature reheat.

Example 8.5 demonstrates the operation of this cycle. More accurate simulations will be shown in Chapter 11.

Example 8.5

Determine the efficiency of the oxy-combustion water power plant shown in Figure E8.5a using the operating conditions shown in Table E8.5a. Assume that the fuel composition is 80% CH_4 and 20% CO_2 (vol.). The isentropic efficiencies of compressors, turbines, and circulating water pump are 80%, 88%, and 75%, respectively.

Table E8.5a

Air compressor at ASU	
Inlet temperature (K)	298
Inlet pressure (bar)	1
Outlet pressure (bar)	6
Compressor 1 / compressor 2	
Inlet temperature (K)	310/310
Inlet pressure (bar)	1/1
Outlet pressure (bar)	40/5
Compressor 3 / compressor 4	
Inlet temperature (K)	298,298
Inlet pressure (bar)	1/1
Outlet pressure (bar)	40/5
CO_2 compressor (C_5)	
Inlet temperature (K)	298
Inlet pressure (bar)	0.05
Outlet pressure (bar)	80
Combustor	
Pressure drop (%)	4
Outlet temperature (K)	873
Reheat	
Upstream pressure (bar)	5
Pressure drop (%)	4
Outlet temperature (K)	1473
Regenerator	
Pressure drop (%)	0
Water outlet temperature (K)	515

Example 8.5 (cont.)

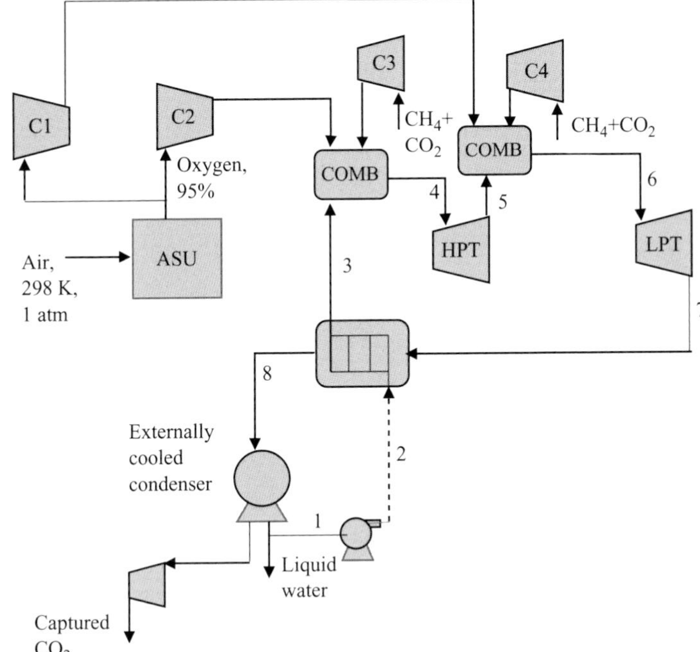

Figure E8.5a

The T–s diagram of this cycle is shown in Figure E8.5b.

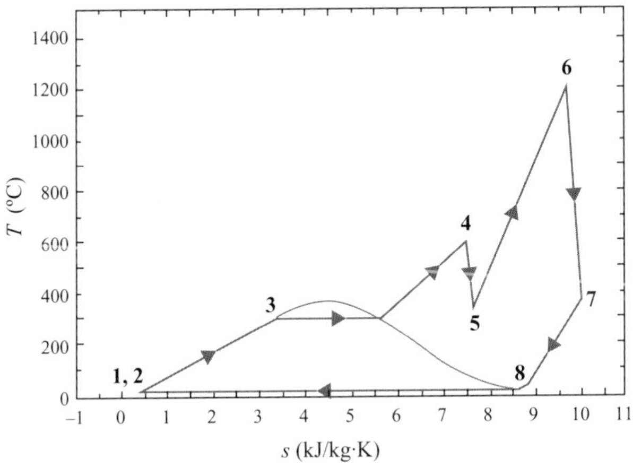

Figure E8.5b The T–s diagram of the cycle shown in Figure E8.5a.

Example 8.5 (cont.)

Solution

Similar to the previous examples, determination of the thermal efficiency of the water oxy-combustion power plant requires one to calculate the work requirement of the compressors (six compressors: C_1–C_5 and the air compressor in the ASU) and the circulating water pump, the work production of the high-pressure and low-pressure turbines, and the total heat input to the cycle by burning the fuel. Notice that the power plant includes a reheat so fuel is burned both in the combustor and the reheat.

Combustor: The reaction equation is

$$(0.8\mathrm{CH_4} + 0.2\mathrm{CO_2}) + a(0.95\mathrm{O_2} + 0.05N_2) + b\mathrm{H_2O}_{(l)} \rightarrow \mathrm{CO_2} + c\mathrm{H_2O}_{(v)} + 0.05a\mathrm{N_2}.$$

Three unknowns, a, b and c, are found from the oxygen and hydrogen balance together with applying the First Law to the combustor. For this, one needs to first determine the outlet temperatures of compressors 1 and 3, which are 930 K and 662 K, respectively. Thus, it can be shown that $a = 1.684$, $b = 11.23$, $c = 12.83$.

These values give a composition of 7.2% CO_2 + 92.2% H_2O + 0.006% N_2. This mixture is expanded in the high-pressure turbine down to 5 bar and 619 K.

Reheat: The combustion reaction equation can be written as

$$(\mathrm{CO_2} + 12.83\mathrm{H_2O} + 0.084\mathrm{N_2}) + a_1(0.95\mathrm{O_2} + 0.05\mathrm{N_2}) + b_1(0.8\mathrm{CH_4} + 0.2\mathrm{CO_2}) \rightarrow c_1\mathrm{CO_2} + d_1\mathrm{H_2O} + 0.05a\mathrm{N_2}.$$

We have four unknowns: a_1, b_1, c_1, and d_1. They can be found by solving the elemental and energy conservation equations. Similar to the combustor analysis, the temperatures of oxidizer and fuel – that is, the outlet temperatures of compressors 2 and 4 – should first be calculated. From the thermodynamic model of compressors 2 and 4, we find 526 K and 448 K as the temperatures of the oxidizer and fuel at the outlet of compressors 2 and 4, respectively.

Solving the energy and C, O, and H conservation equation for the reheat, we find

$$a_1 = 1.576 \qquad b_1 = 0.943 \qquad c_1 = 1.943 \qquad d_1 = 14.34.$$

Note that for each mole of fuel burned in the combustor, 0.943 mole additional fuel needs to be fed into the reheat combustor. The hot gas mixture leaving the reheat at 1473 K and 5 bar is expanded within the low-pressure turbine down to 0.05 bar and 726 K. At this state, the mixture still carries a considerable amount of thermal energy. Within the regenerator, a fraction of this energy is transferred to the pressurized water coming from the pump, which raises its temperature to 515 K. By applying the First Law to the regenerator, the temperature of the exhaust gas is found to be 433 K.

Table E8.5b shows the work consumption of the compressors and pump, and the work production of the turbines.

Example 8.5 (cont.)

Table E8.5b

Compression work (kJ)	340,592
Air compressor	132,617
C_1	33,426
C_2	16,272
C_3	10,275
C_4	5566
C_5	142,436
Pumping work (kJ)	1080
Turbine work (kJ)	687,228
High-pressure turbine	148,385
Low-pressure turbine	538,843
Net work output (kJ)	345,556

The net work production of the power cycle is 345.556 kJ for each mole of fuel fed to the combustor. The thermal efficiency can be calculated as follows:

$$\eta_{th} = \frac{W_{net}}{0.8 \times (1 + b_1)\, M_{CH_4} LHV_{CH_4}} = \frac{345{,}556}{0.8 \times (1 + 0.943) \times 16 \times 50{,}050} = 0.278.$$

The net work output without carbon dioxide compression work is 487.992 kJ, and the thermal efficiency is

$$\eta_{th} = \frac{487{,}992}{0.8 \times (1 + 0.943) \times 16 \times 50{,}050} = 0.392.$$

This reveals the efficiency penalty due to CO_2 compression is 11.4 percentage points.

8.6 Hybrid Fuel Cell–Gas Turbine/CC Cycles

The thermodynamic advantages of hybridization of a fuel cell with a heat engine were discussed in Chapters 4 and 7. It was demonstrated that operating certain types of fuel cells at high temperature to take advantage of lower overpotential loss or other desirable features should not negatively impact the overall efficiency of the plant if a bottoming cycle is used to take advantage of the high-temperature heat rejected by the cell due to the overpotentials and ohmic losses. Depending on the operating temperature and pressure, the high-temperature heat may be used to drive either a steam cycle, a gas turbine cycle, or a CC, all acting as bottoming cycles for the fuel cell. Another advantage of hybridization is that fuel not used by the cell, due to lower fuel utilization efficiency, can also be burned in the bottoming cycle. These hybrid systems can be optimized for carbon (dioxide) capture, either complete or partial, as will be discussed in Chapter 11. Many configurations have been proposed, and several examples are described in the following.

8.6.1 Integrated Gas Turbine and an SOFC Cycle

The concept of integrating a gas turbine cycle and an SOFC has been explored and demonstrated in practice [7,8]. The reported thermal efficiency of hybrid power plants varies depending on the components and system layout. Figure 8.10 illustrates an example of an integrated gas turbine SOFC regenerative power plant; the *T–s* diagram of the gas turbine cycle is depicted in Figure 8.11 [9]. The system consists of a compressor, regenerative heat exchanger, high-temperature SOFC, combustor, gas turbine used to power the compressor (GT), and another to produce power – the power turbine (PT). The total power of the plant

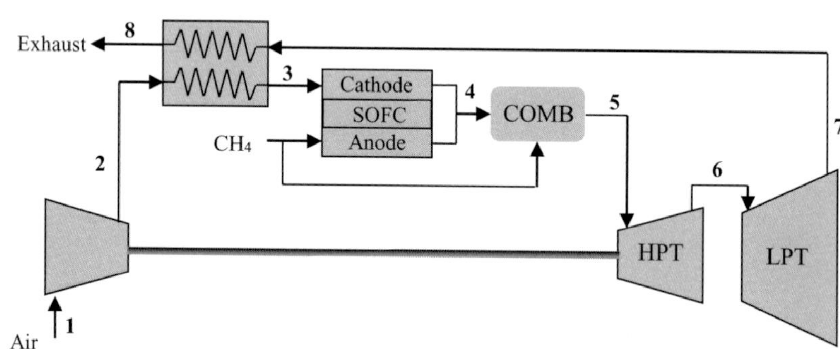

Figure 8.10 The layout of a regenerative gas turbine power cycle combined with a SOFC [9].

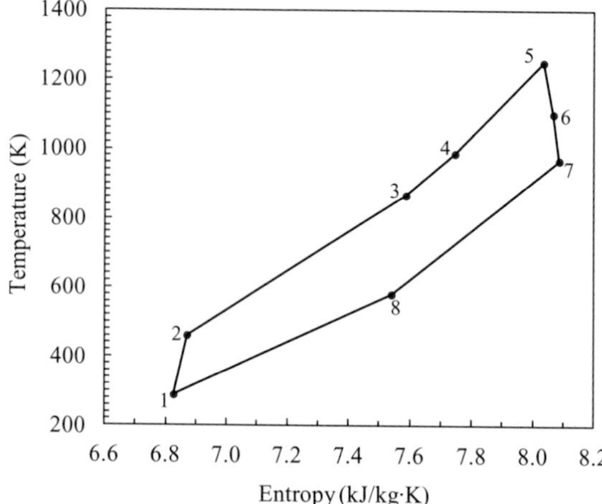

Figure 8.11 The *T–s* diagram of the GT–SOFC power cycle shown in Figure 8.10 at typical operating conditions; pressure ratio: 4, TIT = 1250 K, current density 300 mA/cm^2.

is that produced by the fuel cell plus the PT. Using two turbines, one to power the compressor and one to power the electric generator, is common in gas turbine applications. It offers some hardware flexibility.

Following the compression of air, it is preheated in the regenerator by the hot flue gases leaving the low-pressure (LP) power turbine. The preheated air is sent to the SOFC stack on the cathode side. The fuel, in this case methane (CH_4), is sent to the anode side of the stack. A combination of thermochemical and electrochemical reactions inside the cell leads to the reforming of the fuel and the electrochemical oxidation of the syngas in the fuel cell. Heat is generated thermodynamically and by the different losses (Chapter 7), which raises the temperature of the reaction products up to state 4.

Depending on the design and fuel utilization efficiency, not all the fuel supplied to the fuel cell stack is oxidized within the stack. The reaction products exiting the SOFC could still contain a significant fraction of the original fuel (in this example it is 15%). Thus, the unburned fuel is sent to the combustor. Extra/supplementary fuel can be added to the combustor, where they are burned raising the temperature before expansion (state 5). The hot products enter the HPT where work is produced, reducing the pressure and the temperature to P_6 and T_6. In the LPT the stream expands to atmospheric condition, producing additional power. The flue gases leaving the LPT (state 7) are sent to the hot side of the regenerator to preheat the pressurized air before entering the SOFC. As always in the case of gas turbines where the expansion is limited by the exit pressure, the exhaust gases leaving at state 8 are still hot enough they can be used in other applications (e.g., combined heat and power (CHP) or another bottoming cycle, waste heat recovery, etc.

The net power (work rate) produced by the GT–SOFC plant is the sum of the work production of the PT and the SOFC. Hence,

$$\dot{W}_{net} = \dot{W}_{FC} + \dot{W}_{PT}. \tag{8.17}$$

The total rate of energy input to the system is

$$\dot{Q}_{in} = \dot{Q}_{FC} + \dot{Q}_{Comb}, \tag{8.18}$$

$$\dot{Q}_{FC} = \dot{m}_{f,FC}\, U_f\, LHV, \tag{8.19}$$

$$\dot{Q}_{Comb} = \left[\dot{m}_{f,FC}\left(1 - U_f\right) + \dot{m}_{f,Comb}\right] LHV, \tag{8.20}$$

where U_f is a fuel utilization factor and $\dot{m}_{f,Comb}$ is the mass flow rate of fuel supplied to the combustor, which can be obtained by applying the First Law to the combustor:

$$\left(\dot{m}_3 + U_f\,\dot{m}_{f,FC}\right) h_4 + \dot{Q}_{Comb} = \dot{m}_5 h_5. \tag{8.21}$$

Notice that the mass flow rate at state 5 is the sum of $\dot{m}_{f,Comb}$ and $\dot{m}_4$. The thermal efficiency of the hybrid power plant is found from

$$\eta_{th} = \frac{\dot{W}_{net}}{\dot{Q}_{in}}. \tag{8.22}$$

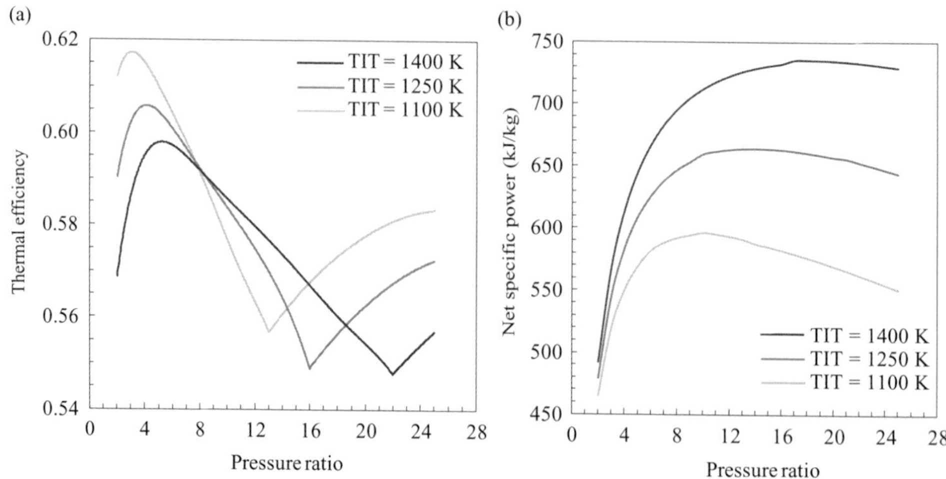

Figure 8.12 Variation of the thermal efficiency and net specific power of the GT–SOFC cycle with the system pressure ratio at three values of the TIT [9].

At a pressure ratio of 4 and a TIT of 1250 K, the thermal efficiency, the specific power produced by the SOFC, and the specific power output of the cycle (SOFC plus PT) are 60.6%, 437.5 kJ/kg, and 583.9 kJ/kg, respectively. The dependence of the thermal efficiency and the net specific power output of the GT–SOFC on the system pressure ratio and TIT are shown in Figure 8.12. The maximum thermal efficiency of the hybrid plant occurs at low pressure ratios in the range of 3–5, depending on TIT. An interesting observation is that, unlike the efficiency–pressure ratio graphs of conventional gas turbines discussed in Chapter 5 in which the thermal efficiency has only one maximum value, a GT–SOFC plant possesses a maximum and a minimum. On the other hand, the net specific power increases with the pressure ratio and reaches a maximum, which is higher at a higher TIT. The optimum pressure ratio giving a maximum power output increases with TIT. The example shows that, similar to many complex power cycles, thermodynamic optimization is necessary to achieve maximum efficiency or power, depending on the design objective. Other studies typically target techno-economic optimization, in which minimizing the cost of electricity is the target.

Example 8.6

A hybrid cycle comprising a regenerative gas turbine engine and a 0.5 MW SOFC is used for power production. Air enters the compressor at 303 K and 1 atm, whose pressure ratio is 4. The regenerator operates with 80% efficiency. The fuel cell stack operates with 0.7 volts. Hydrogen is the fuel that is sent to the fuel cell anode. The fuel utilization factor is 85%, and a combustor is located downstream of the fuel cell stack, where the unconverted part of the fuel is

Example 8.6 (cont.)

completely burned, which raises the temperature of the main stream. The isentropic efficiencies of the compressor and turbine are 80% and 88%, respectively. Assume a dc–ac inverter efficiency of 89%, and an air utilization factor of 0.25. Determine: (a) the mass flow rate of hydrogen consumed in the hybrid cycle; (b) the TIT; (c) the net power production of the system; and (d) the efficiencies of the fuel cell and the hybrid cycle.

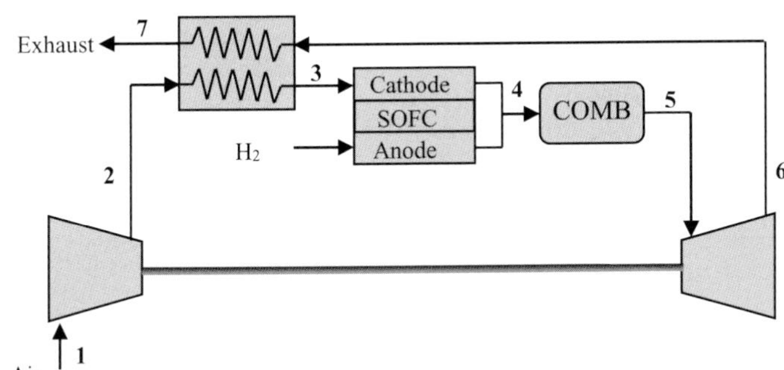

Figure E8.6

Solution

Compressor:

$$T_2 = T_1\left\{1 + \frac{1}{\eta_C}\left[(r_p)^{\frac{\gamma-1}{\gamma}} - 1\right]\right\} = 303\left\{1 + \frac{1}{0.8}\left[4^{\frac{1.394-1}{1.394}} - 1\right]\right\} = 484.7\ \text{K},$$

$$h_1 = 303.3\ \text{J/g, and } h_2 = 487.8\ \text{J/g}.$$

Turbine: To determine the turbine exit temperature, we guess $T_5 = 1373$ K, which will be checked later.

$$T_6 = T_5\left\{1 - \eta_T\left[1 - (r_p)^{\frac{\gamma_T-1}{\gamma_T}}\right]\right\} = 1373 \times \left\{1 - 0.88 \times \left[1 - \left(\frac{1}{4}\right)^{\frac{0.315}{1.315}}\right]\right\} = 1031.6\ \text{K},$$

$$h_6 \approx h_{air}(1031.6\ \text{K}) = 1083\ \text{J/g}.$$

Regenerator:

$$h_3 = h_2 + \eta_{reg}(h_6 - h_2) = 487.8 + 0.8 \times (1083 - 487.8) = 964\ \text{J/g}.$$

SOFC:

The dc power produced by the fuel cell is

$$P_{dc} = \frac{P_{FC}}{\eta_{inv}} = \frac{500{,}000}{0.89} = 561{,}798\ \text{W}.$$

Example 8.6 (cont.)

The molar flow rates of hydrogen and oxygen flowing through the anode and cathode are determined using the results of Example 4.4:

$$\dot{n}_{H_2,in} = \frac{P_{dc}}{2F_aV_{act}}.\frac{1}{U_f} = \frac{561{,}798}{2\times 96{,}485\times 0.7}\times\frac{1}{0.85} = 4.893\,\text{mol/s},$$

$$\dot{n}_{O_2,in} = \frac{P_{dc}}{4F_aV_{act}}\frac{1}{U_a} = \frac{561{,}798}{4\times 96{,}485\times 0.7}\times\frac{1}{0.25} = 8.318\,\text{mol/s},$$

$$\dot{n}_{N_2,in} = \dot{n}_{N_2,out} = \frac{0.79}{0.21}\dot{n}_{O_2} = 31.292\,\text{mol/s},$$

$$\dot{n}_{O_2,out} = \frac{P_{dc}}{4F_aV_{act}}\left(\frac{1}{U_a}-1\right) = \frac{561{,}798}{4\times 96{,}485\times 0.7}\times\left(\frac{1}{0.25}-1\right) = 6.239\,\text{mol/s}.$$

For each mole of hydrogen participating in the cell chemical reaction, one mole of water is produced. So, the molar flow rate of water at the SOFC exit stream is

$$\dot{n}_{H_2O} = U_f\dot{n}_{H_2,in} = 4.159\ \text{mol/s}.$$

a. In terms of mass flow rates, we have

$$\dot{m}_{H_2,in} = 2\dot{n}_{H_2,in} = 9.79\ \text{g/s},$$

$$\dot{m}_{H_2,out} = (1-U_f)\dot{m}_{H_2,in} = 1.47\ \text{g/s},$$

$$\dot{m}_{O_2,in} = 32\dot{n}_{O_2,in} = 266.18\ \text{g/s},$$

$$\dot{m}_{O_2,out} = 32\dot{n}_{O_2,out} = 199.65\ \text{g/s},$$

$$\dot{m}_{N_2,in} = \dot{m}_{N_2,out} = 28\dot{n}_{N_2,in} = 876.18\ \text{g/s},$$

$$\dot{m}_{H_2O} = 18\dot{n}_{H_2O} = 74.86\ \text{g/s}.$$

The mass flow rate of the fuel to be consumed in the cycle is 9.79 g/s.

b. We write the First Law for the fuel cell:

$$\dot{m}_3h_3 + (\dot{m}_{H_2,in}.U_f).LHV = P_{dc} + \dot{m}_4h_4,$$

where $\dot{m}_3 = \dot{m}_{O_2,in} + \dot{m}_{N_2,in} = 266.18 + 876.18 = 1142.36$ g/s,

$$\dot{m}_4 = \dot{m}_{H_2,in} + \dot{m}_3 = 9.79 + 1142.36 = 1152.15\ \text{g/s},$$

$$LHV = 120{,}000\ \text{J/g},$$

$$h_4 = \frac{\dot{m}_3h_3 + \dot{m}_{H_2,in}.U_f.LHV - P_{dc}}{\dot{m}_4} = \frac{1142.36\times 964 + 9.79\times 0.85\times 120{,}000 - 561{,}798}{1152.15}$$
$$= 1334.9\ \text{J/g}.$$

Example 8.6 (cont.)

Combustor:
Noting that $\dot{m}_4 = \dot{m}_5$, the First Law for the combustor is

$$\dot{m}_4 h_4 + \dot{m}_{H_2,in}(1 - U_f)LHV = \dot{m}_5 h_5.$$

Rearranging the above equation yields

$$h_5 = \frac{\dot{m}_4 h_4 + \dot{m}_{H_2,in}(1 - U_f)LHV}{m_5} = \frac{1152.15 \times 1334.9 + 9.79 \times (1 - 0.85) \times 120{,}000}{1152.15}$$

$$= 1487.9 \text{ J/g}.$$

Assuming that the mixture at state 5 can be approximated as air (note that only 6.5% of the mixture is water vapor), the temperature at state 5, corresponding to the above enthalpy, is $T_5 = 1373$ K. This verifies our initial guess for the temperature at state 5. In fact, this is the TIT.

c. The net power output of the gas turbine cycle is

$$P_{Gas} = P_T - P_C = \dot{m}_5(h_5 - h_6) - \dot{m}_1(h_2 - h_1),$$

where: $\dot{m}_5 = \dot{m}_4 = 1152.15$ g/s, $\dot{m}_1 = \dot{m}_3 = 1142.36$ g/s, $h_6 = 1083$ J/g, $h_1 = 303.3$ J/g, and $h_2 = 487.8$ J/g.

Substituting,

$$P_{Gas} = 1152.15 \times (1487.9 - 1083) - 1142.36 \times (487.8 - 303.3) = 255{,}740 \text{ W}.$$

The net power output of the hybrid cycle is therefore

$$P_{hybrid} = P_{SOFC} + P_{Gas} = 500{,}000 + 255{,}740 = 755{,}740 \text{ W}.$$

d. The efficiency of the hybrid cycle is

$$\eta_{th,\text{Hy}} = \frac{P_{hybrid}}{\dot{m}_{\text{H}_2,in} LHV} = \frac{755{,}740}{9.79 \times 120{,}000} = 0.643.$$

The efficiency of the fuel cell is $\eta_{th,\text{Hy}} = \dfrac{P_{SOFC}}{\dot{m}_{\text{H}_2,in} LHV} = \dfrac{500{,}000}{9.79 \times 120{,}000} = 0.426.$

Comparing the two efficiencies, it is clear that combining the high-temperature fuel cell with a gas turbine cycle leads to a 21.7 percentage point increase in the conversion efficiency.

Example 8.7

The proposed power plant shown in Figure E8.7a consists of a hydrogen-fueled SOFC and a closed helium gas turbine cycle. Separate streams of hydrogen and air enter an adiabatic heat exchanger at 298 K, where they are heated by the exhaust stream of the SOFC. The hydrogen

Example 8.7 (cont.)

and air streams are further heated to 1.273 K (1,000 °C) and then supplied to the SOFC, which operates at a steady temperature of 1273 K and pressure 1 atm. The molar flow rate of hydrogen is 100 mol/s. The flow rate of oxygen is such that the SOFC operates stoichiometrically.

Heat is rejected from the SOFC to a closed helium gas turbine cycle. Helium leaves the heat exchanger at 1200 K and enters a turbine of 95% isentropic efficiency and pressure ratio 4:1. From there it is cooled to 320 K and then compressed at 85% isentropic efficiency.

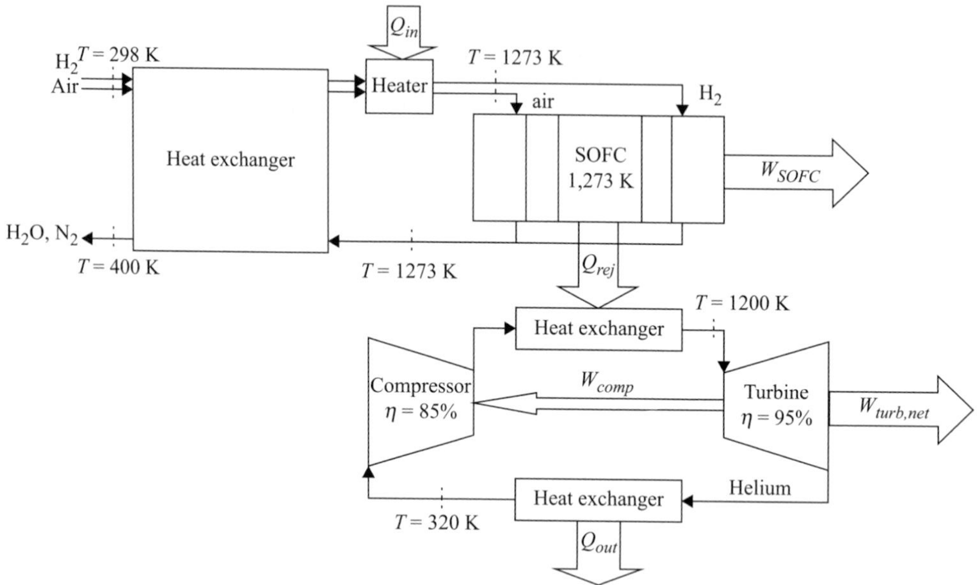

Figure E8.7a

SOFC specifications:

- number of stacks: 20,
- number of cells per stack: 220,
- cell area: 1 m^2,
- exchange current density at the cathode (i_o): 300 mA/cm^2,
- limiting current density (i_{lim}): 900 mA/cm^2,
- charge transfer coefficient (α): 0.5,
- fuel utilization: 100%,
- area-specific resistance for ohmic overpotential (r): 300×10^{-6} kΩ.cm^2,

Example 8.7 (cont.)

Helium gas turbine cycle specifications:

- turbine inlet temperature: 1200 K,
- compressor inlet temperature: 320 K,
- turbine isentropic efficiency: 95%,
- compressor isentropic efficiency: 85%, and
- turbine and compressor pressure ratio: 4:1.

Fuel Cell Voltage and Current

Practical fuel cells combine a number of cells connected in series in a stack in order to produce usable voltages. If each cell has a voltage V_{cell}, and there are n cells in a stack, the total voltage for the stack is nV_{cell}. Many stacks can then be connected in parallel to produce usable current. If each cell (and therefore each stack) produces a current I, and there are N stacks, the total current for the system is NI. The power available from a system of N stacks, each with n cells is $P = NInV_{cell}$.

Useful data

$$\hat{c}_{p,\mathrm{H_2O}} = 38.2\ \mathrm{J/mol{\cdot}K} \quad \hat{h}^o_{f,\mathrm{H_2O(g)}} = -242{,}000\ \mathrm{J/mol}$$

$$\hat{c}_{p,\mathrm{H_2}} = 30.0\ \mathrm{J/mol{\cdot}K} \quad \hat{h}^o_{f,\mathrm{H_2O(l)}} = -286{,}030\ \mathrm{J/mol}$$

$$\hat{c}_{p,\mathrm{O_2}} = 33.4\ \mathrm{J/mol{\cdot}K}$$

$$\hat{c}_{p,\mathrm{N_2}} = 31.1\ \mathrm{J/mol{\cdot}K}$$

$$\hat{c}_{p,\mathrm{He}} = 20.8\ \mathrm{J/mol{\cdot}K}$$

Questions

a. What is the temperature of the reactants leaving the first heat exchanger?
b. What is the rate of additional heat transfer needed (Q_{in}) to increase the temperature of the reactants to 1273 K?
c. What is the open-circuit voltage of the SOFC?
d. Activation and concentration overpotentials (dominant at the cathode), and ohmic overpotentials are present in the SOFC. What is the actual cell voltage of the SOFC?
e. What is the current density in each cell?
f. What is the rate of electrical power produced by the SOFC (W_{SOFC})?
g. What is the required molar flow rate of helium in the gas turbine cycle?
h. What is the total power produced by the entire cycle ($W_{SOFC} + W_{turb,net}$)?

Example 8.7 (cont.)

i. Define an efficiency for the power plant and determine it.
It has been proposed that power plant efficiency can be improved by using a regenerator in the gas turbine cycle. In designing such a regenerator, answer the following questions.

j. What is the new molar flow rate of helium in the gas turbine cycle?

k. What is the new total power produced by the entire cycle ($W_{SOFC} + W_{turb,net}$)?

l. Determine the new efficiency for the power plant.

m. Compare the efficiency of this hybrid SOFC–GT design with that of a design that uses methane to power the simple, helium-based gas turbine. Take the pressure ratio of 10 and maximum and minimum temperatures in the cycle of 1200 K and 320 K, with the same turbine and compressor isentropic efficiencies. To simplify analysis, you may assume that all of the heating value of methane is available to heat the helium. Take the efficiency of producing hydrogen from methane as 85%.

Solution

a. Apply the First Law to a control volume around the first heat exchanger:

$$\dot{n}_{H_2}\hat{h}_{H_2,in} + \dot{n}_{air}\hat{h}_{air,in} + \dot{n}_{exhaust}\hat{h}_{exhaust,in} = \dot{n}_{H_2}\hat{h}_{H_2,out} + \dot{n}_{air}\hat{h}_{air,out} + \dot{n}_{exhaust}\hat{h}_{exhaust,out}.$$

Air consists of 21% O_2 and 79% N_2, and is supplied at a stoichiometric ratio. The following is also known:

$$\dot{n}_{H_2,in} = \dot{n}_{H_2,out} = \dot{n}_{H_2} = 100\ \text{mol/s}$$
$$\dot{n}_{O_2,in} = \dot{n}_{O_2,out} = \dot{n}_{O_2} = 50\ \text{mol/s}$$
$$\dot{n}_{N_2,in} = \dot{n}_{N_2,out} = \dot{n}_{N_2} = 188\ \text{mol/s}$$
$$\dot{n}_{exhaust,in} = \dot{n}_{exhaust,out} = \dot{n}_{H_2O} + \dot{n}_{N_2} = \dot{n}_{H_2} + \dot{n}_{N_2} = 288\ \text{mol/s}.$$
$$T_{H_2,in} = T_{O_2,in} = T_{N_2,in} = T_{cold,in} = 298\ \text{K}$$
$$T_{exhaust,in} = 1273\ \text{K}$$
$$T_{exhaust,out} = 400\ \text{K}.$$

Solving the First Law, we find

$$\boxed{T_{H_2,out} = T_{O_2,out} = T_{N_2,out} = T_{cold,out} = 1100.4\ \text{K}}.$$

b. Now apply the First Law to a control volume around the reactant heater.

$$\dot{Q}_{in} = \left(\dot{n}_{H_2}\hat{h}_{H_2,out} + \dot{n}_{air}\hat{h}_{air,out}\right) - \left(\dot{n}_{H_2}\hat{h}_{H_2,in} + \dot{n}_{air}\hat{h}_{air,in}\right)$$
$$\text{where}: T_{in} = 1100.4\,\text{K}, \quad T_{out} = 1273\,\text{K}.$$

$$\text{Thus}: \boxed{\dot{Q}_{in} = 1.81\ \text{MW}}.$$

Example 8.7 (cont.)

c. The SOFC operates at atmospheric pressure, so the expression for open-circuit voltage is

$$\Delta\varepsilon_{OC} = -\frac{\Delta G_r}{n_e\mathfrak{I}_a} + \frac{\Re T}{n_e\mathfrak{I}_a}\ln\left(\frac{X_{H_2,in}X_{O_2,in}^{1/2}}{X_{H_2O,out}}\right) = \frac{\Re T}{n_e\mathfrak{I}_a}\left(\ln(K_P) + \ln\left(\frac{X_{H_2,in}X_{O_2,in}^{1/2}}{X_{H_2O,out}}\right)\right)$$

where: $\Re = 8.314 \text{J/mol·K}$, $T = 1273$ K, $n_e = 2$, $\mathfrak{I}_a = 96,485°\text{C}$,

$K_P \approx 15 \times 10^6$, $X_{H_2,in} = 1$, $X_{O_2,in} = 0.21$, $X_{H_2O,out} = 1$ (anode side).

Therefore, $\boxed{\Delta\varepsilon_{OC} = 0.863 \text{ V}}$.

d. Actual cell voltage can be expressed as follows:

$$V_{cell} = \Delta\varepsilon_{OC} - |\eta_{act}| - |\eta_{conc}| - |\eta_{ohm}|$$

where

$$\eta_{act} = \frac{\Re T}{\alpha n_e\mathfrak{I}_a}\ln\left(\frac{i}{i_o}\right)$$

$$\eta_{conc} = \frac{\Re T}{n_e\mathfrak{I}_a}\left(1 + \frac{1}{\alpha}\right)\ln\left(1 - \frac{i}{i_{\lim}}\right)$$

$$\eta_{ohm} = ir.$$

We first need to find the current density, which is obtained from

$$I = iA = \dot{n}_{fuel}n_e\mathfrak{I}_a.$$

Fuel is consumed by each cell in each stack, such that

$$\dot{n}_{fuel} = \frac{\dot{n}_{H_2}}{N_{cell}N_{stack}} = 0.023 \text{ mol/s, and } A = A_{cell} = 1 \in \text{m}^2$$

Thus, $I = 4386$ A, and $i = 438.6 \text{ mA/cm}^2$.

The overpotentials and cell voltage can now be found:

$$|\eta_{act}| = 0.042 \text{ V}$$

$$|\eta_{conc}| = 0.110 \text{ V}$$

$$|\eta_{ohm}| = 0.132 \text{ V}$$

so

$$\boxed{V_{cell} = 0.580 \text{ V}},$$

Example 8.7 (cont.)

e. The current density for each cell has already been found above:

$$\boxed{i = 438.6\ \text{mA/cm}^2}.$$

f.

$$\dot{W}_{SOFC} = N_{cell} N_{stack} I V_{cell}$$

where, $N_{cell} = 220$, $N_{stack} = 20$, $I = 4386\ \text{A}$, $V_{cell} = 0.580\ \text{V}$

Thus, $\boxed{\dot{W}_{SOFC} = 11.2\ \text{MW}}$.

g. First, we must find the rate of heat transfer from the SOFC to the gas turbine cycle. Performing the First Law on a control volume around the SOFC gives

$$\dot{Q}_{rej} = \dot{W}_{SOFC} + \dot{H}_{out} - \dot{H}_{in} = (11.2 + (-14.8) - 10.3) \times 10^6 = -13.9\ \text{MW}.$$

Now we must consider the flow of helium in the cycle. Writing the First Law for a control volume around the high-temperature heat exchanger of the helium cycle gives

$$0 = -\dot{Q}_{rej} + \dot{n}_{\text{He}} c_{p,\text{He}} (T_{in} - T_{out})$$

or

$$\dot{n}_{\text{He}} = \frac{\dot{Q}_{rej}}{c_{p,\text{He}} (T_{in} - T_{out})}$$

where

$$T_{out} = 1200\ \text{K}.$$

In order to find the inlet temperature of helium, we must analyze the compressor. For a reversible compressor:

$$T_{out,s} = T_{in} \left(\frac{P_{out}}{P_{in}} \right)^{R/c_p} = 557\ \text{K}.$$

For an 85% efficient compressor:

$$T_{out} = T_{in} - \frac{T_{in} - T_{out,s}}{\eta_{comp}}$$

$$T_{out} = 599\ \text{K}.$$

This is equal to the heat exchanger inlet temperature, so

$$\boxed{\dot{n}_{\text{He}} = 1109\ \text{mol/s}}.$$

Example 8.7 (cont.)

h. Performing analysis on the turbine similar to that on the compressor, we find the turbine outlet temperature is 715 K. The power produced by the turbine and the power consumed by the compressor are therefore:

$$\dot{W}_{turb} = \dot{n}_{\text{He}} c_{p,\text{He}} (T_{in} - T_{out}) = 11.2 \text{ MW}$$

$$\dot{W}_{comp} = \dot{n}_{\text{He}} c_{p,\text{He}} (T_{in} - T_{out}) = -6.3 \text{ MW}$$

$$\text{Therefore}: \ \dot{W}_{turb,net} = \dot{W}_{turb} + \dot{W}_{comp} = 4.8 \text{ MW}.$$

The total power produced by the entire cycle is therefore

$$\dot{W}_{total} = \dot{W}_{SOFC} + \dot{W}_{turb,net}$$

$$\boxed{\dot{W}_{total} = 15.9 \text{ MW}}.$$

i. We can define a fuel utilization efficiency for the plant as

$$\eta_{FU} = \frac{\dot{W}_{total}}{\dot{n}_{\text{H}_2} LHV_{\text{H}_2} + \dot{Q}_{in}}$$

$$\boxed{\eta_{FU} = 61.3\%}.$$

This is a comparatively high efficiency, but it does not account for hydrogen or whatever process is used to supply Q_{in}. In theory, the hydrogen production plant may be able to supply Q_{in}.

j. The helium cycle now has the design shown in Figure E8.7b. The rate of heat transfer from the SOFC is the same, as are the turbine and compressor inlet temperatures. Therefore, the turbine and compressor outlet temperatures are the same as the original design.

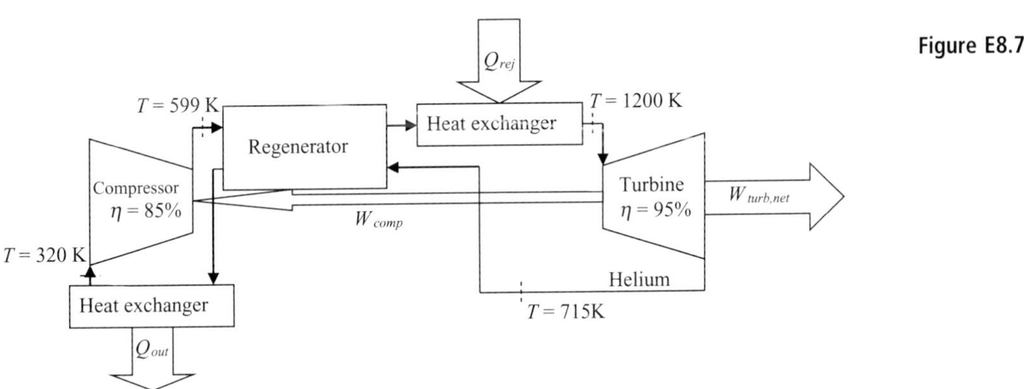

Figure E8.7b

Example 8.7 (cont.)

A temperature difference between hot and cold streams in the regenerator of 10 K is assumed to be reasonable. Therefore, the temperatures entering the hot and cold heat exchangers are

$$T_{hotHX,in} = 705 \text{ K}$$

$$T_{coldHX,in} = 609 \text{ K}.$$

Performing the First Law around the hot heat exchanger, we find

$$\dot{n}_{\text{He}} = \frac{\dot{Q}_{rej}}{c_{p,\text{He}}(T_{in} - T_{out})}$$

$$\boxed{\dot{n}_{\text{He}} = 1345 \text{ mol/s}}.$$

k. The First Law applied to the turbine and compressor gives

$$\dot{W}_{turb} = \dot{n}_{\text{He}} c_{p,\text{He}}(T_{in} - T_{out}) = 13.6 \text{ MW}$$

$$\dot{W}_{comp} = \dot{n}_{\text{He}} c_{p,\text{He}}(T_{in} - T_{out}) = -7.8 \text{ MW}.$$

$$\text{Therefore}: \dot{W}_{turb,net} = \dot{W}_{turb} + \dot{W}_{comp} = 5.8 \text{ MW}.$$

The total power produced by the entire cycle is therefore

$$\boxed{\dot{W}_{total} = 16.9 \text{ MW}}.$$

l. We can define a fuel utilization efficiency for the plant as

$$\eta_{FU} = \frac{\dot{W}_{total}}{\dot{n}_{\text{H}_2} LHV_{\text{H}_2} + \dot{Q}_{in}}$$

$$\boxed{\eta_{FU} = 65.2\%}.$$

m. When we include hydrogen production in the plant design, the fuel utilization efficiency is found to be

$$\eta_{total} = \eta_{H_2prod}\eta_{FU} = 0.85 \times 0.652$$

$$\boxed{\eta_{total} = 55.4\%}.$$

Now let us consider the methane-fired simple gas turbine cycle:

$$\eta_{simple} = \frac{\dot{W}_{net}}{\dot{n}_{\text{CH}_4} LHV_{\text{CH}_4}}.$$

Example 8.7 (cont.)

The differences between this cycle and the previous one are the source of the high-temperature heat transfer and the pressure ratios across the turbine and compressor.

We can perform virtually identical analysis to that for the previous plant to find the following:

$$T_{turb,out} = 514 \text{ K}, \text{ and } T_{comp,out} = 890 \text{ K}$$

$$\dot{W}_{turb}/\dot{n}_{\mathrm{CH_4}} = 1774.9 \text{ kJ/mol}_{\mathrm{CH_4}}, \text{and } \dot{W}_{comp}/\dot{n}_{\mathrm{CH_4}} = -1473.2 \text{ kJ/mol}_{\mathrm{CH_4}}.$$

$$\text{Thus: } \dot{W}_{net}/\dot{n}_{\mathrm{CH_4}} = 301.7 \text{ kJ/mol}_{\mathrm{CH_4}}.$$

Therefore the fuel utilization efficiency of the simple cycle plant is

$$\boxed{\eta_{simple} = 37.6\%}.$$

Combined cycle plants or hybrid plants can be developed to vastly increase the efficiency of power generation.

8.6.2 Integrated Steam Turbine and an SOFC Cycle

An example of a combined steam cycle and SOFC is schematically shown in Figure 8.13 [10]. The fuel used is natural gas, which is preheated in a heat exchanger before entering a reformer. The reformed fuel is heated in a second heat exchanger so that its temperature reaches 650 °C before entering the anode side of the SOFC. The operating temperature of the SOFC is 780 °C. The outlet temperature is the same as the cell temperature. Air is pressurized in a compressor, and then heated up in two heat exchangers to 600 °C before entering the cathode side of the SOFC. Located downstream of the SOFC is a combustor, where the unburned fuel is completely oxidized with the air coming out of the cathode side of the stacks. The high-temperature combustion products flow through a HRSG for producing steam to be used in a steam cycle, which is based on the Rankine cycle with an open feedwater heater (i.e., deaerator). The pressurized water after the feedwater pump is superheated within the HRSG. The steam is expanded in a turbine that generates additional power. The flue gases from the HRSG are used to preheat the air after the compressor. During the steady-state operation of the hybrid cycle, the steam requirement of the reformer is provided by recycling the stream coming from the anode side of the SOFC using an ejector. Note that water is a byproduct of the hydrogen–oxygen reaction taking place within the SOFC.

Table 8.3 lists the operating conditions [10]. The net power output from the SOFC and the steam cycle are 30.42 MW and 8.77 MW, respectively. Thus, the net power output of the hybrid cycle is 39.19 MW. Moreover, the thermal efficiencies of the SOFC, steam cycle, and hybrid cycle are reported to be 44.1%, 35.8%, and 65.8%, respectively.

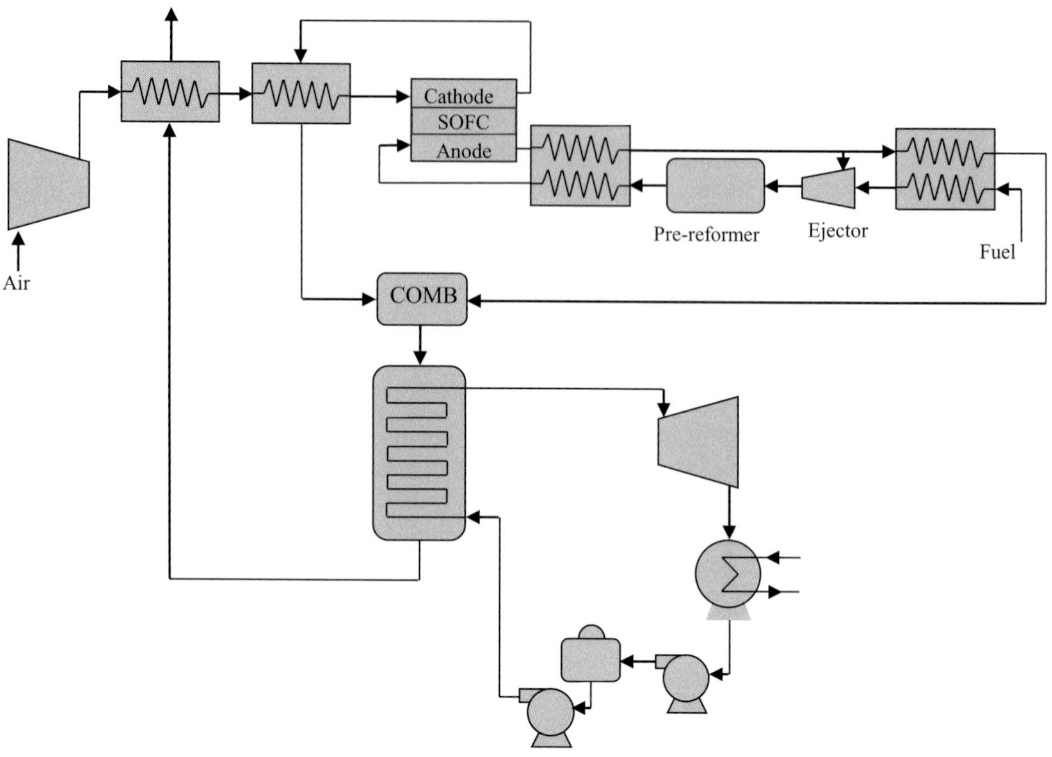

Figure 8.13 The layout of a steam power cycle combined with an SOFC.

Table 8.3 The operating conditions of the hybrid system in Figure 8.13 [10]

Compressor inlet temperature	25 °C
SOFC cathode inlet temperature	600 °C
SOFC anode inlet temperature	650 °C
SOFC operating temperature	780 °C
SOFC utilization factor	0.80
SOFC cathode-side pressure drop	0.05 bar
SOFC anode-side pressure drop	0.01 bar
Heat exchanger pressure drops	0.01 bar
Fuel inlet temperature	25 °C
Pre-reformer inlet temperature	500 °C
HRSG outlet temperature	90 °C
Extraction pressure	2 bar
Condenser pressure	0.05 bar
Steam turbine pressure	60 bar
Steam turbine inlet temperature	490 °C

8.6.3 Integrated MCFC–CC Cycle

This system is illustrated using an example in which carbon dioxide capture is added as part of the design objective of the plant.

Example 8.8

Post-combustion-based approaches to carbon dioxide capture can lead to significant efficiency penalties. For instance, removing 90% of CO_2 from the exhaust of conventional NGCC using amines leads to a near 8-percentage-point reduction in efficiency. Molten carbonate fuel cells (MCFCs) are proposed for production of electricity and the partial separation of CO_2 as described below; when integrated with CCs, the MCFC system can reduce the efficiency penalty of CO_2 capture.

Natural gas, assumed to be 100% methane, is completely burned in air in the combustor of a gas turbine cycle. The equivalence ratio of combustion is 0.4. The turbine exit temperature is 650 °C. The oxygen-rich exhaust of the gas turbine is sent to the cathode of the MCFC. The anode of the MCFC is fed with a syngas mixture (CO, CO_2, H_2, and H_2O) produced by steam reforming of natural gas, as shown in Figure E8.8. The syngas enters the anode at 650 °C, and the fuel cell operates at the same temperature.

The MCFC consumes hydrogen electrochemically, and works according to the following redox reaction:

$$\text{Anode: } H_2 + CO_3^{2-} \rightarrow H_2O + CO_2 + 2e^-,$$

$$\text{Cathode: } \tfrac{1}{2}O_2 + CO_2 + 2e^- \rightarrow CO_3^{2-}.$$

Thus CO_3^{2-} effectively transports some CO_2 from the gas turbine exhaust stream that enters the cathode across the electrolyte to the anode side. At the anode, CO_3^{2-} reacts with hydrogen and produces electrons. The MCFC only consumes hydrogen, leaving CO unburned. The figure shows the layout of the plant and mass flow.

Some of the gases leaving the cathode are used to provide heat for the reformer, and the rest are used in the HRSG of the steam cycle. The gases leaving the anode, which include some hydrogen and the original CO, are completely burned in pure oxygen and used in the HRSG as well.

Assume that for each mole of natural gas entering the reformer, four moles of natural gas are used in the gas turbine cycle. Do all your calculations per four moles of natural gas entering the gas turbine cycle and mole entering the reformer.

Use the following data, including Tables E8.8a and E8.8b for calculation:

1. The efficiency of the gas turbine cycle is 38%, based on the LHV of methane.
2. The efficiency of the steam turbine cycle is 35%.
3. 3.5 moles of H_2O are added to the reformer per one mole of CH_4 entering the reformer.
4. The MCFC is running at a constant temperature of 650 °C and 1 bar.
5. The outside temperature is 25 °C.
6. The LHV of CH_4 = 801.600 J/mol.
7. c_p of the gas turbine exhaust and the cathode exhaust is 31.17 J/mole·K.
8. The enthalpies (J/mol) and entropies (J/mol·K) (for pure species) and the Gibbs free energy (for pure species) are given as in Tables E8.8a and E8.8b.

Example 8.8 (cont.)

Table E8.8a Properties at 650 °C, 1 bar

	H_2	H_2O	CO	CO_2	CH_4	N_2	O_2
h^0	18,360	−218,840	−91,330	−364,060	−41,450	18,950	20,030
s^0	164	229	232	265	242	225	241
g^0	−132,763	−430,521	−305,254	−608,498	−265,185	−189,002	−202,108

Table E8.8b Properties at 25 °C, 1 bar

	H_2	H_2O	CO	CO_2	CH_4	N_2	O_2
h^0	0	−241,700	−110,470	−393,300	−74,560	0	0
s^0	131	189	198	214	186	191	205
g^0	−38,917	−297,933	−169,334	−456,968	−130,063	−57,034	−61,096

The system is shown in Figure E8.8.

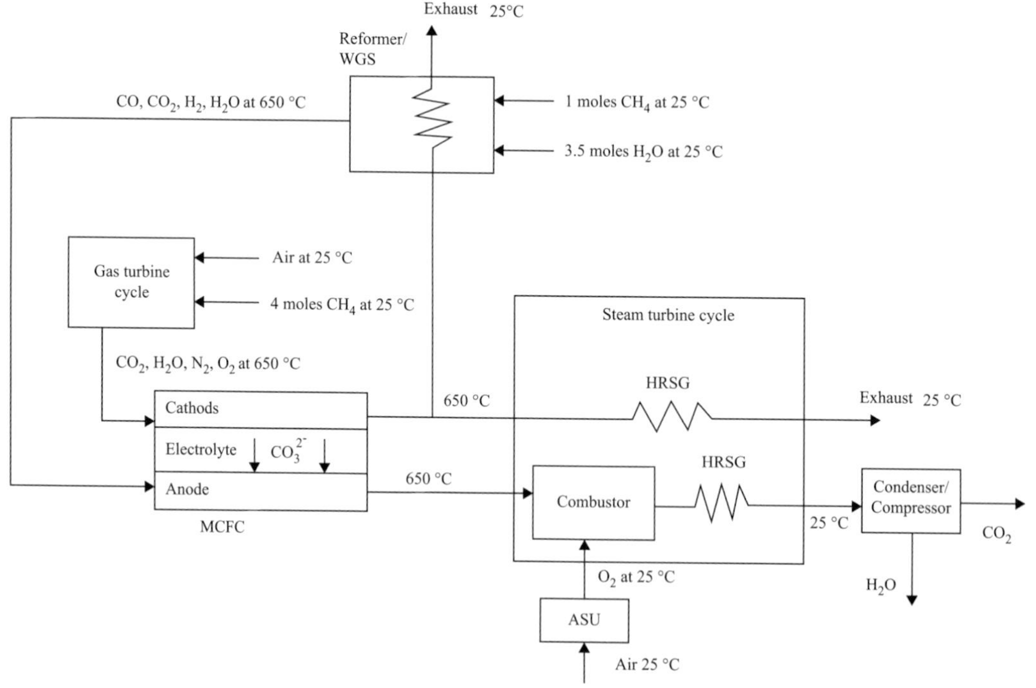

Figure E8.8 The CC with MCFC.

Example 8.8 (cont.)

a. Calculate the composition of the turbine exhaust gases and the work of the gas turbine cycle.

b. Calculate the number of moles of CO and CO_2 at the inlet of the anode, knowing that there are 3.5323 moles of H_2 and 1.9677 moles of H_2O in the mixture (per mole of methane added to the reformer). Assume that all the methane is converted and the mixture of CO, H_2O, H_2, and CO_2 is at equilibrium. The pressure-based equilibrium constant of the water–gas shift reaction at 650 °C is $K_p = 2.0426$.

c. Only 75% of the CO_2 introduced at the cathode participates in the electrochemical reaction (and flows from the cathode side to the anode side). Hence, it can be shown that the mixture composition at the cathode exit is 1 mole of CO_2, 8 moles of H_2O, 10.5 moles of O_2, and 75.2 moles of N_2. Calculate the mixture composition at the exit of the anode.

d. How much work is produced by the electrochemical reaction in the MCFC, when the actual cell voltage is 0.6 V?

e. Calculate the maximum (ideal) cell voltage at the exit of the MCFC.

f. How much of the cathode exhaust should be diverted to the reformer to supply the required heat of reforming?

g. Calculate the work produced by the steam turbine cycle. Heat is provided by the HRSG utilizing the cathode and anode exhaust streams.

h. Calculate the CO_2 recovery rate, defined as the fraction of CO_2 captured of that produced, and the First Law efficiency of the plant, given that ASU work and CO_2 liquefaction work are 1815 J and 77,106 J, respectively. CO_2 is delivered in the liquid form.

Solution

a. The complete combustion proceeds as follows:

$$CH_4 + 2(O_2 + 3.76N_2) \rightarrow CO_2 + 2H_2O + 7.52N_2.$$

When the equivalent ratio is 0.4:

$$CH_4 + 5(O_2 + 3.76N_2) \rightarrow CO_2 + 2H_2O + 3O_2 + 18.8N_2.$$

When four moles of CH_4 are completely combusted, the composition of the turbine exhaust gas is as follows:

$$4CH_4 + 20(O_2 + 3.76N_2) \rightarrow 4CO_2 + 8H_2O + 12O_2 + 75.2N_2.$$

GT cycle: When four moles of CH_4 are utilized, the work from the gas turbine cycle is

$$W_{GT} = \eta_{GT} \times 4 \times LHV_{CH_4} = 1{,}218{,}432 \text{ J}.$$

b. For steam reforming and the water–gas shift reaction, two moles of water are needed per one mole of CH_4 stoichiometrically:

$$CH_4 + 2H_2O \rightarrow 4H_2 + CO_2.$$

Example 8.8 (cont.)

With the abundance of steam (steam to carbon ratio is 3.5) and a temperature of 650 °C, the steam reforming reaction ($CH_4 + H_2O \rightarrow CO + 2H_2$) can proceed completely in the right direction. At equilibrium, the CH_4 mole fraction is around 1%. Therefore, we will assume that at the exit of the reformer, there are only CO, H_2O, H_2, and CO_2 in equilibrium composition. The pressure-based equilibrium constant for the shift reaction ($CO + H_2O \rightarrow H_2 + CO_2$) is $K_p = 2.0426$:

$$CH_4 + 3.5H_2O \rightarrow aH_2 + bH_2O + cCO + dCO_2.$$

Element conservation: $2a + 2b = 11$, $c + d = 1$ and $3.5 = b + c + 2d$.
Equilibrium constant: $K_p = \frac{ad}{bc}$.
a = 3.5323, b = 1.9677, c = 0.4677, $d = 0.5323$.

c. Assume that there is no thermochemical reaction in the MCFC, only electrochemical reactions occurring:
Cathode: $\frac{1}{2}O_2 + CO_2 + 2e^- \rightarrow CO_3^{2-}$.
3 moles of CO_2 (75%) and 1.5 moles of O_2 are removed in the cathode, consuming 6 moles of electrons.
Cathode inlet composition: $4CH_4 + 20(O_2 + 3.76N_2) \rightarrow 4CO_2 + 8H_2O + 12O_2 + 75.2N_2$.
Cathode exit composition: $CO_2 + 8H_2O + 10.5_2 + 75.2N_2$.

Table E8.8c

	CO_2	H_2O	O_2	N_2	Total
Inlet, mole numbers	4	8	12	75.2	99.2
Inlet, mole fractions	0.004032	0.008065	0.012097	0.075806	
Exit, mole numbers	1	8	10.5	75.2	94.7
Exit, mole fractions	0.01056	0.084477	0.110876	0.794087	

Anode: $H_2 + CO_3^{2-} \rightarrow H_2O + CO_2 + 2e^-$.
3 moles of H_2 are consumed and 3 moles of H_2O and CO_2 are produced while producing 6 moles of electrons:
Anode inlet composition: $aH_2 + bH_2O + cCO + dCO_2$.
Anode exit composition: $(a - 3)H_2 + (b + 3)H_2O + cCO + (d + 3)CO_2$.
Mole numbers and mole fractions are as given in Table E8.8d.

Table E8.8d

	H_2	H_2O	CO	CO_2	Total
Inlet, mole numbers	3.5323	1.9677	0.4677	0.5323	6.5
Inlet, mole fractions	0.543431	0.302723	0.071954	0.081892	1
Exit, Mole numbers	0.5323	4.9677	0.4677	3.5323	9.5
Exit, Mole fractions	0.056032	0.522916	0.049232	0.371821	

Example 8.8 (cont.)

d. The current out (per 5 moles of CH_4) is 6 moles of electrons:

Work = V·Q = 0.6 × 6 × 96,485 = 347,346 J.

e. $H_{2,anode} + \frac{1}{2}O_{2,cathode} + CO_{2,cathode} \rightarrow H_2O_{,anode} + CO_{2,anode}$,

$$V_{\max} = \frac{-\Delta g}{2F} - = \frac{1}{2F}\left(g^0_{H_2} + \frac{1}{2}g^0_{O_2} - g^0_{H_2O}\right) + \frac{RT}{2F}\ln\left(\frac{p_{H_2,anode}\left(p_{O_2,cathode}\right)^{1/2}p_{CO_2,cathode}}{p_{H_2O,anode}\,p_{CO_2,anode}}\right).$$

The standard Gibbs free energies for CO_2 cancel out between cathode and anode. As the reaction proceeds, the concentrations along the electrolyte change. Therefore, the inlet and outlet will have different voltages, which are $V_{inlet} = 0.835$ V and $V_{exit} = 0.745$ V. (You might also consider the difference between the inlet and outlet of the cell to get an average value.)

The maximum cell voltage (OCV) at the exit is higher than the operating cell voltage (0.6 V). So, the MCFC can work.

f. The reformer chemical reactions are as follows.

$$\text{Input}: CH_4 + 3.5\,H_2O\ 25\,°C,\ 1\,\text{atm}$$

$$\text{Output}: 3.5323\,H_2 + 1.9677\,H_2O + 0.4677\,CO + 0.5323\,CO_2(650\,°C,\ 1\,\text{atm})$$

The heat should be balanced by the diverted cathode exhaust:

$$\sum_{out} n_i h_i - \sum_{in} n_i h_i = n_{divert} c_{p,cathode}\Delta T,$$

$$\sum_{out} n_i h_i - \sum_{in} n_i h_i = 318{,}247.4\ \text{J}.$$

Total moles at the cathode exit is 94.7 moles.

$n_{divert} = 16.33$ moles of cathode exhaust should be diverted.

g. Heat is provided by the HRSG utilizing the cathode and anode exhaust streams:

$$Q_{in,ST} = Q_{cathode} + Q_{anode},$$

$$Q_{cathode} = (n_{cathode\ exhaust} - n_{divert})c_{p,cathode}\Delta T.$$

The number of moles of cathode exhaust entering the steam turbine cycle = 78.36 moles:

$$Q_{cathode} = 1{,}526{,}627\ \text{J}.$$

The air provided by the ASU should be at the stoichiometric ratio, because the ASU requires substantial work for oxygen separation and to provide more than the stoichiometric ratio wastes energy. However, in this problem, the amount of O_2 above stoichiometry does not change the answer, because the unreacted O_2 is out at 25 °C, the same temperature of incoming O_2 from the ASU. The enthalpy term for the unreacted O_2 is canceled out for the calculation of $Q_{in,\text{anode}}$.

$$Q_{anode} = \sum_{in} n_i h_i - \sum_{out} n_i h_i.$$

Example 8.8 (cont.)

Anode stream combustion and HRSG:

$$\text{Input}: (a-3)H_2 + (b+3)H_2O + cCO + (d+3)CO_2 \text{ at } 650\ ^\circ\text{C and}$$

$$1 \text{ atm, and } \frac{(a+c-3)}{2} O_2 \text{ at } 25\ ^\circ\text{C and } 1 \text{ atm.}$$

$$H_{in} = -2,406,042.619 \text{ J.}$$

$$\text{Output}: (a+b)H_2O + (c+d+3)CO_2 \text{ at } 25\,^\circ\text{C, } 1 \text{ atm,}$$

$$H_{out} = -2,902,550 \text{ J.}$$

$$Q_{anode} = 496{,}507.381 \text{ J.}$$

$$Q_{in,ST} = 2{,}023{,}134.375 \text{ J.}$$

$$W_{ST} = \eta_{ST} Q_{in,ST} = 708{,}097.0313\text{J.}$$

h. The plant efficiency considering the ASU and liquefaction:

$$\eta_{plant} = \frac{W_{GT} + W_{ST} + W_{MCFC} - W_{ASU} - W_{liq}}{5 \times LHV_{CH_4}} = 54.8\%.$$

Therefore, we can design a power plant with almost the same efficiency and with 80% CO_2 capture: 5 moles of CH_4 is used, 1 mole of CO_2 is emitted through the cathode exhaust, and 4 moles of CO_2 are sequestered from the anode exhaust stream. The CO_2 recovery rate is 80%.

8.7 Summary

Simple gas turbine cycles, and modifications to improve their efficiency, were discussed in Chapter 5. It was shown that the largest irreversibility, without significant internal regeneration, is associated with the turbine exhaust (flue gases leaving the turbine at relatively high temperature). The cost of higher-temperature heat exchangers required for regeneration can be too high to justify the benefit. In this chapter, improvements to recycle some of the flue gas energy were introduced, including the humid air cycle and thermochemical recuperation. In both, a fraction of the exhaust enthalpy is captured in the form of steam or chemical energy, respectively, that are added to the gases entering the turbine. On the other hand, the most practical so far has been the CCs in which a Rankine cycle is used as a bottoming cycle for the gas turbine. Combined cycles reach impressive efficiencies exceeding 50% and are used extensively in NGCCs. They have also been used with coal following gasification and gas clean-up (in IGCC plat, which will be discussed in Chapters 12 and 13).

As concerns over CO_2 emissions have grown, "carbon capture" has been suggested as one of the most viable solutions. In one form (which will be discussed in detail in Chapters 10,

11, and 13) oxy-combustion is used, in which pure oxygen, instead of air, is used and the combustion temperature is moderated by recycling part of the CO_2, water, of both. Several solved examples were used in this chapter to introduce alternative forms of oxy-combustion cycles and to determine its impact on cycle efficiency.

In Chapters 4 and 7 it was shown that higher-temperature fuel cells, such as SOFCs, reject high-temperature heat. To improve the efficiency of an FC, it has been suggested to use this heat in a bottoming cycle. Again, several examples were used in this chapter to demonstrate how to construct a hybrid FC–GT cycle or FC–ST cycle, and the impact of the hybridization on the efficiency.

Problems

8.1 Consider the steam injected gas turbine cycle shown in Figure 8.1. Air at 20 °C and 1 atm is pressurized to 12 atm in the compressor, which operates with 80% isentropic efficiency. It is heated to 1373 K in the combustor. The turbine exhaust flows through the HRSG, which produces superheated steam at 500 °C. The exhaust gases leave the HRSG at 100 °C. Water at 20 °C and 1 atm is pumped to the HRSG. The isentropic efficiencies of the gas turbine and pump are 89% and 65%, respectively. The fuel burned in the combustor is methane, with an LHV of 50,050 kJ/kg. The properties of the flow at states 3, 4, and 5 may be assumed to be those of an air–steam mixture. Determine:

a. the amount of steam injection and fuel consumption per unit mass of the air;
b. whether the water content of the exhaust gas condenses;
c. the work output of the cycle; and
d. the thermal efficiency of the cycle.

8.2 A designer investigates the possibility of improving the performance of a steam injected gas turbine cycle by adding a regenerator between the compressor and the combustor. The layout of the proposed cycle is depicted in Figure P8.2. The efficiency of the regenerator is 80%. Take the data given in Problem 8.1 and determine: (a) the work output of the cycle; (b) the thermal efficiency of the cycle; and (c) whether it is recommended to operate the steam injected gas cycle with or without a regenerator.

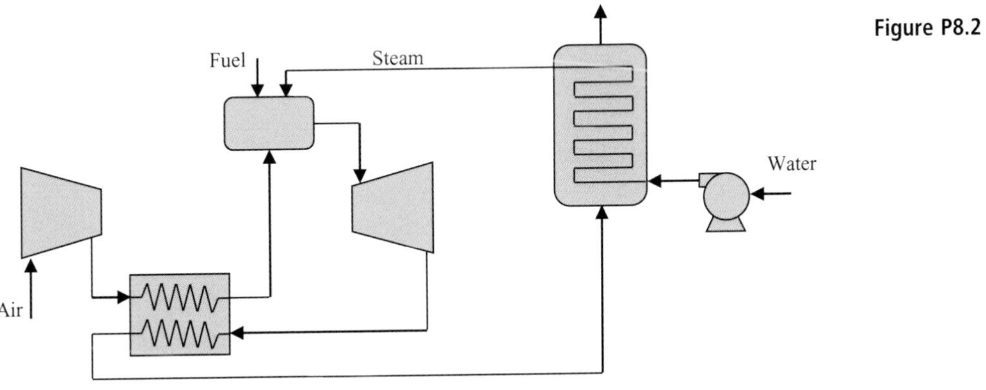

Figure P8.2

8.3 Consider the power cycle shown in Figure 8.2. Air at 20 °C and 1 atm is pressurized within two compressors to 10 atm. Both compressors have an identical pressure ratio, which operates with isentropic efficiency of 80%. The pump, operating with 70% isentropic efficiency, raises the pressure of the water, initially at 20 °C and 1 atm, to 10 atm. For each 1 kg air, 0.2 kg of water is used in the cycle. The intercooler operates with 85% efficiency. The hot stream leaving the combustor at 1000 °C enters the turbine, which operates with an isentropic efficiency of 90%. The efficiencies of the recuperator and the economizer are 80% and 82%, respectively. Determine:

a. the temperature of the air–water mixture at the exit of the mixture;

b. the temperature of the mixture at the entrance of the combustor;

c. the exhaust temperature;

d. the mass of fuel (methane with LHV = 50,050 kJ/kg) required in the cycle;

e. the specific work output of the cycle; and

f. the thermal efficiency of the cycle.

8.4 Consider the CC power plant shown in Figure 8.4. Air at 20 °C and 1 atm enters the compressor, operating with a pressure ratio of 12 and isentropic efficiency of 82%. The inlet temperature of the gas turbine operating with an isentropic efficiency of 92%, is 1200 °C; 85% of the energy content of the turbine exhaust is received in the HRSG. Superheated steam at 400 °C and 10 MPa enters the steam turbine, where it is expanded to 7.5 kPa. The isentropic efficiencies of the steam turbine and pump are 88% and 68%, respectively. The mass of fuel burned in the combustor is negligible compared to that of the air. Determine:

a. the amount of steam produced in the HRSG per unit mass of air;

b. the exhaust temperature;

c. the specific work output of the CC; and

d. the thermal efficiency of the CC.

8.5 A schematic of a gas turbine cycle combined with a regenerative steam cycle is shown in Figure P8.5. Air at 293 K and 100 kPa enters the compressor, where it is pressurized to 1 MPa. It is heated in the combustor to 1550 K, and is expanded in the gas turbine to 100 kPa. The gas turbine's exhaust cools down to 400 K in the HRSG, where a superheated stream at 400 K and 4 MPa is produced. At a certain pressure, 20% (mass basis) of the steam is extracted for preheating the water in an open heat exchanger. The rest of the steam is expanded to 7 kPa. The isentropic efficiencies of the compressor, gas turbine, steam turbine, and the pumps are 80%, 90%, 88%, and 65%, respectively. Assume that the mass of fuel is negligible to that of the air.

a. Determine the fraction of heat that is received in the HRSG.

b. Determine the steam pressure at the extraction point.

c. Calculate the specific work output of the gas cycle.

d. Calculate the specific work output of the steam cycle.

e. Determine the temperature of the water at the entrance of the HRSG.

f. Determine the thermal efficiencies of the gas and steam cycles if they operated individually.

g. Determine the thermal efficiency of the CC.
h. Calculate the amount of heat rejected from the CC.

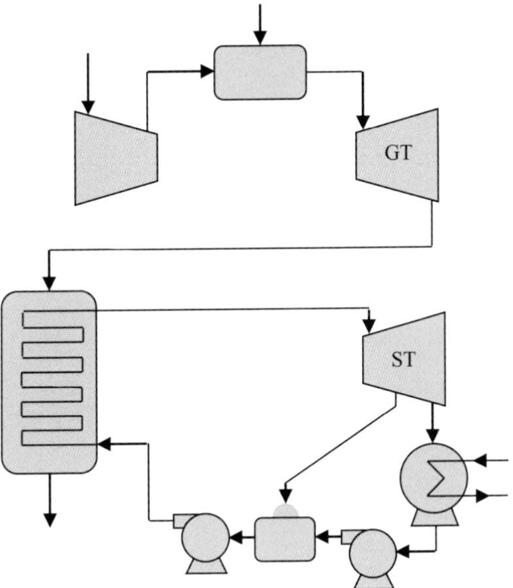

Figure P8.5

8.6 Fuel utilization factor, U_f, is an important design parameter that may significantly impact the performance of a hybrid cycle. To investigate its influence, consider the hybrid cycle analyzed in Example 8.6.

a. Plot the hybrid cycle efficiency versus fuel utilization factor varying between 60% and 90%.
b. Plot the net power output of the hybrid cycle versus fuel utilization factor varying between 60% and 90%.

8.7 A gas turbine engine integrated with an externally reforming SOFC is schematically shown in Figure P8.7. The average operating cell voltage and current density are 0.64 V and 350 mA/cm^2, respectively. The SOFC stack consists of 4000 cells each with a surface area of 800 cm^2. The dc–ac inverter efficiency is 90%. The air and fuel utilization factors are 0.22 and 0.81, respectively. Neglect any pressure drop in the system. The isentropic efficiencies of the compressor, turbine, water pump, and fuel pump are 80%, 89%, 70%, and 75%, respectively. Other process data are given in the figure.

a. Determine the mass flow rate of methane fed to the reformer.
b. Determine whether additional fuel has to be used in the combustor to raise the temperature of the main stream to 1000 °C.
c. Calculate the heat transfer rate in the reformer + shift reactor.
d. Determine if the enthalpy of the exhaust is sufficient to meet the heat requirement for the reforming process.

e. Determine the power output and the efficiency of the SOFC.
f. Determine the net power output of and the thermal efficiency of the hybrid cycle.

Figure P8.7

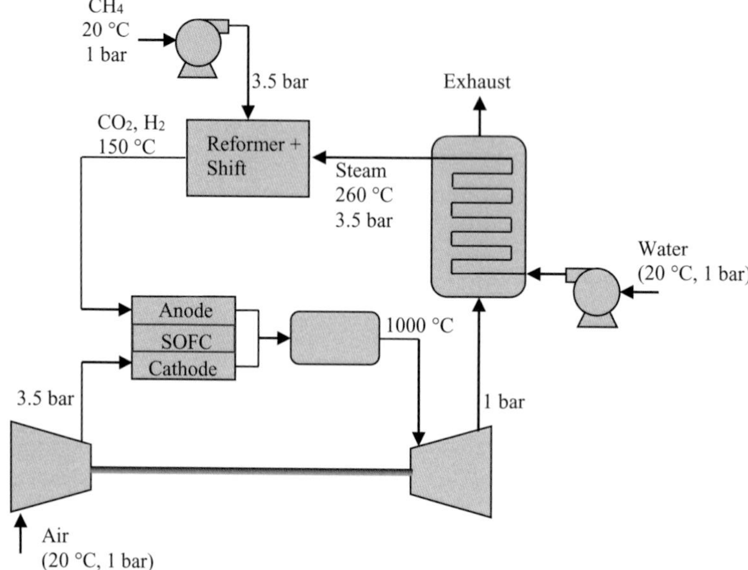

8.8 Consider the combined gas–steam power cycle shown in Figure P8.8. The topping cycle is a simple Brayton cycle that has a pressure ratio of 7. Air enters the compressor at 15 °C at a rate of 10 kg/s. The inlet temperature of the gas turbine is 950 °C. The bottoming cycle is a reheat Rankine cycle between the pressure limits of 6 MPa and 10 kPa. Steam is heated in a HRSG at a rate of 1.15 kg/s by the exhaust gases leaving the gas turbine. The exhaust gases leave the HRSG at 200 °C. Steam leaves the high-pressure turbine at 1 MPa and is reheated to 400 °C in the HRSG before it expands in the low-pressure turbine. Assume 80% isentropic efficiency for all pumps and turbines. Determine:

a. the moisture content at the exit of the low-pressure turbine;
b. the steam temperature at the inlet of the high-pressure turbine;
c. the power corresponding to the turbines, compressor, and pump;
d. the net power output from the combined plant;
e. the net heat input rate; and
f. the thermal efficiency of the combined plant.

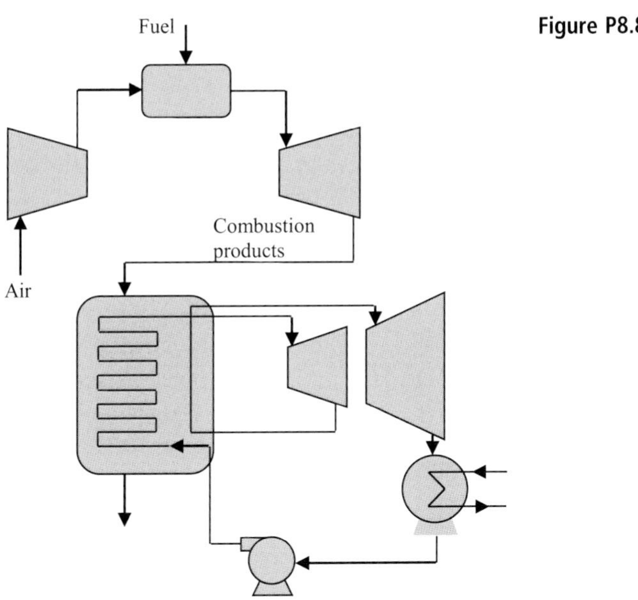

Figure P8.8

8.9 Consider the hypercritical CO_2 cycle examined in Example 6.5 that is now combined with a bottoming steam cycle (Figure P8.9). The rejected heat from the CO_2 cycle is transferred to a steam cycle via a heat exchanger and the steam cycle produces extra work. The CO_2 stream leaves the HRSG at 125 °C. Steam enters the steam turbine with a pressure of 100 bar and temperature of 450 °C, where it is expanded to 0.06 bar. The condenser exit temperature is 20 °C. Assume that the isentropic efficiency of the pump is 85%, and that of the steam turbine is 90%. Use the other parameters given in Example 6.5.

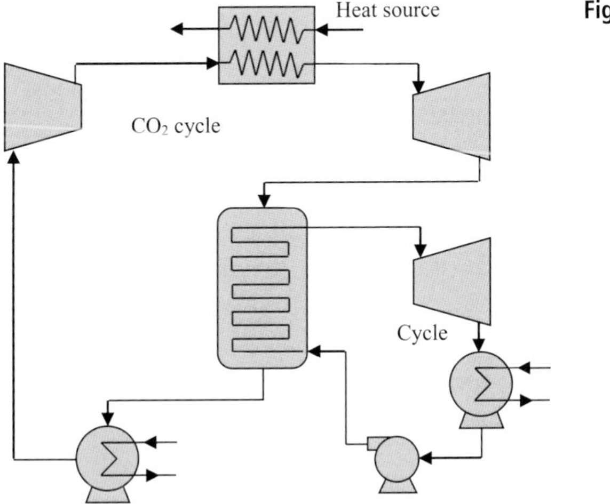

Figure P8.9

a. Calculate the mass flow rate of the steam per unit mass flow of the CO_2.
b. Determine the value of the pinch point in the steam cycle.
c. Calculate the total net work produced per unit mass flow of the CO_2.
d. Determine the thermal efficiency of the CC.

8.10 A hybrid power plant consists of a hydrogen-fueled SOFC and a closed helium gas turbine cycle, shown in Figure P8.10. Separate streams of hydrogen and air enter an adiabatic heat exchanger at 298 K, where they are heated by the exhaust stream of the SOFC. The hydrogen and air streams are further heated to 1273 K and then supplied to the SOFC, which operates at 1273 K and 1 atm. The molar flow rate of hydrogen is 100 mol/s. The flow rate of oxygen is such that the SOFC operates stoichiometrically. Heat is rejected from the SOFC to a closed helium gas turbine cycle. Helium leaves the heat exchanger at 1200 K and enters a turbine. From there it is cooled to 320 K before entering a compressor. The specifications of the SOFC and helium cycle are given in Table P8.10.

Table P8.10

SOFC	
Number of stacks	20
Number of cells per stack	220
Cell area	1 m^2
Exchange current density at the cathode	300 mA/cm^2
Limiting current density	900 mA/cm^2
Charge transfer coefficient (α)	0.5
Fuel utilization	100%
Area-specific ohmic overpotential	300×10^{-6} $k\Omega.cm^2$
Gas turbine cycle	
Turbine inlet temperature	1200 K
Compressor inlet temperature	320 K
Turbine isentropic efficiency	95%
Compressor isentropic efficiency	85%
Pressure ratio	4

a. Determine the temperature of the reactants leaving the first heat exchanger.
b. Determine the rate of additional heat transfer needed, $\dot{Q}_{in}$, to increase the temperature of the reactants to 1273 K.
c. Calculate the open-circuit voltage of the SOFC.
d. Activation and concentration overpotentials (dominant at the cathode) and ohmic overpotentials are present in the SOFC. Determine the actual cell voltage of the SOFC.
e. Determine the current density in each cell.
f. Calculate the rate of electrical power produced by the SOFC.
g. Determine the molar flow rate of the helium in the gas turbine cycle.
h. Determine the total power produced by the entire cycle.
i. Determine the efficiency of this hybrid cycle.

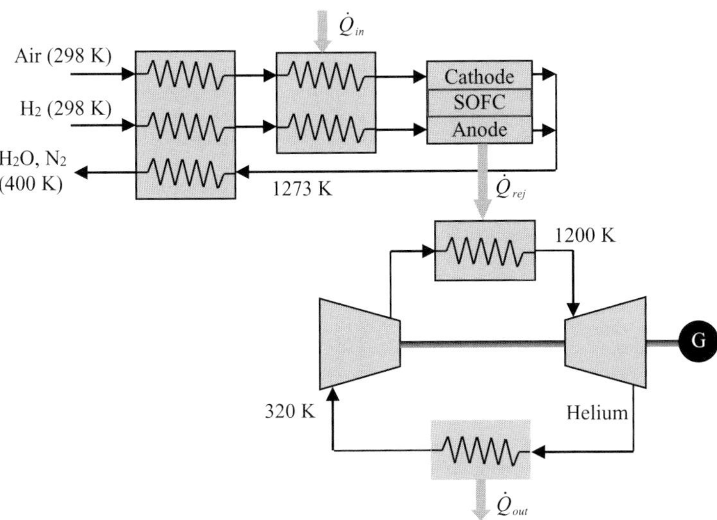

Figure P8.10

8.11 The hybrid power plant of Problem 8.10 is modified in that a regenerator is used downstream of the compressor. For this new configuration, determine:

a. the molar flow rate of helium in the gas turbine cycle;

b. the total power produced by the entire cycle; and

c. the efficiency for the hybrid power plant.

d. Compare the efficiency of this hybrid design with that of a design that uses methane to power a simple, helium-based gas turbine. Take the pressure ratio of 10 and the maximum and minimum temperatures in the cycle of 1200 K and 320 K, with the same turbine and compressor isentropic efficiencies. To simplify analysis, you may assume that all of the heating value of methane is available to heat the helium. Take the efficiency of producing hydrogen from methane as 85%.

8.12 Consider the case in which the SOFC in Problem 4.11 is operated on syngas, which is now integrated into a combined steam–gas cycle, as sketched in Figure P8.12. The energy required for the methane reformer (for production of syngas) and for heating the air to 700 °C before entering the fuel cell, is 414 kJ/mol CH_4, which is fully supplied by the heat rejected from the fuel cell. The total pumping and compressor work required to pressurize the streams entering the fuel cell to 10 atm is 100 kJ/mol CH_4. The mixtures at the outlet of the anode and the cathode are mixed isothermally, and then enter a combustor, where the portion of fuel unconverted in the fuel cell is completely burned stoichiometrically, leaving a mixture of H_2O and CO_2 at the exit of the combustor. The high-temperature mixture is then expanded to 1 atm in a gas turbine. The exhaust of the gas turbine is used directly to supply heat to the bottoming Rankine cycle through a HRSG. The condenser exit temperature is 40 °C. The pressure and temperature at the entrance of the steam turbine are 30 atm and 675 °C. The pinch point temperature difference in the HRSG is 10 °C. The isentropic efficiencies of the air and methane compressors, gas turbine, steam turbine, and pump are 80%, 90%, 87%, and 70%, respectively.

Figure P8.12

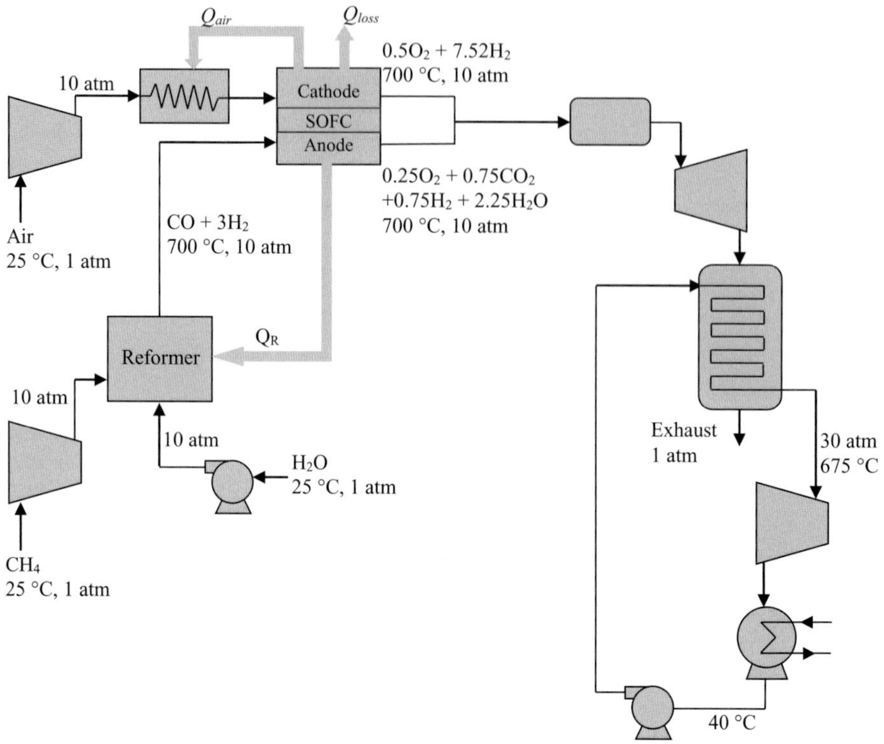

a. Show that the heat rejected from the fuel cell is enough to supply the energy required for the methane reformer and for heating the air entering the fuel cell to 700 °C.

b. Calculate the temperature at the exit of the combustor.

c. Determine the temperature of the mixture leaving the gas turbine.

d. Calculate the work produced by the gas turbine per mole of methane.

e. Calculate the exhaust temperature at the exit of the HRSG. (Hint: Boiling can be assumed to be initiated at the point the water becomes saturated at the heat exchanger pressure, and that point determines the pinch point.)

f. Determine the ratio of the molar flow rate of steam in the Rankine cycle to the molar flow rate of methane entering the reformer.

g. Calculate the net work produced by the Rankine cycle per mole of methane entering the reformer.

h. Calculate the fuel utilization efficiency based on the HHV of methane without the Rankine cycle.

i. Calculate the fuel utilization efficiency of the entire system based on the HHV of methane.

REFERENCES

1. N. V. Kharchenko and V. M. Kharchenko, *Advanced energy systems*. Boca Raton, FL: CRC Press, 2014.
2. W. H. Rao and A. D. Day, "Mitigation of greenhouse gases from gas turbine power plants," *Energy Convers. Mgmt.*, vol. 37, no. 6–8, pp. 909–914, 1996.
3. A. D. Morton, and R. T. Rao, *Perspectives for advanced high efficiency cycles using gas turbines*. Irvine, CA: Flour Daniel Inc., 1989.
4. Westinghouse Electric Co., "Advanced natural gas fired turbine system utilizing thermochemical recuperation and/or partial oxidation for electricity generation, greenfield and repowering applications," 1997.
5. R. Decher, *Energy conversion: systems, flow physics, and engineering*. New York: Oxford University Press, 1994.
6. M. M. El-Wakil, *Powerplant technology*. New York: McGraw-Hill, 1984.
7. J. Larminie and A. Dicks, *Fuel cell systems explained*. Chichester: Wiley, 2003.
8. R. P. O'Hayre, S.-W. Cha, W. G. Colella, and F. B. Prinz, *Fuel cell fundamentals*. Hoboken, NJ: Wiley, 2016.
9. Y. Haseli, I. Dincer, and G. F. Naterer, "Thermodynamic modeling of a gas turbine cycle combined with a solid oxide fuel cell," *Int. J. Hydrogen Energy*, vol. 33, no. 20, pp. 5811–5822, 2008.
10. M. Rokni, "Plant characteristics of an integrated solid oxide fuel cell cycle and a steam cycle," *Energy*, vol. 35, pp. 4691–4699, 2010.

9 Geothermal, Solar Thermal, and Integration

9.1 Introduction

Renewable sources of thermal energy have been used to generate electricity in power plants using power cycles similar to those described in Chapters 5 and 6, although with some modifications to make them compatible for the intermediate- and lower-temperature heat sources. These renewable sources include geothermal energy and concentrated solar thermal energy. The power cycles used for these two types will be discussed in this chapter, starting with general guidelines regarding how to maintain the cycle efficiency as high as possible while using lower-temperature heat sources. For instance, using different working fluids in Rankine cycles can improve the efficiency by operating the cycle as a supercritical cycle even though the source temperature is relatively low.

In the geothermal section, following a classification of source and availability (in the USA), unique power cycles including binary and flash plants are discussed, and a numerical example is used to demonstrate efficiency calculations. Most geothermal sources used today are classified as hydrothermal, and hence are compatible with two-phase cycles that can utilize lower-temperature heat sources. Indirectly heated systems, called binary plants, tend to use organic fluids, as described in Chapter 6, running supercritical cycles to reduce the temperature difference between the source and the working fluid. Another type of plant uses the geofluid as a working fluid, after its pressure is reduced to generate the saturated steam for the turbine, while the liquid is sent back to the well. Combination of both concepts is also possible.

Next, concentrated solar thermal power cycles are presented, starting with source evaluation (worldwide) and the characteristics of solar collectors/concentrators, the temperatures achieved by the working fluid in each case, and the system efficiency. Solar collectors and concentrators are discussed, in particular the efficiency of the solar field and its dependency on the collector temperature and concentration ratio. Matching concentrators with heat engines is important for maintaining high system efficiency. A number of examples are used to demonstrate the operation of different designs.

An important recent trend has been to integrate solar thermal energy with fossil fuel energy to overcome the intermittency of the former and the need for significant storage.

A number of options are described. Different forms of this integration have been proposed and some implemented; in some cases the solar energy is used to heat the working fluid of the cycle, and in others it is used to reform the fuel before combustion using the solar energy. In the latter case, the integration enables storage since the chemical energy of the syngas (produced by reforming) is higher than the chemical energy of the original fuel.

9.2 Working with Lower-Temperature Sources

Solar thermal and geothermal energy provide relatively low-grade heat, 300–600 °C for the former and 150–250 °C for the latter. Rankine power cycles operating on steam work well for solar thermal, but organic Rankine cycles are necessary to achieve reasonable efficiency for geothermal power. The Carnot efficiency is low in these cases, and also the real cycle efficiency, as shown in Figure 9.1. Clearly, gas turbine cycles cannot be used for either. And while the heat source is "free," it is important to collect the thermal energy and deliver it to the power island. Like other forms of renewable energy, the source is distributed with relatively low intensity per unit area.

As shown in Chapter 6, the ideal working fluid for a Rankine cycle has the following characteristics:

1. high critical temperature at relatively low critical pressure to allow higher maximum cycle temperatures without increasing the hardware cost and complexity;
2. large enthalpy of evaporation to minimize the mass flow rate for a given thermal load; and
3. rapidly diverging constant pressure lines on the h–s diagram to improve the effectiveness of reheat.

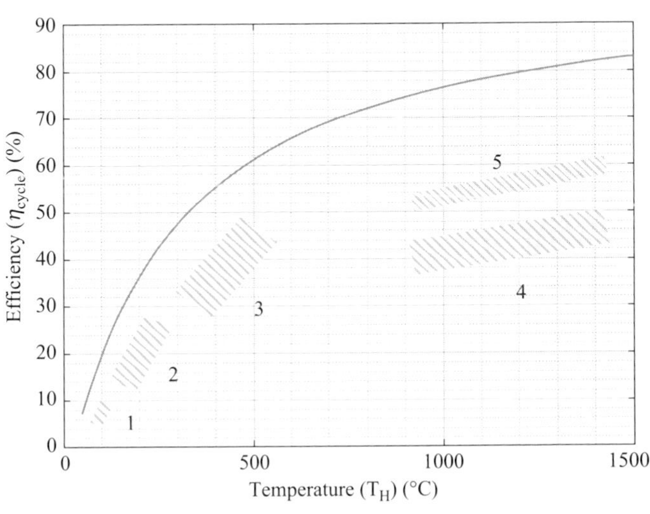

Figure 9.1 The dependence of the ideal Carnot efficiency on temperature, shown as a solid line, and a number of systems according to the source temperature and practical limitations: 1 is for geothermal energy, 2 for concentrated solar power, 3 is for Rankine cycles operating on fossil or nuclear energy, 4 for Brayton cycles, and 5 for combined cycles operating on fossil fuels.

The working fluid should also be non-corrosive, nontoxic, nonflammable, cheap, and widely available. Some exotic fluids, which have reasonable saturation pressures at high temperatures, such as sodium–potassium mixtures, have been used for space power applications.

Fluids with lower-saturation temperatures, close to the atmospheric pressure, work better with renewable energy applications, such as geothermal and solar thermal energy. These include: freon 12, $Cl_{12}F_2$, with $p_{cr} = 40.6$ atm and $T_{cr} = 112\,°C$; propane, C_3H_8, with $p_{cr} = 42.1$ atm and $T_{cr} = 97\,°C$; ammonia, NH_3, with $p_{cr} = 111.3$ atm and $T_{cr} = 132\,°C$; isobutane, isopentane, toluene, and a number of refrigerants. More fluids were listed in Chapter 6.

9.3 Geothermal Energy Applications

Geothermal energy has been used around the world for power generation for decades. It is a renewable energy source that is widely available worldwide, and is largely considered one of the cheapest sources of renewable energy. With an underground temperature gradient of 20–30 K/km, it is possible to provide low-grade heat at 150–200 °C by drilling relatively shallow wells. Figure 9.2 shows sites that have been

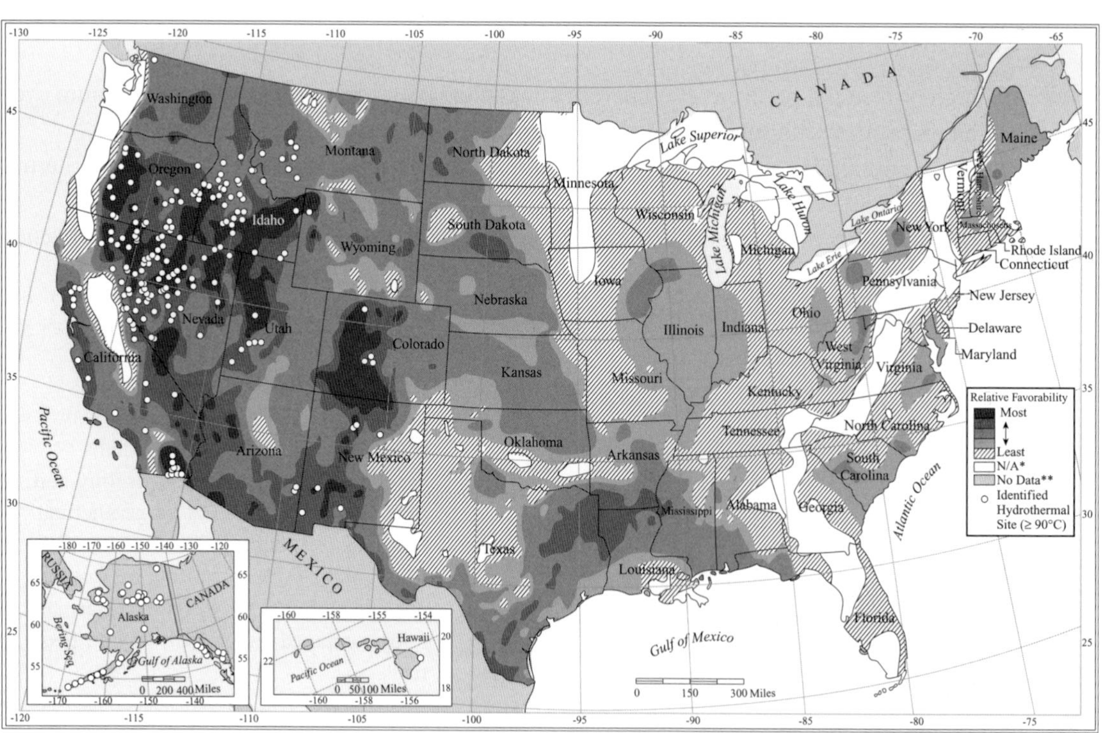

Figure 9.2 The quality of geothermal energy resources in the USA, with darker shades showing the most favorable and lighter shades the least favorable. Circles show sites that have been identified for hydrothermal energy (www.nrel.gov/gis/assets/images/geothermal-identified-hydrothermal-and-egs.jpg).

identified for hydrothermal energy utilization in the continental USA. A fresh geothermal (hydrothermal) well typically lasts for about five years before it needs recharging (which takes 2–3 times as long), but a neighboring well can be drilled and used to supply energy to the same power plant. With a heat transfer rate ~0.05–0.1 W/m^2, multiple wells distributed over $O(1\,km^2)$ are typically required for power plants with capacity 5–20 MW. Geothermal energy is dispatchable – that is, available 24hours/day, all year round – and does not need storage (unlike concentrated solar power, but the latter operates at higher efficiency because of the higher-temperature heat source). When hydrothermal sources (to be defined next) are available, geothermal energy is cost competitive, even with conventional power plants. Drilling deep wells for alternative geothermal technologies, known as enhanced geothermal energy, can raise the cost significantly (and could lead to weak earthquakes in some areas).

The conventional and most widely used form of geothermal energy is **hydrothermal**, in which the geofluid (the fluid coming out of the well) can be liquid and superheated water that includes brines as well, or vapor and dry steam. An alternative form is **geopressure**, which consists of methane, hydro, and thermal energy. An unconventional (and still experimental) type is known as magna or **hot dry rock**, and could be used in enhanced geothermal systems or **EGS**. In the hydrothermal type of geothermal energy, water permeating under its own pressure through underground hot porous rock can reach the surface via a natural well (like geysers) or an artificial well drilled deep enough to reach these rock formations. As the pressure is reduced through the well, partial or complete boiling occurs, forming a mixture of steam and water or pure dry steam. In EGS, natural underground water is not available or not sufficient, and water must be injected deep underground, where hot rock is available. The heated water then permeates though the rock, finding its way to a production well that brings it to the surface.

Two types of geothermal power plants have been used extensively: a **binary plant** using the geothermal fluid as a heat source for the plant working fluid, and a **flash plant** in which the geothermal fluid is flashed to produce steam for the power island. Dry steam has also been used directly in what could be considered an open Rankine cycle – that is, it is fed to the steam turbine directly and, following condensation and pumping, it is sent to an injection well. Binary plants operate reasonably efficiently with low-temperature geofluids in the range of 120–180 °C, since they take advantage of most of the thermal energy available in the well stream. They operate on a closed organic Rankine cycle using an organic fluid as a working fluid, and achieve moderate efficiency. Flash plants are simpler since they operate on the steam of the geofluid, but achieve lower efficiency because they do not take advantage of all the thermal energy available in the well stream (the hot liquid fraction is sent to the injection well directly after flashing the geofluid). Because of the need to flash the original stream, it is necessary to have a higher-temperature well (200–250 °C) with a good flow rate. The operating efficiency of either plant depends strongly on the geofluid temperature. It is possible to integrate flash and binary plants to take advantage of the vapor and liquid components of the geofluid, as will be shown in the following.

9.3.1 Binary Plants

A **binary plant** is a conventional Rankine cycle plant using mostly an organic fluid as the working fluid, which is heated in a heat exchanger using the fluid from the geothermal well (superheated water, brine, and/or steam), as shown in Figure 9.3. Because of the low temperature of the geofluid, it is often necessary to use a working fluid with low supercritical temperature, and for best efficiency, it is best to run a supercritical cycle with multiple stages of regeneration (see Chapter 6). Furthermore, it is best to use two-stage heating, as shown in Chapter 6, to reduce the heat transfer irreversibility in the heat exchanger (or the temperature difference between the geofluid and the working fluid). Depending on the working fluid used and the cycle design, the thermal efficiencies of these plant are ~10–20%. The cool geofluid is pumped back into the well. Organic Rankine cycles used for these plants have been described before, as have the properties of the working fluids. Dry fluids are preferred to avoid condensation inside the turbine.

9.3.2 Flash Plants

An alternative plant design, which is used more frequently because of its simplicity and lower cost, is the **flash plant**. In this design, steam is generating by the adiabatic expansion of the geothermal fluid in a flash chamber. As the pressure of the geothermal fluid is reduced adiabatically, more of it evaporates, producing steam at lower pressure. The well geofluid often comes out at atmospheric pressure and hence it is possible to generate steam by throttling it to a lower pressure, intermediate between atmospheric pressure and the condenser pressure, which is determined by the temperature of the cooling water. More steam flow is generated by flashing to a lower pressure (and temperature), but that also lowers the temperature of the produced steam and hence the work produced per unit mass of steam in the turbine. A general rule of thumb is that the flash temperature is chosen as the arithmetic average (or geometric average) of the reservoir temperature and the

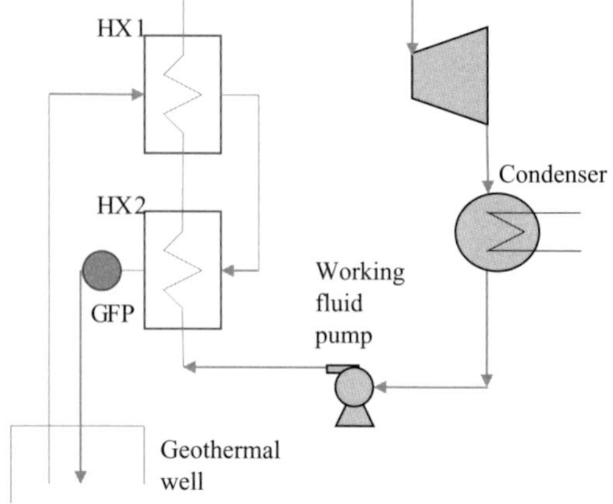

Figure 9.3 A binary geothermal plant using a Rankine cycle, and a two-stage working fluid. HX1 and HX2 are heating units. The working fluid is returned to the well using a geothermal fluid pump (GFP).

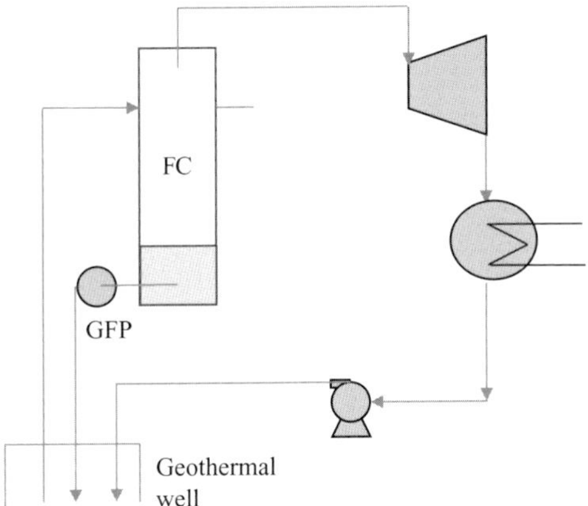

Figure 9.4 A flash plant design. Geofluid from the well is flashed in FC and steam from this flash chamber is sent to the turbine. The expanded stream from the turbine is condensed and pumped back to the well.

condenser temperature. Multiple flash chambers and hence multiple turbines are also used to improve the plant efficiency.

The produced steam from the flash chamber expands in an adiabatic turbine, followed by a condenser and reinjection into the injection well after being pumped. Flash supercritical plants have also been proposed to further raise the efficiency.

To improve the flash plant efficiency, it is possible to add a binary cycle to take advantage of the thermal energy in the liquid, which gets separated from the original geofluid in the flash chamber. Instead of injecting this liquid back into the injection well directly, it can be used in a second power cycle. This liquid is hot, at the same temperature of the steam generated during the flash process, and can be used to heat the working fluid of a binary cycle similar to the one shown in Figure 9.3. Example 9.1 illustrates the working of a **combined/integrated flash–binary plant**.

Example 9.1

Part I

The schematic of a single-flash geothermal power plant is given in Figure E9.1a. The geothermal resource exists as saturated liquid at 230 °C. The liquid is withdrawn from the production well at a rate of 230 kg/s, and is flashed to a pressure of 500 kPa by an isenthalpic flashing process in which the resulting vapor is separated from the liquid in a separator and directed to the turbine. The steam leaves the turbine at 10 kPa with a moisture content of 10% and enters the condenser, where it is condensed and routed to a reinjection well along with the liquid coming off the separator.

Example 9.1 (cont.)

Determine:

a. the mass flow rate of steam through the turbine;
b. the isentropic efficiency of the turbine;
c. the power output of the turbine; and
d. the thermal efficiency of the plant (the ratio of the turbine work output to the energy of the geothermal fluid relative to standard ambient conditions).

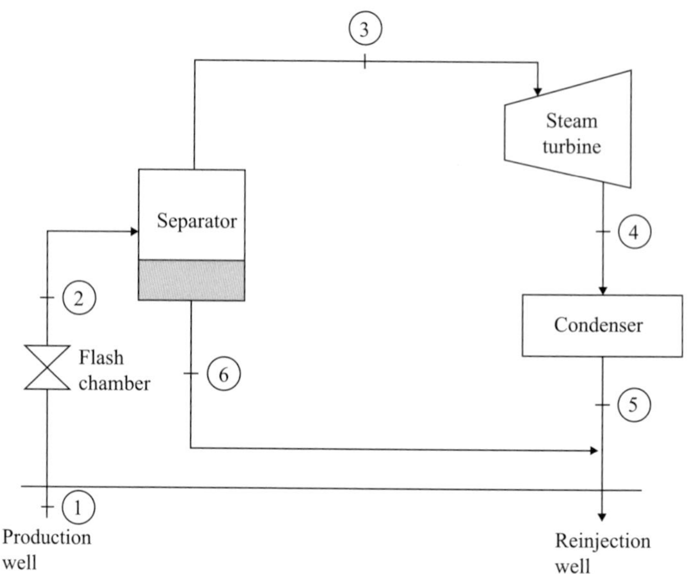

Figure E9.1a Single-flash geothermal power plant.

Part II

It is proposed that the liquid water coming out of the separator be used as the heat source in a binary cycle, with isobutane as the working fluid. Geothermal liquid water leaves the heat exchanger at 90 °C, while isobutane enters the turbine at 3.25 MPa and 145 °C, and leaves at 80 °C and 400 kPa. Isobutane is condensed in an air-cooled condenser and then pumped to the heat exchanger pressure. Assuming an isentropic efficiency of 90% for the pump, determine:

a. the mass flow rate of isobutane in the binary cycle;
b. the net power outputs of both the flashing and the binary sections of the plant; and
c. the thermal efficiencies of the binary cycle and the combined plant.

Example 9.1 (cont.)

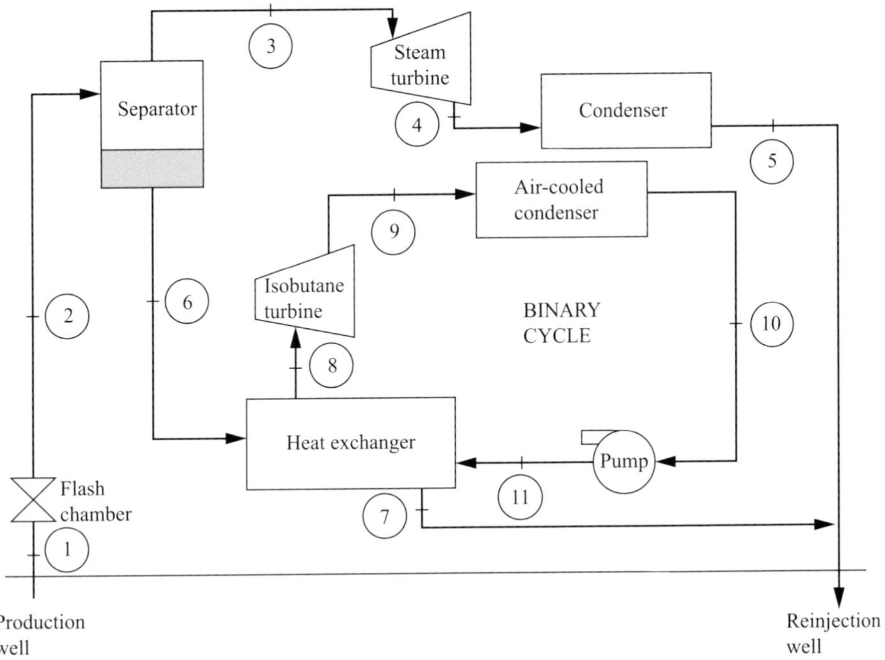

Figure E9.1b The binary cycle with isobutane as the working fluid.

Solution

Part I

We use the properties of water for geothermal water.

$$\left.\begin{array}{l} T_1 = 230\,^\circ\text{C} \\ x_1 = 0 \end{array}\right\} h_1 = 990.14\,\text{kJ/kg},$$

$$\left.\begin{array}{l} P_2 = 500\,\text{kPa} \\ h_2 = h_1 = 990.14\,\text{kJ/kg} \end{array}\right\} x_2 = \frac{h_2 - h_f}{h_{fg}} = 0.1661.$$

a. The mass flow rate of steam through the turbine is $\dot{m}_3 = x_2 \dot{m}_1 = 38.20\,\text{kg/s}$.

b. Turbine:

$$\left.\begin{array}{l} P_3 = 500\,\text{kPa} \\ x_3 = 1 \end{array}\right\} \left\{\begin{array}{l} h_3 = 2748.1\,\text{kJ/kg} \\ s_3 = 6.8207\,\text{kJ/kg·K}, \end{array}\right.$$

Example 9.1 (cont.)

$$\left.\begin{array}{l} P_4 = 10\,\text{kPa} \\ s_4 = s_3 \end{array}\right\} h_{4s} = 2160.3\,\text{kJ/kg},$$

$$\left.\begin{array}{l} P_4 = 10\,\text{kPa} \\ x_4 = 0.90 \end{array}\right\} h_4 = h_f + x_4 h_{fg} = 2344.7\,\text{kJ/kg},$$

$$\eta_T = \frac{h_3 - h_4}{h_3 - h_{4s}} = \frac{2748.1 - 2344.7}{2748.1 - 2160.3} = 0.686.$$

c. The power output from the turbine is:

$$\dot{W}_{T,out} = \dot{m}_3(h_3 - h_4) = 38.20 \times (2748.1 - 2344.7) = 15.41\,\text{MW}.$$

d. We use the saturated liquid state at the standard temperature for the dead state enthalpy:

$$\left.\begin{array}{l} T_0 = 25\,^{\circ}\text{C} \\ x_0 = 0 \end{array}\right\} h_0 = 104.83\,\text{kJ/kg},$$

$$\dot{E}_{in} = \dot{m}_1(h_1 - h_0) = 230 \times (990.14 - 104.83) = 203.622\,\text{MW}.$$

Note that we have used energy here as highlighted in the problem statement (not exergy):

$$\eta_{th} = \frac{\dot{W}_{T,out}}{\dot{E}_{in}} = 0.0757 = 7.6\%.$$

Part II

We use the properties of water for the geothermal water:

$$\left.\begin{array}{l} T_1 = 230\,^{\circ}\text{C} \\ x_1 = 0 \end{array}\right\} h_1 = 990.14\,\text{kJ/kg},$$

$$\left.\begin{array}{l} P_2 = 500\,\text{kPa} \\ h_2 = h_1 = 990.14\,\text{kJ/kg} \end{array}\right\} x_2 = \frac{h_2 - h_f}{h_{fg}} = 0.1661.$$

a. The mass flow rate of steam through the turbine is

$$\dot{m}_3 = x_2 \dot{m}_1 = 0.1661 \times 230 = 38.20\,\text{kg/s},$$

$$\dot{m}_6 = \dot{m}_1 - \dot{m}_3 = 191.80\,\text{kg/s},$$

$$\left.\begin{array}{l} P_3 = 500\,\text{kPa} \\ x_3 = 1 \end{array}\right\} \left\{\begin{array}{l} h_3 = 2748.1\,\text{kJ/kg} \\ s_3 = 6.8207\,\text{kJ/kg·K}, \end{array}\right.$$

$$\left.\begin{array}{l} P_4 = 10\,\text{kPa} \\ x_4 = 0.90 \end{array}\right\} h_4 = h_f + x_4 h_{fg} = 2344.7\,\text{kJ/kg},$$

Example 9.1 (cont.)

$$\left.\begin{array}{l} P_6 = 500\,\text{kPa} \\ x_6 = 0 \end{array}\right\} h_6 = 640.09\,\text{kJ/kg},$$

$$\left.\begin{array}{l} T_7 = 90\,^\circ\text{C} \\ x_7 = 0 \end{array}\right\} h_7 = 377.04\,\text{kJ/kg}.$$

The isobaric properties are obtained from EES:

$$\left.\begin{array}{l} P_8 = 3250\,\text{kPa} \\ T_8 = 145\,^\circ\text{C} \end{array}\right\} h_8 = 755.05\,\text{kJ/kg},$$

$$\left.\begin{array}{l} P_9 = 400\,\text{kPa} \\ T_9 = 80^\circ\text{C} \end{array}\right\} h_9 = 691.01\,\text{kJ/kg},$$

$$\left.\begin{array}{l} P_{10} = 400\,\text{kPa} \\ x_{10} = 0 \end{array}\right\} \left\{\begin{array}{l} h_{10} = 270.83\,\text{kJ/kg} \\ v_{10} = 0.001839\,\text{m}^3/\text{kg}, \end{array}\right.$$

$$w_{p,in} = v_{10}(P_{11} - P_{10})/\eta_P = 0.001819 \times (3250 - 400)/0.90 = 5.82\,\text{kJ/kg},$$

$$h_{11} = h_{10} + w_{p,in} = 276.65\,\text{kJ/kg}.$$

An energy balance for the heat exchanger gives

$$\dot{m}_6(h_6 - h_7) = \dot{m}_{iso}(h_8 - h_{11})$$

$$\Rightarrow \dot{m}_{iso} = 191.81 \times (640.09 - 377.04)/(755.05 - 276.65) = 105.46\,\text{kg/s}.$$

b. The power outputs from the steam turbine and the binary cycle are

$$\dot{W}_{T,steam} = \dot{m}_3(h_3 - h_4) = 38.19 \times (2748.1 - 2344.7) = 15.410\,\text{MW},$$

$$\dot{W}_{T,iso} = \dot{m}_{iso}(h_8 - h_9) = 105.46 \times (755.05 - 691.01) = 6.753\,\text{MW},$$

$$\dot{W}_{net,binary} = \dot{W}_{T,iso} - \dot{m}_{iso} w_{p,in} = 6.139\,\text{MW}.$$

c. The thermal efficiencies of the binary cycle and the combined plant are

$$\dot{Q}_{in,binary} = \dot{m}_{iso}(h_8 - h_{11}) = 50.454\,\text{MW},$$

$$\eta_{th,binary} = \frac{\dot{W}_{net,binary}}{\dot{Q}_{in,binary}} = 12.2\%,$$

$$\left.\begin{array}{l} T_0 = 25\,^\circ\text{C} \\ x_0 = 0 \end{array}\right\} h_0 = 104.83\,\text{kJ/kg},$$

$$\dot{E}_{in} = \dot{m}_1(h_1 - h_0) = 203.622\,\text{MW},$$

$$\eta_{th,plant} = \frac{\dot{W}_{T,steam} + \dot{W}_{net,binary}}{\dot{E}_{in}} = 10.6\%.$$

9.4 Solar Thermal Energy Applications

The use of solar energy has grown significantly over the past few decades, including solar photovoltaics, solar heat to power heat engines, and solar heat for other applications. An interesting implementation of Rankine cycles is in concentrated solar power (CSP), in which solar radiation is collected and concentrated using special collector-concentrators and used as a moderate-temperature heat source for a heat engine. Some of the heat can be stored in different forms (steam, hot salt, bricks, etc.) for use after hours, giving this technology a potential low-cost storage advantage. Figure 9.5 shows a generic schematic for a CSP system in which the **solar field**, which consists of a large number of parabolic solar collectors/concentrators, or **parabolic troughs**, is used to raise the temperature of a **heat transfer fluid** circulating between the solar collectors and the power island [1]. Instead of a boiler, a heat exchanger is used to transfer the heat from the heat transfer fluid to the working fluid of the power cycle, in this case water/steam. Parabolic solar collectors have been used to raise the heat transfer fluid temperature to 300–400 °C. Some of the extra heat can be stored for use when solar energy is low or during evening operation. Moreover, an extra burner can be used when no solar heat is available or to supplement the solar energy, making it a hybrid system that integrates solar and fossil energy. This trough-based solar field has been used extensively around the world.

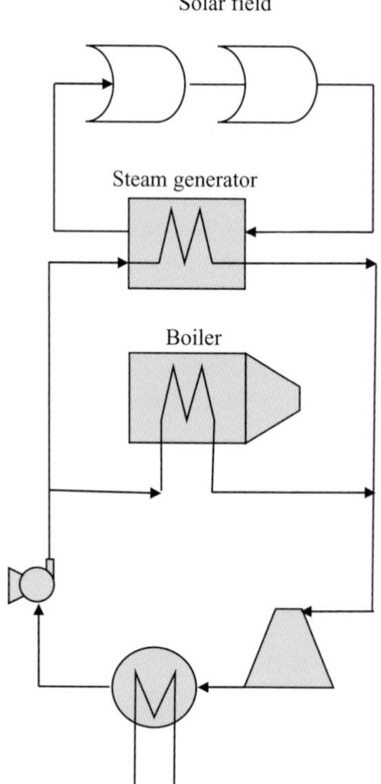

Figure 9.5 A hybrid natural gas–solar thermal electric power plant that uses solar concentrators to heat up the working fluid, and uses natural gas as a backup fuel when solar heat is not sufficient. In the picture shown the collector/concentrator is a parabolic trough; they are arranged in series to heat up a heat transfer fluid. The heat transfer fluid is sent to a steam generator for the power island.

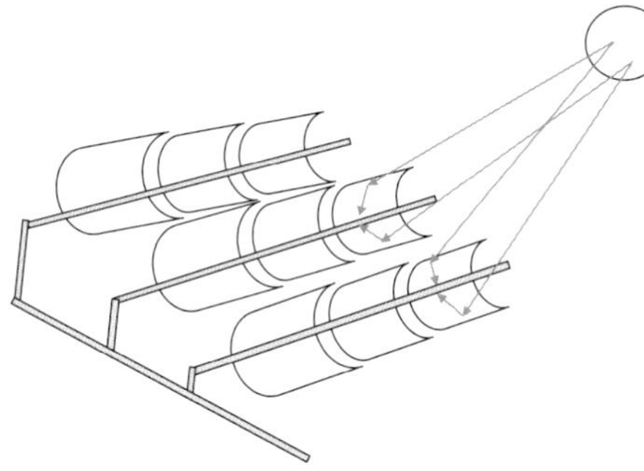

Figure 9.6 A series of parabolic collectors used to heat a heat transfer fluid flowing in a transparent pipe located along the focal area of the concentrator.

A trough collector/concentration is shown in Figure 9.6. It consists of a 2D parabolic mirror/reflector used to concentrate the solar energy onto its line of focus, where a transparent pipe carrying the heat transfer fluid is mounted. The pipe is transparent to the incoming solar radiation, but more opaque to outgoing radiation (at the temperature of the fluid), and in some cases insulated to reduced convective losses (e.g., by incorporating a vacuum space between an inner pipe where the fluid flows and an outer glass pipe). Trough collectors have moderate concentration ratios (defined next) and are surface collectors (in contrast to volume collectors, discussed later) because the surface area of the collector where the heat transfer fluid flows is parallel to the surface area of the reflector. A parabolic trough can track the sun by rotating around the focal line. They can achieve low to moderate fluid temperature because of the moderate concentration and relatively large heat loss surface area of the collector.

The average solar energy density varies significantly by geographic location, as shown in Figure 9.7, where the annual average irradiance is depicted in kWh/m^2. The average in sunny regions is ~0.1–0.3 kW/m^2 (with a maximum of ~2.7 $MWh/m^2/y$ or ~7 $kWh/m^2/day$), which is viable for application of CSP technology. For an overall plant efficiency (defined as the electricity produced per unit of solar energy received) of 20%, the plant generates ~60 MWe/km^2. Indeed, Nevada 1 CSP was built on ~1.3 km^2 for 64 MWe power (with a capacity factor of 25%). The overall plant efficiency is the product of the power cycle efficiency and the solar field efficiency, to be defined next. Given the relatively low-temperature heat available in CSP plants, typically 300-400 °C, the Rankine cycle efficiency of these plants tends to be lower than 30%. The capacity factor (the fraction of time the plant can operate) is limited by the fraction of time sufficient solar irradiance is available, and while it varies by location, it averages ~25%. Like other solar energy technologies, storage or backup power is necessary.

9.4.1 Collector/Concentrator Temperature Analysis

Solar concentrators are used to raise the temperature of the heat transfer fluid; they collect the solar radiation and concentrate it by focusing the radiation toward an area where the heat transfer fluid is flowing. The concentration ratio is defined as $C_R = A_{conc}/A_{col}$ – that is, the

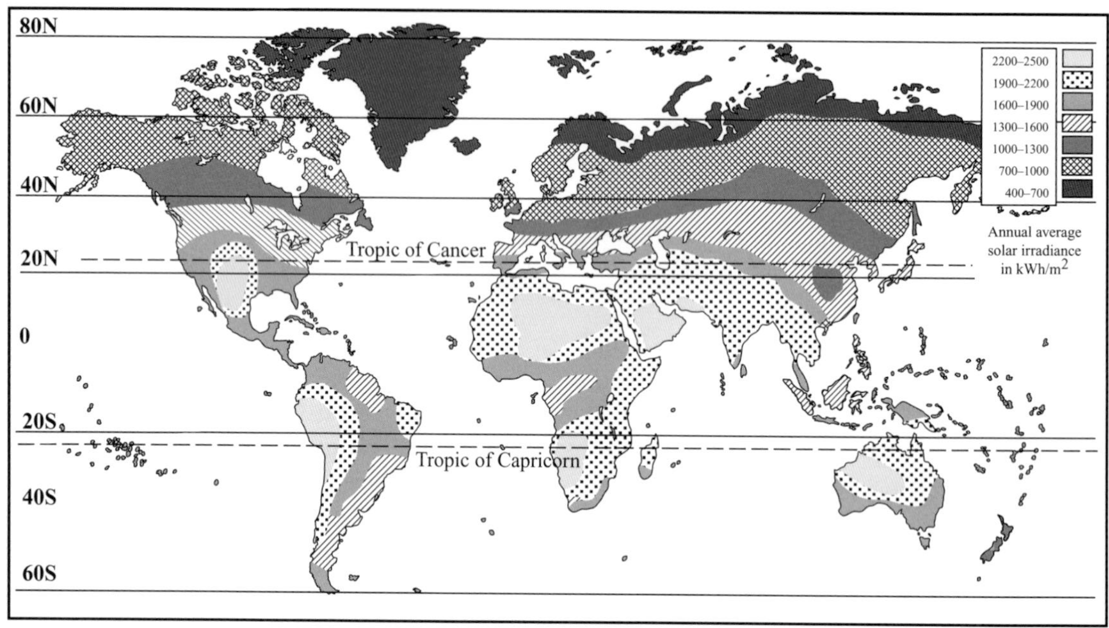

Figure 9.7 Annual average solar irradiance measured in kWh/m^2; data from the National Renewable Energy Laboratory.

receiver/reflector/concentrator area, A_{conc}, divided by the collector/absorber area, A_{col}. C_R ranges between 1 (for flat collectors) to ~10,000 for dish collectors. Achieving higher temperatures is important for raising the plant efficiency and improving the economics. The need for concentration becomes clear by reviewing some relations for the energy/power delivered by the collector/concentrator. The thermal power absorbed by the heat transfer fluid is given by:

$$\text{net absorbed flux}: \quad qA_{col} = \beta A_{conc} I - \hat{h} A_{col}(T_c - T_a),$$

where q is the heat flux received by the collector and absorbed by the fluid, β is a factor that depends on the reflective and transmissive properties of the glass cover of the heat transfer fluid pipe and the absorptive properties of the collector, I is the incident radiation, $\hat{h}$ is the heat transfer coefficient between the collector and the environment accounting approximately for all modes of heat loss (conduction, convection, and radiation), T_c is the temperature of the collector, and T_a is the environment temperature. Collectors are typically designed to be spectrally selective (i.e., can absorb energy without much reflection or emission). They have a low heat transfer coefficient. The maximum temperature of the collector is obtained by setting $q = 0$,

$$(T_c)_{max} = T_a + \frac{\beta I C_R}{\hat{h}}, \tag{9.1}$$

and the collector efficiency is:

$$\eta_{col} = \frac{q}{IC_R} = \beta - \frac{\hat{h}(T_c - T_a)}{IC_R}, \tag{9.2}$$

also known as the **solar field efficiency**, defined as the thermal energy added to the heat transfer fluid per unit of solar energy received. Note how both the maximum temperature and the solar field efficiency increase with the concentration ratio. However, at high temperature the losses increase rapidly, especially for lower-concentration ratio collectors and the efficiency drops very fast. This poses a challenge in selecting a matching power cycle since the power cycle efficiency increases with increases in the high temperature of the cycle [2]. Another collector efficiency model based on radiative heat transfer as the primary heat exchange mechanism will be developed in the next section, and will be used to demonstrate the impact of the collector/concentrator temperature on the collection efficiency and the overall system's efficiency.

Parabolic troughs that can track the sun by rotating around the collector axis have been used widely, but more and more **tower** collectors are being installed [3]. In tower technology, shown schematically in Figure 9.8, a **heliostat** – that is, a system of reflective mirrors that mostly surround the tower – is used to concentrate the solar energy onto the top of a tower, where the heat transfer fluid is heated. These systems are capable of achieving higher concentration ratio and higher field efficiency at temperatures higher than those reached by parabolic troughs. These are essentially volume collectors; the small volume where the heat transfer fluid receives the concentrated solar radiation from nearly the entire heliostat field. The larger the heliostat field, the higher the concentration ratio, but this is limited by the challenge of focusing a large array of mirrors onto a single point, especially while tracking the sun in two directions. The smaller concentration volume at the top of the tower raises the concentration ratio and reduces the heat loss area, contributing to the higher overall efficiency of these collectors and of systems using them in the solar field, leading to increasing use of these towers. While large and tall towers with large-area heliostats have

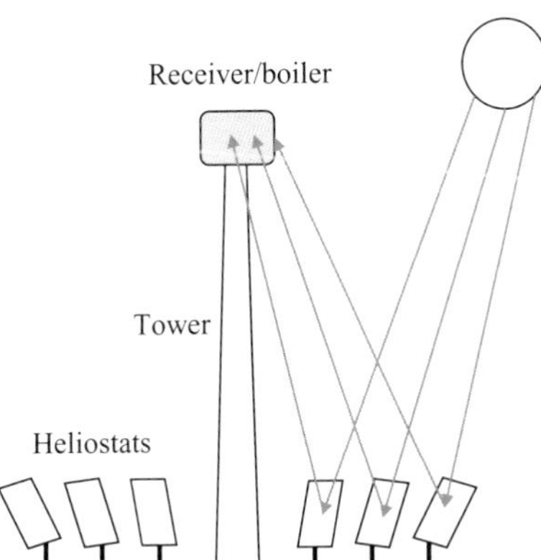

Figure 9.8 A schematic showing a solar tower with a central receiver at the top, and a heliostat field on the ground.

been built and operated for more than 20 MWe, smaller modular systems have also been proposed. The latter could reduce the cost by being fabricated in an assembly line fashion.

Even higher concentration ratio can be achieved using a dish concentrator, shown schematically in Figure 9.9, in which a 3D mirror shaped in the form of a spherical section is used to concentrate the solar radiation onto a single focal point where heat is collected, also tracking the sun in two dimensions. Two-dimensional tracking, higher concentration ratios, and the fixed location of the concentrator (mirror) with respect to the collector (the focal point) improves the efficiency of dish systems. These units, however, have limited capacity because of their size, and have not been scaled beyond demonstration. Dish collectors, because of their design, have been paired with small Stirling engines fixed to the disk focal point. On the other hand, they could be fabricated in large numbers and connected in series for large-scale power generation.

Some characteristics of concentrated solar plants are summarized in Table 9.1, showing the concentration ratio, heat transfer fluid maximum temperature, and, when paired with a heat engine, the overall plant efficiency. One of the largest CSP plants was built in Dry Lake, California is 2015 using power tower technology. The overall capacity (made up of several units) is 392 MW, running at a capacity factor of 28.72%. The plant was built on 4000 acres using 173,500 mirrors.

Table 9.1 Three solar collector/concentrator technologies and their important characteristics, including the overall solar to electricity efficiency or the collectors when paired with heat engines [1]

CSP technology	Concentration ratio	Tracking requirement	Operating temperature	Solar to electric efficiency
Power tower	500–1,000	Two-axis heliostats	400–600 °C	12–18%
Parabolic trough	10–100	One-axis reflector	100–400+ °C	8–12%
Dish – engines	600–3000	Two-axis	600–1500 °C	15–30%

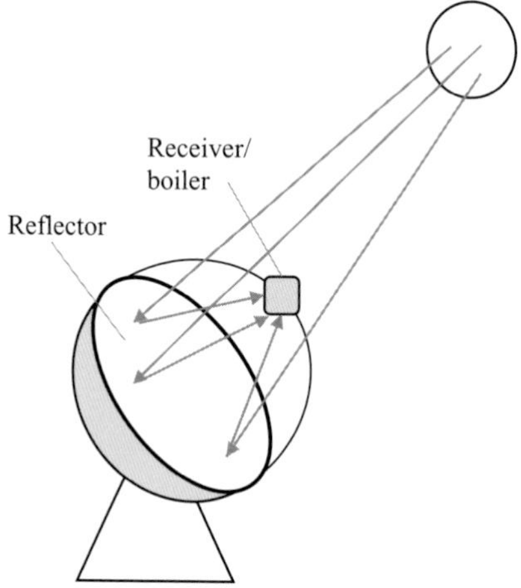

Figure 9.9 A schematic showing a solar dish collector with an engine mounted at the focal point.

Example 9.2

A proposed solar thermal power plant uses parabolic trough collectors to focus the sun's rays onto absorber tubes, which contain the working fluid for a power cycle. The absorber tubes are located at the focal point of the parabolic troughs, such that $D_m = F$, and can be considered isothermal in the axial, radial, and tangential directions. Thermal energy can be lost from the absorber tubes via convective and radiative heat transfer to the environment.

a. Develop an expression for the net flux collected by a solar trough collector as a function of the collector temperature. Assume the convective heat transfer coefficient is temperature-independent. Recall that the rate of radiative flux from a body to its surroundings can be expressed as $\dot{q} = \varepsilon\sigma\left(T^4 - T_{amb}^4\right)$, where ε and σ are the emissivity of the body and the Stefan–Boltzmann constant, respectively.

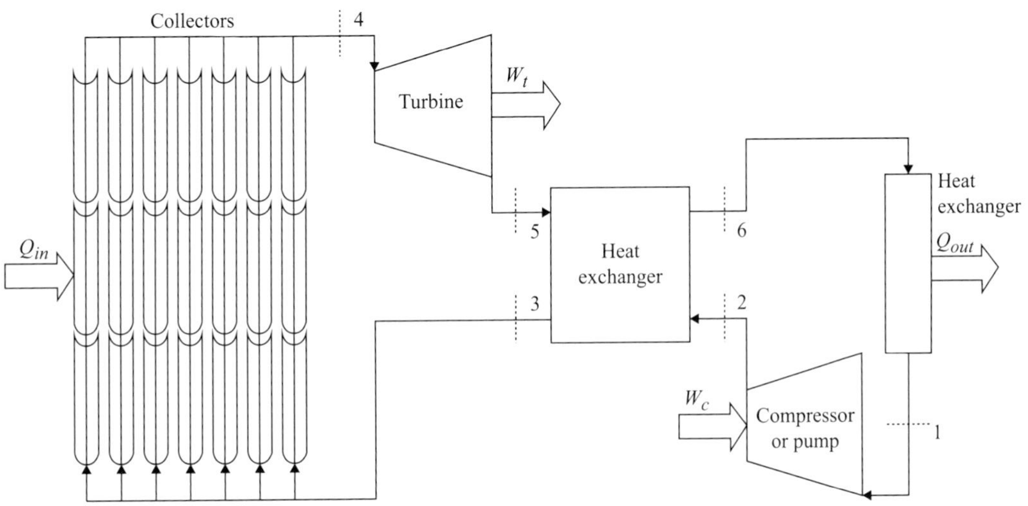

Figure E9.2a

One of the proposed plants is a supercritical CO_2 Brayton cycle, where CO_2 enters the compressor at its critical point. The pressure ratio across the turbomachinery is 3. The hot working fluid exiting the turbine is used to preheat the fluid entering the absorber tubes. The temperature difference between the streams at the "cold end" of this heat exchanger is 15 °C. The isentropic efficiencies of the compressor and turbine are 85% and 95%, respectively. The developers of the plant wish to run the plant at high efficiency, and for this reason are considering three different absorber tube (collector) temperatures: 200 °C, 400 °C, and 600 °C. The working fluid would leave the collectors and enter the turbine at these temperatures.

Example 9.2 (cont.)

b. Which of these collector operating temperatures gives the highest overall power plant efficiency?

c. Draw a *T–s* plot of the most efficient cycle found in part (b).

An alternative proposal operates the collectors at 300 °C and uses a steam Rankine cycle for thermomechanical conversion. Water leaves the collectors as a saturated vapor at 300 °C and enters a turbine, where it is expanded to 10 kPa. After heat exchange in the condenser, it enters the pump as a saturated liquid. The isentropic efficiencies of the pump and turbine are 85% and 95%, respectively.

d. Compare the overall efficiency of this plant with that of the supercritical CO_2 plant.

e. For a 100 MW_e power plant, calculate the mass flow rates of working fluid for the most efficient CO_2 Brayton cycle from part (b) and the H_2O Rankine cycle from part (d).

f. Find the maximum volumetric flow rates of working fluid for the two plants considered in part (e). What do these values suggest about the relative sizes of the turbomachinery needed for the plants?

g. Find the total collector area needed for the two plants considered in part (e).

Useful Data

$\beta = 0.8$, fraction of solar radiation absorbed
$I = 900\,\text{W/m}^2$, solar irradiance
$h_{conv} = 25\,\text{W/m}^2\cdot\text{K}$, convective heat transfer coefficient (independent of temperature)
$T_a = 30\,^\circ\text{C}$, air temperature
$C_R = 107.5 D_m/F$, concentration ratio
$D_m/F = 1$, ratio of mirror dimension (radius) to mirror focal length
$\varepsilon_c = 1$, absorber tube (collector) emissivity
$\sigma = 5.67 \times 10^{-8}\,\text{W/m}^2\cdot\text{K}^4$, Stefan–Boltzmann constant

Thermophysical property data for CO_2 and H_2O can be obtained from the NIST webbook at: http://webbook.nist.gov/chemistry/fluid.

Solution

a. Let us write the First Law for a collector:

$$\dot{q}_{net} = \dot{q}_{solar} - \dot{q}_{conv} - \dot{q}_{rad}$$

where, $\dot{q}_{solar} = \beta I C_R$, $\dot{q}_{conv} = h_{conv}(T_c - T_a)$, $\dot{q}_{rad} = \varepsilon\sigma\left(T_c^4 - T_a^4\right)$.

so

$$\boxed{\dot{q}_{net} = \beta I C_R - h_{conv}(T_c - T_a) + \varepsilon\sigma\left(T_c^4 - T_a^4\right)}\,.$$

Example 9.2 (cont.)

b. Power plant efficiency can be defined as follows: $\eta_{plant} = \eta_{coll}\eta_{cycle}$. We know collector efficiency is

$$\eta_{coll} = \beta - \frac{h_{conv}(T_c - T_a) + \varepsilon\sigma\left(T_c^4 - T_a^4\right)}{C_R I}.$$

From the cycle diagram, we can see that cycle efficiency is

$$\eta_{cycle} = \frac{W_{out}}{Q_{in}} = \frac{W_t + W_c}{Q_{in}}, \text{ or } \eta_{cycle} = \frac{h_4 - h_5 + h_1 - h_2}{h_4 - h_3}.$$

Going through the cycle analysis, we find (from the NIST webbook) at $T_c = 200\,°\text{C} : \eta_{coll} = 73.2\%$:

Table E9.2.1

State	Temperature (°C)	Pressure (bar)	Enthalpy (kJ/kg)	Entropy (J/g·K)
1	31.03	73.8	329.0624	1.422996
2	76.02572	221.4	357.9952	1.435458
3	80.15719	221.4	368.0471	1.464041
4	200	221.4	591.1419	2.015882
5	98.67925	73.8	522.8596	2.02556
6	91.02572	73.8	512.8076	1.99819

Substituting in the cycle efficiency expression: $\eta_{cycle} = 17.6\%$ and $\boxed{\eta_{plant} = 12.9\%}$,

at $T_c = 400\,°\text{C}, \;\; \eta_{call} = 58.9\%$:

Table E9.2.2

State	Temperature (°C)	Pressure (bar)	Enthalpy (kJ/kg)	Entropy (J/g·K)
1	31.03	73.8	329.0624	1.422996
2	76.025724	221.4	357.9952	1.435458
3	188.9161	221.4	575.1346	1.981716
4	400	221.4	847.8825	2.47027
5	278.48171	73.8	729.947	2.4747
6	91.025724	73.8	512.8076	1.99819

Substituting in the cycle efficiency expression: $\eta_{cycle} = 32.6\%$, and $\boxed{\eta_{plant} = 19.2\%}$,

at $T_c = 600\,°\text{C}, \eta_{coll} = 31.7\%$:

Example 9.2 (cont.)

Table E9.2.3

State	Temperature (°C)	Pressure (bar)	Enthalpy (kJ/kg)	Entropy (J/g·K)
1	31.03	73.8	329.0624	1.422996
2	76.02572	221.4	357.9952	1.435458
3	346.158	221.4	781.2209	2.367016
4	600	221.4	1096.2	2.7931
5	458.1172	73.8	936.0333	2.799187
6	91.02572	73.8	512.8076	1.99819

Substituting in the cycle efficiency expression: $\eta_{cycle} = 41.7\%$, and $\boxed{\eta_{plant} = 13.2\%}$.

It is clear, therefore, that the best option is to run the collectors at 400 °C.

c. The T–s diagram is shown in Figure E9.2b.

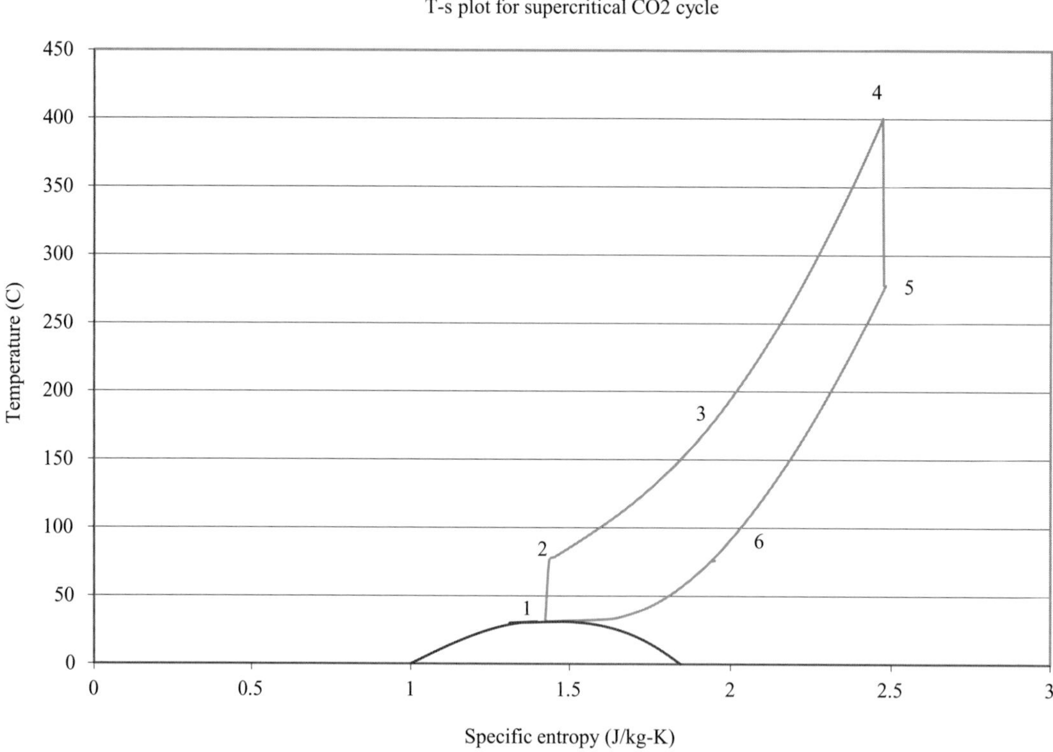

Figure E9.2b

Example 9.2 (cont.)

d. Again, the efficiency of the plant is defined as $\eta_{plant} = \eta_{coll}\eta_{cycle}$. Note that no recuperation is possible with steam in the Rankine cycle as the turbine outlet temperature is lower than the pump outlet temperature.

At $T_c = 300\,°\text{C}$, and $\eta_{coll} = 67.2\%$:

Table E9.2.4

State	P (Pa)	T (C)	h (kJ/kg)	s (kJ/kg·K)	x
1	1.00E+04	45.81	191.81	0.6492	0
2	8.59E+06	46.46	201.9747	0.654	–
3	8.59E+06	300	2749.6	5.7059	1
4	1.00E+04	45.81	1851.952	5.85402	0.694013

Substituting in the cycle efficiency expression: $\eta_{cycle} = 34.8\%$, and $\boxed{\eta_{plant} = 23.4\%}$.

It is clear that in this case, the efficiency of the steam Rankine cycle is higher than the CO_2 Brayton cycle.

e. We can relate the specific enthalpies to the net power output by the following:

$$\dot{m}_{CO_2} = \frac{\dot{W}_{net}}{h_4 - h_5 + h_1 - h_2} \text{ and } \dot{m}_{H_2O} = \frac{\dot{W}_{net}}{h_3 - h_4 + h_1 - h_2},$$

$$\text{therefore, } \boxed{\dot{m}_{CO_2} = 1123.6\,\text{kg/s, and } \dot{m}_{H_2O} = 112.7\,\text{kg/s}}.$$

f. The volumetric flow rate can be found from: $\dot{V} = \dot{m}v$ (where v = specific volume). First, we must determine where the highest specific volumes occur in the two cycles. At low pressures and high temperatures (or enthalpies), fluids have lower densities, and hence higher specific volumes. Therefore, the highest specific volumes are found at the turbine outlets. For the two fluids:

$$v_{max,CO_2} = 0.0137\,\text{m}^3/\text{kg}, \text{ and } v_{max,H_2O} = 10.18\,\text{m}^3/\text{kg}$$

$$\text{Therefore, } \boxed{\dot{V}_{max,CO_2} = 15.41\,\text{m}^3/\text{s}, \text{ and } \dot{V}_{max,H_2O} = 1147\,\text{m}^3/\text{s}}.$$

The required volumetric flow rate of H_2O is almost two orders of magnitude larger than that for CO_2. This suggests that for similar velocities in the turbomachinery, the equipment for the steam Rankine cycle needs to have ~100 times the cross-sectional area. This is a very rough measure of relative equipment size.

Example 9.2 (cont.)

g. We can use the efficiency expression of the plant to find the collector trough area needed:

$$\eta_{plant} = \frac{\dot{W}_{net}}{\dot{Q}_{in}} = \frac{\dot{W}_{net}}{IA_{solar}}, \text{ and } A_{solar} = \frac{\dot{W}_{net}}{I\eta_{plant}}.$$

Solving for the two plant designs, we find:

$$\boxed{A_{Brayton} = 0.578\,\text{km}^2 \text{ and } A_{Rankine} = 0.475\,\text{km}^2}.$$

9.4.2 Overall System Efficiency

It should be noted that the power cycle efficiency, η_{cycle}, increases with the heat source temperature, while that of the solar fuel moves in the opposite direction. Thus, maximizing the product $\eta_{cycle}\eta_{col}$ is necessary. This is illustrated by looking at the efficiency of an ideal absorber (with no convective losses) for which the efficiency, η_{abs}, is given by the balance between the energy received and the energy radiated [4]:

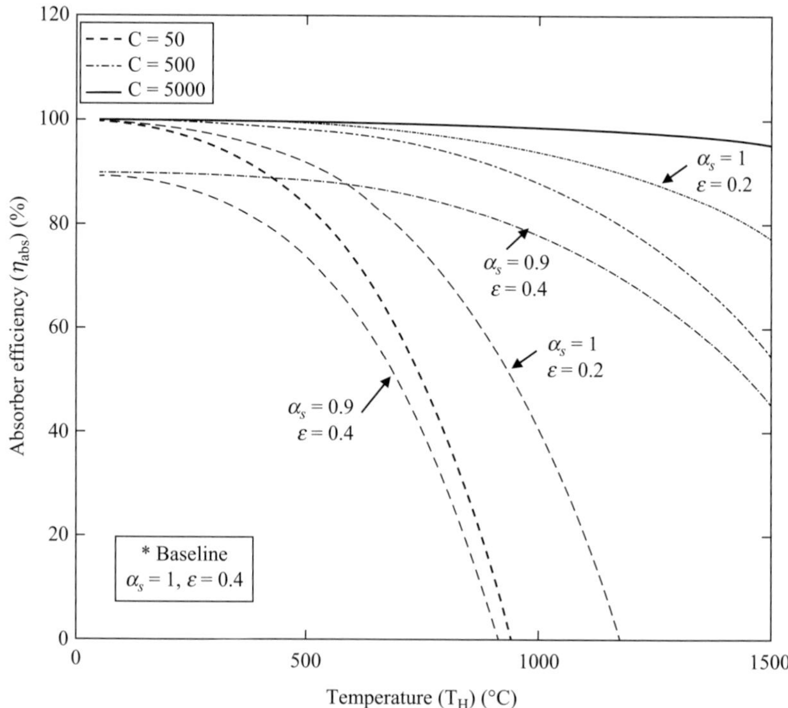

Figure 9.10 The dependence of the receiver/concentrator/absorber efficiency on its temperature for different concentration ratio, absorptance, and emittance. C is the concentration ratio. Unlabeled curve correspond to the baseline.

$$\eta_{abs} = \frac{\alpha_s I C_R - \varepsilon \sigma T_{rec}^4}{I C_R} = \alpha_s - \frac{\varepsilon}{C_R} \frac{\sigma T_{rec}^4}{I}, \tag{9.3}$$

where α_s is the absorptance of the collector (replacing β in (9.2)), ε is the emittance, and T_{rec} is the receiver temperature (assumed to be the same as the working fluid temperature) (Figure 9.10).

The overall efficiency of the system using a Second Law efficiency η_{II} is

$$\eta_{sys} = \eta_{abs} \eta_{II} \left(1 - \frac{T_o}{T_{rec}}\right). \tag{9.4}$$

The system maximum efficiency with respect to the receiver temperature can be obtained by differentiating the efficiency with respect to the receiver temperature and equating to zero to find the receiver temperature that maximizes the system efficiency. Figure 9.11 show the dependency of the total system efficiency on the parameters.

Example 9.3 shows the detailed calculations for the coupled steam Rankine cycle and solar collector, and compares it with a coupled CO_2 cycle and the solar collector. The following examples shows calculations for a dish collector coupled with a Stirling engine.

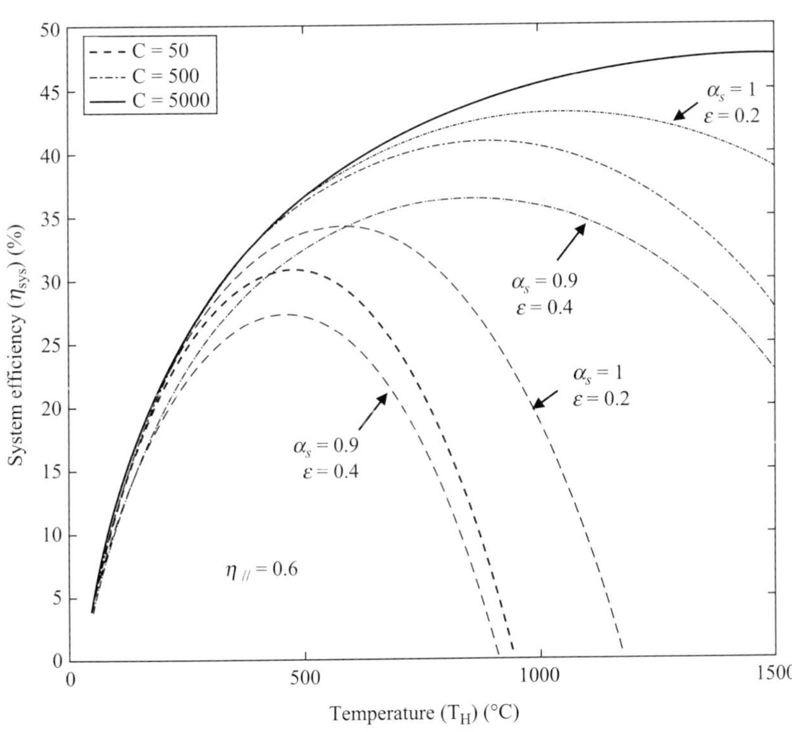

Figure 9.11 The impact of the receiver temperature, T_H, assumed to be the same as the high temperature of the working fluid, on the overall system efficiency, shown for different values of the concentration ratio C, the effective absorptance, and the emittance of the collectors, α_s and ε, respectively. Unlabeled lines correspond to the baseline values $as = 1$, $e = 0.4$.

Example 9.3

A dish solar collector is paired with a Stirling engine for electricity generation. The Stirling engine operates by interacting with a high-temperature heat source, in this case a solar collector, and a low-temperature heat sink, in this case fins that transfer heat to the environment. There is also a regenerator that transfers heat from the hot side to the cold side of the engine, in between the power strokes. The solar irradiation is $I = 300\,\text{W/m}^2$, the collector irradiation efficiency is $\beta = 0.8$. The collector temperature is maintained at $T_c = 1000\,\text{K}$. At this temperature the combined convective/radiative heat transfer coefficient from the collector is $\hat{h} = 100\ \text{W/m}^2\cdot\text{K}$. The temperature of the environment is $T_o = 300\,\text{K}$. The heat transfer flux to the high-temperature reservoir of the Stirling engine is given by: $\mathbf{q} = \beta \times I \times C_R - \hat{h}(T_C - T_0)$, where $C_R = 1200$ is the concentration ratio for dish-type solar collectors, and defined as the ratio between the collector surface area and the receiver surface area. Thus, the total incident flux on the solar dish panel is $I \times C_R$.

Figure E9.3a shows a schematic representation of solar collector–engine system using a Stirling engine.

Figure E9.3a

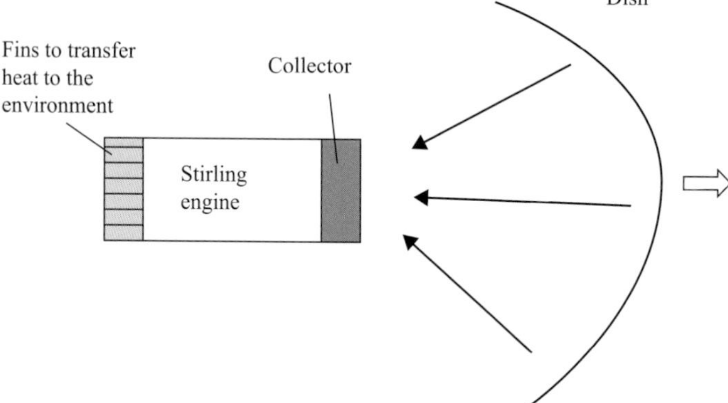

The engine has a compression ratio of $r_C = 2$. The working fluid for this engine is air, assumed to behave as an ideal gas with $R = 0.287\,\text{kJ/kg}\cdot\text{K}$, and $c_v = 0.782\,\text{kJ/kg}\cdot\text{K}$. Under ideal conditions, the working fluid undergoes following four reversible processes:

1. Isothermal compression (process 1–2) at the cycle's lowest temperature.
2. Isochoric (constant volume) heating (process 2–3) via internal regeneration.
3. Isothermal expansion (process 3–4) at the cycle's highest temperature.
4. Isochoric (constant volume) cooling (process 4–1) via internal regeneration.

Part I

a. Calculate the net rate of energy absorbed by the collector (W/m^2). What is the overall collector efficiency, defined as the absorbed flux as a fraction of the collector incident flux?

Example 9.3 (cont.)

b. Derive expressions for heat input/kg-air to the engine and net work done/kg-air by the ideal Stirling engine in terms of r_C, T_H, and/or T_C. Calculate heat input/kg-air and net work done/kg-air.

c. Derive an expression for the ideal Stirling cycle thermal (First Law) efficiency. How does the ideal Stirling cycle efficiency compare with the Carnot cycle efficiency? Briefly explain your comparison.

d. What is the overall efficiency of the solar collector–engine system? If this ideal Stirling engine produces 20 kW, what is the collector area?

Part II

The actual Stirling engine suffers from two forms of losses due to the finite rate of internal heat transfer and the surface areas of the heat exchangers:

1. The highest temperature of the cycle T_{HR} is 20 K below that of the concentrator and the lowest temperature T_{LR} is 30 K above the external temperature.
2. The engine has an imperfect regenerator, and the maximum temperature that can be achieved during the isochoric process is 30 K less than T_{HR}.

Additional heat from the external source is provided to raise the temperature to T_{HR} following regeneration. Similarly, more heat is removed by designing fins for increased heat exchange with the surrounding air that lowers the temperature to T_{LR}.

We will assume that the expansion and compression remain isothermal at T_{HR} and T_{LR}, respectively.

e. Draw this non-ideal Stirling cycle on a T–s and p–v diagram. Label the processes in the diagram.

f. Derive expressions for the total heat input and net work output for the non-ideal Stirling engine. Express your answer in terms of r_C, the regeneration efficiency defined as the actual temperature rise due to regeneration divided by the maximum temperature rise between the low and high temperatures of the cycle, η_R, T_H, T_C, η_{Carnot}, and/or γ (isentropic index of air), and the temperature increments between the working fluid and the receiver and the fins.

g. Calculate for heat input/kg-air to the system and net work done/kg-air. $\gamma = 1.4$ for air.

h. What is the overall efficiency of the solar collector–engine system under these conditions? What is the collector area needed to produce 20 kW?

i. Is the system operating at maximum overall efficiency?

Example 9.3 (cont.)

Solution

Part I

a. $q = \beta \times I \times C_R - \hat{h}(T_C - T_0) = 218{,}000\,\mathrm{W/m^2}$,

$q_i = I \times C_R = 300 \times 1200 = 360{,}000\,\mathrm{W/m^2}$

$\eta_{COL} = q/q_i = 218{,}000/360{,}000 = 0.6056$ or 60.56%.

b. Heat input to the system causes the expansion of fluid during the expansion stroke. Therefore: $q_{IN} = \int_{v_2}^{v_1} p\,dv$.

Using the ideal gas law: $pv = RT \Rightarrow p = \frac{RT}{v}$.

Therefore, heat input to the system: $q_{IN} = \int_{v_2}^{v_1} \frac{RT}{v}\,dv = RT_H \int_{v_2}^{v_1} \frac{dv}{v} = RT_H \ln\left(\frac{v_1}{v_2}\right)$.

Using the definition of the compression ratio $r_C = v_1/v_2$, we get the heat input to the system as

$$q_{IN} = RT_H \ln(r_C) = 287 \times 1000 \times \ln(2) = 198{,}933\,\mathrm{W/kg\text{-}air}.$$

Net work done is equal to the difference in the heat input from the high-temperature reservoir and the heat rejected to the low-temperature reservoir:

$$w_{net} = q_{IN} - q_{OUT}.$$

Following similar derivation for q_{out} as used for q_{in}, we get $q_{out} = RT_L \ln(r_c)$.

Therefore, $w_{net} = R\ln(r_C)(T_H - T_L) = 139{,}253$ W/kg-air.

c. $\eta_{ST} = \frac{w_{net}}{q_{in}} = \frac{R\ln(r_C)(T_H - T_L)}{R\ln(r_C)T_H} = \frac{(T_H - T_L)}{T_H} = 1 - \frac{T_L}{T_H} = \eta_{Carnot} = 0.70$ or 70%.

The ideal Stirling cycle efficiency is equal to the Carnot cycle efficiency for two reasons:

1. The Stirling engine also interacts with heat source and sinks at constant temperature.
2. The regenerator is perfect; that is, the heat interaction between the Stirling engine with the regenerator during the isochoric process is perfect.

d. The overall solar collector–engine efficiency is

$$\eta_{OVERALL} = \frac{W_{OUT}}{Q_{SOLAR}} = \eta_{ST} \times \eta_{COL} = 0.70 \times 0.6056 = 0.4239 \text{ or } \mathbf{42.4\%}.$$

$$\Rightarrow Q_{SOLAR} = 20/0.4388 = 47.18\,\mathrm{kW} = q_i \times A_{COL}.$$

From the solution in (a), we have $q_i = 300 \times 1200\,\mathrm{W/m^2}$. Solving the above equation:

$$\Rightarrow \mathbf{A_{COL}} = \mathbf{0.1311}\ \mathrm{m^2}.$$

Example 9.3 (cont.)

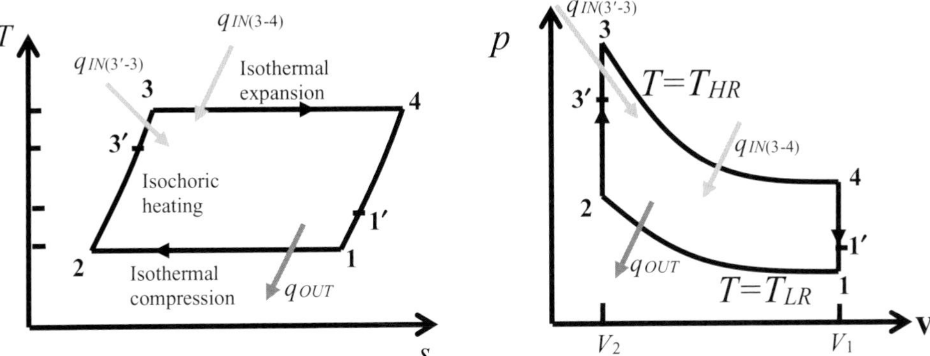

Figure E9.3b

Part II

e. The T–s and p–V diagrams of the engine are shown in Figure E9.3b.

In this cycle, heat is added into the Stirling cycle during process 3′–3 (heat addition from external source) and during isothermal expansion. The regenerator exchanges heat during the process 2–3′ and 4–1′. Additional cooling from 1′–1 is obtained by using a fan for increased heat exchange with the surroundings.

f. The regenerator efficiency is defined as:

$$\eta_R = \frac{(T_{HR,a} - T_{LR})}{(T_{HR} - T_{LR})} = \frac{(T_{HR} - 30 - T_{LR})}{(T_{HR} - T_L R)} = 1 - \frac{30}{(T_{HR} - T_{LR})} = 1 - \frac{30}{650} = 0.9538 = 95.38\%.$$

Heat input:

$$q_{IN} = q_{3'-3} + q_{3-4} = c_V(T_{HR} - T_{HR,a}) + \int_{v_2}^{v_1} \frac{RT}{v} dv = c_V(T_{HR} - T_{HR,a}) + RT_{HR} \int_{v_2}^{v_1} \frac{dv}{v}$$

$$= c_V(T_{HR} - T_{HR,a}) + RT_{HR} \ln\left(\frac{v_1}{v_2}\right)$$

$$\Rightarrow \qquad q_{IN} = c_V(T_{HR} - T_{LR} - (T_{HR,a} - T_{LR})) + RT_{HR} \ln(r_C).$$

Recognizing that $(T_{LR,a} - T_{LR}) = -(T_{HR,a} - T_{HR}) = (T_{HR} - T_{LR} - (T_{HR,a} - T_{LR})) = (T_{HR} - T_{LR})(1 - \eta_R)$

$\Rightarrow q_{IN} = c_V(T_{HR} - T_{LR})(1 - \eta_R) + RT_{HR} \ln(r_C).$

Net work done involves Pdv work and remains the same as calculated in

$$w_{net} = R \ln(r_C)(T_{HR} - T_{LR}),$$

Example 9.3 (cont.)

$$\eta_{ST} = \frac{w_{net}}{q_{in}} = \frac{(T_{HR} - T_{LR})[R\ln(r_C)]}{[c_V(T_{HR} - T_{LR})(1-\eta_R) + RT_{HR}\ln(r_C)]}$$

$$= \frac{R\ln(r_C)}{\left[c_V(1-\eta_R) + R\left[\frac{T_{HR}}{(T_{HR} - T_{LR})}\right]\ln(r_C)\right]},$$

$$\gamma = c_P/c_V = (R - c_V)/c_V = R/c_V - 1$$

$\Rightarrow R/c_V = \gamma - 1$. Also, $\eta_{Carnot,R} = 1 - \frac{T_{LR}}{T_{HR}}$. Using this in the above equation, we have:

$$\eta_{ST} = \frac{(\gamma-1)\ln(r_C)}{\left[(1-\eta_R) + \frac{(\gamma-1)}{\eta_{Carnot,R}}\ln(r_C)\right]} = 0.592 = 59.2\%.$$

g. Using $T_H = 980$ K and $T_L = 330$ K, heat input is given as per the equation above:

$$q_{IN} = c_V(T_{HR} - T_{LR})(1-\eta_R) + RT_{HR}\ln(r_c) = 218415 \text{ W/kg-air}.$$

Net work done is given by the following equation:

$$w_{net} = (T_{HR} - T_{LR})[R\ln(r_c)] = 129{,}307 \text{ W/kg-air}.$$

h. The overall solar collector–engine efficiency is

$$\eta_{OVERALL} = \frac{W_{OUT}}{Q_{SOLAR,actual}} = \eta_{ST} \times \eta_{COL} = 0.592 \times 0.6056 = 0.3585 \text{ or } \mathbf{35.85}\%$$

$$\Rightarrow Q_{SOLAR,actual} = 20/0.3585 = 55.79 \text{ kW} = q_i \times A_{COL}.$$

From the solution in (a), we have $q_i = 300 \times 1200$ W/m^2. Solving the above equation:

$$\Rightarrow \mathbf{A_{COL} = 0.155}\,\text{m}^2.$$

i. The collector efficiency does not change. The Carnot efficiency of the Stirling cycle changes due to the operating maximum and minimum temperatures. Hence, the maximum overall efficiency is

$$\eta_{max,\ overall} = \eta_{CARNOT} \times \eta_{COL} = (1 - 330/980) \times 0.6056,$$

$$\eta_{max,\ overall} = 0.4016 \text{ or } 40.16\% \quad > \eta_{overall} = 35.85\%.$$

The system is not operating at maximum overall efficiency.

9.5 Integrating Solar and Fuel Energy

The temperature of the heat transfer (or the working) fluid in CSP is not sufficiently high to power a gas turbine (unless very high concentration ratios are used that could raise the cost significantly, and hence this concept has not been applied). Plants that use troughs and towers have been built at large scale, and their utilization as central (instead of distributed) power sources that integrate well with existing electricity distribution networks has been successful. They can store thermal energy for short periods of time (in the form of high-pressure steam, molten salt, or potentially hot rock), making it possible to operate that power island (the Rankine cycle plant) when solar energy is not available. However, this is typically for approximately two hours after sunset. Moreover, it takes a few hours in the morning after sunrise to warm up the heat transfer fluid to the required temperature. Short-term storage can also be useful during partly cloudy times.

A backup combustion system can be installed to provide heat for the power island during after-hours operation instead of the more expensive option of storage. Careful integration of fossil sources with the solar energy can provide other advantages, such as better utilization efficiency, longer-term storage, and lower-cost electricity that could encourage more extensive utilization of solar energy. In particular, careful integration of both resources – solar heat and fuel thermal energy – can overcome the storage challenge at an affordable cost [5]. Some approaches to integration are reviewed next, and Figure 9.12 summarizes three of the most typical approaches, namely heating up the air before combustion in a gas turbine (known as solarized gas turbine), using the solar energy to raise more steam for the Rankine cycle in a steam or a combined cycle (CC) power plant, or using solar energy to reform the fuel before burning it in the gas turbine combustor. It is also possible to implement more than one of these approaches in the same plant.

9.5.1 Integration Approaches

The literature discusses three approaches to integrating fossil and solar thermal energy: solarized gas turbines, solar hybrid integrated cycles, and solar reforming, with a detailed review and evaluation given in [7], along with many more references that describe the process and technology. The first is shown schematically in Figure 9.13 in a conceptual drawing of a gas turbine/Brayton cycle plant integrated with a power tower that preheats the compressed air out of the compressor to temperatures in the range of 1000 K. Other designs using parabolic troughs that achieve lower temperatures, ~700 K, have been proposed. The higher the temperature of the gas leaving the solar receiver, the less fuel is used in the combustor to achieve the prescribed turbine inlet temperature (TIT) for the same power output. Moreover, adding the fuel chemical energy at higher temperatures improves the fuel utilization efficiency, as long as there is complete fuel combustion. Higher inlet combustion temperatures may pose some challenges associated with higher NOx (nitric oxides) emissions and combustion stability. Thus, alternatively, the heated air from the receiver could be added post-combustion. In this case, the combustion products are mixed with hot air from

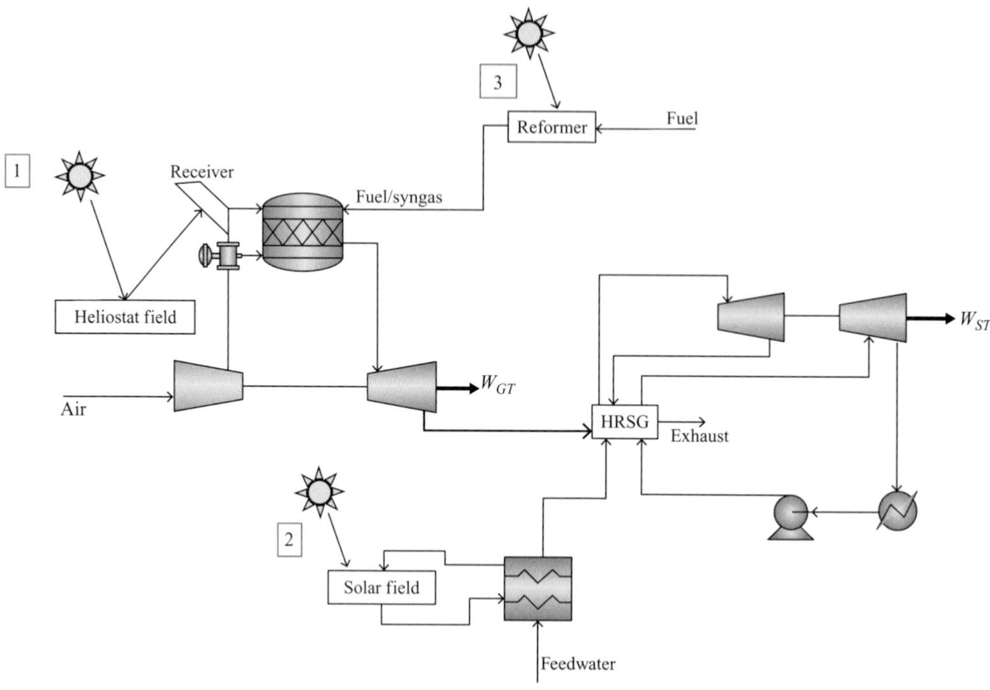

Figure 9.12 Three schemes for integrating solar energy into fossil power plants: a solarized gas turbine in which solar energy is used to heat up the compressed air, a simple Rankine cycle or CC in which the solar energy is used to raise steam, and a cycle in which solar energy is used to reform the fuel before combustion in a gas turbine combustor in a simple or combined cycle [6].

Figure 9.13 Solarized gas turbine power plant [6].

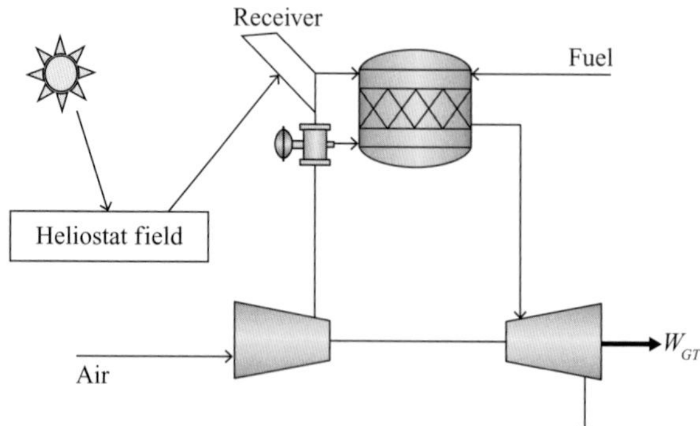

the receiver post-combustion, which is likely to lower the products temperature. The system can also operate with flexibility; the amount of fuel injected into the combustor or balance of pre- and post-combustor air can be adjusted depending on the solar energy available (time of the day, cloudiness, etc.) and load. This reduces the cost of electricity, and economic studies show favorable numbers for solarized gas turbines compared to CSP systems. The overall efficiency of these power plants has been estimated to be in the range of 35–40%, based on the total power output as a fraction of the total primary energy (incident solar energy + fuel chemical energy). These are higher than the efficiencies of conventional CSP systems because of the higher temperature achieved through the integration schemes, and are closer to gas turbine plants efficiency operating on fossil fuel alone. However, care must be taken when evaluating these efficiencies to account for hourly and daily variations of solar energy.

Other solarized gas turbine schemes have been suggested, such as using some or all of the solar energy to raise steam and mix the steam with the combustion air or products before introducing it to the gas turbine. This cycle strongly resembles the humid air cycle described previously, and anticipated efficiencies should be close to those shown for the humid air turbine (HAT) cycle. The plant can also operate using low-temperature collectors (troughs) instead of the higher-temperature volumetric receivers (towers), since raising steam can be done at lower temperatures. However, technological challenges similar to those associated with HAT cycles are anticipated, such as turbine blade corrosion and condensation in the turbine.

9.5.2 Integrated Solar Combined Cycle

Another approach is the hybrid combined cycle, or **integrated solar combined cycle** (ISCC), in which solar heat may be added to the gas turbine combustion air or to the heat recovery steam generator (HRSG), or to both, depending on the design. Most earlier studies showed that adding solar energy to the bottoming cycle (in the ISCC), in parallel (or series depending on the temperature of the solar heat) to the heat from the gas turbine exhaust achieves higher efficiency than other options, potentially because of combustion inefficiencies (incomplete burn) associated with preheating the compressed air before the gas turbine combustor, or the lower isentropic efficiencies of the turbomachinery in the Brayton cycle part of the CC. The differences were small, but plants built based on this concept indeed added solar heat to the bottoming cycle, which could also be the more economical option. Future plants could implement other arrangements, such as adding the solar energy to the compressed gas (before the gas turbine combustion) and take advantage of better gas turbine designs (high TIT and high isentropic efficiency) and better combustors (that achieve complete burn at higher temperatures).

The layout of an operating ISCC plant is shown in Figure 9.14, which was built in Egypt (similar plants have been built in other places around the world). Water from the condenser is heated in two stages (in series) by the solar energy and the gas turbine exhaust to best match the available temperature. Based on 700 W/m^2 direct normal solar irradiation (at noon on March 21), the solar field, made up of 160 parabolic trough collectors with total area of 130,800 m^2, generates thermal power rated at 50 MWt at 393 °C, with a total peak plant power of 125.7 MWe. Without the solar heat (night-time production), natural gas alone generates 103.8 MWe. A heat transfer fluid is heated by the troughs and sent to the low-temperature side of the HRSG to heat the feedwater. It returns to the solar field at 293 °C. The gas turbine is rated at 74.4 MWe, while the steam turbine is at 59.5 MWe.

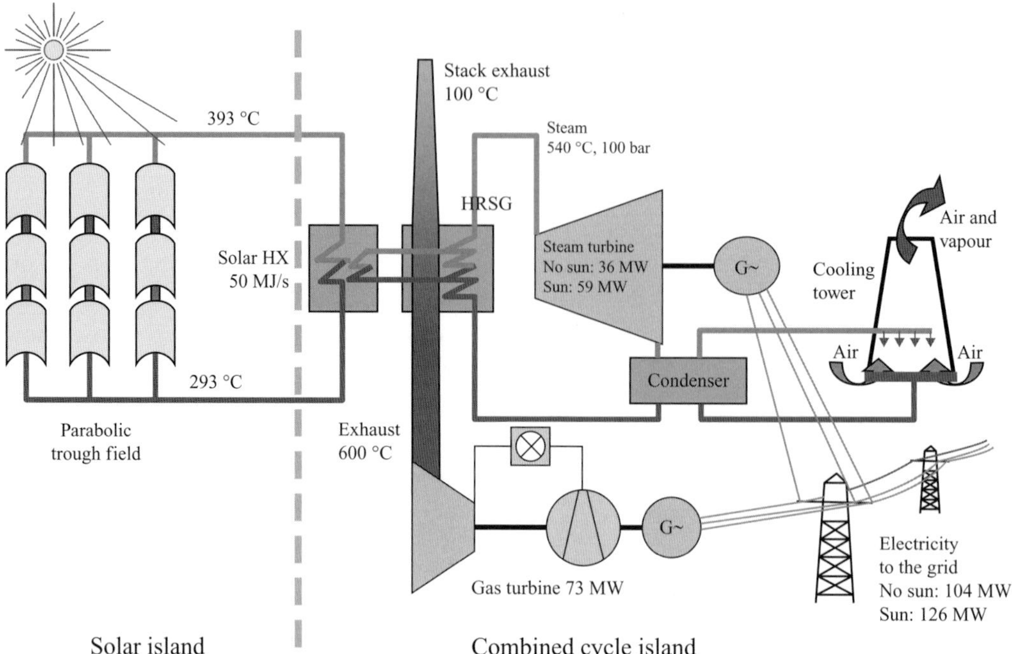

Figure 9.14 Integrated solar combined cycle plant in Kuraymat [8]. In the ISCC, solar thermal energy contributes to the generation of superheated steam for the steam cycle section of the plant. The steam condenser is cooled by water circulating between the condenser and a cooling tower.

Thus, based only on natural gas operation (without the solar part), power is divided to ~70% by the gas turbine and 30% by the steam turbine, which is typical in natural gas combined cycles (NGCCs).

ISCC plant designs could also split the solar energy between the gas turbine (preheating the compressed air before combustion) and the HRSG that supplied energy to the steam turbine. This way, it is possible to raise the overall cycle efficiency by taking advantage of the higher-temperature operation of the gas turbine while keeping the cost down. ISCC plants must implement a carefully designed dynamic control system to manage the intermittency of the solar energy and to respond to the load.

9.5.3 Solar Reforming Combined Cycle

Another concept in integrating solar energy into fossil power plants is **solar reforming**. This concept is similar to the thermochemical recuperation cycles described in Chapters 6–8, but it uses solar energy in steam reforming (or possibly dry reforming using CO_2). In this arrangement, the endothermic energy of reforming is supplied from a renewable solar resource instead of using the gas turbine exhaust energy (as is the case in thermochemical recuperation). The reformed fuel (whose heating value is higher than that of the original

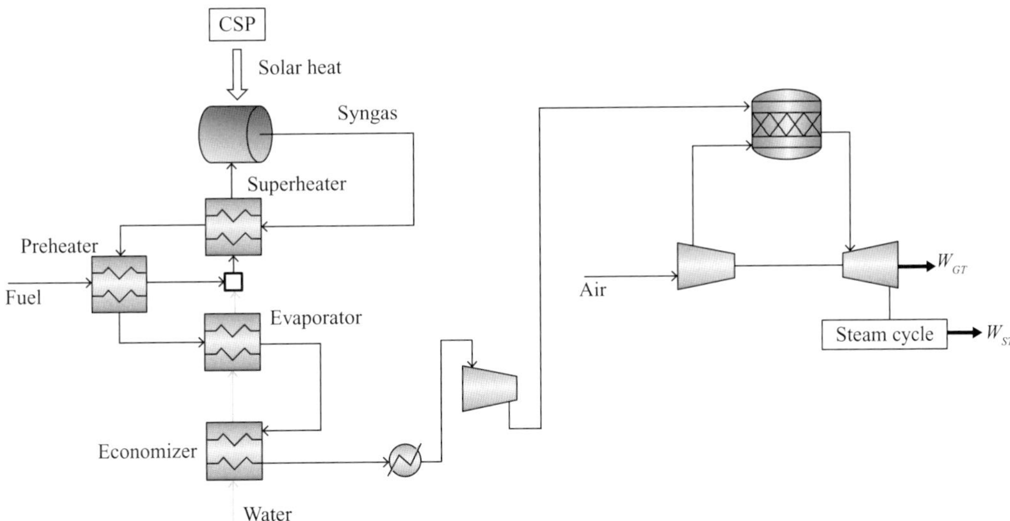

Figure 9.15 An integrated fuel reforming CC plant. Thermal energy from the solar collectors is used in the reforming of methane and production of syngas, which is used in the gas turbine [6].

fuel) consisting mostly of syngas (but with methane if the reforming temperature is relatively low) is burned in the gas turbine combustor. Some of the syngas (produced by reforming the original fuel) can also be stored for later use. Solar thermal reforming, in which solar collectors are used to supply the reforming energy, has been studied extensively [5,10]. The system shown in Figure 9.15 is just one example of how to construct such an integrated cycle. Given the increase in the heating value of the fuel, a potential fuel saving of up to 25% can be achieved using solar reforming. The overall cycle efficiency is expected to be similar to that of the thermochemical recuperation cycles described in Chapter 8.

Another advantage of using solar reforming to produce a "solar fuel" before combustion is the possibility of adding extra storage to the system – that is, storing the solar energy in the form of chemical bond energy (in the reformed fuel), instead of storing the solar heat as thermal energy in steam, oil, or molten salt. As shown before, chemical energy is one of the highest volumetric and gravimetric energy storage approaches, and can also provide longer-term storage with very little loss. Some of the solar reformed syngas can be stored in pressurized tanks for use when solar energy is not available, thus providing a solar share to the plant energy for more hours. The stored solar fuel can also be used to start up the plant in the morning.

One of the challenges of using solar energy for reforming is the relatively low temperature of the heat transfer fluid. Unless the concentration ratio is very high, the maximum temperature of the receiver, and hence the heat transfer fluid and the reformer reactor, is likely to be below 600 °C. At these temperatures, direct steam–methane reforming cannot convert a large fraction of the methane in the mixture, as governed by equilibrium (using a catalyst can speed up the reaction rate, but cannot change the equilibrium conversion).

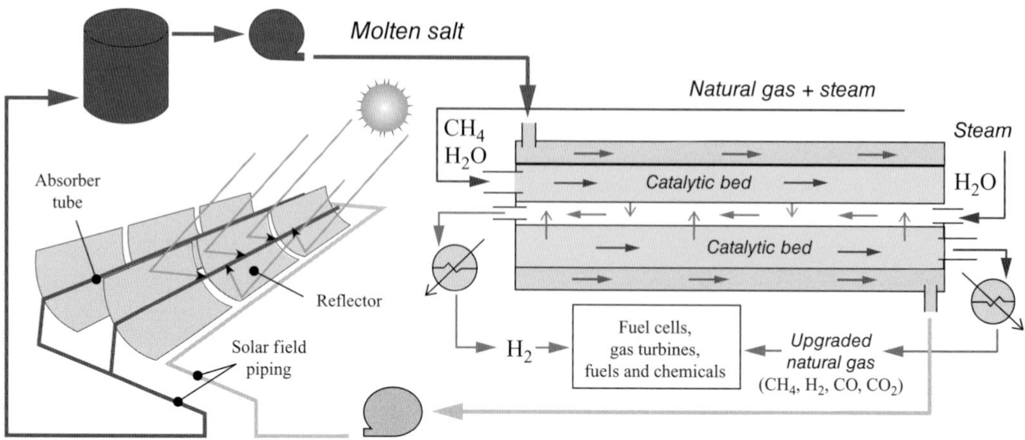

Figure 9.16 A hybrid solar fuel power plant in which a hydrogen-permeable membrane reactor is used to reform methane and produce a stream of pure hydrogen, which can be used to power a fuel cell. The fuel left in the reformed steam is burned in the gas turbine combustor. A heat transfer fluid circulates between the solar collector and the reactor [10].

Low methane conversion at lower temperature limits the contribution of the solar energy to the power plant storage. One way to overcome these equilibrium limitations at lower temperatures is to use redox/chemical looping systems or membrane reactors (see Chapters 10 and 11) to raise the fraction of methane converted at lower temperature.

Using membrane reactors (membranes and membrane reactors will be described in more detail in Chapter 10), as shown in Figure 9.16, it is possible to continuously remove some of the reforming products and hence overcome the equilibrium limitation. This can be shown using the equilibrium concentrations in the mixture:

$$\text{For } CH_4 + H_2O \rightarrow CO + 3H_2$$
$$\text{we have } \frac{p_{CO}p_{H_2}^3}{p_{CH_4}p_{H_2O}} = K_p(T). \tag{9.5}$$

Therefore, reducing the partial pressure of hydrogen in the mixture by continuous removal will lead to reducing the concentration of methane by converting more of it to syngas. In this case, and to provide the thermal energy necessary for reforming, the membrane reactor is heated by the hot molten salt, which is used as a heat transfer fluid circulating between the absorber and the reactor. A mixture of methane and steam are introduced on one side of a metallic hydrogen-permeable membrane (made mostly of palladium alloys, to be described in more detail in Chapter 10). Low-temperature methane steam reforming kinetics on that side of the membrane, known as the feed side, is promoted by a catalytic bed. The hydrogen produced by reforming permeates across the membrane to the other side, known as the sweep or permeate side, and is swept away using a stream of steam. Removing the hydrogen promotes further reforming at relatively low temperature by

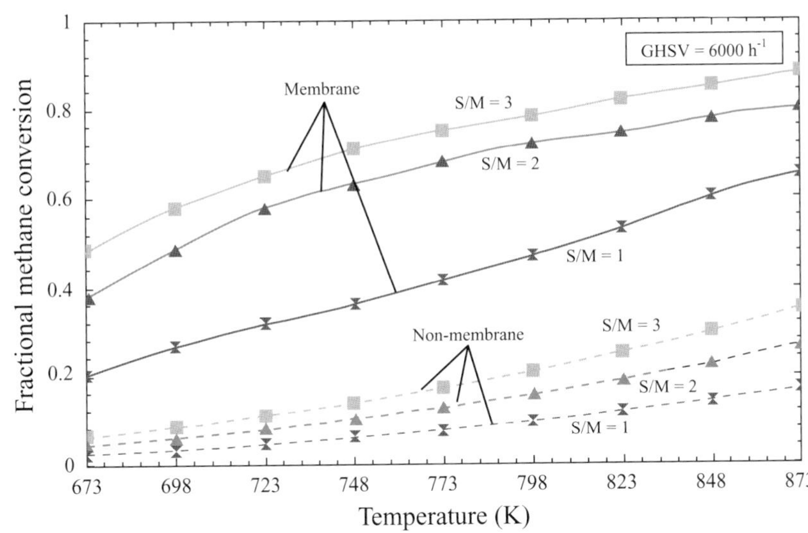

Figure 9.17 Using a hydrogen-permeable membrane reactor enables more conversion of methane in steam–methane reforming at lower temperatures, by removing the hydrogen produced by methane reforming across the membrane. The figure shows the fractional conversion of methane as a function of temperature without and with membrane separation. S/M is the steam to methane ratio in the feed stream [10].

shifting the reforming equilibrium reaction toward the products according to (9.5). It also produces a pure hydrogen stream that can be used in a fuel cell. The reformed gas stream, whose hydrogen has permeated to the sweep side, is mostly CO, and is burned in the gas turbine combustor. Figure 9.17 shows the impact of removing hydrogen across the membrane on the fractional conversion of methane (the fraction of methane introduced into the reactor that is converted within the reactor length) as a function of the reactor temperature. Much higher conversion at lower temperature compatible with solar collectors is possible using the hydrogen-permeable membrane reactor.

Chemical looping can lower the reforming temperature to values compatible with those of the heat transfer fluid in solar collectors. In chemical looping reforming (to be described in more detail in Chapters 10 and 11), water is used to oxidize a metal, producing hydrogen [11]. The metal oxide is then used to partially oxidize methane, producing syngas. The two reactions are called a redox pair, where M and MO are the metal and metal oxide used to extract oxygen from water and partially oxidize methane (note that M can stand for a pure or partially oxidized metal), and are shown by the following reactions:

$$\text{Metal oxidation}: M + H_2O \rightarrow MO + H_2, \tag{9.6}$$

$$\text{Metal reduction}: MO + CH_4 \rightarrow CO + 2H_2 + M, \tag{9.7}$$

$$\text{Overall}: CH_4 + H_2O \rightarrow CO + 3H_2. \tag{9.8}$$

Oxidation and reduction (of the metal) are performed in two different reactors and the metal/metal oxide are circulated between the two reactors (or are fixed to the walls of the micro channels of a rotary reactor). The circulating metal/metal oxide is also known as the oxygen carrier OC. Metal reduction during the fuel partial oxidation step is always endothermic, while metal oxidation in the hydrogen production step can be endothermic or exothermic, depending

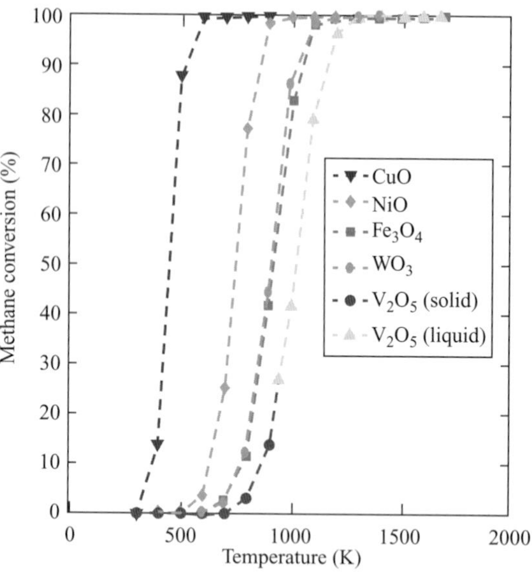

Figure 9.18 Methane conversion during methane partial oxidation according to (9.7) using a number of metals, including copper (CuO), nickel (NiO), iron (magnetite Fe_3O_4), tungsten (WO_3), and vanadium (V_2O_5). In all cases, stoichiometric metal oxide to fuel is assumed, although in practice it is likely that super-stoichiometric mixtures would be used. Mixed metals have also been suggested to improve the performance. Besides conversion, other considerations for choosing the metal include oxygen carrying capacity, price, and toxicity [6].

on the metal. However, the overall process is always endothermic and the heat is provided by the solar collectors. These systems will be described in Chapter 10 and 11.

Figure 9.18 shows methane conversion (partial oxidation according to (9.7)) by metal oxide reduction for a number of metal oxides, demonstrating the high conversion for copper and nickel at relatively lower temperatures compatible with those achieved by solar collectors/concentrators.

A CC incorporating solar redox reforming is shown in Figure 9. 19. Ferrite on alumina is used as the oxygen carrier. Exhaust leaving the HRSG is condensed to produce water for reforming, which is sent to the oxidation reactor after being evaporated in the HRSG. Steam for the Rankine cycle is raised using the thermal energy in the gas turbine exhaust, and more steam is raised in the heat exchanger (HEX) between the two redox reactors. The performance of this and similar cycles is often measured as a function of the solar share X_{solar} defined as the ratio of the heat available to the solar field to the total heat delivered to the solar field and by the fuel:

$$X_{solar} = \frac{\dot{Q}_{solar}}{\dot{Q}_{solar} + \dot{Q}_{fuel}}, \tag{9.9}$$

where $\dot{Q}_{solar} = IA$, and A is the collector's total area. The efficiency of a hybrid cycle is defined as

$$\eta_{cycle} = \frac{\dot{W}_{hybrid}}{\dot{Q}_{fuel} + \dot{Q}_{solar}}. \tag{9.10}$$

The system shown in Figure 9.19 was modeled [7,12]using Aspen, and the computed efficiency is shown in Figure 9.20. As the solar share in the input energy increases, the overall

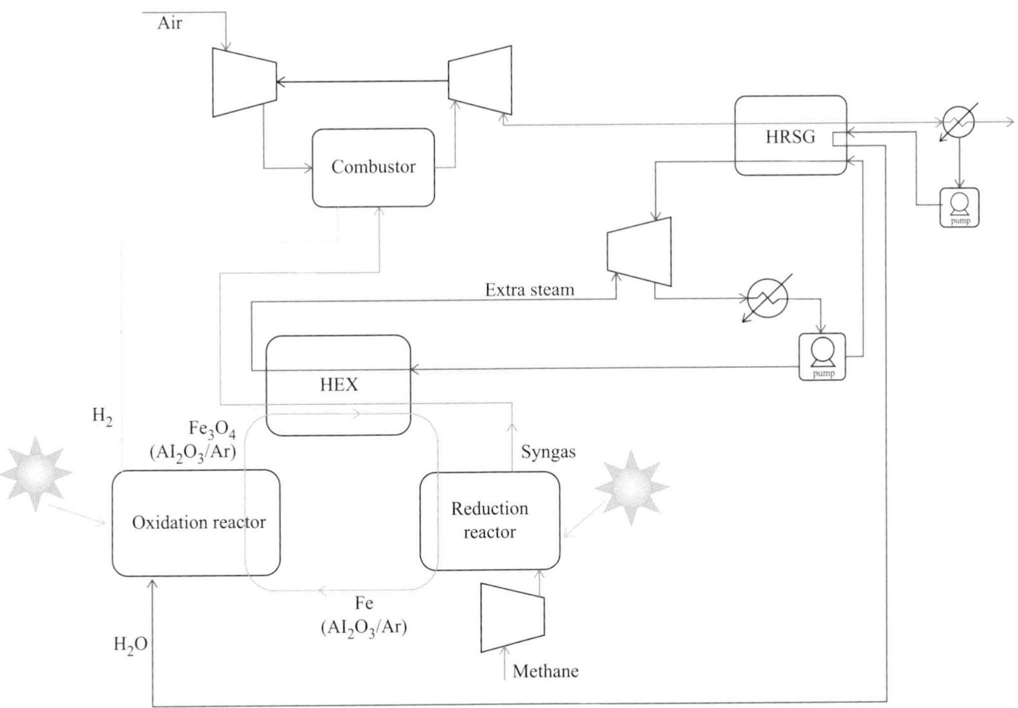

Figure 9.19 A schematic diagram showing the integration of an iron redox reforming into an NGCC. Methane is fed to the reduction reactor along with the metal oxide (in this case ferrite on alumina particles carried by argon). Metal leaving the reduction reactor is transported to the oxidation reactor, where it is oxidized by steam, which is generated in the HRSG. The syngas leaving the reduction reactor (after being cooled in the HEX, along with the metal particles to raise more steam for the bottoming Rankine cycle) and hydrogen leaving the oxidation reduction are sent to the gas turbine combustor.

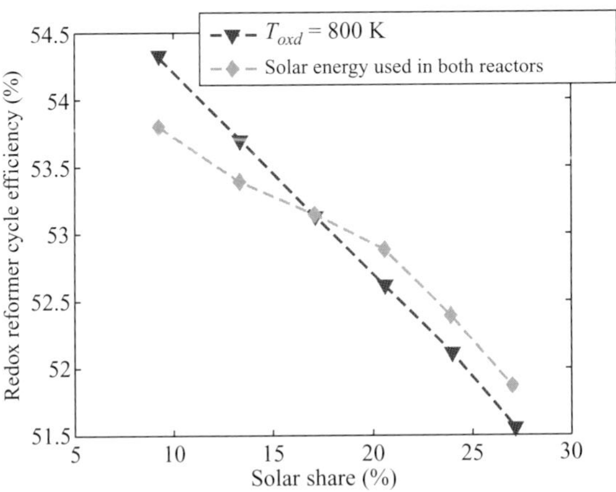

Figure 9.20 The overall efficiency of the hybrid cycle as a function of the solar share. The overall efficiency decreases with increasing solar share because it raises the collector temperature and hence increases the heat losses, decreasing the absorption efficiency [6].

efficiency decreases. Raising the solar share improves methane conversion but reduces the effective efficiency of the solar contribution, since collection efficiency is negatively correlated with temperature.

9.6 Fair Allocation Method

Hybrid plants pose a challenge to regulators in the form of how much, in terms of incentives, to allocate to the fraction of electricity produced by the renewable section of the plant, since in most designs the input energies from the fossil part and the renewable part are added to the same power island in a complex way. Approaches to apportion the output power to the input have been proposed in which the fractions of the power produced are determined using average conversion efficiencies of existing fossil plants and solar plants (see [12]). The calculations are relatively straight forward, although they do not necessarily account for the design of the plant under consideration. Another approach is based on determining the exergy loss in the solar part and fossil part of the plant, hence focusing the analysis on the design and integration of the particular plant being considered. This approach is described in [13].

9.7 Summary

Renewable energy is one of the important approaches for combating climate change, and its utilization has grown recently. Thermal energy from geothermal wells and solar concentrators is used for electricity production utilizing power cycles similar to those applied in fossil fuel power plants, with some modifications related to the fact that renewable sources tend to supply lower-temperature thermal energy. This is especially true for geothermal energy, where organic Rankine cycles make it possible to operate higher-efficiency supercritical cycles. Alternative cycles replace the heat exchanger between the geofluid and working fluid, and use flashing to raise steam for a steam cycle.

Concentrated solar power plants mostly use steam Rankine cycles coupled with trough or tower concentrators, since the working fluid temperature can be raised to levels at which steam cycles achieve reasonable efficiency. Interestingly, the overall system efficiency that accounts for the solar field efficiency (which decreases as the working fluid temperature increases) and power cycle efficiency peaks at intermediate temperatures. A solar dish, which achieves higher temperatures, can be used to power a Stirling engine.

The intermittency of solar energy makes it necessary to utilize storage technologies or to hybridize CSP plants with fossil fuels. Several approaches to hybridize the plants were discussed, including the production of syngas (by reforming the original fuel using solar heat) and storing the gas for later use. Reforming processes were discussed in the previous chapter, and more technologies will be introduced in Chapters 10 and 11.

Problems

9.1 Supercritical Rankine cycles using organic working fluids are used to produce power from low-grade geothermal resources. A cycle using butane is shown in Figure P9.1. The critical point of butane is 425.1 K and 37.96 bar. You may use the NIST database to find the thermodynamic properties of butane at each state.

The isentropic efficiency of the turbine is 85%. You may neglect the pumping power when calculating efficiencies. The cooling water enters the condenser at the same temperature as the environment, $T_o = 15\,°C$.

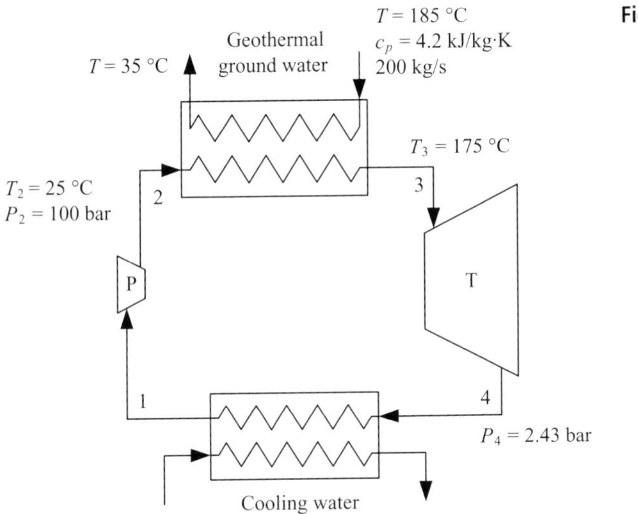

Figure P9.1

a. What is the mass flow rate of butane in the cycle?
b. How much power is produced by the turbine?
c. What is the efficiency (First Law) of the cycle?
d. What is the Second Law efficiency of the *plant*?

9.2 A clever engineer suggests that the efficiency of the cycle in Problem 9.1 could be improved by heating the working fluid in a solar concentrator before feeding it into the turbine. The modified design is shown in Figure P9.2.

Figure P9.2

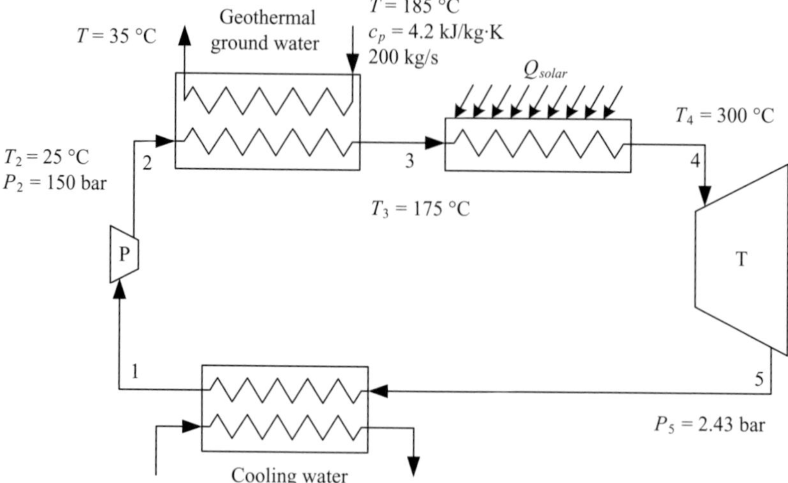

a. What is the required solar heat flux to attain a turbine inlet temperature of 300 °C?
b. How much total power is produced by the turbine in this case?
c. What is the efficiency (First Law) of the cycle?
d. What is the Second Law efficiency of the plant in this case? Consider the solar energy as a heat source at 350 °C.

9.3 As an alternative to integrating the solar concentrator with the geothermal power plant in Problem 9.2, the two heat sources could also be used independently, with the geothermal fluid being used in the original cycle and the solar concentrator being used in the cycle shown in Figure P9.3, with the same solar heat flux as in part (a) of Problem 9.2.

Figure P9.3

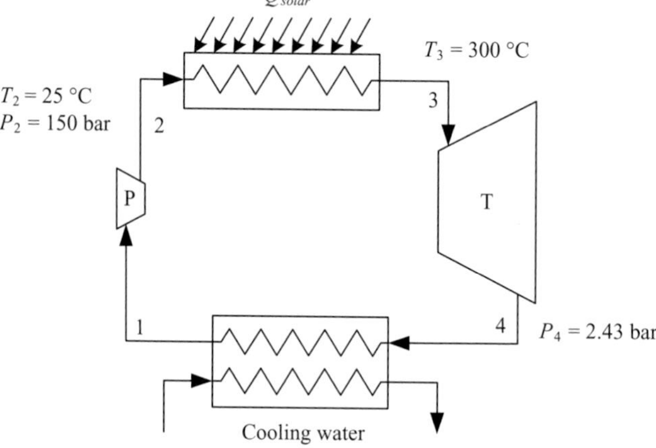

a. What is the mass flow rate of butane through the solar cycle?
b. How much power is produced by the solar cycle turbine?
c. What is the efficiency (First Law) of the solar cycle?
d. Which configuration utilizes the source better?

REFERENCES

1. T. R. Mancini, CSP Overview Presentation. 2004, http://www.eia.doe.gov/cneaf/solar.renewables/page/solarreport/solarsov.pdf.
2. C. J. Winter, R. L. Sizmann, and L. L. Vant-Hull, *Solar power plants: fundamentals, technology, systems, economics*. Berlin: Springer-Verlag, 1991.
3. J. W. Tester et al., *Sustainable energy: choosing among options*. Cambridge, MA: MIT Press, 2012.
4. T. Kodama, "High-temperature solar chemistry for converting solar heat to chemical fuels," *Prog. Energy Combust. Sci.*, vol. 29, no. 6, pp. 567–597, 2003.
5. B. Powell, K. M., Rashid, K., Ellingwood, K., Tuttle, and J. Iverson, "Hybrid concentrated solar thermal power systems: a review," *Renew. Sustain. Energy Rev.*, vol. 80, pp. 215–237, 2017.
6. E. J. Sheu, "A solar reforming system for use in hybrid solar–fossil fuel power generation," Ph.D. thesis, MIT, 2016.
7. A. Q. Sheu, E. J. Mitsos, A. Enter, et al., "A review of hybrid solar–fossil fuel power generation systems and performance metrics," *ASME J. Sol. Energy Eng.*, vol. 134, no. 4, pp. 41006–41017, 2012.
8. G. Brakmann, F. A. Mohammad, M. Dolejsi, and M. Wiemann, "Construction of the ISCC Kuraymat," in *International Solar PACES Conference 2009*, 2009.
9. E. J. Sheu, E. M. A. Mokheimer, and A. F. Ghoniem, "A review of solar methane reforming systems," *Int. J. Hydrogen Energy*, vol. 40, no. 38, pp. 12929–12955, 2015.
10. S. A. M. Said, D. S. A. Simakov, M. Waseeuddin, and Y. Román-Leshkov, "Solar molten salt heated membrane reformer for natural gas upgrading and hydrogen generation: a CFD model," *Sol. Energy*, vol. 124, pp. 163–176, 2016.
11. E. J. Sheu and A. F. Ghoniem, "Redox reforming based, integrated solar-natural gas plants: reforming and thermodynamic cycle efficiency," *Int. J. Hydrogen Energy*, vol. 39, no. 27, pp. 14817–14833, 2014.
12. G. P. Beretta, P. Iora, and A. F. Ghoniem, "Allocating resources and products in multi-hybrid multi-cogeneration: What fractions of heat and power are renewable in hybrid fossil–solar CHP?," *Energy*, vol. 78, pp. 587–603, 2014.
13. G. P. Beretta, P. Iora, and A. F. Ghoniem, "Exergy loss based allocation method for hybrid renewable fossil power plants applied to an integrated solar combined cycle," *Energy*, vol. 173, pp. 893–901, 2019.

10 Gas Separation Processes and Application to Carbon Capture Plants

10.1 Introduction

Carbon dioxide production in electric power plants depends strongly on the fuel and power cycle efficiency. Coal and natural gas generate approximately one mole of CO_2 for each mole of fuel burned, but they differ in their plant conversion efficiency. Table 10.1 shows average data for CO_2 production when using either fuel for electricity generation. The absolute majority of pulverized coal plants operate on steam Rankine cycles. The most efficient natural gas plants operate on combined cycles, although some operate on simple cycles.

Reducing CO_2 production in power generation in fossil fuel plants requires: improvement in energy conversion efficiency; hybridizing with renewable sources; and capturing CO_2 and storing it in deep geologic formations (or reusing some of it). Separating carbon dioxide from the flue gases, or at any other point within the power cycle, and storing it in deep reservoirs is a viable option for continuing the use of fossil fuels in power plants while stabilizing the atmospheric concentration of carbon dioxide at acceptable levels.

Figure 10.1 depicts three different approaches to capturing CO_2 from power plants: post-combustion capture, which can be used in the case of air combustion in a conventional power plant; pre-combustion capture, where reforming or gasification followed by water–gas shift is used to produce hydrogen for the power plant; and oxy-combustion, in which the fuel is burned in a pure, nitrogen-free oxidizing steam. Carbon dioxide separation from flue gases is necessary when conventional air-based combustion systems are used. If reforming or gasification in pure oxygen or pure oxygen and steam/CO_2 are used, it is necessary to separate hydrogen from a mixture of CO_2 and hydrogen. In the case of oxy-combustion, air separation must be performed to produce the oxygen required for combustion. In oxy-combustion, either carbon dioxide must be recycled back to the combustor after the products stream has been cooled in the power equipment, or water must be injected at or immediately downstream of the combustor to control the flame temperature. In the case of CO_2 recycling, 85–95% of the working fluid in the gas turbine is carbon dioxide, depending on the fuel used. In oxy-combustion, water condensation is the only separation process required to purify the CO_2 stream. If the oxidizing stream is not pure O_2, or some air leakage

Table 10.1 Thermal efficiency and CO_2 production in (Rankine cycle, pulverized) coal-fired and (combined cycle) natural gas-fired electricity generation plants, where MJe and MWhe are the electricity generated. Only average values shown

	Pulverized coal (Rankine cycle-based)	Natural gas (combined cycle-based)
Efficiency (%)	35	55
Lower heating value (MJ/kg-fuel)	30	50
CO_2 production (kg/MWhe)	1200	400
CO_2 specific energy (MJe/kg-CO_2)	3	9

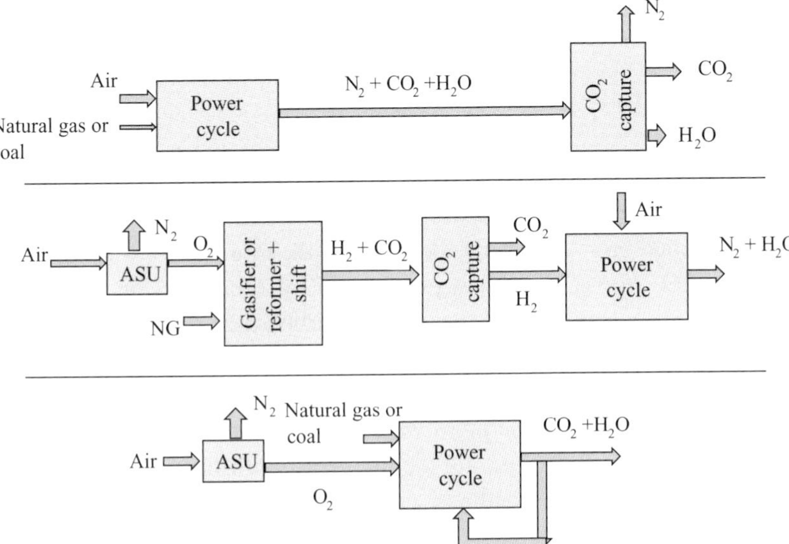

Figure 10.1 Different approaches for CO_2 capture from hydrocarbon fuel-fired electricity generation power plants using post-combustion capture, top; pre-combustion capture, middle; and oxy-combustion capture, bottom (excluding electrochemical methods using high-temperature fuel cells). The diagrams for post-combustion and oxy-combustion capture show the case using steam plants only, while the pre-combustion case shows a combined cycle.

is expected in the intermediate equipment between combustion and CO_2 capture, then separation of N_2 is still necessary. Large air-separation units (ASUs) would be needed. For instance, a 500 MW power plant burning coal (heating value ~30 MJ/kg-coal) and running at 33% efficiency requires almost 50 kg/s of coal and 125 kg/s of oxygen. This translates to 9 Mt/day of oxygen if the plant runs near 80% capacity every day.

A significant energy penalty in these technologies is associated with gas separation, whether it is separating CO_2 from the combustion products (mostly N_2), separating hydrogen from the syngas following water–gas shift, or air separation to produce pure oxygen. Another penalty is associated with compressing/liquefying CO_2 for sequestration (the critical point of CO_2 is 31 °C and 7.39 MPa). The captured carbon dioxide would have to be purified, compressed to pressures greater than 75 bar, and kept below 31 °C to remain in the liquid form for pipeline transmission to the disposal sites. CO_2 drying is important to avoid pipeline damage. This adds to the complexity of the plant and the energy penalty.

Clearly, gas separation technologies are important for the design and operation of CO_2 capture power plants. The fundamentals of some of the most widely used, and some under development, gas separation processes are reviewed in this chapter. The chapter starts by reviewing the exergy loss, or separation energy penalty, in an ideal gas separation system and the implications for air separation and the separation of CO_2 from combustion gases (discussed in Chapter 2). Next, the essentials of gas separation techniques are summarized, including physical and chemical absorption, adsorption, low-temperature distillations, and membrane separation. Some of the related fundamentals of each of these techniques are introduced, along with their implications regarding energy expenditure and compatibility with different CO_2 capture approaches. References on the application of gas separation technologies to CO_2 capture can be found in [1–4], and other texts that will be referenced in the chapter.

10.2 Separation Energy Penalty

As shown in Chapter 2, gas separation processes consume energy (or more precisely exergy). The lower the concentration of the gas to be separated from a mixture, the more energy is required per unit of the separated gas. Modifying the power cycle in a way that reduces the energy required to separate CO_2 from combustion products of hydrocarbons by integrating the power cycle with the separation process, and maximizing carbon dioxide concentration in the stream from which it is separated, are important for reducing the efficiency penalty. The separation process in a particular application depends on the design and operating conditions of the power cycle, and the state of CO_2 at the point where it is separated.

The ideal, isothermal work required to separate a component with molar concentration X_1 from the rest of the mixture, assuming ideal gases and mechanical equilibrium with the environment, evaluated per mole of the mixture and per mole of gas component 1 are, respectively:

$$\hat{w}_{mole\ of\ mixture} = -\Re T_o\left(X_1 \ln\frac{X_1}{1-X_1} + \ln(1-X_1)\right) \tag{10.1}$$

or

$$\hat{w}_{mole\ of\ X_1} = -\Re T_o\left(\ln\frac{X_1}{1-X_1} + \frac{\ln(1-X_1)}{X_1}\right). \tag{10.2}$$

Plots for the impact of the concentration on both values are shown in Figure 10.2. These equations and the figure show that to reduce the separation work, we need to:

1. reduce the number of moles of the gas mixture;
2. separate CO_2 (or any other gas component) at low temperature; and
3. separate CO_2 from a stream with a relatively high concentration of carbon dioxide.

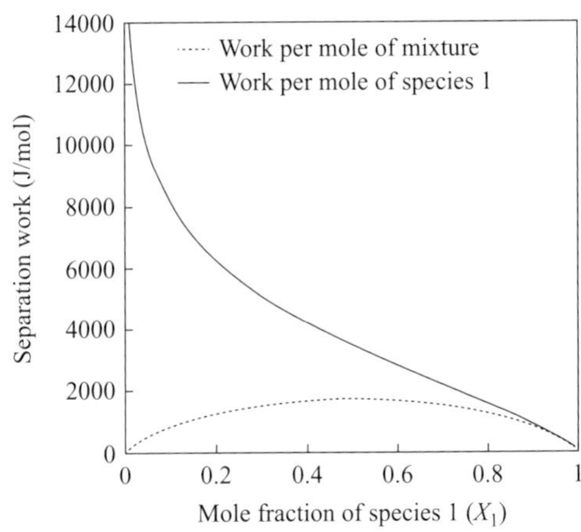

Figure 10.2 The minimum energy required to separate a gas stream into two components, X_1 and $(1 - X_1)$, where X_1 is the mole fraction of the separated gas. The separation energy is shown per mole of mixture (broken line) and per mole of separated gas (solid line).

Separating CO_2 from products of combustion, especially in the case of fuel-lean burning in air, where the mixture composition is dominated by nitrogen and the typical concentration of CO_2 is O (3–5%), may result in a high efficiency penalty. One can increase CO_2 concentration in the flue gases by recycling some of the flue gases back to the combustor, which can also control the burning temperature (recycling some of the CO_2 also reduces nitrogen concentration in the case of air combustion and hence lowers NO_x production in the combustion process). Recycling makes it possible to burn closer to stoichiometry, hence raising the fraction of CO_2 in the products. We note here that different expressions for the work of separation should be used if the output stream is not a gas, or if the separation processes are not isothermal, at constant pressure, or in equilibrium with the environment. Some of these expressions were developed in Chapter 2.

To reduce the total volume of the exhaust gas stream and increase CO_2 concentration, oxygen instead of air may be used in the combustion process – that is, oxy-combustion may be used to replace air combustion. In this case, an ASU is used to produce a stream of oxygen that is sent to the combustor. Oxygen concentration in air is 21%, and air separation for oxy-combustion can be costly and energy-intensive, and care must be taken to carefully integrate the ASU with the rest of the power cycle to minimize the separation efficiency penalty (heat, mass, and energy integration). The separated nitrogen may be used later in the cycle, in other applications, or simply vented back into the atmosphere. Different oxy-combustion cycle layouts will be discussed in the next chapters.

Alternatively, gasification (in the case of coal) or reforming (in the case of natural gas) can be used to produce a stream of synthetic gas (syngas), or a mixture of CO and H_2. This stream can then be reacted with water to produce a stream of CO_2 and H_2, that is water–gas shift (WGS). In both cases, oxygen and steam are used in the first stage – that is, in the endothermic gasification or reforming – and steam in the second. Carbon dioxide can be separated from the CO_2–H_2 mixture. In this case, we still need an ASU, but with a smaller capacity since oxygen is used for the partial oxidation only (with some of the partial oxidation oxygen coming from the steam). Hydrogen produced in this process is burned in air. More on cycle layout and alternative cycles is covered in Chapters 11 and 13.

The selection of the separation techniques depends on the choice of the power cycle, the separation energy penalty, and the cost of associated equipment. It is important to understand some elements of gas separation physics/chemistry in order to understand the origin of energy losses and estimate the efficiency penalty associated with CO_2 capture, and to define which gas separation technique should be used in a particular application. As will be shown, gas separation may require heating and cooling, or heating/cooling and/or work for pumping or compression, or work only for compression through a membrane. The ideal energy of separation can be thought of as an overall minimum penalty on availability/exergy. The actual energy penalty depends on the separation process, where it is executed in the cycle, and the overall "heat, mass, and work" integration in the plant. In the next several sections, we review some existing and promising techniques for gas separation. In Chapters 11 and 13, we discuss how these technologies are integrated into power cycles.

10.3 Gas Separation Processes

Gas separation techniques include chemical or physical absorption (in a liquid sorbent) or adsorption (on the solid surface), cryogenic distillation, and selective membrane separation. In the following, these processes are reviewed in the context of separating CO_2 in power plants; that is, separating CO_2 from flue gases or from a reformed and shifted gas stream. Distillation is discussed in the context of separating oxygen from air, which is needed in:

1. pre-combustion capture – that is, for separating CO_2 from a mixture of hydrogen and carbon dioxide produced by reforming (or gasifying) the original fuel, in which oxy-gasification or reforming is used to produce the syngas; or
2. oxy-combustion applications.

The objective of this introduction to gas separation processes is not to show how to design the process or the necessary equipment, but rather to familiarize the reader with the physics and chemistry of the different processes, and to set the stage for computing the energy penalty associated with removing CO_2 from the products before they are vented into the atmosphere. Thus, the primary objective here is to compute the energy required for separating carbon dioxide from the working fluid or the flue gas in a power plant, or oxygen from air. This energy is needed in the form of higher-temperature heat (higher than the atmosphere), mechanical or electrical work, or both. For a more comprehensive overview of gas separation processes [1].

Chemical and physical absorption processes are shown schematically in Figure 10.3. During absorption, a liquid solvent is used to remove one gaseous component from a gas mixture by bringing a liquid solvent in close and continuous contact with the gas

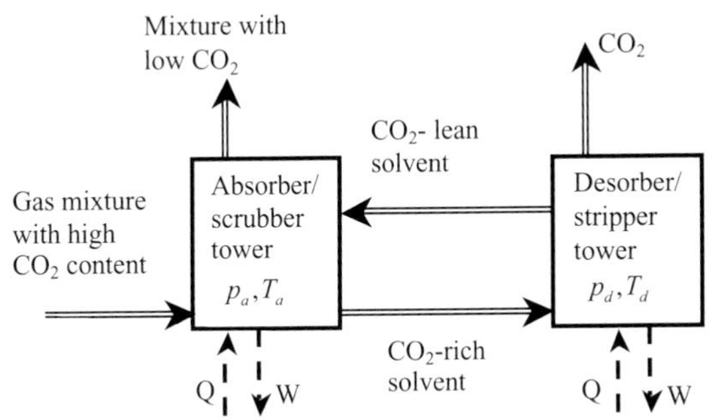

Figure 10.3 A schematic diagram showing physical or chemical separation process in which a solvent is used to remove a gas from its original mixture in an absorber, and the solvent is regenerated by removing the solute in a desorber. The solvent is circulated between the two vessels. The diagram applies mostly to physical and chemical absorption techniques.

mixture. In chemical absorption, the liquid solvent forms chemical bonds with the gas component to be separated. In physical absorption, the bond between the gas component to be separated and the solvent is physical. For the process to operate continuously, an **absorber/scrubber** unit that removes CO_2 (or any other **solute**) from the gas mixture stream using an aqueous **solvent** is employed. The CO_2-rich solvent is then transported to a **desorber/stripper** unit, where CO_2 (the solute) is separated from the solvent, and the solvent is regenerated to its original state. The CO_2-lean solvent is then fed back to the absorber/ scrubber unit for reuse. This looping of the solvent makes the overall process continuous, and requires reversible solvent. In the stripping unit, the solvent is regenerated before it is fed back to the absorber unit.

The pressure and temperature in both vessels, the absorber and the stripper, must be adjusted to create the conditions under which equilibrium favors the desired process – that is absorption or desorption of the gas (CO_2 in this case). Thus, the pressure or temperature or both must be set to favor absorption of CO_2 in the first unit and desorption of carbon dioxide in the second. Depending on the process, energy transferred in the form of heat and/or work may be required in either or both the absorber/stripper unit and desorber/ scrubber unit. These separation processes are performed in a continuous mode by recycling the solvent between the absorber unit/tower and the carbon dioxide stripper unit/tower. The solvent must be regenerable, that is, it must be possible to separate the solute (CO_2) from the solvent almost completely in the desorber unit to bring the solvent to its original state. The reversibility of the reaction between the solvent and the solute guarantees that the solvent is regenerative. (An exception is when the product of the absorption process has a market value, or if the solvent is cheap and the carbon dioxide-rich solvent is easily disposable.) Both physical and chemical absorption have been used in gas separation and can be successfully applied to separating CO_2 from either flue gases or shifted partially oxidized fuels.

In **adsorption processes**, the gas to be separated is adsorbed on the surface of a porous solid as the gas mixture flows through the solid. These processes are often performed in a batch mode, and the adsorption and desorption occur in the same tank, as shown in Figure 10.4. When the solid, or adsorbent, is saturated with the adsorbate (the separated gas), the gas stream is stopped and the conditions in the adsorption tank, that is, the pressure or

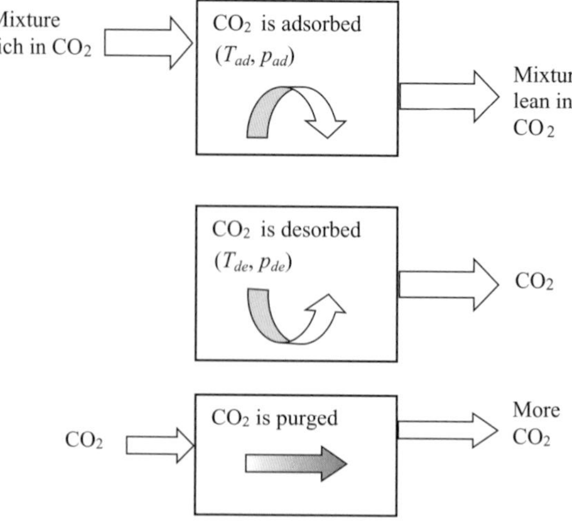

Figure 10.4 A three-step batch process for gas separation by physical adsorption: adsorption, desorption, and purging of adsorbate (CO_2). In the first step, one component of the gas mixture, in this case CO_2, is adsorbed by the solid, while a CO_2-lean gas flows out. When the solid is saturated with CO_2, or when the front reaches the opposite section, the flow is stopped. Conditions are changed to allow the desorption of the gas, by lowering the pressure or raising the temperature, and the gas flows out. In the final step, CO_2 is introduced to purge the remaining CO_2 out of the bed matrix.

temperature or both, are allowed to change to free up the adsorbate from the solid. After flowing out of the solid naturally, the rest of the adsorbate remaining in the tank is flushed out by flowing a stream of similar composition to the adsorbate through the tank; this is the regeneration process. In a continuous separation operation, more than one adsorption tank is required so that the CO_2-rich stream can be diverted between the desorbing tank and the adsorbing tank, while the former is venting its CO_2 or is being purged. The flow through the tank causes a finite-pressure drop, and the mixture must be pumped through the porous medium of the tank. Changing the conditions within the tank to promote adsorption or desorption involves changing the pressure or temperature or both, in what is known as the pressure-swing or temperature-swing separation, respectively. The more energy-demanding process in the series of events taking place during adsorption-based separation processes is associated with the regeneration of the adsorbent through heating in temperature-swing adsorption, or lowering the pressure in pressure-swing adsorption. Alternatively, it is possible that the pressure of the mixture may be raised to increase the amount of the adsorbate on the surface, since solubility of a gas on a surface grows with pressure. Several adsorbers have been suggested and used, such as activated carbon (used mostly in the powder form) activated alumina, silicon gels zeolites, or molecular sieves. Chemical adsorption has also been suggested.

Cryogenic separation or distillation processes are used extensively in the oil-refining industry, and rely on the characteristics of the phase diagram of a mixture of components with different volatilities. In distillation, by reducing the temperature of a gas mixture below the dew point of the mixture, a liquid with higher concentration of the less volatile component forms and flows downwards, and a gas with higher concentration of the more volatile component flows upwards. The fraction of the original mixture-forming liquid is determined by the pressure and temperature of the mixture. As the original gas stream is cooled further, more of the less volatile component separates in the form of a liquid and flows downwards, while a more purified gas flows upwards. This technique is used for oxygen separation from air in cryogenic ASUs, especially the high-capacity ones. New research on applying distillation to separating CO_2 from flue gases has focused on separating carbon dioxide in the solid form (the triple point of CO_2 is -56.6 °C and 518 kPa). Most of the energy required for distillation is in the cooling or cryogenic processes, and in pumping the gas into the distillation columns.

Semipermeable membranes that allow one or more gaseous components to pass through more readily than others are used to separate gases into their components, and to separate dissolved or suspended solids from liquids. Especially in the case of mixtures of components that have different molecular weights, this technique can achieve high purity. Both inorganic mixed ceramic (metal oxides) metal-based membranes and organic polymer membranes have been used for gas separation, and the application temperature usually determines the type and form of membrane used in the separation process. Contrary to the previously described gas separation processes that rely on essentially equilibrium processes for separation, membrane separation relies on diffusive transport across a pressure drop (mostly partial pressure drop for the permeating gases) through the membrane to achieve separation, as shown in Figure 10.5. The nature of the diffusion process through the membrane is determined by its material, as will be shown later. Membranes can be made of metallic, ceramic, or polymeric materials, and can be designed in the non-porous or porous form, wherein the gases are transported by grain boundary diffusion or by gas transport through the pores, respectively. Separation membranes can be unsupported (symmetric) or supported (asymmetric), depending on the material strength. Due to the finite selectivity of some membranes, multistage separation may be necessary to achieve high purity. Ceramic membranes transport charged species – that is, oxygen ions or hydrogen protons – and can be used for the production of 100% pure species.

In the next sections, these separation processes are described in more detail, covering either the thermodynamics or the transport/kinetics, depending on the physics that govern the energy consumption. The implementation of the process in the context of separating CO_2 or oxygen from air is also described.

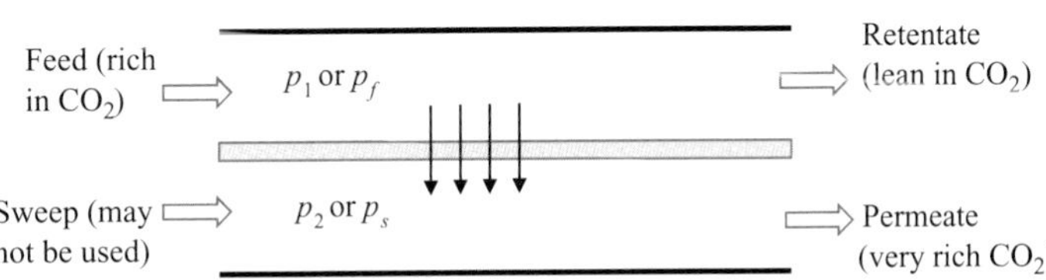

Figure 10.5 Gas separation by membranes. The process is driven by a partial pressure difference of certain gas components across the membrane, which allows the preferential diffusion of one or more of these gas components to permeate to the other side. Depending on the design, a sweep gas may or may not be used.

10.4 Absorption

Absorption refers to the ingestion of a gas by a liquid. Two types of absorption have been used: physical absorption, in which the chemical structure of the gas remains the same in the liquid (i.e., only physical bonds are formed that depend on the solubility of the gas in the liquid); and chemical absorption, in which the gas bonds chemically with the liquid, forming new chemical compounds.

10.4.1 Physical Absorption

In physical absorption, a gas (e.g., CO_2) is absorbed in a liquid solution in the **absorber or scrubber tower** (at lower temperatures and higher pressures), and is released in a series of **flash expansion chambers or drums** under successively lower pressures or higher temperatures. In most cases, physical absorption occurs under higher pressure and/or lower temperature, both of which raise the solubility of gases (CO_2) in the **solvent** – that is, raise the equilibrium concentration of CO_2 in the liquid solvent. Thus, having a high-pressure or low-temperature gas mixture favors using this technique. Following absorption, the gas-saturated mixture is transported to the desorber/stripper tower, where the pressure is lowered and/or the temperature is raised. In these flash or expansion chambers, conditions are set to allow the gas (CO_2) that dissolved in the liquid in the absorber tower to escape. Multiple flashing is used to reduce the pressure to atmospheric or lower than atmospheric pressures to release CO_2 in the desorber tower. Following the gas release from the solvent, the solvent is pumped back to the absorber tower pressure, and the released gas is compressed (to a delivery pressure). Thus, this process requires mechanical energy. In some applications, heat is required to raise the solvent temperature in the desorber to desolve CO_2 and release it from the solvent. The total energy required in this absorption–desorption-based separation process depends on the pressure and temperature differences that must be maintained between the two units, the absorber and the desorber, and these depend on the characteristics of the solvent.

The availability of a high-pressure gas mixture stream facilitates the application of physical absorption for gas separation. Thus, in the context of applying this process for CO_2 separation in power cycles, it may be more suitable for separating CO_2 from a CO_2–H_2 mixture produced in a high-pressure gasification process (IGCC plants that will be discussed in detail in Chapter 13) or in a natural gas reforming process, both followed by a water–gas shift reaction (Chapter 11). The energy used in physical absorption depends on the CO_2 concentration or partial pressure in the gas mixture, the solvent, and the conditions in the absorption and stripper towers. Several physical absorption processes have been developed for gas separation in oil refineries (e.g., Rectisol®, Selexol®, Purisol®) and many achieve high removal efficiency. One of the primary differences among these processes is the choice of solvent. For instance, methanol is used as a CO_2 solvent in the Rectisol® process (its **solubility** at 263 K and $p_{CO2} = 1$ bar is 10 L of CO_2/L of methanol, while the solubility of CO_2 in water at the same conditions is 2.5 L of CO_2/L of water).

10.4.1.1 Solubility and Henry's Law

The **solubility** of a gas in a liquid is defined as the mole fraction of a gas dissolved in a liquid at equilibrium with uniform conditions in both phases. At equilibrium, the mole fraction of the gas dissolved in the liquid is proportional to the gas partial pressure in the gas phase above the liquid (see Figure 10.6). This is expressed by Henry's law, which states that, at equilibrium,

$$X_{L,i} = p_i/\mathrm{He}_i, \tag{10.3}$$

where $p_i = X_{vi} \cdot p$ is the partial pressure of the **solute** (the gas dissolved in the liquid) in the gas stream, $X_{L,i}$ is the mole fractions of the solute in the liquid phase, given in terms of moles of solute per mole of liquid, and He is Henry's constant (or Henry's law constant) for this gas in a dilute solution for a particular solvent. Henry's constant depends on the temperature and is given in several units, but the most correct units are those of pressure (i.e., bar or MPa). Henry's law shows the following:

1. The mole fraction of the dissolved gas increases with increasing total pressure of the gas mixture and the mole fraction of the gas component to be dissolved in the liquid in the gas mixture.
2. Increasing both the gas total pressure and the mole fraction of a particular component raises the fraction of gas dissolved in the liquid.
3. Inversely, lowering the gas pressure or the mole fraction of the solute in the gas allows the gas to escape from the liquid phase.
4. The concentration of a gas dissolved in a liquid at equilibrium is inversely proportional to Henry's constant, and gases with high values of He do not dissolve well in liquids.
5. Higher total pressures promote the dissolution of a gas with a low Henry's constant in the liquid.
6. As the pressure is reduced, the dissolved gas tends to escape (what you experience as you open a can of carbonated soda in which a large fraction of CO_2 has been dissolved).

The choice of solvent is important in reducing the amount of solvent required and the pressure differential across the absorber–desorber (and hence the energy of separation).

Table 10.2 shows the values of Henry's constant for a number of gases dissolving in water at low and moderate pressures. The table shows that oxygen and nitrogen have

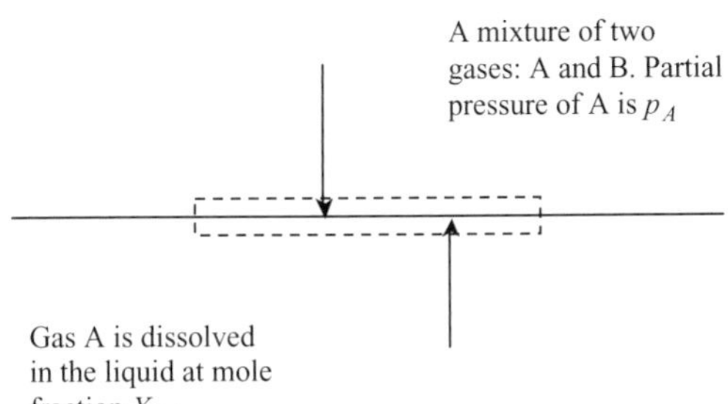
A mixture of two gases: A and B. Partial pressure of A is p_A
Gas A is dissolved in the liquid at mole fraction X_{LA}

Figure 10.6 Equilibrium at the gas–liquid interface; one of the gaseous components is also dissolved in the liquid. The relation between the partial pressure of A in the gas and its molar concentration in the liquid is given by Henry's law.

Table 10.2 Henry's constant (in bars) for some gases in water at low to moderate pressures [5]

gas	290 K	300 K	310 K	320 K	330 K	340 K
H_2S	440	560	700	830	980	1140
CO_2	1280	1710	2170	2720	3220	–
O_2	38,000	45,000	52,000	57,000	61,000	65,000
H_2	67,000	72,000	75,000	76,000	77,000	76,000
CO	51,000	60,000	67,000	74,000	80,000	84,000
Air	62,000	74,000	84,000	92,000	99,000	104,000
N_2	76,000	89,000	101,000	110,000	118,000	124,000

weak solubility in water, while CO_2 and H_2S have high solubility in the same liquid. Clearly, a mixture of carbon dioxide and nitrogen brought into contact with water will lose much more of its CO_2 than its N_2 to the liquid. Henry's constant increases with temperature, and it is possible to increase the concentration of the gas in the liquid by lowering the temperature, or to release the dissolved gas by heating the solvent. Note that while Henry's law applies to the thin film at the interface between the gas and the liquid, at equilibrium the gradients in the liquid are negligible and the gas is dissolved in the volume. In practice, thin liquid films are created by spraying the liquid solvent inside the **absorber tower**, and they remain in contact with the gas mixture until equilibrium is reached. Thus, it is a reasonable approximation to assume that the gas composition within the liquid would remain almost uniform. In these towers, the liquid solvent is sprayed at the top and the gas is injected at the bottom, creating a counter-current flow arrangement and longer contact time between the gas and liquid.

The change in Henry's constant with temperature can be expressed in a similar form to that of the reaction equilibrium constant. Henry's law, which is accurate at low solute concentrations, can be written in a different form, in terms of the equilibrium constant K for the absorption/desorption reaction:

$$\chi_V \rightleftarrows \chi_L, \tag{10.4}$$

where the left-hand side and the right-hand side of this equation stand for the vapor phase and liquid phase, respectively. The equilibrium constant for this reaction can be defined in terms of the activities, or the mole fraction in the liquid phase and the partial pressure in the gas phase, as follows: $K_i = X_{Li}/(pX_{vi})$. Species that have high values of K dissolve in liquids at higher concentrations.

Henry's law of liquid–vapor equilibrium is used when the temperature is above the critical temperature of the gas component of interest – that is, when this component does not condense in the liquid phase. Should conditions be subcritical (i.e., the temperature is below the critical temperature of the participating species), Raoult's law is used:

$$p_i = X_{V,i}p = X_{L,i}p_i^{sat}(T), \tag{10.5}$$

where $p_i^{sat}(T)$ is the saturation pressure of the component at the given temperature. Raoult's law applies to mixtures of ideal gases and ideal liquids – that is, when the temperature is relatively high and the pressure is relatively low (with respect to the critical values) – and when the species are of similar chemical nature and the molecules are of comparable sizes.

Absorption reactions such as those shown in (10.4) are exothermic, and cooling of the absorber towers is often required to keep the solvent at a sufficiently low temperature, at which the solubility of a gas component in the falling liquid is high. Heating can be used in the **desorber tower** to achieve the opposite effect – that is, to allow the gas to dissolve and escape from the liquid solvent. The hot solvent leaving the **desorber or stripper tower** should be cooled before it is reintroduced into the absorber column. A typical design for an absorption tower or column is shown in Figure 10.7. The figure shows a packed column used to improve the contact between the liquid (flowing down) and gas (flowing up). Other designs use tray columns in which horizontal trays with hole patterns are staked along the column. In both cases, a counter-current flow configuration is established between the gas mixture and the liquid solvent. The solvent is introduced through a demister to increase the surface area available for absorption.

Both Henry's law and Raoult's law show that higher partial pressures in the gas lead to higher solubility in the liquid phase (higher value of Henry's constant mean low solubility). Moreover, Henry's constant depends strongly on the temperature, and temperature "swings" could be used between adsorption and desorption processes to facilitate both, in different columns. For instance, for the dissolution of CO_2 in methanol, Henry's constant changes from 170 to 110 MPa/(mol/kg-methanol) as the temperature drops from −40 to −70 °C. Henry's law constants for a number of gases in methanol are shown in Table 10.3.

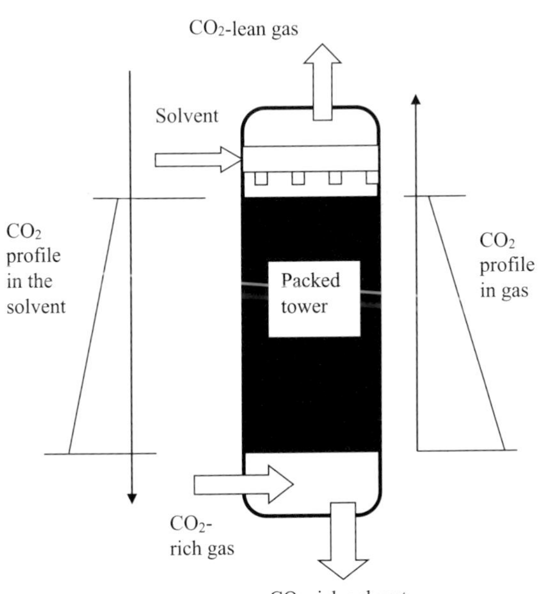

Figure 10.7 Absorption in a packed tower, where a liquid solvent is introduced at the top and a gas mixture is admitted at the bottom. The flow of the liquid downward over a packed bed allows close contact with the gas flowing upwards, and for the absorption to occur over a large surface area. The schematic shows the change in the absorbed gas concentration in the gas and liquid streams within the tower.

Table 10.3 Henry's law constants for the absorption of CO_2, CH_4, and H_2 in methanol, measured in MPa/(mol-gas/kg-methanol) [6]

	Temperature °C	
Gas	−40	−70
Carbon dioxide	1.4	0.31
Methane	70	30
Hydrogen	670	930

Example 10.1

A mixture with 10% CO_2 molar concentration is in equilibrium with a methanol solvent at 10 atm and −70 °C. Determine: (a) the concentration of CO_2 according to (9.3); (b) the number of moles of CO_2 that would be released if the mixture with the same solvent were transported to an environment maintained at 0.1 atm; and (c) the number of moles of CO_2 that would be released if the solvent were transported to an environment maintained at 1 atm and −40 °C.

Solution

a. Henry's law constant at −70 °C is 0.31 MPa/(mol-CO_2/kg-methanol). Thus, from (9.3) the molar concentration of CO_2 at 10 atm (1 MPa) and −70 °C is

$$X_{L,CO_2}(1\ \text{MPa},\ -70\ °\text{C}) = \frac{0.1 \times 1}{0.31} = 0.32\ \text{mol-CO}_2/\text{kg-methanol}.$$

b. If the same solvent with the same CO_2 molar concentration is transported to an environment in which the pressure is reduced to 0.1 atm (0.01 MPa) and in the absence of other gases, then the new equilibrium concentration of CO_2 in the solvent drops to

$$X_{L,CO_2}(0.01\ \text{MPa},\ -70\ °\text{C}) = \frac{0.01 \times 1}{0.31} = 0.032\ \text{mol-CO}_2/\text{kg-methanol}.$$

Therefore, 0.32 − 0.032 = 0.288 moles of CO_2 per mole of solvent is released in the gaseous form. The concentration of CO_2 in the solvent would drop further if some of the solvent evaporates and mixes with the separated CO_2. This is how pressure-swing absorption operates.

c. If the solvent is transported to an environment in which the pressure is lowered to 1 atm (0.1 MPa) and the temperature is raised to −40 °C, then the new equilibrium concentration of CO_2 in the solvent is $X_{L,CO_2}(0.1\ \text{MPa},\ -40\ °\text{C}) = \frac{0.1\times1}{1.4} = 0.071$.

Notice that the Henry's law constant at −40 °C is 1.4 MPa/(mol-CO_2/kg-methanol). So, 0.32 − 0.071 = 0.249 mole of CO_2 is released per mole of solvent. This is how temperature-swing absorption operates.

Clearly, pressure-swing absorption is a good option for separating a component from a gas mixture at high pressures.

10.4.1.2 Energy Requirement in Pressure-Swing Absorption

In pressure-swing absorption (PSA), the solvent must be circulated between the high-pressure absorption tower and the desorber chambers, where pressure reduction is needed to release CO_2 and regenerate the solvent. The energy required in this process is essentially the energy needed to circulate the solvent between the desorption tower and the absorption tower and back (in some case, the energy required to compress the gases leaving the desorber tower is also added). This energy is required to overcome the solvent pressure drop between the absorption tower and the flash (low pressure, desorption) chambers, and through the towers. In the absorption tower, pressure drop is due to the flow of the gas and the solvent through the packed bed. Moreover, pressure must be reduced in the flash chambers to release CO_2 from the solvent. Thus, the total solvent pressure drop can be written as follows:

$$\Delta p_{total} = \Delta p_{between\ tanks} + \Delta p_{across\ absorber} + \Delta p_{across\ flash\ chambers}. \tag{10.6}$$

Pumping power for the solvent can be estimated from the following:

$$\wp_{pump} = \dot{\forall}_{solv} \frac{\Delta p_{total}}{\eta_{pump}}. \tag{10.7}$$

The total pressure drop as the solvent is circulated between the absorption tower and the flash chamber is Δp, the pump efficiency is η_{pump}, and the required solvent volume flow rate is $\dot{\forall}_{volv}$. Extra power is required to pump the original gas mixture through the absorption tower, and to compress the separated CO_2 from the flash chamber to the atmospheric pressure, or the storage pressure.

The solvent flow rate depends on: the fraction of CO_2 in the gas mixture; the total flow rate of CO_2 that needs to be removed; and the solubility of the solute in the solvent. The flow rate of the solvent can be calculated as follows. Taking an overall mass balance for CO_2 between an arbitrary section along the tower – where the molar concentration of CO_2 in the gaseous and liquid streams are, respectively, X_v and X_L, and the molar flow rates of the gas stream and the liquid streams are $\dot{n}_v$ and $\dot{n}_L$ – and the section at the top of the tower, where the gas mixture leaves the tower (see Figure 10.8), we write:

$$\dot{n}_V X_V + \dot{n}_L X_{L,in} = \dot{n}_{V,out} X_{V,out} + \dot{n}_L X_L. \tag{10.8}$$

In this equation, subscripts "in" and "out" refer to the streams entering and exiting the tower, respectively. In writing the mass conservation equation, uniformity is assumed across the cross-section of the tower. Because of the transport of CO_2 from the gaseous stream to the liquid stream, the total molar flow rates of the liquid and gas streams change along the tower, as well as the mole fraction of CO_2 in both streams.

The same equation can be written across the entire tower, between the bottom and top sections, as follows:

$$\dot{n}_{V,in} X_{V,in} + \dot{n}_L X_{L,in} = \dot{n}_{V,out} X_{V,out} + \dot{n}_{L,out} X_{L,out}. \tag{10.9}$$

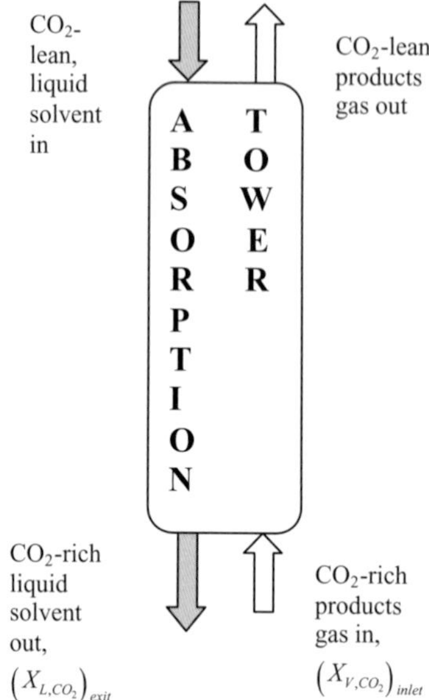

Figure 10.8 Mass balance for the physical absorption process described in the text. Liquid solvent is used to absorb CO_2 from the products gases.

If we assume ideal conditions in which $X_{v,out} \approx X_{L,in} \approx 0$, we can write:

$$\dot{n}_{V,in} X_{V,in} = \dot{n}_{L,out} X_{L,out}. \tag{10.10}$$

Thus, the required molar flow rate of the liquid solvent is the molar flow rate of CO_2 entering the tower, $\dot{n}_v X_{v,CO_2}$, divided by CO_2 solubility in the liquid at the given conditions, X_{L,CO_2}, and the required solvent mass flow rate of

$$\dot{n}_L = \frac{(\dot{n}_V X_{V,CO_2})_{in}}{X_{L,CO_2}}. \tag{10.11}$$

In applying (10.11) to the case shown in Figure 10.8, the parameters are as follows: $\dot{n}_L$ is the molar flow rate of the liquid solvent, $\dot{n}_V$ is the molar flow rate of the product gases, X_{V,CO_2} is the mole fraction of CO_2 in the products gases as they enter the absorption tower, and X_{L,CO_2} is the mole fraction of CO_2 in the liquid solvent leaving the tower. Assuming equilibrium at the tower exit, we can write, using Henry's law:

$$(X_{L,CO_2})_{exit} = \frac{\left(pX_{V,CO_2}\right)_{inlet}}{\text{He}}. \tag{10.12}$$

Henry's constant in this equation is evaluated at the bottom of the tower. The value of $\left(pX_{V,CO_2}\right)_{inlet}$ is the partial pressure of CO_2 in the incoming stream. Substituting (10.12) into (10.11), we get the required solvent molar flow rate:

$$\dot{n}_{sol} = \frac{\text{He}}{p} \dot{n}_{prod}. \tag{10.13}$$

Using a solvent with lower Henry's constant and operating at a higher pressure makes it possible to use lower solvent flow rate. The solvent molar flow rate determines the pumping power using (10.7). Equation (10.13) shows that, under these optimal conditions, the molar flow rate of the solvent depends on the molar flow rate of the gas mixture, the pressure, and the properties of the solvent, irrespective of the CO_2 concentration.

Pressure drop through the tower depends on the flow rate of the solvent and the design of the tower. More detailed analysis is required to determine the size of the absorber tower, which takes into consideration absorption kinetics – that is, the rate at which the gas is absorbed by the liquid solvent. Non-equilibrium conditions between the gaseous and liquid streams must exist inside the tower to promote the mass transfer of CO_2 between the gaseous and liquid phase. That is, at any section, the partial pressure of carbon dioxide in the gas phase must be higher than that corresponding to equilibrium of CO_2 in the gaseous and the liquid phases in order to generate the necessary flux toward the liquid surface.

Figure 10.9 shows the layout of a PSA system used for separating CO_2 from combustion gases. One absorption tower is employed, in which a cool CO_2-lean solvent is sprayed from the top, while the CO_2-rich gas is introduced from the bottom. The CO_2-rich liquid leaves the tower from the bottom and is expanded in two different stages, where the pressure is lowered to allow more CO_2 to be released from the solvent. The released carbon dioxide is compressed to the desired high pressure. The liquid solvent collected at the last drum is cooled and pumped back to the absorber tower.

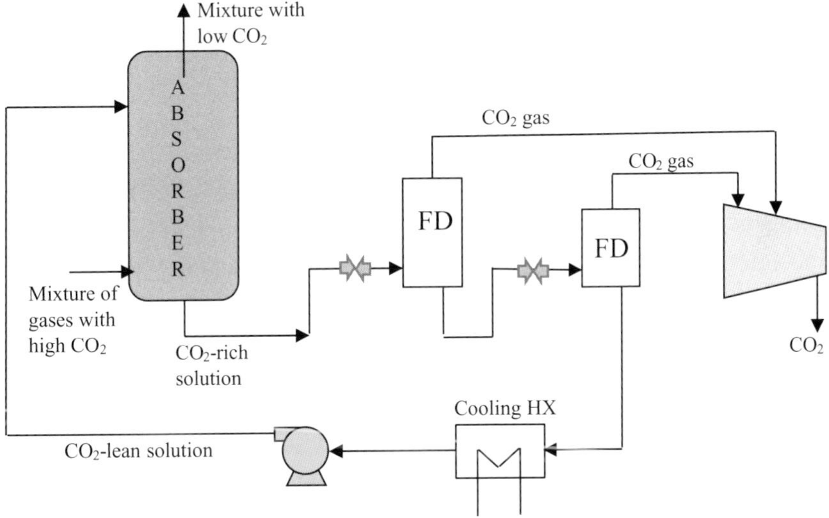

Figure 10.9 A pressure-swing absorption process used to separate CO_2 from combustion products of combustion using a circulating solvent. In the absorber tower, the liquid solvent absorbs CO_2 at atmospheric pressure. The CO_2-rich solvent then flows through flash drums (FD), where CO_2 is released as the pressure is lowered. Pressure is lowered before the FD using a throttle valve. The CO_2-lean solvent is cooled and pumped back to the absorption tower pressure. CO_2 is compressed to the desired transport/storage pressure.

10.4.1.3 Temperature-Swing Absorption

In temperature-swing absorption (TSA), desorption is carried out in a tower similar to the absorption tower, with steam used to heat the solvent. Steam is injected from the bottom and the CO_2-rich liquid solvent is introduced though a distributor at the tower top. The distributer produces a shower of fine mist. The separated CO_2 gas flows to the top of the tower, often with some of the steam, and the lean solvent continues flowing to the bottom of the tower. Steam is condensed outside the tank and separated from CO_2. In TSA, it is important to minimize the amount of steam needed for desorption since the energy of evaporation contributes significantly to the energy penalty of the separation process.

With respect to CO_2 separation, physical absorption has been shown to be compatible with synthetic gas production applications, because the high pressure of these gases reduces the energy requirement for solvent regeneration, both in power and fuel production. The total energy expenditure amounts to the heat required for the regeneration step and the pumping work required to circulate the working fluid within the separation plan. The steam used in the regeneration or desorption tower, in the case of temperature swing, is taken from the steam generator or extracted from the steam turbine at intermediate pressure stages, hence reducing steam turbine power. For synthetic gas applications, this work loss amounts to 0.15–0.25 MJ/kg-CO_2 removed. Physical absorption methods are not suitable for removing carbon dioxide from flue gases in air-combustion applications (post-combustion capture) because of the low concentration of CO_2 in these gases. Chemical scrubbing methods are more suitable for these applications, and are described in the next section.

10.4.1.4 Using Physical Absorption in CO_2 Capture

The availability of a high-pressure gas mixture containing CO_2 favors the application of physical absorption to separate CO_2 in power plants or fuel production plants that use synthetic gas produced in coal gasification, or partial oxidation of natural gas. Gasification and reforming are often performed under higher pressures to reduce the equipment size, promote the chemical kinetics, and produce fuel gas that can be used in the high-pressure combustor of a combined cycle (CC) (or to produce synthetic fuels). The high pressure of the gas promotes the absorption into the solvent. The solvent is chosen such that the relative solubility of other gases, such as H_2, in the solvent is much lower than that of CO_2. In the case of coal gasification (partial oxidation), the sulfur in coal is converted to H_2S as it reacts with water in the reducing environment of the gasifiers (that is, an environment that lacks oxygen), which must be removed before the gas can be used in a gas turbine combustor. When using physical absorption for carbon dioxide capture, it is possible that the same steps and equipment can be used to remove CO_2 and H_2S in the same solvent (the mixture is known as the acid gas). This is an attractive option in coal-fired plants running an IGCC cycle, but this requires special solvents and care must be taken to minimize corrosion. The sulfur compound concentration is much lower than that of CO_2, and carbon dioxide absorption requires much higher solvent flow rate than that required for removing H_2S.

While most of the power required in physical absorption is in circulating the gases and solvent between the desorber lower pressure and the absorber higher pressure, the exothermic absorber reactions require cooling to keep the temperature low and the solubility of CO_2 in the solvent as high as possible. Moreover, because the solubility of most gases is approximately proportional to the partial pressure of the gaseous component in the mixture, higher pressures reduce the size of the equipment and the power expended in the process. (Absorbers operate at high pressures (e.g., 10–50 bar) and low temperatures (e.g., 20 to −20 C)). While the low temperature promotes absorption, it also raises the viscosity, which leads to a higher pressure drop in the equipment, and possibly a lower mass transfer in the tower, thereby increasing the tower length and further contributing to the pressure drop. Gas absorption towers often operate at higher pressures and lower temperatures for another reason: to raise the mass transfer rates between the gas and liquid streams and hence further reduce the overall size of the tower. The driving force of mass transfer is the difference between the partial pressure of CO_2 in the gas stream and that at the gas–liquid interface, and the higher the difference the faster the absorption of the gas into the liquid solvent. However, the exothermicity of the absorption reaction heats up the solvent and may lead to some water evaporation. Thus, makeup water is often required.

The solvents used in physical absorption processes include:

1. methanol, which is used in the **Rectisol® process**;
2. N-methyl-2-pyrrolidone (NMP), used in the **Purisol® process**; and
3. dimethylether polyethylene glycol (DMPEG), used in the **Selexol® process**.

The pressure drop in (10.6) consists of: the pressure drop between the two towers, in the flash chambers, in the absorber towers, and in the pipes that circulate the fluid throughout the system. In order to estimate the pressure drop across the absorption tower, one needs to know the detail of its design. Gas absorption is carried out in packed towers in which a cylindrical vessel is packed either at random or in a structured pattern, with inert material such as clay, porcelain, or plastics, and occasionally metals. The packed bed porosity is controlled by the packing pattern and the spaces between the packing units. Different patterns and corrugations have been used to encourage the formation of thin liquid films, hence increasing the surface area between the liquid and the gas. Direct contact between the liquid and gaseous streams over a long residence time within the tower is necessary for absorption.

The pressure drop across the tower depends on the gas and liquid flow rates. The pressure drop rises almost linearly with increasing the gas mass flow, until the gas flow starts to impede the down flow of the liquid. At high gas flow, the gas may hold up the liquid. At this point the pressure drop rises sharply as the liquid flow rate increases beyond a critical value, called the **loading point**. At even higher values of liquid flow rate, the tower reaches the flooding point, beyond which the liquid phase becomes continuous. Clearly, from the perspective of the size of the tower, higher velocities are desirable, but that would be at the cost of higher pressure drops. However, these pressure drops are often small, and reduction of the tower size is more important. Several generalized correlations for pressure drop have been suggested, one of which gives the pressure drop per unit length as a function of $C_s F_p^{0.5} \nu^{0.05}$ and $G_l/G_x\sqrt{\rho_g/\rho_l}$, where $C_s = V_0\sqrt{\rho_g(\rho_l-\rho_g)}$ is a capacity factor, l and g correspond to the liquid and gas, respectively, V_0 is the superficial gas velocity, F_p is a packing factor which decreases with increasing packing size and/or void fraction, and ν is the kinematic viscosity of the liquid, G is the mass velocity, and ρ is the density. Some of these correlations are given in figures 18.6 and 18.7 in [1]. The following example demonstrates how these correlations are used.

Example 10.2

A process engineer is tasked with designing a packed tower for handling a gaseous mixture of 40% CO_2 and 60% CH_4, using pure methanol as the absorbent. The volumetric flow rate of the mixture is 0.25 m^3/s. The gas mixture is at 400 K and 1 atm, and the methanol is at 293 K and 1 atm. The design pressure drop is set to 817 Pa/m. Calculate: (a) the mass flux of the gas and the absorbent; and (b) the diameter of the tower. Assume $F_p = 92$ and $G_l = 1.4G_g$, and treat the mixture as an ideal gas.

Solution

The mass flux of the gaseous mixture can be determined from $G_g = u_0\rho_g$, where the gas density is obtained using the ideal gas law. Hence,

$$\rho_g = \frac{M_g \times 101.3}{8.314 \times 400},$$

where the average molecular weight is $M_g = (M.X)_{CO_2} + (M.X)_{CH_4} = 44 \times 0.4 + 16 \times 0.6 = 27.2$.

Substituting in the ideal gas law equation, we get. $\rho_g = 0.8285\ \text{kg/m}^3$.

The superficial velocity is determined using figure 18.7 in [1]. The liquid methanol density is $\rho_l = 791.8\ \text{kg/m}^3$. Thus, for $\Delta p = 817\ \text{Pa/m} = 1\ \text{in-H}_2\text{O/ft}$ and

$$\frac{G_l}{G_g}\left(\frac{\rho_g}{\rho_l}\right)^{0.5} = 1.4 \times \left(\frac{0.8285}{791.8}\right)^{0.5} = 0.0453,$$

we find $C_s F_p^{0.5} \nu^{0.05} = 1.65$. With $F_p = 92$ and $\nu_{293\text{K}} = 0.64cSt$, the capacity factor is

$$C_s = \frac{1.65}{92^{0.5} \times 0.64^{0.05}} = 0.176.$$

Hence, $V_0 = C_s\left(\frac{\rho_l - \rho_g}{\rho_l}\right)^{0.5} = 0.176 \times \left(\frac{791.8 - 0.8285}{0.8285}\right)^{0.5} = 5.44\ \text{ft/s} = 1.66\ \text{m/s}.$

The mass flux of the gaseous mixture is now calculated:

$$G_y = 1.66 \times 0.8285 = 1.375\ \text{kg/m}^2\text{·s}.$$

Also, $G_l = 1.4 \times 1.375 = 1.925\ \text{kg/m}^2\text{·s}.$

Example 10.2 (cont.)

b. The cross-sectional area of the tower is obtained from

$$A_{tower} = \frac{\dot{V}_g}{V_o} = \frac{0.25\,[\mathrm{m^3/s}]}{1.66[\mathrm{m/s}]} = 0.151\,\mathrm{m}^2.$$

Thus, the required diameter of the tower is

$$d_{tower} = \left(\frac{4 \times 0.151}{\pi}\right)^{0.5} = 0.439\ \mathrm{m}.$$

10.4.2 Chemical Absorption

Chemical separation of a component from a gas mixture, for instance carbon dioxide from a mixture of flue gases (CO_2, H_2O, N_2), starts with forming a chemical bond between the **absorbate** (e.g., CO_2) and a liquid solvent in an **absorber/scrubber tower**. The liquid solvent is an aqueous solution of a chemical with strong affinity for CO_2, such as an alkanolamine, in short amine, in water. The result of the chemical reaction between CO_2 and the absorbing chemical is the formation of a CO_2-rich molecule in the aqueous medium [7].

$$\mathrm{CO_2 + amine \rightleftarrows COO - R\ + \cdots}, \tag{10.14}$$

where COO – R stands for a CO_2-rich amine. The mixture on the right-hand side of (10.14) is then transported away from the absorber tower and into an environment in which this bond is broken in a desorber tower. In the desorber/stripper tower, CO_2 is separated from the chemical in the reverse reaction. The CO_2-lean mixture that forms in the stripper tower is recycled back to the absorber tower to complete the cycle, while CO_2 is released into a holding tank for further processing (e.g., either compression or liquefaction). The conditions in the absorber tower and the stripper towers are chosen such that the equilibrium concentration of CO_2 in the mixture is as high as possible in the former and as low as possible in the latter, according to (10.14). The forward reaction is exothermic, and low temperature favors the forward reaction – that is, a higher concentration of CO_2 in the amine solution.

The higher the CO_2 concentration in the gas mixture, the more attractive chemical absorption is for gas separation. Mass concentrations of CO_2 higher than 12% in the gas mixture are often recommended when chemical separation is used. Strong dilution of combustion products in lean-burn gas turbine power plants operating on natural gas or syngas reduces the effectiveness of chemical absorption separation processes (and flue gas recirculation could be used to increase CO_2 concentration in the gas turbine working fluid and in the exhaust).

The following reaction shows the process symbolically:

$$\text{CO}_2\text{-lean solvent} + \text{CO}_2 \underset{\text{desorber@high T}}{\overset{\text{absorber@low T}}{\rightleftharpoons}} \text{CO}_2\text{-rich solvent.} \tag{10.15}$$

Increasing the concentration of carbon dioxide in the mixture helps to reduce the size of the equipment – that is, the absorber tower – because it increases the kinetic rates of the reactions. The sizes of these towers depend primarily on the volumetric gas flow rates of the flue gases and the solvent. Operating at higher pressures is also beneficial, since the equilibrium concentration of CO_2 in the solvent and the mass transfer rate between the gas mixture and the solvent rise with pressure. However, higher pressure requires more compression work for the gas and for the solvent. Other solvent characteristics are also important, besides a strong affinity to CO_2. The solvent should not be toxic at concentrations of the amine in water that make it useful in absorbing CO_2 and, most importantly, it should be easily and readily regenerable in the desorber/scrubber at low-energy requirement – that is, the reaction in (10.15) should be reversible. To be regenerable, it is important that changing the conditions can promote the reverse reaction, in which CO_2 is released and the original solvent is brought back to its original state.

A typical arrangement used to strip CO_2 from flue gases is shown in Figure 10.10. Similar to physical absorption, the gas mixture comes into contact with the solvent in a counter-flow configuration within the absorber tower. The design of the absorber tower ensures direct contact between the gas stream and the solvent. Here also, packed towers are used for both the absorber and the desorber. First the gas mixture, the flue gases in this case, is cooled to low temperatures, 30–50 °C, and its pressure is raised to overcome the pressure losses in the

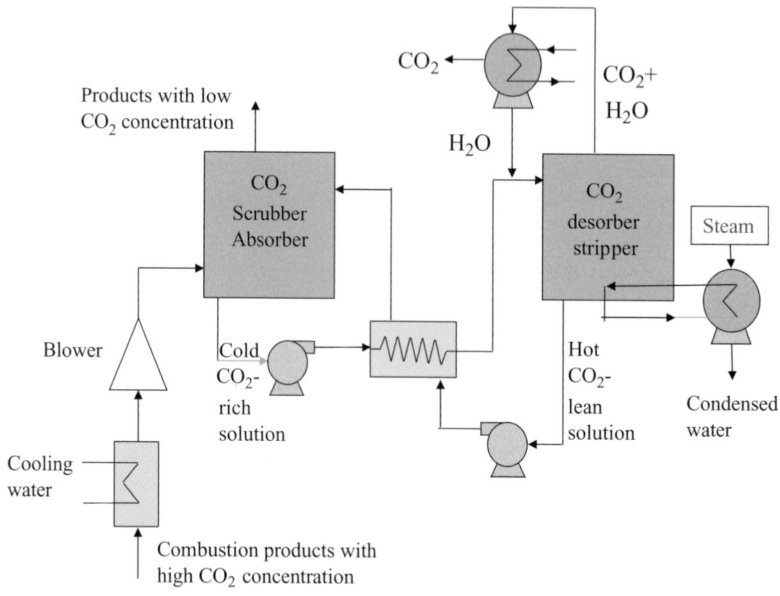

Figure 10.10 Schematic diagram and flow sheet for the recovery of CO_2 from a gas mixture using chemical absorption. After cooling, the combustion gases are blown into the absorber tower, where they contact the solvent and lose most of their CO_2. The CO_2-rich solvent is warmed up in a heat exchanger, then blown into the desorber tower, where it is heated by steam and releases the CO_2. The CO_2-lean solvent is pumped back to the absorber tower after cooling in the heat exchanger.

absorber tower. Flowing upwards, the flue gases come into contact with the descending liquid absorber within the packed tower. During this process, carbon dioxide is removed from the flue gas mixture and absorbed into the solvent. At the top of the tower, the CO_2-lean mixture leaves, while the CO_2-rich liquid mixture is collected and leaves at the bottom of the tower. The rich mixture is pumped through a heat exchanger/recuperator, in which it is heated by the liquid CO_2-lean absorber flowing back from the desorber tower. The regenerator is used between the two columns for heating the CO_2-rich solvent and cooling the CO_2-lean liquid returning to the absorber tower. The warm CO_2-rich liquid enters the desorber tower at the top, and flows downwards while getting further heated by a steam reboiler. The steam reboiler is used in the stripper to provide the necessary thermal energy to maintain the CO_2-rich liquid mixture at sufficiently high temperature for the CO_2 separation reaction to proceed at sufficiently high rate, 100–180 °C. Carbon dioxide gas leaves at the top of the desorber tower, with some water and other solvent vapor. This water/solvent is condensed and returned to the desorber tower. The CO_2-lean liquid is collected at the bottom of the tower and pumped into the heat exchangers to heat up the CO_2-rich gas coming from the absorber tower. The cool liquid then flows to the top of the absorber tower.

Flue gases should be purified from other contaminants, such as sulfur compounds, nitric oxides, and oxygen before being treated by amine-based solvents, since these contaminants can react with amines and form salts, thus causing solvent loss. Chemical absorption-based separation is recommended for post-combustion separation, in which coal or natural gas sulfur is converted to SO_2 during combustion. NO_x levels below 20 ppm are also recommended, which is currently above the levels that most power plants must satisfy to meet air-quality regulations. On the other hand, there is a trade-off between the cost of flue gas desulfurization and that associated with the solvent required to compensate sulfur compound-related degradation. For monoenthanolamine (MEA), a SO_x level of less than 10 ppm is desired for the Fluor Daniel Econamine^{TM} process. On the other hand, a level of 50–100 ppm can be tolerated by the KEPCO/MHI process [8,9]. SO_x can form corrosive sulfuric acid/sulfurous acid aerosols in wet scrubbers, and a mist eliminator or a wet electrostatic precipitator is required in the flue gas desulphurization unit to reduce its concentration to safe levels. NO_x and SO_x removal of up to 98.6% have been demonstrated (down to 6.5 ppm). Oxygen can oxidize the absorbent, and additives must be used to allow for high MEA concentration. Some of the compounds that form in this process can be environmentally harmful. While flue gas clean-up before CO_2 scrubbing should minimize amine loss, one should nevertheless expect some loss and a continuous replenishment of the solvent.

Aqueous solutions of MEA have been used for CO_2 separation in several applications, such as in removing CO_2 from natural gas at the wellhead. The overall chemical reaction for the MEA is:

$$CO_2 + 2R\text{-}NH_2 \leftrightarrow R\text{-}NH_3^+ + R\text{-}NH\text{-}COO^-, \tag{10.16}$$

where $R \equiv (\text{-}CH_2CH_2OH)$. The enthalpy of reaction of the forward reaction is -1.919 MJ/kg-CO_2. Since the absorption reaction is exothermic, extra cooling is required in the absorber tower to maintain favorable conditions for the forward reaction in (10.16).

Table 10.4 The enthalpy of the absorption reaction of CO_2 in some amines

		MJ/kg-CO_2
Monoenthanolamine (MEA)	$R \equiv (\text{-}CH_2CH_2OH)$	−1.919
Diethanolamine (DEA)	$R = (\text{-}CH_2CH_2OH)_2$	−1.519
Triethanolamine (TEA),		−0.989
Methyldiethanolamine (MDEA),		−1.105

Other solvents have been proposed to raise the amine content in the absorber solution, and hence reduce the heat required for regeneration below that required by MEA. Several solvents [10] and their enthalpies of reaction measured per kg-CO_2 released are shown in Table 10.4.

10.4.2.1 Energy Requirements

There is extensive experience in using amines as scrubbers in refineries to remove H_2S, and in gas purification for liquid petroleum gas (LPG) production. High removal efficiency, up to 85%, has been achieved in large-scale applications of CO_2 separation from other gases. One large-scale example is the STATOIL Sleipner T project in Norway, in which CO_2 from sub-sea oil/gas reservoirs is separated from the oil/gas, and injected back underground.

The total thermal energy or heat required in chemical scrubbing can be expressed using the following equation [2]:

$$\begin{aligned}\text{energy for separation} = {} & \text{enthalpy of the desorption reaction} \\ & + \text{thermal enthaply of solution} \\ & + \text{enthalpy of evaporation of vaporized water} \\ & + \text{enthalpy of evaporation of vaporized absorbent} \\ & + \text{enthalpy of desolution of CO2 from water} \\ & + \textit{pumping work}.\end{aligned} \tag{10.17}$$

Note that to convert the pumping work to heat one needs to assume a value for the plant efficiency, and the same to convert the heat required for separation to work/exergy for separation. The exergic efficiency of heat is $\left(1 - \frac{T_o}{T_{heat}}\right)$, where T_{heat} is the temperature at which this heat is required.

Heat is used to raise the solution temperature in the stripper to the value required for the desorption reaction to proceed, and as enthalpy of reaction of the desorption reaction. Since this temperature is higher than 100 °C and the desorption tower operates at atmospheric pressure, some of the thermal energy is "wasted" in evaporating the solvent, in this case water. If conditions are chosen carefully (e.g., higher concentration of amine in the solution) to reduce the solvent flow, the energy penalty can be minimized. The mass of steam required to raise the temperature in the heater of the desorber can be estimated from the following thermal energy balance [2,3]:

$$Q_{recovery} = n_{solution}\hat{c}_{solution}\Delta T + n_{water}\Delta\hat{h}_{evap} + n_{sorbent}\left(\Delta\hat{h}_{desorption} + \Delta\hat{h}_{CO_2,sol}\right), \tag{10.18}$$

$$n_{sorbent} = \frac{n_{CO_2}}{\bar{X}_{CO_2}}, \tag{10.19}$$

$$n_{solution}\frac{n_{sorbent}}{X_{sorb}} \tag{10.20}$$

where: $Q_{recovery}$ is the thermal energy required to remove CO_2; $n_{sorbent}$ and $n_{solution}$ are the number of moles of the sorbent and the solution, respectively; $\bar{X}_{CO_2}$ is the CO_2 solubility in the sorbent (number of moles of CO_2 per mole of sorbent); X_{sorb} is the sorbent concentration in the solution, ΔT is the temperature difference between the scrubber and the stripper; n_{water} is the number of moles of water evaporated during scrubbing; $\Delta\hat{h}_{evap}$ is the enthalpy of evaporation of water inside the stripper; $\hat{c}$ is the solvent specific heat; $\Delta\hat{h}_{desorption}$ is the enthalpy of desorption, and $\Delta\hat{h}_{CO_2,sol}$ is the enthalpy of dissolving CO_2 from water.

The following observations are important in evaluating the thermal energy required in chemical absorption processes for removing carbon dioxide:

1. It is reasonable to assume that the number of moles of water evaporated during stripping is the same as the number of moles of CO_2 leaving the solution, $n_{CO_2} = n_{water}$. Since water is condensed outside the desorber tower, the enthalpy of condensation is lost.
2. The temperature difference between the two towers is about 100 °C. A heat exchanger between the two towers can be used in a heat recovery process, hence reducing the energy required to raise the temperature of the solvent as it flows into the scrubber tower.
3. Under these conditions, most of the thermal energy required in the scrubber is the enthalpy of the desorption reaction, as well as the enthalpy of evaporation of water in the desorber (which is unavoidable since the temperature is sufficiently high).
4. The flow rates in both towers are not equal; the flow leaving the scrubber and entering the stripper is larger than that leaving the stripper and entering the scrubber. The difference is equal to the amount of CO_2 removed in the stripper.
5. The sorbent concentration is often 30% for MEA.
6. Two moles of sorbent are required for each mole of CO_2 absorbed (according to the (10.16)).
7. It is difficult to realize 100% sorbent regeneration efficiency, which increases the amount of sorbent used over the ideal value.

Calculations show that MEA requires about 2–5 MJ/kg-CO_2 of thermal energy, depending on the concentration of MEA in the aqueous solution. Typical MEA concentration in water ranges between 10% and 30% by mass, although mixtures with 50% MEA have been attempted. Higher concentrations of the sorbent in the solution reduce the energy required in the separation process since it reduces the volume flow rate of the solvent, the amount of thermal energy required to raise the solvent temperature in the stripper tower, and the steam evaporation in the desorption process. However, higher sorbent concentration can lead to corrosion and practical limits of ~30% have been used. Other factors affecting the separation energy include the percentage of CO_2 removed from the gases and the original concentration of carbon dioxide in the flue gases

Typical values for the variables in these expressions are now given using a coal combustion example. Overall, the total energy required in physical separation is of the same order of magnitude as that in chemical separation.

Example 10.3

Carbon dioxide produced from burning a high-rank coal stoichiometrically is captured using the chemical absorption process. The mole concentration of the sorbent in the solvent is 10%. The heating value of the coal is 30 MJ/kg. Calculate the heat requirement of the reboiler.

Solution

Burning one mole of coal stoichiometrically generates one mole of CO_2, and 360 MJ of thermal energy (high-rank coal with 30 MJ/kg-coal). The reaction energy of desorption is 1.919 MJ/kg-CO_2, or 84.5 MJ for one mole of CO_2. The enthalpy of evaporation of water is nearly 2.25 MJ/kg-water. Assuming one mole of water evaporating for each mole of CO_2 recovered (see above), then one needs ~40 MJ of heat. Thus, the required thermal energy in the reboiler is about 125 MJ. This is a substantial fraction of the enthalpy of reaction of the fuel (360 MJ). This does not account for the heat required to raise the solvent temperature between the absorber and stripper towers. Regeneration between the two towers provides a fraction of this sensible energy. The impact of this energy on the plant efficiency is moderated by the lower temperatures needed for separation. It should be noted that these numbers are merely illustrative and more detailed calculations should be used to estimate the energy required by the separation plant, taking into consideration how it is integrated with the power plant.

The thermal energy used in the separation process can be converted into equivalent work loss by multiplying by the thermal efficiency of the cycle. However, this is only an approximate estimate since the lower-temperature heat required in the stripper would have been used in the lower-pressure part of the steam cycle, where the conversion efficiency is lower than that of the overall cycle. The work loss is:

$$W_{CO_2,removal} = Q_{recovery}(\eta_{cycle})_{lowerT}, \tag{10.21}$$

where $(\eta_{cycle})_{lowerT}$ is the conversion efficiency discounted according to the lower temperature of the steam required in the reboiler, or the exergic efficiency of the thermal energy used in the separation process (see Chapter 11). The steam extracted from the power cycle to

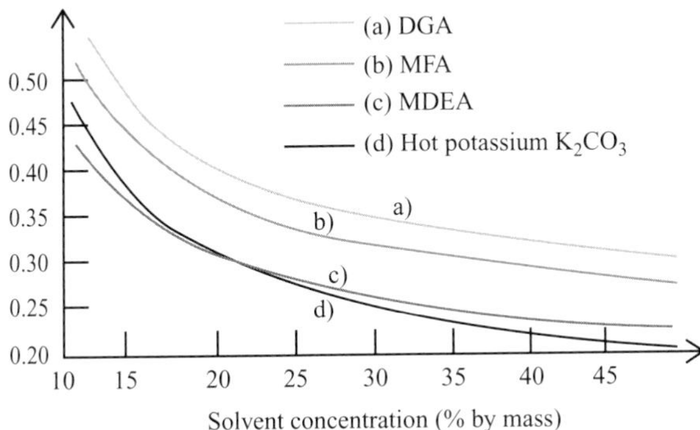

Figure 10.11 The energy penalty for CO_2 separation using chemical absorption expressed as work (exergy) or reduction of electricity production in kWh/kg-CO_2. Increasing the concentration of the sorbent in the solution reduces the energy required for CO_2 removal. With higher sorbent concentration, more energy is expended in the desorption reaction enthalpy than in heating and evaporating the solvent. However, higher sorbent concentration can be corrosive, and mass concentrations below 30% have been recommended (kWh = 3.6 MJ). DGA is known as Econamine (*n*-diglycolamine in aqueous solution), MDEA is 2*n*-methyldiethanolamine.

provide this heat, whose temperature must be above that required in the reboiler (the heater of the desorber) – that is, above 150 °C – leaves the reboiler as condensed water at the same temperature, and is added to the water leaving the condenser (after being throttled to the condenser pressure). This saves some of the fuel required to heat the water for the steam cycle, and hence (10.21) may represent a pessimistic estimate for the work of separation using chemical absorption.

The estimated energy for removing CO_2 from flue gases is in the range of 0.8–1.5 MJ/kg-CO_2 for the pumping power and 2–5 MJ/kg-CO_2 of the heat for solvent regeneration. The actual value depends on the fuel and plant design (IGCC, pulverized coal steam plant, etc.). Figure 10.11 shows estimates of the energy penalty for separating CO_2 using chemical absorption with different absorbents, based on exergic efficiency of 19% in (10.21). The energy decreases nonlinearly with the solvent concentration in the aqueous mixture. For the total energy penalty associated with carbon dioxide capture, we should add the CO_2 compression and liquefaction.

Other solvents have been considered for chemical separation of CO_2, including aqua ammonia. Ammonia solutions have high capture capacity for CO_2, according to the following chemical reactions:

$$NH_{3(aq)} + H_2O + CO_{2(g)} \leftrightarrow NH_4HCO_{3(aq)},$$

$$2NH_{3(aq)} + H_2O + CO_{2(g)} \leftrightarrow (NH_4)_2CO_{3(aq)},$$

$$(NH_4)_2CO_{3(aq)} + H_2O + CO_{2(g)} \leftrightarrow 2NH_4HCO_{3(aq)}.$$

Adding these reactions shows that three moles of ammonia capture four moles of CO_2, while two moles of amines are required for capturing one mole of CO_2. Aqua ammonia can also be used to remove SO_2 and NO_x from the gas, an additional advantage when coal is used as a fuel. The enthalpy of reaction is -1.455 MJ/kg-CO_2, and the scrubber and stripper temperatures are 27 and 82 °C, respectively. The low temperature of regeneration lowers the energy penalty. While these reactions are reversible (i.e., the solvent can be regenerated in the desorber), the products of the forward reactions can potentially be used as fertilizers directly if so desired. One disadvantage of using ammonia is that these reactions require substantial amounts of water.

An important advantage of chemical absorption over physical absorption is the fast mass transfer rates at the gas–liquid interface associated with the depletion of the solute at the interface due to the reaction with the solvent. The mass transfer rate within the gas phase depends on the difference between the partial pressure of CO_2 in the bulk and at the interface. With CO_2 consumed by the reaction at the interface, its partial pressure drops, creating more favorable mass transfer conditions. Chemical absorption is currently the favorable method for separating CO_2 from flue gases, as it requires the lowest energy. Chemical absorption achieves high purity.

Chemical scrubbing using an aqueous solution with 20% by weight of MEA was used at the Lubbock power plant in Texas, where natural gas was fired in a 50 MW plant, and in a steam generator in Carlsbad, NM. The first plant produced ~1000 t/day of CO_2, while the second produced ~113 t/day. In both cases, CO_2 was used for enhanced oil recovery (EOR) in nearby oil fields. CO_2 scrubbing with 15–20% MEA solutions in the 300 MW Shady Point Combined Heat and Power Plant in Oklahoma has been producing nearly 400 t/day CO_2, which is used in the food industry and in EOR. A similar operation is done in a Botswana plant burning coal. The Norway Sleipner Vest gas field separates CO_2 from the recovered natural gas to reduce CO_2 concentration in the produced gas from 95% to 2.5%. The separated CO_2 is injected back into a 250 m deep aquifer located 800 m below the ocean surface.

Example 10.4

A coal-fired power plant employs a CO_2 capture system as shown in Figure 10.4.1. Supercritical steam enters the high-pressure turbine (HPT) at 580 °C and 29 MPa. After leaving the HPT at a pressure of 2.2 MPa, the steam is reheated to 580 °C and fed into the intermediate pressure turbine (IPT). A fraction f of the working fluid is bled off at a pressure of 600 kPa and used as a heat source in the CO_2 capture system. The remaining fluid is expanded in the low-pressure turbine (LPT) to a pressure of 3.16 kPa. The fluid that was extracted for the CO_2 capture system exits the desorber as a saturated liquid at 600 kPa and passes through a throttle before mixing with the rest of the working fluid. The working fluid exits the condenser as a saturated liquid at 3.16 kPa, and is then pumped up to the HPT inlet pressure of 29 MPa before entering the steam generator. The coal used has a heating value of 29 MJ/kg, and its chemical formula is $C_{1.0}H_{0.8}O_{0.2}$. Assume that the entire heating value of the coal is available to heat the working fluid in the steam generator and reheater. The pump has an isentropic efficiency of 70%, and all turbines have isentropic efficiencies of 90%. Table E10.4.1 gives the states of the steam cycle.

Example 10.4 (cont.)

Table E10.4.1

State	P (kPa)	T (°C)
1	3.16	
2	29,000	
3	29,000	580
4	2200	
5	2200	580
6	600	
7	3.16	
8	3.16	
9	600	
10	3.16	

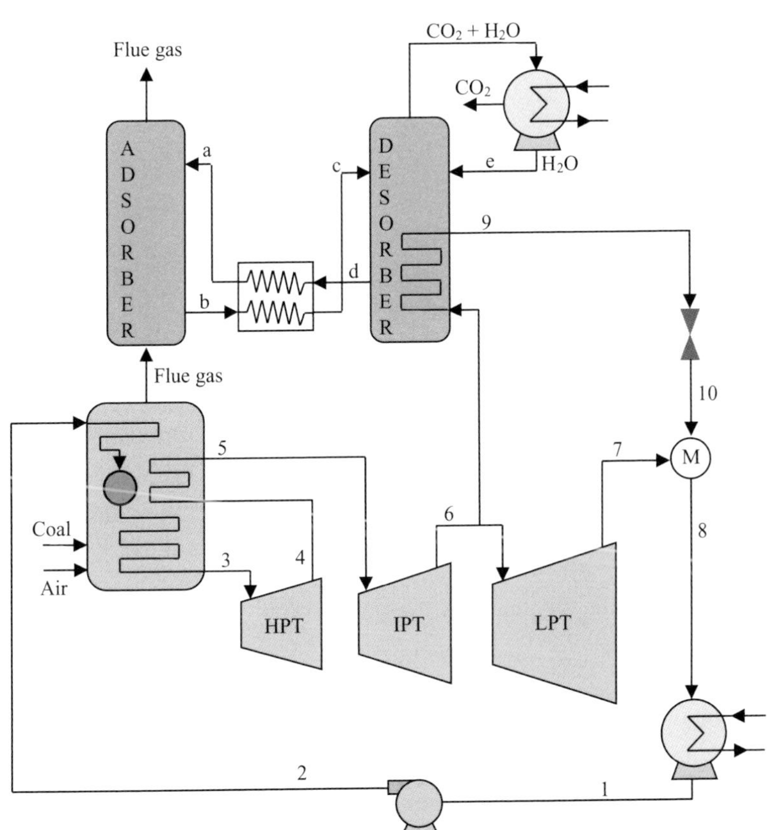

Figure E10.4 Power cycle with post-combustion separation of CO_2.

Example 10.4 (cont.)

The CO_2-rich flue gas enters the absorber at 125 °C, where it comes into contact with a mixture of water and MEA (formula $NH_2CH_2CH_2OH$). The amine mixture is 40% MEA by mass. Two moles of MEA react with each mole of CO_2 captured; 90% of the CO_2 in the flue gas reacts this way. After leaving the absorber at 25 °C, the rich amine solution is preheated by the lean solution leaving the desorber. Steam extracted from the power cycle supplies heat to raise the temperature of the amine solution and break the CO_2–amine bonds. Additionally, one mole of H_2O evaporates for every mole of CO_2 captured. The enthalpy of the stripping reaction is 1.919 MJ per 1 kg of CO_2 captured. The lean amine solution leaves the desorber as a liquid at 100 °C. The CO_2 and water vapor exit the desorber at 150 °C. The water that evaporated in the desorber is condensed and returned to the desorber at 90 °C, and the CO_2 leaves the system as pure stream. The lean amine solution is cooled by the rich solution in a regenerator and reintroduced to the absorber. The temperature difference between the streams of the regenerator is 15 °C. The heat capacity of MEA is 2.95 kJ/kg·K. Both the absorber and desorber operate at atmospheric pressure. The states of the MEA–water mixture are summarized in Table E10.4.2.

Table E10.4.2

State	T (°C)	Note
a		Lean mixture
b	25	Rich mixture
c		Rich mixture
d	100	Lean mixture
e	90	Water

Assume that the rich amine solution consists of MEA and CO_2 as separate components with individual C_p values. (a) Calculate the efficiency of this power plant without CO_2 capture. (b) If the net power output is 1000 MW, find the mass of coal consumed and the mass of CO_2 emitted annually. The capacity factor of the plant is 75%. (c) Determine the fraction of the Rankine cycle working fluid that must be bled off to supply heat for the CO_2 capture system. (d) Calculate the net power output and the efficiency of the power plant.

Solution

a. To calculate the efficiency of the plant, we need to calculate the work consumed by the pump, the work produced in each turbine stage, and the heat transfer in the steam generator. For convenience, all calculations are carried out assuming a unit mass flow rate of the working fluid. The specific work consumed by the pump is $w_p = h_1 - h_2$,

Example 10.4 (cont.)

where $h_1 = h_{sat}^{3.16\text{ kPa}} = 104.5\text{ kJ/kg}$ and $h_{2s} = h_{s=s_1}^{29,000\text{ kPa}} = 133.4\text{ kJ/kg}$.

Hence, $h_2 = h_1 + \dfrac{h_{2s} - h_1}{\eta_P} = 104.5 + \dfrac{133.4 - 104.5}{0.7} = 145.8\text{ kJ/kg}$.

The specific work requirement of the pump is therefore

$$W_P = h_2 - h_1 = 145.8 - 104.5 = 41.3\text{ kJ/kg}.$$

State 4 can be determined by using the isentropic efficiency of the HPT:

$$h_3 = h_{29,000\text{ kPa}}^{580°\text{C}} = 3389\text{ kJ/kg and } h_{4s} = h_{s=s_3}^{2200\text{ kPa}} = 2741\text{ kJ/kg},$$

$$h_4 = h_3 - \eta_T(h_3 - h_{4s}) = 3389 - 0.9 \times (3389 - 2741) = 2805.8\text{ kJ/kg}.$$

Similarly, for the IPT, we have

$$h_5 = h_{2200\text{ kPa}}^{580°\text{C}} = 3644\text{ kJ/kg and } h_{6s} = h_{s=s_5}^{600\text{ k}Pa} = 3203\text{ kJ/kg},$$

$$h_6 = h_5 - \eta_T(h_5 - h_{6s}) = 3644 - 0.9 \times (3644 - 3203) = 3247.1\text{ kJ/kg},$$

and for the LPT:

$$h_{7s} = h_{s=s_6}^{3.16\text{ kPa}} = 2283\text{ kJ/kg}$$
$$h_7 = h_6 - \eta_T(h_6 - h_{7s}) = 3247.1 - 0.9 \times (3247.1 - 2283) = 2379.4\text{ kJ/kg}.$$

The specific work production in each turbine stage is

$$W_{HPT} = h_3 - h_4 = 3389 - 2805.8 = 583.2\text{ kJ/kg}$$
$$W_{HPT} = h_5 - h_6 = 3644 - 3247.1 = 396.9\text{ kJ/kg}$$
$$W_{LPT} = h_6 - h_7 = 3247.1 - 2379.4 = 867.7\text{ kJ/kg}.$$

The heat transfer in the steam generator is

$$q_H = (h_3 - h_2) + (h_5 - h_4) = (3389 - 145.8) + (3644 - 2805.8) = 4081.4\text{ kJ/kg}.$$

So, the efficiency of the power plant without carbon dioxide capture is

$$\eta_{th} = \frac{W_{HPT} + W_{IPT} + W_{LPT} - W_P}{q_H} = \frac{583.2 + 396.9 + 867.7 - 41.3}{4081.4} = 0.443.$$

b. The heat rate required to produce 1000 MW is $\dot{Q}_H = \dfrac{\dot{W}_{net}}{\eta_{th}} = \dfrac{1000}{0.443} = 2257.3\text{ MW}$.

The rate at which coal is consumed is: $\dot{m}_{coal} = \dfrac{\dot{Q}_H}{HV_{coal}} = \dfrac{2257.3}{29} = 77.8\text{ kg/s}$.

The rate at which CO_2 is generated is then

$$\dot{m}_{CO_2} = \frac{W_{CO_2}}{W_{coal}}\dot{m}_{coal} = \frac{44}{16} \times 77.8 = 213.95\text{ kg/s},$$

Example 10.4 (cont.)

and the annual amount of CO_2 released is

$$m_{CO_2} = \dot{m}_{CO_2} \times \text{capacity factor} \times 1\text{year} = 213.95 \times 0.75 \times 365 \times 24 \times 3600$$
$$= 5,060,345,400\,\text{kg} \approx 5.06 \times 10^9\,\text{kg}.$$

Thus, the annual amount of coal consumption is 1.84×10^9 kg.

c. To find the separation work, we consider an energy balance for the desorber. There are several processes occurring in the desorber, which require an enthalpy transfer from the steam. We can divide the total enthalpy transfer into the following parts: heating the rich amine solution coming out of the regenerator to 100 °C; stripping the CO_2 from the amine solution; evaporating some of the water from the amine solution; heating the water vapor and CO_2 to 150 °C; and heating the water returning from the condenser to 100 °C. We assume the rich amine solution to consist of MEA and CO_2 as separate components with individual C_p values.

The molar flow rate of MEA is twice the molar flow rate of CO_2. For each 1 kg of CO_2 absorbed, the amount of MEA that must flow through the desorber is

$$m_{MEA} = W_{MEA} \times \frac{2}{W_{CO_2}} = 61 \times \frac{2}{44} = 2.77\,\text{kg}.$$

Since the MEA solution is 40% MEA + 60% water, the amount of water that flows through the desorber (per 1 kg of CO_2 absorbed) is: $m_{H_2O} = m_{MEA} \times \frac{0.6}{0.4} = 4.159\,\text{kg}$.

We have the energy equation across the desorber as

$$\dot{m}_9\,(h_6 - h_9) = \dot{m}_{MEA}c_{p,MEA}(100 - 85) + 0.90\dot{m}_{CO_2}\left[c_{p,CO_2}(150 - 85) + h_{strip}\right] +$$
$$\dot{m}_{H_2O}xc_{p,H_2O(l)}(100 - 85) + \dot{m}'_{H_2O}\left[h_{fg} + c_{p,H_2O(g)}(150 - 100) + c_{p,H_2O(l)}(100 - 90)\right],$$

$$\dot{m}_9\,(h_6 - h_9) = \dot{m}_{MEA}c_{p,MEA}(100 - 85) + 0.90\dot{m}_{CO_2}\left[c_{p,CO_2}(150 - 85) + h_{strip}\right] +$$
$$\dot{m}_{H_2O}xc_{p,H_2O(l)}(100 - 85) + \dot{m}'_{H_2O}\left[h_{fg} + c_{p,H_2O(g)}(150 - 100) + c_{p,H_2O(l)}(100 - 90)\right],$$

where

$$\dot{m}_{MEA} = W_{MEA} \times \frac{2}{W_{CO_2}} \times 0.90\dot{m}_{CO_2},$$

$$\dot{m}_{H_2O} = \dot{m}_{MEA} \times \frac{0.6}{0.4} = W_{MEA} \times \frac{2}{W_{CO_2}} \times 0.90 \times \dot{m}_{CO_2} \times \frac{0.6}{0.4},$$

$$\dot{m}'_{H_2O} = W_{H_2O} \times \frac{1}{W_{CO_2}} \times 0.90 \times \dot{m}_{CO_2},$$

$$\Rightarrow \dot{m}_9(h_6 - h_9) = 0.9\dot{m}_{CO_2}\Delta h_{desorber}.$$

Example 10.4 (cont.)

Substituting, we get:

$$h_{desorber} = W_{MEA} \times \frac{2}{W_{CO_2}} c_{p,MEA}(100 - 85) + \left[c_{p,CO_2}(150 - 85) + h_{strip}\right] + W_{MEA} \times \frac{2}{W_{CO_2}} \times \frac{0.6}{0.4} c_{p,H_2O(l)}(100 - 85) + W_{H_2O} \times \frac{1}{W_{CO_2}} \left[h_{fg} + c_{p,H_2O(g)}(150 - 100) + c_{p,H_2O(l)}(100 - 90)\right],$$

$$h_{desorber} = 61 \times \frac{2}{44} \times 2.95 \times (100 - 85) + 1.23(150 - 85) + 1919 + 61 \times \frac{2}{44} \times \frac{0.6}{0.4} \times 1.87(100 - 85) + 18 \times \frac{1}{44} \times [2260 + 2.27(150 - 100) + 1.87(100 - 90)]$$

$$\Rightarrow \Delta h_{desorber} = 3217 \text{ kJ/kg}.$$

Water at state 9 is saturated liquid, so the specific heat transferred within the desorber is

$$h_6 - h_9 = 3247.1 - 670.7 = 2576.4 \text{ kJ/kg}.$$

The required steam flow rate through the desorber is then

$$\dot{m}_9 = 0.9 \frac{\dot{m}_{CO_2} \Delta h_{desorber}}{h_6 - h_9} = \frac{0.9 \times 213.95 \times 3217}{2576.4} = 240.4 \text{ kg/s}.$$

The total flow rate of steam through the cycle is:

$$\dot{m}_1 = \frac{\dot{Q}_H}{q_H} = \frac{2,257,300}{4080.4} = 553.2 \text{ kg/s}.$$

The fraction of steam diverted through the desorber is then $f = \frac{\dot{m}_9}{\dot{m}_1} = \frac{240.4}{533.2} = 0.435.$

d. All of the states that define the plant are the same, only the mass flow rate through the LPT has changed. The net work produced by the plant is now

$$\dot{W}_{net} = \dot{m}_1(w_{HPT} + w_{IPT} - w_P) + \dot{m}_9 w_{LPT} = 553.2 \times (583.2 + 396.9 - 41.3) + 240.4 \times 867.7 = 727,939.2 \text{ kW}.$$

The thermal efficiency with CO_2 capture is: $\eta_{th} = \frac{727,939.2}{2,257,300} = 0.322.$

Comparing with the thermal efficiency of the power plant without CO_2 capture calculated in part (a), the energy penalty due to the capturing CO_2 is $0.443 - 0.322 = 0.121$, or 12.1 percentage points reduction in the thermal efficiency of the power plant. On the other hand, the net power output of the plant decreases from 1000 MW to 727.94 MW. The CO_2 capture system reduces power by about 272 MW. Therefore, for each 1 kg of CO_2 captured about 1.4 MJ (= 272 MW/(0.9 × 213.95 kg/s)) work is required.

10.5 Adsorption

Adsorption is the binding of a gas molecule to a solid surface, and under equilibrium, the gas is in the solid bulk as well. The bulk can be porous and the gas attaches to the pore surface. In chemical adsorption, the gas molecules form chemical bonds with the solid. Weak binding forces, such as electrostatic forces, characterize physical adsorption, or **physisorption**, while chemical binding is associated with stronger **chemisorption**.

10.5.1 Physical Adsorption

In adsorption, one or more components in a gas mixture bond preferentially to a solid surface as the gas stream flows over the surface. Molecules of different gases have different tendencies to attach to a given solid surface; and the tendency depends on the bonding forces or bonding energy between the molecule and the surface. Thus, a gaseous mixture flowing over a surface or through a porous solid with a large surface area could lose a fraction of one or more species to the surface. The adsorbed molecules will remain attached to the surface until conditions change, allowing them to leave the surface. The solid, or **adsorbent** must have a large surface area to be effective. Porous surfaces, such as those of activated carbon, alumina, silica gels, and zeolites that contain many small pores, down to nanometer scales, are ideal for this application. The surface area within the pores contains many active sites for gaseous molecules to attach to the surface. Zeolites, or molecular sieves, made of natural or synthetic aluminosilicates, have regular fine pore structure, and are effective adsorbents. Molecular sieves are materials with pores that admit small molecules but not large molecules. The adsorbed gas, or **adsorbate**, may stay on the surface or move along the surface – that is, become stationary or mobile. As conditions change, the adsorbate can leave the surface, and the surface may get regenerated and become ready for repeating the adsorption/desorption cycle [10].

When used to adsorb a component from a mixture, the effectiveness of separating the adsorbed gas is defined by the **separation factor** $\tilde{\alpha}_{ij}$, which is the ratio of the mole fraction of a component in the solid phase to the same in the gas phase divided by the same for another component:

$$\tilde{\alpha}_{ij} = \frac{\left(X_{ad}/X_g\right)_i}{\left(X_{ad}/X_g\right)_j}. \tag{10.22}$$

In this equation, X_{ad} and X_g are the equilibrium mole fractions of the species in the adsorbent and in the gas, and i and j refer to two different molecules. The separation factor depends on the adsorber and the adsorbed species; the larger the separation factor, the more efficient the separation process is, the less adsorbing material is needed, and the less energy is required in the process.

The energy associated with the process is the **enthalpy of adsorption**, and it is, in general, higher for chemisorption, O (10–100 kcal/gmol) than in physisorption O (1–10 kcal/gmol).

Besides its application in the separation of gases (and liquids), adsorption is important in catalytic and heterogeneous reactions. Adsorption on carbon surfaces is used to remove contaminants such as H_2S from flue gases of coal power plants, and in purifying air passing through air-conditioning systems. Carbon canisters are used in automobiles to prevent gasoline vapor from escaping. In large-scale applications, adsorption is often done in packed beds in semi-batch processes.

The adsorptive capacity of a solid adsorbent depends on the surface area available for molecules to attach to, or the **active sites**. The number of molecules that can attach to a surface depends on the pressure of the gas and the molar concentration of the adsorbing component. Higher pressures promote more extensive **surface coverage**. In some cases, when the surface is fully covered – that is, following the formation of the top surface **monolayer** – with further increase in pressure, multilayer coverage or **multilayer** adsorption occurs.

10.5.1.1 Adsorption Equilibrium

Adsorption equilibrium is defined by the **adsorption isotherm** – that is, the constant-temperature relationship between the partial pressure of the species in the gas phase and its concentration in the solid phase, at equilibrium. The concentration of the adsorbed species, the adsorbate, in the solid, the adsorbent, is defined in terms of mass ratio, or mole ratio. There are several theories and approaches for deriving the relation between the surface concentration of the adsorbed species and their partial pressure in the gas phase, or the equilibrium isotherm. The most widely used is the **Langmuir isotherm**. The form of this relation can be determined from thermodynamic consideration or from the equilibrium between the rate of adsorption reaction and the rate of desorption reaction. Given the total number of available sites on a surface, N_s, and the number occupied by gas molecules, N_g, the fraction of occupied sites is defined by $\theta = N_g/N_s$. The rate of adsorption, which is controlled by the collision rate between the gas molecules and the surface, depends on the gas partial pressure, p_i, and the number of available sites. If only one gas is being adsorbed, the number of available sites is $(N_s - N_g)$, and the adsorption rate is $k_{ads}p_i(N_s - N_g)$, where k_{ads} is the adsorption reaction rate constant. The rate of desorption depends on the number of adsorbed gas molecules, $k_{des}N_g$, where k_{des} is the reaction rate constant of the desorption reaction. At equilibrium, these two rates are equal, $k_{ads}p_i(N_s - N_g) = k_{des}N_g$. Assuming that only a single species is being adsorbed, the fraction of occupied site is given by:

$$\theta = \frac{p_i}{1/K + p_i}. \tag{10.23}$$

The (Langmuir) equilibrium constant for adsorption is $K_i = (k_{ads}/k_{des})_i$, and it is a function of the Gibbs free energy of adsorption of this component in a form similar to other reaction equilibrium constants. At low concentration or partial pressure p_i, the isotherm reduces to the linear form $\theta_i = K_i p_i$ (similar to Henry's law for gas–liquid equilibrium, where $K = 1/\mathrm{He}$). At high p, the surface is saturated and $\theta_i \to 1$ (in the absence of other gas that

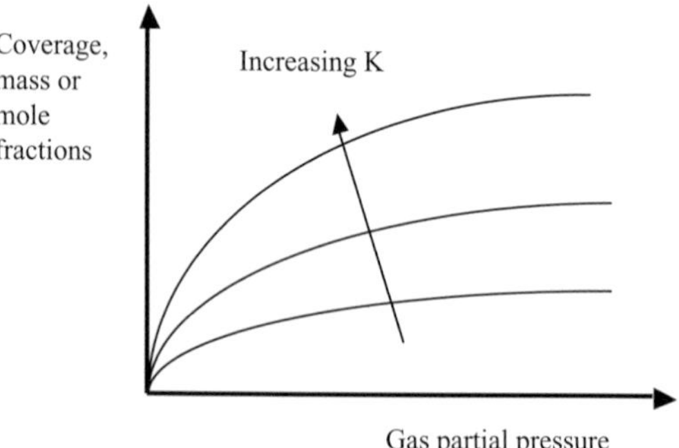

Figure 10.12 The dependence of the surface coverage on the gas partial pressure for different values of the equilibrium constant, as expressed by the Langmuir isotherm. Shown is only a schematic of a generic adsorber.

might adsorb on the surface and compete for sites). The amount of gas adsorbed is directly proportional to the gas partial pressure, and K is the "solubility." At higher pressures, the surface saturates (see Figure 10.12). Langmuir isotherms provide good models for estimating the necessary solid surface area, or solid volume if the surface area per unit volume is known, for adsorption.

The same form can be used to describe the mass fraction of the gas adsorbed in the solid. In this case, θ in (10.23) is replaced with the mole or mass fraction, X or Y, while the right-hand side of the equation remains the same. Given the number of active sites per unit area of the solid surface, $\tilde{\Gamma}$, the surface concentration of the adsorbate is $\theta\tilde{\Gamma}$, measured in molecules per unit surface or $\phi\tilde{\Gamma}/N_a$ in units of mole/area, where N_a is Avogadro's number. To convert to mole concentration, that is, in units of mol/volume, this quantity is multiplied by the surface per unit volume, $(A_s/\forall)$, yielding $C_i = (A_s/\forall)\phi_i\tilde{\Gamma}/N_a = a_s\phi_i$, measured in mole/vol. The Langmuir isotherm in terms of the molar concentration of the adsorbed gas is:

$$C_i = \frac{\alpha_s p_i}{1/K(T) + p_i}. \tag{10.24}$$

The mole fraction of the adsorbate in the solid is C_i/C_{solid}, where C_{solid} is the molar density of the solid. The surface area per unit volume of the solid depends on the morphology (i.e., the pore structure).

Higher values of $K(T)$ correspond to an isotherm with a high equilibrium constant, and a relatively low partial gas pressure would lead to a higher concentration in the solid phase. Adsorption isotherms for a number of adsorbates on zeolite, defined here as moles per unit mass of the solid, as a function of the gas pressure, is shown in Figure 10.13. Zeolite is a very good dehumidifier, and it is indeed used to remove humidity from gases in several applications. Moreover, in a shifted syngas stream of hydrogen and carbon dioxide, zeolite would adsorb (and hence remove) significantly more CO_2 than H_2, but the ratio depends on the initial concentrations, and 100% removal is difficult to achieve. On the other hand, in products of combustion, dominated by nitrogen and carbon dioxide, adsorption may not be very effective for separating CO_2 since the zeolite tends to adsorb both (and other adsorbents should be used for this purpose).

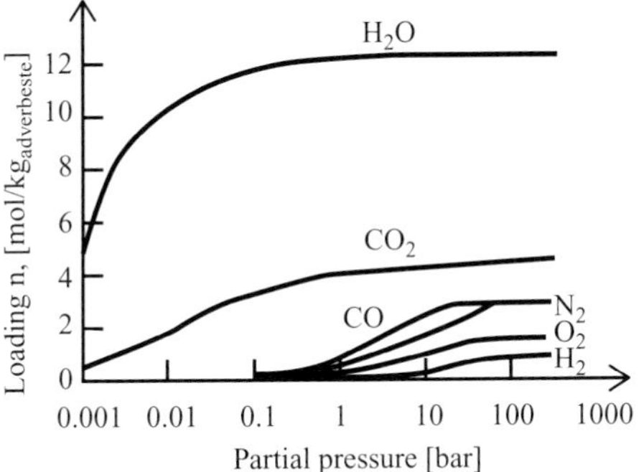

Figure 10.13 Adsorption isotherms for the pure component for zeolite 5A molecular sieve at 20 °C [2] showing the amount of gas (water, carbon dioxide, carbon monoxide, nitrogen, oxygen, and hydrogen) adsorbed per kilogram of the solid as a function of the gas partial pressure at constant temperature.

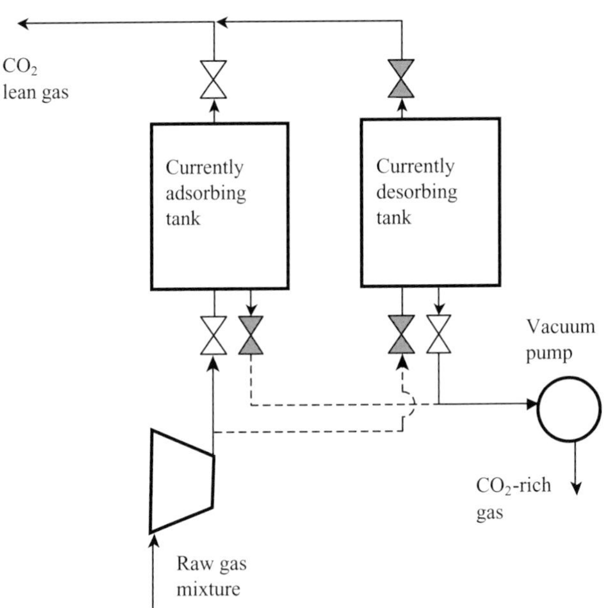

Figure 10.14 A schematic diagram of an adsorption-based separation process using two tanks while switching the gas stream into each tank on and off, depending on the state of the absorbing medium inside the tank. Using two tanks makes it possible to continue feeding gas continuously to the system, while switching the feed stream between the two tanks, and producing two continuous streams of the separated gases. The case shown is for a raw feed gas mixture containing CO_2, and two separated streams: one is CO_2-rich and the other is CO_2-lean. In the figure, light valves are currently open while dark valves are closed.

Pressure-Swing Adsorption Pressure-swing adsorption is used in many applications. In this technique, shown schematically in Figure 10.14, a gas mixture is introduced into an adsorption tank from the bottom side, typically at a higher pressure. As the gas stream advances, it loses one of its components, which is adsorbed on the solid surface of the packing material. The concentration profile of the adsorbed component in the solid resembles a wave front moving from one side to the other. When the advancing wave reaches the opposite end of the tank, the gas stream is switched off. Next, the tank is depressurized (by connecting it to a vacuum) for desorption and is opened to let the desorbed gas out. At the end of desorption, the tank may be purged using a gas stream of the same composition as that being desorbed (not shown in the figure). To achieve continuous operation, two packed-bed reactors, packed with molecular sieves, one adsorbing and the other regenerating, are used. When

PSA is used in small ASUs, nearly pure oxygen can be produced. Nitrogen, at the same concentration, is adsorbed nearly four times as much as oxygen.

Adsorption processes are exothermic; molecules give up energy as they attach to the surface, and the adsorbing tank must be cooled to remain effective. Conversely, desorption is endothermic, in which case energy is provided for the molecules to escape the surface, and the adsorption tank must be heated.

In these semi-batch processes, in which a gas stream is flown through a packed bed of adsorbent, the time required to remove a certain amount of adsorbate can be estimated as follows. Assume that the superficial velocity of the gas stream, that is, the apparent convective velocity of the gas measured as the volume flow rate per unit bed cross-section, is V_o, and the solute density in the gas is ρ_i, then [1]

$$t^* = \frac{L_R \rho_b \left(\tilde{Y}^e - \tilde{Y}_o\right)}{V_o \rho_i}. \tag{10.25}$$

In this equation, L_R is the tank length, ρ_b is the bulk or average density of the reactor bed material, $\tilde{Y}^e$ is the solute mass fraction at equilibrium, measured as a fraction of the bulk density, and $\tilde{Y}_o$ is the solute mass fraction at the inlet of the reactor; $\tilde{Y}$ is also known as the absolute loading. In writing this equation, it is assumed that the adsorption rate is much faster than the flow and the dispersion of the adsorbate in the flow direction, and hence the distribution of the solute concentration in the gas is a Heaviside function. The assumption is an approximation that is violated by finite streamwise dispersion by the finite mass transfer rate between the gas and the bed material. Soon before reaching the actual breakthrough time, t^*, the supply or feed mixture is switched off and the tank is depressurized for desorption.

In applying PSA for CO_2 separation from a gas mixture, CO_2 first adsorbs into a solid sorbent, mostly under pressure (and/or lower temperature). Regeneration, and the release or desorption of CO_2, is accomplished by reducing the pressure (and/or increasing the temperature) of the solid. This process takes advantage of the different solubility of CO_2 and other gases in the original mixture in certain solids, and the change in the solubility with pressure (and temperature). This is shown using schematics of equilibrium isotherms (an isotherm is a constant-temperature line showing the concentration of the mole fraction of a species in the solid as the gas pressure is varied), at different temperatures in Figure 10.15. Along an isotherm the mole fraction of CO_2 changes with pressure. If the process of adsorption and desorption is controlled by the temperature, this is called a temperature-swing adsorption process, and the state of the adsorbent jumps between two isotherms (as shown in the figure). Some applications involve both pressure and temperature changes between adsorption and desorption (PTSA). Activated carbon or coke, carbon molecular sieves, zeolite molecular sieves (natural or synthetic aluminosilicates with regular fine pore structure), or activated aluminum can be used for regenerative PSA. All these highly porous substances adsorb CO_2 more readily than N_2 or O_2 – that is, they are selective adsorbents (where selectivity is defined as the relative adsorptivity of two different gases). The production of a high-purity gas can be achieved using physical adsorption. The separation energy varies widely depending on the gas mixture and conditions.

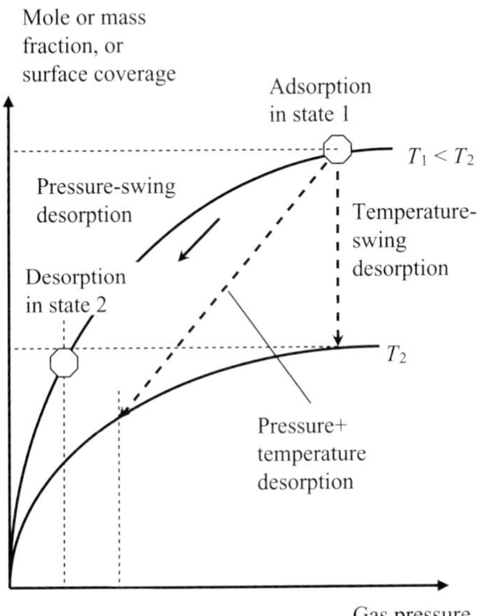

Figure 10.15 A representation of PSA and TSA on the equilibrium isotherms. In PSA, adsorption occurs at high pressure, p, where the surface coverage is high; desorption occurs at low pressure, p_2, to release the adsorbate. In TSA, adsorption occurs at low temperature, T_1, and the temperature is raised to T_2 to release the adsorbate. In PTSA, the pressure is reduced and the temperature is raised for desorption.

Energy Requirements Energy required in physical adsorption depends on whether the process is pressure- or temperature-driven. To calculate the energy required for separation, one must decide on the layout of the separation plant, including the number of stages and dimension of each tank, adsorbent, adsorption pressure, the packing density, and how much adsorbent is needed to achieve the desired result (e.g., the fraction of CO_2 removed). Estimating this separation energy follows steps similar to the those used in estimating the energy required for separation when using physical absorption. For a pressure-swing process, this energy is required for:

- raising the raw gas pressure to the adsorption pressure to increase the solubility of CO_2 in the solid, according to the relevant adsorption isotherm. This process could use work produced by the expanding clean gas;
- running a vacuum pump to reduce the desorber pressure; and
- overcoming the pressure changes in the absorber and desorber.

Pressure drop in a packed-bed reactor is given by the Ergun equation [1]:

$$\frac{dp}{dz} = -\frac{V_0}{\Phi_s d_p}\left(\frac{1-\varepsilon}{\varepsilon^3}\right)\left[\frac{150(1-\varepsilon)\mu}{\Phi_s d_p} + 1.75(\rho V_0)\right], \tag{10.26}$$

where z is measured along the flow direction in the bed, V_0 is the superficial velocity (the volumetric flow rate divided by the bed cross-sectional area); ρ is the gas density; μ is the gas viscosity; d_p is the bed particle diameter; ε is the porosity, defined as the void volume divided by the total bed volume; and $\Phi_s = \frac{6d_p}{A_p/\forall_p}$ is known as the sphericity, A_p is the particle surface area, and $\forall_p$ is the particle volume. In most processes, the pressure drop due to the gas flow

Table 10.5 The separation work and efficiency penalty for CO_2 removal (CO_2 separation factor 80%) using zeolite 5A [2]

	CO_2 mass fraction (%)	Separation work (Wh/kg-CO_2)	Efficiency penalty $\Delta\eta$ (%)
CO_2 from (CO_2 + N_2) in:			
Gas-fired CC	4.94	1.28	7.8
Integrated coal gasification plant	11.70	0.72	21.4
Coal-fired power plant	17.14	0.42	13.9
CO_2 from (CO_2 + H_2) in IGCC after shift	82.02	0.21	6.0

through the bed is smaller than the pressure changes between the two towers. Simplified expressions for the pressure drop within each tower can be obtained.

Estimates for the energy required for physical adsorption separation of CO_2 from flue gases are in the range 1.5–5.0 MJ/kg-CO_2 removed, depending on whether it is separated from: (1) the flue gases of a pulverized coal or a natural gas-fired plant; or (2) the syngas (mostly after water–gas shift) produced by coal gasification or natural gas reforming, and whether the gasification/reforming process is oxygen- or air-based. The energy spent in separating CO_2 depends also on the desired purity of the product gas. Table 10.5 shows the efficiency penalty resulting from CO_2 removal from the flue gases in gas-fired CC, IGCC, and coal-fired pulverized steam plants, and from coal-fired IGCC in a pre-combustion arrangement (i.e., from the shifted syngas). The separation work per unit mass of CO_2 is higher for a gas-fired plant than for a coal-fired cycle because of the lower CO_2 concentration in the exhaust in the gas plant. However, the efficiency penalty is lowest because NGCC has the highest overall efficiency, and the amount of CO_2/unit energy is lower. IGCC flue gases contain less CO_2 since lean burning is necessary to produce low temperatures for gas turbines. On the other hand, regular coal-fired plants operate close to stoichiometry, with higher CO_2 concentrations in the flue gases. This explains why the efficiency penalty in IGCC is higher than that for coal-fired plants. In the IGCC plants, in which CO_2 is separated from the syngas after shift and before combustion to produce a pure hydrogen stream, the separation factor is high because of the large disparity of the molecular weights between CO_2 and H_2. The IGCC plant, after syngas shift, produces the highest concentration of CO_2 in a stream, leading to the lowest energy of separation per kilogram of CO_2 and lowest efficiency penalty. These numbers are discussed further in Chapters 11 and 13.

Continuous physical adsorption processes have been developed for water purification and for applications in the chemical industry. In these processes, two towers are used, an adsorber and a desorber, and the solid material is circulated or looped between the two towers, as shown in Figure 10.16. In the adsorber, the feed gas flows in counter-current to the solid bed material, and the solute is removed. In the desorber, a desorbent is also flowed counter-current to purge the solute from the solid bed. Pressure and temperature in the adsorber and desorber are adjusted to promote the thermodynamics and kinetics.

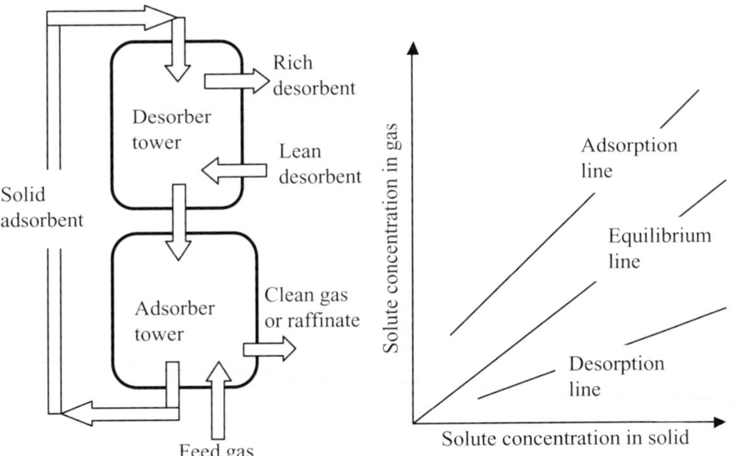

Figure 10.16 Schematic of physical adsorption flow process and the concentration of solute in the gas and solid phases. The solid adsorbent is looped continuously through the adsorber and desorber towers. In the adsorber, CO_2-rich gas in introduced and clean gas is removed. In the desorber, lean desorbent is introduced and rich desorbent is removed. The process takes advantage of the deviation of the sorbent CO_2 concentration from its equilibrium value with respect to the flowing gas, as shown on the right.

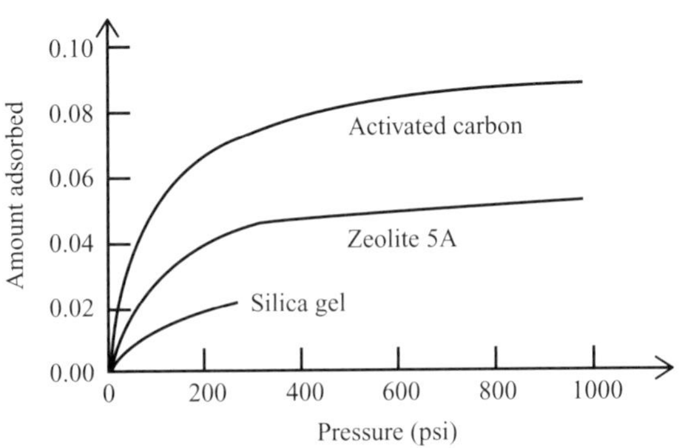

Figure 10.17 Equilibrium isotherms for methane at 25 °C on silica gel, zeolite 5A, and activated carbon [12].

The throughput of this looping process should be higher than batch processes. More discussion on "chemical looping" processes is presented in Chapters 11 and 13, in the context of CO_2 capture.

Physical Adsorbents Adsorbent selection for gas separation applications is based on a number of factors, including its capacity to adsorb a particular gas; the method of regeneration and whether it is temperature- or pressure-swing dependent; and the product purity. The product purity depends on the adsorbent selectivity, and its equilibrium isotherm [11]. The equilibrium isotherm of methane (as a function of its partial pressure in the gas phase) is shown in Figure 10.17 for three adsorbents: activated silica, zeolite 5A, and silica gel. Moreover, the total void space in the bed material where gas may accumulate should be considered. In the following, some adsorbents are described along with their important characteristics and applications.

Activated carbons, which are manufactured from carbonaceous matter such as coal, petroleum coke, or wood, are used in the powder form as the adsorbent. Activation involves carbonization – that is, elimination of volatile matter from the original material – and partial gasification to develop the porosity and enlarge the surface area. The heat of adsorption of activated carbon is generally lower than that of other adsorbents, and stripping the adsorbate for regeneration requires less energy, which makes them efficient gas separation material. Activated carbon is nearly hydrophobic (water has a low tendency to stick to its surface) to low-humidity mixtures, and hence can be used in separating humid gas mixtures, such as combustion or gasification/reforming products. The pore surface area per unit mass is high, 300–2500 m^2/g, the largest among all known sorbents. The pore diameter in activated carbon is 10–35 Å.

Molecular-sieve carbon (MSC) is produced by the carbonization of polymers such as cellulose and sugar, and other methods. The pore structure and pore diameter of this material depend on the original material and the carbonization temperature, and diameters in the range 2–8 Å have been demonstrated. These MSCs are used in ASUs and in other applications in the chemical industry. Carbon (3%) deposited into the pores of lignite char by cracking methane at 855 °C produces MSC that shows significant molecular sieving between CO_2 (admitted) and N_2 (hindered). (CO_2 is a linear molecule with a diameter of 3.7 Å.)

Activated alumina is dehydrated or partially dehydrated alumina hydrated, both crystalline and amorphous, with high surface area. It is strongly hydrophilic and used in drying gases. It has a surface area in the range of 250–350 m^2/g, with pore diameters larger than 50 Å.

Silica gel is a synthetic amorphous silica produced by mixing sodium silicate solution and mineral acid, and the reaction product is activated after being dried. The surface area ranges from 750–850 m^2/g with pore diameter of 22–26 Å for regular gel, or 300–350 m^2/g with pore diameter 100–150 Å for activated gel. Silica gels have relatively low heat of adsorption for water. Silica gels are used for drying; the enthalpy of adsorption of water to silica gel surface is low, it is close to the liquefaction energy of water (11 kcal/mol), and hence it is easy to regenerate the adsorbent at relatively low temperatures of around 150 °C.

Zeolites are crystalline aluminosilicates of alkali or alkali earth elements (such as sodium, potassium, and calcium). They are found naturally or can be synthesized. The basic structure of zeolite is an aluminosilicate skeleton, with a "window aperture" of 3–10 Å. Sorption into zeolites can occur with high selectivity because of the size of the window. They can also act as molecular sieves. Zeolites have been used in hydrogen purification and air separation. Although they can act as molecular sieves, most of their applications are based on their ability to adsorb different amounts when exposed to a mixture of gases with different equilibrium adsorption isotherms. They have been used in ASUs, but the purity of oxygen in the separated steam is relatively low. The equilibrium isotherms for nitrogen and oxygen are shown in Figure 10.18.

10.5.2 Chemical Adsorption

Solid sorbents can also be used to capture CO_2 chemically – that is, by forming a metal carbonate compound when it is exposed to the CO_2-rich stream. The metal carbonate solid is

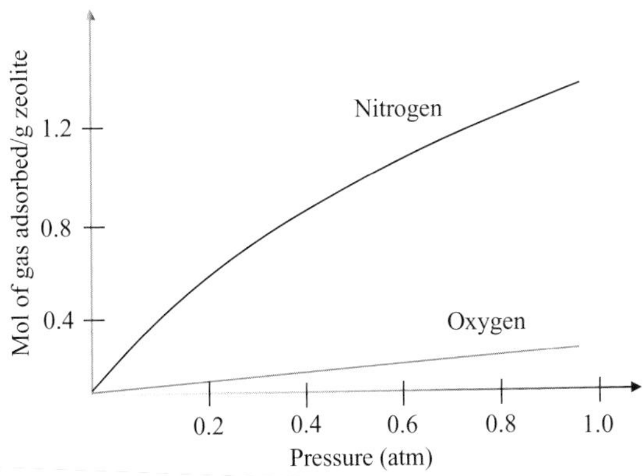

Figure 10.18 Adsorption isotherms of 5A zeolite ($Li_{94.5}Na_{1.5}$-X) at 350 °C. The selectivity ratio for N_2/O_2 declines at higher pressure [13]. The argon isotherm is very close and slightly lower than that of oxygen.

then transported to another reactor in which it is heated to release the CO_2. For instance, the following heterogeneous reactions can be used:

$$X_2CO_{3(S)} + CO_{2(g)} + H_2O \rightleftarrows 2XHCO_3, \tag{10.27}$$

where X stands for lithium (Li), sodium (Na), or potassium (K). The energy required for these reactions is higher than that for MEA or ammonia, being 2.92 and 3.3 MJ/kg-CO_2 for Na and K, respectively. These are low-temperature absorption reactions. These processes can be carried out in circulating fluidized-bed transport reactors or multiple fixed-bed reactors (descriptions of these reactors can be found in Chapters 13–14). Alternatively, higher-temperature absorption reactions using metal oxides that do not require as much cooling of the gas stream have also been proposed. These processes strongly resemble chemical looping processes, which will be discussed in Chapter 11, in the context of CO_2 capture in natural gas and coal plants.

10.6 Cryogenic Distillation

Cryogenic separation exploits the properties of the phase diagram of a two-component mixture. An example for this diagram is shown schematically in Figure 10.19, plotted at constant pressure. The diagram shows the dew or condensation line and the bubble or evaporation line of a mixture of two components, A and B, with A being the more volatile component (with lower boiling/evaporation temperature). The diagram shows the equilibrium state of the mixture, that is the concentrations of A and B in the liquid phase and in the gas phase, as a function of temperature. For instance, starting with state 1, with vapor which is much richer in B, and cooling the mixture down to the temperature of the bubble line, state 2, produces a liquid mixture much richer in B and vapor much richer in A. Further cooling starting with state 3 to state 4 produces a liquid richer in B, and vapor richer in A, and so on. The repeated cooling of the vapor can produce an almost pure vapor of A.

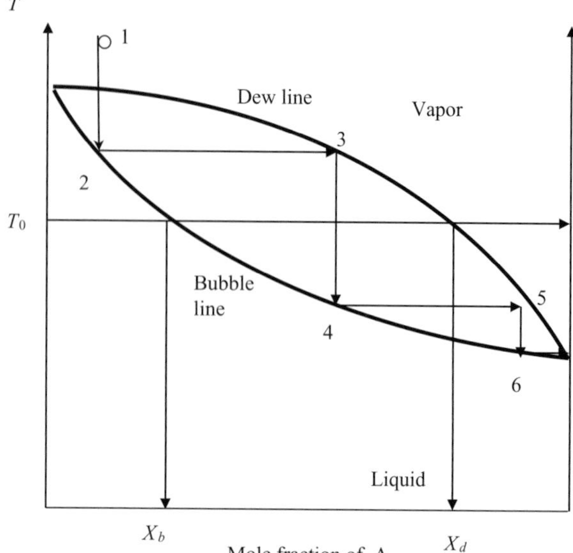

Figure 10.19 Constant-pressure phase diagram of a two-component mixture. Above the dew line, the mixture is in the vapor state; below the bubble line, it is in the liquid state. A has a lower boiling temperature, it is more volatile than B. Between the two lines, a fraction of the mixture is in the liquid state and a fraction is in the vapor state. At T_0, the composition of the liquid is given by X_b and that of the gas is given by X_d.

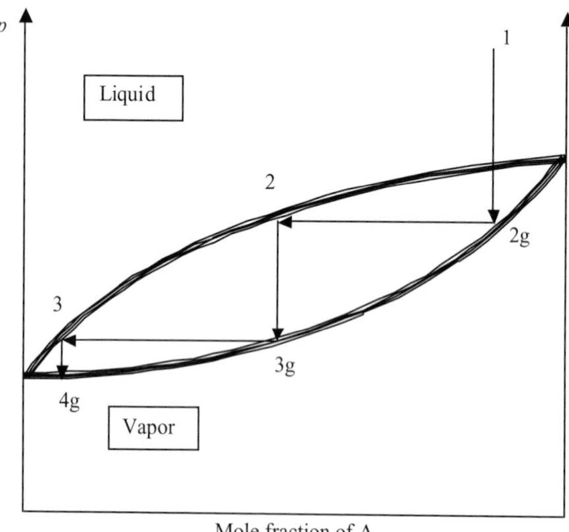

Figure 10.20 Constant-temperature phase diagram of a two-component mixture, showing a trajectory for producing a pure gas stream which is rich in one of the two components.

Alternatively, one can construct a similar phase diagram at constant temperature by varying the pressure of the mixture. The properties of this diagram and that of the constant-temperature diagram can be used to separate the component through distillation or pressure reduction/flashing of a two-component mixture, as shown in Figure 10.20. Starting with a cool liquid state, defined as state 1 in the figure, and reducing the pressure to that corresponding to state 2 produces a liquid much richer in B and vapor that is much richer in A. Flashing the liquid in state 2 down to the pressure of state 3 further increases the concentration of B in the liquid and concentration of A in the vapor, and so on. The repeated flashing on the liquid state produces a liquid that is almost pure in B.

10.6.1 Air Separation Using Cryogenic Units

In the case of air, cooling the mixture at constant pressure raises the fraction of nitrogen, which is the more volatile component in the mixture (upon heating, it evaporates first), in the vapor phase. The boiling temperatures of oxygen and nitrogen at atmospheric pressure are −183 °C (90 K) and −196 °C (77 K), respectively. Cryogenic distillation (or rectification) is used extensively in air separation, shown schematically in Figure 10.21. The system consists of a compressor train to raise air pressure, followed by a heat exchanger to cool the compressed air regeneratively using the two cold separated gas streams of almost pure nitrogen and almost pure oxygen. Air is liquefied in a refrigeration unit. It is cooled further in the liquid pool, made up of mostly oxygen, at the bottom of the rectification column. The critical temperature of air is −140 °C (133 K); its boiling temperature at 1 atm is −195 °C (78 K). Cold liquefied air is throttled in a Joule–Thompson valve to drop the pressure (and in this case the temperature too) of the mixture to near atmospheric conditions. The low-pressure mixture enters the distillation column and separates into a gas mixture that flows

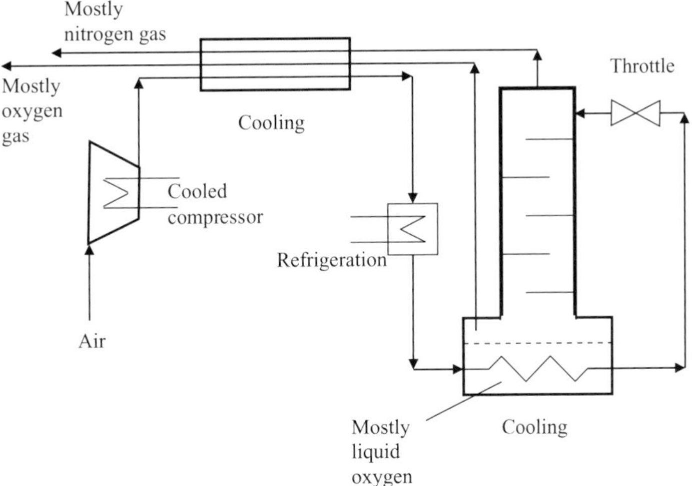

Figure 10.21 A column distillation cycle for air separation. An air compressor is used to raise the air pressure, with the final pressure depending on the desired purity (higher-purity separated gases require higher pressures). A regeneration unit is used to cool the compressed air stream using the cool products (N_2 and O_2 are separate streams). A refrigeration unit is then utilized to reduce the air temperature, with further cooling of the air in the oxygen pool at the bottom of the distillation column. A Joule–Thompson valve lowers the pressure and temperature of the air, flashing it into the distillation column. There, liquid, mostly oxygen, flows downwards and gas, mostly nitrogen, flows upwards. The process can be repeated again in other distillation columns to further separate the component with the smaller concentration and deliver higher-purity streams. Two columns are required if purity higher than 90–95% is desired.

toward the top of the column, and a liquid mixture that falls downwards to the bottom of the column. The more volatile component (i.e., nitrogen) has the higher concentration in the gas mixture, whereas the less volatile component (i.e., oxygen) has the larger concentration in the liquid mixture. The liquid mixture at the bottom of the column is warmed by the liquefied air passing through, and is evaporated into a separate stream. The two cooled separated gas streams, one mostly nitrogen and one mostly oxygen, flow up the column in two separate streams, passing first through a heat exchanger in which they are warmed up by the compressed air entering the system, then to their storage tanks.

In a well-designed, well-integrated distillation plant, the refrigeration load can be reduced as most of the cooling of the incoming air is done using the products streams, while energy is mostly required to compress air. The energy requirement depends strongly on the purity of the products streams (see Figure 10.22). For instance, to achieve 90% oxygen purity (volume fraction) air must be pressurized to 5 bar. Higher-purity requires compression to a higher pressure, up to 10 bar, since more than a single stage of distillation is required. The products in both cases are delivered at pressures and temperatures close to the atmospheric values. The compression work is a large fraction of the energy required, the rest being the energy used in refrigeration, and loss in the distillation column. Adiabatic expansion in a work-producing device could be used in place of the Joule–Thomson valve, and would drop the temperature further, but at a large cost of the hardware. Ideal work for air separation is 0.104 kWh/kg-O_2 or 0.374 MJ/kg-O_2. Air separation Second Law efficiencies in cryogenic plants are low, at 15–30%, depending on the size of the plant and the efficiency of individual components. Detailed calculations of a cryogenic ASU will be discussed more in Chapter 13.

Cryogenic air separation into nitrogen, oxygen, and argon is used widely in large and medium size plants to produce these gases. Production of liquefied products (for storage in tanks) requires almost twice as much energy as required by separation alone. In these applications, air is first dried, and trace species (such as CO_2 and hydrocarbons) are removed to prevent freezing and deposit formation inside the equipment. On average and depending on the local humidity, air contains 5% H_2O (by volume). In modern plants, water separation is achieved using molecular-sieve units discussed previously. The purity of air following this step depends on the application, and may not be as important in

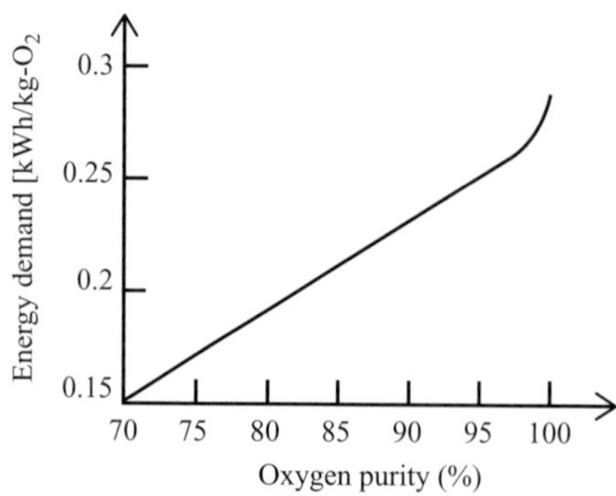

Figure 10.22 Energy expended to produce oxygen in a large-scale double-column cryogenic facility.

power plant applications. The next step is to cool the purified air in a heat exchanger (regenerator) to cryogenic temperatures of −195 °C (78 K). This is done using the cool products (and waste streams) that have already been produced, followed by a refrigeration process. This also helps bring these product gas temperatures back to atmospheric temperatures (unless the objective is to store them in a cold state). Cryogenic air is then introduced into the distillation column.

Because the boiling temperatures of nitrogen and oxygen are close, more than one distillation column is required for high purity. In the first column, an oxygen-rich liquid settles at the bottom, and a nitrogen-rich gas leaves at the top. The nitrogen-rich gas enters the second (top) column, where it is distilled further. The oxygen-rich liquid is also pumped into the top column for further distillation. The separated chilled gases are routed back into the heat exchanger to chill the incoming air and minimize the cooling load of the refrigeration unit. The entire process can be run at higher pressures, which is beneficial when the separated products are used at higher pressure (as in gasification plants). In this case, a compressor at the front end of the plant is used. Moreover, liquid oxygen drawn from the distillation column can be pumped further to a desired high pressure.

10.6.2 Cryogenic Separation of CO_2

Separation of CO_2 from flue gases or shifted syngas is possible using distillation. However, special attention should be paid to the condensation conditions since the thermodynamic characteristics of CO_2 can impact the outcome of the process. The triple point of CO_2 is at −56.6 °C and 518 kPa. Figure 10.23 shows the phase diagram of CO_2. To separate CO_2 in the liquid phase, the temperature and pressure must be higher than the triple point values, falling within the liquid island. For instance, to liquefy 90% of CO_2 in the flue gases (with about 15% CO_2), the gases must be compressed to 35 MPa. Separation of CO_2 from other gases, like methane at natural gas wells, has been used to purify the fuel (many natural gas

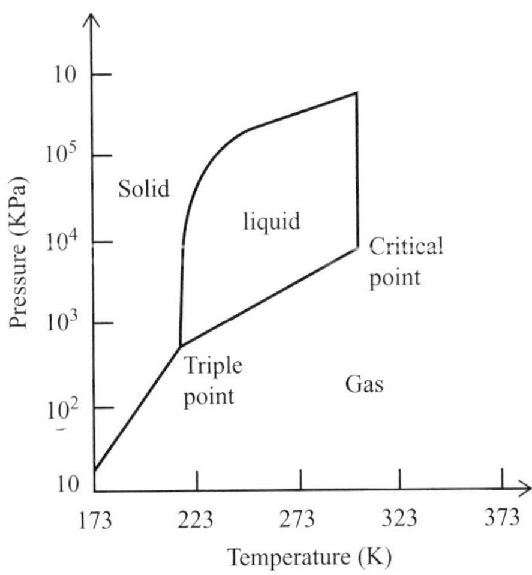

Figure 10.23 The phase diagram of carbon dioxide showing the triple point and critical point.

wells produce a mixture of methane and carbon dioxide, as well as other hydrocarbons but at lower concentrations). Low-temperature distillation of CO_2 in a $CO_2 + N_2$ mixture has not yet been developed commercially, and other methods are still favored for separating CO_2 from combustion products (such as chemical absorption).

Another approach to separate CO_2 is to freeze it out of the mixture; that is, to separate CO_2 in the solid phase by reducing the temperature and raising the pressure to the left of the gas–solid and liquid–solid line in Figure 10.23 (such methods have been developed (by Exxon), and are called the controlled freezing zone). Pressures higher than 3.8 MPa are required, and the precipitate (solid CO_2) is collected in a tray, where it is heated to the liquid phase. It is possible, however, to freeze CO_2 in a mixture at much lower pressures, if the temperature is sufficiently low, again as seen in the phase diagram of CO_2. Reference [3] shows a scheme for removing 90% of CO_2 from a flue gas by cooling it down to −108 °C and 400 kPa. In this case, the power required for separation includes that of compression and refrigeration. Studies show that the energy requirement for freezing CO_2 in a 13% CO_2 flue gas is 0.353 kWh/kg-CO_2 for a carbon dioxide capture ratio of 80%. The energy required grows to 0.3768 kWh/kg-CO_2 for a capture ratio of 90% by cooling to −100 °C and raising the pressure to 4 bar. When applied to a plant with a net efficiency of 41% (without capture), that efficiency is reduced by 11.1 percentage points [3].

10.7 Membrane Separation

Semipermeable membranes that allow one or more fluid (liquid or gas) components to pass through more readily than others – that is, have **selective permeability** – are used to separate gases and liquids from mixtures. Membranes can be made of thin, rigid layers, such as porous glass or sintered metal, or flexible films of synthetic polymers permeable only to certain molecules or ions, ceramic–metal compounds, or pure metal or metal alloys. In a porous membrane, separation depends on the flow of certain gas molecules through the pore. In the non-porous type, separation depends on the difference in the solubility of the gas or liquid into the membrane surface, and its diffusivity through the membrane material. In ceramic membranes, charged species, ionic or protonic, diffuse through the lattice following the adsorption of the original gas and its incorporation in an electrically charged molecule – that is, it is an electrochemical process.

10.7.1 Porous Membranes

Porous membranes are thin layers of rigid materials such as **porous glass or sintered metal/ceramic**. Porous membranes are made by pyrolysis of polymers deposited in the pores of a tubular porous support. Materials used in these membranes are similar to those described for physical adsorption applications. Some of these carbon molecular-sieve membranes have wide variation of permeability of different gases, making them effective separation membranes [1,11,14].

Permeation Flux, Permeability and Selectivity in Porous Membranes: Low-molecular-weight components permeate or diffuse (by flowing through the pores) through a membrane more easily, from the high to the low partial pressure side of the membrane, partial pressure being that of the diffusing component. The molar transport flux within the membrane is governed by Fick's law, which states that the flux is driven by molecular diffusion across a species' molar concentration gradient:

$$J_i = D_i^e \nabla C_i, \tag{10.28}$$

where C_i is the concentration of the fluid in the membrane and D_i^e is the effective diffusion coefficient. When the pore diameter, d_p, is smaller than the mean free path, λ, that is, when the Knudsen number $\mathrm{Kn} = \lambda/d_p$ is larger than 1, gases diffuse independently by the mechanism of **Knudsen diffusion**. In this case, the diffusion coefficient of a gas in the pores is given by the Knudsen diffusivity:

$$D_{Ki} = \sqrt{\frac{8}{9\pi}} d_p \sqrt{\frac{\Re T}{M_i}}, \tag{10.29}$$

where M_i is the molecular weight of the diffusing gas. In cases when the mean free path is on the order of magnitude of the pore diameter, one must account for both Knudsen diffusion and bulk diffusion in the gas mixture. In this case, the total diffusion coefficient is given by:

$$D_i^e = \left(\frac{1}{D_{Ki}} + \frac{1}{D_{Bi}^e} \right)^{-1}, \tag{10.30}$$

and the effective bulk diffusivity, D_{Bi}^e, is given by

$$D_{Bi}^e = \frac{\varepsilon}{\tau} D_{Bi}, \tag{10.31}$$

that is, it is the diffusivity modified by the **porosity**, ε, defined by the ratio of the void volume to the total volume of the porous medium, and **tortuosity**, τ, defined by the ratio of the actual distance traveled by the molecule within the pore to the straight line across the pore, or membrane thickness. Moreover, the diffusivity in the bulk gas is given by the binary diffusivity in the mixture, $D_{Bi} = (1 - X_i)/\sum_{k \neq i} \frac{X_k}{D_{ik}}$.

Because Knudsen diffusion, the dominant mode for separation in molecular sieves, porous ceramic, or glass membranes, depends on the molecular weight of the diffusing gases, it is effective for separating gases with widely different molecular sizes/weights. Bulk diffusion is less sensitive to the molecular mass.

The diffusion flux can be written in terms of the concentration difference across the membrane (the concentrations evaluated at the membrane surface) divided by the membrane thickness. Moreover, if the gas concentration immediately above the membrane surface is the same as the concentration of the absorbed gas at the membrane surface (that is, neglecting any film resistance at the surface or assuming low mass transfer resistance in the gas phase), the concentration can be expressed in terms of the gas partial pressure on both sides, and the resulting expression is:

$$J_i = \frac{D_i^e}{\Re T t}\Delta p_i = \frac{\tilde{p}_i}{t}\Delta p_i = \tilde{P}_i \Delta p_i, \tag{10.32}$$

where t is the membrane thickness, $\tilde{p}_i = D_i^e/(\Re T)$ is the **permeability coefficient**, and $\tilde{P}_i = D_i^e/(\Re T t) = \tilde{p}_i/t$ is the **permeability.** It is the proportionality constant between the flux and the pressure difference across the membrane. The gas component with the higher diffusivity has the larger flux and the higher partial pressure on the lower-pressure side of the membrane, the permeate side. The flux across these porous membranes depends on the permeability.

10.7.2 Polymer Membrane (for CO_2/N_2 and CO_2/H_2 Separation)

Polymers are extremely long chain molecules that consist of chain units, where each unit is similar in dimensions to other regular molecules. Polymers often contain thousands of chain units. When fully stretched, a polymer molecule's length can be close to a micrometer, but polymer molecules are strongly coiled and intertwined (like worms) with a single coil length of the order of 10 nm. The chain units can be small, such as polyethene, whose unit is ethene, which produces thin, flexible polymers. Chain units can also be bulky, such as cellulose, which produces much stiffer polymers (Chapter 14). Apolar chain units, such as ethane or other hydrocarbons and fluorocarbons, produces hydrophobic polymers that do not interact with water, but swell in apolar solvent. Polar polymers such as cellulose contain large numbers of the hydroxyl group and swell strongly in water and may be soluble. Copolymers consist of different chain units.

Cross-linked polymers form 3D networks with high local concentrations and can swell in a solvent, but are not soluble. Cross-linking occurs chemically, via entanglement or as crystallite. Cross-linked polymers are important in separation membranes. Most polymers are partly amorphous – that is, they contain regions where the molecules show little ordering – and partly crystalline – that is, polymer chains are more or less aligned in an organized structure [15]. Crystallinity and the type and shape it takes depends on the polymer type, and may depend on the history, local composition, pressure, and temperature, hence may change in time. The crystalline zone has small dimensions, on the order of 1–10 nm, and can be practically impermeable to many species. Thus, diffusion occurs mainly in the open spaces within amorphous regions.

Polymer properties change abruptly and drastically with temperature. For instance, the modulus of elasticity of a polymer drops by orders of magnitudes as its temperature passes upwards through the glass transition temperature, as shown in Figure 10.24. As it crosses this temperature the polymer changes from a solid, brittle, glassy material to a rubbery, elastic material [16]. At lower temperatures, polymers are rigid and brittle materials that form a glass; at higher temperatures, the polymer becomes soft and elastic, forming a rubber. At even higher temperatures, the polymer goes through another

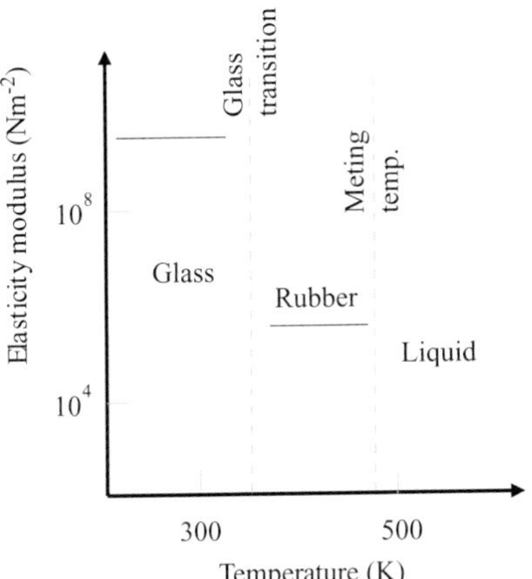

Figure 10.24 The modulus of elasticity showing the glass transition temperature and melting temperature of the polymer.

transition at which it melts, forming a viscous fluid (except for cross-linked polymers). Polystyrene, or ordinary glass, has a glass transition temperature above normal temperatures, and hence are often seen in the solid phase. Silicone rubber glass temperatures are below normal temperatures, and hence we know them in their rubbery forms. These phase transition temperatures may depend on the rates of heating and cooling, and other internal polymer properties may depend on its history. Because of this history dependence and long characteristics time, polymers are not considered to be in thermodynamic equilibrium.

Permeation Flux, Permeability and Selectivity in Non-porous Polymer Membranes The diffusivity of permeates through polymers increases rapidly with temperature, with several orders of magnitude change between the glassy and rubbery phases. Moreover, swelling of polymers by increasing the volume of the permeant in it also increases the diffusivity. The concentration profile across a polymer membrane is shown in Figure 10.25.

Polymer membranes are non-porous, and gas transport through these membranes occurs by a **solution-diffusion** mechanism in which the gas first adsorbs/dissolves into the polymer surface on the high-pressure side, diffuses through the membrane polymer phase, and desorbs into the gas stream on the low-concentration/partial pressure side. In modeling mass transfer across the membrane, it is often assumed that **equilibrium** is maintained on each interface, that is, the partial pressure of the gas on each side of the membrane is proportional to the concentration of the dissolved gas at the membrane surface:

$$C_{is} = p_i S_i, \tag{10.33}$$

where C_{is} is the molar concentration of the dissolved phase at the interface, and S_i is the **solubility** of the gas into the membrane material. (Observe the similarity with gas-liquid equilibrium Henry's constant, $p_i = X_{is}\mathrm{He}_i$.) Equation (10.33) can be written to relate the concentration of the dissolved phase on the surface to the gas partial pressure on both sides of the membrane. The diffusion of the dissolved phase across the membrane is given by

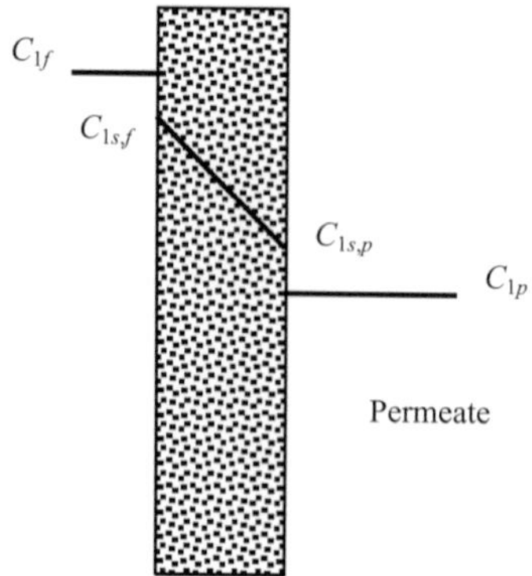

Figure 10.25 Permeation in a non-porous membrane where the concentration on the surface is different than that in the gas phase, immediately above the surface. Gas is first adsorbed on the surface on the feed side, diffuses through the membrane, and then desorbs on the sweep side. There are concentration jumps at the surface associated with the adsorption/desorption or solution/desolution processes at the surfaces. This is different than the case of the porous membrane shown earlier, where there is no jump of concentration on the surface.

Fick's law, which when written for a constant flux/gradient across the membrane, takes the following form:

$$J_i = D_i \frac{\Delta C_{is}}{t}, \tag{10.34}$$

where t is the membrane thickness. The diffusion coefficient, D_i, in this equation is that for the dissolved phase in the membrane material.

Using (10.33) to express the concentration in terms of the gas partial pressure in (10.34), the flux can be written in terms of the pressure difference across the membrane:

$$J_i = \tilde{p}_i \frac{\Delta p_i}{t} = \tilde{P}_i \, \Delta p_i, \tag{10.35}$$

where $\tilde{p}_i = \frac{D_i}{\Re T} S_i$ is the **permeability coefficient**, which is given the unit Barrers (Barrer $= 10^{-10}$cm^3(STP) cm$/(\text{cm}^2\text{s cmHg})$). Note the difference between $\tilde{P}_i$ in the porous membrane case and $\tilde{p}_i$ in the non-porous membrane case, with the former depending only on D while the latter depends on both D and S, and $\tilde{P} = \tilde{p}/t$. Permeability coefficients in selected polymers are given in Table 10.6 [1,17].

The permeability coefficient is much higher for rubber and silicone than for glassy material. The permeability goes to a minimum with increasing weight or size of the gas molecule.

The ratio of the permeabilities of a binary mixture is the membrane **selectivity** with respect to that mixture, or the **ideal separation factor**:

$$\alpha_{ij} = \frac{\tilde{P}_i}{\tilde{P}_j} = \frac{D_i \, S_i}{D_j \, S_j}. \tag{10.36}$$

Table 10.6 Permeability coefficient, $\tilde{p}$, **in Barrers, at 25 °C**

	Gas		
Polymer	H_2	N_2	CO_2
Natural rubber	49	8.7	134
Polysulfone	14	0.25	5.6
Cellulose acetate	24	0.33	10
Polyimide	2.3	0.018	0.41

Thus, high selectivity can be arranged by selecting membranes with high diffusivity ratios (and high fluxes) or high solubility ratios (and high separation). The diffusivities of a solute may vary with its concentration, especially when appreciable swelling of the membrane occurs. In general, membrane selectivity decreases with temperature. For instance, the ideal separation factor for H_2/CO is 40 for cellulose acetate and 76 for polyimide. For H_2/CO_2, they are 2.4 and 2.5, respectively [3].

The **separation factor** is the ratio between the concentration of gases being separated on the feed side and the permeate side. For instance, when separating H_2 from a mixture of CO_2 and H_2, the separation factor can be written as:

$$\tilde{\alpha}_{H_2/CO_2} = \left(\frac{X_{H_2}}{X_{CO_2}}\right)_{permeate} \Big/ \left(\frac{X_{H_2}}{X_{CO_2}}\right)_{feed}. \tag{10.37}$$

Given the wide range of diffusivity and solubility of different gases, it is possible to design membrane units with high selectivity. For example, for silicone rubber, the selectivity of CO_2/H_2 is 4.9, and CO_2/O_2 is 5. For Kapton, an aromatic polyether diimide and a glassy polymer, the selectivity of CO_2/H_2 and CO_2/O_2 are 0.18 and 3.1, respectively, but the permeabilities are 3–5 orders of magnitude lower than those for silicone rubber. For most gases, the permeability increases with temperature according to $Q = A\exp(-E/\Re T)$, because the increase in diffusivity overcomes the decrease in solubility, with the activation energy 1–5 kcal/mol. Operating temperatures must be selected to balance the need for high flux (diffusivity) and high selectivity. It is possible to develop "composite" membranes with a high solubility outside layer and higher diffusivity bulk layer. Membrane separation technology has been used in small ASUs [1], achieving relatively high purity of nitrogen, approaching 99% (necessary for inert gas applications).

Example 10.5

An ASU consisting of a blower, a membrane, and an ejector, is shown in Figure E10.5. Atmospheric air at 30 °C and 1 atm is sucked through the blower and fed to the membrane at 1.1 atm. The oxygen mole fraction at the permeate side is 35%. The ejector maintains a pressure of 0.3 atm on the permeate side. The membrane is made of a special silicon rubber

Example 10.5 (cont.)

with 1.5 μm thickness and 10 m^2 area. The permeability coefficient of oxygen is 500 Barrer (at STP). The behavior of gas in all parts of the system can be approximated as an ideal gas. (a) Assuming perfect mixing at the feed and the permeate sides, determine the daily production of the oxygen-enriched stream. (b) Take an isentropic efficiency of 75% for the blower and ejector, and that $\dot{m}_f = 18\dot{m}_p$; calculate the energy requirement of this unit.

Figure E10.5

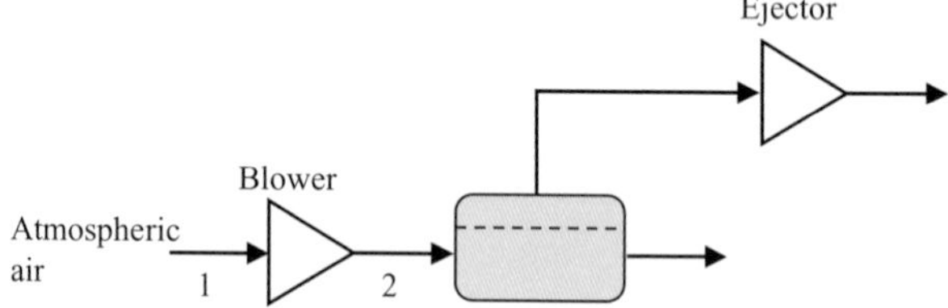

Solution

a. The volumetric flow rate of the oxygen-enriched stream (permeate) can be determined using the following relationship: $\dot{V}_p = \frac{J_{O_2} A}{X_{p,O_2}}$, where the flux of oxygen is obtained, assuming uniform conditions on both sides, from $J_{O_2} = \frac{\tilde{p}_{O_2}}{t_m}\left(X_{f,O_2}p_f - X_{p,O_2}p_p\right)$.

Noting that 1 Barrer = 7.5005 × 10^{-18}m^2/(s·Pa), and substituting the values, we get

$$J_{O_2} = \frac{500 \times 7.5005 \times 10^{-18}\left[\frac{\text{m}^2}{\text{s·Pa}}\right]}{1.5 \times 10^{-6}[\text{m}]}\{0.21 \times (1.1 \times 101,300)[\text{Pa}] - 0.35 \times (0.3 \times 101,300)[\text{Pa}]\}$$

$$= 3.191 \times 10^{-5}\frac{\text{m}^3}{\text{m}^2\text{·s}}.$$

Thus, $\dot{V}_p = \frac{3.191 \times 10^{-5} \times 10}{0.35} = 0.000912\frac{\text{m}^3}{\text{s}}$,

or $\dot{m}_p = \left(X_{p,O_2}M_{O_2} + X_{p,N_2}M_{N_2}\right)\frac{p\dot{V}_p}{RT} = (0.35 \times 32 + 0.65 \times 28)\frac{101.3 \times 0.000912}{8.314 \times 303}$

$$= 0.001078 \text{ kg/s}.$$

Oxygen-enriched stream production per day = 0.001078 × 3600 × 24 = 93.2 kg.

b. The operation of the ASU needs to supply the work requirement by the ejector and the blower. The thermodynamic model for the blower and ejector is similar to that of a compressor. Hence,

$$\dot{W}_{blower} = \frac{\dot{m}_f c_p (T_{2s} - T_1)}{\eta_{blower}}, \text{ where } \dot{m}_f = 20 \times 0.00178 = 0.02156\frac{\text{kg}}{\text{s}}.$$

Example 10.5 (cont.)

Taking $\gamma = 1.4$ (corresponding to $c_p = 1000\,\mathrm{J/kg{\cdot}K}$), T_{2s} is calculated as follows:

$$T_{2s} = T_1\left(\frac{p_2}{p_1}\right)^{\frac{k-1}{k}} = (30 + 273)\left(\frac{1.1}{1}\right)^{0.2857} = 311.4\ \mathrm{K}.$$

Thus, the power requirement for the blower is $\dot{W}_{blower} = \dfrac{0.02156 \times 1000 \times (311.4 - 303)}{0.75} = 241.5\ \mathrm{W}.$

The power requirement of the ejector is calculated using a similar procedure:

$T_{4s} = T_3\left(\dfrac{p_4}{p_3}\right)^{\frac{k-1}{k}} = 311.4\left(\dfrac{1}{0.3}\right)^{0.2857} = 439.2\ \mathrm{K}$. Note that we assumed $T_3 = T_2$.

$$\dot{W}_{ejector} = \frac{0.001078 \times 1000 \times (439.2 - 311.4)}{0.75} = 183.7\ \mathrm{W}.$$

The total power requirement for the operation of the ASU is

$$\dot{W}_{total} = 241.5 + 183.7 = 425.2\ \mathrm{W}.$$

Separation of CO_2/N_2 Using Polymer Membranes The flue gas composition of a pulverized coal plant, after desulfurization, is typically 75% N_2 and 13% CO_2, and the remaining is water vapor (oxygen originally in the gas is removed in the desulfurization process). To use a membrane unit for separating CO_2 from these combustion products, the gases must first be compressed; a pressure of 1–3 MPa is typically used. Multistage compression with intercooling is necessary to reduce the compression work. Finned copper tube heat exchangers using cooling water to reduce the gas temperature between the stages can be used.

The low selectivity of polymer membranes to CO_2/N_2 mixtures means that a single-stage separation produces a CO_2-lean retentate stream and a CO_2-rich permeate stream. Following the first stage, the permeate stream is still diluted with nitrogen, which is undesirable for storage purposes. This is because while CO_2 condenses under 80 MPa pressure, nitrogen remains as a gas (compression and liquefaction of CO_2 is necessary for transportation and storage). Compressing the permeate of the first stage to 80 MPa, most CO_2 will condense, except for a small fraction that separates in the form of vapor and leaves. To further purify the permeate stream from nitrogen, a second-stage membrane may be used, after compressing the permeate again, as shown in Figure 10.26. The permeate of the second stage is much richer in CO_2. The retentate streams can be expanded to recover some of the work required for compressing the feed stream.

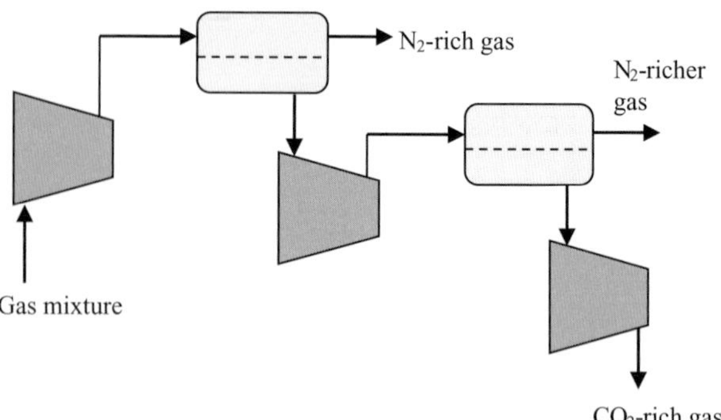

Figure 10.26 A two-stage membrane separation of a mixture of CO_2 + N_2 requiring compressors on the feed side of the membrane units, and more units to compress the CO_2-rich stream toward the storage side. The membranes used in this unit are selective toward CO_2. The N_2-rich streams out of the two units are vented into the atmosphere at atmospheric pressure.

Organic membrane units for CO_2/N_2 separation are based on material like cellulose derivatives, polysulfone, polyamide, or polyimide. The membrane material is supported by porous nonselective high-strength, highly permeable layer, fabricated mostly as hollow fiber with outer diameter of 500 μm and high surface-to-volume ratio (2000 m^2/m^3). A number of manufacturers produce membranes for CO_2/N_2 separation, such as Ube's polyimide membrane with $\alpha_{CO_2/N2} = 43$, Delair's PPO with $\alpha_{CO_2/N2} = 19$, and Monsanto's polysulfone with $\alpha_{CO_2/N2} = 31$ [3].

10.7.3 Design Calculations for Porous and Polymer Membranes

Gas separation membrane units can use vacuum on the permeate side and compression on the feed side to create a pressure gradient, and hence partial pressure gradients, to derive gas permeation across the membrane. The gas flux is often a mixture of gases from the feed side, whose permeability is sufficiently high. The higher the selectivity, the more of one particular gas dominates the total flux.

Permeate Composition under Well-Mixed Conditions The composition of the permeate (gas crossing the membrane) and the retentate (residue in the original stream) depends on several variables, including the pressure difference across the membrane, the flow rates, and compositions of the feed and sweep gases, the permeability of the different components, the total membrane area, and the arrangement of the two streams or the flow pattern. Flux expressions shown above are often used to calculate the membrane area needed to separate a gas, given the flow rates and composition, based on how the streams are arranged and operating conditions. Accurate calculations often involve integrating differential equations describing the variation of the composition along the membrane [1,14]. In the following, two approximate models are shown, in which explicit or semi-explicit expressions of the membrane area and the retentate gas composition can be written. While approximate, they provide insight into the dependency of the process on the operating conditions. Figure 10.27 shows a schematic of a unit with cross-permeate flow; conditions on the permeate side are assumed uniform along the membrane.

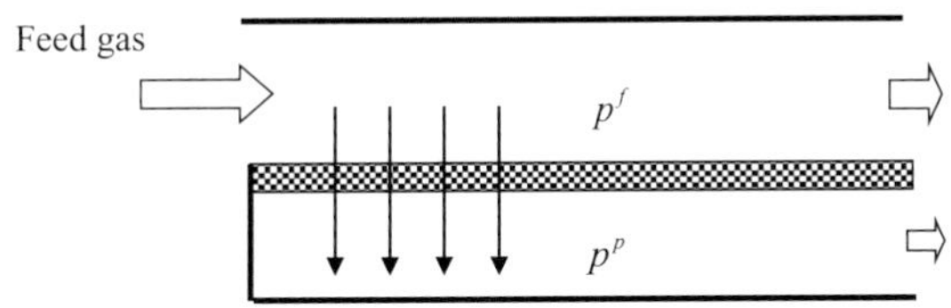

Figure 10.27 Model for calculating composition at the exits of permeate and retentate streams assuming perfect mixing on both sides.

Analytical expressions can be derived for the case when both streams inside the membrane separation unit are assumed to remain fully mixed, at all locations, during the permeation process. This model is similar to the "well-stirred reactor" model used in chemical reactors including combustion, and essentially assumes that the mixing rate between the permeating gas and the permeate or sweep stream is much faster than the flux. Thus, the concentrations are independent of location. Moreover, the conditions inside the unit are the same as those at the exit. Assume only two permeating gases, 1 and 2. For both, the feed and permeate are denoted by f and p. The total flux across the membrane (at any location) is J, and the fluxes of the two components across the membrane are $J_{1,2} = X_{1,2}J$:

$$\begin{aligned} J_{1,2} &= \tilde{P}_{1,2}\left(p^f_{1,2} - p^p_{1,2}\right) \\ &= \tilde{P}_{1,2}\left(X^f_{1,2}p^f - X^p_{1,2}p^p\right), \end{aligned} \tag{10.38}$$

where $\sum_i X_i^{f,p} = 1$. In this case we assume perfect mixing on each side of the membrane; the retentate composition (on the higher-pressure feed side) is the same along the membrane and the same as at the exit (but not the same as the feed inlet); the permeate composition (on the lower-pressure sweep side) is the same everywhere and the same as the exit; and the molar flow rates, N_p, across are [1,14]:

$$\dot{N}_i^p = J_i A_m,$$

where for simplicity we assume that the inlet flow on the permeate side is zero, A_m is the total membrane surface area, and $\dot{N}_i^p = X_i^p \dot{N}^p$. Mass conservation shows that:

$$\dot{N}^f_{i,in} = \dot{N}^f_{i,out} + \dot{N}^p_i$$

or

$$X^f_{i,in} = (1-\varphi)X^f_{i,out} + \varphi X^p_i,$$

where $\dot{N}^f_{i,in/out} = X^f_{i,in/out}\dot{N}^f_{in/out}$ and $\varphi = \dot{N}^p/\dot{N}^f_{in}$, known as the **stage cut**. Substituting in the permeate flow rate expression and using the flux expressions, and solving for the mole fractions and membrane area:

$$X_i^p = \frac{p_i^f}{(1-\varphi)\left(\dfrac{\dot{N}^p}{A_m\tilde{P}_i} + \dfrac{\varphi}{1-\varphi}p^f + p^p\right)}$$

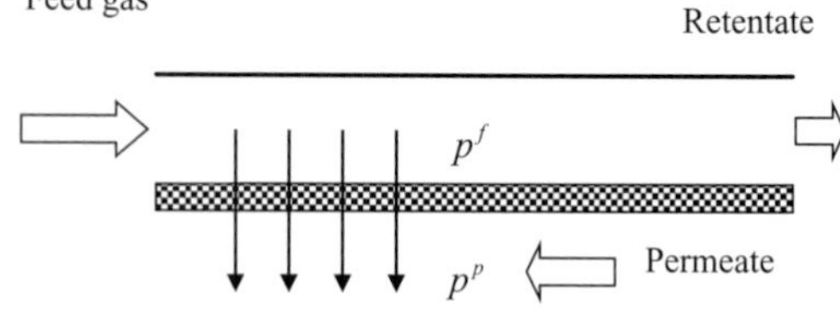

Figure 10.28 Model for calculating composition at the exit of the permeate and retentate streams in a counter-flow separation unit with length-dependent flux.

and

$$A_t = \frac{\dot{N}_{1p}}{\tilde{P}_1\left(\frac{p^f}{1-\varphi}\left(X^f_{1,in} - \varphi X^p_1\right) - p^p X^p_1\right)}.$$

These equations should be solved iteratively in the two unknowns: the permeate gas composition and the membrane area required to separate the stage cut φ from the feed gas.

Another design is the counter-flow arrangement in the unit, as shown in Figure 10.28.

Approximate analysis [1] in which the local flux is assumed to vary along the membrane length depending on location (and local concentrations) can be done to estimate the required membrane length and concentrations. The analysis distinguishes between the local concentration associated with the local flux at a certain point,

$$\frac{X^p_1}{X^p_2} = \frac{J_1}{J_2} \text{ and hence } X^p_1 = \frac{J_1}{J_1 + J_2},$$

and the average concentration along the membrane, $\bar{X}$, which is taken as the average value along the membrane (in order to avoid solving the differential equations that describe accurately the variation of the concentration). The flux expressions are evaluated using the average concentrations at this point, $J_{1,2} = \tilde{P}_{1,2}\left(\bar{X}^f_{1,2}p^f - \bar{X}^p_{1,2}p^p\right)$. Substituting for the fluxes for the expressions above, the local concentration associated with the flux is given by

$$X^p_1 = \frac{\bar{X}^f_1 - R\bar{X}^p_1}{\bar{X}^f_1 - R\bar{X}^p_1 + \left(1 - \bar{X}^f_1 - R\left(1 - \bar{X}^p_1\right)\right)/\alpha},$$

where $R = p^p/p^f$ is the total pressure ratio across the membrane and $\alpha = \tilde{P}_1/\tilde{P}_2$ is the **permeability ratio** or **selectivity** of the membrane. In a counter-flow separation unit, the concentration on the permeate side at the end of the unit, where the first permeation takes place, is the same as the average permeate concentration at the same location, $X^p_{1,0} = \bar{X}^p_{1,0}$ and $\bar{X}^f_1 = \bar{X}^f_{1,out}$. Substituting in the equation above:

$$(\alpha - 1)\left(X^p_{1,0}\right)^2 + \left\{(1-\alpha) - \frac{1}{R}\left[1 + \bar{X}^f_{1,out}(\alpha - 1)\right]\right\}\left(X^p_{1,0}\right)^2 + \frac{\alpha\bar{X}^f_{1,out}}{R} = 0.$$

This **quadratic equation** can be solved for the permeate concentration at this point (the exit of the separation unit) by assuming that the concentrations on the feed side remain almost constant (when the gas being separated has a small concentration in the feed stream). In this approximate solution, the stage cut can be used to evaluate the feed stream concentration at the same point (the end of the unit) using conservation, and the permeate stream concentration at the beginning of the reactor, also using conservation. The concentrations at the beginning and end of the unit are used to evaluate the fluxes at the terminal points and average flux values are used to estimate the membrane area.

These equations show that

- increasing the feed concentration, X_1, raises its concentration in the permeate gas stream;
- reducing the pressure ratio across the membrane increases the concentration of X_{1p} in the permeate, but there is a terminal value;
- the permeate side can be purged continuously with an inert to reduce the partial pressures of the gases on that side, and hence to increase the fluxes and separation in the permeate gas. This technique can reduce the separation work since an overall lower pressure difference across the membrane can be utilized; and
- separation improves as the selectivity increases, but to a limit that is determined by equal partial pressures for X_1 in the feed and the permeate.

In case the feed stream remains almost uniform in concentrations, and its concentrations remain almost the same as that at the inlet, and the concentrations in the permeate stream stay closer to the conditions at the exit (in the counter-flow case), an approximate estimate of the membrane area required is given by

$$A_m = \frac{\dot{N}^p \bar{X}_1^p}{\tilde{P}_1 \left(p^f \bar{X}_1^f - p^p \bar{X}_1^p \right)},$$

where the concentrations are taken as the averages of the inlet and outlet sections.

Example 10.6

A mixture of 40%CO_2/60%N_2 is separated by a Delair-made membrane. The pressures at the feed and permeate sides are 3 bar and 0.6 bar, respectively. The volumetric flow rate at the permeate side is 7.6 m^3/h. Assuming perfect mixing at feed and permeate sides, determine: (a) the permeate composition; (b) the flux of carbon dioxide; and (c) the membrane area. Use the following data: the selectivity of this membrane is $\alpha_{CO_2/N_2} = 19$, and the permeability of carbon dioxide is 2750×10^{-12} m^3/(m^2 Pa s).

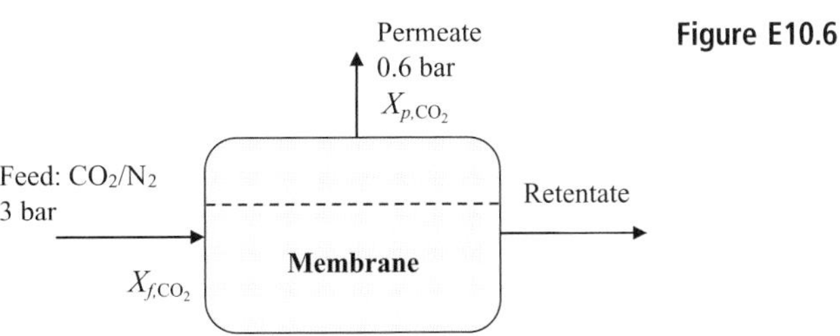

Figure E10.6

Example 10.6 (cont.)

Solution

a. The mole fraction of carbon dioxide at the permeate side is obtained by solving the quadratic equation on page 580:

$$X_{p,CO_2} = \frac{59 - \sqrt{59^2 - 4(19-1) \times 19 \times (3/0.6) \times 0.4}}{2(19-1)} = 0.88.$$

b. We can now calculate the carbon dioxide flux using (10.38):

$$J_{CO_2} = 2750 \times 10^{-12} \left[\frac{\text{m}^3}{\text{m}^2 \cdot \text{Pa} \cdot \text{s}}\right] \times \left(X_{f,CO_2} p_f - X_{p,CO_2} p_p\right) [\text{Pa}]$$

$$= 2750 \times 10^{-12} \times \left(0.4 \times 3 \times 10^5 - 0.88 \times 0.6 \times 10^5\right)$$

$$= 1.85 \times 10^{-4} \frac{\text{m}^3}{\text{m}^2 \cdot \text{s}}.$$

c. The required membrane surface area is

$$A = \frac{\dot{V} \times X_{p,CO_2}}{J_{CO_2}} = \frac{7.6 \left[\frac{\text{m}^3}{\text{h}}\right] \times \frac{1}{3600} \left[\frac{\text{h}}{\text{s}}\right] \times 0.88}{1.85 \times 10^{-4} \left[\frac{\text{m}^3}{\text{m}^2 \cdot \text{s}}\right]} = 10.04 \text{ m}^2.$$

Figure 10.29 shows three different commonly used arrangements in membrane separation: counter-flow, co-flow, and radial cross-flow. The last is used with hollow fibers. In a typical practical separation unit, millions of these fibers are arranged in a shell, as shown in the figure. In the counter-flow arrangement, the permeate rate starts at zero at the closed end and builds up toward the open end, while in the co-flow case, the permeate flow rate is zero in the middle of the fiber. None of the arrangements shown are pure; in all cases mixed flow are obtained.

The energy consumption during the separation process can be estimated from the power required to move the gas through the unit, on both sides – that is, the power required to overcome the pressure drop in the channels of the membrane unit, as well as the power required to pump the gas back to the pressure in the next units in the plant. These calculations can only be done after the separation unit has been designed (i.e., knowing the dimensions of the channel and their length, and the conditions of the neighboring units before and after the separation unit). The total power consumption required for compression, $\wp_c$, is given by:

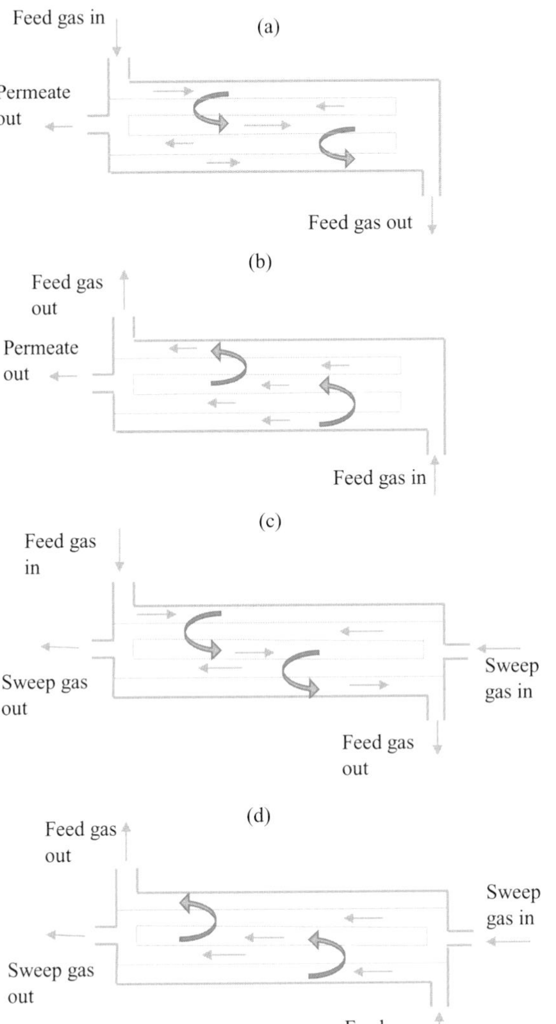

Figure 10.29 Membrane separation units (a) counter-current flow without sweep, (b) co-current flow without sweep, (c) counter-current flow with sweep, (d) co-current flow with sweep. Thick lines show the unit boundaries and thin lines show the membrane units. Curved arrows show feed gas flow around the membrane elements, which can be cylindrical or rectangular cross-sections.

$$\wp_c = J_c A \frac{n\Re T_{in}}{\frac{k-1}{k}\eta_c}\left[\left(\frac{p_{out}}{p_{in}}\right)^{\frac{k-1}{k}} - 1\right], \tag{10.39}$$

where J_c is the molar flux through the compressor, n is the number of compression stages, $\Re = 8.314$ J/mol·K, T_{in} is the temperature before each compression stage, p_{out} and p_{in} are the pressures after and before the stage, k is the isentropic index of the gas through the stage, and η_c is the isentropic efficiency of the compressor.

Separating CO/H_2, or CO_2/H_2 Using Polymer Membranes Membrane separation of CO_2 from flue gases is not very efficient, and chemical absorption techniques are preferable

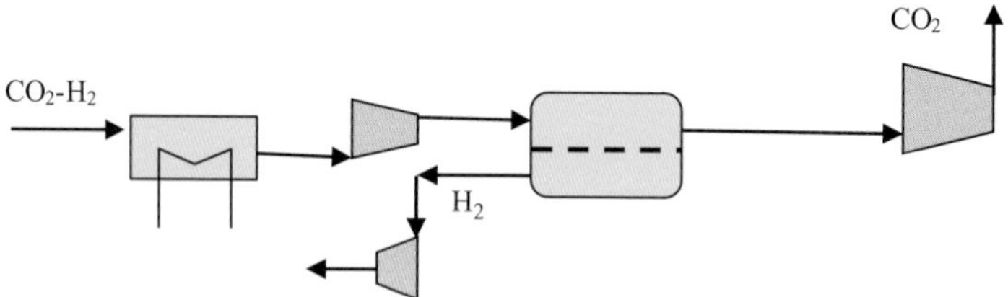

Figure 10.30 A layout showing the use of a membrane to separate H_2 from a mixture of CO_2 and H_2. A feed gas stream, in this case syngas after WGS, is first cooled and compressed to higher pressure. It is introduced on the feed side, where hydrogen diffuses through the membrane toward the permeate (sweep) side driven by the partial pressure differential across the membrane. The permeate is removed using a stream or purge gas (nitrogen or steam). Hydrogen is compressed as it leaves the membrane unit for use in the gas turbine. The retentate stream, CO_2, leaves the membrane unit and is compressed for storage.

for this purpose. On the other hand, it is possible to use membranes to separate H_2 from the products of gasification/reforming, leaving behind $CO + CO_2$ in the retentate gas stream. If WGS is used, then H_2 is separated from a stream of $H_2 + CO_2$, leaving only CO_2 in the original stream. The unit is shown schematically in Figure 10.30. Hollow fiber polyimide membranes can be used for this purpose. The permeability of H_2, CO, and CO_2 are 1374, 23, and 461 [m^3(STP)m^{-2} Pa^{-1} s^{-1}] for the polyimide Ube membrane. The selectivity of H_2 with respect to CO_2 is 60 (this is lower than the ideal value because of some plasticization that occurs as the permeating gas exhibits high chemical affinity for the polymer). Polymer membranes operate at relatively low temperature, around 100 °C.

The actual design of the membrane, namely its surface area, the pressure ratio across it, and the feed rate, should be chosen to satisfy a number of criteria, including minimizing the power requirements and reducing the equipment size [3]. Figures 10.31 and 10.32 show the dependence of the energy consumption on membrane selectivity and pressure ratio (feed/permeate) in the case of H_2+CO_2 and H_2+CO mixtures, respectively. Clearly, as the membrane selectivity improves, the required pressure ratio across the membrane decreases and the separation energy is lowered. However, the difference in the total energy consumption of separation between the high selectivity and very high selectivity membranes is not that significant for higher selectivity membranes. Note that the optimum conditions for high selectivity membranes correspond to moderate pressure ratios of 10.

As expected, reducing the pressure on the permeate side increases the concentration of both gases, but also the separation ratio. The lower the pressure on the permeate side, the higher the percentage of H_2/CO and the less of the same gas is left in the feed side. A larger membrane supports more gas transport across the membrane. However, as the area increases, the partial pressure of H_2/CO on the permeate side increases, reducing the driving force of the flux. Reducing the total pressure on the permeate side delays this effect (at the expense of creating a vacuum). However, reducing the pressure to improve the separation

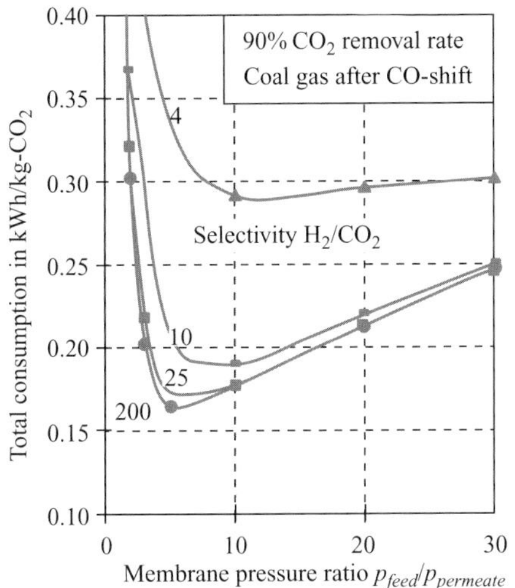

Figure 10.31 Total energy required for CO_2/H_2 separation; the syngas is produced from coal gasification and CO shift for the first case. The separation work is evaluated for different values of membrane selectivity with respect to H_2/CO_2 [2].

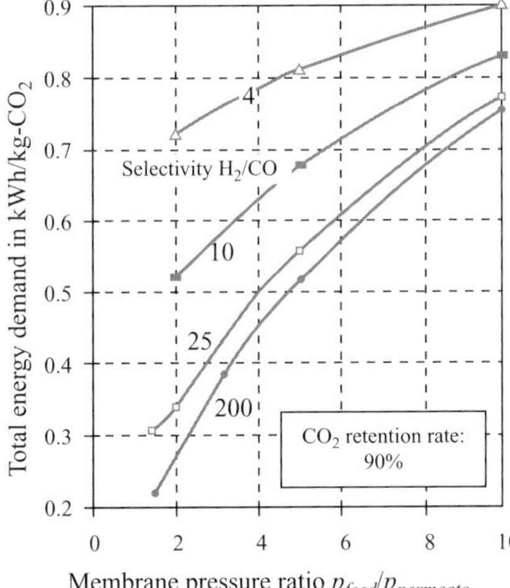

Figure 10.32 Total energy required for CO/H_2 separation; the syngas is produced from coal gasification. The separation work is evaluated for different values of membrane selectivity with respect to H_2/CO [2].

ratio results in a low-pressure hydrogen/CO stream that must be compressed back to the gas turbine combustor pressure. The composition of the gas following separation is discussed in Chapters 11 and 13.

Polymer membranes do not produce very high-purity gases. Often multistage purification is necessary, with either several membrane units arranged in series or another separation process, such as PSA, used if high purity is needed. Membrane separation has been used to remove CO_2 from natural gas at the wellhead. At 35 °C and 40 atm, CO_2/CH_4 have selectivity of 20–30 for polycarbonate, polysulfonate, and cellulose acetate membranes.

10.8 Dense Metallic Membranes

Dense metallic membranes allow one component of a gas mixture flowing on their feed side to permeate through and hence are capable of producing pure gases on the permeate or sweep side. These include mixed conducting ceramic membranes that transport charged species through the lattice and metallic membranes that allow only hydrogen atoms to diffuse.

10.8.1 Mixed Conducting Membranes for O_2 and H_2 Separation

Mixed ionic electronic conducting ceramic membranes (MIEC) have been proposed for gas separation. At high temperature, in the range 800–1000 °C, these membranes can be designed to separate oxygen from air or H_2 from syngas (before or after shift). These are dense membranes that conduct charged species only. For gas separation using thermal energy only, the membrane material must conduct both positive and negative charges (contrary to membranes used in solid oxide fuel cells, which conduct oxygen ions only), as shown in Figure 10.33. Mixed ionic electronic conducting ceramic membranes are dense membranes (non-porous) and are capable of producing high-purity gas – that is, only oxygen (or hydrogen) permeates and flows on the sweep side or mixes with the sweep gas flow (barring leaks through sealing). If the sweep gas includes a reactive component, such as a fuel in the case of oxygen-permeable membranes, it may react with it to form products (complete or partial oxidation depending on the oxygen flux and fuel concentration and sweep flow rate). Ceramics (mixed oxides) such as perovskites or fluorites are often used for these membranes, given their relatively high ionic and electronic conductivities at intermediate temperatures and good surface catalytic activities.

10.8.2 MIEC for Oxygen Separation

Most MIEC membranes for oxygen permeation are made of perovskites. These are mixed metal oxides with oxygen defects (or vacancies). Similar to solid-oxide fuel cell electrolytes, oxygen permeates through these membranes as oxygen ions. Meanwhile electrons (or electron holes) must also be conducted through the same material and hence both electronic and ionic conductivities are important for performance. Under conditions of finite oxygen

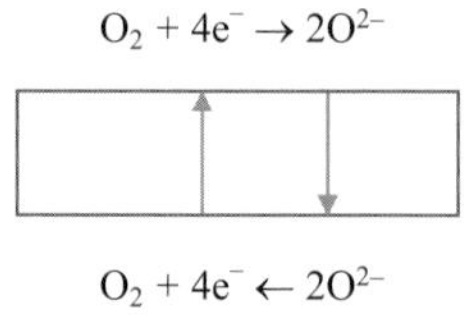

Figure 10.33 In an ion-transport membranes (ITM), oxygen, an electrochemical reaction occurs on both gas–solid interfaces, and both oxygen ions and electrons are conducted (in opposite directions) through the membrane.

partial pressure gradient across the membrane, oxygen can flow from the high to the low partial pressure side while maintaining overall charge neutrality without the need to polarize (applying external finite voltage to) the membrane [18,19]. The permeation rate, or flux, depends on the oxygen partial pressure on both sides, temperature, conductivity, and surface activities on both sides of the membrane, as will be shown.

The ideal stoichiometric structure of perovskites is represented by $ABO_{3-\delta}$, where the A site is often occupied by a large alkali earth metal such as lanthanum (La), strontium (St), or barium (Ba), and the B site is occupied by a smaller transition metal such as iron (Fe), cobalt (Co), or cerium (Ce). The lattice structure of these material is orthorhombic. Examples of oxygen-conducting membranes include lanthanum-calcium-iron (LCF), lanthanum-strontium-cobalt-iron (LSCF), and LGX, where G stands for gallium and X can be nickel, magnesium, or cobalt (or a combination). The associated charge distribution in the lattice depends on the elements and can be $A^{1+}B^{5+}O_3$, $A^{2+}B^{5+}O_3$, or $A^{3+}B^{3+}O_3$ (note that oxygen is O^{2-}). This ideal structure, however, is not a good conductor, and a certain number of defects, leading to oxygen non-stoichiometry, must be present to improve the conductivity of the material. These defects are known as oxygen vacancies, or vacant lattice sites, but they also take different forms. Ion vacancy indicates the absence of an oxygen ion where it should have been. Electron defects can take the form of an ion whose charge is different than the ion that would normally occupy the lattice site, which can also be an electron hole. These vacancies and electron holes are where ions and electrons move (or hop) during permeation.

10.8.3 Defect Chemistry

Defects can be created by doping the material – that is substituting some of the A or B sites with other atoms – homogeneously throughout the material during the manufacturing process. By substituting some of the A site or B site atoms with atoms with similar radius to the original atom but with a different valency (charge number), oxygen vacancies or defects are formed to maintain overall charge neutrality. Take the example of lanthanum ferrite, $L^{3+}F^{3+}O_3$. Substituting some of the La with calcium (Ca^{2+}), the charge is compensated by partial oxidation of Fe^{3+} to Fe^{4+} while changing the oxygen stoichiometry according to:

$$La^{3+}Fe^{3+}O_3 + xCa^{2+}\text{-}xLa^{3+} \rightarrow La_{1\text{-}x}Ca_xFe_y^{3+}Fe_{1\text{-}y}^{4+}O_{3\text{-}d} + \frac{\delta}{2}O_2. \tag{10.40}$$

The value of δ depends on the doping level and temperature, as well as oxygen partial pressure in the gas phase (or other reducing agents, such as H_2). Representative values of the vacancy concentration in LCF as a function of the dopant concentration of Ca are shown in Table 10.7 [20]. Increasing Ca raises the fraction of oxidized iron concentration (Fe^{4+}/Fe_{tot}) in order to compensate for the charge difference between La (3+) and Ca (2+). Meanwhile, the ratio Fe^{4+}/Ca^{2+} (which remains almost constant irrespective of x) is not sufficient to maintain charge neutrality in the material, leading to the formation of an oxygen defect with a concentration of δ. Note that the fraction of oxygen vacancy remains rather small, less than 0.1.

Table 10.7 Oxygen defect concentration in LCF at different doping levels, all measured under the same conditions [20]

Sample	Chemical formula	Fe^{4+}/Fe_{tot}	Fe^{4+}/Ca^{2+}
$x = 0.0$	$LaFe_{0.013}{}^{4+}Fe_{0.987}{}^{3+}O_{3+0.0065}$	1.3	–
$x = 0.1$	$La_{0.9}{}^{3+}Ca_{0.1}{}^{2+}Fe_{0.068}{}^{4+}Fe_{0.932}{}^{3+}O_{3-0.016}$	6.8	0.68
$x = 0.2$	$La_{0.8}{}^{3+}Ca_{0.2}{}^{2+}Fe_{0.142}{}^{4+}Fe_{0.858}{}^{3+}O_{3-0.029}$	14.2	0.71
$x = 0.3$	$La_{0.7}{}^{3+}Ca_{0.3}{}^{2+}Fe_{0.208}{}^{4+}Fe_{0.792}{}^{3+}O_{3-0.046}$	20.8	0.69
$x = 0.4$	$La_{0.6}{}^{3+}Ca_{0.4}{}^{2+}Fe_{0.262}{}^{4+}Fe_{0.738}{}^{3+}O_{3-0.069}$	26.2	0.65
$x = 0.5$	$La_{0.5}{}^{3+}Ca_{0.5}{}^{2+}Fe_{0.310}{}^{4+}Fe_{0.690}{}^{3+}O_{3-0.095}$	31.0	0.62

The actual defect concentration in the material depends on the temperature and oxygen partial pressure at the material surface. For LCF, at equilibrium, the defect concentration is governed by the following two reactions:

$$\frac{1}{2}O_2(g) + V_O^{\bullet\bullet} + 2Fe_{Fe}^{x} \rightleftarrows O_O^{x} + 2Fe_{Fe}^{\bullet},$$

$$2Fe_{Fe}^{x} \rightleftarrows Fe_{Fe}^{'} + Fe_{Fe}^{\bullet},$$

where $V_O^{\bullet\bullet}$ is a double-charged oxygen vacancy, where the defect charge is written as a dot for a positive excess charge (and a prime for a negative excess charge), $Fe_{Fe}^{x}{}^{\circ}Fe^{3+}$, $Fe_{Fe}^{\bullet}{}^{\circ}Fe^{4+}$, and $Fe_{Fe}^{'}{}^{\circ}Fe^{2+}$ are the different iron states in the Fe^{3+} site (neutral, oxidized, or reduced, respectively), and O_O^{x} is the oxygen ion in an oxygen site, $\delta = \left[V_O^{\bullet\bullet}\right]$ where [ ] is the unit formula concentration and $C_i = [X_i]/V_m$ and V_m is the molar volume. The first reaction is the oxygen incorporation reaction where oxygen occupies a vacancy and an iron atom is oxidized; the second is the iron disproportionation reaction. Electron defects in this case appear as (iron) ions with different charge number than their normal value (also called small polarons). Along with the iron site conservation and charge neutrality described by the two equations below, these equations can be solved to determine the vacancy concentration:

$$\left[Fe_{Fe}^{x}\right] + \left[Fe_{Fe}^{'}\right] + \left[Fe_{Fe}^{\bullet}\right] = 1,$$

$$\left[Ca_{La}^{'}\right] + \left[Fe_{Fe}^{'}\right] = 2\left[V_O^{\bullet\bullet}\right] + \left[Fe_{Fe}^{\bullet}\right].$$

Oxygen site conservation is also used in the analysis. The Gibbs energies of these two reactions were obtained from experimental measurements [21], and Figure 10.34 shows the defect concentration at different temperatures and oxygen partial pressures. The figure also shows the change in the iron states. More oxygen defects are generated at higher temperatures and with lower oxygen partial pressures. Moreover, the electronic conductivity

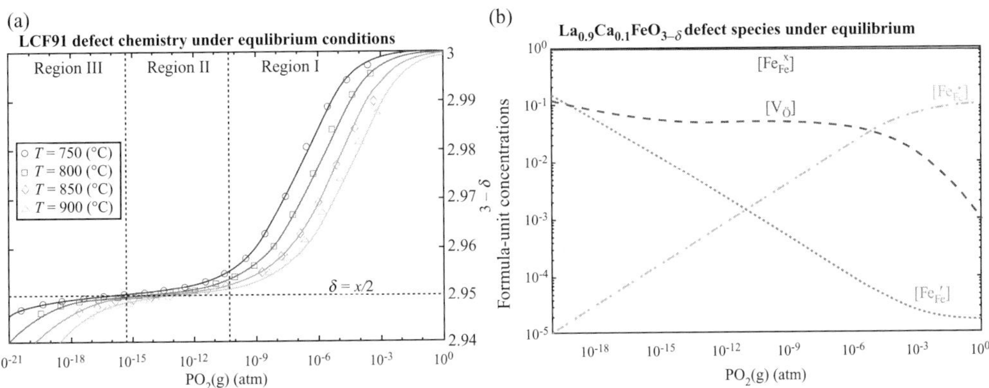

Figure 10.34 (a) Measured and predicted equilibrium non-stoichiometry of $La_{0.9}Ca_{0.1}FeO_{3-\delta}$ (LCF) as a function of temperature and oxygen partial pressure; solid curves calculated from the defect model. (b) Unit formula concentration of defects, and (left) the three iron states, at 950 °C. Data reproduced from [19,21].

is dominated by the electron holes at high p_{O2} and by electrons at lower, p_{O2}, both localized (polarons) on the Fe site, reaching a minimum in between.

The ionic and electronic conductivities depend on the defect species concentration:

$$\sigma_k = \frac{z_k^2 F^2 C_k D_k}{\Re T},$$

where z_k is the carrier charge and D_k is the diffusion coefficient of the carrier. Since C_k depends on the partial pressure of oxygen in the gas phase, the conductivity is often approximated as $\sigma_k = \sigma_k^o p_{O_2}^{-n}$.

Another example of creating defects in a common perovskite is in the case of substituting the lower valence Sr^{2+} ion for some of La^{3+} , in LCF ($La^{3+}Co^{3+}O_{3-\delta}$) [22]. The amount of oxygen vacancy defects and the oxidation state of cobalt can be adjusted through the relation:

$$La^{3+}Co^{3+}O_3 + xSt^{2+} - xLa^{3+} \rightarrow La_{1\text{-}x}St_xCo_y^{3+}Co_{1\text{-}y}^{4+}O_{3\text{-}\delta} + \frac{\delta}{2}O_2,$$

where x is the amount of Sr^{2+} and y is the amount of Co^{3+} in $La_{1-x}Sr_xCoO_{3-\delta}$, commonly referred to as LSCO(1-x)x (that is, LSCO28 for $La_{0.2}Sr_{0.8}CoO_{3\text{-}\delta}$). The effect of Sr^{2+} substitution on oxygen vacancy concentrations was determined experimentally [22]. The results are shown in Table 10.8. There is an increase in the oxidation state of Co as well as in the concentration of oxygen vacancies as more Sr^{2+} is substituted for La^{3+}. Processing conditions can also affect the oxygen content and oxidation state of cobalt significantly.

It is also possible to dope both A and B sites, $A_{1-n}A'_nB_{1-m}B'_mO_{3-\delta}$ to enhance the material conductivity and improve its chemical stability.

Table 10.8 Oxygen vacancy concentration, δ, and cobalt oxidation state, y, as functions of the material stoichiometry; x is the strontium fraction in $La_{1-x}Sr_xCoO_{3-\delta}$ [19]

x	δ	y
0	−0.01 ± 0.01	3.01 ± 0.01
0.2	0.01 ± 0.01	3.18 ± 0.02
0.4	0.05 ± 0.04	3.30 ± 0.08
0.6	0.09 ± 0.01	3.43 ± 0.01
0.8	0.16 ± 0.01	3.48 ± 0.02
1.0	0.30 ± 0.03	3.40 ± 0.06

Figure 10.35 shows representative values of the oxygen permeation flux dependence on the temperature for a number of oxygen permeation membranes. While permeability is high for barium- and cobalt-containing perovskite, these can suffer from chemical stability (especially in a reducing environment – that is, in the presence of a fuel or CO_2).

10.8.4 Flux Models

The oxygen permeation flux depends on the charged species conductivity as well as the surface chemistry processes, including adsorption of oxygen on the feed side and desorption on the sweep side. For oxygen separation from air, oxygen must first be adsorbed on the surface of the feed side and incorporated in the lattice. This is followed by the diffusion of the charged species toward the sweep side, where oxygen partial pressure is kept lower. On the sweep side, oxygen ions are recombined and desorb into the gas stream (see Figure 10.36). The following description of the processes is used to develop an analytical model for the oxygen flux as a function of the operating conditions. It is based on a number of simplifications that enable the development of an analytical expression of the flux in terms of some membrane properties, the temperature, and oxygen partial pressures. While approximate, it is still a useful tool in reactor modeling applications [14,25].

The oxygen surface reaction is written in terms of gas phase oxygen concentration, surface vacancy concentration, lattice oxygen, and electron holes. The reaction is written as a reversible reaction, with forward and backward reactions to describe the process on both sides of the membrane:

$$\frac{1}{2}O_2(g) + V_O^{\bullet\bullet} \rightleftarrows O_O^X + 2h^{\bullet}.$$

The forward reaction rate constant k_f, and the backward reaction, with rate constant k_r, are used to express the flux on the feed side, f, and sweep side, s, as follows:

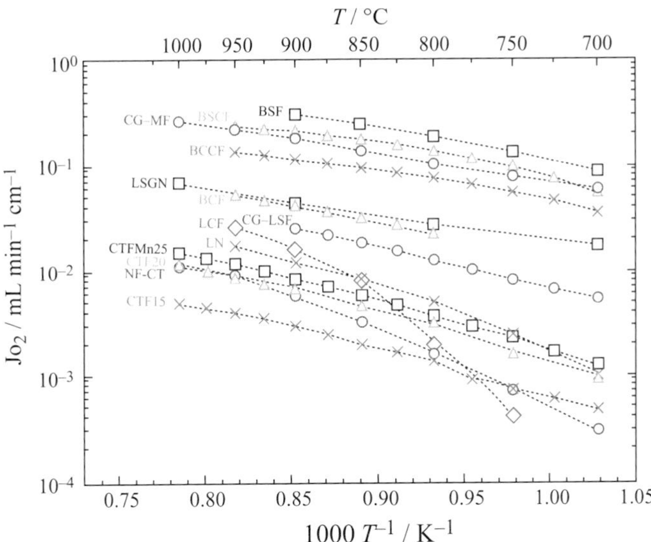

Figure 10.35 Dependence of the permeation flux on temperature for a number of membranes, using air on the feed side. The flux is given per unit membrane thickness; that is, it is flux divided by the membrane thickness [23,24]. The different membranes are:

$Ca_{1-x}Sr_xTi_{0.6}Fe_{0.15}Mn_{0.25}O_{3-\delta}$ (CTFMn25), $Ba_{0.3}Sr_{0.7}FeO_{3-\delta}$ (BSF)
$Ba_{0.5}Sr_{0.5}Co_{0.8}Fe_{0.2}O_{3-\delta}$ (BSCF), $Ce_{0.8}Gd_{0.2}O_{2-\delta}$ containing 15 vol% MnFe2O4 (CG-MF)
$BaCe_{0.1}Co_{0.4}Fe_{0.5}O_{3-\delta}$ (BCCF), $La_{0.8}Sr_{0.2}Ga_{0.7}Ni_{0.3}O_{3-\delta}$ (LSGN)
$BaCe_{0.15}Fe_{0.85}O_{3-\delta}$ (BCF), $Ce_{0.9}Gd_{0.1}O_{2-\delta}$ containing La0.6Sr0.4FeO$_{3-\delta}$ (CG-LSF)
$La_{0.6}Ca_{0.4}FeO_{3-\delta}$ (LCF), $La_2NiO_{4+\delta}$ (LN), $CaTi_{0.8}Fe_{0.2}O_{3-\delta}$ (CTF20)
$NiFe_2O_4$ containing 40 vol% $Ce_{0.8}Tb_{0.2}O_{2-\delta}$ (NF-CT), $CaTi_{0.85}Fe_{0.15}O_{3-\delta}$ (CTF15).

$$j_{O_2,f} = k_f \sqrt{p_{O_{2f}}} C_{V_{O,f}^{\bullet\bullet}} - k_b C_{O_{O,f}^x} C_{h_f^\bullet}^2$$
$$j_{O_2,s} = k_r C_{O_{O,s}^x} C_{h_s^\bullet}^2 - k_f \sqrt{p_{O_{2s}}} C_{V_{O,s}^{\bullet\bullet}}.$$

Oxygen ions are incorporated into the lattice, where they diffuse through the lattice toward the permeate side. Assuming that diffusion is dominated by vacancy diffusion, since it is typically the case that electronic conductivity is higher than ionic conductivity, then

$$j_{V_O^{\bullet\bullet}} = -D_{V_O^{\bullet\bullet}} \frac{dC_{V_O^{\bullet\bullet}}}{dx} = D_{V_O^{\bullet\bullet}} \frac{C_{V_{O,s}^{\bullet\bullet}} - C_{V_{O,f}^{\bullet\bullet}}}{t}.$$

Now we assume that the concentrations of the electron holes and the oxygen in the lattice are almost constant and uniform throughout. The first assumption is based on the high electron conductivity, and the second on the number of defects being rather small compared to the total oxygen concentration in the lattice. Using these assumptions, the constant concentrations are absorbed in k_r in the reaction rate expressions. Since the flux is the same throughout, these three expressions can be combined to obtain the following expression:

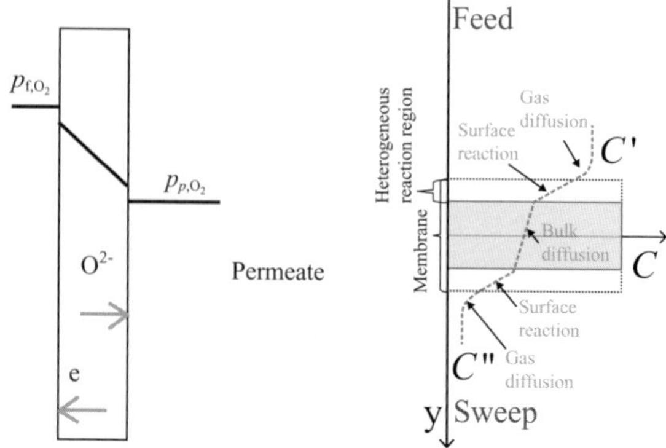

Figure 10.36 Oxygen permeation through a mixed ionic electronic conducting membrane showing on the left side, the flow of oxygen ions and electrons through the membrane from the high to the low-pressure side of oxygen, and the corresponding change in the oxygen partial pressure. Note the sudden finite changes of oxygen concentration at the surfaces associated with the finite rate of the adsorption/incorporation and desorption reactions. On the right side, more detail including the effect of mass transfer through the boundary layers on both sides, besides the changes on the surfaces and in the membrane bulk.

$$j_{O_2} = \frac{D_v}{2t} \frac{k_f\left(\sqrt{p_{O_2f}} - \sqrt{p_{O_2s}}\right)}{k_f\sqrt{p_{O_2f}}\sqrt{p_{O_2s}} + \frac{D_V}{2t}\left(\sqrt{p_{O_2f}} - \sqrt{p_{O_2s}}\right)},$$

where t is the membrane thickness. For relatively thicker membranes and at higher temperatures (where overall transport is limited and there is more charged species diffusion) this expression is reduced to:

$$J_{O_2} = \frac{D_V}{2tK_V}\left(\frac{1}{\sqrt{p_{O_2s}}} - \frac{1}{\sqrt{p_{O_2f}}}\right),$$

where $K_V = k_f/k_r$ is the equilibrium constant. The prefactor in this expression is a strong function of temperature. The MIEC ceramic membranes are O ($t \sim 1$ mm), and typical values for the flux are 0.1–1.0 μmol/cm^2·s at 800 °C $< T <$ 1000 °C, depending on the material and conditions on the permeate side. While not shown explicitly in the expression because it is a function of the flow rates, mass transfer in the gas phase on both sides of the membrane can impact the flux as well, especially if the flow rates are low, since the concentrations in the bulk flow and at the membrane surface would be different [26]. Experimental measurements of the permeation flux and the oxygen partial pressures have been used along with the above expression to determine the reaction rate constants [27]. More elaborate models utilize multistep rate equations [28], and expressions and data have been obtained for these models [29,30]. It is also common to approximate the flux expression as follows:

$$j_{O_2} = A \exp\left(-\frac{E}{\Re T}\right)\left(p^n_{O_2f} - p^n_{O_2s}\right),$$

where the Arrhenius expression combines the effects of surface chemistry and material bulk diffusion (for a particular membrane thickness), and A, E, and n are obtained by curve-fitting this expression to experimental data (Table 10.9).

Oxygen permeation fluxes for most membranes are rather low at moderate temperatures (<900 °C) when using nonreactive or inert sweep gases. Creating a vacuum on the sweep side

Table 10.9 Semi-empirical oxygen flux curve-fits [31]

type	n	A (mol Pa^n/m^2·s)	E/R (K)	T (K)	p_{O2} (bar)
LSCF	0.5	26.75	16,510	1023–1233	6 10^{-3}–1
LNO	0.25	2.011	10,240	1123–1273	8 10^{-4}–2.1

is necessary to raise this flux, which adds to the energy penalty. However, care should be taken with most perovskites, especially those with relatively higher flux, since the material tends to be unstable, especially at higher temperature/higher flux when higher defect concentrations are generated. These conditions can also lead to higher thermal and mechanical stresses. Also, especially in oxy-combustion when CO_2 is the most likely sweep gas, surface reactions and carbonate ($CO_{3\_}$) formation on the surface can lead to significant degradation. This is especially true for perovskites containing Ba and Co, which tend to have high flux. Hydrogen sulfides, which are present in most coal syngas, can also poison the surface even at low concentration.

10.8.5 Membrane Separation Unit Design and Energy Requirements

The energy requirement of a membrane-based ASU is important in determining the efficiency of a power plant utilizing this unit (in oxy-combustion or gasification/reforming applications). Calculating this energy requirement accurately can only be done using detailed models based on mass and energy balances inside the unit and the permeation flux, as well as how it is integrated with other units in the system and plant that exchange heat and mass with the membrane unit. In Chapters 11 and 13, examples of such units and their integrated analyses will be discussed using advanced modeling techniques.

Example 10.7

In this example, a model is described that focuses first on the membrane unit analysis, then adding necessary "peripherals" that are important for its operation [32]. The model is based on mass and energy balances in the feed and sweep streams, with transfer of mass and heat across the membrane, while accounting for the pressure drop in both channels. Two configurations are considered: co-current and counter-current sweep stream flow, both with respect to the feed stream flow direction. Integrating the governing equations analytically is not possible, and the conservation equations are written in discrete forms across a finite cell, Δx, with the positive direction being that of the feed stream flow. For this purpose, the domain, that is the membrane unit, is divided into a number of n cells with $\Delta x = x_{j+1} - x_j$, where j =1, 2, 3, ... n + 1, and j increasing in the direction of the feed stream flow. All other discrete changes across the cells, such as the molar flow, temperature, pressure, etc., are written in similar forms. For instance, the change in the molar flow rate of oxygen across the cell is $\Delta\dot{N}^f_{O_2,j} = \dot{N}^f_{O_2,j+1} - \dot{N}^f_{O_2,j}$.

Example 10.7 (cont.)

Mass balance

Feed side: $\Delta \dot{N}^{f}_{O_2,j} = -\dot{N}^{t}_{O_2}$

Sweep side, co-flow: $\Delta \dot{N}^{s}_{O_2,j} = \dot{N}^{t}_{O_2}$

Sweep side, counter-flow: $\Delta \dot{N}^{s}_{O_2,j} = -\dot{N}^{t}_{O_2}$

Moreover, the oxygen transported across the membrane: $\dot{N}^{t}_{O_2} = J_{O_2}\Delta x \ell^{f}_{per}$, where $\dot{N}^{f}_{O_2,j}$ is the oxygen molar flow rate in the feed stream, $\dot{N}^{t}_{O_2,j}$ is the oxygen molar flow rate transported across the membrane, $\dot{N}^{s}_{O_2,j}$ is the oxygen molar flow rate in the sweep stream, J_{O_2} is the oxygen flux across, and ℓ^{f}_{per} is the perimeter for the flux from the feed side (for a monolith with square channel, channel with w, $\ell^{f}_{per} = 4w$).

Energy balance:

In the feed stream $-\Delta\left(\dot{N}^{f}_{O_2}\hat{h}^{f}_{O_2} + \dot{N}^{f}_{bl}\Delta\hat{h}^{f}_{bl}\right) - \dot{N}^{t}_{O_2}\hat{h}^{t}_{O_2} - q^{f}\Delta x\ \ell^{f}_{per} = 0,$

where subscript bl corresponds to the balance of the material in the stream besides oxygen. Using the approximation, $\hat{h}^{t}_{O_2} = \frac{1}{2}\left(\hat{h}^{f}_{O_2,j+1} + \hat{h}^{f}_{O_2,j}\right) \approx \hat{h}^{f}_{O_2,j} + \frac{1}{2}\Delta\hat{h}^{f}_{O_2,j}$, and the mass balance, we get:

$$\frac{\Delta\dot{N}^{f}_{O_2,j}\Delta\hat{h}^{f}_{O_2}}{2} + \dot{N}^{f}_{O_2,j}\Delta\hat{h}^{f}_{O_2} + \dot{N}^{f}_{bl,j}\Delta\hat{h}^{f}_{bl} + q^{f}\Delta x\ \ell^{f}_{per} = 0.$$

Similarly:

Sweep stream, co-current feed:

$$\Delta\dot{N}^{f}_{O_2,j}\left(\hat{h}^{m}_{O_2} - \hat{h}^{s}_{O_2}\right) - \dot{N}^{s}_{O_2,j}\Delta\hat{h}^{s}_{O_2} - \dot{N}^{s}_{bl,j}\Delta\hat{h}^{s}_{bl} + q^{s}\Delta x\ \ell^{s}_{per} = 0.$$

Sweep stream, counter-current feed:

$$\Delta\dot{N}^{f}_{O_2,j}\left(-\hat{h}^{m}_{O_2} = \hat{h}^{s}_{O_2}\right) + \dot{N}^{s}_{O_2,j}\Delta\hat{h}^{s}_{O_2} + \dot{N}^{s}_{bl,j}\Delta\hat{h}^{s}_{bl} + q^{s}\Delta x\ \ell^{s}_{per} = 0,$$

where $\hat{h}^{m}_{O_2}$ is evaluated at the membrane temperature T^{m}, ℓ^{s}_{per} is the perimeter between the membrane and the sweep streams, (for a monolith, it is $4w$). Convective heat transfer into and out of the membrane (to the free streams) is given by

$$q^{f} = h^{f}\left(T^{f} - T^{m}\right)$$
$$q^{s} = h^{s}(T^{s} - T^{m}),$$

and expressions for the heat transfer coefficients were given in Nusselt number correlations. To solve the problem (i.e., find the flow rates of both streams, and the temperatures of the two streams and the membrane, at all the cell boundaries), five equations are written for

Example 10.7 (cont.)

each cell (two mass balance and three energy balance), and the system of equations is solved simultaneously.

The pressure drop across each cell is given by: $\Delta p = -f\left(\frac{1}{2}\rho V^2\right)\frac{\Delta x}{D_h}$, where D_h is the hydraulic mean diameter of the channel, and f is the friction factor which is a function of the local Reynolds number. The velocity is calculated from the flow rate.

For the purpose of comparing the two configurations and evaluating the ASU's performance, the data in Table E10.7 are used. Note that air is compressed to 3 bar and preheated to 800 K, while the sweep stream is steam at 385 K and 1 bar. Also, the feed flow rate is much higher than the sweep. These numbers are selected for illustration only.

Table E10.7 Data used in the membrane ASU modeling. Note that the initial temperatures of both streams are different, and both are impacted by the heat transfer across the membrane. Also the membrane parameters: A, E, and n are used in the flux expression: $j_{O_2} = A\ \exp\left(-\frac{E}{\Re T}\right)\left(p^n_{O_2f} - p^n_{O_2s}\right)$

Feed stream	Heat air
Initial temperature	800 K
Average pressure	3 bar
Flow rate	0.1 kmol/s
Oxygen mole fraction	0.18
Sweep stream	Steam
Initial temperature	385 K
Average pressure	1 bar
Flow rate	0.01
Oxygen mole fraction	0.0001
Flux pre-exponential term	$A = 6.7\text{e-}3/0.02$ kmol/m^3·s
Activation temperature	$B = 3855$ K
Pressure index	$n = 0.5$
Membrane channel structure	Square monolith
Membrane side length	0.1 m
OTM length	11.5 m

The algebraic equation is solved numerically. Figure 10.7a shows the change in the oxygen partial pressure in both streams for the two cases, co-flow and counter-flow. In the co-flow case, the partial pressure difference, which is the driving force for oxygen permeation, decreases along the membrane, as oxygen moves across from the feed (air) side to the sweep (steam) side. On the other hand, the difference remains almost constant in the counter-flow case. In the meantime, the

Example 10.7 (cont.)

temperature equalizes rapidly near the unit entrance in the co-flow case, and remains constant afterward. The opposite is seen in the counter-flow case, where the temperature equalizes at the exit side. Overall, the temperature is higher in the counter-flow case.

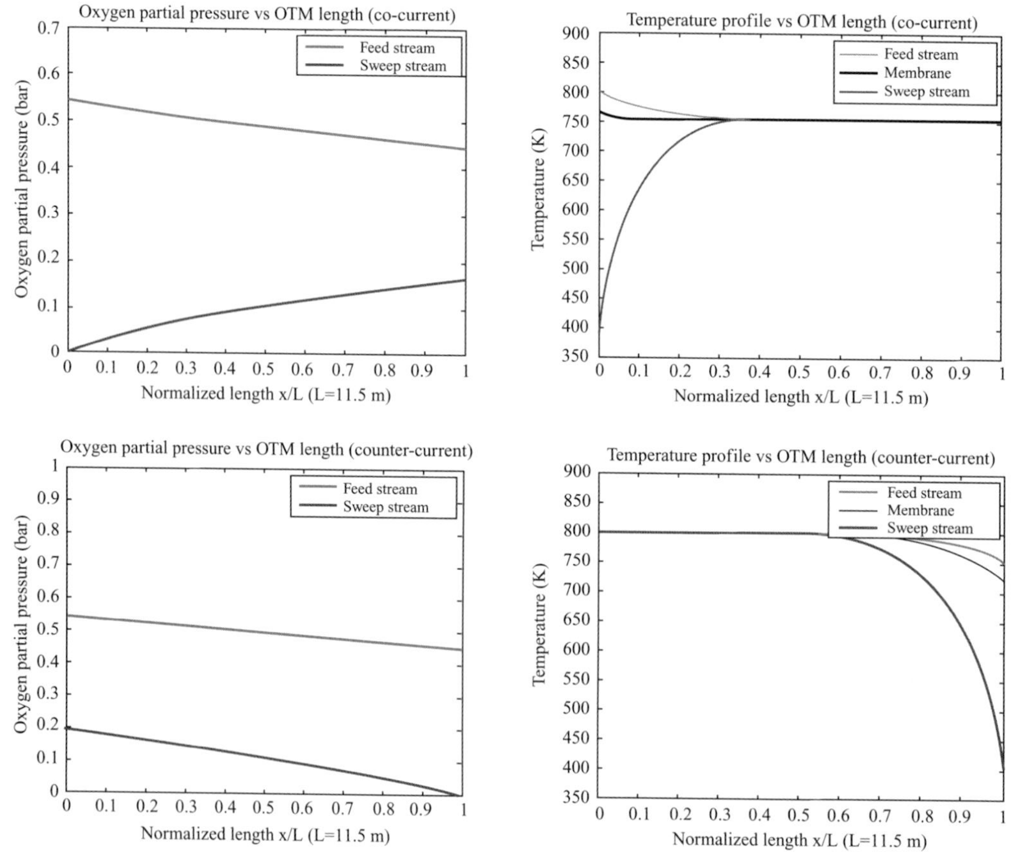

Figure E10.7a The oxygen partial pressure (left) and temperature along the membrane (right) in the case of co-flow (top) and counter-flow (bottom) in the ASU membrane.

Figure E10.7b shows the variation of the oxygen flux along the membrane in both cases. The flux decreases monotonically in the co-flow case, as the partial pressure difference decreases. In the counter-flow case, it increases slowly, reaching a maximum near the unit exit. The maximum can be explained by a combination of the oxygen partial pressure difference, which remains almost constant along the length, and the temperature profile. The membrane temperature is higher in the counter-flow case by ~50 °C, and since the flux is Arrhenius in T_m, this effect is significant.

In order to evaluate the energy consumption of air separation using this unit, a system must be built around it that provides the thermal energy for heating the streams to the temperature required for sustaining high flux, and the work for compressing the gases (air before the unit

Example 10.7 (cont.)

and oxygen produced in the unit up to 1 bar). Some of the heating can be done regeneratively. A possible system is shown in Figure E10.7c in which air is compressed and combustion is used to heat it up after preheating in a recuperator. After leaving the membrane unit, the oxygen-depleted air is cooled in the same recuperator. Extra fuel is used to reheat the stream in order to raise steam (the sweep gas) in a boiler. The oxygen-depleted stream is then expanded in a gas turbine to recover some work. Steam is sent to the separation unit to act as a sweep gas. After leaving the membrane unit it is cooled to condense the water for recycling.

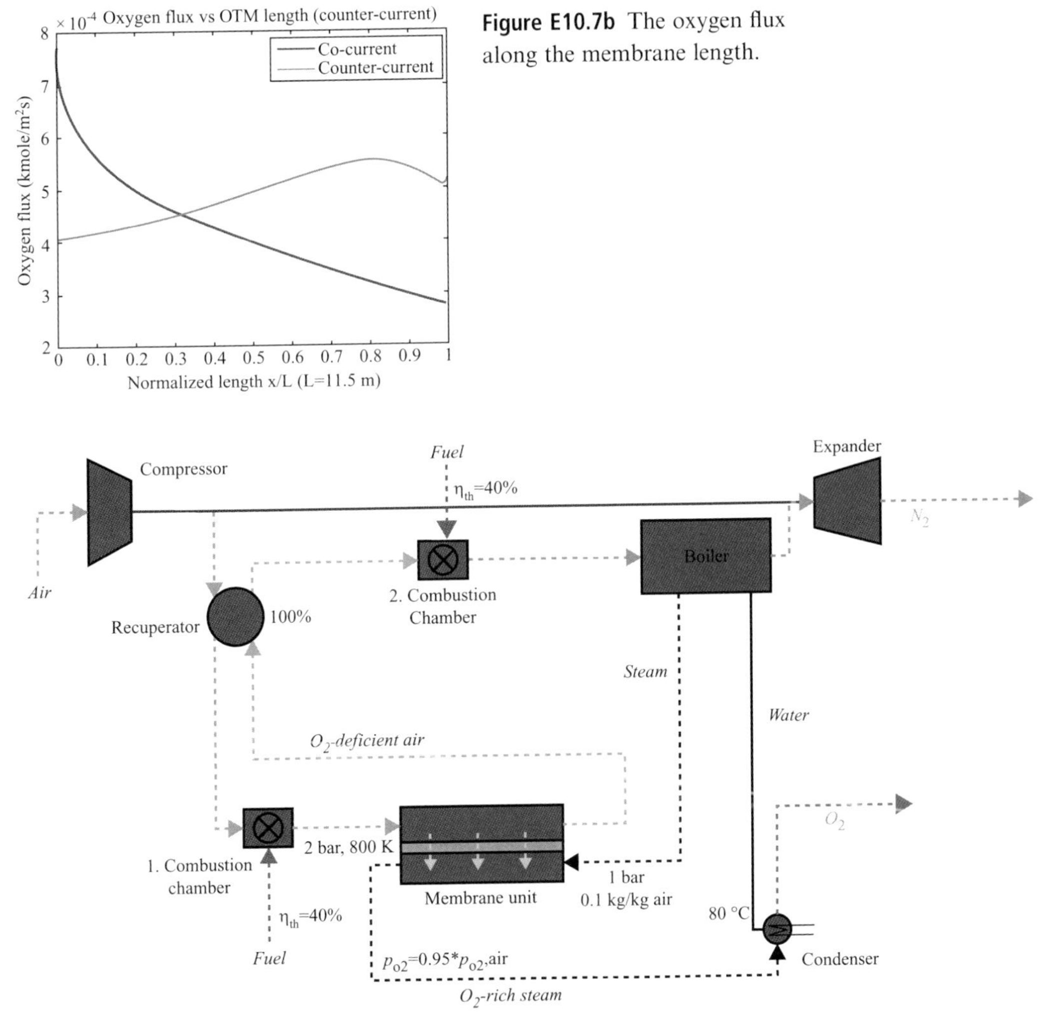

Figure E10.7b The oxygen flux along the membrane length.

Figure E10.7c The system used to evaluate the energy requirement for the membrane ASU [27].

Example 10.7 (cont.)

An Aspen® model was constructed to evaluate the energy required by the different units, the compressor work, and fuel/equivalent thermal energy for the combustion/heating units. To compare against cryogenic ASUs, which report the energy of separation in terms of work/kg-O_2, the thermal energy part was converted to work using a First Law efficiency of 40%. Calculations show that the base case energy requirements of this unit was 1.35 MJe/kg-O_2. Of this, 60% was used in the form of heat and hence it offered good potential for integrating the unit in a power cycle where relatively low-temperature heat could be available at a lower efficiency penalty. Parametric studies showed that the energy penalty could be reduced by increasing the sweep flow rate and the feed pressure, which lead to higher flux in a smaller unit, hence reducing the pressure drop and associated work requirements to below 1.00 MJe/kg-O_2.

Ceramic membrane units are capable of supplying oxygen at 100% purity, barring leakage around the membrane or other joints. Since most of the energy requirement is in the form of heat at intermediate temperatures, it is possible to integrate ceramic membrane separation units with power cycles (IGCC and oxy-combustion, which will be described in the next chapters).

10.8.6 MIEC for O_2 Separation Using Reactive Sweep

Besides using a sweep gas under nonreactive conditions to collect oxygen separating from air, it is possible to use a reactive sweep gas to consume oxygen as it reaches the membrane surface and hence lower the partial pressure of oxygen and enhance the flux. In this case, using syngas or methane in the sweep gas, the membrane unit becomes a membrane reactor unit for the partial or complete oxidation of the fuel. Consuming oxygen at the sweep surface reduces its partial pressure without incurring an extra energy penalty, but the associated exothermicity/heat release must be carefully managed in order to keep the membrane at a safe temperature. Figure 10.37 shows the significant increase in the oxygen flux as higher concentrations of H_2 or CO are added to the sweep side streams (with inert Ar as the carrier gas). Hydrogen is more reactive and leads to higher oxygen flux, but both fuels achieve more than an order of magnitude increase over the nonreactive case (with pure argon as a sweep gas). It is also interesting to note that the flux has not saturated yet even at 40% reactive gas concentration, and values above 1 μmol/cm^2·s could be achieved.

Using a methane–argon sweep raises the oxygen flux as well, especially at lower methane concentration, but saturates at methane concentration lower than H_2 or CO because of the formation of carbon via methane pyrolysis on the sweep side, as shown in Figure 10.38. Carbon deposition on the membrane surface can poison the surface and covers the active sites. Adding CO_2 to the sweep side gases, on the other hand, avoids flux saturation and carbon formation, and the flux continues to rise as more CO_2 is added. In this case, dry reforming of methane forms syngas that reacts with oxygen at the membrane surface, raising

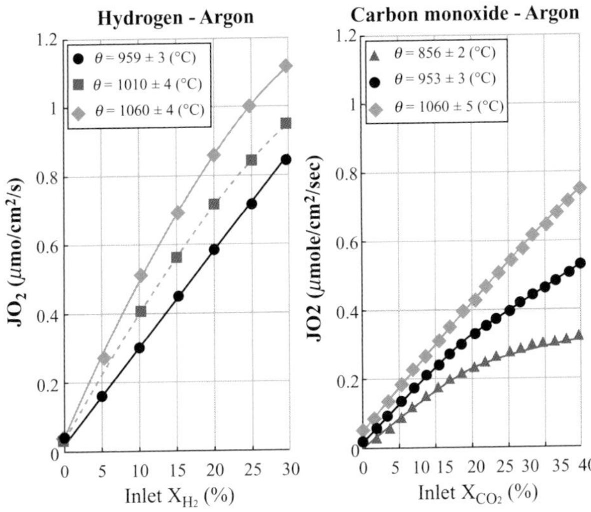

Figure 10.37 Impact of fuel concentration on the sweep side on the oxygen flux, with air on the feed side. Both sides are at 1 bar. Inlet concentration refers to local concentration near the membrane surface [19].

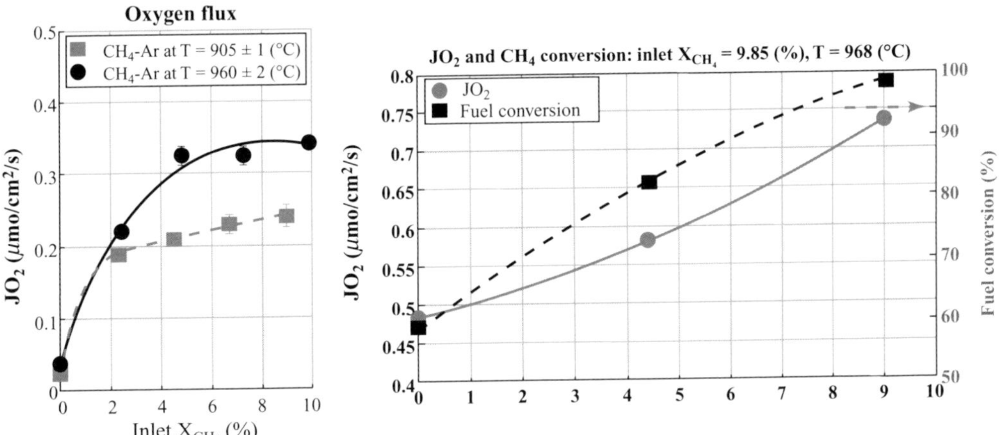

Figure 10.38 Adding CO_2 instead of Ar reduces carbon formation because of dry reforming, and the flux continues to increase as more CO_2 is added. The balance sweep gas was Ar [19].

the flux in ways similar to what is shown in Figure 10.37 (with either H_2 or CO only). Carbon dioxide also reforms the carbon forming by methane pyrolysis (or suppresses carbon formation by pyrolysis in dry reforming) and avoids surface poisoning and the saturation observed in the absence of CO_2. Similar improvements can be achieved by adding water instead of CO_2 to the methane stream; in this case steam reforming of methane or carbon forms the syngas. Having CO_2 and/or H_2O in the sweep gases is consistent with oxy-combustion power cycles in which some of the combustion products are recycled to the combustor to control the flame temperature. It is also consistent with gasification/reforming-based plant (pre-combustion capture is described in detail in Chapters 11 and 13) in which a fraction of the syngas is used in the membrane-based ASUs. These power cycles will be discussed in detail in the following chapters.

Further improvements have been achieved by using porous support on one or both sides of the membrane to increase the surface area for adsorption/desorption, and/or application of extra catalysts to further accelerate the surface reactions [33]. When using thin membranes, with $t < 1$ mm, surface kinetics are more likely to be the rate-limiting step in oxygen permeation, and improved kinetics can be used to overcome these limitations and increase the flux. Both approaches have shown marked improvements in the oxygen flux, especially when using a reactive sweep gas. When adding porous support layers to either or both sides of the membrane, care must be taken to match the thermal expansion of the membrane material and the support layer material to avoid thermal stresses. The porous layer can be made of the same material as that of the membrane, and hence avoid the problem of differential thermal expansion. A transition metal catalyst can be applied directly on the membrane surface or on the porous support layer to accelerate partial or full oxidation kinetics. A porous support is also useful when a very thin membrane is necessary for enhancing the permeation flux – that is, when conditions are such that the overall process rate is limited by the diffusion across the membrane thickness [33]. In this case, besides improving the surface kinetics, the porous layer provides mechanical support.

10.8.7 Mixed Conductors (MPEC) for H_2 Separation

Mixed protonic electronic conducting (MPEC) membranes, mostly perovskite, have also been used for hydrogen separation from syngas, as shown in Table 10.10. In this case, a mixed proton–electronic (electron/electron hole) conducting membrane is used. According to the following reaction, hydrogen is conducted as positive hydroxyl [18]:

$$H_2 + 2O_O^x + 2h^{\bullet} \rightleftarrows 2OH_O^{\bullet}.$$

Pervoskites such as $AB_{1-x}M_xO_{3-d}$, with A being strontium, calcium, or barium, B being cerium or zirconium, and M selected from the group of titanium, chromium, manganese, cobalt, nickel, copper, gallium, or yttrium, have been used at $600\,°C < T < 800\,°C$, including $BaCe_{0.825}Y_{0.1}Ru_{0.075}O_{3-\delta}$ and $SrCe_{0.95}Yb_{0.05}O_{3-\delta}$. Hydrogen fluxes of the order of 1 $\mu mol/cm^2{\cdot}s$ have been measured using these membranes [14]. Similar to oxygen-conducting membranes, the hydrogen flux depends on the temperature, partial pressures of hydrogen on both sides, and the membrane thickness. The hydrogen fluxes in proton-conducting membranes are relatively low, and the material can be poisoned by gases such as CO_2 and H_2S.

Table 10.10 also shows a comparison between different membranes covered in this chapter and used in energy applications. Dual-phase membranes, in which more than one perovskite, or a perovskite and a fluorite, are homogeneously mixed, have shown very promising trends in increasing the fluxes and raising the overall reactivity as well.

Table 10.10 Comparison between three permselective membranes [34]

	MIEC	Metallic	MPEC
Temperature (°C)	700–1000	300–600	700–1000
Conduction	$V_O^{\bullet\bullet}$, e′, or h$^{\bullet}$	Hydrogen atom, H	OH$^{\bullet}$, $V_O^{\bullet\bullet}$, e′, or h$^{\bullet}$
Materials	Perovskite/fluorite/dual-phase	Pd/Pd-alloy	$ACeO_3$, A = La, Sr, Ba
Flux	0.1–5 μmol/cm^2·s	~10 μmol/cm^2·s	~1 μmol/cm^2·s
Stability	Perovskites, such as Sr- and Ba-based, can be poisoned by CO_2, H_2O, H_2S, SO_2	H_2S, SO_2, and, unsaturated hydrocarbons, contaminate Pd membranes	$BaCeO_3$-derived compounds are unstable in CO_2 or H_2O environments
Main disadvantages	Higher permeability, more active materials are more often less stable. Co-based materials are unstable in reducing environments	Use of precious metal (amount of Pd can be lowered by using alloys)	Low permeability; degraded in H_2O and CO_2

$V_O^{\bullet\bullet}$ is oxygen vacancy, e′ is electron, h$^{\bullet}$ is electron hole, H is a hydrogen atom, and OH$^{\bullet}$ is a hydroxyl proton.

10.8.8 Metallic Membranes for H_2 Separation

Yet another type of hydrogen-permeable dense membrane is metallic membranes, composed primarily of palladium (Pd). Hydrogen atoms are small and can diffuse through dense metals, hence producing almost pure gas, similar to ion or proton transport membranes described above [35]. Similar to MIEC/MPEC membranes, the overall transport mechanism consists of hydrogen adsorption on the high hydrogen partial pressure surface, followed by dissociation and (Fick's) diffusion of the atoms through the material bulk. Hydrogen atoms recombine on the low hydrogen partial pressure side and desorb into the gas. Similar to polymer membranes, it is common to express the flux in terms of the permeability coefficient, $\tilde{p}_i$:

$$j_{H_2} = \frac{\tilde{p}_{H_2}}{t}\left(p^n_{H_2f} - p^n_{H_2s}\right),$$

with $0.5 < n < 1.0$, and where $\tilde{p}_{H_2} = \tilde{p}^0_{H_2}\exp\left(-\frac{\tilde{E}}{\Re T}\right)$ is the permeability coefficient for the material. Palladium has high permeability, especially at moderately low temperatures, $O(10^{-8}$ mol/s·m·Pa$^{1/2})$ of around 400 °C, and has hence has been used for these metallic membranes. Hydrogen permeability in a number of metals can be found in Reference [35]. Pd can suffer from hydrogen embrittlement, forming cracks after long exposure to the gas. It can also be adversely affected by sulfur compounds, carbon, CO, and H_2O. Sulfur can react chemically with Pd and change its permeation characteristics. CO and H_2O can block hydrogen from adsorbing onto the surface. Some of these problems can be partially avoided by using alloys of Pd and metals such as silver, copper, and cerium. Besides their resistance to embrittlement, in some cases palladium alloys can have higher permeability than the pure metal.

Hydrogen-permeable membranes can be used effectively to separate hydrogen in a number of applications. For instance, starting with methane, steam methane reforming can be performed on the feed side of the membrane, producing $CO + H_2$. Methane conversion by steam reforming is limited by equilibrium, and higher temperatures are needed to achieve full conversion. More methane conversion can be achieved at lower temperatures if hydrogen (product) is removed by the membrane, since the continuous separation of hydrogen would drive (shift) the reaction equilibrium toward the products side. Moreover, the two processes, reforming and separation, can be conducted in the same reactor, instead of using one reactor for reforming followed by separation. The same is the case for the WGS, which is conventionally done in two separate reactors at two different temperatures to achieve complete conversion of CO to CO_2. If the mixture of CO and H_2O is introduced on the feed side of a hydrogen-permeable membrane, the separation of the produced hydrogen would continue to shift the WGS reaction equilibrium toward the products side (more $CO_2 + H_2$), while simultaneously producing a pure stream of H_2 on the permeate side. In all these applications, the membrane material must be resistant to CO, CO_2, and H_2O poisoning.

Problems

10.1 Consider a packed tower with a diameter of 400 mm that treats a mixture of 70% CO_2/10% CH_4/20% ammonia, which enters the tower at 200 °C and 1 atm. The pressure drop is 409 Pa/m. Methanol at 25 °C and 1 atm is used as the absorbent. Assuming $F_p = 92$ and $G_x = 1.5G_y$, and treating the gas mixture like an ideal gas, determine (a) the superficial velocity; and (b) the mass flow rate of the mixture. (Use figure 18.7 in [1] in this problem and the next two problems.)

10.2 A tower packed with 38.1 mm ceramic Intalox saddles treats 1000 m^3/h of an air stream at 20 °C and 1 atm that contains 6% (vol.) acetone. The packing factor is 52, and water (at 20 °C and 1 atm) is used as the absorbing liquid. If the pressure drop is 0.5 kPa/m, and the superficial velocity is 1.5 m/s, determine the mass flow rate of the water.

10.3 A tower packed with 25.4 mm metal Pall rings (packing factor: 56) is designed to process a mixture of 60% H_2/40% CO_2 at 180 °C and 1 atm, which is fed at the rate of 0.35 kg/s. The height and diameter of the tower are 4 m and 0.6 m, respectively. The absorbing liquid (methanol) enters the tower at 25 °C and 1 atm. The superficial velocity is 1.4 m/s, and the ratio of water flow to gas flow is 1.4. Calculate the total pressure drop in this tower.

10.4 A membrane manufactured by Monsanto is used to separate carbon dioxide from a mixture of 30% CO_2/70% N_2. The volumetric flow rate at the permeate side is 15 m^3/h. The total pressure at the feed and permeate sides is 4 bar and 0.8 bar, respectively. The permeability of CO_2 is 450 Barrers and selectivity is 31. Assuming perfect mixing and uniform conditions on the feed and permeate sides, determine: (a) the flux of the carbon dioxide; and (b) the membrane area. (Hint: You may assume that the flux ratio across the membrane is the same as the molar concentration ratio on the permeate side, and use equation (10.38) to express the fluxes in terms of the conditions on both sides.)

10.5 An equimolar mixture of gas A and gas B is supplied to a 1 mm-thick polymer membrane with a surface area of $12\,m^2$ to produce $40\,m^3/h$ permeate with 90% gas A. The selectivity (of A with respect to B) is 12.3, and the permeability of gas A is 500 Barrers.

a. Determine the feed side and the permeate side pressures.

b. Identify the equipment required to maintain the calculated pressures at the feed and permeate streams.

10.6 A mixture of 60% H_2 + 40% CO_2 (by volume), initially at 298 K and 1 atm, is separated using a 1 mm-thick membrane. The mixture is fed to the membrane at a rate of $100\,m^3/h$. The membrane, with surface area $20\,m^2$, produces $25\,m^3/h$ H_2-enriched permeate with 95% purity. The permeability of the hydrogen is 300 Barrers, and the selectivity is 16.

a. Determine the pressure at the feed and permeate sides.

b. Select the equipment to maintain the pressures at the feed and permeate sides as calculated in part (a).

c. Calculate the power requirement for this separation process assuming an isentropic efficiency of 70% for the items identified in part (b).

10.7 A gas separation membrane operates under cross-flow conditions. The concentration at the feed side changes gradually from an inlet to the exit. Perfect mixing can be assumed on the permeate side with mole fraction X_p. Because of the large change in concentrations on the permeate side, X_f in (10.38) is replaced with an average molar fraction between the feed side (inlet) and the retentate side (exit) in terms of the log mean mole fraction as follows:

$$X_{f,ave} = \frac{X_{fi} - X_{ri}}{\ln\left(\frac{X_{fi}}{X_{ri}}\right)}.$$

This cross-flow membrane is used for production of $20\,m^3/h$ of 88% N_2 + 12% O_2 (vol.) mixture (see Figure P10.7). Atmospheric air is pressurized in a compressor to 8 atm before entering the membrane, whose thickness is 1.5 mm. The pressure at the permeate side is 1 atm. The mole fraction of oxygen at the retentate stream is 12%. The selectivity of the membrane is 5, the oxygen permeability is 50 Barrer, and the isentropic efficiency of the compressor is 80%. Determine: (a) the membrane area; (b) the power requirement of the compressor; and (c) the stage cut. (Use the same hint as in Problem 10.4.)

Figure P10.7

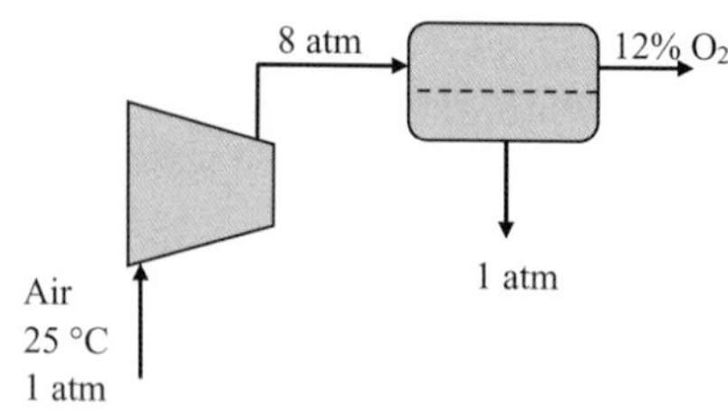

10.8 A membrane is operated for 40% H_2/60% CH_4 separation, giving a permeate composition of 85% H_2 and a residue of 30% H_2. The feed and permeate pressures are 5 and 1 atm. The permeate flow is 25% of the feed flow. Determine: (a) the membrane selectivity; and (b) the permeate composition if a vacuum is used on the downstream side.

10.9 A 1 µm-thick polysulfone membrane is used for separating CO_2 from a mixture of 30% CO_2 and 70% CH_4 by volume. The mixture initially at 25 °C and 1 atm enters a compressor at the rate of 100 m^3/h. The pressure at the feed and permeate sides is 5 atm and 0.5 atm, respectively. The methane concentration at the retentate side is 92%. The system is shown in Figure P10.9. The permeability coefficients for CO_2 and CH_4, respectively, are 5.6 and 0.27 Barrers.

a. Determine the methane flux in the permeate stream.

b. Calculate the power requirement of the compressor and the ejector. The isentropic efficiency of both is 75%.

Figure P10.9

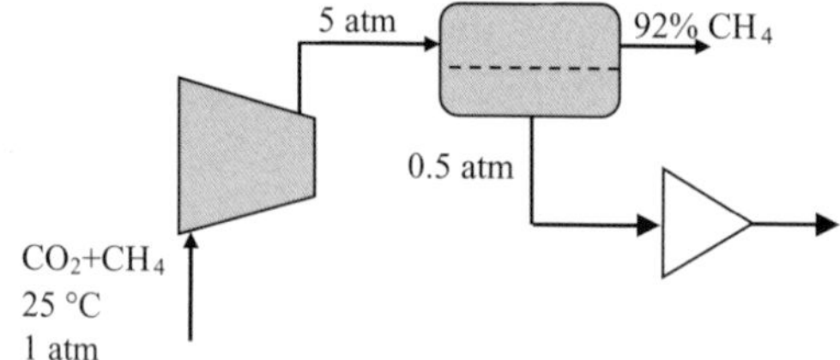

10.10 An ASU is built for the production of a nitrogen-enriched stream with 98% (vol.) purity. A blower is used to flow air at 300 K and 1 atm at a rate of 100 m^3/h to a 1.5 µm-thick silicone (PDMS) membrane. The pressure at the permeate side is maintained at 0.2 atm using an ejector. The pressure at the feed side is 10% above the atmospheric pressure. The system is shown in Figure P10.10. The permeability coefficients of O_2 and N_2, respectively, are 930 and 440 Barrers.

a. Calculate the required membrane area using the log mean mole fraction described in Problem 10.7.

b. The log mean mole fraction method provides a reasonable result if $X_r/X_f > 0.5$; otherwise the error may be too high. If $X_r/X_f < 0.5$, it is better to use a stepwise calculation satisfying $X_r/X_f > 0.5$ in each step. In this problem, $X_f = 0.21$ and $X_r = 0.02$. Use three steps (three membranes operating in series) and calculate the membrane area. Compare the result with that obtained in part (a).

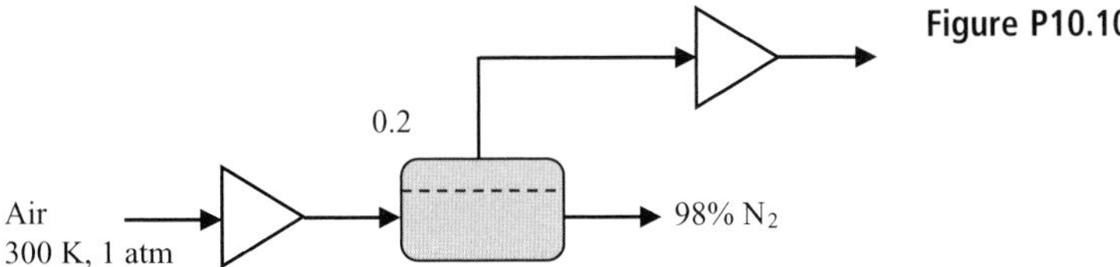

Figure P10.10

10.11 A two-stage polycarbonate membrane unit is employed for separation of CO_2 from a mixture of 30% CO_2/70% N_2, initially at 300 K and 1 bar, as shown in Figure 10.11. The thickness of both membranes is 1.5 μm. The gas mixture is pressurized in a compressor up to 20 bar. It is then directed to the first membrane, where the composition of the retentate stream is 16% CO_2/84% N_2. The permeate stream at 1 bar is directed to another compressor to increase the pressure of the stream to 10 bar, after which it enters the second membrane. The CO_2 mole fraction at the retentate of the second membrane is 8%. Two gas turbines located downstream of the retentate streams are used to recover part of the power requirement of the compressors. The exit flows of the turbine are exhausted into the atmosphere. The isentropic efficiencies of the compressors and turbines are 75% and 90%, respectively. The permeability coefficients of CO_2 and N_2, respectively, are 6.5 and 0.018 Barrers. Calculate: (a) the composition at the permeate side of each membrane; and (b) the net power requirement of this separation plant per unit mass of the initial mixture.

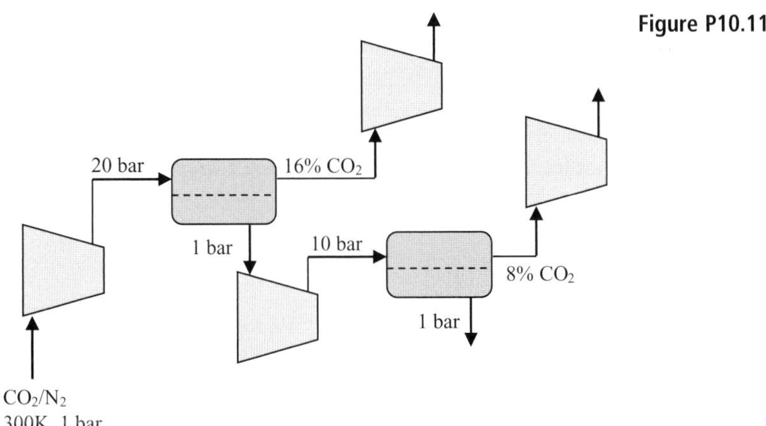

Figure P10.11

REFERENCES

1. W. L. McCabe, J. C. Smith, and P. Harriott, *Unit operations of chemical engineering*. Boston, MA: McGraw-Hill, 2005.
2. G. Gottlicher, *The energetics of carbon dioxide capture in power plants*. Washington, DC: U.S. Department of Energy, Office of Fossil Energy, NETL, 2004.

3. C. Hendricks, *Carbon dioxide removal from coal fired power plants*. Dordrecht: Kluwer Academic Press, 1994.
4. B. Smit, J. A. Reimer, C. M. Oldenburg, and I. C. Bourg, *Introduction to carbon capture and sequestration*. London: Imperial College Press, 2014.
5. Y. A. Çengel and M. A. Boles, *Thermodynamics: an engineering approach*. New York: McGraw-Hill Education, 2015.
6. R. F. Probstein and R. E. Hicks, *Synthetic fuels*. Mineola, NY: Dover Publications, 2006.
7. M. Halmann and M. M. Steinberg, *Greenhouse gas carbon dioxide mitigation*. London: Lewis Publishers, CRC Press, 1999.
8. D. E. A. Clarke, "CO_2 capture and storage." Essen, Germany: VGB PowerTech e.V.
9. E. B. Betz, O. Davidson, H. C. de Coninck, M. Loos, and L. A. Meyer. *IPCC special report on carbon dioxide capture and storage*, Cambridge: Cambridge University Press, 2005.
10. M. Aresta (ed.), *Carbon dioxide recovery and utilization*. Dordrecht: Kluwer Academic Publisher, 2003.
11. R. T. Yang, *Gas separation by adsorption processes*. Singapore: Imperial College Press, 2006.
12. J. Saunders, "No Title," State University of New York, Buffalo, 1982.
13. R. Hutson, N. Zajic, S. Rege, and S. Yang, "Air separation by pressure swing adsorption using superior adsorbents," Technical Report, 2001. Available at: www.osti.gov/biblio/789503
14. K. Li, *Ceramic membranes for separation and reaction*. Chichester: Wiley, 2007.
15. M. Mulder, *Basic principles of membrane technology*. Dordrecht: Kluwer Academic Press, 1996.
16. J. A. Wesselingh and R. Krishna, *Mass transfer in multicomponent mixtures*. Delft: Delft University Press, 2000.
17. A. I. Teplyakov, V. V., Shalygin, M. G., Kozlova, et al., "Membrane technology in bioconversion of lignocellulose to motor fuel components," *Pet. Chem.*, vol. 57, no. 9, pp. 747–762, 2017.
18. Y. Liu, X. Tan, and K. Li, "Mixed conducting ceramics for catalytic membrane processing," *Catal. Rev.*, vol. 48, no. 2, pp. 145–198, 2006.
19. G. T. Dimitrakopoulos, "Experimental study and modeling analysis of ion transport membranes for methane partial oxidation and oxyfuel combustion," Massachusetts Institute of Technology, Ph.D. thesis, 2017. Available at: https://dspace.mit.edu/handle/1721.1/108949.
20. P. Ciambelli , S. Cimino, L. Lisi, et al., "La, Ca and Fe oxide perovskites: preparation, characterization and catalytic properties for methane combustion," *Appl. Catal. B Environ.*, vol. 33, no. 3, pp. 193–203, 2001.
21. T. C. Geary and S. B. Adler, "Oxygen nonstoichiometry and defect chemistry of the mixed conductor $La_{0.9}Ca_{0.1}FeO_{3-\delta}$ at low oxygen partial pressure," *Solid State Ionics*, vol. 253, pp. 88–93, 2013.
22. K. J. Tyler Mefford, J. Rong, X. Abakumov, et al., "Water electrolysis on $La_{1-x}St_xCoO_{3-d}$ perovskite electrocatalysts," *Nat. Commun.*, vol. 7, p. 11053, 2016.
23. J. M. Polfus, W. Xing, G. Pećanac, et al., "Oxygen permeation and creep behavior of $Ca_{1-x}Sr_xTi_{0.6}Fe_{0.15}Mn_{0.25}O_{3-\delta}$ ($x = 0, 0.5$) membrane materials," *J. Memb. Sci.*, vol. 499, pp. 172–178, 2016.
24. X. Wu, "Membrane-supported hydrogen/syngas production using reactive H_2O/CO_2 splitting for energy storage," Massachusetts Institute of Technology, Ph.D. thesis, 2017. Available at: https://dspace.mit.edu/handle/1721.1/111696
25. A. Hunt, "Experimental investigation of oxygen-separating ion transport membranes for clean fuel synthesis," Massachusetts Institute of Technology, Ph.D. thesis, 2015. Available at: https://dspace.mit.edu/handle/1721.1/100059

26. J. Hong, P. Kirchen, and A. F. Ghoniem, "Numerical simulation of ion transport membrane reactors: oxygen permeation and transport and fuel conversion," *J. Memb. Sci.*, vols. 407–408, pp. 71–85, 2012.
27. A. Hunt, G. Dimitrakopoulos, and A. F. Ghoniem, "Surface oxygen vacancy and oxygen permeation flux limits of perovskite ion transport membranes," *J. Memb. Sci.*, vol. 489, pp. 248–257, 2015.
28. J. R. Adler, S. B. Chen, and X. Y. Wilson, "Mechanisms and rate laws for oxygen exchange on mixed-conduction oxide surfaces," *J. Catal.*, vol. 245, pp. 91–109, 2007.
29. G. Dimitrakopoulos and A. F. Ghoniem, "A two-step surface exchange mechanism and detailed defect transport to model oxygen permeation through the $La_{0.9}Ca_{0.1}FeO_{3-\delta}$ mixed-conductor," *J. Memb. Sci.*, vol. 510, pp. 209–219, 2016.
30. G. Dimitrakopoulos and A. F. Ghoniem, "Developing a multistep surface reaction mechanism to model the impact of H_2 and CO on the performance and defect chemistry of $La_{0.9}Ca_{0.1}FeO_{3-\delta}$ mixed-conductors," *J. Memb. Sci.*, vol. 529, pp. 114–132, 2017.
31. A. Mancini, and N. D. Mitsos, "Ion transport membrane reactors for oxy-combustion: Part I – intermediate fidelity modeling," *Energy*, vol. 36, pp. 4701–4720, 2011.
32. C. Zebian, and H. Botero, "Air separation technologies: evaluation of a stand alone oxygen transfer membrane unit," 2.62 class term project, MIT, 2010.
33. X. Y. Wu, A. F. Ghoniem, and M. Uddi, "Enhancing co-production of H_2 and syngas via water splitting and POM on surface modified oxygen permeable membranes," *AICHE J.*, vol. 62, no. 12, pp. 4427–4435, 2016.
34. X. Wu and A. F. Ghoniem, "Mixed ionic electronic conducting membranes for thermochemical reduction of CO_2: a review," *Prog. Energy Combust. Sci. Combust. Sci.*, vol. 74, pp. 1–30, 2019.
35. M. Basile, A. Lulianeli, A. Longo, et al., "Pd-based selective membrane state of the art," in M. De Falco, L. Marrelli, and G. Iaquaniello (eds.), *Membrane reactors for hydrogen production processes*, New York: Springer, 2011, pp. 21–55.

11 Low-to-Zero CO_2 Cycles

Natural Gas

11.1 Introduction

In this chapter, methods proposed for integrating CO_2 capture into power cycles are presented. While "carbon capture" is used to describe this technology, the word refers to producing a separate stream of pure CO_2 at the "tailpipe" of the power plant. Moreover, while the term is sometimes used to describe separating CO_2 from the combustion products of air combustion-based power plants (or any air combustion process), it is actually meant to refer to any technology that uses hydrocarbon fuels in electricity generation power plants while producing a pure stream of CO_2 for storage (or reuse). These technologies include three broad categories: post-combustion capture, pre-combustion capture, and oxy-combustion capture. It is also mostly assumed that in a carbon capture power plant, CO_2 is delivered in the liquid phase, at pressures above 75 bar and temperature around 32 °C.

The objectives of the coverage are to describe the processes involved in some of these cycles, and to evaluate their efficiency and how much the capture process contributes toward derating the power production in the plant. As a general rule, capture should only be applied in high-efficiency plants since it is inevitable that an efficiency penalty would be encountered. Thus, combined cycle (CC) plants are better candidates for capture than simple cycles. Some form of gas separation is always required in a capture technology (e.g., CO_2 separation from flue gases, oxygen separation from air, or H_2 separation following natural gas reforming). Since the energy consumption of separation processes depends on the amount of gas being separated, thermodynamic conditions of the gas and the percentage of component to be separated, and cycle design, are important to minimizing that energy penalty. Thus, the cycle integration with the capture process is important.

The choice of cycle depends strongly on the fuel used. While this is also the case for conventional power cycles, as described in Chapters 5, 6, 8, and 9, it is more critical when one of the design objectives is to produce separate CO_2. For instance, natural gas can be used directly in combustion cycles, but coal and biomass can be used only in Rankine cycles, unless they are gasified first and the syngas in cleaned up (removing particulates and sulfur compounds). Moreover, CO_2 emission levels depend on the cycle and the fuel. To put things in perspective, consider the following numbers, which give order of magnitude estimates. A power plant burning coal in a steam Rankine cycle at 35% efficiency produces 1 mole of CO_2 for each mole of coal. Taking 32 MJ/kg as the heating value of a high-rank coal, like anthracite, the plant produces $(0.35 \times 32 \times 12/44) = 3.05$ MJ/kg-CO_2 or about 1180 kg-CO_2/MWh. Meanwhile, a natural gas plant operating at 55% efficiency, also producing 1 mole CO_2 for each mole CH_4 consumed, but with 45 MJ/kg-CH_4 heating value, generates $(0.55 \times 45 \times 16/44) = 9$ MJ/kg-CO_2 or about 400 kg-CO_2/MWh. Thus, using natural gas instead of coal cuts down the amount of CO_2 generated by a factor of 2–3 (depending on the natural gas plant efficiency), even without implementing CO_2 capture technologies. Even if a simple gas turbine cycle is used with natural gas, due to the higher efficiency of the cycle, the higher heating value of natural gas and the higher heating value per CO_2 produced, the CO_2 emissions of the gas turbine power plant would be half that of a plant operating with coal.

The chapter starts revisiting the work of separation generically and offers insight into best practices. Next, technologies that have been proposed for power plant designs with CO_2 capture are summarized. Post-combustion capture is discussed in detail, including a simplified model that can be used to estimate the energy and efficiency penalty in a well-integrated plant using chemical absorption for separating CO_2 from the combustion gases. Oxy-combustion approaches are reviewed, with a focus on the associated penalties, while using air separation to produce the oxygen stream for combustion. In this case, several alternatives have been proposed and analyzed, including the water cycle, combined cycle, supercritical and Graz cycles, and more. The need to separate oxygen from air in these oxy-combustion plants, and integrate air separation with combustion, has been addressed by applying innovations in chemical looping and high-temperature membranes. Chemical looping is discussed in some detail, given its demonstrated high efficiency. Membrane reactors can achieve similar efficiency, but they are only summarized. Post-combustion cycles incorporating reforming are introduced, along with a simplified model to estimate the associated penalties. Finally, the different approaches are compared.

11.2 CO_2 Capture Requirements

Separating CO_2 from the exhaust gases, or at any other point within the power cycle, and storing it in deep underground reservoirs is considered a viable option for continuing to use hydrocarbon fuels while stabilizing the atmospheric concentration of CO_2. As shown in Chapter 2, separating a component from a gas mixture requires work (exergy), with lower concentration for the component being separated requiring more work per unit mass of that component. When CO_2 is the component being separated, modifying the power cycle in a way that reduces the required separation work by maximizing CO_2 concentration where it is separated is important for reducing the efficiency penalty. In the following, we review some developments in power cycle designs that enable efficient CO_2 capture. Separation processes were reviewed in Chapter 10, and they include absorption, adsorption, distillation or freezing, and membrane separation.

The ideal isothermal work required to separate the component with molar concentration X_1 from the rest of the mixture, assuming ideal gases, per mole of the mixture was given in previous chapters. To reduce the separation work, we need to reduce the number of moles of the original mixture, and carry out separation at a low temperature from a stream with a relatively high concentration of the gas to be separated, as indicated by the following expression:

$$\hat{w}_i = \frac{\hat{w}}{X_1} = -\Re T_o\left(\ln X_1 + \left(\frac{1-X_1}{X_1}\right)\ln\left(1-X_1\right)\right).$$

This expression suggests several strategies, that is, "dos" and "do-nots," for minimizing the energy penalty of separation. For instance, removing CO_2 from air, in which its current concentration is ~400 ppmv or 0.0004, is inefficient. Separating CO_2 from the products of lean burning of methane, where the mixture composition is dominated by nitrogen and the typical concentration of CO_2 is O (3–5%), is also rather inefficient. To reduce the total volume of the stream and increase CO_2 concentration, oxygen–based combustion can be used. In this case, an air separation unit (ASU) must be used to produce oxygen that is sent to the combustor along with a diluent such as recycled CO_2 or steam to maintain the products temperature at an acceptable level. Oxygen concentration in air is 21%, and air separation was discussed in Chapter 10. Oxyfuel combustion reaches higher temperatures than air-fuel combustion, much higher than necessary in power applications and higher than allowable in heat exchangers or for the turbine inlet temperature. To lower the temperature in the combustor, the mixture is diluted by recycling some of the CO_2 from the turbine exhaust, after H_2O has been separated, back to the combustor. Water injection is also used to lower the combustion temperature. Oxy-combustion may also have the added positive effect of reducing NO_x formation. Another approach is to separate CO_2 from the "fuel" stream after steam reforming of natural gas and water–gas shifting (WGS) the syngas to convert all carbon in the fuel to CO_2. This is pre-combustion separation.

11.3 Low CO_2 Emission Power Cycle Layout

Three broad categories for removing CO_2 from power plant exhaust have been suggested: (1) separation from flue gases; (2) oxy-combustion; and (3) reforming and separation before combustion, all shown schematically in Figure 11.1. Some approaches are more compatible with retrofit (i.e., modifying existing plant for CO_2 capture), and some require total redesign (i.e., new construction or greenfield design). There are many variations on each theme, and the exact cycle layout varies. But distinctions can be made, and the method of separation depends on the category of capture technology. For instance, in the second and third approaches, it is always necessary to use an air separation technology since oxygen is used in combustion and/or reforming. (Air separation can be implicit, i.e., integrated with combustion, if membrane reactors or chemical looping reactors are used.

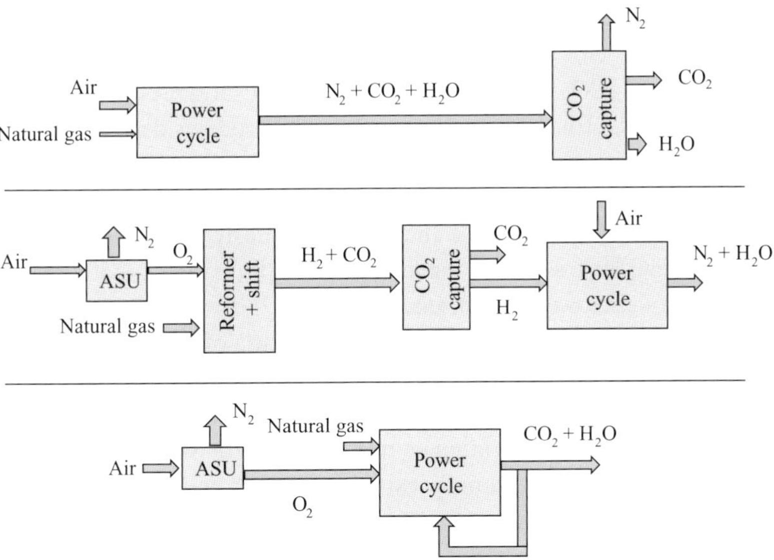

Figure 11.1 Layout for different carbon (CO_2) power plant technologies, including post-combustion capture (which can also be used for industrial processes) in which CO_2 is separated from air combustion flue gases; pre-combustion capture in which CO_2 is separated from the products of steam reforming and WGS; and oxy-combustion capture in which oxygen, separated from air, is used to burn the fuel in the presence of a diluent (CO_2 or steam or a mixture of both). Technologies incorporating fuel cells in pre-combustion capture have also been suggested.

In both cases, starting with separate streams of air and fuel, it is possible to produce separate streams of depleted air and CO_2. These technologies will be described in this chapter.) In some cases, as in oxy-combustion, separation is performed before the power island, and the power island produces an almost pure stream of CO_2 at the "tailpipe" of the plant. In other cases, it is necessary to use another separation process in different places. For instance, in pre-combustion capture, an ASU provides pure oxygen for reforming. Following reforming (and the WGS) another separation process is used to separate H_2 and CO_2. In the latter case, pure H_2 is used as the fuel for a gas turbine cycle or a fuel cell. Pre-combustion capture is also suitable for plants designed to co-produce power and synfuels, since both start with syngas. While the discussion in this section applies to natural gas, syngas can also be used.

Carbon dioxide concentration in a gas stream leaving conventional power plants varies by technology. The concentration of CO_2 depends on the combustion stoichiometry. For instance, in a conventional gas turbine, combustion is lean and hence the low concentration of CO_2. In natural gas-fired boilers, combustion stoichiometry is closer to unity and CO_2 concentration in the products is higher. The concentration, on a dry basis, is 7–10% by volume when fired in a boiler (for a steam cycle), while for a direct-fired gas turbine it is 3–4% by volume [1].

11.4 Post-Combustion Capture

In this approach, CO_2 is separated from the combustion products mixture, which includes N_2, CO_2, H_2O, and O_2 (the fraction of oxygen in the products depends on the stoichiometry of the combustion process and the fuel), following the power cycle in Figure 11.2. Chemical absorption can be used for separating CO_2, with an energy penalty/efficiency loss that depends on the percentage of CO_2 removed and the absorbent. Early studies showed that the energy, mostly thermal energy, requirements for chemical absorption removal of CO_2 from the flue gases (see Chapter 10) might lower the efficiency by 50% of the original (reference plant) efficiency due to the thermal energy requirement for the regeneration of the absorbent. Later, it was confirmed that this efficiency penalty could be reduced substantially if the thermal energy is drawn from the steam turbine at intermediate pressures (between the steam turbine inlet and exit pressures) since the absorbent regeneration temperature is low compared to the steam turbine inlet temperature, instead of burning extra fuel to provide the requisite thermal energy [2]. Studies show that if low-pressure steam from intermediate stages in the power cycle (drawn by extracting partially expanded steam from intermediate stages of the steam turbine) is used instead of using a separate boiler to generate the steam required to regenerate the absorber, the efficiency loss could be reduced to ~10–15 percentage points. Further reduction in the energy penalty of separation has been suggested if different CO_2 solvents are used, and if the operating conditions are optimized, such as the absorbent concentration and the temperature difference between the absorber and desorber towers.

Alkanolamines (MEA) have been suggested as the most likely candidate for the absorber, but other chemicals and mixtures of chemicals are also under investigation. When using these amines, most impurities, including sulfur oxides, nitrogen oxides, and particulates (as well as oxygen), must be removed from the flue gases, since amines form stable salts when reacting with the oxides, and oxygen causes oxidation of the absorbent. Moreover, flue gases should not contain more than 10 ppm of SO_2, even though some processes can tolerate up to 100 ppm sulfur.

The efficiency penalty associated with CO_2 capture is often measured and reported as the reduction in percentage points from the original efficiency (without capture) – that is, if the original efficiency, without capture, is η, and the new efficiency, with capture, is $\eta_{w/cap}$, then the efficiency penalty due to CO_2 capture is:

$$\Delta\eta_{cap} = \eta - \eta_{w/cap}. \tag{11.1}$$

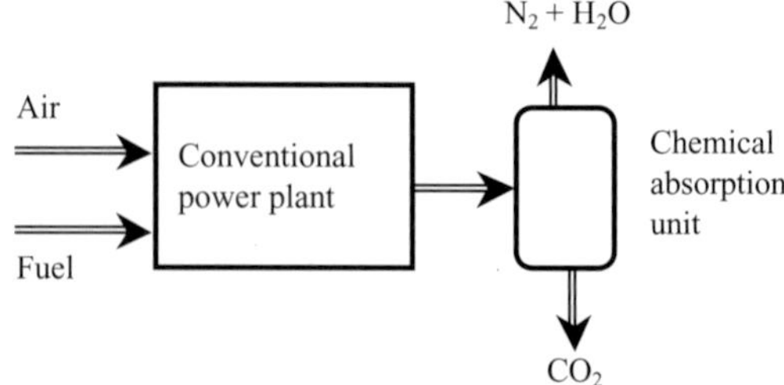

Figure 11.2 Post-combustion CO_2 separation from the power plant flue gases.

Both the efficiency without capture and with capture are measured as useful work divided by the fuel heating value (i.e., for the same denominator), and hence the difference accounts for the work equivalent of the total energy used in the separation of CO_2. Calculating this equivalent work depends on the design and the integration of the plant. In absorption plants, most of the separation energy is in the form of heat, usually "low quality" heat (i.e., at low temperature).

Another penalty associated with CO_2 capture is the work required to liquefy and compress it to 100 bar. This is because, for underground injection, CO_2 must be in the liquid form. The compression work required to liquefy CO_2 by raising its pressure to ~100 bar, W_{comp}, is reported as an efficiency reduction measured in percentage points:

$$\Delta\eta_{comp} = \frac{W_{comp}}{\text{FHV}}, \tag{11.2}$$

where FHV is the heating value of the fuel, and W_{comp} is the compression work of CO_2 per unit mass of the fuel, and hence it is evaluated for the total mass of CO_2 produced by a unit mass of fuel burned in the cycle (note that the heating value is also measured per unit mass of the fuel). Thus, compression work is measured as a fraction of the fuel heating value.

Most of the CO_2 separation energy in the case of chemical absorption is required in the form of heat (for absorbent regeneration), and mostly low-temperature heat around 150–200 °C. Estimating the impact of the separation energy on the efficiency requires more information with regard to where the heat is extracted. Recall from Chapter 10 that the enthalpy of reaction for releasing CO_2 from amines is 1.1–1.9 MJ/kg-CO_2 at temperatures close to 150 °C. This energy is supplied by low-temperature steam extracted from the low-pressure section of the gas turbine, hence degrading the power produced by the steam turbine.

11.4.1 A Model for CH_4

The CO_2 scrubbing system using liquid absorbents was described in detail in Chapter 10, and is shown schematically in Figure 11.3 and Example 10.4. Similar systems are used for coal and gas, as shown in Chapter 10, but more flue gas clean-up is needed in the case of coal.

An approximate model to estimate the energy requirement and the efficiency penalty associated with CO_2 capture from the flue gases of natural gas-fired power plants using chemical absorption was suggested in Reference [3]. While calculations using detailed models yield more accurate results, this approach is illustrative. The model largely neglects interactions among components in the system and attributes the overall loss in efficiency to loss of work potential in individual components as the carbon capture strategy is implemented. The net efficiency of the plant accounting for the impact of the CO_2 separation process is written as follows:

$$\eta_{w/cap} = \eta - \eta_{pump} - \Delta\eta_{heatloss} - \Delta\eta_{CO_2comp}. \tag{11.3}$$

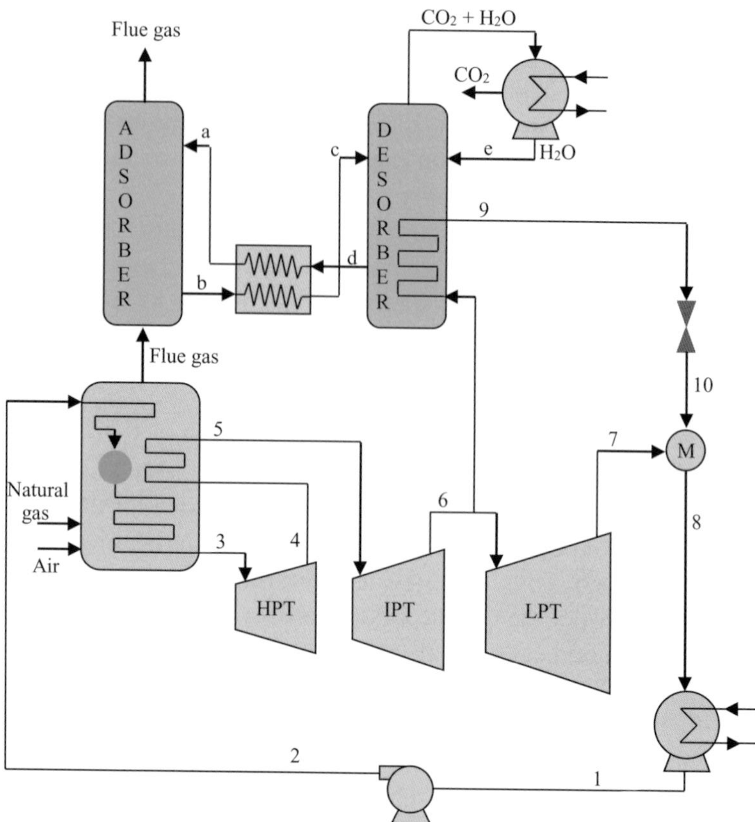

Figure 11.3 Post-combustion capture in a natural gas steam power plant using chemical absorption. The flue gases out of the boiler are flown through the absorber tower, where an amine solution is used to remove CO_2. The solution is then moved to the desorber, where it is heated up using steam extracted from an intermediate stage of the turbine, releasing its CO_2 and some steam. The released gases are cooled to condense water and CO_2 is compressed and liquefied. A recuperator is used between the two towers to reduce the heat duty for amine regeneration; not shown are pumps used to circulate the amine solution between the two towers, and CO_2 compression train. HPT, IPT, and LPT: high-, intermediate, and low-pressure turbine, respectively.

The first term on the right-hand side is the efficiency of the plant without capture. The second term is the penalty associated with the pumping work, W_{pump}, used to circulate the amine solution in the absorption plant and the flue gases through the tower, measured per mole of fuel:

$$\Delta\eta_{pump} = \frac{W_{pump}}{\text{FHV}}. \tag{11.4}$$

The fuel heating value is taken here as the lower heating value. The pumping work overcomes the pressure drop throughout the chemical absorption plant, including that for the flue gas stream through the absorption tower, and for the aqueous solution used for CO_2 removal through the desorption tower and the pipes between the two towers. Given the flow rate of the flue gases, the pressure drop associated with their flow through the absorption tower is much higher than that due to the flow of the solvent (see Chapter 10 for more detail on chemical absorption for CO_2 separation).

The mechanical pumping work depends on the volumetric flow rate of the products and the pressure drop, Δp, and can be estimated as follows: For 1 kg of CO_2, the pumping power is: $W_{pump}^{CO_2} = \frac{\forall_p \Delta p}{m_{CO_2}}$, where $\forall_p$ is the volume of the products and m_{CO_2} is the mass of CO_2 in the products. Using the mole fraction of CO_2 in the products, $X_{p,CO_2} = (n_{CO_2}/n)_p$, which is determined by the overall stoichiometry of combustion, the molecular weight of CO_2, M_{CO_2}, and the ideal gas law: $\forall_p = n_p \Re T/p$, the pumping work per unit mass of CO_2 is $W^{CO_2}_{pump} = \frac{1}{M_{CO_2}} \frac{\Re T}{p} \frac{\Delta p}{X_{p,CO_2}}$, and the total pumping work of the products through the adsorption plant is $\tilde{m}_{CO_2} W^{CO_2}_{pump}$,

$$W^{CO_2}_{pump} = \tilde{m}_{CO_2} \frac{1}{M_{CO_2}} \frac{\Re T}{p} \frac{\Delta p}{X_{p,CO_2}}, \tag{11.5}$$

where $\tilde{m}_{CO_2}$ is the mass of CO_2 produced per 1 kg of fuel. Thus, for a generic hydrocarbon, C_mH_n, the relation between the stoichiometric CO_2 produced and the fuel composition is given by $\tilde{m}_{CO_2} = 44m/(12m + n)$. The typical pressure drop across the tower, while varying with the packing and overall design, is ~150 mbar, leading to 0.34 MJ/kg-CO_2 work of separation. Additional work is consumed to overcome other pressure losses, but it is much smaller, and is taken to be 0.05 MJ/kg-CO_2. Note that the pumping work is independent of the fraction of CO_2 removed in the stripper. Furthermore, the products pumping work is linearly proportional to the pressure drop in the tower, and inversely proportional to the CO_2 mole fraction in the products, as shown in (11.5) [3]. Clearly, raising the concentration of CO_2 in the products by recycling a fraction of the flue gases reduces the work of separation.

The third term in (11.3) is the efficiency penalty associated with using some of the fuel thermal energy, in the form of steam drawn from an intermediate pressure stage in the steam turbine, at the temperature needed for effective stripping of CO_2 from the absorbent. To account for this loss in terms of efficiency penalty, the thermal energy of the extracted steam should be converted to work lost, by multiplying it by a "degraded efficiency," α or $\eta^{deg}_{CO_2}$, since the heat is drawn at a temperature lower than the maximum cycle efficiency. The degraded efficiency is less than the reference plant efficiency, $\alpha < \eta$ (the reference plant is without CO_2 capture). The higher the temperature at which the thermal energy is drawn, and the higher the corresponding saturation pressure, the greater the degraded efficiency, and the higher the efficiency penalty. The thermal energy required for stripping consists of the energy required by the desorption reaction; that is, the enthalpy of reaction for desorption, and the energy needed to raise the temperature of the solvent to a temperature at which the desorption reaction proceeds at a reasonable rate. This temperature is ~150 °C for alkanolamines. The total amount of energy required is in the order of $\Delta h_{CO_2,strip} = 3\text{–}5$ MJ/kg-CO_2 for these solvents. If the fraction of CO_2 removed is f_{CO_2}, the efficiency penalty for stripping is given by:

$$\Delta \eta_{strip} = \frac{\alpha f_{CO_2} \Delta h_{CO_2,strip}}{\text{FHV}}. \tag{11.6}$$

Estimating α is not simple and it requires detailed cycle analysis to evaluate the impact of removing some of the steam at a different temperature (and pressure) during the expansion process on the turbine output. A model for calculating this efficiency penalty for different values of extraction saturation temperature was suggested in Reference [3]. The model is

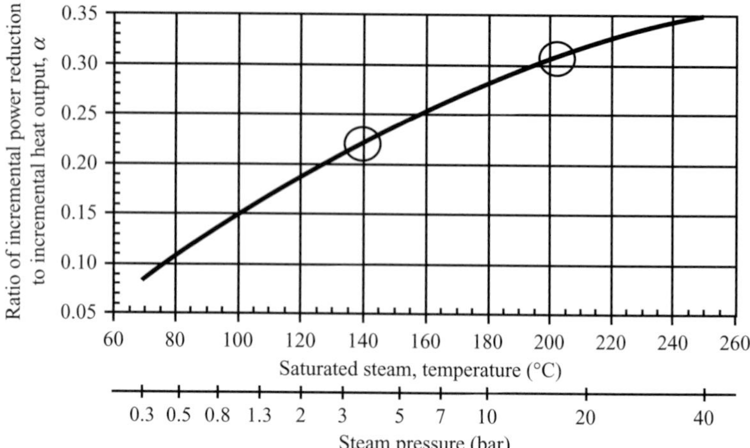

Figure 11.4 The "degraded efficiency" or the incremental power reduction to the incremental thermal energy removed by extracting steam at a given temperature. A high-efficiency CC was used in the calculations with triple pressure reheat and a condenser pressure of 0.04 bar. The extracted steam is delivered in the form of saturated steam, and the condensate is returned at 70 °C [3].

based on a plant with triple pressure reheat, and a condenser pressure of 0.04 bar, while assuming that the condensate is returned to the cycle at 70 °C. The incremental work reduction as a ratio of the incremental heat reduction, that is, the degraded efficiency, for this example is shown in Figure 11.4. The degraded efficiency increases with the temperature (and pressure) of the steam drawn from the cycle.

The fourth term in (11.3) accounts for the efficiency penalty due to the compression of CO_2, nominally to 100 bar:

$$\Delta\eta_{CO_2,comp} = \frac{\tilde{m}_{CO_2} f_{CO_2} W_{CO_2,comp}}{HV_{fuel}}. \tag{11.7}$$

On average, the work required for CO_2 compression to 100 bar is 0.33 MJ/kg-CO_2. While the compression work penalty may be considered as an extra penalty and not part of the cycle performance, it is critical for comparing the performance of different CO_2 capture cycles that the performance is evaluated for a fixed state of the CO_2 at the plant exit. This is because, in some cases, such as supercritical CO_2 cycles, the separated CO_2 may be naturally at a higher pressure than in others, making it less energy-intensive to compress and liquefy CO_2 to the final state of, say, 100 bar and 32 °C. A case of higher-pressure separation of CO_2 will be discussed in the pre-combustion capture section.

For a natural gas combined cycle (NGCC) with 58% efficiency without capture, and using $f_{CO_2} = 0.9$, $\eta_{w/cap} = 49.6\%$ [3].

Example 11.1

The efficiency of an NGCC without CO_2 capture is 55%. The flue gases, which contain 4% CO_2, are sent to a CO_2 separation unit that uses chemical absorption by amine solution, which captures 90% of the CO_2 and compresses it to 100 bar. The flue gases pressure drop in the absorption tower is 150 mbar. The thermal energy required in the stripper is 4 MJ/kg, which is

Example 11.1 (cont.)

provided by steam at 140 °C. Determine the efficiency of the CC with CO_2 capture. Treat the natural gas as methane.

Solution

The efficiency penalty due to the three factors in (11.3) needs to be determined. The specific fan work to overcome a pressure drop of 150 mbar on the flow path of the exhaust that contains 4% CO_2 is 0.31 MJ/kg (according to (11.5)). For each kilogram of fuel (assumed methane with lower heating value = 50.05 MJ/kg), 44/16 = 2.75 kg CO_2 is produced. Hence,

$$\eta_{pump} = \frac{0.31 \times 2.75}{50.050} = 0.0170.$$

From Figure 11.4, the degraded efficiency due to the extraction of steam at 140 °C is around $\eta_{CO_2}^{deg} = 0.22$. The heat requirement for the stripping process is $\Delta h_{CO_2,strip} = 4\ \mathrm{MJ/kg}$. Since only 90% of the CO_2 emitted is captured, $f_{CO_2} = 0.9$, and $\eta_{strip} = \frac{0.22 \times 0.9 \times 4 \times 2.75}{50.050} = 0.0435$.

The work required for compression of CO_2 captured to 100 bar is 0.335 MJ/kg. Hence, $\eta_{CO_2,comp} = \frac{0.335 \times 2.75}{50.050} = 0.0184$.

Thus, the efficiency of the power plant with capture is $\eta_{w/cap} = 0.471$.

For NGCC, the efficiency reduction has been estimated in the range of 5–10 percentage points, using chemical absorption with MEA. Raising the pressure of the separated CO_2 to 110 bar costs about 2 percentage points. A schematic of a CC power plant with post-combustion CO_2 capture is presented in Figure 11.5.

In lean-burning gas turbine plants using flue gas CO_2 separation or post-combustion capture, it is possible to rely on recycling a percentage of the exhaust gases to increase the CO_2 concentration in the stream from which CO_2 is actually removed. Assuming 33% stoichiometry, and complete combustion, CO_2 in the flue gases is close to 5%, according to

$$CH_4 + 2 \times 3(O_2 + 3.76N_2) \rightarrow CO_2 + 2H_2O + 4O_2 + 6 \times 3.76N_2, \tag{11.8}$$

depending on how much water is left in the flue gases. Because of the limitations on the turbine inlet temperature in a CC power plant, and the heating in the compressor, natural gas is burned closer to an equivalence ratio of 0.2, leading to smaller concentrations in the products. Figure 11.6 shows a schematic of a plant with flue gas recycling.

Recycling cools the combustion temperature and makes it possible to burn a stoichiometric mixture in the combustor without violating the limits on the inlet gas turbine temperature. With stoichiometric burning, the CO_2 concentration increases in the products, and raising CO_2 partial pressure in the flue gases reduces the energy penalty of CO_2 separation.

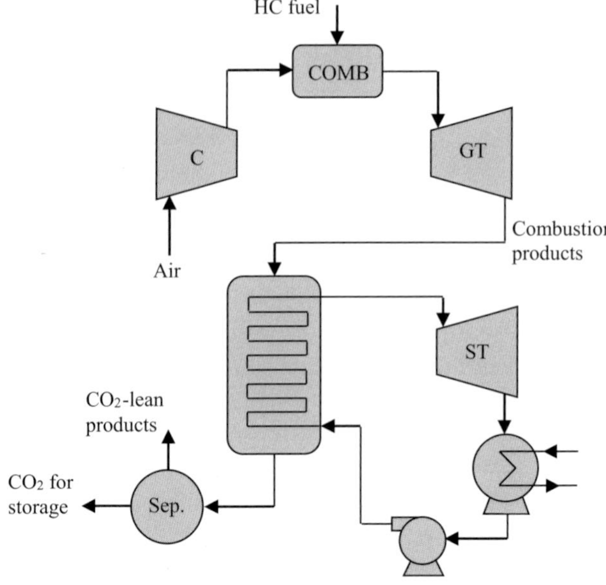

Figure 11.5 Layout of an NGCC power plant with post-combustion CO_2 capture, using chemical absorption. The estimated reduction in efficiency and natural gas is 5-10 percentage points. This amounts to increasing the fuel use by 10–22%. GT, gas turbine; ST, steam turbine.

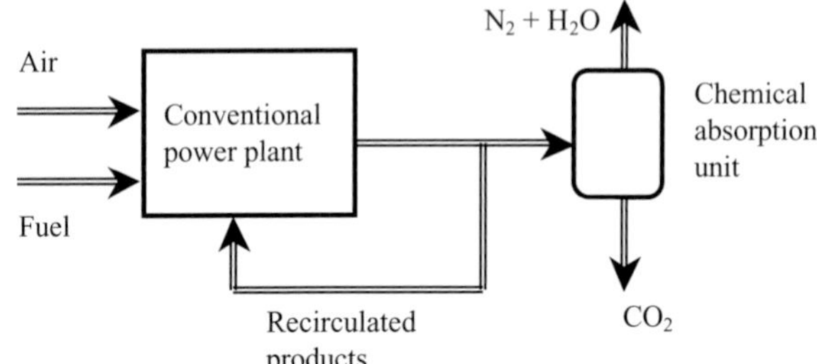

Figure 11.6 CO_2 separation from flue gases in a post-combustion approach, with some flue gas recirculation to increase CO_2 concentration.

However, doubling CO_2 concentration from 3% to 6% by volume, the level expected in most natural gas plants, has a minor impact on reducing the efficiency penalty of capture, and further increase in CO_2 concentration is necessary [2].

Cycles that recycle some of the flue gases are known as semi-closed gas turbine cycles (SCGT), often implemented in a CC plant. In these cycles a large fraction of the gas turbine exhaust, after raising steam for the steam cycle part of the CC, is cooled and recirculated back to the combustor. Semi-closed cycles require extra hardware to cool the working fluid, and gas turbines compatible with a high percentage of CO_2 in the working fluid. The advantage, however, is reduction of the total volume of the gas in the separation unit, and reducing the energy required for separation by increasing the fraction of CO_2 in the flue gas.

11.5 Oxy-Combustion

In oxy-combustion, an ASU is used to supply pure or almost pure oxygen to the combustion process, and the fuel is burned in a mixture of oxygen and a recycled stream composed of

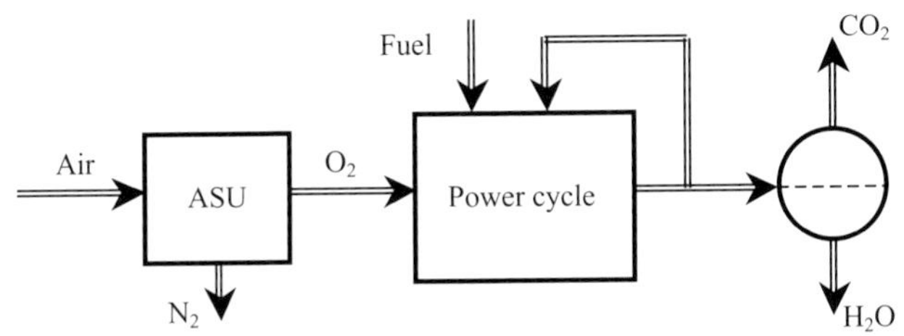

Figure 11.7 A schematic for oxyfuel combustion in a power cycle using combustion products ($CO_2 + H_2O$) recirculation to reduce the combustion gas temperature before the gas turbine.

CO_2, H_2O, or a mixture of both (Figure 11.7). The products of combustion, CO_2 and H_2O, are used in the power cycle in ways that depend on the cycle configuration. Beyond the power island (e.g., a CC), water is condensed and CO_2 is captured. Because nitrogen is removed before combustion, CO_2 concentration in the working fluid is high, and this may have an impact on the operation of the heat exchanger or the gas turbine. To moderate the combustion temperature, a high percentage flue gas recirculation, CO_2 or $CO_2 + H_2O$, is necessary. The primary gas separation process in this case is the separation of oxygen in an ASU. Large-scale ASUs use cryogenic distillation, but other technologies are also possible (see Chapter 10). The working fluid is these systems has a concentration of CO_2 running, impacting the heat exchangers and the gas turbines.

In oxy-combustion, and in order to reduce the energy penalty in air separation, burning is always stoichiometric (or slightly rich in some cases). The fraction of the gas recirculated back to the combustor, which acts as a diluent replacing nitrogen, is determined by the maximum allowable temperature for the heat exchangers or the gas turbine. Oxy-combustion allows for maximum CO_2 removal, theoretically 100%. Air leakage in different equipment, and CO_2 dissolved in water during other processes such as H_2S removal, accounts for most of the less-than-perfect performance.

Air separation is the largest CO_2 capture-related energy penalty in this approach. The most common technology for air separation is the cryogenic approach using the Linde liquefaction and rectification in a double column [4] (also see Chapter 10). Air is fed to the distillation column at high pressure, above 5 bar, resulting in oxygen production at nearly atmospheric pressure, but higher pressures compatible with those desired in the cycle can be reached by raising the distillation pressure. For instance, starting with air at 16 bar, the resulting oxygen is at 6 bar. Pressure-swing adsorption (PSA) and temperature-swing adsorption (TSA) methods can also be used in small ASUs. Membranes have been in development for the same purpose, but have not been used at scale yet. The ideal separation work of O_2 from air (0.21 O_2 by volume) is 0.104 kWh/kg-O_2. Actual work, for near 100% purity, is closer to 0.3 kWh/kg-O_2 or nearly 1 MJ/kg-O_2.

11.5.1 A Simplified Model for CH_4

A simplified model to calculate the efficiency of an oxy-combustion CC using a SCGT in now described, similar to the one proposed in Reference [3]. The layout of the plant in shown in Figure 11.8. Near-stoichiometric combustion of CH_4 in oxygen (97% purity) generated in

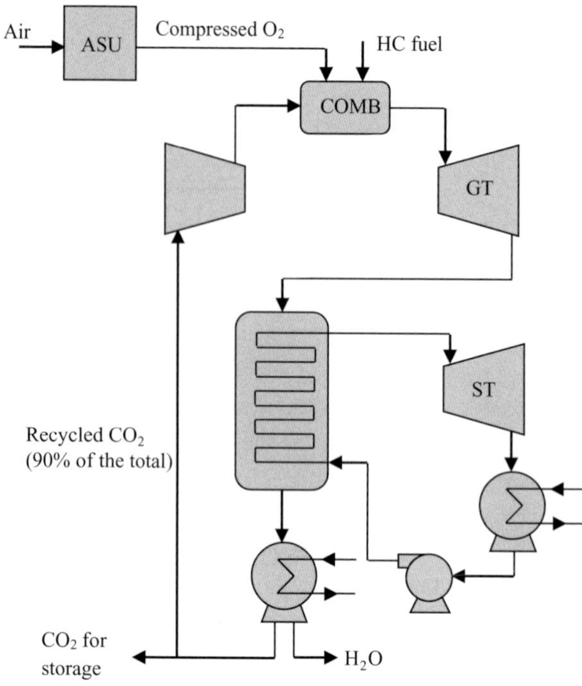

Figure 11.8 Layout for an oxy-combustion cycle, in which an ASU is used to deliver oxygen to the gas turbine combustion chamber.

an ASU is assumed. The combustion temperature is controlled by recycling a large fraction of CO_2. Thus, the working fluid for the gas turbine is CO_2 and some water. The starting point in evaluating the efficiency with CO_2 capture is a cycle efficiency in which CO_2 is used as a working fluid without capture. Because of the different properties of CO_2, the optimal pressure ratio when using it as a working fluid is different, and the chosen cycle is assumed to run at the optimal pressure ratio for CO_2. As shown before, for the same gas turbine exit temperature, the pressure ratio in CO_2 gas turbine must be higher than that in an air combustion turbine (dominated by nitrogen). A pressure ratio of 35 (versus 15 for conventional gas turbines) has been suggested. When oxygen is supplied at 200 °C (from the ASU), and at the highest pressure of the cycle, the CO_2 cycle efficiency without accounting for the energy penalty associated with the ASU is ~ 61%. This efficiency must then be reduced by the work required for air separation and CO_2 compression:

$$\eta_{oxy/cap} = \eta_{oxy} - \Delta\eta_{ASU} - \Delta\eta_{CO_{2comp}}. \tag{11.9}$$

The efficiency penalty incurred by the ASU, η_{ASU}, is evaluated as follows. The work required to produce a unit mass of oxygen at atmospheric pressure (with high purity, around 97%) is $W_{O_2} \sim 1$ MJ/kg-O_2. Moreover, oxygen compression work from atmospheric pressure to the cycle highest pressure is ~0.54 MJ/kg-CO_2 (since oxygen is injected directly into the combustor, along with the fuel). Thus, $W_{O_2} = 1.54$ MJ/kg-O_2. To express that per kilogram of CO_2, we note that for each kilomole of methane burned, two kilomoles of O_2 are needed, and 1 kilomole of CO_2 is produced. Thus, for 1 kmol of CH_4, we need 64 kg of O_2, and produce 44 kg of CO_2, or $r_{CO_2:O_2} = 0.69$ kg-CO_2/kg-O_2. Thus, the efficiency penalty is:

$$\Delta\eta_{ASU} = \frac{W_{O_2}\tilde{m}_{O_2}}{r_{CO_2:O_2}\mathrm{FHV}}. \tag{11.10}$$

The compression work for CO_2 is calculated as before, but with $f_{CO2} = 1$. Calculations show the efficiency of this cycle is reduced to 47% (from 61%). While this efficiency is within the range predicted by more accurate models, better cycle designs that integrate the components better, which will be discussed later, could have higher efficiency.

11.5.2 Combined Cycle and Water Cycle

Here, we describe these two cycles and discuss their performance based on higher-fidelity models that enable the use of more accurate equations of state for the working fluid, because conditions are such that it is no longer possible to do so using ideal gas analysis or pure substances. Moreover, analysis of these cycles requires models of unconventional components such as a cryogenic ASU (ASU) and a CO_2 purification and liquefaction unit (CPU). The modeling platform used here is Aspen© [5]. In this application, we use the Peng–Robinson (PR) cubic equation of state with the Boston–Mathias alpha function for all the working fluids.

11.5.2.1 The Combined Cycle

This cycle strongly resembles the air combustion CC except for the addition of an ASU to produce oxygen and a CPU to clean the flue gases. The fuel is burned stoichiometrically. In the following we review results of cycle efficiency computations using Aspen© to model the cycle, components, and working fluid properties. This cycle is the same as the semi-closed oxy-combustion combined cycle (SSOC-CC) [6].

In the modeled cycle [7,8], shown in Figure 11.9, the turbine inlet temperature is fixed at 1300 °C, with gas composition of 78% CO_2 and 14% H_2O by volume. Entering the compressor is a recycled stream with 88% CO_2 and 12% H_2O by volume, compressed to 40 bar and mixed with O_2 with 95% purity (remaining gases are nitrogen and argon) and methane in the combustor. The exit pressure of the gas turbine is slightly above 1 bar. In the steam cycle, the intermediate pressure is 6 bar, and maximum temperature of both stages is 560 °C. The exit pressure of the steam turbine is 0.1 bar. Following the condenser, 93% of the fluid stream is returned to the combustor. The CPU produces liquid CO_2 at 110 bar. More conditions are shown in Table 11.1.

The computed cycle efficiency is 45.9%, accounting for the energy penalties in the ASU, consuming 11% of the total turbine power (equivalent to 0.225 kWh/kg-O_2 in the ASU, followed by compression to the combustor pressure), and the CPU consuming 3.3%. Raising the turbine inlet temperature (TIT) and isentropic efficiencies of the turbomachinery has been shown to raise the cycle efficiency to 47-49% [6,9].

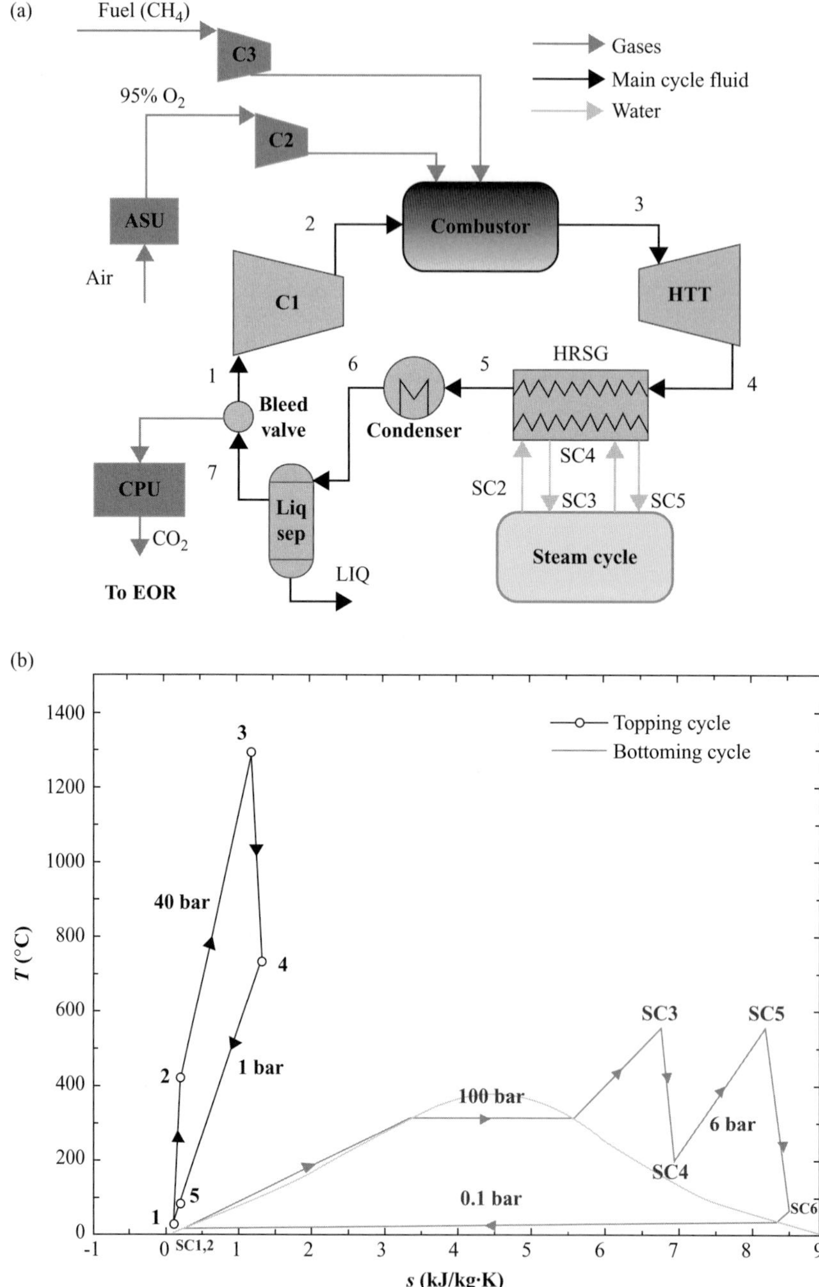

Figure 11.9 Methane oxy-combustion CC (a) and the corresponding *T*–*s* diagram (b) [7]. HTT, high-temperature turbine; HRSG, heat recovery steam generator.

Table 11.1 Conditions used in the oxy-combustion cycle presented in Figure 11.9 [7]

	Combined cycle	Water cycle
Fuel, lower heating value (LHV)	100% CH_4, 50.1 MJ/kg	
Combustors		
Operating pressures (bar)	40	100, and 15 (main and reheater)
Pressure drops (%)	5	10, and 6 (main and reheater)
Turbines		
TITs (°C)	1300	HPT, 600; LPT, 1200
Isentropic efficiencies	85%	HPT, 87%; LPT, 90%
Isentropic efficiency	Compressor 85%	Pump 75%
Heat exchangers		
MITA[a] (°C)	20	20
Pressure drops (%)	5	5
Steam cycle		
HRSG pressures (bar)	100 and 6	N/A
TITs (°C)	560	N/A
Turbine efficiencies (%)	90	N/A
Condenser pressure (bar)	0.05	N/A
Pump efficiency	75	N/A
ASU specific power	0.225 kWh/kg-O_2	
O_2 stream composition	95% O_2, 4.2% Ar, 0.8% N_2 by volume	
O_2 compression		
Isentropic efficiency	84	84
Compression stages	3	3
Intercooler temperatures (°C)	25, 50	25, 50
Max exit temperatures (°C)	200	200
CPU O_2 delivery pressure	110 bar	
Stream composition	>99% CO_2 by volume (EOR ready)	

[a] Minimal internal temperature approach.

11.5.2.2 The Water Cycle

This cycle is essentially a Rankine cycle with reheat, as shown in Figure 11.10, whose working fluid is 95% H_2O and 5% CO_2, operating with the first TIT of 600 °C, between 100 and 15 bar. Following expansion in the HPT, the working fluid is reheated to 1200 °C, and expands in the LPT to 0.1 bar. The high pressure and temperature of the second turbine exit allows for regeneration. Following regeneration, water is condensed at 25 °C, and 83% of it is recycled back to the combustor, with some excess water removed. Carbon dioxide is removed as a gas following the regenerator, and is sent to the CPU.

More of the model assumptions used in the analysis are shown in Table 11.1. The ASU consumes 17.2% of the total turbine power, while the CPU consumes 8.6%. The efficiency of this cycle is 41.4%. Clearly the CC efficiency is higher, at the expense of adding more equipment.

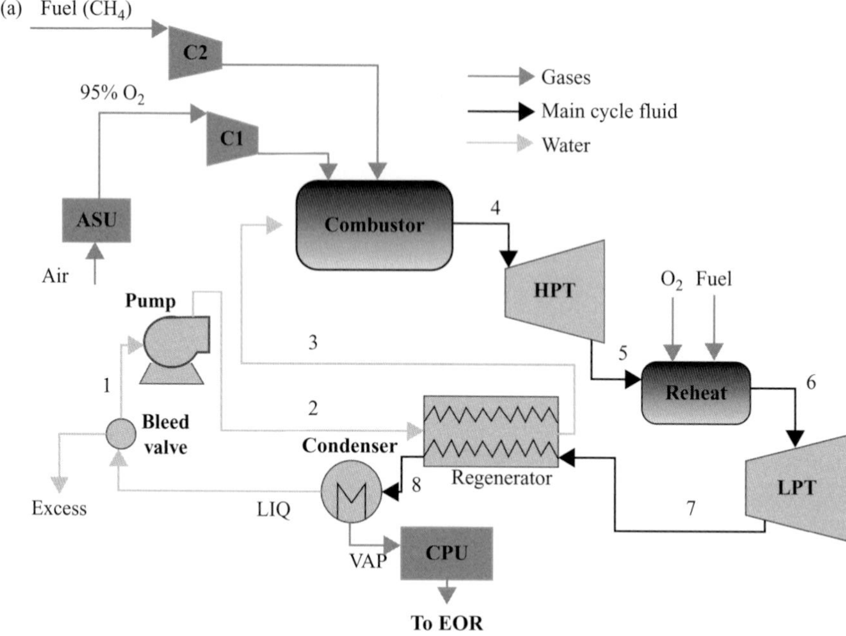

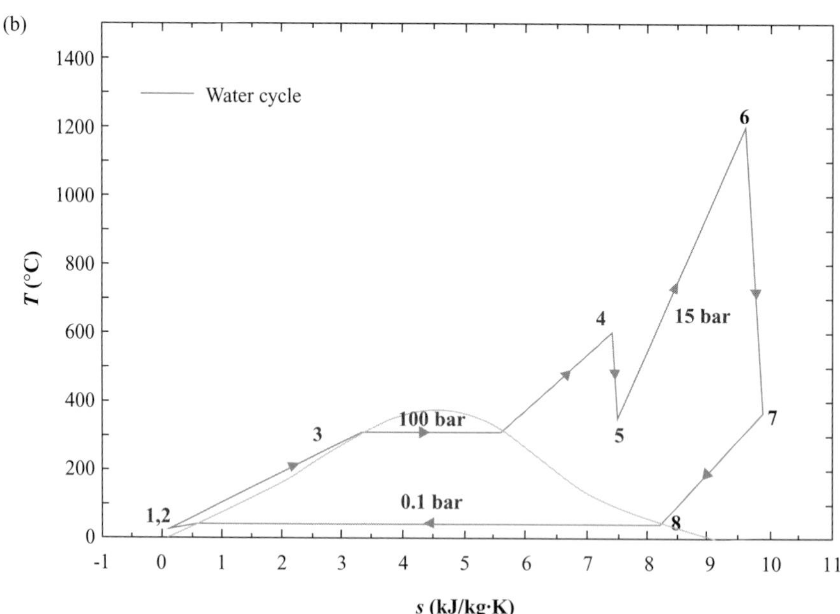

Figure 11.10 Methane oxy-combustion water cycle and the corresponding *T–s* diagram [7].

11.5.3 Supercritical Cycles

Brayton cycles operate at temperatures much higher than the critical temperature, and pressures mostly lower than the critical pressure of the working fluid, in a regime in which the working fluid behaves essentially as an ideal gas. That is, they operate to the far right of the critical point and the vapor dome on the *T–s* diagram. Gas cycles in general are not called supercritical cycles; this designation is reserved for two-phase cycles whose maximum pressure exceeds the critical pressure, while their lowest pressure is well below that.

Gas-based cycles that utilize CO_2 as a working fluid have been proposed for carbon capture applications, in which the lower pressure of the cycle is closer to or slightly lower than the critical pressure, and hence one can take advantage of the low compression work of working fluids when their pressures are close to or below the critical pressure (some of these cycles were discussed in Chapter 6, and Figure 6.20 shows the *T–s* diagram of CO_2). These cycles operate at higher pressures, and the working fluid during most of the cycle is in the supercritical regime, except during heat rejection and the early compression part, where the working fluid conditions are close to the critical point, on either side of that point. The low compression work improves the efficiency of these cycles over those achieved by the Brayton cycle, but operating at high pressures in the heat-addition process and during expansion poses material challenges. These "mixed" cycles, which have some of the characteristics of supercritical vapor cycles, are sometimes called "hypercritical" cycles because they operate largely above the critical point.

Another advantage of these cycles is that the high density of the working fluid near the critical point improves the heat transfer characteristics during heat rejection. Since these mixed cycles already operate at high pressures, the pressure ratio across the compression and expansion stages is relatively low. This improves the opportunity for regeneration. Two "simple" hypercritical cycles are generally quoted in the literature. In the Gohstjejn cycle, CO_2 condenses during heat rejection, $p_{min} < p_{cri}$, and the compression starts with a liquid. The other is known as the Feher cycle, where the lowest pressure is just above the critical pressure. Both were discussed in more detail in Chapter 6, including their *T–s* diagrams. Both cycles can take advantage of reheating and intercooling to reduce the compression work and increase the turbine work, as well as to improve their efficiency. Furthermore, to improve the heat exchange between the turbine exhaust and compressed fluid, two fluids with different densities/heat capacities, **partial or split compression**, may be used. In partial compression cycles, similar to the one shown schematically in Figure 11.11 on a *T–s* diagram, the turbine exhaust stream at state 6 is used in a high-temperature regenerator to heat up the total stream to state 4. Following the low-temperature regenerator, the stream is split into two parts at state 8: part I gets compressed to the high pressure of the cycle (state 3); and part II is cooled further in the heat-rejection part of the cycle to state 1, before being compressed to state 2. Part II is then heated in the low-temperature regenerator, to state 3, and mixed with the stream that has been split and compressed (without rejecting heat). In the low-temperature regenerator, the turbine exhaust is cooled from state 7 to state 6, while part II is heated from state 2 to state 3. While the total stream flows between states 3–4–5–6–7–8,

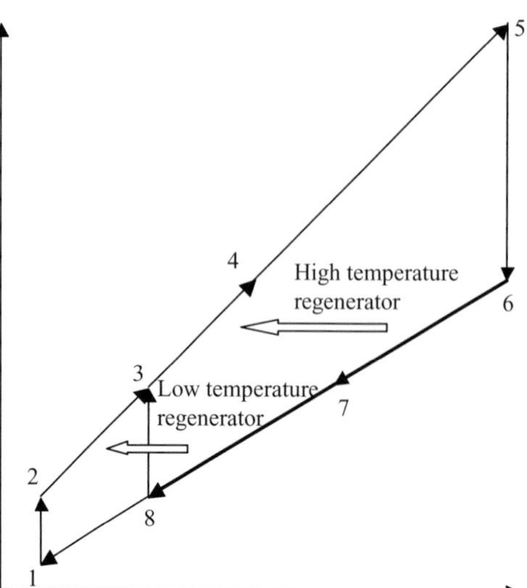

Figure 11.11 The T–s diagram of a split compression or pre-compression gas cycle. The gas leaving the turbine at state 6 is cooled in a high-temperature regenerator to state 8. Only a fraction of the stream is compressed back to the turbine high pressure in process 8–3, while the rest is cooled in a low-temperature regenerator to state 1. It is then compressed to the turbine high pressure and regeneratively heated in 2–3, then mixed with the rest of the gas at 3.

only part I is compressed in process 8–3, and only part II is cooled, compressed, and heated by regeneration in processes 8–1–2–3. This cycle rejects less heat than conventional Brayton cycles, and balances the heat exchange process within the regenerator better. The Sulzer cycle is a Gohstjejn cycle with partial compression, and the Schabert cycle is the Feher cycle with partial compression. Cycles with condensation, such as the Gohstjejn and Sulzer cycles, require high flow rates of low-temperature coolant to be available to remove the large enthalpy of condensation.

Another cycle that employs many of these features – that is, compression near the critical conditions, multistage intercooling, reheating, and regeneration – while using a working fluid composed essentially of CO_2 is the MATIANT cycle. The Graz cycle uses CO_2 as a working fluid, but with extra H_2O injection in the gas turbines (i.e., it is humidified). The Graz cycle is described next.

11.5.4 The Graz Cycle

The **Graz cycle** is another example of a semi-closed humid CO_2 cycle that uses pure oxygen in the combustion process [6]. Contrary to the MATIANT cycle, in the Graz cycle a fraction of the water in the exhaust is pumped back and used in the combustor and gas turbine during expansion. The Graz cycle can be regarded as a well-integrated cycle that includes a humid gas turbine (where the gas is CO_2 instead of the classical humid air) with separate compression stages for the recycled CO_2 and H_2O, and a steam cycle. Thus, while some of CO_2 and H_2O are captured and removed every cycle, most of both are actually recycled after separation in the condenser. Clearly, according to mass conservation, the amount of CO_2 and H_2O captured in every cycle is the same as that produced in the single-stage

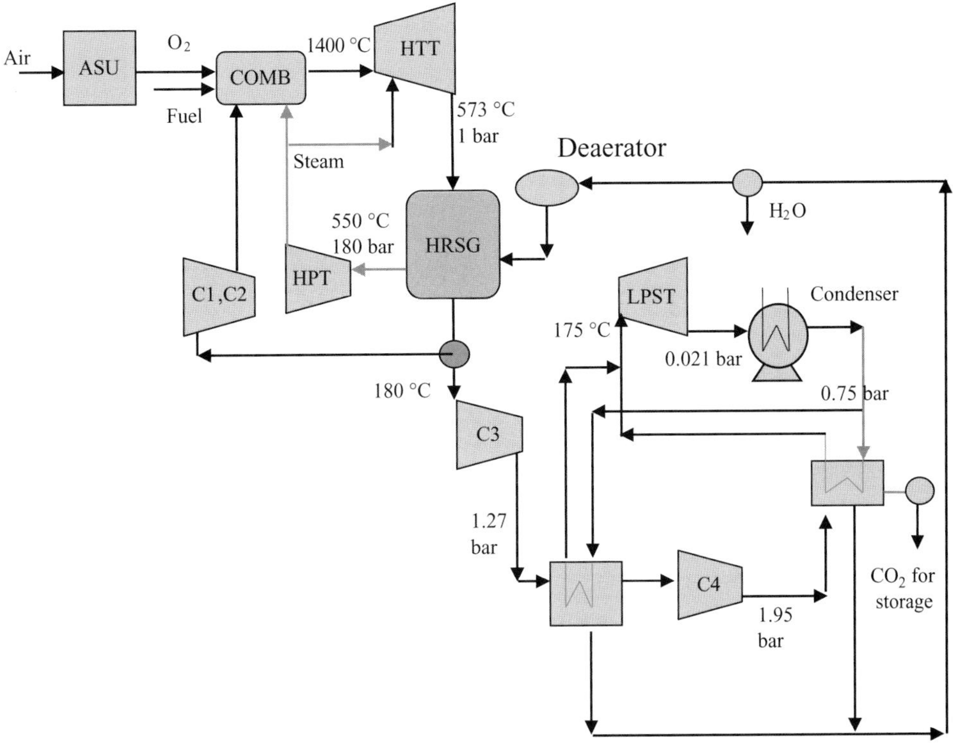

Figure 11.12 A schematic of an optimized Graz cycle [10]. The cycle fluid is 97% water and the rest is carbon dioxide. HPT, high-pressure steam turbine; LPST: low-pressure steam turbine; C1,C2, C3, and C4 are cycle compressors.

oxy-combustion process. Gaseous fuel, natural gas or syngas, oxygen, recycled steam, and recycled CO_2 are burned in a combustor.

Figure 11.12 shows an optimized design of the Graz cycle as presented in Reference [6]. This cycle can be considered as a high-temperature semi-open cycle, with a high-temperature turbine (HTT) that expands humid CO_2 stream, and a high-pressure steam turbine (HPT), followed by low-temperature condensation of the water in the primary stream, and a bottoming closed steam cycle with a low-pressure steam turbine (LPST) – that is, it is a complex CC with a humid gas turbine. The high-temperature cycle includes the HTT (expanding a stream of mostly CO_2 followed by the same stream after being humidified by the addition of steam) and HPT (expanding steam), while the bottoming steam cycle employs the LPST in a closed cycle. The combustor operates at a modest 40 bar, with oxygen, fuel, and recycled gases from the HHT and some steam from the HPT to control the temperature. Gases leave the combustor at ~1400 °C, typically with 75% CO_2 and 25% H_2O (by mass), expand in the HHT to 1.06 bar and 573 °C, with steam (drawn from the HPT) added to the turbine at an intermediate pressure of 10 bar and 330 °C. Adding the steam to the HHT raises its working fluid water content to 79% (by mass) and enables significantly more power. The exit (primary) stream from the HHT passes through a HRSG to raise steam for the HPT, leaving at 180 °C.

Beyond the HRSG, the (primary) stream is split in (almost) two halves; one returns to the compressors (C1 and C2) and combustor, and the other is compressed in C3 to 1.27 bar, cooled by the (closed loop) condensate from the LPST, and some water (~63%) is removed. Further compression to 1.95 bar in C4 (and further cooling using the LPST condensate) allows more water to be removed (25%) from that stream. The water removed in these two condensation steps is collected and sent to the HRSG to be heated by the exit stream from the HTT before expanding in the HPT (this is a steam turbine). Carbon dioxide is removed following the second condensation step from what remains of the primary stream (after two-stage water separation). The pure water in the low-pressure steam cycle (where the LPST operates) is sent to the LPST for expansion, entering at around 175 °C and expanding to 0.021 bar. The expanded steam from the LPST is then condensed (by external cooling) and the condensate is pumped up to 0.75 bar. This condensate is used in the two stages to lower the temperature of the primary ($CO_2 + H_2O$) stream after C3 and C4. The overall efficiency of this cycle, including the work used by the ASU and compressing CO_2 (to 100 bar) is 53%. Optimization studies of similar cycle design were conducted in Reference [11].

Contrary to the MATIANT cycle, the pressure of the working fluid in the Graz cycle remains rather low, but the HHT operates in a humid gas (not air in this case) form. The high pressure, high temperature, and high steam concentration in the HHT increases its power output significantly, but also poses some technological challenges.

11.6 Chemical Looping

As shown in previous sections, a large fraction of the oxy-combustion energy penalty is spent on air separation. Chemical-looping combustion (CLC) is an oxy-combustion process that does not require an ASU. A metal oxide is used to oxidize the fuel in one reactor (called the reduction reactor, where reduction refers to the metal oxide), and the reduced metal oxide is re-oxidized in air in a separate reactor (called the oxidation reactor). Thus, the system is composed of two reactors (or a single reactor capable of performing both functions): an air or oxidation reactor and a fuel or reduction reactor, used to perform the redox reaction pair depicted schematically in Figure 11.13. Fuel is introduced into the fuel reactor, which contains metal oxide, Me_xO_y. The fuel and the metal oxide react according to reaction:

$$(2n+m)Me_xO_y + C_nH_{2m} \rightarrow (2n+m)Me_xO_{y-1} + mH_2O + nCO_2. \tag{11.11}$$

The gas stream leaving this reactor contains only CO_2 and water.

The reduced metal oxide, Me_xO_{y-1}, is then transferred to the air reactor, where it is oxidized. The oxidation reaction proceeds according to the following reaction:

$$Me_xO_{y-1} + \text{air} \rightarrow Me_xO_y + N_2 + \text{unreacted } O_2. \tag{11.12}$$

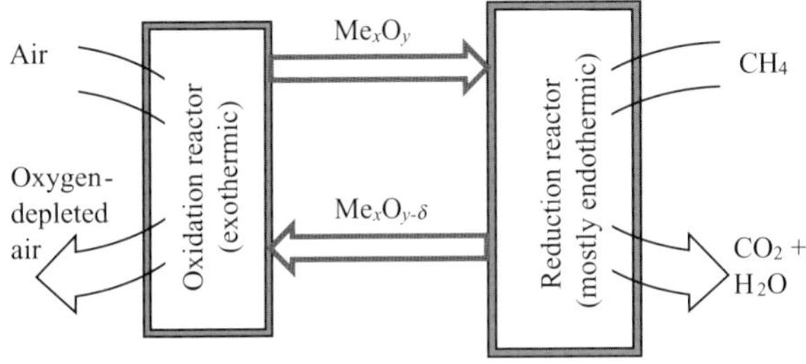

Figure 11.13 Schematic representation of the CLC process. The curved arrows indicate the pathways of the different streams, and oxidation and reduction are performed in separate reactors. Metal oxidation is exothermic. Metal oxide reduction depends on the fuel used, and can be endothermic or mildly exothermic. The metal is partially oxidized.

Table 11.2 Properties of metal oxides used in CLC with CH_4 used as the fuel. The table shows enthalpy of reaction for the reduction reaction, and the oxidation reaction for copper, nickel, and iron oxide (cycling between two states of oxidation). Also shown is the equilibrium conversion as percentage of complete combustion at typical operating temperatures. The oxygen transport capacity is defined as $(m_{ox}-m_{red})/m_{ox}$ where *m* is the mass and subscripts *ox* and *red* correspond to the oxidized and reduced state of the metal [12]

	CuO/Cu	NiO/Ni	Fe_2O_3/Fe_3O_4
Reduction ΔH_0	−178.1 kJ/mol	156.5 kJ/mol	136.6 kJ/mol
Oxidation ΔH_0	−312.1 kJ/mol	−479.4 kJ/mol	−469.4 kJ/mol
Methane conversion (800/1000/1200°C)	100%/100%/100%	100%/99%/98%	100%/100%/100%
Oxygen transport capacity	0.20	0.21	0.03
Material cost	Intermediate	High	Low
Carbon deposition	Low	High	Low

The gas stream leaving the air reactor contains nitrogen and unreacted oxygen. The metal reduction is often endothermic (depending on the metal) if methane is used, whereas it is exothermic if syngas is used. The metal oxidation reaction is highly exothermic. The enthalpy of reaction depends on fuel type and on the metal oxide used the as oxygen carrier (OC). Some examples are shown in Table 11.2.

The total energy released from these two reactions is the same as that obtained from direct combustion:

$$C_nH_{2m} + \text{air} \rightarrow mH_2O + nCO_2 + N_2 + \text{unreacted } O_2. \tag{11.13}$$

The temperature at which complete combustion is achieved is governed by the equilibrium of the methane oxidation reaction, where (s) indicate a solid phase:

$$CH_4 + Me_xO_y(s) \rightarrow CO_2 + 2H_2O + Me_xO_{y-4}(s), \tag{11.14}$$

with

$$K_p = \frac{p_{CO_2}p_{H_2O}^2}{p_{CH_4}} = \exp\left[\frac{-\hat{g}^o_{CO_2} + \hat{g}^o_{H_2O} + \hat{g}^o_{Me_xO_{y-4(s)}} - \hat{g}^o_{CH_4} - \hat{g}^o_{Me_xO_{y(s)}}}{\Re T}\right]. \tag{11.15}$$

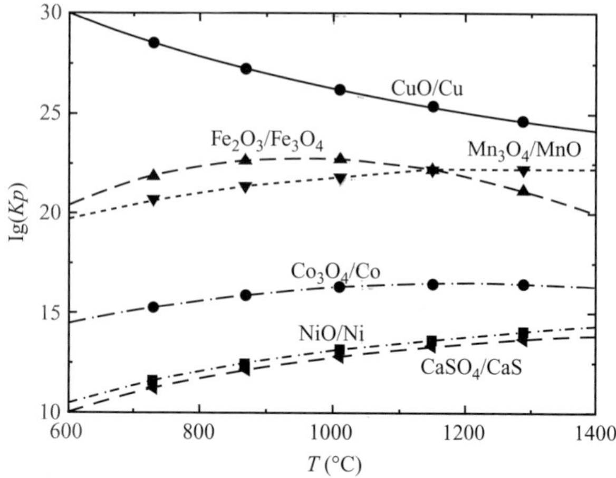

Figure 11.14 The equilibrium constant for methane metal oxide reduction reactions leading to complete oxidation of methane [12].

The equilibrium constants for a number of OCs are shown in Figure 11.14.

While originally proposed for hydrogen production, CLC was later suggested for reducing the exergy penalty associated with conventional combustion in power plants [13], and much later for lowering the energy penalty in carbon capture and sequestration (CCS) plants [14]. Splitting the combustion process into two parts, and operating at temperatures lower than those corresponding to conventional combustion temperatures but closer to the redox reaction equilibrium temperature, reduces reaction entropy generation. Also, operating the reactors closer to temperatures tolerated by the power equipment reduces heat transfer exergy losses. In the case of CCS plants, and since the reduction reactor produces a pure stream of CO_2 and water (similar to oxy-combustion but without an ASU), carbon capture is accomplished without the typical energy penalty associated with an ASU (however, there is still an inherent penalty associated with the fact that leaving the system are two separate gas streams instead of one). Also similar to conventional (gas phase) oxy-combustion, CLC reduces (or eliminates) NO_x production because of the lower reaction temperatures.

The looping medium in CLC acts as an OC, transporting oxygen while it flows between the two reactors. The selection of the OC is a very important component in CLC. Depending on the design of the reactor pair, thermal energy can be transported from the highly exothermic air reactor to the less exothermic (or at times endothermic) fuel reactor with the circulating OC, or by other means such as spinning a single reactor between the oxidizing and reducing streams [15]. Basic properties for a good candidate for an OC include:

1. *Reactivity*: Reactivity characterizes the reaction kinetics. Higher oxidation and reduction reactivity result in shorter residence time, smaller reactor, and hence lower capital cost.
2. *Selectivity*: Some OCs have high selectivity toward complete combustion products, CO_2 and H_2O, while others may be selective toward CO or H_2.
3. *Oxygen capacity*: This describes the maximum amount of oxygen that can be transported by the OC. Higher oxygen-carrying capacity increases the thermal capacity of the reactor.

4. *Fuel conversion:* Fuel conversion is restricted by the thermodynamic equilibrium of the fuel and the OC, but can be impacted by the residence time, reactor design, and solid–gas contact.
5. *Stability*: This requires that the OC circulate continuously without significant changes in its characteristics, while resisting agglomeration, attrition, and contamination.
6. *Costs and safety*: The OC material should be abundant and cheap, and safe to use and to dispose of when necessary.

Chemical-looping processes have been proposed and used for other applications beyond oxy-combustion [16].

11.6.1 CLC Cycle Analysis

Various system designs for CLC-based power generation have been reported in the literature. One of the most common designs is a combined cycle with a combustor replaced by a CLC reactor pair (CLC-CC). Figure 11.15 shows a simplified layout of the CLC-CC. The high-temperature gases from the air/oxidation reaction, essentially oxygen-depleted air that was used to oxidize the metal/metal oxide, is used to drive the gas turbine, as in conventional CC. This is because the metal oxidation reactions are the more exothermic of the two redox reactions, and hence the gas can reach the highest temperature in the cycle. A secondary CO_2 gas/steam turbine (humid gas turbine) can be used to extract the energy from the gases from the fuel/reduction reactor, which although less exothermic than the oxidation reaction or even endothermic, is heated by the circulating OC. The gases leaving the gas turbine of the oxidation reactor are used to drive the bottoming steam cycle, while the gases leaving the lower-temperature reduction reactor turbine can be utilized to preheat the fuel.

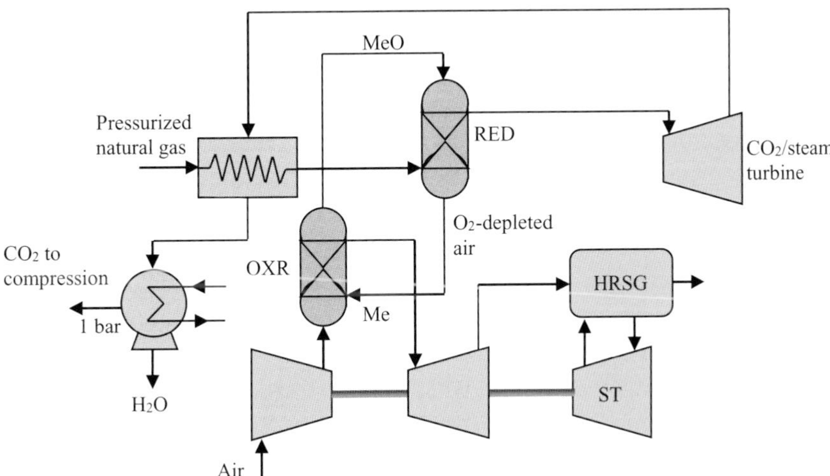

Figure 11.15 The flow diagram of CLC-CC using two separate reactors. OXR, oxidation reactor; RED, reduction reactor.

Several variations of the CC design have been investigated. For instance, the lower-temperature flue gas from the reduction reactor can be used to drive the bottoming steam cycle, called a CLC-CC(s). Another option is to use that gas in a recuperator to preheat the compressed air for the oxidation reactor (the CLCL-CC(r)), such that the CO_2 stream from that reactor is not expanded, and the work of CO_2 compression is reduced. The combined cycle with CO_2 gas/steam turbine, that is the CLC-CC, offers the highest efficiency, while the characteristics of the other two (CC with recuperation, and CC with fuel reactor flue gas driving the steam cycle) are almost the same [12].

The oxidation reactor exit temperature is important for the overall cycle efficiency, as in all other combined cycles. The higher the metal oxidation temperature, the higher the gas TIT. Modern gas turbines can operate at TITs higher than 1400 °C, but other limitations associated with the oxidation reactor temperature must be considered, such as the increased risk of agglomeration and attrition of OC particles, melting, and other material degradation concerns. Other options for improving the efficiency at lower oxidation reaction temperatures include reheat in the gas turbines and using a humid air turbine, as previously shown in other cycles. Supplementary firing in the oxygen-depleted stream is also possible, as described in a later section, but this produces extra CO_2. The presence of two different reactors, possibly operating at two different temperatures, offers more possibilities for integration with different components in the cycle otherwise not available, which could boost the overall efficiency. For instance, when intercoolers are used to reduce the compression work, the low-temperature heat is often wasted. In a CLC cycle, this heat could be used to warm the fuel prior to injecting it in the fuel reactor.

As shown in Figure 11.16, single-stage reheat can achieve efficiencies around 51% with T_{ox} of only 1000 °C, whereas an oxidation temperature above 1200 °C is necessary to obtain similar efficiency for CLC without reheat. At the oxidation temperature of 1200 °C, single-stage reheat shows above 53% net plant efficiency. The double-reheat CLC, however, does not exhibit substantial efficiency improvement compared to the single reheat cycle. The efficiency of the conventional CC with CCS (reference plant in this case) does not exceed 49%.

Another challenge in CLC power generation is that as air flows through the air reactor, part of its oxygen is adsorbed by the OC, and hence only a portion of the compressed air is used to drive the gas turbine. In order to deliver certain specific work output, more compressed air is needed to make up for the fraction lost in metal oxidation. With more air, compression work increases, which lowers the efficiency. In order to overcome this concern, CLC cycles operating on a humid air cycle could be utilized to enhance the total mass flow rate in the turbine without increasing the compressor work. In contrast to a conventional humid air cycle in which the impurities from the exhaust (NO_x, CO_2) cause corrosion of the turbine blade, corrosion in the CLC humid air cycle is less of a concern since fuel is not present in the air reactor. Compared to the conventional CC, the humid air CLC has similar or even lower compression power requirements [17]. The overall efficiency of the humid air CLC is reported to be over 55%, making it 2% higher than the multistage CC.

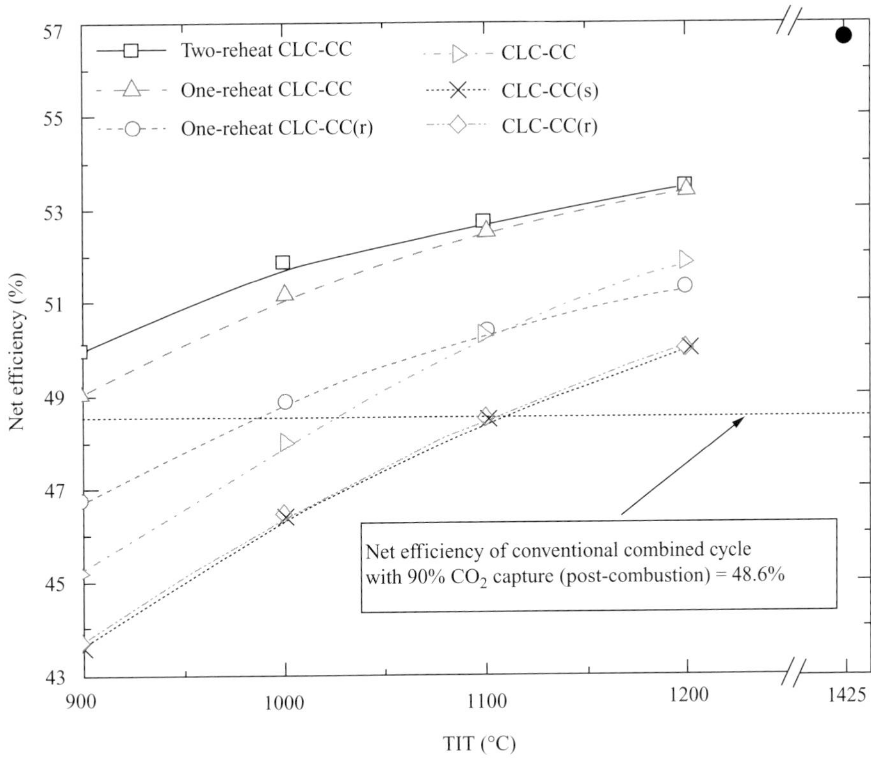

Figure 11.16 Comparison of multistage CC designs. The solid circle on the top right-hand corner is for the CC without CCS. CLC-CC with no reheat, one or two reheats. CLC-CC(r) is the CLC-CC with fuel reactor flue gas recuperation (no reheat and a single reheat); CLC-CC(s) is the CLC-CC with fuel reactor flue gas powering a bottoming steam cycle. The TIT plays a very important role in determining the efficiency [12].

Example 11.2

A chemical-looping-based power cycle using Ni (metal) and NiO (metal oxide) and methane is schematically presented in Figure E11.2. The dotted line represents the chemical loop of the OCs (NiO/Ni). In the oxygen reactor, the highly exothermic metal oxidation reaction heats up the air stream, which is utilized to run turbine 1. In the metal reduction reactor, the endothermic reduction reaction of NiO is used to oxidize methane to CO_2 and H_2O. While these two reactions proceed, the OCs (Ni/NiO) circulate the chemical loop (dotted line). When the NiO particles are prepared, YSZ (yttria-stabilized ZrO_2) is added to NiO to improve the reactivity (adding YSZ improves the porosity of the solid particles and raises the oxygen content). The mass ratio of NiO/YSZ is 3:2 when the mixture consists of only NiO and YSZ. To prevent carbon deposition on the Ni surface, 3 moles of H_2O is added for each mole of CH_4 in the

Example 11.2 (cont.)

metal oxidation reaction. The degree of reaction, X, is defined at the exit stream of each of the two reactors as follows:

$$X = \frac{m - m_{red}}{m_{ox} - m_{red}}, \tag{11.16}$$

where m_{red} is the mass of the metal when it is fully reduced and m_{ox} is its mass when it is fully oxidized. m stands for the mass of the mixture of NiO and Ni. From experiments, it is shown that $X_{ox} = 1$ at the exit of the oxidation reactor and $X_{red} = 0.3$ at the exit of the reduction reactor.

Figure E11.2

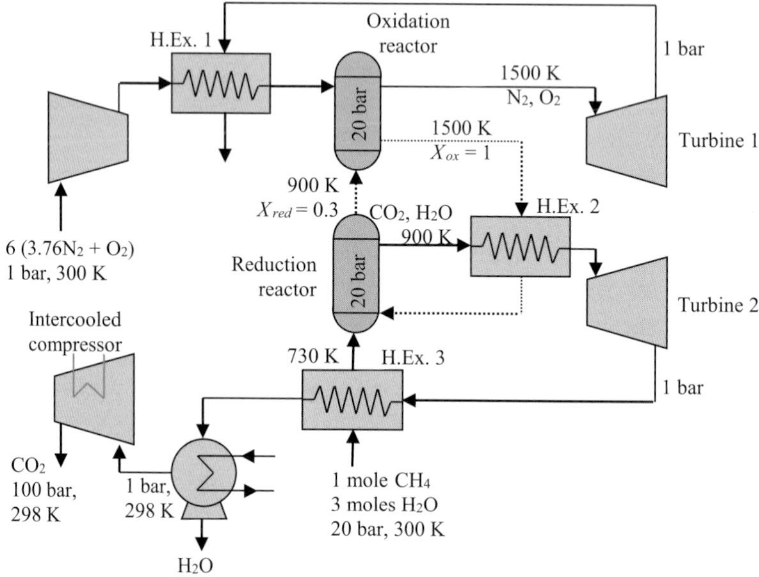

The oxidation reactor and the reduction reactor are at 1500 K and 900 K, respectively. Both reactors are at 20 bar. Only solid particles (a mixture of Ni, NiO, and YSZ) circulate through the chemical loop. Assume that YSZ is ZrO_2 with molar weight 123.2 kg/kmol. The enthalpy of reaction of metal oxidation reaction $[Ni_{(s)} + 0.5O_2 \rightarrow NiO_{(s)}]$ is $\Delta h_{Ox} = -233.11$ kJ/mol-Ni at 1500 K. The enthalpy of reaction of metal reduction reaction $[CH_4 + 4NiO_{(s)} \rightarrow CO_2 + 2H_2O + 4Ni_{(s)}]$ is $\Delta h_{Red} = 141.38$ kJ/mol-methane at 900 K. The isentropic efficiencies of the compressor and turbines are 80% and 90%, respectively. Methane is supplied at 20 bar, 300 K, and completely oxidized with NiO. The temperature of CH_4 at inlet of the reduction reactor is 730 K. For each mole of CH_4, 6 moles of air are supplied.

Example 11.2 (cont.)

Determine: (a) the composition of the chemical loop and the fuel exhaust stream at the exit of the reduction reactor; (b) the composition of the chemical loop and the air stream at the exit of the oxidation reactor; (c) the temperature of air at the inlet of the oxidation reactor; (d) the temperature of the chemical loop at the inlet of the reduction reactor; (e) the temperature at the exit of heat exchanger 2; and (f) the thermal efficiency of the plant with 100% CO_2 recovery rate, assuming a liquefaction work of 20 kJ per mole of methane.

Assume constant values for the specific heats as given in Table E11.2.

Table E11.2

Species	O_2	N_2	CO_2	H_2O	CH_4	Ni	NiO	ZrO_2
C_p (J/mol·K)	34.35	32.09	52.99	39.99	46.35	33.93	54.96	76.74

Solution

a. The chemical reaction within the reduction reactor is

$$CH_4 + 3H_2O + 4NiO_{(s)} \rightarrow CO_2 + 5H_2O + 4Ni_{(s)},$$

so, the fuel exhaust gas stream is $CO_2 + 5H_2O$.

Let's denote "a" as the mole number of NiO entering the oxidation reactor:

$$CH_4 + 3H_2O + aNiO_{(s)} \rightarrow CO_2 + 5H_2O + (a-4)NiO_{(s)} + 4Ni_{(s)},$$

$$X_{Red} = \frac{m - m_{Red}}{m_{Ox} - m_{Red}} = \frac{[(a-4)M_{\text{NiO}} + 4M_{\text{Ni}}] - aM_{\text{Ni}}}{aM_{\text{NiO}} - aM_{\text{Ni}}} = 0.3.$$

Noting that $M_{\text{NiO}} = 74.7$ kg/kmol and $M_{\text{Ni}} = 58.7$ kg/kmol, and solving for a, we find that $a = 5.71$. In other words, 70% of NiO entering the reduction reactor is reacted with 1 mole of CH_4. So, 5.71 moles of NiO enters the reduction reactor and 1.71 moles of NiO leaves unreacted.

The mass ratio of NiO to YSZ is 3:2 when $X = 1$.

$$\frac{(nM)_{\text{NiO}}}{(nM)_{\text{YSZ}}} = \frac{3}{2} \Rightarrow \frac{n_{\text{YSZ}}}{n_{\text{NiO}}} = \frac{2M_{\text{NiO}}}{3M_{\text{YSZ}}} = 0.4.$$

The mole number of ZrO_2 is $0.4 \times 5.71 = 2.284$. So, the chemical loop composition is $1.71NiO_{(s)} + 4Ni_{(s)} + 2.284ZrO_{2(s)}$.

b. The composition of the chemical loop into the reduction is NiO because $X_{ox} = 1$. All Ni is oxidized to NiO in the oxidation reactor: $5.71NiO_{(s)} + 2.284ZrO_{2(s)}$.

For the air stream, $4Ni_{(s)} + 2O_2 \rightarrow 4NiO_{(s)}$.

Two moles of O_2 are used to oxidize $Ni_{(s)}$. The composition of the exit air stream is $4O_2 + 22.56N_2$.

Example 11.2 (cont.)

c. Applying the First Law to the oxidation reactor, we have

$$\left(6h_{O_2}^{T}+22.56h_{N_2}^{T}\right)+\left(1.71h_{NiO}^{900K}+4h_{Ni}^{900K}+2.284h_{ZrO_2}^{900K}\right)$$
$$=\left(4h_{O_2}^{1500K}+22.56h_{N_2}^{1500K}\right)+\left(5.71h_{NiO}^{1500K}+2.284h_{ZrO_2}^{1500K}\right),$$

$$4\Delta h_{Ox}^{1500K}=\left(6c_{p,O_2}+22.56c_{p,N_2}\right)(1500-T)$$
$$+\left(1.71c_{p,NiO}+4c_{p,Ni}+2.284c_{p,ZrO_2}\right)(1500-900).$$

The exothermic enthalpy of the reaction is used to heat up the reactants to the oxidation reactor temperature, 1500 K. A solution of the above equation yields $T=749$ K.

d. Applying the First Law to the reduction reactor, we have

$$\left(h_{CH_4}^{730K}+3_{H_2O}^{730K}\right)+\left(5.71h_{NiO}^{T}+2.284h_{ZrO_2}^{T}\right)=\left(1.71h_{NiO}^{900K}+4h_{Ni}^{900K}+2.284h_{ZrO_2}^{900K}\right)$$
$$+\left(h_{CO_2}^{900K}+5_{H_2O}^{900K}\right)\Rightarrow$$

$$\Delta h_{Red}^{900K}+\left(c_{p,CH_4}+3c_{p,H_2O}\right)(900-730)=\left(5.71c_{p,NiO}+2.284c_{p,ZrO_2}\right)(T-900).$$

The inlet stream is cooled down to 900 K by the endothermic reduction reaction and by heating up the fuel stream to 900 K. Solving the above equation, we find $T=1246.6$ K.

e. Applying the First Law to heat exchanger 2 gives

$$\left(c_{p,CO_2}+5c_{p,H_2O}\right)(T-900)=\left(5.71c_{p,NiO}+2.284c_{p,ZrO_2}\right)(1500-1246.6).$$

Solving for the exit temperature yields $T=1390$ K.

f. The work production of the turbines is obtained from

$W_{turbine}=n_{moles}\eta_{turbine}c_pT_{in}\left(1-\gamma^{-\frac{R}{c_p}}\right)$, where γ is the pressure ratio.

For turbine 1, we have $n_{moles}=26.56$, $c_{p,ave}=32.43$ J/mol·K, $\gamma=20$. Hence, $W_{turbinerl}=623{,}393$ J.

For turbine 2, we have $n_{moles}=6$, $c_{p,ave}=42.15$ J/mol·K, $\gamma=20$. Hence,

$$W_{turbiner2}=141{,}134 \text{ J}.$$

The work required by the compressor is determined as follows:

$$W_{compressor}=\frac{n_{moles}c_pT_{in}}{\eta_{compressor}}\left(\gamma^{\frac{R}{c_p}}-1\right).$$

With $n_{moles}=28.56$, $c_{p,ave}=32.56$ J/mol·K, $\gamma=20$, and $T=300$ K, we get

$$W_{compressor}=400{,}533 \text{ J}.$$

Example 11.2 (cont.)

The thermal efficiency of the power plant is now obtained as follows.

$$\eta_{th} = \frac{W_{turbine1} + W_{turbine2} - W_{compressor} - W_{liquefaction}}{LHV_{CH_4}}$$
$$= \frac{623{,}393 + 141{,}134 - 400{,}533 - 20{,}000}{16 \times 50{,}050} = 0.43.$$

11.6.2 Impact of the Reactor Temperature Ratio

An interesting consideration in CLC technology is the impact of the reactors' temperatures, in particular whether both reactors should operate at the same temperature or at different temperatures. Heat transfer between the hotter oxidation reactor, always operating exothermally with large negative enthalpy of reaction, and the reduction reactor, which operates either exothermically with small negative enthalpy of reaction or endothermically, can be detrimental to the cycle efficiency if the temperature difference between the two reactors is large. The following analysis was developed to guide the process of choosing the temperatures using progressively higher fidelity thermodynamics models [18].

In an ideal CLC system, processes are assumed to be reversible in order to minimize entropy generation associated with heat transfer and chemical reactions, and hence reactions proceed isothermally at their equilibrium temperature while exchanging heat using a reversible engine. The equilibrium temperature is determined by [19,20]

$$\Delta G_{rxn} = 0 = \Delta H_{rxn} - T_{eq}\Delta S_{rxn}, \tag{11.17}$$

where ΔG_{rxn} is the reaction Gibbs free energy, ΔH_{rxn} is the reaction enthalpy, ΔS_{rxn} is the reaction entropy, and T_{eq} is the reaction equilibrium temperature. In CLC, the oxidation reaction is exothermic ($\Delta H_{ox} < 0$), while the reduction reaction is typically endothermic ($\Delta H_{red} > 0$), or weakly exothermic, depending on the OC and the fuel, with heat from the oxidation reactor used to support the reduction reactor. Moreover, the equilibrium temperatures of the two reactions are such that $T_{ox} \gg T_{re}$. Table 11.3 shows the equilibrium temperatures for a number of redox reactions.

Using a reversible engine interacting between the two reactors and the environment, as shown in Figure 11. 17, the maximum work can be obtained using the First and Second Laws [13]; $-W_{MAX} = (Q_{ox} + Q_{red}) + Q_0$ and $\left(\frac{Q_{ox}}{T_{ox}} + \frac{Q_{red}}{T_{red}}\right) + \frac{Q_0}{T_0} = 0$. Eliminating Q_0, we get an expression for the maximum work $-W_{MAX} = Q\left(1 - \left(\frac{T_0}{T_{ox}}\right)\right) - Q_{red}\left(\frac{T_0}{T_{ox}}\left(\frac{T_{ox} - T_{red}}{T_{red}}\right)\right)$, where $Q = Q_{ox} + Q_{red}$, T_{ox} is the oxidation reaction temperature, T_{red} is the reduction reaction temperature, T_0 is the environment temperature, $-W_{MAX}$ is the net work output of

Table 11.3 Oxygen carriers for CLC, the oxidation and reduction (using methane or hydrogen) reactions, their enthalpy of reaction in MJ/kmol, and the equilibrium temperatures of the two reactions in K [18]

		Oxidation			Reduction reaction		
OC	$T_{melting}$	Reaction	ΔH_{ox}	T_{eq}	Reaction	T_{eq}	ΔH_{red}
Ni/NiO	1728	$O_2+2Ni=2NiO$	−479	2542	$CH_4+4NiO=CO_2+2H_2O+4Ni$	420	156
					$H_2+NiO=Ni+H_2O$	NA	−2
Cu/CuO	1358	$O_2+2Cu=2CuO$	−312	1676	$CH_4+4CuO=CO_2+2H_2O+4Cu$	NA	−179
					$H_2+CuO=Cu+H_2O$	NA	−86
Fe_2O_3/ Fe_3O_4	1811	$O_2+4Fe_3O_4=6Fe_2O_3$	−464	1751	$CH_4+12Fe_2O_3=$ $CO_2+2H_2O+8Fe_3O_4$	241	126
					$H_2+3Fe_2O_3=2Fe_3O_4+H_2O$	NA	−10
Mn_2O_3/ Mn_3O_4	1161	$O_2+4Mn_3O_4=6Mn_2O_3$	−190	1153	$CH_4+12Mn_2O_3=$ $CO_2+2H_2O+8Mn_3O_4$	NA	−422
					$H_2+3Mn_2O_3=2Mn_3O_4+H_2O$	NA	−147

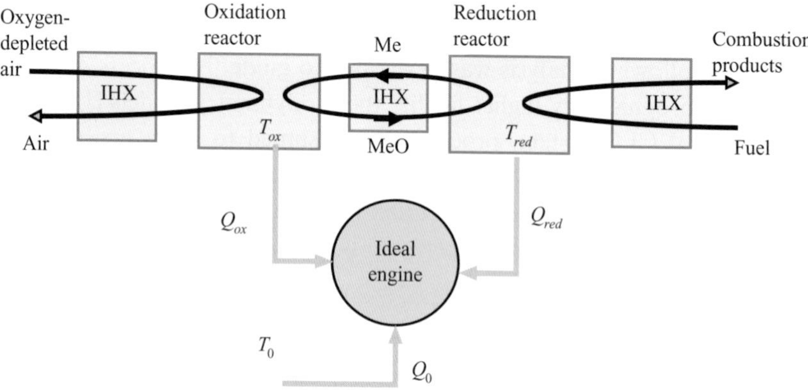

Figure 11.17 Idealized CLC system with two reactors and an engine interacting with the environment, used to develop expressions for the maximum work from this system and the impact of the reactor temperatures on the maximum work. The OC circulates between the two reactors and exchanges the corresponding thermal energy. Ideal heat exchangers are used to recover the thermal energy in the different gas streams and between the two reactors.

the system and the net reaction enthalpy, ΔH, is the sum of the oxidation and the reduction reaction enthalpies, given by $\Delta H = \Delta H_{ox} + \Delta H_{red}$, where $Q_{ox} = -\Delta H_{ox}$ and $Q_{red} = -\Delta H_{red}$.

For an exothermic reduction reaction, $Q_{red} = |\Delta H_{red}|$ and $-W_{MAX} = |\Delta H|\left(1 - \left(\frac{T_0}{T_{ox}}\right)\right) - \left(|\Delta H_{red}|\frac{T_0}{T_{red}}\right)\left(1 - \frac{T_{red}}{T_{ox}}\right)$.

For an endothermic reduction reaction, $Q_{red} = - \mid \Delta H_{red} \mid$ and $-W_{MAX} = |\Delta H|\left(1 - \left(\frac{T_0}{T_{ox}}\right)\right) + \left(|\Delta H_{red}| \frac{T_0}{T_{red}}\right)\left(1 - \frac{T_{red}}{T_{ox}}\right)$. Where ΔH_{red} is the reduction reaction enthalpy, ΔH_{ox} is the oxidation reaction enthalpy. In both expressions, the first term on the right-hand side is the work of a Carnot engine operating between the oxidation reactor and the environment, and supplied with the total enthalpy of reaction at that high temperature. The second term is an additional availability that depends on $(T_{ox} - T_{red})$ [20,21]. While T_{ox} is determined by the equilibrium of the oxidation reaction, T_{red} can take values between T_{ox} and its own equilibrium value (because of heat transfer between the two reactors).

For an exothermic reduction reaction, $T_{red} \leq 0$, the maximizing work output corresponds to $T_{ox} = T_{red}$ – that is, the thermally balanced CLC system is given by

$$-W_{MAX} = |\Delta H|\left(1 - \left(\frac{T_0}{T_{ox}}\right)\right). \tag{11.18}$$

For an endothermic reduction reaction, $T_{red} > 0$, and extra work can be obtained by operating an engine between the two reactors with the heat input equal to the (small) enthalpy of the reduction reaction discounted by T_0/T_{red}. This is difficult to perform in practice, although a solution was suggested in Reference [20]. In the absence of such an arrangement, it is best to transfer the heat between the two reactors (using, e.g., the circulating OC). To minimize the loss due to heat transfer, it is best to keep the two reactors at the same temperature (i.e., thermally balanced reactor configuration $T_{ox} = T_{red}$). Under these conditions, the maximum work is the same in the endothermic reduction case.

A similar analysis can be applied to a gas turbine Brayton cycle with CLC reactors to demonstrate the applicability of this result to more complex designs. A schematic of the regenerative CLC cycle is shown in Figure 11.18. The corresponding *T–s* diagram for an ideal regenerative cycle was shown in Chapter 5. Two cycles are constructed on the fuel and the air sides (denoted by the subscripts 'f' and 'a' respectively). The CLC reactor in the figure comprises both the oxidation and the reduction reactors. Since the control volume is placed around the CLC reactor, it only captures the net heat release, represented by Q. The air and

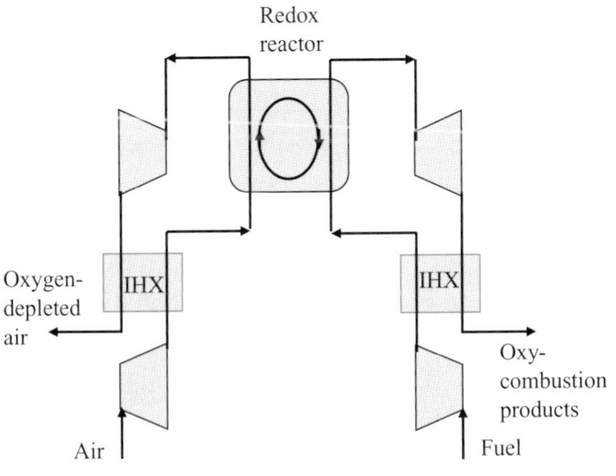

Figure 11.18 An ideal regenerative gas turbine Brayton CLC cycle. The two reactors are represented by a single two-sided reactor. The air and fuel sides are interacting with their own power islands [18].

fuel side pressure ratios are equal and the turbines and compressors are isentropic. In this analysis the work W and heat Q are defined as positive into the control volume, the fuel flow rate m_f is fixed, and therefore the net heat release Q in the reactor is constant, the air side reactor exhaust temperature is fixed, and the air flow rate m_a is varied to control the fuel side reactor exhaust temperature.

Applying the First and Second Laws on either side and assuming ideal regeneration, we get

$$W_{Net_a} = m_a c_{pa} T_0 \left(\pi^{\alpha} - \frac{T_{ox}}{T_0} \right) (1 - \pi^{-\alpha})$$

and

$$W_{Net_f} = m_f c_{pf} T_0 \left(\pi^{\alpha} - \frac{T_{red}}{T_0} \right) (1 - \pi^{-\alpha}).$$

The energy balance of the reactor gives: $Q = -\Delta H = \left(m_f c_{pf} T_{red} + m_a c_{pa} T_{ox} \right) (1 - \pi^{-\alpha})$, where π is the pressure ratio. Combining these equations, the cycle efficiency is

$$\eta = 1 - \Psi_1 + \Psi_2 \left(\frac{T_{red}}{T_{ox}} \right),$$

where

$$\Psi_1 = \frac{\left(\left(\frac{Q}{(1-\pi^{-\alpha}) c_{pa} T_{ox}} \right) c_{pa} \pi^{\alpha} \right) + \left(m_f c_{pf} \pi^{\alpha} \right)}{\left(\left(\frac{Q}{(1-\pi^{-\alpha}) c_{pa} T_{ox}} \right) c_{pa} \frac{T_{ox}}{T_0} \right)} > 0, \text{ and } \Psi_2 = \frac{\left(\left(\frac{m_f c_{pf}}{c_{pa}} \right) \pi^{\alpha} \right)}{\left(\left(\frac{Q}{(1-\pi^{-\alpha}) c_{pa} T_{ox}} \right) \frac{T_{ox}}{T_0} \right)} > 0,$$

showing that the efficiency is positively correlated with the reduction/oxidation reactor temperature ratio. For either exothermic or endothermic reduction, the maximum value is $T_{red} = T_{ox}$, and

$$\eta_{thermal} = 1 - \pi^{\alpha} / (T_{ox} / T_o). \tag{11.19}$$

This is the same expression for a conventional (ideal) regenerative Brayton cycle operating over the same temperature range. The benefit of using the CLC system in this case is the production of a separate stream of CO_2.

Both the Carnot cycle and Brayton cycle models arrive at the same result, that isothermal operation in which both reactors operate at the same temperature is needed to achieve maximum efficiency. However, both were based on significant idealizations. A high-fidelity Aspen© model for a recuperative gas turbine cycle and a CC, as shown in Figure 11.19 as a CLC regenerative cycle, was used to examine the same. The operating conditions and other specifications are shown in Table 11.4. Air is compressed (in a multistage process with intercooling), preheated in the regenerator, and sent to the oxidation reactor side. The oxidation reactor exit stream is a mixture of oxygen-depleted

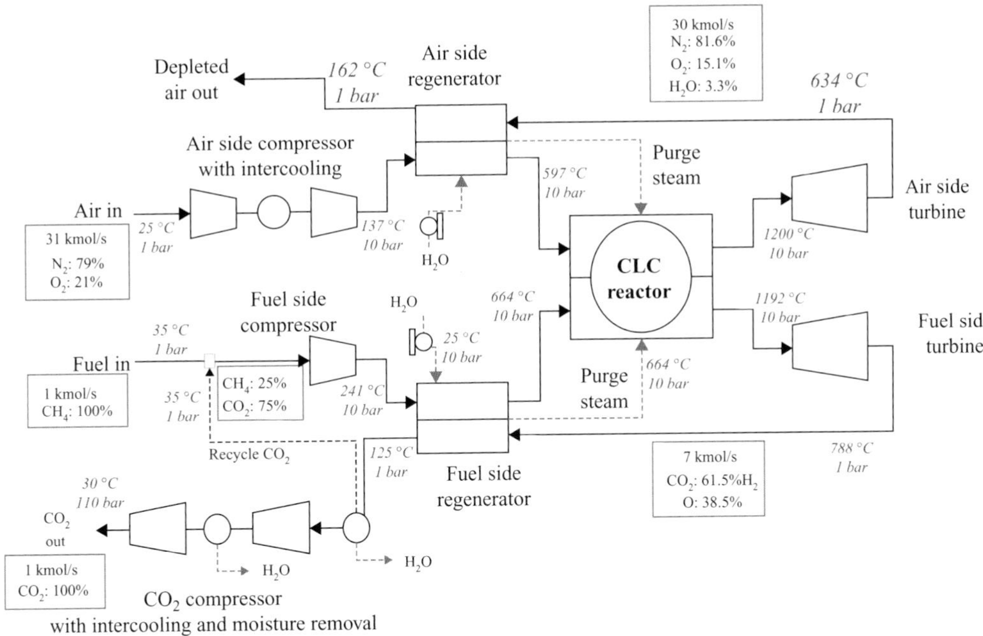

Figure 11.19 Schematic of the Aspen© model for the regenerative CLC cycle (with base case temperature/pressure/flow/composition data). The CLC reactor is shown as one block with two sides, oxidation on the top and reduction on the bottom [18].

Table 11.4 The base case model of the regenerative CLC Brayton cycle. Oxygen carrier is Ni/NiO supported on boron nitride. Air flow is varied to control T_{ox}. Carrier circulation is varied to control T_{red} [18]

Oxidation reactor temperature	1200 °C
Ambient temperature	25 °C
Reactor/operating pressure	10 bar
Ambient pressure	1 bar
Compressor isentropic efficiency	90%
Turbine isentropic efficiency	90%
Sequestration CO_2 compression pressure	110 bar
Regenerative minimum pinch	25
Fuel	Methane
Inlet fuel flow rate	1 kmol/s
Lower heating value (LHV) fuel	50 MJ/kg
Inlet air N_2 mole fraction	0.79
Inlet air O_2 mole fraction	0.21
Recycled CO_2/CH_4 in inlet stream	3:1
OC (Ni + NiO) / fuel (CH_4) mole ratio	6:1
Fuel side purge steam	0.5 kmol/s
Air side purge steam	1 kmol/s

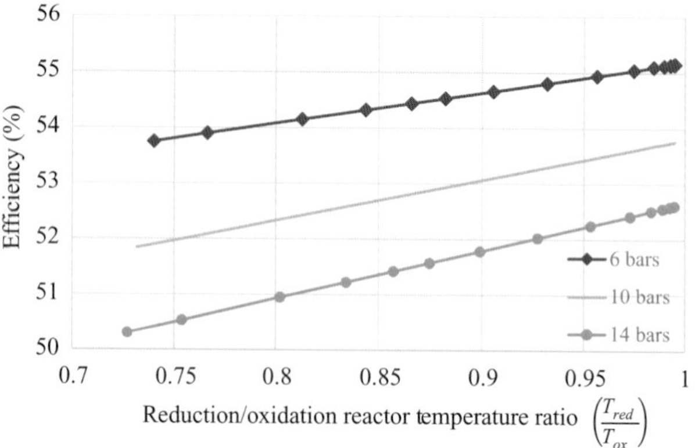

Figure 11.20 Regenerative CLC cycle efficiency is linear in the reduction/oxidation reactor temperature ratio T_{red}/T_{ox}. Maximum efficiency corresponds to thermally balanced reactor operation [18].

air and "purge" steam used to remove any left-over air from the rotary reactor; it expands in the air turbine and is used for heat recovery in the air side regenerator. The reduction reactor exit stream contains the combustion products (CO_2 and H_2O) and "purge" steam used to remove left-over fuel from the rotary reactor; it expands in the "fuel turbine" and is then used in the regenerator. Some of the CO_2 from the cool regenerator exhaust stream is recycled to the reduction reactor, where it serves as carrier gas/diluent for the fuel. The rest of the CO_2 is compressed up to 110 bar, where staged compression with intercooling enables the condensation and removal of water vapor. In the case of the rotary CLC reactor[1] described later in this chapter, the purge steam in the two sides of the CLC reactor is used to clear the channels off of one stream (air or fuel) before the next stream (fuel or air) is introduced. Thus, the turbines resemble humid air or humid CO_2 turbines.

The model was used to compute the efficiency of the CLC cycle for different reduction reactor and oxidation reactor temperatures. Results show ~54% efficiency for the thermally balanced reactor compared to 52% for the imbalanced design (with different air and fuel reactor temperatures). The latter case is characterized by a higher reduction reactor temperature, which reduces the air flow required for temperature regulation, and increases the fuel side turbine output, while the lower air flow rate reduces both the turbine output and the compressor power requirement on the air side, such that the overall effect is a smaller net reduction in air side work output. A close examination of the component contributions to the net system work output shows that the increase in fuel side turbine output is larger than the corresponding decrease on the air side, and hence the net effect of thermally balanced reactor operation is an increase in system efficiency. Figure 11.20 shows the dependence of the cycle efficiency on T_{red}/T_{ox}. The efficiency is almost a linear function of the temperature

[1] The rotary reactor is a monolith that consists of many microchannels covered with the OC. As the reactor spins between the air stream and fuel stream, the OC on the walls of the channels are oxidized and reduced periodically, thus performing the functions of the oxidation reactor and reduction reactor. The design will be discussed more later.

ratio, with a pressure ratio-dependent slope, similar to the analytical result of the ideal regenerative CLC cycle model.

Cycle efficiencies reported in the literature for reactor configurations with different degrees of thermal imbalance (T_{red} ranging from 800 °C to 1100 °C), using methane and assuming complete CO_2 separation, range from 47% to 53.5% for CLC-CC systems. Efficiencies up to 53% for nickel-based humid air CLC cycles were reported (with the CO_2 compression work subtracted from the net work of the cycle) [14]. Including a solid-to-gas heat exchanger between the oxidation and reduction reactors to raise the temperature of the fuel reactor exhaust stream and minimize reactor exergy loss was shown to raise the efficiency to 54% [19]. In the absence of internal thermal coupling, installing a heat exchanger between the two reactors is a good option for improving system availability, though implementing it currently remains technically challenging.

11.6.3 CLC Reactor Designs

A CLC reactor system is built around interconnected fuel and air reactors. Several options have been proposed for the design of these reactors, including interconnected fluidized-bed reactors, moving-bed reactors, packed-bed reactors, and rotating-bed reactors. Two are shown in Figure 11.21 and briefly described next [12].

11.6.3.1 Interconnected Fluidized-Bed Reactors

This reactor set, shown in Figure 11.21a, is the most commonly proposed system. It consists of an air reactor, a fuel reactor, and a cyclone. The air reactor is a fast fluidized bed with concurrent pneumatic transport (for more on fluidized-bed reactors, see Chapter 12 on coal utilization and Chapter 14 on biomass energy). The top part of the air reactor operates as a riser where the oxidized particles are entrained out. The OC is subsequently separated from the carrying flue gases in the cyclone, the remaining gas leaving the cyclone as oxygen-depleted air. The particles fall from the cyclone into the fuel reactor, which is fluidized with a gaseous fuel injected from the bottom. The fuel reactor is a bubbling fluidized bed or a turbulent fluidized bed. The reduced OC particles are recirculated to the air reactor by gravity through a loop-seal for regeneration. Particle loop-seals located between the air and fuel reactors are used to support the pressure difference between the two reactors, and avoid mixing of the flue gases in their respective exhausts.

11.6.3.2 Packed-Bed Reactor

A packed-bed reactor system was described in Chapter 11. Figure 11.21b shows this system as applied to CLC. The OC particles are packed into the reactor and are alternately exposed to reducing and oxidizing conditions via periodic switching of the gas feed streams. Two reactors working in parallel are used alternately to ensure a continuous high-temperature gas stream supply to the downstream gas turbine. The process consists of

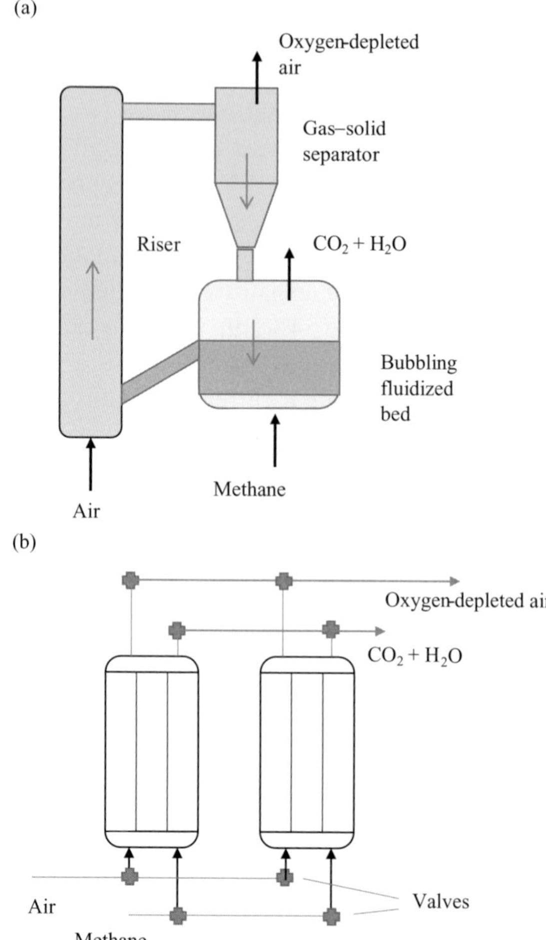

Figure 11.21 Some reactor systems for CLC: (a) interconnected fluidized-bed reactors in which the tall riser is used to oxidize the metal particles in an air stream, and the shorter bubbling fluidized bed is used to reduce the particles in methane; and (b) packed-bed reactors in which each packed metal bed is periodically exposed to air and methane for oxidation and reduction.

alternate oxidation and reduction cycles in two separate reactors, intermittently alternated with short periods of mild fluidization of the bed after each cycle to level off temperature and concentration profiles.

The primary advantages of packed-bed reactors are that the separation of the gas and the particles is intrinsically avoided, and the reactor design is much more compact. Furthermore, since the operation is in stationary beds, no extra energy is needed for circulation. However, the packed particles can be subjected to a large temperature fluctuation within the reactor. Other disadvantages include the fuel slip during the switching period and the necessity to use a high temperature and a high-flow gas switching system.

An alternative design that replaces a two-reactor system with circulating OC is the rotary reactor, in which micro-channels within a rotating wheel are exposed alternately to fuel and air streams [22]. As shown in Figure 11.22, this is a rotating wheel made of a monolith and communicating with two stationary channels at the top and four channels at the bottom

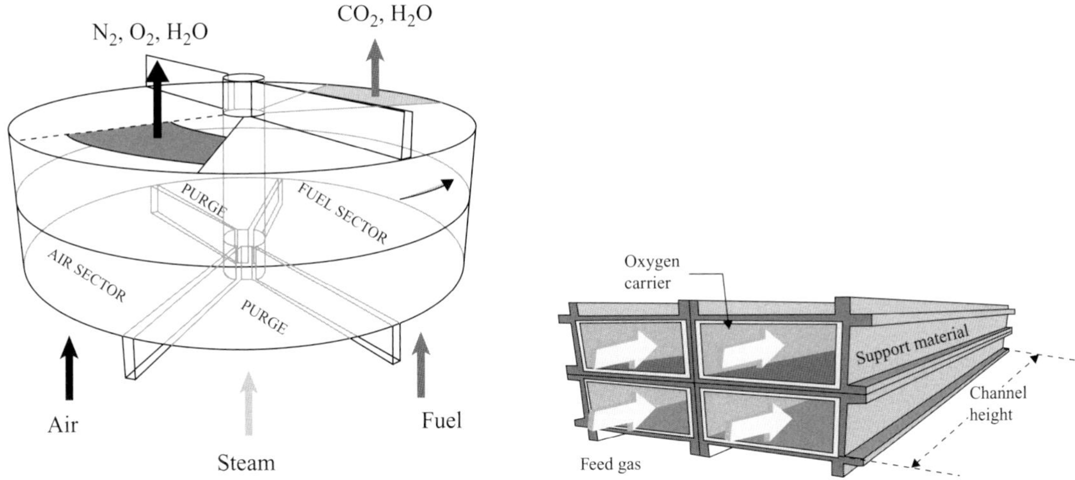

Figure 11.22 Rotary reactor for CLC and the micro-channel system [12].

side, with gas flow from bottom to top. The four feeding stationary channels flow air, purge steam, fuel, and purge steam from the bottom side of the wheel. The two receiving channels at the top side receive combustion products mixed with steam and oxygen-depleted air mixed with steam. The rotating wheel is a matrix/monolith that consists of many micro-channels oriented along its axis, with the OC coated onto their porous wall. The channel wall has two solid layers: the highly porous OC layer that hosts the OC, and a bulk dense ceramic layer with high thermal inertia and thermal conductivity. The OC reacts with the fuel and air alternately as the wheel rotates, exposing the micro-channels to the four streams alternately. The inert ceramic substrate helps maintain the pore structures and chemical reactivity of the OC. Two purging sectors are used between the fuel and the air sectors to sweep the residual gases out of the reactor. The high thermal conductivity redistributes the heat generated during oxidation across the wheel. This design avoids the circulation of the OC and related challenges.

11.7 Membrane-Based Oxy-Combustion Cycles

Membranes for air separation were described in Chapter 10. These mixed ionic electronic membranes operate at moderately high temperatures, ~900 °C, and can be integrated into power cycles. The oxygen flux is O(1 μmol/cm^2·s) and is driven by the oxygen partial pressure difference across the membrane. Air is flown on the feed side, and oxygen is collected on the sweep side. Depending on the configuration, air is compressed on the feed side, or vacuum is applied to the sweep side to enhance the oxygen permeation flux. Otherwise, combustion products are used to sweep the oxygen and transport it to the combustor. In this case the membrane unit is an ASU connected to a combustor (on the

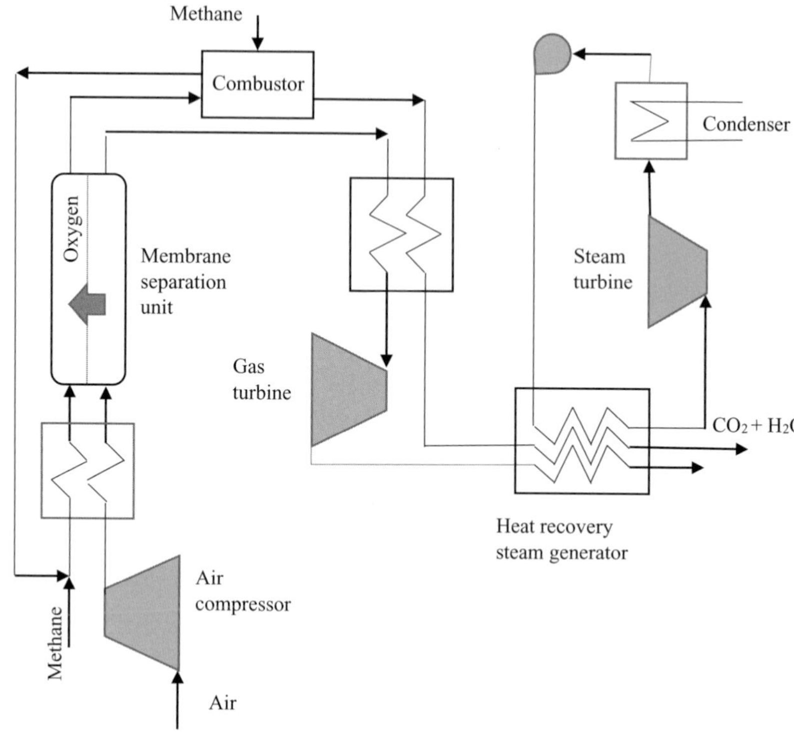

Figure 11.23 Oxy-combustion cycle in which a membrane ASU is used for air separation and possibly partial oxidation of methane. The sweep gases in the membrane unit are recycled combustion products, which can be mixed with a fraction of methane. More methane is burned in the combustor. Combustion products heat up oxygen-depleted air, which expands in the gas turbine. Following expansion, the oxygen-depleted air and the products of combustion are used in the HRSG.

sweep side) and a heat exchanger (on the feed side), as shown in Figure 11.23. The hot products from the combustor are sent to the sweep side of the membrane unit and to a heat exchanger to heat up the feed air stream. Oxygen-carrying products are sent to the combustor, where methane is injected. The products leaving the heat exchanger are sent to the HRSG. High-pressure, hot, oxygen-depleted air expands in the gas turbine, and is then sent to the HRSG.

Reactive operation is also possible, in which fuel is added at the sweep side to further reduce the oxygen partial pressure there. In this case, the membrane unit is actually a reactor/combustor. Depending on the membrane material, the maximum temperature inside the membrane unit is limited to temperatures typically lower than allowed by high-efficiency gas turbines, and post-firing may be necessary to boost the efficiency at the expense of reducing the capture ratio.

11.8 Pre-Combustion Capture

In pre-combustion capture, the hydrocarbon, in this case natural gas or methane, is first reformed into a mixture of H_2 and CO_2. The CO_2 is separated and the H_2 is burned in a CC. In the first step in reforming, partial oxidation (which is exothermic) or steam reforming (highly endothermic) converts methane into CO and H_2. This is followed by a WGS reaction

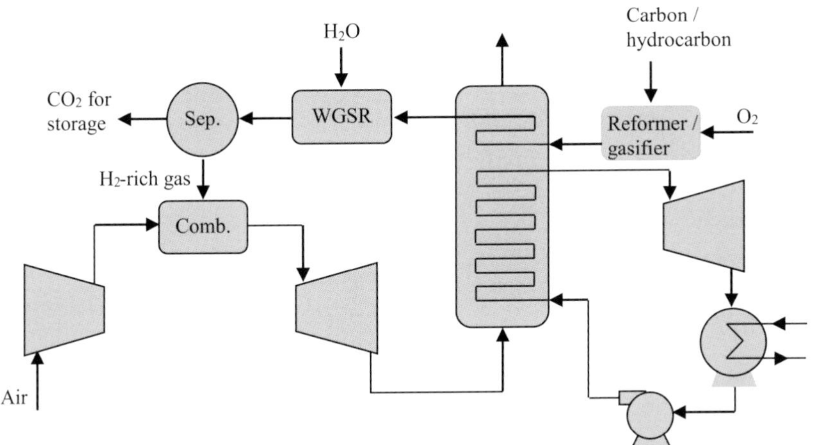

Figure 11.24 Layout for pre-combustion capture. Natural gas is first reformed then shifted, then CO_2 separated from H_2, with the latter burned in air. Shift reactors use extra steam. In this implementation, an ASU is needed for the partial oxidation of natural gas.

to convert the CO into CO_2 while forming more H_2. After separating CO_2, hydrogen is burned in air in the gas turbine combustor. Separation can be performed using several techniques, including physical absorption or membranes; some were described in Chapter 10. A schematic diagram for the integration of methane steam reforming, CO_2 separation, and the power island in shown in Figure 11.24.

In direct reforming, partial oxidation,

$$CH_4 + H_2O_{(g)} \rightarrow CO + 3H_2 \qquad \Delta H_r = 206.28\,\text{kJ/mol}, \tag{11.20}$$

is followed by the WGS, $CH_4 + 2H_2O_{(g)} \rightarrow CO_2 + 4H_2$, with $\Delta H_r = 139\,\text{kJ/mol}$. The process is performed in multiple reactors, at different temperatures, with cooling and possibly with intermediate separation of CO_2 between the reactors. The first stage, partial oxidation, takes place around 700–1000 °C. The WGS takes place in two reactors; the high-temperature shift reactor operates at 300–500 °C, with nickel–chromium oxide catalysts, followed by the low-temperature shift reactor operating at 180–270 °C to achieve near-complete conversion (according to equilibrium) with copper and zirconium oxide catalyst. Direct reforming can be done in oxygen or in air. In the first case, an ASU is added, while in the second case, CO_2 is separated from a mixture on $H_2 + N_2$ (see Chapter 3 for more detail).

The thermal energy removed between the different reforming stages can be used in other parts of the cycle, such as: to reheat the same stream after separation (mostly the hydrogen); to heat air going into the hydrogen combustor; or in the HRSG of the CC. Clearly, "heat integration" among components is important in order to reduce the efficiency penalty associated with cooling the gases between the reforming reactors.

Another penalty in pre-combustion capture cycles is in the separation of CO_2 following steam reforming. Depending on the separation method used, the gas may have to be cooled further to a much lower temperature prior to separation (e.g., to 40 °C for chemical absorption processes). The "low-temperature heat" removed during this process is unlikely to be economically useful in other parts of the cycle, and hence represents an energy loss (note that adding heat exchangers for heat transfer to other components adds to the capital cost of the plant). Higher-temperature gas separation processes, that is, separating H_2 from a $H_2 + CO_2$ mixture without cooling, would raise the overall efficiency of the integrated cycle since this loss would be avoided, but the penalty in this case is in the form of pressure drop across the membrane. See Chapter 10 for hydrogen separation membrane development at relatively higher temperatures.

In steam reforming, the gas turbine exhaust can be used to provide the necessary heat for this endothermic reaction, hence possibly reducing some of the efficiency penalty associated with the ASU or with having to heat up the extra nitrogen if air is used in the partial oxidation system. To ensure near-complete conversion, extra water should be available, which contributes to some extra energy losses (enthalpy of vaporization of H_2O). Nearly twice the stoichiometric requirement for water is often needed to complete methane reforming. Methane steam reforming is done in Cr/Ni catalytic tubular fixed reactor.

Example 11.3

A gas turbine power plant operates on a pre-combustion CO_2 capture cycle, as shown in the Figure E11.3. Air at 298 K and 1 bar is used for reforming methane (partial oxidation), and to supply the gas turbine cycle compressor. The reformer operates at 1073 K. The reformate mixture (made of only N_2, CO, and H_2) passes through a heat exchanger, where it preheats the compressed air (at pressure 10 bar), so its temperature reduces to 673 K before entering a shift reactor. In that reactor, superheated steam at 573 K generated in the HRSG reacts with the reformate to produce additional hydrogen (and form CO_2). The products of the shift reactor (N_2, H_2, and CO_2) leave at 308 K, and enter a membrane separation unit where CO_2 is separated. The unit is composed of a compressor that raises the pressure of the gaseous mixture on the feed side to 10 bar. The retentate on the feed side, a mixture of hydrogen and nitrogen, leaves the separation unit at 10 bar and flows into the combustor, while the permeate, a stream of CO_2, leaves at 1 bar. The temperature of the combustion products is 1473 K. The isentropic efficiencies of all the compressors and the gas turbine are 75% and 88%, respectively. Gases leave the gas turbine at 1 bar, entering the HRSG to raise steam (for the shift reactor), the shift reactor, then the reformer. Determine the thermal efficiency of the cycle, and the outlet temperature of the turbine exhaust gas.

Use the following data in the analysis:

$$\hat{h}^o_{f,H_2O(g)} = -242 \text{ MJ/kmol}, \hat{h}^o_{f,H_2O(l)} = -286 \text{ MJ/kmol}, \hat{h}^o_{f,CO} = -110.6 \text{ MJ/kmol}.$$

$$\hat{h}^o_{f,CO_2} = -393.9 \text{ MJ/kmol}, \hat{h}^o_{f,CH_4} = -74.9 \text{ MJ/kmol}.$$

$$\hat{c}_{p,O_2} = 33.4\,\text{kJ/mol.K},\ \hat{c}_{p,N_2} = 31.3\,\text{kJ/mol·K},\ \hat{c}_{p,CO_2} = 50.6\,\text{kJ/mol·K},$$

$$\hat{c}_{p,H_2} = 30.0\,\text{kJ/mol.K},\ \hat{c}_{p,CO} = 29.3\,\text{kJ/mol·K},\ \hat{c}_{p,H_2O} = 38.2\,\text{kJ/mol·K},\ \hat{c}_{p,CH_4} = 46.35\,\text{kJ/mol·K}.$$

Example 11.3 (cont.)

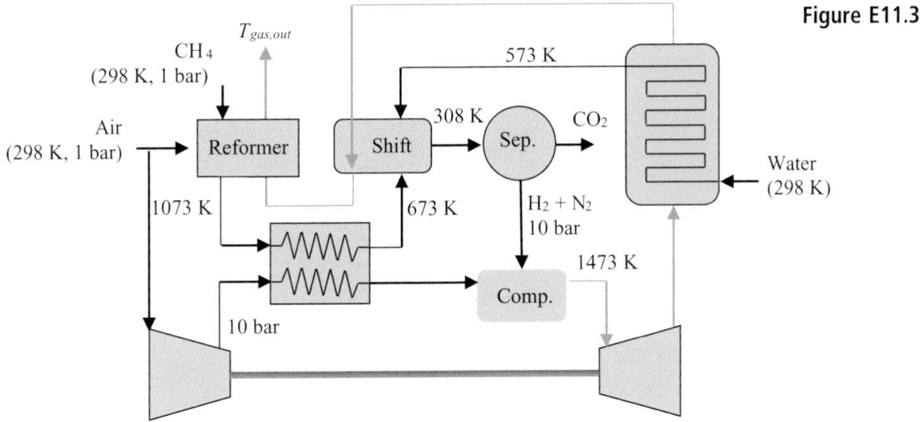

Figure E11.3

Solution

Reformer:

$$CH_4 + 0.5(O_2 + 3.76N_2) \rightarrow CO + 2H_2 + 1.88N_2$$

$$Q_R = \hat{h}_{CO}^{1073K} + 2\hat{h}_{H_2}^{1073K} + 1.88\hat{h}_{N_2}^{1073K} - \left(\hat{h}_{CH_4}^{298K} + 0.5\hat{h}_{O_2}^{298K} + 1.88\hat{h}_{N_2}^{298K}\right)$$
$$= 78.82 \text{ MJ/kmol of methane.}$$

For each kmol of methane fed to the reformer, 78,820 kJ should be supplied to reform methane with air and raise the reformate temperature (by the products stream).

Air compressor exit temperature:

$$T_{AC,o} = T_{AC,in}\left[1 + \frac{\left(\frac{p_o}{p_{in}}\right)^{\frac{k_{air}-1}{k_{air}}} - 1}{\eta_c}\right] = 298 \times \left[1 + \frac{10^{\frac{1.388-1}{1.388}} - 1}{0.75}\right] = 657.3 \text{ K.}$$

Heat exchanger, on the reformate side:

$$Q_{HEx} = \left(\hat{h}_{CO}^{1073} - \hat{h}_{CO}^{673}\right) + 2\left(\hat{h}_{H_2}^{1073} - \hat{h}_{H_2}^{673}\right) + 1.88\left(\hat{h}_{N_2}^{1073} - \hat{h}_{N_2}^{673}\right)$$

$$= 59.107 \text{ MJ/kmol of methane.}$$

Example 11.3 (cont.)

On the air side, where a is the number of moles of oxygen per mole of methane:

$$Q_{HEx} = a\left(\hat{h}_{O_2}^{T_{HEX,o}} - \hat{h}_{O_2}^{T_{AC,o}}\right) + 3.76a\left(\hat{h}_{N_2}^{T_{HEX,o}} - \hat{h}_{N_2}^{T_{AC,o}}\right).$$

Shift reactor: $CO + 2H_2 + 1.88N_2 + H_2O \rightarrow CO_2 + 4H_2 + 1.88N_2$.

The heat requirement in the shift reactor:

$$\begin{aligned} Q_S &= \hat{h}_{CO_2}^{308K} + 4\hat{h}_{H_2}^{308K} + 1.88\hat{h}_{N_2}^{308K} - \left(\hat{h}_{CO}^{673K} + 2\hat{h}_{H_2}^{673K} + 1.88\hat{h}_{N_2}^{673K} + \hat{h}_{H_2O}^{673K}\right) \\ &= -105.13\,\text{MJ/kmol of methane.} \end{aligned}$$

This heat is thus added to the products stream.

Separator compressor:

$$\hat{c}_{p,mix} = \frac{1}{6.88}\hat{c}_{p,CO_2} + \frac{4}{6.88}\hat{c}_{p,H_2} + \frac{1.88}{6.88}\hat{c}_{p,N_2},$$

$$k_{mix} = \frac{\hat{c}_{p,mix}}{\hat{c}_{p,mix} - 8.314} = 1.33,$$

$$T_{Comp,o} = T_{Comp,in}\left[1 + \frac{(p_o/p_{in})^{\frac{k_{mix}-1}{k_{mix}}} - 1}{\eta_c}\right] = 308\left[1 + \frac{10^{\frac{1.33-1}{1.33}} - 1}{0.75}\right] = 624.5\text{ K}.$$

Work requirement for the separation unit compressor:

$$\begin{aligned} W_{SC} &= \left(\hat{h}_{CO_2}^{624.5K} - \hat{h}_{CO_2}^{308}\right) + 3\left(\hat{h}_{H_2}^{624.5K} - \hat{h}_{H_2}^{308}\right) + 1.88\left(\hat{h}_{N_2}^{624.5K} - \hat{h}_{N_2}^{308}\right) \\ &= 63.005\text{ MJ/kmol of methane.} \end{aligned}$$

Combustor reaction: $3H_2 + 1.88N_2 + a(O_2 + 3.76N_2) \rightarrow 3H_2O + (a-1.5)O_2 + (1.88 + 3.76a)N_2$.

From the heat exchanger heat balance equation and the combustor heat balance equation, we get

$$\begin{aligned} &3\hat{h}_{H_2}^{624.5K} + 1.88\hat{h}_{N_2}^{624.5K} + a\left(\hat{h}_{O_2}^{T_{HEX,0}} + 3.76\hat{h}_{N_2}^{T_{HEX,0}}\right) \\ &\quad = 3\hat{h}_{H_2O}^{1473K} + (1.88 + 3.76a)\hat{h}_{N_2}^{1473K} + (a-1.5)\hat{h}_{O_2}^{1473K}. \end{aligned}$$

Solving for a yields: $a = 5.238$.

Thus, from the heat balance equation of the heat exchanger, the temperature of air at the heat exchanger exit is $T_{HEx,o} = 732.1$ K.

Gas turbine:

$$\hat{c}_{p,gas} = \frac{2}{19.62}\hat{c}_{p,H_2O} + \frac{2.517}{19.62}\hat{c}_{p,O_2} + \frac{15.1}{19.62}\hat{c}_{p,N_2} = 32.11\text{ kJ/kmol·K},$$

Example 11.3 (cont.)

$$k_{gas} = \frac{\hat{c}_{p,gas}}{\hat{c}_{p,gas} - 8.314} = 1.35.$$

Turbine exit temperature: $TOT = 1473 \times \left[1 - 0.88 \times \left(1 - 0.1^{\frac{1.35-1}{1.35}}\right)\right] = 890.3\ \text{K}.$

Work:

$$W_T = 3\left(h_{H_2O}^{1473K} - h_{H_2O}^{890.3K}\right) + 3.738\left(h_{O_2}^{1473K} - h_{O_2}^{890.3K}\right) + 21.57\left(h_{N_2}^{1473K} - h_{N_2}^{890.3K}\right)$$
$$= 530.52\ \text{MJ/kmol of methane.}$$

The net work production of the power cycle can be determined from

$$W_{net} = W_T - W_{AC} - W_{SC},$$

where

$$W_{AC} = 5.238\left(h_{O_2}^{657.3K} - h_{O_2}^{298K}\right) + 19.69\left(h_{N_2}^{657.3K} - h_{N_2}^{298K}\right) = 282.71\ \text{MJ/kmol of methane.}$$

Hence, $W_{net} = 530.52 - 282.71 - 63 = 184.81$ MJ/kmol of methane.

The thermal efficiency of the power cycle is therefore obtained as

$$\eta_{th} = \frac{W_{net}}{M_{CH_4}\text{LHV}} = \frac{184.81}{16 \times 500.50} = 23.1\%.$$

To determine the turbine exhaust temperature, we write energy balance around the HRSG and along the products stream to get the following equation:

$$3\left(h_{H_2O}^{890.3K} - h_{H_2O}^{T_{gas,out}}\right) + 3.738\left(h_{O_2}^{890.3K} - h_{O_2}^{T_{gas,out}}\right) + 21.57\left(h_{N_2}^{890.3K} - h_{N_2}^{T_{gas,out}}\right)$$
$$= Q_R + Q_S + Q_B,$$

where:

$$Q_B = \left(h_{H_2O}^{573K} - h_{H_2O}^{298K}\right) = 0.075.3(373 - 298) + 40.65 + 0.0382(573 - 273)$$
$$= 57.76\ \text{MJ/kmol of methane.}$$

Substituting into the above equation, and solving, we find: $T_{gas,out} = 856.4$ K.

The thermal efficiency is low because of CO_2 separation, the relatively low pressure ratio of the gas turbine cycle, and the high products temperature at the exit. Adding a steam cycle as a bottoming cycle can raise the efficiency significantly.

11.8.1 A Model for Natural Gas [3]

A typical layout for a pre-combustion capture NGCC is shown in Figure 11.24. Air is used for autothermal reforming, which comprises partial oxidation followed by WGS reaction to

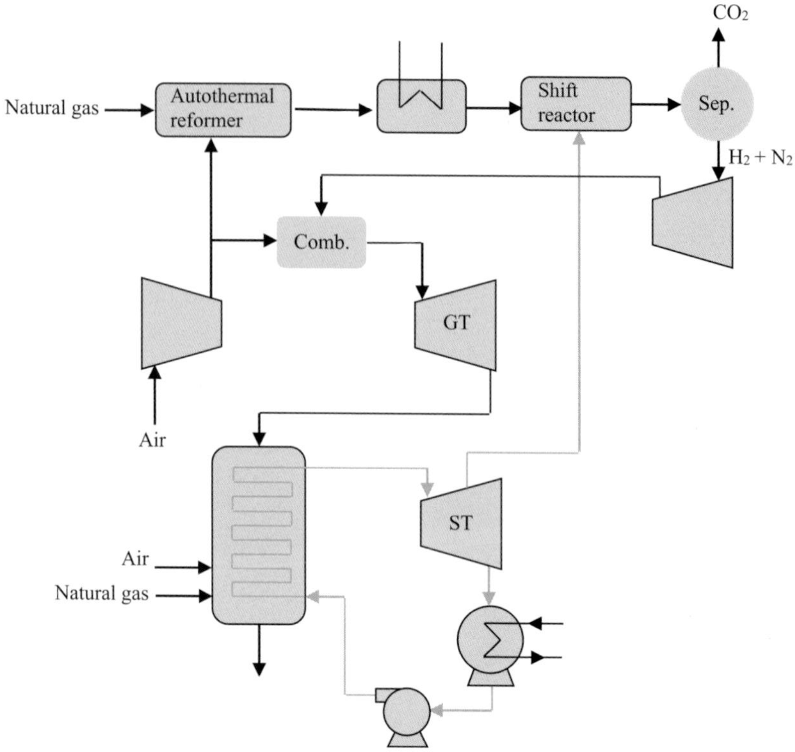

Figure 11.25 Pre-combustion CO_2 capture in natural gas-fired CC. Supplementary firing could be used in the HRSG to raise the steam temperature, making the cycle partial capture only. Steam is extracted at an intermediate pressure from the turbine for natural gas reforming and WGS.

produce a mixture of $N_2 + H_2O + CO_2 + H_2$. Next, water is condensed and a chemical absorption plant is used to remove CO_2, and the $H_2 + N_2$ fuel stream is pressurized back to the compressor pressure, and burned in air in the gas turbine combustor. Compressing the hydrogen stream is necessary since a substantial pressure drop may occur in the gas separation process. Air originally supplied by the compressor is split into two streams, with nearly 15% used to reform methane, and the rest used in the gas turbine combustor. The reformed gas before separation and the thermal energy is supplied to the HRSG. Extra fuel is burned in the HRSG to supply sufficient heat for steam generation and thermal reforming (which generates extra CO_2 that is not captured, hence reducing the effective capture ratio of the plant). The pressure and temperature in the desorption tower are sufficient to separate CO_2 as the pressure is reduced to atmospheric conditions – that is, there is no need for extra heat to separate CO_2 in the stripping process (unlike the case with CO_2 removal from the flue gases). The thermal energy in the steam supplied to the shift reactor provides the energy needed in the separation plant.

Efficiency losses in this cycle are associated with several aspects. (1) The H_2 CC, which is fed with a fuel (H_2) that has less chemical energy than the original fuel (the reforming efficiency is less than 100%, typically ~85%). (2) The need to add extra fuel in the HRSG to supply enough energy for raising steam for the steam turbine and the fuel reforming. (3) Extracting medium-pressure steam from the steam turbine for the WGS, and compressing the hydrogen fuel back to the gas turbine combustor pressure. (4) CO_2 compression to 100 bars (this penalty is lower than in the separation of CO_2 from flue gas case since CO_2 concentration is higher). There is an efficiency credit due to the thermal energy supplied from the gas cooler to the CC. Starting with the efficiency of the H_2 CC plant, the overall efficiency with capture is [3]:

$$\eta_{pre\_comb} = \tilde{\eta}_{H_2} - \Delta\eta_{steam} + \Delta\eta_{cooler} - \Delta\eta_{CO_2comp} - \Delta\eta_{H_2comp}. \qquad (11.21)$$

The energy of separation is neglected in this analysis. All the terms are evaluated per 1 mole of fuel entering the reformer. The efficiency of the hydrogen CC plant, $\tilde{\eta}_{\mathrm{H_2}}$, is based on the chemical energy of the fuel produced in the reforming step. This efficiency can be related to the efficiency of an H_2 CC plant calculated on the basis of the fuel fed to this cycle, $\eta_{\mathrm{H_2}}$, as follows. If the chemical energy of the produced hydrogen is a fraction ζ_{H_2} of the chemical energy of the original fuel, then the hydrogen cycle efficiency based on the original fuel chemical energy is $\eta_{\mathrm{H_2}}\zeta_{\mathrm{H_2}}$. Typical methane to hydrogen reforming efficiency, that is the chemical energy in the reformate as a fraction of the chemical energy in the methane, is 85–95%. This efficiency is further reduced by the chemical energy in the extra fuel used in the HRSG. If ζ_{HRSG} is extra fuel is used for each mole of fuel used in the reformer, then

$$\tilde{\eta}_{\mathrm{H_2}} = \frac{\zeta_{\mathrm{H_2}}}{1 + \zeta_{HRSG}} \eta_{\mathrm{H_2}}. \qquad (11.22)$$

The second term in the efficiency stands for the efficiency loss because of the steam removed from the steam turbine. The energy associated with the extracted steam is the number of moles of steam extracted from the turbine times its enthalpy. It is convenient to express the number of moles of steam extracted in terms of the number of moles of carbon/mole of fuel, $n_{\mathrm{C},fuel}$, times the number of moles of H_2O/mole of carbon, $n_{\mathrm{H_2O,C}}$. The number of moles of steam required for reforming, $n_{\mathrm{H_2O,C}}$, depends on the number of carbon atoms in the fuel. Since most of the steam is used for the WGS, the smallest number for $n_{\mathrm{H_2O},C}$ is 1, but in most cases it is ~2. The resulting expression for the efficiency loss due to steam extraction is:

$$\Delta\eta_{strip} = \frac{n_{\mathrm{H_2O,C}} n_{C,fuel} h_{st}}{\mathrm{FHV}} \eta_{\mathrm{H_2}}^{\mathrm{deg}}. \qquad (11.23)$$

The steam enthalpy is h_{st}. Similar to the analysis of CO_2 removal from flue gases in Section 11.4.1, and the use of (11.6) and η^{deg} to account for the drop in the plant efficiency as some intermediate pressure steam is drawn from the turbine to heat up the solvent, we use $\eta_{\mathrm{H_2}}^{\mathrm{deg}}$ to account for the drop in plant efficiency due to the extraction of some intermediate pressure steam in the reforming process in (11.23). The value of $\eta_{\mathrm{H_2}}^{\mathrm{deg}}$ can be obtained from Figure 11.5.

The third term in the net efficiency expression stands for the efficiency *gain* by transferring heat from the cooled reformate back to the HRSG (note that loss and gain are with respect to the efficiency of the H_2 cycle). This term depends on the flow of reformate at this stage and the drop of their enthalpy in the cooler, that is:

$$\Delta\eta_{cooler} = \frac{m_{reformate}\, c_p \Delta T_{cooler}}{\mathrm{FHV}} \eta_{HP,steam}^{\deg}.$$

The mass of reformate per mole of fuel into the autothermal reactor (ATR) is $m_{reformate}$, the reformate specific heat is c_p, the temperature drop across the cooler is ΔT_{cooler}, and $\eta_{HP,steam}^{\deg}$ corresponds to the efficiency at which this heat is used in the power cycle, and can be obtained again from Figure 11.5. The mass of reformate per mole of fuel into the ATR is given by the following mass balance:

$$\begin{aligned} m_{reformate} &= \frac{\text{mass of fuel} + \text{mass of air into ATR} + \text{mass of steam}}{1 \text{ mole of fuel}} \\ &= M_{fuel} + n_{O_2,C} M_{air} + n_{H_2O,C} M_{H_2O}. \end{aligned} \tag{11.24}$$

The molecular weights of the fuel, air, and H_2O are, respectively, M_{fuel}, M_{O_2}, M_{H_2O}. $n_{O_2,C}$ is the number of moles of oxygen per mole of carbon in the ATR, and it is taken to be slightly greater than 1.

Finally, the mass flow leaving the absorber plant is given by:

$$\begin{aligned} m_{abs,ex} &= \frac{1}{\text{fuel mole}} \left\{ \begin{array}{l} \text{mass of reformate} - \text{mass of } CO_2 \text{ capture} \\ -\text{mass of water condensed} \end{array} \right\} \\ &= m_{reformer} - \tilde{m}_{CO_2} f_{CO_2} - \frac{2\dfrac{n_{H_2O,C}}{n_{O_2,C}} n_{C,fuel} - \dfrac{4}{2}}{\dfrac{n_{H_2O_2,C}}{n_{O_2,C}}} M_{H_2O}. \end{aligned} \tag{11.25}$$

The "4" in this equation is the number of hydrogen atoms in the natural gas fuel. The second term is the mass of CO_2 capture (per mole of original fuel) and the third term is the water remaining after the reforming process, which is condensed to a large extent before the CO_2 capture process.

The last term in the efficiency is the compression work to raise the hydrogen fuel stream pressure to the combustor pressure, and is given by:

$$\Delta\eta_{H_2comp} = \frac{W_{comp}^{H_2stream} m_{abs\_ex}}{\mathrm{FHV}}, \tag{11.26}$$

where $W_{comp}^{H_2stream}$ is the work required per unit mass of that stream. Applying this model [3] shows a loss of ~13 percentage points from the reference case, and the new cycle efficiency is reduced to ~45%. Note that many of these figures are approximate and can change if more accurate calculation methods are used.

11.9 Comparison

The performance of the different cycle/plant designs presented in this chapter depends strongly on the performance of the different units and critically on the energy and mass integration among these units. Progress is being made in the design of more efficient gas

Table 11.5 Efficiency and CO_2 reduction of a number of capture technologies [9]

	Efficiency (%)	CO_2 captured g/kWh$_e$	CO_2 avoided (g/kWh$_e$)
Base case	57		
Post-combustion	48	390	323
Pre-combustion	47	401	325
Oxy-combustion CC	47	485	367
Oxy-combustion water	45	471	367
CLC	51	405	367
Ion-transport membrane-based oxy-combustion	50	405	367

separation units for separating CO_2 and N_2, CO_2 and H_2, or air; more efficient turbomachinery units; and high-temperature heat exchangers. This progress is bound to change the relative performance of different plants. Meanwhile, results of a comparison between many of the systems was published in Reference [9], which used similar assumptions regarding the performance of common equipment and software for modeling the different systems. The results are shown in Table 11.5, compared with the base CC without capture, with uncertainty in the different estimates in the range of 1–2% percentage points [9]. Amine-based post-combustion shows a significant drop in efficiency because of the energy required to regenerate the solvent. Conventional oxy-combustion cycles are penalized by the energy of separation using cryogenic technology, with the Graz cycle showing the best performance because of the better integration among the different units. The AZEP and CLC cycles show better performance, although the maximum temperatures of the cycles are limited by relative lower TIT, and hence the 85% (capture) AZEP cycle shows higher efficiency because of the post-firing.

While plant efficiency is an important metric for measuring the technology performance, other metrics include CO_2 emitted per kWhe and CO_2 avoided per kWh$_e$. CO_2 emission per kWh$_e$ was defined before for a regular cycle (i.e., without any form of CO_2 reduction):

$$CO_2 \text{ emission} = \frac{\nu_{CO_2} M_{CO_2}}{\eta_e M_f \left|\Delta \hat{h}_{R,f}\right|} \quad [\text{kg-}CO_2/\text{MJ}_e]. \tag{11.27}$$

For the same fuel, it depends strongly on the cycle efficiency (including the mechanical energy to electricity efficiency). This formula does not apply for CO_2 capture cycles since CO_2 is removed deliberately at some point, and it should be calculated from the cycle analysis. The CO_2 avoided per kWh$_e$ is the difference between CO_2 emission in a plant with capture and the reference plant (i.e., a plant with the same power island design using the same fuel and operating under the same conditions but without the capture process implemented). The captured CO_2 in kg-CO_2/kWh$_e$ is the amount of CO_2 captured in the plant (difference between CO_2 produced in a particular capture cycle and the amount actually emitted after capture in the same plant), noting that adding capture technology increases fuel consumption per kWh_e and hence the CO_2 produced. The CO_2 captured is always more than the CO_2 avoided except in the solid oxide fuel cell/gas turbine case, since the solid oxide fuel cell is inherently a capture technology. The captured and avoided emissions for several cases are shown in Table 11.5.

Problems

11.1 A power plant operating on the Graz cycle (see Figure 11.12) produces a net power of 50 MW. Methane at 200 °C and 40 bar, and oxygen at 250 °C and 40 bar are fed to the combustor. The isentropic efficiencies of turbines, compressors, and pumps are 92%, 88%, and 70%, respectively. Table P11.1 provides additional data for various states of the cycle. Calculate the mass flow rates at different states and the thermal efficiency of the power plant.

Table P11.1

State	Fluid	Temperature (°C)	Pressure (bar)
1	H_2O		0.25
2	H_2O		5
3	H_2O		5
4	H_2O	140.7	5
5	H_2O		185
6	H_2O	567	180
7	H_2O		40
8	H_2O+CO_2	1400	40
9	H_2O+CO_2	642	1
10	H_2O+CO_2	160	1
11	H_2O+CO_2		0.25
12	CO_2		0.25
13	CO_2		1
14	CO_2		2.7
15	CO_2	25	2.7
16	CO_2		40

11.2 A 100 MW gas turbine power plant operates on pre-combustion CO_2 capture as shown in Figure P11.2. Hydrogen is separated from reformed methane and used in the gas turbine combustor.

In the compressor, air at 25 °C and 1 bar (state 9) is compressed to 12 bar. Water at 25 °C and 12 bar (state 1) is evaporated and superheated to state 14 in a HRSG to state 14. Methane, initially at 25 °C and 12 bar, is reformed using the superheated steam in the reformer, where the ratio of water to methane is 5 to 1 (molar ratio). The reforming reaction can be assumed to progress to completion (no CH_4 in the products; only CO, H_2, and H_2O in the products). The products of the reforming reaction leave the reformer at 700 °C, and pass through a heat exchanger, preheating the pressurized air coming from the compressor.

The temperature of the gas mixture drops to 450 °C at the heat exchanger outlet (point 4 in the figure), after which it undergoes a water–gas shift reaction (all CO is converted to CO_2). The products of the shift reactor leave at 40 °C (this process is not adiabatic). Water is then separated from the product stream (assume that this occurs without work transfer), and the mixture of CO_2 and H_2O are directed to a membrane separation unit. Pure CO_2 and H_2 streams exit the separation unit. Assume the pressures of streams 1–8 are all 12 bar (with no pressure drop through these processes). Assume also that the membrane separator operates isothermally, but not adiabatically (heat is rejected). The Second Law efficiency of the membrane separator is 25%. The isentropic efficiencies of the compressors and the gas turbine are 88% and 94%, respectively.

The stream of hydrogen flows into the combustor and is burned to completion (but not necessarily stoichiometrically). The combustion products temperature is 1450 °C. The exhaust of the gas turbine is at 1 bar, which provides the heat requirement of the reformer and steam production.

a. Determine the power consumed by the membrane separation unit.

b. Determine the thermal efficiency of the cycle (use HHV). Do not consider the energy required to pressurize the initial streams of methane and water.

c. Find the molar flow rates of the methane, air, and superheated steam.

d. Find the thermal energy in/out of the reformer and membrane separation unit.

e. Find T_{15}.

f. Propose two ways to make the system more efficient that do not include increasing the isentropic efficiencies of the components.

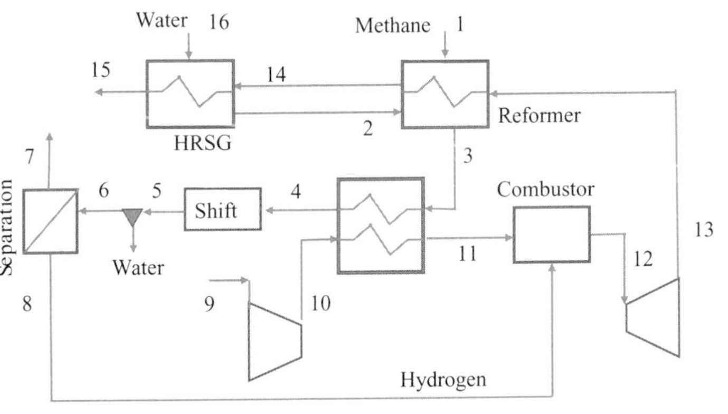

Figure P11.2

REFERENCES

1. O. Bolland and S. Sæther, "New concepts for natural gas fired power plants which simplify the recovery of carbon dioxide," *Energy Convers. Manag.*, vol. 33, no. 5, pp. 467–475, 1992.
2. G. Gottlicher, *The energetics of carbon dioxide capture in power plants*. Washington, DC: U.S. Department of Energy, Office of Fossil Energy, NETL, 2004.
3. H. Bolland and O. Undrum, "A novel methodology for comparing CO_2 capture options for natural gas fired combined cycle plants," *Adv. Environ. Res.*, vol. 7, pp. 901–911, 2003.
4. R. F. Probstein and R. E. Hicks, *Synthetic fuels*. Mineola, NY: Dover Publications, 2006.
5. AspenTech Ltd., "Aspen Plus." 2012 [software].
6. W. Sanz, H. Jericha, B. Bauer, and E. Göttlich, "Qualitative and quantitative comparison of two promising oxy-fuel power cycles for CO_2 capture," *J. Eng. Gas Turbines Power*, vol. 130, no. 3, pp. 31702–31711, 2008.
7. N. W. Chakroun, "Techno-economic analysis of sour gas oxy-fuel combustion power cycles for carbon capture and sequestration," MIT, M.Sc. thesis, 2014.
8. O. Bolland and P. Mathieu, "Comparison of two CO_2 removal options in combined cycle power plants," *Energy Convers. Manag.*, vol. 39, no. 16, pp. 1653–1663, 1998.
9. O. Kvamsdal, H. M., Jordal, and K. Bolland, "A quantitative comparison of gas turbine cycles with capture," *Energy*, vol. 32, no. 1, pp. 10–24, 2007.
10. F. Heitmeir, H. Jericha, and W. Sanz, "Graz cycle: a zero emission power plant of highest efficiency," in *Gas turbine handbook*, R. Dennis (ed). Washington, DC: US DOE, 2006.
11. B. R. Alexander and A. F. Ghoniem, "Analysis and optimization of the coupled parameters of the GRAZ cycle," in *ASME Turbo Expo 2008: Power for Land, Sea and AIr GT2008, GT2008–50588*, 2008, p. 10.
12. Z. L. Zhao, "Rotary bed reactor for chemical looping combustion with carbon capture," MIT, M.Sc. thesis, 2012.
13. K. F. Ritchter, and J. R. Knoche, "Reversibility of combustion processes," *ACS Symposium Series*, 1983, pp. 71–85.
14. H. Ishida, and M. Jin, "A new advanced power generation system using chemical looping combustion," *Energy*, vol. 19, pp. 415–422, 1994.
15. Z. Zhao, C. O. Iloeje, T. Chen, and A. F. Ghoniem, "Design of a rotary reactor for chemical-looping combustion. Part 1: Fundamentals and design methodology," *Fuel*, vol. 121, pp. 327–343, 2014.
16. L. S. Fan, *Chemical looping systems for fossil energy conversions*. Chichester: Wiley, 2010.
17. M. Jin, and H. Ishida, "A novel gas turbine cycle with hydrogen fueled chemical looping combustion," *Int. J. Hydrogen Energy*, vol. 25, pp. 1209–15, 2000.
18. C. O. Iloeje, "Rotary (redox) reactor-based oxy combustion chemical looping power cycles for CO_2 capture: analysis and optimization," MIT, Ph.D. thesis, 2016.
19. C. Iloeje, Z. Zhao, and A. F. Ghoniem, "Analysis of thermally coupled chemical looping combustion-based power plants with carbon capture," *Int. J. Greenh. Gas Control*, vol. 35, pp. 56–70, 2015.
20. N. R. McGlashan, "Chemical looping combustion, a thermodynamic study," *Proc. Inst. Mech. Eng.*, vol. 222, pp. 1005–1019, 2008.
21. V. K. Chakravarthy, C. S. Daw, and J. A. Pihl, "Thermodynamic analysis of alternative approaches to chemical looping combustion," *Energy Fuels*, vol. 25, no. 2, pp. 656–669, 2011.
22. Z. Zhao, T. Chen, and A. F. Ghoniem, "Rotary bed reactor for chemical-looping combustion with carbon capture. part 1: reactor design and model development," *Energy Fuels*, vol. 27, no. 1, pp. 327–343, 2013.

12 Coal, Power Cycles, Gasification, and Synfuels

12.1 Introduction

Coal is a widely available cheap fuel that has been used extensively in heating, electricity generation, and industrial processes. Coal reserves and resources are the largest among other known fossil fuel reserves and resources. Besides carbon, hydrogen, and some oxygen, raw coal contains, among other things, sulfur, metallic compounds, mercury, and nitrogen. Technologies have been developed to utilize coal while limiting the emissions of "criteria" pollutants, including sulfur compounds, nitric oxides, mercury, and fine particulates. While increasing the cost of electricity by raising the plant capital cost and lowering its efficiency, these technologies made it possible to continue to expand the use of coal without negatively affecting air quality. More recently, coal use has accelerated significantly in developing economies. This and the fact that coal produces the largest amount of CO_2 per unit of useful energy has intensified the effort to improve the overall efficiency of coal power plant and to develop technologies for CO_2 capture from these plants.

In this chapter, we summarize the properties of coal as they pertain to energy conversion, and discuss fundamentals of pollutants clean-up for exhaust gases in power plants operating on pulverized coal boilers. Developments in fluidized-bed combustors have allowed for lower emissions and for the implementation of some forms of combined cycles (CC), although not integrated cycles, to improve the thermal efficiency of the plant. For a true CC, it is necessary to thoroughly clean-up the fuel, and this is doable using gasification. Gasification is used to convert coal to a synthetic gas in partial and steam oxidation, and systems used for this purpose are briefly discussed. The environmental characteristics of these plants are superior.

12.2 Resources and Consumption

Vast amounts of coal exist, in the form of reserves and resources of hydrocarbons, as well as "additional occurrences." Broadly, **reserves** are resources that have been discovered and are

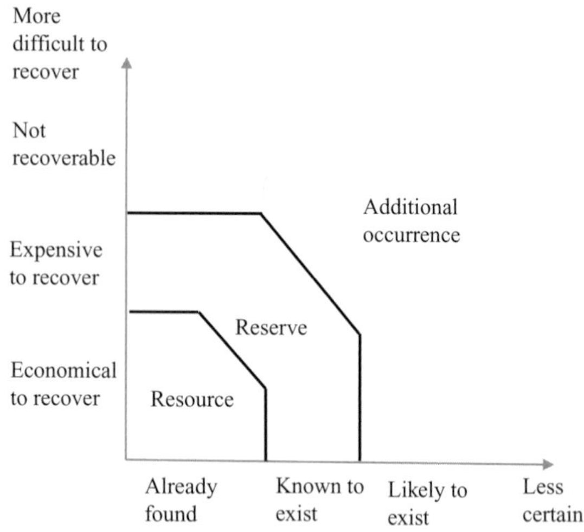

Figure 12.1. A schematic, "McKelvey"-like diagram showing the three categories of natural resource and the relation between the source, its economic value, and its availability.

economically recoverable, as shown in Figure 12.1. **Resources** are resources thought to exist and which might become economically recoverable as technologies are developed to recover these resources. **Additional occurrences** are resources that might be discovered in the future (with reasonably strong evidence for their existence) [1]. By far the largest reserves and resource bases are of coal, and the same is true for the additional occurrences (if methane clathrates or hydrates are discounted since little is known about how these might be recovered or the impact of such recovery on the ocean environment). A schematic representation of the resource value and its utilization is shown in Figure 12.1. Coal has remained one of the cheapest sources of energy worldwide among hydrocarbons, and given the estimates for the resources, it is likely to remain the same for many years to come. Technology for using coal for electricity generation have also evolved such that the price of electricity from coal has remained competitive with other technologies.

Coal resources are distributed worldwide, and many developing countries, like China and India, have sufficient domestic resources to contribute significantly to their growing needs. Furthermore, while coal contains many contaminants, such as sulfur, nitrogen, mercury, and ash, exhaust gas clean-up technologies for the removal of sulfur compounds, nitrogen compounds, particulates, and mercury, have been developed to meet tightening regulations. The application of these environmental technologies raises the price of electricity, but their application has expanded. Finally, as will be discussed in Chapter 13, technologies to capture CO_2 from coal power plants are being developed and some have been piloted already.

Recently, coal has been consumed at the rate of about 100 EJ/year (exajoule = 10^{18} J), measured by the thermal or primary energy produced by burning coal. **Proven reserves** – that is, those known to exist with reasonable certainty and which can be extracted economically with current technology – are estimated to have a total heating value of about 25,000 EJ. With a growth consumption rate of nearly 0.8% per year, this reserve will last close to

150 years. On the other hand, it is estimated that coal **resources**, that is, those that may exist but require different technologies and economic conditions to extract, exceed 150,000 EJ. Among all fossil fuels, coal reserves and resources are the largest when measured by total heating value (larger than oil and gas reserves and resources combined). Coal has been used directly to fuel steam power plants, and more recently it has been gasified to power CC power plants. Coal gasification has also been used to produce gaseous and liquid fuels (indirect liquefaction). Direct coal liquefaction has also been used to produce gasoline and diesel-like fuels.

12.3 Plant Efficiency

The efficiency of coal power plants has improved over the years. Figure 12.2 shows a comparison between efficiencies achieved using different technologies based on pulverized coal combustion in subcritical steam cycles, supercritical steam cycles, and ultra-supercritical steam cycles (USC); all are computed based on the lower heating value (LHV) of coal. Also, shown are the efficiencies achieved using pressurized fluidized-bed combustors (PFBCs) and integrated gasification combined cycle (IGCC) plants. The PFBC plants have lower emissions, and can achieve better efficiency if the plant is partially integrated with a topping gas turbine cycle, as will be described in detail in this chapter. In partially or fully integrated combined cycle plants, the overall efficiency depends on the gas turbine inlet temperature. While the efficiency of USC using pulverized coal is currently the highest achievable value, PFBC and IGCC have other advantages, such as low emissions and more compatibility with CO_2 capture technologies. In the next sections, we describe some of these cycles in more detail and discuss their components.

Table 12.1 shows the operating conditions of subcritical and supercritical pulverized coal power plants, and their efficiencies based on the higher heating value (HHV). The table

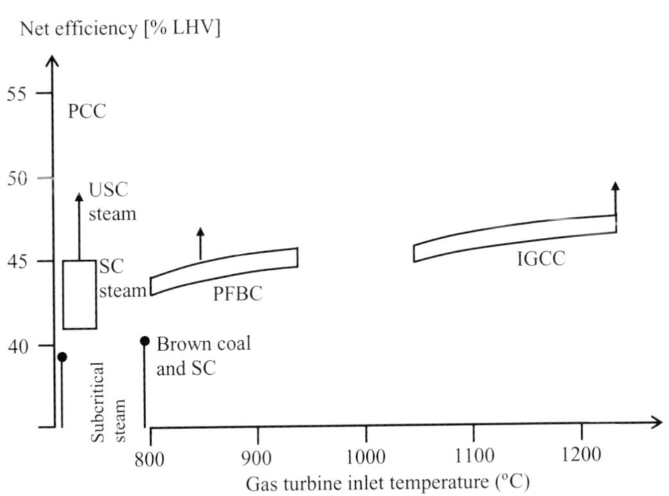

Figure 12.2 Net conversion efficiency based on the LHV for different coal-burning technologies, including pulverized coal burning to power a subcritical, supercritical, or ultra-supercritical steam cycle, PFBCs to power a gas turbine and a steam turbine, and an IGCC plant [2]. The efficiency of the last two options depends on the inlet gas turbine temperature. In all cases, the range of efficiencies depends on the coal type used and the level of gas clean-up technologies used in the plant. In general, lower-rank coals and applying stricter products gas clean-up degrades the plant efficiency.

Table 12.1 Efficiency of coal-burning power plants, classified by the state of steam in the first stages of the turbine. In all cases, multiple reheat is used

Description	Cycle	Efficiency (HHV-based)
Subcritical	16.8 MPa/558 °C/538 °C	37
Supercritical	24.5 MPa/565 °C/565 °C/565 °C	39–40
Supercritical – advanced	31.5 MPa/593 °C/593 °C/593 °C	40–42
Ultra-supercritical	38.5 Mpa/760 °C/760 °C/760 °C	47.5–48

shows the number of reheat stages and the reheat temperatures used to achieve the target efficiency. The impact of the maximum turbine pressure and temperature are clearly seen in the table, and the target values of more than 47% for the USC (pulverized coal-based) is shown. (The standard in Europe is to evaluate the efficiency of power plants on the basis of the LHV, while the convention used in the USA is to use the HHV.)

12.4 Coal Ranks

Coal is a solid fuel formed of carbon and hydrogen with nontrivial amounts of sulfur, oxygen, nitrogen, moisture, and ash (non-combustible minerals of different types, including sodium, silica, aluminum, calcium, magnesium, and iron). It is often represented chemically as $CH_m, m < 1$, or $CH_mO_n, m < 1$ and $n < m$. Coal is a sedimentary rock formed of fossilized vegetation of different types, with a bit of animal content, forming deep underground or close to the surface (under varying levels of pressure and temperature). The chemical characteristics of coal, that is, its carbon and hydrogen percentage, and heating value, as well as its physical properties, that is, hardness, porosity, volatile and moisture contents, etc., vary widely, even within each type.

Coal is defined by its **rank**: lignite, subbituminous, bituminous, and anthracite. The rank refers to how young coal is, with lignite being the youngest and anthracites being the oldest. The lowest-rank coal, lignite, has the least carbon content, about 33% by mass, and highest volatile contents, near 27% (volatiles are essentially oxygen and hydrogen but also some light hydrocarbons), with moisture content of about 33%. All percentages here are by mass. Lignite has the lowest heating value at 12–14 MJ/kg. Bituminous and anthracite sit on the opposite side of the scale, with heating values in the range 31–35 MJ/kg (see Reference [3] for more detail), with subbituminous ~23 MJ/kg. Table 12.2 shows data from the proximate analysis (the percentages of fixed carbon, volatile matter, and moisture) and ranges of heating values of different coal ranks. It is estimated that half of the coal available worldwide is bituminous and anthracite, while the other half is subbituminous and lignite.

Sulfur content can be significant, up to 4% or more, and it poses a significant challenge in burning coal in power plants. Some coal sulfur can be washed away prior to combustion, but most of it forms sulfur oxides or hydrogen sulfides during combustion or gasification,

Table 12.2 The proximate analysis, defined in terms of the fixed carbon, volatiles, and moisture content, all on an ash-free basis and measured as mass percentage, and the heating value of different ranks of coal

	Proximate analysis (ash-free)			
Classification	Fixed carbon	Volatile matter	Moisture	Heating value (MJ/kg)
Lignite	30%	25%	45%	19
Subbituminous	35-40%	30–35%	20–30%	21–30
Bituminous	45-70%	20–35%	5–20%	30–36
Anthracite	85-90%	2–5%	5–8%	31–35

respectively. These must be scrubbed away from the products. The sulfur compounds forming during coal conversion depend strongly on whether coal is combusted or gasified. In coal combustion, most sulfur forms sulfur oxides because of the availability of oxygen. In gasification, and because of oxygen deficiency, sulfur forms H_2S. For sulfur removal from combustion or from gasification products, limestone is used, but other techniques for using these "acid gases" are also available (see Chapter 10), such as amine scrubbing.

"**Ash**" is a term used to describe the inorganic residue left after combustion or gasification. The chemical composition of ash residues may be different in combustion or gasification because of the oxidation/reduction environment in the two cases. In combustion, more metallic oxides are expected than in the reducing environment of gasification. Under high-temperature conditions, above the ash fusion temperature, ash melts, forming slag that flows down the reactor walls. In general, operating a coal reactor (pulverized coal combustor, fluidized-bed combustor, fluidized-bed or entrained-flow gasifiers) is impacted strongly by the ash/slag removal method.

The mass percentages of coal constituents are often given in two forms. In the "**proximate analysis**," moisture, ash, volatile matter, and fixed carbon are reported on a mass basis. It also possible to report the proximate analysis on an ash-free basis. In the "**ultimate analysis,**" the "elemental" composition is reported, including that of the moisture, carbon, hydrogen, nitrogen, sulfur, oxygen, and ash. If either or both are given on an "as received" (ar) basis, the moisture is given separately as a percentage of the original coal, while the rest is given on a dry basis, as shown in Table 12.3. In the ultimate analysis, moisture is given separately from the other hydrogen and oxygen gas contents. Percentages of these constituents can vary even within the same rank, but carbon content is approximately the same within the same rank. Volatile matter and carbon contribute the most to the heating value, while the mineral matter forms ash and/or slag (slag is molten ash if coal is burned or gasified at high temperature, above the ash fusion temperature). The ultimate analysis is used in estimating the HHV of coal, also called the gross calorific value, by applying the Dulong formula:

$$\mathrm{HHV} = 33.83\,Y_C + 144.3\left(Y_{\mathrm{H}_2} - \frac{Y_{\mathrm{O}_2}}{8}\right) + 9.42\,Y_S \quad \text{in MJ/kg}, \tag{12.1}$$

Table 12.3 Proximate and ultimate analysis (on a dry basis) of a number of American coals. Moisture is shown as received (ar). Values given here are averages.

	Rank			
	Bituminous	*Subbituminous*		*Lignite*
Seam	Illinois # 6	PRB	Beulah-Zap	North Dakota
	Proximate analysis (weight %)			
Moisture	11.12	28.09	32.24	36.08
Ash	10.91	8.77	9.72	15.43
Volatile matter	39.37	44.73	44.94	41.49
Fixed carbon	49.72	45.87	44.54	43.09
HHV (MJ/kg)	30.51	27.25	25.59	24.25
	Ultimate analysis (weight %)			
C	71.72	68.43	65.85	61.88
H	5.06	4.88	4.36	4.29
O	7.75	16.24	18.19	16.44
N	1.41	1.02	1.04	0.98
Cl	0.33	0.03	0.04	0.0
S	2.82	0.63	0.80	0.98
Ash	10.91	8.77	9.72	15.43

where Y_i is the mass fraction in the ultimate analysis. The corresponding LHV is $\text{LHV} = \text{HHV} - 21.6 Y_{H_2}$ in MJ/kg.

12.5 Coal Burning in Boilers

The majority of coal power plants run on steam Rankine cycles. They burn pulverized coal in atmospheric (or slightly subatmospheric) pressure boilers (burning in a slight vacuum ensures that the products do not leak out). Boilers are large structures, used to burn coal and to transfer heat to the working fluid of the thermodynamic cycle. The lower section of the boiler constitutes the combustion zone, in which a number of burners are installed around the periphery. Each burner introduces a stream of pulverized coal, with coal particle size O (50–100 μm), surrounded by one or several streams of air. Mixing and combustion of the concentric jets of a burner produce a hot flame, or multiple flames, and interaction between neighboring flames forms a "fire zone" at the bottom section of the boiler. Fine coal particles are burned in air, with excess (or overfire) air used in most cases. The extra air is used to reduce the products gas temperature to values tolerable by the heat transfer equipment, guarantee complete combustion, and control NO_x emissions. Figure 12.3 shows schematically a typical pulverized coal boiler and its integration with the rest of the plant equipment [2]. Given the boiler size and the nature of coal and its combustion, the primary heat transfer

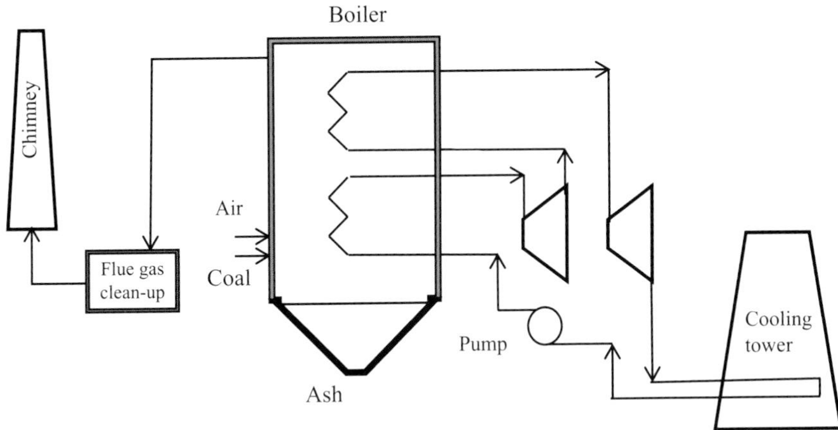

Figure 12.3 Layout of a pulverized coal steam power plant, showing the boiler, the flue gas cleaning equipment on the left-hand side of the boiler, and the power island and the condenser on the right-hand side.

mode in these systems is radiation from the fire zone in the bottom to the steam boiler, superheater, and preheater at the top side.

Before burning, coal is pulverized, and it may also be dried to remove some of its moisture content. In other cases, pulverized coal may be mixed with water and introduced into the burner as a stream of slurry. Slurry is easier to transport and to pressurize, but it requires special pumps for pressurization. Fine moist coal particles are dried within the flame zone, followed by devolatilization, during which the volatiles are released from the particle pores and burned. This is followed by the combustion of the carbon char, leaving behind the non-combustible mineral matter or ash/slag. The state of the mineral matter, whether it is ash or slag, depends on the temperature and combustion environment. Fine particulates or flyash forming during combustion are blown upwards with the combustion products. The rest of the ash settles and is collected at the bottom of the boiler. Besides the flyash, the boiler exhaust contains sulfur oxides and sulfuric acid, nitrogen oxides, and small amounts of carbon monoxides, besides the nitrogen.

Controlling the combustion temperature is important for the fate of the ash; higher temperatures can cause the ash to melt and form molten slag. Typically, in these large coal boilers, a fraction of the thermal energy is transferred from the burning pulverized coal flames, or the fire zone, to the working fluid by radiation and some by convection. Coal flames are highly luminous, forming strong radiative fluxes that distribute its energy within the boiler cavity. Combustion products flow upward by the combined effects of natural and forced convection, and are cooled by the working fluid in the evaporator and the economizer sections of the boiler. The temperature of the flame is high, $>1200\,^{\circ}C$, a necessary condition for fast kinetics, complete coal consumption, and complete combustion, but the high temperature increases the rate of NO_x formation and sulfur dioxide formation in the coal combustion zone. Despite the use of overfire air, the flame temperature is much higher than the highest temperature achieved by the Rankine cycle working fluid. While this results in loss of availability efficiency, the temperature difference between the combustion gases and the working fluid generates higher heat transfer rates and reduces the overall size of the boiler. Metals often found in coal may be oxidized during burning, forming alkali compounds. Boiler designs have been optimized to ensure near-complete combustion of the fuel, and to deliver very high combustion efficiency (defined by the ratio of the thermal energy delivered to the steam to the chemical energy originally in the fuel). More overfire air reduces the temperature and NO_x emissions, but it may lower the overall system efficiency since more air carries unused enthalpy out of the stalk.

12.5.1 Exhaust Gas Clean-Up in Boilers

Modern power plants utilize complex exhaust clean-up technology to reduce the emission of harmful regulated (also called criteria) pollutants. These technologies include wet scrubbers to remove sulfur oxides, ammonia scrubbers to reduce NO_x, and precipitators to remove most of the flyash and fine particulates from the exhaust stream before it is sent to the chimney. Flue gas clean-up is done after the gases have been cooled inside the boiler, and further cooling may be necessary to improve the efficiency of the removal processes for different pollutants. Incorporating exhaust clean-up technologies adds to the cost of the power plant significantly, and penalizes its efficiency, but it is essential for maintaining air quality. The energy required in the scrubbing of the pollutants and the regeneration of the absorbents used in these clean-up processes reduces the energy available for the steam turbines and hence the power output of the plant. Flue gases in the chimney must be hot enough for natural convection to disperse the plume outside.

Exhaust gas clean-up is carried out in steps. First, fine particulates and flyash are removed using electrostatic precipitators or fabric filters. Most of the mineral matter in coal leaves the boiler in the form of flyash, which includes toxic metals such as arsenic, selenium, cadmium, manganese, chromium, lead, and mercury, and nonvolatile organic matter, or soot, which also contains polycyclic aromatic hydrocarbon (PAH). Electrostatic precipitators operate by charging particles negatively by a corona discharge and attracting the charged particles to a grounded plate, where they are collected for disposal. Electrostatic precipitators work better in collecting larger-diameters particles ($>1\ \mu m$), but are not as effective with submicron diameter particles. Another, less effective approach uses fabric filters. A fabric filter device resembles a baghouse, and operates according to the same principal as a vacuum cleaner, utilizing a fabric membrane that is capable of passing air through while holding particles of certain diameter behind (see Figure 12.4). Particulate removal is energy intensive due to operating the compressors or suction pumps.

Some of the sulfur in raw coal is washed away by floating crushed coal in water to allow the heavier mineral contents to settle before combustion. However, some sulfur remains attached to the carbon molecules and forms sulfur oxides during combustion. Sulfur oxides (SO_2) are removed from the flue gases using a sorbent such as limestone ($CaCO_3$) or calcium oxide (CaO) in a wet or dry scrubber. The most effective flue gas treatment approach for

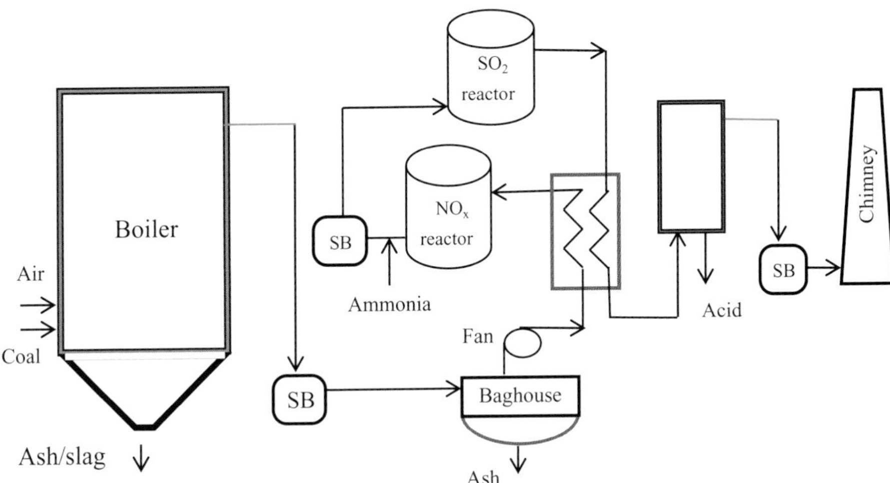

Figure 12.4 Typical pulverized coal power plant exhaust gas clean-up technologies, including a baghouse to remove particulates and ash, a catalytic reactor to remove NO_x using ammonia, and another catalytic reactor to remove SO_x compounds form the exhaust gases. The flue gases must be reheated to regain their buoyancy. SB: support burner to heat the gas.

sulfur removal is wet scrubbing, shown schematically in Figure 12.5. Flue gases enter an absorption tower in which they are sprayed with water slurry of the sorbent, which undergoes the following two reactions to remove SO_2 from the exhaust:

$$CaCO_3 + \frac{1}{2}H_2O + SO_2 \rightarrow CaSO_3.\frac{1}{2}H_2O + CO_2, \tag{12.2}$$

$$CaSO_3.\frac{1}{2}H_2O + \frac{3}{2}H_2O + \frac{1}{2}O_2 \rightarrow CaSO_4.2H_2O. \tag{12.3}$$

The final product of the process, hydrated calcium sulfate, $CaSO_4{\cdot}2H_2O$, is similar to natural gypsum. This sulfate falls to the bottom of the tower, where it is removed, dried, and disposed of. The energy required for this process can amount to 2–3% of the plant available energy, and is used to drive the pumps and to reheat the flue gases exiting the scrubber (gases are cooled as they are sprayed with water, and must be reheated to regain their buoyancy in the chimney). Wet scrubbing can remove up to 90–99% of sulfur oxides in the flue gases. Dry scrubbing, in which dry sintered limestone or calcium oxide is injected directly into the flue gases at the boiler exit, is also used, but the removal efficiency in this case is much lower than in wet scrubbing (the reactions are listed in Section 12.6.1).

Nitrogen oxides are formed during coal combustion via two mechanisms: (1) The oxidation of nitrogen from air at the high combustion temperatures inside the boiler,

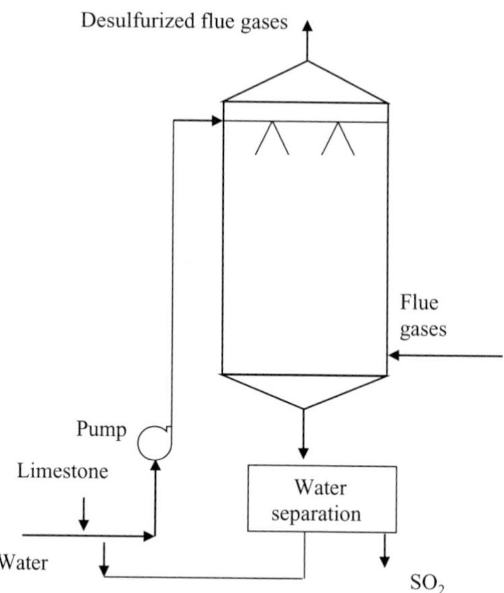

Figure 12.5 Wet limestone scrubber for removing SO_2 from flue gases.

according to the thermal and prompt mechanisms; and (2) the oxidation of coal-bound nitrogen during the same processes. More nitric oxides are formed at higher temperatures. This is because NO formation reactions are endothermic and are extremely temperature-dependent (and hence it is called thermal NO_x, although other prompt NO_x also forms). Modern coal power plants utilize low-NO_x burners in which air staging is used to keep the flame temperature low and hence reduce the formation rates of NO_x due to both mechanisms, thermal and fuel-bound NO. In the first stage of these burners, only a fraction of the stoichiometric combustion air is introduced with the fuel, to force combustion to occur under fuel-rich conditions. Under these oxygen-starved and low-temperature conditions, NO formation rates are low, and fuel-bound nitrogen is released in the form of nitrogen without being oxidized. Downstream, and on the outer periphery of the first-stage rich flame, more air is introduced such that the overall stoichiometry of the first and second stages remains lower than unity. In the second stage, the remaining fuel, especially CO, burns under low-temperature lean conditions. The impact of staging is to reduce the overall NO_x formation rates in these flames. A schematic diagram showing an air-staged coal burner is shown in Figure 12.6.

If combustion modification is not sufficient to meet modern stringent NO_x emission limits, chemical processes are used to remove NO_x compounds from the exhaust stream. One such process uses ammonia in the following **selective catalytic reduction** process:

$$4NO + 4NH_3 + O_2 \rightarrow 4N_2 + 6H_2O. \tag{12.4}$$

This process must be carried out at elevated temperatures, 300–400 °C, in the presence of a catalyst, mostly titanium and vanadium. Hence it requires thermal energy, besides ammonia. Other processes, which do not need a catalyst, are also used. In one such

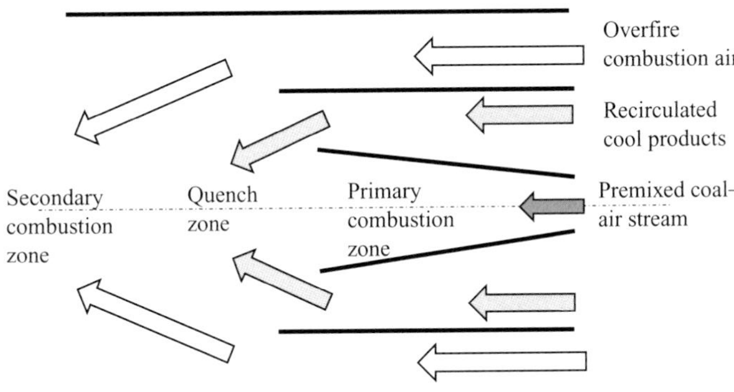

Figure 12.6 An air-staged coal burner utilizing an inner premixed coal–air stream, an intermediate steam of recirculated products to cool/quench the reactions, and an outer air stream used to burn the fuel left over from the first stage. Swirl vanes (not shown) are used to improve mixing between streams.

selective non-catalytic reduction process (SNCR), an aqueous solution of urea is used at temperatures of 900–1000 °C to reduce NO in the following reaction:

$$4NO + 4CO(NH_2)_{2(aq)} + O_2 \rightarrow 4N_2 + 4CO_2 + 2H_2O. \tag{12.5}$$

Both technologies can achieve up to 90% reduction of NO. The power requirement is greater in the SNCR since it operates at a higher temperature, reaching up to 4% of the plant output power.

12.6 Gasification-Based Combined Cycles

Combustion products from atmospheric boilers cannot be used in CC plants. This is because a gas turbine needs high-pressure, high-temperature combustion products to produce power. Moreover, these products must be free of acid gases and particulates that harm the gas turbine blades. One approach to adopt CC in coal plants is to use closed cycle gas turbines. In closed cycles, the working fluid of the gas turbine is heated in the coal boiler. High-efficiency gas turbine cycles must operate at high temperature and hence a high-temperature heat exchanger is needed to raise the working fluid temperature in the coal boiler. Following expansion in the gas turbine, the working fluid is then cooled to raise steam for the steam cycles. These cycles require a high-temperature heat exchanger to heat the gas turbine working fluid, and another heat exchanger to cool the same stream before it is compressed back to the high pressure.

There are two ways to use coal in CCs that employ an open cycle gas turbine [4]:

1. burning coal in a **PFBC**; or
2. partially oxidizing coal in a **gasifier** in an **IGCC**.

Both technologies are described briefly next to show how a high-efficiency CC that uses coal can be constructed. The first technology uses pressurized fluidized-bed reactors (PFBRs) to partially burn coal, as shown in Figure 12.7. A PFBR (in contrast to an atmospheric-pressure boiler) provides a high-pressure, high-temperature clean-gas products stream for

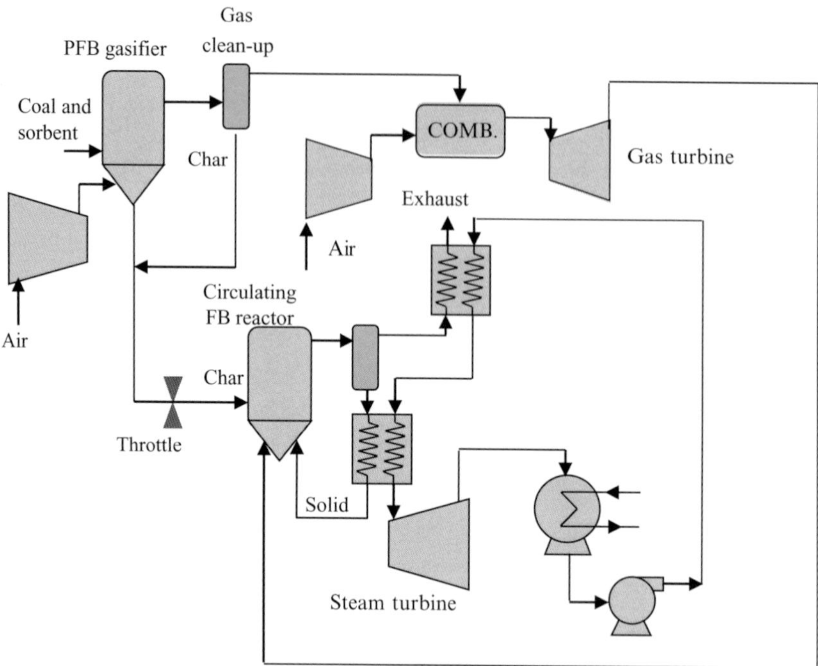

Figure 12.7 Low-temperature gas turbine integrated with PFBC. Coal is devolatilized and partially oxidized in the first pressurized-bed reactor to produce high (higher than atmospheric) pressure gaseous fuel for the gas turbine, which is burned in the combustor (COMB.). The char from this gasifier is burned in the second circulating reactor to raise steam for the steam cycle. The exhaust of the gas turbine is used as the oxidizer for the circulating-bed reactor.

the gas turbine. The proposed design utilizes a PFBR along with a circulating atmospheric fluidized-bed reactor. Coal is *partially gasified* in the first of these two reactors, under rich conditions, in which the volatiles and some of the carbon char are partially oxidized. The gases produced from this partial burning process are used in the gas turbine combustor after removing all the sulfur compounds and the solids including the remaining carbon. The unburned carbon (char) and ash are fed to the second circulating-bed reactor. The gas produced by the partial combustion of coal in the PFBR, which contains CO_2, CO, H_2, and H_2O, is burned in the gas turbine combustor and expanded in the gas turbine. The exhaust of the gas turbine, exiting at atmospheric pressure but at a higher temperature, is fed into the second circulating-bed reactor, where char is burned. The gas turbine exhaust is oxygen-rich, and hence is used as the oxidizer in the circulating reactor.

The unburned char leaving the fluidized bed of the first reactor, and the solids removed during gas clean-up, are fed to the circulating atmospheric fluidized-bed combustor. Circulating-bed reactors are designed to continuously circulate the remaining solids back to the reactor until they are burned. During each pass, more carbon is burned, and the remaining solid is separated from the combustion gases, cooled, and fed back. The gas turbine exhaust is used as an oxidizing stream for the circulating fluidized-bed reactor.

Steam raised in this second reactor is used in the steam cycle. The cycle is not fully integrated since not all the coal combustion products are used in the gas turbine; only some of the coal is partially burned in the first reactor to generate the fuel for the gas turbine. Thus, the gas turbine is likely to operate at relatively lower temperature, limiting its efficiency and the power that can be produced at high temperature. More discussions of the design and operation of the fluidized-bed reactor are presented in the next section.

In the second technology, coal is *fully gasified* in a mixture of steam and oxygen (or air) to produce synthetic gas. This synthetic gas, composed mostly of CO, H_2, CH_4, and CO_2, and some steam is then purified thoroughly to remove all the sulfur compounds and the particulates. The cleaned synthetic gas is then burned in the gas turbine combustor, and the combustion products are expanded in the gas turbine. The gas turbine hot exhaust is used to raise steam in a heat recovery steam generator (HRSG). This is a fully integrated cycle in which all the coal energy is used to produce synthetic gaseous fuel, and all the synthetic fuel is burned in the gas turbine combustor, similar to the natural gas integrated combined cycle described before.

12.6.1 Partially Integrated Combined Cycles Using Fluidized-Bed Combustors

Fluidized-bed combustors have been introduced to coal combustion because of their superior emissions and better heat transfer characteristics. In these reactors, a mixture of crushed coal and fine refractory material, which constitutes the primary component of the bed material, is fluidized. In **fluidization**, the granular material of the bed is forced to become suspended in gas by injection of a (relatively) high-velocity gas though the grated base of the reactor. The degree of fluidization depends on the velocity of the injected gas, which is either air or oxygen. Two types of fluidization have been observed in fluidized-bed reactors:

1. When a low injection gas velocity is used, gas bubbling occurs inside the bed and the bubbles burst at the surface, and the reactor is called a **bubbling reactor**.
2. When a higher injection velocity in used, the reactor material moves vigorously, eventually leaving the reactor with the gas, and the reactor is called a **recirculating bed reactor**.

In the bubbling reactor the motion of the suspended material is relatively weak, while in the circulating reactor the suspended material motion is much more vigorous. Because of the vigorous motion of the bed material, a fraction of the char forming in the bed leaves the reactor with the gas, and must be separated in a cyclone and reinjected. The fraction of coal to the fluidized fine refractory material is small, 2–3% of the total volume. Crushed limestone or dolomite is also added to the bed with the coal for the purpose of desulfurization. Limestone chemically captures a large fraction of the sulfur oxide forming during the combustion of coal, forming calcium sulfate. Burned ash and calcium sulfate settle to the bottom of the bed and are removed. The motion of the gas in the voids of the suspension greatly improves heat transfer and combustion within the bed. In contrast to coal burners that use pulverized coal, fluidized-bed combustion uses crushed coal, 6–12 mm in size, along with limestone, both introduced at the top of the bed. The subject of fluidized-bed reactors (combustors and gasifiers) will be covered more in Chapters 13 and 14.

Combustion products leaving a fluidized-bed reactor often carry fine particulates in the form of unburned ash, which are removed in single or multiple hot cyclone separators, and are fed back to the reactor bed for complete burning. The temperature within the reactor bed is kept uniform because of the large thermal inertia of the bed material and the effective heat transfer throughout the fluidized fine grain of the bed material. Low combustion temperature lowers NO_x emissions, but can produce more CO. Higher bed temperature increases the rate of heat transfer within the bed and makes the overall reactor more compact. Part of the combustion exothermic energy is removed by embedding heat exchanger tubes within the fluidized bed itself, and some is removed from the exhaust gases as they exit the reactor. Combustion temperature within the fluidized bed depends on the reactor type, but is kept around 800–900 °C. The bed temperature is controlled in part by the circulating feedwater tubes within the bed material and in part by adjusting the fuel and air flow rates. Both pressurized and atmospheric fluidized-bed reactors have been designed and used in simple steam cycles instead of boilers. The latter includes the atmospheric bubbling fluidized-bed reactor (AFBC) and the recirculating fluidized-bed reactor (CFBC).

Fluidized-bed combustion is used with all types of coal: anthracite, bituminous, subbituminous and lignite, as well as other solid fuels such as **biomass**, including wood chips, agriculture refuse, etc. Fluidized-bed reactors achieve:

- high carbon burnout because of the thorough mixing within the bed materials;
- low NO_x emissions because of the lower uniform temperature; and
- low sulfur emissions because of the presence of limestone.

Fluidized-bed reactors are more compact than conventional coal boilers because of the improved heat transfer conditions in the fluid bed. As mentioned before, sulfur is removed using the added limestone. First, limestone breaks down into calcium oxide in an exothermic reaction, followed by the formation of calcium sulfate in the second reaction:

$$CaCO_3 \rightarrow CaO + CO_2, \tag{12.6}$$

$$SO_2 + CaO + \frac{1}{2}O_2 \rightarrow CaSO_4. \tag{12.7}$$

CaO is an effective absorber of sulfur oxide in the presence of oxygen.

Both bubbling and recirculating type reactors can be operated at higher pressures, 6–12 bar, producing a pressurized products gas stream, known as **pressured fluidized-bed reactors**. Pressurizing raises the density and temperature of air, and produces higher-density and higher-pressure products of combustion. The advantages of pressurizing are:

1. the reduction of the reactor size, as pressurized air is used, which improves the burning and heat transfer rates; and
2. improving the overall cycle efficiency by using the combustion products to run a gas turbine – that is, in a CC.

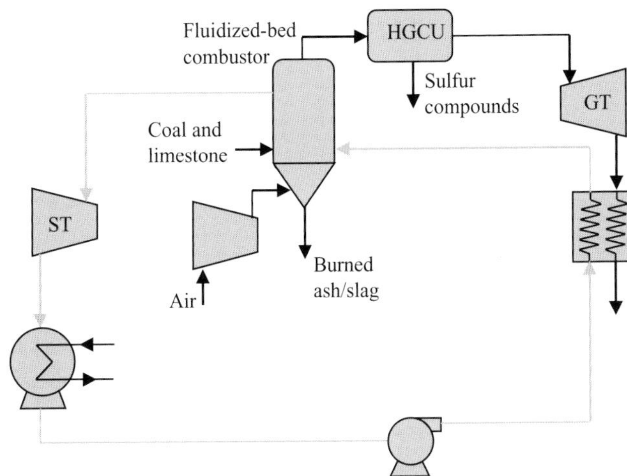

Figure 12.8 A partially integrated combined steam–gas cycle using a PFBC to burn coal. Steam is raised at the combustion section of the fluidized bed, and the hot gases leaving the reactor are cleaned up before they are admitted into the gas turbine (GT). The gas turbine exhaust is used to heat the feedwater of the steam cycle. Combustion products leaving the combustor are expanded in the gas turbine, without the need for a gas turbine combustor.

In a CC plant burning coal in a PFBR, steam is generated in the fluidized-bed reactor by running the feedwater tubes at different sections of the reactor, within the bed and on top of the bed, as shown in Figure 12.8. Feedwater heating, evaporation, and superheating take place within these tubes, instead of the boiler cavity as in the pulverized coal boiler case. The steam generated is used in the steam cycle while cooling the reactor. Moreover, the products gases leaving the reactor are expanded in the gas turbine, after particulates and other turbine-harming gases are removed. The steam turbine is used to generate most of the power, ~80%. Hot combustion products must be cleaned thoroughly by removing the particulates, sulfur, and metal compounds at a relatively high temperature, above 600 °C. The clean-up process at these high temperatures, or the hot gas clean-up (HGCU), often requires several hot cyclones to remove the particulates. Furthermore, since combustion products are used to raise steam in the PFBC, the products temperature entering the gas turbine is relatively low. In some designs, air staging is used – that is, only part of the combustion air used to burn the coal is introduced into the reactor bed and the remaining air is injected into the combustion products above the bed to burn the partially oxidized coal combustion products. In other designs, the partially oxidized products of the fluidized-bed combustor are burned in a **topping combustor** before the gas turbine [2,5,6].

The gas turbine exhaust temperature is still higher than the condenser temperature, and some of the thermal energy in the gas turbine exhaust is used to heat the steam turbine feedwater. Thus, this is a CC, although it is not as well integrated as the natural gas CC described earlier, or the gasification-based coal CC described next. Pressurizing the fluidized-bed reactor leads to slightly higher combustion temperatures, which speeds up the reaction and reduces the overall size of the reactor. It also helps reduce CO in the products. But, the high pressure within the fluidized-bed combustor can also lead to higher NO_x formation. The attraction of this cycle compared to IGCC is its simplicity. Fluidized-bed reactor beds are maintained at 800–900 °C to ensure CO burn out, with pressurized reactors running near the higher end of the limit. Gases leaving the reactor bed are kept around 550 °C. In part because of the pressurization, fluidized-bed combustors are relatively compact and can achieve high capacity.

12.6.2 Integrated Gasification Combined Cycles

In a fully integrated coal-burning CC, coal is first gasified, in air or oxygen and water, producing a stream of high-pressure synthetic gas made of CO, CO_2, H_2, H_2O, and CH_4. The exact composition of the synthetic gas depends on the feedstock composition, that is, the ratio of coal to air/oxygen to water, the gasification pressure and temperature, and the gasifier type. It also depends on the rank of the coal. Syngas produced in air-fired gasifiers is an LHV fuel (because of the presence of nitrogen), while that produced by oxygen-blown gasifiers is a HHV fuel, both measured volumetrically. In both cases, the distinction refers to the enthalpy of reaction per unit volume of the fuel gas. Some of the chemical energy in the original fuel (coal) is used in the gasification process to raise the temperature of the feed stream to values sufficient to favor the shift of the equilibrium toward the gasification products. Higher temperatures are also required to speed up the reaction kinetics and hence reduce the gasifier volume. The overall gasification reaction can be written symbolically as

$$\text{Coal} + \text{oxygen} + \text{steam} \xrightarrow{Heat} CO, CO_2, H_2, H_2O, CH_4, \text{ash/slag}. \tag{12.8}$$

A number of reactions occur during gasification. In its most elementary form, coal can be partially oxidized or combusted in oxygen according to the following two exothermic heterogeneous reactions, respectively:

$$C + \frac{1}{2} O_2 \rightarrow CO \qquad -123\,\text{MJ/kmol} \tag{12.9}$$

and

$$C + O_2 \rightarrow CO_2 \qquad -393\,\text{MJ/kmol}. \tag{12.10}$$

Thus, by gasifying carbon to carbon monoxide, one converts only 31% of the original chemical energy in the carbon to thermal energy. The theoretical "cold gas efficiency" of gasification, defined as

$$\text{cold gas efficiency} = \frac{\text{chemical energy or heating value of products}}{\text{chemical energy or heating value of reactants}}, \tag{12.11}$$

is 69%. On the other hand, if water is added during the gasification process, a synthetic gas (syngas), or syngas plus methane is produced, depending on the composition of the feedstock and the pressure and temperature of gasification. For instance, the following endothermic heterogeneous shift reaction gasifies one mole of coal to one mole of CO and one mole of H_2:

$$C + H_2O \rightarrow CO + H_2 \qquad 118\,\text{MJ/kmol}. \tag{12.12}$$

Thus, one mole of carbon produces two moles of fuel, while consuming nearly the same energy produced in the carbon partial oxidation reaction (12.9). Reactions (12.9) and (12.12) combined form a nearly **thermally neutral** reaction:

$$2C + \frac{1}{2}O_2 + H_2O \rightarrow 2CO + H_2 \qquad - 5\,\text{MJ}/2\,\text{kmol}\cdot\text{C}. \tag{12.13}$$

Adding more water to the gasification process allows more of the chemical energy in the original fuel, coal, to be retained in the form of chemical energy in the products gases, $CO + H_2$. Because of equilibrium constraints, some of the products gases are CO_2 and H_2O. These and other considerations limit gasification efficiency to about 85%. Another often-quoted efficiency, known as the carbon conversion efficiency, is defined as

$$\text{carbon conversion efficiency} = 1 - \frac{\text{carbon in flyash}}{\text{carbon in coal}}. \tag{12.14}$$

In the IGCC plant, the synthetic gas is purified by removing particulates, sulfur, and metal compounds and other chemicals that can harm the gas turbine, after being cooled to lower temperatures. Copl gas clean-up (CGCU) technology is currently more effective than HGCU technology in removing these species. Moreover, because of the higher gas density at the lower temperature, it requires a smaller volume of gas clean-up equipment. As shown in Figure 12.9, first the syngas is cooled to temperatures below 600 °C in a heat exchanger while raising the feedwater temperature for the steam cycle. Next, either an HGCU is performed, or the syngas is cooled further to a temperature below 50 °C for CGCU. In the second case, the thermal energy is recycled back to the gas after it leaves the clean-up device. Next, the syngas (CO and H_2) are burned in a topping combustor, and the combustion products are expanded in the gas turbine. The gas turbine exhaust is used in a HRSG to raise steam for the steam cycle.

Gasification is used to produce high-quality "fuels" such as syngas and hydrogen from coal. These fuels can be consumed in a high-efficiency power plant (e.g., CC or CC integrated with fuel cells) to raise the overall cycle efficiency, or can be purified and sold as a fuel. Better efficiencies should be achieved as higher turbine inlet temperature (TIT) is used in the gas turbine cycle, but the overall performance will also depend on the type of gasifier used. Different gasifiers enable better levels of cycle integration. Higher efficiencies depend critically on heat and mass integration among components, as described next.

An alternative layout of an integrated cycle is shown in Figure 12.10, where an air separation unit (ASU) is used to generate pure oxygen for the gasifier. Oxy-gasification requires a smaller gasifier and reduces the volume of the gas stream produced by the gasifier, and hence reduces the size of the gas cooling and clean-up equipment. High-pressure oxy-gasifiers running at pressures up to 80 bar are more compact, and favored for IGCC applications. Oxy-gasification reduces NO_x formation since nitrogen is removed from the oxidizer stream (coal-bound nitrogen may produce a small amount of nitrogen oxides). Following the gas clean-up step, the syngas is burned in air in the gas turbine combustor, and the nitrogen that was originally separated in the ASU may be added to control the

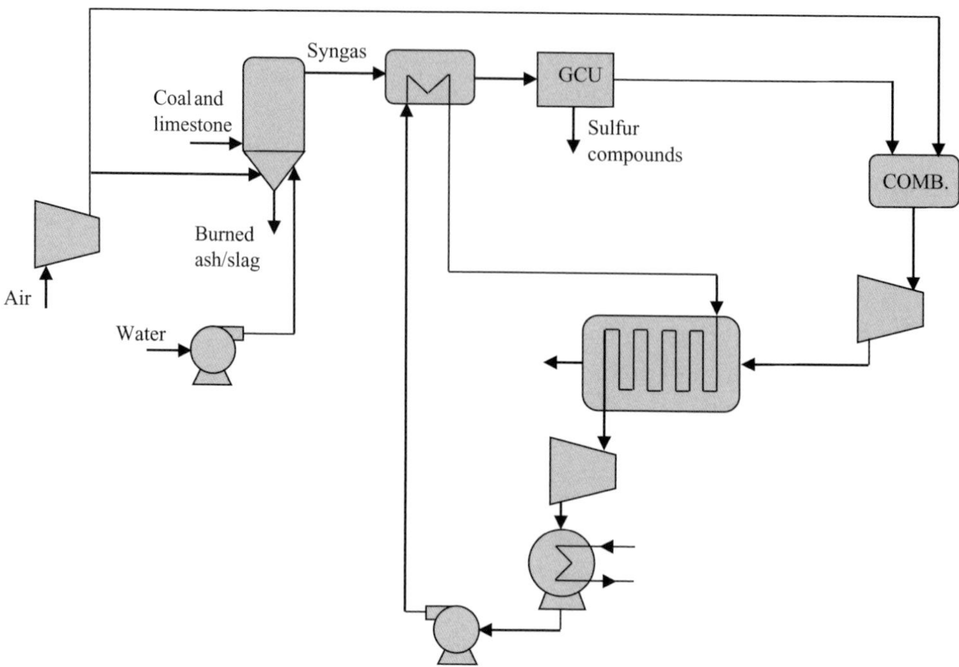

Figure 12.9 IGCC using a high-pressure coal gasifier, where coal, air, and steam are used to produce syngas. The gasifier gas is cooled by the steam-cycle feedwater stream, and sulfur compounds are separated in a gas clean-up unit. Following the gas clean-up unit, the syngas is burned in the gas turbine combustor (COMB.) using extra air from the compressor. The exhaust of the gas turbine is used in a HRSG to raise steam for the steam cycle. The steam leaving the steam turbine is condensed and pumped up to the steam cycle high pressure.

temperature. Nitrogen reduces the products temperature to values compatible with the gas turbine. Following expansion in the gas turbine, the hot exhaust is used to raise steam, as in all other CC plants. Oxy-coal gasification is more widely used than air gasification.

In most gasification plants, CGCU processes are used because they are more effective. In CGCU, the syngas is first cooled, and the heat removed before the clean-up step is recycled back to the same stream after gas clean-up, as shown in Figure 12.11. Prior to the CGCU, some of the gas stream thermal energy is used to heat the feedwater for the steam generator and/or to raise steam. In regenerative cooling of the syngas for the CGCU step, heat is removed from the gas between the gas cooler and the CGCU equipment, and added back to the same stream before the syngas is burned in the gas turbine combustor.

12.6.3 Efficiency and Environmental Performance

Reported IGCC plant efficiencies are below 45%, which is better than most pulverized coal counterpart (with similar steam power island). Natural gas combined cycle plant thermal

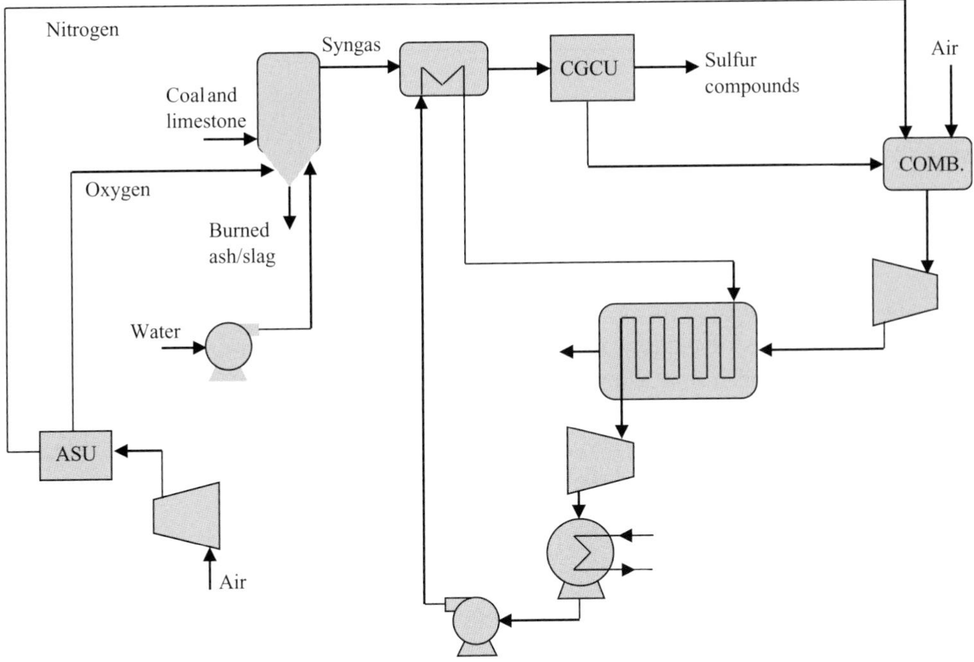

Figure 12.10 IGCC using an ASU and a pressurized entrained-flow or fluidized-bed gasifier to partially oxidize coal in oxygen and steam. The syngas is cooled by the feedwater of the steam cycle. The cooled gas is then cleaned by removing particulates and sulfur compounds and then burned in the gas turbine combustor. Nitrogen separated in the ASU is used to cool the combustion products in the combustor. The products are expanded in the gas turbine, and the gas turbine exhaust is used to raise steam in a HRSG.

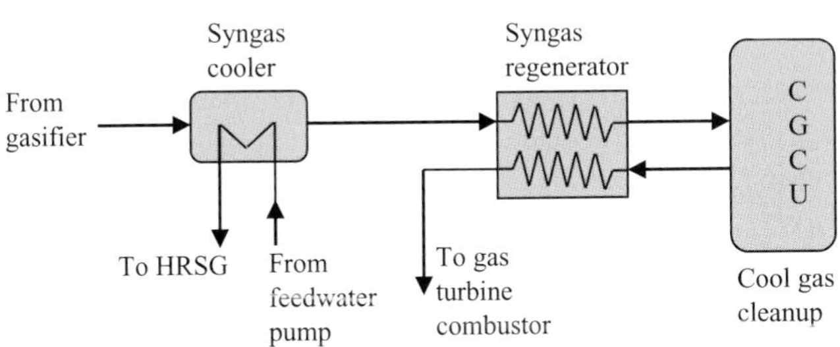

Figure 12.11 Syngas cooling (the heat is used to raise steam in the HRSG), and regenerative cooling before CGCU. The syngas is reheated in the regenerator before it is sent to the gas turbine combustor.

efficiency is 55% and above. The lower efficiency of IGCC plant is because of the losses associated with gasification, extensive gas clean-up, and the complexity of the overall plant integration. The efficiency of CC plants depends strongly on the inlet temperature to the gas turbine, the steam turbine cycle, whether it is subcritical or supercritical, the efficiencies of components such as compressors, the effectiveness of the different heat exchanger units, and the gas clean-up unit. Minimizing heat loss and limiting the temperature difference in the

Table 12.4 Coal plant efficiency

Technology	Steam conditions	Efficiency (%, LHV)
Pulverized coal (desulfurized)	Subcritical	35–37
Pulverized coal (desulfurized)	Supercritical 25–30 MPa, 550–600 °C	45–48
IGCC	Subcritical	43
PFBC, gasification	Supercritical, 25 MPa	42–45

Table 12.5 The environmental performance of different coal-burning power generation technologies in terms of sulfur oxides, nitric oxides, particulate matter, CO_2, water use, and total solids, including pulverized coal with advanced emissions controls technologies, AFBC with SNCR technology, PFBC without SNCR, and IGCC plants [7]

Pollutants, CO_2, water, and solids	Pulverized coal plant (advanced control)	AFBC (with SNCR)	PFBC (without SNCR)	IGCC plant
SO_2 (lb/MWh)	2.0	3.9	1.8	0.7
NO_x (lb/MWh)	<1.6	1.0	1.7–2.6	0.8
PM10 (lb/MWh)	<0.3	0.12	0.13–0.26	<0.14
CO_2 (lb/kWh)	2.0	1.92	1.76	1.76
Water usage gallons/MWh	1750	1700	1555	750–1100
Total solids generated, (lb/MWh)	367: ash, gypsum	494: ash, spent sorbent	450: ash, spent sorbent	175: slag, sulfur

heat transfer components are important for raising the overall efficiency of the plant. Better integration of the plant components and advances in HGCU technologies could improve the efficiency. Because of its higher efficiency, gasification-based coal technology reduces CO_2 emission, and it lends itself better to a high-efficiency CO_2 capture technology (to be discussed in Chapter 13). Table 12.4 shows the efficiency of different coal power plants.

The table shows that alternative coal technologies can achieve similar or higher efficiencies (depending on the design of the Rankine cycle). However, the superior emission characteristics of these plants are currently their most significant selling points, as shown in Table 12.5, where they are compared with pulverized coal power plants equipped with advanced pollution controls, atmospheric fluidized-bed combustion (AFBC) equipped with SNCR, PFBC without SNCR, and IGCC plants. IGCC plants belong to the "clean coal" category.

However, because of the requirements for gasification, the complex power island, and the gas clean-up processes, this technology is currently more expensive than pulverized coal technology.

IGCC plants have been built in the USA and around the world. Their entrained-flow gasifier, using bituminous and subbituminous coal, will be described next. They incorporate a cryogenic ASU to deliver 95–99% pure oxygen to the gasification unit. Intermediate- and

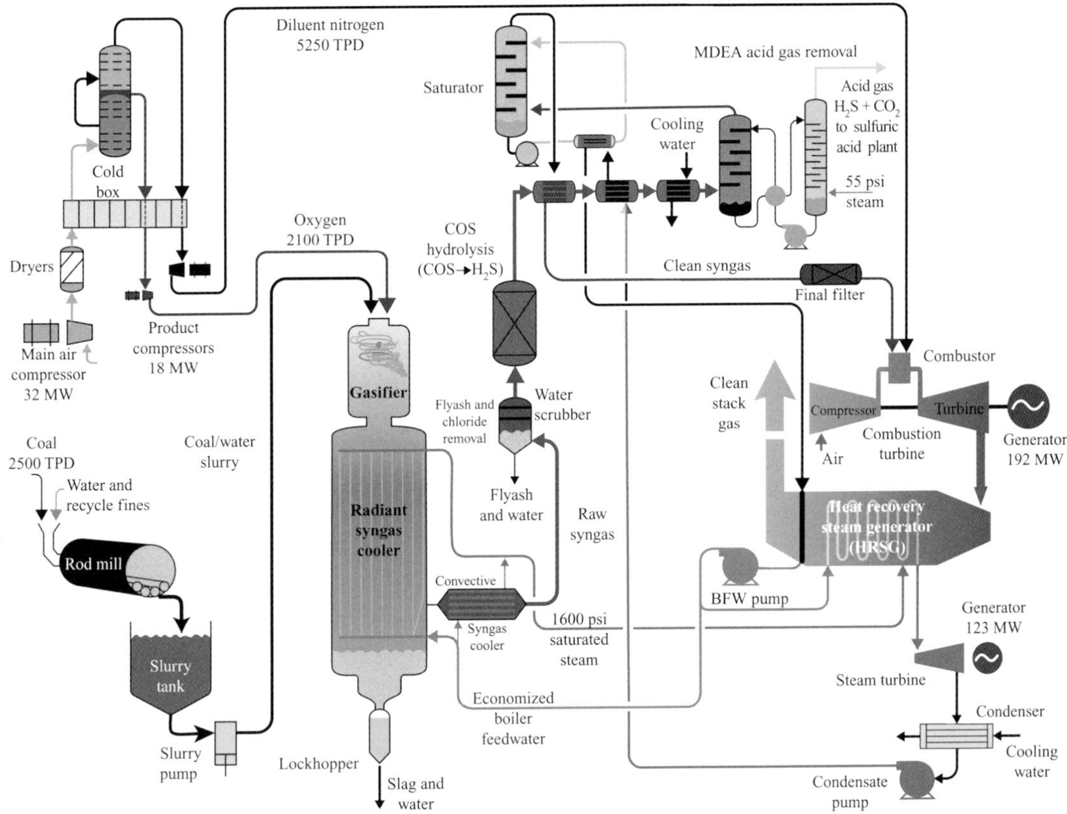

Figure 12.12 The Tampa Electric Polk IGCC power plant [8].

high-sulfur coals have been used (taking advantage of their superior gas clean-up capability). Most of these plants separate hydrogen sulfides that form during gasification (H_2S) using amine scrubbers, with up to 99% sulfur recovery. NO_x control is accomplished using nitrogen and steam dilution in the gas turbine combustors. Other compounds such as ammonia and chlorides are removed via water scrubber.

Figure 12.12 shows a schematic of Tampa Electric's 250 MW_e Polk Power Station (Tampa Electric Integrated Gasification Combined Cycle CCT Project), located near Lakeland, Polk County, Florida. The plant was built in 1996 and runs at 35.3% efficiency. It uses 2500 tons of coal per day and produces 200 tons of sulfuric acid.

Example 12.1

A schematic diagram of an IGCC plant is shown in Figure E12.1. The gasifier operates at 1100 °C and 30 bar, within which each mole of coal (LHV = 29 MJ/kg), represented as

Example 12.1 (cont.)

$CH_{0.85}O_{0.2}$, reacts with 2.5 moles of water and 0.76 moles of oxygen produced using an ASU. The reactants enter the gasifier at 25 °C and 30 bar. The syngas is cooled down to 400 °C within a heat exchanger, after which it enters the combustor of the gas turbine, where it mixes and reacts with the pressurized air coming from the air compressor at stochiometric ratio. A fraction of nitrogen produced at the ASU is fed to the combustor to keep the gas TIT at 1150 °C. The combustion products expand in the gas turbine and leave at 1 bar. The hot gaseous mixture leaving the gas turbine is cooled within a HRSG. The mixture is finally discharged to the atmosphere at 125 °C. The inlet temperature and pressure of the steam turbine are 530 °C and 10 MPa. The saturated steam condenses within a condenser operating at 8 kPa. The isentropic efficiencies of compressor, gas turbine, steam turbine, pump 1, and pump 2 are 80%, 89%, 92%, 75%, and 70%, respectively.

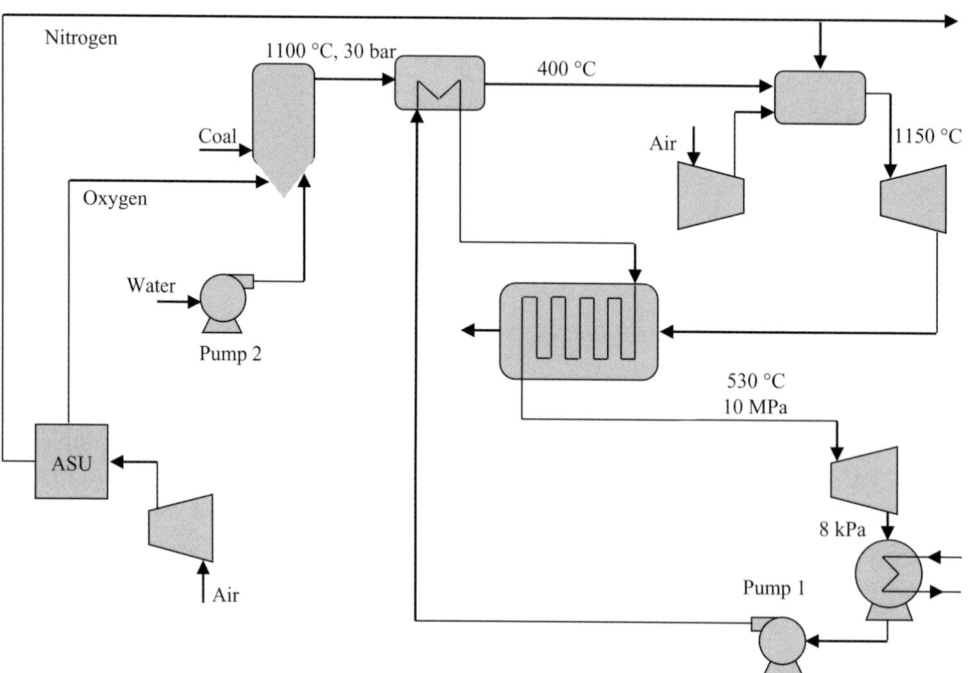

Figure E12.1

a. Determine the composition of the syngas leaving the gasifier, and confirm that concentration of CH_4 in the syngas is negligible.

b. Determine whether the gasifier operates adiabatically.

c. Calculate the number of moles of nitrogen per mole of coal added to the combustor and the number of moles of nitrogen that exit the power plant. Assume that the combustor is adiabatic.

Example 12.1 (cont.)

d. Determine the number of moles of water in the steam cycle and the temperature at the exit of the syngas cooler, assuming no heat losses in the cooler and the HRSG.
e. Calculate the work production of the combined cycle per mole of coal.
f. Calculate the thermal efficiency of the IGCC power plant.

Solution

a. The chemical reaction equation of the gasification process can be written as

$$CH_{0.85}O_{0.2} + 2.5H_2O + 0.76O_2 \rightarrow aH_2O + bH_2 + cCO + dCO_2 + eCH_4,$$

$$\text{C Balance}: \quad c + d + e = 1$$

$$\text{H Balance}: \quad 2a + 2b + 4e = 5.85,$$

$$\text{O Balance}: \quad a + c + 2d = 4.22.$$

Two additional equations can be established using the equilibrium reactions (11.22) and (11.23) for the products of the gasification reaction. For equilibrium reaction (11.22), we have

$$\frac{a.c}{b.d} = K_p(1373\,\text{K}) = 2.046.$$

The value of $K_p(1373\text{ K})$ is obtained from Table 3.4 by interpolation. For equilibrium reaction (11.23), we need to calculate the corresponding value of K_p by first determining the Gibbs free energy of the reaction at 1373 K (see Chapter 3). Therefore,

$$\Delta G(1373\,\text{K}) = 169{,}938\,\text{kJ/kmol},$$

$$K'_p(1373\,\text{K}) = \exp\left(-\frac{\Delta G}{R.T_g}\right) = \exp\left(-\frac{169{,}938}{8.314 \times 1373}\right) = 3.425 \times 10^{-7}$$

$$\frac{X_{H_2O}.X_{CH_4}}{X_{CO}.X^3_{H_2}} = p^2 K'_p(1373\,\text{K}) \Rightarrow \frac{a.e}{c.b^3}(a+b+c+d+e)^2 = p^2 K'_p(1373\,\text{K}).$$

A solution of the above five equations (three from elemental balance, two from equilibrium) yields

$$a = 2.486 \quad b = 0.4394 \quad c = 0.2656 \quad d = 0.7344 \quad e = 0.2 \times 10^{-12}.$$

It can be seen that the content of methane in the syngas is negligible.

b. To determine whether the gasifier operates adiabatically, we calculate the net heat of gasification reaction. For this, we need to first determine the enthalpy of formation of the coal. The LHV of the coal is given as 29 MJ/kg. This is equal to the heat of complete combustion of coal with oxygen at 298 K, with water appearing in the form of vapor. Hence,

$$CH_{0.85}O_{0.2} + 1.1125O_2 \rightarrow CO_2 + 0.425H_2O,$$

Example 12.1 (cont.)

$$\hat{h}^{298K}_{CO_2} + 0.425\hat{h}^{298K}_{H_2O} - \hat{h}^{298K}_{CH_{0.85}O_{0.2}} = -\text{LHV}_{CH_{0.85}O_{0.2}} = -29 \times 10^3[\text{J/g}] \times 16.05[\text{g/mol}]$$
$$= 465{,}450 \text{ J/mol}.$$

The sign of the LHV is negative because the reaction is exothermic. Rearranging, we get

$$\hat{h}^{298K}_{CH_{0.85}O_{0.2}} = -393{,}800 - 0.425 \times 242{,}000 + 465{,}450 = -312 \text{ kJ/mol}.$$

Applying the First Law to the gasifier, we have

$$Q_{Gasifier} = \sum_P n_i\hat{h}_i - \sum_R n_j\hat{h}_j,$$

$$a = 2.486b = 0.4394c = 0.2656d = 0.7344e = 0.2 \times 10^{-12}$$

$$\sum_P n_i\hat{h}_i = 2.486\hat{h}^{1373K}_{H_2O} + 0.4394\hat{h}^{1373K}_{H_2} + 0.2656\hat{h}^{1373K}_{CO} + 0.7344\hat{h}^{1373K}_{CO_2} = -747.467 \text{ MJ},$$

$$\sum_R n_j\hat{h}_j = \hat{h}^{298K}_{CH_{0.85}O_{0.2}} + 2.5\hat{h}^{1298}_{H_2O(1)} + 0.5\hat{h}^{298K}_{O_2} = -746.2 \text{ MJ}.$$

Substituting, we find $Q_{Gasifier} = -747{,}467 - 746{,}200 = -1267$ kJ.

So, for each mole of coal gasified, 1267 kJ of heat is lost through the walls of the gasifier.

c. Noting that the syngas is burned in air at stoichiometric ratio, the combustion reaction equation is

$$\underbrace{2.486H_2O + 0.4394H_2 + 0.2656CO + 0.7344CO_2}_{\text{syngas at 400 °C}} + n_{air,sto}(O_2 + 3.76N_2) + n_{N_2}N_2$$
$$\rightarrow \underbrace{a'H_2O + d'CO_2 + (3.76n_{air,sto} + n_{N_2})\,N_2}_{\text{products at 1150 °C}}.$$

By applying the elemental balance, $n_{air,sto}$, a', d' are determined as follows.

$$\text{C Balance}: \quad d' = 0.2656 + 0.7344 = 1,$$
$$H_2 \text{ Balance}: \quad a' = 2.486 + 0.4394 = 2.9254,$$
$$\text{O Balance}: \quad 2.486 + 0.2656 + 2 \times 0.7344 + 2n_{air,sto} = a' + 2d' \Rightarrow n_{air,sto} = 0.3525.$$

The number of moles of nitrogen to be added to the combustor is obtained by applying the conservation of energy to the combustor. To do this, the temperature of air coming from the compressor should be known. With the compressor inlet temperature and pressure of 298 K and 1 bar, and isentropic efficiency of 80%, the outlet temperature of the compressor is 871.4 K.

Example 12.1 (cont.)

Assuming an adiabatic combustion process, the conservation of energy gives

$$\left(2.486\hat{h}_{H_2O}^{673K} + 0.4394\hat{h}_{H_2}^{673K} + 0.2656\hat{h}_{CO}^{673K} + 0.7344\hat{h}_{CO_2}^{673K}\right) + 0.3525\left(\hat{h}_{O_2}^{871.4K} + 3.76\hat{h}_{N_2}^{871.4K}\right) + n_{N_2}\hat{h}_{N_2}^{298K}$$
$$= 2.9254\hat{h}_{H_2O}^{1423K} + \hat{h}_{CO_2}^{1423K} + (1.3254 + n_{N_2})\hat{h}_{N_2}^{1423K}.$$

Solving the above equation results in $n_{N_2} = 2.29$.

Since 0.76 moles of oxygen is fed to the gasifier, the number of moles of nitrogen produced at the ASU is $0.76 \times 3.76 = 2.8576$. From this, 2.29 moles are sent to the gas turbine combustor. So, 0.5676 ($= 2.8576 - 2.29$) moles of nitrogen leave the plant.

d. The number of water moles in the steam cycle and the temperature at the syngas cooler exit can be determined by analyzing the HRSG and the syngas cooler.

The inlet temperature of the combustion products entering the HRSG is obtained from the thermodynamic model of the gas turbine. With the inlet temperature and pressure of 1423 K and 30 bar, an exit pressure of 1 bar, and an isentropic efficiency of 89%, the exit temperature is calculated as 780.4 K. We now apply the First Law to the HRSG. Hence,

$$n_{water}\left(h_{10MP_a}^{803K} - h_{water}\right) = 2.925\left(h_{H_2O}^{780.4K} - h_{H_2O}^{398K}\right) + \left(h_{CO_2}^{780.4K} - h_{CO_2}^{398K}\right) + 3.6154\left(h_{N_2}^{780.4K} - h_{N_2}^{398K}\right).$$

The unknowns in this equation are n_{water} and h_{water}. The second equation is obtained by applying the First Law to the syngas cooler. The enthalpy of water at the inlet of the syngas cooler is determined from the thermodynamic model of the pump. The water on the upstream of the pump is at 0.08 bar and saturated:

$$h_{Pump1,o} = h_{Pump1,in} + \frac{h_{Pump1,os} - h_{Pump1,in}}{\eta_p} = 3119 + \frac{3300 - 3119}{0.75} = 3360.3 \text{ kJ/kmol},$$

where $h_{Pump1,os} = h_{s=s_{in}}^{10\text{ MPa}}$.

Applying the First Law to the cooler, we get

$$n_{water}\left(h_{water} - h_{Pump1,o}\right) = 2.486\left(h_{H_2O}^{1373K} - h_{H_2O}^{673K}\right) + 0.4394\left(h_{H_2}^{1373K} - h_{H_2}^{673K}\right)$$
$$+ 0.2656\left(h_{CO}^{1373K} - h_{CO}^{673K}\right) + 0.7344\left(h_{CO_2}^{1373K} - h_{CO_2}^{673K}\right).$$

Likewise, the only unknowns in the equations are n_{water} and h_{water}. Solving the two equations obtained by applying the First Law to the HRSG and syngas cooler, we find

$$n_{water} = 2.998 \quad \text{and} \quad h_{water} = 28{,}624 \text{ kJ/kmol}.$$

From water/steam thermodynamic tables, it can be inferred that $h_f^{10\text{ MPa}} < h_{water} < h_g^{10\text{ MPa}}$. Thus, the water leaving the cooler is a mixture of liquid and vapor with a quality of $x = 0.138$. The corresponding temperature is $T_{water}^{sat}(10\text{ MPa}) = 584.2$ K.

Example 12.1 (cont.)

e. The net work production of the IGCC plant is obtained from:

$$W_{net} = W_{GT} + W_{ST} - W_C - W_{Pump1} - W_{Pump2}$$

$$W_{GT} = 2.9254\left(h_{30\text{ bar}}^{1423\text{K}} - h_{1\text{ bar}}^{780.4\text{K}}\right)_{H_2O} + \left(h_{30\text{ bar}}^{1423\text{K}} - h_{1\text{ bar}}^{780.4\text{K}}\right)_{CO_2} + 3.6154\left(h_{30\text{ bar}}^{1423\text{K}} - h_{1\text{ bar}}^{780.4\text{K}}\right)_{N_2}$$
$$= 192{,}317\text{ kJ}.$$

$$W_{ST} = n_{water}\left(h_{100\text{ bar}}^{803\text{K}} - h_{0.08\text{bar}}^{sat}\right)_{water} = 67{,}412\text{ kJ}$$

$$W_C = n_{air}\left(c_{p,\text{O}_2} + 3.76c_{p,\text{N}_2}\right)(871.4 - 298) = 29{,}261\text{ kJ}$$

$$W_{Pump1} = n_{water}\left(h_{Pump1,o} - h_{Pump1,in}\right) = 723.4\text{ kJ}$$

$$W_{Pump2} = n_{\text{H}_2\text{O}}\left(h_{Pump2,o} - h_{Pump2,in}\right) = 192.8\text{ kJ}.$$

$$h_{Pump2,o} = h_{Pump2,in} + \frac{h_{Pump2,os} - h_{Pump2,in}}{n_{P_2}} = 1876 + \frac{1930 - 1876}{0.7} = 1953.1\text{ kJ/kmol}.$$

Hence, $W_{net} = 192{,}317 + 67{,}412 - 29{,}261 - 723.4 - 192.8 = 229{,}551.8\text{ kJ}$.

f. We can now calculate the thermal efficiency of the IGCC plant as follows.

$$\eta_{IGCC} = \frac{W_{net}}{\text{LHV}_{Coal}} = \frac{22{,}951.8}{465{,}450} = 0.493.$$

Note that the thermal efficiency of the IGCC plant would be lower than 49.3% if we had accounted for the energy requirement of the ASU.

12.7 Gasification

Gasification converts solid fuels, such as coal, biomass, and petcoke (the heavy residues of oil refining) to gaseous by partial oxidation and/or steam reforming. Other heavy liquid fuels, with high carbon content, can also be gasified to produce lighter gaseous fuels [3]. The products of gasification are determined by the carbonaceous feedstock, percentages of steam and oxygen added, and the gasifier pressure and temperature. Gasifiers are operated at sufficiently high temperature that carbon is converted to carbon oxides and some methane; although some tars may form (especially in low-temperature conditions) and some carbon may escape with other solid compounds (ash). Tars are condensable light hydrocarbons. The products of gasification, for properly designed gasifiers with good mixing and sufficient resident time, satisfy thermodynamic equilibrium at the exit conditions. Gasification is particularly desirable when using high-carbon "dirty" fuels since it is easier to clean-up the synthetic gas leaving the gasifier (especially after cooling the gas) than to clean-up combustion products, and for the production of gaseous fuels (for combustion or electrochemical conversion) and easy distribution.

Several gasifier designs have been used to convert coal to synthetic gas. The simplest and oldest is a fixed- or moving-bed gasifier. In a moving-bed gasifier with counter-flow of coal and gas, coal enters the gasifier at the top while steam and oxygen (or air) enter from the gasifier bottom. Coal undergoes a series of processes sequentially, including drying, pyrolysis, gasification, and combustion, as it descends downwards. (In other types, coal particles are likely to go through the four stages along their trajectory, although the different states may not be as distinct.) In the first stage, drying, moisture trapped in the pores of the solid particles is released. The amount of moisture in coal depends on the coal rank and whether it is pre-dried. (Biomass has much higher moisture content and is often pre-dried before gasification.) Devolatilization and pyrolysis refer to thermal decomposition of the fuel in an oxygen-starved environment. Pyrolysis of complex hydrocarbons often leads to the release of smaller hydrocarbon components, while devolatilization is mostly the release of volatiles trapped in the pores of the solid. During this process, volatile matter, which tends to be hydrogen-rich, is distilled, and mineral matter and carbon-rich residues are left in the solid form. Kinetic rates of coal pyrolysis peak around 400–500 °C, and drop off sharply outside this temperature range. Because of the dependence of the heating time on the coal particle size, the actual conversion time is a function of the temperature and particle size (see Chapter 14 for more detail).

The energetics and equilibrium of gasification thermodynamics were discussed in Chapter 3. As shown in Figure 12.13, following devolatilization, the left-over char is then partially or fully oxidized. Some of the volatiles and carbon are burned in the combustion stage at the bottom of the gasifier, to provide the thermal energy for many of the endothermic gasification reactions. Oxygen in the feed is used during the gasification and combustion stages. Left-over solids in the form of ash and char settle to the bottom of the gasifier and leave.

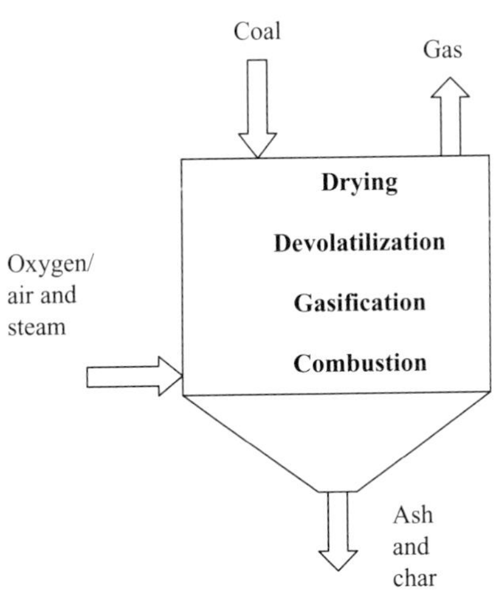

Figure 12.13 The different stages of coal transformation in a moving-bed gasifier. Coal is introduced at the top, and oxygen/air mixed with steam are introduced at the bottom. Gas, essentially $H_2 + CO + CO_2 + CH_4 + H_2O$, leaves at the top, while left-over solids accumulate at the bottom. The same stages of coal transformation can be thought of as taking place in other gasifier types, but the sequence will be more of a time sequence along the coal particle trajectory.

Table 12.6 Some coal gasification reactions, and the enthalpy of reaction (at standard conditions). Hydrogen oxidation is included since most coal contains volatile matter that contains hydrogen and other hydrocarbons. Depending on *T* and *p*, other reactions can be important, such as the methane forming reaction, $CO + 3H_2 \leftrightarrow H_2O + CH_4 - 205$ MJ/kmol

Reaction type	Reaction	ΔH_R MJ/kmol
Partial oxidation	$C + \frac{1}{2}O_2 \leftrightarrow CO$	−123
Combustion	$C + O_2 \leftrightarrow CO_2$	−393
Boudouard reaction	$C + CO_2 \leftrightarrow 2CO$	160
Heterogeneous shift reaction	$C + H_2O \leftrightarrow CO + H_2$	118
Hydrogasification or methanation	$C + 2H_2 \leftrightarrow CH_4$	−88
Hydrogen oxidation	$H_2 + \frac{1}{2}O_2 \leftrightarrow H_2O$	−242

Other types of gasifiers include: (1) fluidized-bed gasifiers in which coal particles, steam, oxygen, and other additives are introduced into a large bed of fine refractory material similar to that used in fluidized-bed combustors; and (2) entrained-flow gasifiers, both dry-fed – where coal is mixed with gas in the feed stream – or water–slurry fed – where coal is mixed with water. In entrained-flow systems, coal is introduced as pulverized fine particles.

Table 12.6 shows some gasification reactions and their enthalpies of reaction. The relative contribution of each reaction to the overall process depends on the gasification medium, temperature, and pressure (the equilibrium constants should be given in Chapter 3). In most cases, given sufficient residence time for the coal inside the gasifier, equilibrium defines the outcome of gasification and the gasifier products. In such cases, a limited number of reactions can be selected to define the gasification products.

The overall gasification process may be endothermic or exothermic, depending on the feedstock, the gasification medium, and the fraction of oxygen. Gasifiers are either heated externally, called indirect gasification, or by some of the exothermic partial oxidation reactions or full oxidation reactions that occur during gasification. Gasifiers are also cooled to maintain their temperature at the desired value by running steam generation pipes through the gasification zone. Even for coal or residues of petroleum refineries, such as petcoke, we are dealing with a hydrocarbon whose composition, or carbon to hydrogen ratio, depends on the source. In general, one can write an overall oxy-gasification reaction as

$$C_nH_m + \frac{n}{2}O_2 \rightarrow nCO + \frac{m}{2}H_2, \tag{12.15}$$

where for coal, $n \sim m \sim 1$, for oil, $n \sim 1$ and $m \sim 2$, and for natural gas, $n \sim 1$ and $m \sim 4$.

Predicting the outcome of the overall gasification process, under equilibrium conditions, was discussed in Chapter 3. The calculations involve elemental mass balance for hydrogen, oxygen, and carbon, and energy balance across the gasifier. More equations are obtained by assuming equilibrium for a number of reactions that define the state of the mixture (e.g., some of (12.16)–(12.21)). One can use an equilibrium computer code to perform these calculations. Using a code, it is easy to use as many species as desired in the calculations over a wide range of conditions. The primary purpose of gasification is to convert carbon to gas (CO, CO_2, CH_4, etc.), and hence gasifiers are run at temperatures that ensure almost complete conversion of all the carbon, and one need not include C in the products species.

Because the feedstock is essentially a carbon-rich fuel, water, and oxygen or air, and under the typical gasification temperatures of 700–1500 °C, the equilibrium mixture is expected to consist of CO, CO_2, H_2, H_2O, and CH_4 (methane is present if the temperature is sufficiently low). Mass balance is used to write three relations between the concentrations of these species, conserving H, O, and C. Performing the calculations using the Law of Mass Action, two more relations are obtained by satisfying the equilibrium conditions for the homogeneous gas shift reaction and the methane forming reaction, respectively:

$$CO + H_2O \rightarrow CO_2 + H_2 \qquad -42\ \text{MJ/kmol} \tag{12.16}$$

and

$$CO + 3H_2 \rightarrow CH_4 + H_2O \qquad -206\ \text{MJ/kmol}. \tag{12.17}$$

Note that these two equations are obtained by algebraic manipulations of (12.16)–(12.20). The conditions of equilibrium for these two reactions show the dependence of the equilibrium molar fractions, X_i, on the pressure and temperature as follows:

$$\frac{X_{CO_2} X_{H_2}}{X_{CO} X_{H_2O}} = K_{CO+H_2O \rightarrow CO_2+H_2}\left(T_{gas}\right) \tag{12.18}$$

and

$$\frac{X_{CH_4} X_{H_2O}}{X_{CO}\left(X_{H_2}\right)^3} = p^{-2} K_{CO+3H_2 \rightarrow CH_4+H_2O}\left(T_{gas}\right). \tag{12.19}$$

The dependence of the equilibrium constants on temperature was discussed before. Given the pressure and temperature, the solution of the five equations (three elemental mass conservation and two equilibrium) yields the composition of the mixture leaving the gasifier. In addition, the energy equation can be used to calculate the heat exchange during gasification,

$$\sum_{in} n_i \hat{h}_i(T_{i,in}) + Q_{in} = \sum_{out} n_i \hat{h}_i\left(T_{gas}\right) + H_{char}. \tag{12.20}$$

In this equation, Q_{in} is the heat transfer to the gasifier, and H_{char} is the enthalpy of the ash/char leaving the gasifier. The latter can be captured and used to heat, for example, the feed stream. The problem can also be solved using one of the equilibrium codes widely available in thermochemical software libraries.

Figures 12.14 and 12.15 show the temperature, cold gas efficiency (CGE), and H_2/CO ratio as functions of the O_2/C ratio and H_2O/C ratio in the gasifier. General trends for the dependence of the gas composition on the feed composition are as follows:

1. The gasification temperature rises with increasing oxygen ratio in the feed as more of the fuel is combusted. More water reduces the temperature because of evaporation.

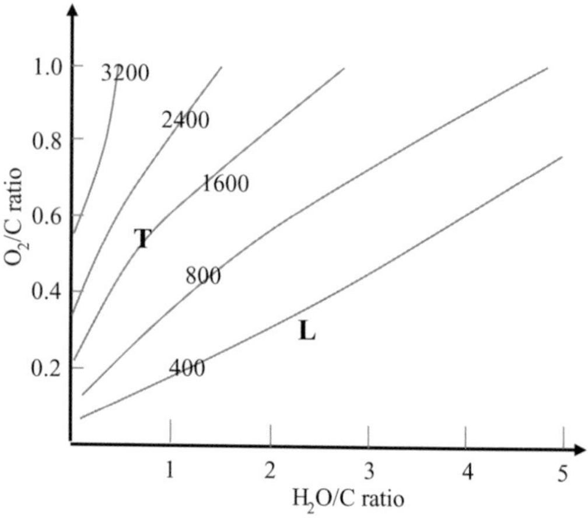

Figure 12.14 The gasification temperature in °C for different water/carbon ratio and oxygen/carbon ratio. T and L are for high temperature and low temperature operation.

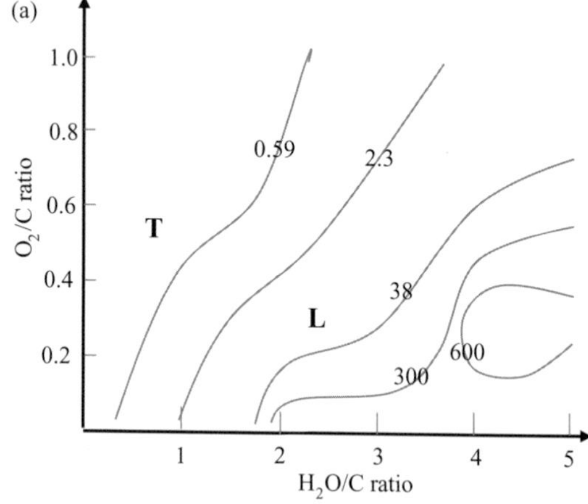

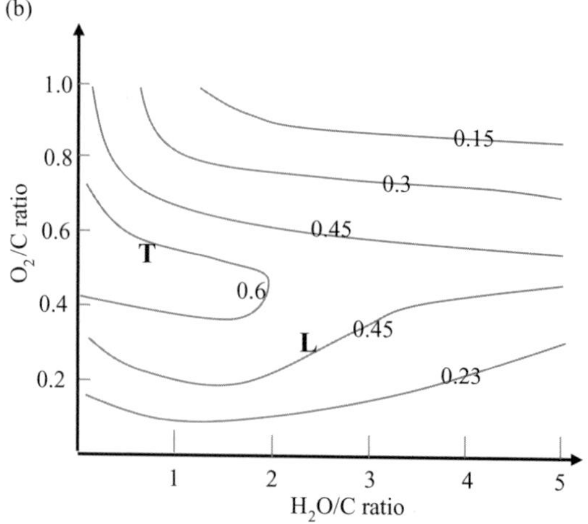

Figure 12.15 The CGE (a) and hydrogen/CO ratio (b) as a function of the oxygen/carbon ratio and water/carbon ratio.

Water–slurry-fed gasifiers require more oxygen to reach the same gasification temperatures than dry-fed gasifiers.

2. More oxygen in the feed increases the concentrations of carbon dioxide and water because of the rise in the oxidation level.
3. Hydrogen increases and carbon monoxide decreases as more steam is added in the feed, but only to a certain extent, as the presence of higher concentration of water encourages the water–gas shift (WGS) reaction and hence increases the concentration of carbon dioxide.
4. The concentrations of hydrogen and carbon monoxide rise with increasing temperature and decreasing pressure. Higher temperatures lead to water and CO_2 dissociation.
5. Methane production is favored at low temperature and high pressure. Low-temperature gasification, below 700 °C, is favored in methane production, as shown in Figure 12.16.
6. Carbon dioxide concentration increases at low temperature, high pressure, and high oxygen concentration.
7. Raising the water level in the feed negatively impacts the CGE, and dry-fed gasifiers have higher CGE than water–slurry-fed ones.

The following observations can be made from the equilibrium analysis and other discussions:

1. Recycling some of the carbon dioxide from the cooler flue gases can be used to change the condition within the gasifier – for instance, to reduce the temperature, especially in the case of oxy-gasification, and to control the concentrations of the different species in the syngas.
2. Gasification without H_2O, also called dry gasification, is slow since the reactions involved in this case are slow, and high temperatures and pressures are required to speed up the reactions, while long residence times are required to reach equilibrium.

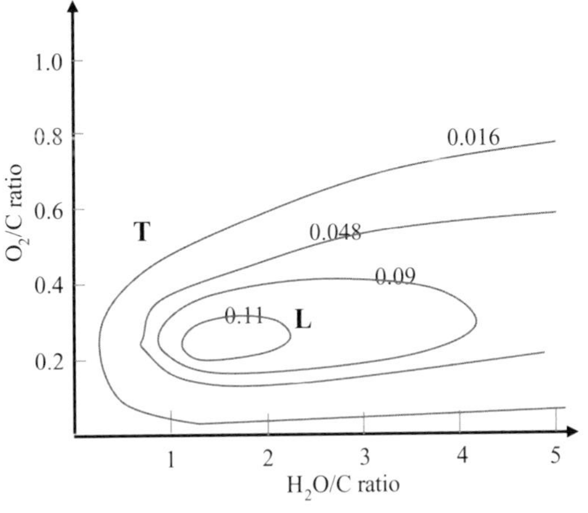

Figure 12.16 The methane mole fraction in the syngas as a function of oxygen/carbon ratio and water/carbon ratio.

3. Steam is added to increase the hydrogen content in the syngas, although it may contribute to lowering the temperature.
4. Increasing the steam fraction promotes the WGS and the methanation reactions.

Products of gasification are characterized by their volumetric heating value, with the volume measured at standard pressure and temperature (STP), and are classified as follows:

- high heating value gas is in the range 35–39 MJ/m^3;
- medium heating value gas is in the range 11–17 MJ/m^3; and
- low heating value gas is in the range 5–8 MJ/m^3.

The volumetric heating value depends on the relative percentages of methane, carbon monoxide, hydrogen, nitrogen, and carbon dioxide. As shown above, the gas composition depends on the pressure and temperature of gasification, the gasification medium (air or oxygen), and the amount of steam added to the gasifier. At STP, CH_4 has a volumetric heating value of 37 MJ/m^3, while H_2 and CO have values around 12 MJ/m^3.

Coal gasification enables better gas clean-up (in part because of the smaller volume of gas) and produces lower emissions of nitric oxides and sulfur compounds than coal combustion. Most gas clean-up processes are performed at low temperature, much lower than those prevailing inside the gasifier. Without careful integration of the overall cycle, a large efficiency penalty can result from cooling the syngas prior to clean-up unless careful integration is done. Integration requires transferring heat from the syngas leaving the gasifier to the gas turbine, the HRSG, or the same gas stream after having passed through the gas clean-up equipment. Multiple heat exchangers operating at high temperatures are required, which would add to the cost of the plant initially and during operation.

Energy losses are incurred during syngas clean-up because of the separation energy required to remove a relatively small concentration gases from the mixture. A number of definitions have been proposed for the overall gasification efficiency. One such definition is the total chemical and thermal energy in the clean synfuel following gas clean-up plus the thermal energy extracted from the gasifier (e.g., to raise steam) as a fraction of the chemical energy in the original fuel:

$$\eta_{gas} = \frac{\text{syngas (chemical + thermal) energy + heat transfer to steam generator}}{\text{chemical energy in original fuel}}. \quad (12.21)$$

Gasification efficiency is strongly dependent on the type of coal used and the integration between components in the combined cycle. High-ash coal (reaching up to 30%) can reduce the gasification efficiency if the coal is gasified in a high-temperature slagging gasifier, since ash melts, consuming a fraction of the original chemical energy in the coal. Heat loss in the gasifiers, in the gas clean-up equipment, and in the different heat exchange processes, lowers the gasification efficiency. Pressure drop in these devices also contributes to the efficiency loss (measured by the availability loss) in the overall cycle.

Example 12.2

Coal can be converted to a syngas, which consists primarily of carbon monoxide and hydrogen. This is done by gasification of coal with steam at high temperature and pressure. Modeling the coal as carbon, the hydrogasification reaction may be written as

$$C + bH_2O \rightarrow \alpha CO + \beta H_2 + \gamma CO_2 + \delta H_2O,$$

where the concentration of each species in the products is determined by the equilibrium of the WGS reaction. We use the following expression for that equilibrium constant:

$$\ln K = -3.821 + \frac{4209}{T}.$$

The gasifier operates at 30 bar. The coal (carbon) and water enter at 298 K. The gasification reaction is endothermic, so an external heat source is required to provide the enthalpy of reaction and to increase the temperature of the gas. Plot, as a function of the outlet temperature between 1200 K and 1800 K, and for $b = 1$, 1.5, and 2:

a. the equilibrium mole fractions of CO and H_2;
b. the heat input per mole of carbon; and
c. the efficiency of the gasification process.

Solution

a. We can write three equations from elemental balance:

$$\begin{aligned} \text{C balance}: &\ \alpha + \gamma = 1 \\ \text{H balance}: &\ \beta + \delta = b \\ \text{O balance}: &\ \alpha + 2\gamma + \delta = b. \end{aligned}$$

Rearranging in terms of α, we get

$$\begin{aligned} \gamma &= 1 - \alpha \\ \beta &= 2 - \alpha \\ \delta &= b + \alpha - 2. \end{aligned}$$

The equilibrium expression for the WGS reaction in terms of the mole fractions is

$$K = \frac{X_{CO_2} X_{H_2}}{X_{CO} X_{H_2O}} = \frac{(1-\alpha)(2-\alpha)}{\alpha(\alpha + b - 2)}.$$

The solution of the above equation gives α as a function of K and the number of moles of water, b.

$$\alpha = \frac{\sqrt{(Kb - 2K + 3)^2 + 8(K-1)} - (Kb - 2K + 3)}{2(K-1)}.$$

Example 12.2 (cont.)

This equation can be used to determine α for given values of K (function of temperature) and b. The mole fractions of carbon monoxide and hydrogen are obtained from

$$X_{CO} = \frac{\alpha}{b+1}$$

$$X_{H_2} = \frac{2-\alpha}{b+1}.$$

The solutions are shown in Figure E12.2a.

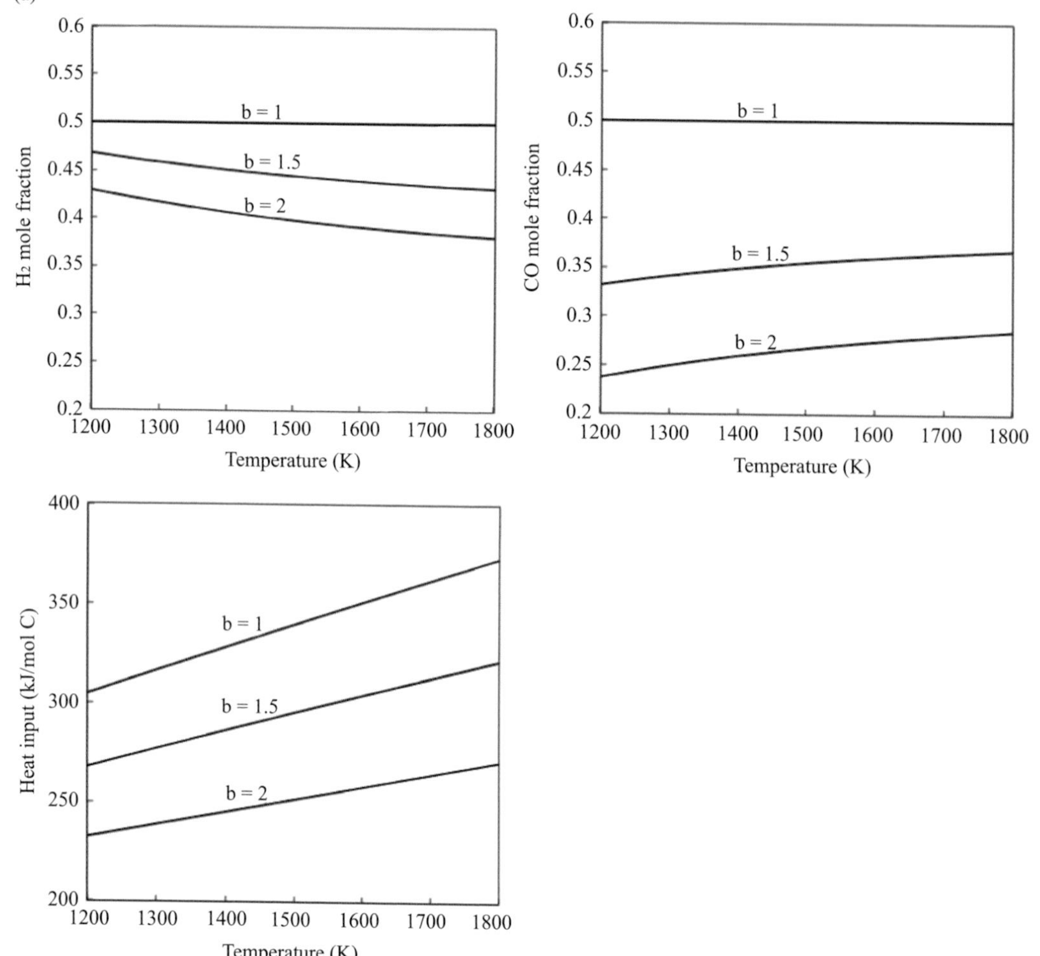

Figure E12.2a

Example 12.2 (cont.)

b. The heat input required for the gasification process is obtained by applying the First Law:

$$Q = \sum_{product} n_i \hat{h}_i - \sum_{reactant} n_j \hat{h}_j = \alpha\left[\hat{h}_{f,\mathrm{CO}} + c_{\mathrm{p,CO}}(T - T_0)\right] + (2-\alpha)\left[\hat{h}_{f,\mathrm{H_2}} + c_{p,\mathrm{H_2}}(T - T_0)\right]$$
$$+ (1-\alpha)\left[\hat{h}_{f,\mathrm{CO_2}} + c_{p,\mathrm{CO_2}}(T - T_0)\right]$$
$$+ (\alpha + b - 2)\left[\hat{h}_{f,\mathrm{H_2O(g)}} + c_{p,\mathrm{H_2O}}(T - T_0)\right] - b\hat{h}_{f,\mathrm{H_2O(l)}}.$$

Substituting the values of enthalpy of formation and constant pressure specific heat for the species involved in the gasification process, we get

$$Q = \alpha[-110,530 + 33.18 \times (T - 298)] + 30.21 \times (2 - \alpha)(T - 298)$$
$$+(1-\alpha)[-393,800 + 54.3 \times (T - 298)]$$
$$+ (\alpha + b - 2)[-242,000 + 41.27 \times (T - T_0)] + 286{,}000b.$$

The heat required the gasification process is shown in Figure E12.2a.

c. The efficiency of the gasification process is obtained from

$$\eta = \frac{\sum_{product} n_i \mathrm{LHV}_i}{\mathrm{LHV_C} + Q} = \frac{\alpha \mathrm{LHV_{CO}} + \beta \mathrm{LHV_{H_2}}}{\mathrm{LHV_C} + Q},$$

where $\mathrm{LHV_{CO}} = 283.27\,\mathrm{kJ/mol}$, $\mathrm{LHV_{H_2}} = 242\,\mathrm{kJ/kmol}$, and

$$\mathrm{C + O_2 \rightarrow CO_2 \Rightarrow LHV_C} = -\hat{h}_{f,\mathrm{CO_2}} = 393.8\,\mathrm{kJ/mol}.$$

Figure E12.2b illustrates the gasification efficiency as a function of temperature and b.

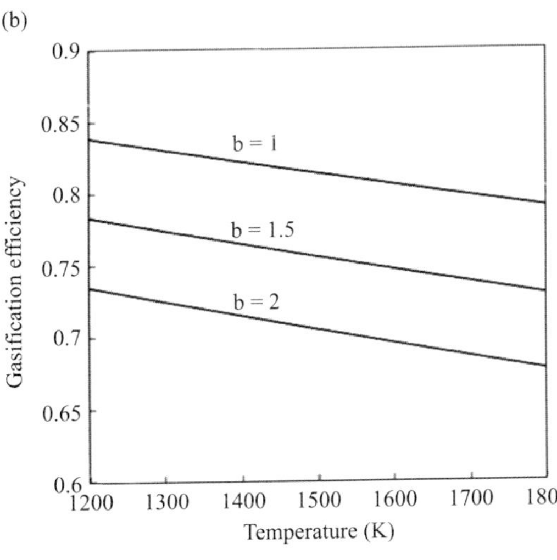

Figure E12.2b

12.7.1 Gasifier Types

Several types of gasifier have been used to produce syngas, including **fixed-/moving-bed gasifiers**, **fluidized-bed** gasifiers, and **entrained-flow gasifiers**. In the moving-bed and fluidized-bed gasifiers, coal and oxygen are introduced in counter-current flow (some moving-bed gasifiers use co-flow or cross-flow currents, and will be discussed in Chapter 14), but the temperature distributions are very different. In the fixed/moving bed, the combustion zone is at the bottom of the gasifier, where oxygen is available at sufficient quantities to completely oxidize the char and provide heat for the gasification reactions (the confusing terminology of fixed/moving bed applies to the same design; the bed actually moves downwards under gravity, slowly, and it "fixed" under steady state). The heat generated at the bottom and the remaining oxygen rises upwards to partially oxidize the coal. In the fluidized-bed system, and because of the thorough mixing associated with fluidization, the temperature is nearly uniform. Coal particle size differs significantly among these types, with the finest required for the entrained-flow gasifiers and the coarsest used in the fixed-bed types. The temperature also differs significantly, with the lowest used in the fluidized-bed type and the highest in the entrained-flow type. While low temperatures impact the gasification kinetics, the better mixing and smaller particle size in the fluidized bed lead to shorter residence time in the fluidized bed than in the fixed bed. The fastest conversion rates are achieved in the entrained-flow system because the temperature is significantly higher and the particle size is smaller. The gasification temperature determines the fate of the ash; gasification above the ash fusion temperatures in a fixed-bed or entrained-flow system is necessary if ash is to be melted and flowed out as slag.

Fixed-bed, also called "moving-bed", gasifiers, like the Lurgi and BG (British Gas) gasifiers, use counter-flows of coal and gas, with coal moving downward and gas moving upward. These gasifiers produce gas at temperature in the range of 800–1200 °C. Because of the relatively low temperature, the syngas may contain some methane, and the ash is separated in the form of solid. Moving-bed gasifiers use larger coal particles and do not require significant grinding or drying of the coal; particles of several centimeters in size are typical for these applications. Oxygen consumption is relatively low, and the temperature of the syngas exiting the gasifier at the top is also low since the hot syngas produced in the bottom part acts as a heating and drying medium for the coal at the top part. Low-temperature gasification and the fact that the coal is devolatilized toward the exit of the gas stream mean that the syngas may contain a fraction of volatiles and pyrolysis products, which can be undesirable. The relatively higher temperature in the bottom half of the vessel means that ash can be separated in the form of slag, but if the temperature is kept low enough it can be separated as ash that leaves at the bottom of the vessel. The lower temperatures and large coal particles lead to a high residence time inside the gasifier for complete conversion. More discussions of moving-bed gasifiers, especially counter- and cross-flow gasifiers, as applied to biomass, can be found in Chapter 14.

Fluidized-bed gasifiers are similar to PFBCs, but they use less air/oxygen and more water to produce the desired syngas composition, instead of completely oxidized products [9].

Similar to their combustion counterpart, they provide a nearly uniform temperature internally, but fluidized-bed gasifiers run at 800–1000 °C, or lower if a catalyst is used to enable fast kinetics at low temperatures. Low-temperature operation reduces the oxygen requirements, but the very good heat and mass transfer of the bed and the uniform temperature reduces the overall residence time required for carbon conversion. Typically, coal is introduced at the top of the bed while oxygen and steam are introduced from ports located below a grate and used as the fluidizing medium. Medium-size coal particles are used, up to 1 cm. Fluidized-bed gasifiers must operate below the ash melting temperature to avoid the formation of molten slag inside the bed, which can lead to agglomeration, and ash is collected at the bottom. Examples of fluidized-bed gasifiers are the Winkler gasifier, the high-temperature Winkler gasifier (up to 1200 °C), and the KRW (Kellogg–Rust–Westinghouse) gasifiers. Fluidized-bed gasifiers are more fuel-flexible than moving-bed gasifiers, and have been used for high-moisture coal and biomass. Desulfurization in these systems is accomplished by adding limestone to the feed. Some fluidized-bed gasifiers, the recirculating bed type in which the fluidization velocity is sufficiently high so that some of the char is carried out with the syngas, recycle some of the char to ensure complete carbon conversion, especially the types that utilize circulating beds. Fluidized-bed gasifiers are discussed more in Chapter 14, along with their application to biomass gasification.

Figure 12.17 shows schematically the three types of gasifiers, the size of the coal particles used in each case, and the temperature range and distribution within each gasifier. Table 12.7 shows a comparison between the three categories of gasifiers in terms of their operating conditions, residence time, gas clean-up technology, range of applications, and more [10].

Entrained-flow gasifiers produce syngas at higher temperatures, 1500–1900 °C. In this type, air is injected at a high velocity, creating a turbulent environment in which vigorous mixing speeds up the burning process. The solid fuel is finely ground, with typical size ~0.1 mm, and is introduced dry or mixed with water in the form of slurry. Moreover, because of the high-temperature environment, non-combustible compounds such as metal residues leave in the liquid phase. To maintain the high temperature, these gasifiers consume relatively high amounts of oxygen. Also, coal should be reasonably dry, and low-quality coals are not used much in this application. Because of the fast mixing and the high temperature, the kinetics are fast and the residence time of the mixture is brief O(1–5 s). Most IGCC plants use this entrained-flow type. Entrained-flow gasifiers operate at high pressure, making them suitable for IGCC plants as well as fuel production plants.

Three types of entrained-flow gasifiers are used extensively in the chemical industry for power and for syngas production. The GE/Texaco gasifier, shown schematically in Figure 12.18, is a slurry-fed down-fired, oxygen-blown gasifier. It is known as a single-stage gasifier since coal is introduce from a single port at the top, and combustion and gasification occur in the same space at the top end of the gasifier column. The gasifier is lined with refractory material to protect the walls against the intense radiation, and the syngas is internally cooled by radiation and convection, with the relative contributions determined by the design of the cooling section and the application. Because of the high temperature, ash

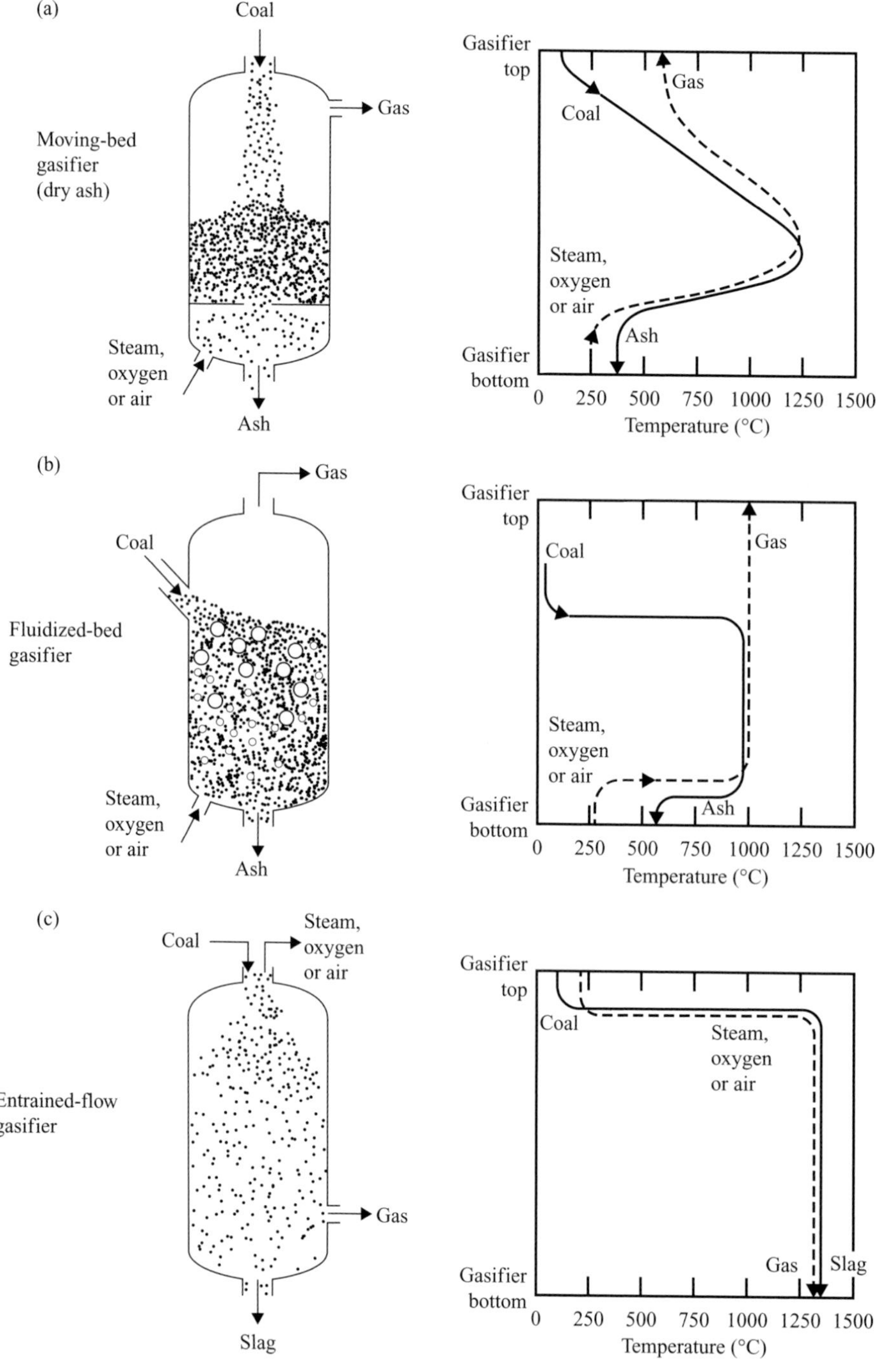

Figure 12.17 Three types of gasifiers and the idealized temperature distribution of the coal (solid line) and gas (broken line) inside each one: (a) fixed/moving bed, (b) fluidized-bed gasifier, (c) entrained-flow gasifier [11]. The temperature distributions are idealized and only detailed measurements or models can

Table 12.7 Gasifier types, their operating temperatures and pressures, feedstock delivery systems and particle size, required residence time for gasification, and sulfur removal technology [10].

	Fixed/moving bed	Fluidized bed	Entrained flow
Maximum temperature (K)	1420	1200	1640–1920
Pressure (atm)	1–27	1–68	1–82
Feedstock particle size (mm)	5–50	1–5	<0.1
Oxidant	Air or oxygen	Air or oxygen	Air or oxygen
Ash condition	Dry or slagging	Agglomerating	Slagging
Residence time	>1 hour	0.5–1 hour	~1–5 seconds
Sulfur removal	Downstream	Limestone/dolomite	Downstream
Syngas HHV (MJ/m^3)	11–14	5.5 (air-blown)	11–13
Advantages	Simple	High productivity, low-quality coal,	Higher productivity and conversion
Disadvantages	High tar	Complex, tar formation	Needs high-quality coal
Application	Synfuels and chemicals	IGCC	IGCC, synfuels and chemicals

leaves at the bottom of the gasifier in the form of slag. Gas can be cooled partially by water quenching. This reduces the size of the cooling (bottom section) and the cost, and contributes to the shift of the produced syngas toward higher hydrogen concentration. This gasifier uses a slurry pump to raise the coal–water slurry and inject the coal, and can operate at very high pressures.

Another type is the Conoco-Philips or E-gas gasifier, shown schematically in Figure 12.19 This gasifier is different because it has two stages; the first is essentially a combustion stage in which coal slurry and oxygen burn to provide heat for gasification. Downstream (upward of the first stage), another coal slurry stream is injected for gasification. Char that leaves with the syngas is captured in a cyclone and recirculated back to the combustion section of the gasifier. The gas is cooled after the second stage. This is also an oxygen-blown system, lined with refractory material and producing molten slag.

A dry-fed gasifier, similar to the Shell gasifier, is shown in Figure 12.20. This dry-fed, up-fired, oxygen-blown system is more suitable for high-moisture coal, and uses lock hoppers to pressurize the pulverized coal, limiting the pressure inside the gasifier to ~35 bar.

Figure 12.17 (*cont.*) show the actual distributions. In each diagram, the size of the coal particles in the feed are shown, as are the locations where the coal, oxygen, and steam are introduced. Note that, at the bottom of the gasifier, different types of solids leave, depending on the temperature and other operating conditions in the gasifier. With a low temperature at the bottom of a fixed-bed gasifier, ash leaves. At higher temperature and with low oxygen concentration at the bottom of the fluidized bed, it leaves in the form of char. At the highest operating temperature in the entrained-flow system, slag leaves.

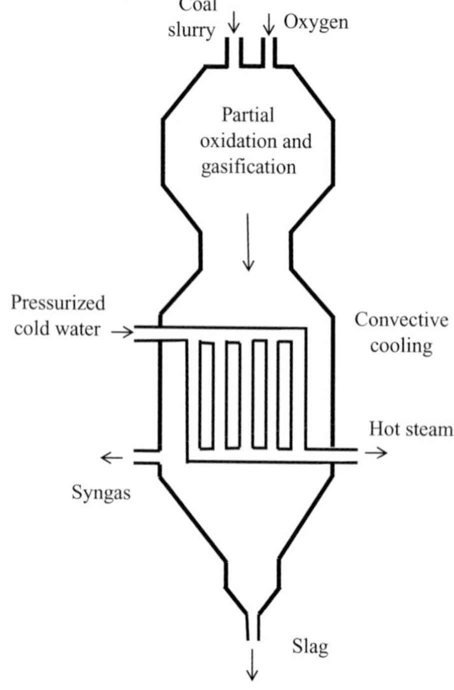

Figure 12.18 The GE (Texaco) entrained-flow gasifier with radiant cooling used to raise steam for a power cycle. The gasifier is fed with coal–water slurry and oxygen. It is a single-stage system that maintains sufficiently high temperature by simultaneous partial and completer oxidation of some of the coal and gasifying the rest.

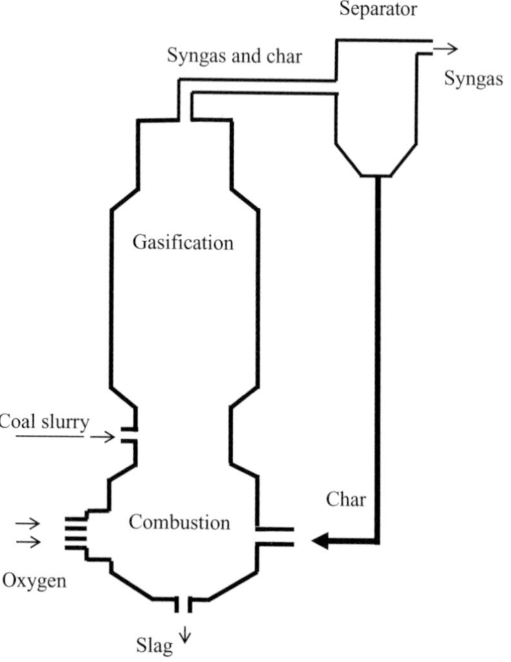

Figure 12.19 The E-gas entrained-flow gasifier. Fed with slurry and oxygen from the bottom, it maintains a high-temperature combustion zone there. Secondary feeding of more coal slurry is for gasification. Some char leaves with the syngas and is captured using a separator and recycled back to the combustion zone. The high temperature at the bottom ensures the flow of hot slag,

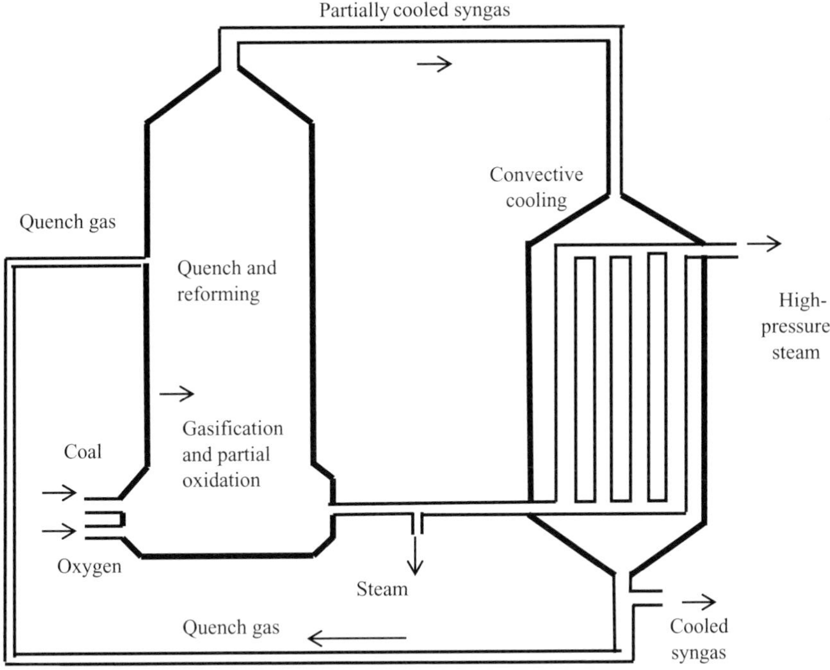

Figure 12.20 The Shell entrained-flow gasifier and gas cooler. Coal, oxygen, and steam are injected for gasification. Quench cooling follows gasification, aided by injecting cooler gas and potential more water/steam. Further cooling of the syngas is done convectively in the gas cooler while raising steam for the power cycle (or other applications).

Gasification and combustion occur concurrently at the bottom of the gasifier column, and the gases are cooled partially inside the gasifier and partially in a separate syngas cooler. The walls are lined with a pipe network that circulates water to cool the gas (known as a membrane wall). Heat transfer takes place by radiation and convection, and ash leaves in the form of slag. However, gas cooling is done partially by quenching by recirculating cool syngas to the gasification products inside the gasifier column.

12.8 Fuel Synthesis Using Syngas

Coal gasification has been used in the production of gaseous and liquid fuels as well as chemicals, in what is known as coal-to-gas, coal-to-chemical, and coal-to-liquid (CTL) (see Figure 12.21). In all cases, the process starts with the production of syngas in one of the gasifiers described in the previous sections. Processes that start with syngas are called "indirect" since they involved an intermediate process, namely gasification (direct processing of coal to fuel has been developed but is much less practiced and will not be covered, with the exception of direct coal to methane, which will be mentioned briefly). Syngas can be used

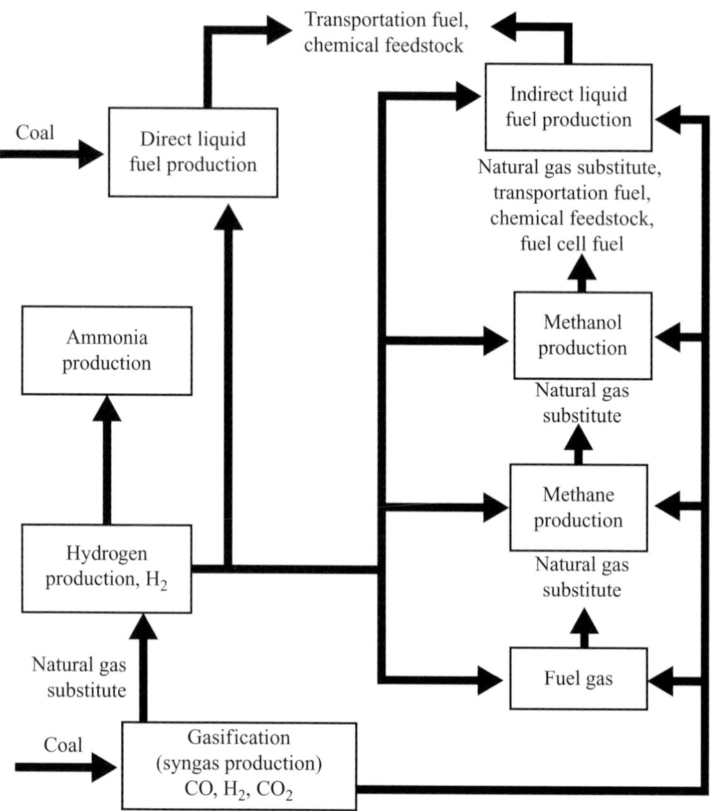

Figure 12.21 The role of gasification as a process to convert solid hydrocarbons, including coal, to a variety of products ranging from gaseous fuel mixtures, to pure hydrogen or liquid transportation fuels, to other chemicals such as ammonia [4].

directly as a fuel (and was used extensively before natural gas). The synthetic gas, with different ratios of H_2 to CO (and some CO_2 in same cases) is used in a catalytic reactor to produce gaseous fuel like methane, or liquid fuels like methanol or synthetic diesel-like fuel [3,4]. Figure 12.21 shows different ways in which coal gasification products can be used as for hydrogen production, for methane (as a fuel) and methanol (used as a fuel and for chemicals) production, or for the production of liquid fuels using Fischer–Tropsch synthesis. In the case of hydrogen, WGS reactors are used to convert CO in the syngas to CO_2 and more hydrogen (using steam) followed by separating CO_2 (using, e.g., pressure-swing adsorption [PSA]). Hydrogen can be reacted with nitrogen to produce ammonia (for many applications such as the production of fertilizers).

12.8.1 Methane

Low-temperature gasifiers, operating below 600 °C, can be used to maximize the percentage of methane in the syngas. At low temperature, it is necessary to use a catalyst to speed up the gasification process. This is a special application of the "direct" gasification of coal into fuel – that is, without using a synthesis process to combine the syngas components into methane or other higher hydrocarbons. Otherwise, the syngas produced using a high-pressure gasifier is used in a synthesis process to produce a hydrocarbon. For instance, **methane** can be synthesized via one or both of the following exothermic reactions in a methanation reactor:

$$CO + 3H_2 \rightarrow CH_4 + H_2O \quad -206 \text{ MJ/kmol}, \tag{12.22}$$

$$2CO + 2H_2 \rightarrow CH_4 + CO_2 \quad -247 \text{ MJ/kmol}. \tag{12.23}$$

These are the reverse steam and dry reforming reactions, respectively. The stoichiometry of the syngas is different in both cases, with significantly more hydrogen used in the first reaction, a fraction of which is rejected in the form of water (or steam). In the second reaction, on the other hand, some of the carbon is rejected in the form of carbon dioxide. In both cases, methane must be separated from the products gases (by water condensation and other gas separation processes; see Chapter 10). The exothermicity is different in both reactions, and operating the second reaction requires more cooling to maintain the reactor temperature at the optimal level. Lower temperature favors methane formation according to equilibrium (although kinetics would suffer). Depending on the temperature at which the reactor is operated, the heat rejected can be at sufficiently high temperature that it is used in raising steam for a power cycle (this is an example of coproduction of fuel and power from a single gasification process). The first reaction is less efficient because it rejects less heat per mole of methane produced, but water separation is easier, and it is the more popular approach for generating methane from syngas. If CO_2 is captured for storage, the second reaction reduces the CO_2 emitted overall. Both reactions are exothermic, and equilibrium favors products formation at lower temperature (as well as higher pressures).

The exothermicity of the methane formation reactions means that heat is rejected during the synthesis process. Depending on the composition of the syngas and reaction pathways, it is 20–25% of the enthalpy of reaction of the syngas, and the "fuel conversion efficiency," that is the chemical energy in the product (methane) as a fraction of the chemical energy of the reactants (carbon monoxide + hydrogen at the given ratio) is significantly lower than unity. The heat rejected at the methanation reactor temperature can be used somewhere else in the plant, such as in the gasification island to raise steam and provide some of the endothermic energy of gasification. Such integration is necessary to keep the overall efficiency of coal to synthetic natural gas high. Values around 60–65% have been reported, depending on the gasifier and level of plant integration. Heat from the methanation reactor can also be used in a power cycle compatible with the reactor temperature, with the electricity used as a product (coproduction of fuel and power) or used in part to power the oxygen production plant, pumps, and compressors, etc., in the plant.

Direct conversion of coal to methane in the gasifier is also possible. The fraction of methane produced using steam gasification of coal is shown in Figure 12.22. As shown, low-temperature and low-steam fractions favor methane formation (over hydrogen and CO), but the kinetics can be slow at these temperatures. Thus, it is important to use a catalyst to speed

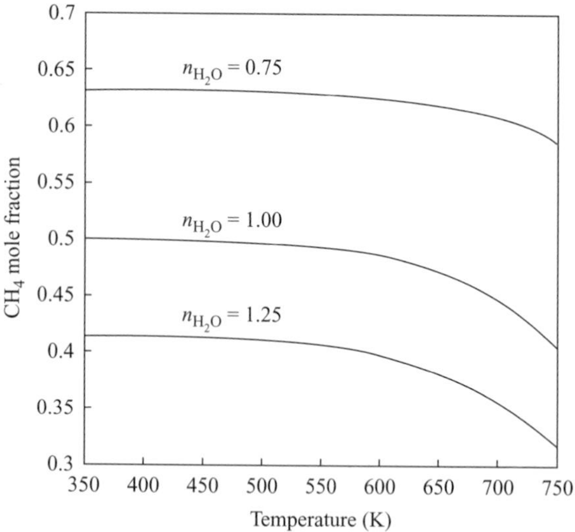

Figure 12.22 Mole fraction of methane in the products of coal (assumed to be $CH_{0.8}$) gasification as a function of temperature, at three values of number of water moles used in the process.

up the reactions at the low temperature favoring the formation of methane. Nickel is used extensively for this purpose, at around 300–400 °C. Other catalysts, such as ruthenium and rhodium, are also effective but significantly more expensive. Moreover, the methane fraction in the products of the gasification process is still low, with the remaining carbon forming CO and CO_2. For utilization of most of the carbon in the production of methane, syngas production at higher temperature is used, followed by "methanation" as described above.

12.8.2 Methanol

Methanol is used as a liquid fuel or fuel additive, solvent, antifreeze, or for the production of other chemicals such as formaldehyde, which itself is used in the production of paints, plastics, etc. Methanol can be produced by combining 1 mole of CO and 2 moles of hydrogen at moderate pressure (5–10 atm), and temperatures in the range of 250–400 °C. Under these conditions, the following reaction is performed over a metal catalyst (copper, zinc oxide, or alumina) [3]:

$$CO + 2H_2 \rightarrow CH_3OH_{(l)} \quad -128\,\text{MJ/kmol}. \tag{12.24}$$

The reaction is exothermic (but less exothermic than the methane synthesis reactions) and low temperatures (and higher pressures) favor the formation of the product. Another reaction that utilizes the CO_2 in the mixture of the syngas combines it with more hydrogen according to the reaction

$$CO_2 + 3H_2 \rightarrow CH_3OH_{(g)} + H_2O_{(g)} \quad -49.5\,\text{MJ/kmol}, \tag{12.25}$$

in which more hydrogen is used and less heat is rejected. Lower temperature is again required to shift the equilibrium toward the products side and minimize the production of

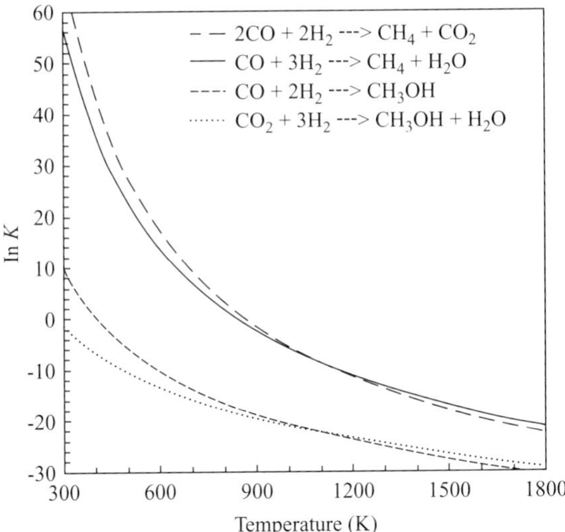

Figure 12.23 The equilibrium constants for methane and methanol production using syngas. The lines follow the same order as the reactions listed in the figure, with methane formation reactions at the top, followed by methanol formation reactions. Because of the exothermicity of the reactions, low temperatures favor the formation of products.

unwanted compounds. Well-integrated gasification-based methanol plants operate with ~50–60% efficiency (coal to methanol), with a significant fraction of the losses in the condensation of methanol.

Figure 12.23 compares the equilibrium constants of methane and methanol production reactions using synthesis fuel. In both cases, as mentioned before, lower temperatures are need for significant formation of products.

A schematic diagram of a low-pressure methanol production plant using synthesis gas is shown in Figure 12.24. Syngas is compressed (5–10 bar), then regeneratively heated by the products of the synthesis reactor. An external heater is used to bring the temperature to the requisite temperature of the reactor (250–300 °C), where methanol is formed. The products are cooled, methanol condensed and separated, and the remaining syngas recycled back to the plant.

12.8.3 Fischer–Tropsch Liquids

Syngas is also used to produce other **liquid fuels**, some of which are similar to gasoline and diesel fuels conventionally produced from oil refining (see also Chapters 3 and 14), in the Fischer–Tropsch (FT) process. For the production of gasoline-like liquids (branched alkenes), the following exothermic reaction applies:

$$CO + 2H_2 \rightarrow (-CH_2-)_{(l)} + H_2O \quad -231\,\text{MJ/kmol}, \tag{12.26}$$

where the enthaply of reaction is calculated assuming that the products is 1/8 of a 1-octene (C_8H_{16}) molecule. The alernative reation is

$$2CH + H_2 \rightarrow (-CH_2-) + CO_2 \quad -228\,\text{MJ/kmol}. \tag{12.27}$$

The ideal ratio of hydrogen to CO required for the FT process is 2, but water forming in the first reaction can react with excess CO to form more H_2, which drives the second reaction.

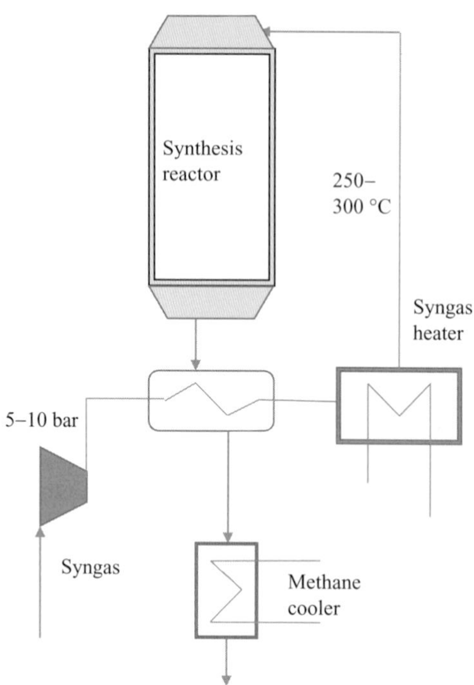

Figure 12.24 A schematic diagram showing methanol synthesis from syngas.

FT processes operate at temperatures in the range 225–365 °C and pressures in the range 0.5–4 MPa. Iron, cobalt, and nickel-based catalysts can be used, with iron used most extensively in practice. The catalyst, besides speeding up the kinetics, plays a key role in the selectivity – that is, determining which products form faster and with higher concentrations. Both the material of the catalyst and how it is fabricated impact the selectivity, as well as its state of oxidation and the presence of other promoters and impurities. Other factors include the residence time and the type of reactor. The temperature of the reactor plays an important role in the selectivity, with lower temperatures promoting the production of higher molar mass compounds. For instance, the production of gasoline (relatively lower molar mass) occurs at higher temperatures. Another factor is the actual CO/H_2 ratio used and the amount of CO_2 and water available. Higher partial pressure of CO_2 in the syngas promotes the formation of liquid fuels. Other compounds, primarily oxygenated hydrocarbons (alcohols) form, as well as solid waxes whose carbon numbers are higher than 35. Product selectivity is ultimately governed by kinetics instead of thermodynamic equilibrium and gas recycling is used to increase the yield.

Figure 12.25 shows a schematic diagram of an FT reactor [3]. The necessary catalyst is formulated in the form of particles that are circulated between a cooled reactor and a separator. While the reactants are sent to the reactor at relatively lower temperature, ~160 °C, the circulating catalyst is introduced at higher temperature, ~350 °C. The exothermicity of the conversion reactions raise the reactor temperature further and significant

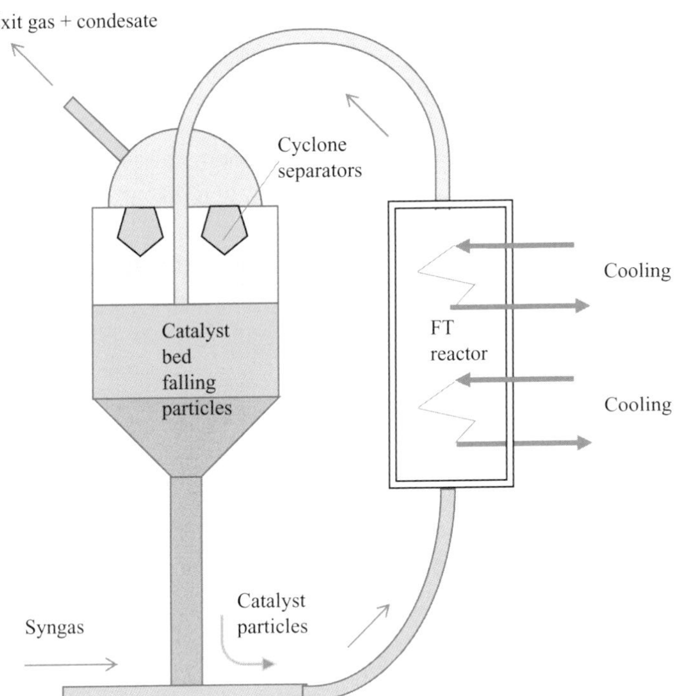

Figure 12.25 Schematic of the Synthol FT fluidized-bed reactor. The catalyst particles are circulated between the cooled FT reactor on the right and a gas–solid separator on the left, along with the syngas as they enter the FT reactor, and with the reformed gas as they exit the FT reactor. The exit gas from the solid separator is cooled to separate the different components and the unconsumed syngas is recycled back. The FT reactor is cooled to remove the excess heat generated by the exothermic reactions.

cooling is needed to keep the temperature sufficiently low for high conversion. The gas–solid mixture leaving the reactor is sent to a cyclone in order to separate the solid catalyst for recycling from the off-gas, which is cleaned and cooled to separate the light hydrocarbons, the product, from heavier ones such as wax. Some of the gas is recycled to the reactor for further processing. The overall efficiency of the process is estimated to be ~55–60%.

Besides coal, most solid organics can be gasified, including biomass, refinery residues such as petcoke, and industrial carbonaceous waste. It is also possible to "gasify" methane through the reforming process described in Chapter 3. The same processes described above for the use of syngas to produce methane, methanol, etc., can be used with syngas produced using these other feedstocks. Biomass gasification will be discussed in detail in Chapter 14 including conversion and reactors.

Example 12.3

A schematic diagram of a coal gasification plant for hydrogen production is shown in Figure E12.3. The chemical composition of coal used in this plant is $CH_{0.8}$, whose enthalpy of formation is 170.6 kJ/mol. For each mole of coal, 1 mole of water and n_{O_2} moles of oxygen enters the gasifier. Oxygen is supplied by an ASU. The reactants are supplied at 298 K and 40 bar and the gasifier operates adiabatically at constant pressure. Inside the gasifier, coal

Example 12.3 (cont.)

reacts with water and oxygen to form a mixture of H_2O, H_2, CO, and CO_2 (other contaminants are neglected). The gas exits the gasifier at 1200 °C and 40 bar. It then enters a high-temperature shift reactor (HTSR), where 0.5 mole of liquid water at 298 K and 40 bar is added to raise the hydrogen concentration in the mixture. The new mixture leaves the HTSR at 600 °C, entering a separation unit to remove all the CO_2 in the mixture at this stage. The mixture of H_2O, H_2, and CO enters a low-temperature shift temperature reactor (LTSR), where an extra 0.5 mole of liquid water at 298 K and 40 bar is added. The gas exits the LTSR at 200 °C. The newly formed CO_2 is again removed from the mixture.

a. Determine the number of moles of oxygen per mole of coal needed in the gasification process, and the mixture composition leaving the gasifier, assuming the gas mixture exiting the gasifier is at equilibrium.

b. Calculate the mixture composition leaving the HTSR and the heat transfer across the reactor. Assume that the mixture is at equilibrium, and it consists of H_2O, H_2, CO, and CO_2.

c. Calculate the mixture composition leaving the LTSR and the heat transfer across the reactor. Assume that the mixture is at equilibrium, and it consists of H_2O, H_2, CO, and CO_2.

d. Determine the coal gas efficiency.

Figure E12.3

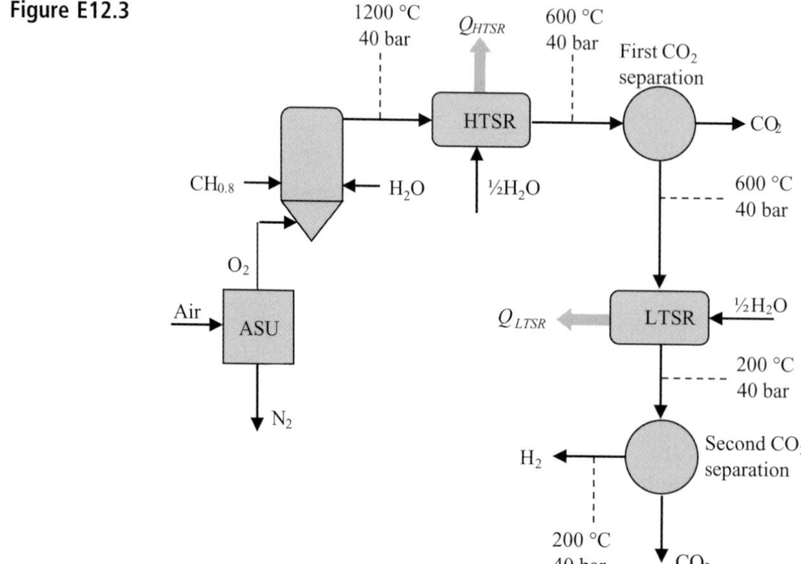

Example 12.3 (cont.)

Solution

a. The chemical reaction equation of the gasification process can be written as

$$CH_{0.8} + H_2O + n_{O_2}O_2 \rightarrow aH_2O + bH_2 + cCO + dCO_2,$$

$$\text{C Balance}: \quad 1 = c + d,$$

$$\text{H Balance}: \quad 0.8 + 2 = 2a + 2b,$$

$$\text{O Balance}: \quad 1 + 2n_{O_2} = a + c + 2d.$$

We have five unknowns and three equations. Two additional equations are required to find the composition. One equation can be established from the equilibrium reaction for the products of the gasification reaction above. Hence,

$$K_p(1473\,\text{K}) = \frac{X_{H_2O}X_{CO}}{X_{H_2}X_{CO_2}} = \frac{a.c}{b.d}.$$

The value of $K_p(1473\ \text{K})$ is obtained from Table 3.4 by interpolation. Hence,

$$K_p(1473\,\text{K}) = 2.454.$$

Another equation is obtained by applying the conservation of energy for the adiabatic gasifier. Hence,

$$\hat{h}^{298\text{K}}_{CH_{0.8}} + \hat{h}^{298\text{K}}_{H_2O(1)} + n_{O_2}\hat{h}^{298\text{K}}_{O_2} = a\left[\hat{h}^{298\text{K}}_{H_2O(g)} + c_{p,H_2O(g)}(T_G - 298)\right] + b\left[\hat{h}^{298\text{K}}_{H_2} + c_{p,H_2}(T_G - 298)\right]$$

$$+c\left[\hat{h}^{298\text{K}}_{CO} + c_{p,CO}(T_G - 298)\right] + d\left[\hat{h}^{298\text{K}}_{CO_2} + c_{p,CO_2}(T_G - 298)\right].$$

Note that the specific heats are evaluated at an average temperature of $(1473 + 298)/2 = 885.5\ \text{K}$. Solving the above five equations yields

$$a = 1.151 \quad b = 0.2486 \quad c = 0.3464 \quad d = 0.6536 \quad n_{O_2} = 0.9025.$$

b. The chemical reaction equation of the HTSR can be written as

$$\left(\underbrace{1.15H_2O + 0.2486H_2 + 0.3464CO + 0.6536CO_2}_{\text{at }1200\,^\circ\text{C}}\right) + 0.5H_2O_{(1)} \rightarrow \underbrace{a'H_2O + b'H_2 + c'CO + d'CO_2}_{\text{at }600\,^\circ\text{C}},$$

$$\text{C Balance}: \quad 0.364 + 0.6536 = c' + d',$$

$$\text{H}_2\,\text{Balance}: \quad 1.151 + 0.2486 + 0.5 = a' + b',$$

$$\text{O Balance}: \quad 1.151 + 0.3464 + 2 \times 0.6536 + 0.5 = a' + c' + 2d'.$$

Example 12.3 (cont.)

We need one more equation to close the system of equations. Assuming the products mixture is in equilibrium, we have, $K_p(873\mathrm{K}) = \frac{d'.c'}{b'.d'}$ where $K_p(873\mathrm{K}) = 0.2672$ is obtained from Table 3.4. Thus, a solution of the above four equations gives:

$$a' = 1.394 \quad b' = 0.5064 \quad c' = 0.0885 \quad d' = 0.9115.$$

To determine the amount of heat transfer from the HTSR, we apply the First Law as follows.

$$Q_{HTSR} = \sum_P n_i \hat{h}_i - \sum_R n_j \hat{h}_j,$$

where

$$\begin{aligned}\sum_P n_i \hat{h}_i &= 1.394 \times \left[-242{,}000 + c_{P,\mathrm{H_2O(g)}}(873 - 298)\right] + 0.5064 c_{p,\mathrm{H_2}}(873 - 298)\\ &\quad + 0.0885 \times \left[-110,530 + c_{p,\mathrm{CO}}(873 - 298)\right]\\ &\quad + 0.9115 \times \left[-393{,}800 + c_{p,\mathrm{CO_2}}(873 - 298)\right]\\ &= -631{,}282\ \mathrm{kJ}.\end{aligned}$$

The enthalpy of the reactants at the inlet of the HTSR is the same as the enthalpy of reactants at the gasifier inlet (due to the adiabatic gasification process) plus the enthalpy of the liquid water. Hence,

$$\begin{aligned}\sum_R n_j \hat{h}_j &= \left(\hat{h}^{298\mathrm{K}}_{\mathrm{CH_{0.8}}} + \hat{h}^{298\mathrm{K}}_{\mathrm{H_2O(1)}} + n_{\mathrm{O_2}} \hat{h}^{298\mathrm{K}}_{\mathrm{O_2}}\right) + 0.5 \hat{h}^{298\mathrm{K}}_{\mathrm{H_2O(1)}} = (-170{,}600 - 286{,}000 + 0) - 0.5 \times 286{,}000\\ &= -599,600\ \mathrm{kJ}.\end{aligned}$$

Thus, $Q_{HTSR} = -631{,}282 - (-599{,}600) = -31,682\ \mathrm{kJ}$.

The amount of heat transferred out of the HTSR is 31,682 kJ per mole of coal.

c. The chemical reaction equation of the LTSR can be written as

$$\left(\underbrace{1.394 H_2O + 0.5064 H_2 + 0.0885 CO}_{at\ \ 600^\circ C}\right) + 0.5 H_2O_{(1)} \to \underbrace{a'' H_2O + b'' H_2 + c'' CO + d'' CO_2}_{at\ \ 200^\circ C}.$$

Note that CO_2 of the mixture coming out of the HTSR is separated in the first CO_2 separation unit:

$$\begin{aligned}\text{C Balance}: &\quad 0.0885 = c'' + d'',\\ \text{H}_2\ \text{Balance}: &\quad 1.394 + 0.5064 + 0.5 = a'' + b'',\\ \text{O Balance}: &\quad 1.394 + 0.0885 + 0.5 = a'' + c'' + 2d'',\end{aligned}$$

Example 12.3 (cont.)

$$\frac{a''.c''}{b''.d''} = K_p(473\,\text{K}) = 0.006292.$$

Solving these four equations yields

$$a'' = 1.805 \qquad b'' = 0.5948 \qquad c'' = 0.000183 \qquad d'' = 0.0883.$$

To determine the amount of heat transfer from the LTSR, we apply the First Law as follows:

$$Q_{LTSR} = \sum_P n_i\hat{h}_i - \sum_R n_j\hat{h}_j,$$

where

$$\sum_P n_i\hat{h}_i = 1.805 \times \left[-286{,}000 + c_{p,\text{H}_2\text{O(l)}}(473-298)\right] + 0.5948c_{p,\text{H}_2}(473-298) + 0.000183$$
$$\times \left[-110{,}530 + c_{p,\text{CO}}(473-298)\right] + 0.0883 \times \left[-393{,}800 + c_{p,\text{CO}_2}(473-298)\right]$$
$$= -523276\,\text{kJ},$$

$$\sum_R n_j\hat{h}_j = 1.394 \times \left[-242{,}000 + c_{p,\text{H}_2\text{O(g)}}(873-298)\right] + 0.5064c_{p,\text{H}_2}(873-298)$$
$$+ 0.0885 \times \left[-110{,}530 + c_{p,\text{CO}}(873-298)\right]$$
$$= -297{,}744\,\text{kJ},$$

$$Q_{LTSR} = -523{,}276 - (-297{,}744) = -225{,}532\,\text{kJ}.$$

The amount of heat to be transferred out of the LTSR is 225,532 kJ per mole of coal. Note that the water content of the product of LTSR is in the form of liquid.

d. The CGE is determined as follows:

$$\eta_{cold} = \frac{b''\text{LHV}_{\text{H}_2}}{\text{LHV}_{coal}} = \frac{0.5948 \times 242{,}000}{320{,}000} = 0.45.$$

Note that the LHV of coal is equivalent to the heat of complete combustion of coal at 298 K.

$$\text{CH}_{0.8} + 1.2\text{O}_2 \rightarrow \text{CO}_2 + 0.4\text{H}_2\text{O}$$

$$\text{LHV}_{coal} = \hat{h}_{\text{CH}_{08}} - \left(\hat{h}_{\text{CO}_2} + 0.4\hat{h}_{\text{H}_2\text{O}}\right) = -170{,}600 - (-393{,}800 - 0.4 \times 242{,}000)$$
$$= 320{,}000\,\text{kJ/mol}.$$

Problems

12.1 The coal gasification for syngas production in Example 12.1 is modified to operate autothermally by adding oxygen in the feed at 298 K. So, the heat requirement of the gasification process is internally supplied by partial combustion of coal modeled as carbon. Assume that the product consists of CO, H_2, CO_2, and H_2O.

a. Show that the number of moles of carbon monoxide, α, can be obtained from

$$(K-1)\left(\hat{h}_{CO_2}-\hat{h}_{CO}\right)\alpha^2+\left[K\left(b\hat{h}_{H_2O(l)}-\hat{h}_{CO_2}-b\hat{h}_{H_2}\right)-\left(b\hat{h}_{H_2O(l)}-2\hat{h}_{CO_2}-b\hat{h}_{H_2O(g)}+\hat{h}_{CO}\right)\right]\alpha$$
$$+\left(b\hat{h}_{H_2O(l)}-\hat{h}_{CO_2}-b\hat{h}_{H_2O(g)}\right)=0.$$

b. Plot the number of moles of oxygen versus temperature (from 1000 K to 1800 K) for $b = 1.5, 2, 2.5$.

c. Plot the mole fractions of CO and H_2 versus temperature (from 1000 K to 1800 K) for $b = 1.5, 2, 2.5$.

d. Plot the CGE versus temperature (from 1000 K to 1800 K) for $b = 1.5, 2. 2.5$.

12.2 The coal gasification process is employed for production of methane. Each mole of coal represented by $CH_{0.8}$ reacts with 0.7 mole of water and 0.1 mole of oxygen in a gasifier operating at a constant pressure of 1 bar. The reactants enter the gasifier at 25 °C, whereas the products exit at 230 °C. The enthalpy of formation of coal is −170,600 kJ/kmol.

a. Determine the composition of the product gas.

b. Calculate the heat requirement of the gasification process.

c. How many separation stages are required for production of a methane-rich stream?

12.3 In a coal-fired power plant with CO_2 capture, coal is combusted under oxyfuel conditions in the furnace of a boiler operating at atmospheric pressure. For each mole of coal represented by the chemical formula $CH_{0.9}O_{0.1}$, 0.6 mole of oxygen is fed to the combustor. A fraction of the CO_2 captured is recycled to the furnace so the combustion products are at 1273 K. The reactants enter the furnace at 298 K and 1 atm. The heat loss from the furnace is 2%.

a. Determine the LHV of the coal using (12.1).

b. Determine the composition of the combustion products.

c. Show that the concentration of methane in the combustion products is negligible.

d. Determine the number of moles of CO_2 recycled to the furnace.

12.4 A gasifier operating at a constant pressure of 10 bar is used for production of syngas. Coal represented by the chemical formula $CH_{0.84}O_{0.1}$ is gasified with water and oxygen. The gasification products leave at 1673 K. The reactants are fed at 298 K and 10 bar.

a. Determine the LHV of the coal using (12.1).
b. Show that the gasification products mainly consist of hydrogen and carbon monoxide if 0.7 mole of water and 0.1 mole of oxygen are fed to the gasifier, and determine the products concentrations.
c. If the number of moles of water is decreased to 0.5, show that the products would consists of hydrogen, carbon monoxide, and methane, and calculate their concentrations.
d. Determine the LHVs of the syngas fuels of parts (b) and (c).

12.5 Coal at 25 °C and 1 bar is fed to an adiabatic constant-pressure gasifier, where it reacts with oxygen and superheated steam. For each mole of coal ($CH_{0.84}O_{0.1}$), 0.9 mole of steam is used in the gasification process. Oxygen is produced using an ASU. A fraction of syngas leaving the gasifier at 1000 °C is fed to the furnace of the boiler, where it burns in air. The heat of combustion is used for production of the required steam for the gasification process. Other process parameters are given in **Figure P12.5**.

a. Determine the composition of the syngas.
b. Calculate the number of moles of oxygen required for the gasification process.
c. Determine the number of moles of syngas to be burned in the furnace of the boiler.
d. Calculate the efficiency of the gasification process, assuming that the power requirement of the compressor is provided from a power plant operating at 40% efficiency.

Figure P12.5

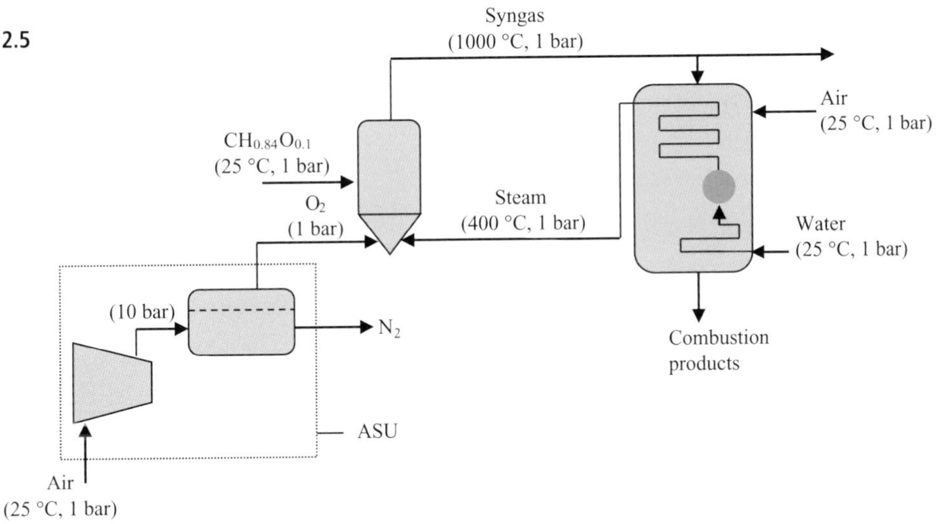

12.6 A hydrogen production plant using coal gasification is shown in Figure P12.6. It includes a gasifier, a HTSR, a CO_2 separation unit downstream of the HTSR, a LTSR, and another CO_2 separation unit after the LTSR. The gasifier operates adiabatically and at a constant pressure of 30 bar. One mole of coal ($CH_{0.84}O_{0.06}$) is gasified with 1.2 moles of water and n_{O_2} moles of oxygen, which all enter the gasifer at 25 °C. The gasification products leaving at 1250 °C react with 0.8 mole of water within the HTSR. The new products leaving 700 °C are sent to the first separation unit, where the CO_2 content of the mixture is separated. The mixture is then sent to the LTSR and mixed with n_{H_2O} moles of water. The products of the second shift reaction exit at 70 °C. At 70 °C and 30 bar, the water content of the mixture is in liquid phase, so it is separated from the mixture. The CO_2 content is separated in the second separation unit.

a. Calculate the LHV of the coal.

b. Determine the composition of the gas mixture leaving the gasifier, and the number of moles of oxygen (n_{O_2}) required for the gasification process.

c. Find the composition of the gas mixture leaving the HTSR.

d. Calculate the number of moles of water (n_{H_2O}) to be added in the LTSR, if the net heat requirement of the HTSR and LTSR is neutral.

e. Find the composition of the mixture at the outlet of the LTSR.

f. If this plant produces 10 kg hydrogen per day, determine the rate of coal feed assuming a steady-state operation.

Figure P12.6

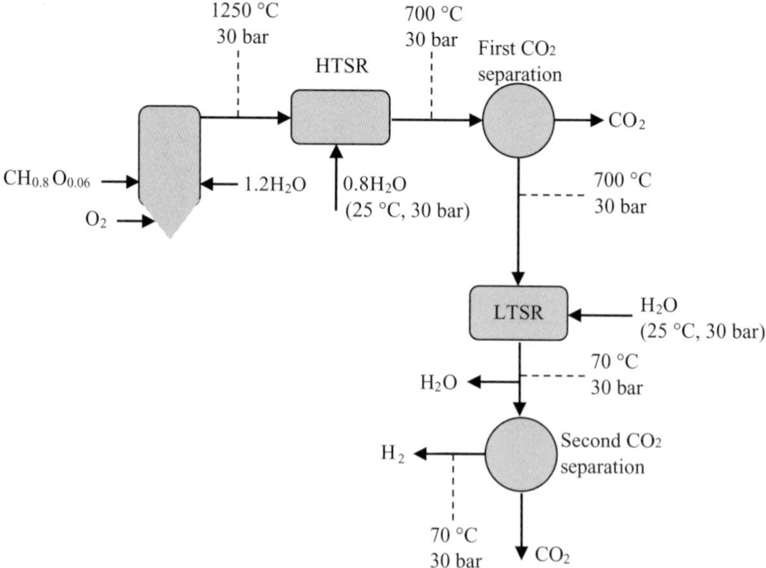

12.7 An IGCC power plant consists of a hydrogen production plant (described in Problem 12.6), a gas turbine cycle, and a bottoming steam cycle, as depicted in **Figure P12.7**. The process data of both gas and steam cycles are shown in the figure. The steam cycle includes a condensate pump, which raises the pressure of the condensate to 1 MPa, and a deaerator. A fraction of the superheated steam is used in the deaerator and the water coming out of it is saturated. For each mole of coal consumed, determine the net work production and the thermal efficiency of the IGCC plant. The pump isentropic efficiency is 85%, the compressor isentropic efficiency is 90%, the GT and ST efficiency is 93%.

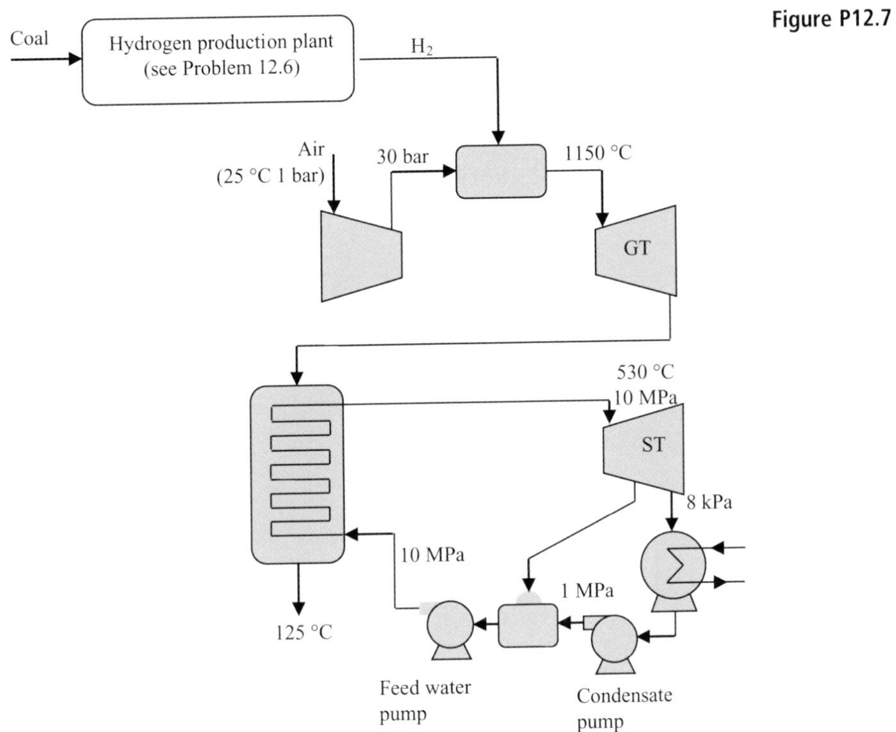

Figure P12.7

12.8 A schematic of an IGCC power plant is shown in Figure P12.8. The gasifier operates adiabatically at a constant pressure of 40 bar. Each mole of coal at 25 °C is gasified with 2 moles of water and n_{O_2} moles of oxygen. The produced syngas is cooled down in a heat exchanger before entering the gas turbine combustor, where it undergoes a complete combustion after being mixed with pressurized air coming from the air compressor. The combustion products leaving the combustor at 1200 °C expand in the gas turbine down to 1 bar. The flue gas leaving the gas turbine is sent to the HRSG for production of superheated steam at 540 °C and 10 MPa. A fraction of steam is used in a deaerator so the pressurized condensate stream at the outlet of the deaerator is saturated. The rest of the steam is expanded in a steam turbine, after which it enters a condenser operating at 6 kPa. A feed water pump further increases the pressure of the feed water coming from the deaerator. The pressurized water is heated in the syngas cooler by exchanging heat between the feed water and the hot stream of the syngas. Use the isentropic efficiencies of the pump, compressor, GT, and ST given in the previous problem.

a. Calculate the number of moles of oxygen required for the gasification process.
b. Determine the composition of the syngas.
c. Calculate the number of moles of air participating in the combustion reaction.
d. Determine the number of moles of water in the steam cycle.
e. Determine the fraction of steam used in the deaerator.
f. Calculate the net work produced by the IGCC plant per mole of coal.
g. Determine the thermal efficiency of this power plant.
h. Find the daily consumption of coal if the net power output of the IGCC plant is 100 MW.
i. Calculate the CO_2 emission of the IGCC power plant per MWh.

Figure P12.8

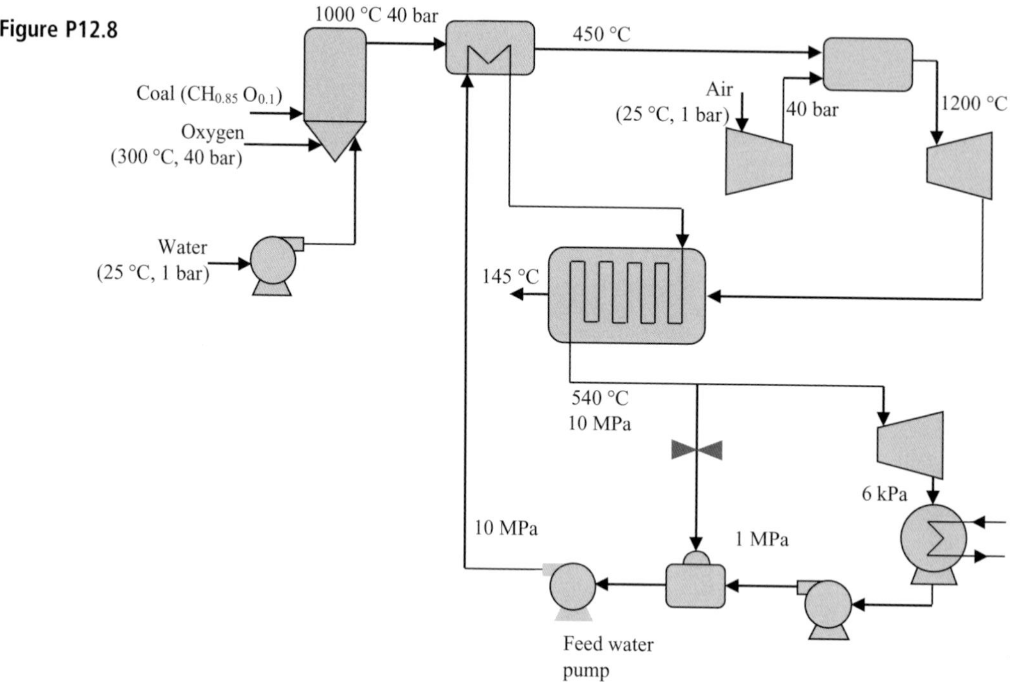

12.9 A schematic of a gas turbine power plant operating with oxyfuel combustion and syngas produced by gasification of coal is shown in Figure P12.9. Each mole of coal ($CH_{0.84}O_{0.1}$), initially at 298 K and 33 atm, is gasified with 0.7 mole of water and 0.1 mole of oxygen. An ASU is employed for production of the oxygen required for gasification and oxy-combustion processes. Atmospheric air at 298 K is pressurized to 9 atm within the ASU producing oxygen at 1 atm. Oxygen is then cooled to 308 K before entering an oxygen compressor, which raises its pressure to 33 atm.

The pressurized oxygen at the outlet of the compressor is divided into two streams; one stream goes to the gasifier, and another is sent to the gas combustor, where it mixes and reacts with the syngas and the recycled carbon dioxide. The combustion process is complete and adiabatic. The products of the combustion at 1573 K and 32 atm expand in a gas turbine down to 1 atm. The flue gas leaving the gas turbine is sent to a condenser, where the water content of the gas condenses and it is therefore separated. The only gas coming out of the condenser is carbon dioxide, a fraction of which is captured and the rest is pressurized using another compressor to the desired pressure. The isentropic efficiencies of the air, oxygen, and carbon dioxide compressors are 85%, 80%, and 81%, respectively. Also, the gas turbine and the pump have isentropic efficiencies of 89% and 75%, respectively.

a. Determine the composition of the syngas.
b. Calculate the heat requirement of the gasification process.
c. Calculate the number of moles of carbon dioxide recycled.
d. Determine the thermal efficiency of the power plant.

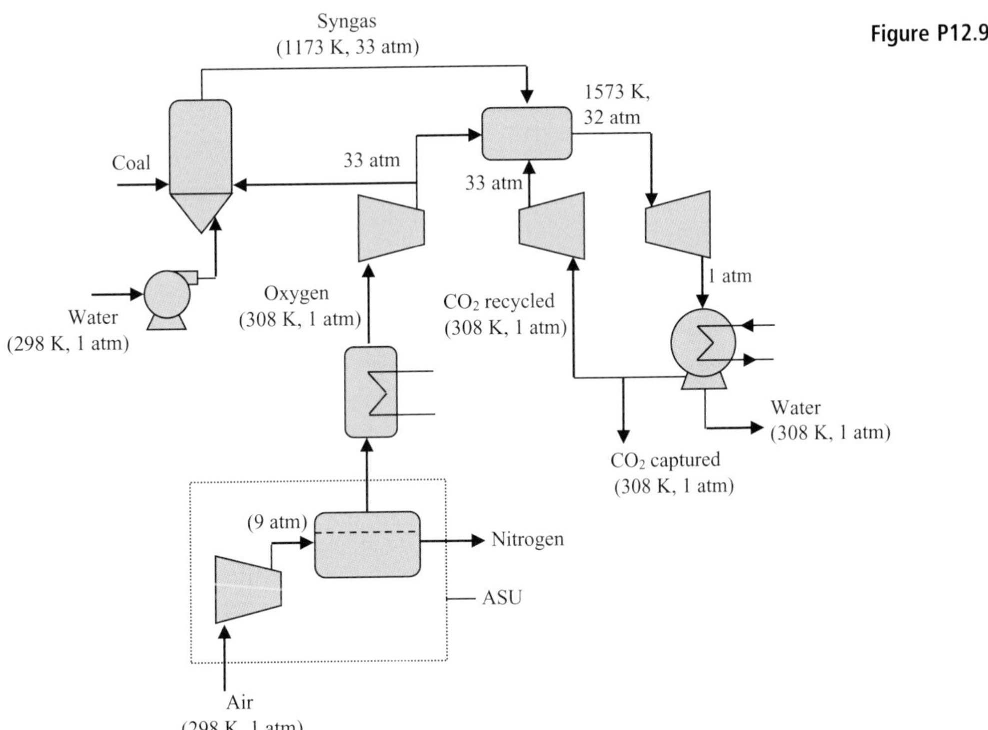

Figure P12.9

REFERENCES

1. H.-H. Rogner, "An assessment of world hydrocarbon resources," *Annu. Rev. Energy Environ.*, vol. 22, no. 1, pp. 217–262, 1997.
2. J. M. Beér, "High efficiency electric power generation: the environmental role," *Prog. Energy Combust. Sci.*, vol. 33, no. 2, pp. 107–134, 2007.
3. R. F. Probstein and R. E. Hicks, *Synthetic fuels*. Mineola, NY: Dover Publications, 2006.
4. The National Coal Council, "Coal-related greenhouse gas management issues," 2003. Available at: www.nationalcoalcouncil.org/page-NCC-Studies.html.
5. J. P. Longwell, E. S. Rubin, and J. Wilson, "Coal: energy for the future," *Prog. Energy Combust. Sci.*, vol. 21, no. 4, pp. 269–360, 1995.
6. N.V Kharchenko and V. M. Kharchenko, *Advanced energy systems*. Boca Raton, FL: CRC Press, 2014.
7. J. Ratafia-Brown, L.M. Manfredo, J.W. Hoffman, et al., "An environmental asssessment of IGCC power systems," 2002.
8. Tampa Electric Company (TECO), "Tampa Electric Polk Power Station integrated gasification combined cycle project: final technical report," 2002.
9. C. Higman and M. van der Burgt, *Gasification*. Amsterdam: Gulf Professional Publishing, 2003.
10. R. F. D. Monaghan, "Dynamic reduced order modeling of entrained flow gasifiers," MIT, Ph.D. thesis, 2010.
11. R. L. Simbeck, D. R., Korens, D. R. M Biasca, et al. *Coal gasification guidelines: status, application and technologies*. Palo Alto, CA: EPRI, 1993.

13 Carbon Capture Cycles

Coal

13.1 Introduction

In this chapter, some proposed power cycles for CO_2 capture in coal power plants are presented. These, as discussed in Chapter 11, include post-combustion capture, pre-combustion capture, and oxy-combustion capture cycles. The objective of the coverage is, besides describing the processes in some of these cycles, to evaluate their efficiency and contribution of the capture process toward derating the power production. While similar in principle to those designed for use with natural gas, coal (and other solid fuels) differs in fundamental ways from natural gas because it is consumed in the solid phase and it is often contaminated with sulfur, nitrogen, and ash, among other undesirable substances. Both the nature of the fuel and the contaminants make coal more challenging to use in energy production, and more so when carbon capture is implemented. However, and given its wide availability, lower cost, and higher CO_2 emission per unit of useful energy produced, it is imperative to develop this technology.

Many of the gas separation technologies used in carbon capture coal power plant design have been described in detail in Chapter 10, and will not be repeated in this chapter. Only reference to their components and energy needs will be made.

After a brief review of CO_2 concentration in different streams in coal power plants, and a brief overview of the three primary categories of carbon capture technologies, we will describe them in detail in three sections, starting with post-capture. Oxy-combustion will follow, then pre-combustion capture.

13.2 Cycles for CO_2 Capture

Table 13.1 shows CO_2 concentration in the products stream leaving coal power plants, using boilers, or in an integrated gasification combined cycle (IGCC), as well as in the syngas in the IGCC case before combustion. If a boiler is used, combustion is under atmospheric conditions, and so are the products. In the case of an IGCC plant, the conditions are those

Table 13.1 CO_2 partial pressure in flue gases of different coal combustion systems. Boilers burn pulverized coal under atmospheric conditions with some overfire air, and products are cooled to raise steam. IGCC plants gasify coal under pressure but the syngas is burned in gas turbine combustors followed by expansion to atmospheric pressure in the gas turbine and cooling in the HRSG. The dry concentrations are high because water resulting from combustion is condensed. In the last row, CO_2 concentration is given in the syngas [1]

Flue gas	CO_2 concentration (%vol [dry])	Pressure of gas stream (MPa)	CO_2 partial pressure (MPa)
Coal-fired boilers	12–14	0.1	0.012–0.014
IGCC after combustion	12–14	0.1	0.012–0.014
IGCC syngas	8–20	2–7	0.16–1.4

following expansion in the gas turbine and cooling in the heart recovery steam generator (HRSG). The concentration of CO_2 depends on combustion stoichiometry. In a conventional gas turbine, combustion is lean and CO_2 is low. The highest concentration in combustion-based plants is in coal-fired plants because of the higher concentration of carbon in the fuel and because combustion stoichiometry is closer to unity. In the case of an IGCC, CO_2 is determined by the gasification medium and the degree of shift before separation. The highest CO_2 concentration in the syngas is when all the CO is shifted to CO_2.

Post-combustion capture can be applied to conventional power plants where CO_2 is separated from the products stream. In oxy-combustion, coal is burned in a mixture of oxygen and recirculated flue gases composed of $CO_2 + H_2O$. In pre-combustion capture, such as in IGCC, CO_2 is separated from the shifted syngas ($H_2 + CO_2$) before burning hydrogen in the gas turbine combustor. Table 13.2 shows a representative comparison between the three technologies in terms of their efficiencies. The efficiency penalty is broken down to its individual contributions. Numbers are averages based on modeling studies, and may change depending on the components used in the plant and assumptions regarding the integration, especially regarding the steam cycle and gas cycle designs, heat integration, type of coal used, design of the air separation unit (ASU), gasifier design in the IGCC, and so on.

In the rest of this chapter, the different technologies are described in some detail; other developments beyond these three technologies are also introduced.

13.3 Post-Combustion Capture

In post-combustion capture, CO_2 is removed from the flue gases after cooling in the HRSG. While several of the gas separation techniques discussed in Chapter 10 can be used to separate CO_2 from nitrogen after condensing water, chemical absorption has been the most widely considered. Alternatives are under development, such as membrane technology or electrochemical separation. Alkanolamines have been suggested as a likely candidate for the

Table 13.2 Representative performance for the three main capture technologies applied to coal-fired plants [2]. The data in the table were obtained for specific plant design including the power island, which consists of steam plant in the pulverized coal and pulverized coal oxyfuel and a combined cycle for the IGCC. The efficiency in all cases is strongly impacted by the assumed operating conditions of the steam and gas cycles, the ASU efficiency, and the CO_2 recovery unit.

	Supercritical pulverized coal		Supercritical pulverized coal oxyfuel	IGCC	
Performance	Without capture	With capture	With capture	Without capture	With capture
Efficiency (%)	38.5	29.3	30.6	38.4	31.2
Efficiency penalty (%)	CO_2 capture (H) −5%		Boiler/flue gas desulfurization 3%	WGS −4.2%	
	CO_2 comp. −3.5%		ASU −6.4%	CO_2 comp. −2.1%	
	CO_2 capture (P) −0.7%		CO_2 comp. −3.5%	CO_2 recovery −0.9%	

H, heat; P, power; comp., compression; WGS, water–gas shift.

absorber fluid, but other chemicals and mixtures of chemicals are also under investigation. Most impurities, including sulfur oxides, nitrogen oxides, and particulates (as well as oxygen) must be removed from the flue gases before chemical separation of CO_2, since amines form stable salts when reacting with oxides, and oxygen causes oxidation of the absorbent. According to most studies, flue gases should not contain more than 10 ppm SO_2, even though some processes can tolerate up to 100 ppm sulfur.

As shown in Table 13.2 and estimates below, most studies show that the reduction of coal-fired cycles efficiency is 8–12 percentage points in steam power plants using pulverized coal in boilers, as compared to 8–10 percentage points in IGCC and pressurized fluidized-bed combustion with topping cycles, all using chemical absorption for CO_2 removal. Raising the pressure of the separated CO_2 to 110 bar reduces the efficiency by 3–4 more percentage points.

Similar to natural gas, one way to increase CO_2 concentration in the flue gases is to recycle some of the exhaust back to the combustor. Recycling cools the combustion temperature and makes it possible to burn a stoichiometric mixture in the combustor without leading to excessive temperatures. With stoichiometric burning, the CO_2 concentration increases in the products. Raising CO_2 partial pressure in the flue gases reduces the energy penalty of CO_2 separation. However, recycling consumes energy because of the associated pressure drop.

13.3.1 Cycle Modifications

It is important to integrate the chemical separation plant with power cycle in order to minimize the efficiency penalty associated with CO_2 capture (for instance, raising steam specifically for the absorber regeneration incurs a much higher energy penalty than

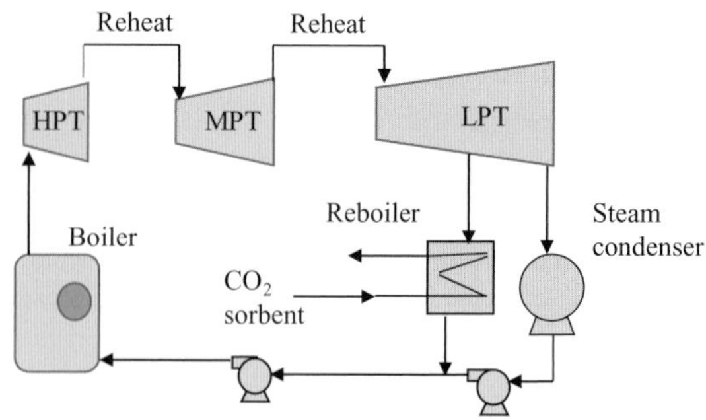

Figure 13.1 Integration of a Rankine cycle with three turbines, two-stage reheat, and a reboiler to heat the sorbent in a chemical absorption CO_2 capture plant. The solvent is heated in the reboiler using low-pressure steam extracted from the low-pressure turbine. HPT, MPT, LPT are high-, medium, and low-pressure turbines, respectively.

extracting steam from the low-pressure turbine). To supply the heat for absorber regeneration, it is important to extract steam from the steam turbine at temperatures slightly higher than those required in the CO_2 desorption tower (to allow for sufficient ΔT for heat transfer), but not at much higher temperatures, as this will further reduce the overall efficiency of the plant (see Chapter 11 for the impact of the steam extraction temperature on the degraded efficiency of the plant in the case of natural gas). Figure 13.1 shows the integration between the chemical absorption plant, reboiler, and the low-pressure steam turbine where steam is extracted.

Temperature differences in the order of 10–15 °C are used in the CO_2-capture reboiler before the absorbent stream is sent to the regeneration tower. Thus, it is desirable to use an absorbent with a low regeneration temperature. The steam leaving the reboiler is added back to the feedwater stream following the condenser and other (regenerative) feedwater preheaters. Thus, only part of the enthalpy in the extracted steam from the turbine is lost for solvent regeneration. Since the solvent leaves the desorption tower at higher temperatures, around 100 °C, steam is fed back to the feedwater stream close to the same temperature. The mass of steam necessary to extract CO_2 was discussed in Chapter 10, and estimates for the power required for this process were also presented. As an example of the efficiency penalty calculations, consider a typical coal-fired 600 MW_e plant, with flow rate of desulfurized combustion products of 500 m^3/s at pressure 1 bar with 75% N_2, 13% CO_2, and 12% H_2O leaving the HRSG [3]. The plant without capture has efficiency of 41%. Aspen modeling of the same plant shows that for 90% CO_2 removal, an amine solution flow rate of 2.4 m^3/s (with CO_2 loading changes between 0.24 and 0.48 mol/mol defined as CO_2/amine in the solution), reboiler heat duty of 565 MW, and extra pumping and blower work of 5.7 MW are required. The overall efficiency loss, including the compression work of CO_2, is 11.3 percentage points.

Reducing the temperatures of the streams within the absorption tower lowers the reboiler heat rate. This can be done by reducing the flue gas temperature and the temperature of the lean solvent as they enter the absorber. Most of the power penalty associated with CO_2 scrubbing is because of steam extraction from the low-pressure side of the steam turbine. Thus, minimizing the amount of steam and lowering its temperature of regeneration are necessary to reduce this power. This is because as the flue gas temperature increases, the solubility of CO_2 decreases and a higher mass flow rate of the solvent would be required. With a larger solvent mass flow, the reboiler duty (heat required) is higher.

13.3.2 CO_2 Separation Using Polymer Membranes

The use of membranes for gas separation was discussed in Chapters 10 and 11. The energy penalty in this case is the compression work required to overcome the pressure drop along the membrane, and to raise the pressure of the separated gas on the low-pressure side of the membrane. To achieve a high-purity CO_2 gas, A multistage membrane system has been suggested, where the CO_2-rich permeate of the first stage is compressed again and allowed to permeate through a second-stage membrane [3,4]. Alternatively, the CO_2-rich permeate from the first stage is compressed to a high pressure, near 80 bar, and cooled to 25 °C to allow CO_2 to condense out while nitrogen remains in the vapor phase. For the same 600 MW_e coal plant discussed in the previous section, in which nearly 500 m^3/s (at standard pressure and temperature [STP]) flue gases are produced, the gases must first be compressed to 9–30 bar for the first-stage membrane. Multistage compression with intercooling is used to reduce compression power, with regenerative heating of the separated gas. Intercooling using the separated gas is not sufficient since the separated gas volume is smaller than the original gas volume, and extra cooling using available cooling water must be used. Figure 13.2 shows the flow diagram of the proposed separation cycle using a membrane.

Organic membranes that can be used to separate N_2 and CO_2 are currently available (see Chapter 10). Hollow fiber modules have a high surface to volume ratio, 1500–2000 m^2/m^3. In these membranes, the feed stream flows outside the fiber and the permeate is collected inside. Inorganic membranes are more suitable for high-pressure applications, including porous ceramic and metal oxide membranes. However, these are not as effective as organic membranes because the difference in molecular weights between N_2 and CO_2 is small.

Along with the membrane units, compressors before and expanders after each membrane unit are used to compress the feed and expand the retentate gas (depending on the feed pressure). The expander recovers a fraction of the compression energy. Compressors use some of the work produced in the power island, and hence incur a relatively high efficiency penalty. Moreover, several heat exchangers are used between the different gas streams to

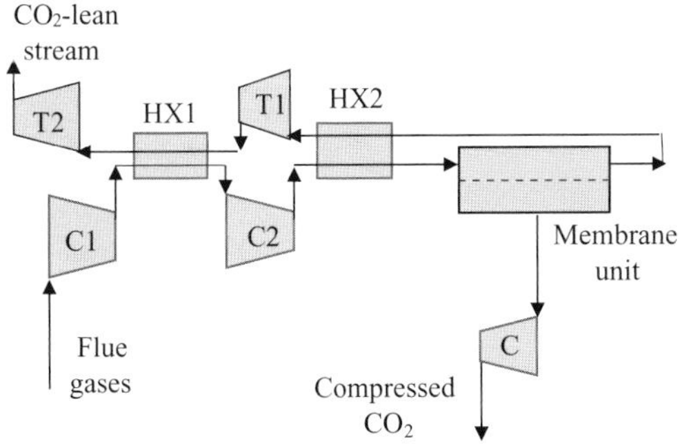

Figure 13.2 Schematic diagram of a membrane unit used to separate CO_2 from flue gases. Two-stage compression C1 and C2 is used to raise the pressure of the gases before the membrane separation unit, with regenerative cooling in HX1 and HX2. The CO_2-free stream leaves the separation unit and undergoes a two-stage expansion in T1 and T2 for energy recovery, assisted by regenerative heating using the original flue gas stream. Further cooling of the flue gas stream is often used before compression to reduce the compression work. The CO_2 stream is compressed for storage.

cool the compressed feed and warm up the separated gas, and for external cooling of the flue gas stream between the different compression stages. Membranes for separating CO_2 from a stream of $CO_2 + N_2$ were described in Chapter 10.

13.4 Oxy-Combustion

In oxy-combustion, coal is burned in oxygen, producing a CO_2-rich stream that is ready for sequestration after condensing water and removing other impurities. Various oxy-coal combustion systems have been proposed, including atmospheric pressure combustion in which flue gases are partially recycled to moderate the flame temperatures, and others in which steam is injected to control the flame temperature. Pressurized combustion has been proposed for oxy-coal combustion with recycled flue gases, and oxy-syngas combustion in combination with gasification. Oxy-combustion allows for maximum CO_2 removal, theoretically 100%. Air leakage in different equipment, and CO_2 dissolved in water, accounts for most of the less-than-perfect performance.

In conventional air-fired power plants, nitrogen accounts for 79% of the feed air. Stoichiometric combustion in pure or nearly pure oxygen results in very high temperatures (~3500 °C for coal) not easily tolerated by the power island, and recirculation of flue gases is required. To match the adiabatic temperature to that of air-fired combustion, nearly 60% (on a mass basis) of the flue gases should be recycled. Moreover, similar flame characteristics, including a temperature and heat flux comparable to those of air-fired combustion, can be achieved by ~30% oxygen concentration in the feed gas, with the rest being the recycled flue gases, with the exact ratio depending on the coal type. Figure 13.3 shows the dependency of the adiabatic flame temperature of coal-oxy combustion on the oxygen mole fraction in the gas at the burner inlet, comparing pure dry (with CO_2 only) recycle and pure wet (with H_2O only) recycle.

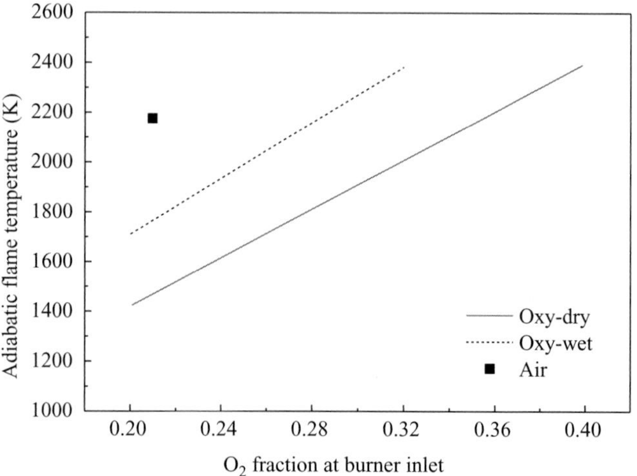

Figure 13.3 The O_2 mole fraction required at the burner inlet to achieve a given adiabatic flame temperature for wet and dry flue gas recycle (residual O_2 mole fraction in the flue gas fixed at 3.3%) [5]. The symbol ■ indicates the air flame temperature (AFT) of air–coal combustion, the solid line and broken line indicate the AFT of oxy-coal combustion with dry and wet flue gas recycle, respectively.

Oxy-combustion has been applied in coal-fired plants [6,7]. Recycled CO_2 is used in the boiler to manage the combustion temperature, with CO_2 captured following the HRSG. Estimates for efficiency penalty in this case are 5–7 percentage points for the ASU and 4 percentage points for recycling CO_2 [3,4]. Reported net efficiency is 28–34% for optimized steam cycles with oxy-combustion (without the liquefaction energy for CO_2). These numbers are consistent with estimates for ~10 MJ/kg coal energy required for air separation (or ~0.27 kWh/kg-O_2). Since 1 kg coal requires $32/12 = 2.67$ kg O_2, and air separation requires about 1 MJ/kg-O_2, the available energy is reduced by about 27% or about 10 percentage points. The origin of these estimates will become clear as we review detailed models of oxy-coal combustion cycles.

13.4.1 Atmospheric and Pressurized Oxy-Combustion

Atmospheric oxy-coal combustion systems, shown schematically in Figure 13.4, were first conceived as a carbon capture and sequestration (CCS) retrofit solution for existing plant. In this technology, recycled flue gases are used to control the flame temperature and hence the flue gas consists primarily of steam, which is later removed through condensation, and CO_2, which is purified, compressed, and liquified for storage. Equipment required in oxy-combustion, similar to that described for the case of oxy-combustion of natural gas except for the purification unit, is described below and in the following section, along with detailed analysis of the energy requirements.

13.4.1.1 Air Separation Unit

Cryogenic distillation, in which air is compressed and cooled prior to being introduced into the distillation column, where it is separated into an oxygen-rich stream and a nitrogen-rich

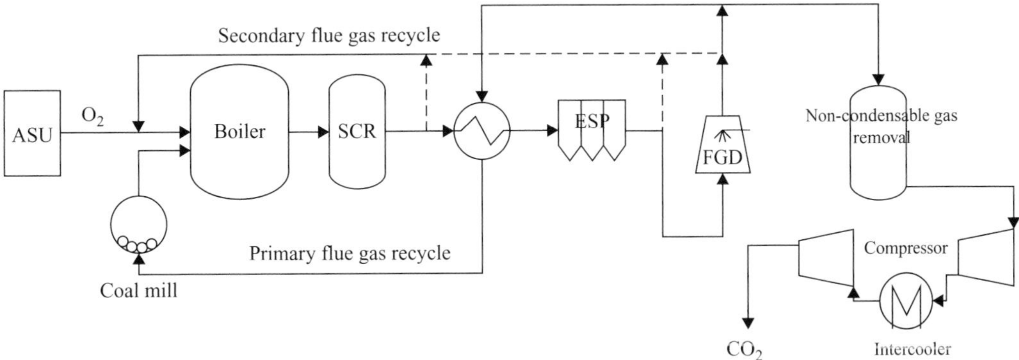

Figure 13.4 Atmospheric oxy-coal combustion system with flue gas recycling [5]. An ASU delivers pure oxygen to the boiler, along with coal and recycled products. The selective catalytic reduction unit after the boiler is used for de-NO_x. The gas is cooled (by the recycled stream) and sent to an ash filter unit (ESP) and a flue gas desulfurization unit (FGD). Some of the purified gas is recycled back to the boiler via the coal mill (for transporting the pulverized coal) and the rest is compressed for storage. Multipoint recycling can be used to improve the efficiency.

stream, is used extensively (see Chapter 10). Cryogenic air separation is energy-intensive, consuming ~0.24 kWh/kg-O_2 for 95% oxygen purity. These units can consume more than 15% of the gross power output of the power plant, leading to ~4–6 percentage point drop in efficiency. A 300 MW oxy-combustion plant requires more than 5000 metric tons per day of oxygen, and existing scalable technology is based on cryogenic distillation. Conventional adsorption methods and membrane separation are also commercially available, but have not been applied at scale yet. Advanced technologies include ion-transport membranes (ITM) and chemical-looping air separation (CLAS), discussed in Chapter 10. In ITM, oxygen is separated by exploiting the oxygen partial pressure gradient across dense mixed-conducting, ceramic membranes at 800–900 °C. If successfully implemented [8], the overall plant efficiency is expected to increase by a few percentage points beyond what is possible with cryogenic separation. Chemical-looping air separation utilizes circulating metal oxide particles between an oxidation reactor and a reduction reactor, a chemical principle similar to that used in chemical-looping combustion (CLC) described in Chapters 10 and 11.

13.4.1.2 Carbon Dioxide Purification Unit

The carbon dioxide purification unit (CPU) removes water, particulate matter, and other pollutant gases from the flue gas before compression. Selective catalytic reduction, electrostatic precipitator (ESP), and flue gas desulphurization (FGD) are typically used for NO_x, particulate matter, and SO_x removal from the flue gases. In oxyfuel combustion, NO_x emissions are reduced to less than one-third of the emissions from air-fired combustion [7]. The lower concentration of nitrogen within the combustor reduces thermal NO_x formation, but most coals contain some nitrogen. On the other hand, the SO_x emission rate does not change considerably, only from 91% in the air-fired case to 64% in oxy-combustion. Non-condensables, such as O_2, can cause corrosion, while others, such as N_2 and Ar, would raise the compression energy, and should also be removed.

13.4.1.3 Flue Gas Recycle System

Flue gases can be recycled at different locations downstream of the heat transfer units (or between these units) in the form of wet or dry recycling. To reduce the efficiency penalty associated with mixing streams with much different temperatures, multipoint recycling should be done, as shown next. The recirculation of the flue gases is associated with a pressure drop. Overall loss because of flue gas recycling accounts for ~2–4% extra penalty in efficiency.

Pressurized oxy-combustion systems have been proposed with the objective of improving the energy efficiency of the cycle by recovering the latent heat of steam in the flue gas at higher pressures and temperatures. Rankine cycles operating on pressurized oxy-combustion of coal, in which a boiler operates at pressures higher than atmospheric, have also been suggested [9,10]. Analysis shows that these cycles can achieve overall higher efficiency because they enable heat recovery at higher temperature (better availability efficiency) and

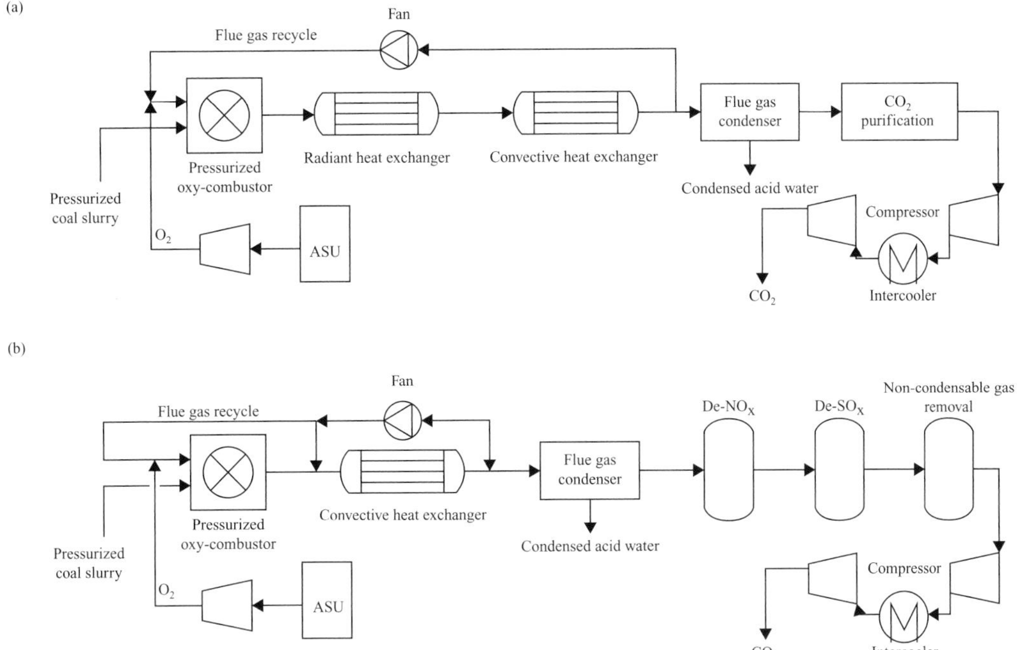

Figure 13.5 Pressurized oxy-coal combustion systems for carbon capture in coal power plants: (a) schematic of the TIPS; (b) the system proposed by ENEL [5].

because CO_2 is delivered to the CO_2 liquefaction unit at higher pressure, hence saving some of the liquefaction energy. The flue gas volume is reduced under elevated pressure, resulting in smaller components and possible reductions in capital cost for the same power output. Several studies have reported on the technical and economic feasibility of this process [11,12], showing improvements in efficiency. Other potential advantages include the reduction of auxiliary power consumption and the elimination of air ingress into the system.

Figure 13.5a shows the pressurized oxy-combustion systems proposed by ThermoEnergy Integrated Power System (TIPS) [13]. It uses a pressurized combustor, heat exchangers, and a flue gas condenser (FGC). Downstream of the radiative boiler and convective heat exchangers, steam in the flue gases is condensed in the FGC, where most of the latent heat in the flue gas is recovered by the feedwater in the steam cycle. The rest of the flue gas, which is essentially CO_2, is purified and compressed for sequestration. Another pressurized system [12,14] will be described in the next section. In pressurized combustion systems, coal is fed in the form of coal–water slurry (CWS).

For pressurized oxyfuel power plants, higher efficiency desulphurization and de-NO_x removal solutions have also been proposed using lead chamber chemistry at elevated pressures [15].

The suggested optimal pressure for this cycle was ~20 bar. In the TIPS system, a temperature difference greater than 100 °C was maintained between the flue gas and the feedwater in the FGC, with the flue gas temperature at the condenser outlet being 123–262 °C. The high temperature difference ensures a sufficiently high heat flux and a relatively small heat exchanger size. The amount of recoverable latent heat and the net efficiency increase monotonically with the operating pressure.

Another pressurized oxy-combustion concept involves gasification. Contaminants, including H_2S and particulates, are removed from the syngas before it is combusted in oxygen diluted by recycled CO_2 and/or steam. Combustion products, consisting mainly of CO_2 and/or steam, are used as a working fluid in a gas turbine, resembling that in natural gas cycles that use oxy-combustion. In this "Oxy-Fuel Zero Emission Power Plant" (O-F ZEPP), the overall efficiency achieved was 30–34% based on the lower heating value (LHV), after taking the ASU and CO_2 compression work into account [16].

Efficiency estimates show a loss of about 10–15 percentage points when post-combustion capture is added to the base case pulverized coal power plant. Atmospheric oxyfuel combustion shows improvement of 1–5 percentage points over post-combustion capture; while the pressurized system gains ~3 percentage points in efficiency over atmospheric oxy-combustion. Although the ASU power consumption is higher in the pressurized oxy-combustion system, the power savings in the CO_2 compression unit and in the recycled flue gas compressor are even higher, resulting in better overall efficiency.

Techno-economic analysis of oxy-combustion of coal has been the subject of many studies [14,17]. Table 13.3 shows a summary of some of these results. Conventional oxy-combustion of coal in boilers was also considered [4]. The cycle efficiency is reduced by ~5–7 percentage points because of the ASUs. In addition, the recirculation of the flue gases to moderate the combustion flame temperature consumes power associated with pressure drops, which leads to another ~4 percentage point penalty in efficiency, for a total reduction of ~9–13 percentage points from the no-capture cycle after accounting for CO_2 compression.

13.4.2 Pressurized Oxyfuel Combustion-Based Power Plants

As shown above, recent studies show that pressurized oxyfuel combustion systems have the potential to reduce the efficiency penalty [17,19]. Because of the higher dew point and the higher-temperature latent enthalpy in the flue gases, a pressurized oxyfuel system can

Table 13.3 Net efficiency of power plants [18] for supercritical (SC) and ultra-supercritical (USC) coal cycles with no capture, post-combustion capture, and oxy-combustion capture

	No capture SC/air	No capture USC/air	Post-combustion 90% capture SC/air	Post-combustion 90% capture USC/air	Oxy-combustion 97% capture SC/oxy	Oxy-combustion 97% capture USC/oxy
Efficiency	39.5	44.6	27.2	32.1	28.3	33

recover more thermal energy from the flue gases. Moreover, oxy-combustion at high pressures may increase the burning rate of char and the heat transfer rates in the convective sections of the heat transfer equipment. Furthermore, the pressurized flue gases utilize smaller equipment within the power plants, and hence the pressurized oxyfuel system may save in capital investment. Figure 13.5b shows a layout for a pressurized oxy-combustion coal cycle that consists of five primary units: (1) an ASU; (2) a pressurized coal combustor; (3) a steam generation unit; (4) a power island; and (5) a CO_2 purification and compression unit.

Figure 13.6 shows a schematic diagram of the plant and the different states of the working fluid. The steam condensate leaves at state 1 and is compressed by the first feedwater pump. The pressurized condensate enters the acid condenser where most of the latent enthalpy in the flue gases is recovered while the flue exhaust stream is cooled from state 20 to state 21. The condensate stream recuperates more thermal energy by cooling the combustor walls before entering the deaerator at 10 bar. After the deaerator, the feedwater stream at state 5 is pumped to the supercritical state.

After leaving the second feedwater pump, the feedwater is heated regeneratively to state 6. Next, the feedwater enters a HRSG, where it is heated to 600 °C at 250 bar (state 7). Across the power island and the HRSG there are two reheat streams, states

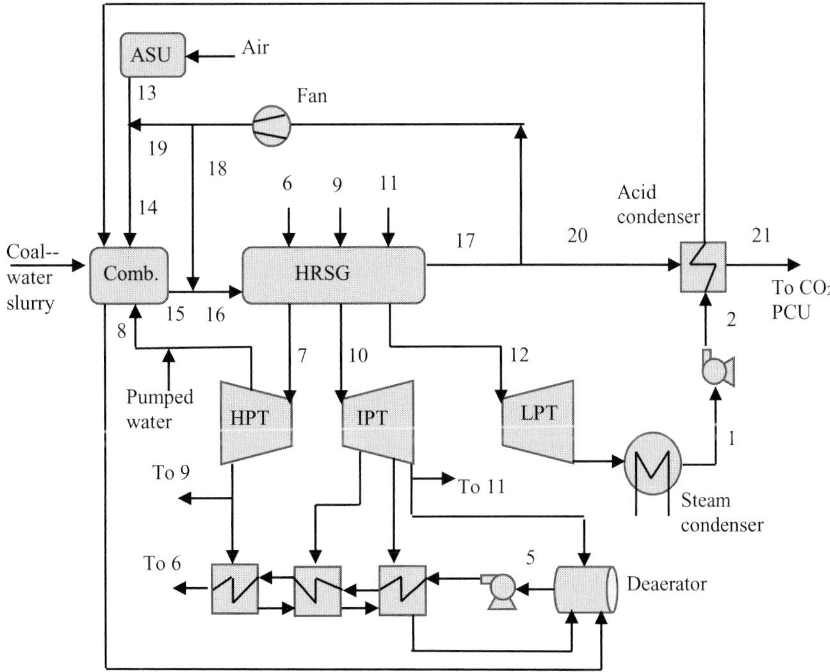

Figure 13.6 A schematic flowchart of the pressurized oxy-combustion power cycle utilizing a pressurized coal combustor [17].

9 and 11, to the intermediate-pressure turbines and the low-pressure turbines, respectively. These two streams are superheated to 620 °C. Steam is bled from the high-pressure turbine to be injected into the pressurized combustor (state 8) in order to atomize the slurry.

The oxygen stream from the ASU (state 13) is mixed with recycled flue gases (state 19) and injected into the pressurized combustor at state 14. Flue gases leave at about 1550 °C (state 15) and are cooled to 800 °C by recycled flue gases (state 18). They then enter the HRSG at state 16 and transfer thermal energy to the steam while being cooled to state 17. A fraction of the flue gases is recycled after the HRSG, and the rest passes through the acid condenser. Next, the flue gas stream enters the purification and compression unit.

Typical design variables of a pressurized oxyfuel combustion power cycle are shown in Table 13.4. These design variables represent commercially available technologies or processes in an advanced development stage.

To operate a pressurized oxy-combustion power plant, a high-pressure deaerator and a flue gas acid condenser are required. In addition, it is necessary to purify and compress the concentrated stream to 110 bar. The purification process includes de-SO_x, de-NO_x, and a low-temperature flash unit to prepare the CO_2 stream for transportation to an enhanced oil recovery (EOR) or a sequestration site.

Table 13.4 Base case design variables of the pressurized oxy-combustion plant analyzed in this section [17]

1. Air separation unit	
Oxygen purity (mol %)	95%
Oxygen in the flue gases (mol %)	3%
Oxygen delivery temperature	200 °C
2. Pressurized coal combustor	
Combustor pressure	10 bar
Combustor temperature	1550 °C
Combustor thermal energy loss	2%
Slurry water (wt %)	35%
Steam injection (wt %)	10%
3. Steam generation	
Inlet temperature of the HRSG	800 °C
Outlet temperature of the HRSG	260 °C
4. Power island	
Turbine inlet pressure	250 bar
Turbine inlet temperature	600 °C
Reheat temperature	620 °C
Deaerator pressure	10 bar
Condenser pressure	0.05 bar
5. Carbon dioxide purification and compression	
CO_2 compression pressure	110 bar

For simulations, Thermoflex® was used to model the steam cycle using the built-in libraries for standard and commercially available components, and Aspen Plus®[1] was used to model the ASU and the flue gas units, including the CO_2 purification and compression unit. In the following, the design and analysis results of the primary units in the plant are described.

13.4.2.1 Air Separation Unit

Cryogenic distillation was described in Chapter 10. Using a two-distillation column system, the ASU delivers an oxygen stream with 95% purity (by volume) at power consumption of 0.245 kWh/kg-O_2 or 0.244 kWh/kg-O_2. Feed air is compressed to 5.5 bar in two stages, necessary to reach the requisite oxygen concentration in the low-pressure column since the heat from nitrogen must be used to boil oxygen. Heat integration is applied between the bottom of the low-pressure column and the top of the high-pressure column.

The pressurized air passes through a regenerator to eliminate impurities, including water and CO_2 that must be removed before entering cryogenic distillation columns. Steam-heated nitrogen is required for regeneration, but 4% of the feed oxygen is assumed to be lost via bed switchover and adsorption/desorption. Through the two distillation columns, an oxygen-enriched stream with 95% oxygen, 4% argon, and 1% nitrogen is produced at 1.24 bar. A two-stage oxygen compressor is used to deliver the stream at 10 bar to the combustor.

13.4.2.2 Pressurized Coal Combustor

Coal is supplied in the form of a water slurry that contains 0.35 kg of water per 1 kg of its total weight. Steam is also injected into the pressurized combustor to atomize the slurry particles. The overall cycle efficiency is raised by extracting the atomization steam from the steam turbines. The high concentration of triatomic molecules, water and carbon dioxide, leads to high emissivity and radiative heat transfer for fast heating of the reactants. At the high temperatures of oxy-combustion, 1400–1600 °C, most of the ash melts and flows down toward the bottom bath. A high combustor wall temperature allows the molten ash to drain out from the combustor. More than 97% of the ash is removed [20]. With high diluent concentration, flameless combustion is maintained (without a diluent, oxy-coal combustion can reach above 3000 °C). An equivalence ratio of 0.989 is maintained and 26.1 percent (by mass) of the flue gases are recycled to maintain the combustion temperature at 1550 °C.

13.4.2.3 Steam Generation Unit

To generate steam for the power island, two superheaters, a once-through boiler, and an economizer, are used in the HRSG, producing supercritical steam at 600 °C and

[1] Thermoflex® and Aspen Plus® are registered trademarks of Thermoflow Ltd and Aspen Technology, Inc., respectively.

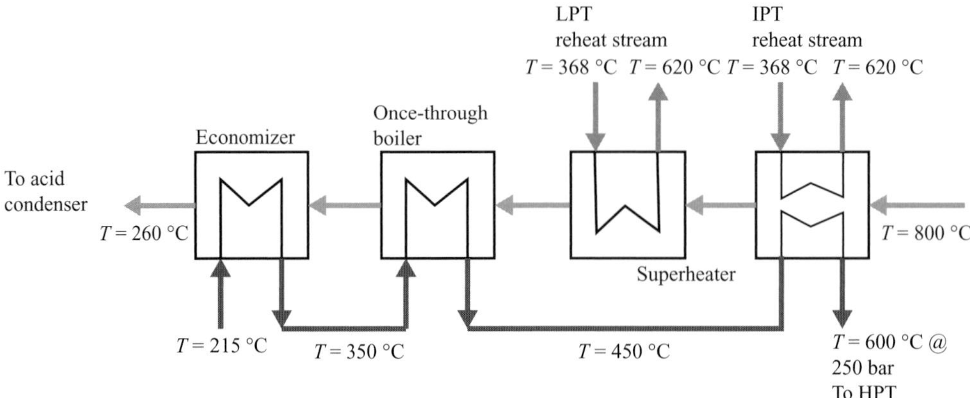

Figure 13.7 Steam generation within the HRSG, starting with water at 215 °C in the far left and moving through the economizer, once-through boiler, and a superheater. Combustion gases enter the superheater at the far right at 800 °C, moving through the superheaters, boiler, and economizer before entering the acid condenser. The diagram also shows the reheat streams [17].

250 bar. The superheaters support two reheat subcritical steam flows, each at 620 °C, as shown in Figure 13.7.

To minimize hot corrosion and oxidation, the inlet temperature of the pressurized flue gases into the HRSG is cooled down to 800 °C by secondary recycled flue gases. Because of the flue gases high pressure, the "acid" dew point is higher than the acid dew point at atmospheric pressure (the presence of acid gases [i.e., SO_x and NO_x] raises the dew point of the flue gases relative to pure water condensation). An acid condenser is added at the end of the steam generation unit to cope with acid condensation and to recover more thermal energy from the pressurized flue gases, which has a higher latent enthalpy.

13.4.2.4 Power Island

At the inlet of the HPT, the steam conditions are 250 bar and 600 °C, and the power island has two reheat steam flows at 620 °C. Steam is condensed at 0.05 bar. Steam bleeding from the high-pressure and the low-pressure turbines is replaced by a high-pressure flue gas thermal energy recovery system, requiring a high-pressure deaerator. Thermal energy recovery from the pressurized flue gases depends on the terminal temperature difference at the hot side of the acid condenser, and the saturation condition inside the deaerator. The feedwater out of the acid condenser (state 3) goes into the deaerator after recovering the thermal energy loss from the combustor. Increasing the exit temperature of the feedwater at the hot side of the acid condenser (state 3) to recover more thermal energy from the flue gases implies a higher inlet temperature for the feedwater entering the deaerator (state 4). According to the saturation condition of the deaerator, the design pressure fixes the exit temperature of the water leaving the deaerator (state 5).

13.4.2.5 Carbon Dioxide Purification and Compression Unit

The carbon dioxide purification and compression unit uses two successive water-wash columns, at 15 bar and at 30 bar, respectively, to remove sulfur and nitrogen oxides. NO and NO_2 are removed as HNO_3, and SO_2 is removed as H_2SO_4. Due to the complete NO_2-catalytic conversion of SO_2 to sulfuric acid with increasing pressure, the SO_x removal process operates in the water-wash column at 15 bar. The column separates all the SO_2 and SO_3 as sulfuric acid, as well as almost half of the remaining water content in the flue gases. After the de-SO_x unit, the flue gases are compressed to 30 bar and introduced to the next water-wash column. Here, more than 90% (by mass) of NO_x is removed as nitric acid. Most NO_x produced from the high-temperature combustion is in the form of NO and must be converted to NO_2 in order to remove it as nitric acid. Because the reaction rate of NO to NO_2 increases at higher pressure and lower temperature, the water-wash column is at 30 bar.

Next, the CO_2 concentrated stream is sent to the low-temperature processing unit, where the remaining impurities, oxygen, argon, and nitrogen, are removed. The CO_2 stream is cooled down to about −54 °C, which is close to the CO_2 triple point, −56 °C. This cooling process produces two different streams, a capture-ready CO_2 stream with 96.5% (molar basis) CO_2 concentration, and an exhaust stream consisting mostly of inert gases. The capture-ready stream is compressed to 110 bar for transportation to a sequestration or an EOR site.

13.4.2.6 Flue Gas Thermal Energy Recovery

Operating at high pressure in the combustor allows for recovering more thermal energy from the flue gases, at higher temperature. Whereas an air-fired combustion system produces a small concentration of water of ~8.7% (by volume), nearly half of oxyfuel combustion power cycle flue gases is composed of water. In addition, the saturation temperature of the water increases with increasing the operating pressure. While the flue gases begin to condense at about 80 °C in the atmospheric oxyfuel system, condensation begins to occur at about 150 °C in the pressurized system. These two facts enable the pressurized system to recover more thermal energy from the flue gas stream. The latent enthalpy of condensation is used to heat up the condensate leaving the first feedwater pump, from state 2 to state 3.

The incremental improvement in the thermal energy recovery is achieved in the acid condenser, which recuperates the latent enthalpy of water. This can be explained through the outlet temperature of the acid condenser represented as stream 21 in Figure 13.10. With the same inlet thermal energy sources (the HRSG outlet stream at 260 °C), the acid condenser of the atmospheric case yields an outlet temperature of 73.19 °C, which is higher than that of the pressurized case, 60.49 °C. With the acid condenser, recovery of ~80% more thermal energy from the flue gases is achievable. The thermal energy recovery from the flue gases is sufficient to replace the thermal energy duty of the feedwater heating system from the high-pressure and the low-pressure steam turbines, and no steam

Table 13.5 Overall performance of the atmospheric oxyfuel power cycle, third column, and the pressurized oxyfuel power cycle, fourth column. All values are based on the higher heating value (HHV) of coal. In both cases, fuel flow is 30 kg/s and oxygen flow is 73.52 kg/s

	Atmospheric	Pressurized
Thermal energy input (MW_{th})	874.6	874.6
Gross power output (MW_e)	388.0	404.5
Net power output (MW_e)	264.3	292.6
Net efficiency (%)	30.2	33.5
Steam demand (kg/s)	211.9	210.0
Flue gas flow rate into the HRSG (kg/s)	1011.1	1004.7
Flue gas flow rate into the purification and compression unit (kg/s)	107.7	87.7

bleeding from those turbines is required, increasing the power generation from the turbines. The overall steam bleeding drops from 32 kg/s in the atmospheric case to 11.9 kg/s in the pressurized system.

Table 13.5 shows the overall performance of the pressurized and atmospheric oxy-combustion power cycles. The atmospheric system is based on 1.1 bar combustion pressure. In this system, the oxygen stream (state 13) is not compressed and thermal energy sources are not sufficient to heat this stream to 200 °C. As a result, the same oxygen delivery temperature target as the pressurized oxyfuel power cycle cannot be achieved. However, because the acid dew point of the atmospheric pressure system is considerably lower than that in the pressurized combustion case, a lower oxygen delivery temperature can be used to avoid acid condensation when it is mixed with the recycled flue gas stream (state 19). As such, the atmospheric oxyfuel power cycle is based on a 100 °C oxygen delivery temperature. Moreover, in the atmospheric pressure system, steam from the low-pressure turbines is used to heat the feedwater leaving the second feedwater pump.

The ASU consumes nearly 20% of gross power output. The pressurized system needs more energy than the atmospheric pressure unit, by 4 percentage points. The carbon dioxide purification and compression unit power demand is 3.7% of gross power output, as compared to the atmospheric system's 9.5%, because the flue gases are already pressurized. Table 13.5 shows that the oxygen demand is lower than the flue gas flow rate into the purification and compression train. In addition, the flow rate of the flue gases undergoing compression in the pressurized system (87.7 kg/s) is smaller than in the atmospheric system (107.7 kg/s). The higher thermal energy recovery increases water condensation and reduces the flue gas flow into the purification and compression unit.

The energy requirement for the fan needed to recycle gas drops by 7.6 MW in the pressurized cycle. This is because of the extent of the pressure drop across the steam generation units and the flue gas recirculation pipe, and the corresponding pressure ratio across the fan. Based on the estimated pressure drop, the fan in the atmospheric pressure case compresses the recycled flue gas stream from 0.98 bar to 1.1 bar, whereas that of the pressurized system compresses it from 9.35 bar to 10 bar. With almost the same mass flow rate as the recycled flue gases, the pressurized system has a smaller pressure ratio across the fan, and thus requires less fan compression work.

The pressurized oxy-combustion system achieves higher net efficiency, 33.5% (HHV) or 34.9% (LHV), compared to the atmospheric combustion cycle because of the high-pressure flue gas thermal energy recovery, the increased power output, and the lower overall compression work.

13.5 Pre-Combustion Capture

Pre-combustion capture cycles for natural gas were described in Chapter 11. In the case of coal (or other solid fuels), and instead of reforming, gasification is used to produce syngas, similar to what is done in conventional IGCC plants. Gasification in oxygen and steam produces a mixture of $CO + H_2$, with an ASU used for oxygen production. Following syngas clean-up, the water–gas shift (WGS) reaction is used to oxidize CO to CO_2 and reduce water to hydrogen. Next, H_2 is separated and CO_2 is captured. Air is used later in the cycle to burn H_2, along with the nitrogen leaving the ASU, which is used as a diluent in the gas turbine combustor. A schematic diagram for the IGCC plant with capture is shown in Figure 13.8. Alternative cycles have been proposed, in which instead of the WGS, the syngas is separated into H_2 and CO, which are sent to two different gas turbines, with the first using air as an oxidizer and the second using oxygen [3]. In the following we will focus on the cycle with WGS and hydrogen separation. The plant layout and detail depend on the technology used to separate hydrogen from carbon dioxide, and in the following, physical separation and membrane-based separation plants are discussed.

13.5.1 IGCC with Physical Absorption Separation

This technology is discussed in the context of an IGCC plant that uses different coal ranks and different syngas cooling technologies following gasification, and different coal feeding systems to pump the pulverized coal into a high-pressure entrained-flow gasifier. Syngas cooling is necessary before its clean-up and its separation into H_2 and CO_2. While bituminous and subbituminous coals have been mostly used in gasification, using lower-rank coals could reduce costs significantly, and hence approaches to using low-rank coal in gasification-based plants are being explored.

13.5.1.1 Water and CO_2 Slurry Feeding

Similar to conventional IGCC plants described in detail in Chapter 12, an entrained-flow gasifier is used, running close to 1400 °C to increase the syngas yield and avoid tar production, and at pressures higher than 30 bar. To feed coal into this high-pressure environment, it

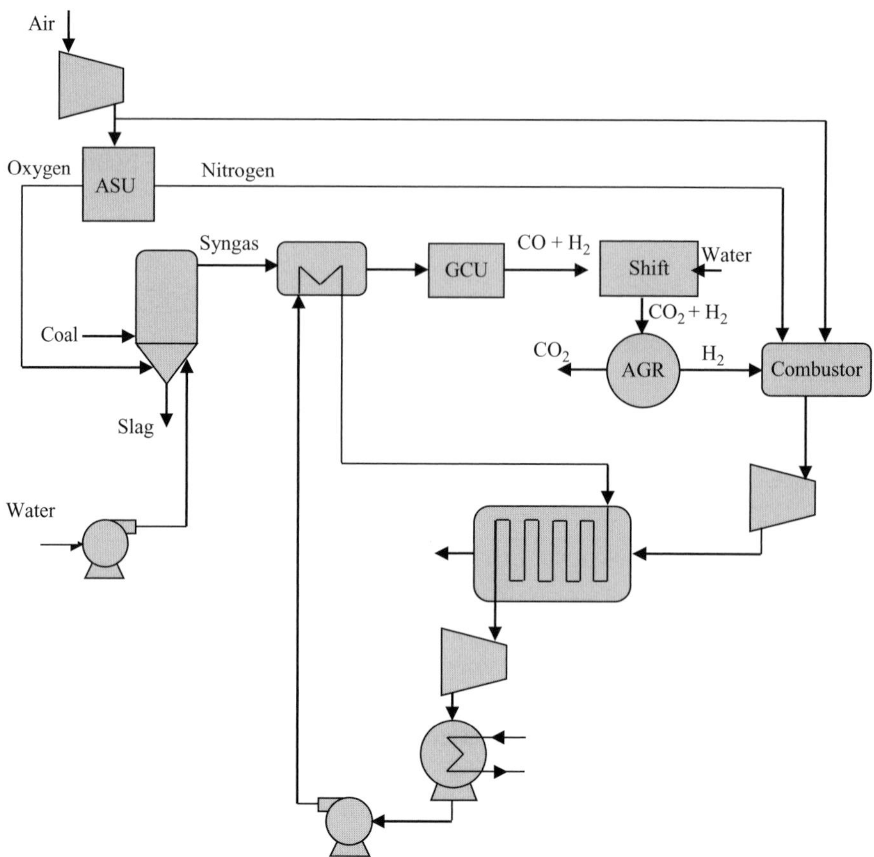

Figure 13.8 Layout of an IGCC plant with pre-combustion capture. Coal gasification uses oxygen and steam. The syngas cooler is used before the gas clean-up unit, and to raise steam for the steam turbine (in the case of radiant coolers). AGR is the "acid gas removal" responsible for separating CO_2 and H_2S (e.g., the Selexol® process), while GCU is the "cold gas clean-up unit" for removing particulates and chlorides. The hydrogen fed to the gas turbines combustion is burned in air, and the gas turbine exhaust is used in the HRSG. CO_2 is collected for compression and storage.

is necessary to start with a coal–water slurry stream and pump this slurry into the gasifier (for lower-pressure gasification systems, lock hoppers are used to pump coal in the dense phase with the help of nitrogen; the maximum pressure achieved when using these lock hoppers, however, is relatively low at ~30 bar). The coal–water slurry steam typically contains much more water than required for gasification, and this water must be vaporized inside the gasifier, which tends to reduce the cold gas efficiency (CGE) to undesirable values around 69–77%. An alternative in a carbon capture plant is to use liquid CO_2 as the slurry medium. Given the availability of liquid CO_2 in these plants, and the lower enthalpy of evaporation of CO_2, the CGE of this "dry feed" system has been shown to be higher, in the range 78–83% (Table 13.6). Since a significant fraction of the energy in the gasifier is

Table 13.6 Properties of water and CO_2 relevant to and at entrained-flow gasification pressure. Properties at average liquid and vapor phase temperatures of 150 °C and 840 °C for H_2O and 18 °C and 710 °C for CO_2, respectively [22]

	H_2O	CO_2
Critical temperature	374 °C	31 °C
Critical pressure	221 bar	74 bar
Slurry (25 °C, 72 bar)		
Liquid viscosity	0.89 cP	0.06 cP
Liquid density	1000 kg/m^3	752 kg/m^3
Gasifier (55 bar)		
Heat capacity of liquid	4.6 kJ/kg·K	4.2 kJ/kg·K
Vaporization enthalpy	1,605 KJ/kg)	146 kJ/kg
Heat capacity of vapor	3.6 kJ/kg·K	2.4 kJ/kg·K

consumed in evaporating the water content of the slurry, replacing water with CO_2 reflects positively on the overall plant efficiency as well, raising it by ~3 percentage points [21], as will be shown in the analysis below. Using CO_2 slurry also makes it possible to use low-quality coal (with higher moisture content) in gasification, which would have been very inefficient otherwise. This concept has been examined in detail [22,23] and the corresponding analysis is summarized here.

Oxygen consumption in the gasifier is determined by the gasification temperature, the amount of water needed for steam gasification, the moisture content of the coal, and the endothermicity of the gasification reaction, which is 131 and 172 MJ/kmol for water and carbon dioxide-based gasification. Dry gasification is about 30% more endothermic, but water evaporation needs more thermal energy. Other energy penalties in IGCC plant are associated with the energy consumption in the ASU unit, and the more oxygen needed for partial oxidation, the higher this energy penalty becomes. The impact of the coal type is important, and the compositions of three different coals used in the simulations are shown in Table 13.7.

Simulations were performed using Aspen [24]. The plant layout is shown schematically in Figure 13.9 for two feeding options for different coals: water slurry or CO_2 slurry. The dry solid loadings are shown in Table 13.8. Two types of gasifiers are used, a radiant cooler in which syngas is cooled radiatively to raise steam for the steam cycle, and a direct quench cooler in which water is sprayed in the gas until it is saturated. The first type, which is more expensive, should lead to overall higher plant efficiency.

The IGCC plant used in the simulations is very similar to Case 2 in [25,26], which was also analyzed in Reference [27] (with modification to account for $CO_{2(l)}$ recirculation for the feed system). The key assumptions and process variables of the IGCC plants are summarized in Table 13.9. In the following, the syngas is followed from the gasifier until the CO_2 compression unit and the gas turbine flue gases.

Table 13.7 Coal composition, measured on a dry basis, and given in terms of the proximate and ultimate analyses, as well as its LHV

Rank	Bituminous	Subbituminous	Lignite
		Proximate analyses (weight %)	
Moisture	11.12	28.09	32.24
Ash	10.91	8.77	9.72
Volatile matter	36.55	44.73	44.94
Sulfur	2.82	0.63	0.80
Fixed carbon	49.72	45.87	44.54
LHV (kJ/kg)	29,544	26,176	24,625
		Ultimate analyses (weight %)	
Moisture	11.12	28.09	32.24
Carbon	71.72	68.43	65.85
Hydrogen	5.06	4.88	4.36
Nitrogen	1.41	1.02	1.04
Chlorine	0.33	0.03	0.04
Sulfur	2.82	0.63	0.80
Ash	10.91	8.77	9.72
Oxygen	7.75	16.24	18.19

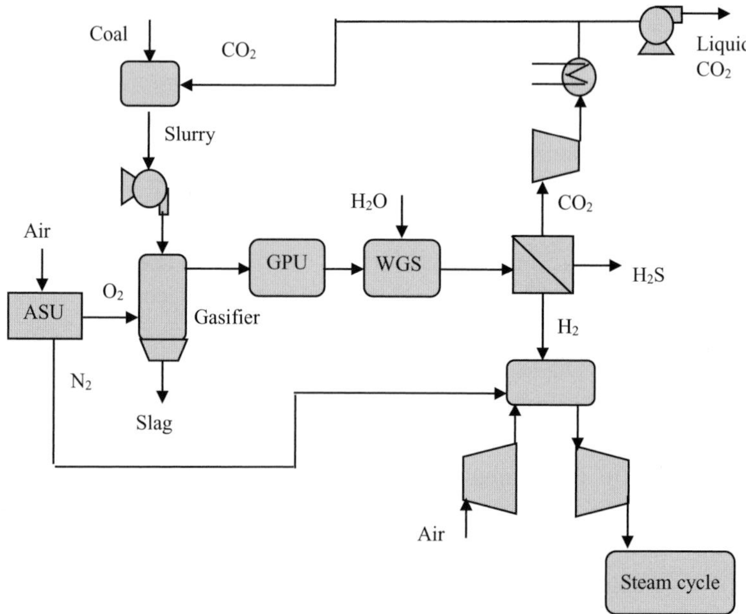

Figure 13.9 The IGCC plant with liquid CO_2 slurry for coal feeding [22].

Table 13.8 Dry solid loading (%wt) in water slurry and CO_2 coal slurry for three coal types, with as received coal moisture. X_m is the mass fraction of moisture in the coal, X_{SM} is the mass fraction of the slurry medium in the coal slurry [22]

		Bituminous	Subbituminous	Lignite
	X_{SM}	(X_m = 11%)	(X_m = 28%)	(X_m = 32%)
Water slurry	29%	**63%**	**51%**	**48%**
CO_2 slurry	20%	**71%**	**58%**	**54%**

13.5.1.2 Syngas Cooling

The gasifier produces a mixture of CO, H_2, CO_2, and H_2O, and small concentrations of HCl, NH_3, H_2S, and COS. Syngas coolers are radiant cooling, reducing the gas temperature to ~600 °C and raising intermediate-pressure steam, or quench cooling, producing saturated gas.

The residence times inside entrained-flow gasifiers are determined by their length and the gas velocity. They have high operating temperatures and, with the exception of the WGS reaction (which is favored at low temperatures), most homogeneous gasification reactions reach equilibrium. These reactions slow down significantly when the gas is quenched. In the analysis, the WGS departure from equilibrium was modeled using the "temperature approach to equilibrium" used in Aspen Plus®. In this method the equilibrium constants are calculated at a temperature that deviates from the gasification temperature by ΔT_{eq}.

13.5.1.3 Plant Layout

The layout of the IGCC plant with liquid CO_2 slurry feed is shown schematically in Figure 13.9. In the gas clean-up unit (GCU), water-wash removes chlorides while ash particulates are assumed to leave with the gasifier slag. In the WGS, a two-stage catalytic reactor is used with a H_2O:CO molar ratio of 2:1 using steam extracted from the steam turbine. The amount is dependent on the humidity content of the syngas. This gas is cooled to ~40 °C before the acid gas removal (AGR) unit.

The AGR unit separates CO_2 and H_2S from the syngas using the Selexol® process (see Chapter 10). In a two-stage process, the H_2S-loaded solvent from the first absorption stage is regenerated thermally in a stripper, and the CO_2-loaded solvent from the second stage is regenerated by reducing the pressure in multiple flash drums (see Chapter 10). High-pressure CO_2 is compressed and recycled to the CO_2 absorber in order to minimize H_2 losses. CO_2 from the intermediate and low-pressure drums of the flash units are sent to the CO_2 compressor. Sulfur recovery uses a Claus unit, in which the exothermic oxidation reaction $2H_2S + SO_2 = 2H_2O + 3S$ produces more heat.

In the power island the decarbonized syngas from the AGR is reheated and expanded to ~30 bar (from the gasifier pressure) before the gas turbine. The expanded syngas is diluted

Table 13.9 Process conditions used in the analysis of the IGCC plant with different coal feed medium, two types of syngas cooling, and three types of coal. The ASU was not integrated with the plant. The water shift reaction (WSR) adds 2 moles of H_2O for each mole of CO, and the Selexol® unit removes 99.6% of the H_2S and captures 90% of the CO_2 [22]. The temperature used for computing the WGS equilibrium deviates from the gasification temperature by ΔT_{eq}. An F class gas turbine is used

Feedstock	
Coal type	Bituminous, subbituminous, or lignite, as received
Coal flow	5450 tonne/day
Feeding system	Coal slurry at 72 bar, with loading described in Table 13.8
Gasifier	
Type	Entrained-flow, single stage at 1370 °C and 57 bar
Oxidant	95% vol. O_2 at 68 bar
Oxidant supply	ASU: 1370 kJ_e/kg
Carbon conversion	98%
Heat losses	1% of HHV
Syngas cooling	Radiant–quench ($\Delta T_{eq.} = -200$ K) or full-quench ($\Delta T_{eq.} = -10$ K)
Power island	
Steam turbine	12.4 MPa /538 °C/538 °C
Condenser pressure	51 mmH_g
Syngas diluent	N_2 (with extra steam if not enough N_2 is available)
Fuel gas HHV	4.8 MJ/N_m^3
CO_2 compression	
Intercooler temperature	30 °C
Compressor pressure	80 bar
Pump pressure	153 bar

with N_2 from the ASU to achieve the heating value shown in Table 13.9. Dilution is necessary in order to moderate the gas turbine combustion temperature to keep NO_x emissions at acceptable levels, with steam extracted from the bottoming cycle used if N_2 from the ASU is not sufficient. The gas turbine exhaust, together with heat available from other units such as the WSR, is used in the bottoming Rankine cycle. Heat integration in an IGCC plant is a trade-off between capital costs and plant efficiency. Hot water and steam are produced at different temperatures from the hot syngas, including from the radiant syngas cooler and cooling prior to the shift reactors.

A multistage compression with intercooling raises the CO_2 pressure to 80 bar and 30 °C, forming a dense phase. A fraction of this liquid CO_2 is recirculated to form the coal–CO_2 slurry, while the rest is pumped to the final pressure required for transportation to the storage site.

A summary of the results are shown in Table 13.10. The specific oxygen consumption increases with lower-rank coals (with higher moisture content). On average, CO_2 slurry feeding consumes 15% less oxygen. Extra oxygen is needed to burn more to provide energy

Table 13.10 Summary of simulation results for the IGCC plant with CC using two different slurry feeding options, water and CO_2, and three types of coal: bituminous subbituminous and lignite. The oxygen requirements are shown per coal thermal energy based on the HHV [22]

	Bituminous		Subbituminous		Lignite	
	Water	CO_2	Water	CO_2	Water	CO_2
O_2 in $kg/kJ_{coal*}10^6$	31.1	26.6	39.4	33.6	41.8	35
Cold gas efficiency	75.2	82	62	72.5	60.5	71.5
Efficiency, radiant cooling	32.7	32.6	27	29.6	24.1	28.6
Efficiency, quench cooling	30	30.9	22.5	26.9	20.2	25.2

to evaporate the liquid contents, besides raising the gasifier temperature (1370 °C) and supplying the endothermic energy for gasification. This reduces the power consumption of the ASU and the capital cost.

The CGE decreases with the coal rank. With CO_2 slurry feeding, higher CGE is achieved. This is especially apparent in the case of low-rank coal, which has lower efficiency. Operation with CO_2 slurry leads to a significant improvement in the CGE, which ranges from 7 percentage points for bituminous coal to more than 11 percentage points for subbituminous coal and lignite slurry.

In most water slurry cases, excess water is present in the syngas and no steam extraction from the turbines is needed. While radiant cooling raises the H_2:CO ratio, its low moisture means more shift steam extraction. Thus, full-quench cooling can reduce the high shift steam requirement in the case of CO_2 slurry feed as it introduces moisture while cooling the gas through direct contact with water. The shift steam extraction in plants with CO_2 slurry is reduced by almost half for bituminous coal, and is unnecessary for subbituminous coal and lignite if full-quench cooling is used. Water slurry feeding raises the H_2:CO ratio, and low total (slurry medium plus coal moisture) water in the feed reduces the H_2:CO ratio of the syngas. Higher CO content in the syngas produced using CO_2 slurry feed raises the shift steam requirement for the WGS at the expense of the steam turbine output. A gasifier with radiant–quench cooling produces syngas with higher H_2:CO ratio because of the slower cooling rate, which allows the WGS reaction to proceed further before its rate freezes at lower temperatures.

The power consumption in the ASU, AGR, and CO_2 compression unit are significant sources of losses in CO_2 capture technology. They are directly affected by the slurry medium; CO_2 slurry reduces the total by 4%. The Selexol® unit and CO_2 compression train, on average, consume 15% more power with CO_2 slurry feed because of CO_2 recirculation. The ASU power reduction is much higher, with ~10% less power consumed with CO_2 slurry feed as a result of the reduced oxygen requirement.

To calculate the efficiency, results are normalized with respect to coal energy input in the HHV. For comparison, a very similar plant was modeled without CC, and its efficiency was estimated to be 39% when using a GE Energy gasifier and radiant cooler, while that with a Shell dry-feed gasifier was 42.1% [25,26]. The same report shows that the efficiency of both plants with CC was 32.6% and 32.2%, respectively. The net efficiency of the plant is higher with CO_2 slurry feed (except for the case of bituminous coal with radiant–quench cooling). Almost 10% higher efficiency is predicted for sub-bituminous coal with radiant–quench cooling (similar to other predictions [21]), and up to 25% higher efficiency is achieved with lignite. While the efficiency is always lower for the lower-rank coal, the reduction is less with CO_2 slurry feed. Moreover, it is also always higher with radiant–quench cooling. However, this comes at the cost of a more expensive gas cooler.

13.5.2 IGCC with CO_2-Gas Feed and Membrane Separation

Other separation technologies following WGS have been considered, especially those using hydrogen-permeable membranes, which were described in Chapter 10. A case using gaseous CO_2 to dry feed coal into an intermediate-pressure gasifier was analyzed in Reference [28]. High-pressure lock hoppers are used for feeding and the gasifier pressure was limited to less than 45 bar. For separation, multiple hydrogen membrane units were used to separate most of the hydrogen. Furthermore, extra firing in the HRSG using left-over hydrogen in the original stream and pure oxygen was used to raise steam. The layout of the IGCC plant, which uses a Shell gasifier, is shown in Figure 13.10. In the analysis, the gasifier model used a kinetics-based model that did not assume equilibrium at the vessel exit [29]. This is because CO_2 gasification kinetics are slower than their H_2O counterpart, and depending on the design and operating conditions, equilibrium may not always be achieved at the gasifier exit. Because of the low membrane tolerance to sulfur, the membrane separation units were placed downstream of the AGR, which uses a Rectisol® process to remove H_2S.

The gasification pressure was 44 bar (this choice is a trade-off between efficiency, which benefits from lower pressure, and the gasifier size). Coal was fed to the gasifier using CO_2 as the carrier gas, at 80 °C and 48 bar, provided by the CO_2 compressor after capture. Effective heat and mass integration were applied in this plant design. The gas turbine compressor supplied 50% of the air required by the ASU distillation column with an expander between the compressor and the ASU to recover part of the compression work. Nitrogen from the ASU was used as a sweep gas in the membrane units in order to reduce the hydrogen partial pressure at the permeate side and hence reduce the required surface area (see Chapter 10 for more detail on membrane separation units). Nitrogen from the ASU was added to the gas turbine combustor to lower the combustion temperature and reduce NO_x formation rate.

Hot syngas from the gasifier was quenched to 900 °C by cold syngas recycling (after the syngas cooler). It was cooled further to 300 °C in the syngas cooler and the recovered thermal energy was used to produce high pressure (HP in the figure) and intermediate-pressure (IP) steam. The syngas was scrubbed to remove flyash, solids, and soluble contaminants, and further cooled to ambient temperature. The thermal energy recovered

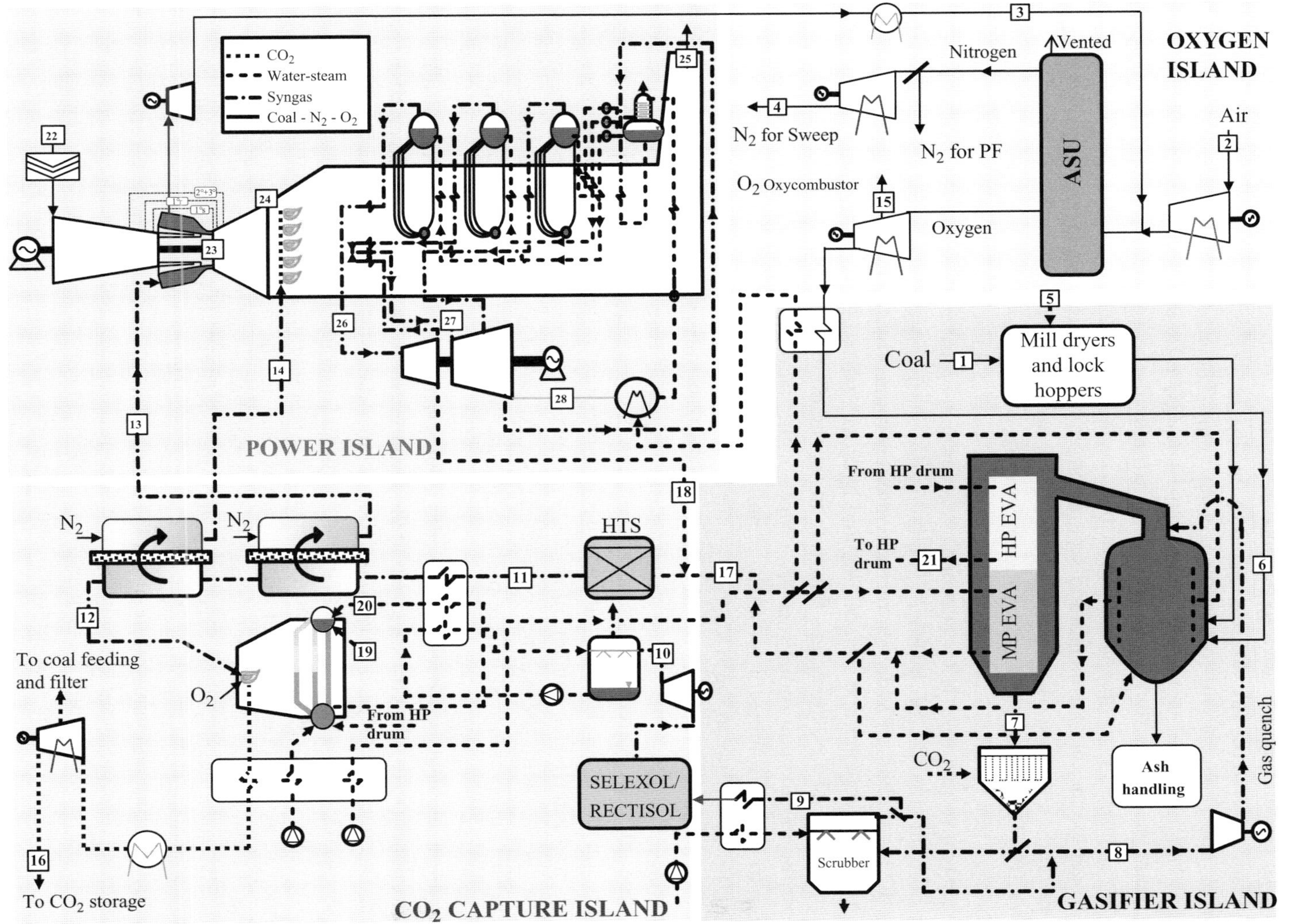

Figure 13.10 The layout of an IGCC with carbon capture using a non-sulfur-tolerant hydrogen membrane. The diagram is divided into quadrants showing the oxygen production island, the gasification island, the power island, and the CO_2 capture island [28]. Sulfur removal from the cooled syngas is based on the Rectisol® process, followed by WGS. HTS is the high-temperature (water–gas) shift unit. Hydrogen is separated in the membrane units using nitrogen sweep gas. Left-over hydrogen energy recovery after membrane separation is carried out with oxy-combustion in the post-firing unit (shown in the middle of the CO_2 capture island). Separated hydrogen is sent to the HRSG.

in that last cooling step was used to produce hot water for the syngas saturator. H_2S and COS were removed in the AGR section by means of a Rectisol® unit using chilled methanol, and H_2S was sent to a sulfur recovery unit. The syngas was then compressed to 54 bar and additional steam added to achieve a H_2/CO ratio equal to 2.0 for the WGS. That steam is drawn from the IP steam generated in the gasification island and, if necessary, from the high-pressure steam turbine.

Hydrogen-permeable Pd membrane separation units were used to separate hydrogen from the shifted syngas, as shown in the CO_2 capture island in Figure 13.10 (see also Chapter 10). The maximum membrane temperature was set at 400 °C, requiring a waste heat boiler after the WGS. Three different values were assumed in the analysis for the hydrogen recovery factors (HRF), defined as the amount of hydrogen separated divided by the maximum amount of hydrogen that could be separated: 90%, 95%, and 98% (Table 13.11). Limits on hydrogen recovery are determined by the membrane separation unit – that is, the membrane surface area and hydrogen partial pressure difference across the membranes. The N_2 sweep gas flow rate in the membrane units was set such that H_2 concentration at the unit outlet was 40%. Most of the hydrogen was separated at 25 bar and sent to the GT combustor, while the remainder (10%, 5%, or 2%) at ambient pressure was sent for post-firing in the HRSG.

After hydrogen separation, the retentate stream, which consisted mainly of CO_2, unconverted H_2, and some CO, was burned in oxygen to recover their heating value. Next, the products stream was cooled to ambient temperature, producing high-pressure steam for the HRSG and intermediate-pressure water for the economizers. At 35 °C and after water condensation, CO_2 molar concentration by volume on a dry basis was 96.2%, with the balance being inert N_2, Ar, and O_2. Conditions in terms of temperature and pressure at different stages for the case with HRF = 90% are shown Table 13.12.

Mass and energy balances over different components were estimated using Gas and Steam software, developed by the GECoS group in the Department of Energy of the Politecnico di Milano. The model uses a gasification reduced-order model in which the gasification kinetics are introduced via a reactor network coupling the mixing and reactions in different parts of the reactor [30]. For the Rectisol® and cryogenic CO_2 separation, detailed simulations were performed in Aspen Plus®. Special attention was paid to the membrane unit modeling because of its significant impact on the results. The membrane surface area was determined using a 2D model developed by SINTEF within the CACHET-II project (details can be found in Reference [31]).

The fuel pressure and temperature at the GT combustor inlet affect the system performance. A 5 bar overpressure above the air pressure is assumed, and hence the sweep gas pressure in the membrane unit is taken as 25 bar. The fuel temperature of 350 °C is used as a reference. (Fuel preheating is commonly used to increase the plant efficiency; a fuel temperature of 400 C raises the efficiency by 0.1 percentage point).

The results of the plant simulations were presented in terms of the net LHV-based (thermal to electrical) efficiency, η_{el}; CO_2 capture ratio (CCR); and the energy penalty

Table 13.11 Assumptions used in modeling the primary components

Gas turbine	
Pressure ratio	18.1
TIT	1360 °C
Steam cycle	
Pressure levels, bar	144, 54, 4
Maximum steam temperature	565 °C
Pinch, subcooling, approach ΔT	10/5/25 °C
Condensing pressure	0.048 bar (32 °C)
Turbine Isentropic efficiency, HP/IP/LP	92/94/88 %
Pump efficiency	70%
Air separation unit	
Oxygen purity	95%
Nitrogen purity	99%
Oxygen outlet temperature	20 °C
Oxygen temperature to the gasifier	180 °C
Oxygen pressure entering the gasifier	48 bar
Gasification section	
Gasifier outlet pressure	44 bar
Gasifier outlet temperature	1550 °C
Coal conversion	99.3%
O/C ratio	0.44
Dry quench exit temperature	900 °C
Scrubber inlet temperature	298 °C
CO_2 separation and compression	
Final delivery pressure	110 bar
Compressor isentropic efficiency	85%
Temperature for CO_2 liquefaction	25 °C
Pressure drop for intercoolers and dryer	1.0%
Pump efficiency	75%
CO_2 purity	>96%

related to CO_2 capture, which is given by the specific primary energy consumption for CO_2 avoided (SPECCA), defined respectively by:

$$\eta_{el} = \frac{\text{Net power}}{\text{thermal power input (LHV}_{\text{NG}})},$$

$$\text{CCR} = \frac{\text{CO}_2\,\text{captured}}{\text{max.amount CO}_2\text{ produced from fuel used}},$$

$$\text{SPECCA} = \frac{HR - HR_{REF}}{E_{REF} - E} = \frac{3600 \cdot \left(\frac{1}{\eta} - \frac{1}{\eta_{REF}}\right)}{E_{REF} - E},$$

Table 13.12 The temperature and pressure for the points labeled in Figure 13.10 with 90% HRF

Point	Temperature (°C)	Pressure (bar)
1	15	44.0
2	15	1.01
3	30	5.8
4	252	25.0
5	80	48.3
6	180	48.0
7	300	41.1
8	291	41.1
9	164	41.1
10	118	54.0
11	507	52.9
12	400	50.3
13	311	25.0
14	362	1.2
15	37	49.3
16	28	110
17	300	54.0
18	417	55.9
19	335	144.0
20	339	144.0
21	339	144.0
22	15	1.0
23	1438	17.6
24	593	1.0
25	90	1.0
26	559	133.9
27	559	44.3
28	32.17	0.048

where *HR* is the heat rate of the plant, expressed in kJ_{LHV}/kWh_e, *E* is the specific CO_2 emission rate, expressed in kg-CO_2/kWh_{el}, and *REF* is the reference case for electricity production without carbon capture. The efficiency is based on the LHV (according to European standards); it should be multiplied by 0.96 to get the value based on the HHV (according to US standards).

Simulations were performed for the reference plant with no capture, the plant with capture but using a Selexol® CO_2 separation process, and the plant with a membrane unit for H_2 separation with three values of the HRF. The thermal energy input for the capture plant was raised to keep the power output almost the same (except for the Selexol® case). The efficiency of the reference plant with no CO_2 capture was 47.1%. The net efficiency of plants with CO_2 capture was 36% for the Selexol® process, and ranged from 38.9% to 39.3% with the membrane separation unit. Higher HRF is better because more energy from the hydrogen is used in the CC rather than in the steam cycle only. However, this comes at the expense of more membrane surface area (i.e., higher cost). The SPECCA for plants with the membrane unit was in the range of 3.3–3.6 MJ/kg-CO_2, depending on the HRF. In comparison, the SPECCA for the reference IGCC with CO_2 capture using Selexol® is 3.71 MJ/kg-CO_2, showing that Pd-based membrane separation could lead to significant improvements in IGCC CO_2 capture technologies.

13.6 Chemical-Looping Combustion

Chemical-looping combustion of natural gas, in which two redox reactions in two separate reactors (or a rotary reactor) are used to produce a separate stream of CO_2 without using an ASU to produce oxygen for oxy-combustion, was described in Chapter 11. A similar concept has been proposed for coal, and several small-scale units have been built to examine its performance. In this case, however, direct contact of the solid oxygen carrier (OC) and coal/char particles is not sufficient for fast reduction of the OC, and other approaches have been proposed. The first is to gasify coal and use the syngas in chemical-looping cycles similar to those proposed for natural gas, the *syngas-CLC* [32]. The other approach is to feed coal directly to the reduction reactor but with a gasifying agent ($H_2O + CO_2$) so that coal is gasified and simultaneously the syngas is burned while reacting with the OC, the *iG-CLC* (*in-situ* gasification with CLC). Another, similar, approach, *chemical-looping with oxygen uncoupling* (*CLOU*), uses an OC that can reduce and release its oxygen under the condition of the fuel reactor, and the released oxygen burns the coal [33]. In this case, CO_2 can be used as the fluidizing medium for the fuel reactor. While the latter cases have the advantage of not requiring a separate gasifier, they require special OCs, and coal contaminants such as ash and sulfur can pose further challenges to the metal oxides. The three concepts are shown schematically in Figure 13.11.

13.6.1 Coal CLC Types

In syngas-CLC (Figure 13.11a) autothermal oxy-gasification (partial oxidation + steam reforming) is used to produce nitrogen-free syngas before the CLC process. An alternative approach is indirect steam gasification, where the enthalpy of the reaction in the gasifier is supplied from the air reactor of the CLC system. Thermal energy can be transferred with the OC and/or using heat exchanger tubes or heat pipes specially designed for this hot environment. The reactivity of natural gas and syngas, and their enthalpy of reaction with different OCs, are not the same, but several OC materials such as Ni, Cu, or Fe supported on alumina can be used, as shown in Chapter 11. Syngas clean-up is necessary in order to condition it for the fuel reactor, such as the removal of ash and sulfur compounds.

In *in-situ* gasification or iG-CLC (Figure 13.11b), the solid fuel, which can be coal, biomass, petcoke, or other pulverized carbonaceous solid fuels, is introduced into the fuel

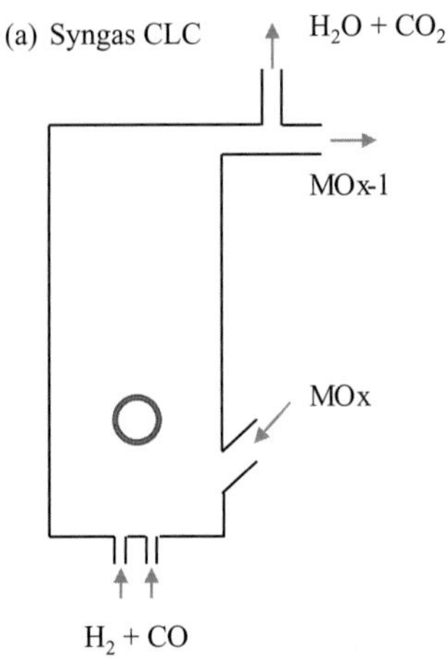

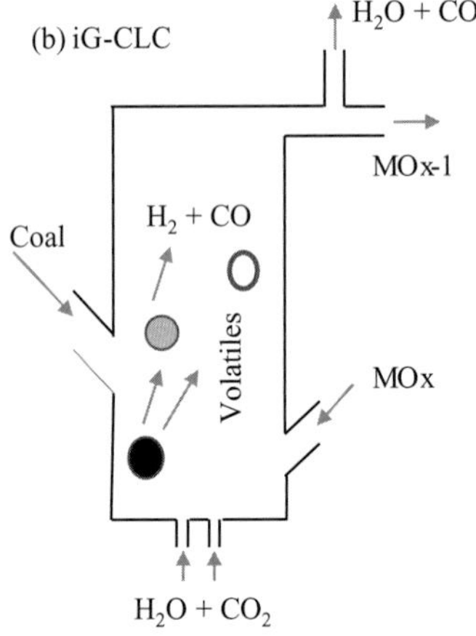

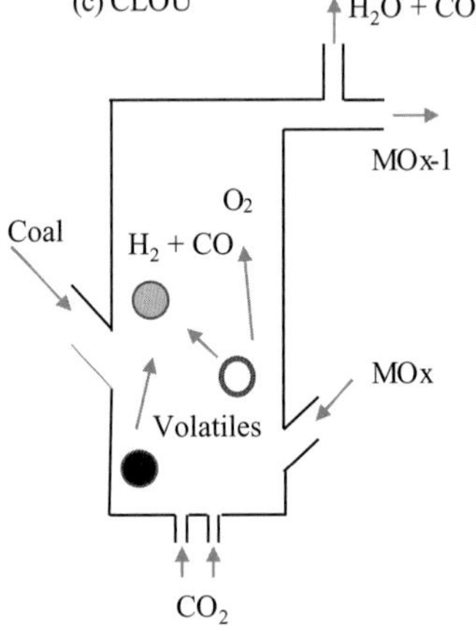

Figure 13.11 Reactions in the fuel fluidized-bed reactor, between the OC, shown as an open circle, syngas, and coal, shown as a dark circle, in the three different processes: (a) syngas-CLC, (b) iG-CLC, and (c) CLOU. The gray circle identify char, which forms as coal releases its volatiles.

reactor along with the hot OC (from the air reactor), and a fluidizing gasifying agent consisting of steam and recirculated CO_2. The process starts with the devolatilization of coal and the oxidation of the released gases, in parallel with the gasification of the char and the oxidation of the syngas. This may require multiple reactors in series to ensure full gasification of all the char and the combustion of the produced gas. Also, a "carbon stripper" may be used downstream of the fuel reactor to separate unburned char from the products and circulate it back to the fuel reactor, otherwise it would reach the air reactor and reduce the CO_2 capture efficiency of the plant. A highly reactive OC, moderately higher temperatures, and longer residence times in the fuel reactor are necessary to achieve complete conversion. Also, higher ratios of OC to coal are required to raise the temperature quickly and provide sufficient energy for the endothermic processes. Otherwise, besides using the stripper, some oxygen addition toward the downstream end of the reactor may be necessary. The process may also be more suitable for highly reactive coal, such as lignite and subbituminous.

In the CLOU process (Figure 13.11c), the OC dissociates, releasing oxygen first, in parallel with the release of volatiles from coal and possibly char gasification. Next, volatile, syngas, and char oxidation take place (the sequence depends on the temperature and reactivity of the coal and OC). A gasifying medium, CO_2, is used to fluidize the mixture of coal and OC and gasify the char. A special metal–metal oxide pair that can spontaneously release oxygen in a sufficient amount and at a sufficient rate under the fuel reactor conditions must be used. The following pairs have been identified as suitable candidates: Cu/Cu_2O, Mn_2O_3/Mn_3O_4, and Co_3O_4/CoO. All reactions are strongly endothermic:

$$\begin{aligned} &4\,CuO_4 \rightarrow 2\,Cu_2O + O_{2(g)} && \Delta H_{850} = 263.2\,kJ/mol\text{-}O_2, \\ &6\,Mn_2O_3 \rightarrow 4\,Mn_3O_4 + O_{2(g)} && \Delta H_{850} = 193.9\,kJ/mol\text{-}O_2, \\ &2\,Co_3O_4 \rightarrow 6\,CoO + O_{2(g)} && \Delta H_{850} = 408.2\,kJ/mol\text{-}O_2, \end{aligned}$$

with oxygen carrying capacity of 0.1, 0.03, and 0.066, respectively. The equilibrium partial pressure of oxygen in the gas phase (both metal oxides are in the solid phase) is shown in Figure 13.12 [32]. Therefore, for stable operation of this CLC system, careful control of the fuel rector temperature, circulation rate of the OC, ratio of OC to coal in the fuel reactor, and the air reactor temperature is needed. It is interesting to note that the reaction of carbon with these oxides can be exothermic or endothermic, which contributes to determining the reactor temperature, as shown in the following metal oxide reduction/carbon oxidation reactions:

$$\begin{aligned} &C + 4\,CuO \rightarrow 2\,Cu_2O + CO_2 && \Delta H_{850} = -133.8\ kJ/mol{\cdot}C, \\ &C + 6\,Mn_2O_3 \rightarrow 4\,Mn_3O_4 + CO_2 && \Delta H_{850} = -203.1\ kJ/mol{\cdot}C, \\ &C + 2\,Co_3O_4 \rightarrow 6\,CoO + CO_2 && \Delta H_{850} = 11.2\ kJ/mol{\cdot}C. \end{aligned}$$

This, as will be shown in the cycle analysis, can impact the integration strategy between components (even when CLOU is not used directly) and the efficiency of the overall system.

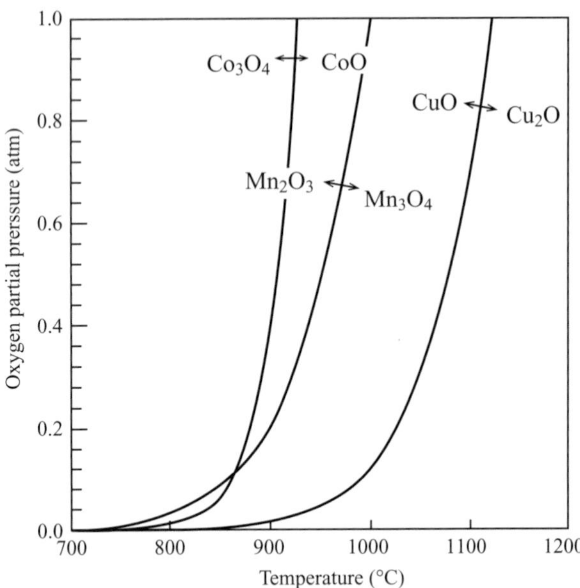

Figure 13.12 Oxygen partial pressure over metal oxide systems of Cu/Cu_2O, Mn_2O_3/Mn_3O_4, and Co_3O_4/CoO at equilibrium [32].

13.6.2 A Case Study for Syngas-CLC

A syngas-CLC plant design case is described in this section [34], and the results of a simulation to evaluate the efficiency of such a power plant are shown. These results are compared with those of a pre-combustion capture IGCC plant using Selexol® to separate H_2 and CO_2. In the IGCC plant, a cryogenic ASU provides oxygen at a purity of 95% to a Shell gasifier (using lock hoppers to feed the coal). A cyclone and a scrubber system are used to remove particulates from the syngas, and an AGR process extracts most of the H_2S and COS from the syngas. Heat recovered from the syngas before the AGR is used in the HRSG. South African bituminous coal (LHV = 25.18 MJ/kg) supplies 1088 MW_{th}. Steam and oxygen flow to the gasifier are ~10 and 126 tonne/hour (t/h), respectively. Gasification pressure is 36 bar. The gas turbine compression ratio is 1:17, with 89% isentropic efficiency. The isentropic efficiency of the compressor is 91%. The gas turbine TIT is 1235 °C and outlet mass flow is 843 kg/s at 596 °C. The steam cycle reaches temperatures above 540 °C at more than 109 bar. The resulting stack temperature is calculated at around 97 °C. The reference plant power flow and efficiency are shown in Table 13.13.

The same IGCC plant with CO_2 capture using a Selexol® absorption train was also modeled. A WGS reactor was placed between the COS hydrolysis and the sulfur removal unit. The former converts carbonyl sulfide to hydrogen sulfide. In the sulfur recovery unit, H_2S is stripped from the syngas using methyl-diethanolamine (MDEA) and sent to the Claus unit for sulfur recovery. Syngas containing ~48% hydrogen, 32 mol% CO_2, 18% water and non-condensables (by mole) was sent to a Selexol® unit to remove CO_2. Following separation, CO_2 was compressed to 110 bar. Hydrogen was burned in the gas turbine combustor with TIT ~1235 °C. The gas turbine exhaust temperature, 587 °C, and flow, 744 kg/s, are

Table 13.13 Power output and input for different units and efficiency for four cases analyzed in Reference [34], where column 1 is an IGCC plant without capture, column 2 is IGCC with pre-combustion capture using a Selexol® process, and column 3 is for IGCC with CLC using two CLC units

	IGCC – no capture	IGCC with pre-combustion capture	IGCC with CLC units
GT output (MW)	663.3	620.6	762.0
ST output (MW)	204.6	155.7	192.9
Comp. power (MW)	316.2	305.1	469.3
Pump (MW)	2.8	2.8	3.4
ASU (MW)	34.6	34.6	34.6
CO_2 removal (MW)		2.5	2.0
CO_2 compression (MW)		24.8	26.0
Auxiliary (MW)	8.6	8.7	10.4
Mech. loss (MW)	2.3	2.2	2.2
Generator loss (MW)	9.6	8.8	9.1
Net electricity (MW)	494.0	386.9	398.1
Plant efficiency (%) (LHV)	45.4	35.6	36.6

below the base case, which explains the reduction of the steam turbine power output compared to the reference case. The temperature and the pressure of the steam cycle were similar to the base case. The stack temperature was around 97 °C. The power flow and efficiency are shown in Table 13.13.

A plant that uses the same gasification unit to produce syngas and CLC unit to burn the syngas and capture CO_2 was analyzed using two designs for the CLC process. The first design used a single CLC unit composed of two fluidized-bed reactors. The OC was nickel stabilized on alumina (NiO/Al_2O_3). No WGS reactor or acid gas removal plant was used. Oxidation of H_2S in the fuel reactor was assumed to be fast enough to form SO_2, which was removed along with CO_2. Compressed air and the OC entered the air reactor (residence time 4–11 s), where nickel was oxidized. Oxygen-depleted air left the cyclone at ~1000 °C, and expanded in a turbine then used in the HRSG. In the FR (residence time up to 60 s), the OC was reduced by the syngas. The CO_2-rich gas leaving the FR expanded in a second gas turbine and was then sent to the HRSG.

In contrast to natural gas-fired CLC units, syngas oxidized with NiO is exothermic, resulting in a slightly exothermic reaction in the FR, where the temperature rises over 1200 °C, increasing the risk of sintering and agglomeration of the OC. (The standard enthalpy of reaction of NiO/Ni with CH_4, H_2, CO, C, and O_2 is, respectively, 156.5, −2.1, −43.3, 85.9, −479.4 MJ/kmol.) To avoid this risk, OC leaving the cyclone of the AR is cooled, reducing the system efficiency. For this reason, a two-stage CLC reactor concept was devised. This system is shown in Figure 13.13. Higher air flow is used in the first AR, but with only a fraction of the OC. The partially oxygen-depleted air expands in

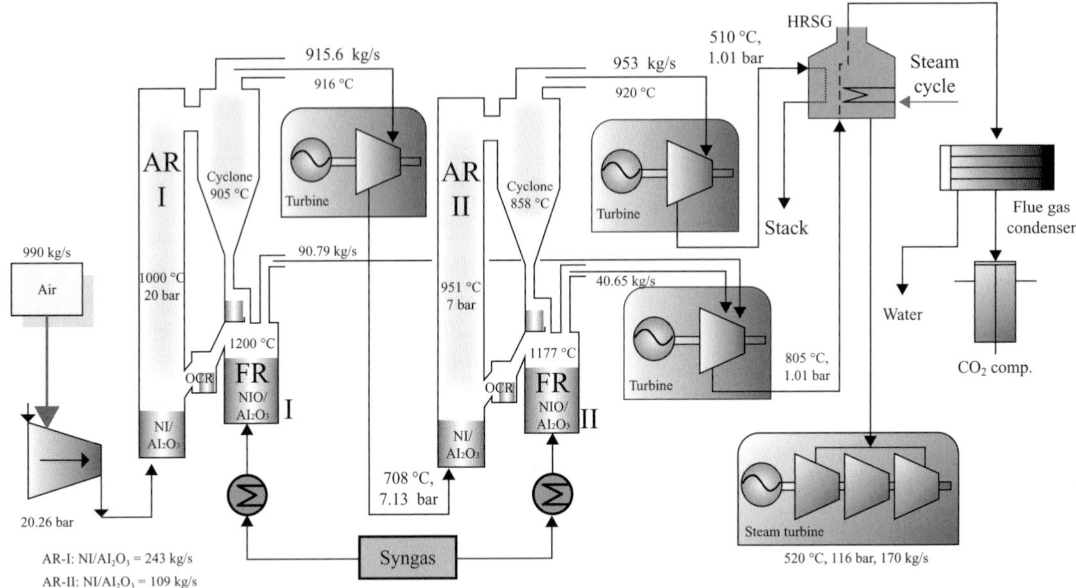

Figure 13.13 IGCC with a two-stage CLC reactor that burns syngas and captures CO_2 [34].

a turbine from 20 bar to 7 bar before it enters the second AR, where it reacts with the remaining OC. The oxygen-depleted air from the second OR expands through the second GT, and is then sent to the HRSG. The syngas supply is distributed between the two FRs. The combustion products from both fuel reactors (mainly CO_2 and H_2O) expand in a third gas turbine, and are then sent to the HRSG. The power flow and efficiency are shown in Table 13.13.

REFERENCES

1. O. Bolland and S. Sæther, "New concepts for natural gas fired power plants which simplify the recovery of carbon dioxide," *Energy Convers. Manag.*, vol. 33, no. 5, pp. 467–475, 1992.
2. MIT, *The future of coal: options for a carbon constrained world.* Cambridge, MA: MIT, 2007.
3. C. Hendricks, *Carbon dioxide removal from coal fired power plants.* Dordrecht: Kluwer Academic Press, 1994.
4. G. Gottlicher, *The energetics of carbon dioxide capture in power plants.* Washington, DC: U.S. Department of Energy, Office of Fossil Energy, NETL, 2004.
5. L. Chen, "Computational fluid dynamics simulations of oxy-coal combustion for carbon capture at atmospheric and elevated pressures," MIT, Ph.D. thesis, 2013.
6. B. J. P. Buhre, L. K. Elliott, C. D. Sheng, R. P. Gupta, and T. F. Wall, "Oxy-fuel combustion technology for coal-fired power generation," *Prog. Energy Combust. Sci.*, vol. 31, no. 4, pp. 283–307, 2005.

7. L. Chen, S. Z. Yong, and A. F. Ghoniem, "Oxy-fuel combustion of pulverized coal: characterization, fundamentals, stabilization and CFD modeling," *Prog. Energy Combust. Sci.*, vol. 38, no. 2, pp. 156–214, 2012.
8. D. M. Dyer, P. N. Richards, R. E. Russek, and S. L. Taylor, "Ion transport membrane technology for oxygen separation and syngas production," *Solid State Ionics*, vol. 134, pp. 21–33, 2000.
9. A. Fassbender, "A pressurized oxy-fuel combustion for multi-pollutant capture," in the *30th International Technical Conference on Coal Utilization and Fuel Systems*, Clearwater, FL, USA, 2005.
10. E. Benelli, G. Cumbo, D. Gazzino, and M. Morgani, "Pressurized oxy-combustion of coal with flue gas recirculation: pilot scale demonstration," in *PowerGen Europe Conference and Exhibition*, Milan, Italy 2008.
11. B. Yang, T. Hunt, P. Lisauskas, and R. Vitalis, "Pressurized oxy combustion carbon capture power systems," in *35th International Technical Conference on Clean Coal & Fuel Systems*, Clearwater, FL, USA, 2010.
12. J. Hong, R. Field, M. Gazzino, and A. F. Ghoniem, "Operating pressure dependence of the pressurized oxy-fuel combustion power cycle," *Energy*, vol. 35, no. 12, pp. 5391–5399, 2010.
13. T. Zheng, L. Pomalis, R. Clements, and B. Herage, "Optimization of a high pressure oxy fuel combustion process for power generation and CO_2 capture," in *44th International Technical Conference of Clean Coal & Fuel Systems*, Clearwater, FL, USA, 2019.
14. J. Hong, G. Chaudhry, J. G. Brisson, et al. "Analysis of oxy-fuel combustion power cycle utilizing a pressurized coal combustor," *Energy*, vol. 34, no. 9, pp. 1332–1340, 2009.
15. C. Iloeje, R. Field, and A. F. Ghoniem, "Modeling and parametric analysis of nitrogen and sulfur oxide removal from oxy-combustion flue gas using a single column absorber," *Fuel*, vol. 160, pp. 178–188, 2015.
16. A. Anderson, R. E. MacAdam, S. Viteri, et al. "Adapting gas turbines to zero emission oxy-fuel power plants," in *ASME Turbo Expo 2008, Power for Land, Sea and Air*, Berlin, Germany, 2008.
17. J. S. Hong, "Techno-economic analysis of pressurized oxy-fuel combustion power cycle for CO_2 capture," MIT, M.Sc. thesis, 2009.
18. DOE/NETL, *Pulverized coal oxy combustion power plants, Vol 1: bituminous coal to electricity*. Washington, DC: DOE/NETL, 2007.
19. M. Hong, J. S. Ghoniem, A. F. Field, and R. Gazzino, "Techno-economic evaluation of pressurized oxy-fuel combustion systems," in *Proceedings of the ASME 2010 International Mechanical Engineering Congress and Exposition 2010*, Vancouver, BC, 2010.
20. E. Benelli, G. Girardi, G. Malavasi, and M. Saponaro, "ISOTHERM: a new oxy-combustion process to match the zero emission challenge in power generation," in *the 7th High-Temperature Air Combustion and Gasification International Symposium*, Phuket, Thailand, 2008.
21. J. P. Dooher, "Physico-chemical properties of low rank coal/liquid CO_2 slurry as gasifier feedstocks," in *34th International Technical Conference on Coal Utilization and Fuel Systems*, Clearwater, FL, USA, 2009.
22. C. Botero, "The phase inversion-based coal-CO_2 slurry (PHICCOS) feeding system: design, coupled multiscale analysis, and techno-economic assessment," MIT, Ph.D. thesis, 2014.
23. C. Botero, R. P. Field, R. D. Brasington, H. J. Herzog, and A. F. Ghoniem, "Performance of an IGCC plant with carbon capture and coal-CO_2-slurry feed: impact of coal rank, slurry loading, and syngas cooling technology," *Ind. Eng. Chem. Res.*, vol. 51, no. 36, pp. 11778–11790, 2012.
24. AspenTech Inc., "Aspen Plus," 2012 [software].
25. L. L. Woods, M. C. Capicotto, P. J. Haslbeck, et al., "Cost and performance baseline for fossil energy plants," Technical Report DOE/NETL-2007/1281. Prepared by Research and

Development Solutions, LLC(RSD) for the U.S. Department of Energy, National Energy Technology Laboratory, 2007.

26. NETL and DOE, *Cost and performance baseline for fossil energy plants. Volume 1: bituminous coal and natural gas to electricity final report, Revision 2*. Washington, DC: NETL and DOE, 2010.
27. R. P. Field and R. Brasington, "Baseline flowsheet model for IGCC with carbon capture," *Ind. Eng. Chem. Res.*, vol. 50, no. 19, pp. 11306–11312, 2011.
28. M. Gazzani, D. M. Turi, A. F. Ghoniem, E. Macchi, and G. Manzolini, "Techno-economic assessment of two novel feeding systems for a dry-feed gasifier in an IGCC plant with Pd-membranes for CO_2 capture," *Int. J. Greenh. Gas Control*, vol. 25, pp. 62–78, 2014.
29. M. Gazzani, G. Manzolini, E. Macchi, and A. F. Ghoniem, "Reduced order modeling of the Shell-Prenflo entrained flow gasifier," *Fuel*, vol. 104, pp. 822–837, 2013.
30. R. Monaghan and A. Ghoniem, "A dynamic reduced order model for simulating entrained flow gasifier. Part 1: model development and description," *Fuel*, vol. 91, pp. 61–80, 2012.
31. M. Gazzani, D. M. Turi, and G. Manzolini, "Techno-economic assessment of hydrogen selective membranes for CO_2 capture in integrated gasification combined cycle," *Int. J. Greenh. Gas Control*, vol. 20, pp. 293–309, 2014.
32. L. F. Adanez, J. Abad, A. Garcia-Lobiano, F. Gayan, and P. de Diego, "Progress in chemical-looping combustion and reforming technologies," *Prog. Energy Combust. Sci.*, vol. 38, pp. 215–282, 2011.
33. H. Mattisson, T. Lyngfelt, and A. Leion, "Chemical looping with oxygen uncoupling for combustion of solid fuels," *Int. J. Greenh. Gas Control*, vol. 3, pp. 11–19, 2009.
34. S. Rezvani, Y. Huang, D. McIlveen-Wright, N. Hewitt, and J. D. Mondol, "Comparative assessment of coal fired IGCC systems with CO_2 capture using physical absorption, membrane reactors and chemical looping," *Fuel*, vol. 88, no. 12, pp. 2463–2472, 2009.

14 Biomass, Bio-energy, and Biofuels

14.1 Introduction

Until the mid-1800s, when fossil fuels became the major source of energy for heating, electricity production, and transportation, biomass derived from plants and animal products was the primary source of energy (heating and lighting by fires and candles). The explosive growth in the use of fossil fuels powered the Industrial Revolution and provided enormous improvements in the standards of living wherever it became the dominant source of primary energy. However, biomass still supplies nearly 10% of the primary energy worldwide, and is a major source in rural communities. Although only a small percentage of available biomass resources are used for energy production, the total potential exceeds 4.5 EJ (exajoules; 10^{18} joules). Early use of biomass was likely to have been for fires. More recently, it has been for heating by combustion for domestic use; electricity production by combustion in power plants; and for the production of biofuels such as ethanol and biodiesel by bioconversion or thermochemically. The contribution of biomass to primary energy consumption, both in quantity and form and as a fraction or absolute amount, depends strongly on the level of economic development and geographic location. In 2003, it was 26% of the total consumption in the developing world, but only 4% in the developed world. During 2010–2012, the US consumed 62 terawatt-hour/year (TWh/y) of bio-energy, followed by Germany, which consumed 37 TWh/y. During the same period, China and India consumed 27 and 3.4 TWh/y, respectively. The majority of bio-energy, close to 90% of the total, is produced using solid feedstock in the form of energy crops like sugar and corn, or lignocellulosic material, with the rest in the form of gaseous fuels produced from landfills and bio-digesters.

Interest in reducing CO_2 emissions has intensified the effort to use more biomass for energy production, especially using it for the production of liquid fuels for transportation, and blending it with coal for electricity production. Direct combustion of biomass for heating or power production is considered to be CO_2-neutral, as plants consume the same amount of CO_2 while growing, and hence recycling the emitted amount. The same may not always be true in the case of using biofuels. This is because fossil fuels are used for the production of biofuels, including agriculture, harvesting, and conversion to fuels. The actual

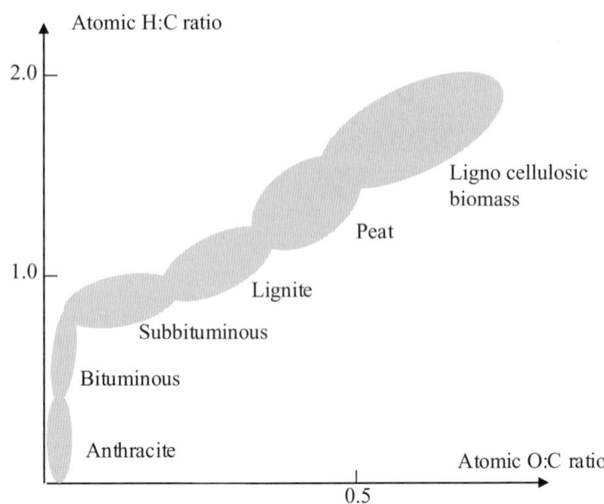

Figure 14.1 A schematic showing H/C and O/C atom ratios for coal and biomass. Coal is plant material that aged under high pressure and temperature. Both ratios are low for coal, with the highest quality having the smallest ratios. The higher heating value increases with increasing H/C ratio and decreasing O/C ratio.

saving in CO_2 emissions depends strongly on the biomass source and conversion technology, and careful analysis is needed to evaluate the advantages and disadvantages of the technology. Other concerns include land and water use if bio-crops are planted, the use of fertilizers, and competition with food production. An example of a biofuel produced from an energy crop is ethanol, which is currently produced from corn (in the USA) or sugar cane (in Brazil). On the other hand, much biomass is produced as "agricultural waste" – that is, what is left of plant after fruits and edibles are harvested.

Biomass in its raw form includes significant fractions of water and volatiles – that is, gaseous and condensed species. An important difference between biomass and solid fossil fuels is the large fractions of oxygen and hydrogen in its molecular structure, as shown in the van Krevelen diagram in Figure 14.1. This partially oxidized state lowers its heating value below that of coal. While coal (an extremely aged plant material) contains relatively small amounts of oxygen, with anthracite having the least and lignite the most, raw biomass contains significantly more, with oxygen to carbon ratios reaching 1:1. Biomass contains even more hydrogen, with H:C ratios approaching 2, while those for coal are significantly lower.

This chapter starts with a review of some elementary organic chemistry to familiarize the reader with the molecular structure of some important classes of hydrocarbons.

14.2 Complex Hydrocarbons

Some familiarity with hydrocarbon molecules and their structures is useful before discussing biomass conversion and fuel production. Hydrocarbon molecules consist of an "inner" arrangement of carbon molecules surrounded by hydrogen atoms. A carbon atom possesses four bonds that it uses to attach to other carbon atoms or to hydrogen atoms, and to oxygen atoms in the case of oxygenated hydrocarbons. The carbon atom uses one, two, or three bonds to attach to a neighboring carbon or oxygen atom, but only a single bond to attach to a hydrogen atom.

14.2.1 Classification of Hydrocarbons

Hydrocarbon families are divided between those that share only a single bond between carbon atoms, the C–C bonds; a single double bond between two carbon atoms, the C=C bond and more single bonds; or a single triple bond between two carbon bonds, the $C \equiv C$ bond and more single bonds. Another classification is made according to whether the molecule is **open chain** or forms **a ring structure**. Chain compounds are also known as aliphatics. This is the structure that binds the inner carbon molecules, in the form of a straight chain of carbon atoms or carbon atoms attached around a ring. **Alkanes** (paraffins) C_nH_{2n+2}, **alkenes** (olefins) C_nH_{2n}, and **alkynes** (acetylenes) C_nH_{2n-2}, are open-chain molecules, with all single bonds, one double bond, and one triple bond, respectively. **Cyclanes** (cycloalkanes) C_nH_{2n} or $(CH_2)_n$, and **aromatics** (benzene family) C_nH_{2n-6} are ring structures.

Open-chain structures use the following prefix to indicate the number of carbon atoms in the molecule: 1 – meth, 2 – eth, 3 – prop, 4 – but, 5 – pent, 6 – hex, 7 – hept, 8 – oct, 9 – non, 10 – dec, 16 – hexadec. Examples of the open-chain alkanes, which as we mentioned before are straight chain molecules since all the carbon bonds are arranged in a chain of (C–C), are methane (CH_4), ethane (C_2H_6), propane (C_3H_8), and butane (C_4H_{10}), which are shown in the Figure 14.2. These are also called saturated molecules since no more hydrogen atoms can be added to the molecule; all the carbon bonds available have been used in a single-bond fashion. When an H atom is extracted from these alkanes, the resulting radical is an alkyl radical (e.g., methyl, ethyl, propyl radical).

Alkenes are open unsaturated chain structures with a single double carbon bond. Examples of alkenes are ethene (C_2H_4), propene (C_3H_6), and butene (C_4H_8) (Figure 14.3).

```
    H                      H   H
    |                      |
H - C - H   Methane    H - C - C - H    Ethane
    |                      |
    H                      H   H

    H   H   H                  H   H   H   H
    |   |   |                  |   |   |
H - C - C - C - H  Propane H - C - C - C - C - H  Butane
    |   |   |                  |   |   |
    H   H   H                  H   H   H   H
```

Figure 14.2 Methane, ethane, propane and butane.

```
                                  H   H
H\      /H                        |   |
  C = C      Ethene        H \    C = C - C - H      Propene
H/      \H                 H /            |
                                          H
```

Figure 14.3 Ethene and propene.

Figure 14.4 Propyne.

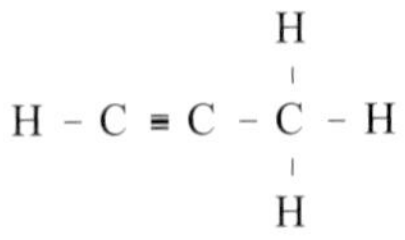

```
        H
        |
H - C ≡ C - C - H
        |
        H
```

They are classified as unsaturated because the double carbon bond can be broken and more hydrogen atoms can be added to the molecule. The first two molecules are shown next.

Alkynes have an open-chain structure and a single triple carbon–carbon bond (C_nH_{2n-2}), such as ethyne (C_2H_2) and propyne (C_3H_4) (Figure 14.4).

Adding hydrogen atoms to an alkene or an alkyne by breaking the single double or triple carbon–carbon bond, respectively, and rearranging the carbon atoms saturates them to an alkane.

Compounds with simple straight chains are often labeled as **normal** and designated with an n in front (e.g., n-pentane C_5H_{12}). This is to distinguish them from the branched-chain molecules described next. Alkanes occur naturally in petroleum, and are not generally reactive at ambient conditions. Alkenes are obtained in the cracking of petroleum and are more reactive due to the presence of the double bond. Alkynes are highly reactive, and are produced in petroleum refineries.

Many higher hydrocarbons, with three or more carbon atoms, appear as **branched**-chain molecules in which one or more C–H bonds is replaced with a branched C–C bond, with the outer carbon atom acting as a host for three hydrogen atoms. Note that the same chemical formula can be used to characterize two different compounds whose chemical structures are different because of branching. This is known as structural **isomerism**, and the compound with branched chains gets the designation iso- in front. These branched molecules are also known as **isomers** (e.g., the isooctane molecule shown in Figure 14.5). Two examples are shown: 2-methyl butane and 2,2,4-trimethyl pentane, or isooctane. The numeric nomenclature attached to the organic name is meant to show where the branches are attached.

Ring structures are molecules in which the carbon atoms are organized around a ring, with single or double bonds between the carbon atoms. Molecules with ring structures in which all carbon atoms share a single bond are known as **cyclanes**, while those in which some carbon atoms share more than one bond belong to the aromatic or benzene family. Examples of cyclanes are shown in Figure 14.6.

The aromatics are ring-structured **benzenes** (C_6H_6), where every other carbon–carbon bond is a double bond, and organized as 6CH pairs (Figure 14.7)

Benzene rings can be combined to form polycyclic aromatics, and side chains can replace the hydrogen atoms with one or more extra radicals; for example, toluene (ϕCH_3) and phenol (ϕOH), where the ϕ stands for the original molecule without a single hydrogen atom (C_6H_5–).

Another important family of hydrocarbons is the **alcohols** family. These are compounds of an alkane radical attached to a hydroxyl radical, such as methanol (CH_3OH) and ethanol (C_2H_5OH), which are shown in Figure 14.8. The hydroxyl radical is also known as a functional group. These alcohols are often written as ROH.

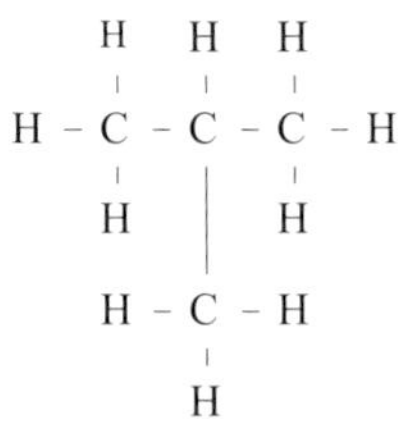

```
    H   H   H
    |   |   |
H - C - C - C - H
    |   |   |
    H   |   H
        |
    H - C - H
        |
        H
```

2-methyl butane

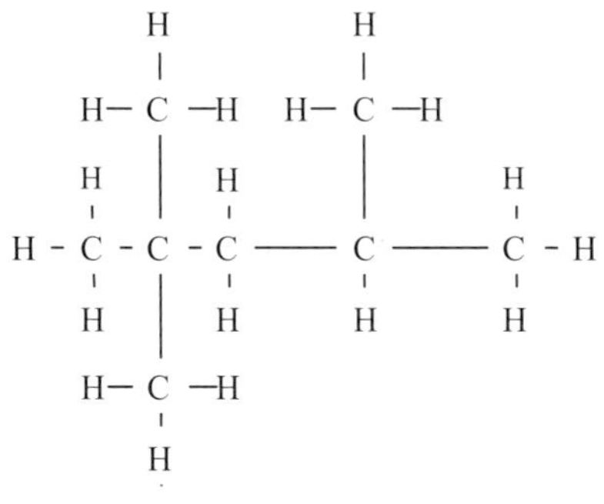

```
        H           H
        |           |
    H - C - H   H - C - H
    H   |   H       |          H
    |   |   |       |          |
H - C - C - C ----- C -------- C - H
    |   |   |       |          |
    H   |   H       H          H
        |
    H - C - H
        |
        H
```

2,2,4 trimethyl pentane or isooctane C_8H_{18}

Figure 14.5 2-methyl butane and 2,2,4-trimethyl pentane.

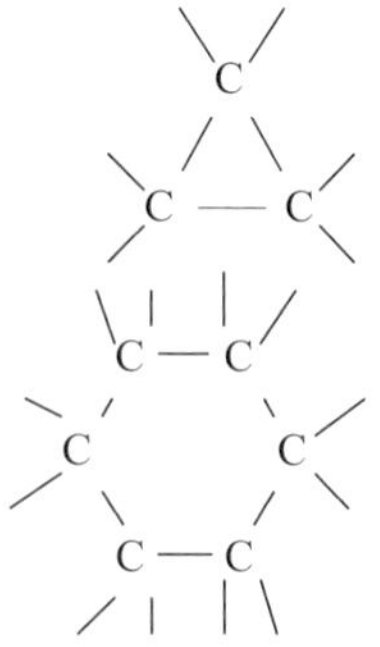

```
      \ /
       C
      / \
   \ /   \ /
    C --- C
   /       \
```

Cyclopropane C_3H_6; the hydrogen atoms are not shown

```
     \ |   | /
       C - C
  \   /     \   /
    C         C
  /   \     /   \
       C - C
     / |   | \
```

Cyclohexane C_6H_{12}; the hydrogen atoms are not shown

Figure 14.6 Cyclanes.

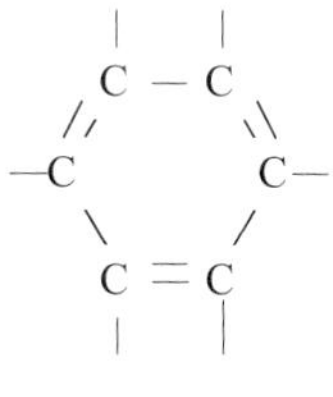

```
        |   |
        C - C
      //     \\
    -C         C-
      \       /
        C = C
        |   |
```

Benzene; hydrogen atoms are not shown

Figure 14.7 Benzene.

These alcohols are fuels – that is, they have finite heating values. Because of the presence of oxygen in the molecules, the heating values of alcohols are lower than those of non-oxygenated compounds (see Chapter 4). For instance, the absolute values of the enthalpy of reaction (corresponding to the higher heating value [HHV]) of methane is 55.5 MJ/kg, while that of methanol is 22.7 MJ/kg. The corresponding values for ethane

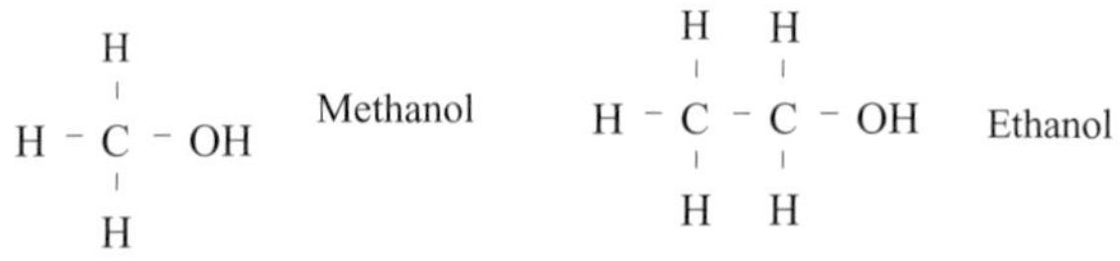

Figure 14.8 Methanol and ethanol.

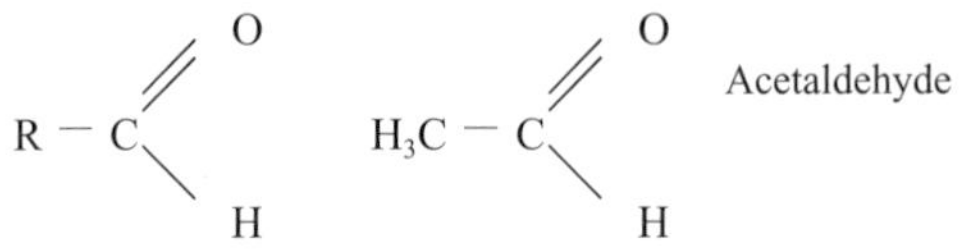

Figure 14.9 Acetaldehyde.

and ethanol are 52 and 29.7 MJ/kg, respectively. Despite their lower heating values, alcohols, especially ethanol, are added to gasoline in different proportions to improve the fuel mixture combustion characteristics and to extend the fuel supply beyond petroleum products. As will be shown in this chapter, ethanol can be produced as a product of the fermentation of starch or sugars (carbohydrates) found in grains and carbohydrates found in biomass. Carbohydrates include cellulose, starch, and sugars, and the process of converting them to ethanol depends on the particular carbohydrate. For instance, starting with starch, $(C_6H_{10}O_5)n$, the hydrolysis (reactions with water) of starch forms sugar maltose ($C_{12}H_{22}O_{11}$), the hydrolysis of maltose forms glucose ($C_6H_{12}O_6$), and the fermentation of glucose forms ethanol (C_2H_5OH).

In other compounds, known as **ketones**, the radical group is CO; in organic acids, the radical group is COOH. Thus, in addition to carbon and hydrogen, organic compounds can contain other elements, such as oxygen, nitrogen, and sulfur. These elements are found in the "**functional groups**." These groups determine the characteristics of the compound, and are used to classify the organic chemicals.

Other compounds can be formed from an organic radical R and another group. Examples include aldehydes, in which the radical is HCO, and the compounds are R–CHO. Examples of aldehydes are formaldehyde, where R is H, and acetaldehyde, in which the radical is H_3C (Figure 14.9).

More complex hydrocarbons, such as carbohydrates, will be described in the discussion of biomass and its conversion to simpler compounds.

Another important group of compounds, which contain nitrogen, is known generally as **amines**. Important examples of these compounds are: (1) methylamine (CH_3NH_2) (the functional group in these compounds is amino ($-NH_2$); (2) dimethylamine ($(CH_3)_2NH$), whose functional group is ($=NH$); and (3) trimethylamine ($(CH_3)_3N$), with the functional group ($\equiv N$).

14.3 Photosynthesis and Biomass Composition

Biomass is derived from plant life, although some is derived from animals (fats, residues, etc.). Biomass includes energy crops (crops grown specifically to provide energy), farm and agricultural waste, and municipal waste. Some energy crops are also food sources, such as corn and sugar cane. Animal waste is another form of biomass (which starts also as plant life). Forests are considered as a renewable source that can be regenerated in under 100 years.

Sources of biomass form in plant by the **photosynthesis** of carbon dioxide and water in the presence of **chlorophyll** according to the following reaction:

$$n\mathrm{CO_2} + m\mathrm{H_2O} \xrightarrow[chlorophyll]{sunlight} \mathrm{C}_n(\mathrm{H_2O})_m + n\mathrm{O_2}. \tag{14.1}$$

The enthalpy of reaction of photosynthesis is 470 MJ/kmol. The sunlight provides the reaction energy while the chlorophyll acts as the catalyst. While absorbing CO_2, the reaction releases a significant amount of oxygen. The organic compounds forming in this reaction are **carbohydrates**; they include **sugars**, **starch**, and **cellulose**. Photosynthesis efficiency is low (0.1–3.0%). Thus, biomass energy is a form of solar energy that recycles CO_2 back into a fuel form. Besides fossil fuel combustion (and before their extensive use), CO_2 is released into the atmosphere via respiration (of humans and animals), biological degradation (rotting), and fires.

14.3.1 Composition and Properties

Organic woody plant materials include **carbohydrates** and **lignin**, while grains and fruits are mostly starch and other sugars. Carbohydrates are **saccharides**: sugars or polymers of sugars. Sugars include **sucrose** ($C_{12}H_{22}O_{11}$; found in sugar cane and beet) and **glucose** ($C_6H_{12}O_6$; found in corn, grapes, and other fruits). Isomers of glucose include mannose and fructose, shown symbolically in Figure 14.10. An important property of sugar, which is relevant to the **bioconversion** of biomass, is whether it is **fermentable** or not. Fermentable sugars are D-glucose, D-mannose, D-fructose, D-galactose, and maltose [1].

Other sugars can be converted to fermentable sugars by **hydrolysis** (reactions with water), usually in the presence of an acid or an **enzyme**. For instance, sucrose (a disaccharide;

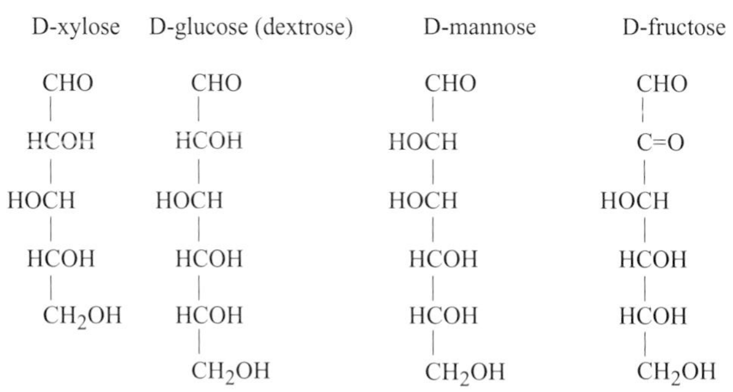

Figure 14.10 The molecular structure of a number of carbohydrates.

$C_{12}H_{22}O_{11}$) can be hydrolyzed by the enzyme **invertase** (part of yeast) into two monosaccharides, glucose and fructose, both having the same chemical formula, $C_6H_{12}O_6$:

$$\underset{\text{sucrose}}{C_{12}H_{22}O_{11}} + H_2O \xrightarrow{\text{invertase}} \underset{\text{glucose}}{C_6H_{12}O_6} + \underset{\text{fructose}}{C_6H_{12}O_6}. \tag{14.2}$$

14.3.2 Woody Material

The primary carbohydrates in plant material are **cellulose** and **hemicellulose**, besides the starch and sugars in roots, grains, and fruits. These carbohydrates are polymers of sugars, also known as **polysaccharides**. Cellulose is the earth's most common organic compound. It is the main constituent of plant cell walls, and it is a fibrous polysaccharide. Cellulose is a long linear chain polymer formed by 10,000–15,000 glucose units linked by glycosidic bonds (Figure 14.11), with hydroxyl groups attached to the sides of the cellulose chain. The long cellulose molecules forms microfibrils 10–500 μm in diameter. These orderly bundles of fibrils contribute to the mechanical strength of the plant cells. Cellulose is extremely insoluble and chemically inert, and it is resistant to acidic and enzymatic hydrolysis.

Sugars and lignocellulosic compounds are found in trees, grasses, legumes, grains, sugar crops, and aquatic plants. Dry wood is almost 66% **holocellulose** (a combination of cellulose and hemicellulose) and 25% lignin; the rest is resins, gums, tannins, and waxes (ash and extractives). About 25% of the cellulose is hemicellulose, $(-C_5H_8O_4-)$, with molecular weight 10,000–35,000, which is easily hydrolyzed to fermentable sugars; the rest is cellulose, $(-C_6H_{10}O_5-)_n$, with molecular weight >100,000, and lignin (molecular weight

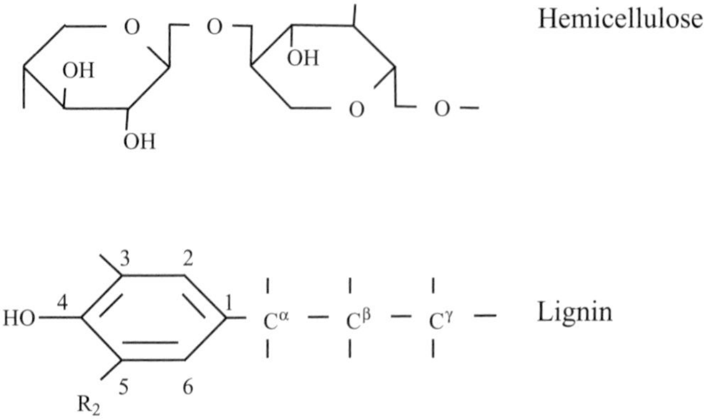

Figure 14.11 The molecular structure of the three components of lignocellulosic biomass: cellulose, hemicellulose, and lignin. R1 and R2 in lignin depend on the molecule.

5000–10,000), which resist hydrolysis. Lignin has the highest heating value of the three components of woody biomass, and it can then be combusted separately if pre-removed from the plant material.

Hemicelluloses are amorphous, low molar mass polysaccharides, also known as xylan. Hemicellulose is more soluble than cellulose, and can be dissolved in dilute alkaline solutions, and hydrolyzed to fermentable sugars. Most hemicelluloses contain several sugars, including D-glucose, xylose, and mannose. Unlike cellulose, which is a homopolymer, hemicelluloses are branched mixtures of polysaccharides [2]. The degree of polymerization ranges from 150 to 250. Different forms of hemicellulose polymers exist in different types of biomass. While herbaceous biomass contains primarily arabinoxylan; deciduous woods contain primarily xylan (80–90% weight); and coniferous woods contain 60–70% glucomannan and 15–30% arabinogalactan. Along with pectins and glycoproteins, hemicellulose binds together the cellulose microfibrils in a cross-linked matrix.

Lignin, a major constituent of plant material, forms the walls of woody cells and acts as the natural glue that provides the plant with strength. It is a polymer of single benzene rings linked with aliphatic chains. Lignin is amorphous and more soluble than cellulose, but it is totally resistant to hydrolysis and to microbial degradation. However, it can be separated from the cellulosic material by steaming or using solvents. Lignin consists of phenylpropane units linked through ether and carbon–carbon linkages. The strength of these carbon–carbon linkages is what provides lignin with high resistance to thermal and chemical degradation. Deciduous woods tend to contain guaiacylpropane units, while coniferous woods contain the guaiacylpropane and syringylpropane units. Lignin is found primarily in the middle lamella and binds together adjacent cells. By encasing the hemicellulose and cellulose components, it protects the plant from enzymatic and microbial attack. The molecular structure of some components of lignin is shown in Figure 14.12.

Along with these components, biomass contains **extractives**, **moisture**, and **ash**. Extractives are nonstructural compounds including proteins, oils, starches, and sugars. They provide plants with odor, color, and durability, and can be extracted by hot water or other solvents. Bark contains 4–5 times more extractives than wood. Ash is inorganic solid residue remaining after a fuel undergoes complete combustion. It often contains

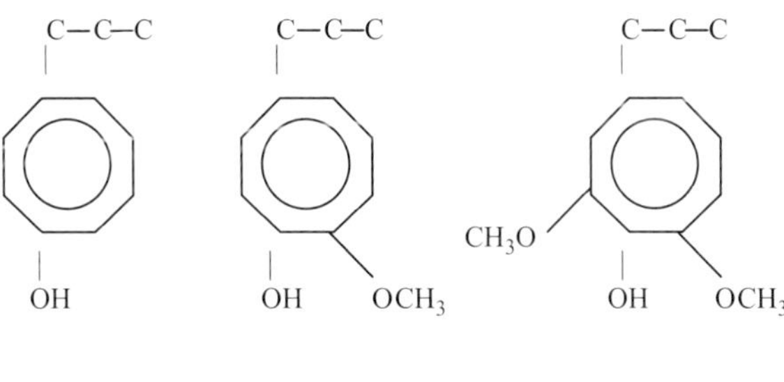

Figure 14.12 A monomer group found in the composition of lignin, all made of a single benzene ring linked with aliphatic chains: (1) p-hydroxyphenolpropne; (2) guaiacylpropane; (3) syringylpropane.

carbonates, phosphates, and sulfates of silica, calcium, magnesium, sodium, and potassium. Some mineral components may not be inherently contained in the biomass; they may actually be from dirt and other impurities picked up during the collection process. Woodchips contain less ash (~8.6 g/kg-fuel) compared to straw (~35 g/kg-fuel), which is characterized by high potassium and silicon content.

Water plays an important role in plant transpiration, photosynthesis, and the transport of other fluids throughout its structure. Raw biomass contains a high moisture fraction. In plants, moisture is classified as free (external or imbibition) and inherent (bound or saturation) moisture. The former is moisture above the fiber saturation point, and generally resides outside the cell walls in cavities. Inherent moisture resides within the cell walls and is a function of relative humidity and air temperature [3]. Moisture content is measured on a **wet** or **dry** basis, depending on whether the moisture mass is divided by the original wet mass or the final dry mass, respectively. For very wet biomass like manure, moisture content on a dry basis exceeds 1. On a wet basis, rice husk contains less than 10% moisture while rise straw contains more than 60%. Table 14.1 shows the composition of some biomass in terms of their primary components.

Table 14.1 Composition of different types of biomass in terms of their primary components, in mass fractions [4]

Biomass	Cellulose	Hemicellulose	Lignin	Ash	Extractives
Softwood					
Japanese cedar	38.6	23.1	33.8	0.3	4
Fir	45	22	30	0.5	2.6
Spruce	43	29.4	27.6	0.6	1.7
Pine	46.9	20.3	27.3	0.3	5.1
Hardwood					
Alder	45.5	20.6	23.3	0.7	9.8
Aspen	52.7	21.7	19.5	0.3	5.7
Beech	44.2	33.5	21.8	0.5	2.6
Poplar	49	24	20	1	5.9
Willow	41.7	16.7	29.3	2.5	9.7
Herbaceous plants					
Bamboo	39.8	19.5	20.8	1.2	6.7
Miscanthus	34.4	25.4	22.8	5.5	11.9
Switchgrass	40-45	31-35	6-12	5-6	5-11
Agricultural residue					
Corn leaves	26.9	13.3	15.2	10.95	22.01
Corn straw	42.7	23.2	17.5	6.8	9.8
Rice straw	34.5	18.4	20.2	13.3	10.1
Rice husk	37	23.4	24.7	7.27	3.19
Wheat straw	37.5	18.2	20.2	3.74	4.05

14.3.3 Starch and Sugar

Other than woody biomass, roots, grains, fruits, "agri-waste," and animal waste are important ingredients/feedstock for bio-energy. **Starch**, $(-C_6H_{10}O_5-)_n$, with molecular weight 35,000–90,000, is a granular polysaccharide found in the storage organs of plants such as seed, tubers, and roots. It is an important constituent of corn, potato, and rice. Starch consists of 10–20% α-amylose, which is water-soluble, and the rest is amylopectin, which is insoluble. Both elements are polymers of D-glucose. Starch is readily hydrolyzed to fermentable sugars by dilute acids and enzymes. Grain crops include corn, wheat, rice, barley, and other cereals. These plant products have high starch contents, which can be hydrolyzed to fermentable sugar. They are grown on agricultural land and consume significant amounts of water and nutrients, because they are primary contributors to the food supply for humans and animals. Using these products for energy production (mostly to produce biofuels) could be considered as a threat to this supply.

Sugar crops include sugar cane, beet, and sweet sorghum. Sugar cane produces nearly twice the energy of beet, measured as unit energy produced/land area. However, its growth is restricted to warm climate, good soil, and where plenty of water is available.

Other sources of plant biomass include **crop residue** – that is, the material left over after harvest. Crop residue is typically low in sugar and starch, but high in lignocellulosic material. Thus, it may be more suitable for thermochemical conversion (gasification and torrefaction) and combustion. The same is likely to be true for switch grass. Other agricultural waste, which is left over after harvesting and processing crops, includes sugar cane bagasse and cotton gin trash. Sold as animal feed, it can also be burned or used in thermal conversion processes.

Aquatic plants include ocean kelp, algae, and buckweed. They are more difficult to harvest than other types of biomass. However, these plants can be encouraged to grow faster by supplying nutrients such as CO_2 to the medium where they are planted.

Animal waste is another source or organic biomass, which does not compete with food production, but its supplies are limited. Municipal solid waste includes cellulosic material, and it may not work well with biochemical conversion processes. Waste lends itself well to combustion and thermochemical conversion.

Extracting energy from biomass can be done through several routes, including: (1) thermal by direct combustion; (2) thermochemical by gasification (to produce gaseous fuels such as syngas and methane) or torrefaction (to produce solid fuels such as bio-char); and (3) biochemical by fermentation into liquid fuels. Thermal processes may generally be preferable for recovering the energy of woody biomass, given its resistance to bioconversion processes. Combustion is the easiest process for getting the thermal energy from woody biomass, but given the low heating value of wood, huge amounts are needed to power a typical power plant. Wood has an open pore structure and high moisture content, and in most cases must be dried to increase the energy density and to make it possible to sustain a combustion process. Raw wood, because of the moisture content and the fibrous structure, requires a significant amount of energy to reduce its size for use in conventional boilers, and fixed-bed or fluidized-bed combustors are used if only larger particles are available.

14.3.4 Analysis and Heating Value

The bulk density of biomass is low, 100–200 kg/m^3 for bales of hay or straw and up to 300–400 kg/m^3 for staked logs of wood (compared to 800 kg/m^3 for coal), depending on the moisture content and the degree of drying.

Similar to coal, biomass is characterized by its ultimate and proximate analysis [5]. The ultimate analysis describes the weight percent of (total) carbon, hydrogen, oxygen nitrogen, and sulfur, reported as received (ar), dry basis (db), or dry ash free (daf). The ultimate analysis and heating values of several biomass materials are shown in Table 14.2. In most biomass, the fraction of carbon is below 50%, with most of the remaining being oxygen, which contributes to its lower HHV, mostly below 20 MJ/kg. Among plant-based biomass, and except for rice husk and rice straw, which contain a significant quantity of ash, most biomass is low in ash.

The proximate analysis describes the moisture content, volatile matter, and fixed carbon. The total carbon reported in the ultimate analysis is the carbon in the volatiles plus the fixed carbon. The ultimate and proximate analysis of willow (woody) and straw (herbaceous) are compared in Table 14.3 [6]. Herbaceous plants (straw, grass, and hay) tend to have more

Table 14.2 Ultimate analysis and HHV of some plant biomass measured on a dry basis

Fuel	C (%)	H (%)	O (%)	N (%)	Ash (%)	HHV (MJ/kg)
Douglas fir	52.3	6.3	40.5	9.1	0.8	21.05
Redwood	53.5	5.9	40.3	0.1	0.2	21.03
Maple wood	50.6	6.0	41.7	0.3	1.4	19.96
Sawdust	47.2	6.5	45.4	0	1.0	20.50
Rice husk	38.5	5.7	39.8	0.5	15.5	15.38

Table 14.3 Ultimate and proximate analysis of willow and straw

	Willow	Straw
Ultimate analysis		
C	49	46
H	6.25	5.9
O	43	43
N	0.5	0.5
S	0.06	0.125
Cl	0.03	0.555
Proximate analysis		
Moisture (ar)	30	30
Volatiles (db)	82	74.5
Fixed carbon (db)	14.55	18.15
Ash (db)	0.87	3.56
HHV (MJ/kg)	51	15

chlorine (in the ash) than woody plants. Also, biomass has much more volatiles than coal, which contributes to their reactivity, and much less fixed carbon.

The higher volatile (75–82 %wt db) in biomass as compared to 36 %wt db in coal, and metal fraction of biomass make it more reactive than coal, and makes it easier to decompose its different components to small fractions at lower temperatures, as will be shown later. Meanwhile, its fixed carbon is low (10–20% wt db) versus coal (55 % wt db). As the temperature is raised, some of the volatiles form as the different fractions decompose. The presence of metals in biomass catalyzes its lower temperature decomposition (pyrolysis) as well as its gasification reactions. As the volatiles escape the biomass, it leaves more porous char, which gasifies and burns faster. The higher reactivity of biomass is manifested in its lower ignition temperatures, in the range of 220–285 °C, as compared to coal's 670–930 °C, with the higher values for the higher-ranked coals.

14.3.4.1 HHV and LHV

The lower heating value (LHV) of dry hemicellulose, cellulose, and lignin are ~16, 17.5, and 26, respectively. On the same dry basis, herbaceous biomass LHV is ~17 MJ/kg, while woody feedstock is 19 MJ/kg [7]. The difference is due to the difference in cellulosic, ash, and extractives content. The LHV of coal can range from 20.6 MJ/kg for lignite to 29.7 MJ/kg for anthracite.

Several correlations have been proposed to estimate the HHV of biomass based on their proximate or ultimate analyses. The following unified correlation for the HHV of solid (e.g., biomass and coal), liquid, and gaseous fuels has been proposed (with 1.45% error limit) [8]:

$$\mathrm{HHV} = 0.3491\,Y_C + 1.1783\,Y_H + 0.1005\,Y_S - 0.1034\,Y_O - 0.0151\,Y_N - 0.0211\,Y_{ash}. \tag{14.3}$$

This expression should be used only under the following conditions:

$$4.745\,\mathrm{MJ/kg} \le \mathrm{HHV} \le 55.345\,\mathrm{MJ/kg}$$

and

$$Y_C \le 92.25\%;\; 0.43\% \le Y_H \le 25.15\%;\; Y_O \le 50\%;\; Y_N \le 5.6\%;\; Y_S \le 94.08\%;\; Y_{ash} \le 71.4\%.$$

Several expressions have been derived for biomass; for instance, the following relationship between the HHV (MJ/kg) and the elemental composition of a biomass fuel [9]:

$$\mathrm{HHV} = -1.3675 + 0.3137\,Y_C + 0.7009\,Y_H + 0.0318(1 - Y_C - Y_H - Y_{ash}), \tag{14.4}$$

where Y represents the dry-basis mass fraction. The predicted HHV using this equation is within ±5% of measured values. It should be mentioned that the moisture content of biomass varies widely, as well as seasonally, and hence impacts the drying requirements (heat) and the resulting composition.

Figure 14.13 The effect of moisture content measured on a wet basis on the LHV of wood in MJ/kg.

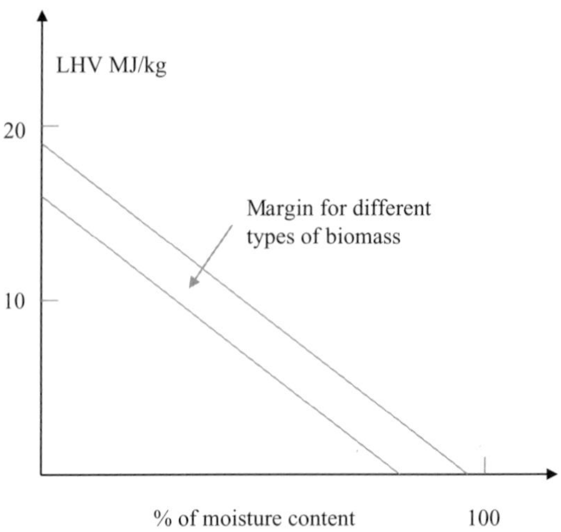

The enthalpy of evaporation of water at 298 K is −2.442 MJ/kg. The LHV of biomass can be calculated by subtracting the evaporation heat of water from the HHV. Hence,

$$LHV = HHV(1 - MC) - 2.442MC, \tag{14.5}$$

where MC is the wet-basis moisture content of biomass. The dependence of the heating value on the moisture contents is shown in Figure 14.13.

Another important property is the bulk energy density, defined as the energy per unit volume of the biomass. This is the product of the bulk density and the heating value, and depends on how the biomass is packed. Values of 4.5–7 GJ/m^3 are often quoted for wood logs, ~3 GJ/m^3 for woodchips, and ~2 GJ/m^3 for hay and straw bales. These low values contribute to the high cost of biomass transportation.

14.4 Conversion

Biomass can be converted to solid, liquid, and gaseous fuels through biochemical processes, namely fermentation or anaerobic digestion, or thermochemical processes, namely torrefaction, gasification, or pyrolysis (Figure 14.14). In all of these, some of the chemical energy originally stored in the biomass material is converted into chemical energy stored in other, simpler molecules. Biomass can also be readily combusted to generate heat, and currently this is the most widely used conversion process for biomass. In the biochemical routes, fermentation produces mainly liquid fuels, while bio-digestion produces gaseous fuels. Thermochemical pyrolysis produces gaseous, liquid, and solid fuels; the ratio of the three depends on the heating rate, final temperature, and environment. Gasification produces gaseous fuels, with possibly some solid (char), although as shown before, most of the solid is recycled back for complete conversion. Gasification can be used with almost any type or form of biomass, while other processes tend to work best with certain types and forms.

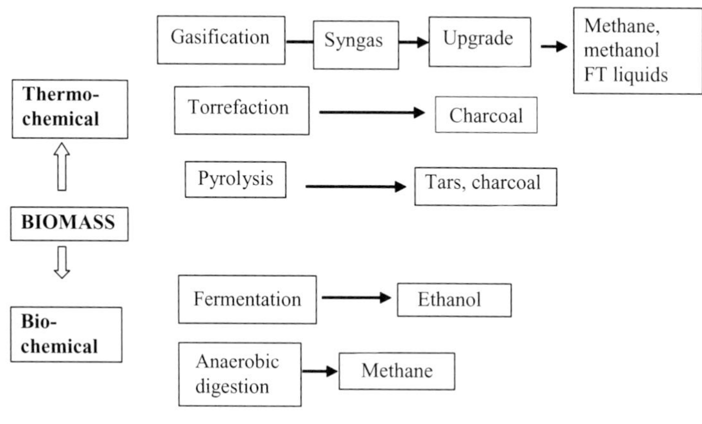

Figure 14.14 Biomass conversion to fuels pathways, including thermochemical and biochemical.

14.5 Bioconversion

Fermentation and anaerobic digestion are used to convert biomass biochemically to mostly liquid and gaseous fuels, respectively, without oxygen and in liquid medium. Fermentation refers to the anaerobic decomposition of carbohydrates to alcohols in the presence of enzymes, with ethanol being the primary product. Digestion refers to the decomposition of organic matter by the metabolic action of bacteria, in the absence of oxygen, with methane and carbon dioxide being the principal products. Bacterial digestion is accomplished by their own enzymes (and hence the distinction between the two methods may be artificial).

Enzymes are catalysts for biological processes. They are complex proteinaceous compounds, produced by living cells that catalyze hydrolysis and oxidation reactions. Because they occur at a lower temperature, biochemical reactions catalyzed by enzymes tend to be slow, one to two orders of magnitude slower than thermal reactions.

14.5.1 Fermentation of Sugar Cane

Ethanol is derived from sugar cane, in large quantities in Brazil and in smaller quantities in other countries. What is delivered to the fermentation plant is called "burned and cropped" (b&c), which represents almost three-quarters of the original raw cane. Leaves are removed and burned in the field (mostly), while the roots are left in the soil to sprout out in the next year (cane is a perennial plant), hence the name burned and cropped. Average b&c production is ~58 t/hectare/year (t is for tonne = 1000 kg). Each tonne yields ~750 kg of juice, made up of 135 kg of sucrose and the rest is water. The residue of the juicing process is wet bagasse, which when dried yields 130 kg of dry bagasse. Bagasse is a valuable fuel with a HHV of 19.7 MJ/kg that can be extracted by combustion. Sucrose's HHV is 16.5 MJ/kg, and hence the total heating value of a tonne of b&c is 4.7 GJ (when adding that produced by burning bagasse). Thus, per hectare per year, the total biomass energy is 270 GJ, which is equivalent to 0.86 W/m^2. Taking the average solar insolation as 225 W/m^2, the photosynthesis efficiency of sugar cane to ethanol is 0.38% [1].

Sucrose in the juice is broken down to glucose and fructose by hydrolysis (in water) with the aid of acids or enzymes. The resulting monosaccharaides are converted to ethanol:

$$C_6H_{12}O_6 \rightarrow 2C_2H_5OH + 2CO_2. \tag{14.6}$$

Thus, two moles of ethanol are produced for each mole of glucose consumed. The heat of reaction of glucose is 15.6 MJ/kg, or 2.81 GJ/kmol, while the energy in ethanol is $2 \times 29.7 \times 46 = 2.73$ GJ. Thus, the theoretical efficiency of conversion is 97.5%. The actual efficiency, as shown next, is lower.

The conversion flow chart is shown schematically on the right-hand side of Figure 14.15. The juice, suitably diluted to encourage the metabolic activity of the enzyme for fermentation to proceed, is placed in large vats, and the yeast cure is added (saccharomyces cerevisiae). The liquid is maintained at 37 °C. The fermentation process, during which CO_2 is released (and could be collected) lasts for 36 hours. At the end of the process, the liquid, called "wine," contains almost 10% alcohol, which is transferred to the distillery. Ethanol is miscible in water, forming an azeotrope whose boiling point is 78.3 °C at atmospheric pressure. Ethanol's boiling point is 78.5 °C. Ethanol is distilled to a mixture with 95.6% purity. Hydrated alcohols contain too much water to add to gasoline, although it is adequate to be used as a fuel by itself. To produce anhydrous alcohol, the mixture must be rectified in a vacuum distillation process or by extracting water. These distillation and dehydration processes add to the energy requirements for ethanol production.

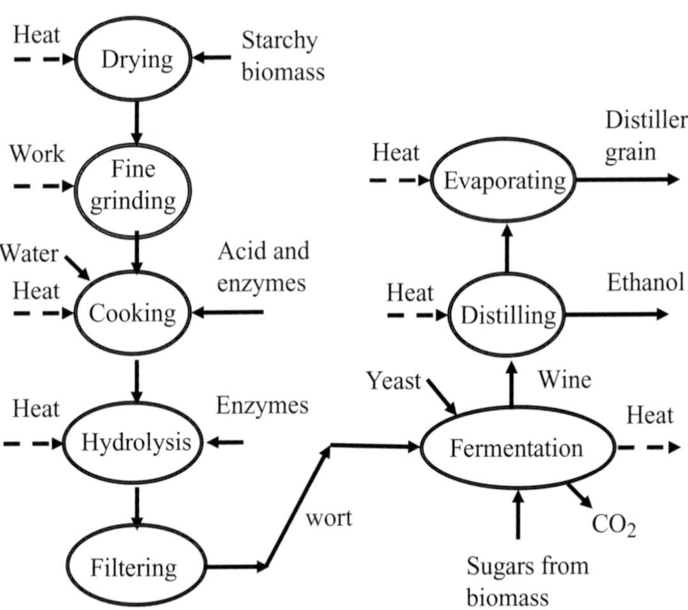

Figure 14.15 Flow diagram of fermentation of corn (left) and sugars (right).

14.5.2 Fermentation of Corn Starch

Simple sugars like mono- and disaccharides can be fermented directly using enzymes. Complex polysaccharides need a pretreatment step before fermentation, in which they are broken down into simple sugars by enzymes or acids in what is known as hydrolysis or saccharification. Hydrolysis is the decomposition of macromolecules by reaction with water. The following description applies to corn starch fermentation to ethanol, the most widely used type of starch in ethanol production. Figure 14.15 shows the corresponding flow chart, starting on the left-hand side.

1. **Hydrolysis**: Corn grain is dried and finely ground then suspended in water. The work required for dry milling is ~1.2 MJ/L of ethanol produced. The suspension is boiled to free the starch molecules. Cooking lasts about an hour. The enzyme α-amylase is added to prevent the regression or settling back of the gelatinized starch. Other enzymes, including α-amylase and β-amylase, also called diastose, are then added to catalyze the next step, the hydrolysis or saccharification. These enzymes are produced by the germination of barley grains into malt. Alternatively, glucoamylase can be used. This process is carried out at 50–60 °C and converts starch to disaccharide. Acids can be added, especially when some cellulosic biomass is used, such as 0.5% sulfuric acid, at 140–190 °C, or 70–80% sulfuric acid at 45 °C.
2. **Fermentation**: Following hydrolysis, or when simple mono- or disaccharide sugars are used in the production of ethanol, the mixture, called wort, is cooled below 30 °C and its pH is adjusted to ~5. Yeast is added to consume the sugars over a period of 2–3 days. Yeast is a microscopic simple cell organism that supplies the necessary enzymes to catalyze the process. Fermentation reactions are strongly exothermic and heat is removed to maintain the temperature around 30 °C. Based on fermentation enthalpy of reaction of –105 kJ/mol-ethanol formed (29.7 MJ/kg), heat dissipation is ~8% of the HHV of ethanol.

Sugar concentration in the feed is 14–18% by mass in an aqueous solution, and the reaction continues until the alcohol concentration is 14%. Higher concentration of alcohol inhibits the metabolic activities of yeast. Finally, alcohol is separated from the liquid by distillation. Estimates show that for 99.5% purity of ethanol, 30–45% of the HHV of ethanol is needed in this step. Membrane separation may lead to a lower energy penalty.

The following equations describe the conversion of starch to maltose, to glucose, to ethanol in three steps involving different types of enzymes:

$$2(-C_6H_{10}O_5-) + H_2O \xrightarrow{diastase} C_{12}H_{22}O_{11}., \tag{14.7}$$

$$(C_{12}H_{22}O_{11}) + H_2O \xrightarrow{maltase} 2C_6H_{12}O_6, \tag{14.8}$$

$$2C_6H_{12}O_6 \xrightarrow{zymase} 4C_2H_5OH + 4CO_2. \tag{14.9}$$

Using these equations to establish the mass balances, one can show that for 324 kg of starch, 184 kg of ethanol is produced. In practice, ~10% of the starch is converted into other byproducts, such as higher alcohols, glycerin, and ethers. Assuming corn is 61% starch and correcting for the 10% conversion to byproducts, shows that 1 kg of ethanol requires 3.2 kg of corn, or a liter of ethanol requires 2.6 kg corn.

Given the HHV of corn as 14.1 MJ/kg and ethanol as 29.7 MJ/kg, and subtracting the energy for milling, cooking, distillation, and recovery of byproducts, an overall thermal efficiency, defined as the ratio between the heating value of the ethanol products divided by the sum of the heating value of corn plus the energy used during the conversion process, is 46%. Thus, energy input is 65% of that of the enthalpy produced. This does not account for the energy used to cultivate and harvest the corn crop, estimated to be 42% of the energy of the ethanol produced, leading to a negative 7% energy overall. These estimates are sensitive to the energy expenditure in each step and the energy and mass integration of the different steps. For the production of ethanol from corn grain to be energy-positive, crop residues and fermentation byproducts must be used to supply some of the heat required by the process.

14.5.3 Anaerobic Digestion

Anaerobic digestion is the decomposition of complex organic matter to methane and carbon dioxide, or bio-gas, in the absence of oxygen [1]. Depending on the application, bio-gas can be used directly, or methane can be purified and used separately. Digestion occurs in three steps, as shown schematically in the flow chart in Figure 14.16.

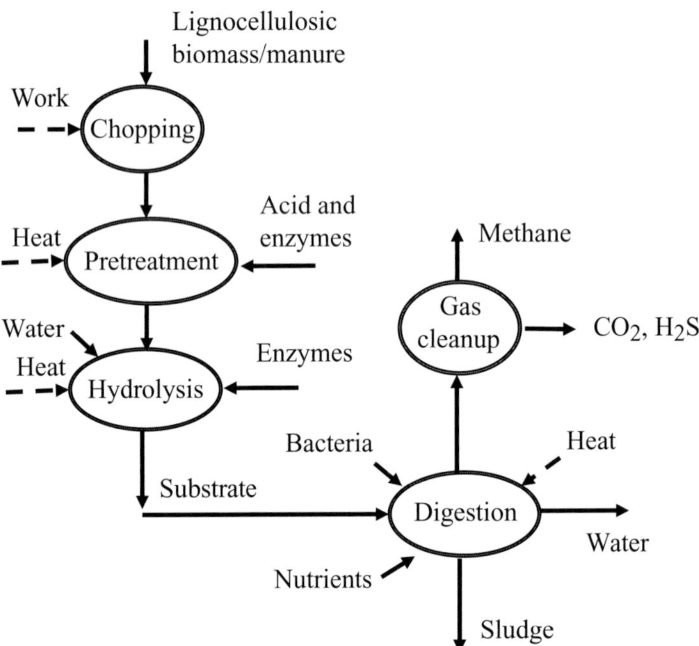

Figure 14.16 Flow diagram of anaerobic digestion.

1. Hydrolysis is used to convert insoluble carbohydrates, proteins, and oils into soluble substances, including sugars and alcohols, in the presence of hydrolytic bacteria.
2. Conversion of the soluble molecules into fatty acids, esters, carbon dioxide, and hydrogen. This step requires hydrogen-producing acetogenic bacteria, and homoacetogenic bacteria.
3. Methanogenesis, during which methane forms, requiring methanogenic bacteria.

These three steps require different bacterial groups, which need different conditions such as temperature and pH levels, to operate optimally. Thus, ideally, the three steps should be operated in different vessels or reactors. In practice, however, a single vessel is used.

Before the hydrolysis, pretreatment may be necessary when using lignocellulosic crop residue biomass and/or similar material. Acids and enzyme are added to improve the gas yield from the biomass.

In the digester, where oxygen concentration is low and biogas forms, bacteria propagate within the dilute organic mixture, or the substrate. Bacteria is naturally present in the biomass, but more can be added or recycled from previous batches. Bacteria best suited to metabolize the feedstock will dominate by natural selection, a process known as acclimation. For acclimation, the proper temperature and pH level, the presence of nutrients and the absence of toxicants are necessary. These are described in some detail next:

1. **Concentration:** To limit toxicity, concentration of biomass in the substrate is limited to 7–12%, thus requiring substantial supplies of water. For instance, higher concentration of ammonia (>3 g/L) or acetate (>2 g/L) can arrest (inhibit) bacterial activity.
2. **Temperature:** Digestion is mildly exothermic and the substrate in the digester may have to be heated to achieve the optimum operating temperature. Bacterial groups are classified according to their ideal thriving temperature: psychophilic bacteria, $0 < T < 20\,^{\circ}\mathrm{C}$; mesophilic bacteria, $20 < T < 45\,^{\circ}\mathrm{C}$; and thermophilic bacteria, $45 < T < 65\,^{\circ}\mathrm{C}$. The reactor design and operating temperature range determine the bacteria that can thrive and the residence time required for significant conversion. Higher temperatures, needed by thermophilic bacteria improves the conversion rate and reduce the residence time required for high conversion level, but at an energy cost. Typical residence time for thermophilic bacteria is 5–10 days, and 10–15 days for mesophilic bacteria. Conversion rates associated with psychrophilic bacteria are too slow to be practical. On the other hand, mesophilic bacteria may achieve higher conversion levels in some types of biomass.
3. **pH and nutrients:** The pH should be kept at 7–7.2 and not allowed to go beyond 6.6–7.6 (except for some special feedstock). Nutrients such as nitrogen, phosphorous, and alkali metals may be needed to improve conversion. (Neutral solutions have a pH of 7, acidic solutions have lower values and alkaline of basic solution have high pH.)

Batch and continuous reactor/digesters have been designed and are commercially available in several sizes, depending on the application. Mixing within the digester may be necessary to improve the conversion rate, at the cost of an energy penalty. It is important to achieve optimal size for the biomass feedstock and continue to reduce individual biomass (mechanically or by cavitation). Multistage digester systems are preferred; most practical systems are either single- or two-stage.

Typical bio-gas composition, by volume, may include 50–65% methane, 35–50% carbon dioxide, up to 1% ammonia, and 0.1–1.0% hydrogen sulfide. Moisture content depends on the temperature; if saturated at atmospheric pressure it is ~5–20% for temperatures in the range 35–60 °C. Bio-gas can be used directly, or can be purified by removing ammonia, hydrogen sulfide, and water. If synthetic natural gas is the desired product, carbon dioxide should be removed using physical or chemical separation techniques, and possibly compressed for storage and transporation.

The HHVs for bio-gas are 20–26 MJ/m^3, or 15–22 MJ/kg. Because of the build-up of toxicants in the digester, only a small fraction of the water used in the digester can be recycled, while the rest is disposed of or used for irrigation. The remaining sludge could be used as a soil nutrient because of its high nitrogen content. For both water and sludge, some treatment may be necessary for purification, depending on the feedstock of the digester.

To calculate the energy efficiency of this process, the mass and energy balances need to be established. The chemical equation for the conversion of cellulose, where in the following equation the total mass of the reacting components is shown below the chemical symbol, is

$$\underset{\text{cellulose, } 162}{(-C_6H_{10}O_5-)} + \underset{18}{H_2O} \xrightarrow{\text{bacteria}} \underset{132}{3CO_2} + \underset{48}{3CH_4}. \tag{14.10}$$

Thus, 0.3 kg of methane (or 0.4 m^3) is produced from 1 kg of cellulosic biomass. Practically, only 25–50% of this amount is produced, leading to a conversion ratio of 0.07–0.15 kg-CH_4/kg-biomass, or 0.1–0.2 m^3-CH_4/kg-biomass. Improving conversion by allowing the bacteria better access to the feedstock can be done with mixing and fragmentation of biomass before and during conversion (including possible use of cavitation).

Energy requirements for anaerobic digestion include heating the substrate to maintain the bacterial activity at the desired level. Some heating for the pretreatment phase may also be needed. For thermophilic processes operating at 55 °C, the heating requirement is significant, nearly 40% of the methane energy produced. Heat recovery (regeneration) from water and the substrate is essential to improve the overall efficiency. Given the small temperature difference between the digester and the environment (between 55 °C and 10–15 °C), large and expensive heat exchangers are needed for regeneration. The **thermal efficiency** of the process is defined as the HHV of methane produced as a fraction of the HHV of the feedstock. With 50% heat recovery, the efficiency (calculated for dry cattle manure with heating value of ~13.4 MJ/kg) is estimated to be near 53%.

One advantage of anaerobic digestion over fermentation is that bio-gas separates naturally, and hence the distillation penalty associated with separating ethanol is avoided. On the other hand, the gas separates at atmospheric pressures and the digester temperature, and may need to be compressed (and purified).

14.6 Thermochemical Conversion

Thermochemical conversion of biomass includes pyrolysis (essentially heating in an inert medium) and gasification. Pyrolysis can be used to produce mainly solid fuels, known as torrefaction, or mainly liquid fuels, known as flash pyrolysis. In all cases, most of the chemical energy in the original feedstock is retained as chemical energy in the products, while the balance is converted to thermal energy. Thermochemical processes can be used to convert lignocellulosic biomass that is not easily fermentable, and hence they are more flexible and versatile. However, they are generally more expensive when used to produce fuels because of the capital cost of the plant, and do not always scale down easily. Dedicated energy crops that do not compete with the fuel supply, such as miscanthus, willow, and switchgrass, can be used for steady supply for thermochemical conversion facilities. It has been estimated that the productivity of these energy crops is 30–240 boe/ha·y (that is barrel of oil equivalent per hectare per year, which is equivalent to 1–6 W/m^2).

Woody and herbaceous biomass consists of cellulose, hemicellulose, and lignin, as well as ash and extractives (such as water and lighter oils). At higher temperatures, following the release of the volatiles (water and lighter oils), these three components decompose chemically to form more volatiles (in the form of gases), condensibles (tars) and liquids, and solids (char). Hemicellulose decomposes first at 225–325 °C, followed by cellulose at 325–375 °C, while lignin decomposes at 250–500 °C. This is shown in Figure 14.17, in which the weight-loss curves of the three individual components of biomass are depicted, as measured in

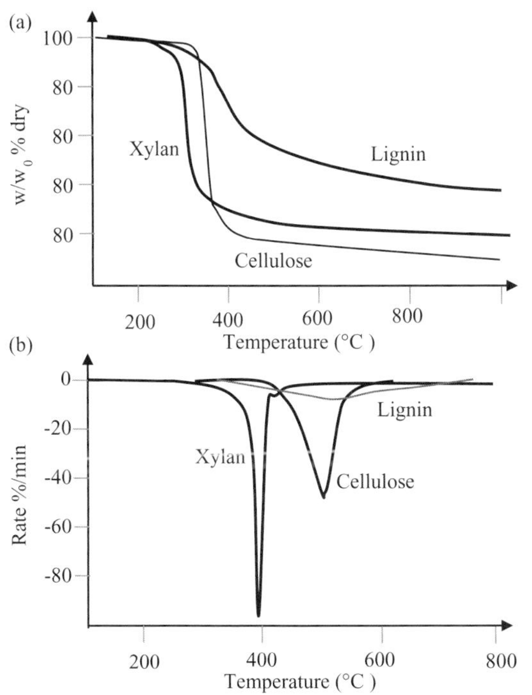

Figure 14.17 (a) Thermogravimetric analysis weight-loss curves for the three primary components of biomass [10]. Xylan is the same as hemicellulose, and it is the most reactive component and decomposes first at 200–300 °C and in a very narrow temperature window (in torrefaction, xylan decomposition is the primary mechanism for the carbonization of the biomass). Cellulose decomposes in the 300–400 °C range, while lignin continues to pyrolyze till >600 °C. (b) Thermogravimetric analysis weight-loss curves for the three primary components of biomass, with the same components as shown in (a). The vertical axis shows the normalized rate of mass loss (per unit weight of the original sample) per minute. Note the fast hemicellulose decomposition, followed by that of cellulose then the slow lignin [10].

thermogravimetric analysis at different temperatures. In this device, biomass decomposes in an inert atmosphere (such as nitrogen). The fast weight loss of xylan (hemicellulose) and cellulose at relatively lower temperature is an indicator of their volatility (or reactivity). Few solids are left at the end of the pyrolysis of both components. Lignin is less volatile, decomposing at higher temperature, and more solid is left at the end of the process.

Temperature controls the rate at which these components decompose, with hemicellulose decomposing the fastest and lignin the slowest, as shown in Figure 14.17. The data in both figures, for the weight loss and corresponding rates, are used in matching the conversion process conditions to the desired product, and in determining the necessary residence time in the corresponding reactor.

Pyrolysis proceeds in an oxygen-starved environment, at temperatures mostly between 200 °C and 600 °C, and at lower rates than combustion and gasification. In practice, low-temperature pyrolysis of woody biomass, known as torrefaction, proceeds at the lowest end of this temperature range, typically around 200–300 °C, with residence time of 20–55 min. It produces mostly solid products, that is char, of up to 80%; gas, 15–20%; and some liquid (tar), at 0–5%. It is the slower type of pyrolysis. Higher-temperature pyrolysis, on the other hand, proceeds at around 500 °C, and at higher rates (and so is called flash pyrolysis), with residence time around 1 s. It results in mostly liquids, at 75%, with the remainder divided almost equally between solid and gaseous products. Slower pyrolysis, proceeding around 400 °C at residence times of hours, produces almost equal amounts of solids, liquids, and gases. Gasification proceeds in oxygen or air, mostly with steam (or CO_2), under fuel-rich conditions, at 700–1000 °C, and at faster rates. It produces mostly gaseous products, around 85%, and some liquids (tars) comprising nearly 5%, and solid (chars) at around 10%. Combustion mostly occurs at temperatures higher than 1200 °C, depending on the moisture content of the biomass [11]. The relative amount of solid, liquid, and gaseous pyrolysis products depends on the feedstock, heating rate, residence time, and final temperature. Hydro-pyrolysis – that is, pyrolysis in the presence of hydrogen – increases the liquid fuel yield. In both cases, heat is needed to raise the temperature to the required value. Heat can be supplied externally or from the partial oxidation of the pyrolysis products.

In gasification, air or oxygen, steam and/or hydrogen are added to partially oxidize the carbon and produce gaseous fuels. Higher steam concentration and higher temperatures increase hydrogen yield in the products, while lower temperatures favor the formation of methane. The stoichiometry of gasification is adjusted to convert carbon to CO, and the water shift reactions convert some or all of the carbon monoxide to CO_2 and produce more H_2. Gasification thermodynamics was described in detail in Chapter 3, and coal gasification and its products in Chapter 12. Biomass gasification will be covered in more detail in a later section of this chapter.

Biomass, compared to other carbon feedstock like coal, has higher moisture, oxygen, and volatile matter, leading to more lighter hydrocarbons in the products of pyrolysis and gasification. Besides the evaporation of its water, more water forms as the biomass decomposes and hydrogen is oxidized by oxygen in the biomass. Most biomass can be pyrolyzed or gasified, although pre-drying may be necessary in some cases.

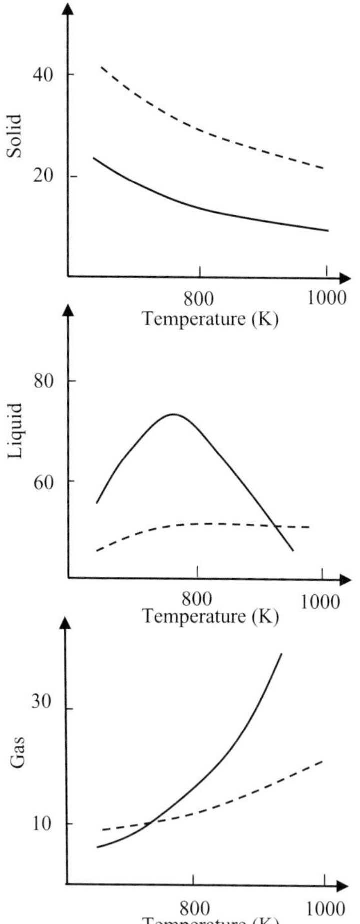

Figure 14.18 Solid (char), liquid (tars), and gas yields of many woody-type biomasses at different temperatures. Only averages are shown, with the broken line for slow pyrolysis and continuous line for fast pyrolysis [11].

Figure 14.18 shows the results of slow and fast pyrolysis in terms of char, liquid, and gaseous composition in the products, and their dependence on temperature. Slow and low-temperature heating leads to higher char yield, while the opposite produces more gaseous products. Fast and flash pyrolysis yields the mostly liquid at intermediate temperatures, while slow pyrolysis yields mostly solids at low temperature.

For the primary constituents of woody biomass – lignin, cellulose, and hemicellulose – the decomposition rate and products differ. Lignin decomposes slowly and produces mostly char and some aromatic liquids. Cellulose decomposes faster and produces the least char. Hemicellulose is the most reactive, decomposing the fastest. Other elements evolving during wood pyrolysis include sulfur, chlorides, nitrogen, oxygen, and ash. Heavier tars are produced at higher pyrolysis temperatures, and contain aromatic hydrocarbon, aliphatics, and higher-molecular-weight phenols. Gaseous products of wood pyrolysis include CO, H_2, CH_4, and some higher hydrocarbons. At lower temperatures, more CO_2 is formed, while at higher temperatures, fast pyrolysis favors the formation of CO and H_2 as gasification reactions (with raw biomass, water, and oxygen) become faster.

As an example, dry sawdust pyrolysis at 825 °C leads to 43.8% CO, 12.5% O_2, 29.3% H_2, 11% CH_4, and 3.1% C_2H_2 (by volume). The HHV of this mixture is 15.9 MJ/m^3. Biomass decomposition at this temperature is almost thermally neutral, with the only energy required being that needed to raise the temperature to the pyrolysis temperature (this can be done partially by regeneration). This energy is estimated to be ~10% of the HHV of the sawdust. With partial heat recovery, the efficiency of this process is 90%. Most pyrolysis reactors (retorts) are fluidized-bed reactors.

In the following three subsections, three different conversion processes for the production of useful fuel products are described. The processes operate within different temperature ranges that take advantage of the natural pyrolysis of biomass described above. At intermediate temperatures, around 500 °C, mostly liquid fuels are produced, that is bio-oils. At lower temperatures, around 250 °C, mostly solid fuel is produced, that is bio-char. At higher temperatures, around 800 °C, gaseous fuels are produced, that is bio-gas.

14.6.1 Pyrolysis: Intermediate-Temperature Conversion

At intermediate temperatures, centered around 500 °C, more condensibles, or bio-oils, are produced from biomass pyrolysis. These are also known as tars, a black viscous fluid consisting of phenolic compounds, water, and oxygenated hydrocarbons with an LHV of 13–18 MJ/kg (obtained for tars produced from biomass with an LHV of 19.5–21 MJ/kg db). These bio-oils break down (secondary pyrolysis) into smaller molecules at higher temperatures exceeding 1000 °C. Under intermediate-temperature pyrolysis, while more liquid is formed, some gases and some solids, or chars, are also formed in a "pyrolyzer." Typically, the liquid/tars and solids/char, with some gases, form a single stream, or an emulsion. Char can be separated using a cyclone, while the gases and liquids are separated by cooling the stream leaving the cyclone. The liquids in the bio-oil contain hydroxyaldahydes, hydroxyketones, sugars and dehydrosugars, carboxylic acids, and phenolic compounds [5]. The fractions of gases depend on the pyrolysis temperature and residence time inside the reactor, which can continue to change the composition via secondary reactions. The chemical reaction representing biomass pyrolysis can be written as follows:

$$C_xH_yO_z + \text{heat} \rightarrow \text{light gases } (CO, CO_2, H_2, CH_4, H_2O) + \text{tar} + \text{char}. \tag{14.11}$$

The main factors influencing the relative amounts of the pyrolysis products include heating rate, reactor temperature, and residence time. As shown in the figures, for maximizing char formation, low temperature (200–300 °C), slow heating rate, and relatively longer residence (30 min) is needed. This is known as torrefaction. For maximizing liquids (tars), intermediate temperature (300–500 °C) is required, along with relatively high heating rate and short residence time (O(1 min)). This is known as flash pyrolysis. For maximizing gaseous products, higher temperature (above 600 °C) with low heating rate and longer

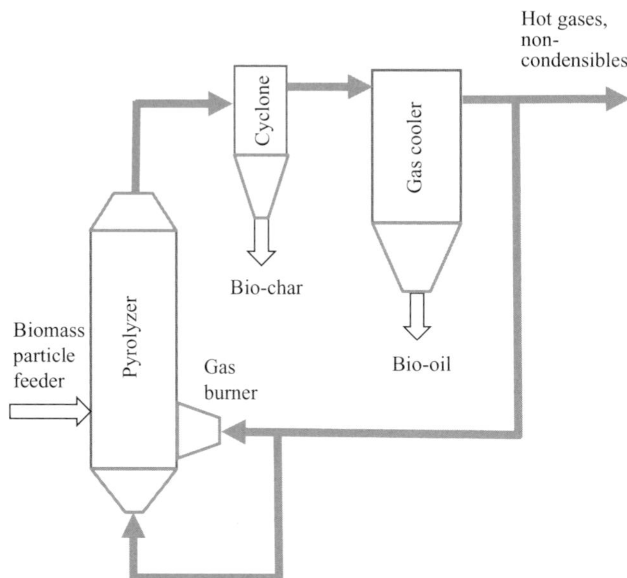

Figure 14.19 Typical layout of a pyrolysis plant in which biomass is fed into a fluidized-bed reactor using a screw feeder, and heated by burning some of the produced gas and char. The products of pyrolysis are separated into char, condensibles (tar), and gases in two stages.

residence times are needed. In this case, steam, hydrogen, and/or some oxygen are usually used for products selectivity.

A typical layout of a flash pyrolysis plant is shown in Figure 14.19. Chopped biomass is fed into the pyrolyzer. While different types have been suggested, a fluidized bed is best suited for the purpose, given the rather uniform temperature conditions and thorough mixing achieved in the reactor. Fluidization is supported by the products gases circulated after the char and condensibles are separated. The bed in the case shown is "externally" heated by burning some of the gases (and possibly the char) produced in the process. In most cases, biomass pyrolysis is not by itself autothermal and external heat must be supplied to bring the biomass to the requisite temperature. How much heat is required depends on the biomass, its moisture content, and the pyrolyzer. Alternatively, the thermal energy is supplied by burning other fuels, which is also necessary for starting up the system. The products of pyrolysis – that is the bio-oil (or condensable tars), gas, and char – are separated into their components using a cyclone followed by a gas condenser. The char is removed at the bottom of the cyclone and the tars are condensed by cooling the products. Some of the non-condensible gas is recirculated back to the gas burner to provide some of the required heat, along with the some of the char.

Discussions in the previous paragraph highlight some of the challenges in determining the energy balances of the pyrolyzer. While some of the reactions are endothermic, such as the depolymerization and secondary cracking, others are exothermic, such as the pyrolysis of hemicellulose and lignin. The enthalpy of reaction of some of these steps depend on the temperature as well. In general, the net heating requirement of the feedstock and the overall heat of reaction in the reactor is supported by burning some of the products (and possibly extra fuels).

14.6.2 Torrefaction: Low-Temperature Conversion

Lignocellulosic biomass suffers from low bulk energy density (energy per unit volume) resulting from high moisture content, volatiles, oxygen, and other non-combustibles. This low density, and the tendency of biomass to decay by micro-biological activity, makes it rather difficult to store, which needs to be overcome, given the seasonal nature of biomass availability. Enzymes and mesophilic organisms contribute to self-heating up to about 60 °C, while thermophilic bacterial activity at 75–80 °C (mentioned previously in the context of anaerobic digestion) causes further decay [7]. The rate of decay is strongly affected by the moisture content. Even dry biomass can suffer from high moisture uptake because of its hydrophilic nature (related to the abundance of hydroxyl groups in cellulose and hemicellulose, which form in hydrogen bonds with water molecules [12]). Torrefaction has been suggested to overcome many of these difficulties.

Cellulosic material can be carbonized, or torrefied, if pyrolysis temperature is kept at 200–300 °C according to the following reaction:

$$(-C_6H_{10}O_5-) \xrightarrow{\text{slow, low-temperature pyrolysis}} 6C + 5H_2O, \tag{14.12}$$

with $\Delta H^o = -463\ \mathrm{kJ/mol}$. This overall reaction shows that the outcome of the process is pure carbon and water. The enthalpy of the overall exothermic reaction is almost 15% of the heating value of cellulose. In practice, if the biomass is externally heated this energy is essentially lost during low-temperature pyrolysis in the form of chemical energy of volatile gases. It is possible to collect the volatiles and warm gaseous products leaving the torrefaction reactor, burn the volatiles, and recycle the thermal energy to warm the raw biomass. According to the reaction, what is left is pure carbon. However, during the process, coking (formation of larger carbon molecules) and cracking reactions lead to the formation of char that contains oxygen and hydrogen – that is, the solid torrefied products.

The slow pyrolysis process is carried out at a reactor temperature in the range 200–300 °C (see Figure 14.18). During this process, moisture evaporates, some volatiles (containing C, O, H, and other components) are released, raising the carbon fraction of the solid left behind. Released volatiles include furfural, formic acid, methanol, lactic acid, and phenol, while the gaseous products contain mostly CO_2, with ~20% CO and a trace of H_2 and methane (the exact fractions depend on the temperature). The remaining solid is more brittle than the original raw biomass. Typical volatiles produced during torrefaction include acetic acid ($C_2H_4O_2$), formic acid (HCO_2H), lactic acid ($C_3H_6O_3$), furfural ($C_5H_4O_2$), and methanol, with heating values in the range 15–22 MJ/kg.

The energy and mass contents of the solid part of the biomass decrease during torrefaction, with the reduction of mass being greater because of the loss of water and volatiles, so that the energy density of the resulting solid increases beyond that of the raw source. Table 14.4 shows an example of the effect of torrefaction on sawdust when subjected to three different temperatures, for 1 hour, at a reactor temperature of 250 °C, 270 °C, and 300 °C. The reduction in the energy content of the solid varies between 27.5% and 44.86%.

Table 14.4 The effect of reactor temperature on the properties of sawdust after torrefaction for one hour. The HHV of the raw material is 18.14 MJ/kg. The proximate analysis of the raw material is: moisture content, 15.43%; volatile content, 70.89%; ash content, 1.48% [13]

	Temperature (°C)		
Property	250	270	300
Energy yield (% of original)	72.5	67.15	55.12
Mass yield (% of original)	67.25	59.5	42.00
Energy density	1.08	1.12	1.31
HHV (MJ/kg)	19.55	20.47	23.8
HHV increase (% of original)	7.8	12.8	31.2
Moisture content (%)	4.4	3.56	3.31
Volatile content (%)	68.34	59.27	40.11
Ash content (%)	2.12	2.25	3.61

The mass reduction, on the other hand, is between 32.75% and 58%, raising the energy density between 108% to 131% over the raw solid material. It is also evident that the heating value, measured per unit mass of the solid, increases by 7.8–31.2% following torrefaction, respectively.

Raw woody biomass has a fibrous structure that requires significant energy to break into smaller particles, especially when moist. Torrefaction reduces the hemicellulose content in the biomass significantly (it is first to decompose, as seen before). This polymer binds the cellulose fibers of the plant cell walls, and with further cellulose depolymerization and reduction of the fiber length, the remaining biomass is more brittle and easier to grind into finer particles. Torrefied biomass is dry and brittle (with a charcoal-like texture). Grinding raw biomass to sizes smaller than 1 mm is infeasible; the grinding energy is a very significant fraction of its HHV. For example, to achieve a 0.5 mm willow particle size, 2.6 MJ/kg or 700 kWh_e/ton is required. This is more than 13% of the dry heating value of the fuel. Assuming a thermal to electric conversion efficiency of 33%, this corresponds to more than 39% of the thermal energy in the fuel [6]. For a 0.2 mm willow particle size, the grinding energy requirement exceeds the heating value of the fuel. Grinding the biomass is necessary to use the fuel in conventional boilers in steam power plants as the primary fuel or as a blend with coal. Moreover, the higher energy density of the torrefied wood makes it possible to achieve higher combustion temperatures when burning it. It is also necessary for entrained-flow gasification (see Chapter 12 in which the pulverized feedstock must be O (100 μm) to meet the gasifier requirements).

Torrefaction creates a more homogeneous product from different sources of biomass, and reduces the seasonal influence on the quality of the feedstock. Reducing the moisture content and depolymerizing the fibrous chains produces a hydrophobic solid product that is more resistant to decay. The hydrophobicity improves as the torrefaction temperature increases,

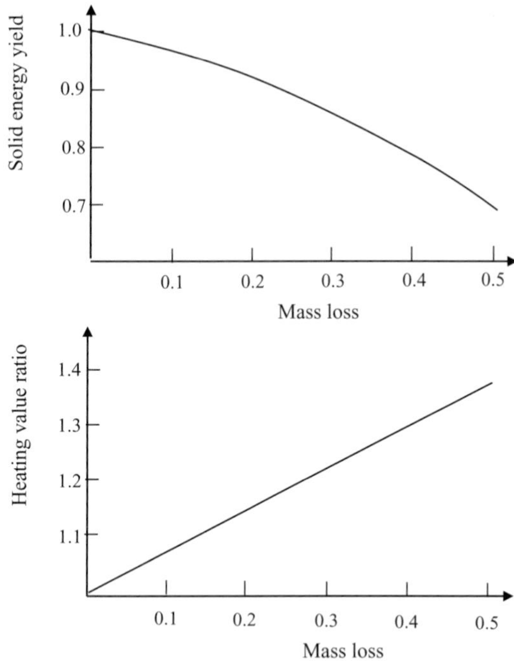

Figure 14.20 Solid energy yield, $\eta_s = Y_s R_{HHV}$, and energy densification, $R_{HHV} = \frac{HHV_{tor}}{HHV_0}$, as a function of mass loss [14].

making it much easier to store the fuel for a longer time in more compact form. The solid energy yield, η_s, is defined as

$$\eta_s = Y_s R_{HHV}, \tag{14.13}$$

where Y_s is the total solid mass yield, or the solid remaining after torrefaction divided by the initial raw biomass. Energy densification ratio R_{HHV} is defined as

$$R_{HHV} = \frac{HHV_{tor}}{HHV_0}, \tag{14.14}$$

where HHV_{tor} is the heating value of the torrefied solid product, and HHV_0 is the initial heating value of the biomass. Figure 14.20 shows the estimated solid energy yield and energy intensification as the mass loss $(1 - Y_s)$ increases. While more of the original biomass energy is lost with torrefaction, the energy intensification increases.

Torrefaction reactors are designed to be indirectly heated (i.e., pyrolysis occurs in an inter-environment with heat supplied externally through the reactor wall or by flowing hot gases through a reactor bed of biomass) or directly heated by flowing air that can burn the volatiles and potentially a fraction of the char to raise the temperature of the raw biomass.

14.6.3 Gasification: Higher-Temperature Conversion

An important difference between gasification and combustion is that gasification allows transfer of energy from the feedstock into the chemical bonds of the product gas, whereas combustion releases the energy content of the fuel in the form of thermal energy by breaking the chemical bonds. A gasifying medium such as oxygen/air or steam (or even CO_2) is required during the gasification process, which reacts with the solid feedstock to form gaseous fuel products such as carbon monoxide, hydrogen, and methane. The gaseous medium required for combustion is either air or oxygen, and the products are mainly carbon dioxide and water.

Increasing the gaseous products and reducing the solids and tars over that produced by pyrolysis alone is achievable via gasification. Moreover, by adjusting the stoichiometry of the input, it is possible to change the H_2/CO ratio in the products. As with other feedstocks (e.g., coal), air gasification of biomass results in a low heating value gas that can be used in electricity generation if combusted in a simple or combined cycle, or in the production of liquid fuels via Fischer–Tropsch (FT) processes. While raw (or dried) biomass can be directly combusted to produce heat for a steam cycle, gasification leads to lower overall emissions and overall better efficiency at these lower levels of emissions. This is in part because the volume of the produced syngas is smaller than that of the products of combustion, and hence less energy is used in the gas clean-up process. Oxy-gasification produces a medium heating value gas, and pre-drying the biomass may be necessary to produce it.

High cold gas efficiency (CGE), up to 90%, can be achieved using biomass, which is higher than that achieved in coal gasification, especially when the gasifier is cooled to raise steam for a power cycle. One reason is that biomass gasification is often done in fluidized-bed gasifiers that operate at temperatures lower than those often used in entrained-flow gasifiers, and with larger particle size. However, drying the biomass can consume some energy. The metal content of biomass acts as a catalyst, speeding up the gasification kinetics even at lower temperature. Because of the oxygen present in biomass, less oxygen is needed in biomass gasification than for coal gasification.

Biomass gasification consists of several subprocesses, including drying (in the case of wet biomass), pyrolysis of the biomass into volatiles and char, partial oxidation of some of the pyrolysis products, and char gasification. Depending on the operating conditions, these subprocesses may take place sequentially or overlap. When moist biomass is fed into a gasifier, it first undergoes a preheating stage until its temperature reaches the evaporation temperature of water. By further heating the biomass, its moisture content evaporates, leading to a dry solid. Heat is transferred to the fuel through convection and radiation. Soon after the biomass temperature reaches a pyrolysis temperature, it begins to undergo thermal decomposition yielding light gases (e.g., CO, CO_2, H_2, CH_4), tar, and char. The pyrolysis products react with the gasifying medium (they may also react with each other), forming the final products of gasification.

In most practical applications, the heat required for drying, pyrolysis, and endothermic gasification processes is supplied by some exothermic oxidation reactions in the gasifier. This is the case of **autothermal** or direct gasification, in which heat is released by partial oxidation of some of the pyrolysis products. Thus, autothermal gasification requires some pure oxygen or air. The heating value of the product gas is in the range 4–7 MJ/m^3 for the case of air gasification, while oxygen gasification produces gaseous products with a heating value in the

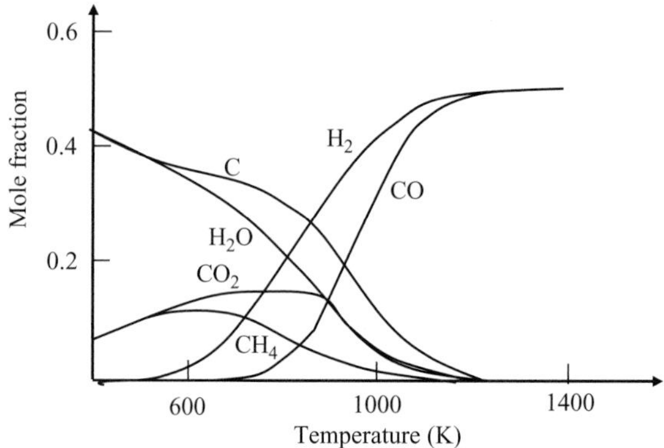

Figure 14.21 The effect of temperature on the composition of equilibrium gasification products of cellulose using steam as the gasification medium. The amount of steam fed to the gasifier is 0.17 mole per mole of biomass.[1]

range 10–18 MJ/m^3. The other case of gasification is **allothermal** or indirect gasification, in which steam is used as a gasifying medium and heat for the endothermic reactions is supplied from an external source. Allothermal gasification produce a hydrogen-rich product gas with a heating value of 14–18 MJ/Nm^3, which depends on the biomass and moisture content.

The gasification temperature, typically in the range 700–1100 °C, and medium, mostly steam but can also be CO_2, influence the composition of products. Gasification reactions are not expected to reach complete conversion to syngas until higher temperatures (see Chapter 3), especially in the case of relatively lower temperatures. Moreover, as the temperature increases, biomass tends to release its volatiles first, forming char, which is then gasified. These are reaction kinetics-limited processes that cannot be modeled using equilibrium. A model has been proposed that assumes the decomposition of the original biomass to some basics elements first, then seeks the equilibrium of these elements, for estimating the composition of gasification products dependence on temperature [15]. For instance, starting with cellulose, the equilibrium products were assumed to satisfy the following reaction:

$$CH_{1.66}O_{0.83} + 0.17H_2O \rightarrow \alpha_1 C_{(s)} + \alpha_2 CH_4 + \alpha_3 CO_2 + \alpha_4 CO + \alpha_5 H_2 + \alpha_6 H_2O.$$

The model fractions α_i are calculated by mass balance between the two sides of the equation and assuming that the species on the right-hand side are at equilibrium. This allows for an approximate estimate of the char and methane in the products. It is not possible to estimate the composition of tar products using this model; kinetic models are needed for this purpose.

Figure 14.21 shows the equilibrium products composition of steam gasification of cellulose, $CH_{1.66}O_{0.83}$. Hydrogen and carbon monoxide increase, while methane decreases, reaching almost zero beyond 1100 K. the equilibrium products of dry gasification of the same material is shown in Figure 14.22. Steam gasification produces more H_2. It should be noted that the stoichiometric amount of the gasification medium is used in these calculations. In practice, this is used in order to minimize energy loss in heating up extra gases.

[1] The author wishes to thank Professor A. Gomez Barea for bring this figure to his attention.

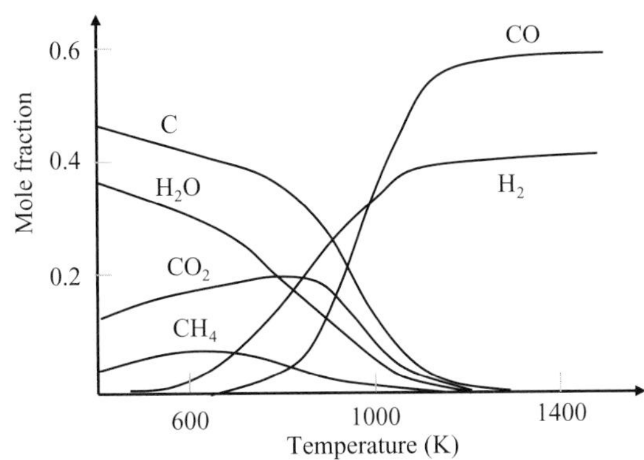

Figure 14.22 The effect of temperature on the equilibrium composition of cellulose gasification products using CO_2 as the gasification medium, with 0.17 mole of carbon dioxide used per mole of cellulose.

14.6.3.1 Types of Gasifiers

Coal gasifiers were described in Chapter 12. Similar designs have been modified to match the needs of biomass, and they are summarized here to highlight their performance [16]. In biomass (and other) gasifiers, the solid feedstock (biomass particles of different sizes, depending on the gasifier type) and the gasifying agent (steam, carbon dioxide, air/oxygen) are mixed and gasified (or in the presence of other solid particles in the case of a fluidized-bed gasifier). Three types of gasifiers have been used with solid fuels (see also Chapter 12 and 13): (1) moving-bed; (2) fluidized-bed; and (3) entrained-flow gasifiers. In moving-bed gasifiers, the solid move downwards slowly under the effect of gravity (in some places they could be classified as "fixed" bed because of the slow movement, but moving is more descriptive of the process). The speed of the bed is determined by the gasification rate. Each type of gasifier is further subdivided into several types, as shown next.

Several moving-bed gasifiers have been designed in which the gas flow and (slow) solid flow are in the same or opposite direction. Because of the relatively slow conversion rates, moving-bed gasifiers are often employed for small-scale applications, with a thermal capacity between 10 kWh and 10 MWh (the thermal capacity is defined as the rate of production of syngas multiplied by its heating value), although larger gasifiers have been used. These gasifiers can operate with larger biomass particles, O (10 cm), because of the relatively longer residence time, O (1 hour), at moderate temperatures, O (700–1000 °C). Because of the moderate temperature, they have relatively high CGE of ~80–85%, but the tar content can be high, at O (50 g/Nm^3), in the updraft type since the volatiles are not exposed to the higher combustion temperatures. Downdraft gasifiers produce a couple of orders of magnitudes lower tars. Moving beds operate at atmospheric pressure and relatively low capacity because of the high residence time.

Fluidized-bed gasifiers are typically used for intermediate-scale applications of 5–100 MWh. Fluidized-bed systems are characterized by higher heat and mass transfer rates than moving-bed systems, and hence faster conversion and shorter residence times than in the moving-bed case. The residence time is defined as the mass of the biomass inside the gasifier (the gasification vessel volume multiplied by the biomass density) divided by the biomass mass flow rate through the gasifier. Residence time is O (1 hour), but it depends strongly on the operating temperature, O (700–950 °C), and particle size, O (50 mm). The tar content of the gas is still high at O (5–10 g/Nm^3) because of the lower temperatures (some have been operated at higher than 1000 °C to reduce the tar content of the syngas). Fluidized beds can accommodate moist biomass, but in this case require longer residence time and operate at lower efficiency. Most fluidized beds operate at atmospheric pressure, but some can be pressurized. Several fluidized-bed gasifier types have been used, such as the bubbling bed and circulating fluidized-bed gasifiers, in which the fluidization regimes are different. Circulating beds operate at higher conversion rates but they are more complex, as will be shown next.

Entrained-flow reactors are appropriate for large-scale systems (>400 MW_{th}) because of their small residence time requirement, O (1 s), but they cannot be used with raw or even dried biomass. They operate on finely pulverized feedstock of O (100 μm), which is not possible to achieve even with dry biomass. It is possible to pulverize torrefied biomass then use entrained-flow systems in a two-step design. However, this may prove uneconomical and complex to operate. Because of the very high-temperature operation, the residence time is much lower, the CGE is relatively low, but the tar content is also very low. Table 14.5 shows a comparison of the specifications of three types of gasifiers in terms of the biomass particle size, operating temperatures, CGE, and technical challenges. The three types are described in some detail next, with more detail on moving- and fluidized-bed systems, which are used in biomass gasification. All systems considered operate under autothermal conditions.

Moving-Bed Gasifier In a moving-bed gasifier, medium-size solid fuel particles (forming the bed) move slowly downward inside the reactor, under gravity and at a rate that depends on the gasification rate. The gasification rate depends on the temperature and the feed rate of the gasification agent. Particles rest on a grate at the bottom of the gasification chamber and dry ash falls off to the chamber below. Moving-bed gasifiers can operate as updraft, down-draft, or cross-draft, depending on the relative flow direction of the gases and the solid. The exit gas temperature is lowest in the updraft type since the products are cooled by the biomass entering the gasifier from the top. Moving-bed gasifiers have been built in many sizes; they are easily scalable and relatively easy to operate. On the other hand, they lack good internal mixing and develop inhomogeneity as they become large. The gasification

Table 14.5 Some characteristics of biomass gasifiers and their operating conditions

Parameter	Moving bed	Fluidized bed	Entrained bed
Feed particle size (mm)	~100	~50	~0.5
Operating temperature (°C)	700–1000	700–950	1100–1400
Operating pressure (bar)	1.0	1–10	20–50
Exit syngas temperature (°C)	~500	~900	~1250
Biomass feedstock conditions	Dry	Moist or dry	Torrefied

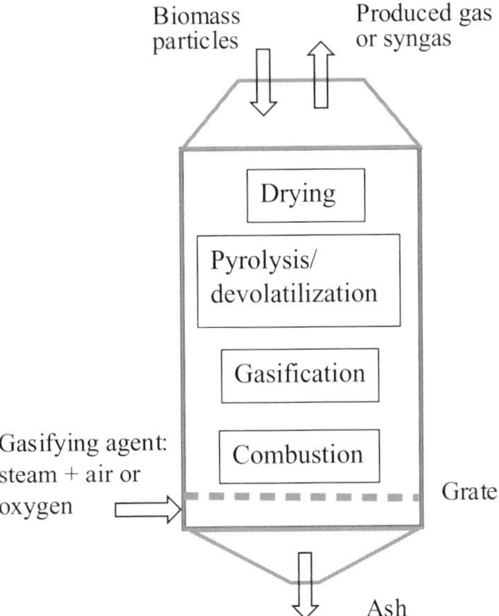

Figure 14.23 Schematic of an autothermal updraft gasifier and the conversion processes taking place inside the gasification vessel. At the bottom of the gasifier, air or oxygen is injected, where char left after gasification is burned to produce heat and combustion products. Both convect upward, where they meet the products of pyrolysis, namely char and volatiles. Both are gasified there. Hot products continue to flow upward, where they heat the descending biomass, causing pyrolysis and drying. The biomass undergoes these processes in reverse as it flows downward.

efficiency is rather high, especially for the updraft. These gasifiers operate autothermally, consuming a fraction of the feedstock to generate heat for the endothermic reactions. In all types, spatially different parts of the bed (the biomass particles) undergo conversion such as drying, pyrolysis, combustion, etc., as described next.

Figure 14.23 depicts a schematic of an autothermal **updraft** gasifier in which biomass in continuously fed from the top and the gasifying agents (steam + air/oxygen) are fed from the bottom through a distributer plate. As the biomass bed moves downwards under gravity, it undergoes drying, pyrolysis and the release of the volatiles, gasification of the gas and char, and combustion of the remaining char. The oxygen is consumed in the lower part of the vessel, where it is introduced. The heat produced there convects upward with the flowing gases. The products of combustion gasify the volatiles and char as they flow upwards (driven by forced and natural convection). Hot syngas (also called produced or producer gas) from the gasification zone contributes to drying the incoming biomass at the top of the vessel. Ash is collected at the bottom.

The hot gas entering the gasification zone provides the necessary heat for the endothermic char gasification reactions with CO_2 and steam, leading to increased concentration of hydrogen and CO. Because of the endothermicity of the gasification reactions, the temperature of the gas mixture declines as it passes the pyrolysis and drying zones. However, the mixture is still hot enough to provide the heat required for drying and pyrolysis of the descending biomass. The volatiles released due to biomass pyrolysis mix with the gas coming from the bottom and the final gas product leaves the bed from the top. The char produced by pyrolysis and the ash fall down. This improves the hot gas efficiency (HGE) to values close to 95%, and produces gas with an LHV of 5–6 MJ/Nm^3, but at the relatively low temperature of 300–400 °C. These gasifiers can be used with ash content up to 25%.

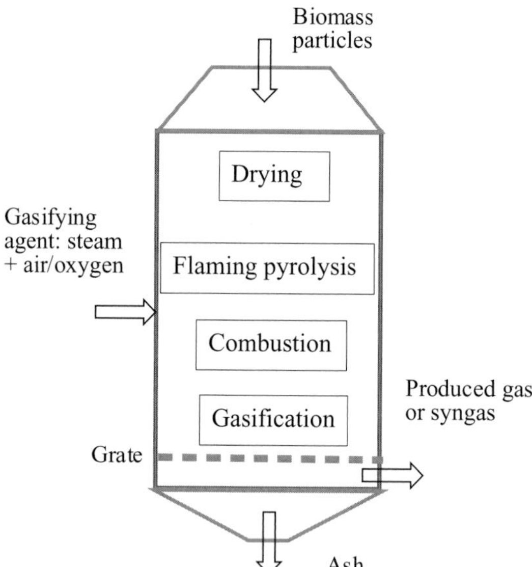

Figure 14.24 Schematic of an autothermal downdraft gasifier and the conversion processes taking place inside the gasification vessel.

The gaseous mixture produced in updraft gasifiers has high tar content, up to 150 g/Nm^3. Updraft gasifiers work well with moist biomass, with moisture up to 60% (which is dried by the exiting hot syngas) and are relatively insensitive to the particle size, using particles up to 10 cm [15].

In the **downdraft** gasifier, shown schematically in Figure 14.24, biomass particles and the gasifying medium move in the same direction, downward. The produced gas leaves the bed at the bottom, following the combustion and gasification zones, near the hottest zone, at around 700 °C. This has the advantage of cracking some of the tar released during gasification into smaller molecules, reducing the concentration of tar in the syngas relative to that in other moving-bed gasifiers to values lower than 3 g/Nm^3. Updraft gasifiers are typically limited to low-MWh application; given the difficulty of achieving cross-sectional uniformity they are limited to smaller cross-sectional areas. As shown in Figure 14.24, more of the volatiles may be partially oxidized in the same pyrolysis location where they are released, close to where air is injected, releasing heat during pyrolysis. More of the combustion of the char occurs further down (as the pyrolyzed biomass falls further downward), also close to where air is injected. As the products of combustion move downward with the remaining char, they gasify that char. The syngas leaves at the bottom of the bed. Ash accumulates at the bottom as well. These gasifiers are typically used with lower-ash biomass (<6%) because of the concern over ash fusion. The HGE of these gasifiers is <90%, and the gas LHV is 4–5 MJ/Nm^3. A more detailed schematic of a (coal) downdraft gasifier was shown in Chapter 12. Similar designs but at smaller scales have been used for biomass.

In the **cross-draft** gasifier shown schematically in Figure 14.25, biomass is fed from the top, the gasifying agent is injected from one side, and the syngas leaves from the opposite side. Both are positioned closer to the bottom of the gasifier. Combustion, as always, takes place close to where air is injected, where volatiles and char are consumed,

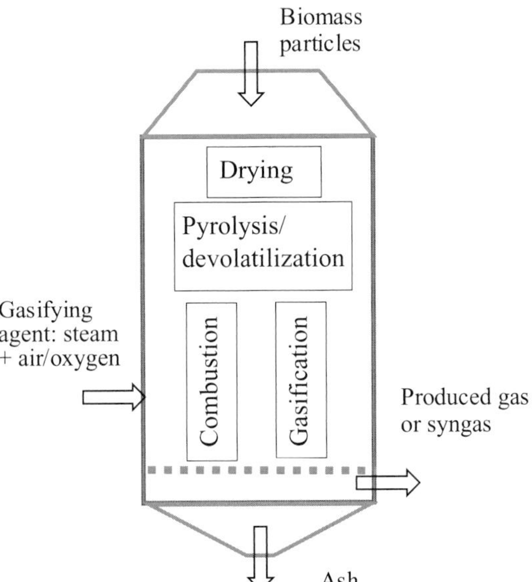

Figure 14.25 Schematic of an autothermal cross-draft gasifier and the conversion processes taking place inside the gasification vessel.

producing combustion products that gasify the remaining char and the volatiles as it flows out of the combustion zone. Heat diffusing out of the combustion and gasification zones contributes to the pyrolysis of the biomass moving downwards. Ash accumulates in the bottom. These systems operate at relatively higher temperatures, up to 1200 °C, lowering the tar in the produced gas to <1 g/Nm^3, but also lowering the HGE. Because of the high temperature, it is advantageous to use smaller particles and low moisture ($<20\%$) and ash content ($<1\%$) biomass.

Fluidized-Bed Gasifiers As shown in Chapter 12, in a fluidized-bed reactor, solid particles are suspended by the flow of a fluidizing agent entering from the bottom through a distributor. The bed material (Geldart B particles) is composed of inert particles (e.g., silica or alumina) of diameter 0.3–0.7 mm and density 1500–5000 kg/m^3. Figure 14.26 shows a time sequence of fluidized-bed dynamics illustrating the motion of bubbles through the solid. The biomass particle, with larger diameters and lower density, typically constitute a small fraction of the total bed material, and hence the bed hydrodynamics are governed primarily by the properties of the inert particles, endowing these reactors with excellent feedstock flexibility [17,18]. During fluidization, gas injected into the bed organizes in the form of bubbles that rise upwards, resulting in superior gas–solid heat and mass transfer. The improved gas–solid mixing and temperature uniformity make these gasifiers less sensitive to fuel quality (moisture content and reactivity). The solid fuel particles are typically smaller than those used in the fixed-/moving-bed gasifiers (but larger than in the case of entrained-flow gasifiers). This is advantageous in the case of biomass, which is rather difficult to grind.

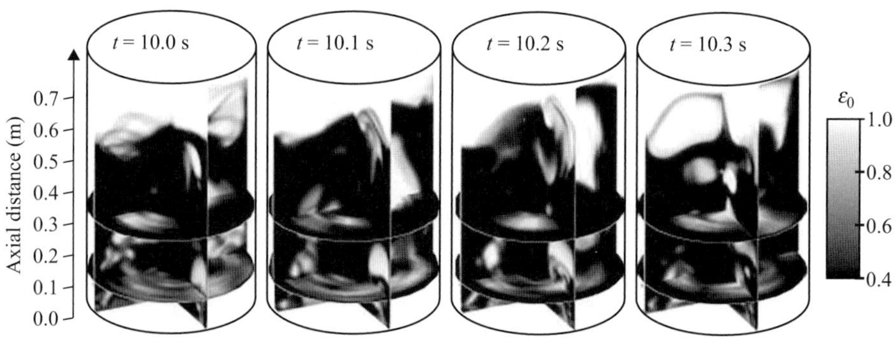

Figure 14.26 Computational fluid dynamics simulations of bubbling bed fluidization of Geldart B particles in a 0.5 m diameter reactor at $u_0/u_{mf} = 3$ (initial bed height is 0.5 m) showing the voidage ε_g filled by gas [17]. Here u_0 is the superficial gas velocity and u_{mf} is the minimum fluidization velocity. Fluidization transitions through multiple regimes. In the packed bed regime, $u_0/u_{mf} < 1$, the gas flows between the particles and no solid motion is observed. In the bubbling bed regime, $1 < u_0/u_{mf} < 7$, the gas organizes in the form of rising bubbles, causing the solid to circulate within the bed. In the slugging regime, $7 < u_0/u_{mf} < 13$, the size of the bubbles is on the same order as the bed diameter and a significant fraction of the gas bypasses the bed. For $u_0/u_{mf} > 13$, a turbulent regime is observed in which the solids organize in the form of clusters.

Because mixing is much better, the capacity of these reactors can be a lot higher than fixed-bed gasifiers. Also because of the smaller particles, the residence time is shorter. Fluidized-bed biomass gasifiers operate at intermediate temperatures and produce less tar (with an average value of 10 g/m^3) than some moving-bed gasifiers (but more tar than that is produced in the higher-temperature entrained-flow systems where tar undergoes secondary reactions).

The operating temperature of a fluidized-bed reactor is kept between 700 °C and 1000 °C to prevent ash melting and particle agglomeration at the higher temperatures that degrades the quality of fluidization. Depending on the residence time, the solid particles leaving the reactor may contain some partially gasified char that should be captured and recycled back to the bed for oxidation/reforming. In these gasifiers, primary gasification reactions occur on the surface and within the pores of solid particles, but secondary reactions continue in the gas phase. This is shown schematically in Figure 14.27 as the biomass particles are heated, volatiles (constituting a significant fraction of the particle) are released, and char forms. Next, char particles are partially or completely oxidized, and the remainder may be fragmented and elutriated by the mechanical interactions with the bed materials, forming finer particles. The gaseous products continue to react in the gasifier free board, above the bed, leading to further reforming of the volatiles, formation of more complex products such as polycyclic aromatics (PAHs) and tars, and water–gas shift reactions. Some of the fines could be carried out by the gas, leaving the reactor and must be separated and, in many cases, recycled back to the bed to improve the overall carbon conversion efficiency.

Two types of fluidized-bed gasifiers have been used: bubbling fluidized bed and circulating fluidized bed. Bubbling fluidized-bed gasifiers operate at a lower superficial gas velocity,

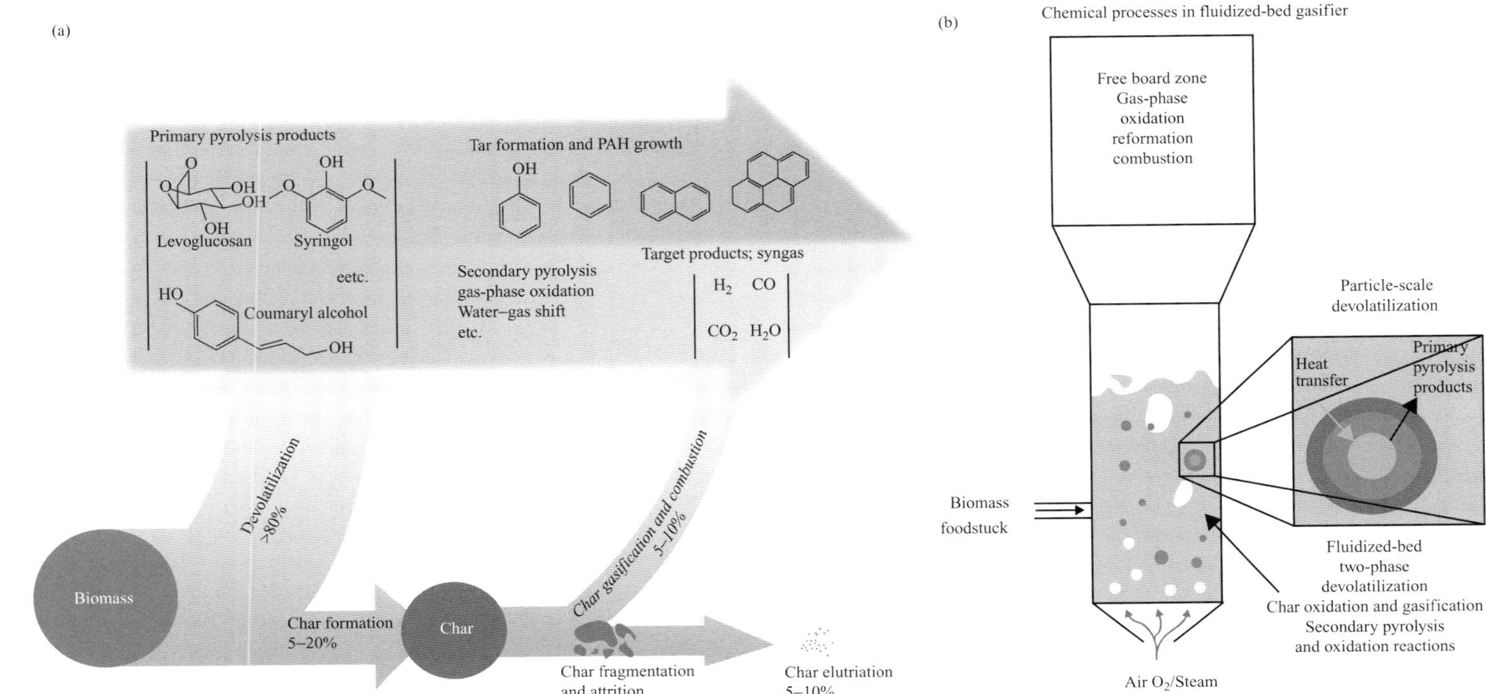

Figure 14.27 (a) A schematic representation of the sequence of events that a biomass particle undergoes in a fluidized-bed gasifier, starting with the release of the volatiles and formation of char particle, followed by the oxidation and gasification of the particles and formation of finer char particles. In the gas phase, the primary pyrolysis products undergo further reforming, oxidation, and formation of tar and PAHs. (b) A schematic representation of the processes occurring in the fluidized bed and the gasifier vessel free board [19].

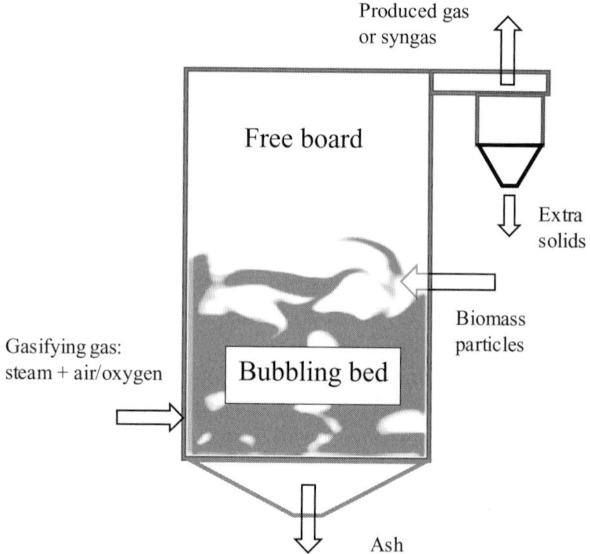

Figure 14.28 Schematic of a bubbling fluidized-bed gasifier, showing the gasification vessel, a cyclone separator for fine particles, a biomass feeder into the bed, and the fluidization agent blowing fan. Ash separates as a solid, falls to the bottom, and is continuously removed.

below the minimum velocity required to lift bed material and flow it with the gas, known as the minimum fluidization velocity. It is typically in the range 0.5–3 m/s, depending on the particles used for the bed. In circulating fluidized-bed gasifiers, the superficial velocity of the fluidizing gas is in the range 4–10 m/s, sufficient to carry some of the fine bed particles with the gas leaving the gasifier. These fine particles exiting with the gas must be capture using a cyclone and recirculated back to the bed.

In bubbling bed gasifiers, shown schematically in Figure 14.28, the bed material and biomass mostly remain in the gasification vessel. While the bed expands as gas bubbles form and flow through the bed, the gas flow velocity is not sufficiently high to carry a significant fraction of the solids as the gas leaves the bed. The injected gas forms bubbles that, while rising, cause the bed material to circulate internally within the bed itself without leaving the vessel, although very fine particles that form as the biomass is consumed may fly out. This may cause incomplete conversion of char and a "return leg" following a cyclone separator could be used to capture these fines and send them back to the bubbling bed. Moreover, some gas bypass can occur – that is, gas that flows in internal channels that does not experience contact with biomass particles. Fluidized-bed gasifiers can operate at higher pressures while keeping the temperature below the ash fusion temperature (e.g., the Winkler gasifier). Char conversion in bubbling beds can drop significantly below 100% if fines (biomass particles that have been broken by a combination of the mechanical action inside the hot bed and thermal cracking of the brittle char) are carried out with the gas leaving the vessel, especially in the absence of a return duct, as used in circulating beds.

Circulating fluidized-bed gasifiers operate at higher superficial gas velocity and hence can levitate the particles and flow them to the exit of the gasification vessel. Because of the higher gas velocity, which improves the mixing within the bed, the residence time can

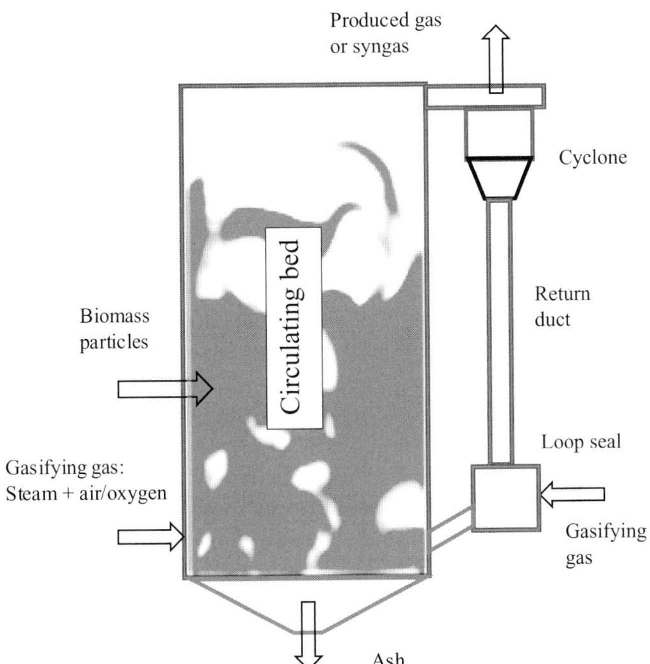

Figure 14.29 Schematic of a circulating fluidized-bed gasifier. The gasification vessel is taller to allow for the flow of the circulating particles internally. These particles are separated in a cyclone and returned to the fluidized bed through a return leg. The loop seal prevents gas and solid from the bed from flowing to the return leg. In gasifiers that operate at pressure higher than 1 bar, lock hoppers (not shown) are used to feed the biomass and remove the ash.

become too low for gasification and char conversion, and hence the gasification vessel must be made taller. The improved mixing within the bed allows more variability in biomass particle distribution. In order to keep the inventory of solid particles in the bed constant, it is important to capture the departing particles with the gas using a cyclone, and to return all these particles to the bed. This same process captures unconverted char particles and returns them to the bed to ensure near-complete conversion of the biomass carbon. As shown in Figure 14.29, the gasifier vessel is tall, sometimes called a riser, and a cyclone is connected to the bed with a return leg. The loop seal prevents bed material from flowing into the return duct. Gasification products leave from the return duct side, above the cyclone. The solid particles, including the unconverted char and bed material, are captured in the cyclone and returned to the bed [15]. A fan is used to feed the returning material to the bed. These systems can be operated at higher pressure to raise the capacity, and in this case lock hoppers are used to feed the biomass (raise the feed pressure) and remove the ash (reduce its pressure).

Because of the relatively lower gasification temperature, the methane content in the produced gas can be high, 4–12% by volume db. Methane has higher volumetric energy density (LHV = 802 kJ/mol) compared to hydrogen and carbon monoxide (LHV = 240 and 280 kJ/mol, respectively), and hence it can account for up to 30–40% of the chemical energy of the syngas. Depending on the application, this can be advantageous (if the produced gas is used as a fuel) or a disadvantage (if it is used in the catalytic FT processes that operate on hydrogen and carbon monoxide only).

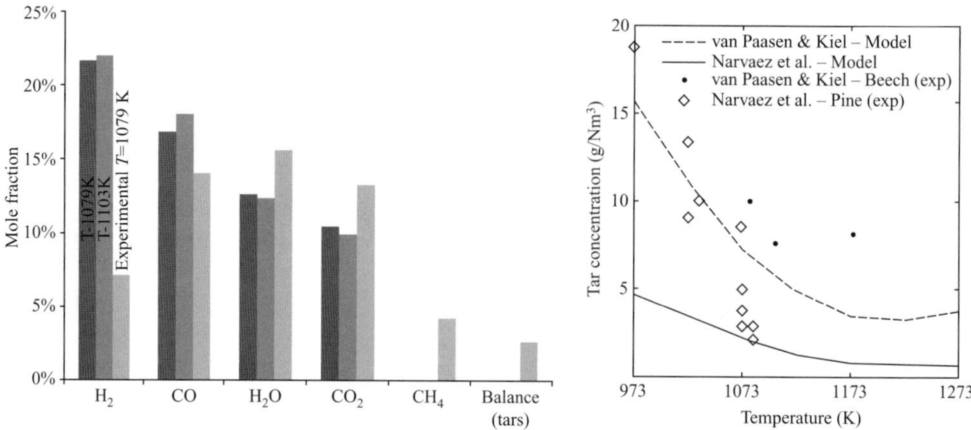

Figure 14.30 (a) Comparison between equilibrium concentrations of gasification products of beech wood in an air-blown fluidized-bed gasifier, at 1079 K (first bar) and 1103 K (second bar) and experimentally measured values at 1079 K (third bar), showing the impact of non-equilibrium effects (kinetics and transport) on the formation of methane and tars (both with near-zero equilibrium concentration). (b) Model prediction (continuous lines) and experimental measurements of tar concentration for beech at equivalence ratio 0.25 (dotted line and black dots) and pine at equivalence ration 0.3 (solid line and black diamonds) wood, at different temperatures. In the predictions, tars are takes to be species with C_{6+}. On average, beech produces more tar than pine, and tar decomposes with temperature. Predicting tar concentration depends strongly on chemical kinetic models coupled with particle conversion models that describe the coupling between chemistry and transport at that scale, and reactor mixing models in the fluidized bed as well as the free board [19].

Tars form in the produced gases of fluidized-bed gasifiers, and constitute a technology challenge for applications (and technical challenge for modeling) [20]. Tars are condensable organic compounds that can cause fouling, plugging, and corrosion in downstream equipment (heat exchangers, gas cleaning equipment, combustors, and turbines), and do not decompose to lighter gases until temperatures ~ 1000–1300 °C, above allowable limits of fluidized-bed gasifiers. Typical tar contents in these systems are 6–12 g/Nm^3, while desirable values are <0.05 g/Nm^3. Higher bed temperature reduces the outlet concentration of tars, at the risk of bed agglomeration (by ash melting). This is shown in Figure 14.30 in terms of total tar concentration and its dependence on temperature for a couple of wood species. Shown also is the measured concentration of methane, and the significant departure from equilibrium (as their formation is kinetics dependent) for methane and tar, and their impact on the concentrations of hydrogen and CO. Non-equilibrium models are necessary to improve the predictions and an example using these models is described in the next section. Tar components can represent a nontrivial fraction of the chemical energy in the gasification products (Table 14.6).

Entrained-Flow Gasifiers Entrained-flow gasifiers, described in detail in Chapter 12, operate at higher temperatures and pressures than fluidized-bed and moving-bed

Table 14.6 Some important tar species produced during biomass gasification

	Formula	Molar mass (g/mol)	HHV (kJ/mol)
Benzene	C_6H_6	78.11	3291.7
Toluene	C_7H_8	92.14	3934.8
Phenol	C_6H_5OH	94.11	3112.3
Cresol	C_7H_8O	108.14	3756.0
Naphthalene	$C_{10}H_8$	128.17	5216.0
Anthracene, phenanthrene	$C_{14}H_{10}$	178.23	7131.0
Pyrene	$C_{16}H_{10}$	202.25	7939.8

gasifiers. They do not use beds of solid particles, and instead create a highly turbulent environment inside the gasification vessel that vigorously mixes the gasifying agent with the particles. Solid biomass particles must be a lot finer, typically O (100–200 μm), and the residence time inside the vessel is much shorter O (1 s). To achieve such short residence time and complete conversion, the operating temperature must be high, ~1200–1400 °C. These gasifiers typically operate at higher pressures of 20–50 bar, and hence have high capacity of ~500 MW_{th}. Because of the higher operating temperature, ash leaves in the form of slag, and tar is mostly fully reformed or partially oxidized. The ash melting temperature of most biomass is lower than that of coal, and biomass reactivity is higher, and hence it is possible to achieve complete conversion and slagging operation at lower temperatures, using less oxygen than in the entrained-flow gasification of coal. However, entrained-flow reactors have not been used for raw biomass gasification because the moisture content of biomass is high, making it impossible to reach the requisite temperature (and significantly reducing the gasification efficiency). Moreover, the short residence time (~1 s) makes it necessary to feed very fine particles, which is difficult with moist and fibrous biomass. Torrefaction can make it possible to reduce the moisture content and make the material more brittle, hence making pulverizing the feedstock easier. However, a two-step conversion process, torrefaction then gasification, is complicated and may not be economical. For more on the design and operation of entrained-flow gasifiers, especially for coal applications, see Chapter 12.

14.7 Modeling Gasification and Combustion

Modeling biomass gasification is important fundamentally and for the design of gasification. Fundamentally the process is multiscale – that is, it occurs simultaneously on the microscale of the particle and the macroscale of the reactor. At the particle scale, individual particles are first heated, leading to drying, pyrolysis, and the release of gases and vapors, and formation of solid char. These transformations and the mechanical interaction with the bed material, especially in the case of fluidized beds, leads to fragmentation and formation of fine particles. The internal mixing within the bed induced by the fluidization ensures a nearly homogeneous environment, but eventually the gases mix with some of the gasifying medium and flow into the freeboard, above the bed. The char continues to gasify while in the bed, although most of the fines flow out (elutriate) with the gases, without much conversion. Thus, chemical reactions occur in the gas phase, as well as on the solid surface of the char. In this section, char particle models are reviewed, focusing on the case of a fluidized-bed reactor.

14.7.1 Char Particle Conversion

The **char particle conversion model** describes the physical and chemical processes contributing to char consumption and formation of fines that get elutriated out of the gasifier. Char particles produced following the drying and devolatilization of biomass particles are highly porous and experience internal as well as external mass and heat transfer, and chemical reactions on their outer surface and internal porous surfaces. The model described next considers two classes of particles: mother particles produced by devolatilization and primary fragmentation; and fines produced by the abrasion/attrition of mother particles [21]. Figure 14.31 describes this process.

Characteristic time scale analysis comparing the rates of the different chemical and physical processes is often used to identify the most relevant processes (or rate-limiting processes) affecting char conversion, and simplifying the analysis (see Reference [21] for detail of this analysis). Small char particles, O (1 mm), thermally equilibrate with their environment in a fluidized bed almost instantaneously, and hence can be considered as the same temperature as the bed. They remain thermally homogeneous throughout the conversion process. The bed material thermal inertia is high and the fraction of biomass is low, such that the bed remains at the same temperature (spatially and temporally) during conversion.

Gasification reactions, because of the relatively lower temperature (as compared to combustion reactions) are generally slower than internal and external mass transfer processes, especially for smaller particles, and hence these reactions are assumed to occur

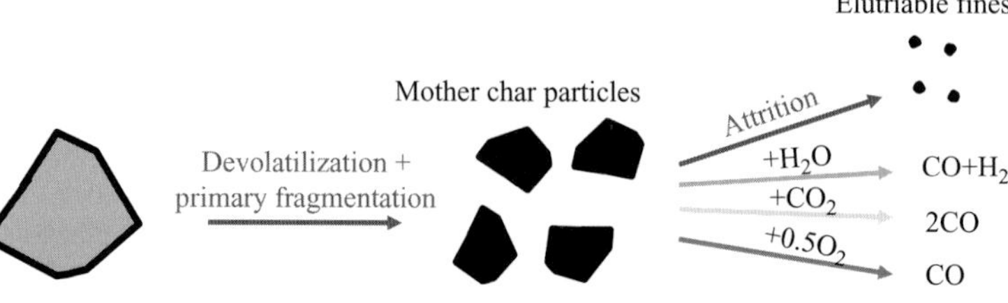

Figure 14.31 Schematic showing the primary physical and chemical processes in char conversion model: pyrolysis and fragmentation, formation of mother char particles and their attrition and formation of fines, and gasification [22].

uniformly within the particle. In a uniform conversion model (UCM), shrinkage/expansion of the particle is also often neglected. Abrasion at the exterior surface may reduce their average diameter and produce fines (by attrition), which are elutriated as soon as they are produced (because the residence time of the fines is small, in the order of a few seconds, compared to the time scale of their chemical reaction). Thus, fine particles produced through elutriation contribute to loss of unconverted carbon (the free board of a fluidized-bed gasifier is typically free of oxygen, and hence combustion of fines is not possible unless extra air or oxygen in injected there) [15]. The process of particle shrinkage via elutriation is described by the shrinking unreactive particle model. Even though the fines are much more reactive compared to the mother particles due to their smaller size, they hardly react with oxygen inside the bed because of their short residence time.

Combustion reactions, enabled by the presence of oxygen and the higher temperatures, are much faster than gasification reactions, and hence, contrary to gasification, internal and external mass transfer limitations play important roles in combustion and should be considered in the model if oxygen is present. Moreover, combustion reactions act on the particle surface and decrease the mass and size of the particle without significantly affecting its density. In both combustion and gasification cases, intra-particle temperature gradients are neglected, and the heat transfer from the bed material to the active particles is sufficiently fast that no temperature difference exists between the particle surface and the boundary layer.

14.7.1.1 Particle Characteristics

When raw fuel particles are introduced into a reactor they rapidly undergo devolatilization and primary fragmentation, resulting in a distribution of mother char particles of smaller size. The initial average char diameter $d_{ch,0}$ forming from an initial raw particle diameter $d_{f,0}$ is given by

$$d_{ch,0} = d_{f,0}/(\varphi_s n_1)^{1/3}, \tag{14.15}$$

where φ_s and n_1 are the shrinkage and primary fragmentation factor, respectively. The primary fragmentation factor is the total number of fragments generated by a single biomass particle. Fragmentation occurs in the early stages, during devolatilization. The volatiles, gases, and condensable vapors mix with the gasifying medium and form bubbles that rise in the bed, causing solids circulation and maintaining a well-mixed environment.

An initial charge of mother particles is formed by devolatilization and primary fragmentation. The initial number of particles $n_{ch,0}$ is computed from their average initial diameter, $d_{ch,0}$, initial density, $\rho_{ch,0}$, and initial mass, $m_{ch,0}$ (the mass left in the solid phase following pyrolysis):

$$n_{ch,0} = 6m_{ch,0}/\left(\pi d_{ch,0}^3 \rho_{ch,0}\right). \tag{14.16}$$

Biomass particles can have irregular shapes, so the average char diameter used throughout this section refers to the volume/surface mean diameter – the diameter of a sphere with equivalent volume to surface area ratio to the particle.

14.7.1.2 Conversion Rate

The carbonaceous portion of char particles is consumed by gasification, combustion, and attrition, and its mass changes according to

$$\frac{dm_{ch}}{dt} = -\left(R_{gasif} + R_{att} + R_{comb}\right)m_{ch}. \tag{14.17}$$

The attrition, gasification, and combustion rates, R_{att}, R_{gasif}, R_{comb}, respectively, with units of 1/s, will be discussed next. Ash is assumed to detach from the particle during combustion, the ash fraction of the char is consumed by attrition and combustion on the surface:

$$\frac{dm_{ash}}{dt} = -(R_{att} + R_{comb})m_{ash}. \tag{14.18}$$

Meanwhile, gasification reactions diminish the density of the char,

$$\frac{d\rho_{ch}}{dt} = -R_{gcif}\rho_{ch}. \tag{14.19}$$

Non-dimensional char conversion parameters, $X_\rho = \left(\rho_{ch,0} - \rho_{ch}\right)/\rho_{ch,0}$, $X_m = (m_{ch,0} - m_{ch})/m_{ch,0}$, and $X_g = X_m - m_{att}/m_{ch,0}$, are used to track its evolution. Assuming the mother particles remain spherical and their number, n_{ch}, remains constant until they are completely converted, the average diameter of the char particles, d_{ch}, can be computed from

$$d_{ch} = \left(\frac{6}{\pi}\frac{m_{ch}}{n_{ch}\rho_{ch}}\right)^{1/3} = d_{ch,0}\left(\frac{1 - X_m}{1 - X_\rho}\right)^{1/3}. \tag{14.20}$$

Attrition A char particle converted solely through gasification exhibits a constant average char diameter, and its average density eventually diminishes to 0. A char particle converted solely through attrition or combustion maintains constant char density, while the average char diameter eventually shrinks to 0. Secondary fragmentation is neglected. The attrition rate of mother char particles, R_{att} (1/s), is modeled as [23]:

$$R_{att} = K_{att,0}\left(u_0 - u_{mf}\right)F_{att}/d_{ch}, \tag{14.21}$$

where $K_{att,0}$ is a dimensionless attrition constant ranging from 10^{-7} to 10^{-8}. It is a strong function of the hardness and shape of the particles. Since hardness itself is not measured or modeled, the density is used as a proxy variable. u_{mf} is the minimum fluidization velocity of the bed, and F_{att} is a dimensionless structural attrition profile,

$$F_{att}\left(X_\rho\right) = \left(1 - X_\rho\right)^{-q} = \left(\rho_{ch}/\rho_{ch,0}\right)^{-q}, \tag{14.22}$$

where q is a structural attrition parameter. Attrition plays an important role in carbon conversion since attrited particles are elutriated before they are gasified.

Gasification The gasification reactivity (expressed here as rates) of char R_j ($j = CO_2$, H_2O, or O_2), has units of 1/s and is defined as the conversion rate per unit mass remaining,

$$R_j = -\frac{1}{m_{ch}}\frac{dm_{ch}}{dt} = r_j \; F_j(X_\rho). \tag{14.23}$$

The reactivity of chars is the chemical kinetics rate, r_j, multiplied by a dimensionless structural profile $F_j(X_\rho)$ that represents a normalized surface area and/or concentration of active sites for heterogeneous reactions.

The chemical kinetics rate, r_j, in a pure gasifying agent is either an Arrhenius expression or more generally a Langmuir–Hinchelwood (LH) expression. The following equations show examples for both, for two different types of wood, where R_g is the universal gas constant, T is temperature in kelvin, and p is partial pressure in bar. For beechwood:

$$r_{CO_2} = k_{CO_2} p_{CO_2}^{0.55}, \tag{14.24}$$

with $k_{CO_2} = 5.518 \times 10^4 \exp\left(-154,000/R_g T\right)$, for $1123 < T < 1273$ K and $0.1 < p_{CO_2} < 0.3$ bar.

The LH kinetics for pinewood are:

$$r_{CO_2} = \frac{k_{CO_2} p_{CO_2}}{1 + \kappa_{CO_2} p_{CO_2} + \kappa_{CO} p_{CO}}, \tag{14.25}$$

with $k_{CO_2} = 5.94 \times 10^7 \exp\left(-180,000/R_g T\right)$, $\kappa_{CO_2} = 4.64 \times 10^{-1} \exp\left(45{,}000/R_g T\right)$, and $\kappa_{CO} = 4.296 \times 10^{-9} \exp\left(213{,}000/R_g T\right)$, and $1023 < T < 1123$ K, $0.8 < p_{CO_2} < 1$ bar, $0 < p_{CO} < 0.2$ bar.

$F_j(X)$ is taken as invariant. The total gasification rate is the sum of all heterogeneous rates, which can be adjusted by a fitting parameter, ψ,

$$R_{gasif} = \psi\left(r_{CO_2} F_{CO_2} + r_{H_2O} F_{H_2O}\right). \tag{14.26}$$

Combustion Because of the high reactivity of biomass chars, the overall combustion rate depends on the kinetics as well as internal and external diffusion rates. Char combustion models can be apparent (or overall, accounting for chemistry and diffusion combined) or incorporate intrinsic kinetics separately. When the latter is used, the particle reaction rate depends on intrinsic reactivity, pore surface area, the concentration (or mass density) of carbon in the char, and the effective diffusivity of reactants through the porous char matrix. A rigorous and more universal model of biomass combustion utilizes the intrinsic approach. In this case, the effective combustion rate R_{comb} (1/s) is a single expression incorporating the three limiting processes,

$$R_{comb} = R_{O_2} \eta_{int} \eta_{ext}, \tag{14.27}$$

where R_{O_2} is the combustion reactivity under kinetically limited conditions, η_{int} is the internal effectiveness factor, and η_{ext} is the external effectiveness factor. The reactivity of biomass char to oxygen, R_{O_2} (1/s) is the kinetic rate, r_{O_2} with units of 1/s multiplied by a dimensionless structural profile, F_{O_2},

$$R_{O_2} = r_{O_2} F_{O_2}. \tag{14.28}$$

The kinetics is often expressed as first order in oxygen concentration, C_{O2},

$$r_{O_2} = k_{O_2} C_{O2}, \tag{14.29}$$

while the structural profile is $F_{O_2} = (1 - X_\rho)^{1.2}$.

Internal diffusion is governed by the structural properties, including the concentration of carbon C_c, char void fraction, $\varepsilon_{g,ch}$, and the effective diffusivity of the reactant (oxygen) through the pores. Char particles are composed entirely of carbon, ash, and void $\varepsilon_g = 1 - \varepsilon_{ch} - \varepsilon_{ash}$. All hydrogen and oxygen contained in the biomass is typically released during devolatilization. The solid volume fractions ε_j is inferred from their apparent densities ρ_j and skeletal densities $\rho_{wall,j}$. Moreover, $\varepsilon_{bio} = \rho_{bio}/\rho_{wall,bio}$, $\varepsilon_{ch} = \rho_{ch}/\rho_{wall,ch}$, and $\varepsilon_{ash} = \rho_{ash}/\rho_{wall,ash}$.

The char yield, Y_{ch} produced from devolatilization is typically the same as the fixed carbon content of the original biomass, and all the ash in the original biomass is retained in the char. Thus, the initial apparent carbon density and apparent ash density are $\rho_{ch,0} = \rho_{bio} Y_{ch} \varphi_s$ and $\rho_{ash,0} = \rho_{bio} Y_{ash} \varphi_s$, where φ_s is a feedstock-dependent, dimensionless shrinkage parameter, and Y_{ash} is the ash content of the biomass. The molar concentration of carbon is $C_c = \rho_{ch}/MW_c$ where MW_c is the atomic weight of carbon. As shown in Table 14.7, the initial carbon concentrations can be different for different woods; for instance, for pine wood char (PWC) and spruce wood pellets (SWP) they are 10,200 and 24,300 mol_C/m^3. The estimated initial void fractions are 0.935 for PWC and 0.85 for SWP. And although a significant amount of volumetric shrinkage occurs during devolatilization, most of the carbon in the original biomass is released in the volatiles, and thus the resultant chars tend to be *more* porous than their raw biomass particles.

Diffusion through the char is often described by Fick's law modified with an effective diffusion coefficient. The binary diffusivity of oxygen D_{binary} (units of m^2/s) in air is shown in Table 14.7. However, the diffusion within the char is reduced due to constrictions and the nonlinear path the molecules must travel. An effective diffusion coefficient which takes this into account is often introduced $D_{eff} = D_{binary}(\varepsilon_g \sigma/\tau_p)$, where σ is the dimensionless constriction factor and τ_p is the tortuosity. The range of these values is $(0.01 < \varepsilon_g \sigma/\tau_p < 0.1)$ or $\sigma/\tau_p = 6$, and generally $\sigma/\tau_p = 0.2\varepsilon_g^3$. Around 800 °C and 1 atm, the average effective diffusivities are $3.25 \times 10^{-5}\ \text{m}^2/\text{s}$ for PWC and $2.6 \times 10^{-5}\ \text{m}^2/\text{s}$ for SWP due to its lower void fraction (higher apparent char density).

The internal effectiveness factor for a spherically equivalent geometry is the actual reaction rate divided by the rate if a uniform reactant concentration existed across the particle:

$$\eta_{int} = \left(\frac{1}{\tanh(3\phi)} - \frac{1}{3\phi} \right), \tag{14.30}$$

where the Thiele modulus is $\phi = \frac{d_{ch}}{6} \sqrt{\frac{k_{O_2} C_c \lambda}{D_{eff}}}$, and the stoichiometric coefficient, λ is $\text{mol}_{O2}/\text{mol}_C$. This approach is used to estimate the initial properties of the char and through conversion.

Table 14.7 Transport properties used in modeling the gasification and combustion of pine wood chip PWC and SWP char

	Quantity	PWC	SWP
Char structural properties			
Biomass skeletal (wall) density		$\rho_{wall,bio} = 1500\,\text{kg/m}^3$	
Char skeletal (wall) density		$\rho_{wall,ch} = 1888\,\text{kg/m}^3$	
Ash skeletal density		$\rho_{wall,ash} = 2600\,\text{kg/m}^3$	
Apparent biomass particle density	ρ_{bio} kg/m^3	550	1300
Char yield (fixed carbon) (db)	Y_{ch}	0.133	0.171
Ash yield Proximate analysis (db)	Y_{ash}	0.00307	0.00327
Devolatilization shrinkage factor	φ_s	1.68	1.31
Initial char void fraction	$\varepsilon_{g,0}$	0.935	0.845
Initial char carbon concentration	$C_{c,0}$ mol$_C$/m^3	10.232	24.273
Mass transfer properties correlations			
Binary diffusivity		$D_{binary} = 1.5815 \times 10^{-4}\left(\frac{T}{1000}\right)^{1.75}\frac{101{,}325}{P}\ (\text{m}^2/\text{s})$	
Effective diffusivity		$D_{eff} = D_{binary}0.2\varepsilon_g^3\ (\text{m}^2/\text{s})$	
Dynamic viscosity		$\mu_f = 1.98 \times 10^{-5}\left(\frac{T}{300}\right)^{\frac{2}{3}}$ (Pa-s)	
Sherwood number		$Sh = 2\varepsilon_{mf} + 0.7\left(Re_p/\varepsilon_{mf}\right)^{1/2}Sc^{1/3}$	
Reynolds number		$Re_p = \rho_f u_{mf} d_{ch}/\mu_f$	
Schmidt number		$Sc = \mu_f/\left(\rho_f D_{binary}\right)$	
Voidage at minimum fluidization		$\varepsilon_{mf} = (14\Psi)^{-1/3}$, Ψ =0.9 (silica sand)	

External transport of oxygen from the bulk through the boundary layer to the particle surface is a significant limitation during combustion of millimeter-size biomass char particles. The mass transfer coefficient (in m/s) depends on the Sherwood number for the particle as well as the diffusivity of oxygen in the boundary layer, $h_m = ShD_{binary}/d_{ch}$. The Sherwood number correlation is shown in Table 14.7, developed specifically for reacting spherical particles under fluidized-bed conditions at elevated temperatures (723 K) and atmospheric pressure and fitted to their experimental data within ±10%. The external effectiveness factor is

$$\eta_{ext} = \frac{1}{\eta_{int}\phi_{ext} + 1}, \tag{14.31}$$

where the external Thiele modulus, $\phi_{ext} = \dfrac{d_{ch}k_{O_2}C_c\lambda}{6h_m}$.

To integrate the transient particle model we need initial char conditions (size, density, etc.) and reactor specifications (size, temperature, pressure, and gas phase concentrations). The ordinary differential equations are integrated until the mass of char remaining in the reactor is 0. Examples of the application of the model are shown next.

Table 14.8 Experimental conditions for the gasification/combustion and attrition study in References [23] and [24] used in the modeling analysis.

Bed material size, density, mass: d_p, ρ_p, m_{bed}	350 μm, 2600 kg/m^3, 0.18 kg	
Fluidization, superficial velocity: u_{mf}, u_0	0.0438, 0.8 m/s	
Bed temperature, bed diameter T_{bed}, d_{bed}	800 °C, 0.04 m	
Inlet gas composition, by volume		
Gasification conditions	60% CO_2, 40% N_2	
Combustion conditions	4.5% O_2, 95.5% N_2	
Initial char loading, $m_{ch,0}$	2 grams	
Feedstock properties		
Feedstock	SWP	PWC
Initial biomass diameter, $d_{f,0}$	$d_{f,cyl,0} = 6$ mm	$d_{f,slb,0} = 10.4$ mm
Reported mean diameter, $d_{ch,0}$	$d_{ch,cyl,0} = 4.9$ mm	$d_{ch,slb,0} = 5.3$ mm
Reported fragmentation factor, n_1	1.4	4.5
Particle geometry	Cylinder	Flake
Aspect ratio	$AR_{cyl} = 20/6$	$AR_{slb} = 2$ (assumed)
Fitted parameters (this study)		
Initial attrition constant $K_{att,0} \cdot 10^{-7}$	0.8	5.5
Gasification reactivity factor, ψ	0.184	0.24
Attrition structural profile constant, q	2.5	2

Char particles of spruce wood and pine wood were gasified in a fluidized-bed, and the elutriation rare and gasification measured [23]. The model parameters were determined by fitting the model to the experimental data, and the values obtained are shown in Table 14.8. The initial attrition constant shows the strongest dependence on the type of wood used in producing the char, likely because of the difference in the density or hardiness between them, and the palletization of spruce.

Figure 14.32 compares the model results and experimental data for spruce pellet gasification. The attrition rate first increases rapidly then decreases at an even faster rate; the first is because of the reduction in the wood strength because of the decrease in the char density during gasification, and the decrease in the diameter which increases the surface to volume ratio. As the char inventory in the bed decreases, the rate of formation of fines drops rapidly. The production of fines reduces the conversion significantly, as shown in the figure, with only ~70% of the carbon being gasified and the remainder leaving in the form of fines.

The results of char combustion of the same wood pellets are shown in Figure 14.33, using the same elutriation rate model parameters as in gasification. In this case, the attrition falls rapidly because of the fast decrease in inventory with combustion. As mentioned before, combustion occurs essentially on the outer diameter of the particles, hardly affecting the char density/hardiness and hence its attrition rate. Almost complete char conversion is achieved by combustion.

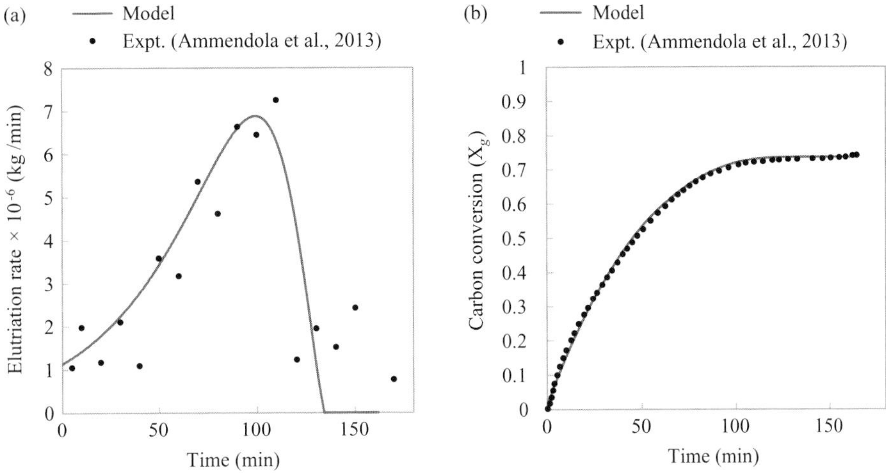

Figure 14.32 (a) Elutriation rate, and (b) carbon conversion (X_g) for SWP char gasification [22].

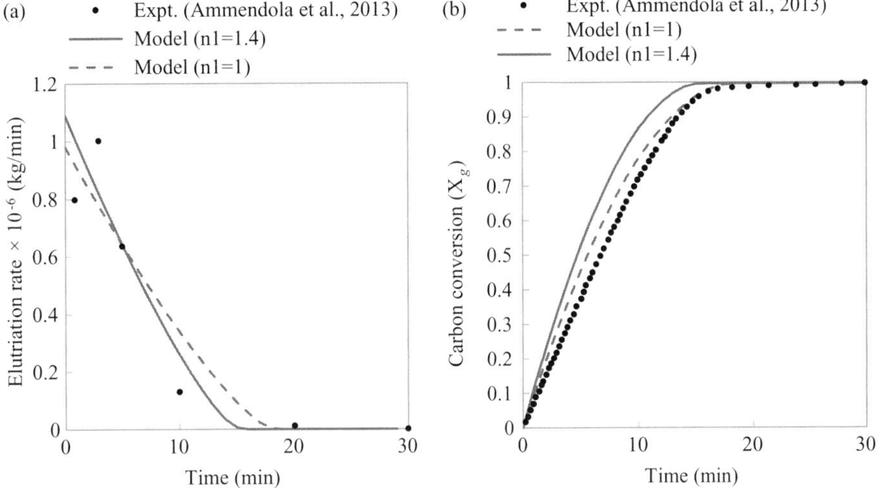

Figure 14.33 (a) Elutriation rate (kg min $^{-1}$ × 10^{-6}) versus time, and (b) carbon conversion (X_g) versus time in minutes for SWP char combustion. Data show the dependency on n_1, the fragmentation factor [22].

14.8 Gasification for Power and Fuel Production

In many cases in which gasification is used for power and fuel production, it is better to produce a fuel gas (syngas when it is $CO + H_2$, and producer gas when it is $CO + H_2 + CH_4$) that is free (or almost free) of nitrogen. A nitrogen-free gas is easier to purify after cooling, and it is better when used in synthesis or for combustion in engines or turbines. It is also easier to store and transport. For the purpose of producing fuel gas that is free of nitrogen, and instead of using an air-separation unit to produce pure oxygen for oxy-gasification, a duel-reactor bed can be used. In a dual-bed system, a separate fluidized-bed reactor is used to gasify the biomass into steam and a heated solid, and another is used to burn the char produced in the first reactor in air and reheat the solid. The char produced during the gasification and some of the solid bed material of the gasifier leave the first reactor and are separated in a cyclone. The gas from this reactor is cooled and cleaned, and used in a combustor (for power production) or a catalytic reactor to produce liquid fuels. The solids – that is the bed material and char leaving the cyclone – are combusted in air in the second reactor. The heated solid bed material is then conveyed to the first reactor, while the hot gaseous products are used to raise steam for gasification, and ash is separated. Thus, the bed material is used, besides fluidization, as a "heat" carrier between the two reactors. This "thermal looping" process resembles the chemical looping process described previously, except in this case only heat is transported between the two reactors. This concept is shown in Figure 14.34.

This dual-bed fluidized reactor was used in the gasification plant in the town of Güssing, Austria, to product electricity and heat using wood chips collected from the nearby forests [25]. The producer gas was cooled and cleaned, and used in an internal combustion gas engine to spin an electric generator. Biomass chips were transported from a hopper to a metering bin and fed into the fluidized-bed reactor via a rotary valve system and a screw

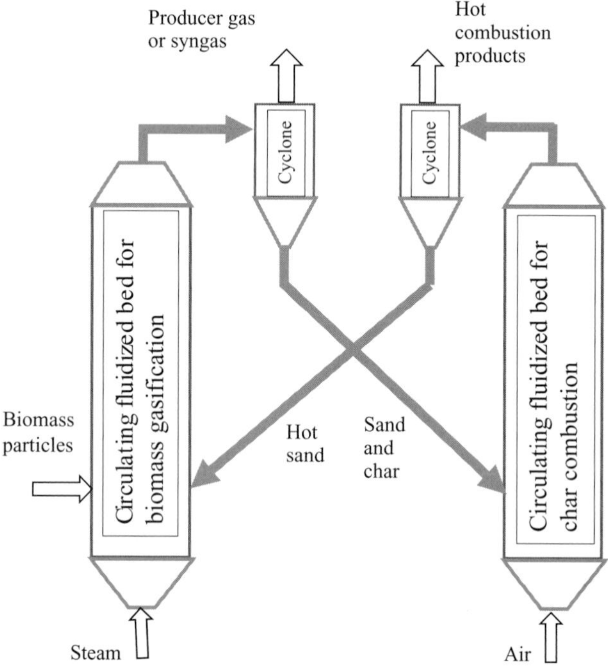

Figure 14.34 Dual circulating fluidized-bed gasifier for the production of nearly nitrogen-free syngas. On the left-hand side reactor, biomass is steam gasified with the help of heated sand particles. The char and cool sand are circulated into the right-hand side reactor, where air is used to burn the char and reheat the particles.

feeder. A water-cooled heat exchanger reduced the temperature of the producer gas from 850–900 °C to about 160–180 °C. The first stage of the cleaning system was a fabric filter to separate the particles and some of the tar from the producer gas. These particles were returned to the combustion zone of the gasifier. In a second stage, tar was removed using a scrubber. The spent scrubber liquid saturated with tar and condensate was vaporized and introduced into the combustion zone of the gasifier. The scrubber was used to reduce the temperature of the clean producer gas to about 40 °C, which was necessary for the gas engine that powers an electric generator, and whose was used for district heating. When the engine was idle, the producer gas could be burned in the boiler to produce heat. The exhaust gas of the gas engine was catalytically oxidized to reduce CO emissions. The "waste" thermal energy in the engine exhaust was used for preheating of the air, superheating the steam, and also to deliver heat to the district heating system. A gas filter separated the particles before the flue gas is released via a stack to the environment.

The thermal capacity of the gasifier was 8 MW_{th}, producing 2 MW_e and 4.5 MW of heat. The heating value of the producer gas was 12 MW/Nm^3, and it was, by volume, made of 35–45% H_2, 20–30% CO, 15–25% CO_2, 8–12% CH_4, and 3–5% N_2. It left the gasifier at 850–900 °C and was cooled to 150–180 °C for removal of impurities such as ammonia (1000–2000 ppm), tar (1500–4500 mg/Nm^3), particles (5000–10,000 mg/Nm^3), and H_2S in the raw gas.

Problems

14.1 One mole of biomass char modeled as pure carbon reacts with steam and oxygen in a gasifier. For each mole of char, 1.2 moles of steam is added. The products of gasification leaving the gasifier at 1273 K consist of carbon dioxide, carbon monoxide, steam, and hydrogen, and are in chemical equilibrium. Determine the number of moles of oxygen required in this process in order to achieve a hydrogen mole fraction of 40% in the products.

14.2 One mole of a woody biomass whose LHV = 21 MJ/kg, $\hat{h}_f = -1267\,\text{kJ/mol}$, $\hat{c}_p = 75.6\,\text{kJ/kmol·K}$, and chemical formula $C_6H_{12}O_6$ reacts with 8 moles of steam at 1073 K in a gasifier/reformer. The products (CO_2, CO, H_2, and H_2O) leave at 773 K. CO in the mixture is converted in a water–gas shift reactor to CO_2 using 1 mole of steam per mole of CO in the mixture. The mixture leaves the shift reactor at 373 K. The process is shown in Figure P14.2.

a. Determine the equilibrium composition of the gaseous mixture leaving the gasifier/reformer.

b. Calculate the heat required in the steam reformer.

c. Find the hydrogen concentration in the gas leaving the shift reactor.

d. Determine the overall process efficiency, including the reforming and shift processes, if no waste heat is recuperated.

Figure P14.2

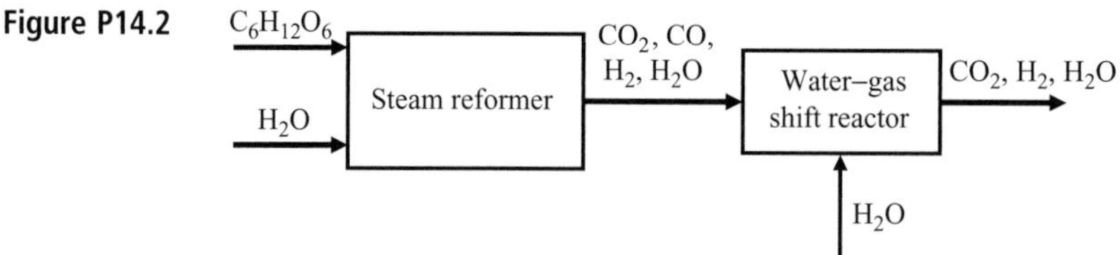

14.3 Consider complete combustion of a woody biomass ($C_6H_{12}O_6$) in air, at a stoichiometric ratio in an adiabatic furnace of a boiler in a steam power plant, where both biomass and air are supplied at 298 K and 1 atm. The temperature of the steam leaving the boiler is 823 K. The combustion products are cooled down to 323 K. Assume a Second Law efficiency of 50% for the steam cycle. Calculate: (a) the adiabatic temperature of the combustion products; (b) the maximum efficiency of the steam cycle; (c) the overall fuel utilization efficiency; and (d) the work produced by the power plant per mole of biomass.

REFERENCES

1. R. F. Probstein and R. E. Hicks, *Synthetic fuels*. Mineola, NY: Dover Publications, 2006.
2. A. Nag, *Biosystems engineering*. New York: McGraw Hill, 2010.
3. P. Basu, *Biomass gasification and pyrolysis*. New York: Elsevier, 2013.
4. Z. Wang, S. Dai, G. Yang, and H. Luo, "Ligno cellulosic biomass pyrolysis mechanism: A state of the art review," *Prog. Energy Combust. Sci.*, vol. 62, pp. 33–86, 2017.
5. P. Basu, *Biomass gasification, pyrolysis and torrefaction: practical design and theory*. Amsterdam: Academic Press, 2013.
6. R. Bates, "Modeling the coupled effects of heat transfer, thermochemistry and kinetics during biomass torrefaction," MIT, M.Sc. thesis, 2012.
7. L. Francecato, V. Antonini, and E. Bergomi, *Wood fuels handbook*. Legnaro: AIEL. 2008.
8. P. P. Channiwala, and S.A. Parikh, "A unified correlation for estimating HHV of solid, liquid and gaseous fuels," *Fuel*, vol. 35, pp. 1051–1063, 2002.
9. J. L. T. Sheng, and C. Azevedo, "Estimating the higher heating value of biomass fuels from basis analysis data," *Biomass Bioenergy*, vol. 28, pp.. 499–507, 2005.
10. E. Biagini, F. Barontini, and L. Tognotti, "Devolatilization of biomass fuels and biomass components studied by TG/FTIR technique," *Ind. Eng. Chem. Res.*, vol. 45, no. 13, pp. 4486–4493, 2006.
11. C. Di Blasi, "Combustion and gasification rates of lignocellulosic chars," *Prog. Energy Combust. Sci.*, vol. 35, no. 2, pp. 121–140, 2009.
12. J. S. Tumuluru S. Sokhansanj, C. T. Wright, and R. D. Boardman, *Biomass torrefaction process review and moving bed torrefaction system model development*, Idaho Falls: Idaho National Laboratory, 2010.
13. A. Pimchuai, A. Dutta, and P. Basu, "Torrefaction of agriculture residue to enhance combustible properties," *Energy & Fuels*, vol. 24, no. 9, pp. 4638–4645, 2010.

14. R. B. Bates, "Modeling the coupled effects of heat transfer, thermochemistry and kinetics during biomass torrefaction," MIT, M.Sc. thesis, 2012.
15. A. Gómez-Barea and B. Leckner, "Modeling of biomass gasification in fluidized bed," *Prog. Energy Combust. Sci.*, vol. 36, no. 4, pp. 444–509, 2010.
16. C. Higman and M. van der Burgt, *Gasification.* Amsterdam: Gulf Professional Publishing, 2003.
17. A. Bakshi, "Multi scale continuum simulations of fluidization: bubbles, mixing dynamics and rector scaling," MIT, Ph.D. thesis, 2017.
18. A. Bakshi, C. Altantzis, L. R. Glicksman, and A. F. Ghoniem, "Gas-flow distribution in bubbling fluidized beds: CFD-based analysis and impact of operating conditions," *Powder Technol.*, vol. 316, pp. 500–511, 2017.
19. A. K. Stark, "Multiscale chemistry modeling of the thermochemical conversion of biomass in a fluidized bed gasifier," MIT, Ph.D. thesis, 2015.
20. A. K. Stark, R. B. Bates, Z. Zhao, and A. F. Ghoniem, "Prediction and validation of major gas and tar species from a reactor network model of Air-blown fluidized bed biomass gasification," *Energy Fuels*, vol. 29, no. 4, pp. 2437–2452, 2015.
21. R. B. Bates, C. Altantzis, and A. F. Ghoniem, "Modeling of biomass char gasification, combustion, and attrition kinetics in fluidized beds," *Energy Fuels*, vol. 30, no. 1, pp. 360–376, 2016.
22. R. B. Bates, "Multiscale simulation of methane assisted fluidized bed biomass gasification," MIT, Ph.D. thesis, 2016.
23. P. Ammendola, R. Chirone, F. Miccio, G. Ruoppolo, and F. Scala, "Devolatilization and attrition behavior of fuel pellets during fluidized-bed gasification," *Energy Fuels*, vol. 25, no. 3, pp. 1260–1266, 2011.
24. F. Scala, R. Chirone, and P. Salatino, "Combustion and attrition of biomass chars in a fluidized bed," *Energy Fuels*, vol. 20, no. 1, pp. 91–102, 2006.
25. H. Hofbauer, R. Rauch, B. Klaus, K. Reinhard, and C. Aichernig "Biomass CHP plant Gussing, a success story," *Conf. Proc. Pyrolysis and Gasification of Biomass and Waste*, Strasbourg, France, 2003, pp. 527–536.

Index